NEUROANATOMY RESEARCH AT THE LEADING EDGE

HUMAN NEUROPHYSIOLOGY

DEVELOPMENT AND REPAIR OF THE HUMAN CENTRAL NERVOUS SYSTEM

Neuroanatomy Research at the Leading Edge

Additional books in this series can be found on Nova's website under the Series tab.

Additional e-books in this series can be found on Nova's website under the e-books tab.

Neuroscience Research Progress

Additional books in this series can be found on Nova's website under the Series tab.

Additional e-books in this series can be found on Nova's website under the e-books tab.

NEUROANATOMY RESEARCH AT THE LEADING EDGE

HUMAN NEUROPHYSIOLOGY

DEVELOPMENT AND REPAIR OF THE HUMAN CENTRAL NERVOUS SYSTEM

GISELHER SCHALOW

NOVA BIOMEDICAL

New York

For permission to use material from this book please contact us:
Telephone 631-231-7269; Fax 631-231-8175
Web Site: http://www.novapublishers.com

Library of Congress Cataloging-in-Publication Data

ISBN 978-1-62257-457-5

Library of Congress Control Number: 2012941954

Published by Nova Science Publishers, Inc. † New York

Contents

Preface

To repair the injured, malformed or degenerating human central nervous system, we have to understand the functioning of the healthy human CNS on cellular and integrative levels to learn what changed in the neural networks. Since there is some similarity between development and repair, also the development of the human CNS has to be studied. Especially the strategy of movement-based learning of the CNS to eliminate occurring errors during development is of interest because it can also be used during repair.

Based on an accurate classification scheme of human peripheral nerve fibres and anatomy, skin, urinary bladder and muscle spindle receptors are analysed in detail and their communication with the neuronal networks of the CNS upon physiologic stimulation. By using the new measuring technique of recording single-nerve fibre action potentials from nerve roots, the natural impulse traffic between the periphery and the CNS is measured simultaneously in several single afferent and efferent fibres in brain-dead humans (rather physiologic) and patients with CNS injury. Additionally in patients surface electromyography (sEMG) and coordination patterns dynamics (CD) measurements are performed. Single-motor unit sEMG recordings, obtained non-invasively, are compared with those of single-nerve fibre action potential recordings, obtained invasively during operations in patients and in brain-dead humans, to get more safety and knowledge about natural impulse patterns in the human nervous system. EMG displays impaired motor unit recruitment in patients and gives information for repair. CNS functioning in patients, pupils and athletes is measured by the CD upon exercising on a special coordination dynamics therapy and recording device. Based on the System Theory of Pattern Formation, CNS functioning can be measured non-invasively by pattern change, which is achieved when exercising on the special device. The computer quantifies the rhythmicity of exercising upon pattern change of arm and leg movements during turning on the special device.

By comparing natural impulse pattern under physiologic and patho-physiologic conditions it is found that the self-organisation of the neural networks becomes impaired following injury, malformation or degeneration. Especially the coordinated firing of neurons and the (motor) patterns become impaired and have to be repaired. For the repair of the impaired neural network organisation the strategy of the development is used, namely movement-based learning. In the book it is analysed what movement patters have the highest rate of re-learning movement and higher mental functions. The therapy using these efficient movement-based learning movements is called Coordination Dynamics Therapy (CDT).

Based on measurements at the neural and integrative levels, learning transfer from movement to other movements, autonomic and higher mental functions is analysed. In the clinical part it is shown that urinary bladder functions could be repaired in a spinal cord injury patient and speech could be induced in a cerebral palsy child upon administered CDT. Reading, speech and other higher mental functions can only be studied and repaired upon medical (human) research.

Four-cell communication is studied in Chapter I during development and repair in the frog and consequences are drawn for the repair of the human CNS. In Chapter II, clinical studies are documented for the partial repair of the human CNS in groups of patients and in single patients up to therapy duration of 11 years. The human development is studied from birth to 18 years of age. It is also shown how athletes could improve their physical performance and how in aging one can live longer with a higher quality of life. In Chapters III and V, based on human neurophysiology, the scientific for the movement-based learning therapy is developed. Chapter IV and I inform about how animal data can be of benefit for patients (translational medicine). Chapter VI shows scientifically that stem-cell therapy is unlikely to be effective since stem/progenitor cells cannot be integrated in the adult neural networks. The dream of repair is to influence the epigenom. In the supplement it is show that medical (human) research is lacking organisation and that it is an adventure to perform it. Interesting is that qualified medical research was possible across the Berlin Wall.

Chapter I

Competitive Interaction between Two Kinds of Motoneurons for the Innervations of Two Types of Muscle Fibres during Development and Repair in Frog, Discussed with Respect to Axon Guidance and Endogenous Stem Cell Supported Structural Repair by Motor Learning in Human Cervical Spinal Cord Injury

Abstract

During development and regeneration the neuromuscular innervation, change of slow muscle fibres from one motoneuron population to another has been measured electrophysiologically and morphologically in the common frog (rana temporaria). Target invasion, selection, and selection change are discussed with respect to long- and short-range attraction and repulsion of possible guidance molecules and receptors for guidance of axon grows cones and synapse profiles. The contact attraction and repulsion over a distance of 0.1μm and shorter is discussed with respect to competitive interactions and electromagnetic field guidance of the two types of motoneuron populations with respect to the specific innervation pattern of the two kinds of target muscle cells.

The invasion of a target muscle by two successive waves of two types of axons and the recognition and specific selection of two types of muscle fibres for innervation during development and regeneration in the frog are used to understand the regeneration of the human spinal cord following severe C5/6 spinal cord injury (SCI) upon movement-based learning. Based on the location of spinal cord tracts in the white matter and premotor networks in the grey matter and specific regeneration, it is found that the regeneration in human is starting from the spared naturally activated nerve fibres at the injury site. The building of new motoneurons from endogenous stem cell reserves for the activation of the index finger and their functional integration started from the activated neuronal networks surrounding the injury site.

In spite of the progress in molecular biology, cell biology, and genetics, it will probably take at least another 100 years of faithful research to make a pure exogenous stem cell therapy work. To partly repeat neural development in humans may be as difficult as achieving eternal life. However, with a faithful translation of animal to human data and administration of Coordination Dynamics Therapy (CDT), movement-based learning, to human patients, the injured or malfunctioning human CNS can already partly be repaired.

Data Summary

1. The neural control of twitch and slow muscle fibres by two kinds of motoneurons in frog during development (metamorphosis) is quantified by electrical changes of membrane properties, changes of contractile properties, and morphologic changes of axons and synapses made visible by electron microscopy (EM). Twitch (similar to FF-type muscle fibres in human) and slow muscle fibres (similar to S-type muscle fibres in human) are innervated in adult frogs by thick, fast conducting and thin, slowly conducting motor axons, respectively. Slow conducting axons grow and regenerate slower than the thick, fast conducting axons, causing neural control changes during development and regeneration (following injury) with competitive interactions between the two kinds of motoneurons for the innervation of the two kinds of muscle fibres. In a first wave, fast conducting axons invade the pyriformis muscle and successfully innervate twitch and slow muscle fibres non-selectively. In a second wave, slowly conducting axons invade the already innervated pyriformis muscle and innervate selectively the slow muscle fibres. The successfully established endplates of the fast conducting axons on the (wrong) slow muscle fibres are repelled. This four-cell-communication model for specific cell organization gives insight into the complexity of neuronal network organization during development and repair both in animals and humans.
2. In the tadpole stage (tail length = 20mm), both slow and twitch muscle fibres are present, as can be seen from EM pictures. All muscle fibres are innervated by the same motoneuron population, namely those that innervate the twitch muscle fibres in adult frogs. Innervated by the wrong fast conducting axons, as in the early stage of reinnervation after denervation, the slow muscle fibres are not able to respond with a sustained contracture but are capable of generating action potentials (APs).
3. With the progress of metamorphosis (tail length = 10mm), before the slowly growing and slowly conducting axons have established neuromuscular transmission with their own slow muscle fibres, the slow muscle fibres respond to direct stimulation only with an AP of reduced size and acquired the ability to respond to acetylcholine application with a maintained contracture (indication of long-range communication).
4. With the nearly finished metamorphosis, the slow muscle fibres of the small frog (no tail remaining) become electrically innervated by their own slowly conducting motor axons and show their adult characteristic properties: the ability to respond with a maintained contracture and the inability to generate APs. The neural control of the

slow muscle fibres has fully been taken over by the small motoneurons with the thin, slowly conducting axons. The twitch muscle fibres kept their control by the large motoneurons with the thick, fast conducting axons, even though a transient reduced innervation indicates that the slowly conducting axons communicated also with the twitch muscle fibres but could not take control over them. This four-cell-communication, including competitive interaction between motoneurons, indicates that attractive and repulsive factors are necessary to generate a specific neuro-muscular innervation pattern.

5. To clarify morphologically the generation of the adult innervation pattern and especially the distances of action of short-range attraction and repulsion, morphology was performed, mainly by electron microscopy. In spite of an over-innervation of the twitch muscle fibres during metamorphosis, no degenerating twitch fibre synapses were found. However, in the small frog stage, degenerating axons were found at the nerve entry to the muscle.
6. Slow muscle fibres in the tadpole stage (20mm tail length), which had only electrical contact with fast conducting axons, showed two kinds of axon contact structures:

   a. Extended synapses with no synaptic folding, which were the synapses from fast conducting axons, transiently established on slow muscle fibres.
   b. Three kinds of axon endings, namely axons with dilatations, exploring looking like synapses with dilatations on slow muscle fibres, and pre-synaptic profiles containing sometimes synaptic vesicles. These axon profiles, making contact with the slow muscle fibres, seem to be the first occurrence of the slowly conducting axons during target invasion at a developmental stage. Here, they start to exert a trophic influence onto the slow muscle fibres but have not established synaptic transmission.

7. In the small frog stage (just no tail remained), slow muscle fibres received contacts from different adjacent synaptic axon profiles. In some cases, one synaptic profile was a part of the already functioning synapse of a slowly conducting axon, whereas the other profile was the remained synapse of a fast conducting axon.
8. The morphology of target invasion and selection indicates that for the competitive interplay of the two kinds of motoneurons with their target cells, the two kinds of muscle fibres, contact attraction and repulsion took place between the axon endings of the two types of motoneuron and the target muscle fibre. The distance of action of attractive and repulsive factors was 0.1μm or less.
9. If the human nervous system is as complex as the frog nervous system, we can expect that for a learning-based repair of the injury sites, the human CNS needs some kind of scaffold along which growing fibres can extend. The effective short-range (contact) distances of action of growing and inhibiting substances are probably also in the range of 0.1μm. For the building of specific innervation patterns, precise geographical landscapes of concentrations of interacting substances are needed.
10. A 17-year-old patient suffered a severe (motoric complete) cervical (C5/6) SCI. Upon 2.5 years of movement-based learning therapy (CDT), the sensitivity improved, some motor functions returned below injury level, and urinary bladder functions, breathing, and cardiovascular performance were cured. This indicates that

some regeneration of the cord had occurred. It also indicates that some propriospinal circuitry was repaired at the injury site, including the building of new motoneurons, since some finger functions returned upon 3.5 years of CDT.

11. Firstly, pressure and touch returned below the injury level, and later on pain and temperature. The later return of pain and temperature feeling can be understood on the basis of functional anatomy. The tractus spinothalamicus anterior, carrying pressure and touch information, was mainly spared by the injury. With regeneration, parts of the tractus spinothalamicus lateralis, carrying pain and temperature information and sited adjacent to the tractus spinothalamicus anterior, was partly repaired. It seems that the activated spared and newly regenerated fibres of the tractus spinothalamicus anterior (carrying pressure and touch information) attracted new fibre growth of the tractus spinothalamicus lateralis (pain and temperature). Target-derived growth factors seemed to have been released from the over-activated fibres and stimulated growth cones to extend along the surface of the axons of spared tract fibres.
12. Upon 3.5 years of optimal CDT, dorsal flexion of the index finger reappeared, indicating that some motoneurons were built anew. The dorsal flexion of the index finger is activated by the musculus extensor indicis, innervated by the radial nerve, ramus profundus, from the spinal segments C8-Th1, which were destroyed by the injury. According to the somatotopic ordering of the motoneurons in the ventral horn, in which the extensor motoneurons are sited ventrally, it seems that the activity in the tractus spinothalamicus anterior and lateralis, activated by CDT, attracted endogenous stem/progenitor cells for homing.
13. It is important that the patient activates all spared and newly formed fibres and premotor networks, especially those circumferential to the injury site for contact attraction and repulsion. This aids in preparing the milieu for the integration of new cells and growing of axonal and dendritic arbors by performing all possible movements, especially the coordinated and integrative ones.
14. Movement-based learning induced adult neurogenesis for repair in a human patient. It seems possible to assist endogenous stem cell reserves by exogenous stem cells.

1. Introduction

1. General Introduction

A movement-based learning therapy, called coordination dynamics therapy (CDT), has been developed and shown to be able to improve central nervous system (CNS) functioning after stroke [56], traumatic brain injury [57,66], hypoxic brain injury [65], cerebellar injury [67-69], spinal cord injury (SCI) [58,59,70-80], in cerebral palsy [63], and in Parkinson's disease [60-62].

The current research in basic clinical neurophysiology and human repair neurophysiology is aimed at developing treatment for CNS injury and malformation and to improve movements and autonomic and higher mental functions.

In a previous publication [72] (see Chapter II), it was shown that if 50% (according to MRI) of the matter of a cervical spinal cord was spared following injury, the patient could relearn walking, running, and jumping and could become continent within two to three years of therapy. The administered CDT could near totally repair this incomplete SCI at C5/6 levels mainly by a functional reorganization. On the other hand, it was shown that if the SCI was ten times more severe, meaning that 95% of the spinal matter was damaged, only partial spinal cord repair was possible. A functional reorganization was no longer sufficient for a complete repair, and the structural repair was too limited. The question is now, how can the regeneration over the injury site and the building of adult-born neurons be enhanced? Can an additionally applied exogenous stem cell therapy safely improve CDT? It seems that more specific in-vivo knowledge is needed of development and repair in animals and humans. The simple application of growth factors or stem cells may not be helpful to improve the repair of the nervous system in humans.

In this article, the neural development of muscle innervation is compared with the repair following nerve injury in the frog and the mechanisms of building a specific innervation pattern are studied. It then attempts to translate the animal data to human data by analyzing how in a patient with a severe cervical SCI, a partial repair took place and attempts to find mechanisms that are conductive for repair. As will be shown, distances of action of target-derived growth factors and electromagnetic fields of natural action potential (AP) impulse traffic from the spared fibres may be of crucial importance for a regeneration and building of new neurons. Because of the necessary close relationship between repair and specific activation of the spared spinal cord tissue (circumferential to the injury site) for regeneration and building, engraftment, and integration of adult-born neurons, an exogenous stem cell therapy can only be successful if it is administered together with a movement-based learning therapy, which stimulates already neuronal network repair, including regeneration and building of new neurons from endogenous stem cells. The exogenous stem cell therapy would just supply more stem/progenitor or other cells for the repair process by offering more cells for repair since the available amount of endogenous stem cells for repair seems to be very limited. The occurring instabilities, due to the changed network connectivity, can probably be managed by movement-based relearning of integrative movement patterns [73, 78]. The training of integrative movement patterns will help to avoid the building of spontaneously firing subnetworks, which may be the triggering points for pathologic patterns like convulsions.

The comparison between development and repair in an animal model and in a human spinal cord may lead to a better understanding of the repair mechanisms at the cellular level. In Chapter II, the necessary integrative mechanisms of repair will be analyzed further by comparing the neural development [64] and repair in humans.

2. Specific Introduction

Frog skeletal muscles possess two distinct types of muscle fibres: twitch muscle fibres, which are innervated by large fast conducting (and growing) motor axons, and slow muscle fibres, which are innervated by small slowly conducting (and growing) motor axons [19,20,42].

While twitch fibres respond with an action potential (AP) to direct stimulation, slow fibres are normally unable to do so [5]. Furthermore, slow fibres are able to maintain a sustained contracture in response to depolarizing drugs in contrast to twitch fibres, which produce tension only transiently [21].

Both the electrical and the mechanical properties of the slow fibres are neurally controlled, and after denervation, slow muscle fibres acquire the AP mechanism [25] but preserve the maintained contracture. Because the thick fast conducting motor axons regenerate more quickly than the thin slowly conducting axons following nerve injury, twitch and slow muscle fibres become first reinnervated non-selectively by the fast conducting axons [36, 37]. Being innervated by the (wrong) fast conducting axons, the slow muscle fibres loose their maintained contracture [8, 9, 40]. This phase is transient. The slow fibres lose the AP mechanism and become able to generate sustained contractures again, when slowly conducting axons re-establish normal synaptic contacts [9, 23]. The time course of contracture properties is controlled by Ca^{2+} transients [26].

Changes of functional properties seem to be a general feature during the ontogenetic differentiation of cells and cell properties [12, 13, 16]. In the newborn cat, for example, all muscles show long contraction times. Within six weeks, some of these muscles become fast (for example the flexor digitorum longus muscle), whilst other ones stay slow (soleus muscle). This differentiation into fast and slow muscle fibres is guided by the nervous system [4]. Under the influence of the innervating nervous system, the differentiation of the contractile and membrane properties of the muscle fibres in the latissimus dorsi anterior and posterior muscles in chicken follows a similar way [14, 15]. In chick embryos the "neurotrophin hypotesis" was put forward by showing the target-dependent nature of embryonic pruning of neuritic branches and cell death [46, 47, 49].

As is shown in this article, the generation of the specific neuromuscular innervation pattern is not only a target-dependent property, but is also influenced by the innervating motoneuron types. In human skeletal muscles, there are three kinds of muscle fibres (fast fatigue, fatigue resistant, and slow muscle fibres), which are innervated by three kinds of motoneurons; α_1, α_2, and α_3. These are characterized by a group conduction velocity, a group nerve fibre diameter of their axons [52, 74] and their rhythmic firing property for high excitation [54]. The developmental pattern of innervation has not been studied so far. But during repair, the establishment of innervation patterns may be partly understood by recording electromyographic activity from the different kinds of muscle fibres (see also "Indication for network and not tract fibre repair" in Discussion).

The main aim of the present investigation in the frog is to demonstrate that the neural control of electrical and mechanical properties of slow (and twitch) muscle fibres during development is similar to that following nerve injury (denervation), and that the intricate neural control changes of two kinds of muscle cells by two kinds of motoneurons cannot be extracted alone from the morphologic data at different stages of development. Functional measurements are also needed. To study in detail this four-cell communication, including competitive interactions between the two kinds of motoneurons, morphologic and electrophysiologic measurements are needed to have a full understanding of this period of frog neural development. The generation of specific innervation patterns during development better expresses the cooperative and competitive interactions among the two nerve and two muscle cells than during regeneration following injury, since mainly no disturbing degenerating nerve fibres and synapses (motor endplates) disturb the morphological interpretation. The main difficulty in this animal study was the electrophysiology. To impale two glass micropipettes into a muscle fibre of approximately 10μm diameter of a muscle of 1mm length and 0.15mm diameter, without (or only little) damaging the muscle fibres, is quite a manual task.

2. Methods

Experiments were performed on pyriformis muscles of tadpoles and small frogs (Rana temporaria) from a breeding-pond. In the course of their development, the tadpoles grow to a maximum size with a maximum tail length of approximately 27 ± 1mm. At this stage, the experiments were begun. During the following stages of metamorphosis, the length of the tails decreased, while the lengths of the bodies remained fairly constant (12 ± 1mm from head to coccyx). The forelimbs broke through the skin at about 10-15 mm tail length, and at about 20 to 15mm tail length the hind limbs had grown to a size that changed only marginally until the tail had disappeared. When the small frog had no tail remaining, the experiments were discontinued, because there no longer remained an obvious external criterion to distinguish the further stages of development. The time for the developmental period from the tail length of 27mm to 0.5mm was measured in two tadpoles to be approximately six days at 18°C. Using the comparable relationship between the reduction of the tadpole tail length and the age of the tadpole from Moser [27], an approximate age scale for 18°C, corresponding to the tail length, could be drawn (Figures 3 and 4).

1. Intracellular Recording

The pyriformis muscles were removed together with the spinal nerves IX and X [7] and a part of the spinal cord. To avoid twitching, the muscles were mounted on a polyethylene rod and stretched to about 150% of their in-situ length (for further details, see 40). Nerves and spinal cord were carefully sealed into a Perspex chamber, which contained a pair of platinum electrodes 1.5mm apart. The distance between stimulating cathode and the site of entry of the nerve into the muscle was approximately 5mm. Square pulses of 0.1ms duration were applied.

Conventional intracellular techniques for potential recording and current application were used [10]. Micropipettes were filled with 3M KCl; their tip resistance was about 100 MΩ.

The current passing microelectrode was placed at a distance of 50-200μm from the recording electrode. All muscle fibres were polarized to a resting potential of -90mV; hyperpolarizing and depolarizing current pulses were of 100ms duration. Approximate values for the membrane time constant were obtained by measuring the time of the electronic voltage deflection to reach 85% of its final value.

The temperature of the nerve-muscle bath was held between 7 and 9°C. Ringer's solution had the following composition (mM): NaCl 100, KCl 2.5, $CaCl_2$ 16; the ph was 7.3 with Tris buffer (5mM). The high calcium concentration helped to identify the slow fibres, because their effective resistance increases more than that of twitch fibres in calcium-rich media [39, 41].

2. Mechanical Recording

The isometric tension of pyriformis muscles were recorded with a UF2 mechano-electric transducer and a two-channel recorder at room temperature. Normal saline solution contained (mM): NaCl 110.4, KCl 2.5, $NaHCO_3$ 2.4, $CaCl_2$ 7.2. To elicit contracture, acetylcholine was added to reach a concentration 10^{-4} g/ml. High potassium solution (100mM) gave the same result.

3. Electron Microscopy

For electron-microscopic examination, the muscles were fixed in 2% glutaraldehyd in phosphate buffer (266m osmol/l) for three hours at room temperature. After washing with the same phosphate buffer (24 hours at room temperature) fixing with 1% osmium tetroxide in phosphate buffer and dehydration in ethanol the muscles, sectioned into three parts, were transferred through propylene oxide into liquid epon 812. Ultrathin sections were cut with a Reichert ultramicronome OMU 3, stained with uranyl acetate and lead citrate, and then investigated in the electron microscope (Siemens Elmiskop 101).

3. Results

Identification and Some Properties of Twitch and Slow Muscle Fibres and Their Innervating Motoneurons

Thirty-two pyriformis muscle fibres were successfully examined to determine certain properties of slow and twitch muscle fibres and their innervating motor axons at different stages of development (metamorphosis). The main difficulties were the softness of the tissue and the smallness of the muscle fibres. In adult frogs, twitch and slow muscle fibres can be distinguished by their membrane time constant (which differ by a factor of approximately 50) and their input resistance (which differ by a factor of approximately 6) [41].

The diameter of the muscle fibres (in the developmental range examined) varied between 2 and 20μm (Figure 6). The diameters were roughly measured under the dissecting microscope during the experiments and confirmed with measurements under the electron microscope. The fibres of the small frogs were only slightly larger than those of the tadpoles. The slow fibres and the small twitch fibres had an average diameter of 5μm in the slow fibre region of the muscle and the large twitch fibres had an average diameter of 15μm. In adult frogs, the muscle fibre diameters are between 30 and 100μm in the pyriformis muscle. The damage, therefore, produced by impaling the micropipettes into the small muscle fibres, may have been considerable, and the values summarized in Table 1 should be taken only as a rough approximation. By comparing the slow fibres with the small twitch fibres in Table 1, it is clear that the identification was not always possible from the passive membrane properties alone. Slow fibres were, therefore, discriminated from twitch fibres by the absence of full action potentials (APs) or by a time constant longer than 40ms. A few unidentified fibres were discarded.

Table 1. Membrane properties of twitch and slow muscle fibres of tadpoles during different stages of development. Measurements were performed in "Ringer" solution with 16mM Ca. By hyperpolarizing current application, the fibres were held to a resting potential of -90mV. In fibre group III only slow muscle fibres are contained, which could not generate an action potential

Fibre type	Mean fibre diameter [μm]	Initial resting potential after impalement of electrode [mV]	Membrane resistance [MΩ]	Membrane time constant [ms]	Number of fibres / Number of muscles	Tale length [mm]
I Large twitch fibres	15	-70 to -100	3 to 10	5 to 36	33/12	12 to 27
II Small twitch fibres	5	-35 to -65	3 to 18	9 to 22	10/9	2.5 to 25
III Slow fibres	5	-30 to -70	10 to 60	21 to 400	13/11	1 to 24

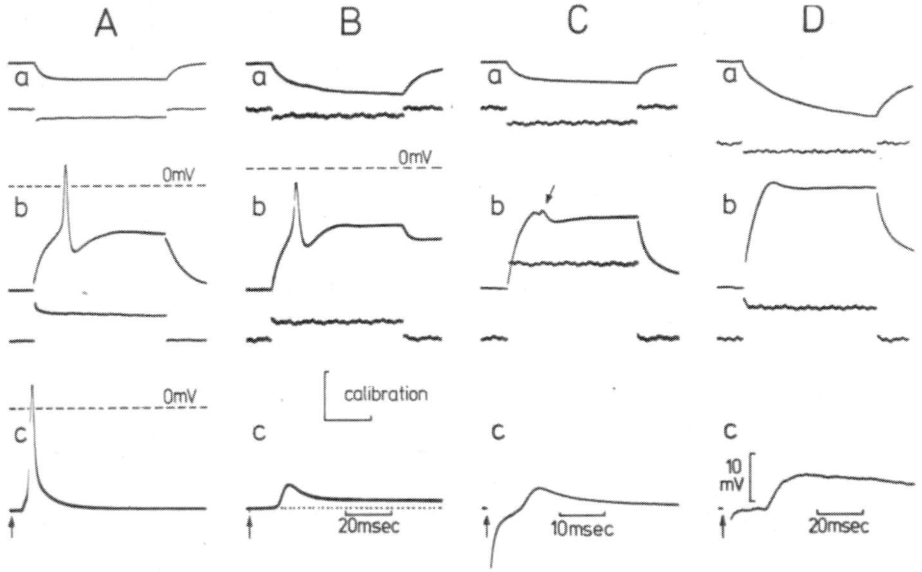

Figure 1. Intracellular recordings from a twitch muscle fibre of a tadpole with 19.5mm tail length (A) and from slow muscle fibres at three stages of tadpole development (B, C, D). a: hyperpolarizing electronic potentials (upper trace; recorded with the voltage electrode) induced by hyperpolarizing current pulses (lower trace; generated with the current applying glass micropipette) to identify the muscle fibre type; b; depolarizing current pulses (lower records), triggering action potentials (APs) in A, B, and C. c: endplate potentials elicited by nerve stimulation (arrows indicate time of stimulation) and made visible with the voltage recording glass electrode). B. Records obtained from a slow muscle fibre with 19.5mm tail length, innervated by a fast conducting axon, and responding with an AP (same muscle as records in "A"). C. Slow muscle fibre of a more developed tadpole (tail length = 11.5mm). The fibre was still innervated by a twitch fibre axon and responding with an AP of reduced amplitude, marked by an arrow. D. Slow muscle fibre of a young frog with no tail, which is innervated by a slowly conducting axon and does not respond with an AP (normal situation as in adult frogs). In record B,b (upper trace), the recording electrode slipped out of the muscle fibre after the electronic potential. Calibration: voltage 40mV; time 40ms; current 20nA for "A," and 4nA for B, C, and D; exceptions are indicated.

3.1. Electrophysiology

1. Muscle Fibres with and without Action Potentials

In tadpoles with tail lengths larger than 18mm, all twitch muscle fibres (Figure 1A) and most slow muscle fibres (Figure 1B) responded with an action potential (AP) to direct electrical stimulation.

The endplate potentials in slow fibres, however, were unable to trigger the AP. The latencies of the endplate potentials and stimulation thresholds were of the same order of magnitude for both kinds of muscle fibres. The innervating axons were therefore regarded as belonging to the same population of motoneurons. As will be clarified below, these were the fast conducting axons, which in adult frogs innervate the twitch muscle fibres only. Later in development (tail length ≈ 12mm), the slow muscle fibres responded with an AP of reduced size (Figure 1C) but were still innervated by the fast axon population of large motoneurons. In small frogs (tail length ≈ 0.5mm), the slow muscle fibres no longer produced APs (Figure 1D). The latencies of their endplate potentials were much longer, and the stimulation thresholds were at least twice as high as those for the axons innervating twitch muscle fibres or slow muscle fibres investigated at earlier developmental stages. Obviously, another population of motor axons had established synaptic contacts with the slow muscle fibres. As will be shown later, these were the slowly conducting axons of small motoneurons, which in adult frogs innervate the slow muscle fibres.

2. Twitch and Slow Muscle Fibres and Their Innervating Motor Axons during Development

The sciatic nerve of tadpoles and small frogs was too short for measuring conduction velocities.

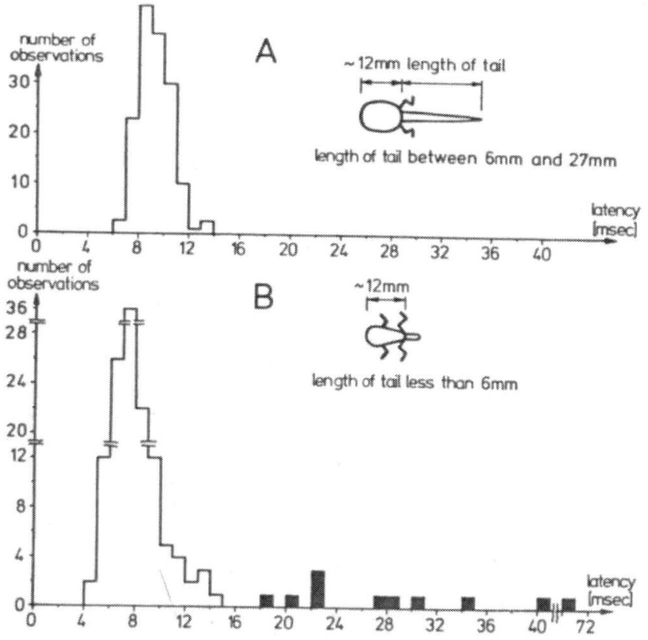

Figure 2. (Continued)

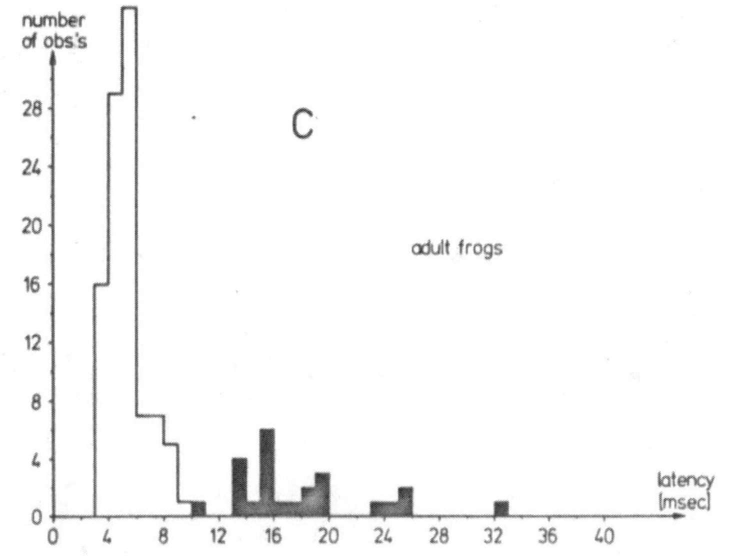

Figure 2. Latencies of endplate potentials of fast conducting axons (open columns) and slowly conducting axons (filled columns) recorded from 19 tadpoles (A), 14 young frogs with tail length shorter than 6mm axons (B), and adult frogs (C) [36].

Slow and fast conducting motor axons were therefore identified by the latencies of the endplate potentials and by the stimulation thresholds, which were approximately twice as high for slowly conducting axons as for fast conducting axons [36]. Figure 2 shows that slow and twitch muscle fibres in the tadpole are innervated by fast conducting axons (Figure 2A), which innervate only twitch muscle fibres in the adult frog. When the tadpole had transformed into a small frog (tail length shorter than 6mm), slowly conducting axons also started to make synaptic contacts (Figure 2B) but only with slow muscle fibres. The latencies of the two axon populations decreased with further development to reach the adult values (Figure 2C).

3. AP Amplitude of Slow Muscle Fibres in Dependence on the Types of Motoneuron Innervation

Figure 3A relates the amplitude of APs, recorded from slow fibres, to the tail length. It shows that the slow fibres do not respond anymore with an action potential for tail lengths shorter than 10mm, while slowly conducting motor axons were found only in tadpoles with tail lengths shorter than 2.5mm.

This means that the AP mechanism disappeared prior to the establishment of neuromuscular transmission by slow motor axons. Figure 4A shows the percentage of slow muscle fibres innervated by fast or slowly conducting axons.

Initially, the percentage of slow muscle fibres, innervated by fast conducting axons, increased with development but then decreased before the slowly conducting axons established neuromuscular transmission. Transient simultaneous reinnervation by fast and slowly conducting axons of slow muscle fibres was observed after denervation [36]. Such slow fibres may have escaped detection in this work due to technical difficulties or because the number of observations was too small.

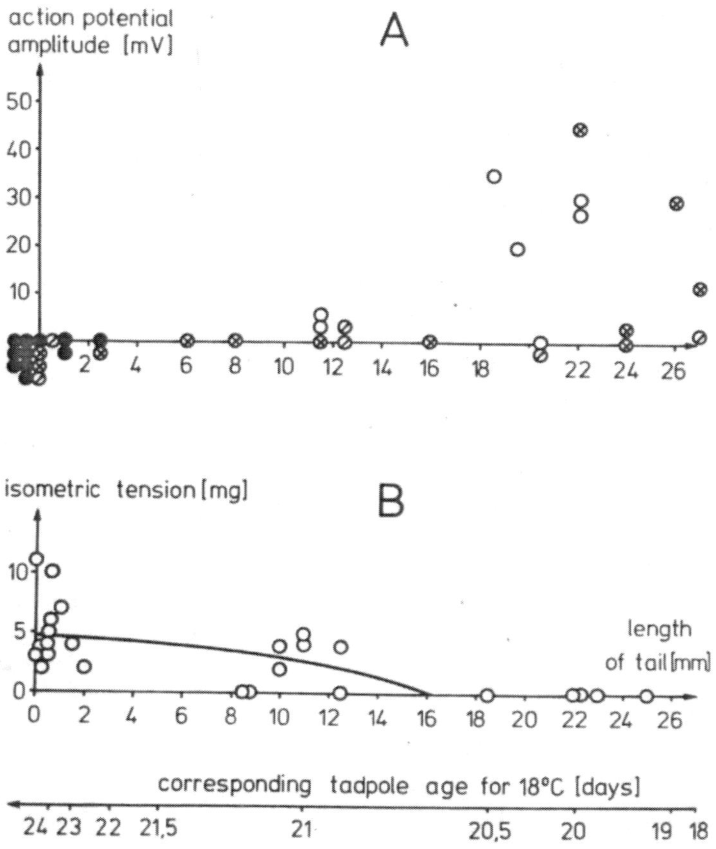

Figure 3. A. Action potential amplitude (ordinate) of slow muscle fibres as a function of tadpole tail length (abscissa). Open circle (○), muscle fibres innervated by fast axons; open circle with diagonal line (Ø), muscle fibre innervation not checked; open circle with a cross, muscle fibres that were not innervated; filled circle (●), muscle fibres innervated by slowly conducting axons (note, slowly conducting axons made electrical contact with the slow muscle fibres at tail lengths shorter than 2.5mm). B. isometric tension of pyriformis muscles measured five min after application of 10^{-4}g/ml acetylcholine. Abscissa as in "A." Line drawn by eye. Lower part of "B": Age of the tadpoles in days corresponding to the length of the tail.

4. Double Innervation of Twitch Muscle Fibres

In a high percentage of recordings from twitch muscle fibres, double innervation could be observed. In most cases, the endplate potential component with the lower stimulation threshold was too small to trigger an AP. In a few fibres, the latency of the endplate potentials with the higher threshold was shorter. In the "pure" twitch fibres region of the muscle, all muscle fibres were innervated and double innervation was most frequent during the developmental stage characterized by tail lengths of 15mm. With further development, the percentage of double innervated twitch muscle fibres decreased to a minimum value of around 10% before it increased again. This period during which few twitch fibres were only double innervated (tail length ≈ 3mm) coincides with the establishment of neuromuscular transmission of slowly conducting axons onto slow muscle fibres (Figure 4A).

This transient reduction of double innervation may, therefore, be interpreted as an unsuccessful try of the slowly conducting axons (innervating in adult frogs only slow muscle fibres) to innervate also the twitch muscle fibres.

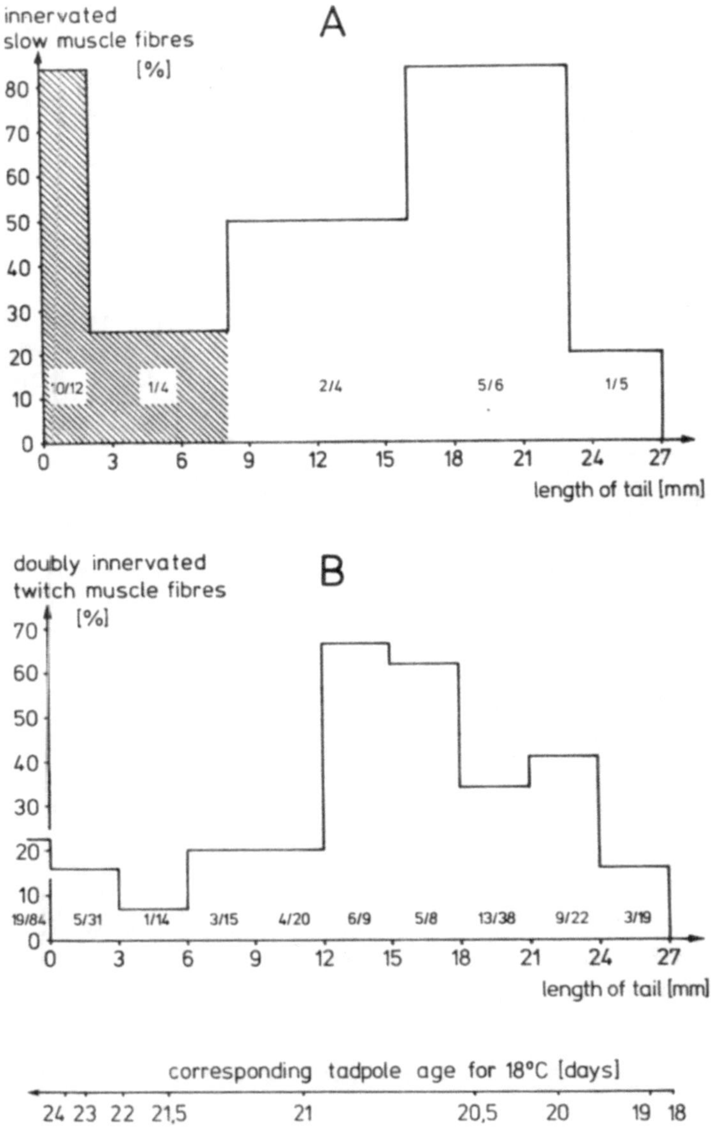

Figure 4. A. Innervation of slow muscle fibres as a function of tail length. Open columns represent muscle fibres innervated by fast conducting axons and hatched columns represent fibres innervated by slowly conducting axons. The first number in a column gives the number of innervated fibres, and the second number gives the total number of measured muscle fibres. B. Double innervation of twitch muscle fibres in the twitch fibre region of the pyriformis muscle in relation to tail length. The first number in the column gives the number of double innervated fibres, and the second gives the total number of measured fibres.

3.2. Isometric Tension

It was of great interest to establish the relationship between the ability of the muscles to respond with a maintained contracture and their functional innervation by slowly conducting motor axons. Figure 5 shows the isometric tension of pyriformis muscles elicited by acetylcholine application at different stages of development.

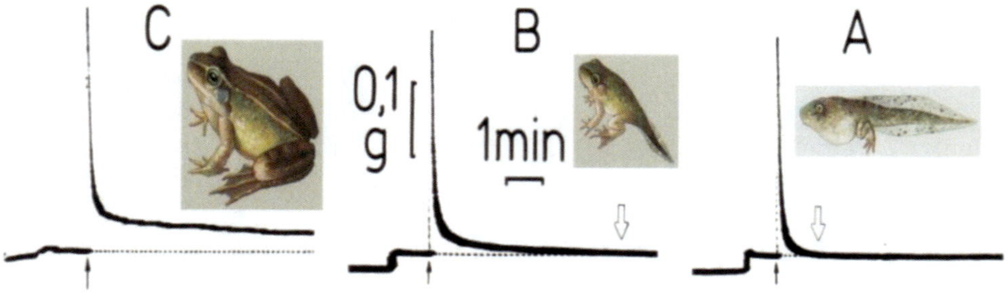

Figure 5. Isometric tension developed by a muscles of a tadpole (A) (tail length 23mm), a very young frog (B) (tail length 10mm), and a small frog (C) (no tail, body length 15mm; a stage far beyond the examined developmental period) upon application of Ach 10^{-4}g/ml (1.8mM Ca in Ringer). Small arrows indicate the time of application of Ach, open arrows the end of tension.

The muscle in Figure 5A (tail length = 23mm) developed tension only transiently, whereas the muscle in Figure 5B (obtained from a tadpole with a shorter tail length (10mm)) showed a small maintained contracture. In Figure 5C, a maintained contracture is shown from a muscle at a later stage of development. In Figure 3B, the isometric tension measured 5min after addition of acetylcholine was plotted against tail length. It was observed that the muscles started to respond with a maintained contracture when the tail length had decreased to below 16mm; this corresponds to the developmental stage when the slow fibres started to loose the AP mechanism (Figure 3A). Thus the inability of AP generation and the ability of maintaining tension developed simultaneously and prior to the establishment of synaptic transmission by slowly conducting motor axons.

Conclusions from the Electrophysiological Results

The experiments demonstrate that during metamorphosis of the frog, fast conducting axons establish neuromuscular transmission non-selectively with both twitch and slow muscle fibres as occurs during nerve regeneration [36]. At a tail length of around 18mm, twitch muscle fibres became doubly innervated at the same time when the slow muscle fibres became innervated by fast axons (Figure 4). It seems, therefore, that the fast conducting axons still have a preference to innervate the twitch muscle fibres. At a later period, at tail lengths less than 5mm, the slow muscle fibres became innervated by slowly conducting axons (Figures 3A, 4A). Twitch muscle fibres showed a transient reduction in the percentage of double innervation at that stage (Figure 4B). This observation may show that the later arriving slowly conducting (and growing) axons also tried to innervate the twitch muscle fibres but lost the competition with the fast conducting axons.

Slow muscle fibres developed a maintained contracture (Figure 3B) and began to lose the AP mechanism (Figure 3A), at a period (developmental stage) before their innervation changed from fast to slowly conducting axons. The neural control of contracture and AP mechanism of slowly conducting axons on slow muscle fibre is, therefore, exerted independently of muscle activity, because during metamorphosis, the slow muscle fibres are mostly innervated by fast conducting axons. This finding is in full agreement with the neural control of contracture, during reinnervation, which is also independent of muscle activity [9].

In slow muscle fibres of tadpoles with long tails, the endplate potential did not trigger the AP (Figure 1Bc). This can be understood if it is assumed that in that developmental stage the AP mechanism is only present in certain areas of the slow muscle fibre membrane and the endplate potential is not large enough at these regions to reach the threshold. In support of this view, it was found that in slow fibres in which the AP has been induced at localized areas by partial denervation, the endplate potential generally failed to trigger an AP [33, 35].

3.3. Morphology (Primarily Electron Microscopy)

1. Four-Cell Communication to Derive Action Distances of Grows and Inhibition Factors for the Repair of the Injured Human CNS

The electrophysiological measurements showed a functional four-cell communication during development. Two motoneuron populations (with fast and slowly conducting axons) competed for the innervation of two muscle fibre types (twitch and slow muscle fibres). Since the large motoneurons with the fast conducting axons grow faster than the small motoneurons with the slowly conducting axons, the fast conducting axons reach the pyriformis muscle first and innervate their own twitch muscle fibres. They also innervate the slow muscle fibres, which in adult frogs are innervated by the slowly conducting axons of small motoneurons.

At that stage of development, tadpoles have a long tail and forelimbs are still under the skin, but the tadpole is using the hind limbs. Tadpoles without hind limbs are just lying on the ground when they rest. However, tadpoles having already hind limbs are using them. The positioning of the legs is somehow similar to that of adult frogs. Therefore during metamorphoses, the animal is using what is possible according to the developmental stage. It uses the muscle fibres that are functioning already, even though the slow muscle fibres are innervated (for holding contracture in some similarity to the red slow muscle fibres in humans) by the fast axons and have similar properties to the twitch muscle fibres. The slow muscle fibres can morphologically be identified by their structure, which does not change during neural control changes (see below). Therefore, without functional measurements (electrophysiology, tension measurements), it would not be apparent that the slow muscle fibres have at that stage of development fast properties similar to twitch muscle fibres.

With further development (metamorphoses), the slowly conducting axons reach and invade the pyriformis muscle and communicate with the slow muscle fibres, the twitch muscle fibres, and probably the fast conducting axons for target selection. They take over control of the slow muscle fibres before their synapses make electrical contact; the slow muscle fibres are reducing their membrane excitability (decrease of AP amplitude) and start to generate a sustained contracture. The neural control of the fast conducting axons is terminated when the slow axons have built a fully functioning motor endplate (including the generation of endplate potentials) and have fully taken control. The slowly conducting axons also communicate with the twitch muscle fibres, quantified by the transient reduction of the double innervation of the twitch muscle fibres but cannot take control over of the "foreign" twitch muscle fibres.

An important question now is how the slowly conducting axons take over the control morphologically. This will give further insight into the mechanisms of cell communication. Over what distances are communication substances (including target-derived grows factors) working? Are the axons of different kinds of motoneurons (with specific intrinsic states during development) attracted by target-derived growth factors from the muscle fibres and form their synapses, or do the two kinds of motoneuron axons also compete with one another for the target (the slow muscle fibres)? In denervation experiments, transient simultaneous reinnervation of slow muscle fibres by both fast and slowly conducting axons was observed. Following denervation, the regenerating axons innervate the area of the former motor endplate, and it is therefore likely that the slowly conducting axons will try to form their motor endplates at the place where the fast conducting axons have positioned their motor endplate. Since the twitch and slow muscle fibres have a different morphologic structure and the fast and slowly conducting axons are forming different kinds of motor endplates (see below), deep functional and structural insight of this four-cell communication will be obtained. The approximate distances of growth factor actions will be measured.

The distances of action of growth and inhibition factors may give information on the structural repair during endogenous and exogenous stem cell therapy, because the regeneration follows a similar schedule. A better understanding of the homing of adult-born neurons and the specific growth of axons and dendrites may show us how to improve Coordination Dynamics Therapy (CDT) to repair more efficiently the injured, malfunctioning, or degenerating human CNS.

2. Cross section of the Pyriformis Muscle

Figure 6 shows an EM picture of two-thirds of the cross section of the pyriformis muscle of a small frog with no tail. The muscle fibre diameters vary approximately between 5 and 30μm. In the upper part, a small nerve branch can be seen with some myelinated nerve fibres. Several motor endplates are indicated by arrows.

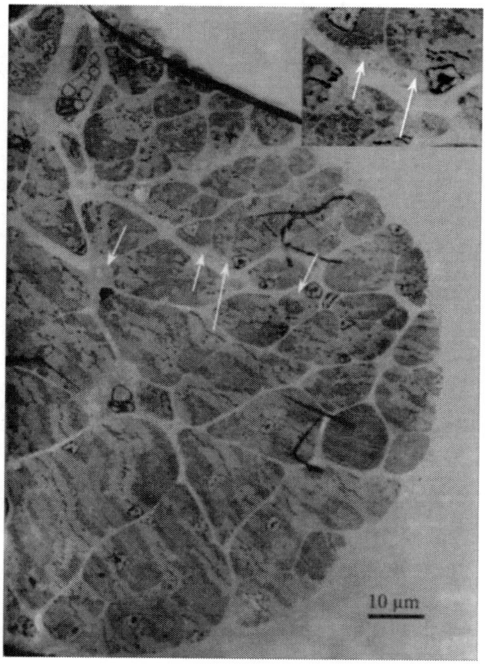

Figure 6. EM cross section of 2/3 of a pyriformis muscle of a young frog (tail length ≈ 0.5mm). The small muscle fibres (average diameter ≈ 5μm) lie in the area (upper right) where the slow muscle fibres are mainly sited. Many of the cell nuclei have a position intermediate between edge and centre of the muscle fibres. A cross section of the pyriformis muscle of a tadpole looks very similar.

At this amplification, not much information can be obtained concerning the type of muscle fibres (slow or twitch) and the kind of motor endplates (from fast or slowly conducting axons).

3. Twitch and Slow Muscle Fibres at the Tadpole Stage

In tadpoles with a tail length of 20mm, two types of muscle fibres could clearly be distinguished. One muscle fibre type was characterized by containing many mitochondria; the Z-lines are straight and there is existence of an M-line in the middle of the A-band (Figure 7). This muscle fibre type matched the criteria of twitch muscle fibres of the adult frog and was therefore considered to be a twitch muscle fibre.

In a second kind of muscle fibre, the Z-line had mostly no straight-line shape, typical M-line was missing (Figure 8), and mitochondria could only seldom be found. This muscle fibre type corresponds to the slow muscle fibres in adult frogs. In cross sections, the twitch and slow muscle fibres cannot clearly be distinguished by the form of the myofibrils [30].

4. Twitch Muscle Fibre Synapses in the Tadpole Stage

In longitudinal sections of tadpole muscles in Figure 7, two twitch muscle fibres can be seen with one synapse each. The motor endplates were longer than 60μm and can be judged as extended synaptic profiles, even though shorter endplates were found at twitch muscle fibres. Synaptic folding was often opposite to electron-dense membrane areas, at which synaptic vesicles accumulated (these are the active zones according to [6]). A comparison with the synapses of adult frogs [2, 31] shows that the twitch muscle fibre synapses had already the shape of matured ones, even though their extension was ten times shorter.

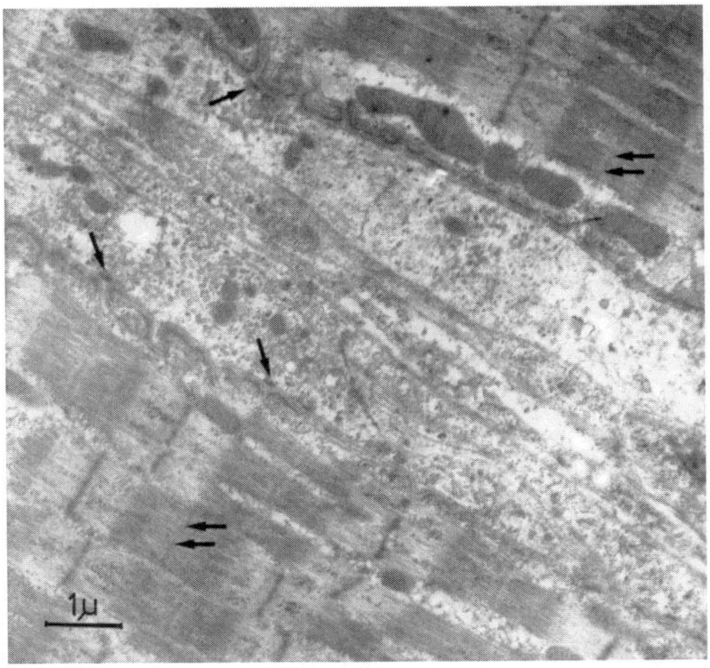

Figure 7. Electron microscopic longitudinal section of two twitch muscle fibres with one synapse each; tadpole with 20mm tail length. The synapses were over 60μm long. Three active zones are marked with arrows; two M-lines are marked with double arrows.

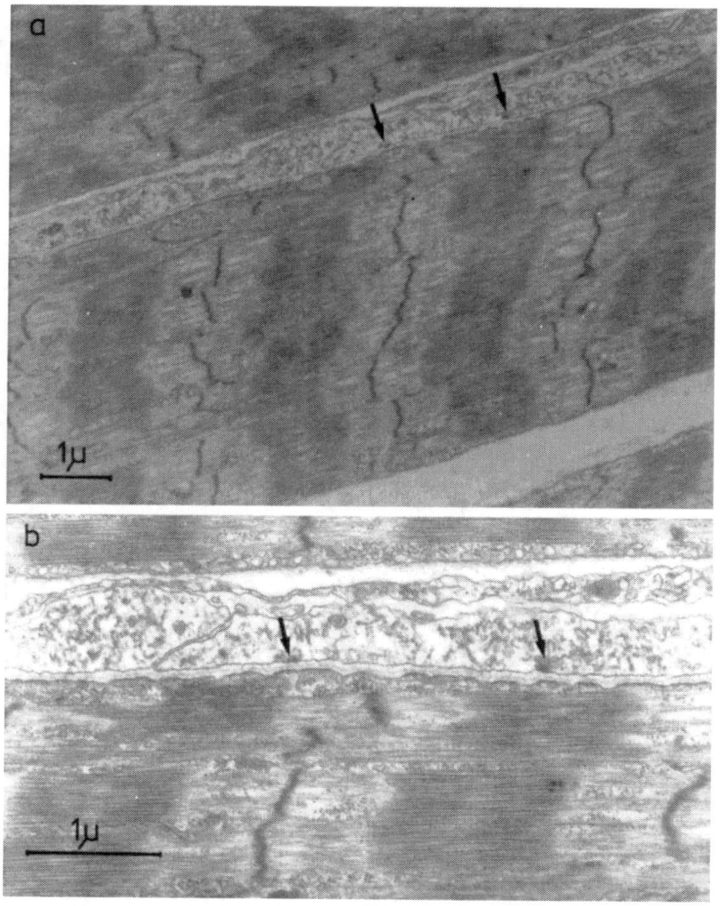

Figure 8. Electron microscopic picture of a slow muscle fibre synapse of a tadpole (tail length 20mm). The motor endplate is the nerve ending of a fast conducting axon since the axon profiles are long and have active zones, marked with arrows. No synaptic foldings. a. Nearly central longitudinal section of the synapse; some synaptic parts are still outside the picture. b. A part of "a" in a higher magnification. Note that for slow muscle fibres, typical missing of the M-line in the A-band and the not linear course of the Z-line.

5. Criteria to Identify the Synapses of Fast and Slowly Conducting Axons Innervating Slow Muscle Fibres

The large synapses shown in Figure 7 and the active zones of the pre-synaptic membranes are characteristic for synapses that fast conducting axons form on twitch muscle fibres in the adult frog. The synaptic endings of fast conducting axons can, therefore, be identified by these properties. The subsynaptic folding, on the other hand, seems to be a property that is determined alone by the muscle fibre type, since they are conserved following cutting and degeneration of the innervating motor axon [3].

The "en grappe" form of slow muscle fibre synapses (11, 22, 28, 29; Figure 9a and b) and the missing active zones are obviously typical for slow muscle fibre synapses in the adult frog. Since slowly conducting axons are building these synapses, the "en grappe" synapse form and the lack of active zones will be used in the following for the identification of nerve endings of slowly conducting axons.

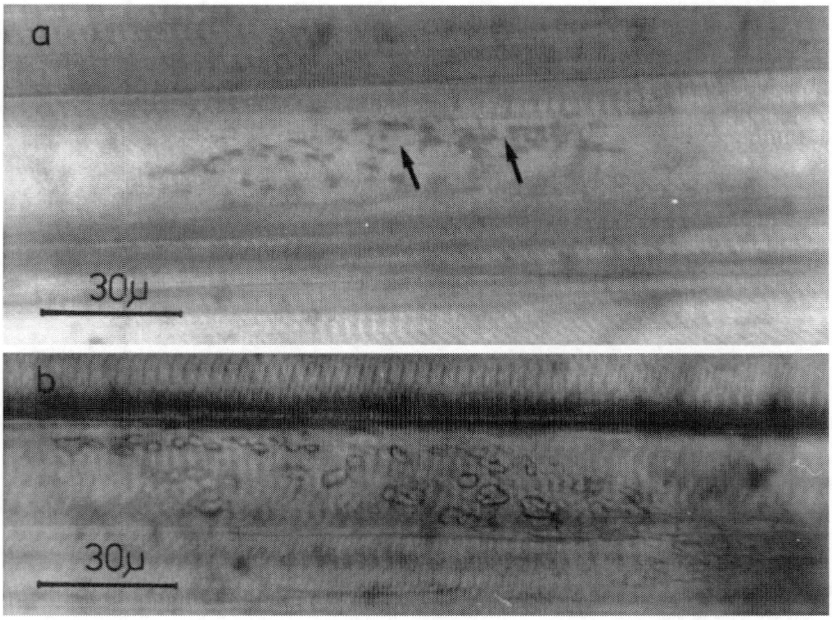

Figure 9. Light microscope pictures of slow muscle fibre synapses (nerve endings of slowly conducting axons) of pyriformis muscles of an adult frog. a. methylene blue staining; the "en-grappe" form of the nerve ending can be recognized (dark profiles can be seen of 2 to 3µm extension, which are connected by nerve fibres (marked with arrows)). b. Karnovski-staining (18); the areas of increased cholinesterase activity have "en-grappe" form and are partly larger than 3µm.

6. Slow Muscle Fibre Synapses in the Tadpole Stage of Development

In the adult frog, the synapses of slow muscle fibres are of "en grappe" type (Figure 9); they have different forms and consist of single a few µm long synaptic contacts, which are arranged in rows and complexes (28). EM examinations show that in adult frogs one to three axon profiles can be aligned with an extension of 1.5 to 4µm each [28, 29]; but not more than two axon profiles were found to be close together.

In the tadpole stage (tail length ≈ 20mm), the existing slow muscle fibre synapses differed in several characteristics to those in adult frogs. Eight slow fibre synapses were examined in longitudinal sections to get information concerning form, extension, and existence of active zones.

The synapses were longer than 10µm and showed up to five compact synapse profiles, of which at least one profile was longer than 4µm (Figure 8). In nearly all synapse profiles, active zones could be found. As Figure 10b and c show, the active zones could not be found in each slice; for it to be apparent, several slices had to be examined. The large synapse extension (longer than 10µm) and the existence of several synapse profiles with active zones allow the conclusion that these slow muscle fibre synapses were formed from the nerve endings of fast conducting axons.

In six synapses the diameters of the pre-terminal axons were between 0.15 and 0.30µm (Figure 10b and c). In centrally cut axons, microtubules can be seen with diameters of 16nm (Figure 10). In four slow fibre synapses one to two axon profiles were found, which were filled with vesicles with larger sizes than those close to active zones (Figure 10a).

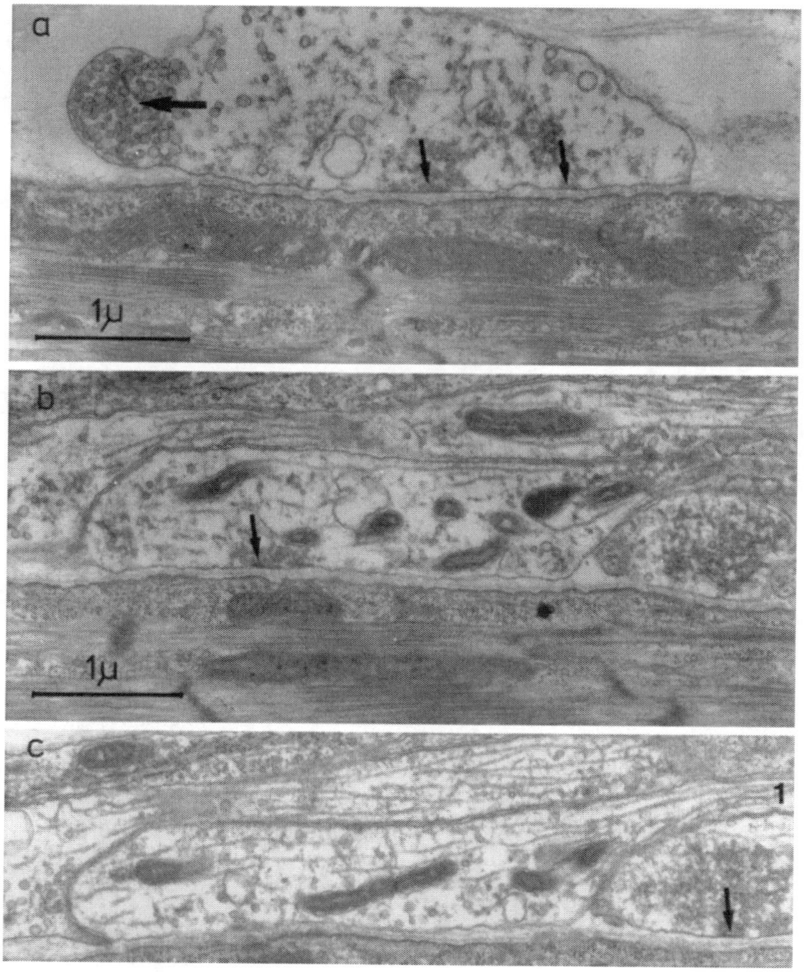

Figure 10. Electron microscopy of a part of an extended slow muscle fibre synapse of a tadpole (tail length 20mm). It is the motor endplate of a fast-conducting axon since active zones (thin arrows) are present and the profiles are quite extended. No synaptic folding can be found. a. Synapse profile with a protrusion (thick arrow) filled with large vesicles, which had no synaptic contact with the slow muscle fibre membrane. b,c. Other part of the same synapse; three axon endings from two different sections can be seen. Not in every section were active zones present. A pre-terminal axon is marked with "1."

Such vesicle structure resembled those found in degenerating synapses with vesicle aggregation following nerve section [3]. Some of the axons had in the area of vesicle aggregation contact with the muscle service, but others did not (Figure 10a, thick arrow).

Besides synaptic nerve endings, elliptic structures were found, which were surrounded by an external lamina, which contained mitochondria and formed with the slow muscle fibre a threefold membrane that is typical for synapses (Figure 11). Most likely, Figure 11 shows an axon with dilatations, which was growing along a slow muscle fibre or retracted [38, 43] and had, via the dilatations, close contact with the muscle fibre. In combination with the electrophysiological results it concluded that the muscle fibres in the tadpole stage were innervated and neurally controlled by the fast conducting axons. Furthermore, slowly conducting axons reached and invaded the muscle and started to communicate with the muscle fibres via axon dilatations through a threefold membrane for target selection.

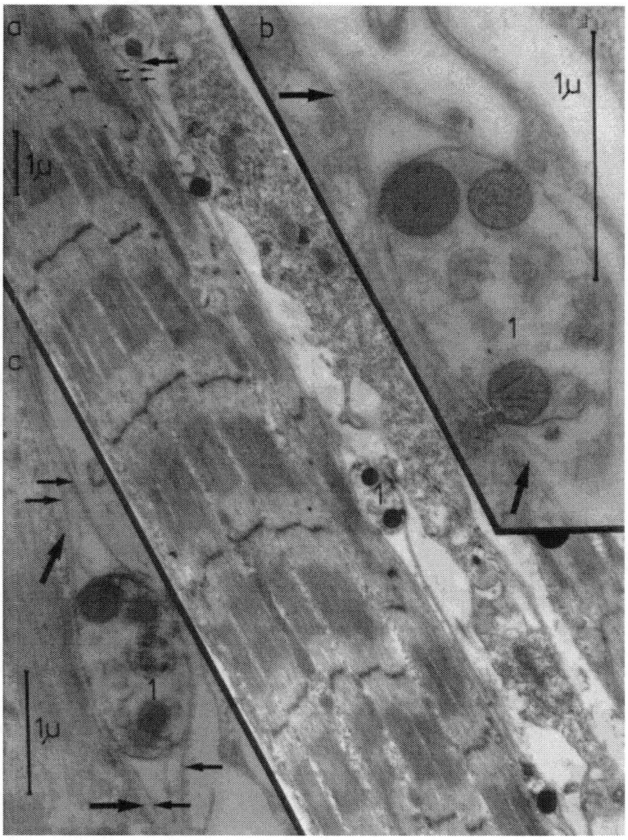

Figure 11. a. Electron microscopy of a exploring like growing axon; the axon dilatations are attached to a slow muscle fibre from a tadpole with 20mm tail length. b,c. Axon dilatation "1" of "a" in different sections at higher magnification. The axon dilatations show no microtubule but plenty of mitochondria. The basal membrane of the muscle fibre is marked with thick arrows, the lamina around the axon dilatations with thin arrows. A part of the poorly fixed axon is marked with small double arrows.

7. Slow Muscle Fibre Synapse Complexes in Small Frogs: Competition of Fast and Slowly Conducting Axons for Specific Target Selection to Build Up Adult Innervation Pattern

In pyriformis muscles of three young frogs, three (tail length = 0.5mm) slow muscle fibre synapses could be found, which were very similar in their structure. One of these synapses is shown in Figure 12. This synapse consists of three parts: an extended nerve ending, a nerve ending of "en-grappe" form, and further axon dilatations. The partly shown extended nerve ending (Figure 12b and e; marked with "1") was approximately 40µm long and had active zones. The "en-grappe" nerve ending had no active zones (Figure 12a and c; marked with "2") and was connected by a thin axon with a vesicle containing axon dilatation (Figure 12a; marked with "3").

It is concluded that the extended synapse part with the active zones was the nerve ending of a fast conducting axon, whereas the synapse profiles of "en-grappe" form without active zones was the nerve ending of a slowly conducting axon. The pre-terminal diameter of this slowly conducting axon was with 0.08 to 0.15µm (three measurements), clearly thinner than the pre-terminal diameter of the fast conducting axon at the earlier tadpole stage of development (0.15 to 0.30µm; see section: Slow fibre synapses in the tadpole stage).

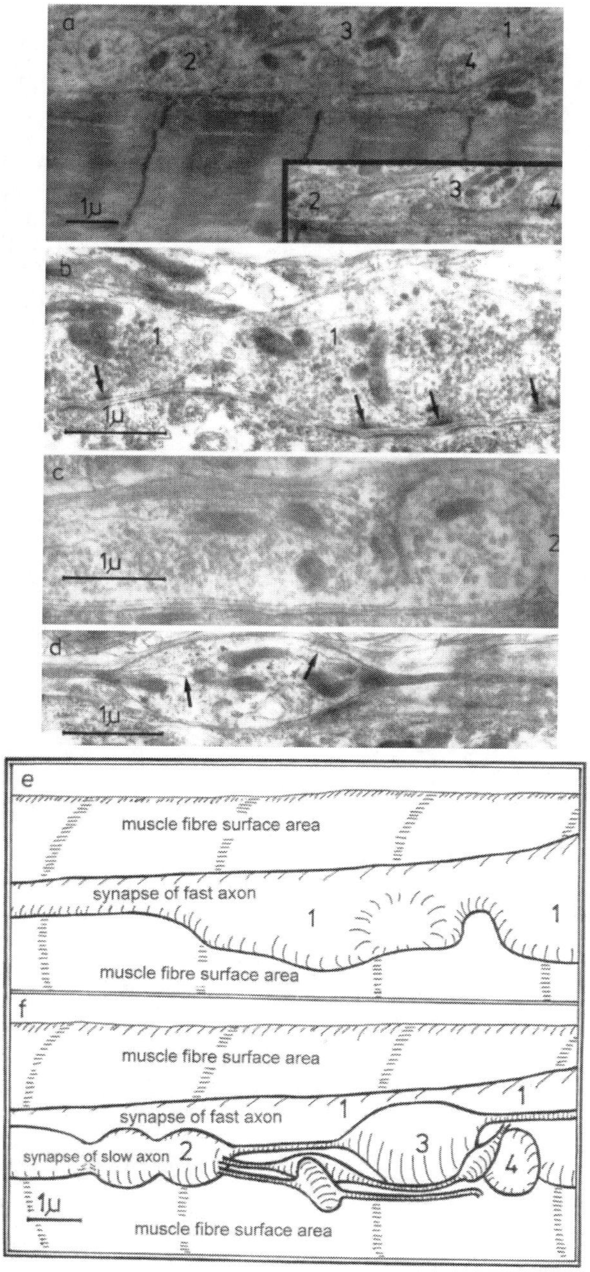

Figure 12. Electron microscopy (EM) of a slow muscle fibre with synapses of a fast (marked with "1") and a slowly conducting axon (marked with "2-4"); pyriformis muscle of a frog at the end of the metamorphosis from the tadpole to the small frog (tail length 0.5mm). By serial sections, including the sections "a" through "d," a picture of a part of the slow muscle fibre with the two synapses could be drawn ("e," "f"). a. Filaments of the slow muscle fibre show no M-line. Endplates of the fast and slowly conducting axons are strongly intermingled; both have contacts with the slow muscle fibre membrane. b. Nerve ending only from the fast axon; active zones are marked with arrows; no synaptic folding opposite to the active zones. c. Synapse profile of the slowly conducting axon; no active zones and no synaptic folding. d. Axon enlargement with marked microtubules. Three-dimensional reconstruction of the synapses of the fast and slowly conducting axons attached to the slow muscle fibre; obtained from EM serial sections, including the ones from Figure 3a-d; in "e" only the synapse of the fast axon is shown; in "f" the synapses of both the slowly and the fast conducting axons are shown.

To recognize the nerve endings of the different axons more easily of this very complex slow muscle fibre synapse with many profiles (Figure 12), a three-dimensional picture of the synapse was drawn from several serial sections (Figure 12e and f).

From the picture of the reconstructed synapse of this young frog, it can clearly be seen that this complex synapse consisted of the nerve ending of a fast conducting axon ("1") and a nerve ending of a slowly conducting axon ("2" in connection with "3").

Some axon profiles of this synapse could not be identified. Other parts were similarly complex, especially as the number of vesicle containing axon dilatations was large. Within the area of the synapse, the axon dilatations contained synaptic vesicles (Figure 12a; "3" and "4").

The axon dilatation of Figure 12d was sited approximately 100μm away from the synapse, but its axon was leading to it. It had elliptic form but was without synaptic vesicle and no adjacent axon dilatations. Further axon dilatations were found further away from the synapse; some had close contact with the slow muscle fibre (with or a without basal membrane) and some did not.

4. Discussion

1. Distances of Communication between Motor Axons and Muscle Fibres

With respect to axon guidance and stem cell therapy for repair, it is of interest to measure the distance of communication between cells. Based on communication distances, one can estimate then the distances of action of target derived factors influencing the growth and retraction of axons during the building of innervation patterns. Attractive and repulsive factors between competing neurons for target innervation will contribute to the generation of specific innervation patterns.

Geographical landscapes of diffusing growth and inhibiting factors should be simulated in realistic culture experiments and in administered exogenous stem cell therapies in which stem/progenitor cells are applied and a homing of nerve and other cells (integration of exogenously applied cells into existing neuronal networks) is intended. For generating a specific innervation pattern, the internal state of the neurons is also involved. Substances have to be transported in microtubules retrogradely to the nucleus in the soma for gene expression change, and there in the soma produced substances have to be transported actively antegradely in microtubules down to the nerve ending for axon growing, synapse formation, and functioning.

The communication distance between the nerve endings of the fast (Figure 12b) and slowly conducting axons (Figure 12c) with the slow muscle fibre was approximately 0.1μm; the motor endplates formed a threefold membrane. The distance of action between the target (slow muscle fibre) competing slowly and fast conducting axons of the two motoneuron populations was less than 0.1μm (no external membrane between the two axons) (Figure 12a and f). The communication distances between nerve and muscle cells are extremely short for this contact attraction and repulsion.

2. Early Phase of Innervation of the Pyriformis Muscle

The electrophysiological measurements demonstrate that during metamorphosis of the frog, it was the fast conducting axons of large motoneurons that established neuromuscular transmission first non-selectively with both twitch (in some similarity to FF-type (white) muscle fibres in human) and slow muscle fibres (in some similarity to S-type (red) muscle fibres in human) (Figure 2A) as occurs during nerve regeneration [36]. Both muscle fibre types responded with an action potential (AP) to direct (electrical) stimulation (Figure 1A, B) and had fast (twitch) contracture properties (Figure 5A for slow fibres). At tail lengths of approximately 18mm, twitch muscle fibres became doubly innervated (Figure 4B) at a time when the slow muscle fibres became innervated by the fast conducting axons (Figure 4A). The fast conducting axons, therefore, still seemed to have a preference to innervate the twitch muscle fibres. The type of muscle fibre was identified by the electrical time course upon hyperpolarizing current pulses.

3. Neural Control without Neuromuscular Transmission

At a later period of development, slow muscle fibres developed a maintained contracture (Figures 3B, 5B) and began to loose the AP mechanism (Figure 3A) before the (synaptic) innervation changed from fast to slowly conducting axons. Therefore, the neural control of contracture and AP mechanism of slowly conducting axons exerted onto slow muscle fibres was independent of muscle activity and neuromuscular transmission, because during metamorphosis, the slow muscle fibres were mostly innervated by fast conducting axons. This finding is in full agreement with the neural control of contracture during reinnervation, which is also independent of muscle activity [9].

The rate of reduction of the AP amplitude was in the range of 40mV per day at 18°C. The rate of development of the AP amplitude after denervation is also approximately 40mV per day at 18°C in frog slow muscle fibres [34]. Thus the rate of reduction of the AP amplitude, when the slow muscle fibres are getting under neural control of the slowly conducting axons and the rate of development of the action potential (AP) amplitude, when their neural control is lost, are similar.

This similarity may be accidental or may have its basis in the similarity of speeds for in- and ex-corporation or unblocking and blocking of Na-channels in the slow muscle fibre membrane. The acetylcholine sensitivity of muscle fibre membranes is also under neural control [24]. But there are no data available to compare the speed of spread of acetylcholine sensitivity with that of restriction.

4. Localized Changes of Membrane Properties

In slow muscle fibres of tadpoles with long tails, the endplate potential did not trigger the AP (Figure 1Bc, Cc). This can be understood if it is assumed that in that developmental stage, the AP mechanism is only present in certain areas of the slow muscle fibre membrane and the endplate potential is not large enough in these regions to reach the threshold. In support of this view, it was found that in slow fibres of adult frogs in which the AP mechanism has been incorporated at localized areas by partial denervation, the endplate potential generally failed to trigger an AP [33, 35].

5. Competition of two Types of Motoneurons for the Innervation of Two Kinds of Target Cells

At a later period of development (tail lengths less than 6mm), the slow muscle fibres became fully innervated by slowly conducting axons of small motoneurons (Figures 4A, 2B); the small motoneurons were therefore successful in fully taking neural control. The fast conducting axons of large motoneurons lost this competition against the slowly conducting axons of small motoneurons even though they had already established successfully working synapses on the slow muscle fibres.

At the developmental stage, when the slowly conducting axons took full control of slow muscle fibre, the twitch muscle fibres showed a transient reduction in the percentage of double innervation (Figure 4B). This observation indicates a communication of the later arriving slowly conducting axons with the twitch muscle fibres and a competition with the fast conducting axons for twitch muscle fibre innervation change. But the slowly conducting axons lost that competition and could not replace the extended motor endplate of fast conducting axons on twitch muscle fibres by their "en-grappe" type synapse. Therefore, growth and guidance are insufficient to explain fully the generation of this specific innervation pattern. The potentiation of some connections and the inhibition of others most likely include the changing responsiveness to target-derived grows factors over time and their reversibility [82].

6. Four-Cell Communication

It is shown that the innervation pattern is not simply formed by the growing of axons of one kind of motoneurons. When the axons of the second motoneuron population were reaching the target cells, the innervation pattern changed completely and the formation of the innervation pattern was not a simple competition; it depended on the inner state of four cells, their gene expression, and their communication via attractive and repulsive substances or molecules and receptor types. Since attraction and repulsion were exerted simultaneously among these four cells, the responsiveness of growth factors may change from attraction to repulsion (and vice versa) by additional factors. Changing electromagnetic fields probably also modulated growth, guidance, and innervation pattern.

If axons of a third population of motoneurons would have appeared at the target area and searched for their targets (as it may be the case during the generation of a neuromuscular innervation pattern in human muscles (three types of muscle fibres are innervated by three kinds of motoneurons)), the communication between the different cells would probably have been even more complicated.

7. Distance of Action of Growth and Inhibiting Factors in Frog

The question now is how this four-cell communication is taking place to build up the specific innervation pattern during development and what can we learn from it for the repair of human CNS injury. By knowing the functions of these four cells during this period of development, simultaneous performed morphology will give us further insight into the occurring structural changes. An important question with respect to repair is how close does the slowly conducting axons have to be to the muscle fibre to take control over and can the axons exert action over longer distances?

From partial denervation experiments, it is known that slow muscle fibres respond with an endplate potential to nerve stimulation, which sometimes triggers an action potential (AP) [33, 35]. This means one can have denervation and innervation properties in the same cell (muscle fibre), and these properties are localized. The endplate potential is electronically conducted from the innervated muscle fibre part to the denervated one and is there generating an AP [35]. From regeneration experiments, it is known that transiently the slow muscle fibre in rana temporaria can simultaneously be innervated by slow and fast conducting axons [36]. The gene expressions can, therefore, probably be different in the nuclei of the corresponding muscle fibre parts. But in the long term, one motoneuron type takes full control over its (own) corresponding muscle cell. These findings already indicate that the slowly conducting axons have to be quite close to the slow muscle fibre to exert their neural control and that there may be even communication among the nuclei in the target muscle cell.

The morphology made here, in combination with the electrophysiology (measuring function), offers more exact data on the distance of action of growth and inhibiting factors.

During development, when axons make initial contact with muscle cells, the muscle fibre membrane becomes refractory to further synapse formation in the surrounding of that synapse [1]. In present experiments when the slowly conducting axons reached the pyriformis muscle, they probably grew along the different muscle fibres (contact attraction), communicated with them (in similarity to the axon in Figure 11), tried to exert neural control, and searched for membrane areas of muscle fibres to build functional synapses. For communication, the axons may have formed dilatation and communicated with the muscle fibre by building trilaminar membranes like in Figure 11, where an exploring like axon with dilatations can be seen. For building a synapse with full neuromuscular transmission on slow muscle fibres, the slowly conducting axons had in the majority of cases to build the synapse at the same place where the synapse of the fast conducting axon was sited and compete with it for the target muscle cell. For the cooperative interplay of the slowly conducting axon with the slow muscle fibre, a trilaminar membrane structure seems to be built of 0.1μm thickness. For the competitive interplay with the synapse of the fast conducting axon (no basal membrane was build), the contact distance was less than 0.1μm (Figures 10-12). In conclusion, in this in vivo experiment, the distance of effective action of factors for attraction and repulsion was in the range of 0.1μm.

8. Culture Experiments (*In Vitro*) Cannot Simulate *In Vivo* Experiments

In Xenopus culture experiments, a distance of action of growth factors of 100μm from the centre of the growth cone was used [82]. Such distance is 1,000 times larger than those obtained from the in vivo experiments (rana temporaria) of this article. Obviously, the culture experiments for demonstrating nerve growth regulation by attractive and repulsive factors were with respect to the distance of action far away from these frog measurements and probably very far away from the human reality for the regeneration of the human spinal cord (see below).

9. Strong Similarity between Development and Repair in Frog, But Not in Humans

It has been shown here that in the frog, during a small period, the development of the nervous system is similar to the repair following nerve injury, even though the repair is never complete. Regenerating axons of the two types of motoneurons recapitulate developmental guidance for the specific innervation pattern of twitch and slow muscle fibres in the pyriformis muscle. Below, it will be shown that following cervical spinal cord injury in humans, the repair cannot recapitulate the development spontaneously.

CNS development and neuronal network generation have shared many similarities throughout evolution. But after the intricate process of development is completed, the CNS response to injury diverges widely among different species. Therefore the repair strategies also have to be different for different species. Following peripheral nerve injury in frogs, the repair occurs mainly spontaneously. If the sciatic nerve is cut and the proximal end put in direction of the head, many axons grow through the whole of the frog body to find its corresponding muscle. In humans, the power of regenerating fibres is much smaller. Proximal and distal nerve ends have to be adapted for possible regeneration. The mismatch [53] of nerve fibres at the nerve injury side (growing along wrongly chosen "Bügners bands") has to be compensated for by a reorganisation in the CNS, namely by learning.

Several fish species retain the ability to regenerate transacted spinal cords in the adulthood, whereas human patients with spinal cord injury remain permanently paralysed unless proper learning therapy is administered. Since such movement-based learning therapy needs to be intensive (four hours training per day) and has to be administered for a few years, the repair strategies have to be understood and optimized.

Since many experiments cannot be performed in humans, animal experiments have to be performed, and these data have to be translated into human reality as much as possible. Since the human nervous system is outstanding among different species with respect to learning and higher mental functions, we can expect tremendous complexity and variability in the human CNS.

10. Establishment of Innervation Patterns

During development, a surplus of axons initially reaches their targets. In our case, the fast conducting axons grow into the pyriformis muscle and innervate slow and twitch muscle fibres. Many preliminary synapses form only to retract after (Figure 10a probably shows a retraction bulb). Winners of this game are determined by competition for limited target-derived growth factors. Motoneurons with fast conducting axons that lack victorious nerve terminals undergo apoptosis. Degenerating axons were found in the nerve branch to the pyriformis muscle.

With further development (the developmental period considered here), the axons (slowly conducting) of a second population of motoneurons reached the muscle and competed among their own motoneuron population for synapse formation and competed with the established synapses (motor endplates) of the other type of motoneurons (with fast conducting axons) for innervation change. Losers of this four-cell competition will undergo apoptosis or will reduce the number of muscle fibres innervated by one motoneuron (reduction of motor unit size). This four-cell communication will depend on the intrinsic state of the two types of motoneurons and the intrinsic states of the slow and twitch muscle fibres (or parts of it) and activity-dependent plasticity. As Figures 10 to 12 show, the communication distance between different synaptic profiles for attraction and repulsion among these four cells is in the range of 0.1 μm.

Already, the establishment of the innervation pattern of the pyriformis muscle with twitch and slow muscle fibres shows high complexity. We can expect a tremendous complexity for the generation of the neuronal networks in the human CNS. It will be difficult to recapitulate for repair the precise growth, guidance and flexibility of the developing CNS.

11. Reduction of Double Innervation

Double innervation of twitch muscle fibres occurred more often in the considered tadpole stage of development than in the later frog stage. Such transient over-innervation has also been observed during the innervation of other muscles [1, 17, 32]. In the phase of reduction of double innervation, no safe signs for degeneration were found; the reduction of double innervation (pruning of innervation pattern) points more towards a retraction of nerve endings than a degeneration [32, 36].

Following injury, additionally degenerating nerve endings and synapses have to be resolved. The repair of an innervation pattern may be therefore more difficult than establishing an innervation pattern during development; recapitulating the development (if possible) is not sufficient for repair.

Understanding details of the building of an innervation pattern (of motoneurons) is not a theoretical game. Here is an example: Following incomplete spinal cord injury, a human patient needs at least 50% of the motoneurons for innervating a muscle to generate sufficient muscle power, because the motor unit size can approximately only be increased by 50% (by sprouting). A rat can increase the motor unit size by a factor of 10. That means, frankly speaking, rats can relearn walking only with a few spared motoneurons (with poor coordination), human patients, however, cannot. Animal experiments have very few consequences for treatment of humans.

12. Principles of Axon Guidance during Development

The features of CNS neuronal network organization and the communication with the environment are determined to a large extent by the intricate network of connections between nerve cells. In adult humans, each of over a trillion neurons makes connections with, on average, over a thousand target cells. How can this intricate neuronal network be repaired, following an injury, to generate the complexity of patterns again with the necessary precision, reliability, and stability? Since there is some similarity between development and repair, it is of interest to study the development and the repair of the nervous system, in animals and humans, to get further ideas how to repair the human CNS.

Neuronal connections form during embryonic development when each differentiating neuron sends out an axon, tipped at its leading end by the growth cone, which migrates through the embryonic environment to its synaptic targets. The axons extend to the vicinity of their appropriate target regions in a highly stereotyped and directed manner, making very few errors of navigation. They do so by detecting molecular guidance cues presented by cells in the environment. The cellular interactions between growth cones and their surroundings that direct the path are the coordinated action of multiple guidance forces that are mediated by mechanically and evolutionary conserved ligand-receptor systems.

The precise pattern of neuronal connections during development appears to involve the sequential operation of two broad sets of mechanisms: those that require activity in neurons (activity-dependent) and those that do not (activity-independent). The events of growth cone guidance and target recognition may not only be activity-independent. When axons extend along the surface of active cells (along other axons or muscle cells), the guidance may be influenced by the natural activity of the successfully established cells. Axons appear to be guided through the combined operation of four guidance mechanisms (short- (contact attraction) and long-range attraction (chemo attraction), and short- and long-range repulsion), and that the outcome of any particular guidance decision appears to reflect the balance of attraction and repulsion at the decision point. These mechanisms act simultaneously and in a coordinated manner.

The complex task of reaching a distant target is reduced to the simpler task of navigating through individual segments and choice points. The axon growth appears to be characterized by at least two types of cellular behaviours: simple linear growth along "highways," punctuated by more complex decision-making behaviours at intermediate targets (choice points), as axons switch from one highway to another.

Further, the wiring of the nervous system occurs in a stepwise manner. The first axons that develop navigate through an axon-free environment when the embryo is still relatively small, but most axons face an expanded environment criss-crossed by a scaffold of earlier projecting axons. Many later developing axons travel along pre-existing axon tracts (or fascicles) for at least some of their trajectory, switching from one fascicle to another at specific choice points. This "selective fasciculation" strategy simplifies the assembly of large nervous systems like that of humans, in which axons extend to their targets in successive waves over a period of several months.

Several guidance molecules are bi-functional – attractive to some axons and repulsive to others. Such responses are presumably dependent on the receptors expressed by the growth cones. The netrins appear to function as both long-range chemo attractants and chemo repellents for distinct classes of axons. A precise spatial distribution of netrins is important for correct directional growth in vivo. Target recognition involves selection of target region, topographic location, and discrete termination site. The secreted protein Beat, made by the motoneuron itself, might function to selectively decrease the attractiveness of some axons to others.

For the establishment of a certain pattern, there are three models. The first involves the recognition of positional information. The axon has a unique label that is complementary to another unique label on its appropriate target cell. Both because of the implausible large number of labels that would be required and because this model does not provide a mechanism for each axon to find its target, except by wandering aimlessly around, this model is not so likely, even though after cutting the sciatic nerve in the frog the growing axons do wander around in nearly the whole body of the frog. The positional information might instead be encoded in the form of gradients of signalling molecules along both the anterior-posterior (AP) and dorso-ventral (DV) axes of the target, and these gradients could be detected by complementary gradients of receptors on the axon. Positional information could thus be specified with a small number of molecules, and all axons could read positional information. In principle, topographic projections could be directed by just one ligand gradient and one receptor gradient (along each of the AP and DV axes). This mechanism requires, however, that each axon seeks out a specific concentration of ligand (a "set point," determined by the level of receptor expression) and migrates down-gradient at higher concentrations and up-gradient at lower concentrations to reach the set point. In this set-point model, the ligand acts sometimes as an attractant for the axon and sometimes as a repellent. An alternative class of models makes use of the antagonistic effects of two ligand gradients (along each axis). For example, an axon that is exposed only to an attractant gradient along a particular axis will tend to migrate all the way up the gradient, but if it is simultaneously exposed to a repellent gradient that starts shallow but becomes steep, it will migrate to that point along the axis where the repulsion precisely balances out the attraction [83].

13. Axon Guidance and Specific Neuromuscular Recognition in the Frog Experiments

The four guidance mechanisms (short- and long-range attraction and repulsion) seem to operate in all types of decisions – linear growth, sharp turns, axon fasciculation and de-fasciculation, and target invasion and selection. The present experiments to study frog development were started when the fast conducting (and fast growing) axons had reached and already invaded the target muscle (pyriformis muscle) and had innervated the twitch and slow muscle fibres mainly non-selectively; a slight preference was given by the fast conducting axons to ("their own") twitch muscle fibres.

The slowly growing (and slowly conducting) axons faced a different environment for target invasion and target selection because all muscle fibres were already innervated ("engaged"). First, they were probably attracted by the slow muscle fibres (long-range attraction) and especially to the muscle fibre surface below and around the endplate of the fast conducting axons. The partial changes of slow muscle fibre properties (see above) before electrical contact was made support the long-range attraction. But by short-range (contact) attraction and repulsion (including gene expression changes), they "pushed off" the motor endplates of the fast conduction axons and established their own "en grappe" type motor endplate. The axon branch of the fast conducting axon retracted. The slowly conducting axon and the slow muscle fibre formed a trilaminar membrane structure (as motor endplates have) (Figure 12) for contact attraction. But between the nerve endings of the slowly and fast conducting axons, there was no basement membrane (Figure 12); the contact was closer. If there was repulsion between the two motoneurons for specific target selection, then the contact distance for repulsion was shorter. The distance of contact attraction and repulsion was therefore 0.1μm or shorter. As the transient disturbance of the innervation of twitch muscle fibres by fast conducting axons indicates, the slowly conducting axon also tried unsuccessfully to replace the extended motor endplates on twitch muscle fibres by their "en grappe" type motor endplate. The inner state of the two kinds of muscle fibres and the two kinds of motoneurons was most likely different, so that finally (in the adult frog) the fast conducting axons innervate the twitch muscle fibres and the slowly conducting axons the slow muscle fibres. That the two kinds of axons had a unique label that were complementary to other unique labels on its appropriate muscle fibre types is unlikely, since after partial denervation of slow muscle fibres (in those fibres which are innervated by slowly conducting axons), the membrane properties changed only in the denervated parts of the muscle fibre. More likely is that the gene expression changed in the muscle cells according to molecules secreted from the growth cones of the two kinds of axons. This view is supported again by the partial denervation experiments of slow muscle fibres. Muscle fibres have several nuclei (controlling parts of the muscle fibres), which are partly under control of the innervating axon.

14. Modulation of Guidance Molecules and Receptors by Electromagnetic Fields Generated by Action Potentials

Netrins, semaphorins, and neurotrophins may have been involved in this highly specific neuromuscular selection process. The short distances of contact attraction and repulsion of 0.1μm raises the possibility that electromagnetic fields, changing dynamically in space and time, contributed to the generation of the specific neuromuscular innervation pattern.

It has been reported that direct-current electric fields are required for normal development, dramatically influence the rate and direction of nerve grows in vitro, and promote nerve regeneration in vivo [51]. Electric fields may work as an extracellular guidance cue. There are cell-specific responses at a given strength of electric fields. Mammalian dendrites are attracted to cathodes, whereas mammalian axons are not [20]. But the information transported in dendrites is different from that in axons. In dendrites, the depolarization is transported electrotonically (passively) from the subsynaptic membrane to the axon hillock, whereas in axons, the depolarization (the action potential) is actively conducted from the axon hillock to the endplate. Similarly, in adult slow muscle fibres, the depolarization is conducted electronically and in twitch muscle fibres actively by action potentials. It could therefore be that slowly changing electric fields contribute to the guidance of dendritic growth cones and fast-changing electric fields (electromagnetic fields) to the guidance of axons. The contribution of electromagnetic fields to the guidance of growth cones could directly and/or indirectly modulate guidance molecules and/or receptors. The electromagnetic fields may also have affected bi-functional properties of guidance molecules, namely a change from attraction to repulsion or vice versa. Expression changes of guidance molecules by the electrical activity of naturally active nerve fibres could explain why the growing or regenerating fibres extend along active axons. Activity-dependent repair needs probably not only naturally activated nerve cells but at least some kind of scaffold for the extension of nerve fibres and the homing of adult-born cells (see below).

15. Factors Promoting Regeneration

1. Growing axons seem to need mechanical support. In Figure 11, it is the muscle fibre. In the CNS, the axons, dendrites and neurons may grow or migrate along processes formed by astrocytes that have de-differentiated into radial glia or other neurons.
2. In incomplete spinal cord injury or brain injury, the surviving neurons (especially when activated) release signals that promote growing of axons and dendrites and the engraftment of replacement neurons. The short-range signals have a distance of action of 0.1μm. This short distance of cell communication is conductive for a fast diffusion of attractive and repellent neurotrophins and may give rise to sufficient strengths of electromagnetic fields, generated by APs conducted along spared axons, to attract or repel neuritis or change molecule expression.
3. Emulating the geographically precise endogenous signalling cues during development, spared nerve fibres and neurons following injury have to be triggered to re-express endogenous neurotrophins or electromagnetic fields at appropriate levels and locations to be conductive for repair.
4. Physiologic ongoing pre- and postsynaptic activity in the injured area plays an important role in guiding the axons and dendrites that extend from surviving and adult-born neurons. To generate physiologic synaptic and nerve fibre activity with their coordinated timing, physiologic and integrative movement (and other) patterns have to be trained with the patients, such as walking, jumping, and exercising on special devices (Figure 14).
5. Whereas in animal repair research it is tried to re-express developmental guidance cues to guide regenerating axons and dendrites and substitute stem/progenitor cells, in human repair research, the learning is used for functional CNS reorganization and the stimulation of endogenous stem cell reserves to partly replace lost tissue. Whether endogenous stem cells can partly replace adult neurons outside the hippocampal dentate gyrus and olfactory bulb is unclear. However, as the human example of the partial repair of the human spinal cord of below shows, some neurons can be built anew in the human CNS and integrated in the spinal cord. This finding is in accordance with the identification of radial glia as a multi-potent progenitor cell type that gives rise to both neurons and glia in the subventricular zone and other regions of the CNS, including the spinal cord.

5. Translation to Human Data

1. Limited Regeneration and Neurogenesis in the Human Spinal Cord upon Coordination Dynamics Therapy (CDT)

In the case of the repair of the human cervical spinal cord, the distance of attraction and repulsion for generating the specific neuromuscular innervation pattern in the frog and the functional anatomy-based analysis of regeneration and building of new motoneurons are used to understand the induction of index finger function upon CDT.

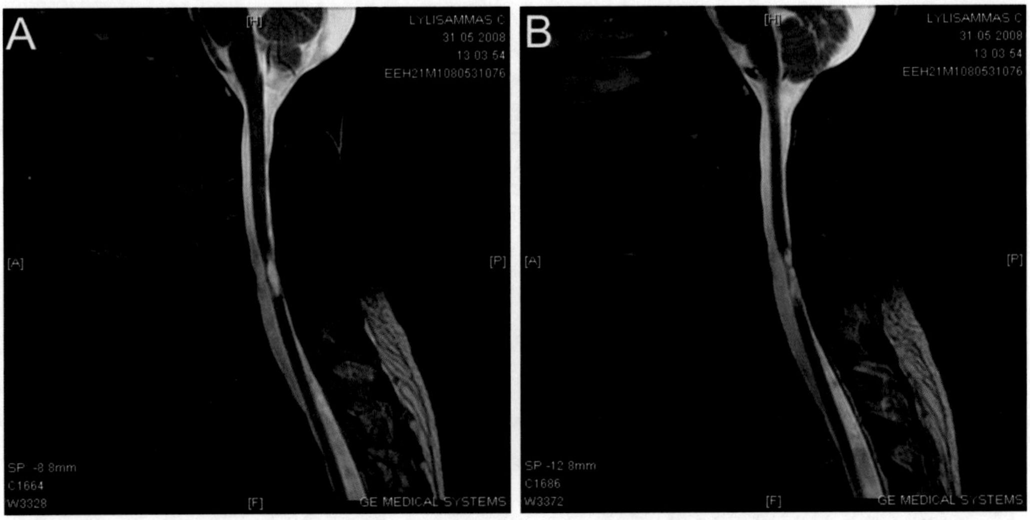

Figure 13. Magnetic resonance images of a severe spinal cord injury at C5/6 levels of a 20-year-old female patient after three years of optimal coordination dynamics therapy. Both sagittal T2-weighted images seem to show connection between the rostral and caudal spinal cord.

It is most likely that endogenous stem cell reserves were activated upon CDT and the migration of stem/progenitor cells guided to the appropriate target site in the spinal cord and stimulated to proliferate into motoneurons (and other cells) and to be integrated into the spared neural networks. Supported treadmill walking, jumping on springboard, and exercising on a special CDT-device (Figure 14) were mainly administered for the limited repair of the human spinal cord.

Case report: A 17-year-old female patient suffered a severe cervical spinal cord injury in a car accident. No motor functions remained below the injury level of C5/6, but the patient had impaired feelings. From the magnetic resonance images (Figure 13), I estimated that approximately 5% of the spinal cord cross sectional area was spared. Two months after the accident, CDT was started. Upon 2.5 years of CDT, the sensitivity improved, and some motor functions returned below the injury level, indicating that some regeneration of the spinal cord had occurred.

The connectivity over the injury site, quantified by magnetic resonance imaging (MRI) may have increased to 8%. In previous publications, I reported about the recovery of some motor functions [73] due to the limited regeneration of the spinal cord and the cure of urinary bladder functions [74-78], breathing, and cardiovascular performance [73, 78].

Here, I will analyze the repair of circuitry at the injury site to show that the strategies for understanding repair mechanisms are quite different in animal and humans and that simultaneous research in animals and humans is needed, possible and beneficial to improve treatment of human patients. The repair of propriospinal circuitry is of special interest in cervical C5/6 levels, since those motoneurons destroyed importantly innervate lower arm, hand, and finger functions. Further, to identify the repair mechanisms in the human spinal cord by the novel CDT, it is of high importance to further improve the efficiency of treatment. The repair of finger (and urinary bladder) functions has not been tackled by animal and biological research so far, as can be seen from recent review articles on the repair of the human spinal cord [45, 48, 50, 84].

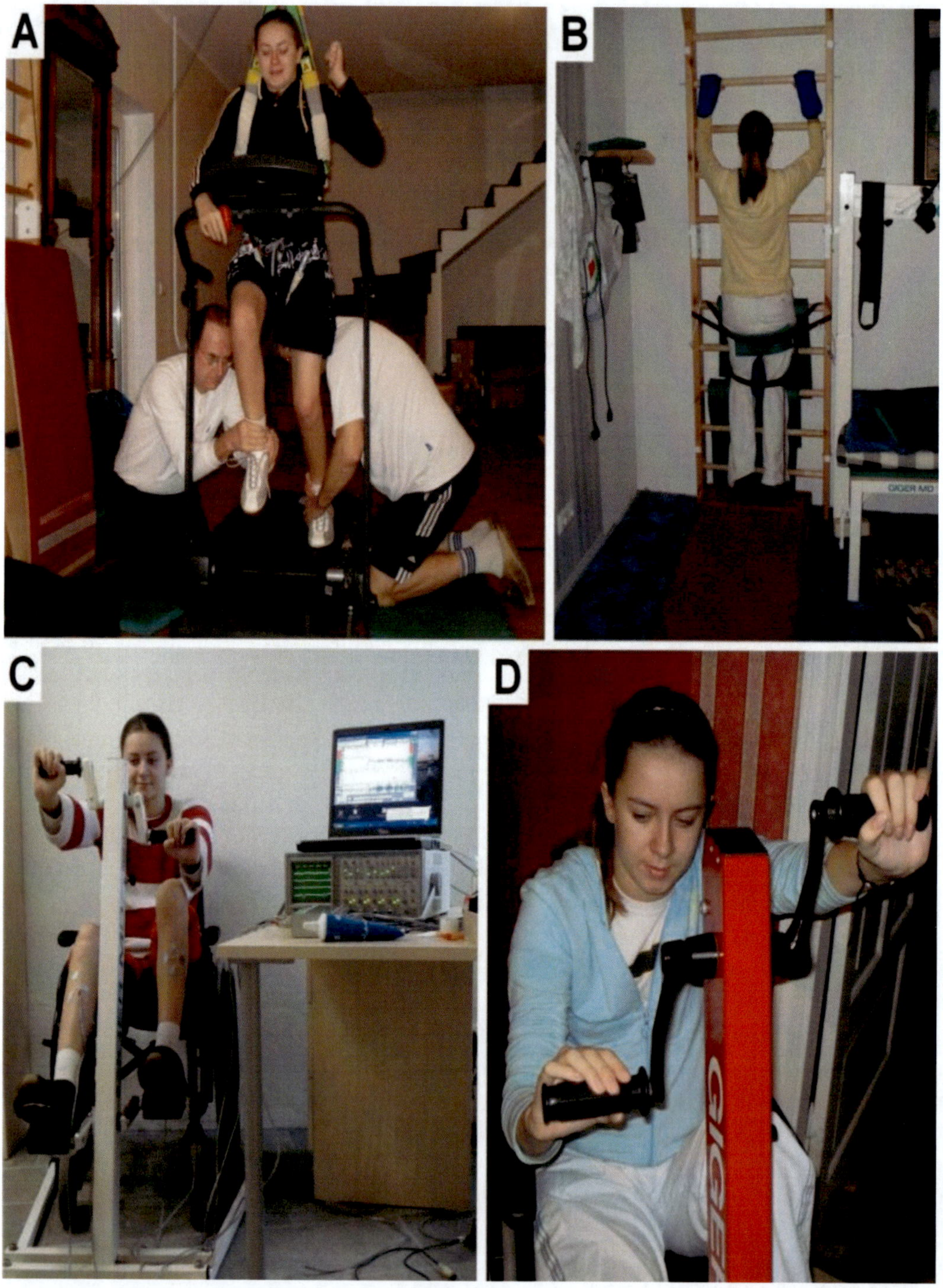

Figure 14. Supported treadmill walking (A), jumping on springboard upon no weight support (B) and exercising on a special coordination dynamics therapy device (C, D) of a 20-year-old patient with a severe spinal cord injury at C5/6 levels after three years of optimal coordination dynamics therapy. In "D" the patient is exercising finger functions besides coordinated arm and leg movements.

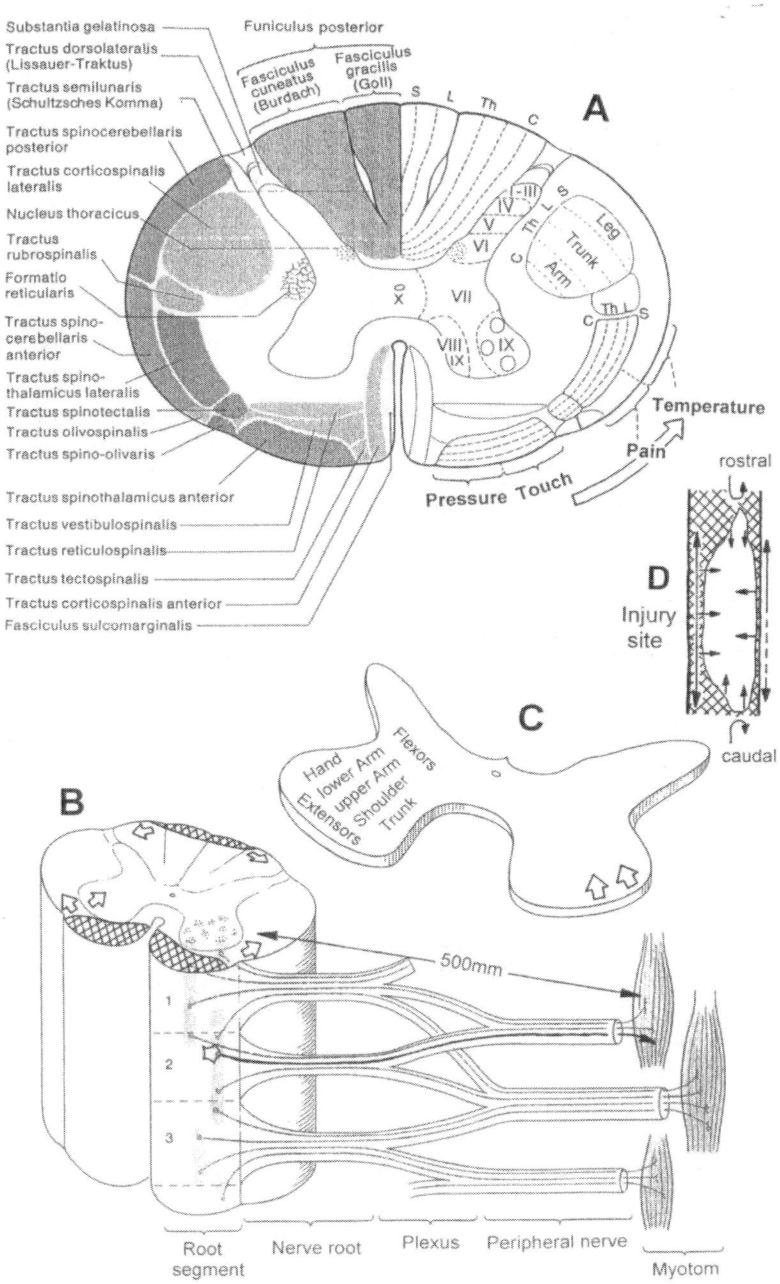

Figure 15. Schematic drawings of the human spinal cord at the injury site. A. Spinal cord cross section with fibre tracts and grey matter. Note the regaining of sensitivity upon CDT from pressure to touch, pain, and temperature is indicated by an open arrow. B. Spinal cord section with suggested spared matter (crosshatched). Open arrows indicate direction of structural repair. The axonal outgrow of a newly built motoneuron is indicated by a bended and long arrow. C. Motoneuron sites for serving different functions are indicated. Note the extensor motoneuron cell bodies are sited more adjacent to the spared spinal cord matter then the flexors and should be generated first in this case, indicated by open arrows. D. Injury site of the patient. Small short arrows indicate direction of structural activity-dependent repair. The structural repair starts from the activated spared matter into the cavity. The ascending and descending tract activity is indicated by long arrows, and the tract and network activity rostral and caudal to the injury site is indicated by bended arrows.

2. Recovery and Repair of Sensory Function in Human

Upon three years of CDT, motor functions reappeared in the rostral-caudal direction, indicating a limited regeneration of the human spinal cord. The recovery of sensory functions started in the caudal-rostral direction in the spinal shock phase, which lasted approximately three to four weeks. Two weeks after the accident, the patient felt first the Achilles tendons followed by feelings in the toes. After the termination of spinal shock, the reoccurrence of the sensitivity was rather complicated with respect to topography because of spared spot wise distribution and reoccurrence. The last location in which the patient got sensitivity back was the abdomen, at a time when the urinary bladder function was cured, two years after the accident [73].

The first sensation the patient experienced after the spinal shock was pressure and then touch. Later on, pain and temperature sensation reoccurred. The later reappearance of pain and temperature feeling can be understood on the basis of functional anatomy. In Figure 15B, the spared white matter is crosshatched pictured. The distribution of spared white matter is taken from the magnetic resonance image (MRI) of Figure 13. The later reappearance of pain and temperature may mean that new sensory fibres were crossing the injury site adjacent to the spared sensory fibres, indicated in Figure 15B by open arrows.

Looking at the cross section of the human spinal cord (Figure 15A), it seems that the tractus spinothalamicus anterior, carrying pressure and touch information, was mainly spared by the injury. With regeneration parts of the tractus spinothalamicus lateralis, carrying pain and temperature information was repaired. Growing of nerve fibres (and possibly homing of adult born cells) were probably attracted and guided by the natural activity in the spared fibres crossing the injury site. The funiculus posterior also seemed to have improved in its functioning. The spared white matter at the dorsal part of the cord cannot be seen well in the T2-weighted MR images in Figure 13. However, on T1-weighted images (not shown), some spared connections can be seen between the proximal and distal spinal cord parts.

The growing of nerve fibres across the injury site adjacent to the activated fibres is pictured in Figure 15D by short arrows pointing into the white cavity adjacent to the longitudinal long arrow in and beside the spared tract fibres. It seems that at the limit, activated spared and newly regenerated fibres attracted new fibre growth. Target-derived growth factors seemed to have been released from the over-activated fibres in the rostral and caudal direction upon exercising for repair adapted movement patterns at the limit (70). Based on the reoccurrence of the reported functions, the mechanism of sensory and motor repair can partly be understood.

3. Partial Repair of Index Finger Function

Upon 3.5 years of optimal CDT, a completely different mechanism of repair became visible: the index fingers started to function minimally. This was most likely achieved by the building of new motoneurons. The only hand functions the patient had regained within the first year following the accident (mainly spontaneous recovery) were the dorsal flexion and a weak plantar flexion.

The dorsal flexion was achieved by the extensor digitorum muscle, innervated by the nervus radialis (ramus profundus) and having its motoneuron cell bodies in the spinal segments C6-C8, which is fitting with the spinal cord injury level at C5/6 levels. The propriospinal premotor network for activating the extensor digitorum is sited just above the injury cavity (Figure 15D) in the grey matter (butterfly). The weak plantar flexion is probably caused by the flexor digitorum superficialis, innervated by the median nerve, having its motoneuron cell bodies in the spinal segments C7-Th1. The very weak muscle power is reasonable, since most motoneurons will have been destroyed by the injury.

But after 3.5 years of therapy, the patient could rather suddenly repeatedly lift the index finger first on the right and shortly later also on the left side. No other fingers could be moved separately. The dorsal flexion of the index finger was probably activated by the musculus extensor indicis, innervated by the radial nerve, ramus profundus, from the spinal segments C8-Th1. But these are the segments, which were destroyed and are now forming the cord cavity (Figures 13, 15D). Why did the extensor start to work but not another hand flexor (or the flexor digitorum superficialis got stronger), as, for example, the musculus digitorum profundus, innervated by the median nerve and ulna nerve from the cord segments C6-Th1? If we look at the somatotopic ordering of motoneurons in the ventral horn of the grey matter, we can see that the extensors are sited most ventrally, whereas the flexors are sited more in the direction of the dorsal horn (Figure 15C). When the patient is performing coordinated leg and arm move-ments on a special device, including the fingers (Figure 14D), then the tractus spino-thalamicus anterior and posterior are activated alongside the cavity (Figure 15D) and the working parts of the funiculus posterior (ascending activation) and some regenerated axons, activating trunk and some leg muscles (descending direction) across the injury site.

Because the patient is holding the handle with the fingers during the coordinated arm and leg movements (Figure 14D, right hand), the extensor digitorum is also activated and sensory input is reaching the cord from the fingers and hands. Growth factors could have been secreted from the sensory fibres in the tractus spinothalamicus anterior/posterior and the premotor network of the extensor digitorum to attract stem/progenitor cells for homing (integrated functioning with the spared neuronal networks). The attraction for engraftment is indicated by open (Figure 15B, C) and small arrows (Figure 15D). The flexors, on the other hand, are far away from supposed target-derived growth factor concentrations (Figure 15C); and further, in the middle of the spinal cord cavity, there was no scaffold where growth cones and cells could get mechanical support for growing and migration. According to this reasoning, the flexors can only come much later with further building of new propriospinal circuitry. It is therefore important that the patient and the therapist understand the therapy (including functional anatomy) and that the patient is activating all spared by the injury and newly formed fibres and premotor networks circumferential to the injury site (the cavity, the cyst) (Figure 15D). This is needed in order to prepare the milieu for the integration of new cells and growing of axonal and dendritic arbors by performing all possible movements, especially the coordinated and integrative ones (according to the System Theory of Pattern Formation [68, 73, 77] and the coincidence principle [62]).

4. Is the Index Finger Function a Separated Function of the Extensor Digitorum Muscle?

The possibility exists that the repaired index finger function is due to a separation of function from the digitorum muscle. But first, the index finger started to function suddenly.

Second, no other finger could perform a dorsal flexion separately. Third, since the dorsal flexion of the hand is generated by the last working motoneurons rostral to the injury site (in direction of caudal), the premotor network of the extensor digitorum will also have been damaged, and with a rather simple connectivity, it may not be possible to separate single finger functions from this muscle, which is probably possible in healthy humans.

5. Time Delay of 3.5 Years to Regain Index Finger Function

Why did it take 3.5 years of optimal CDT to regain some finger functions back? From brachial plexus injuries, it is known that functions start to reappear after two years. This time interval results to a regeneration speed in the PNS of less than 1mm/day (v = s/t = 500mm /700days ≈ 0.7 mm/day) for the thickest, fastest regenerating axons. To get index finger function back, new motoneurons must be built and their axons must find their way to the rootlets, may be guided by activated axons leading to the extensor digitorum, and grow along the Bügners bands to the extensor indicis and there form functional motor endplates (Figure 15B). The muscle fibres of the extensor indicis were most likely atrophied and needed another two to three months of regaining full power.

The building of new motoneurons from endogenous stem cell reserves, therefore, needed approximately 1 to 1.5 years. A further 2.5 years were needed to get function back because of the slow growing speed and the long growing distance (0.5m, Figure 15B). If some of the motoneurons innervating the index finger had survived the injury and only their axons were injured close to the soma (and the motoneurons would not have died), then axon growing times would be in the range of two years. If axons of pyramidal tract neurons (innervating arm, hand, or leg motoneurons) were injured, then already the time for the retrograde axonal transport along microtubules to "inform" the nucleus in the soma (in motor cortex) for gene expression change would need a few months (see below).

Some physicians make patients with CNS injury believe that following exogenous stem cell therapy, they can get functions back within a few days or weeks. Such hope, generated in spinal cord victims, is not supported by the data of human repair physiology.

6. Very Limited Building of Functionally Successful Motoneurons

So far only few motoneurons were successfully built anew to functionally replace injured ones. This is for a number of reasons. Firstly, only a few motoneurons seem to be built anew following injury, which is in accordance with the measured very limited neurogenesis in humans (84). Secondly, as known from treatment of spinal cord cysts, the lost spinal matter is replaced by a cyst or cavity (mostly filled with cerebrospinal fluid), with no scaffold for growing of axons and dendrites and migration of cells. Replacement of lost neurons and regeneration and guidance of projections over the entire distance of the formally injured area (and the building of migratory scaffolds from there) seem to be only encouraged along spared and regenerated tract fibres in some similarity to the extending of axon branches along muscle fibres (Figure 11).

Thirdly, there is mostly no proper guidance to the rootlet. Fourthly, there is probably mismatch [53] to find the specific Bügners bands, in this case, to the muscles activating the dorsal flections of the index fingers. In our patient, it appeared that the dorsal flections of the index finger were not the same as in healthy persons.

7. Surface Electromyography (EMG) of the Extensor Indicis and Tibialis Anterior Muscles

To further understand the repair pattern of the spinal cord in this patient, EMG was performed from the extensor indicis (rostral to the injury site) and the tibialis anterior muscle (caudal to the injury site) (Figure 16).

Single motor unit action potentials could be recorded from the extensor indicis and the tibialis anterior muscles. Even though the extensor indicis is lying below the extensor carpi ulnaris and the tibialis anterior is sited directly under the skin (80), the largest motor unit action potentials of the extensor indices had with 400μV a double as high amplitude than those of the tibialis anterior (200μV).

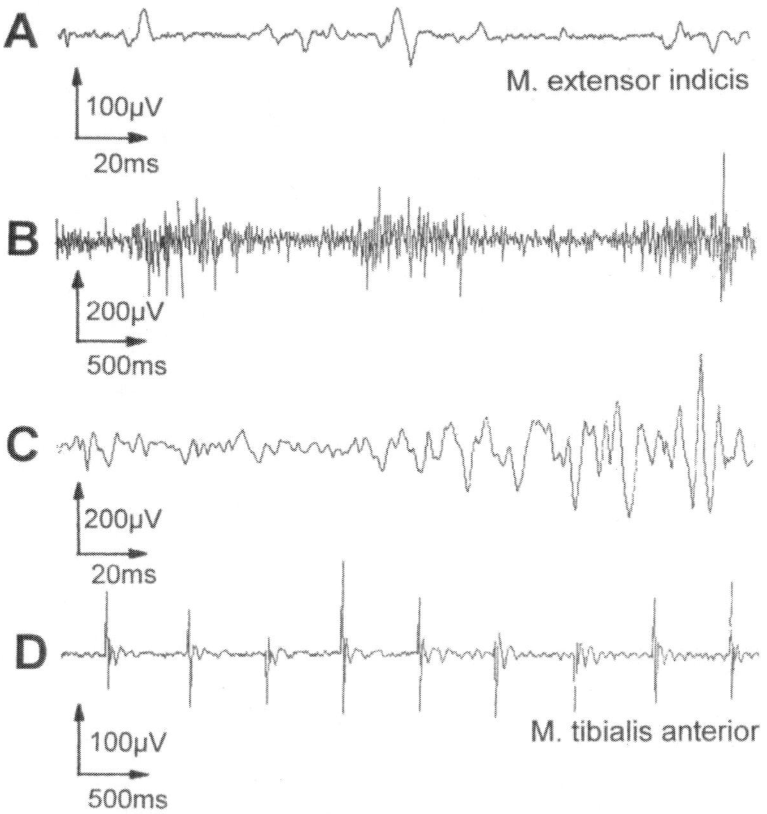

Figure 16. Surface EMG recordings from the extensor indicis (A,B,C) and tibialis anterior (D). Note the different calibrations and that the activity of the extensor indicis (B,C) has a higher amplitude than that of the tibialis anterior (D). The motor unit in the tibialis anterior fires at a frequency of 1.8 Hz; this frequency falls into the frequency range of oscillatory firing α_3-motoneurons innervating under physiologic conditions slow (red) muscle fibres.

Physiologically, a motoneuron innervates approximately 20 muscle fibres in the extensor indicis and approximately 200 in the tibialis anterior. The single motor unit action potentials recorded from the tibialis anterior muscle should therefore be much larger than those from the extensor indicis, namely over 1,000μV (Figures 4 and 5 of [79]); but they were not. This unphysiologic small amplitude of the motor unit action potentials of the tibialis anterior and the earlier start of functioning of the tibialis anterior give further insight into the regeneration process taking place in the spinal cord.

8. Indication for Network and Not Tract Fibre Repair

For the volitional activation of the tibialis anterior, some regeneration of the spinal cord was necessary. Some axons had regenerated over the injury site to reconnect the distal spinal cord part. The very limited regeneration of the spinal cord was faster than the building of new motoneurons and the growing of their axons from the grey matter to the rootlets and then via the Bügners bands in the median nerve to reinnervate the musculus indicis (Figure 15B) to induce finger function. But why did the recordings from the tibialis anterior muscle show only small motor unit action potentials, which cannot be explained only by atrophied muscle fibres?

Human muscles have FF, FR, and S-type muscle fibres, which are innervated by α_1, α_2, and α_3-motoneurons, respectively [55]. The large, fast conducting α_1-motoneurons innervate the FF-type muscle fibres, show large motor unit action potentials and are integrated only in small networks. The thick tract axons from the cortical pyramid cells will preferentially innervate these α_1-motoneurons to generate high muscle power quickly. The FR-type motor units generate smaller motor unit potentials than the FF-type motor units and the innervating α_2-motoneurons are integrated in larger networks and have specific premotor networks, generating for high activation stable oscillations. The S-type motor unit action potentials are most likely small; the innervating α_3-motoneurons can be activated poly-modal and are integrated in large neural networks.

The small motor unit action potentials, recorded from the tibialis anterior, indicates, therefore, that the recorded activity was mainly generated by the activated smaller α_3, and α_2-motoneurons, integrated in larger and more complex networks. An oscillation frequency of 1.8 Hz of a motor unit action potential in the tibialis anterior muscle (Figure 16D) supports this view that α_3-motoneurons fire oscillatory in the range of 1.8Hz [54]. (A change of the commitment of motoneurons, due to a partial denervation, cannot be excluded). This means that the regenerating axons over the injury site were not giving rise to a partial regeneration of the pyramidal tract but contributed to a reinnervation of the neuronal networks caudal to the injury site. So far, after 3.5 years of optimal CDT, the repair was probably mainly achieved by network repair (and not by the repair of tracts), including the building of a few neurons, and a substantial functional reorganization for generating physiologic functioning. Further, a regeneration of the axons of the pyramid cells over the injury site down to the motoneurons of the tibialis anterior, sited in the lumbar-sacral cord segments, would probably take much more regeneration time (if possible at all) if one assumes a regeneration length of 150mm from the cervical to the lumbar-sacral spinal cord. Already the retrograde information transport in the microtubules in the axons for stimulation gene expression in the DNA of the nucleus would need a few months, if a retrograde transport velocity of 1.5mm per day is assumed, and the proximal axon length would be 150mm (t = s/v = 150mm/1.5mm/day = 100days). For the repair of the spinal circuitry, the distances for information transport and regeneration are much shorter (only the neurogenesis times may be long), which makes it likely that primarily networks were repaired for improved functioning.

The regeneration of the motoric complete cervical SCI (sensoric incomplete) was so far achieved by the growing of some axons over the injury site (including the building of new neurons and forming and activating new pathways), combined with a substantial functional reorganization to regain physiologic functions. At this research stage of spinal cord repair in humans, it is unclear whether the regeneration-associated genes have the possibility to partly restore original spinal cord structure or not. But for regaining finger functions, new motoneurons are needed; a network repair by growing of axon and dendrite arbores cannot or can only partly substitute for lost motoneurons.

9. Physiologic Distribution of Factors for CNS Repair

The presented details of four-cell communication during frog development demonstrate how intricate the geographical landscapes of factors are, and probably have to be structured, to guide growing axons to generate physiologic innervation patterns. Distances of action in the range of 0.1μm to achieve physiologic concentrations of attractive and repulsive factors (secreted from target and growing cells) and/or effective strength of electromagnetic fields cannot simply be simulated by adding grows factors during or after an operation.

Also the inhibition of the inhibition of regeneration [81] to enhance repair in SCI has nothing to do with the building of a physiologic environment at the injury site to be conductive for the generation of physiologic innervation patterns. Repellent neurotrophins and inhibiting pathways are necessary to generate physiologically structured innervation patterns and physiologic functioning. Mass contractions are not helpful for patients.

In the upper frog model, blocking the repellence of fast conducting axons from slow muscle fibres would probably result in the generation of chaotic patterns of multi-innervated slow and twitch muscle fibres. Missing inhibition in Parkinson's patients results in tremor [61], trunk spasticity and other deficits. The blocking of inhibitions will result in mass contraction, as, for example, some kind of spasticity in human CNS injury.

10. Altered Signalling Environment Following Injury in Comparison to Development

If it would be possible to recapitulate development for repair, the milieu for repair at the starting point of recapitulation would be different from that for development. During normal neuro-development, there are also variations of signalling cues, but those variations are probably small in comparison to the changes of the environment following severe injury or malformation. Also in the frog, the quality of repair to reinnervate the pyriformis muscle depends on how severely the sciatic nerve was injured. Cutting the sciatic nerve and placing it in the direction of the brain will result in a much lesser repair than following only squeezing the sciatic nerve. The repair of the CNS in lizards following disconnection of the tail from the rest of the body, when trapped by the tail in dangerous situations, is incomplete, even though the separation of the tail when caught is an innate escape mechanism.

The repair by movement-based learning (CDT) also depends on the kind of injury, the lost CNS parts, and the severances of the injury in general. Therefore, during relearning, functional aspects of the anatomy have to be taken into consideration to adapt the therapy to the kind of injury to enhance the rate of learning. For the repair of the human nervous system, human research is needed or the animal data have to be translated first to human (as is tried here) before using them in repair treatment for human patients. Partial re-establishing of the developmental pattern of guidance molecule expression could contribute to the regeneration of the human CNS. But the proper geographical landscape of attractive and repulsive guidance molecules or neurotrophins cannot be achieved by simply adding growth factors as is sometimes done after plastic surgery.

11. Adult Neurogenesis for Stimulating Plasticity

Movement-based learning has been shown to induce the building of new motoneurons in the human spinal cord. It is likely that stem/progenitor cells also differentiated into other neurons and supporting cells to improve propriospinal circuitry. Adult neurogenesis is important in general because the newly built neurons are, at least temporarily, immature and probably more plastic to support learning. The immature neurons are not that much differentiated and can form more easily new connections, in a given period of time, than mature neurons. By being structurally more plastic, the new neurons are highly susceptible to changes in the environment and to different movement-based learning protocols. Since new neurons are only immature and strongly plastic in a given period of time, CDT should not be discontinued for longer periods of time. Practical experience is that with a longer break of CDT than three days the performance of learned movements decreases again. The site of origin of adult-generated cells in the CNS is unclear. One possibility is the subventricular zone, which is the source of new olfactory bulb neurons in the monkey. Also the migration ways are unknown. However, movement-based learning enhances the rate of neurogenesis and the survival of new neurons.

12. Stem Cell Therapy

Since exogenously applied adult-born stem/progenitor cells are similar or the same as endogenous stem/progenitor cells, it is likely that movement-based learning therapy can control the differentiation also of the added cells into different kinds of neurons and glia cells and can improve the impaired stability of integrative neural networks during functional and structural repair. Stem cell and coordination dynamics therapy of a patient with a complete spinal cord injury was not successful [71], probably because the spinal cord injury was complete (no spared tract fibres and scaffold for regeneration and homing of cells) and the CDT was not optimal before and after the stem cell therapy. But most spinal cord injuries are incomplete. In brain and brain stem injuries (especially in young children), the repair does not depend that much on spared fibres and tissue as in spinal cord injury.

It is unclear whether CDT can control neurogenesis from embryonic stem cells, since repair cannot recapitulate development in humans and cannot generate the necessary environment for physiologic growing.

13. Angiogenesis

Brain and spinal cord pressure do not only damage nervous tissue directly but also indirectly via reduced blood (oxygen) supply. Blocked or destroyed blood vessels result in an infarct of parts of the CNS (brain, brain stem, spinal cord) if sufficient blood cannot be delivered through anastomosis [89]. The repair and functioning of the CNS needs sufficient blood supply for the microcirculation; it needs angiogenesis. In animals, lack of exercise-induced angiogenesis is not a rate-limiting factor for neurogenesis (86). Exercise increases hippocampal neurogenesis and improves learning.

In the dentate gyrus, new cells are clustered close to blood vessels and proliferate in response to vascular growth factors. In young adult animals, exercise increases endothelial cell proliferation, vascular growth factor levels, and angiogenesis throughout the brain. In the dentate gyrus, the effects of exercise on the vasculature may also be important for enhancing neurogenesis. However, enhanced fitness in runners is probably not the main reason for improved learning. Movement performance is more indicative of recall ability than fitness of the subject. Stressful experience decreases the number of new neurons in the dentate gyrus [85], which means that the vegetative nervous system is also involved in the regulation of cell proliferation. There appears to be no simple relationship between the number of new neurons and learning [86].

In the spinal cord injury patient of above, there was a repair of the cardiovascular performance: the skin was no longer vulnerable, and no pressure ulcers occurred. CDT is a human science-based learning therapy, which includes the powerful learning tools of improving the coordinated firing of neurons and learning transfer [77]. Fitness is a by-product of this intensive training and offers the possibility for the patients to train at their limits (range of overreaching [70]) to induce all repair mechanisms. That enhanced fitness in animals is not the main reason for learning and that there is no simple relationship between the number of new neurons and movement-based learning is supported by the human data of this report, because different learning strategies were used (partly shown in Figure 14) to induce neurogenesis and angiogenesis in the spinal cord by learning (which cannot even be simulated by animal research).

14. Link between Adult Neurogenesis and Learning

During the past several years, evidence has accumulated suggesting a relationship between newly born cells in the hippocampus and various types of learning and memory [87]. The involvement of the hippocampus in learning and memory has long been recognised. It is usually assumed that synaptic plasticity within the hippocampal formation contributes to the acquisition and retention of memories, but the exact mechanism remains unknown. It seems that the dentate gyrus of the adult hippocampus produces new neurons in substantial numbers and does so in a wide range of mammalian species, including humans [88]. Some studies have shown that learning tasks do not require the hippocampus but nonetheless activate or engage it and do not change the number of new granule neurons in the dentate gyrus of the hippocampus [90].

In this article, evidence is given that adult neurogenesis in the human spinal cord is achieved by motor learning in which the hippocampus is not directly involved. The acquisition and retention of movement patterns is so strong that it is also supported by the building of motoneurons and probably other neurons. Learning-induced structural repair can probably be achieved throughout the human CNS. The specificity (finger function) and integrativity (coordinated arm, leg, and trunk movements; Figure 14D) of the performed movements combined with functional anatomy-based activation of the environment of the injury site (Figure 15) gave rise to neurogenesis and acquisition and retention of the learned movement. It seems therefore that the relearning of the index finger function induced the building of new neurons (motoneurons) in those networks in the spinal cord that were directly required for the movement, in some similarity to the learning in the hippocampus (see above). The effect of various learning tasks on the repair of certain CNS injuries and network structures has to be further explored.

As has been shown earlier, the learning of a new pattern disturbed the organization of the CNS networks (Figure 3 of [73]), but by continuing CDT, CNS instabilities could be repaired; newly learned and spared movement patterns were not destroyed. That adult neurogenesis is a vestige of development, which, in adulthood, has no functional significance, is extremely unlikely. It seems that by understanding our neuronal network organization (including functional anatomy), which generates learning, memory and movements, we can alter CNS networks – the spirit can rise above physis (the limiting factor is the time).

15. Motor Learning and Memory for Neural Repair

Long-term potentiation (LTP) has been proposed as a model of learning and memory. The sensory cortex is required for the acquisition of new motor skills. LTP can be induced in the motor cortex with stimulation of the sensory cortex or associatively when stimulation was applied to both sensory cortex and thalamus. It was proposed that motor learning involves the formation of loop circuits between the motor cortex and the periphery involving the sensory cortex and the thalamus [91]. Loop formation has also been measured in humans; the loop formation was extracted from coordinated, natural firing patterns of α- and γ-motoneurons and secondary muscle spindle afferents [77]. First, the loop circuits are diffuse but become specific by producing LTP through practice. LTP participates in the execution of learned motor skills by maintaining the efficacy of synaptic transmission in selected circuits. This short-term memory, or working memory, refers to memory formed and retained for a relatively short time, from a few minutes up to several hours. This learning in the short-term memory probably occurs when during CDT, the patient is exercising at intervals (crawling or walking) of one to three minutes. During the intervals for rest, the efficacy is only reduced marginally because of the short-term memory and the patient with CNS injury has some rest for the overloaded muscles and neural networks to recover. With each further crawling, the performance improves until the patient is getting exhausted and the performance is decreasing again due to exhaustion.

The cerebral cortex (together with other brain parts) consists of multiple loop circuits, and the function of the cerebral cortex depends on the interaction of these various loop circuits. Some of these circuits become progressively more specific. With the interaction of hierarchies of loop circuits, we may understand the dynamic organization of a motor behaviour, in which motor learning is generated more or less by all the contributing subsystems of the CNS.

With ongoing therapy over weeks and months, the repeated usage of specific synapses results in proliferation of these synapses; the changed efficacy of synapses becomes permanent. Further structural changes can be achieved by synaptogenesis, which has been demonstrated in the mature brain. For the acquisition of novel skills (such as playing piano) or relearning of motor patterns in patients with injury-deficient neural networks, further structural changes are needed; the establishment of new pathways by growth of terminal dendrictic and axonal arborisations and building of new synapses is needed. Long-term memory may be retained indefinitely. The modification of existing damaged motor pattern requires learning through feedback information. Repetitive activity of the motor cortex changes the excitability of not only the motor cortex but also of the spinal cord. Motor learning is based on increased excitability of selected efferent zones by circulation of impulses in selected loop circuits [91]. If the growth and ramification of axons and dendrites, together with their myelination, and building of new synapses is not sufficient for learning a motor skill or relearning a motor pattern in injury-deficient neural networks, then new neurons are needed. In the patient with the cervical spinal cord injury, the changed excitability of the cervical spinal cord networks was not sufficient: new motoneurons were needed and built. But those motoneurons were built first by motor learning-activated spared networks. In the frog, during metamorphosis, the change of the innervation pattern was accompanied by the movement of the hind limbs (before the forelimbs broke through the skin). When the muscles in the growing hind limbs became innervated by the fast conducting axons, meaning that the legs became operational, the tadpole was using them, as can be seen every year if one takes tadpoles from a breeding pond to an aquarium. Adult-generated cells may have unique properties that increase their impact to more mature neurons during learning and repair. Immature neurons may form more new connections, and their axons and dendrites may grow over longer distances in a given period of time than mature neurons. Is gene expression in new neurons affected by learning and does it differ from mature neurons?

16. Motor Behaviour and Motor Learning Are Products of All Contributing Subsystems

It was proposed that motor learning involves the formation of loop circuits between the motor cortex and the periphery involving the sensory cortex and the thalamus [91]. In a system theory of motor development and repair (Chapter II), sensory and motor inputs are not identified as separate entities but rather as one system having an impact on both input and output. Sensory pathways do not transmit information to the brain, where it is received, acted upon, and sent out as a motor component. Rather, sensory and motor influences work together nested within each other, simultaneously filtering, assessing input, and contributing to the output. Afferent and efferent pathways are made up of both sensory and motor components and cannot be partitioned into separate compartments [92]. This system theory of motor development (and repair) is supported by electrophysiological measurements of phase and frequency coordination among single afferent and efferent nerve fibres (Chapter III) for pattern self-organization [74-76] and learning transfer in humans [77]. When exercising on the special CDT device (Figure 14B,D), sensory and motor influences are trained together for movement-based learning and relearning. This system theory of motor development and repair will be tackled further in Chapter II, when the human neuro-development is compared with the repair of the human CNS.

Conclusion

It has been shown that there is strong similarity between neural development and repair in the frog peripheral nervous system. There is nearly no similarity between development and repair in the human CNS; repair cannot recapitulate development. Therefore, unlike many similarities across species in patterns of neural development, responses to injury differ greatly between frogs and humans. But if animal data are translated into human data (as is tried in this article), human treatment research [52-80] may get substantial support from biological research.

The building of a specific neuromuscular innervation pattern in a frog related to the functional and structural repair in a human patient with a cervical spinal cord injury indicates that the simple adding of growth factors cannot support the repair of the human nervous system. Administration of exogenous stem cells may only be helpful for structural repair (including the building of new neurons) if the patient has previously obtained CDT (or a similar scientifically based movement-based learning therapy) for a year or more to naturally activate all the necessary guidance cues for repair. It could well be that under these neural milieu conditions, exogenously applied stem cells can support the endogenous stem cell reserves. As long as necessary medical research is replaced by pure biological research, it is unlikely that substantial CNS repair can be achieved in human patients – because the human CNS is different from those of animals with respect to repair, learning (neural network complexity and variability), and pattern stability.

References

[1] Bennet, M.R. and Pettigrew, A.G.: The formation of synapses in striated muscle during development. *J. Physiol.*, 241: 515-545, 1974.

[2] Birks, R., Huxley, H.E. and Katz, B.: The fine structure of the neuromuscular junction of the frog. *J. Physiol.*, 150: 134-144, 1960.

[3] Birks, R., Katz, B. and Miledi, R.: Physiological and structural changes at the amphibian myoneural junction, in the course of nerve degeneration. *J. Physiol.*, 150: 145-158, 1960.

[4] Buller, A.J., Eccles, J.C. and Eccles, M.: Differentiation of fast and slow muscles in the cat hind limb. *J. Physiol.*, 150: 399-416, 1960.

[5] Burke, W. and Ginsborg, B.L.: The electrical properties of the slow muscle fibre membrane. *J. Physiol.*, 132: 586-598, 1956.

[6] Couteaux, R. and Pecot-Dechavassine, M.: Vesicules synaptiques et poches au niveau des "zones actives" de la jonction neuromusculaire. *C.R.Acad.Sc.* (Paris), 271: 2346-2349, 1970.

[7] Ecker, A.: Die Anatomie des Frosches. Verlag Friedrich Vieweg, Braunschweig, 1864.

[8] Elul, R., Miledi, R. and Stefani, E.: Neurotrophic control of contracture in slow muscle fibres. *Nature* (London), 217: 1274-1275.

[9] Elul, R., Miledi, E. and Stefani, E.: Neural control of contracture in slow muscle fibres in the frog. *Acta physiol. latinoam.*, 20: 194-226, 1970.

[10] Fatt, P. and Katz, B.: An analysis of the endplate potential recorded with an intracellular electrode. *J. Physiol.*, 115: 320-370, 1951.

[11] Gray, E.G.: The spinal and extrafusal innervation of a frog muscle. *Proc. R. Soc. London, Series B,* 146: 416, 1957.

[12] Guth, L.: Trophic influences of nerve on muscle. *Physiol. Rev.*, 48: 645-487, 1968.

[13] Gutmann, E.: Neurotrophic relations. *Ann. Rev. Physiol.*, 38: 177-216, 1976.

[14] Gordon, T., Purves, R.D. and Vrbora, G.: Differentiation of electrical and contractile properties of slow and fast muscle fibres. *J. Physiol.*, 269: 535-547, 1977.

[15] Gordon, T., Vrbova, G.: The influence of innervation on the differentiation of contractile speeds of developing chick muscles. *Pflügers Arch.*, 360: 199-218, (1975).

[16] Harris, A.J.: Inductive functions of the nervous system. *Ann. Rev. of Physiol.,* 36: 251-305, 1974.

[17] Hughes, A.: Cell degeneration in the larval ventral horn of Xenopus laevis (Daudin). *J. Embryol. Exp. Morph.*, 9: 269-284, 1961.

[18] Karnovsky,M.J: The localisation of cholinesterase activity in rat cardiac muscle by electron microscopy. *J. Cell. Biol.*, 23: 217-232, 1964.

[19] Kuffler, S.W. and Gerald, R.W.: The small nerve motor system to skeletal muscle. *J. Neurophysiol.,* 10: 383-394, 1947.

[20] Kuffler, S.W. and Vaughan-Williams, E.M.: Small nerve junction potentials. The distribution of small motor nerves to frog skeletal muscle and the membrane characteristics of the fibres they innervate. *J. Physiol.*, 121: 289-317, 1953.

[21] Kuffler, S.W. and Vaugham-Williams, E.M.: Properties of the slow skeletal muscle fibres of the frog. *J. Physiol.*, 121: 318-340, 1953.

[22] Lännergren, J. and Smith, R.S.: Types of muscle fibres in toad skeletal muscle. *Acta Physiol. Scand.*, 68: 263-274, 1966.

[23] Lehmann, N. and Schmidt, H.: Contractile responses to direct stimulation of slow muscle fibres before and after denervation. *Pflügers Arch.*, 382: 43-50, 1979.

[24] Miledi, R.: Properties of regenerating neuromuscular synapses in the frog. *J. Physiol.,* 154: 190-205, 1960.

[25] Miledi, R., Stefani, E. and Steinbach, A.B.: Induction of action potential mechanism in slow muscle fibres in the frog. *J. Physiol.*, 217:737-754, 1971.

[26] Miledi, R., Parker, I. and Schalow, G. Calcium transients in normal and denervated slow muscle fibres of the frog. *J. Physiol.* 318: 191-206, 1981.

[27] Moser, H.: Ein Beitrag zur Analyse der Thyroxinwirkung im Kaulquappenversuch und zur Frage nach dem Zustandekommen der Frühbereitschaft des Methamorphose-Reaktionssystems. *Rev. Swisse Zool*, Sept. 1950.

[28] Orkand, M., Orkand, R.K. and Cohen, M.W.: Distribution of acetylcholine receptors on Xenopus slow muscle fibres determined by Bungarotoxin binding. *Neuroscience*, 3: 435-446, 1978.

[29] Page, S.: A comparison of the fine structure of frog slow and twitch muscle fibres. *J. Cell Biol.*, 26: 477-497, 1965.

[30] Peachey, L.D. and Huxley, A.F.: Structural identification of twitch and slow striated muscle fibres of the frog. *J. Cell Biol.*, 13: 177-180,1962.

[31] Peper, K., Dreyer, F., Sandri, C., Akert, K. and Moore, H.: Structure and ultra structure of the frog motor endplate. A freeze-etching study. *Cell Tiss. Res.,* 149: 437-445, 1974.

[32] Riley, A.: Spontaneous elimination of nerve terminals from the endplates of developing skeletal myofibrils. *Brain Res.,* 134: 279-285, 1977.

[33] Schalow, G. and Schmidt, H.: Action potentials induced in slow muscle fibres by partial denervation. *Nature,* 253: 122-123, 1975.

[34] Schalow, G. and Schmidt, H.: Effect of nerve length and temperature on the induction of action potentials in denervated slow muscle fibres of the frog. *Pflügers Arch.*, 372: 17-22, 1975.

[35] Schalow, G. and Schmidt, H.: Local development of action potentials in slow muscle fibres after complete or partial denervation. *Proc. R. Soc. Lond, B* 203: 445-457, 1979.

[36] Schmidt, H. and Stefani, E.: Reinnervation of twitch and slow muscle fibres of the frog after crushing the motor nerves. *J. Physiol.*, 258: 99-123, 1976.

[37] Schmidt, H. and Stefani, E.: Action potentials in slow muscle fibres of the frog during regeneration of motor nerves. *J. Physiol.*, 270: 507-517, 1977.

[38] Speidel, C.C.: Adjustment of nerve endings. In: Harvey lectures, 36: 126-158, 1941.

[39] Stefani, E. and Schmidt, H.: A convenient method for repeated intracellular recording of action potentials from the same muscle fibre without membrane damage. *Pflügers Arch.*, 334: 276-278, 1972.

[40] Stefani, E. and Schmidt, H.: Contractile responses to depolarizing drugs of slow muscle fibres during reinnervation by fast motor axons. *Acta Physiol. Latinoam.*, 20: 320-322, 1973.

[41] Stefani, E. and Steinbach, A.B.: Resting potential properties of frog slow muscle fibres. Effect of different external solutions. *J. Physiol.*, 203: 383-401, 1969.

[42] Tasaki, I. and Mizutani, K.: Comparative studies on the activities of the muscle evoked by two kinds of motor nerve fibres. *Jap. J. med. Sci. Biol.*, 10: 237-244, 1944.

[43] Wernig, A., Pecot-Dechavassine, M. and Stöver, H.: Signs of nerve regression and sprouting in the frog neuromuscular synapse. In: Ontogenesis of functional mechanism of peripheral synapses. INSERM Symposium No 13, Ed. J. Taxi, Elsevier/North Holland Biomedical Press, 1980.

[44] Bhardwj, R.D. et al.: Neocortical neurogenesis in humans is restricted to development. *PNAS,* 103 (33): 12564-12568, 2006.

[45] Bradbury, E.J. and McMahon, S.B. (2006). Spinal cord repair strategies: why do they work? *Nat. Rev. Neurosci.*, 7: 644-653.

[46] Hamburger, V.: The effects of wing bug extirpation in chick embryo on the development of central nervous system. *J. Exp. Zool.*, 68: 449-494 (1934).

[47] Hamburger, V.: Cell death in the development of the lateral motor column of the chick embryo. *J. Comp. Neurol.*, 160: 535-546 (1975).

[48] Harel, N.Y. and Strittmatter, S.M. (2006). Can regenerating axons recapitulate developmental guidance during recovery from spinal cord injury? *Nat. Rev. Neurosci.*, 7: 603-616.

[49] Levi-Montalcini, R. and Levi, G.: Les consequences de la destruction d'un territoire d'innervation peripherique sur le development des centres nerveux correspondants dans l'embryon de poulet. *Arch. Biol.*, 53: 537-545 (1942).

[50] Martino, G. and Pluchino, S. (2006). The therapeutic potential of neural stem cells. *Nature Reviews / Neuroscience*, 7: 395 -406.

[51] McCaig, C.D., Rajnicek, A.M., Song, B. and Zhao, M.: Has electrical growth cone guidance found its potential? *Trend in Neuroscience*, 25: 345-359, 2002.

[52] Schalow, G.: Schalow, G. Conduction velocities and nerve fibre diameters of touch, pain, urinary bladder and anal canal afferents and α and γ-motoneurons in human dorsal sacral roots. *Electromyogr. Clin. Neurophysiol.* 31: 265-296, 1991.

[53] Schalow, G., Aho, A. and Lang, G. Microanatomy and number of nerve fibres of the lower intercostal nerves with respect to a nerve anastomosis. Donor nerve analysis. I (IV). *Electromyogr. Clin. Neurophysiol.* 32: 171-185, 1992.

[54] Schalow, G. (1991). Oscillatory firing of single human sphincteric α_2 and α_3-motoneurons reflexly activated for the continence of urinary bladder and rectum. Restoration of bladder function in paraplegia. *Electromyogr. Clin. Neurophysiol.*, 31: 323-355.

[55] Schalow, G. and Zäch, G.A. (2000): Reorganization of the Human CNS, Neurophysiologic measurements on the coordination dynamics of the lesioned human brain and spinal cord. Theory for modern neurorehabilitation (31 case reports). Gen. *Physiol. Biophys.*, 19, Suppl. 1: 1-244.

[56] Schalow, G.: Stroke recovery induced by coordination dynamic therapy and quantified by the coordination dynamic recording method. *Electromyogr. Clin. Neurophysiol.*, 42: 85-104, 2002.

[57] Schalow, G.: Improvement after traumatic brain injury achieved by coordination dynamic therapy. *Electromyogr. Clin. Neurophysiol.*, 42: 195-203, 2002.

[58] Schalow, G.: Recovery from spinal cord injury achieved by three months of coordination dynamic therapy. *Electromyogr. Clin. Neurophysiol.*, 42: 367-376, 2002.

[59] Schalow, G.: Partial cure of spinal cord injury achieved by 6 to 13 months of coordination dynamic therapy. *Electromyogr. Clin. Neurophysiol.*, 43: 281-292, 2003.

[60] Schalow, G., Pääsuke, M., Ereline, J. and Gapeyeva, H.: Improvement in Parkinson's disease patients achieved by coordination dynamics therapy. *Electromyogr. Clin. Neurophysiol.*, 44: 67-73, 2004.

[61] Schalow, G. Tremor in Parkinson's disease patients can be induced by uncontrolled activation and uninhibited synchronization of α_2-motoneuron firing to which α_1-motoneuron firing synchronizes. *Electromyogr. Clin. Neurophysiol.*, 45: 393-406, 2005.

[62] Schalow, G., Pääsuke, M., Jaigma, P.: Integrative reorganization mechanism for reducing tremor in Parkinson's disease patients. *Electromyogr. Clin. Neurophysiol.*, 45: 407-415, 2005.

[63] Schalow, G., Jaigma, P.: Cerebral palsy improvement achieved by coordination dynamics therapy. *Electromyogr. Clin. Neurophysiol.*, 45: 433-445, 2005.

[64] Schalow, G.: Functional development of the CNS in pupils between 7 and 19 years. *Electromyogr. Clin. Neurophysiol.*, 46: 159-169, 2006.

[65] Schalow, G.: Hypoxic brain injury improvement induced by coordination dynamics therapy in comparison to CNS development. *Electromyogr. Clin. Neurophysiol.*, 46: 171-183, 2006.

[66] Schalow, G. and Jaigma, P.: Improvement in severe traumatic brain injury induced by coordination dynamics therapy in comparison to physiologic CNS development. *Electromyogr. Clin. Neurophysiol.*, 46: 195-209, 2006.

[67] Schalow, G.: Surface EMG- and coordination dynamics measurements-assisted cerebellar diagnosis in a patient with cerebellar injury. *Electromyogr. Clin. Neurophysiol.* 46: 371-384, 2006.

[68] Schalow, G.: Symmetry diagnosis and treatment in coordination dynamics therapy. *Electromyogr. Clin. Neurophysiol.*, 46: 421-431, 2006.

[69] Schalow, G.: Cerebellar injury improvement achieved by coordination dynamics therapy. *Electromyogr. Clin. Neurophysiol.*, 46: 433-439, 2006.

[70] Schalow, G., Vaher, I. and Jaigma, P.: Overreaching in coordination dynamics therapy in an athlete with spinal cord injury. *Electromyogr. Clin. Neurophysiol.*, 48: 83-95, 2008.

[71] Schalow, G.: Stem cell therapy and coordination dynamics therapy improve spinal cord injury. *Electromyogr. Clin. Neurophysiol.*, 48: 233-253, 2008.

[72] Schalow, G.: Near-total functional recovery in partial cervical spinal cord injury (50% injury) after three years of coordination dynamics therapy. *Electromyogr. Clin. Neurophysiol.,* 49: 67-91, 2009.

[73] Schalow, G.: Partial cure achieved in a patient with near-complete cervical spinal cord injury (95% injury) after three years of coordination dynamics therapy. *Electromyogr. Clin. Neurophysiol.*, 49: 199-221, 2009.

[74] Schalow, G.: The classification and identification of human somatic and parasympathetic nerve fibres including urinary bladder afferents and efferents is preserved following spinal cord injury. *Electromyogr. Clin. Neurophysiol.*, 49: 263-286, 2009.

[75] Schalow, G.: Coordination impairment between the somatic and parasympathetic nervous systems in the human sacral micturition centre. *Electromyogr. Clin. Neurophysiol.*, 49: 337-367, 2009.

[76] Schalow, G.: Coordination repair between the somatic and parasympathetic nervous systems in patients with spinal cord injury. *Electromyogr. Clin. Neurophysiol.*, in press.

[77] Schalow, G.: Scientific basis for learning transfer from movements to urinary bladder functions in human. *Electromyogr. Clin. Neurophysiol.*, in press.

[78] Schalow, G.: Cure of urinary bladder function in severe (95%) motoric complete cervical spinal cord injury in human. *Electromyogr. Clin. Neurophysiol.*, in press.

[79] Schalow, G.: Relative Coordination between Neuron firing and Generation of a Motor Program in the human CNS. In: Berkovsky, T.C. (Ed.), *Handbook of Spinal Cord Injuries*, Chapter V, pp. 265-295, Nova Science Publishers, 2010.

[80] Schalow, G.: Building of New Motoneurons in the Human Spinal Cord upon Coordination Dynamics Therapy to Improve Finger Functions in Motoric Complete Cervical Spinal Cord Injury. In: Berkovsky, T.C. (Ed.), *Handbook of Spinal Cord Injuries*, Chapter IV, pp.231-264, Nova Science Publishers, 2010.

[81] Fouad, K., Dietz, V. and Schwab, M.E.: Improving axonal growth and functional recovery after experimental spinal cord injury by neutralizing myelin associated inhibitors. *Brain. Res.*, 36: 204-12, 2001.
[82] Buchli, A.D. and Schwab, M.E.: Inhibition of Nogo: a key strategy to increase regeneration, plasticity and functional recovery of the lesioned central nervous system. *Ann. Med.,* 37: 556-567, 2005.
[83] Song, H. et al.: Conversion of neuronal growths cone responses from repulsion to attraction by cyclic nucleotides. *Science,* 281: 1515-1518, 1998.
[84] Tessier-Lavigne, M. and Goodman, C.S.: The molecular biology of axon guidance. *Science*, 274: 1123-1131, 1996.
[85] Thuret, S., Moon, L.D.F. and Gage, F.H. (2006), Therapeutic interventions after spinal cord injury. *Nat. Rev. Neurosci.*, 7: 628-643.
[86] Gould, E. and Gross, C.G.: Neurogenesis in adult mammals: Some progress and problems. *J. Neurosci.*, 22: 619-623, 2002.
[87] Praag, van H., Shubert, T., Zhao, C. and Gage, F.H.: Exercise enhances learning and hippocampal neurogenesis in aged mice. *J. Neurosci.*, 25: 8680-8685, 2005.
[88] Leuner, B., Gould, E. and Shors, T.J.: Is there a link between adult neurogensis and learning? *Hippocampus* 16: 216-224, 2006.
[89] Eriksson, P.S., Perfilieva, E., Bjork-Eriksson, T., Alborn, A.M., Nordborn, C., Peterson, D.A., Gage, F.H.: Neurogenesis in the adult human hippocampus. *Nat. Med.*, 11: 1313-1317, 1998.
[90] Schalow, G.: Feeder arteries, longitudinal arterial trunks and arterial anastomoses of the lower human spinal cord. *Zbl. Neurochir.*, 51: 181-184, 1990.
[91] Gould, E., Beylin, A., Tanapat, P., Reeves, A. and Shors, T.J.: Learning enhances neurogenesis in the hippocampal formation. *Nat. Neurosci.*, 2: 260-265, 1999.
[92] Asanuma, H. and Pavlides, C.: Neurobiological basis of motor learning in mammals. *NeuroReport*, 8: 1-6, 1997.
[93] Motor assessment of the developing infant. Piper, M.C. and Darrah, J., Chapter 1, *Theories of motor development*, Saunders Company, Philadelphia, 1994.

Chapter II

Development and Repair of the Human CNS

Abstract

Based on the System Theory of Pattern Formation (a) and with the single-nerve fibre action potential recording method measured phase and frequency coordination among neuron firings, the oscillatory firing of motoneurons and the impairment of coordination and oscillatory firing following injury or malformation (b) - the Coordination (Pattern) Dynamics (CD) measurements (by motor pattern change) for quantifying human CNS organization during development and repair and the Coordination Dynamics Therapy (CDT) to repair the injured or malformed human CNS by movement-based learning are developed. During human central nervous system (CNS) development in healthy infants and pupils between 2 and 18 years of age, improvement of pattern performance, stability, and symmetries are evaluated by CD measurements. Complexity, quality, and stability of CNS self-organization developed strongly in the age-range between two (when it became possible to measure) and six years of age. Transient high-frequency exercising to improve the coordinated firing of neurons was measured to occur preferentially between 5 and 11 years of age. CD values were very similar in girls and boys. The ability to do two things at the same time (exercising and speaking, for example) improved with CNS maturing. Movement-based learning was the fastest in young pupils. In groups of patients and case reports, the improvements in severe cervical spinal cord injury (SCI), cerebral palsy, traumatic brain injury, and cerebellar injury, achieved by CDT, were quantified by CD measurements, surface EMG and movement performance. Depending on the duration, intensity, and quality (complexity, coordination, symmetry, and integrativity of movements) of movement-based learning, probably weight, wiring, neural replacement, and coordination plasticity was induced for the repair of physiologic patterns. Understanding the complex epigenetic regulation of CDT-induced nervous system repair, including limited adult neurogenesis, may give further possibilities of increasing the rate of CNS repair. Manual support of the movement improved the movement performance, and via the improved movement-induced afferent input, there were also in the patient generated motor programs. The comparison of motor pattern improvement in severe cervical SCI patients upon 0.5 and 5 years of CDT with the development of motor patterns in the developing infant showed similarity in trunk rotation, unsustained sitting, four-point kneeling and hand-to-knees positioning. But the supported standing and walking with substantial antigravity muscle activation in SCI patients were far behind those of the developing infant. Continence functions of the urinary bladder were repaired in patients earlier than or within the same time period as their establishment in the developing infant. The induced learning transfers from exercised and improved movements to vegetative and higher mental functions (including the induction of speech) upon CDT seem to be the essential progress in the repair of the developing and adult human CNS. The rate of repair may be five times faster in a five-year-old infant than in an adult patient.

1. Introduction

1.1. General Introduction

The current research in basic clinical neurophysiology and human repair neurophysiology is aimed at developing treatment for CNS injury and malformation and to improve movements, autonomic, and higher mental functions. A movement-based learning therapy, called Coordination Dynamics Therapy (CDT), has been developed and shown to be able to improve central nervous system (CNS) functioning after stroke [39], traumatic brain injury [40,52], hypoxic brain injury [51], cerebellar injury [53-55], spinal cord injury (SCI) [41,43,56-64,67,68], in cerebral palsy [49], and in Parkinson's disease [45,48]. To (further) derive treatment strategies and to understand abnormal neural development, normal neural development must be the reference point. A first step to compare normal neural development and repair in human has been done [50].

In Chapter I, normal frog development was compared with its repair at the cellular level; during regeneration, the development could partly recapitulate. It was shown that during development and regeneration, the neuromuscular innervation of slow muscle fibres changed from one motoneuron population to another. Target invasion, selection, and selection change were discussed with respect to long- and short-range attraction and repulsion of possible guidance molecules and receptors for guidance of axon growth cones and synapse profiles. The contact attraction and repulsion over a distance of 0.1µm and shorter were discussed with respect to competitive interactions and electromagnetic field guidance of the two types of motoneuron populations with respect to the formation of the specific neuromuscular innervation patterns on the two kinds of target muscle cells.

The invasion of the target muscle by two successive waves of two axon populations and the recognition and specific selection of the muscle fibres for innervation during development and regeneration in the frog was used to understand the regeneration of the human spinal cord following severe C5/6 spinal cord injury (SCI) upon movement-based learning for repair. Based on the location of spinal cord tracts in the white matter and premotor networks in the grey matter and specific regeneration, it was found that the regeneration in the human spinal cord started from the spared naturally activated nerve fibres at the injury site. The building of new motoneurons from endogenous stem cell reserves, for the activation of the index finger, and their functional integration started from the activated neuronal networks surrounding the injury site.

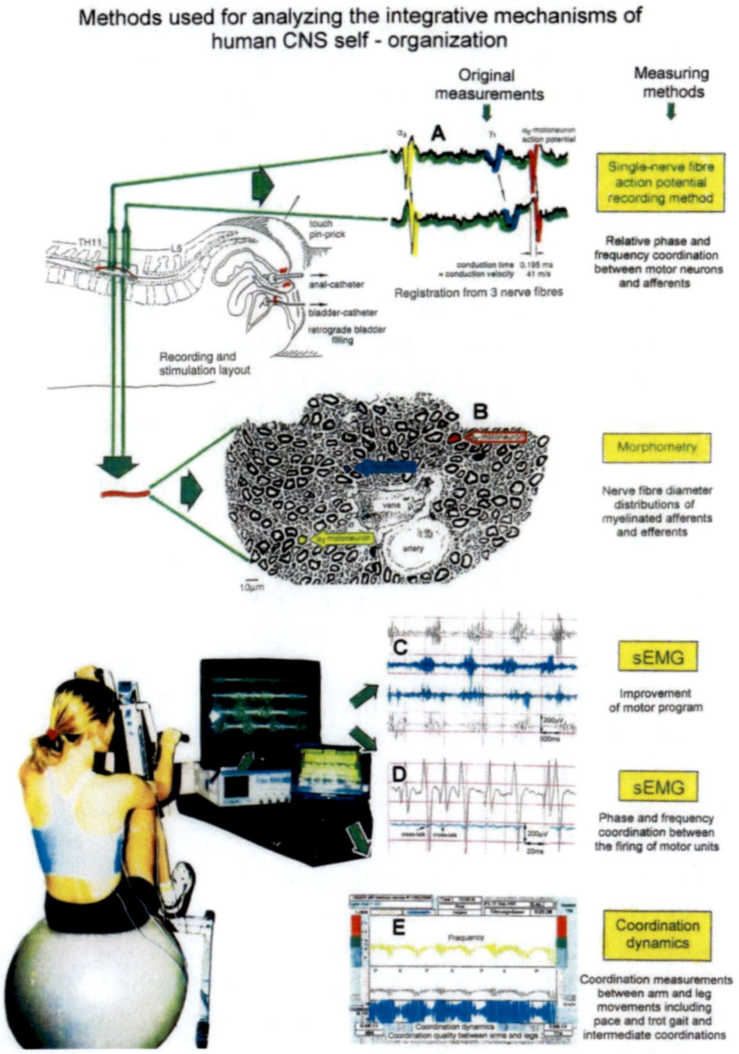

Figure 1. Schematic layout of the measuring methods used to study the self-organization of neuronal networks of the human CNS under physiologic and pathophysiologic conditions. A, B. By recording with two pairs of platinum wire electrodes from sacral nerve roots containing between 200 and 500 myelinated nerve fibres, records were obtained in which single-nerve fibre action potentials (APs) could be identified from motoneuron axons (main AP phase downwards) and afferents (main AP phase upwards). By measuring the conduction times and with the known electrode pair distance (10 mm), conduction velocities could be calculated. Conduction velocity distribution histograms were constructed in which the myelinated nerve fibre groups larger than 4µm could be characterized by group conduction velocity values. Recording was followed by morphometry. Distributions of nerve fibre diameters were constructed, and nerve fibre groups were characterized by the peak values of asymmetrical distributions. By correlating the peak values of the velocity distributions with those of the diameter distributions obtained for the same root, a classification scheme was constructed of the human peripheral nervous system. It became thus possible to record natural impulse patterns simultaneously from identified single afferent and efferent nerve fibres and to analyse self-organizing mechanisms of the human CNS under physiologic and pathologic conditions. C,D. By performing surface electromyography (sEMG) with up to four electrode pairs (two indifferent electrodes, one earth electrode; pre-amplification x1000, four-channel oscilloscope) the changes of motor programs were measured (C). When recording from appropriate patients, natural activation patterns of several single-motor units were obtained (D), and coordination between motor unit firing could be studied. E. Coordination between arms and legs was quantified by the single integrative parameter, arrhythmicity of turning (df/dt or df/dt/f; f = frequency) during exercising on a special coordination dynamics therapy device.

In this chapter, development and repair will be compared at the integrative level in humans to formulate treatment strategies. Abnormal motor development cannot be recognised, correctly interpreted, and partly repaired without solid knowledge of normal development. The neuromaturational model, the system theory of motor development, and learning aspects will be used to understand the systems measurements of CNS organization during development and repair.

To identify the neural development of the normal or deviant infant or older child and to formulate repair strategies, additional information is required to assess the ongoing motor development rather than assessing reflex capabilities and stereotyped patterns of posture and movement. With the "coordination dynamics" (CD) measurements such an assessment has been developed to quantify CNS organization integratively and non-invasively. Integrative means when many parts of the CNS are activated. This is the case when arms, legs, trunk, and head are moving and activated simultaneously in coordination. Upon exercising on the special coordination dynamics therapy (CDT) device for measuring (Figures 1,17,19), the integratively working sensory and motor influences to produce motor output are assessed by motor pattern change imposed by a mechanically exact working device. Since CNS organization is assessed online, dynamic changes of CNS organization can directly followed visually.

1.2. Specific Introduction

1. CNS Assessment during Development by "Infantile Patterns and Reactions" within the Framework of the Maturational Theory

Assessment of neurological development within the framework of the maturational theory during infancy results from the neurological maturation of the CNS. Maturation is characterised by increased neuronal network complexity (increased connectivity (growing of axons and dendrites and formation of synapses with appropriate strength) and myelinisation) and concomitant inhibition of the lower subcortical nuclei of the brain by the higher functioning cerebral cortex.

The cerebral cortex is construed as the organizational centre for controlled movement. Within this model, motor development and changes in motor skills are intrinsically driven, and the impact of the environment plays a secondary role in the emergence of motor skills. At all stages of motor development, the intrinsic influence of the CNS transcends any effect the environment may contribute. Neonate's movement patterns are interpreted as being initially dominated by reflexes. The movements operate in a stimulus-response fashion, with the appropriate stimulus eliciting a predictable and stereotyped response [8]. Primitive reflexes are assumed to represent the dominance of lower levels of the CNS – the subcortical nuclei located in the brain stem.

The integration of these reflexes is perceived to indicate maturing of the CNS and inhibition of the lower centres by the higher functioning cerebral cortex. Viewed from this perspective, primitive reflexes offer a tangible gauge to assess the level of maturation of the CNS.

Evaluation and documentation of the presence or lack of primitive reflexes are used extensively in infant motor assessment to evaluate the level of neurological integrity [6,8]. The persistence of these reflexes past the accepted age of integration is interpreted as abnormal, representing a lack of maturation of the cerebral cortex and subsequent inhibitory control over the lower centres. Movement evolves from an uncontrolled reflexive state to a controlled voluntary state as the cerebral cortex matures. Thus, the refinement of movement is dependent on CNS maturation.

The so-called "infantile patterns and responses" (as, for example, the stepping movements, palmar grasp reflex, and Moro reaction) follow a definite, though variable, developmental course. The "infantile pattern and reaction and responses" decline and are superseded by so-called "voluntary motor behaviour" of increasing complexity. They are related to the degree of maturation of the brain (and not to the moment of the infant's birth). The main significance of the infantile patterns and reactions in the neonatal period is that they are age-specific manifestations of an active brain, demonstrating that the "hardware" of the brain appropriate to that age is operational. This mode of assessment is based on the assumption that reflexes and reactions underlie most volitional movements and that there is a relationship between functional motor achievement and reflex activity in early life [7]. In older children and adults, the level of maturation can no longer be evaluated by these patterns and reactions.

2. Motor Maturation and Motor Performance

Following injury, in adults and children, therapists are trained to evaluate an individual's motor performance in terms of skill – that is, how the individual is able or unable to correctly move in relation to a previously acquired motor skill. Therapy interventions involve helping individuals regain mastery of the performance of everyday tasks. The underlying assumption in both assessment and treatment of motor disorders is that a normal pattern has been impaired or replaced with an abnormal pattern of movement; thus the assessment and resulting treatment are aimed at assisting the individual to overcome the deficiencies of movement by relearning motor programs of performance [11].

The evaluations of the motor skills of a developing infant requires an orientation to motor performance different from that normally applied to older children or adults. The motor assessment of the developing infant entails the evaluation of an evolutionary process or maturation of motor behaviours over time. This unique emphasis on process and maturation differentiates infant motor assessment from the assessment of older children and adults [11]. McGraw [24] defined motor development as the sequential change in specific motor activity with age, thereby often referring to it as motor maturation. Accordingly, the motor skills of the developing infant are not static but rather a complex evolutionary process. Therefore, assessing the reflex capabilities of an infant provides very little information on the development of functional motor skills [15].

The presence and evaluations of primitive reflexes to assess early motor development is based on the assumption that reflexes and reactions underlie most volitional movement and that there is a relationship between functional motor achievement and reflex activity earlier in life [7].

To identify, therefore, the neural development of the normal or deviant infant or child, additional information is required to assess the ongoing motor development than assessing reflex capabilities and stereotyped patterns of posture and movement.

3. CNS Assessment during Development by Coordination Dynamics (CD) Measurements within the System Theory of Pattern Formation

The CD is such additional information to assess the ongoing neuronal network development of the CNS in infancy and childhood and is a different assessment from assessing CNS development by patterns, reactions, and voluntary motor functions.

Neuronal network maturation and quality of organization is measured. CNS development is assessed by the quality of coordinated firing of neurons and complexity of neuronal network self-organization, quantified by the quality of performance of coordinated arm and leg movements during pattern change, imposed by the device when a subject is exercising on the special CDT device. The very complex networks of the human nervous system are the morphological substrate for the functional variability and accuracy of the nervous system. When the child is exercising on the special CDT and recording device, the coordination dynamics can be evaluated non-invasively and online. Because development increases the complexity of the networks formed by the neuronal elements, the infant in the neonatal period is not able to exercise on the special CDT device by himself, because the changing coordination between arm and leg movements are too difficult as more network complexity and support is needed. Within two to three years, the infant becomes able to exercise in the forward direction with poor performance; later on, exercising backwards also becomes possible (see Results). However, supported treatment with the special CDT device to improve CNS functioning in infants with CNS injury or malformation is already possible from one year on. The specific input from coordinated movements serves as a teacher for repair and improved development. Furthermore, the measuring of the arrhythmicity of exercising during continuous pattern change from pace to trot gait and backwards can also be seen as a further development of the assessment of motor skills. Not all patterns are judged or assessed, but the movement adaptation to a continuously changing range of motor patterns is assessed.

4. The System Theory of Pattern Formation of CNS Development: The CNS Is One Network to Which All Subsystems Contribute and Is Self-Correcting "En Route"

The System Theory of Pattern Formation of CNS development encompasses all areas of development and is derived from mathematics, physics, human neurophysiology, clinical research, and developmental psychology. The CNS is considered as one neuronal network. A new behaviour is generated, which is dependent on the input of all subsystems. This behaviour may have a characteristic that could not have been determined by evaluating the contributing behaviours individually [70].

The neuromaturational theory of development evolved from a prescriptive, structural framework in which the instructions controlling movement exist before the motor behaviour emerges. The instructions are encoded in the CNS, and the higher control centres control movement by a "feedback" system; all modifications to a movement must originate from the cerebral cortex – the command centre. In contrast, the systems approach is developed from a functional rather than a structural framework. Within this framework, the behaviour itself can affect and modify the resultant behaviour in contrast to all the commands being issued by the cerebral cortex. This theory exemplifies a "feed-forward system" that is self-correcting en route rather than hardwired from the cerebral cortex. It also implies that all factors, subsystems, or structures contributing to the motor behaviour (or patterns in general) are important and exert an influence on the outcome [70]. If, for example, the thalamus is damaged in an infant by an infarct, then many patterns become abnormal. The higher centres can no longer control movements or other patterns sufficiently. But upon movement-based learning, a self-correction en route may partly compensate for the missing contribution of the subsystem in the way that other subsystems take functions over and/or the thalamus is partly repaired.

1.3. Theoretical Implications for Treatment

1. Treatment Principles Seemingly Embedded in the Neuromaturational Model

It is believed that fundamental treatment principles are embedded in the neuro-maturational model of motor development. The emergence of motor skills in cephalocaudal and proximal-distal directions and the refinement from reflexes to voluntary movements are all beliefs incorporated into the planning of intervention strategies with motor-delayed infants. Head control is facilitated before pelvic control, and proximal stability is achieved before distal mobility is anticipated. This theory of motor developments appears to provide the framework and basis for many treatment strategies [25].

The repair of spinal cord injuries seems also to proceed in cephalo-caudal and proximal-distal directions [59,64,67]. But the repair depends also on what is spared from the injury. Further, from the mechanical point of view, trunk stability is needed during walking and crawling before arms and legs and other movements can be used. But this does not necessarily mean that treatment interventions have to proceed in cephalo-caudal and proximal-distal directions. Bernstein argued already [3] that joints and muscles never work in isolation but rather in a coordinated manner. Muscles are strengthened and recruited in functional patterns.

Therefore, functional patterns also have to be trained for repair. During development, the rate, quality, and sequence of motor skills are genetically predetermined. But variations in (motor) development are probably corrected by the natural movements of the infant (crawling, walking, running, jumping, balance training). Therefore, repair (by learning, see below) can be seen as environmental information, necessary to fulfil the genetically predetermined patterns in the case of disturbance in the generation of the operational hardware. Interventions may not be the antithesis of the hierarchical, cerebral cortex-dominated framework associated with the neuromaturational model, since the genetically predetermined patterns may include some adaptation to the environment to be more successful for the survival of the species.

2. Treatment Principles Embedded in the System Theory of (Motor) Development

The system theory of motor development embraces the role of interventions. Designated boundaries defining treatment may not exist. Because of the influence of many subsystems, many possibilities exist to facilitate (motor) behaviour. All subsystems contribute to the enhancement or delay of (motor) behaviours and not only the motor system. In addition to exercising motor functions, instructions or self-instructions may contribute, for example, to the improvement of motor functions. Questions remain. What degree of deprivation in any subsystem can be compensated by the other systems? Are some subsystems more important than others or is the severity of involvement the restricting factor? It seems that humans are resilient to minor disturbances within any subsystem. The degree of resiliency is unknown. The systems approach challenges therapists to interpret, based on measurements, why treatment techniques are effective.

3. Treatment Principles Following on from Human Electro-Neurophysiology: CNS Strategy to Improve the Coordinated Firing of Neurons During Development

Treatment strategies are not only embedded in theories of neurodevelopment but do also follow from the self-organization of neuronal networks, measured by electrophysiology [63,64]. If phase and frequency coordination is fundamental for CNS self-organization and gets impaired during structural changes as during repair [59], then this coordinated firing of neurons should also get continuously impaired during development, because the neuronal structure is, due to the increasing complexity, changing and maturing all the time. To avoid chaotic organization, the developing CNS (and the structural repairing CNS) should have strategies at hand to improve continuously this coordinated firing of neurons. One of these strategies (or the strategy) seems to be to make the child move very fast for a short time, as, for example, during running. Upon this fast exercising, the CNS can learn a better coordinated firing of neurons from the improved movement induced afferent input generated by the kinetically improved movement performance. This fast moving to improve the coordinated firing of neurons for an improved self-organization can be measured nicely when healthy children are exercising on the special CDT and recording device (Figure 19). As will be measured (Figure 27), such fast exercising is most prominent during a certain period of CNS development.

4. Retarded, Accelerated or Deviant Development of Motor Functions

Many fluctuations occur during normal development. There are accelerations and decelerations in the developmental rate of different functions within one infant. Characteristic for normal development is variability in motor performance. But the impaired nervous system is not able to attain such variability in motor performance because the structure of the CNS is deficient in enabling the infant to use various modes of operation for single performance. When there is serious damage to the CNS, it is even possible that the usual modes cannot be developed at all; the brain has to resort to unusual modes. Distorted motor patterns arise, which fall outside the range of the normal variability and which are stereotyped, due to the deficiency of the "hardware."

Due to the reduced variability in motor performance of the impaired nervous system, patients with CNS injury or malformation have big problems exercising on the special CDT devices because the device imposes complex motor pattern variability. But motor pattern variability can partly be learned from the device. Due to the deficiency of the neuronal structure, the development of motor functions may be retarded, accelerated or deviant. Some functions may not develop at all, while others show only a decrease in variability.

5. Learning for Repair by Recapitulating Development

The mechanisms, or strategies, developed by the newborn, infant, or young child are not peculiarities or movement patterns that occur only in the neonatal or later period. They belong to one phase of the developmental course of the nervous system.

These patterns are elements that have been tested and trained and that can be modified for use in more differentiated and complex motor behaviour during later development. Their integrity must be regarded as a prerequisite for normal development during infancy. Sufficient crawling has to be trained by the infant to have no coordination problems between arms and legs in later life.

The stepping movement of the newborn (Figure 2) (characterized by the emphasized lifting of the knees) is partly reappearing in severe cervical spinal cord injury. Upon partial regeneration [67] (Figure 3), induced upon CDT, the stepping automatism is partly changing again into volitional walking.

The aim of learning is not only to equip the deviant infant or child with just one efficient mode of motion for each motor performance but to lay down a foundation in the CNS for adaptive responses in specific circumstances and to induce learning transfer to other (movement) patterns that cannot be trained. Jumping on springboard and exercising on the special CDT device improves continence in severe cerebral palsy (see repair in cerebral palsy) and spinal cord injury [64] by learning transfer [63]. Exercising walking in the backward direction on treadmill improves the forward walking in stroke patients by learning transfer.

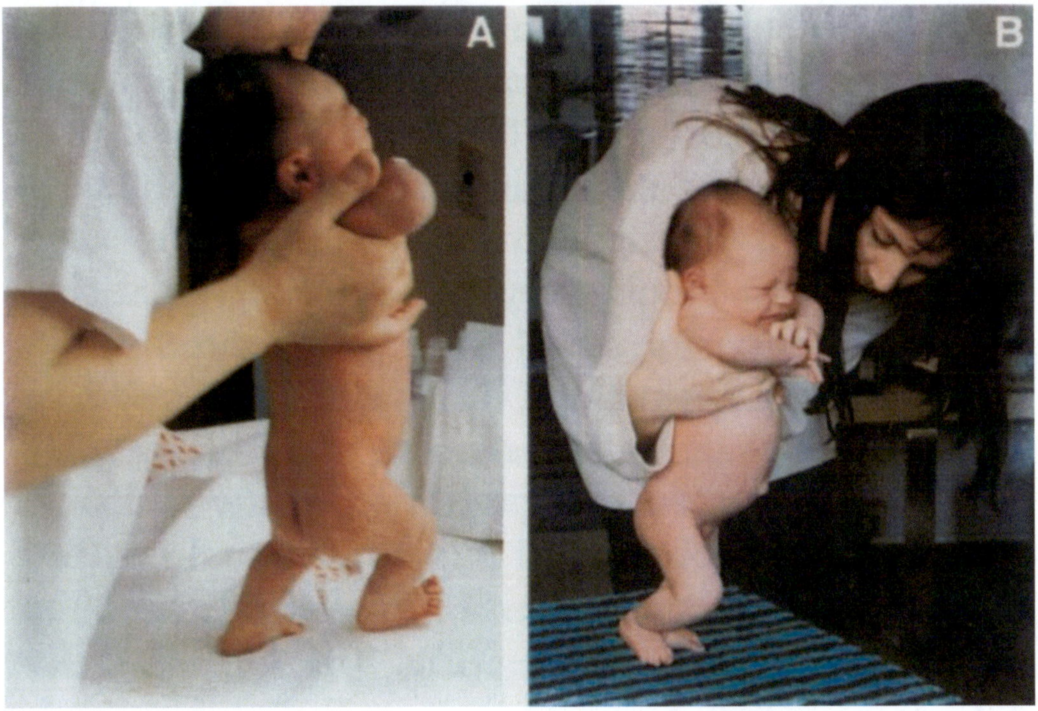

Figure 2. Automatic stepping in a newborn infant. A. A five-day-old infant, Juliane, performing primary automatic stepping; slight backward posture. The heel of the right foot touched the ground first. B. Infant Juliane, eight days old, performing automatic stepping.

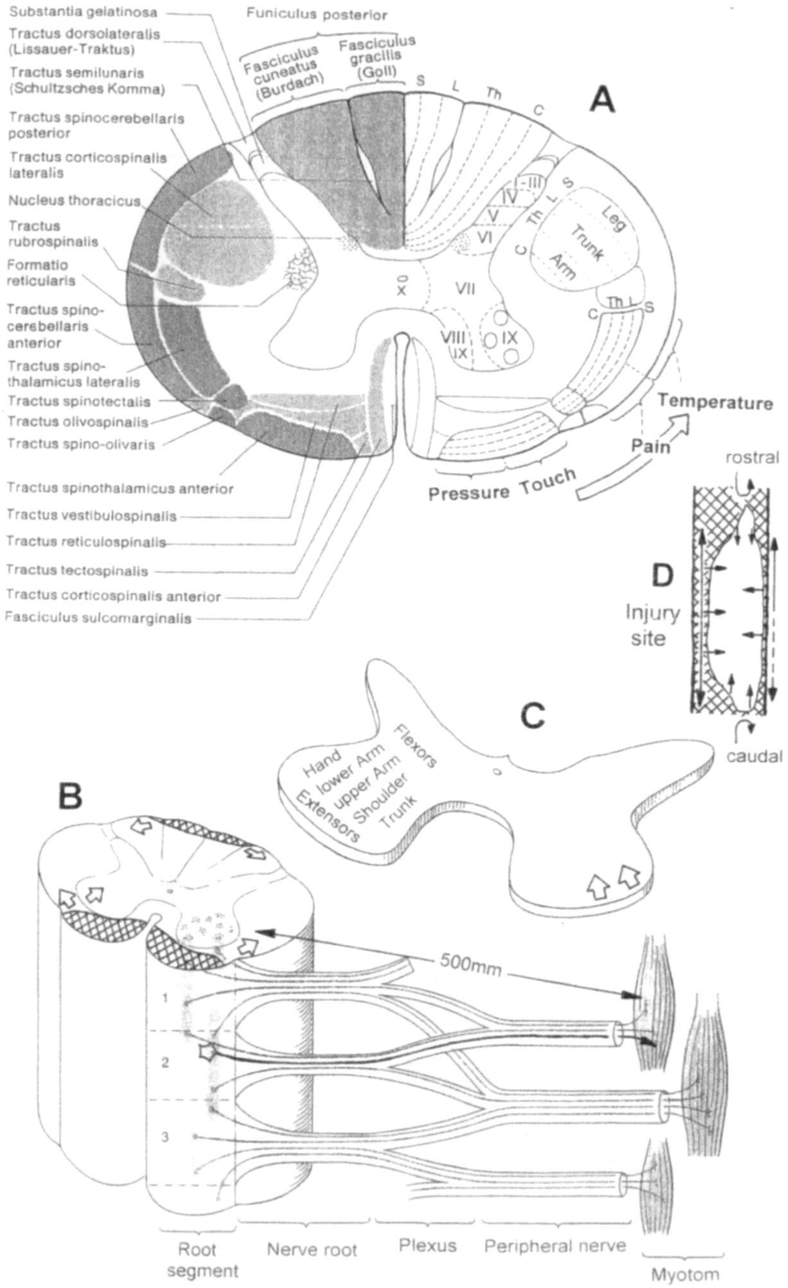

Figure 3. Schematic drawings of the human spinal cord with a cervical injury. A. Spinal cord cross section with fibre tracts and grey matter. Note the regaining of sensitivity upon CDT from pressure to touch, pain, and temperature is indicated by an open arrow. B. Spinal cord section with suggested spared matter (crosshatched). Open arrows indicate direction of structural repair. The axonal outgrow of a newly built motoneuron is indicated by a bended and long arrow. C. Motoneuron sites for serving different functions are indicated. Note the extensor motoneuron cell bodies are sited more adjacent to the spared spinal cord matter than the flexors and should be generated first in this case, indicated by open arrows. D. Injury site of the patient. Small short arrows indicate direction of structural activity-dependent repair. The structural repair starts from the activated spared matter into the cavity. The ascending and descending tract activity is indicated by long arrows, and the tract and network activity rostral and caudal to the injury site is indicated by bended arrows.

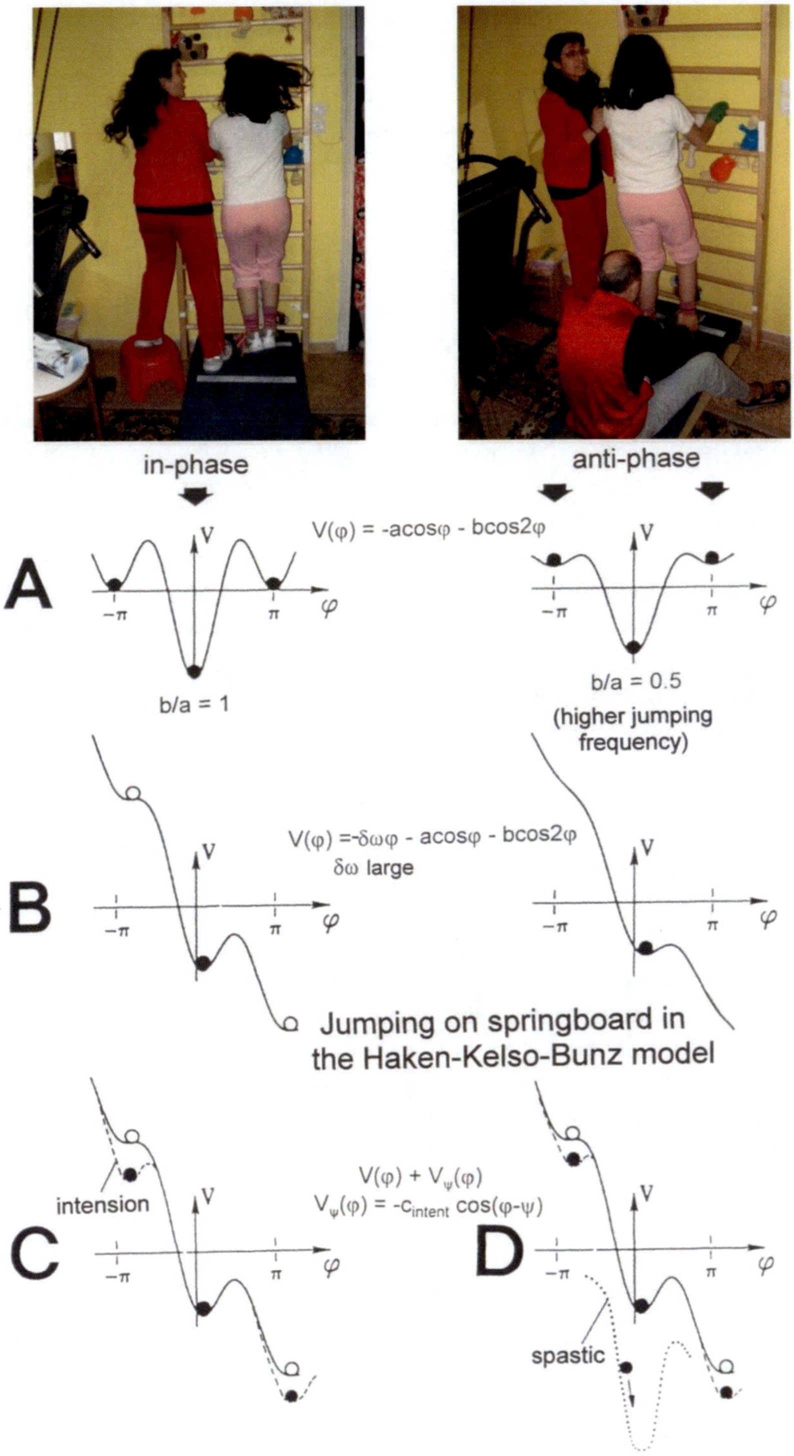

Figure 4. The jumping on springboard in in-phase and in anti-phase, analysed by the Haken-Kelso-Bunz model in the framework of coordination dynamics. The stability of jumping patterns is represented by the potential wells (derived from the formulas) and a ball moving in the potential landscape. Dark ball = stable state (attractor state), white ball = unstable state. In "A," the CNS injury is small; in "B" and "C" the injury is more severe with impaired symmetries.

The learning process may be hampered by a deficiency of the neuronal structure on which they are dependent. Stimulating the patient to use uncommon modes of operation may enable him to achieve results that he cannot achieve spontaneously. If an adult patient or child cannot jump in anti-phase on springboard (Figure 4) continuously, he may learn it by including, for example, a swing in between the jumps. Elements of function also have to be learned, which are necessary for more complex motor patterns. If one recognizes which learning processes are deficient, one may then be able to offer the infant a cue to start other learning processes or help him to eliminate errors and to progress in the right direction. If treatment strategies become more personalized to the needs of individual infants or children, it may become more difficult to evaluate the effectiveness of any one treatment program. But if CDT would be ten times more efficient than conventional treatments, differences in effectiveness can be found. Proper treatment of the handicapped infant or child consists of close cooperation between an active parent and an active therapist.

6. Learning for Repair

Pathology: Because of experienced adverse perinatal events, such as hypoxic ischemic encephalopathy, bronchopulmonary dysplasia, or bacterial meningitis, we may find retarded, accelerated or deviant development of motor and other functions. Some functions may not develop at all, while others show only a decrease in variability. Combinations of these possibilities may occur within single infants and result in individual motor and other patterns. The deviant patterns are characterized by a lack of variability and assume a thoroughly stereotyped appearance.

In most cases, some parts of the brain are damaged more than others, and some parts are not damaged at all. The undamaged parts will try to become operational. In correspondence with the localization, extent and type of the damage, it may happen that processes that in themselves are normal cannot bring about good results because some areas of the brain which are also necessary for the accomplishment of the particular motor function are deficient. For example, certain functions that have developed normally cannot be integrated into larger patterns because other elements of these larger patterns are missing or have not developed sufficiently. In general, both impaired and healthy parts of the brain mature over time and thus lead to increased complexity, which has direct repercussions on the quality of the learning (trial and error-elimination [26], see below) processes.

The well-known symptoms and signs of cerebral palsy in the first year of life (poverty of movements, stereotypy of posture and motility, inability to “discover” new motor possibilities, neglect of one extremity, stereotyped extension of the legs during vertical suspension, head-lag during the traction test or during sitting) can all be traced back to a lack of trial-and-error-elimination processes (learning) as a consequence of deficient brain structure. In the case of an inability to “discover” new motor possibilities, there may be a disturbance in the chain of events because errors are not recognized (or not eliminated), with the result that the processes stop prematurely [73].

Problem solving therapy by learning during (deviant) development: From the therapeutic point of view, trial-and-error-elimination in abnormal infantile development can be of great value in that it may indicate which type of treatment is needed. It is of crucial importance to recognize the deficiency or absence of particular functions and learning processes. If one recognizes which processes are deficient, one may then be able to offer the infant a cue to start other processes or help him to eliminate errors and to progress in the right direction. It may be sensible to "teach" the infant specific modes of operation to avoid wrong postures or wrong movement patterns or to stimulate him to use uncommon modes that may enable him to achieve results that he cannot achieve spontaneously. It has to be assessed which functions are left intact and the ways in which the child can avail himself of these functions to promote further development [73].

Motor learning and problem solving therapy by learning: Motor learning, such as learning to ride a bicycle or play piano, is partly different from the learning for CNS repair or optimizing development. The memory of motor skills (nondeclarative memory) such as riding a bicycle is called procedural memory. Nondeclarative memory is formed in brain regions other than the hippocampus (probably in the sensory-motor cortex and brain stem nuclei). The problem-solving learning tries to repair subnetworks that are necessary for functioning and learning. By inducing trial–and-error-elimination processes in subunits of the normal developing nervous system, an optimal development is achieved [26]. To teach the injured CNS to repair itself by trial–and-error-elimination processes, the CNS has to, similarly to the development, recognize upon CDT which subnetworks, regulation units or subloops are not functioning properly (or are missing) and to repair them by error elimination, including the possibility that other brain parts partly take over functions, and subnetworks build anew to a limited extent.

1.4. Outline

The first part of development is studied longitudinally in an infant using EMG and motor performance assessment. From five years of age on the coordination dynamics data were collected from pupils from the 1st class of a primary school until the 12th class of a high school. Because of inter-individual variability of the cross sectional study certain aspects of sequence of development seem not to come out as pronounced as in a longitudinal study. A general longitudinal study would be better for analysing development, but it is difficult to organize. A correlation of motor performance to higher mental functions (academic level, effort, speed of learning, will, interest, attitude to others) is tried.

For comparing the development with the repair over years, patients with cervical spinal cord injury, a cerebral palsy child, and three children who suffered a traumatic brain injury will be mainly considered. It will turn out that human development cannot be recapitulated by repair. Still, there seem to be some CNS organization strategies (to achieve coordinated firing of neurons, change of motor functions) that are similar during development and repair. Based on electrophysiologic measurements, it will be shown that the natural impulse patterns in many single afferent fibres stimulated by touch and pinprick cannot be simulated by electrostimulation.

As was shown in Chapter I, exogenous stem cell therapy will most likely not work because of not including electrophysiologic data on how to integrate newborn neurons into the existing injured neural networks – it will turn out in this Chapter again that patients get conventionally false treatment and have to suffer, because the treatments are not based on human neurophysiology (especially electrophysiology to measure function) and clinical research. Since the human CNS functions very accurately (non-chaotic) and the dynamic self-organization is very complex, there seems to be no easy way for causal repair.

2. Method

2.1. Theoretical Basis of the Schalow Coordination Dynamics Therapy

If we want to understand the functioning and the repair of the human central nervous system (CNS), we have at least to understand CNS functioning at the neural level, neural assembly level, muscular and movement level. We have to discover or identify the laws or principles of CNS organization at the different levels of organization. The link between neuronal activities (microscopic events) and behaviour (macroscopic events) consists of collective effects (pattern formation) at the microscopic level that create macroscopic order and disorder. At the neural level, communication takes place by cooperation and competition.

To compare within the System Theory of Pattern Formation the motor development with the repair, the same assessment for measuring the systems functions of CNS organization of pupils and patients is used: the coordination pattern dynamics for exercising on a special coordination dynamics therapy (CDT) device.

1. Movement Level: Coordination Dynamics for CNS Assessment during Development and Repair on the Basis of the System Theory of Pattern Formation

By cooperative and competitive interplay, the many billions of neurons of the human CNS generate dynamics in self-organization that cannot be explained by the properties of the single neurons and neuron assemblies alone. Therefore, a dynamic system theory of self-organization and pattern formation of the human CNS was derived from the concepts of synergetic. Synergetics is the theory of self-organization and pattern formation in non-equilibrium systems [95,96]. Since the amount of information necessary to describe individual states of neurons is very large, ways must be found to select relevant variables to compress the amount of information. In synergetics, methods have been found to compress system complexity to only one or a few degrees of freedom, the so-called order parameters, whose dynamics (equations of motion) are low dimensional. This dynamic system theory of pattern formation is applied here also to the injured or malfunctioning CNS (on the macroscopic level) to improve CNS functioning by learning including structural repair.

The organization dynamics of the human CNS, viewed through the window of coordinated movements, are termed coordination (pattern) dynamics. Collective variables or order parameters capture the collective coordinated activity of the neurons of the CNS. The specific equations of motion (the dynamics) of these collective variables generate the time course of organizational states.

The collective variables can be designated by the vector X, and the coordination pattern dynamics can abstractly be formulated by equations of motion [69]:

$$dX/dt = F_{intr}(X) \qquad (1)$$

The functional form of the right-hand side of the equation is called vector field and must be modelled for practical application. The major constraint of such modelling is that stable, reproducible patterns are asymptotically stable solutions (called attractors) of these dynamics. If the constraints do not specify a particular pattern, the coordination dynamics are called intrinsic dynamics (F_{intr}). For a mathematical solution of the intrinsic dynamics (F_{intr}) for the jumping on the springboard in the Haken-Kelso-Bunz model, see below.

A further step on the dynamics of the system theory of coordinated pattern formation is to map the behavioural patterns (observed coordination patterns) onto attractors of the order parameter dynamics. In a complex system like the CNS, patterns can be generated by the system seeking cooperative stability. Stability is what defines collective states. The system has the tendency to slip into the collective states to which it is attracted. Commonly attractive states and attractors are pictured as a ball in a potential well or more generally in an attractor layout (see Figure 70E,G). When an infant crawls, its arms and legs are strongly attracted to the pace and trot gait patterns (attractor states). Changes in CNS functioning are characterized as continuous stabilization and destabilization, over time, of preferred attractor states.

However, coordination patterns are not only determined by the task or biological function. Patterns adjust continuously to requirements from the environment, memory, intention, and support given by the therapist. All the specific requirements are captured by the concept of behavioural information and are made part of the vector field that attracts toward the required patterns. The coordination pattern dynamics, characterized by equations of motion of collective variables, takes now the following form [69]:

$$dX/dt = F_{intr}(X) + \sum c_{inf} F_{inf}(X,t) \qquad (2)$$

where F_{intr} designates the intrinsic dynamics of nervous system. These intrinsic dynamics capture the anatomical (neuronal network structure), physiological, and pathological states of the CNS and its muscular-skeletal elements.

$\sum c_{inf} F_{inf}(X,t)$ represents the sum of external influences ($F_{inf}(X,t)$) with their relative strength (c_{inf}) pertaining to each influence. The so-called behavioural information $F_{inf}(X,t)$ includes cognitive states, emotional states, intentions, motivations, instructions, interpersonal coordination, movement support, etc. While applying therapy to the patient, these extrinsic influences become extremely important, because the intrinsic (pattern) dynamics can be changed with these extrinsic influences by altering the equation of motion. By modulating the behavioural information, the intrinsic dynamics of the neuronal networks can be influenced further, if CDT is no longer efficient anymore in repairing the injured CNS (updating of the therapy).

If the behavioural information includes the exercising of extremely coordinated, integrative movements, like exercising on the special CDT device for turning, then the quality of the CNS self-organization can be enhanced by improving the exactness of self-organization, namely the precision of phase and frequency coordination between neuron and neural assembly firing (see below). By improving the precision of organization of the intrinsic dynamics, that is, the specific variability of the injured networks, certain patterns do reappear.

2. Clinical Implications for Treatment Derived from the Equations of Motion of the Collective Variables (Formula 2)

1. Behavioural requirements F_{inf} (like intention, support, and instruction) affect the whole coordination dynamics, including stability rather than only certain coordination patterns [69]. The change of the whole coordination pattern dynamics of

the CNS by the behavioural information is one scientific basis for learning transfer [63] between different patterns and stability changes of patterns (as, for example, the reduction of spasticity).

2. Intrinsic coordination tendencies captured by the intrinsic dynamics influence the performed pattern systematically because the degree to which intrinsic tendencies conflict or agree with the required patterns determines the variability of the performed coordination pattern [69].

3. Reduction in stability (of movement and other patterns) when intrinsic and informational requirements conflict may lead to loss of stability and abrupt change as behavioural information is changing smoothly [69].

4. The intrinsic dynamics F_{intr} include vegetative and higher mental functions (these are also patterns of the coordination dynamics), which indicate that via exercising coordinated movements with support and instructions (F_{inf}), urinary bladder function, intelligence and speech may partly be repaired following CNS injury or malfo-rmation.

5. When in an injured CNS with a certain set of behavioural information ($\sum c_{inf}F_{inf}$), the intrinsic coordination dynamics (F_{intr}) cannot be changed anymore during coordination dynamics therapy, then this set of behavioural information has to be changed (using different F_{inf}) or balanced differently (using different c_{inf}) to further improve CNS organization dynamics.

6. However, the equations of motion of the coordination pattern dynamics (Formula 2) provide no information about the specific behavioural information (F_{inf}) and training intensity (c_{inf}) with the use of which the CNS can be efficiently repaired in the given patient. We need to have detailed knowledge of the human CNS on the single neuron and neural assembly level [39-38,46,60-63,65,66], besides the knowledge at the integrative level, to find the specific behavioural information for the repair of the human CNS.

7. The first novel step in coordination dynamics therapy is the inference derived from Formula 2 of the equation of motion. It suggests that the movement training not only improves the performance of that particular movement but also improves the other non-trainable functions by transfer of learning [63]. These functions include vegetative functions like bladder control, speech, and higher mental functions. Furthermore, we have an applicable tool at hand with which the stability of physiological network states can be increased (e.g., movements, continence, continuous concentration to certain tasks, speech, etc.). The same tool can then be used to decrease the stability of pathological network states, like spasticity. The coordination (pattern) dynamics therapy partly based on the System Theory of Pattern Formation thus offers us an important theoretical basis and a practical tool to diagnose, quantify and repair the malfunctioning human nervous system at the macroscopic level (see below).

3. Geographical Landscape of Attractors

The drawback of the equation of motion of the order parameters (formula 2) is that it is normally not possible to find a mathematical solution to it. But by defining a potential function and by picturing the attractive states and attractors by a ball in a potential well or rather by a ball moving in a geographical landscape of attractors [69], we get a theoretical basis to measure the stability of certain coordinated movement patterns (i.e., the deepness of the potential well of an attractor) in patients with CNS injury who are receiving ongoing therapy. By studying the pattern change of certain highly coordinated arm and leg movements while a subject is exercising on a special coordination dynamics therapy device, pattern stability can be made visible, and the mean stability per one minute can be measured by the arrhythmicity of exercising. Such value, so-called coordination dynamics value, quantifies CNS functioning objectively, integratively, and non-invasively. The assessment of quality of CNS organization by pattern change is a second novel step in CDT. To make the strategy of pattern formation, pattern stability, pattern assessment, and pattern picturing understandable, the procedure is demonstrated for the simple movement "jumping on springboard," which is used during CDT, especially for the repair of the urinary bladder and training in the upright weight-bearing posture (very important in patients with SCI).

4. Equation of Motion, Potential Function and Attractor Layout for the Movement "Jumping on Springboard"

For the special movement "jumping on springboard" with no behavioural information ($\sum c_{inf} F_{inf}(X,t) = 0$), the equations of motion (formula 2) take the form:

$$d\varphi/dt = f_{intr}(\varphi)$$

where φ is the relative phase between the two moving legs and is the only collective variable of this special movement. The mathematical solution of $d\varphi/dt = f_{intr}(\varphi)$ in the Haken-Kelso-Bunz model [12,19] (for the approximations being made, see Ref.50) gives the equation of motion for jumping on a springboard for the symmetric case [52]:

$$d\varphi/dt = -\,a(t)\sin\varphi - 2b(t)\sin 2\varphi$$

The so-called potential function is defined by

$$d\varphi/dt = -\partial V(\varphi,t)/\partial\varphi \quad (19).$$

By integration, we obtain the potential function for jumping on a springboard:

$$V(\varphi,t) = -\,a(t)\cos\varphi - b(t)\cos 2\varphi$$

The potential function $V(\varphi,t) = -\,a(t)\cos\varphi - b(t)\cos 2\varphi$ can be plotted for different φ, and certain ratios of the parameters a and b and is shown in Figure 4.The potential function shows two attractor states, namely the jumping in in-phase ($\varphi = 0$) and the jumping in anti-phase ($\varphi = \pm\,\pi$). Especially for higher frequencies (smaller b/a), the jumping in-phase has a higher stability (the potential well is deeper) than the jumping in anti-phase. Asymmetry (not tackled mathematically here) strongly changes the stabilities of the attractor states (depths of potential wells) (Figure 4).

The human CNS, seeking for cooperative stability, slips into the collective states to which it is attracted. For jumping on springboard, these attractive states are the jumping in in-phase and in anti-phase. For crawling (not creeping), the attractive states are the pace (in-phase) and in trot gait coordinations (anti-phase). Since such a potential function can no longer be derived from more general movements, especially when the CNS is injured, malformed or degenerating, the temporal stability of different movement patterns for a characterization of CNS functioning has to be measured. This is partly possible by measuring the so-called coordination (pattern) dynamics (see below).

5. Assessment of the Geographical Landscape of Attractors, Pattern Stability and Coordination Dynamics (CD) Values of Patients with CNS Injury

The dynamics of CNS self-organization at the neuron level by phase and frequency coordination (see below) is partly reflected in the temporal stability of coordination patterns, which can be assessed through a process of pattern change. The non-invasive measurement of pattern stability is the third novel step in CDT. When a healthy subject or patient exercises on the special CDT and recording device (Figures 1,17,19), the device is imposing a spectrum of continuously changing coordinations of arm and leg movements between the two attractor states "pace gait" and "trot gait." On the CD trace, the arrhythmicity of turning (df/dt or df/dt/f; f = frequency of turning) is increasing and decreasing and thus forming an attractor layout of states of stability for this range of movement patterns (Figure 70). It is possible to judge whether the attractor states are positioned at the pace and trot gait patterns or shifted away from them because of pathological or suboptimal CNS organization. The mean arrhythmicity of turning (the mean stability) over one minute of exercising is referred to as the CD value and is used to quantify the quality of CNS organization with ongoing therapy. Since the quality of CNS functioning can be quantified by a single value, progress in CNS functioning can be evaluated in percentages and can be used to judge improvements in CNS functioning in patients during therapy or in children during development [50].

6. Symmetry Impairment and Its Repair

In addition to stability, symmetries also structure the state space of a multicomponent system with multiple patterns. Also, embryologists were discovering that the embryo developed in a symmetrical manner. Furthermore, if the symmetry is impaired (if $\delta\omega$ is large in the potential function $V(\varphi)$ in Figure 4B) in a patient with CNS injury, a relatively lower level of fluctuation (due to increased phase and frequency variability (see below) or a perturbation) is required to "kick" the system out of its present state than is required with a symmetric potential well (Figure 4). The CNS organization switches more easily from movement state into spastic state or into another movement state. The movement jumping on a springboard in anti-phase becomes more difficult in the asymmetric case of CNS functioning than in the symmetric case because of stability reduction. The patient switches often into the in-phase jumping during jumping in anti-phase or cannot jump anymore in anti-phase (Figure 4).

During therapy, the movements have to be trained with respect to symmetries in order to repair the different impaired symmetries of CNS functioning to increase the stability of the physiologic network states.

Symmetry in forward-backward movement seems to be an important symmetry to be exercised. The impairment of forward-backward symmetry can be measured (and reduced) when the patient is exercising on the special device in the forward and backward direction.

7. Finding the Behavioural Information (the Treatment) for the Repair of the Intrinsic Dynamics of CNS Self-Organization from Human Neurophysiology and Pathophysiology

The fourth novel step in CDT and the "System Theory of Pattern Formation for Repair" is that we can partly find from human neurophysiology and pathophysiology the behavioural information ($\sum c_{inf} F_{inf}(X,t)$) (the therapy), through which it is possible to change and repair the intrinsic dynamics ($F_{intr}(X)$) of CNS organization.

The learning transfer for bladder repair [60-64] can be achieved by the movement "jumping on a springboard" (Figure 4), even though the behavioural information's "exercising on the special device for turning" (Figures 1,17) and "supported treadmill walking" (Figures 17,37) also contribute to the learning transfer to rebuild an attractor for physiological bladder functioning (Figure 5) following severe cervical spinal cord injury [64].

But because in severe cerebral palsy the continence could be cured (see repair in cerebral palsy, below) before the child learned the in-phase jumping on springboard (Figure 4, upper left, anti-phase jumping was only possible with support (upper right)), it seems that the improvement of the coordinated firing of neurons is more important than stimulating via rhythmic pelvic floor movements the sacral micturition centre for repair [63].

Below, it will be shown that with two different methods, the phase and frequency coordination between neuron firing for CNS self-organization can be measured and that this coordinated firing is getting impaired following CNS injury, degeneration, or malformation. The improvement of the coordinated firing of neurons (if impaired) is of fundamental importance for the physiologic functioning of the human CNS.

8. Relative Coordination between Neuron and Neural Assembly Firing as an Organization Principle of Human CNS Organization (Human Neurophysiology)

Afferent and efferent extracellular action potentials can be recorded from lower human (thin and long) sacral dorsal and ventral nerve roots (Figure 6).

Because of missing epineurium, the action potentials are not electrical shunted as in peripheral nerves and can be detected. The recorded summed natural impulse traffic can be split into sets of afferent and efferent impulse patterns of single-nerve fibres (Figure 6) and schematically drawn (Figure 7).

It can be seen that there is relative phase and frequency coordination between the action potentials of the firing patterns upon natural stimulation. Relative coordination means complicated coordination, especially frequency coordination, which can be nicely seen between spindle afferents and oscillatory firing motoneurons when distribution histograms are constructed (Figure 8).

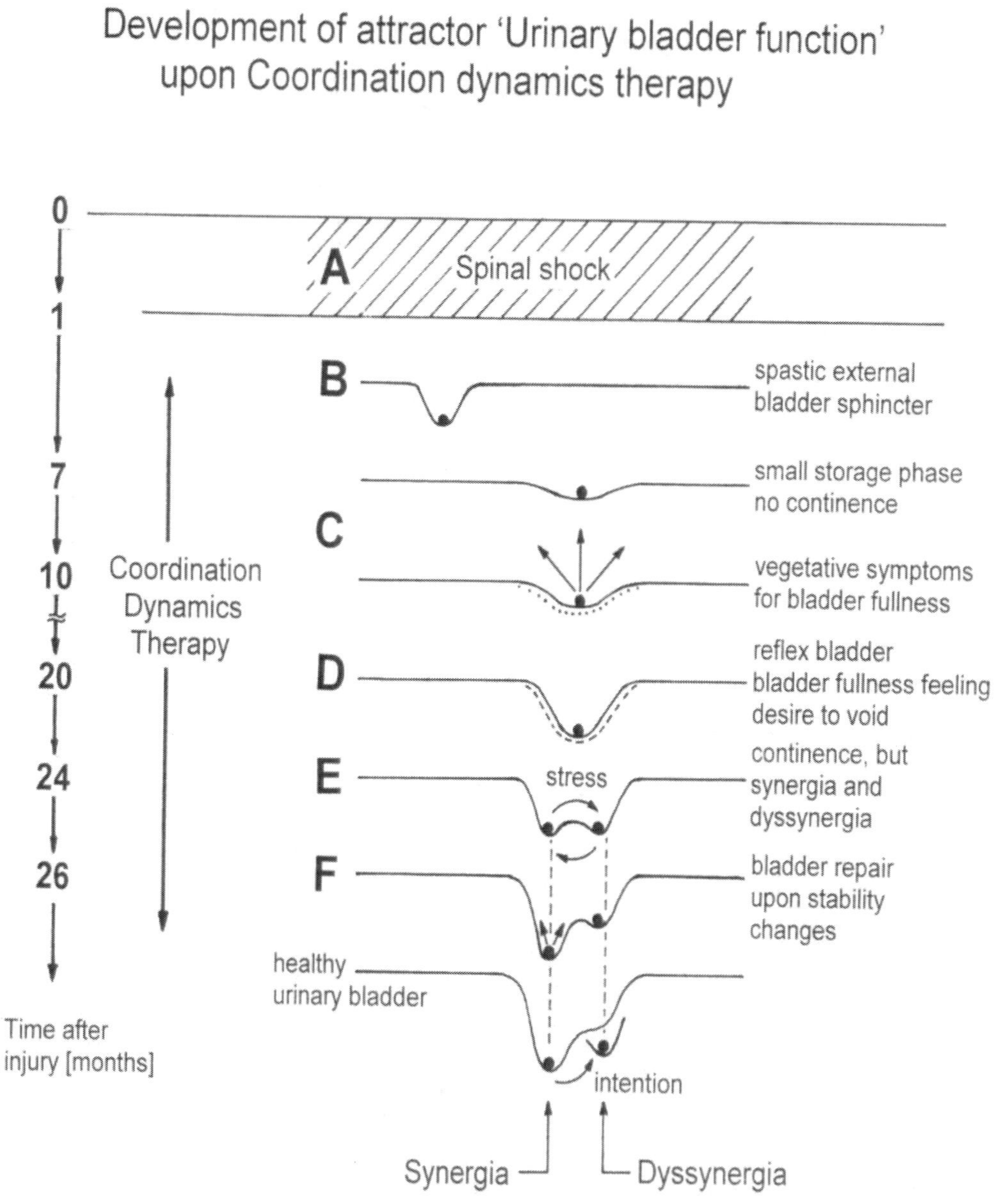

Figure 5. Evolution of the attractor layout of bladder functioning induced by learning transfer from movements to bladder functions upon CDT. The region around each local minimum of the potential landscape acts like a well that weekly traps the system into a coordinated state. Black balls correspond to stable minima of the potential. With learning the pattern "spasticity of the external bladder sphincter" vanishes, and the patterns for bladder functioning ("synergia" and "dyssynergia") appear anew and gain their physiologic stability (physiologic deepness of each basin of attraction). The corresponding attractor layout for physiologic bladder functioning is given. Fluctuation of pattern state (the black ball) (C), and their decrease (F) due to the impairment of phase and frequency coordination of neuron firing, is pictured in "C" and "F" by long and short arrows. Dotted and dashed lines indicate the reoccurrence of bladder sensation. Note that more than two years of optimal CDT were needed for bladder repair.

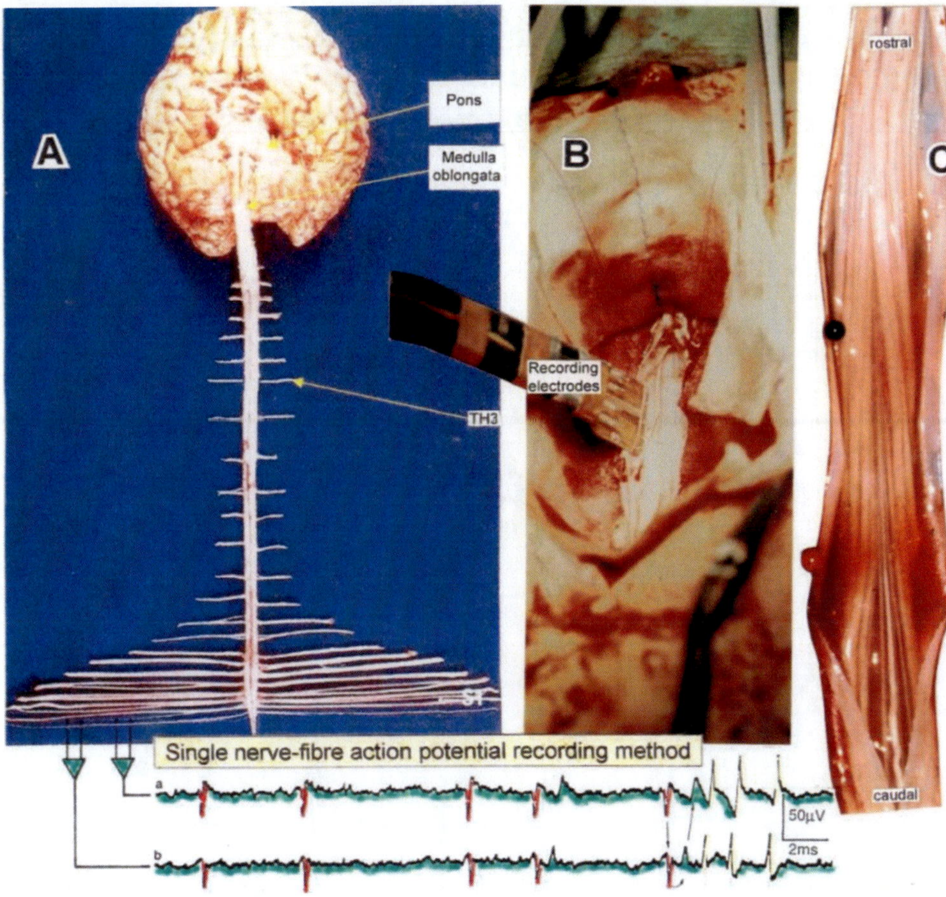

Figure 6. Layout of the recording of single-nerve fibre action potentials to analyse the self-organization of neuronal networks of the human CNS under physiologic and pathophysiologic conditions. A,B,C. By recording with two pairs of platinum wire electrodes (B) from sacral nerve roots (cauda equina, C) containing between 200 and 500 myelinated nerve fibres, records were obtained in which single-nerve fibre action potentials (APs) were identified from motoneurons (main AP phase downwards) and afferents (main AP phase upwards). By measuring the conduction times and with the known electrode pair distance (10 mm), conduction velocity distribution histograms were constructed in which the myelinated nerve fibre groups larger than 4µm could be characterized by group conduction velocity values. After the recording, morphometry was performed. Distributions of nerve fibre diameters were constructed, and nerve fibre groups were characterized by the peak values of asymmetrical distributions. By correlating the peak values of the velocity distributions with those of the diameter distributions obtained for the same root, a classification scheme was constructed of the human peripheral nervous system. It became thus possible to record natural impulse patterns simultaneously from identified single afferent and efferent nerve fibres and to analyse self-organizing mechanisms of the human CNS under physiologic and pathologic conditions. A. Human CNS with the schematic illustration of the recording layout and original record of single-nerve fibre action potentials. Note the time calibration of 2ms. B. Intraoperative recording layout (when implanting a bladder stimulator) with two pairs of wire electrodes and one temperature sensor. A thin nerve root is positioned over the platinum wire electrodes. C. Dissection of the human cauda equina; at the caudal end of the filum terminale with thin nerve roots are seen. D. Schematic splitting of the activity of several nerve fibres into simultaneous impulse patterns of single fibres by comparing waveforms, conduction velocities and reoccurring characteristic impulse patterns (rhythmic firing of sphincteric motoneurons). The different conduction times and waveforms were recognized on an expanded time scale. Stretch receptor and secondary muscle spindle afferents contribute to the drive of sphincteric motoneurons and form, together with other afferents, regulation units.

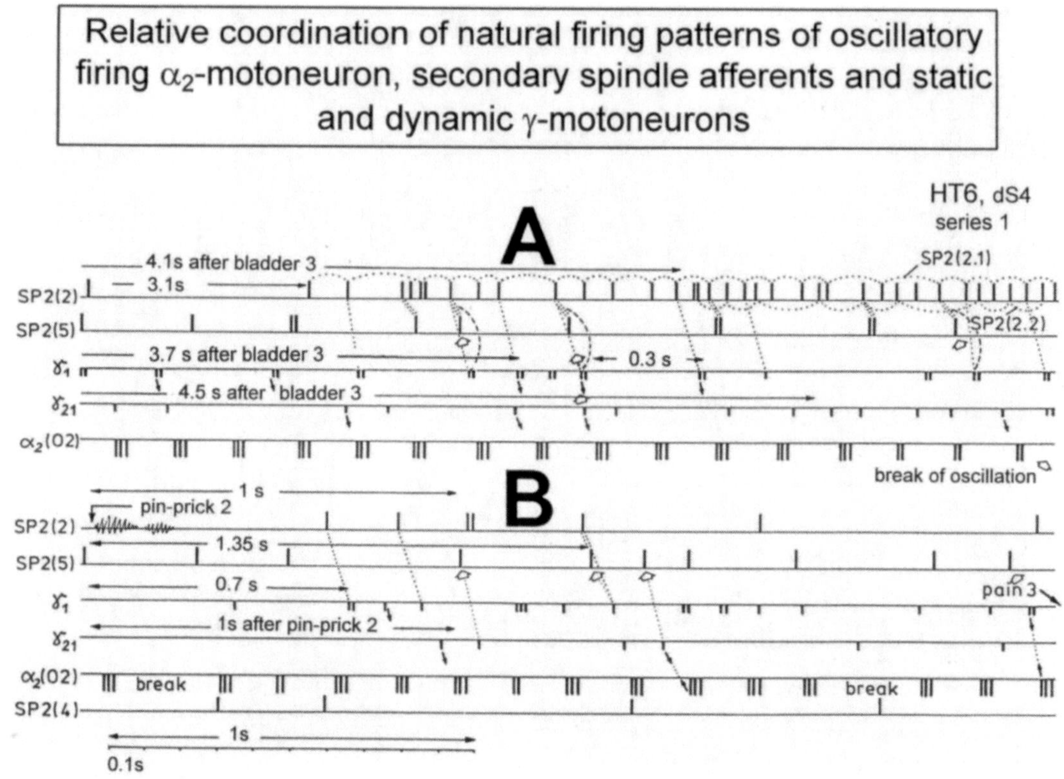

Figure 7. Phase and frequency coordination between the extracellular action potentials (APs) of simultaneously recorded γ-motoneurons (γ_1 and γ_{21}), secondary muscle spindle afferent fibres (SP2(2), SP2(4), SP2(5)) and oscillatory firing α_2-motoneuron O2 following bladder catheter pulling (bladder 3) (A) and pinprick (pinprick 2) (B). B was recorded before A. In A, the impulse patterns of the two encoding sites SP2(2.1) and SP2(2.2) of the single parent fibre SP2(2) are indicated by the dotted curves. Times to activity increases of γ-motoneurons and secondary muscle spindle afferents following stimulation are indicated. Similar time intervals of the occurrence of γ-motoneuron (APs) and SP2(5) fibre APs (phase coordination) are indicated by the open arrows, and the similar time intervals of γ-motoneuron APs and α-motoneuron APs are indicated by the small arrows. Similar time intervals of the APs of fibres SP2(2) and SP2(5) are indicated by the double dotted lines, those of γ_1-APs and the SP2(2) fibre APs by the dotted line, and those of γ_1-APs and the SP2(2)-SP2(5) correlation by the curved dashed line. Brain-dead HT6; dS4-root.

Rhythmic motor unit firing can also be recorded by using surface electromyography (sEMG). By comparing the single-nerve fibre recordings with the sEMG recordings, it can be seen that the invasively obtained motoneuron patterns are the same as the non-invasively obtained sEMG patterns (Figure 9).

Relative phase and frequency coordination can also be followed up when subjects are moving by the non-invasive and not painful sEMG.

By using human patients, to whom CDT has been administered, coordinated firing between several single-motor units can be observed in the same muscle (Figure 10) and also in different muscles (Figure 11).

Relative coordination between neuron firing can, therefore, also be observed between neurons sited in different CNS parts.

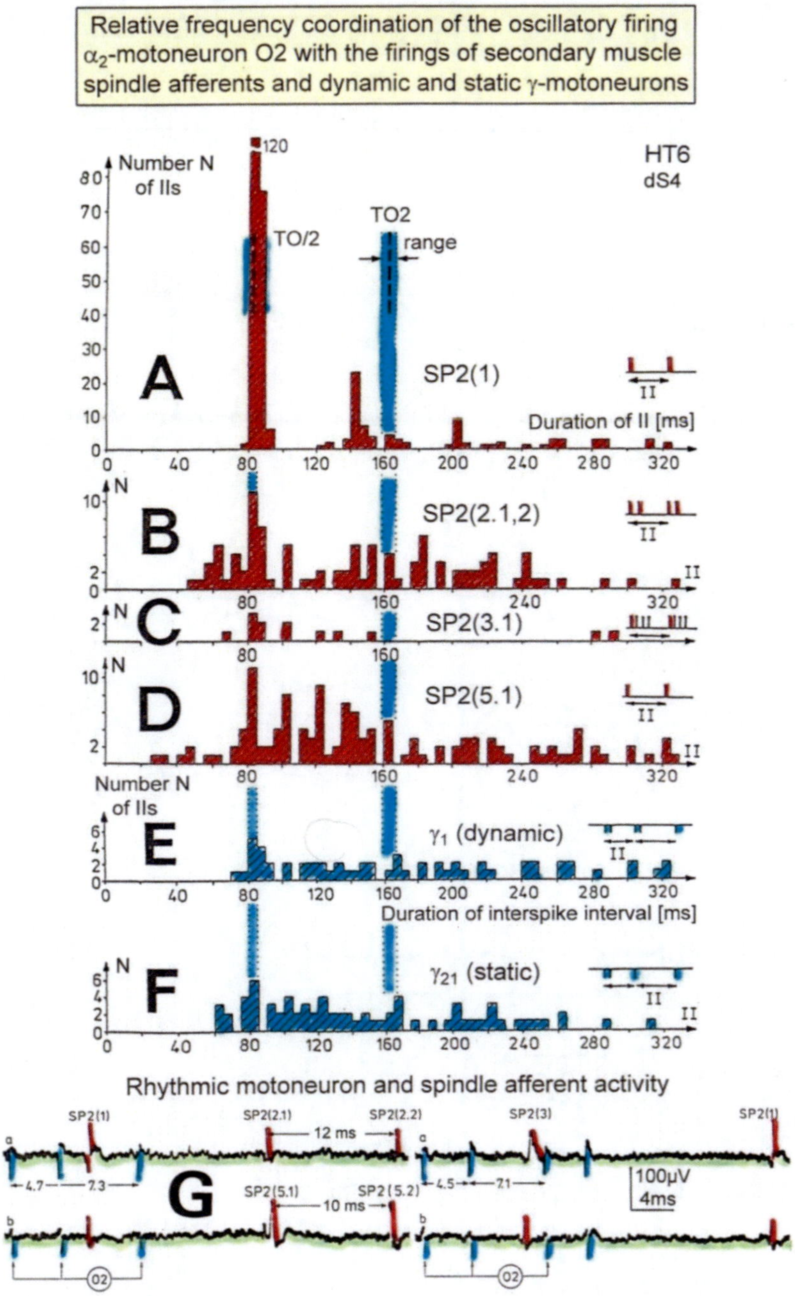

Figure 8. Interspike interval distributions of single endings of four secondary muscle spindle afferents (SP2) and two γ-motoneurons, recorded simultaneously. In A, the oscillation period TO2 (impulse train length = 3 APs) with its range of simultaneously recorded oscillatory firing α_2-motoneuron O2 (see G) is drawn for comparison; also, the halves of the oscillation period TO2/2 are indicated. Note that the interspike interval distributions of spindle afferents and γ-motoneurons have shortest interspike interval, nearly identical to the half of the oscillation period (relative frequency coordination). The schematic impulse pattern in A to F shows the procedure for measuring the interspike intervals. Original records of the firing patterns of α_2-motoneuron O2 and the secondary muscle spindle afferents SP2(1), SP2(2), SP2(3) and SP2(5) are shown in G. Brain-dead human HT6, dS4 root.

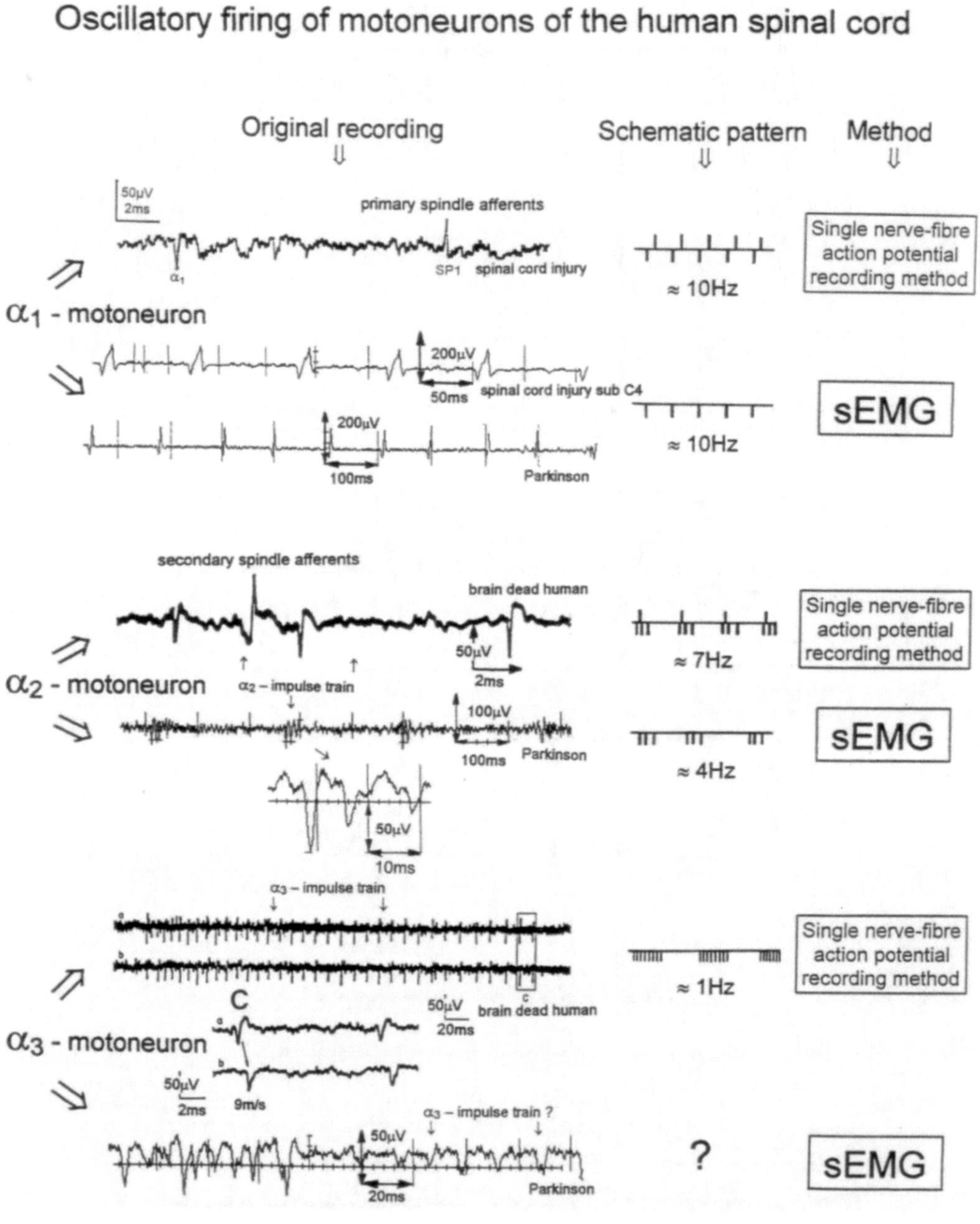

Figure 9. Oscillatory firing patterns of α_1, α_2, and α_3-motoneurons recorded from motoneuron axons with the single-nerve fibre action potential recording method and by surface electromyography (sEMG) from FF, FR, and S-type motor units. The left panel shows original recordings, the middle panel the schematic patterns; the recording methods are indicated on the right side. The recordings were taken from patients with spinal cord injury and Parkinson's disease and from brain-dead humans.

This coordinated firing of single human neurons with respect to time and space can be recorded from patients when performing natural movements. The relative coordinated firing between single neurons of the human CNS is documented by the single-nerve fibre action potential recording method and sEMG.

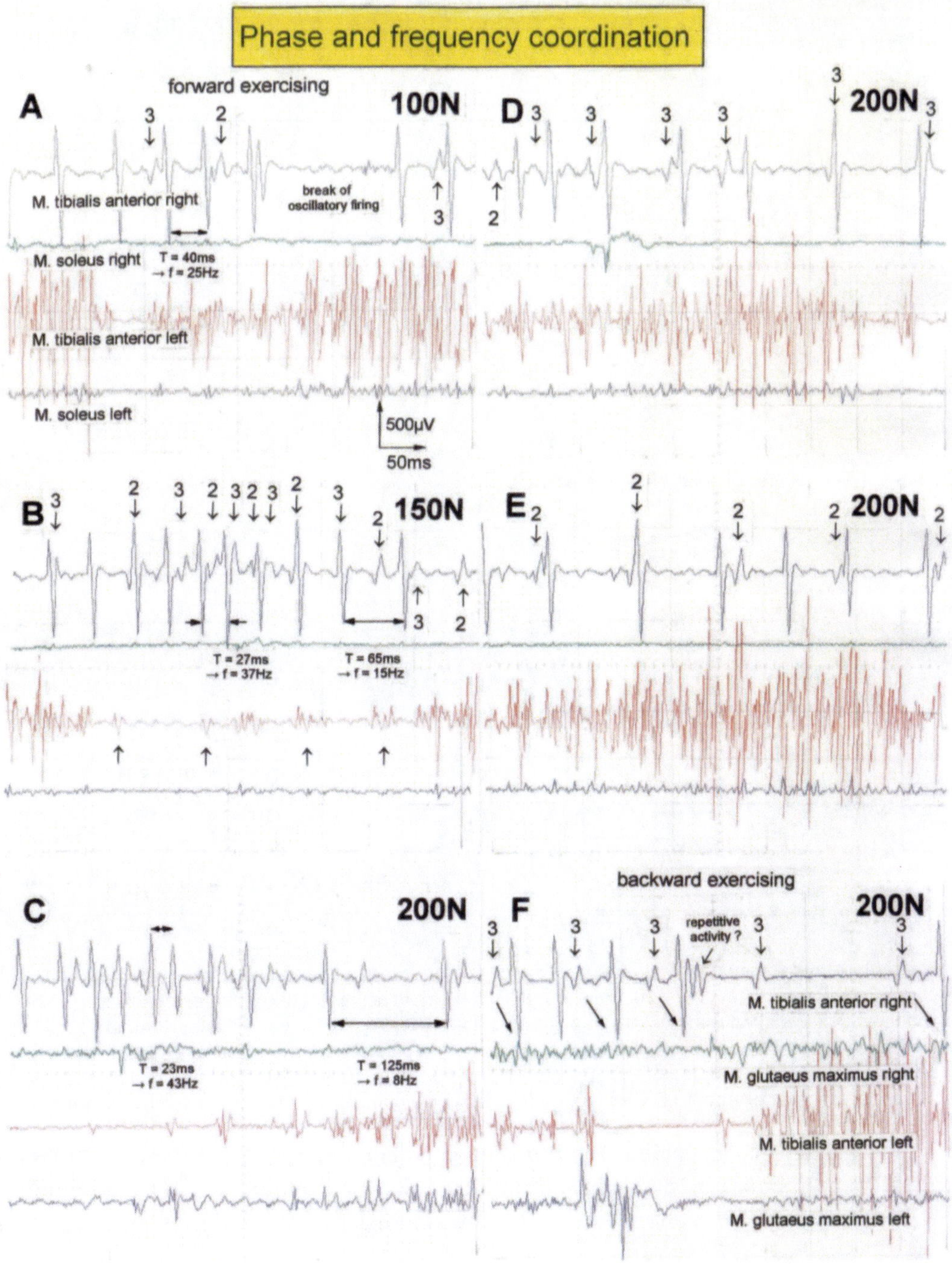

Figure 10. Phase and frequency coordination between oscillatory firing motor units (FF-type) during the generation of a motor program during exercise on the special coordination dynamics therapy device at loads increasing from 100 to 200 Newton. Oscillation periods (T) and oscillation frequencies (f [Hz]) of oscillatory firing motor unit 1 are partly indicated. "C,F" soleus electrodes shifted to gluteus muscles to check early reinnervation upon therapy. The waveforms of the three identified FF-type motor unit potentials "1," "2," and "3" are indicated; motor units "2" and "3" are partly marked. In "F," some coordinations between motor unit "3" and "1" are marked.

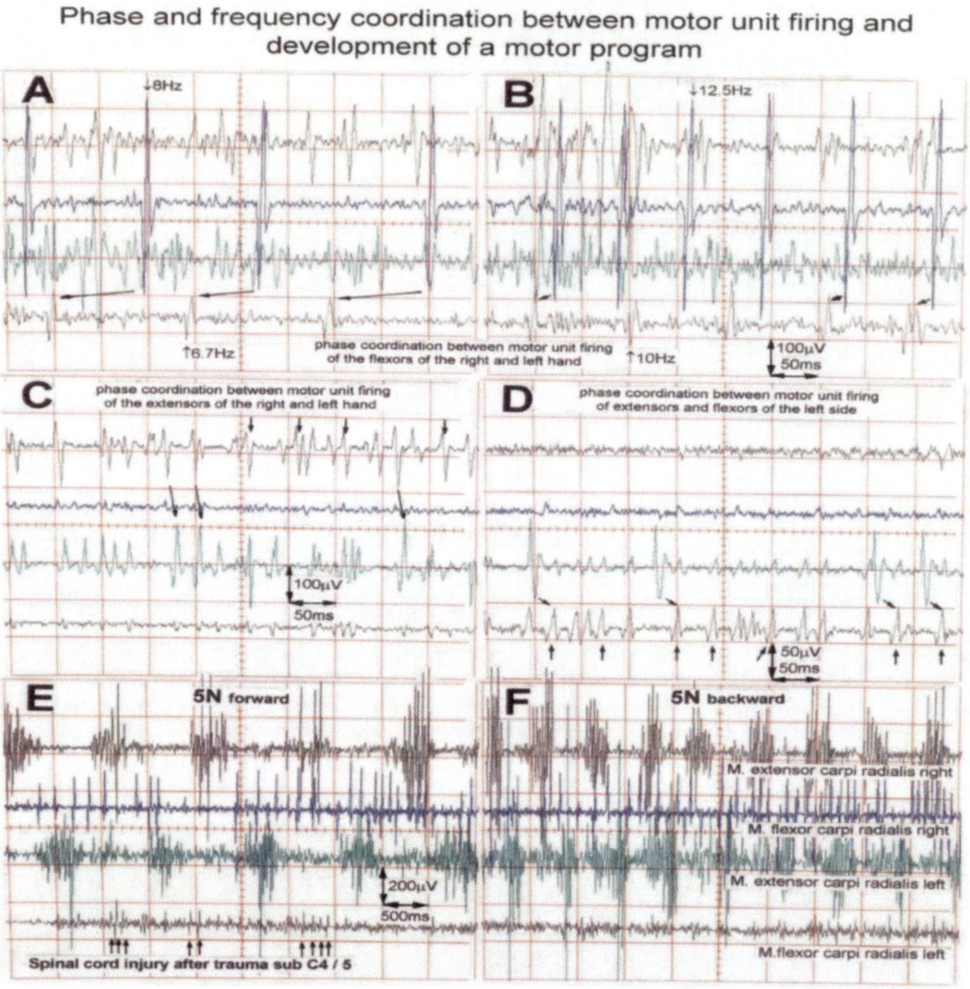

Figure 11. Phase and frequency coordination of single-motor unit firing between different muscles and between different arms. Phase coordinations are indicated by the arrows between motor unit potentials. A-D. The patient with a spinal cord injury was in position at the special coordination dynamics therapy device but activated muscles on volition upon looking onto the oscilloscope screen. E,F. Patient exercising on the special device. Note the beginning of the appearance of a motor program in the right flexor carpi radialis muscle for 5N forward turning. Recordings for a patient with a complete spinal cord injury sub C4/5.

The advantage of the single-nerve fibre action potential recording method is that relative coordination can be measured also between afferent and efferent fibres, and the advantage of sEMG is that the relative coordination can be observed non-invasively during movements.

9. Impaired Coordination (Human Pathophysiology)

Upon injury, malformation, or degeneration, the relative coordination between neuron firing becomes impaired and gives rise to pathologic CNS organization. The impaired coordination can again be assessed with the single-nerve fibre action potential recording method and sEMG. In a brain-dead human, the phase and frequency coordination is rather stable in the spinal cord (Figure 12). Following spinal cord injury, this phase and frequency coordination becomes impaired (Figure 13) in the way that it becomes instable and varies very much. Upon degeneration of the human CNS (in Parkinson patients), the physiologic coordination becomes impaired and motor units fire more synchronized (impaired inhibition among motoneuron firing) (Figure 14) with the consequence of tremor. For further details of the physiologic and pathologic coordinated firing of neurons, see references [46,61-63,65,66].

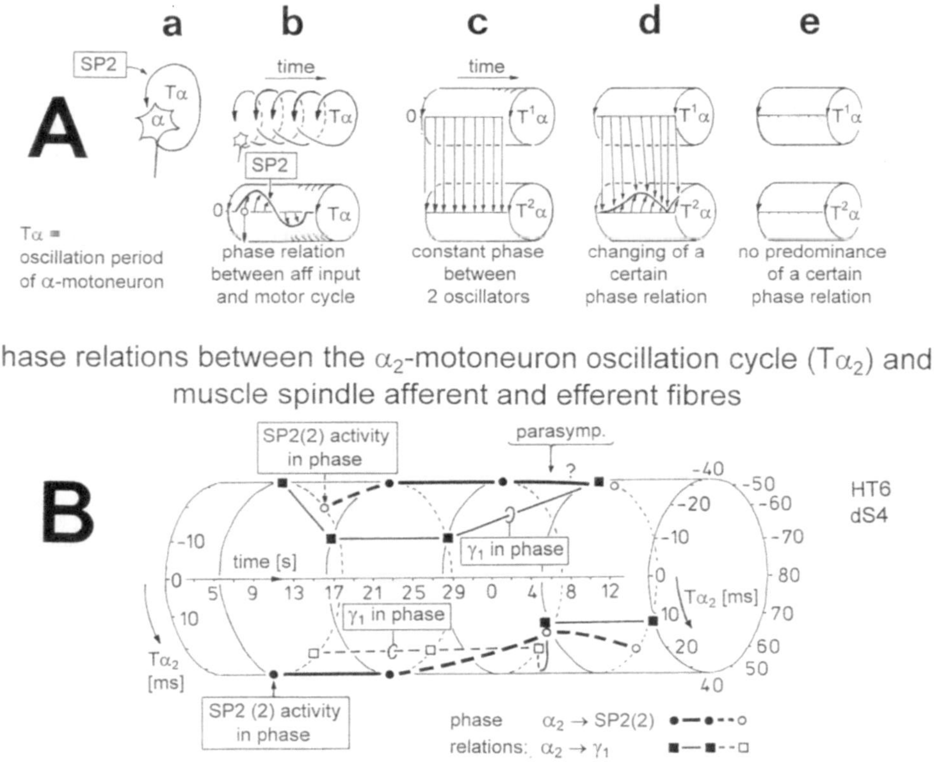

Figure 12. (A) Derivation of the simultaneous description of interspike intervals and phase relations. (a,b) The oscillation period of an oscillatory firing α-motoneuron is schematically characterized by the length of the loop (perimeter). Successive oscillation periods with ongoing time yield a cylinder. Flashing with a stroboscope on such a cylinder with the same frequency as that of the rotation of the cylinder would make a black spot on the turning cylinder not move up or down. If the frequency of the cylinder or the stroboscope changed slowly, the black spot would move up or down. If the black spot moves from left to right with ongoing time, a curve is obtained. By replacing the flashing of the stroboscope by the occurrence of the APs of the spindle afferent fibre (or another oscillatory firing motoneuron) with respect to the APs of the oscillatory firing motoneuron, phase relation changes are made visible in the lower part of "b" for a constant oscillation period (cylinder with no diameter changes). (c) A constant phase between two oscillatory firing motoneurons results to a constant line on the cylinder with ongoing time. (d) A changing phase gives a curve on the cylinder circumference. (e) If there is a loss of predominance of a certain phase between two motoneurons (the black spot gets diffused with ongoing time and is then lost), there is no line or curve. (B) Interspike interval and phase data from the brain-dead human HT6 (root dS4) are plotted in the representation of A. Filled dots and squares represent average phases (phase relations); thick and thin lines connect the dots to show trends. Note that the phase relations change only little; the frequency of the sphincteric α_2-motoneuron ($1/T\alpha_2$) changes only little - the cylinder does not change its diameter.

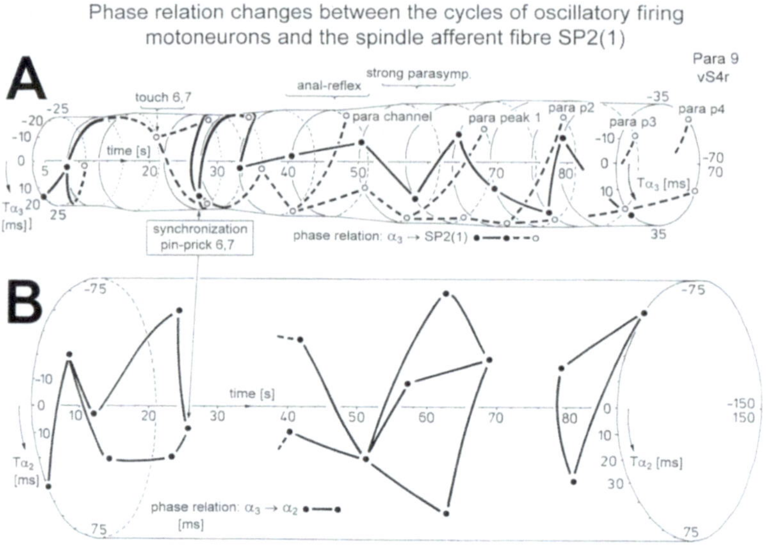

Figure 13. (A) Phase relations between the secondary muscle spindle afferent fibre SP2(1) and the oscillatory firing α_3-motoneuron, taken from Figs.10-12 (and additional data), are plotted on the oscillation period cylinder $T\alpha_3$ (mean oscillation periods are taken from Figs.11A,12A) according to Figure 21A. The cylinder is changing its diameter (perimeter) because the oscillation period changes. Phase changes in ms are scaled on the cylinder circumference. The ongoing time (to the right) is scaled on the axis of the cylinder (time intervals are taken from Figure 10A). Existing phase relations are represented by dots (filled and open (backside)); lines (filled and dashed (backside)) only connect the phase relations to show trends. para peak 1, para p2, para p3, para p4 = activity peaks of the SP2(1) fibre due to parasympathetic activation (see Figure 10A right). (B) Phase relations between the α_3 and α_2-motoneurons plotted onto the oscillation period cylinder of the α_2-motoneuron. Dots represent phase relations, taken from Figs.11B,12B. Note that the phase relations of the paraplegic 9 are much more variable than those of the brain-dead human HT6 (Figure 21B); also, the number of phase relations changes.

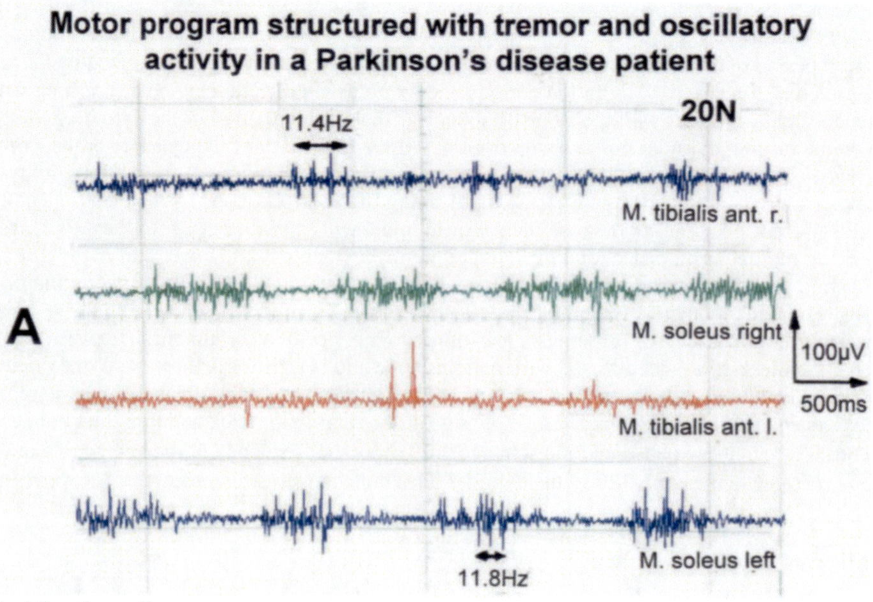

Figure 14. (Continued)

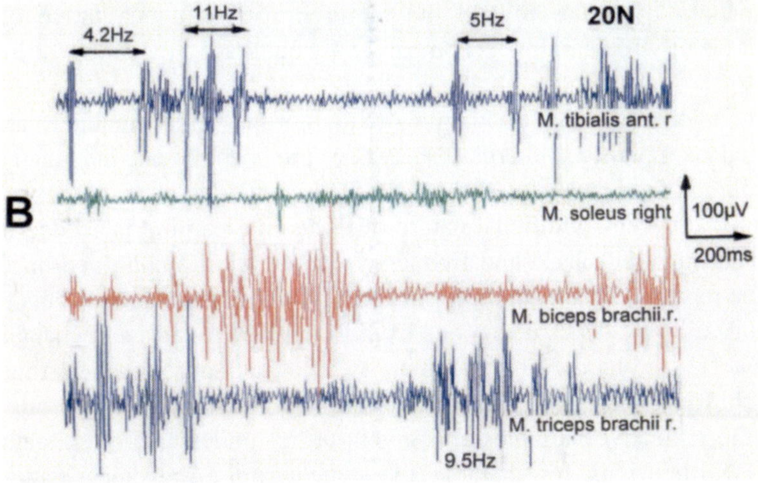

Figure 14. EMG motor programs of a patient with Parkinson's disease during the exercise on the special coordination dynamics therapy device at a low load of 20N. The motor program muscle bursts are more or less structured by rhythmicity (no rhythmic structure in normal motor bursts (Figure 1A)). In A rhythmic activity at a frequency of 11.4 and 11.8Hz is suggested. In B (faster sweep and partly different muscles), low-frequency rhythmicity of 4.2 and 5Hz (frequency range of α_2-motoneuron oscillatory firing) and higher frequency rhythmicity of 11 and 9.5Hz (frequency range of α_1-motoneuron oscillatory firing) are suggested to be seen.

10. Training of Phase and Frequency Coordination as Behavioural Information

Following CNS injury, movements and other functions are lost or impaired and the phase and frequency coordination of CNS self-organization is impaired, the behavioural information to be used for repair has to include (I) the relearning of movements and other CNS functions and (II) the improvement of the time and space-coordinated firing of neurons, which in turn improves the intrinsic dynamics ($F_{intr}(X)$) of CNS organisation (see above). The behavioural information required for improving the phase and frequency coordination in the intrinsic dynamics ($F_{intr}(X)$) is the physiological afferent input generated by the exact coordinated arm, leg, and trunk movements while the patient is exercising on special CDT devices. These types of inputs are also provided when the patient is exercising very coordinated movements like brisk crawling, walking or running.

The System Theory of Pattern Formation for Repair is mainly an abstract theory. But by using real neurophysiologic and pathophysiologic data of natural human CNS functioning, the theory becomes a basis of human CNS organization and reorganization by learning. Every theory claiming that it can explain or describe human CNS functioning must be applied to human reality for proof. The proof that the "System Theory of Pattern Formation for Repair" is applicable to human is given in the Introduction and Result sections. For other behavioural information to improve the intrinsic dynamics, see below.

11. Including the Variability of Phase and Frequency Coordination among Neuron Firing into the Equation of Motion of the Collective Variables

Depending on the relationship between the initial coordination dynamics (so-called intrinsic dynamics, $F_{intr}(X)$, depending strongly on the severity of the injury) and the patterns to be relearned (termed behavioral information, $\sum c_{inf}F_{inf}(X,t)$, which act as attractors of the coordination pattern dynamics toward the required patterns), qualitative changes in the attractor layout occur with learning, accompanied by qualitative evidence for loss (or change) of stability.

The nature of change due to learning (e.g., abrupt versus gradual) arises from the cooperative and competitive interplay between the behavioral information (supported jumping or walking of the patient) and the intrinsic dynamics.

A completely different, additional nature of necessary learning is needed in the repair of CNS injury. The impaired phase and frequency coordination among neuron firing (Figures 13-15) has to be repaired by relearning for proper CNS self-organization. This perturbation of CNS self-organization produces deviations from the attractor states and changes the attractor layout because of altered hardwiring due to injury. In a first approximation, this tremendously increased variability of phase and frequency coordination can be included into the equations of motion of the collective variables and gives further understanding of pattern change in patients with CNS injury as, for example, the switch from a movement pattern to a spastic pattern (Figure 16B).

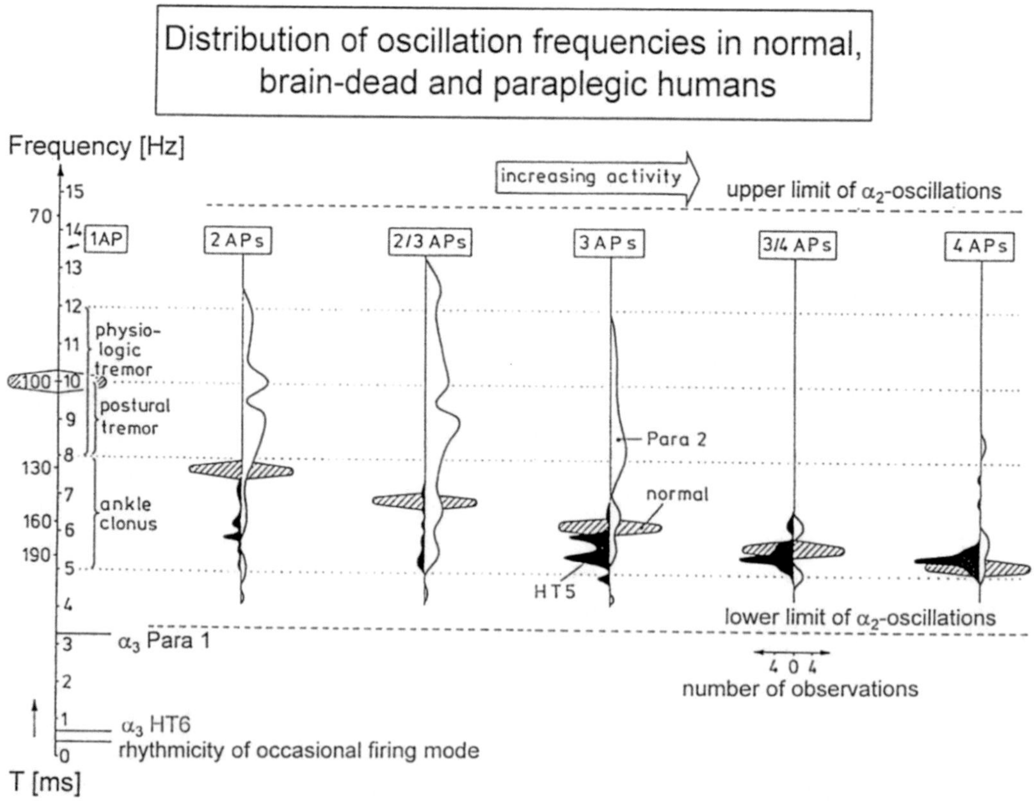

Figure 15. Frequency distributions of oscillation frequencies of continuously oscillatory firing α_2-motoneurons with increasing number of APs per impulse train (increased activity) in paraplegic 2 (open), in brain-dead HT5 (filled), and probably normal human (crosshatched). Frequencies and rhythmic activity changes in the occasional and oscillatory firing mode are indicated. Ranges of physiologic tremor, postural tremor and ankle clonus are also drawn. Note that frequencies for the brain-dead HT5 are too low, and the oscillation frequencies of the spinal cord isolated for a long time (Para 2) are too high and too spread as compared to the theoretically predicted frequency ranges (crosshatched). T = oscillation frequency.

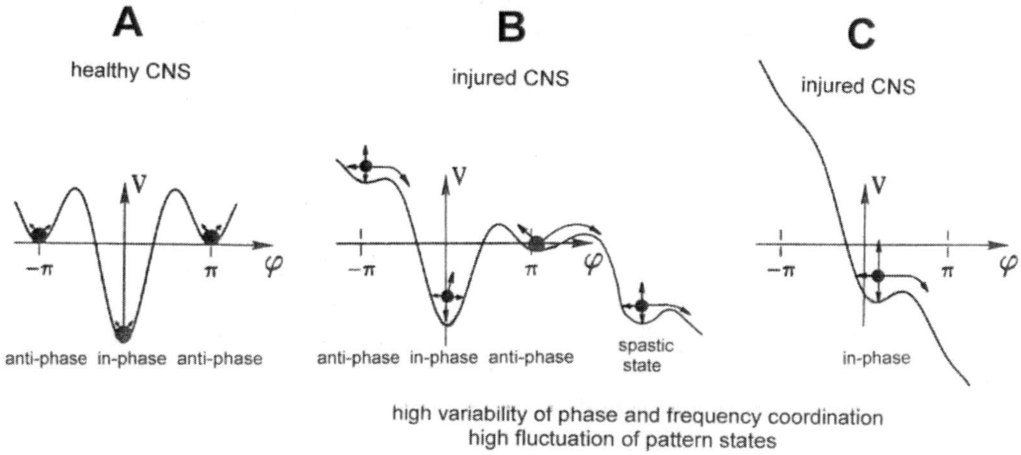

Figure 16. The potential, V(φ), of the coordination dynamics for jumping on springboard of a healthy (A) and injured CNS (B,C). The region around each local minimum acts like a well that weakly traps the system into a coordinated state. Behavioral changes are represented by the over-damped movement of a rolling ball in the potential "landscape." High fluctuations in the stable state, due to the high variability of phase and frequency coordination (in the injured case), will have a greater probability of "kicking" the system out of the basins of attraction (B,C) than for low fluctuations, due to small variability of phase and frequency coordination (in the healthy case (A)). In B, only the in-phase jumping is stable, even though the fluctuation is high. In C, there is only an attractor basin for the in-phase jumping, but the fluctuation is so high that there is high probability that the system is kicked out of the basin of attraction. The patient cannot jump anymore in anti-phase and has problems to jump in-phase. The stability of jumping depends on the motor program (deepness of basin of attraction) and the fluctuation of the pattern state caused by the increased variability of phase and frequency coordination due to the injury.

In the Haken-Kelso-Bunz model, the jumping on springboard (Figure 4) can be described in terms of relative phase between the rhythmically moving legs. Without specific behavioral information, the dynamical description is defined by a vector field (a differential equation) expressing the rate of change in relative phase, dφ/dt, as a function of the derivative of its potential, V(φ):

$$d\varphi/dt = -\, dV(\varphi)/d\varphi + (Q\xi_t)^{1/2} \qquad (3)$$

where $V(\varphi) = -\, a\cos(\varphi) - b\cos(2\varphi)$ and $(Q\xi_t)^{1/2}$ is the phase and frequency variability of strength Q (where ξ_t is Gaussian white noise of unit variance). Zanone and Kelso [75] introduced noise in Equation 3 (from a logic point of view), because all real systems described by low-dimensional dynamics are coupled to many subsystems at a more microscopic level. One may view noise as a continuously applied perturbation that produces deviations from the attractor state. Such fluctuations are conceptionally important in dynamical modeling of phase transition or bifurcation phenomena and are essential in effecting transitions.

I included noise in Equation 3 (from the experimental point of view) because of the measured increased variability of phase and frequency coordination among the coordinated firing of neurons and neural assemblies in the human CNS.

This neuron level measured fluctuation of phase and frequency coordination is giving rise to phase transitions or bifurcation phenomena and is essential in causing transitions among attractor states under physiologic (small fluctuation; Figure 15 (HT5 or normal

Eigenfrequency distributions); Figure 12B) and pathologic conditions (large range of the Eigenfrequency (Figure 15) (Para 2 distribution) and in consequence large variability of phase and frequency coordination (Figure 13)).

The relative stability of an attractor states is, therefore, reflected by the depth of each potential well (I) and the strength Q of the variability of phase and frequency coordination (II), and the attraction of attractor states is reflected by the slope at each point of the potential curve.

The behavioral changes when jumping on springboard (Figure 4) are represented by the over-damped movement of a rolling ball in the potential landscape for the physiologic (Figure 16A, Q small = little fluctuation of phase and frequency coordination) and the pathologic case (Figure 16B,C, Q large = large variability). The increased fluctuation in the rather stable state, due to increased variability of phase and frequency coordination, will have greater probability of "kicking" the system out of attractor the basin (Figure 16B,C), especially in the asymmetric case.

In the healthy CNS, the phase and frequency variability is small (short arrows), and the jumping in-phase and anti-phase is stable (Figure 16A). Following injury, the potential landscape is deformed, and the fluctuation of the network states, generating jumping, is high (Figure 16B). The in-phase jumping is still stable in spite of the increased fluctuation because the basin of attraction is deep. The jumping anti-phase became unstable because the basin of attraction is shallow, and the increased fluctuation in the state has a greater probability of "kicking" the system out of the basin. A switch into a spastic state is also possible. In severe CNS injury or malformation, the patient cannot jump anymore in anti-phase because of the missing of attractors for anti-phase jumping (Figure 16C). Support is needed for anti-phase jumping (Figure 4, upper right). The jumping in-phase is still possible but unstable (Figure 4, upper left).

Upon performing very exact coordinated movements, imposed by devices (as for example shown in Figure 1), the nervous system of the patient learns to reduce the variability of phase and frequency coordination and achieves in this way a small fluctuation of the network states again, as shown in Figure 16A. The progress in treatment (learning) is that the in-phase jumping in Figure 16C and the anti-phase jumping in Figure 16B become stable (Figure 16A) again.

Also the potential landscape will change due to the reduction of the phase and frequency variability. The important consequence for treatment is that when exercising on special devices and reducing in this way the variability of phase and frequency coordination, the patient can induce the formation of patterns again, without having trained them (learning transfer).

Upon improving the coordinated firing of neurons, a cerebral palsy child may become able to speak or may develop social behaviors (see repair in cerebral palsy).

In conclusion, the impairment of phase and frequency coordination, measured at the neuron level in human, can be included in the coordination dynamics at the collective variable level.

The decrease of the variability of phase and frequency coordination (one kind of coordination repair) is an essential part of CNS development and repair, as will be shown in the Results.

2.2. Repair

The System Theory of Pattern Formation provides us with a tool to understand the functioning of the human CNS with ongoing time. With the behavioural information $F_{inf}(X,t)$, we can change intrinsic dynamics $F_{intr}(X)$. However, the equations of motion of the coordination dynamics (formula 2) provides us with no information about the behavioural information (F_{inf}) that can be used to efficiently repair the CNS of the specific patient. In every patient, intrinsic dynamics will be different due to the different nature of the injury, the different history of the CNS, and the different genetics of the CNS.

To find out what behavioural information may change the intrinsic dynamics in the direction of wanted repair, we have to understand how the human CNS is functioning physiologically and what has changed in the functioning due to injury, malformation or degeneration on the neuron level, on the neural assembly level, on the integrative level, and on the functional anatomy level. Repair really requires thorough understanding of the functioning of the human CNS. The behavioural information includes movements to be performed, manual supports, touch input, instructions, neuro-feedback, and other information. Surface electromyography (sEMG) helps to identify the movement patterns for which the motor programs of the critical muscles are the best ones (Figures 55,68,69). Learning or relearning of movements may be understood as change of coordination dynamics [69]. How, then, do we learn on the different levels of CNS self-organization after CNS injury, malformation, or degeneration? Compared to a healthy person, the rate of learning may differ by the factor of 50 [42]. To achieve repair of the CNS, we have to know how the human CNS generates certain functions and what does change upon an injury.

After CNS injury some CNS functions get lost (or functions become pathologic), and the physiologic self-organization of the neuronal networks becomes impaired. Impairment of the self-organization of neuronal networks means impaired phase and frequency coordination between neuron firings, impaired symmetries of CNS functioning, pathologic self-organization of neuronal assemblies (e.g., spinal oscillators), impaired phase and frequency coordination among neuronal assemblies, pathologic recruitment of assemblies, and other impaired organizations.

To repair the impaired phase and frequency coordination, different special coordination dynamics therapy devices have to be used so as to give the injured CNS the opportunity to relearn (from the devices) the exact coordination in self-organization. To improve the functioning of premotor spinal oscillators (self-organization and recruitment), the patient has to exercise rhythmic, dynamic, stereotyped movements. To achieve functional repair, automatisms and certain movements have to be exercised, which make the intrinsic dynamics more physiologic. Also, inborn automatisms like creeping, crawling, walking, running, breathing, swallowing, continence functions, blood circulation, and others will get genetic support for structural repair. By going to the limits (more than 30 hours therapy per week) for more than probably six months, some structural repair may also be induced. Also, movements that the patient wants to relearn have to be exercised. Old-learned movements remain (at least partly) as attractor states in the intrinsic dynamics and should also be exercised.

Further ideas on how to repair injured or malfunctioning CNS may be derived from studying the development of the CNS to see how the developing CNS improves its self-organization (for example, fast by transient moving (nice to be seen when the child exercises on the special CDT device (Figures 21E,F;27)). In the system theory of motor development

[71], a new behaviour is constructed, which is dependent on the input of all the contributing subsystems. We may learn from the development how the “feed-forward system” is self-correcting en route. In healthy infants, there seems to be substantial coordination deficits in the movements of arms and legs [71]. Self-corrections “en route” seem to need substantial corrections of the intrinsic dynamics (changing upon the development). Stability that characterizes the rhythmic movement forms as a whole does not apply to individual, temporal subdivisions (von Holst [97]). This may mean that the splitting of automatisms into submovements may not always be beneficial for repair. It may be stated in general with respect to CNS repair that the timing of neuron firing has to be improved, along with the stability of physiologic movements, whereas stability of the pathologic organizations has to be reduced.

1. Repair of the Intrinsic Dynamics by (Re-)Learning of Movements

For improving development and repair, the phase and frequency coordination for CNS self-organization has to be improved. But repairing an organization principle is not sufficient. Damaged or lost neuronal structures have also to be repaired or substituted. The generation of a motor (or other) pattern can be seen as a dynamic cooperation of the many subsystems in a task-specific context. A higher transmission frequency of human nerve fibres than in animals allows the dynamic cooperation of a higher number of subsystems and therefore a higher dynamic complexity of CNS organization (see also below). Substitution of missing networks, nuclei, or brain parts seems until now impossible in humans. But by training different movements, the CNS may recognize functional deficits and the absence of particular learning processes and may induce missing particular learning processes for repair and functional reorganization so that other brain parts can partly take functions over. Missing of essential neurons (and other cells) or minor but essential small networks may be generated by neurogenesis from endogenous stem cell reserves. There is indication that there is limited neurogenesis in humans [67]. Those movements have to be trained and (re-)learned that are most efficient for repair by functional reorganization or optimizing physiologic development.

Learning programs include:

1. Improvement of phase and frequency coordination.
2. Training of patterns and reactions (stepping automatism, palmar grasp reflex, etc.) to be used as functioning elements of complex movements. The integrity of pattern elements may be a prerequisite for the repair of integrative CNS functions.
3. Training of integrative innate automatisms like crawling, up-righting, walking, and running and learned automatisms like climbing staircases.
4. Training of integrative movements so that the CNS can recognize the mistakes of functioning (due to damage or malformation) by recognizing the different requirements of the innate “blueprint of the organism” (even though it is unclear how the different requirements of the innate blueprint are exerted).
5. Teach the abnormal infant specific or patient modes of operation (backward walking) to avoid or improve pathologic movement patterns (to improve forward walking).
6. Stimulate uncommon modes of operation, which may enable the infant to achieve results, which he cannot achieve spontaneously.
7. Training at the limits (range of overreaching [56]) to induce all possible repair mechanisms (even the unknown ones).

8. Improvement of movement performance to enhance the relearning of physiologic patterns. Support, given by a therapist, improves the movement-induced afferent input and in turn the generated motor program, which can be measured by sEMG (Figure 68).
9. Addition of behavioural information, which may enhance learning: Instructive training, bio-feedback, enhancement of physiologic movement-induced afferent input.
10. Induction of learning processes, which are hampered by the deficiency of the neuronal structure.
11. For the enhancement of structural repair, a safe thoughtful stem cell therapy, combined with CDT, could be beneficial.
12. Modulating the epigenom for enhanced repair.

2. CDT Learning (Treatment) Strategies Included in Important Training Movements for Repair

1. Exercising on a special CDT device:

   A. Training of phase and frequency coordination (necessary for the improvement of movements, urinary bladder and other functions).
   B. Learning of pattern variability from the changing coordination between arm and leg movements (repair in the cephalo-caudal direction). Learning of horizontal motor pattern variability (between arms and hands) (Figure 17B) in addition to vertical pathways.
   C. Training of trunk stability from the performed rotational trunk movements, if the patient is exercising on the special CDT device in the lying position (Figure 17A). From the mechanical point of view, because of the changing coordination between arms and legs, the thoracic and lumbar segments are rotated sequentially, which is important also for reducing the scoliosis of these patients.
   D. Upon exercising with the right hand at the left handle and with the left hand at the right handle, the corpus callosum can substantially be activated and repaired, which is important in corpus callosum injury.

2. Walking and running:

   A. Re-learning of an integrative innate automatism, which may get strong genetical support for repair.
   B. The stepping automatism (= neonatal stepping movement) is a movement element to be integrated (modified, refined and brought under volitional control) in the more integrative pattern walking.
   C. Natural walking (or running) in the forward and backward direction, if necessary with support of therapists (Figure 17C), is the most specific human movement and must therefore be trained, because the CNS may use this movement also for other purposes.
   D. Sky-walking (neuro-walking): Since in severe SCI (C5/6) the physical load for therapists during supported treadmill walking is high and the weight support for

the patient is large, sky-walking on an exactly functioning device helps to relearn walking. Such patient (tetra) with little trunk stability can relearn the trot gait moving without weight-bearing support (Figure 17D).

E. Sky-walking enhances the stability of the trot gait pattern, which may have been weakened by the training of pattern variability and phase and frequency coordination upon exercising on the special CDT device (Figure 17A).
F. The support given by the therapist improves the movement-induced afferent input and in turn the performed motor pattern as can be shown by sEMG (Figure 68).
G. During fast running, if possible, the movement-induced afferent input is higher per time and may be more physiologic, because the kinetic improves the performance. Because of the higher and more physiologic input, the pathologic CNS organisation may be reduced and the performance and learning improved.

3. Jumping or swinging on springboard:

A. Injured premotor spinal oscillators are entrained upon the rhythmic, dynamic, stereotyped movements [63] (partly imposed by the mechanical properties of the springboard or swinging device) to reduce their frequency range and to get a real "Eigenfrequency again" (Figure 15). The frequency of jumping is close to the normal "Eigenfrequency" of the α_3-premotor spinal oscillators ($\approx$ 1Hz) or half of it (in-phase jumping $\approx$ 0.5Hz, anti-phase jumping $\approx$ 1Hz).
B. Induction of learning transfers from movements to urine storing and continence functions of the urinary bladder (similar networks and receptors are activated [63]).
C. Swinging and jumping can also be performed in patients with severe SCI (C5/6) without weight-bearing support in the upright position (Figure 17E).

4. Training of trunk stability:

A. Recapitulating development to repair trunk stability. Trunk stability can be trained by the change of the crawling posture, exercising on the special CDT device in the lying position, and by training patterns and reactions of infancy development (Figures 36,37).

5. Endogenous stem cell therapy:

A. Activating endogenous stem cells in or at the blood vessels, especially when going to the physical limits, will contribute to the adaptation and repair of the blood vessel system in general. When activating the neuronal networks, upon therapy, adjacent to the injury site (in SCI, Figure 3D), blood vessels are generated also specifically at the injury site. Multipotent stem cells, whatever way they migrate or generate, find the environment for homing. The naturally activated neuronal networks generate the electromagnetic fields and guidance molecules for the integration of the different differentiated cells and regenerating nerve fibres (see Chapter 1). The more severe the injury is the less proper environment can be generated for structural and functional repair. This reasoning is speculative, but the nervous system seems to be involved in nearly all functions. The innervation change during development and repair in the frog (Chapter 1) showed that reinnervation is a very complex process in which close cell communication plays an important role.

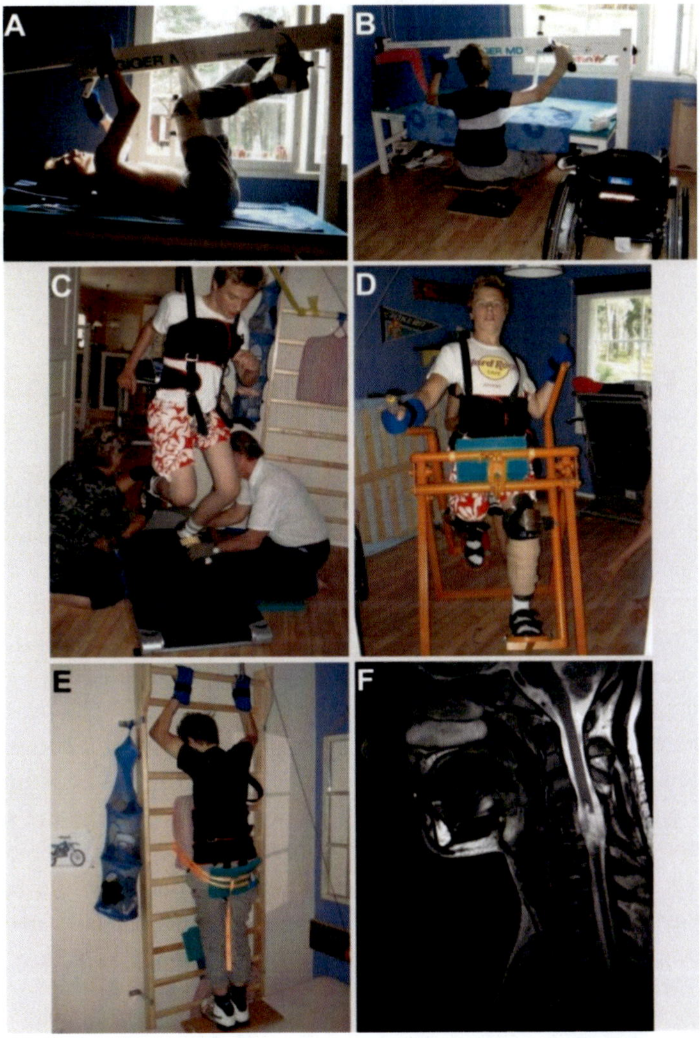

Figure 17. Typical movements to induce movement-based learning for CNS repair, shown in the case of an almost complete cervical spinal cord injury SCI (injury level C5/6, age 16, motor bike accident, one year after the accident). A. Exercising coordinated movements of arms, legs, and trunk (note the trunk rotation) over the injury site (mainly in cephalocaudal direction) and relearning of phase and frequency coordination. B. Exercising exactly coordinated arm, hand, and trunk movements (mainly horizontal plane of spinal cord network activation). C. Treadmill walking in the backward direction (V = 4.5km/h) to induce walking repair in the backward direction and by learning transfer also in the forward direction. Feet and knees are lifted high by the therapists to induce the stepping automatism in the backward direction, which reappeared a bit. D. Sky-walking on an exact self-made device to train trot gait coordination. Note: there is no weight support needed! The harness is only for bringing the patient in the appropriate position. E. Swinging at wall bars on a self-made swinging device. Note: there is no weight support. F. MRI of this SCI patient; probably a few percent of the spinal cord tissue remained at the injury site. Note: even a patient with a practically complete cervical SCI can train integrative scientifically based movements for CNS (mainly spinal cord) repair.

3. Natural Therapy

Hippocrates' ancient wisdom that "natural forces within us are the true healers of disease" is in accordance with the strategy of coordination dynamics therapy to use natural means (mainly movements) for CNS repair. The question remains, however, how to find and stimulate the natural strategies, especially to achieve structural repair. After denervating one leg of a rat, the rat often eats the denervated leg, because the leg has no sensitivity and it hinders the animal's movement. If the rat knew that within some months the leg would be reinnervated and the feeling and the functions would at least partly recover, it would not eat the leg. Humans, on the other hand, have the possibility to find out, understand, and use natural means of regeneration and leave destructive operations and therapies out. By training integrative functions coordinatedly, many body functions will improve and not only those ones that are trained, especially because there is learning transfer in the human CNS (Chapter V).

3. Results: Development and Repair

CNS development is mainly measured in two periods. In the developmental period from 0.5 years up to 5 years of age, the healthy infant Jürgen has been followed up longitudinally. From 5 up to 18 years of age, the CNS development will be assessed by the coordination dynamics of female and male pupils from one school (preschool plus classes 1 through 12). It turns out that the interpersonal variation of CNS development was so large that in spite of large groups, the true development of function and neuronal network complexity became only partly visible. Even though a longitudinal study of motor development of infants may offer more specific data, the evaluation of a developmental period of 20 years for a study is difficult to organize. Lumping the data from several schools together (even though further data were collected) was not improving the results, since the CNS development seems to be affected by many environmental factors, including the time of the year, the social situation of the pupils at home and at school, the variation of the rate of development, and physical activities. In a second part, long-term repair of the injured CNS is studied in several case reports. It is then tried to partly compare CNS development with CNS repair to see whether the repair can recapitulate development and to understand better the development and the repair of the human CNS.

3.1. Development of the Human CNS in the Age Range from 0.5 to 18 Years

1. Postnatal Development and Repair

With respect to CNS repair, the stepping automatism of the postnatal development is of importance. The heal strike induced stepping automatism of the newborn (Figure 2) is partly reappearing following spinal cord injury. The stimulation of the partly reappearing stepping automatism during supported treadmill walking is the start to relearn walking.

2. Intrapersonal Development from 0.5 Years to 5 Years of Age

2.1. EMG Motor Programs

At an age of five months, the infant Jürgen walked more or less in a stepping automatism fashion (Figure 18a). The EMG motor program was not fully developed. The antagonistic action between the right tibialis anterior and right gastrocnemius muscles was not developed (Figure 18b). Three months later, the supported walking was more walking like (Figure 18c), but there was still no antagonicity between the EMG activity of the tibialis anterior and gastrocnemius muscles (Figure 18d). At an age of five years, Jürgen could walk nicely by himself, and the motor programs showed probably antagonistic action between the tibialis anterior and gastrocnemius muscles. Since the EMG motor program varies among different trials, motor development was measured mainly by the coordination dynamics (CD).

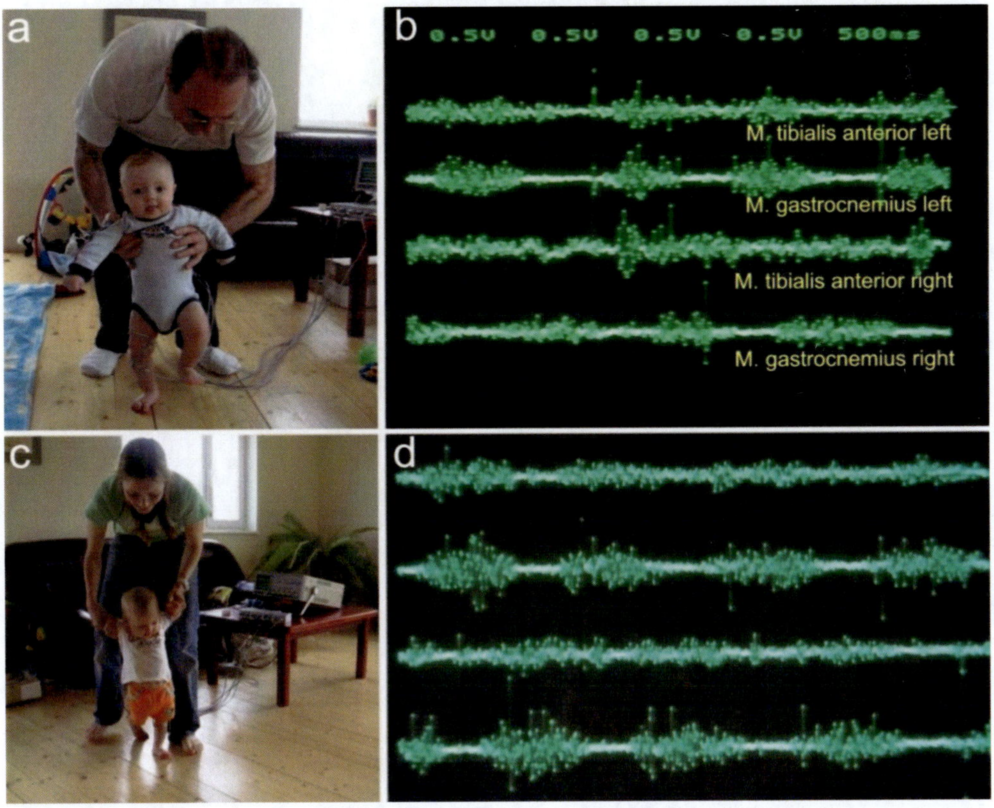

Figure 18. Surface EMG obtained from the healthy five-month-old (a,b) and eight-month-old old "Jürgen" (c,d) during supported walking. a. Walking resembles a bit the stepping automatism, because of the strong lifting of the left knee during walking. The toes of the right foot are plantar flexed, which is not physiologic. b. Surface EMG motor programs of left and right tibialis anterior and gastrocnemius muscles. Note that there is no antagonicity between the tibislis anterior and gastrocnemius muscles. The right tibialis anterior muscle shows no motor program. C. Supported is walking more walking like and not so much automatic stepping like. d. Motor programs a bit better than three months ago (b). Still there exists no antagonicity between the tibialis anterior and gastrocnemius muscles. The activation of the right tibialis anterior muscle is a bit better than three months ago (b). e. Healthy infant Jürgen at an age of five years. Real walking is achieved. f. A physiologic EMG walking pattern has been developed: Antagonicity EMG activity between the tibialis anterior and gastrocnemius muscles can be seen.

2.2. Neuronal Network Complexity Needed for Complicated Coordinations

At an age of six months, the infant Jürgen was too small and not able to turn on the special coordination dynamics therapy (CDT) device. At an age of one year, the infant was made familiar with the device (Figure 19F), and he liked it. With two years, Jürgen was tall enough to exercise on the baby CDT device, and he wanted it strongly. But he was not able to exercise continuously. He got stuck again and again after one to three turns. He solved the problem by himself of not being able to turn with arms and legs continuously by taking the hands or the feet from the device (Figure 19A,B) and in this way avoiding the complicated coordination patterns of arms and legs (with respect to coordination) for which his CNS had not developed so far the needed complexity.

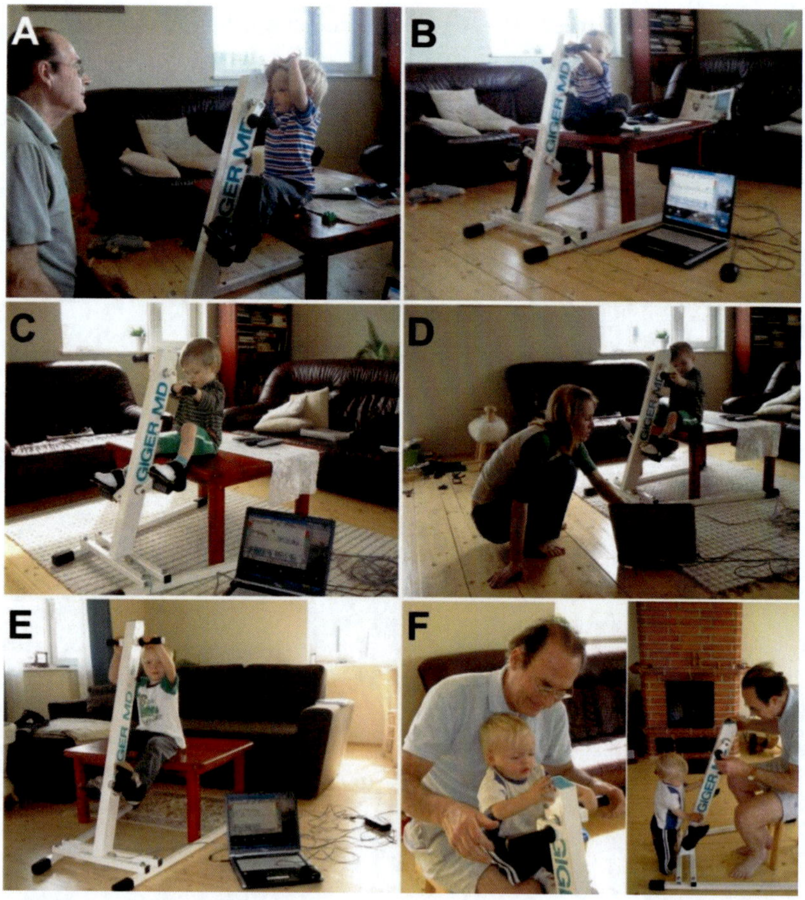

Figure 19. Development of the ability to exercise on the special CDT device. F. Healthy one-year-old Jürgen is made familiar with the special CDT device before he is big enough to exercise on it and his CNS is not developed enough to generate the complicated patterns between pace and trot gate. A, B. At two years, he is now big enough to turn on the special CDT device for babies, but he could manage only a few turns continuously by himself. He solved the problem to be able to exercise continuously by taking the hands (A) or the legs from the instruments (B). C, D. With nearly three years, Jürgen became able to exercise on the baby device. He now wanted his mother to explain the online coordination dynamics measurements (D). E. With five years, Jürgen became able to exercise on the special device for children; the children's device is bigger than the one for babies.

Jürgen thus nicely demonstrates that high complexity of neuronal network organisation is needed to be able to perform coordinated arm and leg movements with changing coordination between arm and leg movements (pattern change). For turning only with the legs (fitness bicycle), no high network complexity is needed. At an age of three years, he became able to exercise on the special CDT baby device (Figure 19C,D); even though the rhythmicity of turning was poor, especially in the backward direction. With five years, Jürgen was able to turn on the larger children's CDT device in the forward and backward direction (Figure 18E). The exercising in the backward direction was less smooth. The assessed CD will give further insight into the maturing of the CNS during development.

2.3. CD Assessment between Two and Five Years of Age:

The generation (increase) of CNS neuronal network complexity during development can be quantified, if the child is performing complicated coordinated arm and leg movements and the quality of performance measured.

The child is made to exercise on the special CDT device, and the arrhythmicity of turning (the quality of performance, the CD) is measured. The improvement of the performance and the reduction of the arrhythmicity of exercising can be nicely followed up in the developing boy Jürgen (Figures 20,21). With two years of development, Jürgen could not turn continuously on the baby device (Figure 20A,B).

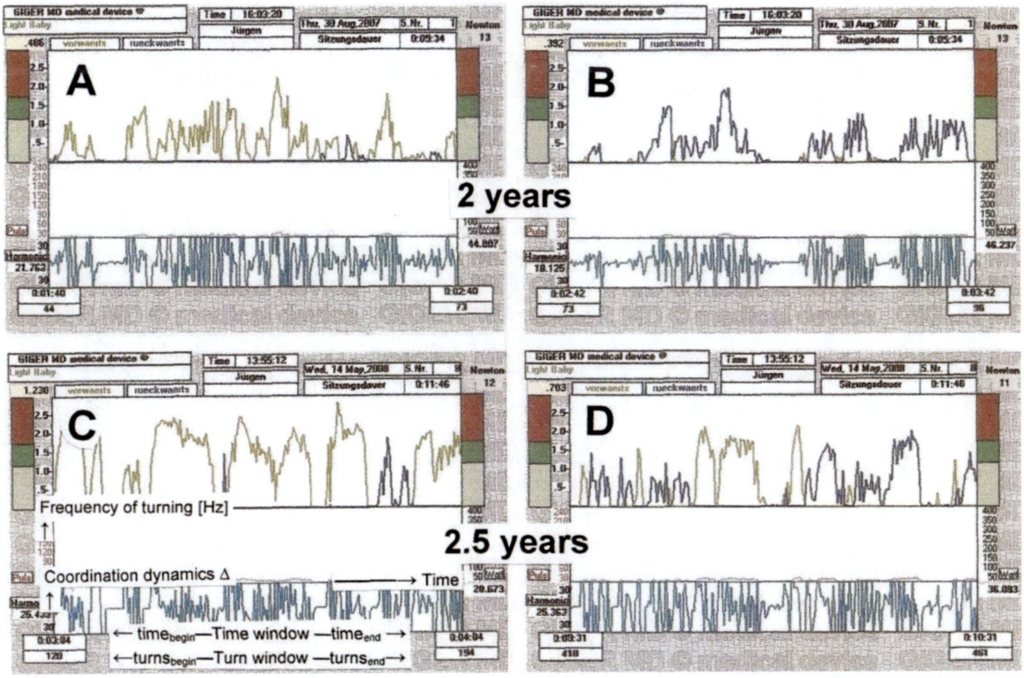

Figure 20. Development of CNS complexity, quantified by coordination dynamics measurements (very early stage of coordination dynamics (CD) assessment during development). Upper traces = exercising frequency; lower traces = arrythmicity of exercising (CD). A,C = forward exercising and B,D = backward exercising. A,B. At an age of two years, healthy Jürgen cannot turn continuously on the baby CDT device: the frequency varies very much and the turning shows many stops. C,D. With 2.5 years, the young child can turn more continuously on the baby device and the frequency of exercising is much higher.

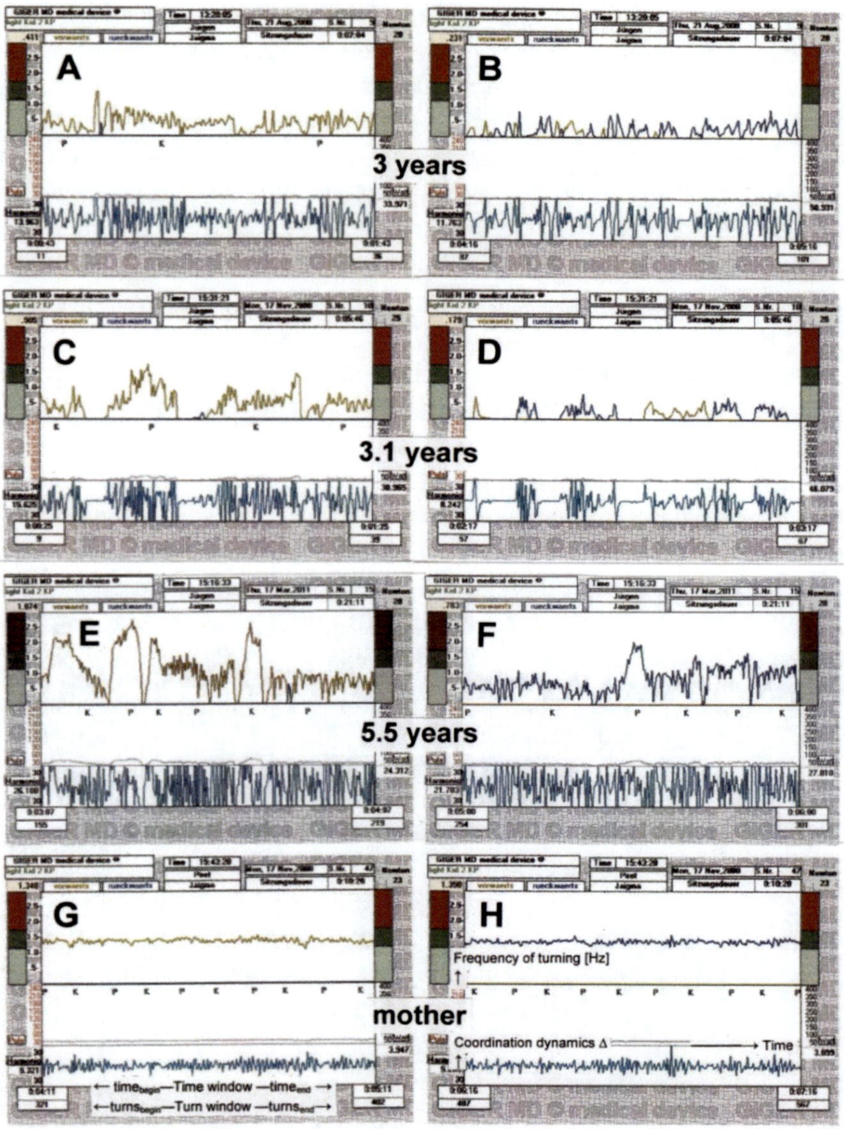

Figure 21. Development of CNS complexity, quantified by coordination dynamics measurements (early stage of coordination dynamics (CD) assessment). Upper traces = exercising frequency; lower traces = arrythmicity of exercising (CD); A,C,E,G = forward exercising and B,D,F,H backward exercising. A,B. With three years, the child Jürgen becomes able to exercise on the children's device. The frequency of exercising is low. The child makes still mistakes in the direction of exercising, especially upon turning in the backward direction (B). C,D. Two months later, he can turn at higher frequency. Now the difficult coordinations between pace (P) and trot gate (K), for which a higher neuronal network complexity is needed, becomes visible (C): he gets stuck or nearly stuck between the patterns pace and trot gait. The backward exercising is still very poor (D); mistakes in the direction of turning still occur. E,F. At an age of 5.5 years, Jürgen can turn more continuously at higher frequencies. The difficult coordinations between pace and trot gate are still visible (reduction of the frequency). For forward exercising, transient high frequencies of turning (four peaks) occurred (E). For exercising in the backward direction, one period of exercising with high frequency occurred (F). G,H. Coordination dynamics of the mother, for comparison. Note, the frequency (f) of exercising is nearly a line (constant frequency), and the arrhythmicity of turning (df/dt /f) is small for forward (G) and backward (H) exercising. Note further, there is a large difference in performance of exercising on the special device of the five-year-old healthy Jürgen and the mother, exercising on the same children device.

The mean frequency was low (Figure 20A,B; upper trace = frequency; no continuous line) and the arrhythmicity of turning (lower trace) was high. The mean arrhythmicity of exercising per minute (the CD value) was with 44.8 for forward exercising (Figure 20A) and 46.2 for backward exercising (Figure 20B) very high. The values would have been even higher if the infant would not have gotten stuck during the turns. At an age of 2.5 years, the frequency of exercising became higher for forward (mean frequency = 1.24Hz, Figure 20C) and backward exercising ($f_{mean,back}$ = 0.71Hz); the corresponding frequencies of turning at an age of two years were $f_{mean,forward}$ = 0.49Hz and $f_{mean,back}$ = 0.39Hz. At an age of 2.5 years, the CD values had decreased (became better) to $\Delta_{forward}$ = 20.8 and $\Delta_{backward}$ = 36.1. As can be seen from Figure 20A through C, the child could not turn continuously in the forward or backward direction. He made mistakes in the direction of forward and backward exercising; he mixed them up a bit. The continuous intention to exercise in one direction was not fully developed; the symmetries of CNS organization were also not fully matured. The exercising in the backward direction was worse in comparison to the exercising in the forward direction.

At an age of three years, Jürgen was taller and became able to turn on the special CDT device for children (Figure 21). This is the size of the device that is used for assessing all later developmental stages. Exercising on the larger CDT device is more difficult, and the coordination dynamics values cannot directly be compared with those obtained from the baby instrument. The CD curves obtained from the children's device (Figure 21A,B) show clearly that it was more difficult to turn on the larger device. The frequency of exercising was smaller ($f_{mean,kid,forward}$ = 0.42Hz (against $f_{mean,baby,forward}$ 1.24Hz); $f_{mean,kid,back}$ = 0.24 (against $f_{mean,baby,forward}$ 0.71Hz) and the coordination dynamics values higher ($\Delta_{forward,kid}$ = 33.6 (against $\Delta_{forward,baby}$ = 20.8); $\Delta_{backward,kid}$ = 51.2 (against $\Delta_{backward,baby}$ = 36.1)). Further, the rhythmicity of turning was still so poor that no preferential turning patterns could be seen (Figure 21A). The exercising in the backward direction (Figure 21B) was worse than the exercising in the forward direction (Figure 21A), because the CD values were bigger (worse) and the child changed un-volitionally more often into the wrong direction of turning. At an age of 3.1 years, the CD improved further. In the forward direction of exercising, Jürgen no longer made mistakes in the direction of exercising, and the backward exercising also became more stable. In the forward direction of exercising, the difficult and easy coordination patterns became visible. For the easy coordination patterns around the pace gait (P) and trot gait coordination (K), Jürgen could turn quite fast (Figure 21C), and among the difficult coordinations between pace and trot gait he got stuck.

At an age of 5.5 years, the coordination patterns of arm and leg movements became much better. The arrythmicity of exercising became smaller ($\Delta_{forward,kid}$ = 24.3, $\Delta_{backward,kid}$ = 27.8) and the frequency of exercising higher ($f_{mean,kid,forward}$ = 1.07Hz, $f_{mean,kid,back}$ = 0.78Hz) (Figure 21E,F). But these values were still far away from those of the adult mother (Figure 21G,H). Her arrhythmicity of exercising was much lower and her frequency of exercising higher ($\Delta_{forward,kid}$ = 4.1, $\Delta_{backward,kid}$ = 3.6; $f_{mean,kid,forward}$ = 1.35Hz, $f_{mean,kid,back}$ = 1.32Hz). The motor maturation from five years on up to the adult stage will be assessed below. Plotting the coordination dynamics and frequency values in the age range from 2 to 5.5 will give further insight into the development of movement patterns of the human CNS.

3. Preference of Function to Symmetry

In general, the frequency of exercising increased with age and the coordination dynamics values (the arrythmicity of exercising) decreased (Figure 22).

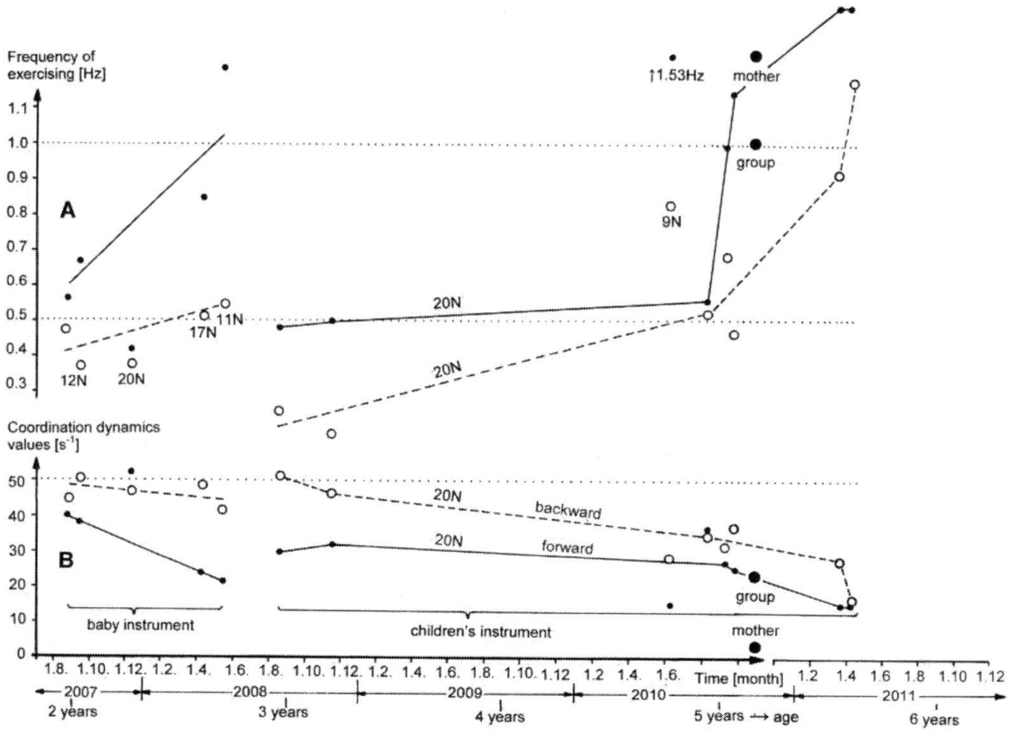

Figure 22. Development of the coordination of movements, quantified by the coordination dynamics (CD) values. Frequency of exercising (A) and coordination dynamics values (B) of a healthy child from two to five years. The values for two to three years are obtained from the baby CDT device and from three to five years from the children's device. Load values in Newton (N) are indicated. CD and frequency values from the mother and a group of five-year-old children are also indicated.

At the beginning during the measuring period between two and three years, upon exercising on the baby device, the performance of the forward exercising of the infant Jürgen became better than those exercising in the backward direction. Then later on, in the age period between three and five years, the performance of exercising in the backward direction improved more than those in the forward direction. That means the exercising frequency in the backward direction improved more than that in the forward direction, and the CD values for backward exercising improved more than those for forward exercising. This means the developing CNS gave first priority to establish the forward moving, probably because the forward moving is more important. Then, at a developmental stage, when this movement pattern was somehow established, the development took care of the symmetry of moving. It seems therefore that during development, function was given priority to symmetry. This movement principle is in accordance with the learning (establishing) of walking. The developing child often learns first the forward walking in the age range between 12 and 18 months and then, three to six months later, also the backward walking. The stepping automatism probably becomes also first operational in the forward direction, because the newborn can only step automatically in the forward direction. When partly reappearing, following spinal cord injury, the stepping automatism can be induced (during supported treadmill walking) in the forward and backward direction. During development, forward crawling becomes also operational before backward crawling.

4. Functions That Need Less Network Complexity Are Developed and Repaired First

As can further be seen from Figure 22, the infant could perform the exercising better on the smaller baby device and better for lower load on both the baby and children's device. Therefore, for smaller movements and exercising with less load (less Newton), less muscle fibre activity is needed and in turn less integrative and complex network activity. That less integrative functions become first operational could nicely be measured during the repair in a patient who suffered an incomplete cervical spinal cord injury. With CDT, the patient relearned first, after six months, the turning for low load (20N) in the forward direction on the special CDT device. For exercising against a load of 100N, he needed 24 months and against 150N more than 36 months (Figure 39). As relearning period, the time was taken when the CD values had reached a plateau of low value and the CD values were in the range of those of healthy persons.

5. The Quality, Complexity, and Stability of CNS Neuronal Networks of a Five-Year-Old Child Is Far Away from Those of the Adult Mother

The plotted curves of the coordination dynamics (CD) values and frequency of exercising (Figure 22) clearly show that the arrythmicity of turning (the CD values) reduced and the frequency of exercising increased. The neuronal network complexity and quality of organization of the CNS had increased strongly during these three years of development. The intention for fast exercising (to improve CNS) was no longer hindered by the undeveloped variability in patterns organization. But the quality, complexity, and stability of CNS functioning of the child, as judged by the CD, was still far away from those of the adult stage of development, as can be seen when comparing the infants CD values with those of the mother (Figure 21G,H;Figure 22). The mother could turn smoothly (with little frequency variation) for all patterns between pace and trot gait and the CD values for forward and backward exercising were small, indicating that her CNS neuronal network complexity was good enough to generate all the necessary patterns for smooth exercising. Good enough network complexity means that the network connectivity was high and that the phase and frequency coordination of neuron firing was high and allowed the generation of difficult movement coordination patterns.

The further maturation during CNS development will now be studied cross sectionally. It will be shown below that with increasing age, the coordination dynamics values continuously decrease until they have reached the adult values, whereas the frequency of exercising shows an increase and decrease.

6. Transient Very Fast Exercising

Before following up the further CNS development, an interesting movement performance phenomenon was already observed in the child Jürgen. When he became able to exercise sufficiently fast, he also sometimes turned transiently very fast (Figure 21E,F), as if somebody was telling him to turn very fast for a short time. It will be analysed below that such transient fast exercising of children may be a strategy of the CNS to improve its coordinated firing of neurons additionally. During repair, the patient can exercise on a special CDT device to improve the injury or malformation impaired phase and frequency coordination of neuron firing. During normal development, the CNS may use the fast performance of different movements, as, for example, fast running or crawling, to improve continuously the phase and frequency coordination of neuron firing, which is necessary, because the developing network (increase of connectivity for example) is changing all the time. Following severe cervical spinal cord injury, it was shown that with every bit of regeneration of the spinal cord (re-appearance of function), the CD values got transiently worse (Figure 2 of Ref.64; Figure 50). Upon exercising on the special CDT device, the transient fast moving during CNS development becomes nicely visible.

7. Development of Coordination Dynamics (CD) Values between 5 and 19 Years of Age

Figure 23 shows the CD values of boys upon exercising in the forward direction (A) and backward direction (B). As can be seen from the curves, the values decreased continuously from 5 to 19 years of age. There may have been a transient increase of CD values at an age of 10 to 11 years. The size of the age groups is given in Figure 23C,D; it varies between 5 and 25 pupils per group. The frequency of exercising increased first strongly and then decreased slowly (Figure 23E,F). The boys turned fastest in the forward direction ($f_{average,forward}$ = 2.02Hz) at an age of nine years. Exercise in the backward direction was fastest also at nine years of age but with a lower frequency (1.59Hz). The values of the infant Jürgen in the age range between 5 and 5.5 years are plotted into Figure 23.

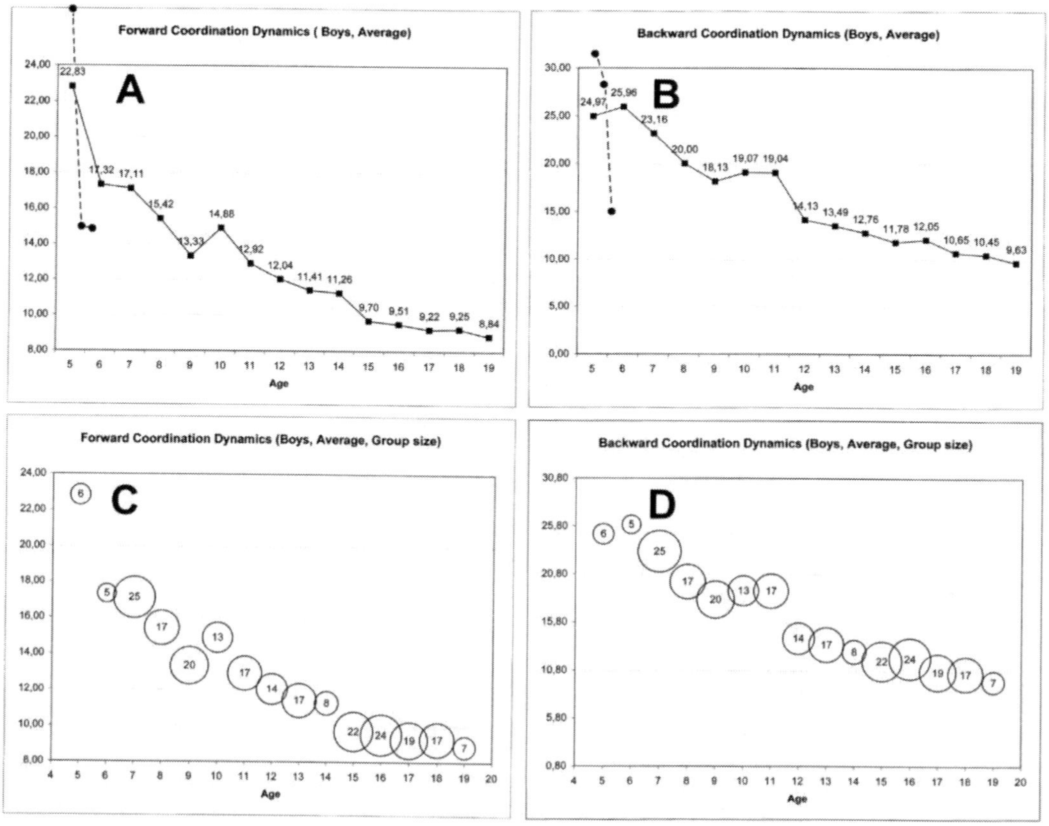

Figure 23.. (Continued)

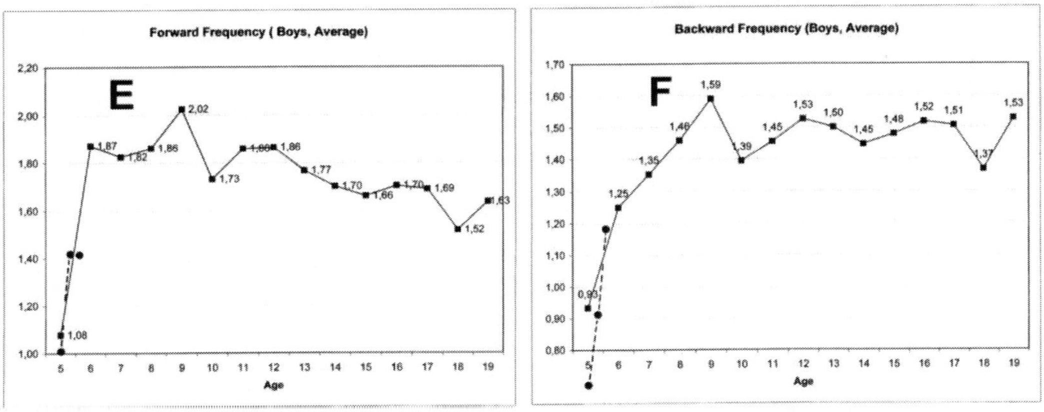

Figure 23. Forward and backward coordination dynamics (CD) values (A,B) and frequency of exercising (E,F) from boys in dependence of age. Note that the CD values decrease (become well) with increasing age, which means that the CNS is functioning better with increasing age. The frequency increases first strongly and then decreases slowly. The group sizes of measured boys are given in the circles in C and D.

8. CD Value of Girls in the Age Range between 5 and 19 Years

As for the boys, the coordination dynamics values of the girls decreased with age (Figure 24A,B). The values for backward exercising were similar to those of the boys; for 17 and 18 years old, the values were higher. The group sizes of assessment were roughly the same as for boys (Figure 24C,D).

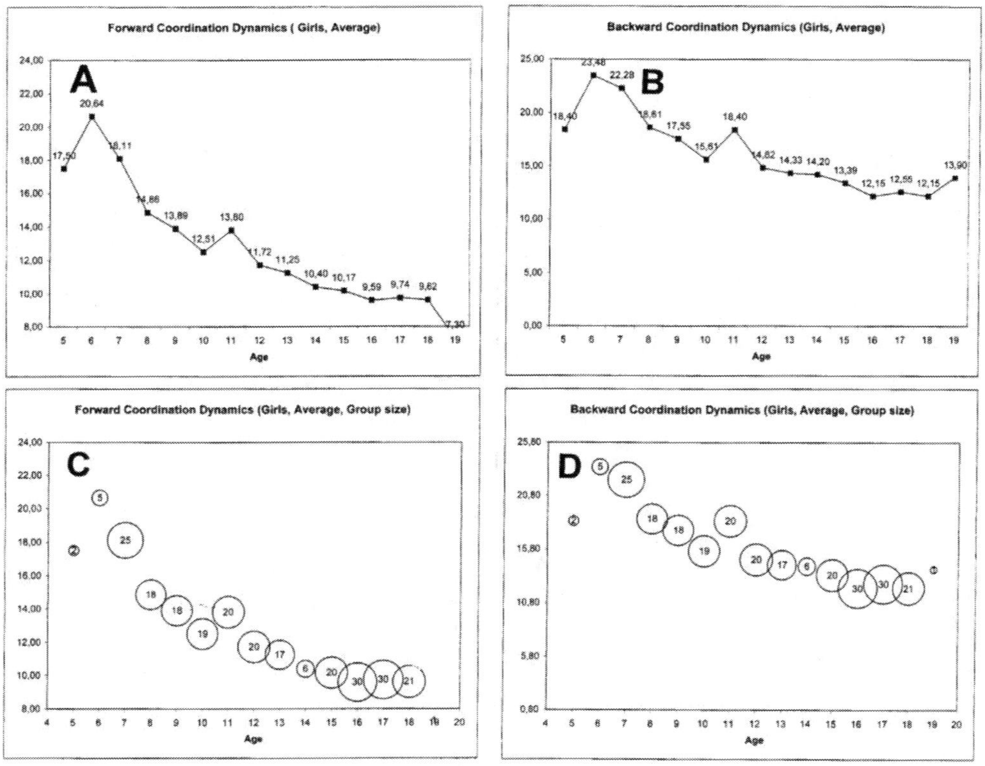

Figure 24. (Continued).

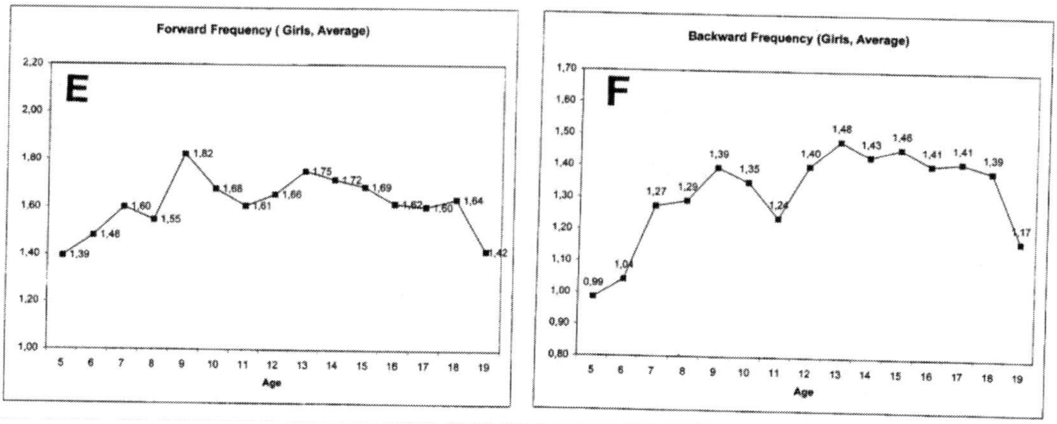

Figure 24. Coordination dynamics values (A,B) and frequency of exercising (E,F) of girls in dependence on age. Group sizes (C,D) are similar to that of boys.

As in the case of boys, the girls turned on average fastest in the forward direction (f = 1.82Hz) at an age of nine years (Figure 24E). For exercising in the backward direction, the girls were fastest at an age of 13 (1.48Hz) (Figure 24F).

9. CD Values for Girls and Boys

Plotting the coordination dynamics values of male and female pupils together, it turns out that the coordination dynamics values and the frequencies of exercising were similar (Figure 25).

The data of boys and girls were therefore gathered together to give the coordination dynamics values (Figure 26A,B) and the frequency of exercising for boys and girls (Figure 26C,D). The most group sizes for each age range now between 30 and 50 (Figure 26E,F). The variations were probably mainly due to the variation of the rate of development and the variation in the environment. Even though the group sizes were on average very large, details of the development seem only partly to become visible; the longitudinal study of the young child between 0.5 and 5 years of age offered more information. The transient increase of coordination dynamics values for exercising in the backward direction (worsening of CNS organization) at an age of 11 years (Figure 26B) may have significance. An expected increase of the CD values during puberty (in the age range of 14 years) cannot be seen. But because these pupils were not very cooperative at that age (as during school lessons), the group size was comparably small. The frequency of exercising was highest at an age of nine years.

At an age of five and six years, boys and girls were not very cooperative. They often were afraid of the assessment situation, even though the measurements were performed in their familiar surroundings.

In Figure 26A,B, the probable dependence of CD values (also taken the data from Jürgen into account) with age is pictured by a dashed arrow. This was the advantage of the infant Jürgen (Figure 29); he was not afraid of the assessment situation. The measurements were performed at home with his mother, and he was used to the instrument. He also forgot from one session to the next one how horribly difficult the exercising on the device was, especially when he was very young.

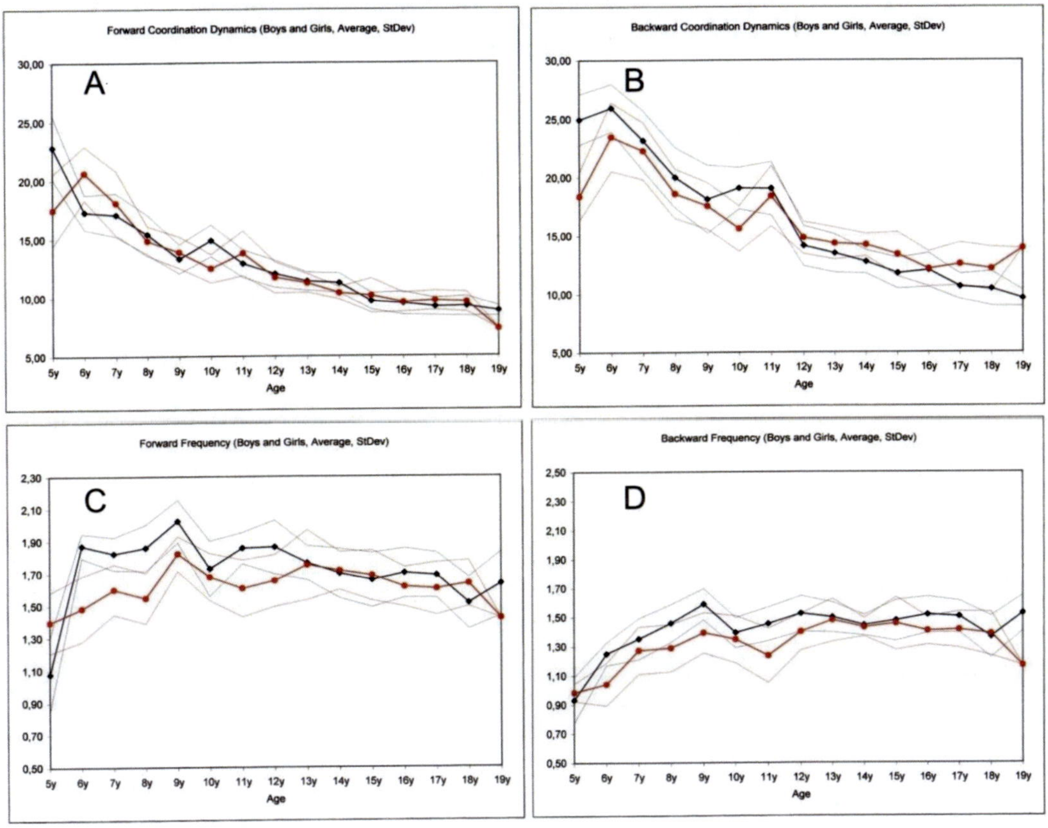

Figure 25. Coordination dynamics values and frequency of exercising of boys and girls in dependence on age. "■" are boys and "●" are girls. Note that the average group values are very similar. The standard deviations overlap much.

In conclusion, the CD values decreased with age that means with ongoing development. The frequency of exercising in the forward direction first increased strongly between five and nine years of age and then deceased slowly. This can be understood upon watching the pupils during exercising.

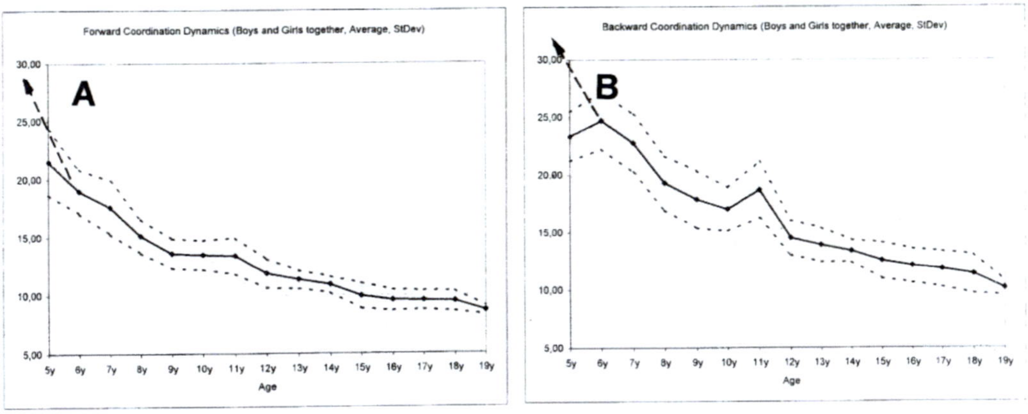

Figure 26. (Continued).

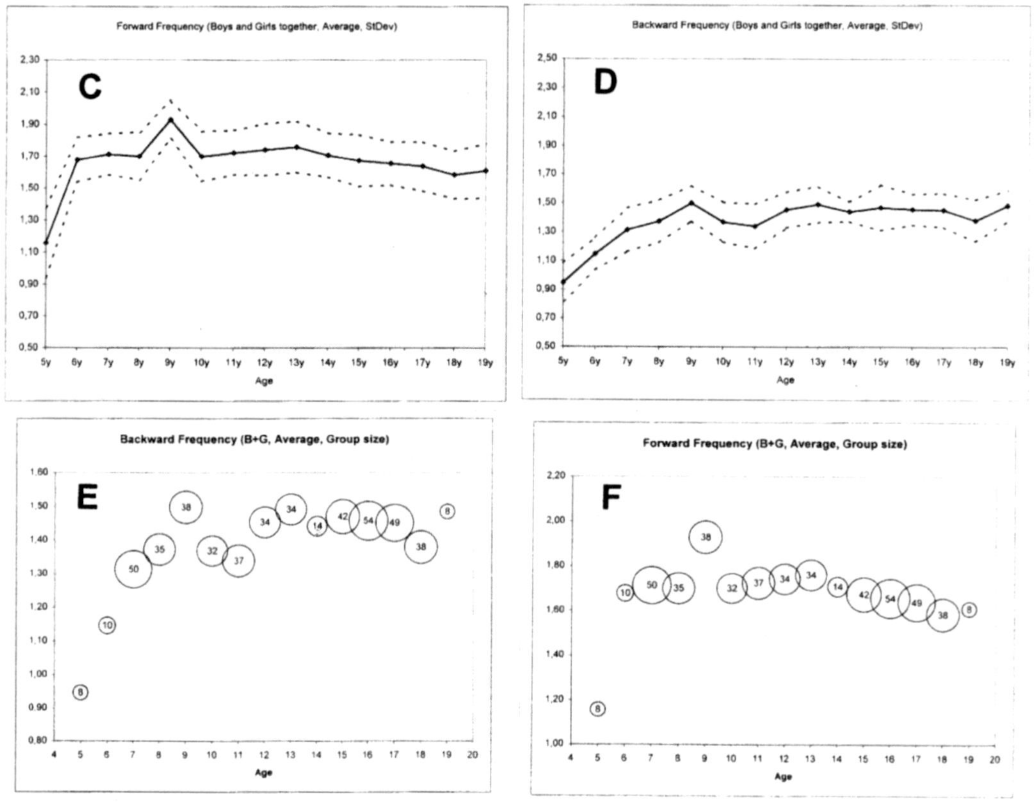

Figure 26. Coordination dynamics values of boys and girls lumped together. Note that the backward coordination dynamics values transiently increase at 11 years (B) and the frequency of exercising is highest at nine years (C,D). Note further the large group sizes (E,F).

The young children wanted to turn fast (sometimes they were even saying it), but they got stuck again and again. The continuous exercising was hindered by the missing network maturation. Especially the complicated coordination patterns of arm and leg movements between pace and trot gait could not be sufficiently generated by their CNS neuronal networks. The network complexity and accuracy were not developed well enough in that developmental period. When Jürgen was older, he even wanted to have the online measurements seen on the screen of the computer explained (Figure 19D).

10. Equality between Women and Men

An important observation, with respect to equality between women and men, is that the girls' CNS was functioning as well as that of the boys with respect to low-load coordination dynamics (CD) values.

Also for high load, the CD values of women and men are similar. The female director of the high school managed better with the high-load assessment than the male sports teacher! But with respect to a high-load assessment, the composition of muscle fibre types (fast and slow) and their upstream fast and slow neuronal subnetworks have to be taken into consideration (Figure 28).

11. Fast Movements to Improve Phase and Frequency Coordination among Neuron Firing for Improving CNS Self-Organization

The slow decrease of the exercising frequency with increasing age, after the fast increase, can also be understood. The high frequency of exercising was used by the CNS for improving its self-organization. The increased movement frequency improves the quality of performance because the kinetics is smoothing the movements, and the improved movement-induced afferent input improves CNS organization by movement-based learning. With ongoing maturing during development, this kind of movement-based learning becomes less important and the frequency of exercising decreased. Fast moving can be observed in infants and children in everyday life. Generally, walking and running (especially when performed fast), jumping, and training balance seem to be necessary for a physiologic development of the nervous system.

12. Occurrence of CNS Instabilities Due to Lack of Sufficient Movement-Based Learning

In reverse, if pupils are not moving sufficiently, they may not get only overweight, but we have to expect that variations, instabilities and small mistakes in CNS development can no longer sufficiently be repaired by movement-based learning. A higher incidence of diseases in children caused by suboptimal CNS development has to be expected. The occurrence of narcolepsy already in children (which normally occurs at an age between 20 and 30 years) may be such a disease caused by suboptimal CNS development.

13. Transient Fast Moving

The learning from fast movements to improve CNS functioning (most likely by improving phase and frequency coordination) seems to be necessary during development because of increasing network complexity. The intention to move fast can nicely be seen by the transient fast exercising of children on the special CDT device, frequently observed in the age range between six and nine years. For a few seconds, the young pupils were turning very fast (Figure 21E,F; Figure 27C,D), just because they liked the feeling, as an older person with a slightly malfunctioning CNS explained. In this way, the CNS is getting the input it needs for proper development. Such transiently fast moving can also be observed in patients with spinal cord injury, cerebral palsy, epilepsy and other CNS diseases.

A six-year-old boy with mild cerebral palsy turned comparably slowly in the forward and backward direction (around 0.5Hz) but turned transiently very fast (up to 2.5Hz) (Figure 27A,B). His healthy and sporty ten-year-old sister turned generally faster (in accordance with the typical age values, Figure 26C,D), but the transient fast turning was comparably not that fast (Figure 27C,D). It seems that the injured or malformed CNS needs especially the fast exercising for improving CNS self-organization, even though this fast exercising is hindered by the impaired self-organization.

Adults with a healthy CNS turn mostly at a frequency in the range between 1 and 2Hz. Athletes turn at higher frequencies, partly determined by the sport they perform. But healthy adults do not turn un-volitionally transiently fast. This transiently fast exercising seems therefore to have something to do with the improvement of the suboptimal functioning CNS. It is probably used by the CNS for improving its phase and frequency coordination for improved self-organization and for a functional CNS repair.

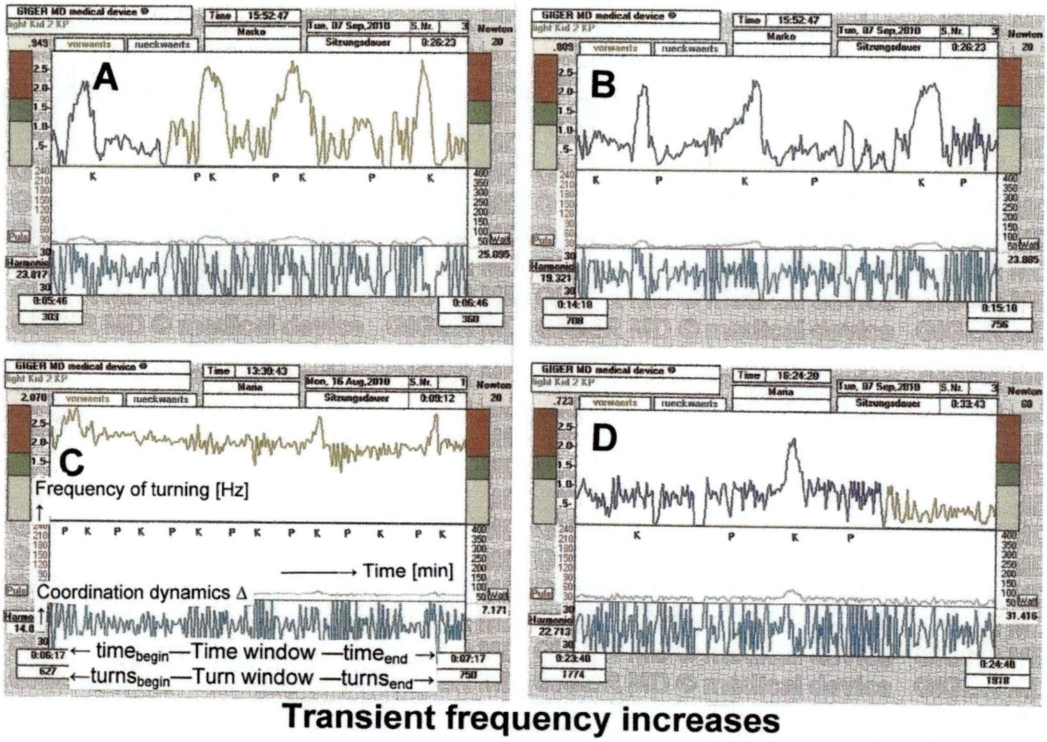

Figure 27. Transient frequency increase of exercising in a six-year-old boy with cerebral palsy (A,B,) and in a healthy ten-year-old girl (sister) (C,D). Note that the boy with a mild cerebral palsy is exercising at a low frequency, but the transient frequency increases are very high in comparison to those of his healthy sister.

14. Learning of Deep Network Complexity by Exercising Coordinated Movements at High Load

To improve CNS functioning, the developing child or the patient has to exercise on the special CDT device in the forward and backward direction to improve symmetries of network organization. By turning fast and slow, the movement performance can be improved via the movement-improving, movement-induced afferent input generated by the slow, medium-fast (fast fatigue resistant) and fast contracting muscle fibres (fast fatigue) and their corresponding upstream neuronal networks including the premotor spinal oscillators (Figure 28).

But exercising against different loads also has to be performed to enhance the integrativity (many subnetworks are activated simultaneously and coordinated) and complexity of neuronal network organization. By increasing the load during longer periods of exercising, the slow muscle fibres (and the upstream networks) (Figure 28) are trained, which, for example, a marathon runner needs. To get sufficient oxygen, to regulate the temperature and the cardiovascular performance, the vegetative nervous system is trained and improved in its functioning in general, especially because of substantial learning transfer to other untrained functions in the injured and uninjured CNS.

In a few pupils, this very healthy high-load exercising could be assessed, even though it was not popular among the pupils because of the hard work. The high-load assessment and training was performed every one, two, or few weeks, to see how the CD improved for high-load exercising.

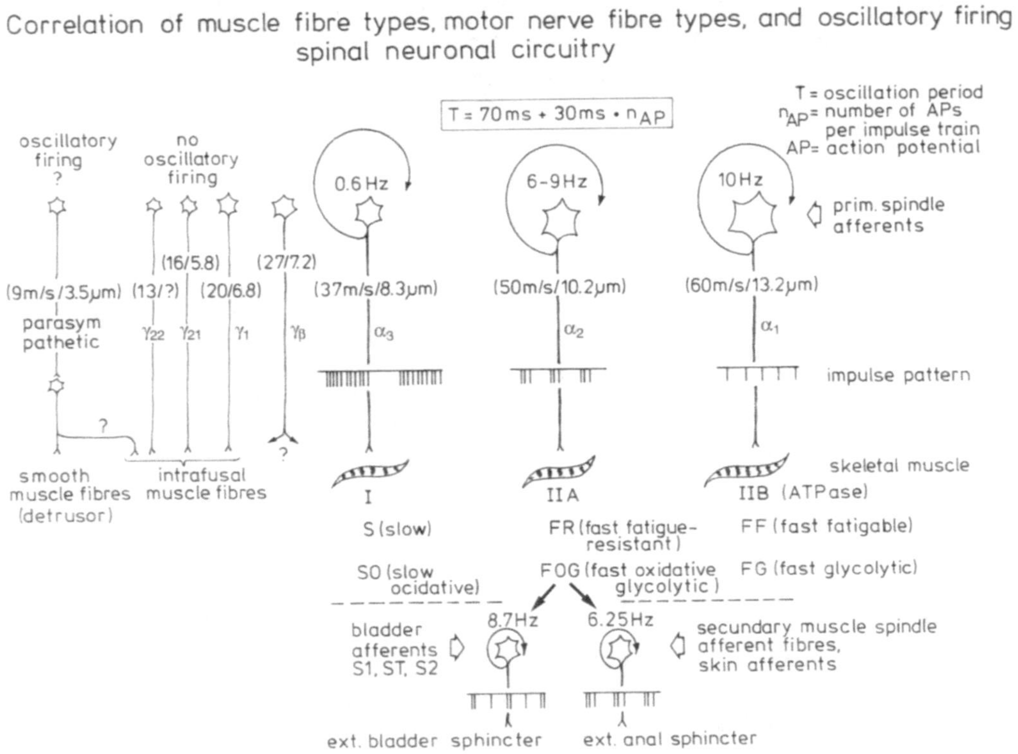

Figure 28. Correlation of muscle fibre types, motor nerve fibre types, and oscillatory firing spinal neuronal networks, based on histochemical, morphological and physiological properties. This Figure provides a simplified correlation between muscle fibre, motoneuron and sacral oscillator types. No additional subtypes have been included. The existence of α_1-motoneuron (FF) oscillators firing at 10 Hz has been predicted (34), and they have been identified in paraplegics (unpublished observation). α = motoneuron, γ_1, γ_2 = dynamic and static fusimotors, parasympathetic = parasympathetic preganglionic motoneuron. S1, ST, S2 = stretch, tension and flow receptor afferents. Different muscle fibres types and premotor spinal oscillators.

The female director of the high school also performed this high-load assessment and training and felt very good in her body afterwards. A sport teacher, on the other hand, of that school failed the test (he gave in at the load of 150N), probably because he had not sufficient fatigable resistant slow or medium-fast muscle fibres (as often 100m runners do not have) and not enough motivation to push himself through the high loads. For this high-load assessment and training, one needs the slow muscle fibres, the medium-fast and fast (fatigable) muscle fibres. The women can therefore perform this high-load exercising as well as the men.

15. Improvement of High-Load CD and Super-Compensation in Healthy Pupils

Figure 29 shows such high-load assessment strategy. The load of exercising was increased from 20 to 200N and then decreased again (Figure 29A). A fit adult or a pupil with a good functioning CNS can turn at the same frequency for all loads (no load escape (Figure 29A)) with low CD values (Figure 29B,C). An unfit subject with suboptimal network organization will reduce the frequency of exercising (or stop exercising) to escape the load, and the CD values will increase more strongly (get worse) for higher load exercising (Figure 29D,E,F).

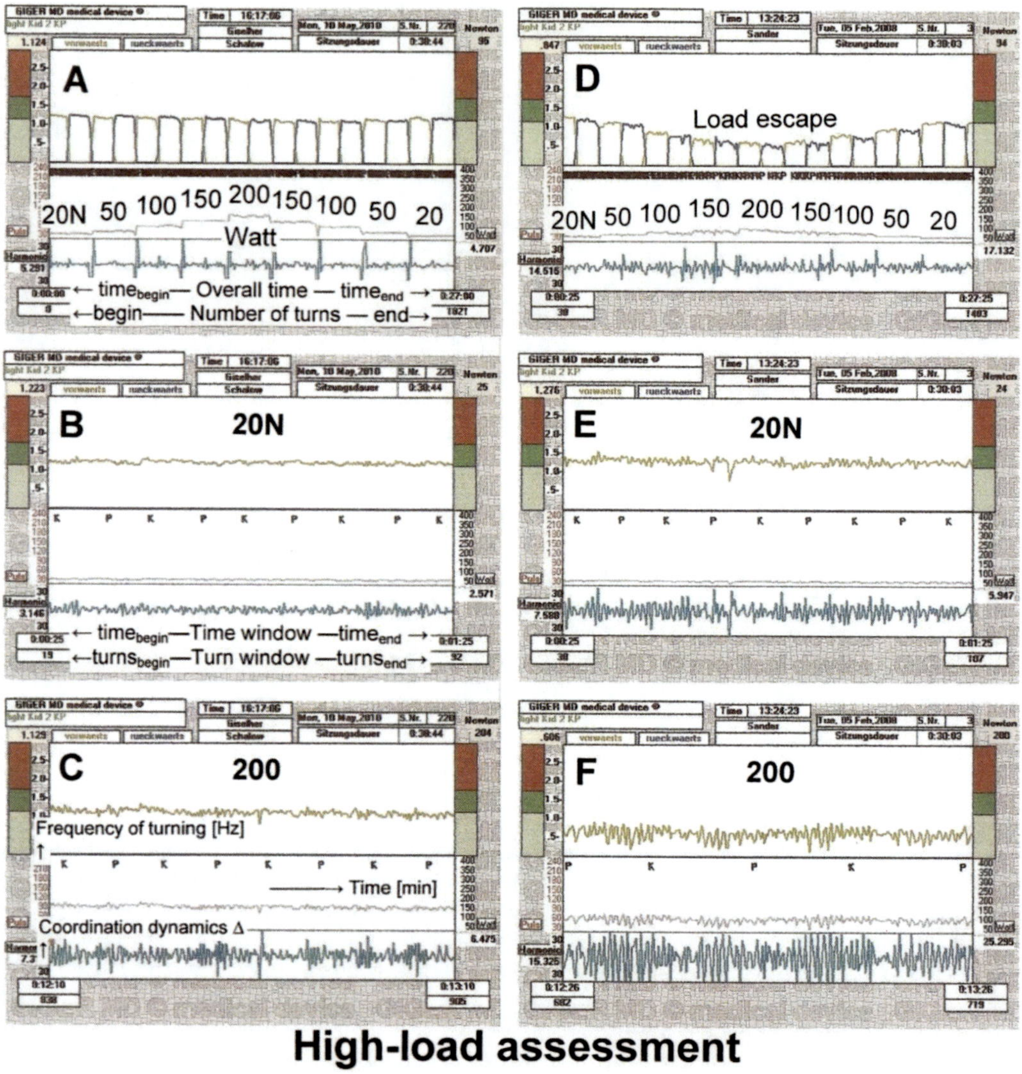

Figure 29. High-load coordination dynamics recordings for a well and balanced functioning CNS of the author (A-C) and a 15-year-old pupil (D-F). In the recordings A through F, the upper traces represent frequency of turning, the middle traces show load changes in Watt, and the lower traces show the coordination dynamics. In A and D, the whole record of 27min duration is shown during force changes in three-min steps (1.5min forward and 1.5min backwards) from 20 to 200N and backwards, and in B,C,E,F, a time window of 1min is shown when the load was approximately 20 and 200N (P and K are the pace and trot gait coordination positions). Note that for force changes between 20 and 200N, the power in Watt changed stepwise in the author (A) but not in the pupil who showed a transient frequency reduction (D). Note further that for the load of 200N, the pupil could exercise quite well on the special coordination dynamics therapy device for the easy coordinations close to pace and trot gait but not for the difficult intermediate coordinations; the frequency of turning decreased strongly, and the coordination dynamics increased strongly (worsened) for the intermediate coordinations, which means that the pupil's CNS had problems to organize the motor control for the difficult intermediate coordinations between arm and leg movements.

In adding up the CD values for the different loads, a value for the quality of high-load performance is obtained. Repeated assessments show the improvement of fitness and CNS organization. The assessment of fit pupils can show the improvement of CNS organization when going by exercising deeply into the network complexity of CNS organization.

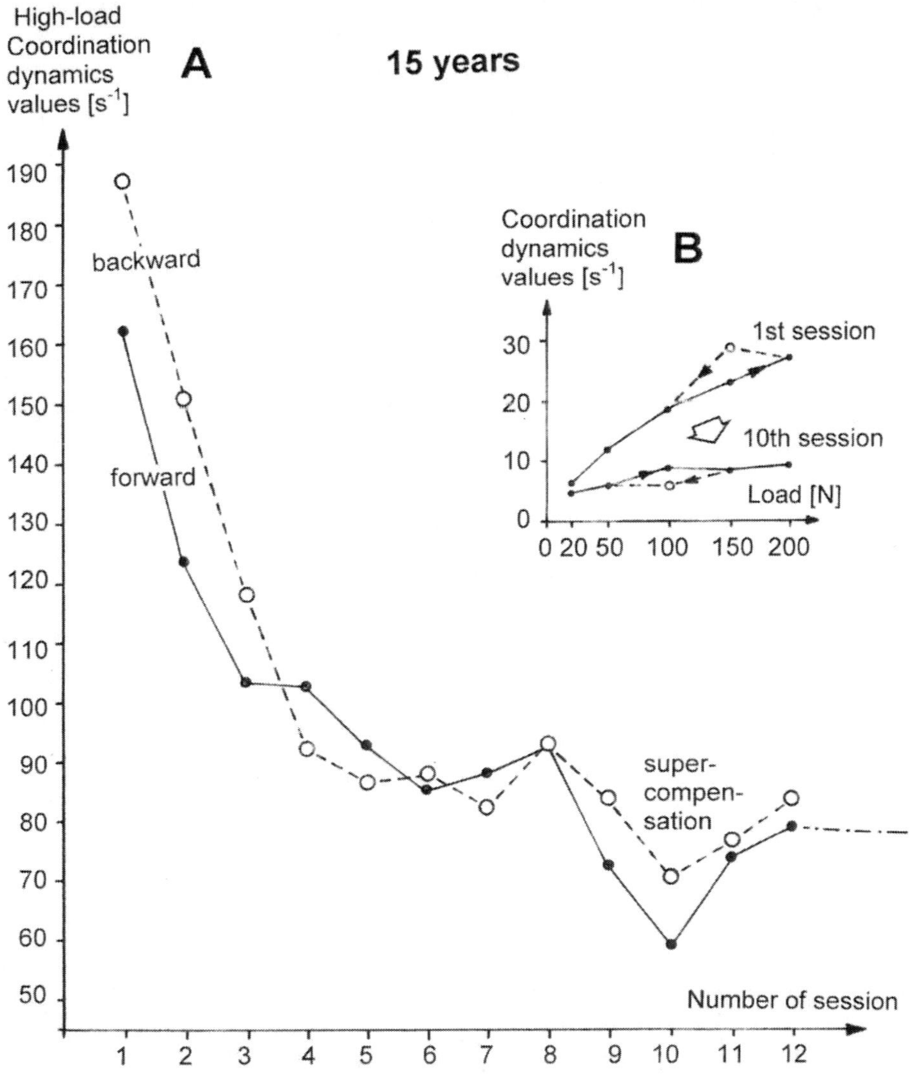

Figure 30. A. Improvement of high-load coordination dynamics (CD) values (for evaluation see B) due to 12 times repeated (low- and) high-load coordination dynamics testing and exercising (low-intensity motor learning) in the forward and backward direction in a 15-year-old pupil. The mean time interval between two sessions was 11 days, and the number of turns per session was approximately 3500. Note the transient improved exercising (super compensation) at the tenth session. The dashed-dotted line suggests the further continuation of the high-load CD values. B (inset). Coordination dynamics values in dependence on load increase (solid line) and decrease (dashed line) for the first and tenth session for exercising in the forward direction. The hysteresis-like curve for the tenth recording shifted to smaller coordination dynamics values. The high-load CD values were obtained by summing up the single CD values, Δ (high-load CD value = Δ20N + Δ50N + 100N + Δ150N + 200N + Δ150N + Δ100N + Δ100N + Δ50N + Δ20N + Δ20N).

Figure 30 shows the high-load CD values with ongoing assessment (and no additional training) of a 15-year-old male pupil. In Figure 30B, the single values of the first and tenth sessions are shown. As can be seen, the hysteresis-like curves improved with ongoing sessions. Plotting the high-load CD values (the sum of the single CD values at different loads) in dependence on the number of sessions, a curve for forward and backward exercising is obtained (Figure 30A). First, the high-load CD values improve strongly (get smaller) and then reach a plateau with a slight decline (Figure 30A). Before the plateau is reached, a phase is passed through in which the performance is best (the CD values are lowest). This phase of super performance is called "super-compensation" and occurred in this case with the tenth session. After this super-compensation phase, most pupils stopped being cooperative, mainly because of lost motivation. As long as they got better, they were motivated to do the hard testing. When they were not getting better anymore, the motivation was lost. The argument that this training is very healthy was seemingly not strong enough. The high-load testing for an 18-year-old pupil is shown in Figure 31.

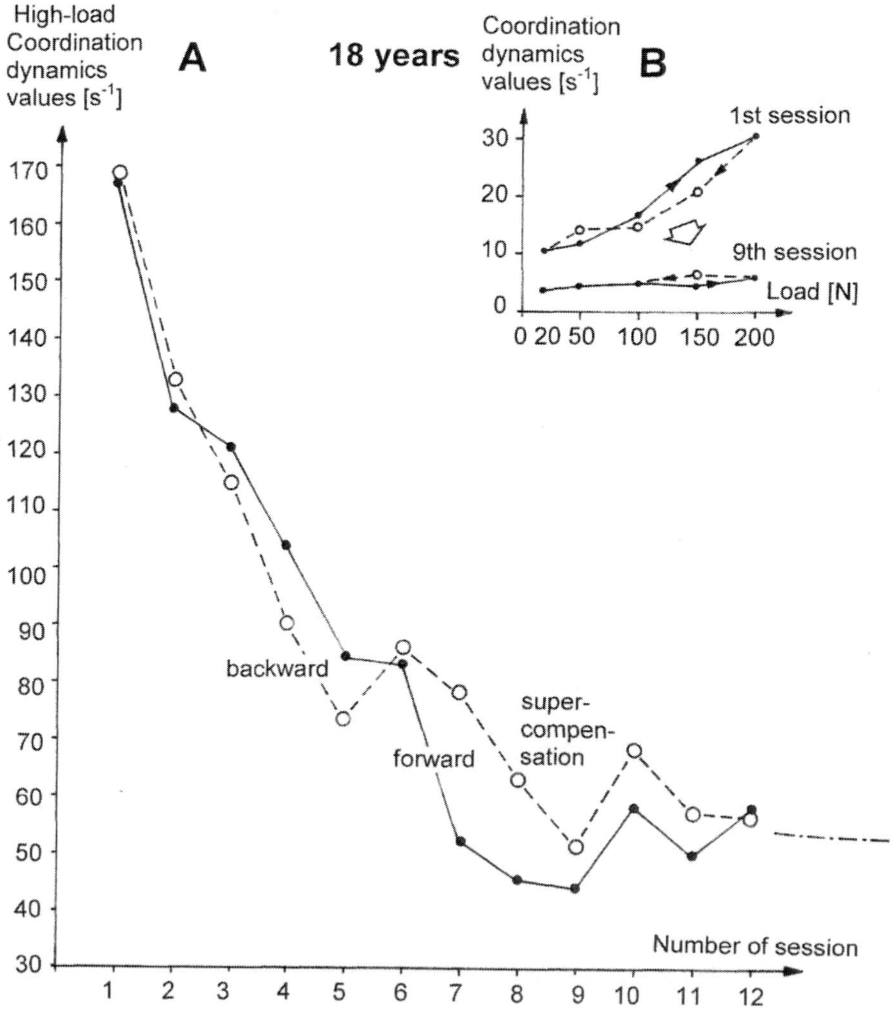

Figure 31. A. Improvement of high-load coordination dynamics (CD) values (for evaluation see B) due to 12 times repeated (low- and) high-load coordination dynamics testing and exercising in the forward and backward direction in an 18-year-old pupil. The mean time interval between two sessions was 31 days, and the number of turns per session was approximately 4,200. Note the transient improved exercising (super compensation) at the ninth session. The dashed-dotted line suggests the further continuation of the high-load CD values. B (inset). Coordination dynamics values in dependence on load increase (solid line) and decrease (dashed line) for the first and ninth session for exercising in the forward direction. The hysteresis-like curve for the tenth recording shifted to smaller coordination dynamics values. The high-load CD values were obtained by summing up the single CD values.

His super-compensation phase was reached with the ninth session. His plateau value was with approximately 60 lower than that of the 15-year-old pupil (approximately $80s^{-1}$). It seems that the high-load CD values also decrease with age (CNS maturing) and not only the low-load values.

16. Improvement of High-Load CD and Super-Compensation in Cerebral Palsy and Epilepsy

A 20-year-old pupil with a mild cerebral palsy showed a similar CD value improvement (Figure 32) than the healthy pupils.

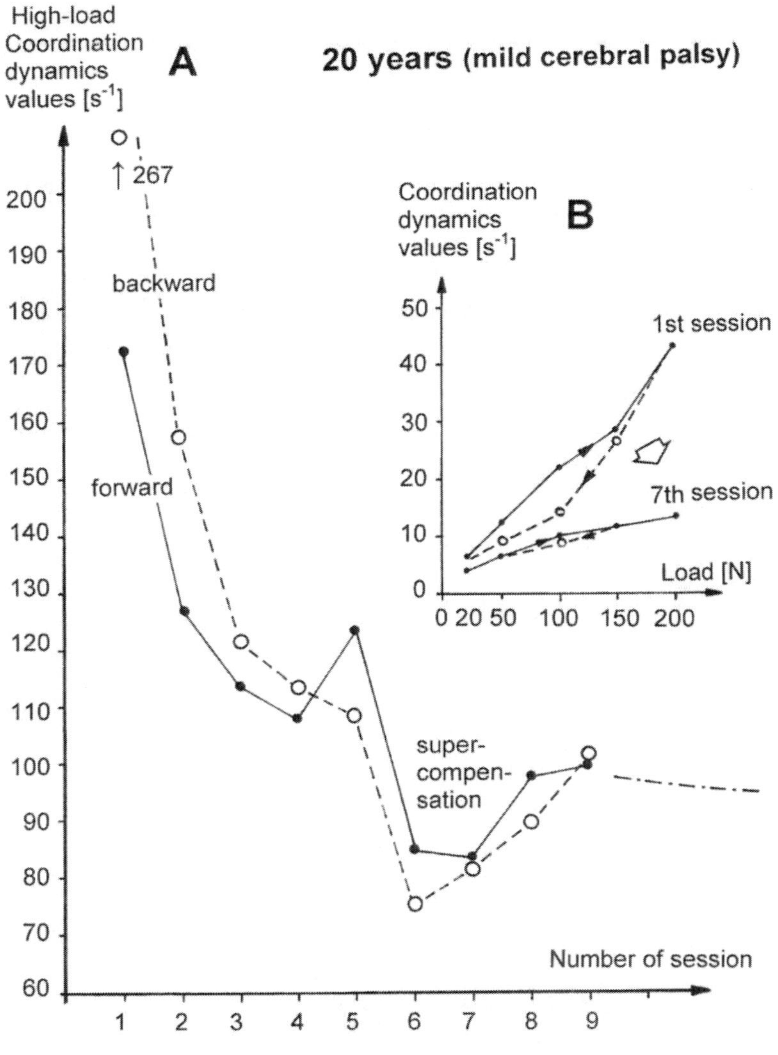

Figure 32. A. Improvement of high-load coordination dynamics (CD) values (for evaluation see B) due to nine times repeated (low- and) high-load coordination dynamics testing and exercising in the forward and backward direction in a 20-year-old pupil with mild cerebral palsy. The mean time interval between two sessions was five days, and the number of turns per session was approximately 4,600. Note the transient improved exercising (super compensation) at the seventh session. The dashed-dotted line suggests the further continuation of the high-load CD values. B. Coordination dynamics values in dependence on load increase (solid line) and decrease (dashed line) for the first and seventh session for exercising in the forward direction.

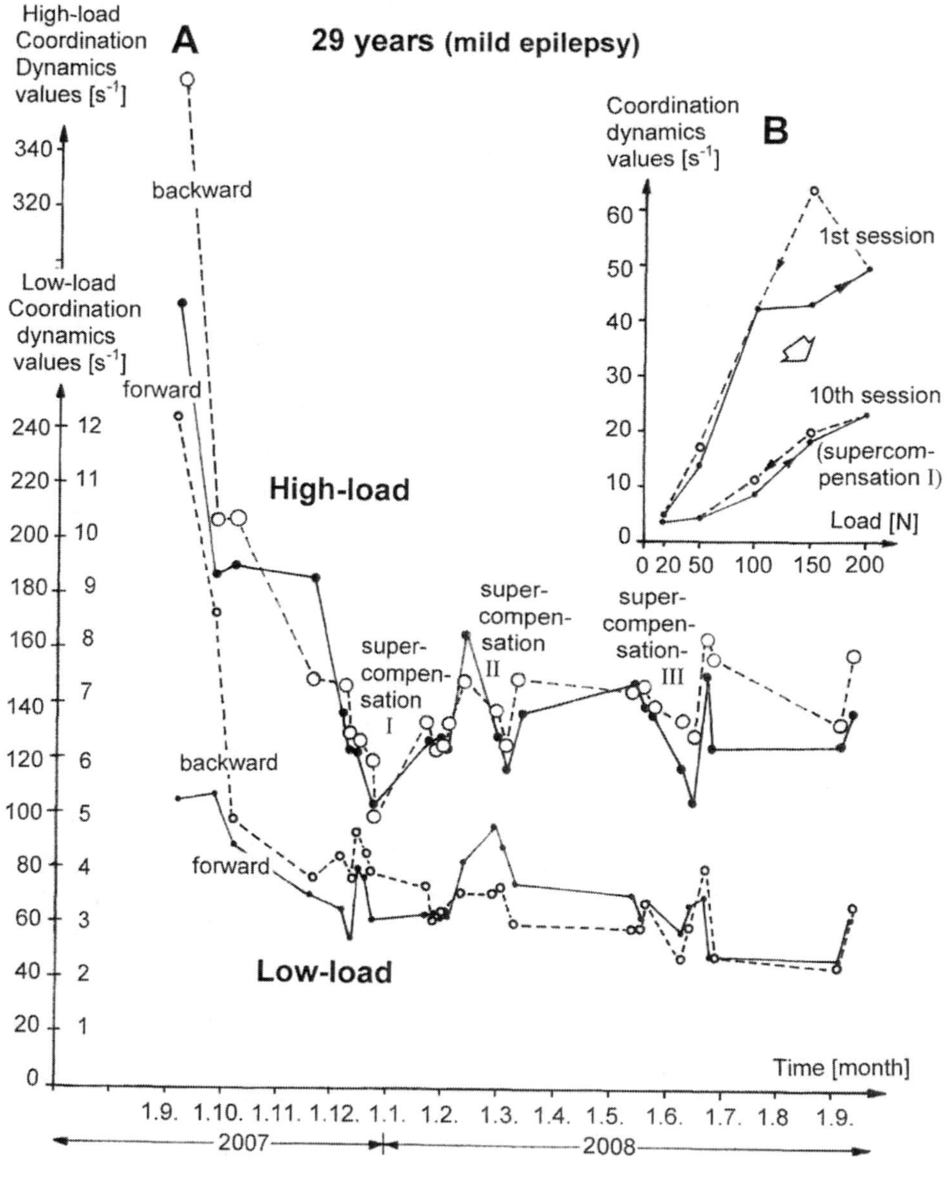

Figure 33. A. Improvement of high-load and low-load coordination dynamics (CD) values upon repeated low- and high-load coordination dynamics testing and exercising in the forward and backward direction in a 29-year-old mother with mild epilepsy. The mean time interval between two sessions was 13 days, and the number of turns per session was approximately 4,300. Note the first transient improved exercising (super compensation I) at the tenth session; with further exercising sessions, further super-compensations (II and III) occurred. Note further that with these 27 exercising sessions, a plateau with slight decline was reached. The low-load coordination dynamics values declined in a similar way but showed no super-compensation. B. High-load coordination dynamics values in dependence on load increase (solid line) and decrease (dashed line) for the first and tenth session for exercising in the forward direction.

His super-compensation phase occurred around the seventh session. His plateau value was approximately at 100. In comparison to the healthy pupils of the same high school, his values were at the beginning worse, and his reached plateau value was also not as good (low) as in the healthy pupils. All three pupils were not overweight and liked moving. The pupil with the mild cerebral palsy (now at university) walked every day for one hour when going to school. This healthy fast walking to school improved his CNS functioning, but the high-load exercising on the special CDT device further improved his CNS functioning.

A sporty 29-year-old mother of a pupil with mild epilepsy performed the high-load exercising for a longer period of time (Figure 33). Her super-compensation phase was reached at the tenth session. But since she went on with the high-load exercising (for improving her disease), further super-compensation periods could be observed (Figure 33A). Her high-load plateau value was around 140; that means much higher than in the pupils. Also, the single high-load CD values started from much higher values (Figure 33B). Even though her high-load CD values were not as good as for the pupils, the disposition to epileptic seizures decreased. Some pupils of the school with mild epilepsy could not be motivated for this demanding training to reduce the tendency of getting a seizure. The organizational effort was too big for them. Movement-based learning was not popular at the high school, even though the head of the school was supporting it. Such CNS and fitness training should be included in school sport lessons. In a patient with an incomplete cervical spinal cord injury (C5/6), the high-load CD values improved even five years after the accident (Figure 39).

17. Super-Compensation May Only Involve High-Load Performance

For the young lady with the mild epilepsy, the low-load (20N) CD values are also plotted in Figure 33A. The low-load CD values also reduced (improved) and reached a plateau around the value of 3 (Figure 33A). But no clear transient improvement of the performance of low-load exercising (super-compensation) can be seen at the super-compensation periods of the high-load assessment. Some changes can be seen on the low-load curve but no clear super-compensation periods. The super-compensation periods seem to involve mainly the performance of exercising at high loads.

18. Importance of Super-Compensation Phases

If the mechanism of super-compensation could be understood, and the time of occurrence predictable, super-compensation could have consequences in sports. To improve the functioning of the CNS of athletes during competition would be fair and an alternative to doping. Understanding of the mechanism of super-compensation could have consequences for CNS repair. Is the termination of a period of super-compensation the start of a structural adaptation to new needs? During limited spinal cord regeneration, following injury, the CD got always transiently worse with the gain of some small function below the injury level (64, Figure 50). This was explained in the way that with getting some more connectivity over the injury site, the CD values got worse and the training (movement-based learning) improved by the slight network enlargement disturbed, coordinated firing of neurons.

19. Movement-Based Learning Is Fastest in Young Children

To convince somebody that exercising on the special CDT device is really improving CNS self-organization, this person has to be made to exercise on the special CDT, because this person will experience then that this turning has nothing to do with a conventional fitness bicycle. This is especially true during backward exercising when the person would feel that his/her CNS has problems to generate such movement patterns, and that the movement performance of these patterns is improving quickly in the short-term memory.

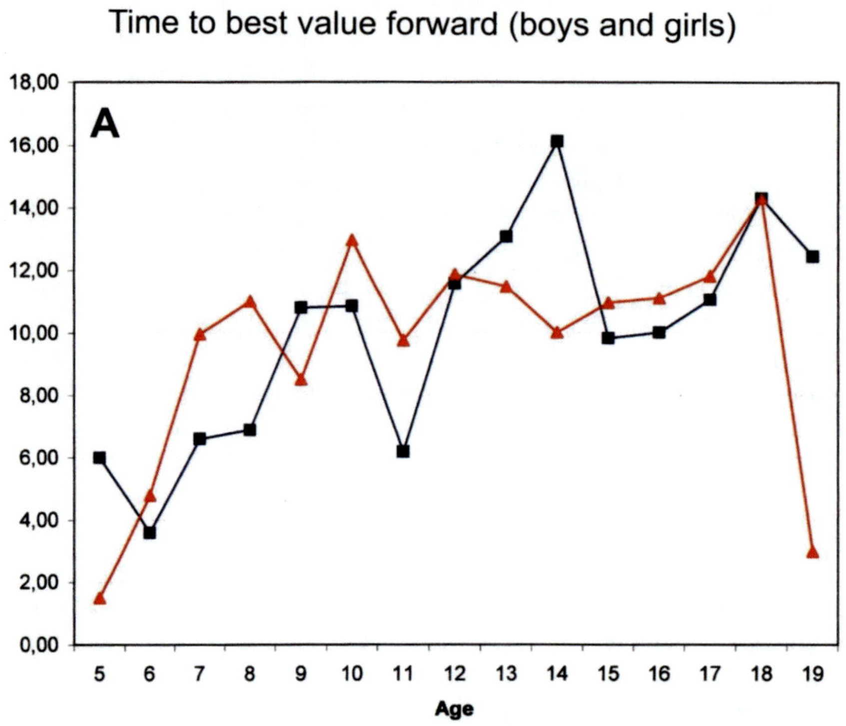

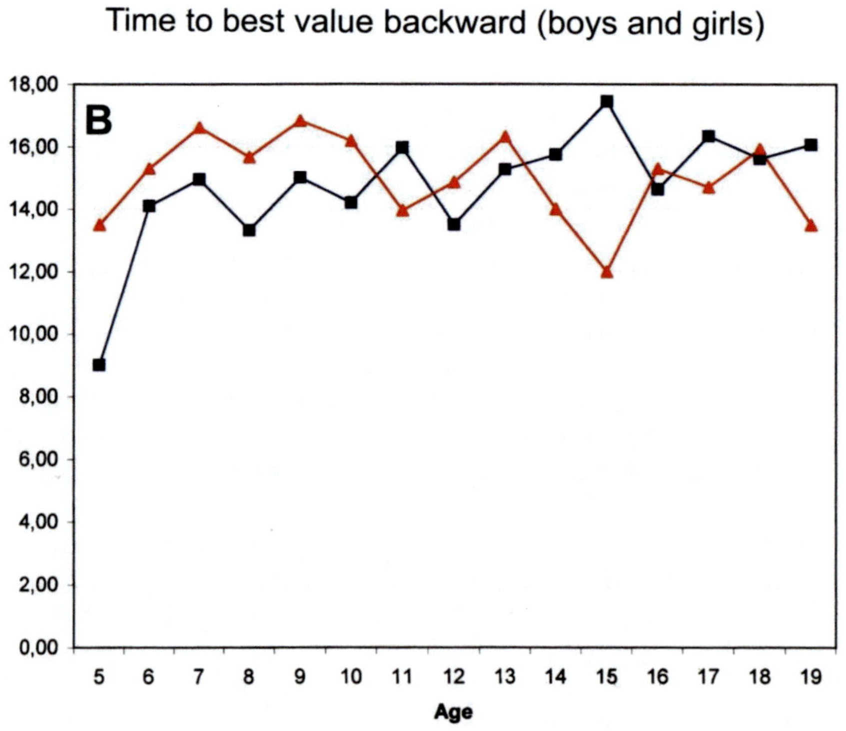

Figure 34. Times until the best CD values are reached during 21min of low-load exercising. Age in years.

It is of interest therefore to measure whether during development the rate of this movement-based learning in the short-term memory is increasing, decreasing or is the same for all ages. The rate of learning was assessed by the time that was needed to achieve the best CD value. As can be seen from Figure 34A, the young children learned fastest. For exercising in the forward direction (Figure 34A), the young children were fastest to reach the best CD value. The time to the best CD value increased (with variations) continuously from 5 to 18 years of age. This means more time was needed for motor learning with increasing age; the rate of motor learning decreased with ongoing development.

For backward exercising (Figure 30B), which is more difficult than to exercise in the forward direction, all ages seem to need the whole exercising time to reach the best value. The shorter times for a five-year-old age group are partly due to the becoming exhausted during exercising time of 21min so that the session was discontinued earlier. It could be expected that younger children learned faster to get their best value for exercising in the forward direction. But that the young pupils had similar problems as the older ones for exercising in the backward direction can also be understood. During development, the CNS is giving more emphasis to the functional more important forward moving than to establish symmetries as in the case of the stepping automatism and walking (see above).

20. The Developing CNS Is Giving More Priority to Forward Movement (Function) Than to Symmetries

After birth, the stepping automatism in the forward direction is present in the newborn, but it is unlikely in the backward direction (see above). This would mean that the symmetry of stepping (in the forward and backward direction) is not present at birth. Also, children can walk first in the forward and then in the backward direction. After severe cervical spinal cord injury, on the other hand, the stepping automatism in older children and adults simultaneously reappears in the forward and backward direction. It seems therefore that during infancy, the forward moving becomes first operational; the development gives preference to the forward moving and not to symmetries.

21. Rate of Learning of Coordinated Firing of Neurons in the Healthy and Injured CNS

It can be expected that movement-based learning is hampered by CNS injury. A six-year-old boy with mild cerebral palsy and his healthy ten-year-old sister trained for three sessions on the same special CDT device. The sporty sister (best in her class) could perform much better ($CD_{forward}$ value = 6) than her brother with cerebral palsy ($CD_{forward}$ value = 25). Mean normal values for an age of six years are 17.5 and for ten years are 13 (Figure 26). The CD value of this boy was worse than the mean $CD_{forward}$ value for the boys group and that of the girl was much better than that of the average girl.

Up until the age of two years, this boy was seemingly normal and developed within one month a pathologic movement pattern (walking on toes), which was not corrected by the further development. The injury of the CNS (cause unknown, although possibly a bacterial meningitis) affected other movements: the whole performance of the feet were pathologic; the boy could not crawl in pace and trot gait (an intermediate stage was used) and could not jump.

With three training sessions on the special device, the $CD_{forward}$ values for the girl did not improve and were around six, and those for the boy improved from 37 to 25 (32%). The transient fast exercising was much more pronounced in the boy (Figure 27A,B) than in the healthy girl (Figure 27C,D), and the rate of learning for improving the coordinated firing of neurons was much higher in the cerebral palsy boy. Obviously, the phase and frequency coordination was already very good in the girl and could not be improved quickly, whereas the coordinated firing of neurons of the boy was impaired by the injury (whatever origin) and could be improved quickly. On the other hand, it can be expected that normal movement patterns can be learned much faster by the healthy sister, because her CNS network will have more network connectivity and accuracy and therefore more network variability. The extremely high transient fast exercising may indicate that the injured CNS tried hard to repair itself. Still, the boy could not relearn to walk properly so far.

A nervous system during development is capable of autonomously modifying the motor skill, depending on the constraints imposed on the system and the level of functioning of each unit in the system. Elements composing the system may mature at different rates, and any factor can act as rate-limiting, delaying the emergence of a new skill. It seems therefore that in that boy, one or several elements of the system were injured so much that the system was unable to autonomously modify the pathological walking, and proper walking was not possible. The nervous system would need help by movement-based learning to recognize what subsystems are no longer working sufficiently and to repair those ones by learning.

22. Correlation of Motor Pattern Variability (Coordination Dynamics) with Higher Mental Functions in Pupils

An old saying is: to a healthy mind belongs also a healthy body and vice versa. It is therefore of interest to try to correlate the CD values of the pupils to their performance of higher mental functions at school. It was shown that a severe cerebral palsy child could learn speaking and improve its higher mental functions upon CDT (see below). But improving CNS self-organization upon CDT including the higher mental functions in normal and abnormal children or adults is a different situation than having good CD values in normal life and correlating them higher mental functions. I have measured good CD values in unsporty, overweight teenagers with seemingly average intelligence and average social behaviour. Top athletes with a very good performance in their sport discipline have mostly very low CD values for low- and high-load exercising but may not be especially intelligent or have an especially good social behaviour.

To have a good CD value is only one parameter of good CNS functioning. The rate of improving the CD values (the rate of improving the CNS self-organization) is a second parameter of good CNS functioning. There are most likely many more parameters to quantify good CNS functioning in general. One has to be extremely careful to draw conclusions of the quality of a person from some parameters of CNS functioning. In spite of all these methodical problems of judging the quality of a person, the low-load CD values were correlated to the performance of the higher mental functions at school to have a first experience whether correlations do exist or not.

Figure 35 shows the correlation of the CD values to the higher mental functions quantified here by academic level, effort, speed of learning, will, interest, attitude to others and other's attitude. The higher mental functions correlated with large CD values mean poor coordination dynamics. This opposite correlation is unclear.

	Abilities	cademic lev	Effort	eed of learn	Will	Interest	itude to oth	thers' attitude
FCD	0,11	0,20	0,25	0,15	0,25	0,28	0,21	0,21
FAVF	0,08	0,15	0,20	0,12	0,19	0,21	0,13	0,12
FF	-0,03	-0,02	-0,03	-0,05	-0,01	0,01	-0,01	0,00
BCD	0,08	0,15	0,23	0,09	0,22	0,26	0,18	0,19
BAVF	0,06	0,14	0,24	0,09	0,22	0,25	0,18	0,18
BF	-0,01	-0,03	-0,12	-0,05	-0,09	-0,06	-0,07	-0,07

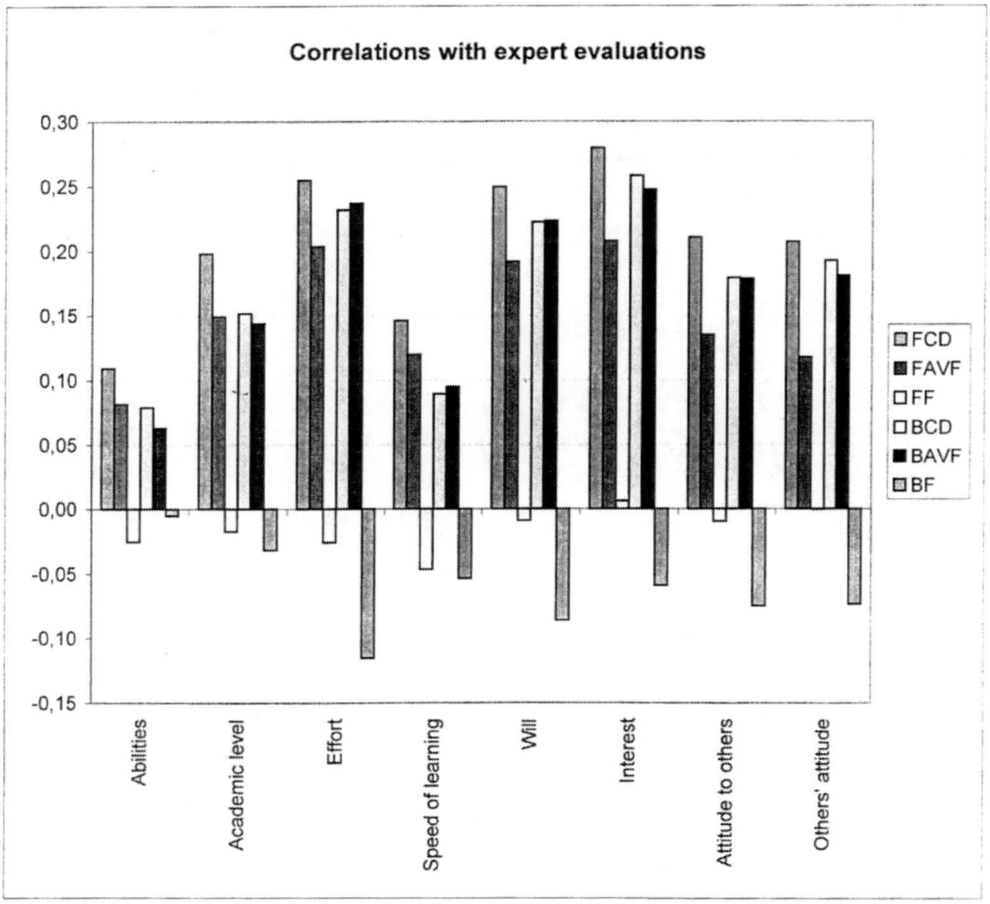

Figure 35. Correlation of the coordination dynamics to higher mental functions measured at school.

23. Liberating Movement Activities by Counteracting Gravity

Gravity has an impact on a SCI patient's improvement of motor functions from the moment of injury, and all movements during the first three years after the injury (and later) involve gaining greater control against gravity. Increasing postural control permits an infant to overcome the forces of gravity; thus lack of gravity control can be viewed as a task-limiting factor preventing the appearance of new motor skills in infants. Therefore, manual trunk support during jumping/swinging on spring-board and sky-walking (Figure 17D,E) may liberate more or new movement activities in SCI and other patients. Under the influence of artificial postural control, more mature skills may emerge.

24. Comparison between Development and Repair

Within the neuromaturational model, motor development progresses in a cephalo-caudal direction. Movement is first controlled proximally and then distally. Neonate's movement patterns are interpreted as being initially dominated by reflexes such as, for example, the plantar and palmar grasp reflex and primary stepping. Primitive reflexes are assumed to represent the dominance of lower levels of the CNS – the subcortical nuclei in the brain stem. The integration of these early reflexes is perceived to indicate maturation of the CNS. Evaluation and documentation of the presence or lack of primitive reflexes are used extensively in infant motor assessments to evaluate the level of neurological integrity. The persistence of these reflexes past the accepted age of integration is interpreted as abnormal, representing a lack of maturation of the cerebral cortex. Movement evolves from an uncontrolled reflexive state to a controlled voluntary state as the cerebral cortex matures. Thus the refinement of movement is dependent on CNS maturation [25].

Within the neuromaturational model, all changes in motor behaviour are viewed as intrinsically driven under the influence of the cerebral cortex; the environment has a minimal impact on evolving motor skills. In contrast, within the systems or dynamic motor theory approach, the behaviour itself can affect and modify the resultant behaviour, motor behaviours are a product of all contributing subsystems, and motivation, the emotional state and other variables influence the motor behaviour.

Yet this is precisely what CDT advocates; motor output can be influenced by more normal sensory and motor cue input. By learning transfer, vegetative and higher mental functions can be improved or repaired. One weakness of CDT is that no designated boundaries defining treatment exist. Because of the influence of many subsystems, and because the cerebral cortex is not viewed as the control centre, many opportunities exist to repair the self-organization of neuronal networks. What degree of deprivation in any system can be compensated for by the other systems? Are some subsystems more important than others or is the severity of involvement the restricting factor?

Humans are resilient to minor disturbances within any subsystem. Following severe injury or malformation, qualified therapy is probably needed for many years. It seems that upon efficient therapy, the repair is of open-ended nature. But the patients with their CNS improvements upon therapy or occurring complications upon no therapy are instructing us about the possibilities and secrets of CNS repair. Only they can feel the repair in their neuronal networks quantified by reoccurring sensitivity and motor and vegetative functions.

3.2. CNS Repair

3.2.1. Repair in Severe (95%) Cervical Spinal Cord Injury (SCI)

1. Repair in a Group of SCI Patients

CDT was applied to 18 patients (average age 31 years) after a SCI between C4/5 and L4/5; the therapy was administered on average five years after the injury for a minimum of three months [41]. The injuries were due to car accidents [5], working accidents [4], jumping into the water [4], motorcycle accidents [2], playing ice hockey [1], a parachute accident [1] and erroneous surgery [1]. All (almost) complete SCI became incomplete, i.e., motor functions improved below the lesion level, including trunk stability and arm, hand and leg functions. The organization of the CNS, quantified by the CD, improved by 42% for forward and by 49% for backward moving when exercising on a special CDT device. The improvements of the CD were 53%, 32% and 48% for injuries in the cervical, thoracic, and lumbar range, respectively. Since the CD did not change substantially prior to CDT—it did further improve with continued therapy and worsened when the therapy was terminated—it is concluded that the improvement of CNS functioning above and below the SCI level was due to the therapy. Since the CNS functioning would further improve if the therapy was administered for 6 to 13 months [43], it is suggested that SCI can partly be cured if coordination dynamic therapy is administered for a few years. It will be tried now to analyze further the mechanism of repair in SCI and to show some similarity to development.

2. SCI Repair in Comparison with Development

In almost complete cervical SCI, the caudal spinal cord is disconnected from the cerebral cortex and brain stem. Following the spinal shock, some reflexes such as the stepping automatism and urinary bladder reflex (similar to those in infants) may reappear, especially when stimulated. Pathologic reflexes or automatisms also appear as, for example, extensor spasticity. With very limited regeneration upon CDT, motor and vegetative functions partly reappear in a cephalo-caudal direction, and movements become controlled first proximally and then distally; spasticity reduces. This cephalo-caudal and proximal to distal scheme only partly holds because the injured CNS uses all existing repair and adaptation mechanisms upon movement-based learning, including plexus connections outside the spinal cord. But with respect to the improvement of trunk stability and breathing, the repair in cephalo-caudal direction seems to hold.

3. Comparison between Development and Repair in Two SCI Patients with Respect to Administered Therapy Time

To improve the efficiency of SCI repair by CDT, it may be helpful to compare the improvement of trunk control and antigravity movements between development and repair. For the comparison, two patients with similar severe cervical SCI are used. The MRIs are shown in Figure 36J,K. Upon 0.5 years of therapy, the boy mainly regained some trunk control and the breathing improved; the urinary bladder function was not repaired so far. To the young lady, five years of CDT were administered. Bladder function was repaired and many movements improved including breathing [59,64]. The MRIs of the patient's SCI are shown in Figure 36J,K, and the two treatment stages are compared with the development in Figures 36 and 37.

Upon 0.5 years of CDT, the patient regained little trunk control and had big problems with the trunk rotation. Arms and legs could not flex for the rotational movement (Figure 36A). Upon five years of CDT, the patient can easily roll from supine to prone position and backwards. The movement performance with slightly flexed arms and legs (Figure 36B) looks much more healthy and similar to those of an infant (Figure 36C). The unsustained sitting (Figure 36F) following 0.5 years of therapy is also very poor. No leg flexion is possible. The back is not straight, and the hand cannot be opened (Figure 36D). Following five years of therapy, the unsustained sitting (Figure 36E) is more similar to the pattern of the infant (Figure 36F).

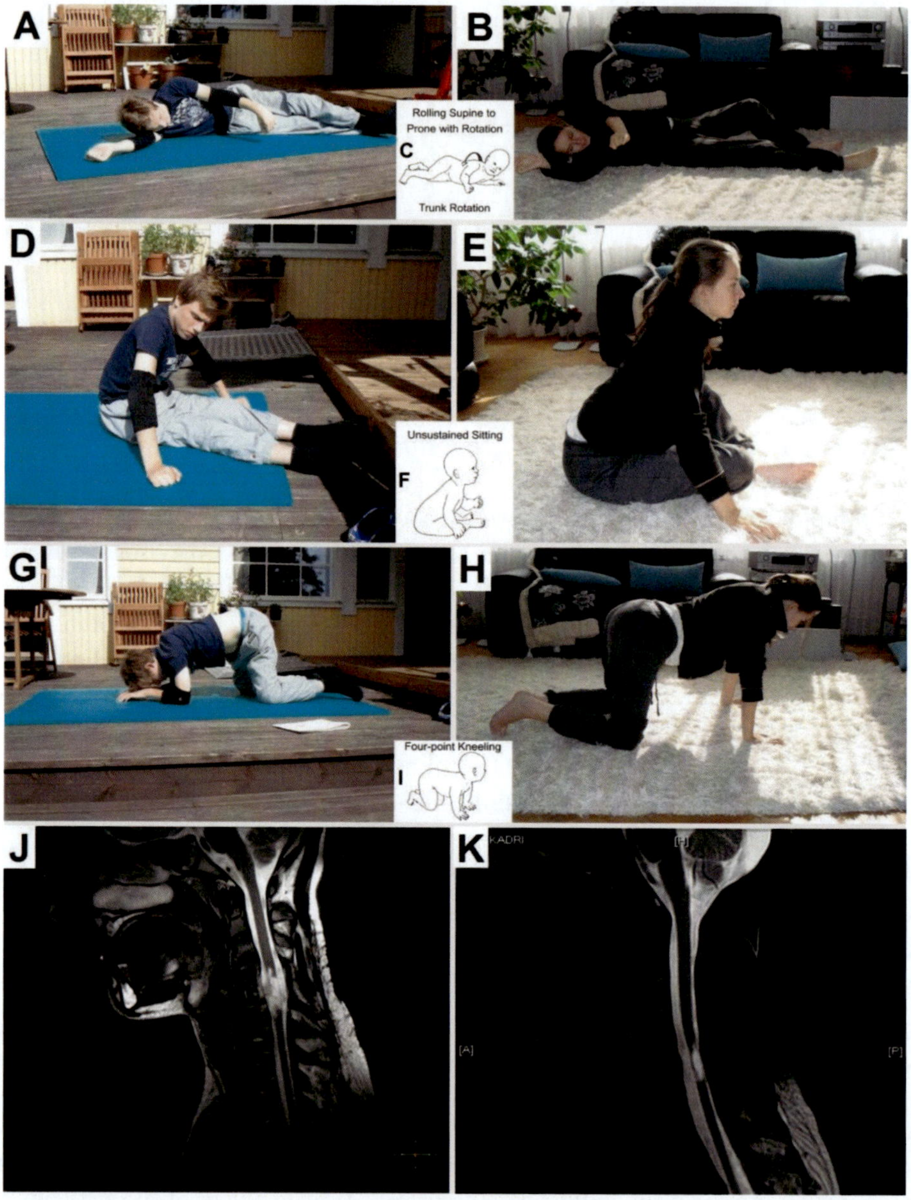

Figure 36. Comparison between development and the repair of an almost complete cervical spinal cord injury upon 0.5 years (A,C,G) (therapy started with 15.5 years) and five years of coordination dynamics therapy (B,E,H) (CDT started with 17 years). The MRI of the SCI of the patient of the left panel is shown in "I" and those one of the right panel in "K." Trunk rotation (A,B,C), unsustained sitting (D,E,F), and four-point kneeling (G,H,I) is compared between 0.5 and five years of CDT and with the positioning of an infant.

The four-point kneeling (Figure 36I) (crawling position) is poor following 0.5 years of treatment (stretched arms are not possible without support) (Figure 36G) and better following five years of treatment (Figure 36H). But the main weight is still on the arms (unsupported crawling is not possible). The feet and leg positioning is much better than in the other patient (36G), and the arms can be stretched.

The hands to knees positioning of the infant (Figure 37C) cannot be achieved in the patient upon 0.5 years of therapy (Figure 37A). The wheelchair is necessary for hands to knees touch. But this position helps the patient to exercise the pace gait coordination and co-movement can be induced in the legs by the synchronous touch of the hands. Upon five years of CDT, the "hands to knees" position can be achieved (Figure 37B) and the abdominal muscles are activated, but very little power is in the legs and there is no (or not enough) dorsal flexion in the feet. The supported standing of the infant with leg movements (Figure 37F,I) is not possible in the patients, even following five years of CDT. The leg muscles have still too little muscle power. But with different supports, both patients can train upright movements with full weight-bearing, such as swinging/jumping on springboard and sky-walking (Figure 37D,E).

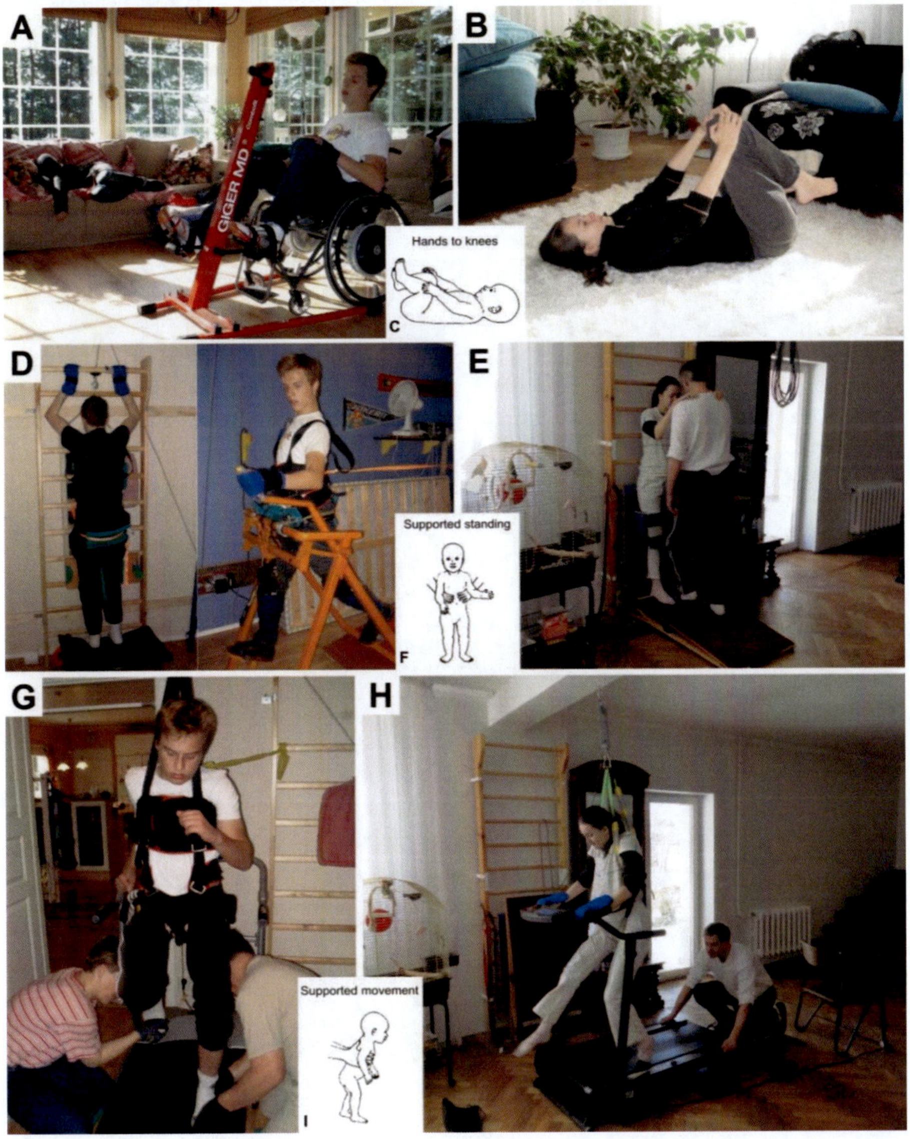

Figure 37. Comparison between the development of an infant and repair of a cervical spinal cord injury upon CDT. Left penal, positioning upon 0.5 years of therapy and right penal after five years of therapy. Same patients as in Figure 36. Reaching of hands to knees, supported standing, and supported walking are compared.

For walking on treadmill upon 0.5 years of therapy, weight-bearing and feet support is needed (Figure 37G). Upon five years of CDT, the patient still needs weight-bearing support, but is able to move the legs volitionally without support (Figure 37), even though the dorsal flexion of the feet is missing. Upon walking barefoot on treadmill, the stepping automatism is stimulated strongly. The supported standing pattern of the infant (Figure 37I) is still a dream for both SCI patients. But development lasts 17 years. Hopefully, the patients can partly catch up on the long-term upon optimal therapy.

3.2.2. Repair in Incomplete (50% Injury) Cervical Spinal Cord Injury (C5/6)

In incomplete cervical SCI (Figure 38), the patients can also relearn running (Figure 40A) and jumping (Figure 40B), besides proper breathing and urinary bladder control and can improve cardiovascular performance [58].

Such repair can be mainly achieved by a functional reorganization, which is quite different from the development. In almost complete cervical SCI, there is nearly no spontaneous recovery. In incomplete cervical SCI, there is some spontaneous recovery, especially the following weeks after the spinal shock.

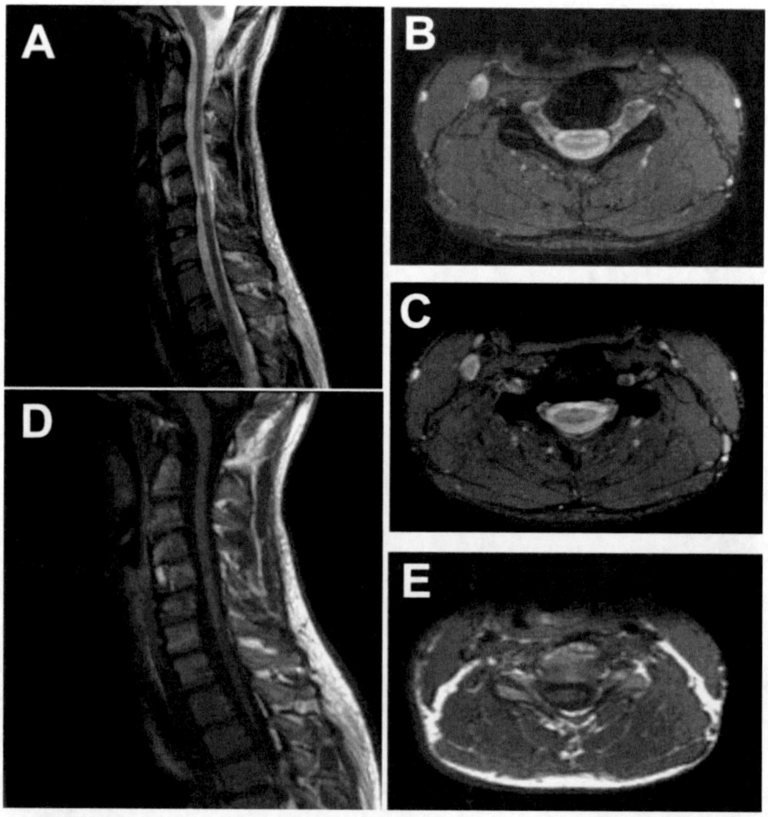

Figure 38. Magnetic resonance images show an incomplete spinal cord injury at C5/6 levels, three years after the accident, from a 20-year-old patient who jumped into shallow water. A,D. Sagittal T2 (A) and T1-weighted (D) images. The region of hyperintense signal within the spinal cord of the T2-weighted image (light) indicates nonhemorrhagic spinal cord damage. B,C,E. Axial T2 and T1-weighted images of the spinal cord showing a more severe injury than in the sagittal plane. The authors estimated an anatomical injury of approximately 50%.

Most spontaneous recovery occurs in the first six months following the injury. One year after the injury, there occurs normally no further spontaneous recovery. Conventional rehabilitation centres, which are concerned with care and not with cure, "sell" the mainly spontaneous recovery as treatment-induced repair. In a case report, it is explained how a 17.5-year-old boy, who suffered an incomplete spinal cord injury (SCI) at the Cervical 5/6 levels, could achieve a near-complete recovery within three years of coordination dynamics therapy (CDT) and improved further in the following three years.

1. Case Report on the Repair of an Incomplete SCI

A 17.5-year-old male patient suffered a cervical spinal cord injury (SCI) at C5-6 levels in a diving accident. The spinal cord injury was incomplete (ASIA grade D). Some sensory-motor functions were preserved below the level of injury. Initially, the patient received conventional rehabilitation treatment and then started coordination dynamics therapy (CDT), five months after the injury. At the time of initiation of the CDT, this patient needed a stick as support for walking, was wearing an orthosis on the left foot and emptied the bladder with intermittent catheterization. During walking, the left heel was not touching the ground. While jumping on the springboard, his feet had to be supported by the therapist. While turning on the special CDT device, the left hand had to be fixed.

He received nearly three years of non-optimal CDT. One year after the injury, the metal support used for spine fixation was removed. At the end of the third year of CDT, he was able to walk, run, and even jump independently and did not need any internal or external support. He is continent again and off of all medications. In that sense, it can be safely said that the patient is almost cured of his quadruparesis. In the following, the clinical aspects and the theoretical concepts behind this recovery will be explained in detail.

2. Quantification of the Anatomical Extent of the Injury with MRI

Magnetic resonance imaging (MRI) of the cervical spine was performed three years after the accident to delineate the anatomical extent of the injury. At the site of the injury, the extent of loss of tissue would indicate the severity of the injury, and the presence of residual tissue would indicate the extent of spared tissue bridging the injury site that was maintaining the connectivity. The T2-weighted images (Figure 38A) highlighted the pathology to be more extensive than what was seen on T1-weighted images (Figure 38D), and axial images delineated a more extensive injury than the sagittal images (Fig 38A-E). By comparing these multiple multi-planar images, the author estimated the spared tissue to be 50% of the cross-sectional area at the site of injury. The grey matter of the central cord seemed to be more damaged than the circumferential white matter. Since in this case, the injury span was less than one vertebral level (relatively modest distance to bridge the injury site), it is possible that a small part of the visible spared tissue was actually from those tract fibres that regenerated during the course of therapy over the three years. It seems that in SCI the grey matter (the neuronal networks) is most vulnerable to pressure and/or lack of sufficient blood supply, whereas the white matter (the tract fibres) survives better.

3. Improvement of Coordination Dynamics (CD) Values Following Therapy

The progressive changes in the CD values (mean pattern stability per minute) for loads of between 20N and 150N (and subsequently 200N) over the course of three years, while undergoing CDT, reflected gradual improvement in CNS functioning (Figure 39). As explained earlier, improvement in CNS functioning was indicated by the lowering of coordination dynamics values, i.e., decreased arrhythmicity of exercising and increased pattern stability. At the beginning of CDT, as the therapy progressed, there was strong improvement in (reduction of) CD values for loads between 20 and 150N for exercising in the forward and backward (measured intermittently) direction. This improvement was achieved in spite of non-optimal/unsupervised therapy. Subsequently, the patient stopped CDT for more than half a year, and then restarted. This break in the therapy resulted in the deterioration in coordination dynamics. Interestingly, when he restarted and continued the therapy, the values improved again, especially for exercising at loads of 100 and 150N. Deterioration of CD values after stopping the therapy and improvement after restarting the therapy indicates that even after a few years of therapy, the functional repair of the CNS is incomplete and not persistent. In other words, to maintain an optimal level of improvement of CNS functioning, the continuation of therapy for longer duration is essential. One year after finishing the therapy, the high-load CD values worsened again in the patient (Figure 39), even though the patient felt more power in arms and legs, and the low-load values (20N) did not increase (got worse) again.

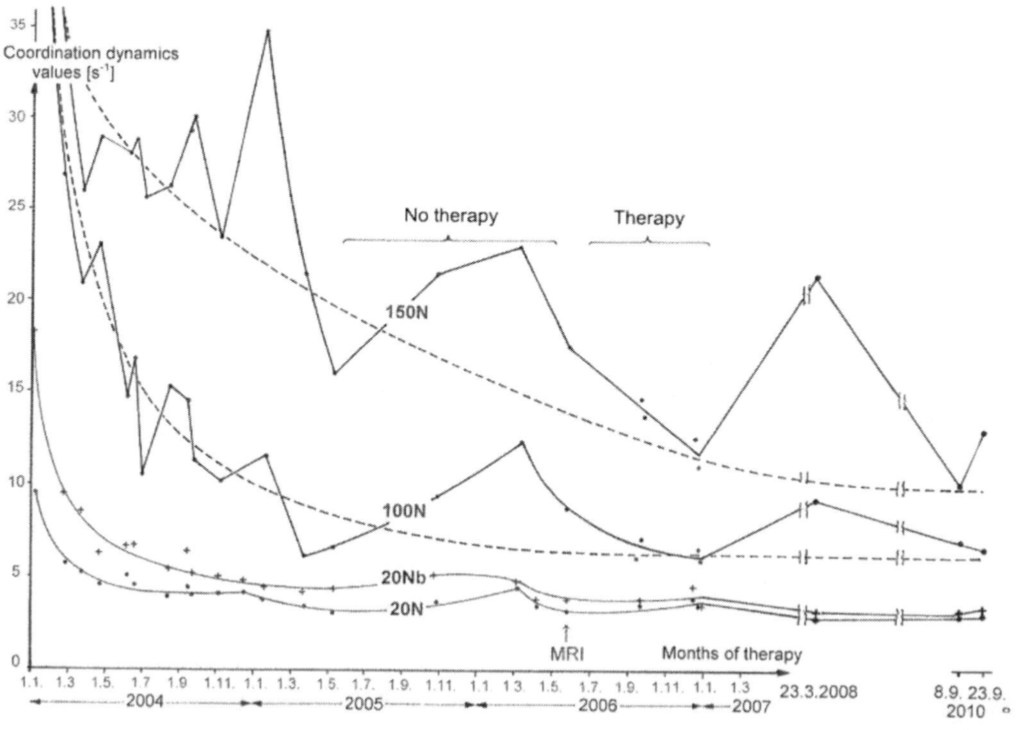

Figure 39. Relation of coordination dynamics values to therapy length for increasing load between 20 and 150N. The loads for forward exercising (dots, 20N, 100N, 150N) are marked at the curves (20Nb = backward exercising (crosses) at 20N). Note that with no therapy, the coordination dynamics values got worse and upon administering therapy again, the values improved again even two years after the accident. After stopping therapy, the coordination dynamics values for 100 and 150N increased again (23.3.2008). When the patient trained himself intensively (including running and jumping), he further improved his high-load CD (8.9. and 23.9.2010).

The reincrease of the high-load values further show that the CNS repair was still incomplete and not fully persistent. The patient would need a maintenance therapy of about an hour per day. When the patient trained further by himself, CNS functioning further improved (see CD values for 150N in 2010 in Figure 39).

4. Learning Transfer and Integrated Functions

It took the patient nearly eight months to achieve the best performance for low-load exercising (at 20N) on the special CDT device. Similar good performance was achieved for high loads of 100N after only 24 months with ongoing therapy and after 36 months for 150N (Figure 39). Exercising at significantly higher loads requires recruitment of more muscle groups of arms and legs and also the trunk. Such recruitment of muscle groups of almost the entire body requires activation of almost the entire central nervous system (CNS). Thus, a widespread activation of downstream and upstream motor neurons and their relevant neuronal networks takes place. Such integrated (more systemic) network activation is important because it not only affects the attractor basin of a particular movement but also changes the entire attractor layout of the state-space. As the whole attractor layout changes, the intrinsic dynamics also change, which causes learning transfer from the movement to other CNS functions, like bladder control. The induction of learning transfer by integrated (systemic) CNS activations is very important, since only those networks that can be can recruited, substantially activated, trained or reached by learning transfer can be repaired. Therefore to achieve learning transfer, exercising at higher loads is required for longer duration.

5. Motor Improvement

According to coordination pattern dynamics theory, if CD values improve, there should also be visible improvement in motor performance. The motor training of the patient (which included crawling, walking and running on treadmill in the forward and backward direction, jumping on spring-board,) was guided by the therapist to correct abnormal movement patterns. At the end of the three years of therapy, the significant improvement of coordination dynamics values (Figure 39) was accompanied by a substantial improvement in the performance of movements like running and jumping with the skipping rope (Figure 40). Quantitatively, the times required for running 60 and 400 meters decreased, and the number of jumps per series increased (Figure 41).

6. Electrophysiology (sEMG)

With ongoing therapy, CD values can be used to quantify improvement in CNS functioning up to a percent (Figure 39). Clinically, this improvement can be correlated with the corresponding improvement in scores of activities of daily life (Figures 40,41). But for knowing and measuring the physiological and pathophysiological changes occurring at neuronal network level in the CNS with ongoing therapy, surface electromyography (sEMG) is needed. Even single neuron activity can sometimes be measured non-invasively by sEMG. The firing pattern of single neuron (motoneurons) can be studied non-invasively in terms of single motor unit potentials.

a) Pathologic patterns of motor activation. Surface EMG was performed in this patient while he was undergoing CDT. Figure 42 shows sEMG recordings of motor patterns while the patient was exercising on the special CDT device at the beginning of therapy (Figure 42A,B) and also after six months of therapy (Figure 42C,D). The motor patterns were completely abnormal at the beginning of therapy (Figure 42A,B). The amplitude of sEMG activity was high in a muscle proximal to the injury (i.e., biceps brachia), while significantly smaller in the muscles distal to the injury site (i.e., tibialis anterior and gastrocnemius muscles). Note the difference in calibration for displaying the amplitude. The motor patterns in the muscles distal to the injury site C5/6 show obvious improvement after six months of CDT (Figure 42C in comparison to Figure 42A,B).

b) Motor pattern stability. At the beginning of therapy, the motor patterns varied significantly while turning on the special CDT device, as can be seen in Figure 42A and 42B. This variation persisted for as many as six months after starting the therapy. In other words, the stability of the motor pattern of the movement's exercising on the special device by turning with arms and legs was very low at the beginning of therapy. This low pattern stability is also reflected in the high CD values (high arrhythmicity of exercising = low pattern stability) (Figure 39). Clinically, this motor pattern variability correlated with the instability of movement performance, observed during walking and running on a treadmill at the beginning of therapy.

c) Impairment of reciprocal relationship of antagonist muscles. The natural reciprocal relationship of agonist and antagonist muscles of the leg, namely the tibialis anterior (dorsal flexor) and gastrocnemius muscle (planter flexor), was reasonably well preserved on the left side (Figure 42B,C) but was significantly impaired on the right side. It persisted for as many as six months in spite of therapy (Figure 42D). This impairment of antagonistic muscle action contributed to the difficulty in gait faced by the patient while walking and running.

d) Measurements of temporal stability of movement patterns by pattern change. As explained above, the sEMG motor patterns were generated and recorded when the patient was exercising on the special CDT and recording device. The mean temporal stability of these motor patterns over one minute (= CD value = mean arrhythmicity value) was used to quantify improvements of CNS functioning with therapy (see above, Figure 39). To measure differences in temporal stability of different movement patterns, such coordinated patterns involving arm and leg movements, need to be generated with accuracy. This is achieved by the high precision mechanics of the special device for turning movements, when the arms turn at a slightly higher frequency than the legs. The relationship of revolution between arm leavers and leg paddles is 19:18. This difference in revolution produces the changing of coordination patterns of arm and leg movements. Such pattern change must also be seen in the sEMG motor patterns.

The way that the muscle activations changed between the left biceps brachii muscle (arm) and left gastrocnemius muscle (leg) can be seen in Figure 42B. During four gastrocnemius activations (four turns), the biceps were activated approximately 4.3 times (4.3 turns). This seemingly simple change of arm and leg movements between pace and trot gait is an extremely difficult task for the CNS to generate. All the muscle patterns of arms and legs have to be changed.

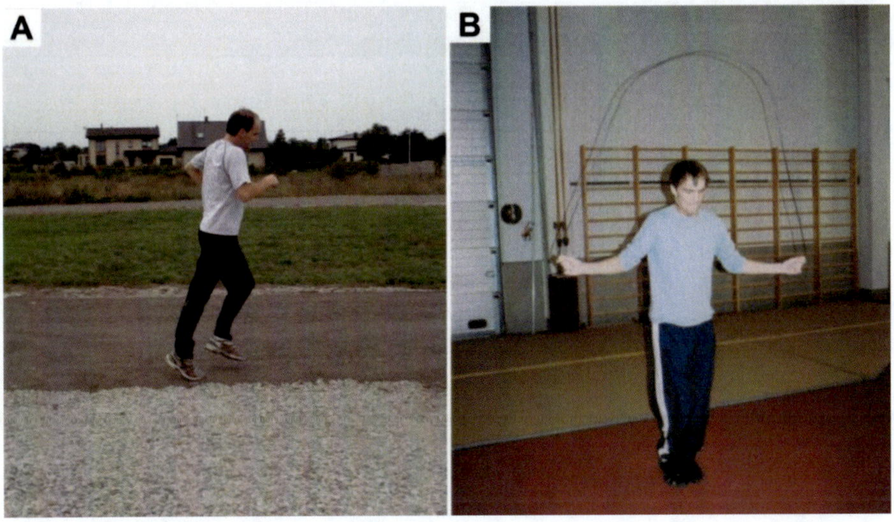

Figure 40. Patient with an incomplete cervical spinal cord injury at C5/6 levels (severity 50%) during running and jumping with a skipping-rope.

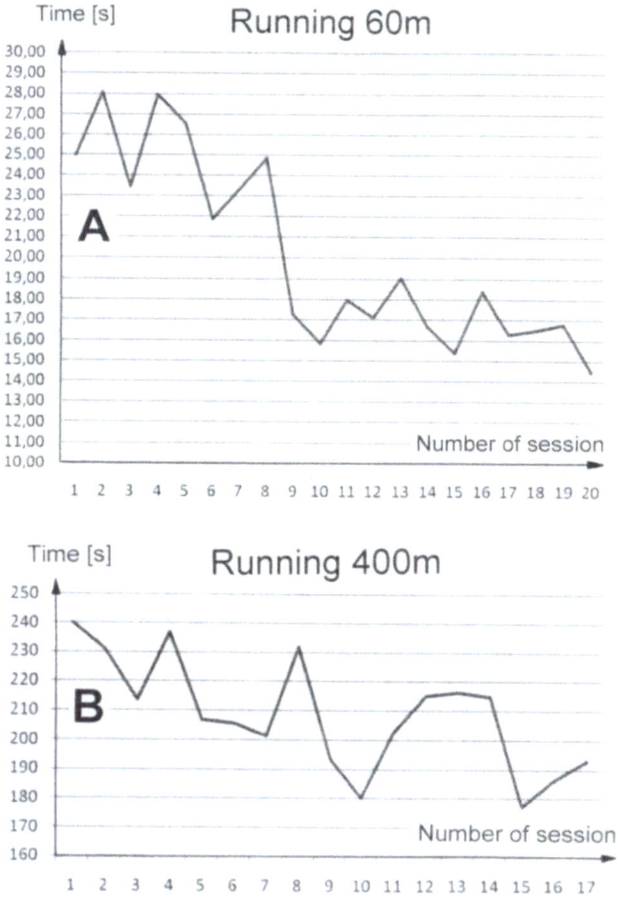

Figure 41. (Continued).

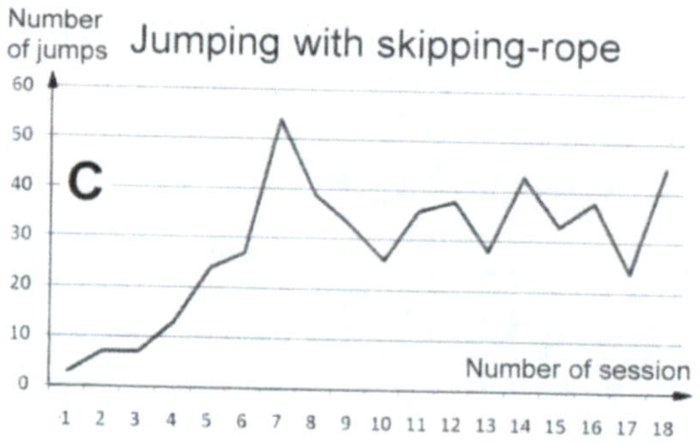

Figure 41. Improvement of times for running 60 and 400m (A,B) and increase of jumps with a skipping-rope per single session (C) in relation to ongoing therapy sessions. Therapy period = second half of 2006 (see Figure 3).

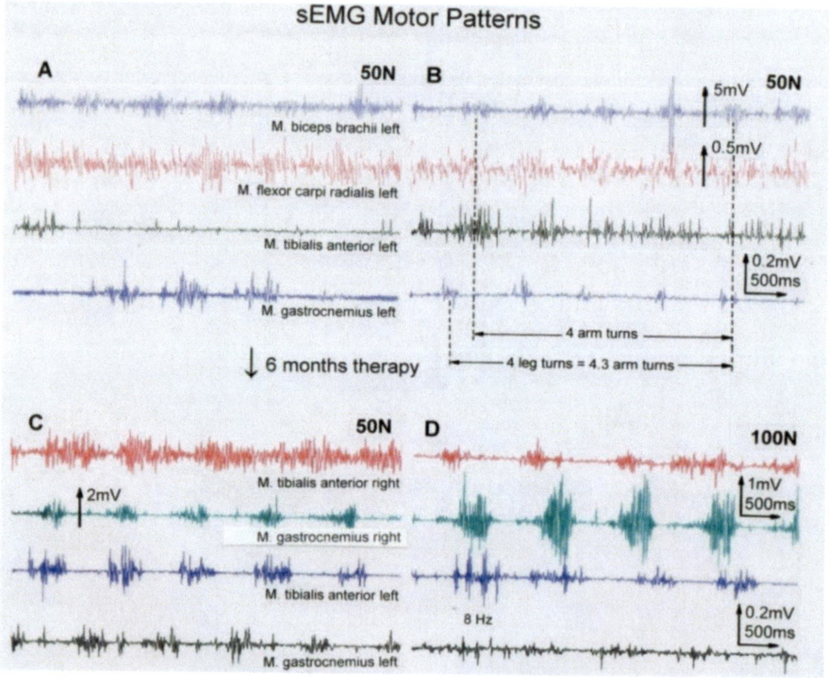

Figure 42. Motor patterns of arm and leg muscles recorded by sEMG for the patient with a 50% spinal cord injury at C5/6 levels at the beginning of the coordination dynamics therapy (A,B) and after six months of therapy (C,D). Note that the motor patterns of the left tibialis anterior and gastrocnemius muscles improved during these six months of therapy (from A,B to C); the patterns varied quite much from recording (A) to recording (B). The sEMG activity has a large amplitude for the biceps brachii muscle and a low amplitude for the tibialis anterior and gastrocnemius muscles (A,B) (note the different calibrations). The patterns were generated when the patient was exercising on the special coordination dynamics therapy device for turning. To measure stability differences of patterns (system theory of pattern formation), the arms turn at a higher frequency than the legs. The revolution relation between arm leavers and leg pedals is 19:18; this relation generates the changing coordination between arms and legs. The different turning frequencies of arms and legs can be seen in the sEMG patterns (B; four leg turns = 4.3 arms turns). Upon increasing the load from 50 (C) to 100N (D) during exercising on the special device for turning, the motor patterns of the left tibialis anterior and gastrocnemius muscles deteriorated (D): specific patterns were lost, antagonicity was lost, pathologic synchronization of premotor spinal α_1-oscillators at 8Hz occurred; this deterioration of motor patterns stability is in accordance with the increase of the coordination dynamics values (= increase of arrythmicity of exercising) at higher loads.

And if the device is positioned further away from the trunk, then the trunk has to also perform rotating movements in coordination. Since this trunk rotation is changing with the coordination position between arms and legs, all the segmental trunk muscles change their coordination in the rostral and caudal direction. If the CNS is functioning physiologically, then the subject can turn rather easily and smoothly without stress. But if the CNS is injured, then such ongoing change of coordination among arm, leg, and trunk muscles becomes a very difficult task for the CNS. Patients are often sweating because of the stress the CNS is facing in generating these complicated movements as the pattern keeps changing. As can be seen from Figure 42B, during the time period of the recording, the patient's CNS could manage quite well the change of the coordination between pace and trot gait. Chaotic activation was not observed on either the biceps brachii or gastrocnemius muscle traces during the movement pattern change.

e) Impairment of pattern formation is revealed with integrated CNS activation. Upon increasing the load of exercising on the special device from 50N (Figure 42C) to 100N (Figure 42D), the motor patterns of the left tibialis anterior and gastrocnemius muscles partly deteriorated (Figure 42D). The specific patterns and their antagonistic relationship were partly lost. Pathologic synchronization of premotor spinal α_1-oscillators occurred at 8 Hz, as can be seen from the trace of the left tibialis anterior muscle (Figure 42D). The increasing impairment of CNS self-organization with increasing load of exercising, made visible by the sEMG recordings, is in accordance with the increase of the arrhythmicity of exercising on the special CDT device for increasing load (Figure 39, increase of coordination dynamics values at higher loads).

7. Increase in Temporal Instability of Movement Patterns While Exercising against High Load

From the above discussion, it is understandable that the increase of the CD values (deterioration of CD) while exercising against high load (Figure 39) is because of the reduction in the temporal stability of movement patterns as shown in sEMG recordings. Kinesiologically, this increased temporal instability of movement pattern formation is apparent when CD traces for 1min are plotted against increasing load (Figure 43). As can be seen from Figure 43, the arrhythmicity of exercising increased with increasing load. Exercising at 20N (A) and 50N (B), in the recording, the arrhythmicity varied without any specific structure. At 100N (C) and 150N (D), however, structure becomes visible in the CD traces. Note in Figure 43, that the CD values increased with increasing load. Upon exercising against a load of 200N in the forward direction (E), it can be seen that there is rhythmic increase and decrease of the arrhythmicity of exercising. At times, when the arrhythmicity is small, the arm and leg movement patterns have a high stability. Different patterns of coordinated arm and leg movements (produced while exercising on the special CDT device) have different degrees of stability. Those patterns that are more stable have small amplitudes of arrhythmicity; they are the attractor states among different movement patterns. The unstable movement patterns (with a large arrhythmicity of exercising) are no attractor states (see Figure 70).

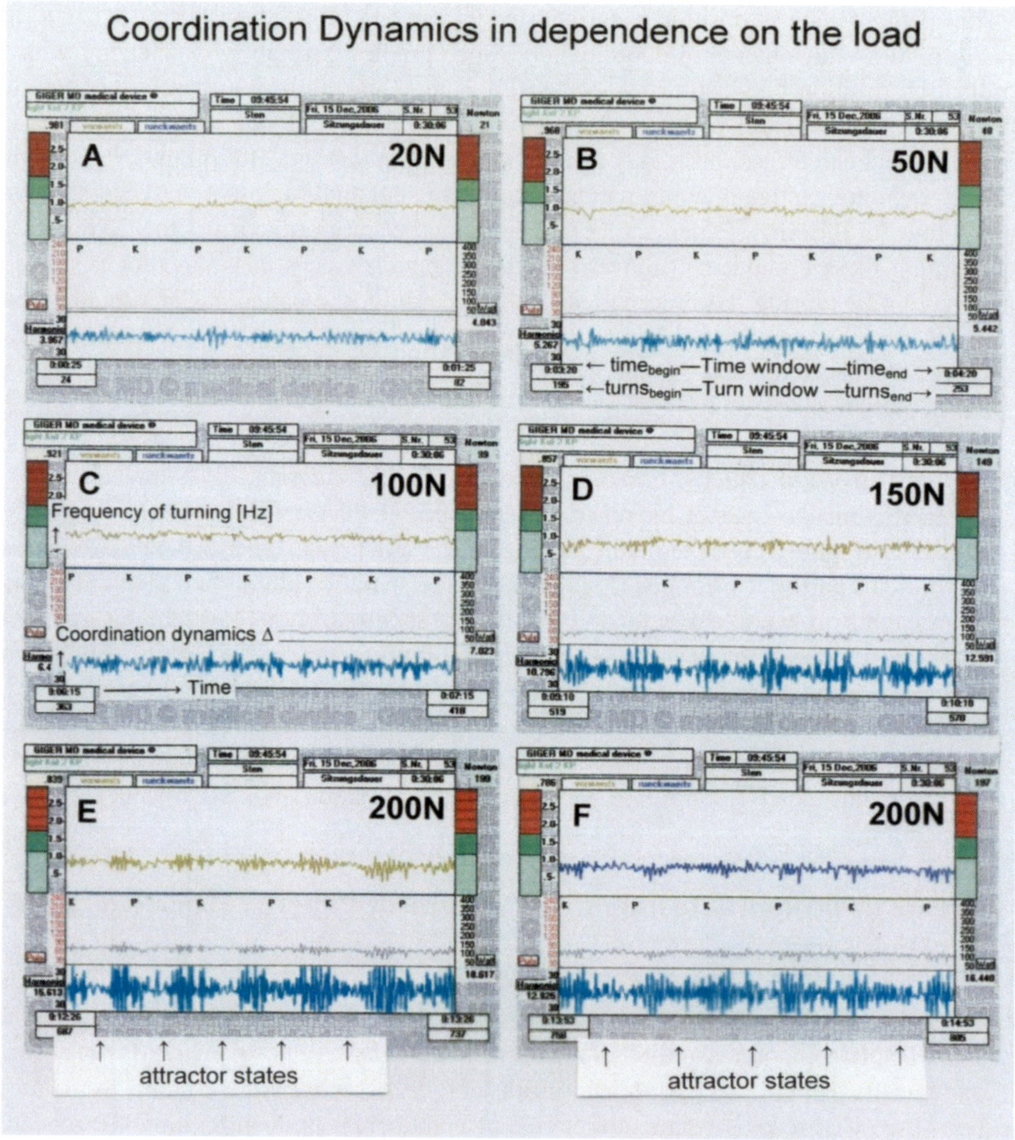

Figure 43. Coordination dynamics measurements in a patient with a 50% spinal cord injury upon exercising on the special coordination dynamics therapy and recording device for forward turning at increasing load from 20 to 200N (A-E). The recording sweeps are 1min long. F. Exercising in the backward direction at 200N. Upper trace = frequency; lower trace = coordination dynamics = arrhythmicity of exercising. P = pace gait coordination, K = trot gait coordination. "P to K" and "K to P" = coordination changes from pace to trot gait and from trot gait to pace gait. Note that for 20N (A) and 50N (B), the coordination dynamics increase is more noise like, whereas from 100N (C) to 200N (E), there are rhythmic increases and decreases of the coordination dynamics with the changes of the coordination. Note further that the lowest arrhythmicity of exercising (highest pattern stability = attractor state) lies for forward exercising at 200N (E) to the right side of the pace (P) and trot gait coordination's (K) and for backward exercising (F) to the left side of the pace and trot gait coordination.

Sudden instabilities of movement patterns are perceived by the patient and healthy subjects as sudden increases in the resistance to smooth turning. These are sudden perturbations of CNS organisation and are felt as a stutter during smooth exercising.

8. Exercising at High Loads Reveals Impairment in the Symmetries of CNS Organisation

While the patient was exercising against a high load of 200N, rhythmic changes of the arrhythmicity of exercising were observed during forward (Figure 43E) as well as backward exercising (Figure 43F). Interestingly, the location of attractor states (patterns of high stability) were seen to be situated at different locations from the calibrated states of "pace" (P) and "trot gait" (K), as shown in Figure 43E,F. For forward turning, the attractor is located on the right side, while for backward turning, the attractor is located on the left side of the pace and trot gait patterns. This impairment of symmetry for forward-backward movements (different attractor states) became visible only with the very integrated activation of the CNS as in exercising against high load.

9. Improvement of Symmetries of CNS Organization Increased Pattern Stability

As the impairment of symmetry is diagnosed with the highly integrated (more systemic) activation of the CNS, in order to improve the symmetry, the high-load training of those movements is also required. As the pattern stability with respect to symmetry increases, movement performance improves (Figure 4). In the case of an asymmetric injury (and subsequent asymmetric CNS organisation, large $\delta\omega$ value in the potential equation in Figure 4B), the movement pattern "jumping in anti-phase" can be so unstable that the patient is unable to perform this movement. In terms of system theory, the potential well for "jumping in anti-phase" in the attractor layout is very shallow or no longer exists (Figure 4B). In this patient, initially, only in-phase jumping was possible without support. The anti-phase jumping needed assistance and strong intention (Figure 4C,D). With training, the patient relearned to jump in anti-phase; the attractor basin for "jumping in anti-phase" became deeper and the movement pattern stable and possible. At the end of therapy, the patient was even able to jump independently with a skipping rope (Figure 40).

10. Repair of Bladder Function

Before World War II, most patients with severe spinal cord injury died because of recurrent urinary tract infections. Needless to say, one of the most important functions to be repaired in patients with spinal cord injury is the control of the urinary bladder. Patients are often reluctant to undergo urodynamic evaluation in a rehabilitation centre because of the fear of contracting urinary tract infection with antibiotic resistant bacteria. Development of urinary tract infection is a major setback to the therapy, as optimal therapy is not possible in the presence of such an infection. If urodynamic assessment is not available, some good practical indicators of improvement in urinary bladder function are the reduction in the incidence of urinary bladder infections and the significant reduction in the volume of post-voidal residual urine in the bladder. Residual urine can be measured by emptying the bladder with a catheter, after the patient has passed urine voluntarily. Residual urine of less than 50 ml indicates good physiological functioning of the bladder.

At the beginning of therapy, our patient was using intermittent catheterisation for emptying the bladder. A few months after starting the therapy, he started emptying the bladder manually. Initially, he had to compress the lower abdomen to empty the full bladder; but later, more physiological bladder emptying on volition became possible. Over six months of therapy, the residual urine reduced from 200 ml to 50 ml. Then the patient refused further measurements to avoid unnecessary catheter trauma to the urethral epithelium. The scientific basis of urinary bladder functioning is given in Chapters III and V. Bowel emptying was not a major problem in the patient. It is usually required to empty the bowel only once a day, and there is no risk of infection as in the case of the urinary bladder. Since the neuronal network for primary sexual functions are located in the same sacral segment S2-S5 as are the centers of bowel and bladder function, improvement in sexual functions is expected to accompany the improvement of urinary bladder functions.

11. Absence of Significant Increase in Muscle Power during Isometric Contraction, in the Face of Improvement of Motor Performance, Indicates That Repair Is Achieved Mainly through Functional Reorganization

Conventional concepts of neuro-rehabilitation focus on improvement of the strength of the muscles to achieve motor improvement. It was tested whether the improvement of motor performance after CDT was due to the increase of muscle power. The maximal power of isometric contraction of quadriceps femoris muscle was measured. In the spring of 2004, the maximal isometric contraction power was 411N on the right side and 378N on the left side (normal values are around 700N; the muscles had therefore only 2/3 of their normal power). In November 2006, the values were 404N on the right and 391N on the left side. In December 2006, at the end of therapy and after an intensive therapy for a few months, the maximal isometric contraction power was 416 N on the right and 386 N on the left side. In short, at the end of therapy, after three years, when the patient had significantly improved and was walking normally, the muscle power of the quadriceps femoris muscles remained more or less the same. Only the speed of power generation increased (improved).

12. Improvement in Hand Grip

The power of hand grip improved considerably during CDT. At the beginning of therapy, after only four turns on the special CDT device, the left hand would slip from the handle and was required to be fixed with a bandage. By the end of therapy, the patient could maintain good hand grip even while turning at a high load of 200N. The improvement of hand grip power was also confirmed when measured with a dynamometer. Unfortunately, subsequent readings could not be made due to logistics issues. At the end of therapy, some residual deficits did remain. Considering the time profile, the author feels that the hand grip improvement was mainly achieved through functional reorganization and not by the building of new motoneurons in the grey matter of injured cervical segments C5/7.

13. Repair in the CNS and PNS

From the MR image of the cervical spine in this patient (Figure 38), it is obvious that it is the grey matter that was mainly damaged. Obviously a part of the white matter was also damaged, considering the clinically apparent long tract deficits in this patient. As the spinal cord is arranged segmentally, the sensory, motor, and autonomic functions of each segment crucially depend on their connections with the supraspinal centres for all conscious, voluntary or automatic actions. Damage to these supra-spinal connections, as happens in white matter tract injury, leaves caudal spinal segments partially or totally isolated. Such de-afferentiation of the cord distal to the injury has devastating consequences. This is especially true in the case of the motoneurons of the lumbo-sacral cord. Adequate activation and control of these neurons from supraspinal centres is required to perform walking movements. In the presence of damage to the pre-existent descending pathways, activation of redundant pathways and reorganisation in spared descending tracts is required for functional repair. As described above in the patient, improvement in walking cannot be attributed to the increased strength of the quadriceps femoris muscle, which in fact remained unchanged. Therefore, the repair is unlikely to be achieved by substantial structural changes like the building of new motoneurons and connections (regeneration). It is most likely that the improved performance was due to functional repair strategies, as, for example, sprouting of corticospinal axons onto the spared, long descending propriospinal tracts that would have re-established the connectivity of supraspinal segments with lumbar motoneurons. In other words, functional repair would encompass plasticity along various levels and subsystems of the CNS like the cortex, brain stem and spinal cord. For example, cortical sensory-motor areas are known to functionally rearrange [80,81]). At subcortical level, the rubrospinal system can reorganise and compensate for the deficits in the corticospinal system [81].

The functional improvement in the upper limbs is probably attributable to both structural repair and functional reorganization. The improvement in the upper limb function, seen as lowering of the functional level of injury, within three years of treatment is more related to recruitment of spared neurons and segmental reorganization of connections. A minor reorganization in the peripheral nervous system could have contributed to this functional recovery.

The grey matter injury at segments C5/6 caused death and loss of neurons, glia and precursor cells. The structural repair in this case would involve mobilization of precursor cells to the injury site, building of new neurons and supporting glia. Even though these neurons may get reinnervated by sprouting or regenerating cortico-spinal axons, for these connections to affect functional recovery, the peripheral axons of these motoneurons would have to grow into the arm up to the hand muscles and innervate them (Figure 3). This would take longer than two years, considering the slow speed of axonal growth (1-2mm/day at its best). The strong improvement observed at the end of the third year of therapy could have been partly due to such structural repair. This also stresses the importance of continuing the therapy beyond two years to achieve clinically relevant structural repair.

14. Importance of Integrated CNS Activation for Learning Transfer

In the treatment of patients with spinal cord injury, learning transfer is of prime importance especially for the training of those vital CNS functions like bladder control and sexual functions, which cannot be trained directly. The repair of the bladder function is in fact of highest priority as that can ensure longevity of the patient by reducing life-threatening urinary tract infections, which are a common cause of mortality in SCI patients.

Neuronal networks of the somatic and autonomic nervous systems are intertwined upstream and downstream and to different degrees. This sharing of networks is of advantage because only then can the autonomic networks be influenced by the training of the somatic networks. But in order to influence and train the autonomic networks, they need to be recruited and adequately activated during movement therapy. Only an integrated (systemic) activation of the CNS, upstream, symmetrical, and across many subsystems can achieve this effect. In this movement therapy, such integrated, highly coordinated activation is achieved while the patient is exercising at high loads (like 200N) on the special CDT device. This exercising is similar to the highly integrative activation during running. It takes longer training time for improving the CD of turning against 150 or 200N than is required for turning at 20N (Figure 39), as large parts of CNS need to be coordinated activated and reorganized.

Within the framework of System Theory of Pattern Formation for Repair (Method), learning transfer can be well understood, going back to the equation of motion. Upon influencing the intrinsic dynamics (F_{intr}) with the behavioural information like running, jumping, and exercising on special devices ($c_{inf}F_{inf}$), CNS self-organization can be modified as a whole integrated entity. Unlike as expected in the pattern generator theory [82], while training movements like running or turning on a special device, not merely a circuitry is activated in the neuronal networks. If it would be so, then we would have to train every specific movement separately.

Only the system theory can easily explain learning transfer. Because of the integrated activation of the entire nervous system, not just a single pattern is formed, but the entire neuronal network is involved, which facilitates learning transfer to other functions; the whole attractor layout is changing.

15. Importance of Symmetries in Impairment and Training

It is well known that asymmetric movements are less stable than symmetric ones (16,54; Figure 4). Symmetries structure the state-space of movement patterns and influence the stability of movement patterns. The impairment of symmetries in CNS functioning, following injury, is uncovered during the integrated activation of the CNS when the patient is exercising at high loads on the special CDT and recording device (Figure 43E,F). For measuring coordination dynamics, the length of levers and pedals has to be short to reduce kinetic and rotational energies interfering with the true measurement of the arrythmicity of exercising. Different symmetries can be improved by training symmetric, integrated, and dynamic movements like brisk walking, running, jumping, and exercising on the special device in the forward and backward direction.

The rotational and/or kinetic energies, generated by fast movements, smooth the movements and make them more symmetrical in the presence of asymmetries. The coordinated and symmetrical afferent inputs generated by these coordinated movements enter the CNS and entrain the neuronal networks, thus correcting the asymmetries. The therapist has to anticipate, observe, support and correct the asymmetries. Such support, given by the therapist with good interpersonal coordination while the patient is jumping on springboard or when he/she is walking on treadmill, increases the symmetry of the movement and in turn the movement induced afferent input.

16. Pathologic Single Neuron Firing as an Important Cause of Impaired CNS Self-Organisation

As shown in the sEMG recordings above (Figure 42D), pathologic synchronization of oscillatory firing motor units occurred intermittently at a frequency of 8Hz. Such pathological synchronisation of pre-motor spinal oscillators occurs when the physiological mutual inhibition of motoneurons is not sufficiently strong.

As analysed and reported earlier in connection with patients with Parkinson's disease [47], such pathological synchronisation starts with the oscillatory firing of α_2-motoneurons. Later, the activated α_1-motoneuron oscillators start to synchronize their firing at the same frequency and thus add to the pathological synchronisation of the neuronal assemblies. Since α_1-motor units have larger amplitudes than α_2-motor units, the synchronisation is observed in sEMG recordings mostly only when the α_1-oscillators are activated and their firing becomes synchronized. This synchronized firing of spinal oscillators indicates that after spinal cord injury, not just motor patterns are impaired, but the physiological cooperative and competitive interplay of CNS neurons is also impaired, which makes the physiologic self-organisation of neuronal networks impossible.

17. Repair of Bladder Function

In this patient, the repair of urinary bladder functions could partly be due to spontaneous recovery (autonomic functions are known to sometimes recover spontaneously, even though much later than the motor functions) and partly due to the learning transfer achieved through integrated activation of the CNS described earlier. Transfer of learning occurred from movement patterns to the patterns of urinary bladder storage and emptying. The rhythmic jumping on springboard (Figure 4) trains the urinary bladder functions most directly. Because of the rhythmic and dynamic load changes exerted onto the pelvic floor during jumping, the external bladder and anal sphincters (somatic muscles), as a part of the pelvic floor, are rhythmically and dynamically activated and relaxed and passively stretched. The inner sphincter (smooth muscle) is probably trained by the pressure changes in the detrusor muscle. This patient had no dyssynergia (co-contraction of the detrusor muscle and the external bladder sphincter) of the urinary bladder, which made it easier to improve bladder functioning.

As explained earlier in the framework of system theory of pattern formation, the behavioural information, in the form of training coordinated, integrated movements, changed the whole attractor layout of pattern formation, thus inducing learning transfer. However, a far deeper understanding of the human neurophysiology of bladder functioning is essential to characterise optimal behavioural information (required training) needed for the bladder repair. A detailed analysis will be given in Chapters III and V.

18. Exercising on the Special CDT Device Repairs only Certain Aspects of CNS Organization As Is Revealed by a Comparison with Healthy Children

Since the patient went on with motor training after three years of therapy and diagnostic was performed, further improvement could be measured by the CD. The low-load coordination dynamics (CD) values (Figure 44C,D) improved in this patient within a few months (Figure 39), whereas the high-load CD needed much more time for recovery. To get good CD values for turning against 100N needed approximately two years (Figure 39) and against 150N more than three years. Upon six years of supervised and only partly supervised treatment, the patient was able to perform the high-load test (Figure 44E,F) as well as healthy pupils (Figure 29D,E,F). Also when exercising on the special CDT device, changing performance difficulties for the different coordinations between pace and trot gait can be seen in the SCI patient (Figure 44F) and the healthy pupil (Figure 29F).

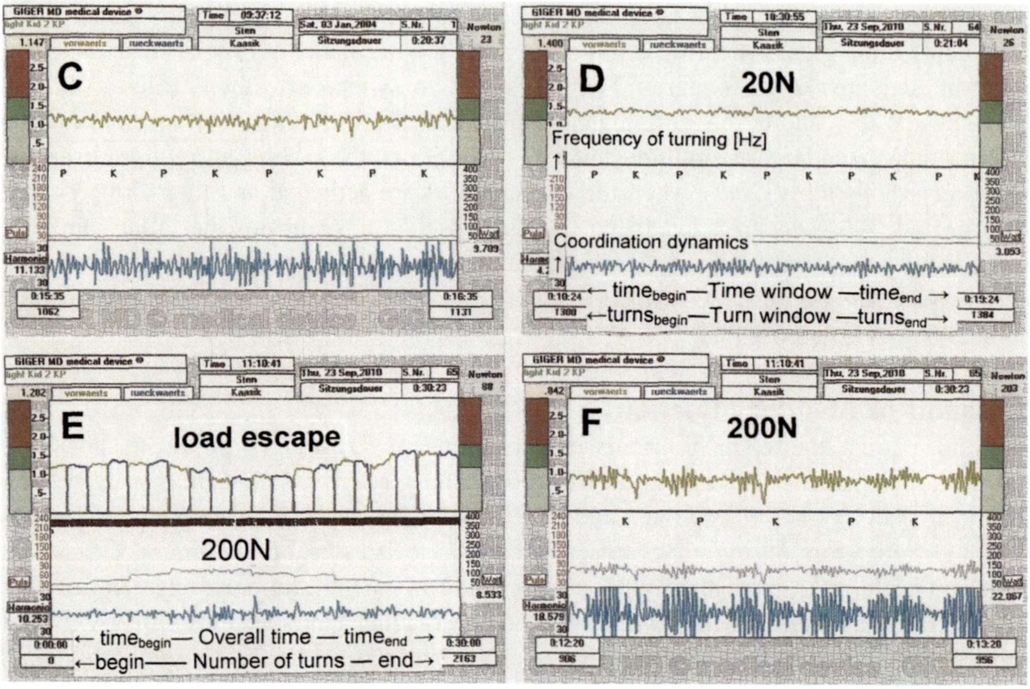

Figure 44. Repair of an incomplete cervical spinal cord injury (SCI) (approximately 50% injury). C. Low-load (20N) coordination dynamics (CD) at the beginning of therapy. D. Low-load CD six years later; notice the improvement (reduction) of CD value from 9.7 to 3.1. E. Overall high-load CD with some load escape; the load increase and decrease is nearly stepwise. In the recording two weeks earlier, the step-wise performance was even better. F. One-minute CD recording at 200N. Note the rhythmic increase and decrease of the arrhythmic turning for the different coordinations.

Being able to perform well on the special device for high loads does therefore not mean that the CNS is fully repaired. First, the performance for high loads depends also on the fitness besides going deeply into the complexity of the neuronal network organization, and second, turning on the special CDT device quantifies only a certain aspect of CNS organization. This patient with the 50% SCI was able to perform on the special device after six years as well as the healthy pupils, but the performance of walking, running and jumping was not as good by far. Even after six years of supervised and not supervised treatment, the CNS was not fully repaired. The patient reported that the training of different movements helped him to still get further on with the repair. CDT is only the beginning part of a treatment. When basic CNS functions are repaired, many different movements have to be trained. Against the advice of the author, this patient was not training sufficiently on the special CDT device. But there was no need to push the patient, because all important functions were repaired, and he was having a normal life. The increase of the 150N CDT value at the end of 2010 (Figure 39) was due to participation in a marathon competition – for normal people – a few days before!

19. The Problem of Muscle Power Generation

In incomplete cervical SCI (50% injury), sufficient muscle power can be developed by the patient to walk, run and jump. In nearly complete cervical SCI (95% injury), the generation of sufficient muscle power is a fundamental problem. Due to limited regeneration, the patient can move different leg muscles and can even walk on treadmill (Figure 37H). But the patient is still far away from walking without weight-bearing support. The problem of muscle power generation can partly be quantified by the maximum load against which a patient can exercise on the special device. In 95% cervical SCI, upon 0.5 years of CDT, the patient can turn on the special device against a maximum load of 30 Newton (Figure 17A) and upon five years of CDT against 50N. The patient with the 50% cervical SCI could exercise already against a load of 200N after two years.

20. Limitations of Basic Science Research

There are many inadequacies in the current research in SCI. Animal research is mainly concerned with structural repair and the regeneration of the spinal cord in animals. The aspect of functional recovery is rarely understood and never analyzed. Even though structural repair is important in SCI, the researchers rarely discuss the difference in the potential of regeneration between animals and humans [4,11,47]. In the goldfish, for example, the spinal cord regenerates spontaneously [3]. But what inferences can be derived from this regeneration that can be relevant and applicable to humans? The presented gross human anatomy may make a casual reader believe that the animal data are indeed applicable to humans. However, the functional recovery is not measured thoroughly. Morphological data at the cellular level, like the degree and extent of growth and regeneration of axons across the injury site, is often not correlated with the functional recovery. The functioning of newly grown axon has to be proven in the context of neuronal network functions and not only by the improved movements of the treated animals, because in animals there also exists spontaneous recovery and functional reorganization, especially if only 50% of the cord is destroyed. Actually, fiber counts in different ascending and descending tracts would be needed to assess what percentage of tract fibers need to be regenerated for meaningful recovery in humans.

21. A Need for Scientific and Clinical Human Approach

The diagnostics at the beginning, during, and at the end of therapy is of particular importance to differentiate between spontaneous recovery and the improvement achieved by the treatment. Since spontaneous recovery may occur for up to one year after the injury, long-term studies are needed. In case of a partial spinal cord injury, most of the recovery occurs between two to six months after injury and to a lesser degree up to one year. In the present patient, the treatment was started six months after the injury and lasted for three years; thus the recovery cannot be attributed to spontaneous recovery.

In the case of complete SCI (as assessed by MRI), only very limited spontaneous recovery occurs, and the potential for functional reorganisation is very low, since the majority of the distal cord is disconnected from the supraspinal control. In such an injury, even the combination treatment of stem cell therapy and coordination dynamics therapy did not bring about significant improvement [56,57]. However, most of the spinal cord injuries are incomplete. Even in the case of a severe injury, a significant part can be salvaged by immediate removal of compression of the spinal cord by surgery and the stabilization of the spine to minimize progressive mechanical injury.

There are many unscientific reports, claiming miraculous recovery in SCI patients, which could very well be attributed to the spontaneous recovery. Most of the rehabilitation centres focus mainly on care and general maintenance of patients and administer inefficient and arbitrary treatment that does not utilise the full potential of recovery in patients with partial SCI. Patients want more choices than the "no hope" approach of conventional rehabilitation centres and the "false hope," given by some physicians and researchers working in the field of animal research and cell and molecular biology.

Over the last 20 years, no significant conceptual or real methodical progress has occurred in the field of neuro-rehabilitation. The current science of neuro-rehabilitation is mostly out-dated, as no real review of human neurophysiology has been done for many decades. The lack of any perceived need for development and in turn the absence of any efforts in that direction has perpetuated the inefficient, out-dated and ineffective treatment modalities in the treatment delivery system. This is particularly relevant considering the fact that the young patients who suffered a spinal cord injury have nearly 50 years of their remaining lifespan that they have to live with a very poor quality of life, while only getting care from the current rehabilitation centres. On the other hand, with optimal and intensive coordination dynamics therapy for a few years after the injury, their quality of life can be enhanced substantially in order to achieve independence in daily activities and most importantly physiological bladder control.

3.2.3. Comparison of the Recovery of 50% SCI between Human and Animal and False Hope Generation

1. Animal Experiments Are Removed from Human Reality

Recently, it was reported on the spontaneous recovery of a SCI in rhesus monkeys and the important role for primate models in translational disease research [27]. A few comparisons will show that animal experiments are far away from human reality and normally cannot contribute to translational medicine or human disease research.

1. In human patients, as can be seen from Figure 38, the injury is mostly in the centre of the spinal cord and not just a half of the spinal cord is damaged as in the monkey experiment (Supplementary Figure 2 of Ref.27). Neurons in the spared grey matter of monkeys may function as relay neurons. The blood supply is probably less impaired in the monkey.
2. Since in 50% SCI the spinal cord can mainly be repaired upon CDT (58, above) and the patient can be brought back to normal everyday life, there is not much sense of after developments in monkeys, unless the experiments are designed to be close to the clinical human setting to obtain parameters that cannot be measured in humans. Is it actually justified to kill monkeys for an after development? Not to cite existing qualified clinical literature, for the repair of a 50% SCI is not convincing, especially when claiming that their research is a contribution to translational medicine.
3. The biggest problems in human SCI are the urinary bladder functions (Refs.60-64, Chapters III and V) and the occurring pressure ulcers due to insufficient blood supply. Before World War II, a majority of SCI patient mortality was due to continuing urinary bladder infections. In India, the life expectancy time of a patient with a severe SCI, after leaving the clinic is one year, as at home bladder and other infections cannot be managed sufficiently. By cutting only one half of the spinal cord in animals, one is not getting urinary bladder and skin nutrition problems. The challenge of the future is to cure the 95% or complete SCI. The neurosurgeon L.W.

Freeman (USA) stated in the 1960s after performing very many spinal cord injury repair experiments on dogs, "it is difficult to find persons who do research in paraplegia and to get money for such research; probably World War II is already too far away." If one cuts the whole spinal cord in animals, then one would have to manually empty the bladder by pressing the urine out. Such work seems to be unpopular among researchers. In human patients, the bladder is, for example, emptied approximately every five hours by intermitted catheterization (and other ways), introduced by Guttmann from Breslau [83] in the Second World War.

4. The improvement of hand function is also very important in cervical SCI; I have not seen animal experiments tackling the hand problem. A "tetra" (tetraplegic patient) cannot empty his urinary bladder by intermittent catheterization without sufficient hand function. Hand functions can be improved in C5/6 SCI upon building of new motoneurons in the human spinal cord, but it needs several years of optimal CDT (67).
5. The repair of the fundamental functions such as breathing, swallowing, and speech in high cervical SCI (severe cerebral palsy (see below) or severe traumatic brain injury) has also not been tackled by animal research. Obviously, animal research is far removed from the patient's real problems.
6. Humans need years for development and repair and not only a few months. Animal experimentalists are therefore working on the wrong time scale with respect to development and repair to translate their data to human reality. Again, it seems not to be popular among researchers to do experiments that last over several years.
7. Translational medicine only makes sense if the researchers have knowledge of animal and human research. A translator can only translate from German to English if he has knowledge of both languages. Also, for peer-reviewed journals in translational medicine, it would be important to have reviewers that are competent in animal and human research. The competence has to be shown by publications in international journals in both fields.
8. When I was for postdoc education with Sir Bernard Katz at University College London, I learned that there should be no more than four authors on genuine research publications. Fourteen authors on a publication is typical for clinical research where clinicians do not have enough time for research and putting their names as much as possible on the publications of friends and assistances to increase the number of their own publications. By avoiding real research, professors in the clinical field may have insufficient understanding of qualified research and are less open to new developments in medicine.
9. In mouse experiments, half of the spinal cord is cut, and treatment success is demonstrated by the staining of neurons and nerve fibres and demonstrating that the mouse can run for example on a horizontal ladder [78]. Whether the regenerated fibres were really functioning is not measured. Something like 50 years ago, there was a treatment for thoracic SCI with which patients learned to move the legs by moving the arms. The information from the upper to the lower part of the body took place most likely by skin traction. In a frog, I could not see from the morphology whether axons and synapses were functioning; when trying to understand regeneration, I used morphology and electrophysiology (Chapter I). Sir Bernard Katz, the specialist for the neuromuscular endplate, explained that sometimes

denervated motor endplates survive very long and still look quite normal (as if they would really still function). In mouse treatment, success was measured by morphology and making the mouse move on a horizontal irregular ladder and arguing then that the progress is coming from animal experiments to human treatment [78,79]. In the upper human patient with the 50% cervical SCI, treatment progress is demonstrated by letting the patient run and jump (Figure 40), measuring the improvement of running and jumping (Figure 41), measuring the coordination pattern dynamics (CD) (Figures 43,44) and measuring motor programs (Figure 42) and single motor unit firing (neuron level) with surface electromyography (sEMG) (Figure 42). And the scientific basis for the repair is the understanding of CNS functioning and repair at the neuron level in human, as can partly be seen from Chapters III and V. For the time being, the progress in SCI is coming from human research. Human data need to be translated into animal data, as partly has been done in this Chapter. Universities and research institutions worldwide seem to have an anxiety with human neurophysiology and qualified clinical research.

2. Consequences of False Hope

In Chapter I, it was shown, based on frog data that exogenous stem cell therapy in its present application is not working. The main problem seems to be that the newborn cells cannot be integrated in the existing neuronal networks for various reasons. Cell communication needs to be better understood. During competition, a motocross athlete suffered a clinically complete SCI at the thoracic 11/12 levels. Six weeks after the accident, the subject began intensive CDT at an up-to-date therapy centre. After six months of therapy, when further improvements were only marginal, the patient opted for haematopoietic stem cell therapy. During two years of stem cell therapy, including four sessions of stem cell application, and ongoing CDT, improvement remained marginal – no more than what would have been achieved with continuing only CDT. This haematopoietic stem cell therapy did not have any beneficial effect on the repair of the spinal cord in this patient [32]. When it turned out that the costly stem cell therapy had no beneficial effect on the regeneration of his spinal cord, the patient lost the belief in SCI treatments and stopped also CDT. He only continued with some training, according to his opinion, to reduce spasticity. Another patient with a probably complete thoracic SCI was sent by a physiotherapist to the author to get advice concerning a stem cell therapy in addition to the administered movement therapy. The patient did not consult the author, because he wanted to have the stem cell therapy administered, whatever the arguments of the author would have been. He even raised the argument that the author may be against progress in SCI. Up until one year after the start of the stem cell therapy, there was no real progress in repair. The patient stopped the therapy in the physiotherapy place and is doing now his own treatment at home. It seems that he had lost the belief in any treatment. During the treatment at the physiotherapy place, no MRI was performed on this patient (messy treatment and diagnostic situation (Switzerland)); it was argued that this is not possible because of the metal fixation. But with a Titan fixation, which is nowadays mostly used, an MRI is giving quite good anatomical information about the severance of the injury of the spinal cord (see Figure 2A of Ref.59).

Instead of raising false hope that the animal data can be easily used to cure diseases of the human CNS, the universities and research institutions should first do their duty, namely organizing qualified human neurophysiologic research, especially treatment research, and also support such research.

3. Ethics of SCI Research, Treatment, and Clinical Trials – False Hope from Animal Treatment Research

"Transplantation of both embryonic stem cells and embryonic stem cell-derived neural (neuronal or glial) progenitors is able to efficiently promote CNS regeneration in preclinical models of stroke, myelin deficiency, acute spinal cord injury [18,23] and Parkinson's disease [22]." This sentence of an article makes the reader believe that the research made already the step from the animal research to the human research and treatment is already administered to humans. Looking up the references, one finds the title: *Human embryonic stem cell-derived oligodendrocyte progenitor cell transplants remyelinate and restore locomotion after spinal cord injury*. Patients with SCI or journalists think immediately that the SCI problem has been already solved! I have learned that it has to be written in the title on what species the research has been performed. In this case, it was the rat. Animal researchers often deliberately choose the words in a way to make the reader believe that this research has already solved the SCI problem in humans. A well-known journal in medicine is even redefining medicine. It makes the reader believe that measurements in mouse are a part of medicine [20]. I have learned in my medical study that medicine has something to do with the cure of diseases in human. Specialist journals and general audience media need to set reasonable expectations of the safety and efficacy of potential therapies to avoid raising and then dashing the hopes of those living with SCI or those in government, those carrying out research, or the general public [72]. High ethical standards are required by researchers, clinicians and journalists to ensure that results are communicated to the general public in a manner that honestly reflects the safety and efficacy of a potential therapy [72].

Medicine is very successful if natural existing repair mechanisms are supported like in surgery or immunology. But if repair mechanisms do not exist, then medicine is not very successful. The repair of the nervous system is very limited and depends on the site of the injury. An exogenous stem cell therapy must be very sophisticated to be better than nature with respect to repair. The integration of newborn cells into the existing neuronal networks is by far not solved. Further, the scientific basis for doing neurobiology is out of date. The understanding of the functioning of the CNS in animals and humans has to be upgraded. In physics, for example, the basis is continuously upgraded. In Chapter I, it was shown how complicated the innervation of two kinds of muscle cells by two kinds of motoneurons is (four-cell communication). The innervation and innervation changes during development and repair are probably much more complicated in the human CNS. It is too simple thinking to believe that an addition of cells can repair the CNS what the CNS cannot do by itself. CDT, on the other hand, rests on learning what is natural for humans. Academically accepted articles are generating false hope by allowing authors to state that stem cells can potentially be used in different CNS diseases including SCI. The authors are stating it to get funding for their animal research, by making believe that their research has direct consequences for the treatment of human patients. No wonder that practitioners also want to participate in getting money. Instead of warning that some clinics in Asia may look more for profit than for qualified stem cell therapy, the editorials should first think over their own research policies, which are giving rise to such unqualified medical treatment. It is not enough to state "The medical promise of stem cells remains real, but largely unrealized for now. The excitement must not be left to dissolve into a muddle of disappointment, frustration and fear because of the practices of a few irresponsible profiteers" [76]. There are principle problems to be solved for stem cell therapy in humans. A few years ago, there was a big propaganda in the TV and other mass media in Switzerland about the suppression of grow-inhibiting properties for the regeneration [78,79] of the spinal cord. It was stated that SCI could be cured in a year's time. It seems to be popular among neurobiologists to raise false hope in society concerning treatment in human patients, as if they have not understood much about treatment of human patients.

If we want to repair function, we also have to measure function in a qualified way, and one powerful tool is the electrophysiology, because the nervous system is mainly functioning by electric currents and potentials. It is difficult to understand why electrophysiology has mainly been destroyed worldwide, even though this tool was very successful in neurophysiology in the past. One reason could be that although the younger generations are capable of handling computers or electronic equipment very well, they do not have the manual skill to perform complicated electrophysiology. Statistics are needed concerning treatment success beyond the placebo effect, including the occurrences of cancer and CNS instabilities (seizure disorders). Hopefully, the globalization did not give rise to coordination in thinking (in German: "Gleichschaltung der Denkweise").

It is depressing for a qualified therapist that "there is an abundance of patients desperate for miracle cures, and one stem cell treatment can bring in tens of thousands of US dollars" [76]. Western scientists and clinicians would argue that controls are necessary to identify unambiguously whether a therapy is safe and effective; some clinicians have claimed that withholding a potential therapy from a patient with SCI is in itself unethical [72]. Many SCI units know about CDT. But the patients are not informed that such a therapy exists! A real change from care to cure seems not to be in the interests of the rehabilitation centres.

There is too much false hope generated by animal physiologists, neurobiologists, and researchers working in the field of genetics. Without detailed knowledge of the human physiology and pathophysiology, the animal knowledge is only of limited help to cure diseases in humans. False hope also stops the patient from fighting for improvements that are needed for everyday life. Why should a patient fight for three years or more with a movement-based learning therapy to get urinary bladder and some motor functions repaired, if in a few years' time walking can be returned with miracle cells or pills? Such false hope is coming from qualified researchers, when they make believe that the animal data can be easily used to repair the human CNS. Qualified human research is needed and has to be organized. Only if there is an overlap between animal and human research is there the possibility of rather safe transfer of knowledge from animals to humans. As long as human research is not organized properly, the patients have to suffer or even to die (for example, Christopher Reeve). Even monkey experiments are far away from human reality (see above). A monkey cannot speak, write or read, and will never be able to solve "differential equations" (mathematics). The power of the human CNS is its learning capacity, which is outstanding among different species. A fly demonstrates with how little brain matter sophisticated fly tasks are possible. Qualified human CNS research, including the single-neuron level, is needed for understanding the functioning and repair of its neuronal networks.

Authors of recent review articles presented approximately 1,000 citations on the repair of the human spinal cord following injury [5,13,22,72]. Less than 10% of the citations were from human research. Interestingly, the author was not cited, even though his work is widely available online. I learned a lot from these brilliant research articles with respect to animal research. But with respect to human research and applicability to human patients, they were out of date by 20 years.

"Miracle" treatments are not the only dangerous for patients and the freedom in research. It is the worldwide research, treatment, and teaching system that do not allow qualified human research, which is urgently needed to cure diseases and make humans live longer with a higher quality of life. And if there are really operations with unbelievable success, then the treatment before those operations was wrong.

For example, a paraplegic patient came by wheelchair from a well-known German rehabilitation centre to a neurosurgery department, was operated on and walked two weeks later out of the neurosurgery department. A second patient was cured in the same way. The reason for this progress was caused by a mistake of the rehabilitation centre. Bones from the spinal canal caused pressure on the spinal cord. The spinal cord stopped working but was damaged only little. A laminectomy freed the spinal cord, and the spinal cord started to work again. Follow up MRIs are needed, which are normally not performed in rehabilitation centres. The argument is that it has no clinical consequence, because the patient is staying in the wheelchair for the rest of his or her life anyway. This argument is wrong since, for example, cysts can build up in the injured spinal cord, which enlarge the spinal cord, and pressure symptoms will occur and the patient will loose further functions. The injury level may rise for example from C6 to C5.

3.2.4. Repair in Cerebral Palsy

1. Treatment Strategies in Cerebral Palsy

In complete or nearly complete cervical SCI, a repair is very difficult because tract fibres have to regenerate and new neurons are needed for hand and arm functions [67]. In brain injury or malformation, on the other hand, quite a bit of repair can be achieved already by a functional reorganization of neuronal networks. In cerebral palsy or brain injuries in children, where we have development and repair at the same time, great progress can be achieved, because trial and error elimination processes are part of optimal development. The trial and error elimination processes have to be enhanced by movement-based learning to repair also severe injuries or malformations.

Within the neuromaturational model, all changes in motor behaviour are viewed as intrinsically driven; the environment has minimal impact on evolving motor and other skills. The rate, quality, and sequence of skills are genetically predetermined since treatment represents a form of environmental manipulation; it cannot be assumed from the neuromaturational model that treatment is capable of either changing a motor behaviour or having an impact on the CNS. It is contrary to the neuromaturational model to surmise that damage to the cerebral cortex can be modified by external factors. Yet this is precisely what CDT advocates; motor output can be influenced by more normal sensory and motor cue input. By learning transfer, vegetative and higher mental functions can also be improved or repaired, as will be shown below. One weakness of CDT is that no designated boundaries defining treatment exist. Treatment strategies are developed from human neurophysiology, the system theory of pattern formation for repair, and the similarity to development. And, last but not least, the patients teach us about the secrets of movement (and other patterns) and its repair. Because of the influence of many subsystems, and because the cerebral cortex is not viewed as the control centre, many opportunities exist to repair the CNS by changing the self-organization of neuronal networks.

What degree of deprivation in any subsystem (localized or distributed) can be compensated for by the other systems? Are some subsystems more important than others or is the severity of involvement the restricting factor? As will be shown below, the severity of involvement and the kind of subsystem involved both seem to be important.

Humans are resilient to minor disturbances within any subsystem. Following severe injury or malformation, excessive qualified therapy is needed for many years to achieve substantial repair. The repair seems to be open-ended (see below under traumatic brain injury).

2. Improvement in a Group of Cerebral Palsy Patients upon CDT

It was shown that in a group of eight cerebral palsy patients (age range 7-27) that all members improved in their motor functions upon three months of CDT [49] as judged by low-load CD values. In one patient, the CDT was further continued, and it turned out that the patient further improved and that vegetative and higher mental functions also improved. The question arises, how much improvement of CNS functioning is possible in severe cerebral palsy if CDT is administered for several years. In a case report, it will be shown that speech, continence and motor functions can be repaired in severe cerebral palsy.

3.2.5. Speech Induction by Learning Transfer from Coordinated Movements in a Patient with Cerebrum, Thalamus, and Corpus Callosum Malformation or Injury

3.2.5.1. Summary

Coordination dynamics therapy (CDT) was administered to a five-year-old cerebral palsy child with a malformation of the right cerebral hemisphere (dominant side) and corpus callosum, an ischemic damage of the left thalamus (Figure 46), and high brain pressure for approximately two years. At the beginning of therapy, the motor functions of the right body side were strongly impaired, most likely mainly caused by the damaged left thalamus and not by the malformed cortex (see below). Her higher mental functions were very low; she could not speak and was not fully continent. Speech therapists had previously tried unsuccessfully for two years to induce speech. Upon an intensive CDT period of three months at an age of six, the speech was induced besides improvements in motor and vegetative functions. The neuronal networks of the cerebrum, thalamus, and cerebellum (and other CNS structures) became able to generate the language pattern with ongoing therapy, most likely because the neural coordination among neurons, neural ensembles, and regional specialized subnetworks improved strongly upon instrument-imposed, coordinated movements (learning transfer from coordinated movements to speech pattern). Within the framework of the System Theory of Pattern Formation for Repair, the change from aphasia to dysarthria can be understood in the way that the stability of the speech pattern increased (and the fluctuation of the pattern decreased) so much that the patient could speak a hundred words at the beginning.

Figure 45. Supported coordinated movements performed by a cerebral palsy child with cerebrum, thalamus, and corpus callosum malformation and/or damage. A. Free coordinated walking with mother and sister. B. Treadmill walking in the backward direction. C. Exercising on the special coordination dynamics therapy device, which imposes very coordinated arm, leg, and trunk movements. Hands and legs are fixed, because the patient is not cooperative. When the patient is looking to the doll, her CNS is not only getting very coordinated movement-induced afferent input from arms, legs, and trunk into her CNS but also coordinated visual input from the movements of the hands. Note, for treadmill walking in the forward direction (D), the mother had to support the right foot and knee for proper performance; for backward treadmill walking (B), no leg support is needed.

With further therapy, the speech, the higher mental functions, and the movements improved further; the patient also became fully continent. The dysarthria, probably due to the impairment of motor coordination of speech muscles, mainly caused by the pathologic processing in the thalamus, is now treated further by CDT and speech therapy. Weight, wiring, neural replacement, and neural coordination improvement and functional network complexity are discussed with respect to memory, pattern stability, and plasticity. The improvement of the timed firing of neurons in the neuronal networks of the CNS, giving rise to the induction of speech, is further analysed with respect to regional specialization of CNS organization (cellular connectionism) and System Theory of Pattern Formation (aggregate field view).

Keywords: Cerebral palsy – Aphasia – Dysarthria – Coordination Dynamics Therapy – Speech repair – Thalamus – Schizencephaly – Regional specialization – Integrative CNS functions – ARAS – Plasticity – Network stability – Membrane transmission frequency.

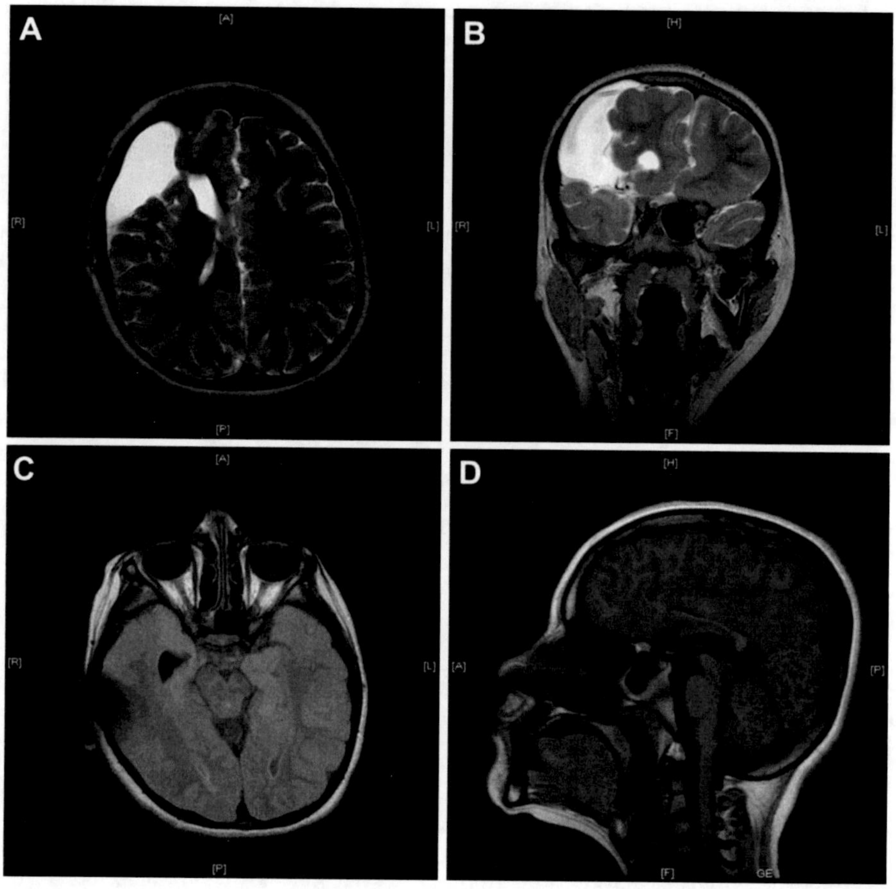

Figure 46. Magnetic resonance images (MRI) of a seven-year-old cerebral palsy child with cerebrum (A,B), thalamus (C), and corpus callosum (D) malformation and/or injury. The cerebrum malformation and/or damage seem to show a schizencephaly. But, as is indicated in the text, the real severe CNS injury with strong deficits in the higher mental functions seems to originate in the damage of the left thalamus (C). The damage of the thalamus may also impair the ascending reticular activating system (ARAS). The third ventricle cannot be seen.

3.2.5.2. Introduction and Additional Method

1. Learning Transfer

By learning transfer (63, Chapter V) not only the trained movements can be improved but also untrained functions like vegetative and higher mental functions. CDT has so far been able to improve by learning transfer autonomic and higher mental functions in cerebral palsy [49] and autonomic functions (cardiovascular performance, breathing, continence and micturition) in patients with spinal cord injury. In this report, it will also be shown that speech can be induced in cerebral palsy by learning transfer from movements. It has long been known that practice of one part of the body in performing a skilled act increases the ability of the bilaterally symmetrical part in the same act [84]. In coordination dynamics theory, it was found that patterns of coordination between symmetrical parts transfer spontaneously. The spontaneous transfer of learning points to an abstract nature of learning [85].

The present knowledge of transfer of learning so far is the following: (a) The amount of transfer seems to be small and positive unless the tasks are practical identical and (b) the amount of transfer depends on the similarity between the tasks [86]. Earlier publications and this report indicate that the amount of transfer of learning has by far been underestimated, at least for patients with CNS injury. Also the tasks or patterns need not to be similar. Jumping on springboard and exercising on a special CDT device, for inducing learning transfer from movements to bladder functions, are quite different network patterns than continence and micturition [63,64]. The speech is also a completely different pattern from exercising on the special CDT device (Figure 45C), which mainly induced the speech in the cerebral palsy child of this report.

For the repair of the urinary bladder [64], the jumping on springboard contributed substantially to the repair by activating simultaneously the neuronal networks for generating the jumping movement and the continence and micturition patterns; even some neurons will have been involved in both network organizations mainly sited in the lumbosacral spinal cord. On the other hand, the neuronal networks for generating movements and speech are different, even though sounds and (self-) instructions can improve rhythmic movements like walking. To a large extent, the learning transfer will have an abstract nature immanent in the self-organization of the human CNS. Since following CNS injury the phase and frequency coordination between neuron firing becomes impaired [46,63], the improvement of this phase and frequency coordination of neuron firing will contribute substantially to the transfer of learning for CNS repair. When a patient performs very coordinated movements on special CDT devices, his CNS can learn from the device-imposed movements, via coordinated arm, leg, and trunk movements, a better coordination between neuron firings for CNS self-organization. This improved phase and frequency coordination is not only giving rise for a better organization of the trained movement but also other patterns will benefit from the improved timed firing of neurons, because during the integrative process of self-organization, most CNS neurons are involved and not only those that directly generate, for example, the pattern of muscle activation (see Method, System Theory of Pattern Formation).

Learning transfer is a powerful tool to repair the human injured, malformed, or degenerating CNS. But to further develop treatment (discipline Neurotherapy), we have to understand the organization changes of the human CNS following injury and during repair. Since the thalamus relays and processes input to the primary sensory areas of the cerebral cortex, as well as information about motor behavior to the motor areas of the cortex, including speech muscles, it is likely and shown that primarily the improvement of functional organization of the left thalamus gave rise to speech induction and to a lesser degree the functional improvements in the Wernicke's and Broca's areas with their connections.

2. Regulatory Circuits and Pathways Involved in the Coordination of Movement, Motor Learning, and Memory That May Be Impaired upon Thalamus Injury

As will be analysed, the most severe CNS damage is caused in this patient by the ischemic injury of the left thalamus and not by the malformation of the cerebrum due to schizencephaly. The cerebellum, involved in the coordination of movement, motor learning, and memory, cannot perform its task properly because the regulatory circuits involving the pontine nuclei project through the thalamus (neocerebellar cortex → dentate nucleus →

dentatothalamic tract → thalamus → thalamocortical tract → cerebral cortex → corticopontine tract → pontine nuclei → pontocerebellar tract → neocerebellar cortex) (Figure 47B) were impaired. The exercising of volitional coordinated movements may therefore only partly be successful for repair.

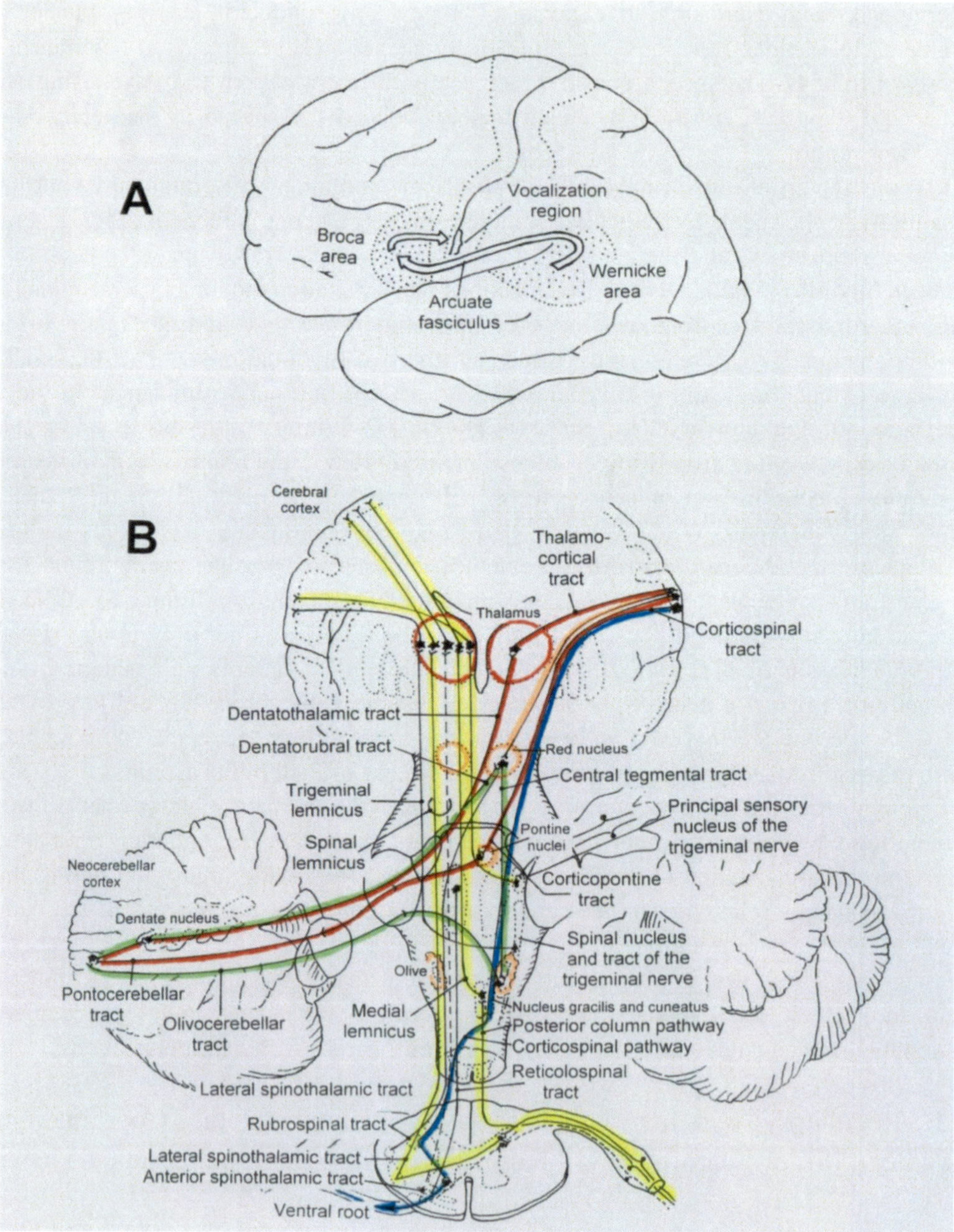

Figure 47. A. Localizations of areas and tracts of the cerebrum involved in speech. B. Regulatory circuits and somatosensory pathways related to the thalamus. Cerebellar regulatory circuits involving the thalamus are marked red and not involving the thalamus green. Somatosensory pathways leading through the thalamus (yellow). Circuits and pathways of the CNS only partly drawn and are not drawn to scale.

But the cerebellar regulatory circuits for motor coordination involving the red nucleus and the olive (triangle of Guillain and Mollaret) (neocerebellar cortex → dentate nucleus → dentatorubral tract → red nucleus → central tegmental tract → olive → olivocerebellar tract → neocerebellar cortex), and not projecting through the thalamus, was probably fully working in this patient (Figure 47B). This motor learning for long-term memory is mainly subconscious and automatic and was stimulated by automatic movements like walking. The regulatory circuits involving the thalamus, the red nucleus, and the olive (neocerebellar cortex → dentate nucleus → dentatothalamic tract → thalamus → thalamocortical tract → cerebral cortex → corticorubral tract → red nucleus → central tegmental tract → olive → olivocerebellar tract → neocerebellar cortex) (Figure 47B) was also impaired, because it projects also through the thalamus.

Upon performing coordinated arm, leg, and trunk movements, the neuronal networks of the intrinsic apparatus of the spinal cord (especially in the cervical (arm movements) and lumbosacral enlargement (leg movements)) and the reticulo-spinal system can be activated, controlled, and coordinated by the corticospinal, the corticorubral, and other tracts.

Damage of the thalamus is not only impairing the function of the cerebellum; it is also impairing the functions of the ascending reticular activation system (ARAS), which is contributing to consciousness. Ascending fibres project from the reticular formation to the thalamus and further on to the cerebrum. The reticular formation includes input from the spinal cord (reticulo-spinal system) and cranial nerves of which only the input from the trigeminal nerve is pictured in Figure 47B. If the level of consciousness is lowered by reduced impulse traffic in the ARAS, then the consciousness network state may switch more easily between healthy and unhealthy consciousness attractors (Figure 49). It may become difficult to keep a patient in a healthy consciousness state. But to keep a patient in a healthy consciousness state or to keep a patient continuously in a certain network state is difficult in many different brain injuries, because in all CNS injuries, the geographical landscape of attractors of patterns will be malformed, the attractor basins are shallower, and the network state fluctuations are larger so that the attractor states are less stable and the network state switches easily between the attractors unintentionally. The lack of continuous concentration to a certain task can be explained by a pathologic geographical landscape of attractors in combination with a high pattern fluctuation (in similarity to and extension of formula 1 in the Method section). Since the thalamus is the largest subcortical collecting point and an important centre for integration and coordination for all exteroceptive and proprioceptive sensory inputs (Figure 47B), somatosensory deficits will occur due to the damage of the left thalamus. The afferent impulses of different modalities, from different regions of the body, are wrongly integrated and false affective coloration may be given. Elementary phenomena such as feeling, pain, pleasure, and well-being may be changed.

Due to the increased brain pressure for a time longer than one year, the reticulo-spinal system may also be slightly damaged. The reticulo-spinal system is the phylogenetically oldest system descending into the spinal cord from the region of the brain stem. This system is contributing strongly to reorganization and has remarkable regulatory effects at segmental level. The reticulo-spinal and the proprio-spinal systems are only partly drawn in Figure 47B, even though they are essential for learning transfer from coordinated movements to speech functions. For the repair of the injured human CNS, therefore, the anatomy of the human CNS has to be taken into consideration. It is not sufficient to think only of a reorganization of neuronal networks or an enhancement of neurogenesis and the growing of nerve fibres.

3. Coordination Dynamics in Speech

Sound emerges as a distinctive and well-known pattern. The many different articulators (tongue, lips, jaw, etc.) whose motion changes the shape of the vocal tract, which in turn creates sounds, allow us to talk and listeners to understand. It takes the coordinated action of about thirty muscles in the tongue, the respirators, and so forth for a baby to say "ba." It was shown that the relative phasing among the movements of the articulators is a collective variable that characterizes stability and change in syllable structure. A key collective variable turned out to be the relative phasing between the opening of the glottis and the closing of the lips [19,87]. Relevant information for perception of dynamic speech patterns lies in the collective variables that characterize the pattern's generation. Even though the relative phasing among articulators is crucial for speech production and improvement, the substantial progress to induce speech, and make further speech therapy possible in the patient of this report, was achieved by enhancing the stability of the speech pattern, probably mainly by improving the timed firing of neurons in the left thalamus within the context of the whole CNS.

4. Non-Cooperative Patient

The primary drawback with the patient of this report was that she didn't understand the need to fight by herself to get healthier upon treatment. The patient learned more or less automatically. The mother had to do a great job to make her training. Often, her hands had to be fixed (Figure 45C,D) to make sure that she was training continuously. Hopefully, her higher mental functions improve further so that she understands that she has to train also by herself and has to become more cooperative for further improvement. EMG recordings would not have been possible in this patient, because she would have destroyed the equipment. The recording of the coordination dynamics was easily possible when she became able to exercise by herself.

3.2.5.3. Results

1. Case Report

Eight months after birth, it was recognised that the patient, called "Popi," was not normal because of the occurrence of strong convulsions. The convulsions were treated. Computer tomography and magnetic resonance imaging (MRI) was performed. The prognosis was that Popi probably would not be able to walk and speak. One year after birth, the right hand and right leg were not functioning as well as on the left side. The patient was behind in reactions in comparison to normal children of her age and could not speak yet. She could only generate vowel sounds. Even though the patient was not like a normal child, she developed further. At an age of two, she could say a few words. But then her functions went backwards. She stopped speaking. The mother asked for a new MRI. The neuro-paediatric refused, because he thought that everything was fine with the child. Upon pressure by the mother, an MRI was performed at an age of 4.5 years. It turned out that the cyst in the cerebrum had enlarged, and there had been probably high brain pressure for one to two years. A shunt was installed. Schizencephaly was diagnosed. The cyst probably generated the high pressure. The mother was advised to have another child. But the mother started to fight for her daughter. For two to three years, licensed speech therapists from Greece and the USA tried to make the child speak, but without success. They concluded that it was impossible that the patient would learn to speak. At an age of six, CDT was started. Speech was induced and motor, higher mental, and vegetative functions improved. When she became able to perform supported jumping on springboard, full continence was achieved.

2. Cerebrum Damage in Relation to the Impairment of Motor and Speech Function

An MRI, done at an age of seven (Figure 46), seems to show a prominent open-lip schizencephaly on the right side (Figure 46A,B) by the connection of an arachnoids cyst to the ventricle system (schizencephalies are clefts of dysplastic grey matter extending from the surface of the brain into a ventricle). At the connection, there is gliosis of the cerebrum from an old ischemic injury. Mental deficiencies correlate well with the presence of frontal schizencephalies. Patients with unilateral clefts, however, as in this case, have a good prognosis for intellectual development but may have seizures. From the damage of the right hemisphere, one expects that the functions of the left arm and leg are more impaired, since each hemisphere is concerned primarily with sensory and motor processes on the contralateral side of the body. But the right side was more affected than the left side. As can be seen from the image of Figure 46C, the left thalamus (including the ventral part) was also injured probably from an ischemic injury of long ago (may be a unilateral thalamus infarction of long ago). Since also the thalamus projects primarily to the contralateral side, the severe impairment of the movement functions of the right arm and leg were primarily caused by the injured left thalamus and not by the damaged right cerebral hemisphere. Compensation between the right and left hemispheres was limited in this patient because of the dysplastic corpus callosum (Figure 46D). In the right-handed patient (only using the left hand more (Figure 45B) because of the impaired motor functions on the right side)the left not (or only little) damaged hemisphere is regarded as being dominant for controlling speech, and the Wernicke's and Broca's areas should be sited there and should not be damaged severely.

Wernicke's area, near the primary auditory cortex (Figure 47A), is important to understand the spoken language. It lies near the angular gyrus, which combines auditory input with input from other senses. The patient seemed to be able to understand the speaking of the mother but could not get the pattern speech activated to answer a question; Wernicke's area seemed to work. The arcuate fasciculus is a fibre tract that connects Wernicke's area to Brocas's area (Figure 47A). Since the essential step of inducing speech was achieved within three months (a time too short for a regeneration of tract fibres), it seems that the arcuate fasciculus of the left hemisphere was also not damaged severely. Broca's area initiates grammatical speech. It, in turn, lies near the vocalization region of the motor area (Figure 47A), which issues the specific commands that cause the mouth, tongue, and palate to form words. Since the patient could not initiate speech, and, when she learned to speak, could only poorly form the words, the motor speech coordination was the main impairment. Therefore, Broca's area and the vocalization region of the motor area were not functioning physiologically. Since no clear structural damage could be found, it is likely that the inability to speak was caused by other structures, which communicate at least with the vocalization area of the motor cortex. Since the injured left thalamus strongly impaired the motor functions of the right arm and right leg, it is likely that the injured left thalamus was also the main cause for the inability to speak.

3. Thalamus Injury as the Main Cause for the Inability to Speak

First of all, the thalamus is the largest subcortical collecting point for all exteroceptive and proprioceptive sensory impulses. Since this collecting point is damaged in the patient, at least on one side, we have to expect an impairment of several functions. The thalamus is a relay station for all impulses arising in cutaneous and visceral sensory receptors, for visual and auditory impulses, and for impulses from the hypothalamus, cerebellum, and brain stem reticular formation, all of which are processed in the thalamus before being transmitted onward to other structures. Through its reciprocal connections (feedback loops) with the motor cortex, some of which pass through the basal ganglia and cerebellum, the thalamus modulates motor functions, including those for speech. If the cerebellum is damaged, dysarthria occurs [53-55]. If the thalamus is damaged, we expect also dysarthria, at least because the impulses from the cerebellum lead through the thalamus (Figure 47B). The patient showed dysarthria when she became able to speak. Further, all sensory impulses (other than olfactory impulses) must pass through the thalamus before they can be consciously perceived. Thus the thalamus was traditionally called "the gateway to consciousness" [88]. Lesion of the left thalamus may produce pain syndromes in addition to other sensory deficits. The patient even seemed to have sometimes "thalamic pain" (pain for which no reason could be found) and had most likely deficits in proper feeling, since even the mother was not allowed to touch her at the beginning of therapy.

The thalamus, however, is not merely a relay station but an important centre for integration and coordination, in which afferent impulses of different modalities, from different regions of the body, are integrated and given an effective coloration. From the damage of the thalamus, we expect not only an impairment of the phase and frequency coordination of neuron firing (due to every CNS injury) but also dyscoordination of the special areas (including the vocalization region of the motor area) necessary for the generation of speech. This dyscoordination may have contributed substantially to the inability to speak. It is conceivable that the improvement of the neural coordination among neurons, neural ensembles, and specialized subnetworks and the improvement of the coordination of the different modalities of sensory inputs lead to the induction of speech in the patient upon exercising on the special CDT device. Some thalamic nuclei (especially the centromedian nucleus) are also components of the ascending reticular activating system (ARAS), a specific arousal system originating in nuclei that are diffusely located throughout the brain stem reticular formation. Activating impulses from the ARAS are relayed by thalamic nuclei to the entire neocortex. An intact ARAS is essential for normal consciousness.

4. Impairment of the Ascending Reticular Activating System (ARAS)

Since the ARAS was impaired at least in the left thalamus, we have to expect impairment of normal consciousness and higher mental functions. The patient showed changing consciousness, dysphoria, and impaired higher mental functions. When the patient became able to speak, she sometimes had a nice, healthy looking face, spoke, and was cooperative, and at other times, her face was looking unhealthy, she was yelling, and was not cooperative. During therapy, it was not possible to keep her continuously in the healthier network state, even though the mother was trying hard (Figure 45C).

5. Dysplastic Corpus Callosum and Right Cerebral Hemisphere in Relation to Emotions in Speech

Affective aspects of language are represented in the right hemisphere (where in this patient the schizencephaly was sited), and their anatomical organization mirrors the organization of the cognitive content of language in the left hemisphere [17]. Damage to the right temporal area homologous to Wernicke's in the left hemisphere leads to disturbances in comprehending the emotional content of language. Although this pattern of localization appears to be inborn, it is by no means completely determined. Young children in whom the left hemisphere is severely damaged early in life can develop an essential normal range of language function [17]. In spite of the malformation of the right hemisphere (Figure 46A,B) and the corpus callosum (Figure 46D), which connects the two hemispheres, the patient had emotions in her speech. This again supports the view that the severe deficits in her higher mental functions did not originate in the two hemispheres and their connections but in the thalamus.

6. Coordination Dynamics Therapy (CDT)

At an age of 5.5, years CDT was started. At the beginning, the therapy was not optimal and not sufficiently intensive. The patient could not turn by herself on the special CDT device. Her exercising had to be supported by the mother. The patient was not cooperative. The mother, therefore, had to make her cooperative by performing a Punch and Judy show (see Figure 45C). The right leg improved, but the patient did not learn to speak. At an age of six years, an intensive therapy for three months was administered with 20 hours per week. When the patient's CNS became improved so much that she could turn by herself on the special device, she started to speak. The repertoire included approx. 100 words. She began to form sentences as taught by the mother. It is concluded that the patient started to speak by learning transfer from coordinated movements (exercising on the special CDT device for turning and supported walking on treadmill).

On the basis of System Theory of Pattern Formation for Repair, this progress can be expressed in the following way. The attractors for walking and exercising on the special CDT device took a more physiologic form. With the deepening of their attractor basins (improving the stability of these motor patterns), an attractor for speech was built anew or its stability enhanced that strong so that the speech pattern could be intentionally recruited by her.

When the patient became able to exercise on the special device by herself, coordination dynamics measurements could be performed (Figure 48A,B). The coordination dynamics (CD) values for forward and backward exercising were with $\Delta(26N) \approx 39$ and 41, respectively, quite poor (Figure 48A,B). A healthy sporty child of her age has values in the range of 9 and 16 (Figure 48C,D). Newer average CD values of healthy children are in the range of 18 and 22 for forward and backward exercising (Figure 24A,B). A healthy grown-up person has for forward and backward exercising approximate CD values of 5 and 10, respectively [42]. A trained person (author) has coordination dynamics values of between 2 and 3 (Figure 48E,F). Even though the patient had plenty of arm and leg power, the CD values improved only slowly, indicating that her CNS improved in its functioning only slowly since large reorganizations had to take place because of the severe damage/malformation and severe pathologic organizations to be repaired.

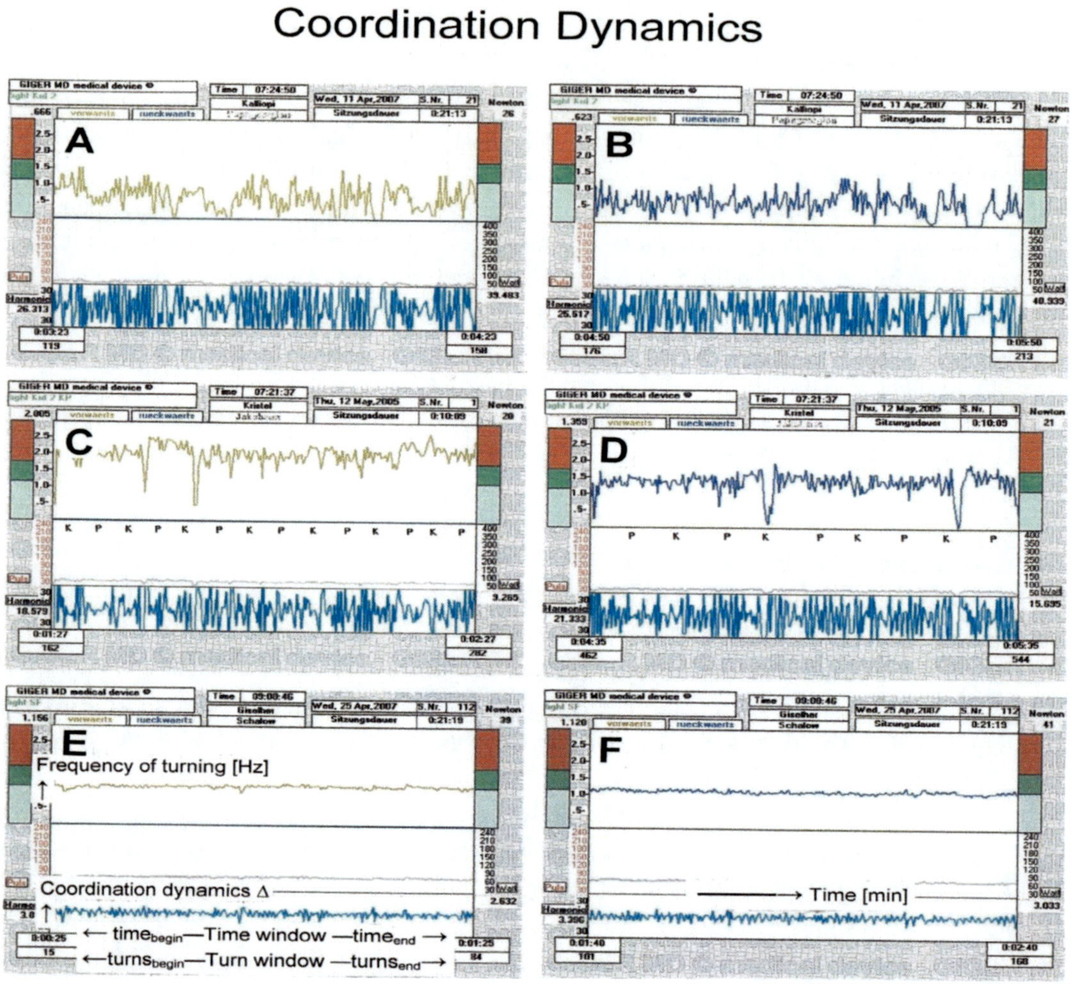

Figure 48. Low-load coordination dynamics upon exercising on the special coordination dynamics therapy (CDT) device in the forward (A,C,E) and backward direction (B,D,F). A,B. Recordings from of a seven-year-old cerebral palsy girl. The frequency of exercising (upper trace) is very irregular, and the arrhythmicity turning (coordination dynamics) (lower trace) is very high (coordination dynamics (Δ) value (mean arrhythmicity per minute) = 39.5 (41(backward turning)). C,D. Coordination dynamics from a healthy girl of the same age. The frequency of turning is more regular and the arrhythmicity of exercising is smaller (Δ = 9.3 (15.7). E,F. Coordination dynamics recordings from the author. The frequency of exercising is regular and the arrhythmicity of exercising is small (Δ = 2,6 (3.0)).

7. Improvement of the Cooperation of the Patient

The biggest problem in this patient was that she was not cooperative. The main load treatment was on the mother. But after six years of treatment (at an age of 12 years), there was indication that the patient became more cooperative. For short times, it became possible that the patient also exercised a bit with the author (Figure 49). The patient tolerated some instructions by laughing (Figure 49A). But the supported exercising, in the way to help the patient to turn more smoothly to improve the movement-induced afferent input for improved movement-based learning, the patient still did not like.

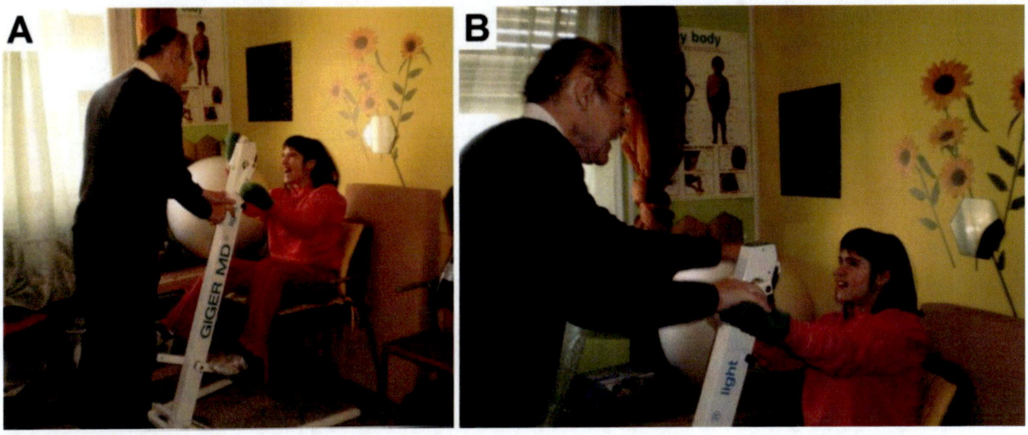

Figure 49. Author during exercising with a 12-year-old cerebral palsy child. When the author is only instructing for improved exercising, the patient is laughing (A). Upon supporting the smoothness of exercising to improve movement-based learning, the patient becomes angry (B).

Her mood changed from laughter to anger, as can be seen from the impression change of her face (Figure 49B). Mostly, cerebral palsy patients are cooperative, because they feel the improvement of their CNS functioning in the short-term memory. Baby patients (if they got used to the equipment and were not anxious anymore) are often more cooperative than healthy young infants. Good mothers can be unbelievable in being patient with their children to improve it and should be included in the treatment.

8. Importance of Induction of Speech

The improvement of the motor functions was a quantitative step in repair, but the induction of speech was a qualitative step in repair. It is very difficult to communicate with a cerebral palsy child (mentally retarded), if the child cannot speak. Also for the further development of its intelligence, the speech is of high importance. The patient speaks quite a bit now, even though the clarity of speech (dysarthria) has to be improved. The speech therapists argue now that the patient Popi will make good progress upon further speech therapy.

3.2.5.4. Discussion

1. Induction of Speech Patterns upon Improving the Neural Coordination

It was shown that a child with severe injury/malformation of the cerebrum, thalamus, corpus callosum, and maybe some other brain (stem) structures could learn to speak by transfer of learning from coordinated movements. This quality step of improvement of CNS functioning is not only of high importance for the quality of life of cerebral palsy children and their parents but also for the understanding of the functioning of the human CNS in general. Since the performance of the walking and sometimes crawling was not especially good, it seems that the main stimulus for the induction of speech originated in the exercising on the special CDT device (Figure 45C). This means that the improvement of the neural coordination (including phase and frequency coordination of firing among neurons, neural ensembles, and specialized subnetworks, see below), which the CNS learned from the device-imposed very coordinated movements, was the main cause to induce speech. Since the three-month period of intensive therapy was too short to induce substantial structural repair in the thalamus and other CNS parts, it seems that the functional reorganization and/or improvement of the neural coordination established the speech pattern or enhanced its stability.

2. Improvement of Speech and Neural Coordination in the Short-Term Memory

An intelligent stroke patient, who spoke four languages before stroke, could still speak after the stroke but could not communicate with other persons, because he mixed up the different languages during speaking. When he once exercised on the special device in the lying position for more than two hours (approximately 6000 turnings), he suddenly could communicate, because he didn't mix up anymore the different languages. When, after a few hours, the improved CNS organization decreased in the short-term memory, he mixed up the languages again. Therefore, the improvement of the neural coordination in the short-term memory influenced the organization of his speech. A patient with mainly a severe cerebellum injury and dysarthria could speak slightly better when he was exercising on the special CDT device. His speech therapy, therefore, was performed while exercising on the special CDT device (Figure 92 of [38]). If the improved timed firing of neurons in the short-term memory is improving speech in the short-term memory, then it is likely that with repeated improvements of the timed firing of neurons and speech in the short-term memory improvements of both properties are also achieved in the long-term memory.

3. Exercising on the Special CDT Device Alone Is Not Sufficient for Improving Substantially the Phase and Frequency Coordination in the Long-Term Memory

Since the arrhythmicity of exercising in the patient of this report (the coordination dynamics) improved only little with further therapy, also other functional and structural changes contributed to the CNS repair, and they are additionally needed to further substantially improve the quality of CNS functioning in the long-term memory. In a ten-year-old boy, who suffered severe brain damage in a car accident, the coordination dynamics only improved further, following a year of therapy, when he exercised additional strongly crawling, walking and running and especially in the backward direction (52, see below under traumatic brain injury). Also in this patient, the right foot was positioned better during backward (Figure 45B) than during forward walking (Figure 45D) on treadmill. For forward walking, the mother had to support the foot and the knee (to avoid overstretching of the knee), whereas during backward walking, the positioning of the right knee and foot was sufficiently good for CNS repair.

There are, therefore, two conclusions. First, even though the training of a single movement, like walking or exercising on the special CDT device, may be very successful for the repair at the beginning of therapy but on the long term, many coordinated movements should be trained, if possible, because the human CNS is an open system that has to be entrained for repair by many patterns. Second, backward exercising (backward walking, crawling, or exercising on the special CDT device) seems to be very successful for repair. This additional backward exercising is more than increasing the ability of the bilaterally symmetrical parts in a skilled act. Exercised forward-backward symmetry may contribute efficiently to the repair of the structure of the state space of patterns. Interestingly, a newborn baby can step automatically only in the forward direction. In severe spinal cord injury, with a few percent of spared white matter (tracts), the stepping automatism transiently reappears during CDT but in the forward and backward direction.

4. Does the Thalamus Contribute to Learning Transfer?

Since the thalamus is a large subcortical collecting point and an important centre for integration and coordination for all exteroceptive and proprioceptive sensory inputs (Figure 47B), there will be a large overlap of networks for the different organizations and integrations. The thalamus may therefore substantially contribute to learning transfer from motor to speech functions in similarity to the learning transfer from movements (jumping on springboard and exercising on the special CDT device) to urinary bladder functions due to the neural network overlap of motor and continence functions in the lumbar-sacral spinal cord segments (63,64, Chapter V).

5. Weight, Wiring, Neural Replacement, and Neural Coordination Plasticity for Learning in the Long-Term Memory to Generate a Physiologic Attractor Layout with Long-Term Stability

The human brain consists of 10^{11} neurons connected by 10^{15} synapses. Even though the patient of this report had fewer neurons and synapses because of malformation, injury and (Valerian) degeneration, the main problem was to achieve, by therapy-induced modification, a rather physiologic functioning of her CNS; a physiologic attractor layout for physiologic pattern formation had to be generated.

Weight-plasticity: Neural learning research has been focused on activity-dependent "weight" changes between previously connected neurons. This mode of plasticity could involve either changes in the efficiencies of existing synapses (LTP) or structural changes that lead to the addition or subtraction of synapses between previously connected pre- and postsynaptic neurons. In either case, the network's connectivity matrix, or wiring diagram, is left unchanged. In the weight-plasticity scenario, the learning capacity lies in the system's ability to increase and decrease the weights on existing connections as a means of encoding learned information. The synaptic efficacy can be modulated in a use-dependent manner to produce weight changes. The therapy-dependent changes (due to weight and other changes) are formulated in the equations of motion of the collective variables $dX/dt = F_{intr}(X) + \sum c_{inf}F_{inf}(X,t)$ (formula 2) by the behavioural information $\sum c_{inf}F_{inf}(X,t)$ (intention, instruction, walking, etc.) (Method).

Wiring-plasticity: In addition to weight changes, learning could involve alterations to the wiring diagram, whereby previously unconnected neurons become connected and vice versa. Unlike weight changes, wiring changes require structural plasticity. In this learning mode, the storage capacity lies in the system's flexibility to choose which presynaptic neurons provide input to each postsynaptic neuron. Weight and wiring changes are not mutually exclusive, and neurons and their synapses might be involved in both forms of learning. Structural changes that would be required to achieve wiring changes, including synaptogenesis and outgrow of axons and dendrites, can occur in the adult brain [89]. In most areas of the brain, only a small fraction of all possible connections between neurons physically exist. In such sparse networks, a capacity to rewire could dramatically increase the number of functionally distinct circuits available to encode learned information. On the other hand, the task of finding appropriate partnerships between pre- and post-synaptic units in a sparsely connected network is a hard combinational search problem and can be accelerated more or less through activity-dependent release of diffusible factors from axons and/or dendrites or through electric fields. If changing electromagnetic fields, due to the natural action potential traffic in activated neuronal networks, are contributing to neurite guidance besides target-derived growth factors, then natural activity would be more efficient for repair than the application of growth factors, even when administered according to a geographical landscape. The structural plasticity in the injured brain, including synapse formation and elimination, and outgrowth and retraction of dendrites and axons, is stimulated in a therapy-dependent manner.

To connect a presynaptic neuron to a postsynaptic neuron through spine/dendrite/axon outgrow, there are two possibilities. First, the population of synapses can be formed without significant growth of axonal or dendritic arbors; new wiring diagrams are within "immediate" reach. Second, the potential population of presynaptic partners consists of those that can be accessed only through growth of new axonal or dendritic branches (over longer distances); longer-range wiring connections are necessary for further changing of wiring diagrams. As the repair of autonomic (urinary bladder function) and motor functions upon CDT in a patient with a severe cervical spinal cord injury indicates [59,64], long-range growth of new axonal or dendritic branches in the human CNS seems possible. Only, it may take two to three years, and the growth of axons and dendrites is mainly in response to movement-based learning. In comparison to development, it also takes two to three years for infants to become continent. Since the repair in severe spinal cord injury extended over one vertebra length, the maximum spatial extent of axonal and dendritic branches may be larger than 10mm.

Neural replacement-plasticity: In a patient with severe cervical (motoric complete) spinal cord injury at C5/6 levels [59,64] (no finger functions remained, because of destroyed motoneurons at the injury site), the index fingers started to work upon three years of intensive CDT [67]. This repair indicates that some motoneurons were built anew in the adult CNS. Therefore neurogenesis exists in the injured human CNS, even though the process seems to be very slow (or starts only after a long latency), is very limited, and has to be stimulated by intensive movement-based learning. A new theory of long-term memory has been proposed, in which there is a relation between neurogenesis in the adult brain and learning in the long-term memory [90]. The replacement of older neurons by new ones that are, in turn, transient and replaceable offers a vehicle for the constant and spontaneous rejuvenation of key brain circuits. The acquisition of a long-term memory for relearned patterns, including neurogenesis, is even more likely in the injured CNS. Wiring and neurogenesis plasticity is also included in the therapy-induced improvements described by the System Theory of Pattern Formation for Repair (formula 2) and should be able to partially repair the integrative functions of the left thalamus, the ARAS, and other damaged CNS structures and tracts in the patient of this report.

Neural coordination-plasticity: It has been shown in our patient that already mainly exercising on the special CDT device induced the speech and improved motor patterns. Therefore, by improving the timed firing of neurons, the stability of the speech pattern increased so much that the patient could recruit it by intention. Before therapy, the patient wanted to speak (one could see it) but could not speak; she could not recruit the speech pattern by intention. Plasticity does not only involve weight and wiring changes or neural replacement, but the exactness of the timed firing of neurons is also important. If waves of excitation do not propagate purely along dedicated lines but can also cross for example in the form of for example synfire chains [1], then a more precise phase and frequency coordination between the firings of neurons offers a higher functional connectivity, since more information can be processed per time interval. By learning, the crossing of excitation waves is possible at higher impulse traffic (higher information processing, which allows more complex CNS self-organization) or gives more safety for physiologic functioning (higher pattern stability that means deeper attractor basins and less pattern fluctuation).

6. Neural Coordination

The measurements so far indicate that a principle of the human CNS self-organization is the neural coordination among neurons, neural ensembles, and specialized subnetworks. This timed firing for communication (neural coordination) consists of specific natural impulse patterns and the coordinated firing by phase and frequency. An injury, malformation, or degeneration of the CNS impairs this complex intricate neural coordination with the consequence that physiologic patterns change or disappear and pathologic patterns appear.

7. Higher Neural Functional Complexity in the Human CNS in Comparison to Animals

Naturally evoked transmission frequencies of nerve fibres in humans can be even higher than 5,000Hz (interspike intervals of single fibres are smaller than 0.2ms (Chapter III)). Transmission frequencies of less than 1,000Hz were expected (91); textbooks for medical student's present values of transmission frequencies of between 500 and 1,000Hz [92]. Absolute refractory periods in the range of 0.2 ms therefore indicate that the membrane properties of secondary muscle spindle afferents in humans differ from those of animals. If axons of other neurons in the CNS also had five times higher transmission frequencies, then CNS self-organization changes in humans could occur five times faster than in conventional laboratory animals. In modelling human CNS functions, one probably has to consider membrane properties typical for human membranes. Transmission frequencies, refractory periods, and times of inactivation of nerve cells are important for the physiologic functioning of the CNS.

If one believes that self-organization and pattern formation of the CNS is not purely constructed of dedicated lines of excitation but also of synfire chains that cross each other, then refractory periods of nerve cells may become of high importance. The reduced refractory periods of membranes in humans may have consequences for the organization of the higher mental functions. A neuron, neural ensemble, or specialized subnetwork can "interview" more other neurons, ensembles, or subnetworks with certain (sub) functions per time; a higher transmission frequency may allow, therefore, a higher complexity in CNS organization. The higher intelligence of humans in comparison to animals may therefore not only be due to the higher structural complexity of the CNS but also due to a higher functional complexity (for further considerations, see below).

8. Patterns as a Means for Long-Term Memory Storage

Learning-related plasticity, mediated by different kinds of functional and structural changes, could underlie long-term information storage in the human CNS. The commonly accepted learning mode based on changes in synaptic weights is not sufficient to explain repair in the human CNS. The physical change(s) encoding a particular long-term memory is called an engram. It could well be that the physical substrate for a certain mode of long-term information storage are learned or relearned neural network patterns.

9. Are Functions Localized to Specific Regions of the Brain (CNS) or Do They Represent a Collective and Emergent Property of the Whole Brain?

One of the frontiers of human neurophysiology is to understand the physiological basis of consciousness and the mental processes by which we receive, act, learn, speak, and remember. What rules relate the anatomy and physiology of the CNS to its specific function in perception, thought or movement? Can these rules be understood better by examining the CNS as a whole, the region, or by studying the individual nerve cells with their connections and membrane properties? Within the framework of the System Theory of Pattern Formation for Repair, the CNS self-organization represents a collective and emergent property of the whole brain. Dynamics in self-organization, pattern stability, limb coordination, and learning transfer can nicely be understood and used to understand CNS repair. But to find strategies for CNS repair, processes localized to specific regions of the CNS and their existing functional and structural subdivision also have to be known. As partly shown above, membrane and nerve cell properties and their functional and structural connectivity also have to be known. A powerful tool to examine human CNS functions is the neural ensemble "premotor spinal oscillator."

Phrenologists had argued that the cortex is a mosaic of specific functions and that even abstract mental attributes are localized to single functionally specific cortical areas. The opposing aggregate field school argued that mental functions are distributed homogenously through the cerebral cortex. This report on the induction of speech indicates that specialized localized areas of the whole CNS are involved, but also integrative organization principles as neural coordination and pattern stability are needed to understand language.

Further, the functions localized to discrete regions in the brain are not complex faculties of mind but elementary operations. More elaborate faculties are constructed from serial and parallel (distributed) functional and structural interconnections of several brain regions. Regional specialization may be a key principle of cortical organization. But these specific areas have their specialized function only in the context of the whole CNS. Extirpation of a specialized area for examination will in large part result in meaningless function, since its functional context is lost. Already a small CNS injury or the building of new nerve cells or connections disturbs integrative functions significantly (see below). Subdivisions of functional areas, in connection with serial and parallel distributed processing of different specialized areas, lead to collective functions of specialized areas. Collective properties are therefore needed anyway for (understanding) pattern organization.

Speech: By improving mainly the phase and frequency of neuron firing of large parts of the CNS (especially of the left thalamus), upon exercising on the special CDT device for three months, the speech was induced in our patient. But when the patient could speak, the injury of the thalamus became obvious: the speech performance was poor (dysarthria). A further partial functional and structural repair of the thalamus seems possible but may take a few years. On the other hand, with a severe damage of the Broca's and Wernicke's areas, the patient most likely would not have learned to speak, at least not in three months. In other words, by improving the complexity of CNS organization (by improving the exactness of timed firing of neurons), the patient became able to speak. In a patient with a severe cerebellum injury (and not injured thalamus), the speech performance was poor but improved within ten years of therapy [53-55]. In that patient with cerebellum injury, the speech and the intelligence improved much more with motor learning than the motor functions, indicating that the higher mental functions are more integrative functions.

Movements: Individual movements are represented in small, quite discrete regions of the cortex, and movements of a limb are produced by stimulating the precentral gyrus in the contralateral motor cortex. Thus, the right hand, commonly used for writing and skilled movements, is controlled by the same hemisphere that controls speech. In most people, therefore, the left hemisphere is regarded as being dominant. The patient of this report is probably right-handed, because for skilled movements she uses the right arm, even though she is trying to write with the left hand. Also for the impairment of the movements of the right arm and right leg, the injured left thalamus is responsible. In the patient with the severe cerebellum injury, the movements (especially the balance) were also strongly impaired [53-55], even though the primary motor cortex was not injured. In conclusion, many specialized areas of the CNS do contribute to the organization of movement and speech and not only those located in the cortex.

As in other CNS injuries, the patient could walk better backward than forward. Are there different specialized areas in the cortex for forward and backward walking or is forward and backward walking a question of integrative functions, for which the symmetries structure the state space of movement patterns? A newborn baby can step automatically (characteristic for the automatic stepping is the strong lifting of the knees). Later on during development, supraspinal centres (including cortex, thalamus, and cerebellum) change the automatic stepping into walking. Following spinal cord injury, the automatic stepping may reappear and is changed again into walking with a partial regeneration of the spinal cord. Therefore a specialized area for walking is at least also located in the lumbosacral enlargement of the spinal cord and not only in the specialized cortex area. The most basic continence function for the urinary bladder is located in the same spinal cord segments (S2-S5) as the continence for the rectum. Therefore, certain specialized areas of the CNS may have more than one function. In conclusion, our patient trained automatically very coordinated arm, leg, and trunk movements and learned in this way to speak by learning transfer.

10. Structural Network Changes Induce Functional Instability

In a patient who suffered a severe cervical spinal cord injury, the spinal cord regenerated very limitedly upon CDT. Motor functions below the injury level reappeared, and autonomic functions, including continence, breathing, and cardiovascular performance, were cured [59,64].

The quality of CNS organization in this patient was measured integratively and continuously by the coordination dynamics values (collective property) and plotted graphically (Figure 50). With each very little step of regeneration of the spinal cord (the regeneration of a few fibres), the coordination dynamics values got transiently worse (seen as peaks in Figure 50). This means, with each little step of structural repair, the CNS organization lost some stability, which was repaired each time again by the ongoing therapy. With each little step of structural change, instability occurred. For extended structural repair, large instabilities have to be expected.

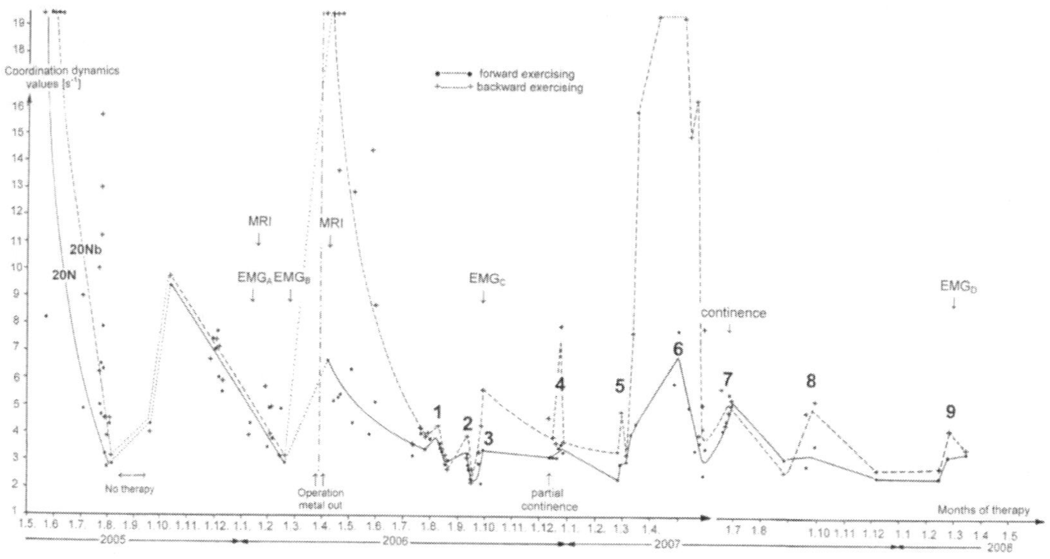

Figure 50. Relation of coordination dynamics values to therapy duration for a load of 20N and for exercising in the forward (lines and dots) and backward direction (20Nb; dashed line and crosses) in a patient with a 95% cervical spinal cord injury (C5/6). Note that with no therapy, the coordination dynamics values got worse (increased), and upon further therapy, they improved again. Upon metal removal, the coordination dynamics values increased strongly. The transient coordination dynamics value increases (peaks, sign of transient CNS network instability) "1" through "8" fall together with the reappearance of certain muscle functions or specific improvements of motor and autonomic functions. Note, this graph clearly shows that with structural network changes of the human CNS (here spinal cord regeneration), network instabilities do occur.

Such occurrence of instabilities due to structural changes is foreign to current brain research in human. Continuous functional measurements of human CNS organization are needed to quantify CNS instabilities scientifically and treat them when occurring.

11. Cerebral Palsy: CNS Instability Induced by Stem-Cell Therapy and Out-of-Date Pharmacological and Orthopaedic Treatments

Two cerebral palsy children (similar ages as the patient of this report) from Greece were sent to Germany for stem-cell therapy to cure or improve CNS functioning. One got cancer and the other one convulsion (network instability). Some orthopaedics still think that they can repair the CNS in cerebral palsy by orthopaedic operations. By changing the position and dimensions of tendons, muscles and bones, legs and feet can be brought into physiologic position, if the CNS is not activated. But the CNS can after such operations still not generate physiologic movement patterns.

The nervous system has to be repaired. Already Bobath (physiotherapist) argued 30 years ago that by changing the length of tendons, the receptors will work in a different range and will not inform the CNS about the real situation in the periphery. The healthy human CNS can adapt quickly to such changes in the periphery but not the severely injured CNS with a very low rate of learning. A movement therapy after such operations would not only have to fight for repair of the malformation and/or injury but would have to additionally to fight for the compensation of the changed afferent input from the periphery due to the operational changes; a boy with cerebral palsy could walk with poor performance before an orthopaedic operation but could not walk anymore after the operation. When our patient was walking in the forward direction, the mother had to support the foot for proper positioning (Figure 45D); it seemed that something was wrong with tendons, muscles and/or bones. But for backward walking, the foot and knee performance was nearly normal (Figure 45B)! A causal therapy has to repair the injured CNS; CDT is a learning therapy, which can partly repair the CNS causally.

It is unlikely that the human CNS can be repaired by drugs. Drugs can shift equilibriums, put CNS parts out of order, dumb systems down and achieve thus beneficial effects for the patient, but they cannot cure nervous system damage causally. Botolinum toxin, for example, cannot solve the problem of spasticity (pathologic pattern); the neuronal networks have to be repaired. For the reduction of spasticity, no drugs are needed; CDT reduces spasticity [93] but without reducing motor power for volitional movements simultaneously.

12. Need for Research on Human for Improvement in Cerebral Palsy

Theories of CNS self-organization derived from the lamprey or rat may not be specific and developed enough to explain learning transfer from movements (like walking, crawling and exercising with hands (and fingers) and feet on special devices) to speech. The enormous learning capacity of the human CNS needs a high variability of the networks with the risk of untreatable instabilities for high rates of structural changes. In lower animals, in which the CNS is more fixedly wired, the repair is more a question of growing of nerve fibres, whereas in humans, the repair is more a question of reorganizing neuronal networks by learning including learning transfer.

For enhancing the very slow structural repair, the endogenous stem cell reserves have to be activated more strongly, probably by exercising more at the limits or maybe by using in addition to CDT thoughtful, safe drugs. Additional administration of (exogenous) stem cells may also help, but the procedures must avoid large risks for the patient; the functional integration of the administered stem cells would be secured by CDT.

The most efficient way to repair the CNS in cerebral palsy children will be to influence the development in the way that physiologic functioning is achieved. The "correction en route" during development has to be used excessively; a treatment has to start therefore with cerebral palsy children as early as possible (when correction, adaptations and the rate of learning and building of new nerve cells are largest), which means with one or two years of age. Human research is needed to compare the development with the repair after injury or malformation. In this way, repair research can learn from the development (the correction en route) how to change and improve the injured, malformed, or degenerating CNS.

3.2.6. Repair of Traumatic Brain Injury in Three Children upon an 11-Year Therapy: An Outcome Comparison between Conventional Neuro-Rehabilitation, Treatment by the Mother, and Coordination Dynamics Therapy (CDT)

3.2.6.1. Repair in a Group of Patients with Traumatic Brain Injury

In six patients, 13 to 47 years old, who suffered traumatic brain injury two to seven years ago, CNS functioning, including motor, vegetative and higher mental functions, could partly be restored by applying CDT for three months [40]. The repair was quantified by the improvement of walking, jumping and hand function and the direct measurement of the CD. The CNS organization (the CD) improved by 69% within three months of therapy. The mean CD value for forward exercising achieved ($5.9s^{-1}$) was slightly higher than those in a population of untrained healthy physiotherapists (coordination dynamics = $5.1 \pm 1.2s^{-1}$). The following longitudinal study reports about the immediate and long-term repair of severe traumatic brain injury in three children.

3.2.6.2. A. Quick Improvement Following Severe Brain Injury in Three Children, When Therapy Is Started in the Vigilant Coma Stage

Three boys, aged 9, 12 and 14 years old (Benjamin, Mario, and Andrej, respectively), suffered severe brain lesions almost at the same time, two in a car and one in a bicycle accident. In the 9- and 14-year-old patients, an intensive CDT was started approximately five to ten weeks after the trauma in vigilant coma stage. The two patients recovered unexpectedly early from the coma and progressed quickly in their recovery of motor functions so that they relearned running after four months of therapy, even though their locomotor functions are still far from normal. The 12-year-old patient Mario did not obtain intensive CDT, only conservative physiotherapy, because of many infections and complicated bone fractures, which were not invasively reconstructed. The patient recovered much later from the coma (six months instead of six weeks) and now has very much extensor spasticity (Figure 51E), shortened tendons, problems with several joints, quite much pain, weakened mental functions, grasp reflex in the right hand (Figure 51F, reappearance of infant automatism) and nearly no useful motor functions in the legs. It will be attempted now to apply the intensive CDT with a delay of five months in comparison with the other two cases. It has to be seen whether the same therapy, even if delayed, can help to reach the same level of reorganization in the end. It is known that in brain injuries (and most likely in all CNS injuries, including SCI), the therapy has to be started as early as possible following the trauma (maybe except for the first week) to avoid pathologic reorganization (for example, spasticity).

For an overall view of the progress seven to nine months after the injury, all three patients are shown together in Figure 51A. The two patients who obtained CDT for six months relearned walking and running (Figure 51A,B), and the mental functions improved. The third patient who obtained no CDT for the first seven months following the trauma could not walk or run but had to be carried from the wheelchair to the training device (Figure 51C). He could hardly stand because of extensor spasticity (Figure 51E) and supination of the feet (Figure 51G). The running improved in the two patients Benjamin and Andrej who received coordination dynamic therapy.

For the 9-year-old Benjamin, the running times over 22 m reduced during eight months from 14.4s to 5.2s and for the 14-year-old Andrej during six months from 18.5s to 6.4s. The jumping in anti-phase in Benjamin increased from 25 per series to over 250 per series, and in Andrej from 7 to 55, with a delay of three months. An assessment of the CD was not available at the beginning of therapy.

The nine-year-old boy Benjamin was running whilst playing in the street and was hit by a car driven at 55 km/h and thrown several meters through the air. The boy was pressed against an iron fence and suffered many injuries, including an open thorax injury on the right side and brain damage of the basal ganglia domain (area = 4 cm x 3 cm) and the frontal lobe. The patient was in a coma for two weeks and started then to react a bit when asked to respond.

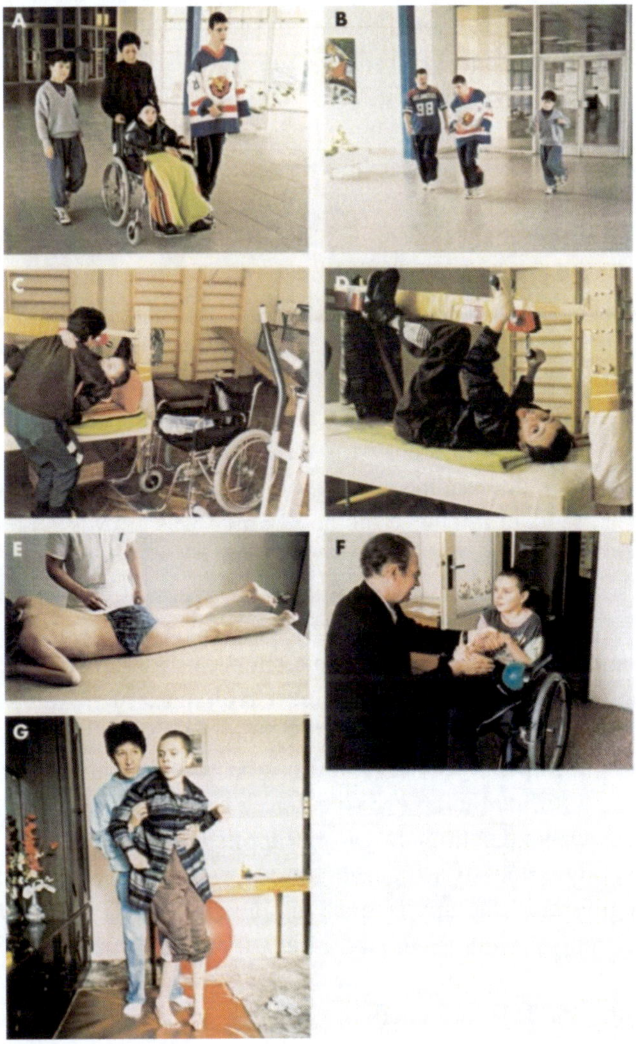

Figure 51. Three patients with severe brain lesions (ten-year-old (A, left), 12-year-old (A centre, in the wheelchair pushed by the mother), 14-year-old (A, right)), seven to nine months following the lesion. The two patients who received coordination dynamic therapy for six months can walk (A) and run again (B), and the third patient who did not receive coordination dynamic therapy for seven months following the lesion is still in wheelchair (A) and has to be carried to the coordination dynamic therapy device by the mother (C). In A, the mother is pushing the wheelchair with her child; the mother of the ten-year-old boy and the father of the 14-year-old boy are in the background. In B, the father of the 14-year-old boy is running, for safety reasons, along with his son and is running in interpersonal coordination. D. The 12-year-old Mario (right-sided hemiparesis; quadruparesis) during exercising on the special coordination dynamic therapy device. The right hand is holding the hand-hold quite strongly, probably because of the induction of the grasp reflex (F) in addition to volitional power. E. Extensor spasticity activated in the 12-year-old Mario; the extensor spasticity is only moderately hindering the therapy. F. During shaking both hands with the author, the patient develops relatively strongly the grasp reflex in the right hand (from the hand reaching position the hand slips into the position of holding the thumb of the author; it could be felt how the network organization slipped from the neuronal network state "hand giving" into the network state "grasp reflex" in similarity to Figure 107). G. Suppination of the right foot during standing (Mario).

The physicians of the intensive care unit forecasted that he would be in vigilant coma for another one to two months. After ten days of vigilant coma (≈4 weeks after the accident), CDT was started in the recumbent position. Especially treadmill walking supported by four persons seemed to frighten the patient. The patient seemed to develop left-sided hemiparesis; the left arm and the left hand seemed to show signs of spasticity. Due to a humerus fracture, the shoulder joint could be moved only little.

Even though the movements in the left arm and left hand seemed to be painful for the patient, he seemed to be cooperative. Because he could not speak, a safe communication was not possible. When the author left the patient after three days of therapy, he was stroked by the patient, which indicated that he was thankful for the therapy, even though the therapy caused much pain. The CDT was continued by a physiotherapist and the mother, herself a physician. After a further three days, speech returned, and the vigilant coma was over. The patient said that he had quite a lot of pain in the left arm and hand during the movements. But because he felt that the therapists and the mother wanted to help him, he helped actively during the arm and hand movements as much as possible, in spite of the pain and being frightened. The recognition of the reality (of the trauma) was still difficult for the patient. Often he said to his mother: "Mum, please wake me up, the dream is so terrible." Depressive thoughts were opposed with an intensive therapy (which was anyway in the line of his neuro-therapy), not to give the patient time to think his situation over, in similarity to the work therapy of neurotic patients. Further, training using the special CDT device seemed to have a positive effect on the patient's mood. As turned out after the end of the vigilant coma, the patient indeed was hemiparetic on the left (with a slight quadruparesis), including spasticity in the left arm, hand and fingers and in the left sternocleidomastoideus muscle (pulling his head to the left side). The organization of the CNS improved continuously with the ongoing intensive therapy (six times per week, several times per day). On the advice of the author, the parents refrained from botuline therapy to selectively reduce spasticity (by a blockade of the neuromuscular endplates in the domain of the application of the toxin for approx. three months), which was suggested by a neurologist. With the early beginning of an efficient intensive neuro-rehabilitation therapy, the development of false organization of the CNS (spasticity) can be strongly reduced or abolished; there was therefore no indication for a botuline therapy.

After three months of CDT, the functions of the paretic arm strongly improved. Thus, the physiologic network states had been strengthened (the corresponding attractor wells had deepened), and the pathologic network states had been weakened (the corresponding attractor wells became more shallow). Only if the left arm can fully be moved in coordination with the other arm and the legs (coupling of the self-organizing network for the right arm and hand movement to the network organization of the other arm and the legs) will the spasticity reduce so much that non-rhythmic volitional movements of the left hand and fingers can fully be trained. By oscillator formation therapy (page 374 of Ref.63; entrainment of injured premotor spinal oscillators as for example upon jumping on springboard) and CDT, the movements of the paretic arm and leg could be improved.

After six months of CDT (seven months after the CNS injury), the paretic arm further improved, the overstretching of the left knee reduced, and the patient relearned running (Figure 51B). It seemed that an essential step forward was achieved in the last three months of therapy, when the patient used the special CDT device in the standing position. At the beginning, the patient had again big problems in the left hand to hold the lever. But six weeks later, the function of the left hand improved. Several months earlier, the patient had big problems holding the lever of an air-walker.

After six months of CDT, the patient needed no support anymore when jumping in anti-phase on springboard, the left knee was not overstretched anymore, the positioning of the left arm, hand, head (reduced spasticity of the musculus sternocleidomastoideus) and trunk were better—hemiparesis had partly subsided. With the improvement of the motor functions the mental functions also improved. His short-term memory became better. The short-term memory was still not normal because he often forgot what sentences he should translate, when translating two to three sentences from English to Slovak. He enjoyed going to school for a few hours per day. The improvement of the mental functions corresponded with the healthy looking face, as judged by the parents and the author.

The 14-year-old patient Andrej was cycling down a steep hill on a school journey, fell on his head and lost consciousness. He suffered many injuries including contusion of the lungs, obstructions of the airways due to facial and mandibular injuries and brain damage in the region of the right thalamus, the right fronto-parietal region, the frontal periventricular region and left basal ganglia. The coma and vigilant coma stage lasted 2.5 months, which is six weeks longer than in the nine-year-old patient, where the therapy was started earlier. CDT was started 2.5 months after the accident in the last phase of the vigilant coma state. When the author first saw the patient, the patient was aggressive, and the author even thought it was a hopeless situation to repair the injured CNS because the patient could hardly perform any coordinated movement, and the speech was strongly impaired. But the parents wanted help, so the author advised them to start CDT using the special CDT device and a treadmill. At the beginning, the patient had to be carried from the bed and the wheelchair to the training device. The progress in the beginning was slow. But with the improvement of motor functions and the vegetative functions (the strong sweating of the hands reduced), the mental functions also improved. The patient became strongly motivated to train hard to get healthy again. He started to like the author and could no longer remember that he had reacted aggressively at the beginning. It seems therefore that with the more integrated functioning of the neuronal networks, the CNS started to work on a mentally higher level. After four months of therapy, the patient could run better than walk. After slipping once during running, the patient became afraid, and his running was not as good anymore. Some virus infections also restricted his therapy. The therapy included also crawling, creeping, and rightening movements.

After six months of therapy, the functioning of the neuronal networks of the patient's CNS was very good in comparison to the following case in which the patient Mario did not obtain any CDT for the first seven months, but was not as good as in the 9-year-old Benjamin because of a later start of the CDT, an illness, a complication with a bone contusion, suboptimal therapy, and of the unnecessary administration of botuline toxin, which disturbs the movement-induced reafferent input.

The improvement of CNS functioning with therapy in this patient, Andrej, could also be judged upon by the reduction of the palmar flexion of the right hand. A good-bye photo after six months of therapy demonstrated that the neuronal network organization generating movement, posture and hand functions was influenced by somatic and mental functions. Even though the right hand and right arm functions were still strongly impaired, as could be seen during crawling and giving both hands on the good-bye photo, the right hand was not palmar flexed. In the nine-year-old Benjamin, the hand function was better during crawling, was not far from normal when giving the right hand, but was a bit spastic on the good-bye photo (Figure 86A of Ref.38). The potential landscape for motivation and emotions after CNS injury [71] was in this emotional situation beneficial for the hand function in the case of Andrej but not for Benjamin. Interestingly, one month later, it was possible to bring the right hand of Andrej in aligned position during handshaking (Figure 86D of Ref.38). The right hand could even be slightly flexed dorsally. Thus, the emotion with the present potential landscape for motivation and emotions [71] showed in this case a progress in the reorganization of the CNS, which was possible without emotion only one month later. For further details and pictures of patients, see Ref.38, case reports 11-13.

The 12-year-old Mario had a car accident when cycling and suffered many bone fractures and brain injuries when he fell under a car. Because of many complications of the bone fractures and the brain injuries (additional subarachnoidal bleeding) and several infections, CDT could be only started seven months after the accident, even though the author was urging all physicians to start the therapy as quickly as possible. The coma and vigilant coma state ended six months after the accident, i.e., much later than in the two other cases where CDT was started earlier. The motor and mental situation was very much delayed in comparison to those of the other two patients, Benjamin and Andrej, nine months after the CNS injury.

The patient had still to be carried from the wheelchair to the therapy device (Figure 51C,D) because of strong extensor spasticity (Figure 51E) and supination of the feet (Figure 51G), and communication was difficult, even though the patient was friendly and cooperative (when the author was present).

After one month of CDT, the motor and mental functions started to improve. It was hoped that the lost early treatment could partly be caught up. It was argued in the literature that there is a time window for early treatment [9]. Even if the suggestion of a time window for early treatment were correct, an early treatment is important for the following reason: if supervised reorganization is efficient and starts very early after the CNS injury, then the reorganization goes directly from the neuronal network organization after the injury towards the functioning with physiologic output. But if such early therapy is not provided, the injured CNS organizes itself in a pathologic way because of missing supervised relearning. Due to the pathologic functioning of the CNS, the mechanical properties of the body become pathologic; tendons shorten, muscles become spastic or atrophic, joints cannot be moved anymore, (pain) and the arms, legs, hands and feet often assume very un-physiologic positions so that it becomes difficult to use them for physiologic movements, and appropriate therapy can only be applied slowly and with many difficulties; then, the therapy is not as efficient anymore.

With these and other difficulties, the pathologically functioning CNS has to be reorganized in the direction of physiologic functioning. In conclusion, following CNS injury, the remnants of formerly physiologic functioning network parts are easier to organize for physiologic output than network parts that started to function pathologically. Even for the nine-year-old Benjamin, it would have been probably better to start CDT already ten days rather than four weeks after the injury. Even if tubes, wires and the necessary apparatus hinder intensive CDT, intensive therapy has to be started as early as possible, best in the intensive care unit. In severe brain injuries, bone and other reconstructions must be designed in a way to allow an early intensive neuro-rehabilitation. Bone fractures, for example, must be reconstructed invasively.

3.2.6.3. B. Stage of Repair in the Three Children One Year after the CNS Injury: Repair Also in the Case of Delayed CDT

The now ten-year-old boy Benjamin obtained a further optimal six month CDT at home. The functions in the left arm and left hand improved further. He could walk and run better after the further treatment. With the left "bad" hand, he could turn over pages of a journal. Benjamin was back to his old school in his old class (!) and was managing well. The strategy to first give priority to the relearning of the motor functions before going back to school and giving in this way priority to the relearning of the basic structures of the CNS with the improvement also of the short-term memory was right in this case. With the improvement of the motor functions, the higher mental functions also improved so that he could better manage at school. Benjamin is not far from normal again. Even though the running of the ten-year-old Benjamin was powerful again one year after the accident, his posture during running was not fully physiologic. On the other hand, when playing with his brother on the beach, his posture was quite normal, and when standing under a natural shower on holiday in Africa during holidays, his posture was symmetrical. It seems therefore that the patient was using the former pathologic posture now at least partly as a habit. The former attractor "pathologic posture with spasticity" of the injured CNS was still somehow existing in the repaired CNS. But the patient had the possibility to switch on volition into the attractor state "physiologic posture." The possibility to have a bi-stability of the two attractors, pathologic posture and physiologic posture, can also be found when tackling idiopathic scoliosis in the case of slight disturbances of CNS functioning. When repairing the urinary bladder upon CDT, two attractors, synergy and dyssynergy of the bladder, do appear transiently (Figure 5E,F) (see also Chapter V). Fifteen months after the accident, Benjamin's CNS had still problems in organizing the trot gait coordination between arms and legs during walking and during exercising on the special CDT device. When falling, he was still not using protection reaction automatisms. The measured coordination between arms and legs was still not physiologic.

The now 15-year-old Andrej received only further conservative treatment for one month. His motor functions did not improve further during this month. The reappearing slight palmar flexion of the right hand may even indicate a loss of regained motor functions. A loss of relearned functions was also found during the therapy of a caudal and cervical SCI (Figure 38G) and cerebellar injury [55] when interrupting the therapy. With restarting CDT, the motor functions in SCI and cerebellar injury started to improve again.

Most interesting was the case of the now 13-year-old Mario. Six to eight months following the brain injury (with no or only little conservative treatment), his motor and higher mental functions were in a seemingly hopeless condition. Because of only little or false organization of the neuronal networks, including different kinds of spasticity, false posture and positioning of leg and feet (Figure 51E,G), he had to be carried by his mother (Figure 51C). Still, an intensive CDT was started with him, including exercising on the special CDT in recumbent (Figure 51D) and sitting position, and treadmill walking. Even though the patient was often crying, especially during treadmill walking, mainly because he was afraid (as he explained later on, when his mental functions had further improved), the neuropediatrician, the physiotherapist, the mother and the author came to the same conclusion, namely that Mario had to exercise intensively, or he had nearly no chance for further essential improvement in the future.

One year after the brain injury (after four months of rather intensive therapy), an essential progress occurred. The patient became able to walk a bit when supported. His hand functions improved, so that he could turn over the leaves of a journal and was able to write again. He was now motivated to exercise, and he had the feeling that his functional state of the CNS was progressing (Figure 86Kh of Ref.38). His higher mental functions also improved substantially. Eight months after the accident, the patient had problems recognizing the author. Twelve months after the accident, he even spoke some words in German to him (his native language is Hungarian). Mario was back to his old school, but two classes lower. The pupils and the teachers were giving him much help so that he could manage. Further progress depends on the intensity of the continuous CDT. When Mario was separated from his mother for a longer period of time, he was so stressed that he had problems speaking. If Mario is therefore treated further at the clinic, then his mother will have to be with him. Seventeen months after the brain injury, he had still problems in organizing trot gait coordination between arms and legs. For pictures and further details, see Ref. 38 (cases 11-13).

The case report of the 13-year-old Mario shows that even when the early CDT is missed, there is still hope that CNS functions can substantially be improved. However, the therapy is more difficult to perform because of the pathologic organizations that have taken place in the CNS in the meantime. The pathologic neuronal network states, which express themselves by spasticity, false positioning of body parts and maybe epileptic seizures, make the reorganization of the CNS more difficult and cause partly pathologic movement-induced reafferent input. A further important point is also that continuous, intensive and early adequate therapy and not only differences in the brain injury are crucial for the outcome of similar severe brain injuries. If the therapy had not been continued eight months after the accident, then the patient Mario would not have made such substantial progress.

3.2.6.4. C. Stage of Repair in the Three Children Six Years after the Injury

Andrej obtained only conventional therapy and did not progress further. Mario was trained at home by the mother, and his motor functions improved. When the staff of his school changed, he did not get further help from the school. Benjamin went back to his original school class as a disabled pupil and was accepted as a normal pupil much later only. He made a substantial step forward in repair upon the administration of one month of optimal CDT in a therapy place with 30 to 40 hours treatment per week. Because of the importance for CNS repair, the case of Benjamin will be reported in more detail.

Benjamin further improved at home, although only little. He was not able to crawl because of the spasticity of his left hand and arm. When the measuring of low-load coordination dynamics (CD) was introduced, it became possible to follow up the evolution of his CNS functioning. With the introduction of high-load exercising on the special CDT device to activate the CNS more integratively (systemically), Benjamin improved further and became able to crawl. After a certain time, his motor functions did not improve any further. The question arose whether Schalow CDT had come to its limits with respect to the potential improvement of CNS functioning or whether the therapy had to be updated and performed more intensively. It will be shown below that the CNS repair was by far not at its limits and that measurements of low-load (Figure 52) and high-load CD (Figure 53) provided further insight into the reorganization changes of the CNS in this patient.

1. Temporary Impairment of Low-Load CD and Exercising Rates

When plotting the values of low-load CD measured over the five years of home therapy (Figure 52Ba), it can clearly be seen that there was a pronounced worsening of CD (higher values). Simultaneously, the exercising rate (frequency) showed a transient strong decrease (Figure 52Aa).

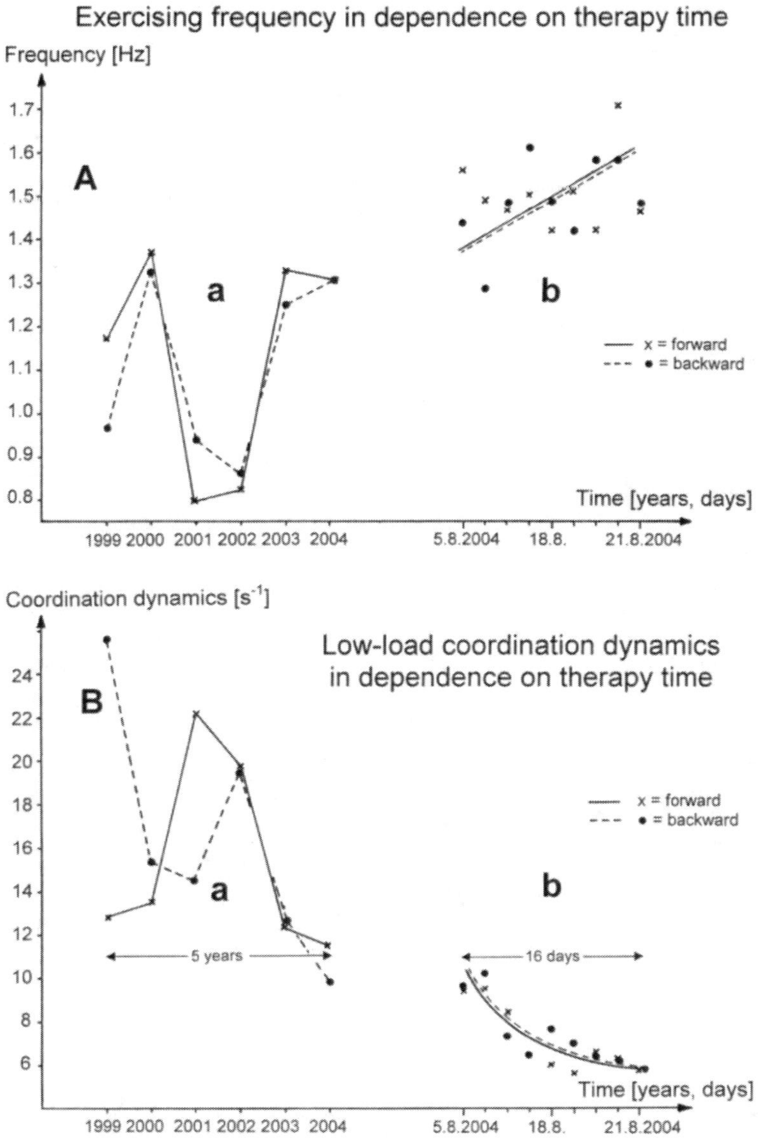

Figure 52. Exercising rate (frequency) and low-load coordination dynamics values with ongoing therapy; patient after severe traumatic brain injury. Solid line = exercising in the forward direction; dashed line = exercising in the backward direction. In "Aa" and "Ba," the rate and the coordination dynamics changes are shown observed within the years 1999-2004 (training at the patient's home). In "Ab" and "Bb," the changes of the rate and coordination dynamics values are shown as observed during 16 days of treatment under professional supervision. Note that the coordination dynamics for exercising in the forward direction (solid line) improved in the 16 days of treatment under professional supervision more than during the five years of home therapy.

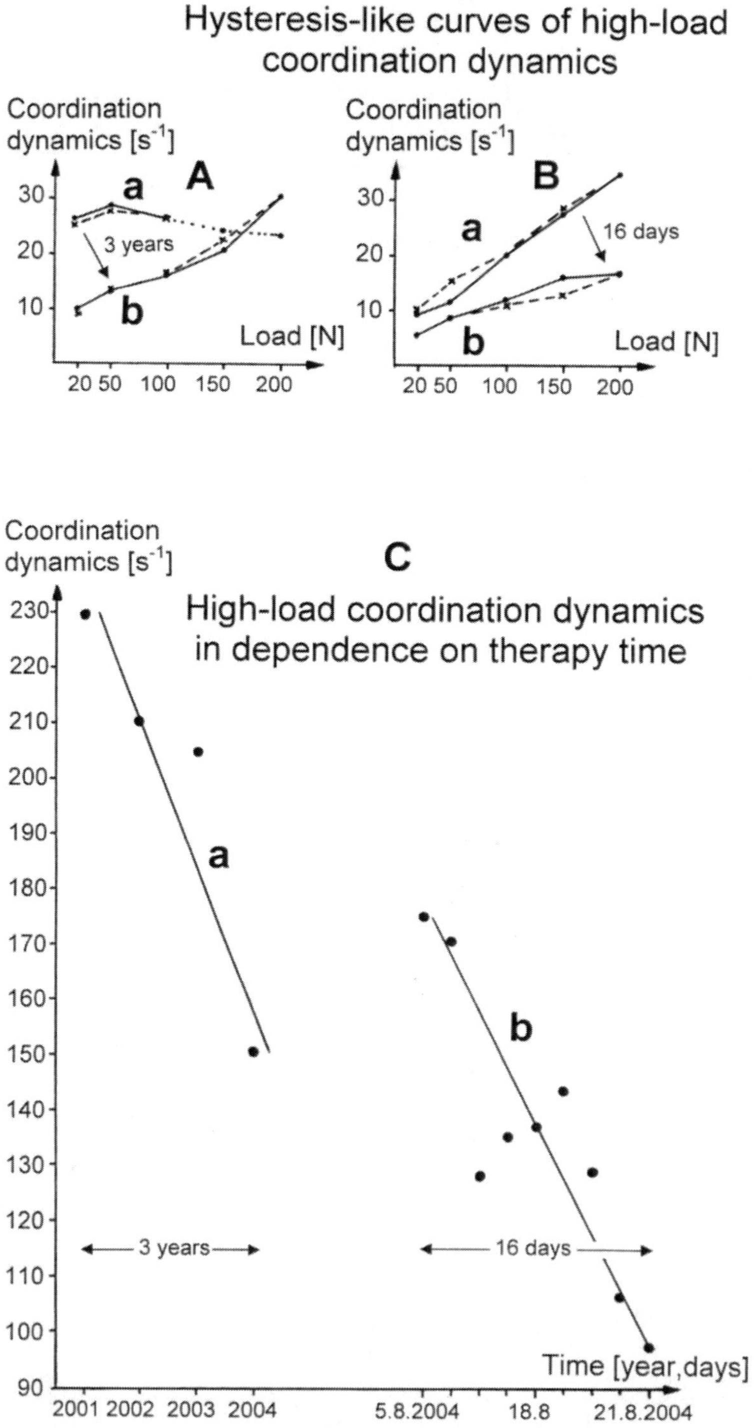

Figure 53. A,B. Improvements (from "a" to "b") of hysteresis-like high-load coordination dynamics values during three years of home training (A) and during 16 days of optimal therapy at a coordination dynamics therapy place (B); patient after severe brain injury. C. High-load coordination dynamics values for increasing and decreasing load ($\sum = (\Delta_{20N} + \Delta_{50N} + \Delta_{100N} + \Delta_{150N} + \Delta_{200N} + \Delta_{150N} + \Delta_{100N} + \Delta_{50N} + \Delta_{20N})s^{-1}$) measured during three years of home training (a) and during 16 days of institutional therapy (b). Note that the high-load coordination dynamics improved during the 16 days more than during the three years before.

The pronounced worsening of the coordination of arm and leg movements is explained by substantial changes of CNS self-organization, probably induced by starting high-load exercising. It was predicted, based on the theory of CD, that CNS organization may markedly worsen when the CNS is changing (due to the treatment) from one organization to another (see Method) (Figure 2 of Ref.64). Such organizational change of the CNS seems to have happened in Benjamin's CNS during the home therapy after the introduction of the high-load exercising. In spite of the principal organizational changes in the CNS, the improvement of low-load coordination dynamics from 12.9 to 11.5 (11%) was comparably small for the duration of the treatment (several years). The low-load CD values (arrhythmicity of exercising) could not be lowered, despite exercising on the special device for at least 12 hours per week. Even the learning effect of these special movements seemed to be very small or non-existent.

2. Improvement of High-Load CD upon Home Therapy

During home training, the high-load CD for increasing and decreasing the load from 20N to 200N improved with 12 hours therapy per week. The hysteresis-like curves for increasing and decreasing load improved from a pathologic (Figure 53Aa) to a physiologic hysteresis curve (Figure 53Ab). The values of the high-load CD improved over the three years of home treatment from 229 to 151 (34%) (Figure 53Ca). This is a small improvement taken to the duration of the therapy (three years).

3. Improvement of the Low-Load CD upon Therapy under Professional Supervision

The improvement achieved in low-load CD under optimal conditions is shown in Figure 52. During the 16 days of quantified therapy, the rate of exercising (frequency) was high and increased slightly (Figure 52Ab). The value of the CD for exercising in the forward direction decreased (improved) from 9.3 to 5.8 (38%) (Figure 52Bb). Normal values measured upon first exercise (healthy persons) are between 5 and $6s^{-1}$ [42].

4. Improvement of High-Load CD upon Therapy under Professional Supervision

High-load CD improved strongly during this short time interval of therapy received at the special therapy place, even though during this time, Benjamin exercised much less on the special CDT device than at home. The hysteresis-like high-load coordination dynamics curves (Figure 53Ba,b) went into the direction of those observed for track and field athletes [44] over the 16 days. The high-load CD improved from 175.3 to 97.3 (44%) (Figure 53Cb).

5. Comparison of Home Therapy and Therapy under Professional Supervision

The values of the low-load CD improved during the five years of home treatment (1,800 days) by 11%, whereas an improvement of 38% was achieved during the 16 days at the therapy place. The values of the high-load coordination dynamics improved during the three years (1,100 days) by 34% and by 44% during the 16 days of institutional therapy. The rate of relearning for low-load coordination dynamics was 0.61%/100days (11%/1,800days = 0.61%/100days) for home therapy and 238% /100days (38%/16days = 237.5%/100days) for institutional therapy place. The efficiency difference in respect of low-load exercising was by a factor of 390 (238%/0.61% = 390). The rate of learning for high-load coordination dynamics was 3.1%/100days (34%/1,100days = 3.09%/100days) for home therapy and 275%/100days (44%/16days = 275%/100days) for institutional therapy. The efficiency difference for high-load exercising between home and institutional therapy was thus by a factor of 89 (275%/3.1%= 89). The efficiency difference in respect of CNS repair, as quantified by low-load (390) and high-load CD [89], between home and institutional therapy thus was approximately by a factor of 100.

6. Improvement of Movements upon Treatment at a Professional Therapy Place

Not only did CD improve strongly during professional therapy, but walking also got much better, too. Upon walking, Benjamin used the left leg much more actively instead of lifting the hip and swinging the leg. The treatment goal could be approached: at a first glance, you could not see that the patient suffered a severe brain injury.

7. Comparison of the Improvement of CNS Functioning After Injury with Physiologic Changes during Individual Development

It was demonstrated that the CNS functioning, as quantified by the low-load CD values, continuously improves during physiologic development within the age range between 5 and 19 years (Figure 23A,B). During the treatment for repair, the values of the coordination dynamics were shown in this report to temporarily worsen, probably as the CNS was changing its self-organization from pathologic to a more physiologic one (Figure 52Ba). Such temporary worsening of CNS self-organization was predicted by the theory of CD. During the three weeks of intensive therapy, the values of the CD improved in this patient continuously, in similarity with physiologic development.

8. Oscillatory Firing FF-Type Muscle Fibres

In this patient after severe brain injury, also the basal ganglia on the left side were probably partly injured. Such partial damage of the basal ganglia may be responsible for the difficulty of functional repair. Rhythmic firing of FF-type muscle fibres was sometimes observed in the left tibialis anterior muscle (Figure 55C), suggesting impaired coordination (including synchronization) between oscillatory firing of motoneurons.

9. Treatment Judgment According to Surface Electromyography (sEMG)

In this case, the optimal Schalow CDT included exercising on the special CDT device to repair the impaired phase and frequency coordination by means of imposing very coordinated arm, leg, and trunk movements in the lying, sitting and standing position by the device. The inborn automatisms creeping, crawling, up-righting, walking and running were exercised indoor or on beach sand outside for optimal adequate skin input. Rhythmic dynamic stereotyped movements, like jumping and swinging on the springboard, were exercised to entrain premotor spinal oscillators as an entrainment of CNS neuronal networks on the ensemble level.

Old learned integrative movements (learned automatisms), like climbing staircases, were also trained. A typical exercising situation at the therapy place is shown in Figure 54. All movements were performed under the aspect of symmetry. To optimally entrain the severely damaged neuronal networks integratively, those integrative movements were exercised most often, during which the injured networks were activated most physiologically, which was quantified by surface electromyography (sEMG). Surface EMG was performed for different integrative movements to see under what conditions critical muscles showed best motor program performance (Figure 55).

Because of an excessive palmar flexion of the left hand, the left brachioradialis and extensor carpi radialis muscles were chosen to judge upon CNS functioning with respect to the arm and hand movements. Because of an excessive plantar flexion of the left foot (and an excessive inward rotation), the left triceps surae and tibialis anterior muscles were selected to record from to judge upon CNS functioning with respect to leg movements. The task was now to get the best sEMG motor programs for all the four muscles simultaneously, to see which integrative rhythmic, dynamic, stereotyped movements entrain the injured CNS optimally in respect of repair.

It is important to improve CNS functioning integratively (systemically). The human CNS is an open system, and if a small subnetwork is being entrained only, the pathologic organization escapes the entrainment by shifting to another CNS neuronal network site, as could nicely be seen in this patient. When the patient was jumping on the springboard and his left hand was in too much palmar flexion, he was instructed to concentrate on the left hand to keep the hand straighter. Then, however, the left foot went into plantar flexion and inward rotation.

Figure 54. Some coordinated movements used at the therapy place. Three patients can be seen who had suffered spinal cord injuries and one patient who had suffered a severe brain injury (on the treadmill, the patient of this case report). The exercising is assisted by two physiotherapists. The patient on the left (incomplete cervical spinal cord injury) is jumping on the springboard to entrain neuronal networks on the assembly level by performing rhythmic, dynamic, stereotyped movements. The patient on the treadmill is exercising the automatism walking. Support is provided by the therapist to arm movements to bring them in coordination with leg movements. The patient in the front (formally complete spinal cord injury Th5/6) is exercising the automatism crawling. Support is provided by a therapist to leg movements. The patient on the right (formally complete spinal cord injury sub Th10/12) is exercising highly coordinated arm and leg movements on the special coordination dynamics therapy device in the lying position; the coordination between arm and leg movements is imposed by the device.

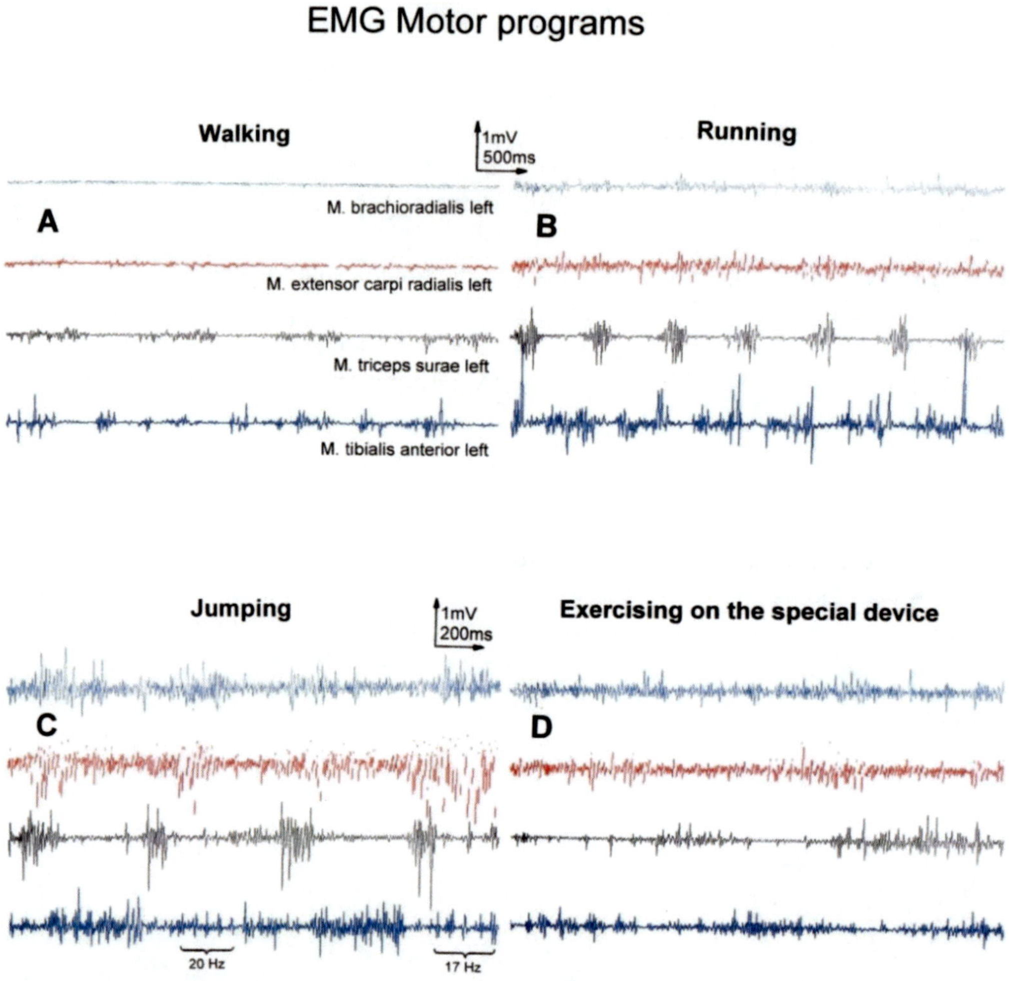

Figure 55. Motor programs registered by surface electromyography (sEMG) during walking (A), running (B), jumping (C), and exercising on the special coordination dynamics therapy and recording device at the load of 100N (D); a patient with formally very severe brain injury. Note that the critical muscles brachioradialis left and extensor carpi radialis left are not activated during walking (A). On the extensor carpi radialis muscle trace, there may be some artefacts on top of the motor program. On the tibialis anterior trace, rhythmic firing of 20 and 17Hz can be seen, which most likely stems from FF-type motor units innervated by α_1-motoneurons.

The patient was then instructed to concentrate on getting the heel down during jumping (to perform more dorsal flexion). He concentrated on the left foot and could improve the position of the foot; but then the left hand went again into pathologic palmar flexion. He was now instructed to concentrate simultaneously on the left hand and the left foot; but then he argued that he was unable to concentrate simultaneously on the arm and the leg. Patients with severe CNS injuries have especially problems concentrating on two things at the same time (as young children). It was therefore important to identify complex movements, which would induce good left arm and left foot positioning, as quantified visually and by good sEMG motor patterns in all the four muscles, mainly not requiring too much patient's concentration.

Figure 55 shows the motor programs of the left brachioradialis, extensor carpi radialis, triceps surae and tibialis anterior muscles during walking (A), running on treadmill (B), jumping on the springboard (C), and exercising on the special coordination dynamics therapy device (D). It can be seen from Figure 55 that the motor programs of the four muscles for running and jumping on the springboard (B,C) were better than those for walking (A) and exercising on the special device (D). In this patient, running and jumping on the springboard thus seemed to be more efficient in respect of CNS repair than walking and exercising on the special device. Improvements of CNS functioning were quantified clinically, mainly by improvements of performance of movements, because Benjamin could already walk and run fast after one year of treatment. Only the performance of the left arm and left leg were poor in nearly all movements. The writing performance was poor. Benjamin was, in general, too slow.

10. Need for Many Different Movements to Reorganize the Injured CNS

During home therapy, Benjamin did not progress anymore, although walking and running every day and exercising nearly every day on the special device for one up to two hours (approx. 5,000 coordinated movements per day). When the patient obtained therapy at the therapy place where he exercised many different movements, his coordination dynamics improved strongly, even though he exercised less on the special device than at home. There are two consequences implied from the strong improvement of the CNS functioning. First, after severe CNS injuries, the learning effect of the special movements exercised on the special device is only small. The special coordination dynamics therapy and measuring device is therefore suitable for the diagnostics of CNS functioning, as improvements of coordinated arm and leg movements are mainly the result of a general improvement of CNS functioning and only little of the learning of these special movements. Secondly, single movements, like exercising on the special device or walking on treadmill will not reorganize the severely injured CNS substantially: Benjamin exercised at home on the special device and was walking and running every day, and he still did not improve further. The therapy has therefore to include many different movements, which have to be adapted to the special CNS injury of the patient with his different pathological CNS organization. Specific complexity of movements and deep knowledge of human CNS organization and repair are needed to achieve CNS repair rather than “miraculous instruments, movements, stem cell therapy” or manipulations, not thoughtful electrostimulation or pure unqualified pharmacotherapy.

11. Need for Optimal Institutional Treatment

The improvement of CNS organization by a factor of approximately 100 shows further that an optimal therapy at a therapy place is needed for efficient repair of the injured or malfunctioning CNS. Home treatment is helpful but cannot substitute for institutional treatment under professional supervision. The difference in the efficiency of the treatment at home and at a therapy place, the latter taking advantage of most recent data on CNS repair from human neurophysiologic research, was approximately by a factor of 100. After these three weeks of treatment at the special therapy place, the author claimed: “Benjamin, if you come here to be treated for one year, in the end you will not differ very much from the other boys at school anymore. You also can exercise at home under my partly supervision, but you may need a 100 years to achieve the same.” Benjamin’s father, a friend of the author, asked also other friends (neurologists and psychiatrists) for their opinions and did not send Benjamin for this one-year-treatment.

3.2.6.5. D. Stage of Repair of the Now Three Young Men 11 Years after the CNS Injury

1. No Further Repair upon Conventional Therapy, Some Progress upon Supervised Therapy at Home, and Back to Normal Life upon CDT

None of the three patients obtained optimal therapy during the 11 years following the injury.

Figure 56. Clinical evaluation of three patients, 21- (Benjamin), 23- (Mario), and 25-years-old (Andrej), who suffered a severe brain injury 11 years ago and obtained conventional therapy for ten years (Andrej), home therapy (Mario), and CDT (Benjamin). A,B. Patient's during crawling. Note the spasticity in both hands of Mario (middle) and the right hand of Andrej (right). C. Patients during walking. Mario (middle) has problems with the spasticity in the hands. D. Mario upon coordination dynamics assessment. Note the strong impression in his face. E. Mario shows proudly the reduction of the hand spasticity, due to the exercising on the special device. F. Eating of the patients after assessment and training. Andrej and Mario are living at home with the mother as unable persons, and Benjamin is having a normal life and is studying languages.

Andrej obtained conventional therapy (against the advice of the author) and did not progress further. He is now 25 years old and living at home with the mother and obtains four weeks of psychiatry treatment every year. He is depressive because he was refused for entering a normal school and because there was no progress with his motor functions. Obviously, the conventional neuro-rehabilitation system is 20 years out-of-date and stole his future.

Mario, now 23 years old, trained with the mother at home, sometimes advised by the author. The help from the school stopped when the staff changed. His motor functions improved further; he relearned how to walk (Figure 56C). His higher mental functions improved but are still far away from normal. He is living as disabled person with his mother as his caretaker. The mother reported that he is sometimes aggressive but not when the author is present (see below).

Benjamin, now 21 years old, obtained CDT, which was still far away from an optimal treatment. He progressed further in motor and higher mental functions. Six years after the accident, he obtained optimal CDT in a therapy place for three weeks and made a big step forward in repairing his CNS (see above). He finished successfully his school (grammar school) and is studying now languages at a university. He leads a normal life and helps to bring better treatment to at least some patients with CNS injury. Upon studying at the university, he neglected the necessary treatment for keeping the level and got a bit worse as he felt it. He pushed the author to help him. To stimulate all three patients to improve their training, the now three young men were brought together to compare them and to motivate them for further training.

2. Clinical Assessment

Figure 56 shows the clinical state of repair. Benjamin, Mario, and Andrej (from left to right) started to crawl (Figure 56A). Benjamin is crawling quite physiologic and leads (Figure 56B). Mario is second, because he has problems with the spasticity in the fingers and hands on the left side. Andrej can hardly crawl at all because the right hand is in a pathologic position (plantar flexed) (Figure 56B). Afterwards, all three patients walked (Figure 56C). Benjamin (left) is walking quite physiologic; only the left arm and hand are not fully in the right position.

Mario (middle) can walk with some balance problems and is fighting with the spasticity of the left hand. Andrej (right) is walking quite physiologic; ten years ago, he could even run. Andrej was astonished that Mario learned to walk in the past ten years.

To compare the CD, all three patients were assessed. Note the strong impression in Mario's face when he was exercising (Figure 56D). When Andrej exercised (Figure 56E) after him, Mario was proudly showing his left hand with not flexed (no spastic) fingers. During crawling (Figure 56B) and during walking (Figure 56C), the left hand was strongly spastic. The flexor spasticity had strongly reduced in the short-term memory upon exercising on the special CDT device. His own special CDT device worked inaccurately, so that his CNS could not learn anymore from the device the coordinated firing of neurons and reduce the spasticity first in the short- and then also in the long-term memory. His special CDT device reduced to a simple muscle training device and could not reduce spasticity anymore. After exercising, the patients had lunch together (with the author, at that moment, the picture was taken) (Figure 56F). Andrej was eating with the left hand, because the right hand was not working properly. To reduce the spasticity of the right hand, he was holding something in his right hand to increase the a+zfferent input from the right hand to reduce spasticity. Mario (middle) could eat quite well with both hands because of the reduced spasticity upon exercising on the special device. Some weight was round his left wrist (doubtful reason) to improve hand and arm performance. Benjamin (right) was eating with both hands, but the left hand and arm were a bit spastic. The muscles activated by spasticity can be recognized in Figure 56F.

3. Comparison of the Coordination Dynamics (CD)

The low-load CD (20N) values improved during approximately ten years in Andrej from 21.3 to 13.4 (37% improvement), in Mario from 23.3 to 21.4 (8%), and in Benjamin from 12.9 to 8.2 (36%). Normal values are between 5 and 6 [16]. The improvement (the lowering) of the CD does not show the whole improvement because at the beginning of therapy, this assessment was not available, and the biggest decrease (improvement) of the CD values is normally at the beginning of therapy. Benjamin had the best low-load CD. But as can be seen from Figure 52Ba, they depend also on the repair state of the neuronal networks. During structural repair periods, the self-organization of the CNS gets transiently worse (see Figure 50). Andrej's CD values improved but did not reach Benjamin's values. This can be understood. Andrej obtained poor treatment. With proper treatment, Andrej could probably have been clinically as good as Benjamin. Mario's CD values for low-load were quite poor in accordance with his clinical state. But his CNS was most likely more severely injured. Interesting is the 20N-recording from Benjamin (made in 1999) in Figure 57D.

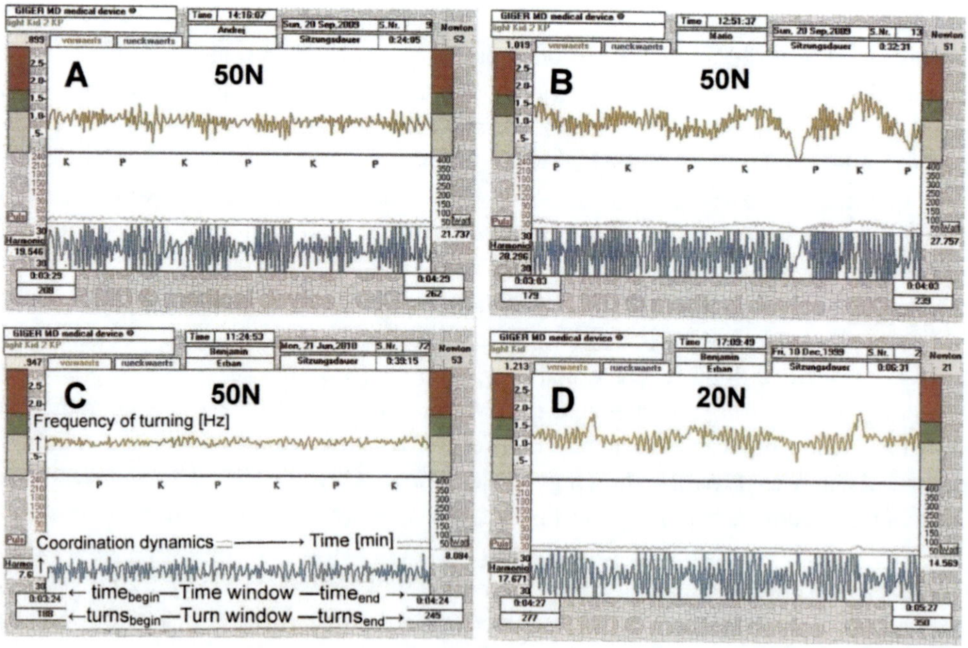

Figure 57. A,B,D. Comparison of low-load coordination dynamics measurements of three patients to whom different treatments were administered for approximately ten years. Upper traces = exercising frequency; lower traces = arrythmicity of exercising (CD); all recordings are for exercising in the forward direction. The CDs in A and B are worse than in C in which the patient was treated by CDT. D. Transiently fast exercising is shown in an 11-year-old patient from the beginning of treatment; same patient as in "C."

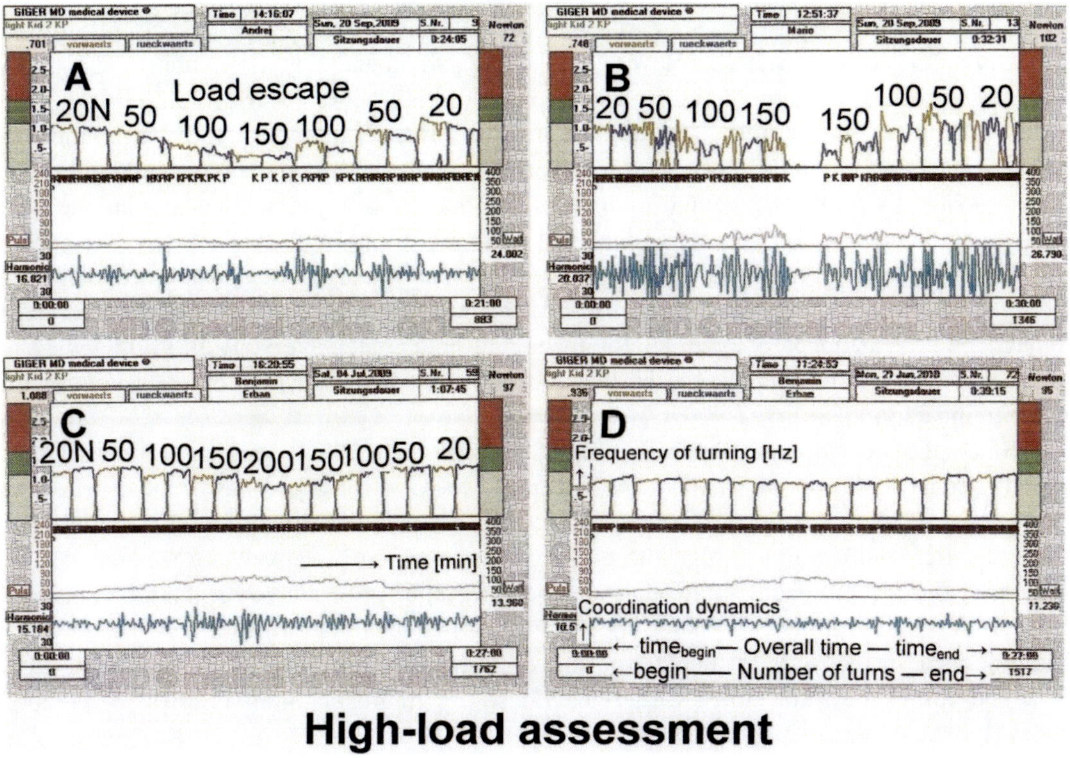

Figure 58. A, B, C. Comparison of high-load coordination dynamics measurements of three patients who suffered severe traumatic brain injuries 11 years ago and to whom different treatments were administered for approximately ten years. Upper traces = frequency of exercising in the forward and backward direction; time periods between 21 and 30min. Note the reduction of the frequency in "A" and "B" (load escape) for loads mainly between 100 and 200N. In "A" and "B" the patients could not manage with the 200N load. Lower traces = arrythmicity of exercising (CD); not the nearly stepwise increase and decrease of the work in "Watt" in "C" and "D," typical for the performance of athletes. The patient in "C," who obtained CDT, shows the best high-load CD. D. After a few months of intensive training, the high-load CD further improved in the patient from "C" (improved stepwise function).

It shows two transient high-frequency turning episodes, which can be observed in healthy pupils in the age range between 5 and 11 years (Figure 27C,D). This recording indicates therefore that Benjamin's CNS improved not only from the movement-based learning therapy, but also the developing CNS helped to repair by trial and error eliminating processes present during normal development.

The 50N-recordings of the three now young men are compared in Figure 57A-C. Again, Benjamin had the best 50N value (8.1), Andrej had the second best (21.7) and Mario the worst one (27.8). By comparing the overall mean high-load CD recordings of Benjamin, Mario, and Andrej in Figure 58A-C, it can be seen that those of Benjamin were the best ones (14.0s^{-1}), followed by those of Andrej (24s^{-1}) and Mario (30s^{-1}); Mario and Andrej did not manage to exercise at the load of 200N. When Benjamin exercised again at the limit under supervision, his mean high-load CD value improved from 14.0 (Figure 58C) to 11.2 (Figure 58D); the Watt-curve (middle trace) shows a nice stepwise function, typical for those of top athletes (Figure 72B).

4. Conclusion

Only CDT could repair the severely injured CNS in the boy (Benjamin). The partly supervised mother at home could repair the injured CNS better (Mario) than conventional neuro-rehabilitation (Andrej). The author predicted these outcomes at the beginning of therapy. In these cases, actually, the neuro-paediatrician of the conventional rehabilitation clinic wanted the progress, wanted to apply CDT, but she was pressurised to continue with out-of-date treatment. Benjamin was included in an EU program to make video films about children with severe brain injury to show how they can manage in everyday life. The film about him won the competition. But to include treatment into the film was not allowed, even though proper treatment is essential for managing better later everyday life. Obviously to think that severe brain injuries can be repaired by proper treatment was out of scope for the EU administration.

Since the developing (and repairing) CNS has difficulties organizing two patterns simultaneously (like playing ball in the street and looking for a coming car, the necessary complexity is not yet achieved), society has to take better care for their children so that they do not suffer injuries and so that the children get adequate therapy if an accident occurs. Courts in Europe still decide that car insurance companies need not pay if a child is hit by a car and the child is at fault for the accident (child-ball-street situation). But how can a normal child be made responsible for being a normal child, whose CNS cannot organize sufficiently two patterns at the same time? Children with severe traumatic brain injuries (or cerebral palsy) will be on the losing side for the next 50 years.

3.2.7. Cerebellum Repair Achieved by Ten-Year CDT

Even though in children movement-based learning is more efficient than in aging, substantial progress can also be achieved in elderly people who suffered CNS injury. Upon ten-year non-optimal therapy, in a cooperative patient with a cerebellum and cerebrum injury, CNS functioning could substantially be improved in its functioning. This CNS improvement can nicely be seen in the face of the patient (Figure 61). The diagnostics performed with the patient gives further insight into the repair mechanism of CDT. As the treatment was administered with increasing and decreasing intensity over ten years, information could also be obtained on the long-term therapy of this condition. The outcome of the long-term treatment will be compared with an optimal therapy approach at an up-to-date therapy place. This optimal therapy in cerebellar injury will further be compared with that in cerebrum injury.

3.2.7.1. Summary

Coordination dynamics therapy of changing intensity was administered at home for ten years (at the beginning at low intensity) to a 58-year-old patient who had suffered a severe injury of the cerebellum (and of the cerebrum). Motor functions improved markedly during the first year of rather intensive therapy (by 42%), but there was but little improvement in the five following years (by 27%). An intensive three-week CDT was administered after four years of quantified therapy at an up-to-date therapy place under optimal conditions. The motor functions improved strongly (by 29% in 13 days). Therapy under professional conditions was shown to be much more efficient (by the factor of 150 as quantified by low-load coordination dynamics values) than that at home. A comparable patient with a brain (cerebral) injury improved slightly more under optimal conditions (by the factor of 390, see above). In the patient with the injury of the cerebellum, the largest improvement concerned speech and higher mental functions.

The strong improvement during the three weeks of optimal therapy may have partly been due to the enhanced training of CNS symmetries. It thus seems that not only phase and frequency coordination between neuron firings has to be improved by therapy but also symmetries in CNS organization. The CNS functioning caused by severe injury of the cerebellum will partly be quantified below by surface electromyography (electrophysiology) and coordination dynamics measurements (kinesiology). When exercising on a special coordination dynamics therapy (CDT) and recording device with different combinations of arms and legs in the forward and backward direction, right-left, rostral-caudal, and forward-backward, symmetry impairments of CNS movement organization are quantified.

3.2.7.2. Results

1. Values of Coordination Dynamics (CD) Measured between 1999 and 2006

In the patient who had suffered a severe injury of the cerebellum, improvements in the CNS functioning were quantified by the CD values for low-load and high-load (between 1999 and 2006, and between 2002 and 2006, respectively) (Figure 59). The administered CDT was not regular and was performed at home under the supervision of the author.

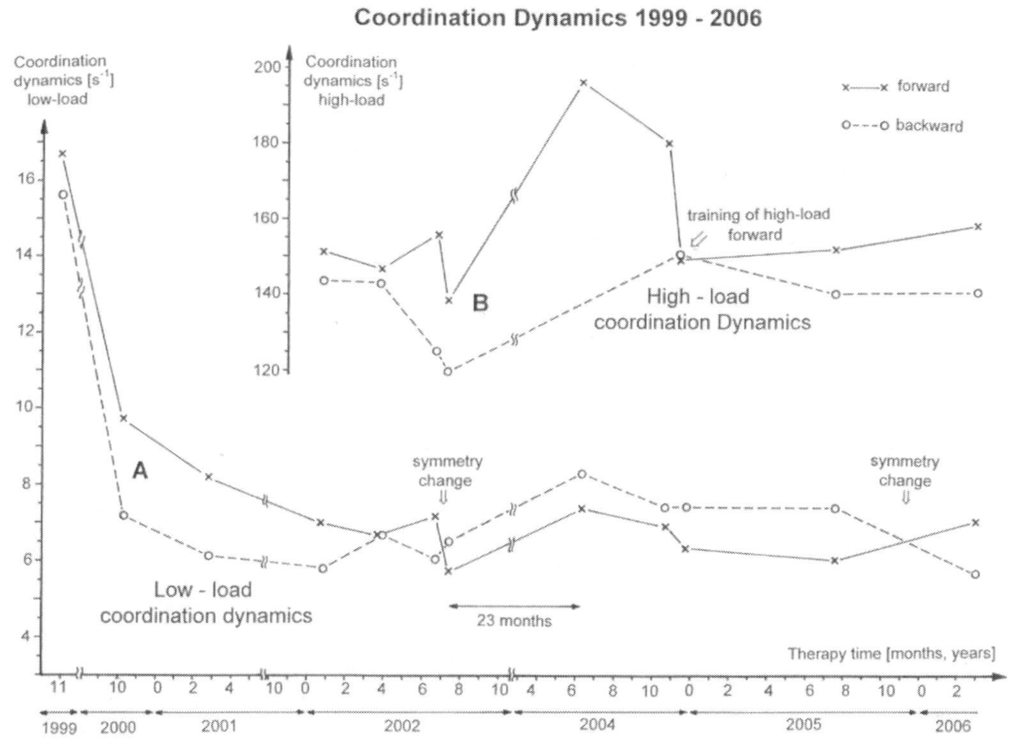

Figure 59. Coordination dynamics values in dependence on therapy time. Low-load coordination dynamics from 1999 to 2006, and high-load coordination dynamics from 2002 to 2006. Crosses and line = forward exercising, open circles and dashed lines = backward exercising.

On occasions, the intensity of the therapy was high for up to one month, but sometimes there was nearly no therapy at all. The patient, however, did exercise nearly regularly on the special CDT device.

As can be seen from Figure 59, at the beginning of the therapy during 1999 to 2001, there was a strong improvement (reduction) in the values of the low-load CD for exercising in the forward and backward direction followed by only little changes over 2002- 2006.

At the beginning of the therapy (until 2002), the values of CD for exercising in backward direction were better than those measured for exercising in forward direction. After high-load exercising on the special CDT device was started and the intensity of the therapy was increased, the symmetry of the CNS organization changed (Figure 59). Better values could be obtained for exercising in the forward than in the backward direction. From July 2002 onwards, the patient exercised only slightly for the next 23 months. The values for low-load and high-load CD worsened (increased) again.

The organization of the patient's CNS worsened with respect to movements. With an optimal therapy at an up-to-date therapy place for three weeks at the end of 2004, movements, low-load and high-load (Figure 60B) CD improved. The progress faded again as soon as the intensity of the therapy diminished. A further symmetry change occurred for the low-load CD between 2005 and 2006 (Figure 59). Between 1999 and 2006, the values of low-load CD improved from 16.6 to 7.0s^{-2} (58%). In the first year, the value improved from 16.6 to 9.6 (42%), and in the following years from 9.6 to 7.0 (27%). The value of high-load coordination dynamics did not improve (Figure 59B), probably because of the interruption of the therapy for 23 months.

In conclusion, the values of CD could not be further reduced substantially from 2002 onwards, probably because the therapy was not sufficiently intensive and continuous. CD improved when the patient received more therapy and worsened again when the patient exercised only little or not at all. With this patient, a principle problem was encountered. Because of a severe problem with keeping balance, the improved movements could only little be used and trained in everyday life. Since motor functions like walking improved only little, there is indication that in this patient after severe injury, the values of CD are a quite good reflection of the quality of the CNS organization with respect to movements.

2. Intensive Therapy under Optimal Conditions

When there was no longer substantial progress, the question arose (as above in severe cerebral injury (52)), whether the patient had reached his limits of improving CNS functioning with respect to CDT or not? To clarify this important therapy issue, a three-week therapy under optimal conditions was administered to the patient at an up-to-date therapy place. The results of this intensive three-week therapy are shown in Figure 60. During the 13 days of optimal therapy, the values of high-load CD improved by 26% (Figure 60B). The value of low-load coordination dynamics (not shown) improved by 29%. When the patient was taking one tablet Neurobion® (containing mainly 100mg Vitamin B_1 + 200mg Vitamin B_2 + 20µg Vitamin B_{12}), the movements and the CD worsened: the value of high-load CD worsened by 8%.

Before this three-week-therapy period, the patient could only crawl a bit forward in trot gait coordination. He could not crawl in pace gait coordination and could not crawl backwards. To enhance the efficiency of the treatment by inducing learning transfer from movements to their symmetry movement counterpart, crawling in pace and trot gait was trained now in forward and backward directions. Figure 60A illustrates the improvements in crawling. The mean crawling times in trot and pace gait coordination for forward and backward movements improved by 48% in 14 days. Administration of Neurobion was associated with a worsening of crawling times by 28%. In conclusion, the functioning of the CNS in this cooperative patient might still be further improved if CDT is administered intensively, over longer periods of time (for several months), and at an up-to-date therapy place.

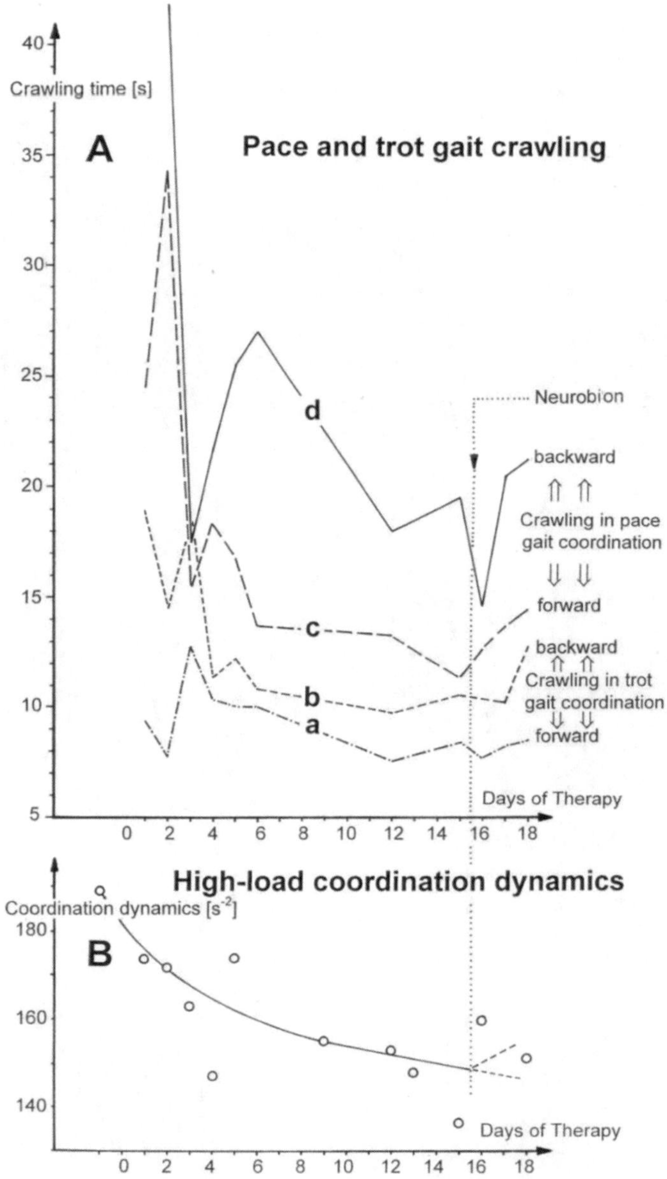

Figure 60. A. Forward and backward crawling in pace and trot gait in dependence on therapy days. Note that with Neuropiron administration, the different kinds of crawling and the high-load coordination dynamics got worse. B. High-load coordination dynamics in dependence on therapy days.

3. Comparison of Home Therapy and Therapy under Professional Supervision

The value of low-load coordination dynamics improved by 58% between 1999 and 2006. The improvement seen between 2000 and 2006, was 27%. During the 13 days of optimal treatment, the value for low-load improved by 29%. The efficiency of optimal and home therapy differs by a factor of 84, when considering the six years (29%/13 days = 2.23%/day; 58%/365•6 = 0.0265%/day; 2.23/0.0265 = 84) and by the factor of 149 when considering the last five years (29%/13days = 2.23%/day; 27%/365•5days = 0.015%/day; 2.23/0.015 = 149).

4. Improvement of Higher Mental Functions, Speech, and Writing

The CDT the patient was receiving was far from optimal. As a consequence, improvement in CD and movements were limited.

Figure 61. Improvement of the impression of the face of a patient with severe cerebellum injury in dependence on therapy: before the accident, after the accident (1995) to 2006.

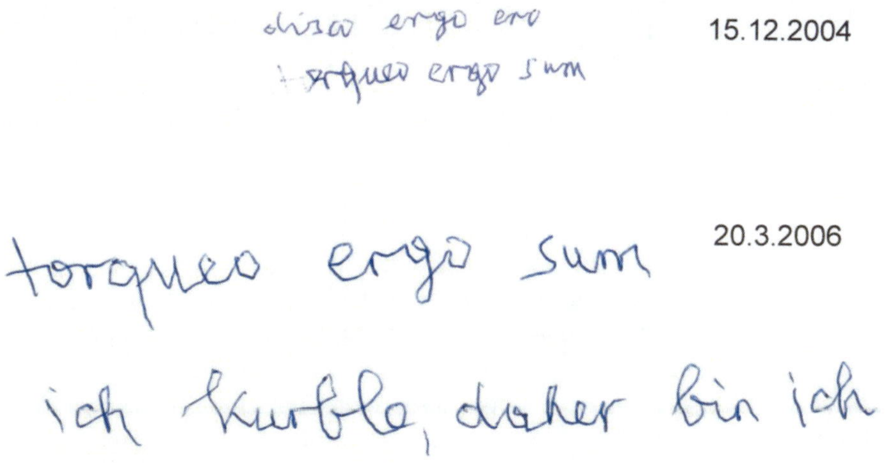

Figure 62. Writings of the patient with severe cerebellum injury as a judgement of goal-oriented movements (task mainly of cerebrum cerebellum).

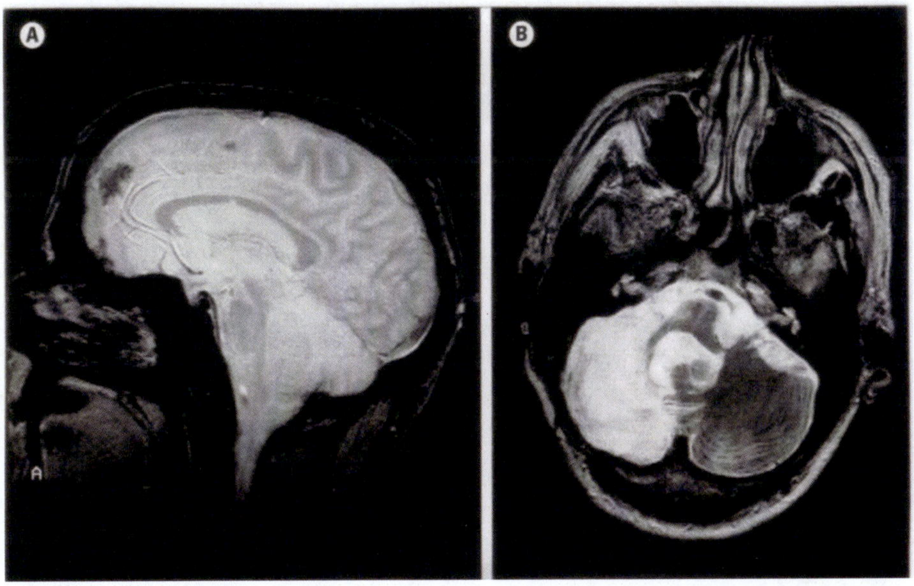

Figure 63. Magnetic resonance imaging (MRI). Sixty-three-year-old patient who suffered a severe cerebellar and cerebral injury. The cerebellum has been destroyed to approximately 60% (B, light parts of the cerebellum). There is a loss of brain tissue in the frontal lobe (A, dark areas of the forebrain).

It is nevertheless sure that the therapy was worthwhile since considerable improvements could be achieved in higher mental functions and the speech. At the beginning, the patient showed scanning speech, and it was difficult to listen to him over longer periods of time. Over the years, a normal discussion became possible. The patient also received speech therapy, mostly during the exercising on the special CDT device. Hyper-salivation subsided already after six months of therapy.

Improvements of higher mental functions became partly reflected in the improved impression of the patient's face (Figure 61). Before the accident (a), the face shows mental strength. Four months after the accident (b), the face seems to be that of another person: as if everything was lost. Seven (c) and ten months after the accident (d), the impression of the face partly recovered, but there is a strong over-emphasizing of the feelings reflected in the face. Nineteen (e) and 20 months after the accident (f), a further recovery in the impression is distinguishable. Still, a slight over-emphasizing of the feelings can be seen. Forty months after the accident (g), some of the old impression of the face is back. Forty-four months after the accident, the face became really similar to that seen before the accident; the face still misses mental strength. Ten years after the accident (i), also some of the mental strength reappeared in the face. In reality and as suggested by his appearance, the patient is still not the old one. His mental strength has not been fully repaired. He himself says that before the injury, he could do two things at the same time: "Listening to somebody and developing at the same time a strategy against the opponent person. This is not possible anymore." In practical life, he has still problems; he additionally suffered a frontal lobe injury (Figure 63).

After the accident that caused the severe cerebral and cerebellar injury, the patient's wife was advised to place the patient in a place where human bodies are kept until they die. Instead of following the advice, the wife was fighting for her husband's future; having learned about CDT, she organized to start it approximately two years after the accident.

Dr. Cwienk's (the patient) intellect recovered so much that the author sometimes asks him for advice, since the patient has much more expertise than the author in certain fields. This is actually a bit of a dream for a research worker who tries to repair the human CNS. One gets a patient after CNS injury, who has been given up by the school medicine, and after treatment, the patient's intellect partly exceeds that of the research worker himself. It was this Dr. Dieter Cwienk (Austria, Graz, "Hofrat") who asked in the past why we don't let patients who suffer severe brain injury die if we don't give them the proper available treatment. To leave the patient to vegetate until the remaining functions of his body die is undignified for humans [51].

The goal-directed movements (cerebrum-cerebellum) in this patient are still limited and were not specially trained. His writing is still poor (Figure 62), and he does not train it. For written communication, he mainly uses the computer. A small improvement in handwriting may have occurred between 2004 and 2006 (Figure 62). In 2006, the letter "o" in "torqueo ergo" has more corners (mostly five) than in 2004 (mostly three) (Figure 62). It thus seems that the handwriting has become smoother, even though the size of the letters is larger in 2006.

3.2.7.3. Discussion

1. Difference between Home Training and Professional Therapy

It could be shown here that motor functions of a patient after severe cerebellar injury improved. However, the improvement was only little when related to the six years of therapy. It was shown further in a three-week trial that the motor functions improved strongly (between 29% and 48%) when the therapy was administered at a professional place. A 6- to 12-month treatment at a professional place would have been needed to substantially improve the CNS functioning in this patient. It was reported above that in severe brain injury (cerebral injury), the efficiency may even differ up to a factor of 390 [52] between home therapy and professional therapy at a modern therapy facility. Professional therapy means that the therapy is based on human neurophysiology data, based on the theory on the repair of the human CNS, and is based on scientific publications in international journals. For the time being, the major problem in relation to CNS repair is not a scientific but an organizational one.

2. Problems of Keeping Balance in Severe Cerebellar Injury

The therapy administered at home was only a part of the problem of the limited progress. Another significant problem was that the patient had a primary severe problem with keeping balance because of the severe cerebellar injury. Due to this, improvements of movements could not be transferred to everyday life's movements and could not be trained in everyday life. In everyday life, essentially balance (vestibulocerebellum) is needed to be able to perform movements, antagonistic muscle activation (spinocerebellum) is needed, and goal-directed movements (cerebrocerebellum) are needed. All these attributes were impaired in this patient.

3. Schalow CDT and Neurogenesis

Schalow CDT claims to be able to help to repair CNS functioning by enhancing functional and structural repair. Structural repair means building of new nerve cells from endogenous stem cell reserves, growing of new connections, and functional integration of the new cells and connections into existing networks. Why, then, did no structural repair did take place in these six or ten years of treatment? So far it seems that substantial structural repair only occurs if intensive therapy lasts for periods exceeding six months. The patient could not receive continuous therapy of such length because of organizational reasons. Also, the cooperative patient (aged about 65 years) argued that he also wanted to live rather than spend his life training only. The argument of the author that the time spent on therapy will provide a longer period of life with a higher quality did not sufficiently impress the patient.

4. Comparison between Cerebral and Cerebellar Injury

Both the patient with predominantly a cerebral injury (above) and the patient with predominantly a cerebellar injury (this case report) improved strongly at the beginning of the therapy. Both patients received optimal therapy at a modern therapy facility after home therapy ceased to induce very much further improvement. Under professional treatment, strong improvements (by approximately the factor of 100) could be noted in both patients. There was a slightly stronger improvement in values of low-load CD in the patient with severe cerebral injury (by the factor of 390) than in the patient with cerebellar injury (by the factor of 150). It seems therefore that injuries of the cerebellum may also impair motor learning [94].

3.2.8. Surface EMG- and Coordination Dynamics (CD) Measurements-Assisted Cerebellar Diagnosis and Improvement in a Patient with Cerebellar Injury

3.2.8.1. Summary

Cerebellar diagnostic was performed in the upper patient who had suffered severe cerebellar injury, using kinesiologic (coordination dynamics (CD)) and electrophysiologic (surface EMG (sEMG)) means. Since the right cerebellum had been completely destroyed, significant coordination problems (especially on the right side) were expected to be found in addition to balance problems. As compared to a healthy person, coordination between arms and legs was 80%, up to 300% worse. Unexpectedly, no pronounced impairment of coordination of the right arm and right leg was found (with respect to the remaining limbs) as a result of the lost right cerebellum. The values of CD measured for the right arm and the right leg for exercising in the forward and backward directions were as good as or even better than those obtained for the left side. However, sEMG disclosed a strong impairment of antagonistic muscle activation (spinocerebellum) between the tibialis anterior and the gastrocnemius muscles on the right side upon exercising the legs only on the special coordination dynamics therapy (CDT) device (similar to a stationary bicycle). The muscles of the right-sided leg worked worse than those of the left-sided leg. But upon exercising both legs and arms on the special device (more integratively), the antagonistic muscle activation improved in the poor right leg by 36% in the short-term memory but not in the good left leg (symmetry improvement). These kinesiological and electrophysiological measurements show that there was no clear ipsilateral correlation between the cerebellum and the laterality of body functions. For different kinds of supported walking, sEMG motor programs worsened with the reduction of the support. Probably the damaged vestibulocerebellum could not sufficiently coordinate balance anymore with walking. When swinging and jumping on the springboard, sEMG motor programs were best for jumping in anti-phase, where maybe more inhibition is activated. Forward-backward movement symmetry of the human CNS was investigated through pattern change during exercising on a special CDT and recording device in the forward and backward direction. A mirror-image symmetry shift was found in the patient with cerebellar injury for the attractor patterns (movement patterns with the highest temporal stability) for exercising on the special CDT device in the forward and backward directions, with respect to pace and trot gait coordinations. Symmetry diagnosis and symmetry learning are of importance in CNS repair because of the learning transfer to the movement symmetry counterpart: training of backward walking improves forward walking or the bad side learns from the good side. More integrative CNS activation and movement support improves CNS functioning and repair. It is concluded that coordination measurements and sEMG are powerful tools to evaluate the outcome of cerebellar injury and to provide a diagnostic tool to quantify improvements in the CNS functioning as a result of CDT.

Keywords: Cerebellum – Injury – Diagnostic – Coordination dynamics – sEMG – Motor program – Symmetry – Antagonicity – Balance.

3.2.8.2. Introduction

To enhance the efficiency of the treatment, diagnostic means are needed to tailor the therapy to the type of the CNS injury of the patient and to the CNS changes associated with the CNS repair during ongoing treatment. Since the cerebellum is the coordination centre for movements and their interplay with gravity and posture, coordination and symmetry measurements in patients suffering severe cerebellar injury may give us an insight into the CNS regulatory mechanisms and may thus help to make CNS repair more effective. It will be shown in this section that surface electromyography (sEMG) and CD measurements can be used to diagnose the function of the cerebellum. It will be shown in the patient who suffered severe cerebellar (and brain) injury that symmetry and balance are impaired. Forward-backward symmetry will be analysed in detail. It will be shown that the anatagonicity impairment between forward and backward moving, quantified by sEMG, can also be measured in terms of CD.

Symmetry of the pattern dynamics seems to be an attribute that also plays an important role in the structuring of the coordination of rhythmic movements [16]. Breaking front-rear symmetry was found to account for the difference between straight and reverse walking [98]. Further, there is learning transfer from a movement to its symmetry counterpart (19, page 179).

Exercising backward walking and crawling may improve forward walking and crawling. Further, symmetry has consequences in terms of transitions between behavioural states. In the asymmetric case, lower level of noise (from variability of phase and frequency coordination) and/or perturbation are required to throw the system out of its present state than is required in the symmetric case (Figures 4,16) [16].

In neurotherapy, the stability of the movement patterns can be experienced during treadmill walking. If a patient with an incomplete spinal cord injury walks on treadmill under weight reduction and leg support given by two physiotherapists, he/she can get better into or stay better in the rhythmic walking pattern, if his injury is more symmetric and if the interpersonal coordination between the two physiotherapists (each supporting one leg) is more symmetric.

Symmetry and stability thus structure the state space of a multicomponent system with multiple patterns (including pathologic patterns like spasticity) so that only a limited set of transitions are observed (99). Symmetry diagnostic and symmetry training with respect to CNS organization are therefore also important during therapy.

3.2.8.3. Method

CD measurements and surface EMG (sEMG) were used to diagnose cerebellar function. CD of arm and leg movements were measured in terms of arrhythmicity of exercising on the special CDT device. The movements were imposed by the device except for the smoothness of exercising. According to the system theory of pattern formation (see method section), arrythmicity of exercising was the collective or order parameter of CNS organization for these special movements. Differential input was used for sEMG (input resistance = 1MΩ, AC coupling). The EMG signals were pre-amplified (1000x, passing frequency 100Hz – 10kHz) and displayed on a four-beam oscilloscope. Figure 64A illustrates the sEMG recording layout for walking with sticks.

By function, the cerebellum is divided into three parts (2), the vestibulocerebellum (archicerebellum), the spinocerebellum (paleocerebellum), and the cerebrocerebellum (neocerebellum).

The vestibulocerebellum is the oldest part and it consists of the floccucolnedullar lobe. It is responsible for keeping the balance, i.e., for coordination of the muscle activations for posture and movements against gravity.

The spinocerebellum mainly consists of the lower vermis, the paraflocculus, and the intermediary part (paravermal zone). It is mainly responsible for the coordinated action of antagonistic muscles and communicates strongly with the spinal cord. The cerebrocerebellum (neocerebellum) consists of the lateral part of the cerebellum and communicates with the cerebrum (the cerebral cortex).

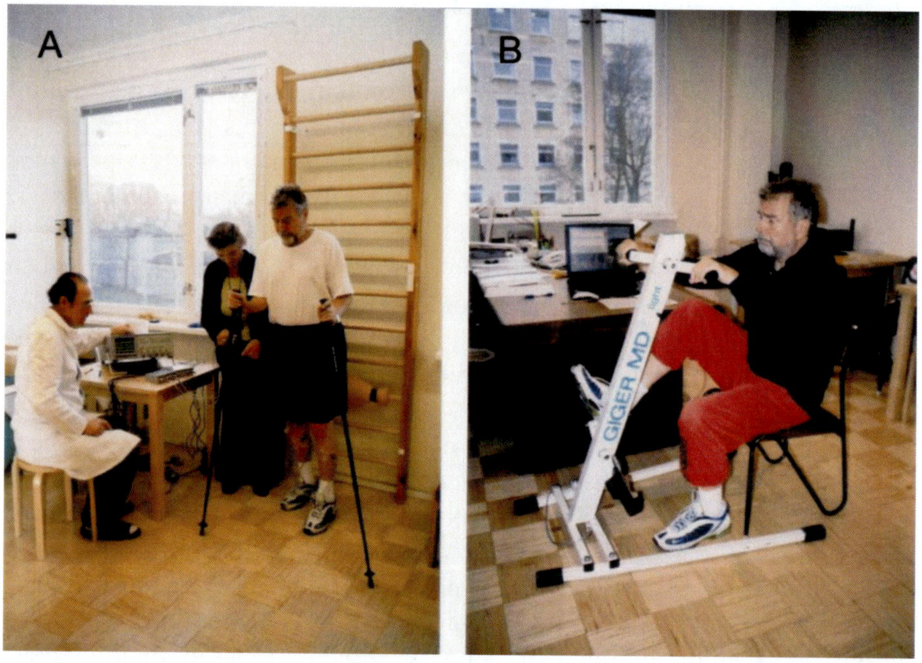

Figure 64. A. Surface EMG recording layout for measuring motor programs during walking with sticks. The author is at the oscilloscope, and the wife of the patient is holding the wires to the surface electrodes to reduce movement-induced artefacts. B. Patient with severe cerebellar injury during exercising with both arms and one leg on the special coordination dynamics therapy device in the sitting position. The coordination dynamics (arrhythmicity of exercising) is recorded and displayed on a laptop.

3.2.8.4. Results

4.1. Functional Anatomy According to the Damage to the Cerebellum

In an accident, a 58-year-old patient suffered cerebral injury and severe cerebellar injury. Low-intensity coordination dynamics therapy was administered. After four years, magnetic resonance imaging (MRI) was performed (Figure 63). Five years later, a further MRI was taken to look for macroscopic changes in the brain. The second MRI looked very similar to the first one. No new brain tissue could safely be identified. On the other hand, there was also no atrophy of brain tissue either distinguishable from the last MRI as compared with the image taken five years earlier. Approximately 60% of the patient's cerebellum was destroyed, as can be partly seen from the light parts of the cerebellum in Figure 63. The right cerebellar hemisphere, the vermis of the cerebellum, and the right middle peduncle of the cerebellum down to the right lateral pons were deficient in substance. The nuclei of the pons may also have been partly damaged. The right cerebellum was nearly completely destroyed (Figure 63). Additionally, there was a small substance deficiency in the right frontal paramedian area (of approximately 1cm in diameter) and a small cortical lesion in the frontal upper gyrus, both due to the injury.

Since the right cerebellum was nearly completely destroyed, strong deficiencies in motor functions on the right side could be expected (according to textbooks [2]), since the efferent fibres of the cerebellar halves are supposed to predominantly influence the ipsilateral body halves. The tractus cerebellorubralis crosses to the other side, and the tractus rubrospinalis crosses back. The fibres of the tractus cerebellothalami cross over to the other side and reach the cortex on the other side via the tractus thalamocorticalis. Via the tractus corticospinalis, the impulses cross back to the ipsilateral spinal cord.

By comparing the CD of the right and left side and of arms and legs and by comparing motor programs of different movements, it will be shown that the severely injured cerebellum had problems coordinating balance with movements (which is the task of the vestibulocerebellum) and coordinating motor control of antagonistic muscles (which is the task of the spinocerebellum). It will further be shown that the recorded symmetries of the motor patterns were not fully in accordance with the assumption of ipsilateral side influence of the cerebellar hemispheres.

4.2. Kinesiology

To measure CNS organization with respect to symmetries and coordination by kinesiologic means, the patient exercised on a special device. The arrhythmicity of exercising over time quantified by CD was measured for different movements. Symmetry differences between right and left, between arms and legs, and between exercising in the forward and backward direction could be analysed by not involving one arm or leg or arms or legs in the exercise.

1. Coordination Dynamics (CD) of Only Arms, Only Legs, and Arms and Legs

To measure movement symmetry differences between arms and legs, the patient exercised with only arms ($\Delta_{arms,forward} = 16.55s^{-1}$ and $\Delta_{arms,backward} = 14.8s^{-1}$) or only legs in the forward direction ($\Delta_{legs,forward} = 9.2$) and backward direction ($\Delta_{legs,backward} = 23.4$) (Figure 65). It can be seen from the CD values that the patient could exercise better in the forward direction (smaller arrhythmicity of exercising = lower CD values) when working with the legs only (Figure 65A; $\Delta = 9.2$), whereas exercising in the backward direction was better when only using the arms. When exercising with arms and legs, the value of CD (arrhythmicity of exercising) was much better (smaller; $\Delta_{forward} = 8.4$, $\Delta_{backward} = 7.0$) as compared with involving legs or arms only. Involving both arms and legs, the patient could exercise better backwards (Figure 65H; $\Delta_{backward} = 7.0$) than in the forward direction (Figure 65G; $\Delta_{forward} = 8.4$). By comparing the values of CD measured for exercising with both arms and legs with those for legs or arms alone, it can be seen that the values for arms and legs individually cannot be derived or extrapolated from the values for involving both arms and legs. With the changing coordination between arms and legs including pace and trot gait and the intermediate coordinations, four-limb movements seem to be completely different from movements performed by arms or legs alone. Exercising with only legs involved is very similar to that on a stationary bicycle.

The corresponding values of CD measured for a healthy sporty 28-year-old lady were much better: $\Delta_{arms,forward} = 4.5s^{-1}$; $\Delta_{arms,backward} = 5.8s^{-1}$; $\Delta_{legs,forward} = 3.5$; $\Delta_{legs,backward} = 5.8$; $\Delta_{forward} = 2.9$; $\Delta_{backward} = 3.9$.

Even though the patient had exercised during at least six years on the special device to improve his CNS functioning, his values of the CD for arms and legs were worse, between 80% and 300% (especially for less integrative movements), than those measured for the healthy person, who only exercised on the device the second time.

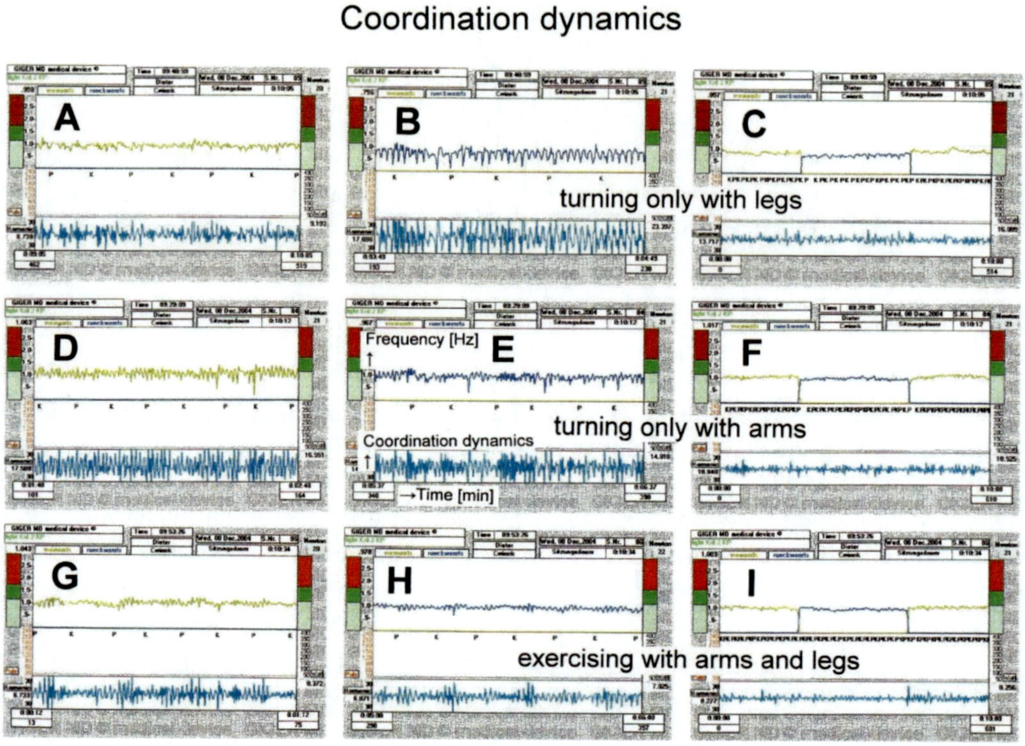

Figure 65. Coordination dynamics measurements in a patient with a severe cerebellar injury upon exercising on the special coordination dynamics therapy and measuring device (at the load of 20N) with only legs (A,B,C), with only arms (D,E,F), and with arms and legs (G,H,I). A,D,G = best 1min record of exercising in the forward direction. B,E,H = best 1min record of exercising in the backward direction. C,F,I = 10min sweep showing exercising in the forward, backward, and forward direction. Upper trace = frequency, f; lower trace = coordination dynamics = arrythmicity of exercising. P = pace gait coordination, K = trot gait coordination."P to K" and "K to P" = coordination changes from pace to trot gait and from trot to pace gait imposed by the device. Note that the legs alone can best do forward movements, and the arms alone can best do backward movements, and arms and legs together can best do backward movements as quantified by the value of coordination dynamics. Note further that upon exercising with arms and legs only, there are rhythmic increases and decreases of the coordination dynamics with the changes of coordination (G,H). For forward movements, "K to P" coordinations are mostly better than "P to K" (G), whereas for backward movements, "P to K "´coordinations are better than "K to P" coordinations (H).

2. Coordination Dynamics (CD) for Involving the Right or the Left Arm Only and Both Legs or for Involving the Right or the Left Leg Only and Both Arms

To measure side differences in movement coordination between arms or between legs, the patient (right-handed) left out one arm or leg during exercising on the device (Figure 64B). Since the value of CD varies with exercising, the patient started to exercise for 2min with his arms and legs and subsequently exercised for 2min with only the right or the left leg (or arm) alternating three times, and the mean value was calculated. At the end of the exercise, the patient involved both arms and legs for 2min as at the beginning of the exercise. Typical best 1min exercising curves are shown in Figure 66 with the corresponding CD values: $\Delta_{left,arm,forward} = 20.1$; $\Delta_{right,arm,forward} = 19.6$; $\Delta_{left,arm,backward} = 22.0$; $\Delta_{right,arm,backward} = 17.6$; $\Delta_{left,leg,forward} = 44.9$; $\Delta_{right,leg,forward} = 44.8$; $\Delta_{left,leg,backward} = 37.7$; $\Delta_{right,leg,backward} = 29.7$. The values of CD (and the curves) suggest that for the exercising in the forward and backward direction, the right arm moved slightly better than the left arm.

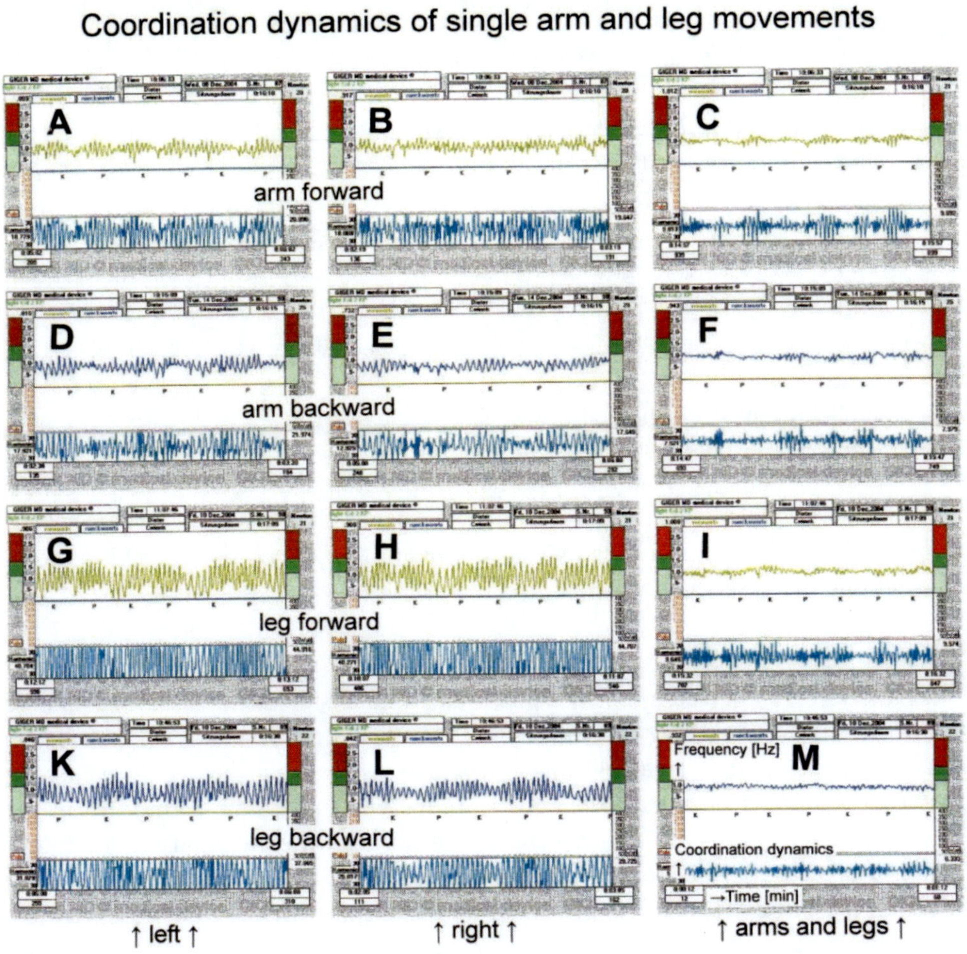

Figure 66. Coordination dynamics measurements upon exercising (at a load of 20N) with only the left arm and both legs in the forward (A) and backward direction (D), with only the right arm and both legs in the forward (B) and backward direction (E), and with only the left leg and both arms forwardly (G) and backwardly (K), and only with the right leg and both arms in the forward (H) and backward direction (L). C,F,I,M. Exercising with arms and legs after using only one arm or one leg. P = pace gait coordination, K = trot gait coordination. "P to K" and "K to P" = coordination changes from pace to trot gait and from trot to pace gait imposed by the device. Note that the right arm and leg can do better movements in the forward and backward direction (B,E,H,L) than the left arm and leg (A,D,G,K), as quantified by the coordination dynamics values. Single arms can move more in a coordinated manner (A,B,D,E) than single legs (G,H,K,L), as quantified by the amplitude of arrhythmicity of exercising. Note further that the more easily performed coordinations are different for single arms and legs and for exercising in the forward and backward direction, as can be seen from the (upper) frequency and the (lower) coordination dynamics traces. For example, less arrhythmicity (A) of "K to P" coordinations was observed for the left arm for exercising in the forward than in the backward direction (D).

Less favourable values were measured for the legs. For exercising in the forward direction, the right and the left leg moved very similarly, as the CD values were very similar (44.8 and 44.9). For exercising in the backward direction, the right leg (29.7) was better than the left leg (37.7). For exercising in the forward direction, the right-sided arm and leg were very similar to the left-sided arm and leg. However, for exercising in the backward direction, the right arm and right leg were better than the left ones. The better (or the same) quality of movements on the right side was unexpected, since poorer movement coordination was expected on the right side due to the destroyed right cerebellum with the ipsilateral impairment of function (see functional anatomy, above).

The corresponding CD values (means) measured for a right-handed healthy sporty 28-year-old physiotherapist (same as above) were (in the brackets values measured for the patient are given for direct comparison): $\Delta_{left,arm,forward}$ = 6.57 (20.1); $\Delta_{right,arm,forward}$ = 5.88 (19.6); $\Delta_{left,arm,backward}$ = 8.57 (22.0); $\Delta_{right,arm,backward}$ = 6.93 (17.6); $\Delta_{left,leg,forward}$ = 21.5 (44.9); $\Delta_{right,leg,forward}$ = 21.5 (44.8); $\Delta_{left,leg,backward}$ = 19.9 (37.7); $\Delta_{right,leg,backward}$ = 21.5 (29.7). The volunteer could coordinate her right hand with her legs slightly better than the left hand for exercising in the forward and backward directions. Moreover, for exercising in the forward and backward directions, she could coordinate her right and left leg very similarly with the arms.

The volunteer could coordinate arm and leg movements much better (lower values) than the patient (values in the brackets), even though the patient had exercised for more than six years on the device and the volunteer had not. The differences in CD values between the patient and the volunteer were bigger between the hands than between the feet. This may suggest a more significant impairment of the coordination of the arms and hands as compared with the legs due to the severe cerebellar (and brain) injury.

4.3. Electrophysiology

After a severe cerebellar injury, the phase and frequency coordination between neuron firing will be impaired. This will have consequences at different levels of CNS self-organization, including the level of motor programs, which can be measured by surface electromyography (sEMG). Motor programs were therefore recorded from different muscles during exercising on the special CDT device and during different kinds of walking (with different amounts of balance needed), running, and jumping on springboard to identify impairments. It will be attempted to allocate functional impairments to the different parts of the cerebellum. When the patient exercised on the special device in the sitting position at 100N, only using legs or using arms and legs (similarly as in Figure 64B), he did not have to keep balance. Deficits in the motor programs will therefore mainly be induced by the injury of the spinocerebellum (responsible for coordination, especially between antagonistic muscles).

1. Symmetry Improvement of Motor Programs of Antagonistic Muscles by Increasing the Integrativity of the Movement (Spinocerebellum)

The surface electrodes were placed on the right and left tibialis anterior and the right and left lateral gastrocnemius muscles. When the patient exercised only with the legs, the left gastrocnemius muscle became activated too late with respect to the left tibialis anterior muscle (Figure 67A, at $\varphi = 200^0$ instead of 180^0 (antagonistic muscles)) for exercising in the forward direction, and it became activated too early for exercising in the backward direction (Figure 67B, at 150^0 instead of 180^0). The antagonicity of the motor programs thus showed pathologic symmetry between exercising in the forward and backward direction, namely $\Delta\varphi_{forward} = 20^0$ against $\Delta\varphi_{backward} = -30^0$. When exercising with arms and legs (Figure 67C,D), this asymmetric coordination did not change in the (better) left leg.

However, an improvement of the symmetry of antagonistic coordination between the gastrocnemius and the tibialis anterior muscles could be observed in the right leg when exercising in the forward or backward direction using arms and legs as compared to only using the legs. Upon exercising in the forward directions only using legs, the right gastrocnemius muscle became activated too late with respect to the right tibialis anterior muscle (Figure 67A, at 230^0 instead of 180^0; $\Delta\varphi_{forward} = 50^0$) and too early for exercising in the backward direction (Figure 67B, at 65^0 instead of 180^0; $\Delta\varphi_{backward} = -115^0$). When exercising in the forward direction with arms and legs, the right gastrocnemius muscle was not activated that late (Figure 67C, only at 210^0 instead of 180^0; $\Delta\varphi_{forward} = 30^0$). When exercising in the backward direction, the right gastrocnemius muscle did not get activated as early with respect to the right tibialis anterior muscle (only at 105^0 instead of 180^0; $\Delta\varphi_{backward} = -75^0$).

Thus, it may be stated that the antagonistic coordination between the right tibialis anterior and the right gastrocnemius muscles improved for exercising in the forward and backward directions by 20^0 and 40^0, respectively, upon exercising with arms and legs as compared to using only the legs. More generally speaking, this antagonicity symmetry between the tibialis anterior and the right gastrocnemius muscles improved by 60^0 for exercising in the forward and backward directions upon exercising with both arms and legs.

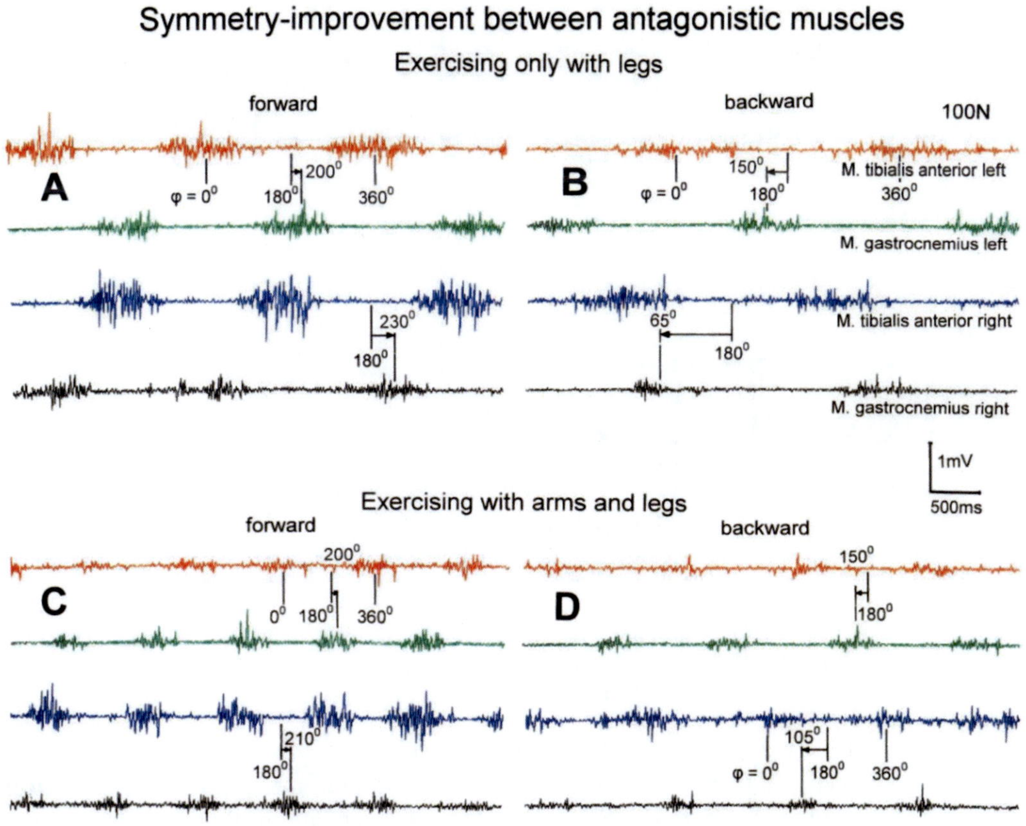

Figure 67. Motor programs of right and left tibialis anterior and gastrocnemius muscles recorded by sEMG for the patient with a severe cerebellar injury upon exercising on the special coordination dynamics therapy device in the forward (A,C) and backward direction (B,D) only using legs (A,B) and with arms and legs (C,D). Note that the gastrocnemius muscle is not activated properly antagonisticly with respect to the tibialis anterior (at 180 degree): it gets activated ealier or later, depending on whether movements are performed in the forward or backward direction and upon whether only legs or arms and legs are used.

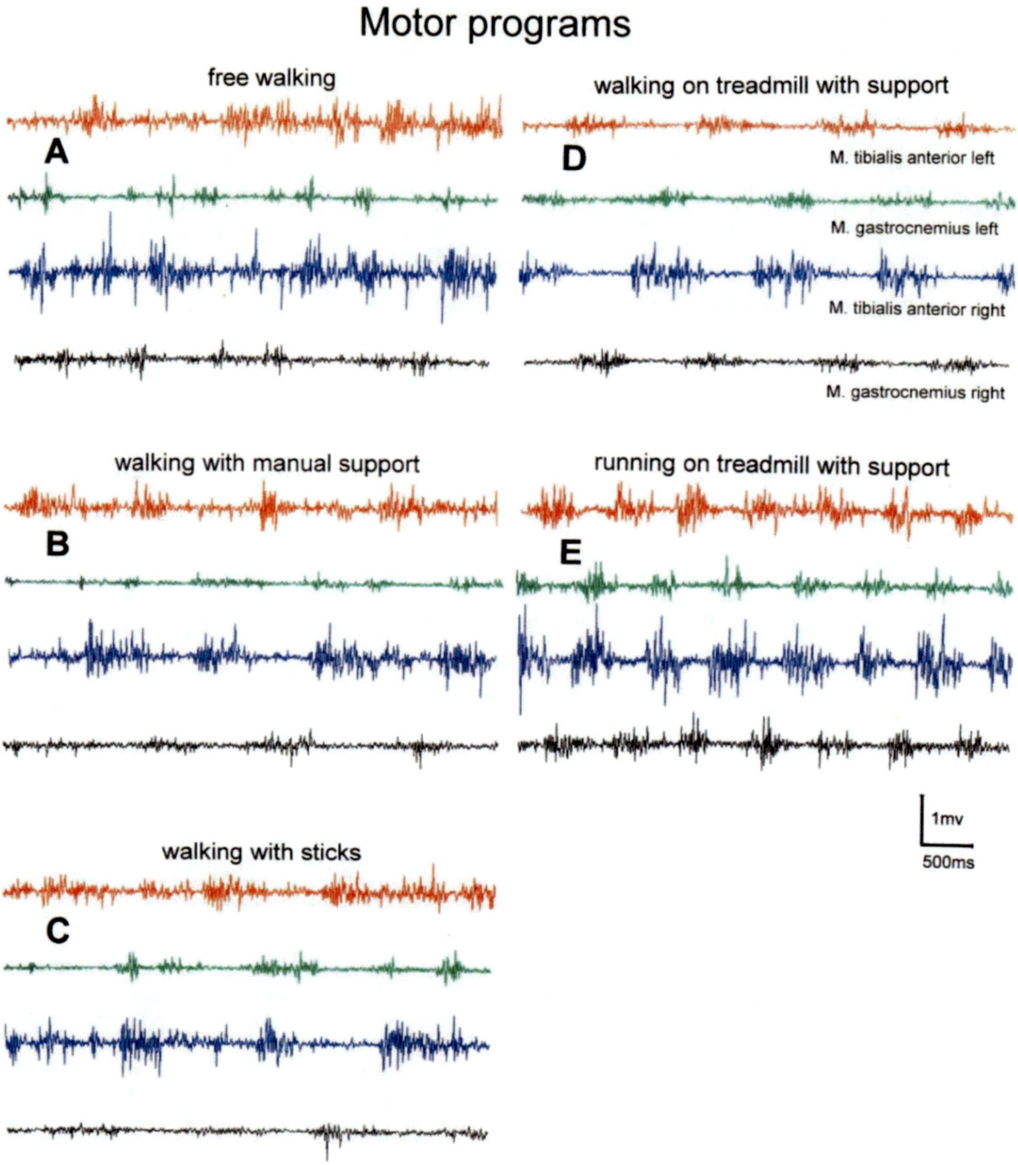

Figure 68. Surface EMG motor program of the right and left tibialis anterior and gastrocnemius muscles for free walking (A), walking with manual support (B), walking with sticks (C), walking (D) and running (E) on the treadmill with support. Note that the motor programs for walking with strong support (D,E) are much better than those for walking with little or no support (A,B,C).

During this more integrative exercising on the special CDT device using both arms and legs, the antagonistic activation of the tibialis anterior and the gastrocnemius muscles improved in the short-term memory. Since it is the spinocerebellum that is supposed to be responsible for the antagonistic muscle coordination, the main improvement in the short-term memory may be located in the spinocerebellum.

2. Learning in the Short-Term Memory from the Better Opposite Side

The better (with respect to antagonistic activation) left side did not improve, whereas the poor right side did improve by 36% ($\Delta\varphi_{forward}$ - $\Delta\varphi_{backward}$ reduced from 165^0 to 105^0). With the integrativity increasing from only coordinated leg movements to coordinated arm and leg movements, the better left side remained unchanged, whereas the poor right side improved by 36%. This one-sided symmetry improvement can be interpreted in the way that the poor right side learned in the short-term memory from the good left side. This learning effect may be sited in the spinocerebellum and/or in the spinal cord (co-movement).

3. Motor Walking Programs and Balance (Vestibulocerebellum)

Subsequently, I studied how the motor program of walking changed when the patient had to keep more balance by himself. Since the vestibulocerebellum was also heavily damaged, increasing motor program deficits for increasing balance needs could have been expected.

Figure 68 shows sEMG motor programs for free walking (Figure 68A) and for supported walking (manual support) (Figure 68B), walking with sticks (Figure 68C), walking on treadmill (Figure 68E), and running on treadmill (Figure 68E).

The motor programs for free walking (A) were worse than those for the different kinds of supported walking (B-E). With the increasing support (from B,C to D,E), the motor programs improved. Obviously, the patient had balance problems; upon spending efforts to keep more balance by himself, his motor programs became worse. The interpretation of this worsening of the motor programs with the increasing balance needs is that upon free walking, the damaged vestibulocerebellum could no longer sufficiently coordinate balance with walking.

4. Motor Programs When Swinging and Jumping on Springboard

To find out what motor programs are best upon performing rhythmic, dynamic, stereotyped movements (functional repair on the neural ensemble level) on springboard, sEMG was recorded for the different kinds of springboard movements.

Motor programs upon swinging and jumping on springboard are shown in Figure 69. For swinging (Figure 69A), the tibialis anterior and gastrocnemius muscles were only slightly rhythmically activated, i.e., no real motor program could be seen. When jumping in in-phase, the leg muscles were more activated, and motor programs could partly be identified (Figure 69B). In the right tibialis anterior muscle, synchronized oscillatory firing of motor units (pathologic) can be seen (see Discussion). For jumping in anti-phase with (Figure 69C) and without support by a therapist (Figure 69D), the motor programs were quite good, even though rhythmic firing of motor units can also clearly be identified during the motor bursts (right gastrocnemius in Figure 69D). The conclusion for the treatment with respect to exercising on springboard is that jumping will entrain the neuronal networks of the patient better than swinging, since real motor programs could only be identified for jumping.

5. Impairment of Functions of the Cerebrocerebellum

The patient had problems with precise and goal-oriented movements like eating, drinking, using keys, playing piano, and solving dexterity tests. These properties of the cerebellum were not analysed.

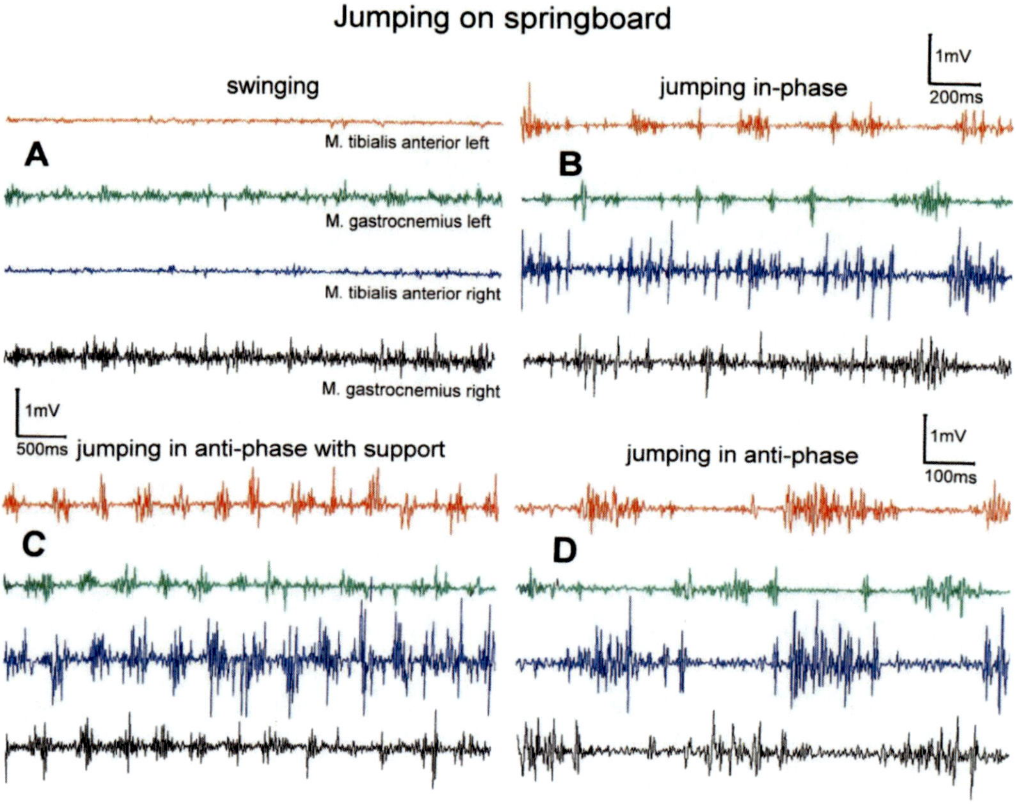

Figure 69. Surface EMG motor program of the right and left tibialis anterior and gastrocnemius muscles for swinging (A) and jumping on springboard in-phase (B), in anti-phase with support (C), and in anti-phase without support (D). Note that the motor programs are better for jumping in anti-phase (C,D) than for jumping in in-phase and swinging. Note further that continuous synchronous oscillatory firing of motor units can be seen on the trace taken from the right tibialis muscle for jumping in in-phase (B).

4.4. Symmetry Diagnosis and Treatment in Coordination Dynamics Therapy (CDT)

Above, it was dealt with cerebellar diagnosis, right-left, rostral-caudal, and forward-backward symmetry diagnosis in a patient with cerebellar injury. The forward-backward movement symmetry of the coordination of the firing of agonist and antagonist muscles was quantified by sEMG. In this section, forward-backward movement symmetry will be analysed with respect to the coordination of firing of antagonistic muscle activation in a patient with cerebellar injury during exercising on the special CDT device, i.e., by coordination dynamics (CD). It will be shown that for high-load exercising, the patterns with the highest temporal stability (attractor states) shifted away in a mirror-image manner from the physiologic position in the coordination pattern dynamics traces, i.e., away from pace and trot gait patterns.

1. Treatment of Symmetry Impairment

Above, it was shown by sEMG that there was antagonicity impairment of CNS functioning with respect to antagonistic muscle activations of the tibialis anterior and gastrocnemius muscles in a patient with severe cerebellar injury. This antagonicity impairment between antagonistic muscles was opposite for exercising in the forward and backward directions (Figure 67). If it were possible to eliminate this opposite antagonicity impairment of antagonistic muscle activation by certain movements, the patient's motor patterns and performance may improve. It was shown further (Figure 67) that increasing integrativity of CNS organization resulted in reduced antagonicity impairment. Upon exercising the legs only on the special CDT device (like on a stationary fitness bicycle)the antagonicity impairment between the tibialis anterior and gastrocnemius muscles was strong (Figure 67A,B). Upon exercising with arms and legs, this antagonicity impairment reduced (Figure 67C,D). In the particular patient, the antagonicity impairment could thus be influenced by exercising highly coordinated arms and legs movements on the special device. Other highly coordinated four-limb movements may probably also result in improvements of the antagonicity of muscle activation.

2. Integrative Forward-Backward Symmetry Impairment in Cerebellar Injury

Antagonicity impairments may occur in many pairs of antagonistic muscles, different for exercising in the forward and backward directions (mirror shape, measured by sEMG). Therefore, a more integrative measure is needed to quantify antagonicity impairment. Such an integrative diagnosis for antagonicity impairment may partly be available if the patient exercises on the special CDT device in the forward and backward directions, and CD traces are recorded.

CD traces (Figure 70) show that in the patient with cerebellar injury, the lack of symmetry between exercising in the forward and backward direction could be made visible and quantified by measuring CD (temporal stability) through pattern change. When exercising at loads between 100N and 200N, the values of exercising frequency and CD increased and decreased rhythmically for exercising in both the forward (Figure 70A) and backward directions (Figure 70B). For exercising in the forward direction, however, the arrhythmicity (the value of CD) was slightly to the right side of the pace (P) and trot gait (K) coordinations (Figure 70A) whereas upon exercising in the backward direction (Figure 70B), the arrhythmicity was small to the left side of the pace (P) and trot gait (K) coordinations. Thus, there seems to be a mirror-image shift of the arrhythmicity between exercising in the forward and backward directions with respect to the pace and trot gait coordinations.

In a healthy person, the smallest arrhythmicity of exercising (the highest pattern stability) was for the pace and trot gait coordinations (Figure 70F). To make this seemingly mirror image change with respect to exercising in forward and backward direction better visible, the CD traces for exercising in the forward and backward directions are shown together in Figure 70C (bottom panel). The recordings were taken one week later to show that patterns of highest stability remained at the same place with respect to the pace and trot gait coordinations to provide evidence that there was no phase drifting. It can clearly be seen in Figure 70C that the periods of smallest arrhythmicity shifted rather symmetrically away from the pace (P) and trot gait (K) coordination position with respect to exercising in the forward and backward directions. When picturing this rather opposite symmetric shift of arrhythmicity of exercising in an attractor layout of pattern formation in the framework of the system theory of pattern formation, the attractor states (patterns with the highest stability) shifted away from the pace and trot gait coordination patterns (Figure 70E).

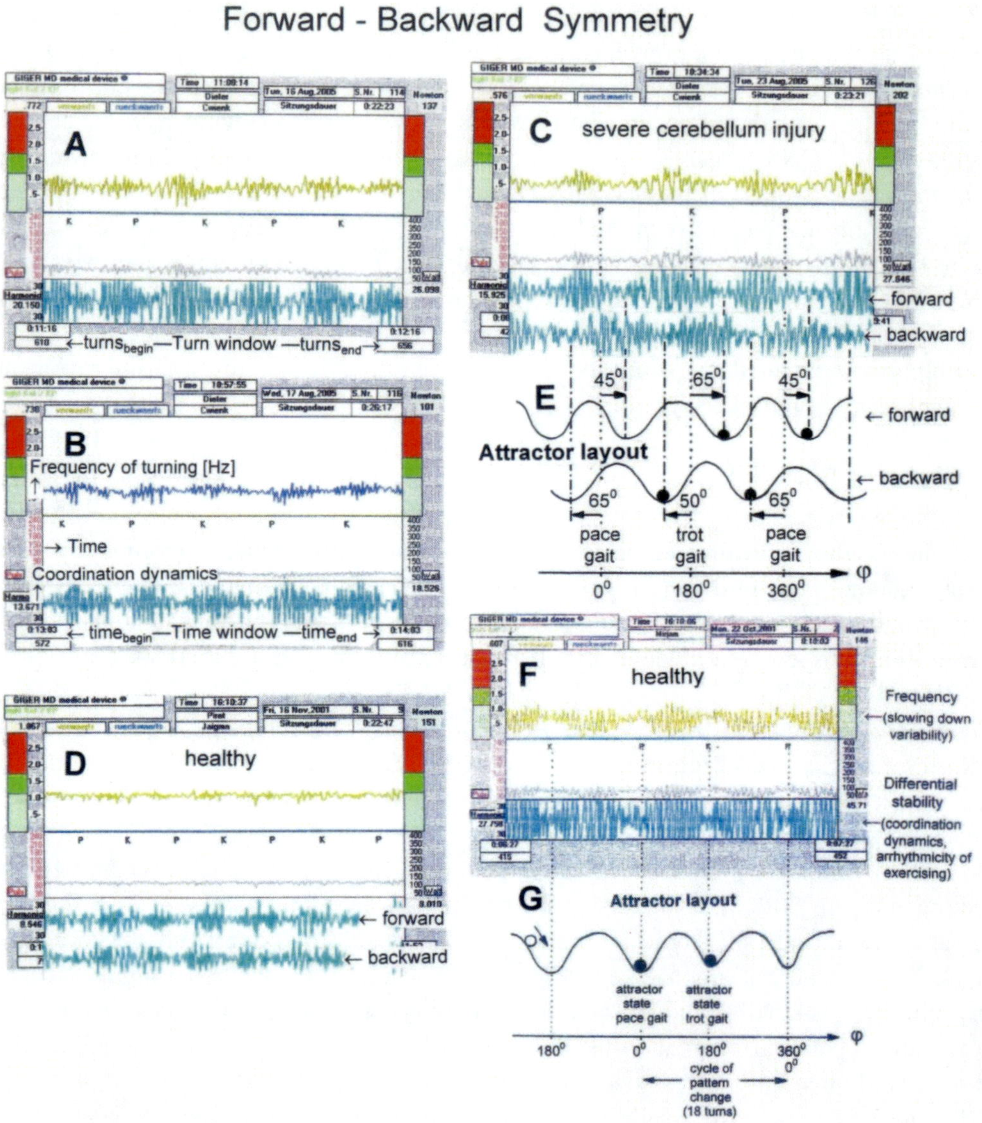

Figure 70. Forward-backward movement symmetry impairment in a patient with severe cerebellum injury for exercising on the special coordination dynamics therapy device (A-C). Coordination dynamics Figures of a healthy person are shown for comparison (D).

In the attractor layout of a healthy sporty girl (Figure 70G), drawn from the CD recording presented in Figure 70F, the stable movement patterns (when exercising on the special device) are at the pace and trot gait coordinations. For the patient with the cerebellar injury, the stable movement patterns (small arrhythmicity of exercising) shifted with respect to the cycle of pattern change (bottom part of Figure 70G) between $\varphi = 45°$ and $65°$ forwardly for exercising in the forward direction and between $\varphi = 50°$ and $65°$ backwardly for exercising in the backward direction (Figure 70E). This rather opposite shift of the attractor states shows similarity to the rather mirror picture change of antagonicity impairment of the antagonistic muscle activation of the tibialis anterior and gastrocnemius muscles, measured by surface EMG (Figure 67). It seems therefore that the special CDT device can be used to measure symmetry impairments of CNS organization with respect to exercising in the forward and backward directions. The symmetry diagnosis performed using the special CDT and recording device, presented above, concerned comparisons between right and left and rostral and caudal. Here, the symmetry diagnosis compared exercising in the forward and backward directions. It remains to be further clarified whether this diagnosis of symmetry CD with respect to exercising in the forward and backward directions (kinesiologic measure) actually measures integratively symmetry impairment of several antagonistic muscle pairs (motor programs measured by sEMG, electrophysiologic measure).

3. Forward-Backward Movement Symmetry Impairment in Healthy Subjects

In healthy subjects, the symmetries in CNS organization may also not be ideal. A forward-backward movement asymmetry seems to exist. For exercising in the forward direction, the value of coordination dynamics obtained for the physiotherapist ($\Delta = 8.0s^{-2}$, trained) was much smaller (better) (Figure 70D) than that for the patient ($27.6s^{-2}$, strongly trained) (Figure 70D). For exercising in the backward direction, the coordination dynamics trace of the physiotherapist is quite good with respect to symmetry (Figure 70). The stronger arrhythmicity (high pattern instability) mostly occurs for the difficult intermediate coordinations between the pace and trot gait coordinations. For exercising in the forward direction (Figure 70D), the stronger arrhythmicity is slightly shifted away from the middle position between the pace and trot gait coordinations. Thus, the symmetry with respect to movements in the forward and backward directions is probably suboptimal even in healthy sporty persons. Also, athletes differed in their coordination dynamics values for low-load exercising in the forward and backward directions [42].

3.2.8.5. Discussion

1. Functional Anatomy

Due to an accident, the right side and part of the left side of the cerebellum of a 57-year-old patient were completely destroyed (Figure 63). According to the functional anatomy described in textbooks [2], an impairment of the function and coordination mainly on the right side of the body can be expected. Measuring of the coordination between one arm and the other arm and the legs or between one leg and the other leg and the arms showed that the coordination between the right arm or right leg and the remaining limbs were still better than the coordination between the left-sided arm or leg and the other limbs, even though the right cerebellum was completely destroyed. The coordination dynamics measurements therefore did not show stronger ipsilateral function impairment.

The electrophysiological measurements however showed a more significant impairment on the right side. For exercising in the forward and backward direction, there was stronger impairment of the antagonicity (deviation from $\Delta\varphi_{forward,backward} = 0^0$) of the antagonistic leg muscles tibialis anterior and gastrocnemius on the right ($\Delta\varphi_{forward} = 50^0$, $\Delta\varphi_{backward} = -115^0$) than on the left side ($\Delta\varphi_{forward} = 20^0$, $\Delta\varphi_{backward} = -30^0$). Other antagonistic muscle pairs were not measured. These kinesiological and electrophysiological measurements show that there was no clear ipsilateral correlation between the cerebellum and the laterality of body functions. A very detailed comparison between the cerebellar injury (anatomy) and the functional impairment is very difficult or not possible at all as the CNS damage cannot be evaluated exactly (including nuclei and tract fibres). There may also be further CNS damage, which may not be detected by the MRI.

2. Impaired Symmetry between Exercising in the Forward and Backward Directions

The special CDT and recording device offers the possibility of comparing forward and backward movements since forward and backward movements are very similar to each other. Right-left differences can also be easily evaluated by not using one leg or one arm during the exercising (Figure 64B). Asymmetries in the CNS functioning can therefore be measured on the motor pattern level (by sEMG) and on the movement level (via CD). On the motor pattern level (by sEMG), the antagonicity impairment between the tibialis anterior and gastrocnemius muscles on the right side were bigger for exercising in the backward ($\Delta\varphi_{backward} = -115^0$) than for the forward direction ($\Delta\varphi_{forward} = 50^0$) (Figure 67). Forward-backward symmetry impairment could also be recorded with the CD when the patient with cerebellar injury exercised at loads between 100N and 200N in the forward and backward direction on the special device. The movement patterns with the highest stability (attractor states) showed a nearly mirror-image shift with respect to the pace and trot gait coordinations for forward and backward exercising (Figure 70).

Many asymmetries were observed on the movement level. For example, when using arms and legs, the CD values were different for the forward ($\Delta_{forward} = 8.4$) and the backward direction ($\Delta_{backward} = 7.0$), even though the patient had exercised for several years on the special device equally in the forward and backward directions. Sometimes, the difference between the forward and backward direction was small; it often was large. Upon exercising with only the arms, the difference was (in this patient) small ($\Delta_{arms,forward} = 16.55s^{-1}$ and $\Delta_{arms,backward} = 14.8s^{-1}$); large differences were seen upon exercising with the legs only ($\Delta_{legs,forward} = 9.2$ and $\Delta_{legs,backward} = 23.4$). When the patient was asked to exercise with arms and legs but without using one leg or one hand (to measure side differences), varying differences between exercising in the forward and backward direction were observed (see above). In this patient, many symmetries of the CNS were strongly impaired.

When studying the different coordinations between arm and leg movements in healthy subjects, differences could also be observed (see above). Symmetry impairment in CNS functioning were even observed in athletes: e.g., the values for low-load coordination dynamics were worse for exercising in the backward (6.9) than in the forward directions (6.0) [42]. In physiotherapists, even larger differences in low-load coordination dynamics values could be observed between exercising in the forward and backward direction ($\Delta_{forward} = 5.9$; $\Delta_{backward} = 9.4s^{-2}$) [42]. The difference in the CNS organization between this patient and a healthy person is that healthy persons can improve their CD between arm and leg movements with respect to symmetries. For patients with severe CNS injuries and especially with severe cerebellar injuries; however, it seems to be not so easy to improve the symmetries in their CNS organization.

3. Implications of Symmetry Learning for the Therapy

Learning transfer from one hand movement to the symmetric one has been reported to occur [19], which means that the symmetry counterpart improves without being trained itself. It has been shown herein (see above) that the antagonicity between the tibialis anterior and gastrocnemius muscles in the poor right leg improved in the short-term memory by 36% when increasing the integrativity of exercising by changing from only leg movements to coordinated arm and leg movements. Assuming that repeated improvements in CNS functioning in the short-term memory repeatedly will slowly also become reflected in the long-term memory would mean that there is learning transfer with respect to the improvement of antagonicity from the left leg to the right leg.

Improvements of symmetries of movements are used in CDT. A certain movement can be also improved by training the symmetry counterpart. If, for example, the walking pattern on the right side has been impaired in a stroke patient for a few years, this pathologic walking pattern has become an old-learned movement and is difficult to improve. The symmetry counterpart, namely backward walking, has, however, not become an old-learned movement after injury. Additional training of backward walking will therefore more efficiently improve forward walking by learning transfer. In conclusion, movements and especially automatisms have to be trained with their symmetry counterparts to enhance the efficiency of CNS repair.

4. Symmetry Improvement

It could be shown that in the patient with cerebellar injury, the symmetries in the arm-leg pattern dynamics were strongly impaired. Symmetry of the pattern dynamics is a property that plays an important role in the structuring of the coordination of rhythmic movements [16]. Front-rear symmetry was found to account for differences between straight and reverse walking (98). Not only could mirror-picture symmetry impairments of antagonistic muscle activation for exercising in the forward and backward directions be seen on the movement level (Figure 70) but also on the motor pattern level (Figure 67). Since after spinal cord injury there is also impairment of the premotor spinal oscillators for activation for the motor programs, neural assemblies are also impaired [35] in their self-organization with respect to symmetry as are their driving muscle spindle afferents. Premotor spinal oscillators have two driving phases from secondary muscle spindle afferents per oscillation period (180° apart (Chapter V, part 3, Figure 5)). In-phase and anti-phase symmetry seems thus to go down to the very single-neuron level as do its impairments and repair.

5. Symmetry and Antagonicity Improvement during Development in Comparison with Those After CNS Injury

CD between arm and leg movements in the forward and backward direction has been shown to substantially improve during CNS development by exercising on the special coordination dynamics therapy device (50) (Figures 22-25). Also, the symmetry between exercising in the forward and backward direction improved (50) (Figure 22B). Besides CD improvement, symmetry improvements may thus also be of importance to improve CNS functioning in the injured as well as normally functioning CNS. It seems worthwhile to further study improvement in CNS organization with respect to coordination and symmetries during individual development [51,52] (Figure 22) to see what can be learned from the improvements of CNS organization during individual development for CNS repair.

The antagonistic coordination between the right tibialis anterior and the right gastrocnemius muscles was impaired in the patient with cerebellar injury and improved for exercising in the forward and backward directions. In a six-month-old infant, nearly no antagonistic action could be measured between the tibialis anterior and gastrocnemius muscles (Figure 18b). With further development, this antagonicity will be established. With respect to antagonicity improvement of antagonistic muscles, there is therefore similarity between development and repair.

6. Giving Emphasis to Training Movements for Which the Motor Programs Are Best

The motor programs of the tibialis anterior and gastrocnemius muscles were shown to be better for jumping than for swinging on springboard (Figure 69). Since the motor programs were better still for jumping in anti-phase than for jumping in in-phase (Figure 69), jumping in anti-phase will be more beneficial for improving neuronal network organization than jumping in in-phase and should be trained more often. Also, the pathologic synchronized oscillatory firing of motor units (see below) seemed to be reduced in the anti-phase jumping mode. Jumping in anti-phase may activate more inhibition. Consequently, jumping in anti-phase should be predominantly trained in this patient at that repair stage.

7. Pathologic Synchronized Oscillatory Firing of Motor Units

Surface EMG (46) and recordings from single human neuron axons (Figure 70, Chapter III) showed that motoneurons and motor units fired coordinated oscillatory so as not to synchronize to avoid rhythmic movements of muscles and tremor. Because of impaired inhibition (lateral field and other inhibitions), motor units in patients with Parkinson's disease synchronize strongly and give rise to strong tremor [47]. There is indication that in patients with Parkinson's disease, oscillatory firing α_2-motoneurons (innervating FR-type muscle fibres) synchronize their firing to which α_1-motoneurons (innervating FF-type muscle fibres) synchronize (47). Synchronized oscillatory firing of motor units was also observed in patients after spinal cord injury (Figure 42) and in the patient with severe cerebellar (and brain) injury reported herein (Figure 69). Pathologic synchronized oscillatory firing seems to frequently occur in the pathologically organizing CNS. There was a suggested less synchronized oscillatory firing upon jumping on springboard in anti-phase (Figure 69D) than in in-phase (Figure 69B). This can be explained by the inhibition acting more during anti-phase than during in-phase jumping. Another or additional possibility is that observed in patients with Parkinson's disease. An integrative reorganization mechanism was induced, which reduced synchronization of oscillatory firing motoneurons [48]. The rhythmic dynamic stereotyped movement of jumping in anti-phase on springboard organized a motor program (which improved in the short-term memory) to which synchronized motor units became coordinated, resulting in reduced synchronization of the motor units.

3.2.9. Health Care in Aging

1. Improvement of CNS Functioning in Aging

CDT was administered for ten years to a 60-year-old patient with cerebellum (and cerebrum) injury. The higher mental functions improved substantially (Figure 61, see above). Also, the health of elderly patients with Parkinson's disease improved upon CDT [45]. For sure, CDT can improve the health in aging. Further, elderly people are often quite open to movement-based health care. Especially the special CDT device is very helpful to exercise at home every time they want.

Figure 71. Exercising on the special CDT device with crossed arms to train enhanced communication between the right and left cerebral hemisphere via the corpus callosum. One hand holds the leaver in suppination and the other hand in pronation for mechanical reasons.

Since exercising on this instrument is not only improving the fitness, but also the functioning of the nervous system, and the nervous system is involved in nearly all body functions including the cardiovascular performance and microcirculation, it seems possible to live longer with a higher quality of life upon performing CDT every day for one to two hours. Continence functions will also improve. Healthy nutrition (including garlic and ginger), training on the CDT device, good sleep, and little stress are beneficial for healthy living. Especially eating garlic and ginger and exercising on the special CDT device seems to be beneficial to improve the blood supply of the body, since the exercising on the special CDT device is activating the arterioles, via the autonomic nervous system, which contribute substantially to the microcirculation. Exercising on the special CDT device is beneficial for sleeping well. Especially when waking up at night and not being able to sleep again, a 30min training will help to start to sleep again. How the mechanism coordinated arm and leg movements can influence the sleeping pattern is unclear. Maybe the stability of the sleeping pattern is improved in the short-term memory.

To train and use both halves of the cerebrum, training of right-left communication as shown in Figure 71 is also beneficial. By throwing a ball from the right to the left hand and backwards, one can feel the improvement of CNS functioning. The right-left communication mainly via the corpus callosum can also be trained for elderly with hand problems upon using the special CDT device and exercising with crossed arms (Figure 71). In stroke patients, one hand may have to be fixed to the leaver. It seems that the turning with crossed arms increases power generation because a certain power generation seems to be easier with crossed arm turning, as if the increased integrative CNS activation increases muscle power generation.

Actually, the cerebral palsy child above with a malformation of the corpus callosum should train on the special CDT device additionally with crossed arms to improve the corpus callosum functioning, which was impaired by the malformation of the brain. The problem so far was that the patient was not cooperative enough to exercise that movement.

A limiting factor to enhance brain health, plasticity and vascularisation in elderly was so far that not enough nerve cells could be built anew and microcirculation stimulated. However, by physical exercise in elderly neurogenesis can most likely also be stimulated, as is shown by animal experiments.

2. Indication of Neurogenesis in Aging upon Movement-Based Learning, Derived from Animal Experiments

Probably, the rate of hippocampal neurogenesis significantly changes on physical activity, stress and aging. It is likely that hippocampal learning and memory formation holds also for other brain parts but to a lesser extent. Stress and aging are among the most potent negative regulators of neurogenesis. Exposure to physical activity and learning, however, increases the generation of new neurons. This leads to the hypothesis that the rate neural circuitry changes of the brain is constantly adapting to the needs [77]. Extrinsic factors of adult hippocampal neurogenesis, such as growth factors and neurotransmitters, secreted by the stem cell niche, diffusing from the nearly vasculature or released from synaptic connections, relay the environmental influence onto progression of neurogensis in the adult hippocampus. Proliferating cells and putative neural progenitors in both subgranular zone of the dentate gyrus and subventricular zone of the lateral ventricles are closely associated with the vasculature, indicating that factors released from the blood vessels may have a direct impact on adult neural progenitors [88]. Hippocampal neural network activity in the physiological context (see Chapter VI), which means, for example, network activity generated by movement-based learning, is a potent stimulator of adult hippocampal neurogenesis [100]. Astroglia is supposed to induce neurogenesis from adult neural stem cells [102]. During stem cell therapy in spinal cord injury (Chapter VI), the astrocytes will not contribute to the stimulation of neurogenesis because outside the spinal cord matter there is no astroglia. But if in elderly only the nerve cells die, then the astrocytes in the neurogenic niche can contribute to the control of neurogenesis.

The brain-derived neurotrophic factor (BDNF) is a neurotrophin that promotes the proliferation of neural precursors and the maturation and integration of newly generated neurons in adult hippocampal neurogenesis. It is normally synthesized in mature neurons in the granular cell layer of the dentate gyrus at much higher concentration than in other neurons of the brain and is localized in their axons [100]. In Chapter I, it was shown that the neurotrophic factors were secreted by the axons from two different kinds of motoneurons to modulate the functions of two kinds of muscle cells. By denervating the muscle fibres (cutting the nerves), the slow muscle fibre properties changed in dependence on the length of the nerve stump. With longer remained nerve (axon) pieces, changes occurred later, indicating that a larger amount of neurotrophic factors remained in the longer nerve stump.

The proliferation of neural stem/progenitor cells is enhanced by physical activity, whereas their fate determination and maturation are strongly impaired in aging and stress [77]. Interestingly, physical activity counteracts some adverse effects of aging on neural maturation [101]. Voluntary running of old animals augments the number of their newborn neurons to the same extent as running of old animals. Despite the running-mediated increase in neurogenesis in old animals, the total number of neurons they generate is still much lower than in non-running young animals. These animal experiments make it likely that movement-based learning in elderly humans induces neurogenesis. But to have the adult-born neurons available at sites where they are needed following cell death, different kinds of movements have to be trained to induce excitation-neurogenesis coupling (see Chapter VI) at sites where repair is needed. There is therefore some similarity between aging and CNS injury. For the repair of the injured CNS, excitation-neurogenesis coupling has to be stimulated at the sites of CNS injury. For the improvement of CNS functioning in elderly people, excitation-neurogensis coupling has to be induced at all the sites of cell death and degeneration.

CDT cannot only repair the injured or aging nervous system but can also improve CNS functioning in general. Movement performance can be improved by this special movement-based learning therapy in sportspeople, musicians and others.

3.2.10. Improvement of Movement Performance in Sportspeople, Musicians, and Other Healthy Adults

1. CD Values in Groups of Healthy Adults

Low-load [42] and high-load CD [44] have been measured in groups of athletes, musicians, physiotherapists, and gymnasts. The musicians had better low-load CD ($5.2s^{-1}$) than athletes ($6.0s^{-1}$). The high-load CD between arms, legs, and trunk movements were, on the other hand, much better in athletes. This can be understood because musicians train primarily the fine control of finger, hand, arm, and trunk movements, whereas athletes mainly train the coordination of movements under power conditions. Surprisingly, the gymnasts did not have an especially good low-load (5.6) and high-load CD as would be expected from their movements with music. Some gymnast teachers (and a sports teacher) were not very good, got angry about this objective assessment, and did not want to be assessed again. Athletes, musicians, and physiotherapists (low-load CD = 5.9) improved their high-load CD substantially upon repeated assessment [44], as was measured in pupils in more detail above (Figure 30-33).

2. Improvement of the Coordination of Movements in Sports and Music

The injured, malformed, or degenerated human CNS can be improved in its functioning by movement-based learning. Since the healthy CNS has a higher capacity in learning, and even the healthy CNS is not functioning optimally, it is obvious that the healthy CNS can also be improved in its functioning by movement-based learning. The improvement of the coordination of finger, hand, arm, trunk, leg, and feet movements, generated by the improvement of the coordinated firing of neurons, can improve the performance in sports and music. Sportsmen will run, jump, and play better with balls and musicians will play better piano or violin if they improve their CNS by movement-based learning. The movement-based learning has to be adapted to the special sport or the music to be performed. An improvement of the performance by 1% could be already important in sports. Whether an athlete runs 100m in 10.0s or 9.9s makes a big difference. Due to first trials, such an improvement seems possible.

In addition to the improvement of the special task, the functioning of the CNS in general has to be improved. An athlete has also to be fit in general and also his CNS has to function optimal in general. A violin player, playing in an orchestra for a few hours, performs in an unhealthy posture and may get posture problems; compensatory movements have to be trained to not overload certain body parts. As we know from sportsmen, musicians, and dancers, the psyche is involved in (movement) performance – too much stress has to be avoided and if occurring, it has to be reduced.

3. Supercompensation, Overcompensation, Load Escape, and CD in Healthy People

It was shown that the high-load CD improved in a healthy pupil upon repeated exercising (Figure 30). At the tenth session of high-load exercising, the high-load CD values were transiently best (smallest). Such transient period of best performance when going to the limits is called super-compensation. Further, the overall CD recordings for high-load testing show a transient reduction of the frequency at high loads to escape the load. Such period of turning at a reduced frequency is called "load escape." As Figures 72D,E,F show, the load escape is smallest during the phase of super-compensation (Figure 72D) with respect to before (Figure 72E) and after (Figure 72F).

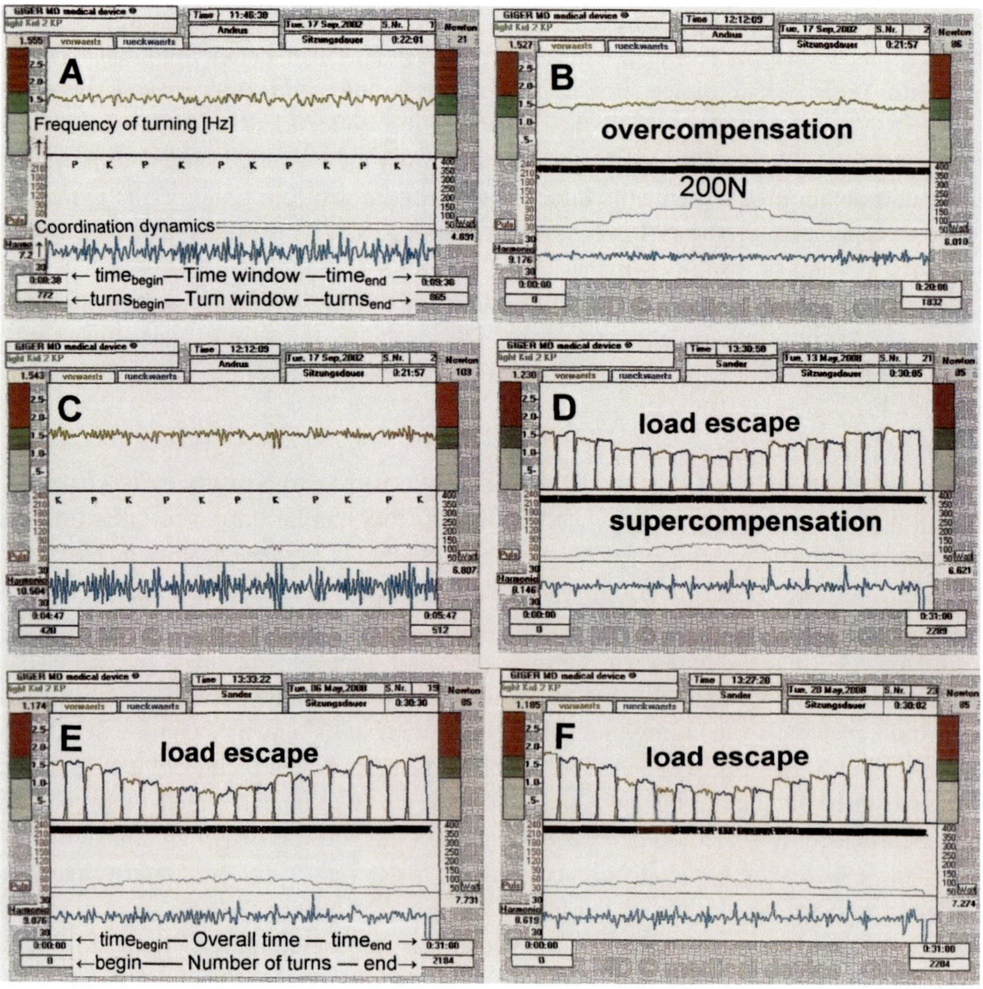

Figure 72. High-load coordination dynamics (CD) assessment in an athlete (A-C) and a pupil (D-F). Note that with increasing load, the athlete increases the frequency of exercising (overcompensation), and the pupil decreases the frequency of exercising to escape the load (load escape). Note further that during supercompensation (D), the load escape is less than before (E) and after (F).

This means during the period of supercompensation, the CNS was working best so that the pupil could develop more coordinated power and could better manage with the high-load (exercised with a higher frequency). A pupil athlete of the same age showed also a reduced load escape during supercompensation. This means a sportsman can better run, jump or play football or tennis during the phase of supercompensation. A fit adult can also turn without load escape, after plenty of exercising on the special CDT device (Figure 29A). A top athlete shows no load escape but overcompensation (Figure 72B). When during high-load assessment the high loads are coming, he is exercising faster. This faster turning is probably not due to a better CNS organization but due to a better motivation. Even though during development, the CNS tried to improve it's functioning by a faster exercising (Figure 27).

If a top athlete exercises with overcompensation, it does not mean that his CNS functions optimally. The author (bcing quitc fit) could exercise on the special device without load escape but did not exercise with overcompensation. But his CD values for low and high load were better (Figure 29B,C) than those of the top athlete (winner of Olympic Games in spear-thrust) (Figure 72A,C). The suboptimal CD values and the rhythmic changes on CD trace (Figure 72C) indicate that this top athlete could probably perform his sport even better if he would have had improved the coordination among arms, legs and trunk, which means the coordinated firing of neurons and cell assemblies, upon CDT.

4. Accuracy of Motor Patterns

To optimize a movement like running or playing piano, it is not sufficient that the muscles work in proper coordination with the needed power. The motor programs of the different muscles (and their coordination) have to be optimized. Bursts of action potentials run along the different kinds of motoneuron axons (Figure 6) down to the muscle fibres of the different motor units and give rise to the different kinds of motor unit action potentials (Figures 9,28). These bursts of motor units have to be optimized in their shape and timing in the way that the muscles work in coordination with each other, especially the antagonistically working muscles, and that the shape and length of the bursts are optimally formed or timed respectively. During early development, these motor programs are far from being optimally arranged (Figure 18); the antagonicity between antagonistic muscles is not fully developed in the child (Figure 18). In the mother, the motor program was much better (Figure 1C). From Figure 10 and 11, it can be seen that the muscle motor burst are generated by the transient firing of the different motor units. Some motor units fire transient oscillatory (for the burst time) and other ones only occasionally. If more power is needed, more motor units fire transiently oscillatory with increasing frequency. In Figure 10A,B, this change from occasionally firing to transiently oscillatory firing can be seen. When the patient exercised at 100N on the special CDT device, motor unit 2 and 3 fired only occasionally in the right tibialis anterior (Figure 10A). When the patient exercised against a load of 150N, motor units 2 and 3 fired rhythmically (transiently) at a low frequency. The change of the firing patterns from occasional to rhythmic firing in motor neuron axons for increasing muscle power has nicely been shown for skeletal continence muscles.

The "on and off" switching of the motor units has to be timed exactly by the neuronal networks in the spinal cord and upstream networks. If, for example, a motor unit was switched on to fire oscillatory but could be switched off, then this motor program of that muscle would be impaired. The relation between contraction and relaxation of the muscle would be impaired. Uncontrolled motor unit firing has been recorded from patients with spinal cord injury (Figure 48 of Ref.38) or patients with Parkinson's disease (Figure 14). The worst cases of uncontrolled neuron firing can be encountered in brain injuries (epilepsy). Uncontrolled continuously activated neurons activate muscles or muscle groups continuously. Such uncontrolled continuous activation is called spasticity. In healthy humans, the optimal activation and relaxation of motor units and whole muscles can be trained by training the neuronal networks generating the motor patters. Muscle-tendon properties also influence the contraction-relaxation properties. If tendons and muscles get too short because of a pathologic motor program like in extensor spasticity in patients with brain injury, then the motor patterns improved by training are ineffective. A too short Achilles tendon in extensor spasticity blocks the walking pattern.

To better understand the functioning of the healthy human CNS, we have to explore the injured human CNS, and this is possible in human patients with injuries, malformation, and degeneration. This research strategy is in similarity to animal research. To study the functions of the CNS in animals, the CNS is injured and the regeneration, adaptation, or response to the injury is studied. The argument that the animal CNS data hold also in humans is wrong. The human is much more intelligent than rats or monkeys. A translation of animal data into human data is only possible if parallel experiments are performed in animals and humans or if similar knowledge is available from humans. An example will be given below that slightly different membrane properties in humans can already result in quite different nervous system functioning.

3.2.11. Different Membrane Properties in Animal and Human with Possible Consequence for Intelligence

It is believed that intelligence increases with the amount of CNS matter; that is how many nerve cells the brain and the spinal cord consist of and the connectivity of these nerve cells. But membrane properties of nerve fibres may also influence the intelligence, as will be analysed below. This thinking is in some similarity to the improvement of nervous system functioning, when nature introduced the myelinisation of nerve fibres for faster communication, because a higher conduction velocity can be achieved if the excitation is jumping from one node of Ranvier to the next one instead of being continuously conducted along very thick unmyelinated nerve fibres, occurring for example in the squid (giant axon).

Medical students learn in physiology lessons that the fastest conduction velocities in thick, thickly myelinated nerve fibres are in the range of 120m/s. For References of old animal classification schemes, see [60]. These measurements were brilliant for their times but out-of-date for modern electrophysiology. The medical students do not learn that the highest conduction velocities in thick, thickly myelinated human nerve fibres are in the range of 60 to 65m/s. An exact classification scheme for human peripheral nerve fibres down to 4μm is known for long [30] and given in Chapter III. This scheme is not used for the teaching of medical students, even though the classification allows the identification of the type of human nerve fibres in natural impulse traffic in human nerve roots (Figure 7) and has therefore consequences for the understanding of the functioning of the human nervous system.

That the peripheral nerve fibres of humans conduct much slower than in animals is astonishing because we would expect in humans higher conduction velocities. The human is larger than most laboratory animals and would need high conduction velocities for quick communication among neurons. The question is, why do human peripheral nerve fibres conduct more slowly? The most likely explanation is that human membrane properties of peripheral nerves are different from those of animals, because the human CNS needs a higher communication speed between cells, cell assemblies and subnetworks for higher intelligence.

Since afferent and efferent nerve fibres are sited inside and outside the CNS, we can expect that the membrane properties of nerve fibres, sited completely inside the CNS, also conduct with a lower velocity. For a higher intelligence, one would expect a higher communication between neurons, neural assemblies, and subnetworks in different CNS nuclei. And probably that is so. But for a complex communication in the CNS, the conduction velocity of nerve fibres is only one parameter. The transmission frequency (limited by the inactivation time of excitable membranes of nerve fibres) is a second important parameter. Firstly, a shorter inactivation time following an action potential allows more complex patterns to be transmitted per time interval. Secondly, if the CNS organization is not only accomplished by the forming of isolated loops between different network parts but also by crossing of excitation streams, then the transmission frequency is very important. Abeles showed that excitation can cross by synfire chains [1]. The crossing of excitation streams depends crucially on the transmission frequency (and the exactness of coordinated firing of neurons). It seems that the transmission frequency in human nerve fibres is much higher than those in animals.

The transmission frequency was measured in humans (Chapter V) and was up to 5,000Hz. Medical students learn that maximum transmission frequencies are in range of 500Hz to 1,000Hz. Most likely, the transmission frequencies will vary among animals. It could well be that the membrane properties of the squid giant axon (diameter up to 1mm; unmyelinated) [14] are different from those of myelinated human nerve fibres. On the other hand, the membrane properties of human nerve fibres will not be completely different than those in animals, because the recorded extracellular single-nerve fibre action potentials (recorded with the same equipment) are similar to those in animals. Actually in a frog, single-nerve fibre action potentials can be recorded also nicely from peripheral nerves [28] (and not only from nerve roots), because in the frog, the nerves have nearly no epineurium shunting the potentials. The frog is therefore still a very suitable animal for demonstrating to students nervous system functioning.

The higher transmission frequency of nerve fibres in human allows the excitation streams to cross more easily (for example by synfire chains). They are not blocked so much by the inactivation times of the nerve fibres. This means that, for example, during 0.1s, a nucleus (or subnetwork) of the human CNS can communicate with more other nuclei or subnetworks than the CNS in animals to generate a movement or other pattern. The functional network complexity in human is higher than that in animals – the human CNS can work on a higher level. A higher intelligence is possible. More complex network organizations can be generated for a given neuronal network connectivity. The exactness of timed firing of neural assemblies and subnetworks (the phase and frequency coordination) is perhaps more important for humans than for animals because of the probably increased communication rate due to the higher transmission frequency.

During human development, the myelinisation is increasing. Nerve fibres can conduct faster. The CNS of the developing child can organize itself faster. A higher complexity of CNS organization becomes possible. The child becomes more intelligent, independent of network complexity, which of course substantially contributes to intelligence. But because the human CNS seems to give priority to the increase of the transmission frequency at the price of a lower conduction velocity, it seems that the crossing of excitation streams for the self-organization of neuronal networks is more important than the speeding up of communication by improving the conduction velocity. If the self-organization of a pattern would be achieved by many formed loops to other subnetworks, then the crossing of loop excitation would be more important than the speed in the loop. This would mean that the crossing of excitation streams is an important part of CNS organization, even though we do not know how the self-organization takes place. If very many self-organized excitation stream loops would be involved, then loop crossing would be an important organizational feature.

A few principle long-range communication loop lines are indicated in Figure 40B. Above, in the case of cerebral palsy, some principle consequences were analysed when the thalamus was injured on one side. It was not discussed that in thousands to millions single loops in and between nuclei there were blocks or delays, mixing the coordinated firing of neurons completely up. It is astonishing that such damaged CNS still somehow works and can even be improved in its functioning by movement-based learning. In conclusion, the price for a higher transmission frequency for more complex subsystem communication in human is the lower conduction velocity. Membrane parameters (like inactivation times), measured by electrophysiological tools, contribute to the understanding of CNS functioning.

3.2.12. Pathologic Pattern Stimulation upon Electro Stimulation – Physiologic Pattern Stimulation from Natural Afferent Impulse Patterns from the Skin

When stimulating the skin (Figure 73G,H), the afferent input from the skin can nicely be recorded (Figure 73A-E). By waveform comparisons on a stretched time scale (Figure 73B-D), sets of natural impulse patterns of different skin mechanoreceptors can be identified (Figure 73Eb,F). The activity patterns (Figure 73A) show a touching and a releasing part of activity as can be seen from the action potential appearance of large amplitude units (T1 (Parcinian Corpuscle) and T2) and from the schematic activity pattern of the $T1_1$ unit (Figures 73Eb,F). All the 4 touch (T1 until T4) and 1 pain groups, recorded here, seem to have their own touching and releasing part but according to their latencies and conduction velocities, at different times. It can be seen from Figure 73 that the touching activity part of the slower conducting T3 and T4 fibres with their smaller action potential amplitudes (Figures 73A,C) started about in the middle of the T1 and T2 activity part and lay mainly between the touching and the releasing parts of the T1 and T2 units. The releasing activity of the T3 and T4 fibres appeared after that of the T1 fibres. The activity of the pain fibres (here the slowest conducting fibres) appeared even later at the recording electrodes and stopped appearing after the end of the pinpricking (recorded at the proximal cauda equina, Figure 6).

The very late pain after pinpricking that one normally also feels is probably conducted by very thin myelinated or non-myelinated fibres and cannot be recorded under such conditions. The different delays and impulse train lengths of the T1 units are interpreted with the spatial distribution of the T1 receptors and the skin indentation upon touch and pinpricking (Figure 73H). The $T1_1$ unit, for example, had its receptor nearest to the needle and responded fastest and longest since the skin indentation first reached the $T1_1$ receptor, gave rise to the largest skin shift and turned away last. The $T1_2$ receptor, being further away from the centre of skin touch, was reached later by the skin indentation; the skin shift was not so large and the skin turned back to the former position quicker.

The action potentials in the Tl_2 unit therefore appeared later, showed only two (extracellular) action potentials in the impulse train, and the release activity part also appeared earlier.

Such natural impulse patterns in many skin afferents inform the CNS about changes in the periphery of the body and give rise to certain network organizations. The neuronal networks of the spinal cord (and supraspinal centres) do self-organize by the natural impulse patterns from the receptors in the periphery (in SCI) and from supraspinal centres (physiologically both contributions). It is difficult to say whether the networks adapted their self-organization to the impulse patterns of the receptors of the periphery or whether, during phylogeny or ontogeny, the receptors adapted their firing to the needs of the neuronal network. Nevertheless, it is obvious that the natural firing patterns of the afferents and their space-time distributions give rise to the self-organization of the neuronal networks and are therefore of utmost importance for the understanding of the functioning of the CNS. Such parameters are difficult to measure.

To make a conclusion on the organization and reorganization of neuronal networks from the wiring of the neurons is risky because one would need to know, for example in humans, the efficacies of the excitatory and inhibitory synapses (LTP, LTD), the membrane properties of local critical areas of neurites, and how membrane properties change with learning and reorganization following CNS injury. Two of the original recordings of simultaneous skin afferent patterns of 14 and 22 receptors are shown in Figure 73. The summed natural firing patterns upon touch (Figure 73Ea) are split and are schematically drawn in Figure 73Eb, and those upon pinprick (Figure 73A-D) can be seen in Figure 73F.

Electrostimulation of nerves (or skin) in patients cannot mimic such spatio-temporal impulse codes elicited by natural stimulation and will give rise to non-natural neuronal network organization of the CNS. When performing electrostimulation repeatedly in patients for repair, it should be therefore clear what network states are about to be activated in the CNS and how the artificial patterns differ from the natural pattern states. In most kinds of electrostimulation, pathologic patterns will be trained in the patient with CNS injury or malformation and the physiologic patterns impaired, because the injured CNS is very vulnerable and instable. In healthy humans, the CNS is much more stable and has many safety margins so that the electrostimulation is probably not doing much harm. Also upon electrical muscle stimulation, the muscles are activated unphysiologically. The oscillatory distributed firing of the many motor units of a muscle with respect to time and space cannot be mimicked by electrostimulation because all the muscle fibres are activated with a certain frequency simultaneously. Even the tremor in Parkinson's patients, where motor units synchronize their oscillatory firing, is more physiologic [47]. The use of simple electrostimulation in neuro-rehabilitation must have other reasons than the improvement of the health of patients. In basic research, it also should be borne in mind that artificial electrostimulation patterns give rise to artificial neuronal network organization of the CNS. Electrostimulation is therefore not suitable to analyse natural integrative functions of the CNS. In Ref.63, the fact is discussed that the spatio-temporal touch afferent patterns (like touch 6 in Figure 73Eb) give rise to a self-organized network state subserving anal continence, whereas additional activated pain afferents (like pinpricking 1 in Figure 73F) stimulate a network state inducing an escape and protection response. For further details, see Chapter V, part 4, Figure 8.

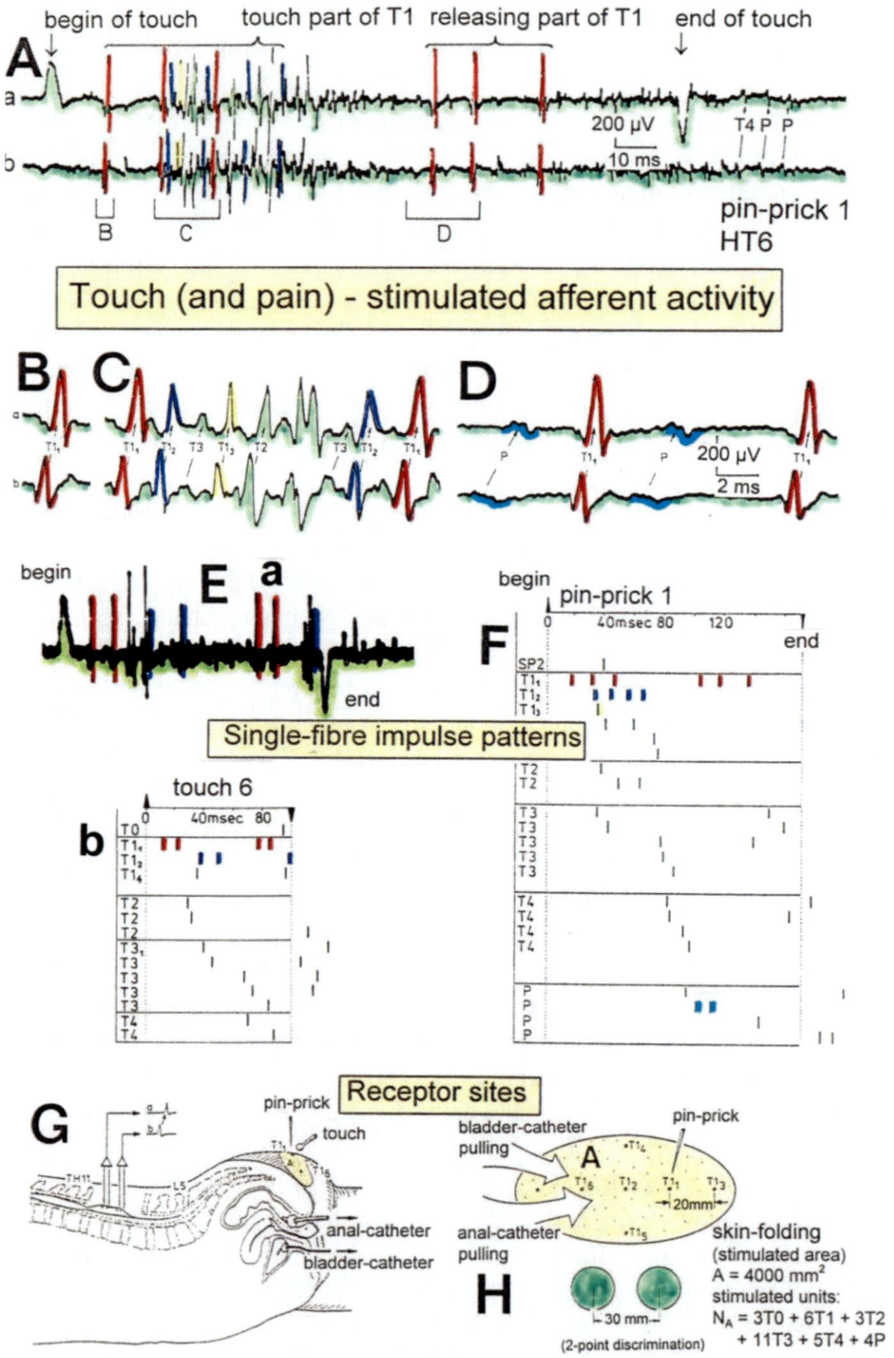

Figure 73. Touch (and pain)-stimulated afferent activity. Touch and pain activity stimulated by pinpricking (A) and touching (Ea) at S5 or Co dermatomes and recorded extracellularly from a dorsal coccygeal root (brain-dead human HT6). T1, T2, T3, T4, P = mark action potentials (APs) from single touch and pain fibres. Subscripts 1, 2, 3 mark single fibres. A. Whole sweep following pinprick 1 shown at a slow time base. The large upward artefact on trace "a" marks electronically the beginning of the pinprick. The large downward artefact on trace "a" marks the end of the pinprick. Note that two intervals of high activity of large APs occur, one after the beginning of the pinprick with one AP in front, and a second before the end of the pinprick; potentials with small amplitude follow potentials of large amplitude. Time intervals B, C and D are shown in

a time-expanded form in Figures B, C and D. B, C, D. Time expanded sweep pieces of A. Identified APs are indicated. Note that the APs from the $T1_1$ touch unit can be safely identified by the waveforms in B, C, D. Eb, F. AP occurrence patterns of single touch and pain fibres following short touch 6 and pinprick 1. No pain afferents are stimulated upon touch 6. Upon pinprick 1, the single-fibre AP activity of the different touch and pain groups is identified by the AP waveforms on traces "a" and "b," and by the conduction times. The single touch afferents of the T1 group are marked with subscripts. One active secondary muscle spindle afferent fibre (SP2) could always be identified in F. Note that for pinprick 1, touch and pain afferents are stimulated whereas for touch 6 only touch afferents. G. Recording and stimulation arrangement for simultaneous recording of several single touch and pain units. A = area stimulated by skin folding, drawn in H in more detail. $T1_1$, $T1_6$ = suggested touch points of the $T1_1$ and $T1_6$-units. H. Drawing of the very approximate skin area stimulated by skin folding. $T1_{1\text{-}6}$ = suggested focal T1 touch points. Two-point discrimination indicated for the sake of comparison. N_A = number of stimulated units in the dorsal coccygeal root. Skin tractions evoked by anal and bladder-catheter pulling are indicated by the large open arrows.

3.2.13. Improvement after CNS Injury in Comparison with Physiologic Development with Respect to Urinary Bladder Functioning

It is difficult to judge whether there is similarity between development and repair of the human CNS. The functioning of the urinary bladder as a complex CNS organization pattern will be used for comparison; the scientific basis for repair has been given in detail [60-64] (Chapter V). Healthy infants become continent within two to three years. The automatism centres for emptying urinary bladder and bowel are sited in the sacral segments S2 to S5 and are operational postnatally. Only the supraspinal centres to volitionally control the bladder and bowel functions need more time to mature because more complex neuronal networks have to be built that do communicate with different subnetworks simultaneously. If the children have stress for different reasons later on, incontinence may reappear. Bed wetting is unvolitional, and the child is suffering itself. Higher mental functions are substantially involved in urinary bladder control.

A boy suffered a thoracic SCI and obtained CDT at an age of six. Upon 0.5 years of therapy, he became continent [43] again, and he valued it. The biggest motivation for him to be cooperative during therapy was the feeling that his bladder functions improved. A seven-year-old cerebral palsy child became continent upon 0.5 years of intensive CDT (cerebral palsy case report, above). Even though her higher mental functions were low, she also valued it to be continent at night. In brain injury or malformation, the patients are only incontinent if the damage is severe. A 17-year-old female patient with a severe cervical SCI (Figure 36B) needed two to three years to be able to control bladder and bowel again [64] (Figure 5). An 18-year-old patient, who suffered an incomplete cervical SCI (58) (Figure 40), needed 0.5 years to become independent from intermittent catheterization. Incontinence in aging probably has different reasons. One unknown reason is that the artery of Adamkiewicz, the main feeding artery to the sacral micturition centre (Chapter III), becomes severely sclerotic during aging [29]. Since motor, vegetative, and higher mental functions are involved in the control of bladder and bowel, the repair needs complex network reorganizations. The repair of bladder and bowel functioning seems to depend on the severity of the CNS injury and the age of the patient. In children, the repair is much faster. A more complex understanding of urinary bladder functioning and repair will be given in Chapters III and V.

3.2.14. Need for Human Cell Neurophysiology

The importance of the system theory of pattern formation for repair is that it gives the therapist a tool at hand to treat the patient and to understand the reason for the treatment.

The property stability gives him/her a goal at hand, which has to be achieved with treatment: The stability of physiologic (motor) patterns has to be increased and the stability of pathologic patterns (for example extensor spasticity) to be decreased. But to understand the stability of patterns according to the geographical landscape of attractors already, the variability of phase and frequency coordination ($(Q\xi_t)^{1/2}$) among neuron firing had to be used in formula 3 (see Method), which means human cell neurophysiology is needed. Further, to partly find the behavioural information ($\sum c_{inf}F_{inf}(X,t)$) (the therapy) through which it is possible to change and repair the intrinsic dynamics ($F_{intr}(X)$) of CNS organization human cell neurophysiology is also needed.

A practical example will demonstrate that human cell neurophysiology is needed to find out what exercising can efficiently contribute to the repair of the patient's CNS. The patient in Figure 74 is exercising on a self-made sky-walker (A) (to make the coordination of the movements more exact) and a seemingly similar device (B) with which it is possible that the patient with severe cervical SCI can perform coordinated trot gait movements. Exercising on the sky-walker is helpful for CNS repair and the other device not so much; it is mainly an instrument to train muscle power because the sky-walker is working more exactly and no weight support is needed.

The human cell neurophysiology of Chapters III and V can also be seen as another approach to understand the functioning and repair of the human CNS, since physiologic and pathophysiologic pattern formation is created by the collective effects of timed cooperative and competitive firing of neurons, neural assemblies and subnetworks.

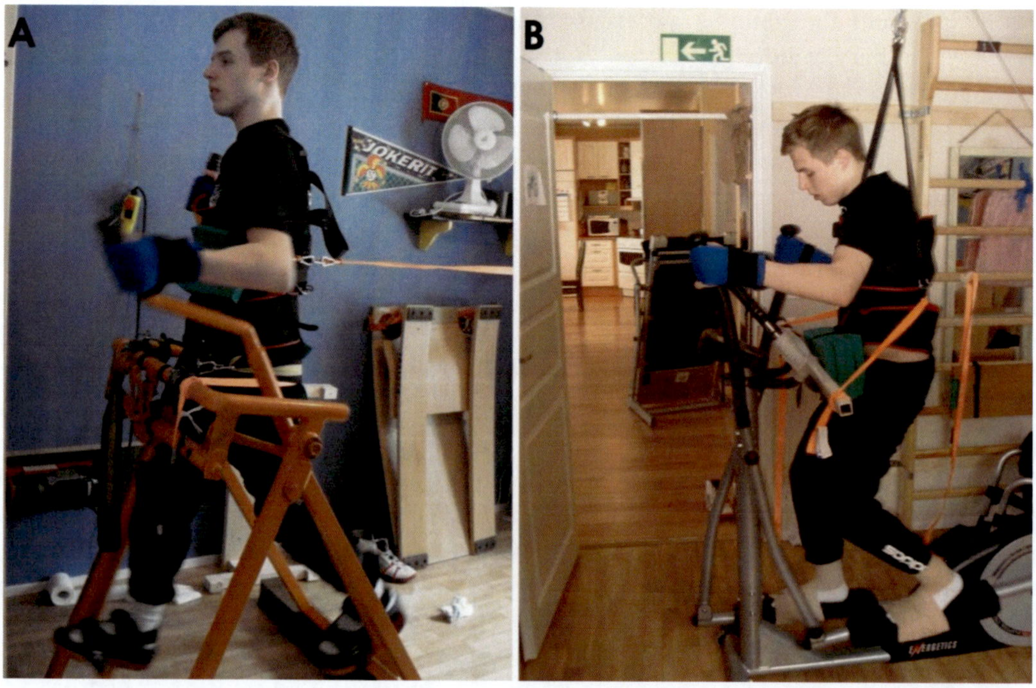

Figure 74. Patient with severe cervical spinal cord injury is exercising very coordinated trot gait movement without weight support (A). The training of this movement is beneficial to repair his injured spinal cord. Exercising trot gait coordination on a similar device (B) is only little beneficial to improve CNS functioning (it is mainly a device to improve muscle power), because the device is not working exact enough to teach his CNS coordinated firing of neurons and also weight support is needed.

References

[1] Abeles M. (1982): *Local cortical circuits: An electrophysiological study.* Berlin: Springer.

[2] Bear, M. and Frotscher, M.: Duus', *Topical Diagnosis in Neurology: Anatomy, Physiology, Signs, Symptoms.* Georg Thieme Verlag, Stuttgart, Germany, 2005. (a very good anatomy book with clinical consequences (a great work of Peter Duus), only the human neurophysiology has not been updated (not available at the times of Peter Duus)).

[3] Bernstein, N.: *The Coordination and Regulation of Movement.* London, Pergamon, 1967.

[4] Bhardwj, R.D. et al.: Neocortical neurogenesis in humans is restricted to development. *PNAS*, 103 (33): 12564-12568, 2006.

[5] Bradbury, E.J. and McMahon, S.B.: Spinal cord repair strategies: why do they work? *Nat. Rev. Neurosci.*, 7: 644-653, 2006.

[6] Capute, A.J., Accardo, P.J., Vining, E.P.G., et al.: *Primitive Reflex Profile. Monographs in Developmental Pediatrics, vol.1* Baltimore, University Park Press, 1978.

[7] Capute, A.J., Shapiro, B.K., Wachtel, R.G., et al.: Motor functions: associated primitive reflex profiles. *Pediatr. Res.*,14: 431, 1980.

[8] Fiorentino, M.R.: *A Basis for Sensorimotor Development – Normal and Abnormal.* Springfield, Carles C Thomas, 1981.

[9] Freund H.-J.(1996): Remapping the brain. *Science* 272: 1754.

[10] Fuchs, A., Jirsa, V.K., Haken, H., Kelso, J.A.S.: Extending the HKB model of coordinated movement to oscillators with different eigenfrequencies. *Biol. Cybern.* 74: 21-30, 1996.

[11] Gentile, A.M.: Skill acquisition, movement, and neuromotor processes. In: Carr, J.H., Sheperd, R.B. (Eds): *Movement Science. Foundation for Physical Therapy in Rehabilitation.* Rockville, M.D., Aspen Publications, 1987, pp 93-154.

[12] Haken, H., Kelso, J.A., and Bunz, H.: A theoretical model of phase transitions in human hand movements. *Biological Cybernetics* 39: 139-156, 1985.

[13] Harel, N.Y. and Strittmatter, S.M.: Can regenerating axons recapitulate developmental guidance during recovery from spinal cord injury? *Nat. Rev. Neurosci.*, 7: 603-616, 2006.

[14] Hodgkin, A.L. and Huxley, A.F.: A quantitative description of membrane current and its application to conduction and excitation in nerve. *J. Physiol.* (London) 117: 500-544, 1952.

[15] Horak, F.B.: Assumption underlying motor control for neurologic rehabilitation. In: Contemporary Management of Motor Control Problems. Proceedings of the II STEP Conference, Alexandria, V.A., *Foundation for Physical Therapy*, 1991, pp 11-27.

[16] Jeka, J.J. and Kelso, J.A.S.: Manipulating symmetry in the coordination dynamics of human movements. *Journal of Experimental Psychology*, 21 (2): 360-374, 1995.

[17] Kandel, E.R., Schwartz, J.H. and Jessell, T.M.: *Principles of neural Science.* Appleton and Lange, East Norwalk, 1991.

[18] Keirstead, H.S. et al.: Human embryonic stem cell derived oligodendrocyte progenitor cell transplants remyelinate and restore locomotion after spinal cord injury. *J.Neurosci.* 25: 4694-4705 (2005). (rat).

[19] Kelso, J.A.S.: Dynamic Patterns. *The Self-Organization of Brain and Behavior.* MIT Press, Cambridge, 1995.

[20] Kerschensteiner, M. et al.: In vivo imaging of axonal degeneration and regeneration in the injured spinal cord. *Nature Med.* 11: 572-577 (2005). (rat).

[21] Maguire, T.O. and Redfern, L.: Theories of motor development. In: *Motor development of the developing infant.* (Eds Piper, M.C. and Darrah, J., W.B. Saunders Company, Philadelphia, 1994.

[22] Martino, G. and Pluchino, S.: The therapeutic potential of neural stem cells. *Nat. Rev. Neurosci.*, 7: 395-406, 2006.

[23] McDonald, J.W. et al.: Transplanted embryonic stem cells survive, differentiate and promote recovery in injured rat spinal cord. *Nature Med.* 5: 1410-1412 (1999).

[24] McGraw, M.B.: *The Neuromuscular Maturation of the Human Infant.* New York, Macmillan, 1945.

[25] Piper, M.C. and Darrah, J.: *Motor Assessment of the Developing Infant.* W.B. Saunders Company, 1994.

[26] Popper, K.R.: *Objective knowledge: An evolutionary approach.* London: O.U.P., 1972.

[27] Rosenzweig, E.S. et al.: Extensive spontaneous plasticity of corticospinal projections after primate spinal cord injury. Nature Neuroscience, Advance online Publication, Nov., 2010.

[28] Schalow, G.: Single unit potential amplitude in relation to the conduction velocity in frog and human. *Zent.bl. Neurochir.* 48: 109-113, 1987.

[29] Schalow, G. Feeder arteries, longitudinal arterial trunks and arterial anastomoses of the lower human spinal cord. *Zent.bl. Neurochir.* 51: 181-184, 1990.

[30] Schalow, G.: Conduction velocities and nerve fibre diameters of touch, pain, urinary bladder and anal canal afferents and α and γ-motoneurons in human dorsal sacral roots, *Electromyogr. Clin. Neurophysiol.*, 31: 265-296, 1991.

[31] Schalow, G.: Oscillatory firing of single human sphincteric α_2 and α_3-motoneurons reflexly activated for the continence of urinary bladder and rectum - restoration of urinary bladder function in paraplegia. *Electromyogr. Clin. Neurophysiol.*, 31: 323-355, 1991.

[32] Schalow, G. and Barth, H.: Group conduction velocities and nerve fibre diameters of α and γ-motoneurons from lower sacral nerve roots of the dog and humans. *Gen. Physiol.Biophys.* 11: 85-99, 1992.

[33] Schalow, G.: Recruitment within the groups of γ_1, α_2 and α_3-motoneurons in dogs and humans following bladder and anal catheter pulling. *Gen. Physiol. Biophys.* 11: 101-121, 1992.

[34] Schalow, G. and Wattig, B.: Recruitment of α and γ-motoneurons in rats, dogs and human. *Electromyogr. Clin. Neurophysiol.* 33: 387-400, 1993.

[35] Schalow, G.: Spinal oscillators in man under normal and pathologic conditions, *Electromyogr. Clin. Neurophysiol.*, 33: 409-426, 1993.

[36] Schalow, G. and Zäch, G.A.: Nerve compound action potentials analysed with the simultaneously measured single-fibre action potentials in humans. *Electromyogr. Clin. Neurophysiol.* 34: 451-465, 1994.
[37] Schalow, G. and Zäch, G.A.: External loops of human premotor spinal oscillators identified by simultaneous measurements of interspike intervals and phase relations. *Gen. Physiol. Biophys.* 15, Suppl.1: 95-119, 1996.
[38] Schalow, G. and Zäch, G.A.: Reorganization of the Human CNS, Neurophysiologic measurements on the coordination dynamics of the lesioned human brain and spinal cord. Theory for modern neurorehabilitation (31 case reports). *Gen. Physiol. Biophys.*, 19, Suppl. 1: 1-244, 2000.
[39] Schalow, G.: Stroke recovery induced by coordination dynamics therapy and quantified by the coordination dynamics recording method. *Electromyogr. Clin. Neurophysiol.* 42: 85-104, 2002.
[40] Schalow, G.: Improvement after traumatic brain injury achieved by coordination dynamic therapy. *Electromyogr. Clin. Neurophysiol.*, 42: 195-203, 2002.
[41] Schalow, G.: Recovery from spinal cord injury achieved by three months of coordination dynamic therapy. *Electromyogr. Clin. Neurophysiol.* 42: 367-376, 2002.
[42] Schalow, G. and Pääsuke, M.: Low-load coordination dynamics in athletes, physiotherapists, gymnasts, musicians and patients with spinal cord injury, after stroke, traumatic brain lesion and with cerebral palsy. *Electromyogr. Clin. Neurophysiol.* 43: 195-201, 2003.
[43] Schalow, G.: Partial cure of spinal cord injury achieved by 6 to 13 months of coordination dynamic therapy. *Electromyogr. Clin. Neurophysiol.*, 43: 281-292, 2003.
[44] Schalow, G., Pääsuke, M. and Kolts, I.: High-load coordination dynamics in athletes, physiotherapists, gymnasts, musicians and patients with CNS injury. *Electromyogr. Clin. Neurophysiol.* 43: 353-365, 2003.
[45] Schalow, G., Pääsuke, M., Ereline, J. and Gapeyeva, H.: Improvement in Parkinson's disease patients achieved by coordination dynamics therapy. *Electromyogr. Clin. Neurophysiol.*, 44: 67-73, 2004.
[46] Schalow, G.: Phase and frequency coordination between neuron firing as an integrative mechanism of human CNS self-organization. *Electromyogr. Clin. Neurophysiol.*, 45: 369-383, 2005.
[47] Schalow, G. Tremor in Parkinson's disease patients can be induced by uncontrolled activation and uninhibited synchronization of α_2-motoneuron firing to which α_1-motoneuron firing synchronizes. *Electromyogr. Clin. Neurophysiol.*, 45: 393-406, 2005.
[48] Schalow, G., Pääsuke, M., Jaigma, P.: Integrative reorganization mechanism for reducing tremor in Parkinson's disease patients. *Electromyogr. Clin. Neurophysiol.*, 45: 407-415, 2005.
[49] Schalow, G., Jaigma, P. Cerebral palsy improvement achieved by coordination dynamics therapy. *Electromyogr. Clin. Neurophysiol.*, 45: 433-445, 2005.
[50] Schalow, G.: Functional development of the CNS in pupils aged 7 to 19 years. *Electromyogr. Clin. Neurophysiol.*, 46: 159-169, 2006.
[51] Schalow, G.: Hypoxic brain injury improvement induced by coordination dynamics therapy in comparison to CNS development. *Electromyogr. Clin. Neurophysiol.*, 46: 171-183, 2006.

[52] Schalow, G. and Jaigma, P.: Improvement after severe traumatic brain injury induced by coordination dynamics therapy: Comparison with physiologic CNS development. *Electromyogr. Clin. Neurophysiol.*, 46: 195-209, 2006.

[53] Schalow, G.: Surface EMG- and coordination dynamics measurements-assisted cerebellar diagnosis in a patient with cerebellar injury. *Electromyogr. Clin. Neurophysiol.* 46: 371-384, 2006.

[54] Schalow, G.: Symmetry diagnosis and treatment in coordination dynamics therapy. *Electromyogr. Clin. Neurophysiol.*, 46: 421-431, 2006.

[55] Schalow, G.: Cerebellar injury improvement achieved by coordination dynamics therapy. *Electromyogr. Clin. Neurophysiol.*, 46: 433-439, 2006.

[56] Schalow, G., Vaher, I. and Jaigma, P.: Overreaching in coordination dynamics therapy in an athlete with a spinal cord injury. *Electromyogr. Clin. Neurophysiol.* 48: 83-95,2008.

[57] Schalow, G.: Stem cell therapy and Coordination dynamics therapy improve Spinal cord injury. *Electromyogr. Clin. Neurophysiol.*, 48: 233-253, 2008.

[58] Schalow, G., Jaigma, P., Belle, V.K.: Near-total functional recovery achieved in partial cervical spinal cord injury (50% injury) after three years of coordination dynamics therapy. *Electromyogr. Clin. Neurophysiol.*, 49: 67-91, 2009.

[59] Schalow, G.: Partial cure achieved in a patient with near-complete cervical spinal cord injury (95% injury) after three years of coordination dynamics therapy. *Electromyogr. Clin. Neurophysiol.*, 49: 199-221, 2009.

[60] Schalow, G.: The classification and identification of human somatic and parasympathetic nerve fibres including urinary bladder afferents and efferents is preserved following spinal cord injury. *Electromyogr. Clin. Neurophysiol.*, 49: 263-286, 2009.

[61] Schalow, G.: Coordination impairment between the somatic and parasympathetic nervous system divisions in the human sacral micturition centre following spinal cord injury. *Electromyogr. Clin. Neurophysiol.*, 49: 337-367, 2009.

[62] Schalow, G.: Phase relation changes between the firings of α and γ-motoneurons and muscle spindle afferents in the sacral micturition centre during continence functions in brain-dead human and patients with spinal cord injury. *Electromyogr. Clin. Neurophysiol.*, 50: 3-27, 2010.

[63] Schalow, G.: Scientific basis for learning transfer from movements to urinary bladder functions for bladder repair in patients with spinal cord injury. *Electromyogr. Clin. Neurophysiol.*, 50: 339-395, 2010.

[64] Schalow, G.: Cure of urinary bladder function achieved in a patient with severe cervical spinal cord injury (95% injury, motoric complete) after three years of coordination dynamics therapy. *Electromyogr. Clin. Neurophysiol.*, 50: 155-179, 2010.

[65] Schalow, G.: Relative Coordination between Neuron firing and Generation of a Motor Program in the human CNS. In: Berkovsky, T.C. (Ed.), *Handbook of Spinal Cord Injuries*, Chapter V. pp 265-295, Nova Science Publishers, 2009.

[66] Schalow, G.: Impaired coordination between oscillatory firing FF and FR-type motor units in Parkinson's disease and patients with spinal cord injury. In: Berkovsky, T.C. (Ed.), *Handbook of Spinal Cord Injuries*, Chapter XV. pp 501-517, Nova Science Publishers, 2009.

[67] Schalow, G.: Building of New Motoneurons in the Human Spinal Cord upon Coordination Dynamics Therapy to Improve Finger Functions in Motoric Complete Cervical Spinal Cord Injury. In: Berkovsky, T.C. (Ed.), *Handbook of Spinal Cord Injuries*, Chapter IV. pp 231-264, Nova Science Publishers, 2009.
[68] Schalow, G.: Competitive interaction between two kinds of motoneurons for the innervation of two types of muscle fibres during development and repair in frog, discussed with respect to axon guidance and endogenous stem cell supported structural repair by motor learning in human cervical spinal cord injury. Chapter I.
[69] Schöner, G., Zanone, P.G. and Kelso, J.A.S.: Learning as change of coordination dynamics. *Journal of Motor Behavior*, 24: 29-48, 1992.
[70] Thelen, E.: The role of motor development in developmental psychology: a view of the past and an agenda for the future. In: Eisenberg, N. (ed.): *Contemporary Topics in Developmental Psychology*. New York, Wiley, 1987, pp 3-33.
[71] Thelen E., Smith L.B. (1994): *A Dynamic Approach to the Development of Cognition and Action*. MIT Press, Cambridge.
[72] Thuret, S., Moon, L.D.F. and Gage, F.H.: Therapeutic interventions after spinal cord injury. *Nat. Rev. Neurosci.*, 7: 628-643, 2006.
[73] Touwen, B.: Variability and stereotypy in normal and deviant development. In: *Care of handicapped child.* Apley, J. (Ed.). Clinics in Developmental Medicine No. 67. Philadelphia, 1978, pp. 99-110.
[74] Zanone, P.G. and Kelso, J.A.S.: Coordination dynamics of Learning and transfer: Collective and component levels. *Journal of Experimental Psychology: Human Perception and Performance*, 23: 1454-1480, 1997.
[75] Zanone, P.G. and Kelso, J.A.S.: Evolution of behavioural attractors with learning: Nonequilibrium phase transition. *Journal of Experimental Psychology: Human perception and Performance*, 18 (2): 403-421, 1992.
[76] Editorials: Order from chaos. *Nature*, 466: 7-8, 2010.
[77] Zhao, C., Deng, W. and Gage, F.H.: Mechanisms and functional implications of adult neurogensis. *Cell*, 132: 645-660, 2008.
[78] Maier, I.C. and Schwab, M.E.: Sprouting and circuit formation in the injured spinal cord: factors and activity. *Philos. Trans. R. Soc. Lond. B Biol. Sci,* 361: 1611-1634, 2006.
[79] Zorner, B. and Schwab, M.E.: Anti – Nogo on the go: from animal models to a clinical trial. *Ann. N.Y. Acad. Sci.*, 1198, Suppl.1: E22-34, 2010.
[80] Bareyre, F.M. et al.: The injured spinal cord spontaneously forms a new intraspinal circuit in adult rats. *Nature Neurosci.* 7: 269-277, 2004.
[81] Raineteau, O. and Schwab, M.E.: Plasticity of motor systems after incomplete spinal cord injury. *Nature Rev. Neurosci.* 2: 263-273, 2001.
[82] Grillner, S.: The motor infrastructure: From ion channels to neuronal networks. *Nature Reviews Neuroscience* 4: 573-586, 2003.
[83] Guttmann, L.: *Spinal cord injuries. Comprehensive management and research.* Blackwell, Oxford, 1973.
[84] Bray, C.W.: Transfer of learning. *Journal of Experimental Psychology*, 11: 443-467, 1928.

[85] Zanone, P.G. and Kelso, J.A.S.: Coordination dynamics of Learning and transfer: Collective and component levels. *Journal of Experimental Psychology: Human Perception and Performance*, 23: 1454-1480, 1997.
[86] Schmidt, R.A.: *Motor control and learning*. Champaign, IJ: Human Kinetics, 1988.
[87] Tuller, B., Case, P., Ding, M. and Kelso, J.A.S.: The nonlinear dynamics of categorical perception. *Journal of Experimental Psychology: Human Perception and Performance*, 20: 3-16, 1994.
[88] Alvarez-Buylla, A. and Lim, D.A.: For the long run: maintaining germinal niches in the adult brain. *Neuron*, 41: 683-686, 2004.
[89] Chklovskii, D.B., Mel, B.W. and Svoboda, K.: Cortical rewiring and information storage. *Nature*, 431: 782-788, 2004.
[90] Nottebohm, F.: Why are some neurons replaced in adult brain? *The Journal of Neuroscience*, 22: 624-628, 2002.
[91] Hodgekin, A.L.: *The conduction of the nervous impulse*, Liverpool University Press, 1964.
[92] Schmidt, R.F. and Thews, G.: *Physiologie des Menschen*, page 16, Springer Verlag, Berlin, 1985.
[93] Schalow, G. Non-drug induced spasticity reduction achieved by coordination dynamic therapy in CNS injury. *Electromyogr. Clin. Neurophysiol.* 42: 281-293, 2002.
[94] Thach, W.T.: On the specific role of the cerebellum in motor learning and cognition: Clues from PET activation and lesion studies in man. *Behavioral and Brain Sciences* 19: 411-431, 1996.
[95] Haken, H.: In: *Synergetics of the Brain* (Haken, H., ed.). Berlin, Springer-Verlag, 1983.
[96] Haken, H. *Information and Self-Organization*. Berlin, Springer-Verlag, 1988.
[97] Von Holst, E.: Relative coordination as a phenomenon and as a method of analysis of central nervous system function. In R. Martin (Ed.), The collected papers of Erich von Holst (pp. 33-135). Coral Gables, Fl: University of Miami. (Original work published in 1939).
[98] Schöner, G., Jiang, W.Y. and Kelso, J.A.S.: A synergetic theory of quadrupedal gaits and gait transitions. *J. Theor. Biol.* 142: 359-391, 1990.
[99] Jeka, J.J., Kelso, J.A.S. and Kiemel, T.: Spontaneous transitions and symmetry: Pattern dynamics in human four-limb coordination. *Human Movement Science* 12: 627-651, 1993.
[100] Covic, M., Karaca, E. and Lie, D.C.: Epigenic regulation of neurogensis in the adult hippocampus. *Heredity,* 105: 122-134, 2010.
[101] van Praag, H., Shubert, T., Zhao, C. and Gage, F.H.: Exercise enhances learning and hippocampal neurogensis in aged mice. *J. Neurosci*., 25: 8680-8685, 2005.
[102] Song, H., Stevens, C.F., Gage, F.H.: Astroglia induce neurogenesis from adult neural stem cells. *Nature*, 417: 39-44, 2002.

Chapter III

Neurophysiology and Pathophysiology of the Neural Networks of the Human Caudal Spinal Cord and Their Communication with Skin, Urinary Bladder and Muscle Spindle Receptor Afferents – A Scientific Basis for the Repair of the Human Central Nervous System

1. General Introduction

The human CNS possesses billions of neurons; each one of which has connections to an average of approximately 4,000 other neurons. New concepts and tools are needed if the inherent complexity of the most complex system of all, the human CNS and its relation to behaviour and thinking, is to be understood and repaired in the case of an injury. Presently, there is a huge void between the knowledge what a single neuron does (which we know a lot of) and what many of them do when they cooperate. The understanding of the principles of organization among large numbers of neurons is of a paramount importance, as this organization lies at the root of the understanding of ourselves, of the world we live in, of how we touch, see, hear, plan, act and think and how we relearn behaviour and thinking in the case of a CNS injury. Such fundamental behavioural functions depend on temporally coherent functional units distributed throughout different regions of the CNS, and standard methods have not the potential to elucidate them. For example, responses to sensory stimuli or activities in relation to motor acts are commonly averaged over successive trials in order to improve the signal-to-noise ratio. This averaging procedure destroys any temporal structure in the activation pattern that is not precisely locked to the stimulus or the motor response. Thus, temporal codes were either ignored or remained undiscovered with the commonly applied methods of single unit analysis [220]. In the human cell neurophysiologic studies of this Chapter, the simultaneous natural firing patterns from several identified single afferent and efferent neurons were therefore recorded and analyzed but not averaged. In this way, it was possible to discover the self-organization of premotor spinal oscillators (functional unit of a motoneuron and interneurons) and phase and frequency coordination between the firings of oscillatory and not oscillatory firing motoneurons, γ-motoneurons and afferents.

By recording single-nerve fibre action potentials from nerve roots, it is possible to analyze simultaneously afferent and efferent impulse patterns and investigate the coupling changes of self-organized premotor spinal oscillators to generate, under physiologic conditions and following a CNS injury, rhythmic and non-rhythmic movements and autonomic functions like continence. These data partly enable the understanding of the integrated functions of the human spinal cord and supraspinal centres; moreover, they allow substantial improvement of locomotor and other functions in patients with CNS injuries. The first paradigm shift in the understanding of the functioning of the CNS concerning the self-organization of neuronal networks gives more reorganization possibilities by the so-called neuronal network plasticity (Chapter II). The second paradigm shift, namely that neurogenesis, may be induced in the human adult spinal cord (new motoneurons for finger function, Chapter II) and adult brain [56], opens further possibilities for repair, regeneration and reorganization of the CNS.

In the last 100 years, many authors have come to suggest that it is the rhythmic firing of human neuronal networks that is responsible for rhythmic movements (trembling, tremor) of the body (trunk) and legs, arms and fingers (for references, see [117,243]). Descartes tackled tremors as early as in 1649 [47]. The conclusions drawn by R. Jung from measurements on tremor and clonus in 1941 [117], were similar to the findings concerning the self-organization of premotor spinal oscillators in recent papers [176, 186, 188, 195]. Only R. Jung did not differentiate between different motoneuron types and analyzed tremor by only using mechanical and electromyographic recordings.

In 1939 and 1950 [93-95], E.v. Holst expressed his disagreement with the common regard of the CNS being only a reflex apparatus producing motor output. His “relative coordination” of different rhythms [93] of the CNS in different species including man is very similar to recent findings of the relative coordination of human spinal oscillators [176, 186, 199]. It was the opinion of the Sherrington school that all reflexory, excitatory and inhibitory influences onto motor output are due to direct interactions at the motoneuron pool itself (reflex theory) [43]. The opinion of R. Jung was that the bottom-level coordination mechanism, at which all impulses run together, is the “Schaltzellenapparat” (neuronal network apparatus) of the spinal cord [117].

The current research on spinal oscillators and rhythm coupling supports the rhythm theory of R. Jung and E.v. Holst that assumes that the neuronal networks with their rhythmic properties driving the motoneurons are the bottom level basic mechanism for coordination and essentially contribute to the coordination (pattern) dynamics of arm and leg movements. At least to understand movements, the coordination dynamics of neuronal networks of the spinal cord must be understood and integrated in theories on human movements, since neuronal networks of the spinal cord show remarkable functions and plasticity, as has been shown by electromyography by the improvement of the motor program in patients with a complete rostral spinal cord injury [48]; brain-dead humans may be able to perform coordinated leg movements when kept in an intensive care unit for several weeks;

anencephalic [155] and healthy newborn babies (Figure 2 of Chapter II) can step automatically; and spinal oscillators self-organize themselves and may fire rather coordinated in patients with a complete spinal cord injury.

In the last 25 years, the understanding of the functioning of the CNS has changed, moving away from the rigid reflex and neuronal response chain theories towards the concept of dynamic self-organization of neuronal networks [4, 120], and this has direct implications for neuro-rehabilitation because of the increased network plasticity including large-scale plasticity [215]. The regulation of neural stem cells and neurogenesis in the intact and damaged adult mammalian and human brain, and probably spinal cord, offers further possibilities with respect to the regeneration and reorganization of the injured human CNS [3, 64, 237, 238]. This chapter is designed to build up a scientific basis from the perspective of human neurophysiology to reorganize the injured human CNS by relearning so that motor, autonomic and higher mental functions of patients can be improved on the same basis. To get more hard data on the functioning of the human nervous system, new precise human neuro-electrophysiologic methods are needed, like single-motor unit electromyography [165, 123], the tungsten electrode method [115, 123, 231] and the single nerve fibre action potential recording method presented in this chapter. This chapter starts with the new human electrophysiologic development, with which it is possible to analyse partly the organization of neuronal networks of the human CNS. As a first step, the firing of motoneurons in the occasional and oscillatory firing mode in relation to the skin and muscle spindle afferent activity is demonstrated; the recordings were obtained from brain-dead individuals (HTs) and partly from patients with spinal cord injury (paraplegics) to measure the CNS changes following injury. The partial loss of phase and frequency coordination following CNS injury was derived from recordings of paraplegics made during surgery for implantation of electrical bladder stimulators. Single-motor unit firing was partly verified electromyographically from leg muscle recordings. The partial loss of phase and frequency coordination prompted the development of coordination dynamics therapy to restore the impaired coordination dynamics of the CNS.

Since there is a gap between animal research, human research and treatment, and comparable measurements in animals and man are missing to translate animal data to man, the current research starts from the scratch with human measurements. But to allow partial approximations of the animal data to man, classification schemes and recruitment of motoneurons are assessed in some animals based on the recording of single-nerve fibre action potentials in the animals with the same recording device and are given for rat and frog in Chapter IV. To improve motor, vegetative and higher mental functions in patients with CNS injury, it is appropriate to observe neuronal network organization in the human lower sacral range, where motor and vegetative (volitional and automatic) functions are partly generated. Self-organized premotor spinal oscillators have been shown to have two driving phases per oscillation cycle [194] when primarily the somatic nervous system is activated (pinpricking of sacral dermatomes to induce escape reaction), but the oscillators have three driving phases [194] when additionally the parasympathetic division is activated (stimulation of urinary bladder filling (vegetative) afferents by bladder catheter pulling). The complexity of neuronal network organization becomes obvious when in addition to somatic also vegetative functions are activated. It is further believed that the vegetative functions are the door to the higher mental functions. The activation of the parasympathetic nervous system division in coordination with the somatic one will be mainly tackled in Chapter V.

2. Specific Introduction

Currently, research in human neurophysiology and clinical neurophysiology is aimed at developing an effective treatment for patients with CNS injury to improve micturition, defecation [169, 170, 192-194], locomotion [194], and higher mental functions (writing, reading, speaking, thinking, social behaviour) in particular. This new development started from scratch by first clarifying the principal aspects of anatomy [169, 170, 177-180] and developing a new basic electrophysiological recording technique suitable for use in humans, the simultaneous recording of single afferent and efferent nerve fibre extracellular action potentials (APs) with two pairs of wire electrodes from undissected nerve roots; this technique can be used for intra-operative diagnosis and research [171, 175, 180, 192]. Further developments in morphometry of nerves (classification of fibre diameters into four ranges of myelin sheath thickness (Figure 19)) made it possible to simultaneously characterize nerve fibre groups by group conduction velocities and group nerve fibre diameters (Figure 1) [175, 181, 192].

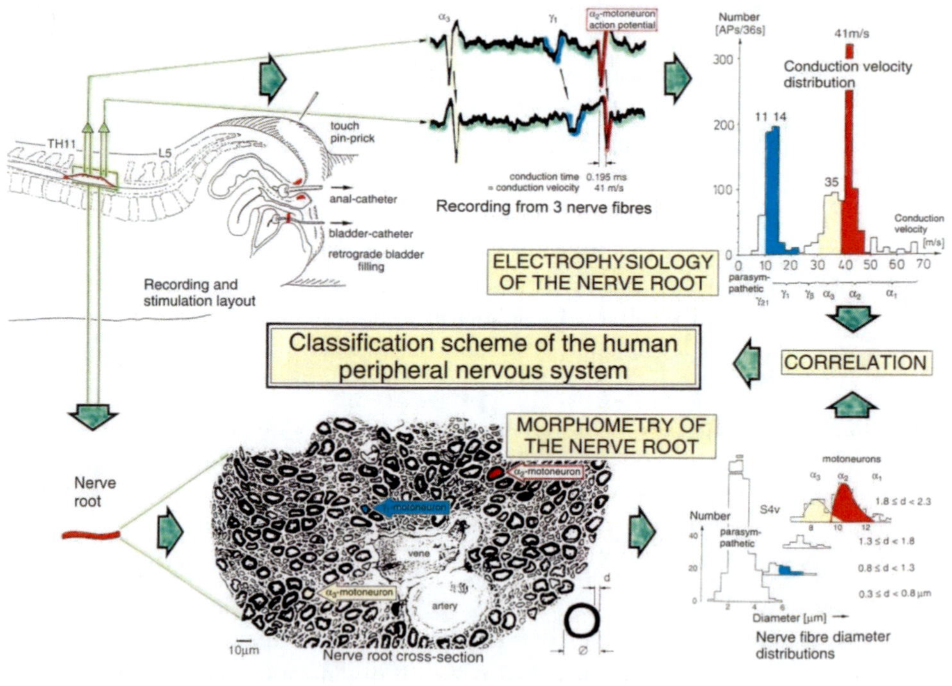

Figure 1. Schematic layout of the classification scheme for the human peripheral nervous system. By recording with two pairs of platinum wire electrodes from a nerve root containing approx. 500 myelinated nerve fibres, a recording is obtained in which three action potentials (APs) from three motoneurons (main AP phase downwards) can be seen. By measuring the conduction times and with the known electrode pair distance (10 mm), a conduction velocity distribution histogram is constructed in which the nerve fibre groups are characterized by ranges of conduction velocity values and peaks in asymmetrical distributions. After recording, the root was removed, fixated, embedded and stained, light microscope cross sections were prepared and used to measure the mean diameter and the myelin sheath thickness (d). Distributions of nerve fibre diameters were constructed for four different ranges of myelin sheath thickness. Nerve fibre groups were characterized by the peak values of asymmetrical distributions. By correlating the peak values of the velocity distributions with those of the diameter distributions obtained from the same root, a classification scheme was constructed of the human peripheral nervous system. Brain-dead human HT6.

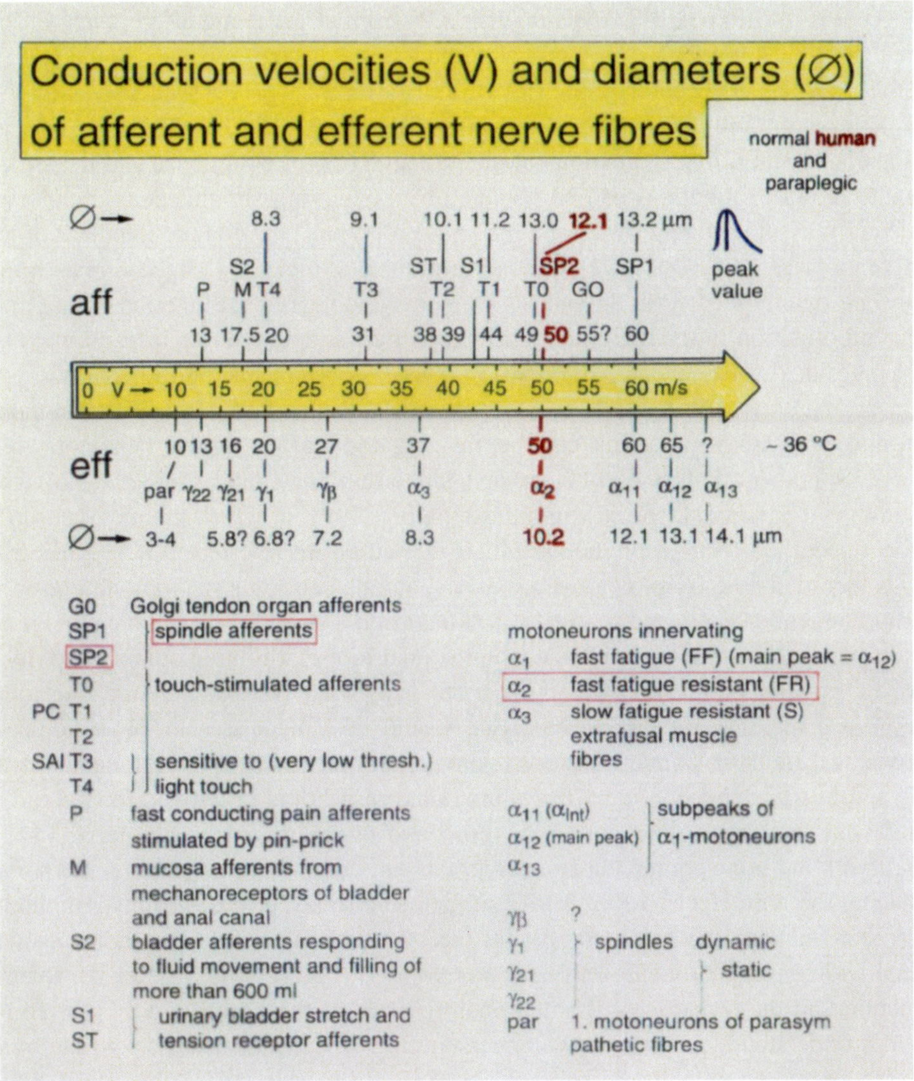

Figure 2. Conduction velocities (V) and nerve fibre diameters (∅) of afferent and efferent nerve fibre groups in normal humans and in patients with a traumatic spinal cord lesion for 0.5 to 6 years. The splitting of the α_1-motoneurons into the three subgroups, α_{11}, α_{12}, α_{13}, has not yet been confirmed.

In this way, an exact classification scheme could be constructed for the human peripheral nervous system (Figure 2). The scheme represents a solid basis for classifying and identifying nerve fibre groups in the human peripheral nervous system (PNS) and analyzing central nervous system (CNS) functions under physiologic and pathophysiologic conditions [192], even though it is incomplete and so far only holds for nerve fibre diameters larger than approximately 3.5µm. Another basic electrophysiological method for use in humans, the recording with tungsten electrodes of single-fibre extracellular APs [115, 231] made it possible to record impulse patterns of single-nerve fibres; the method, however, rests on the classification schemes of animal peripheral nerve fibres, which do not apply to humans.

The investigation of time-dependent conduction velocity distributions of afferent and efferent fibres allowed an analysis of activity level changes within nerve fibre groups [184,

185]. Since it is further possible to distinguish APs from afferent and efferent nerve fibres and to extract from the summed activity of a fibre bunch the discharge patterns of single fibres, it is possible to analyse receptor properties of skin afferents, primary and secondary muscle spindle afferents and urinary bladder afferents [180, 189, 192, 196] and analyze parasympathetic and somatic functions of the human CNS for continence and movements. An insight into the CNS functions can be obtained from studying simultaneous impulse patterns of single afferent and efferent fibres and the phase relations between the impulses of afferents and efferents [186, 188, 189, 191], following natural stimulation. Since human continence functions are mainly located in S3 and S4 segments and segmental functions seem to overlap in their representation in the roots, as indicated by dermatome overlap, especially continence functions are ideal for analysis An extension of the recording method to several roots from different segments will bring further information. With the discovery of the human spinal oscillators [176], CNS functions can be investigated based on the behaviour of spinal oscillators. Spinal oscillators can be divided into premotor spinal oscillators, of which the motoneuron is a part, and of propriospinal oscillators [199], of which the motoneuron is not a part. If not specified further, premotor spinal oscillators are referred to here, since rhythmic activity is recorded here from motoneuron axons, and it seems that the motoneurons are a part of the rhythm generating subnetwork [186]. It has been argued by animal physiologists that it is not surprising that there are spinal and supraspinal loops exhibiting spiking activity. But to improve the quality of life of patients, it is relevant what we can measure and identify in humans. But a big step forward in medicine would take place if neurophysiologists would also cover the field of human neurophysiology by measuring with the same method on animals and humans and proving what animal data also holds in human, to find out what is fact and what is fiction The author has successfully recorded single-nerve fibre action potentials with the same equipment from frog, rat, cat, dog and human nerves and nerve roots. By replacing the wire electrodes by EMG surface electrodes, also electromyographic activity can be recorded with the same equipment (see Chapter II) and single-motor unit activity compared with single motoneuron firing patterns [199] (Figure 9 of Chapter II). Even though the structure and the function of the neuronal network of the human CNS is far from being even partly understood, progress in the understanding of CNS disorder is possible by making use of the premotor spinal oscillators [176, 186, 188, 189, 191, 194, 199]. The spinal oscillators are functional units that organize themselves in preformatted neuronal subnetworks of the CNS by natural adequate impulse patterns of the particular afferents and supraspinal centres (brain stem, motor and sensory cortex, cerebellum). In the isolated spinal cord (complete SCI), only an afferent input from the periphery induces self-organization of the oscillators. These oscillators are characterized by their rhythm of firing, namely by the frequency of repeated firing with impulse trains and the interspike intervals of the impulse train (Figure 3). If the afferent input to the neuronal network is too low to fire oscillatory or if the neurons in the circuitry are inhibited, the motoneurons are integrated in another organization form of the network. Then, the motoneurons fire in the occasional firing mode [182, 184, 185] repeatedly every three seconds approximately, or they fire transiently oscillatory [176, 196]. Figure 3 summarizes what has been known so far about these premotor spinal oscillators, which, under physiologic conditions, fire rhythmically with impulse trains [176, 186, 188, 189, 191]. α_1 (FF), α_2 (FR) and α_3-motoneurons (S), characterized by group conduction velocity and group nerve fibre diameter, are integrated in their own neuronal network to fire oscillatory for high activation by their respective afferent input and to innervate their respective muscle fibre type.

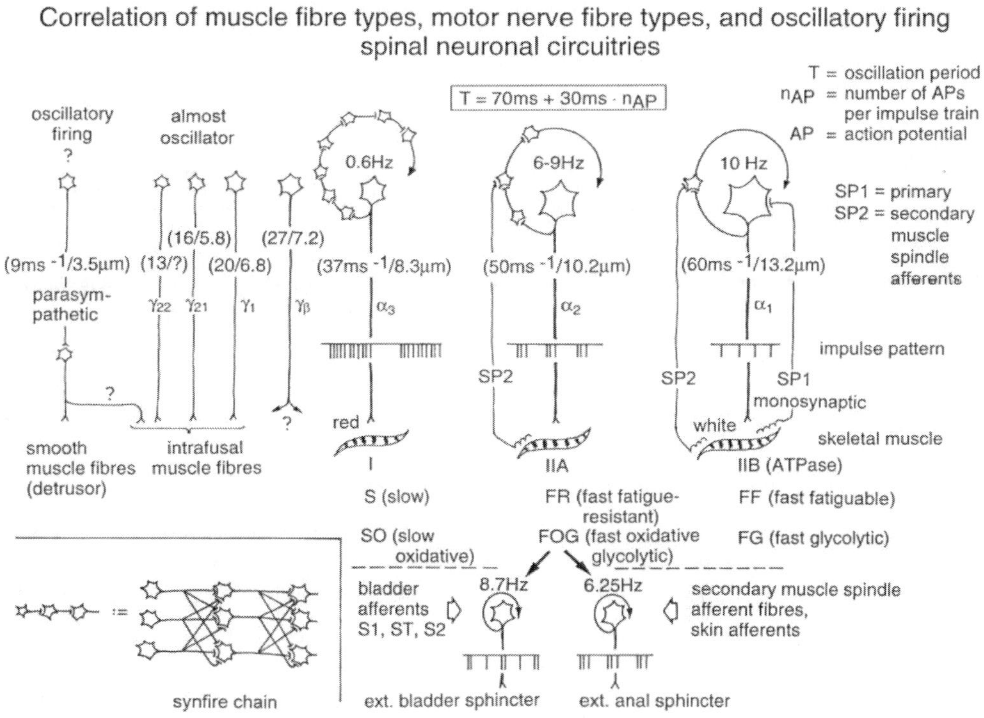

Figure 3. Correlation of muscle fibre types, motor nerve fibre types, and oscillatory firing spinal neuronal networks, based on histochemical, morphological and physiological properties. This Figure provides a simplified correlation between muscle fibre, motoneuron and sacral premotor oscillator types. No additional subtypes have been included. The existence of α_1-motoneuron (FF) oscillators firing at approx. 10 Hz had been predicted [83], and they have been identified in this work [91]. α, motoneuron; γ_1; γ_2, dynamic and static fusimotors; parasympathetic, parasympathetic preganglionic motoneuron; S1, ST, S2, stretch, tension and flow receptor afferents. α_2-Oscillators, innervating the striated ext. anal sphincter, show multifrequency oscillations (for example, f = 5.25, 6.25, 7.7 Hz) [83], as predicted mathematically (page 156 of [60]). The dashed line neurons in the synfire chain, in the left lower part, indicate the functional fringe of subthreshold excitation of the synfire chain.

In man, there are mainly three organization forms of somatic neuronal networks in the lower sacral spinal cord, which drive three kinds of striated muscle fibre (Figure 3). E.g., the α_2-motoneurons that subserve continence functions are integrated in neuronal network organization forms firing at 6 to 9 Hz; they innervate fast fatigue-resistant muscle fibres (FR) of histochemical type IIA, which are fast oxidative glycolytic (FOG) (for Refs. see [194]). The α_2-motoneurons innervating the external (striated) bladder and anal sphincters are mainly specifically driven by oscillators that fire with one to two APs every 110 ms (8.7 Hz) and with three APs every 160 ms (6.25 Hz), respectively (Figure 3). The neuronal network driving the external anal sphincter is channelled to fire oscillatory upon the activity of secondary muscle spindle afferent fibres (probably innervating the spindles of the anal sphincter) and skin and mucosal afferents innervating the anal canal. The oscillation period (T) of the premotor spinal oscillators is roughly related to the number of APs per impulse train (n_{AP}), and this relationship can be expressed as T = 70 ms + 30 ms • n_{AP}. Further properties of the spinal oscillators have been discussed elsewhere [176, 186]. So far, preganglionic parasympathetic neurons could not be observed to fire oscillatory. Fusimotors seemed not to fire oscillatory under physiologic conditions [189] but seemed to fire oscillatory following spinal cord injury (SCI) [194, 198].

Since premotor spinal oscillators are essential organization forms of the neuronal network of the spinal cord (since their functions can be measured following complete SCI), they can be expected to change their properties when the neuronal network of the spinal cord becomes damaged or when there is some pathology in its functioning. Early measurements of the dysfunction of spinal oscillators were used to study principle structures of their neuronal network [186]. Intrinsic oscillatory firing properties of individual cells have been found in humans so far only in the heart. The quick switching on and off of oscillatory firing of motoneurons down to one oscillation period cannot be explained by internal busters. It seems as if premotor spinal oscillators, subserving continence functions, can consist of two halves in some similarity to the lamprey locomotor system [83, 84]. Since however premotor spinal oscillators of leg muscles can change from alternating to symmetrical oscillatory firing [199], which even seems to correlate with phase changes in the synchronization with other spinal oscillators [199], the alternating oscillatory firing has also been attributed to changes in the coupling states of spinal oscillators [199]. More generally, therefore, alternating oscillatory firing oscillators may indicate coupling of non-equal suboscillators, reciprocal inhibition of half-centre oscillators or mutual inhibition of oscillators driving antagonistic muscles.

It was shown [192-194] that the classification scheme of the human peripheral nervous system does not change 0.5 to 6 years following spinal cord injury [192]. Data related to dysfunctions of the urinary bladder after SCI seem to indicate that dyssynergia of the urinary bladder is due to a false organization of the neuronal networks in the sacral micturition centre. The detrusor, driven by the parasympathetic neuronal network, is activated at the same time as is the external bladder sphincter (dyssynergia of the bladder), driven by the somatic neuronal networks. In healthy individuals, the external bladder sphincter is inhibited upon the activation of the detrusor. Dyssynergia of the urinary bladder could be explained by a threefold disorganization of the spinal neuronal networks, if one splits in first approximation spinal networks into interlacing somatic and parasympathetic networks: (i) the somatic neuronal networks organized themselves pathophysiologically, which was judged from the dysfunction of oscillatory firing α-motoneurons. (ii) Also, the parasympathetic neuronal networks organized themselves in a false way, as revealed by the detrusor not receiving sustained activation from preganglionic parasympathetic motoneurons. (iii) The interaction between the somatic and the parasympathetic neuronal networks had changed, as judged from the dys-coordination of the detrusor and the external bladder sphincter (dyssynergia of the bladder). The idea was put forward that the parasympathetic neuronal network was no longer able to fire continuously oscillatory; the transiently oscillatory firing in paraplegics with a spastic bladder gave only rise to undulating or spot-like activation of the bladder smooth muscle. The continence problem encountered by paraplegics with a "spastic" bladder (overactive bladder, nearly no filling volume) seemed to be only better understandable if the disorganization of the somatic neuronal network was better understood, since it is easier to record from the thicker α-motoneuron axons (on average large AP amplitude) than from the thinner parasympathetic preganglionic fibres (on average small AP amplitude). Disorganization of the somatic neuronal networks also causes a dysfunction of premotor spinal oscillators. Since the premotor spinal oscillators are probably organized by the same interneurons that are used for the organization of propriospinal oscillators (rhythmically firing subunits probably recruited for the organization of the spinal patterns for rhythmic movement (198)), the disorganization and therefore dysfunction of premotor spinal oscillators reflect the disorganization of somatic neuronal networks. The disorganization of somatic neuronal networks can therefore be detected by the properties of the premotor spinal oscillators as the resonance frequency band width, the regularity of the interspike intervals of the impulse train (with which the oscillator is firing) and the coupling properties between spinal oscillators or systems of coupled oscillators. For systems of coupled oscillators as models of central pattern generators, see [41, 126, 160]. The disorganization of the somatic neuronal networks in paraplegics may cause spasticity and clonus. In the following, the self-organization of somatic neuronal networks under physiologic and pathophysiologic conditions are under consideration.

In 1939 [93] and 1950 [94], E.v. Holst challenged that the CNS was commonly regarded as a reflex apparatus producing motor output. His “relative coordination” of different rhythms [93] of the CNS in different species including man is very similar to recent findings on the correlation of human spinal oscillators [176, 186, 199]. According to Moshe Abeles [1], the nervous system is essentially a statistical machine [243]. According to von der Malsburg, the information in this statistical machine is encoded not only in the number of active nerve cells and their firing rates (“frequency code”) but also in interactions between neuronal and neuromuscular elements of cell assemblies expressed in correlation of their firing patterns [233]. This author argues that the coordination of discharges has a strong influence on nervous dynamics, since neurons essentially are “coincidence detectors,” because coordinated inputs excite a postsynaptic cell more effectively than do uncorrelated inputs. Correlations could therefore be processed by neuronal networks. The correlation could be initiated in the network from an external source and then be propagated by “synfire chains” (Figure 3), as suggested by Abeles [1]. The functional organization of the common brain stem system and its dynamic changes is a consequence of afferent inputs and the rhythmic properties of the common brain stem system network. The neuronal network reveals different types of functional organization. In the state of functional organization, in which the neurons discharge rhythmically, the phase relations between the different rhythms are essential in the transfer of information [129, 164, 213]. It was the opinion of the Sherrington school that all reflectory, excitatory and inhibitory influences onto motor output are due to direct interactions at the motoneuron pool itself (reflex theory) [43]. The opinion of R. Jung was [117] that the bottom-level coordination mechanism, at which all impulses run together, is the “Schaltzellenapparat” (neuronal network apparatus) of the spinal cord; this concept also admits that there are spinal pathways that bypass the “Schaltzellenapparat” and synapse directly onto the motoneurons. This neuronal network apparatus is the common substrate that controls the volitional and unvolitional motor output. The generation of rhythmic activity, as revealed by physiologic tremor, is one function of this “Schaltzellenapparat” (oscillator theory). The current research on spinal oscillators supports the oscillator theory of R. Jung, which assumes the neuronal network with its rhythmic properties driving the motoneurons being the bottom level basic mechanism for coordination; this theory opposes the reflex theory, which states that the functions of the human CNS can be explained on the basis of reflexes as assumed mostly in clinics. Sir Matthews could show that the so-called long-latency stretch reflex can be explained by the response of secondary muscle spindle afferents in the spinal cord [139]. In his opinion, nevertheless, there is no need for a conceptional revision to be able to understand the function of the spinal cord and its interactions with supraspinal centres. Probably the functioning of the human spinal cord is intermediate, between dynamically preferred connectivity patterns (see system theory of pattern formation in Chapter II) as a basis for self-organization and a rigid network. The monosynaptic reflexes belong to rather rigid network structures, whereas the spinal oscillators are the first step towards self-organization of pre-formatted neuronal network structures.

3. Methods and Subjects Recorded From

3.1. Spinal Cord Injury Patients and Brain-Dead Humans from Which Data Were Recorded

The measurements in humans were collected from seven brain-dead humans (HTs), 11 patients (mean age 27 years) with spinal cord injury (SCI) and a few other patients. The SCI patients had a spastic bladder (hyperactive detrusor (nearly no bladder filling volume)), spastic pelvic floor (including external striated sphincters), dyssynergia of the urinary bladder, and spasticity in general. Electrophysiologic measurements were performed during surgery, with the new method of recording single-nerve fibre action potentials (APs) extracellularly from undamaged nerve roots or nerve root fascicles. The patients underwent procedure to have implanted a sacral anterior root stimulator (according to Brindley [31]) to improve bladder control and to save the kidneys. The strategy of the surgery is to deafferentiate the urinary bladder to increase its storage volume and to subsequently stimulate the motor roots (mainly S3 and S4) to empty the bladder. To deafferentiate the bladder, afferents and efferents in the nerve roots were identified by electrical stimulation and partly by using the single-nerve fibre AP recording method to recognize afferents in certain roots. The surgical procedure involves extirpation of parts of the dorsal roots. The cut dorsal roots, from which recording was performed, were fixated and prepared for morphometry. Light anaesthesia was administered with Propofol. The implantation of anterior root stimulator (a destructive operation) is not justified anymore since urinary bladder function can be repaired nowadays by CDT (movement-based learning) (Chapters II and V).

3.2. Electrophysiology: Single-Nerve Fibre Action Potential Recording Method

Single-nerve fibre action potentials (APs) were recorded extracellularly (Figure 1) from nerve roots with two platinum wire electrode pairs (electrode pair distance (mostly) = 10mm; electrode distance in each pair = 4mm) at two sites, preamplified (x1000), filtered (RC-filter, passing frequency range 100Hz-10kHz) and displayed on a digital storage oscilloscope (Vuko Vks 22-16) and also stored using a PCM-processor (Digital Audio Processor PCM-501ES) and a video recorder. The beginning of a touch or a pinpricking was marked with an upward pulse, and the end with a downward pulse on trace "a" (Figure 4a). These pulses were generated by a markation pulse generator connected to the digital scope, which was switched on and off with a touch sensor working on the basis of resistance changes.

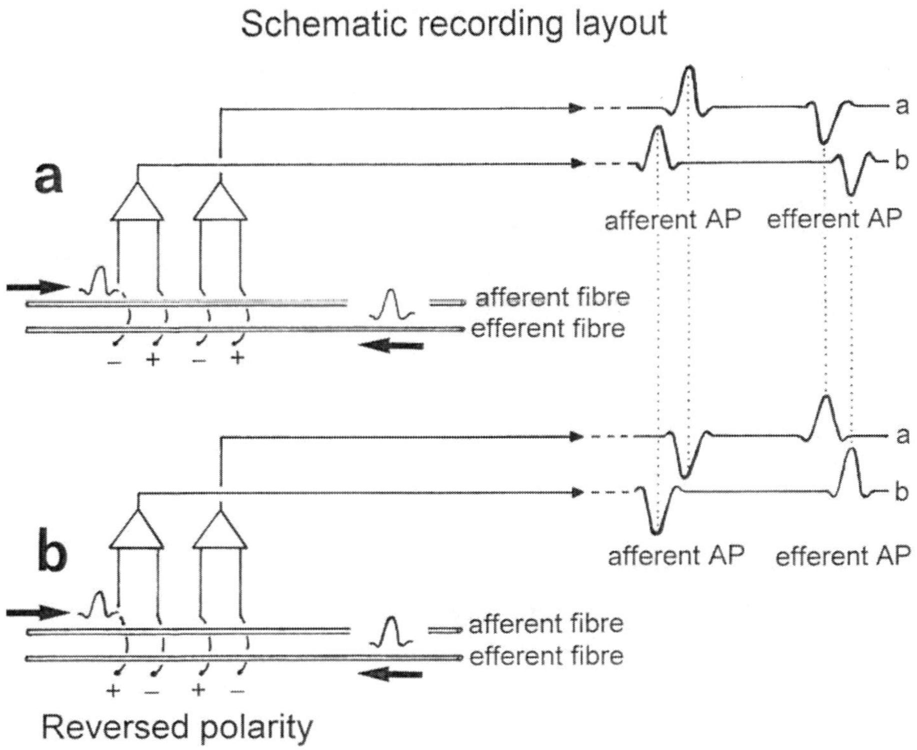

Figure 4. Schematic layout for recording single-nerve fibre action potentials (APs). The reversing of the inputs to both preamplifiers does not change the ability to differentiate between afferent and efferent APs.

Also, the pulling and releasing of anal and bladder catheters were mostly marked with the help of a pull-switch connected to the catheters and working in connection with the same markation pulse generator. Trace "a" was the recording from the proximal electrode pair and trace "b" from the distal pair. Conduction velocities of single-nerve fibres were calculated from the conduction distance (electrode pair distance) and the respective conduction times, the time needed for an AP to cover the conduction distance (time difference between traces "a" and "b" for a particular AP) of 10mm. APs from afferents and efferents could clearly be distinguished since for the used electrode arrangements, the main phase (second phase) from afferent fibres is upwards and that of efferents downwards (Figures 4a). E.g., the AP of a skin afferent fibre reaches a pair of electrodes first as negative and then as positive. According to the electrode setting used, the main phase is upwards. An AP of a motoneuron, coming from the opposite direction, would reach the electrodes in the order positive-negative. The potential changes are therefore opposite, and the main triphasic AP will point downwards. An AP in an afferent fibre reaches first the caudal electrode pair and then the rostral pair, whereas an AP of the efferent fibre reaches first the rostral electrode pair and then the caudal one. The conduction times are therefore also opposite. For further clarity, a change of the inputs to the preamplifiers does not change the ability to differentiate between afferent and efferent APs because the amplitudes of both types of APs change their upward or downward direction (Figure 4). Figure 5 shows schematically how to get from a two-fibre recording (Figure 4) to a recording from many afferent and efferent nerve fibres, which means to the recording of the summed afferent and efferent impulse traffic in a thin root. The APs of the different nerve fibres add up algebraically.

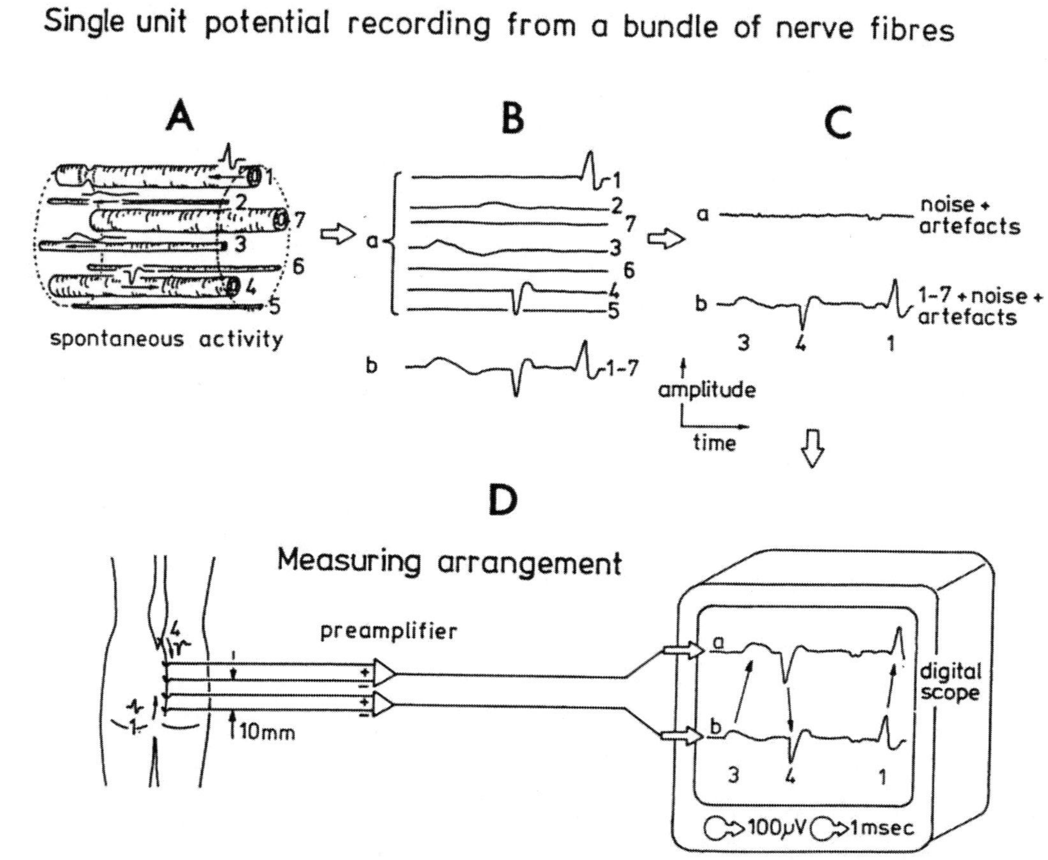

Figure 5. Schematic layout of the recording of extracellular single-nerve fibre action potentials from a bundle of seven nerve fibres. A. The afferent nerve fibres 1, 2 and 3 and the efferent fibre 4 are occasionally active, as indicated by the arrows and the single-unit potentials; fibres 5, 6 and 7 are not active. Large single-nerve fibre potentials from thick fibres are indicated by larger potential changes and small potentials from thin fibres by small potential changes. B. Traces I (1 to 7) show the theoretical recordings of the single fibres. Trace II shows the arithmetic sum of the seven recordings. C. Trace I shows the noise and artefact background levels; trace II is the constructed recording (sum of traces II (B) and I (C)) of three single unit potentials from three nerve fibres in a bundle of seven nerve fibres. The potential of fibre 2 is lost because of low amplitude. D. Stimulation and recording layout. The numbers at the single-nerve fibre potentials correspond to the nerve fibre numbers in A, B and C. Traces a and b are the records from the two sites; downward arrow (4) = efferent, upward arrow (3,1) = afferent. The stimulations used were touch, pinprick, anal and bladder catheter pulling.

In reverse, to get from a multi-unit recording to the simultaneous impulse patterns of the single fibres, the multi-unit recording has to be split up (Figure 6). The splitting is achieved by recognizing the APs from the certain single fibres on the basis of waveform comparisons on the two recording traces and the conduction time that an AP needs to travel from one electrode pair to the other one (10mm) and selecting these APs out, as is done in Figure 6. Being able to measure simultaneously the natural impulse patterns generated by certain single receptors in the periphery, which run into the spinal cord (CNS), and those patterns that leave the cord in single fibres, it becomes possible to analyze integrative properties at the cellular level of the only little changed human CNS and its response to natural stimuli in the periphery. The use of this new electrophysiological recording technique brought the main progress in repairing the human CNS.

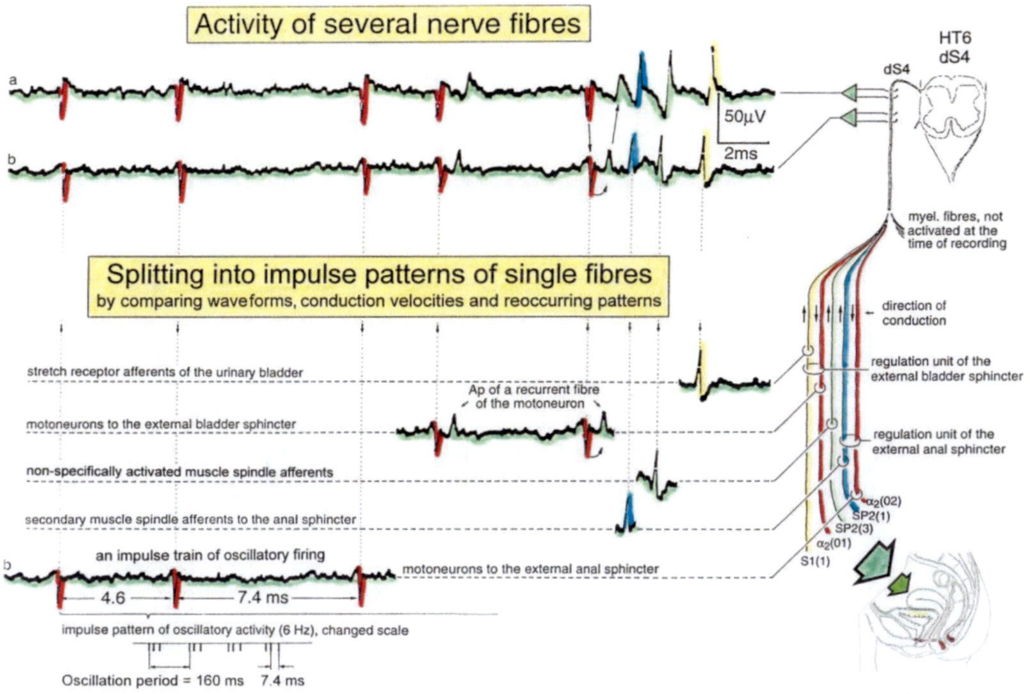

Figure 6. Schematic splitting of the activity of several nerve fibres into simultaneous impulse patterns of single fibres by comparing waveforms, conduction velocities and reoccurring characteristic impulse patterns (rhythmic firing of sphincter motoneurons). The different conduction times and waveforms were recognized on an expanded time scale. For receptor sites and the location of the muscles that are innervated by the sphincter motoneurons, see Figure 52. Stretch receptor and secondary muscle spindle afferents contribute to the drive of sphincter motoneurons and form, together with other afferents, regulation units.

3.3. Relation between Conduction Velocity, Action Potential Amplitude and Action Potential Duration

In accordance with animal data [149], the AP amplitude of single afferent APs decreases, and the AP duration increases with decreasing conduction velocity (increasing conduction time) (Figure 7), if the different nerve fibres had the same distance to the recording electrodes. Similar relationship among conduction velocity, AP amplitude and AP duration hold for efferent APs (Figure 8). With increasing conduction velocity (and increasing diameter), AP amplitude increases and AP duration decreases. The scatter in these relations is large because different fibres in a root are at different distances from the recording electrodes. Also nerve fibres change their diameter slightly from internode to internode (see Figure 29). And as a matter of fact, damaged nerve fibres have strongly reduced conduction velocity and increased AP duration. Poor digitalisation and too strong filtering will reduce large AP amplitudes. Nevertheless, good recordings clearly demonstrate that average AP amplitude decreases and AP duration increases with the reduction of conduction velocity when changing from one nerve fibre group to another.

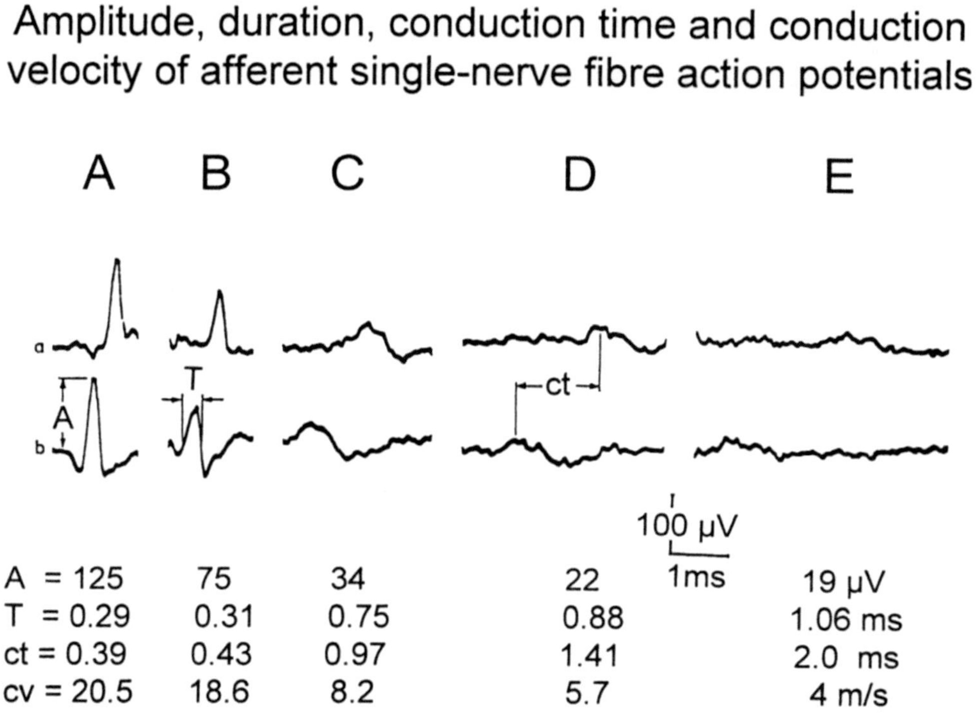

Figure 7. Extracellular AP waveforms of different amplitude (A), duration (T), conduction time (ct) and conduction velocity (cv). The different nerve fibres, in which these APs were conducted, had approx. the same distance to the recording electrodes. The afferent nerve fibres were stimulated by touch, pain, and bladder and anal canal-catheter pulling and releasing. Brain dead individual HT4, 56 years, 32 °C (approximate measuring temperature), recording made from a dorsal S5 root (diameter = 0.17 mm).

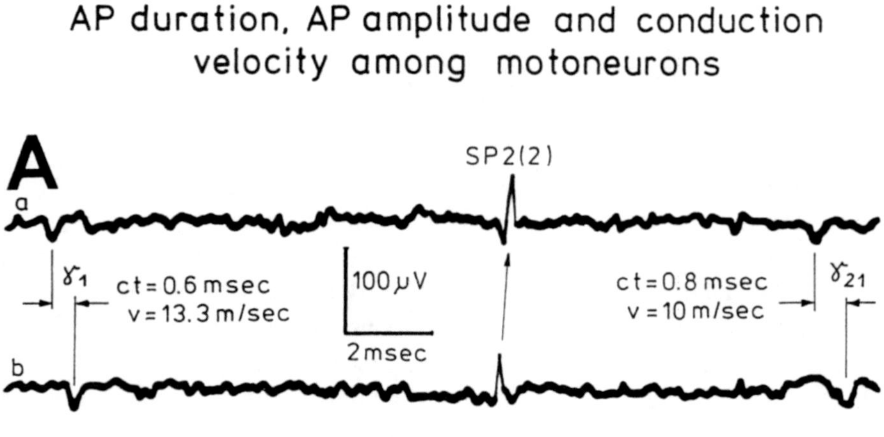

Figure 8. (Continued)

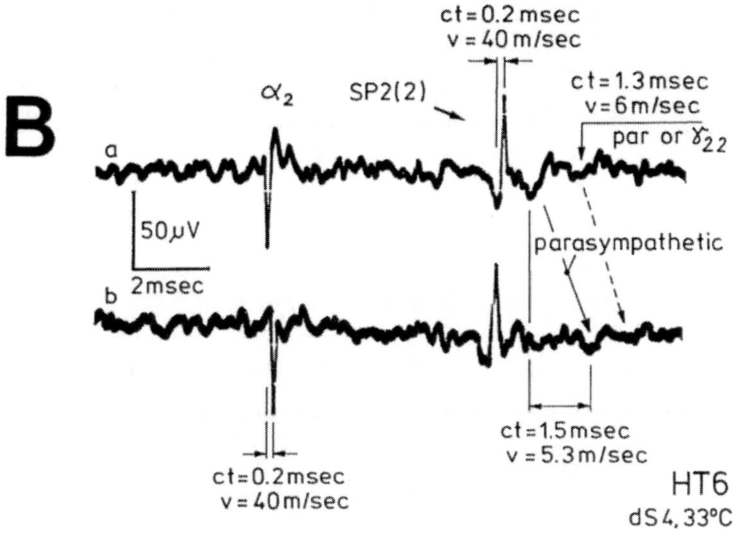

C

AP duration	$\alpha_2 : \gamma_1 : \gamma_{21} : par = 1 : 3 : 4 : 6$
AP amplitude	$\alpha_2 : \gamma_1 : \gamma_{21} : par = 1 : 1/3 : 1/3.5 : 1/8$
conduction velocity	$\alpha_2 : \gamma_1 : \gamma_{21} : par = 6 : 2 : 1.5 : 1$

Figure 8. Recordings of single-nerve fibre action potentials (APs) (A,B). C. Relation between the AP durations, AP amplitudes and conduction velocities of single α_2, γ_1, γ_{21} and parasympathetic motoneurons. HT6, dS4-root, 33°C.

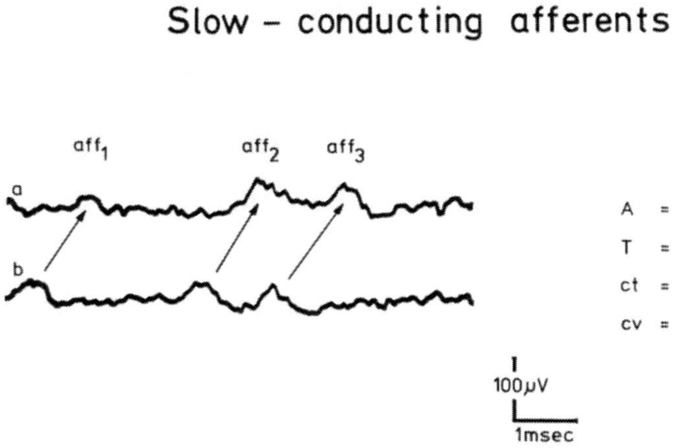

		aff1	aff2	aff3
A	=	28	28	25 µV
T	=	0,63	0,63	0,63 msec
ct	=	0,91	0,94	1,16 msec
cv	=	8,8	8,5	6,9 m/sec

Figure 9. Three extracellular action potentials from slowly conducting touch-stimulated afferents. The potentials are probably from three different fibres. HT4, 56 years, 32°C, S5d, Ø = 0.17mm, same root as in Figure 7. A = amplitude, T = action potential duration, ct = conduction time, cv = conduction velocity.

These relations are of interest, since with known expected conduction times, AP amplitude and AP durations, it is much easier to identify or to pick up certain APs on traces a and b, in particular in the presence of small amplitudes and high levels of noise and artefacts. From Figure 8A and others, it can be calculated that the AP duration increased from α_2

through γ_1 to parasympathetic motoneurons (1:3:6) and the AP amplitude decreased in the same sequence (1:1/3:1/8) (Figure 8B).

As in the case of parasympathetic efferents (Figure 8), AP amplitude, duration, and conduction time can be calculated as long the APs can be recognised (Figure 9). Therefore high-quality recordings are necessary for detecting the APs from thin fibres. They can be obtained from the thin and long lower sacral nerve roots.

3.4. Touch Afferents from the Nearly Hairless Skin

AP activity in response to touching sacral dermatomes is shown in Figure 10 in a ventral S4 root recording. The extracellularly recorded APs are marked with the group (T1, T2, T3, T4); they belong to according to the peak conduction velocity values from the corresponding constructed conduction velocity distribution histogram (not shown). The APs of a certain group could originate from one or several touch units.

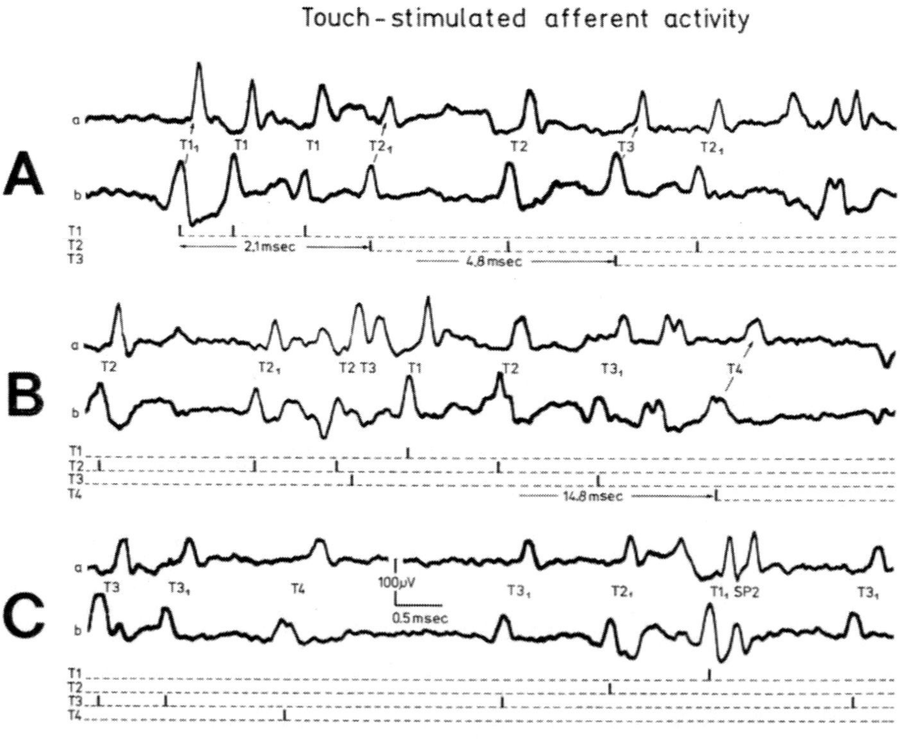

Figure 10. First part of touch stimulated activity from an S4 ventral root recording, HT3. A, B, C are successive sweep pieces. Touching area about 15-20mm (touching with a finger). Extracellular action potentials (APs) are marked with the touch afferents groups (T1, T2, T3, T4) to which they belong according to the conduction velocity distribution histogram (not shown). Subscripts mark APs from single fibres. Traces "a" and "b" in A, B, C are original registrations. The dashed line traces T1, T2, T3, and T4 below traces "b" give the activity of separated touch groups schematically. Each upward deflection marks the occurrence of an AP. Note, the APs of group T2 touch afferents arrive 2.1ms later at the place of measuring than the APs from the T1 group. The APs from the T3 and T4 group arrive 4.8 and 14.8ms later. SP2 marks the AP from a secondary spindle afferent fibre.

By recognizing very similar waveforms on the traces “a” and “b,” a few APs were identified as originating from the same fibre and were marked with an index. Only in good recordings, the waveform identification are safe. The approximate number of stimulated touch units in each class of Figure 10 could be counted and amounted to three in T1, three in T2, three in T3 and one in T4.

The sweep piece of Figure 10 is, therefore, a recording of about ten touch units. The APs with shorter conduction times (higher conduction velocities) have, on average, higher amplitudes in accordance with Figures 7 and 8. In the lower part of Figures 10A and B, the arrival of the first APs from the fastest touch units of the T2, T3 and T4 groups are given in relation to the arrival of the first AP from the fastest unit of the T1 group. With these relative latencies and the group conduction velocities (peak values), it can be calculated how far the place of touching sacral dermatomes was away from the recording electrodes. A few values were chosen, and it was verified that a distance value of 550mm fitted best (for further see Ref. [175]. Below, an analysis will be given when the beginning and end of touch is marked on the recording trace with a markation pulse generator.

3.5. Anal Canal and Urinary Bladder Afferent Fibres

To repair the continence of patients, the functioning of the urinary bladder and anal canal has to be understood in detail. By pulling anal and bladder catheters, afferent (and efferent) natural AP patterns can be measured. The activity from a S5 dorsal root stimulated by pulling and releasing an anal catheter with a balloon in the ampulla recti is shown in Figure 11. Single-unit potentials from afferent fibres (aff) with low amplitude and long duration, probably from mucosal afferents, are indicated by dashed arrows (v = 7.3m/s). Since these potentials start to merge, it becomes difficult to identify them.

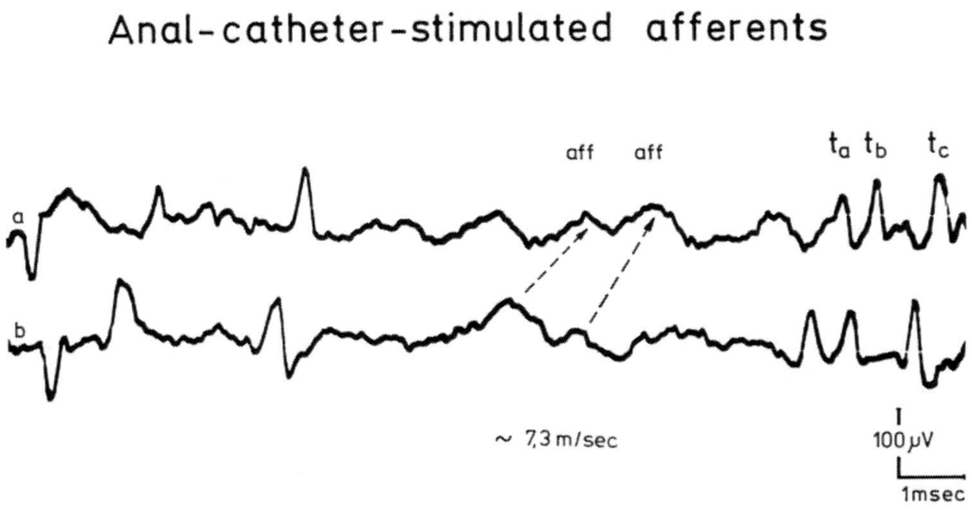

Figure 11. Action potentials from slow-conducting afferent fibres (dashed arrows) among potentials from fast-conducting fibres (marked on the right with t_a, t_b, t_c). HT4, 56 years, 32°C, S5d, Ø = 0.17mm (same root as in Figures 7 and 9).

The larger potentials with shorter durations, marked with ta, tb, and tc in Figure 11, can still be identified. These large potentials are most likely from touch units around the anus, since they were absent in a recording from a S3 dorsal root (the S3 dermatome does not innervate around the anus).

Similar activity can be recorded from the bladder. It becomes obvious that an identification of the natural impulse traffic in nerve roots is only possible if the quality of recording is high and the activity level is low so that the different kinds of APs can be distinguished from each other. The human S4, S5, and Co roots are ideal for recording.

3.6. Ventral Root Afferent and Dorsal Root Efferent Fibres

There has been considerable discussion about the existence of dorsal root efferent and ventral root afferent fibres since Sherrington's first report [218]. Actually, the ventral root afferent and dorsal root efferent fibres in the human lower sacral nerve roots (violation of the Dale's principle) improved the recordings and made it possible to record from a few fibres. The S4 ventral root and the coccygeal root are ideal to record afferent activity, and the S4 dorsal and the coccygeal root are ideal to record natural efferent activity because of low overall root activity.

3.7. Algebraic Adding Up Single Action Potentials

If APs of two efferent fibres meet at one electrode pair at the same moment, the APs add up algebraically. The adding up of many fibres' impulses following electrical stimulation will give rise to a compound AP (see below). If the APs of an afferent and an efferent fibre meet at the same point of time at one electrode pair, they partly or fully abolish each other (subtraction of the AP amplitudes). However, since these APs will not meet each other at the other electrode pair, the afferent and the efferent APs are clearly distinguishable on the trace of the other electrode pair.

3.8. Conduction Velocity Distribution Histograms

The conduction velocities of afferent APs were plotted in an afferent conduction velocity distribution histogram and those from efferent APs in an efferent velocity histogram. Histogram classes were half-closed intervals (≤ and <); the left border belongs to a particular class, the right one does not. Group conduction velocity and nerve fibre diameter values were the peak values of asymmetrical distributions.

3.9. Recording Conditions

Single-fibre APs were always recorded from whole roots or fascicles. I did not try to tease single-nerve fibres. In nerve roots thinner than 0.6 mm in diameter (radial decline of AP amplitude due to volume conductance is approx. by a factor of 1/10 per 0.3 mm; the root flattens when positioned on the wire electrodes), it is possible to record single-nerve fibre APs from all fibres with a diameter larger than approx. 3.5 μm. The radial decline of low-amplitude long-lasting extracellular APs, due to volume conductor embedding, of thin fibres, including preganglionic parasympathetic fibres, is less pronounced than it is for high-amplitude short-lasting APs since the low frequency sinusoidal coefficients will have large amplitudes in a Fourier-expansion. Since nerve roots have no epineurium and nearly no perineurium, the nerve fibres in the roots can easily be damaged when recording with wire electrodes. Pressure and stretch will change the AP waveform or even block conduction (most easily at the node of Ranvier) so that one AP can be recorded from one electrode pair only. Double-peaked APs can occur, probably when a node of Ranvier is blocked. Mechanical stimulation, nerve fibre compression, resistance artefacts and trigger zones may change impulse shape and activity. For further references, see [192-194].

3.10. Doublet Firing of Secondary Muscle Spindle Afferents

It has been argued that the doublet firing of secondary muscle spindle afferents for slight parasympathetic activation [193] is caused by the injury of nerve fibres. Generally, by measuring functions in a system, one changes more or less the functions of the system. By recording with wire electrodes from a bundle of nerve fibres (nerve root (Figure 22)), protected only by a thin sheath of cells, stretch, pressure, membrane injury or drying up of the fibres can occur. Mechanical alterations of nerve fibres can be judged by the appearance of double-peaked APs caused by blocks at the nodes of Ranvier, which were not observed. Ectopic AP generation due to the drying up of nerve roots can be avoided or eliminated by wetting the root. Further, injury potentials of nerve fibres (long membrane cable) are not long lasting. The injury will depolarize the membrane and therefore, at least partly, inactivate the membrane so that the excitability of the injured membranes will be reduced rather soon. In intracellular recordings from a muscle fibre, one therefore additionally injects current to keep the membrane potential or to hyperpolarize it slightly [199]. Without such membrane potential stabilization, the recordings from the thin muscle fibres of tadpoles and young frogs of Chapter I would not have been possible. Actually, the doublet firing of the spindle afferents lasted long, and the firing even increased in that case instead of decreasing. Injury as a cause for the doublet firing is therefore very unlikely. To get more safety concerning the repeatability of natural human firing patterns, further measurements have to be performed.

4. Anatomy

4.1. Anatomy of Urinary Bladder Innervation

4.1.1. Cauda Equina: Unique Anatomical Situation for Measuring Single-Nerve Fibre APs

I dissected more than 50 human cadavers to clarify the necessary anatomy (including morphometry of nerves and nerve roots and urinary bladder innervation) for performing this human electrophysiology. The development of this new human recording technique was possible because of the unique anatomical situation present in the human spinal canal. Because of the ascensus of the spinal cord, the lumbosacral nerve roots have become very long and form the cauda equina (Figure 12).

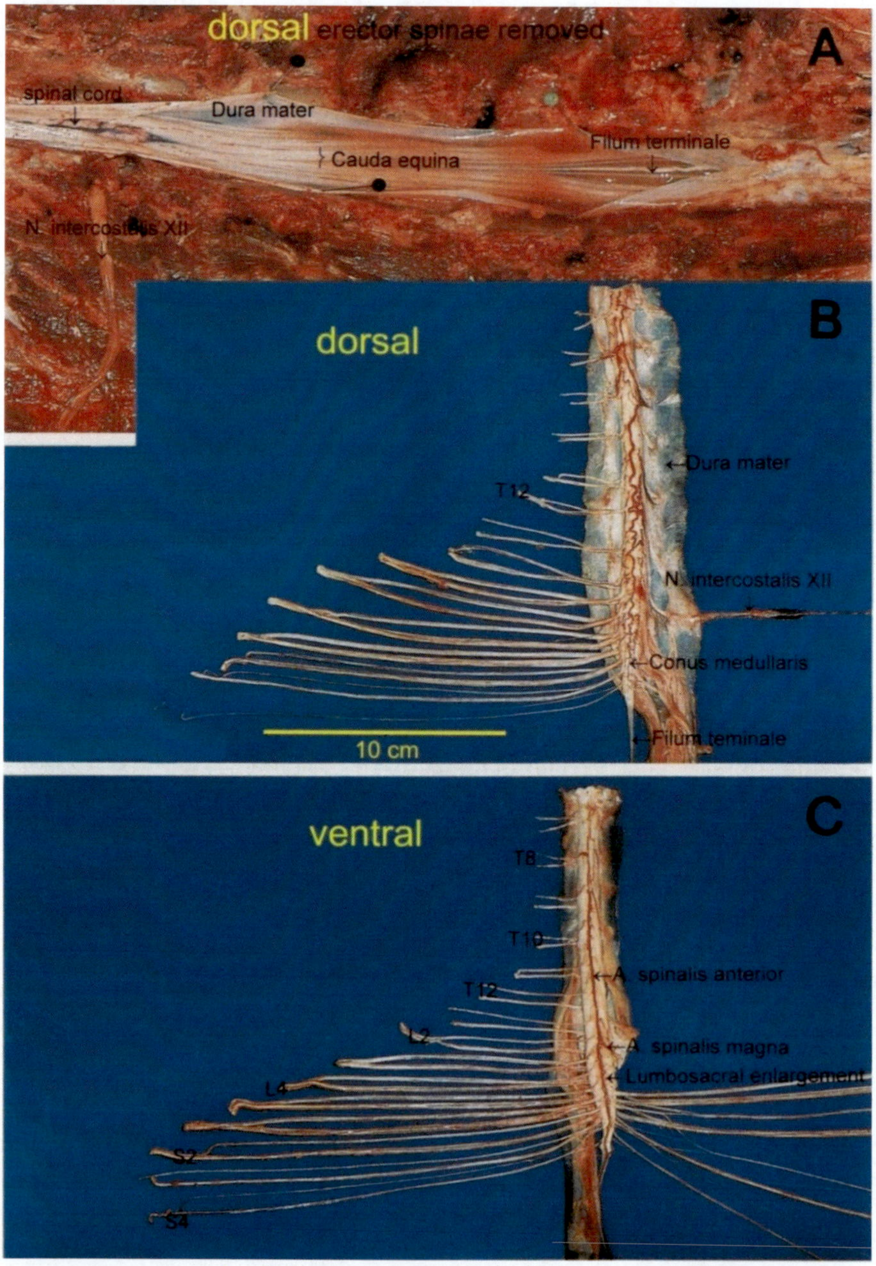

Figure 12. A. Cauda equina and the intercostal nerve XII (subcostalis). Laminectomy, the dura mater spinalis opened and the erector spinae removed. Nerve roots lie in a position similar to a horse's tail (cauda equina). Cadaver dissection. B. The dorsal spinal cord, the cauda equina nerve roots, the dura mater and the intercostal nerve XII removed and split up. The roots are cut at the dura mater. Note that the caudal ventral roots are thinner than the dorsal roots. C. The ventral spinal cord and the cauda equina nerve roots. The passage of the artery spinalis magna is indicated. Note the root interconnections on the left side (right side of the Figure).

Since the caudal sacral nerve roots are very thin (Figure 12B,C) and nerve roots are only sheathed by a thin layer of cells (Figure 22), they are ideal for recording single-fibre APs from undissected nerve roots. It is also possible to record single-nerve fibre APs from thin human [179] (Figure 18) and animal peripheral nerves (Chapter IV), but the epineurium has to be removed and the quality of recording cannot reach that of root recordings, because the perineurium is still shunting AP currents, and tissue parts may give rise to resistance artefacts [167].

4.1.2. Nerve Roots

Since further, humans have no tail, mainly continence (mainly S2 to S5) and sexual functions are located in the conus medullaris only. Those functions are therefore represented in the lower sacral nerve roots, and they don't mix with tail functions as, for example, in rat, cat or dog [181-183].

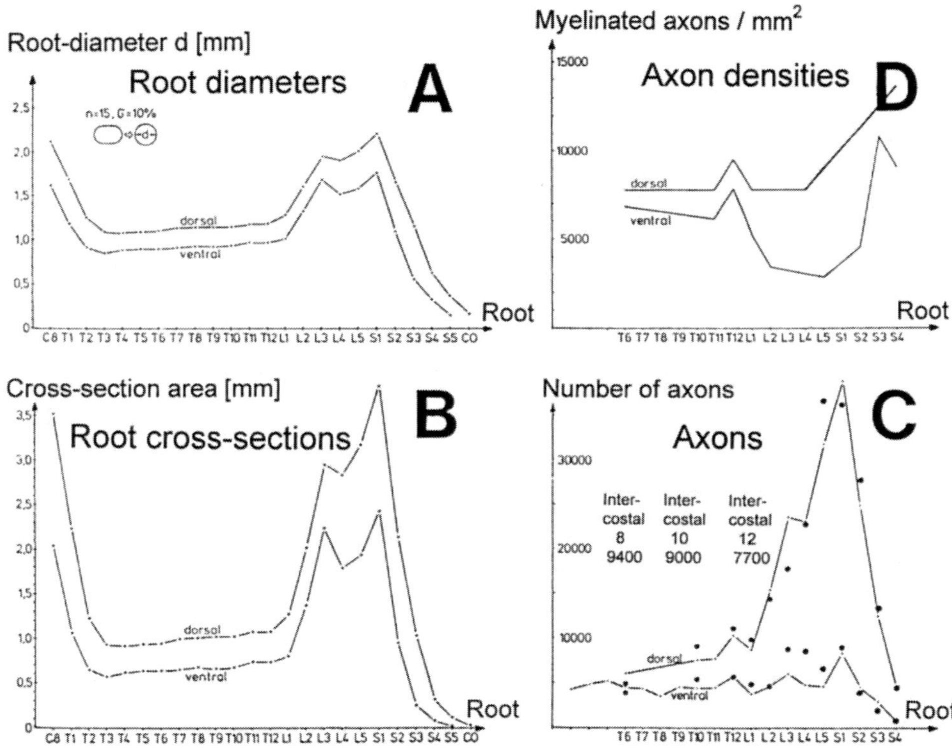

Figure 13. A. Dorsal and ventral mean root diameters of 15 cadavers in relation to the root level. In the case of elliptic-shaped cross sections, the diameters of the circle with the same area were taken. The standard deviation was 8 – 10%; 10% for the smaller roots. B. Dorsal and ventral mean root cross section areas, calculated from A, in relation to the root level. C. Number of myelinated axons of dorsal and ventral roots. The line for dorsal roots is the average of two van Gieson stains and the tuloidin blue stain. The line for the ventral roots is the average of three ventral root counts. The large dots are counts from the tuloidin blue stain; upper dots are the dorsal counts, lower dots are the ventral counts. The deviation of the dots indicates the variation of the axon counts. The inserted axon counts from the intercostals 8, 10, and 12 are from the tuloidin blue stain count. As viewed under the light microscope, the intercostals 8 and 12 had about the same cross section area; the tenth intercostals had about a 10% smaller one. D. Approximate densities of myelinated axons in relation to the root level, calculated from B and C (density = number of axons / cross section area).

A few nerve fibres to leg muscles run through the lower sacral nerve roots (not mentioned in anatomy books), as can easily be verified by electrical and manual nerve root stimulation and the observation of foot or toe movements. Root thicknesses of dorsal and ventral roots are shown in Figure 12. Lower sacral nerve roots show quite a lot of variation, including root interconnections (Figure 12C, right) and ventral root afferents and dorsal root efferents [183]. With respect to electrical nerve root stimulation [31] for urinary bladder control, the functions in the roots have to be identified functionally, including electrical nerve root stimulation and recording of afferent APs, to stimulate efferents in ventral roots after the surgery, most efficiently for bladder control. The implantation of an electrical bladder stimulator is not justified anymore because the urinary bladder can be repaired by CDT non-destructively. Values of nerve root diameters, axon densities, root cross sections and the numbers of nerve fibres per root [170] are given in Figure 13. It can be seen that the dorsal roots are much thicker than the ventral roots in the lower sacral range and that the root diameters strongly decrease in the caudal direction (Figure 13, Figure 12B,C). It was shown that nerve fibre counts of roots obtained from cadavers with the van Gieson method gave similar nerve fibre numbers than those from patients of nerve roots stained with thionin acridine-orange [192]. In one case, it was even possible to obtain similar myelinated nerve fibre numbers with the morphometric method as described below and the electrophysiologic method [180] (see below).

4.1.3. Blood Supply

The blood supply of the caudal spinal cord has been quantified by measuring blood vessel diameters and estimating blood supply contributions [173] (Figures 14,15).

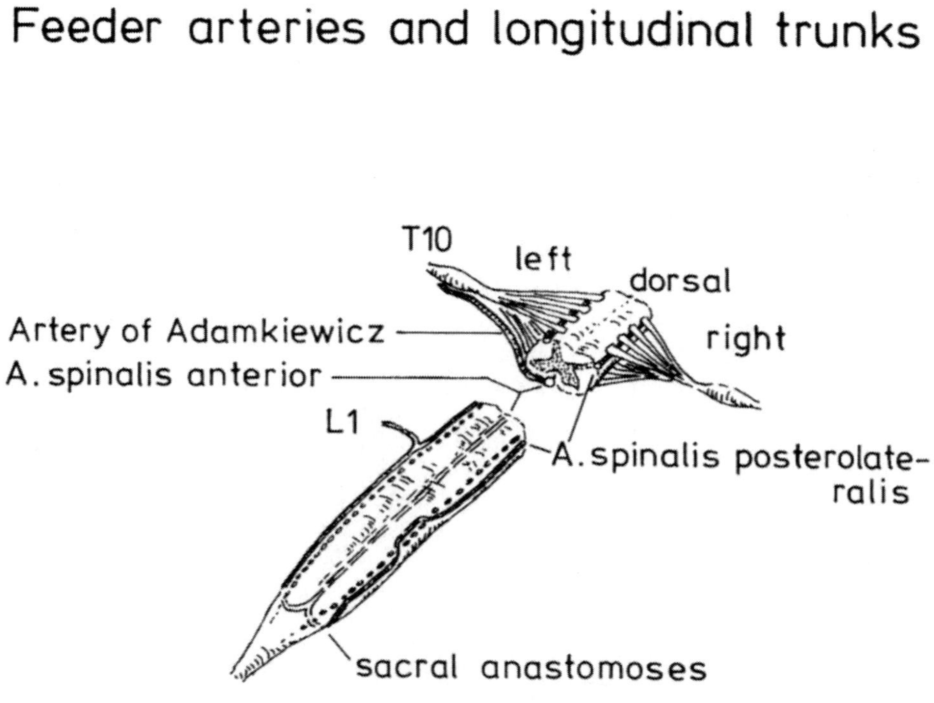

Figure 14. (Continued)

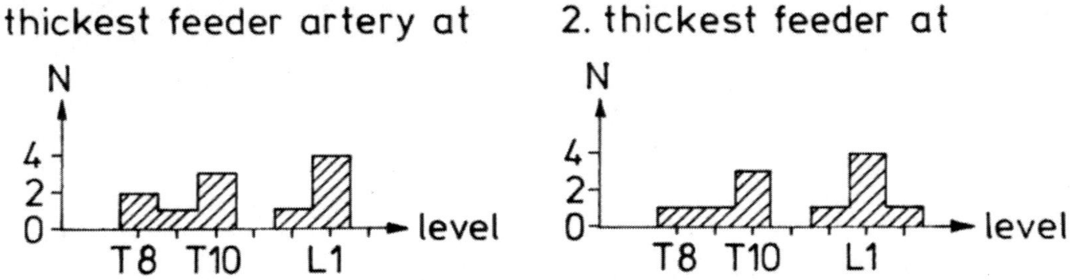

Figure 14. Schematic drawing of the main extra-medullary arteries of the thoraco-lumbar territory of the spinal cord. A ventral (artery of Adamkiewicz) and a dorsal feeder artery, the two dorso-lateral (artery spinalis anterior) and the sacral anastomoses are shown. In the lower part, number (N) of thickest and second thickest feeder arteries in dependence of spinal cord level are given. Measured cases were 11.

Medullary feeder arteries

Cad	arteries schematically l dvd r	thickness order	level	ϕ[mm]	ventral or dorsal	left or right	arterial anastomoses sacral / lumbal
1	T8, T11, L1, L2, L4	1	T8	1.20	v	r	l
		2	L1	0.45	d	l	
		3	L2	0.35	d	l	
		4	T11	0.35	d	r	
		5	L4	0.30	v	l	
2	T10, L1, L1	1	T10	0.50	v	l	1/2s
		2	L1	0.45	d	l	
		3	L1	0.25	d	r	
3	T9, L1	1	T9	0.70	v	l	l
		2	L1	0.45	d	r	s
4	T10, T12	1	T10	0.65	v	l	s(2x0.30; 0.40)
		2	T12	0.28	v	l	
5	T7, T10, L1	1	L1	0.75	v	l	s(0.60; 0.60)
		2	T10	0.45	v	r	
		3	L1	0.35	d	l	
		4	T7	0.35	v	r	
6	T7, T10, L2, L1	1	T10	0.65	v	l	s(0.25;0.25)
		2	L2	0.45	d	l	
		3	L1	0.45	d	r	
		4	T7	0.40	v	l	
		5	T12	0.40	d	l	
		6	T9	0.40	d	l	
		7	L2	0.25	v	l	
7	T7, T11, L1, L3	1	T11	0.60	v	l	s(0.35; 0.35)
		2	L1	0.55	d	l	
		3	T7	0.40	v	l	
		4	L3	0.35	d	r	
		5	T12	0.35	d	r	
		6	T11	0.35	d	l	
		7	T9	0.30	v	l	

Figure 15. Inner diameters of medullary feeder arteries and arterial anastomoses. The column "arteries schematically" shows the extramedullary arteries of the thoraco-lumbar territory projected into plane. Cad = cadaver, l = left, r = right, d = dorsal, v = ventral, T = thoracal feeder artery, L = lumbar feeder, C = cervical feeder, Ø = diameter, l = lumbar anastomosis, s = sacral anastomosis, ½ = anastomosis only to one side, 2X = 2 anastomoses. Thickness order only indicated for the thick feeders of the lower spinal cord.

Since more than 80% of the blood to the thoraco-lumbar arterial territory is supplied by the thickest ventral (A. spinalis magna feeds 68% of the blood from the left ventral side mostly at T10 or L1 levels (Figures 12C,14)), and the thickest dorsal radicular-medullary arteries, these two feeder arteries have to be saved in cauda equina surgery [173].

4.1.4. Number of Myelinated and Unmyelinated Nerve Fibres in Nerve Roots

The number of myelinated nerve fibres supplying the lower human body lies in the range of $0.3x10^6$ (Figure 13). Ingbert counted 650,000 myelinated nerve fibres in all the dorsal roots on one side in a human [107]. Probably, there are more than 10^6 myelinated fibres contained in the nerve roots emerging from the CNS. First measurements indicate that the number of unmyelinated nerve fibres exceeds those of the myelinated fibres in dorsal roots approximately six times (depending on the root) (see also page 71 of [51]). This may explain why patients with a clinically complete SCI (some tract fibres may still be preserved) often report still having deep slowly appearing feelings, since the chance of remaining tract fibres connected to unmyelinated fibres may be much higher than those connected to myelinated fibres; but also, other pathways (sympathetic chain, connection of plexuus) not running through the spinal cord may contribute to the deep slow feeling.

4.1.5. Number of Myelinated Nerve Fibres Innervating the Urinary Bladder [178]

In six human cadavers, the nervus pudendus, the nervi pelvini and the nervus (plexus) hypogastricus were dissected to study the gross anatomy. In further two human cadavers, a few centimetres of the whole nerves of the nervi rectales inferiores and perinealis, the nervi pelvini and the nervus hypogatricus (at the sacral promontorium) were removed.

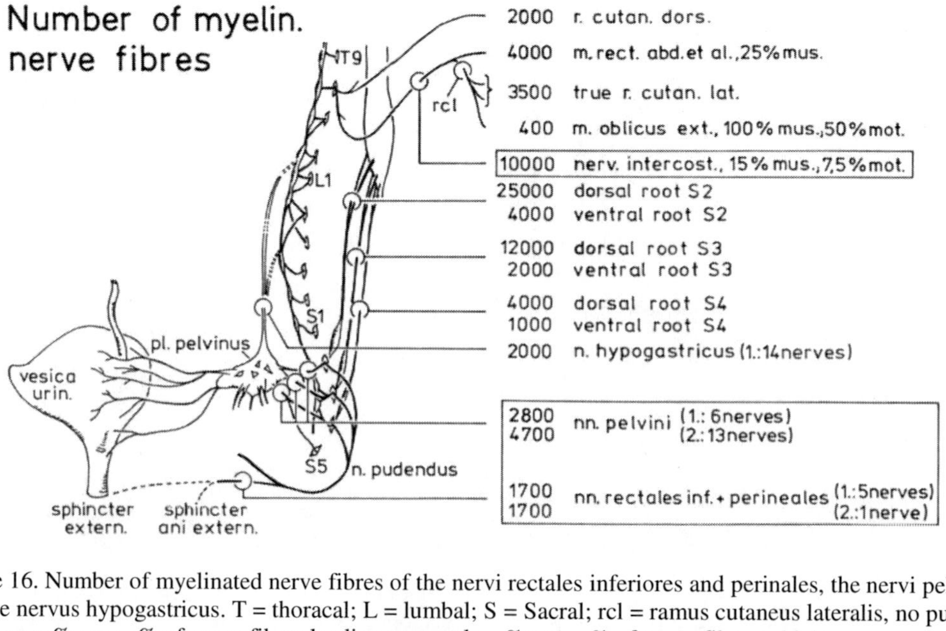

Figure 16. Number of myelinated nerve fibres of the nervi rectales inferiores and perinales, the nervi pelvini and the nervus hypogastricus. T = thoracal; L = lumbal; S = Sacral; rcl = ramus cutaneus lateralis, no pure skin nerve; %mus = % of nerve fibres leading to muscles; %mot = % of nerve fibres which are motoric; 1 and 2 = Nerve fibre counts from cadavers 1 and 2. Number of nerves in the bracket gives the number subnerves of which the nerve consisted. The question mark indicates that the innervation pathway of the external bladder sphincter is unclear.

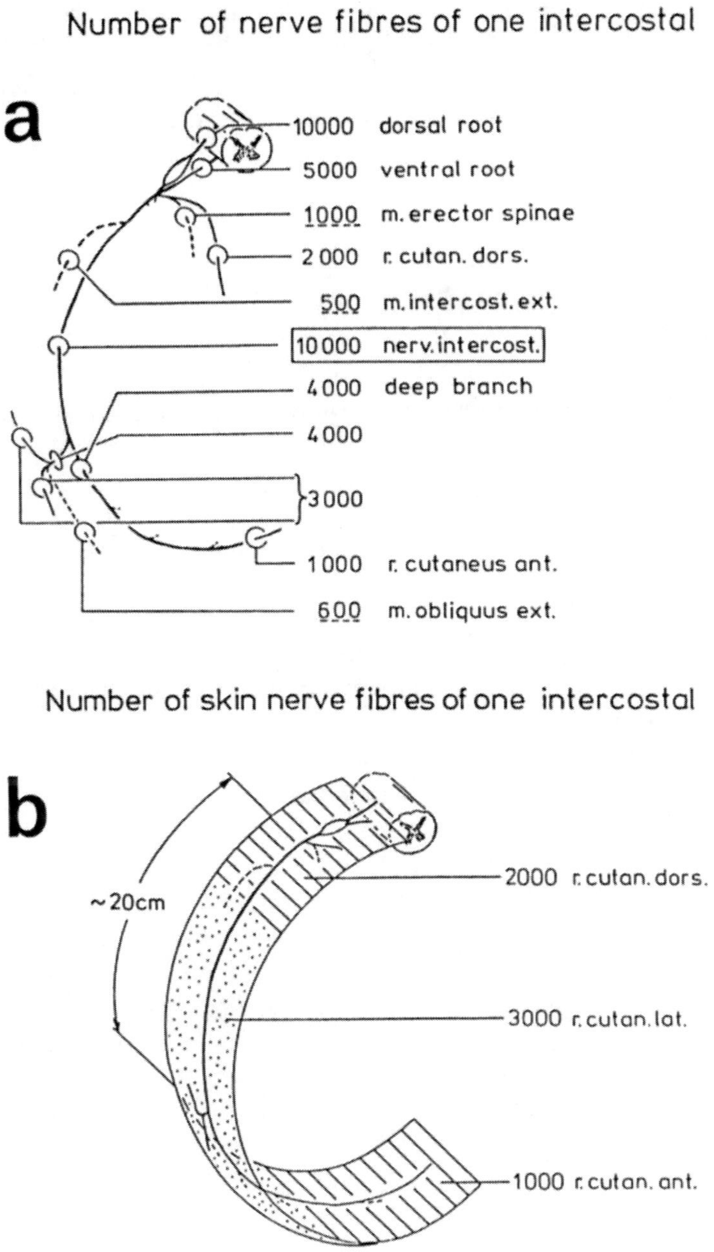

Figure 17. a. Summarized approximate numbers of myelinated nerve fibres of an average intercostal nerve (Th8-Th12) and its branches. Dashed nerve branches indicate pure muscle branches. Dashed underlined numbers indicate numbers of nerve fibres leading to muscles (about 50% motor). The nerve fibres of the rami communicantes have not been counted. b. Summarized numbers of myelinated nerve fibres of the three skin branches of the spinal nerve. Lined or pointed areas indicate approximate area of innervation. Dashed branches indicate muscle branches.

The nerve pieces were fixated, embedded, stained (fixated in 4% glutaraldehyde in cacodylate buffer, after-fixated in 1% OsO_4, embedded in Araldite) and cut, and the myelinated nerve fibres counted. Figure 16 shows the obtained nerve fibre numbers. The nervi pelvini had 2,800 and 4,700 myelinated fibres and consisted of 6 and 13 subnerves. The nervi rectales inferiores and perineales contained 1,700 myelinated fibres and consisted of five and one subnerves. The thick nervus hypogastricus contained only 2,000 myelinated nerve fibres. Most fibres of this vegetative nerve were most likely unmyelinated. The unmyelinated fibres were not counted here. The number of myelinated nerve fibres of an intercostal nerve is taken from Figure 17. From the number of subnerves of the two cadavers of the nervi pelvini (6 and 13) and the nervi rectales inferiors and perineales (5 and 1) (Figure 16), one can expect many variations of these nerves.

4.1.6. Approximate Number of Myelinated Nerve Fibres Leading to the Skin and to the Muscles of the Lower Intercostal Nerves [177]

Twenty human cadavers (cad) were dissected to clarify the gross anatomy of the lower intercostal nerves (Th7-Th12). In further six human cadavers, nerve pieces were removed and myelinated nerve fibres counted in intercostal nerves and its muscle and skin branches, which were identified by following them up to the skin or to the muscles. The spinal nerves split into the dorsal and ventral ramus. The dorsal ramus leads to the musculus erector spinae and the dorsal skin. After the branching off of the rami communicantes, the ventral ramus splits around the angulus costae into the nervus intercostalis externus (inspiration, pure muscle nerve) and the nervus intercostalis internus, normally called the intercostal nerve (Figure 17). The intercostal nerve normally runs under the rib of the same number. A few small branches lead from the intercostal nerve to the musculus intercostalis internus (expiration) and possibly to the musculus serratus posterior inferior. Then the intercostal splits at about the middle or anterior line into the ramus cutaneus lateralis (not pure skin nerve and the mixed ramus (deep branch) to the musculus rectus abdominis and other muscles and the anterior skin (ramus cutaneus anterior). The ramus cutaneus lateralis splits after a few centimetres into a more dorsal skin branch, a more ventral skin branch and a muscle branch to the musculusobliquus externus (Figure 17a). The deep branch to the musculus rectus abdominis can be followed for a few centimetres until it runs deeper into the musculus intercostalis internus. It supplies the musculus obliquus abdominis internus, the musculus abdominis transversus and the musculus rectus abdominis and ends in the ramus cutaneus anterior. The numbers of the counted myelinated nerve fibres of the different skin and muscle branches of the intercostal nerve of cadaver 1 are in Figure 17 (for all cases, see Ref.177). The variation in the number of nerve fibres between the intercostals in each cadaver is large, and the variation from case to case is also large. It seems that the number of nerve fibres increases from the 8th to the 12th intercostal nerve in accordance with the thickness of the average intercostal nerves seen in the dissection and also in accordance with the thickness of intercostal nerve roots and the number of nerve fibres in the intercostal nerve roots. An intercostal nerve has something like 10,000 myelinated nerve fibres. The number may be higher in lower intercostals nerves and in young humans and lower in higher intercostals and in older humans. Since a skin nerve contains mainly sensory fibres and a muscle nerve contains about 50% sensory fibres, an intercostal nerve consists of about 80 to 90% sensory fibres.

4.1.7. Single-Nerve Fibre Action Potential Recording from a Peripheral Nerve in a Human [179]

Single-nerve fibre action potentials (APs) were recorded from the nerve to the musculus obliquus externus abdominis (Figure 18) of the brain-dead human cadaver HT7 with two pairs of wire electrodes in the same way as from nerve roots. Even though the recording quality is low, afferent and efferent APs can be identified in the recording of Figure 18. On average, the afferent activity was much higher than the efferent one, even though about 50% of the fibres of a muscle nerve are motor [21, 177]. Very often, single APs merged. But the skilled eye was able to select the non-merged APs out of the merged ones. Conduction velocity distribution histograms could also be constructed, and the histograms of different recording sweeps compared. Following pinprick of the skin on top of the musculus obliquus externus abdominis, the activity of the secondary muscle spindle afferents in the nerve to the musculus obliquus externus reduced and the activity of the primary spindle afferent fibres disappeared. These afferent changes are interpreted as the result of the defence contraction of the muscle underlying the skin.

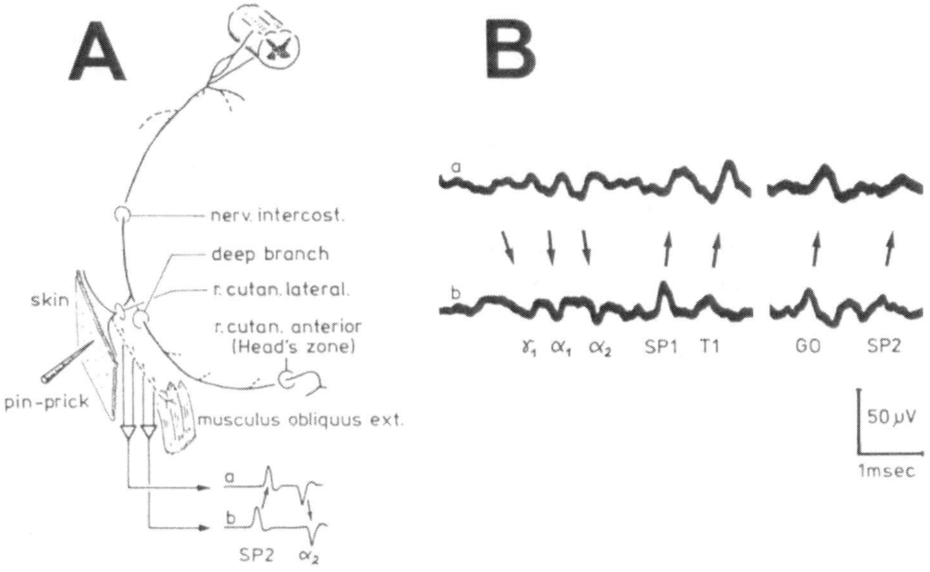

Figure 18. A. Recording arrangement for measuring single extracellular action potentials. Dashed nerve branches are muscle nerves. Electrodes are attached to the nerve innervating the musculus obliquus externus abdominis. As indicated, the ramus cutaneus anterior innervates in the segments T11 to S1 the Head's zone of the urinary bladder. B. Single-nerve fibre action potential recording from the nerve (T11) to the oblique muscle. HT7. Nerve class symbols are explained in Figure 2.

4.1.8. Defence Reaction of the Body

The defence behaviour of the human body following pinprick of the trunk results in a reflex pattern, which tends to induce a movement away from the stimulus as well as a contraction of adjacent muscles for protection of underlying tissue [128]. The withdrawal response consists of the ipsilateral flexion and crossed extension reflex [86, 128]. The stretch reflex, not stimulated here, is distinguished from the skin reflex by its proprioceptive afferent input from the muscle, its shorter latency and synchronized volley [128].

4.1.9. Comparison of Recording Conditions between the Peripheral Nerve and the Nerve Root

The epineurium was removed to a large extent from the nerve to reduce the electrical shunt for the following recording. The perineurium was left in place. Care was taken to leave as many blood vessels as possible intact, so as to not disturb the blood supply of the obliquus externus nerve, so that the nerve fibres do not stop conducting APs after some time. In nerve root recordings, the interruption of the blood supply is not the main problem, because of the good blood supply of the roots. The main problem in root recordings is the mechanical alteration of the nerve root fibres by overstretch and pressure because of the missing of protecting tissue, namely epineurium and perinuerium. As can be seen from Figure 18, it is possible to record from a whole pure muscle nerve. That means, the perineurium is not electrically shunting the AP activity of single fibres too much, so that the recording of single extracellular APs is possible. Since one has to remove the epineurium, which is shunting single-fibre activity strongly, and this is not completely possible without damaging the nerve, bits of epineurium will stick to the nerve and change a bit the AP current spread [167]. This, in turn, can result in a larger variation of the measured conduction velocities or can deform AP waveforms. High-quality recordings can therefore not to be expected from nerve recordings, but they can be expected from nerve root recordings. This quality difference between nerve recordings and nerve-root recordings becomes obvious when comparing the recording of Figure 18B with that of Figure 37. But the advantage of a nerve recording is that functions are measured, which must not be the case in root recordings. The disturbing high activity in the recordings causing the single-nerve fibre APs to merge is also encountered in root recordings. But in the lower sacral roots, the activity is low and afferent activity can be recorded from the ventral root afferents and the efferent activity can be recorded from the dorsal root afferents because of the mixing of afferent and efferent nerve fibres in dorsal and ventral roots (dorsal and ventral roots are close in the conus medullaris). Since humans have no tail, there are mainly continence and sexual functions in the conus medullaris. This is different in most laboratory animals; they have a tail and no ascensus of the spinal cord. Therefore, the high quality of the nerve root recordings in humans, in connection with the anatomical situation that few functions are running through the long and thin caudal sacral roots, mainly implemented the progress in CNS repair upon CDT.

4.2. Functional Morphometry

4.2.1. General Method

Pieces a few cm long were removed from dorsal roots of patients during surgery and from roots of brain-dead humans (HTs) and used to record from, fixated for four hours in 4% glutaraldehyde in cacodylate buffer, after-fixated in 1% OsO_4 for four hours, and dehydrated and embedded in Araldite according to standard techniques. Pictures of semi-thin sections (approximate depth = 1μm) stained with thionin acridine-orange were taken with a light microscope (x1000). Nerve fibre diameters $\varnothing = 1/2(\varnothing_1 + \varnothing_2)$ ($\varnothing_1$ and $\varnothing_2$ are the larger and the smaller diameter of non-round shaped fibres) and the mean myelin sheath thickness "d" were measured by hand. A correction of 8% for shrinkage was allowed. The nerve fibre diameters measured were divided into four classes of myelin sheath thickness: $0.25\mu m \leq d < 0.8$; $0.8 \leq d < 1.3$; $1.3 \leq d < 1.8$; $1.8 \leq d < 2.5\mu m$. For each range of myelin sheath thickness, a diameter distribution histogram was constructed. In the case of damaged fibres (split myelin sheath), the myelin sheath thickness was measured at the most preserved part. Very strongly damaged fibres were not considered. The morphometry was performed manually. Computer-assisted morphometry would have less accuracy, since computer programs cannot handle artefacts and altered nerve fibres as if they were normal fibre.

4.2.2. Relation between Nerve Fibre Diameter and Myelin Sheath Thickness

In order to gain further information about nerve fibre populations innervating the bladder and the anal sphincters, nerve fibre diameter frequency distribution histograms were determined from a cadaver. To identify possible nerve fibre classes in the nerve fibre diameter distribution histograms, the diameter spectra were split according to different myelin sheath thickness ranges. To choose feasible ranges, the relation between nerve fibre diameter and myelin sheath thickness of the nervi pelvini was constructed and shown in Figure 19.

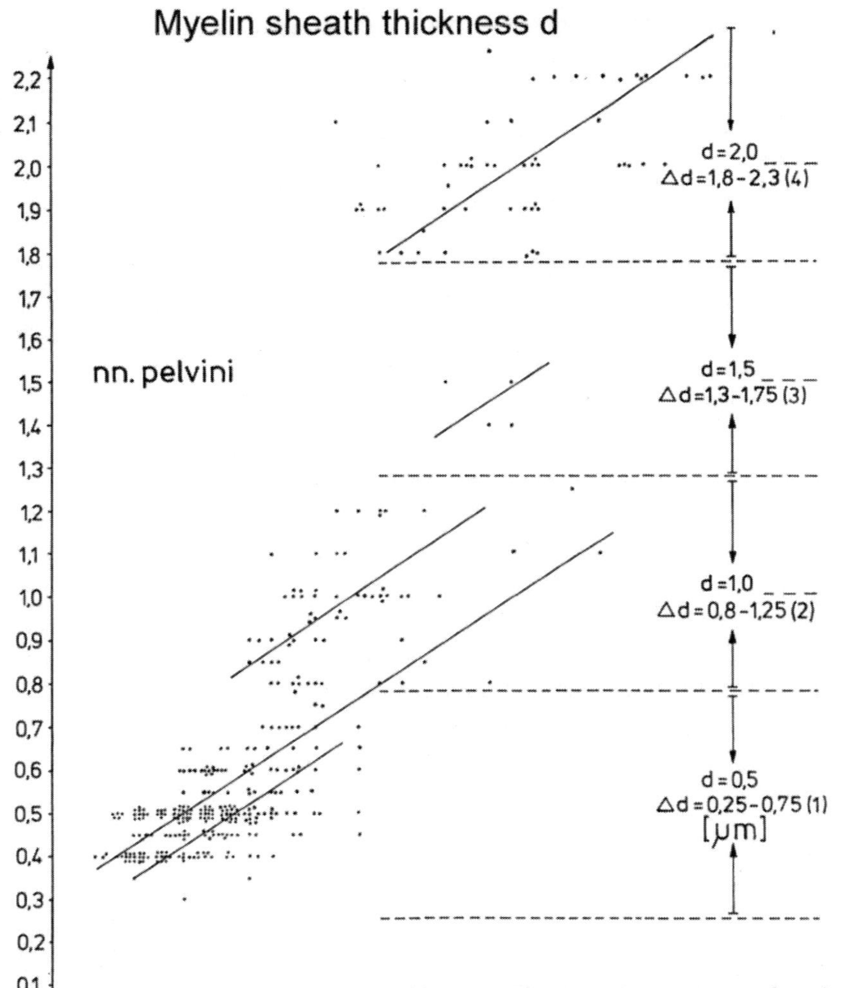

Figure 19. Myelin sheath thickness in relation to the nerve fibre diameter Ø of the nervi pelvini. For definitions of d and Ø, see previous Figures. Lines indicate suggested correlation between d and Ø for different nerve fibre groups. Note, there may be several nerve fibre groups in one class ($0.3 \leq d < 0.8\mu m$), and a nerve fibre group may extend into another class. Correlation dots in class $0.3 \leq d < 0.8\mu m$ are shifted slightly because of closeness.

The myelin sheath thickness roughly increased with the diameter. But the correlation points indicate that there may be different populations of nerve fibres with their own correlation between myelin sheath thickness and diameter. Correlation lines for possible nerve fibre populations are drawn in into Figure 19. It can be seen from the large scatter of the correlation points that it is difficult or impossible to identify different nerve fibre populations on these morphological data alone. The myelin sheath thicknesses were therefore lumped together in four classes characterized by the myelin sheath thickness ranges $0.3 \leq d < 0.8$, $0.8 \leq d < 1.3$, $1.3 \leq d < 1.8$ and $1.8\mu m \leq d < 2.3\mu m$. Nerve fibre diameter distribution histograms were then calculated for these myelin sheath thickness ranges. With the chosen classification ranges, it is still possible that a certain population of nerve fibres covered more than one myelin sheath thickness range, as is indicated in Figure 19. In other nerves, it also seemed as if nerve fibre groups have their own correlation between fibre diameter and myelin sheath thickness.

As Figure 19 indicates, the choice of the myelin sheath thickness ranges was not arbitrary. The split histograms separate different nerve fibre groups better than the whole diameter histogram. Maybe the best method of identifying nerve fibre groups from morphometric data is to correlate the nerve fibre diameter, the myelin sheath thickness (or better the number of myelin sheath lamellae) and the number of occurrences in a three-dimensional space. The method of plotting data in Figure 19 is a first step towards a three-dimensional representation.

4.2.3. Nerve Fibre Diameter Distribution Histograms [178]

4.2.3.1. Nervi rectales inferiores and perineales: Figure 20C shows the nerve fibre diameter distribution histograms of the nervi rectales inferiors and perineales for the four myelin sheath thickness ranges according to Figure 19. In Figure 20Cb, the characteristic cross section area and in Figure 20Ca, the corresponding spectra are shown. To analyze the diameter spectra of the nervi rectales inferiors and perineales, the spectra of a L4 ventral nerve root are shown for comparison in Figure 20Ba, with the identified three α-motoneuron peaks [172, 175] for myelin sheath thicknesses between 1.8 and 2.3μm. The nerve fibres of a L4 dorsal root are distributed differently over the four myelin sheath thickness ranges (Figure 20Aa). The majority of fibres are thin and have a thin myelin sheath and lay therefore in the histogram with sheath thicknesses 0.3 and 0.8μm. Since this spectrum shows two peaks, at least two populations of very thin afferent nerve fibres can be expected to exist in this range. For the mixed (afferent and efferent) nerve rectales inferiors and perineales (Figure 20Ca), the majority of fibres lie in the class with myelin sheath thicknesses between 0.3 and 0.8μm. Several fibre groups can be expected in this histogram.

From the nerve fibres of the histogram with $1.8 \leq d < 2.3\mu m$, an estimation of the number of nerve fibres leading to the sphincter muscle fibres can be obtained. Sphincter muscles have muscle fibres mainly of the slow twitch type (FR) (Figure 3) [134, 205]. These muscle fibres are innervated by α_2-motoneurons [172, 175] (Figure 3) with a nerve fibre diameter range of 9.5 to 12μm (hatched in Figure 20Ca). In these nerve fibres, diameter and myelin sheath thickness ranges are about 4% of the nerve fibres of the nervi rectales inferiors and perinales, namely 68 fibres (0.04 x 1700). This number of nerve fibres contains α_2-motoneurons and afferent fibres with the same nerve fibre diameter and myelin sheath thickness ranges.

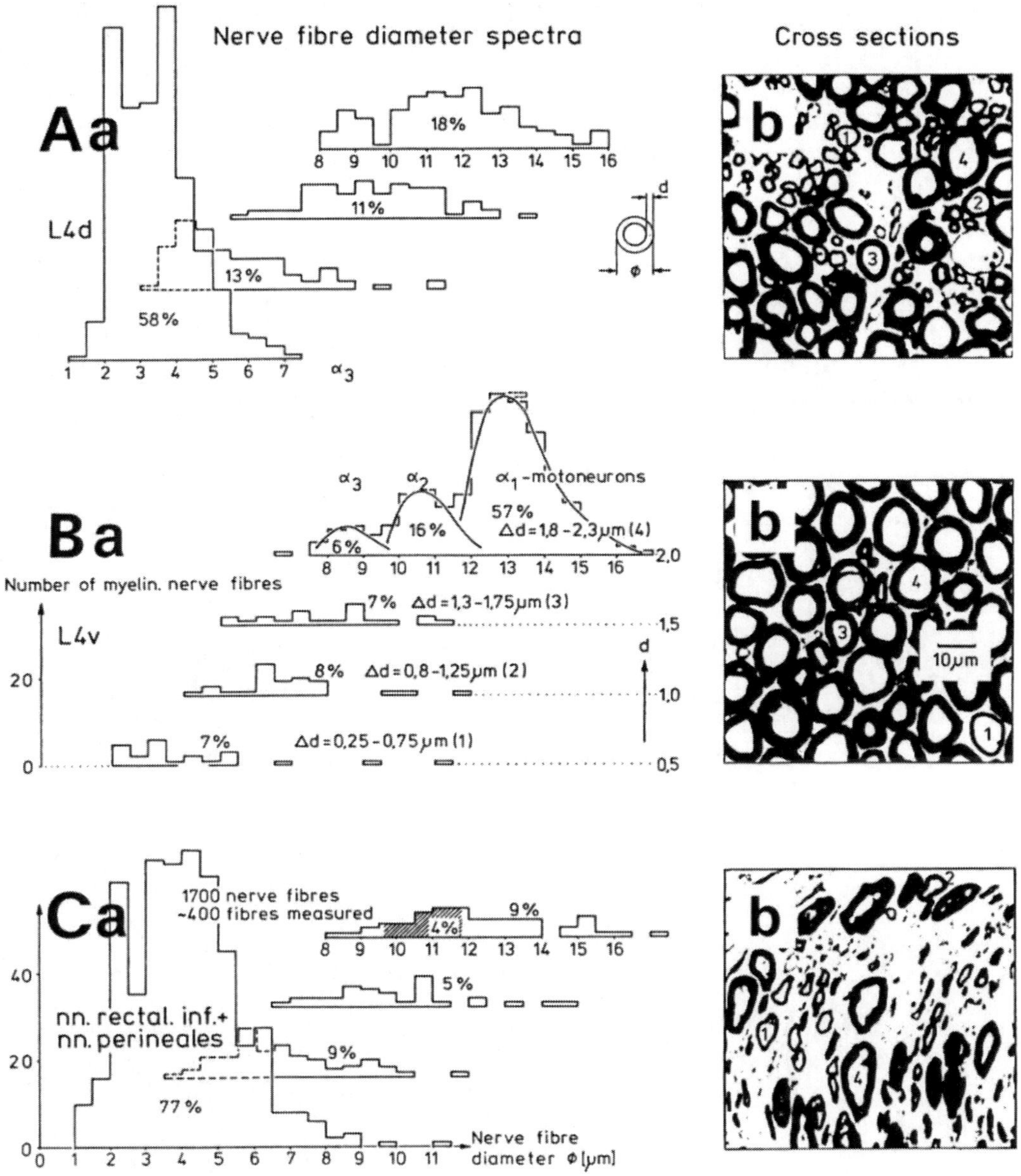

Figure 20. a. Nerve fibre diameter distribution histograms classified by four classes of myelin sheath thicknesses as indicated in Ba. % indicates percentage of fibres in classes or subgroups. b. Corresponding characteristic cross sections. A few fibres are numbered by their myelin sheath thickness range to which they belong. Dimension scale for A, B, and C is drawn in Bb. 8% shrinkage correction. For the definition of fibre diameter Ø and myelin sheath thickness d, see insertion in Aa. – A. Nerve fibre diameter spectrum of a fourth dorsal lumbar root of a 47-year-old female human cadaver removed two to five hours after death, 660 fibres were measured. B. Spectra of a fourth ventral lumbar root (same case as in A), 320 fibres were measured. In the myelin sheath thickness range $1.8 \leq d < 2.3\mu m$, the distribution curves of the three α-motoneuron classes are drawn into the histogram. C. Spectra of the nervi rectales inferiors and perineales. Note, the majority of fibres (77%) has a very thin myelin sheath ($0.3 \leq d < 0.8\mu m$) with a relatively large amount of thick fibres. In the histogram of very thick myelin sheaths ($1.8 \leq d < 2.3\mu m$), the diameter range of α_2-motoneurons, to which sphincter motoneurons belong, is cross hatched (4% of the fibre).

If one assumes a similar distribution of afferent and efferent fibres in this myelin sheath thickness range ($1.8\mu m \leq d < 2.3\mu m$), then one can expect to have an upper limit of sphincter

α_2-motoneurons leading to the anal sphincter and possibly to the external bladder sphincter through the S3 to S5 ventral and dorsal roots [176] of 30 fibres on one side. The number of additional necessary γ-motoneurons will not exceed 30 [175] since there are no or only few spindles in the sphincter muscles [73]. The α-motoneurons to the pelvic floor muscles normally do not lead through the pudendal nerves [131, 156].

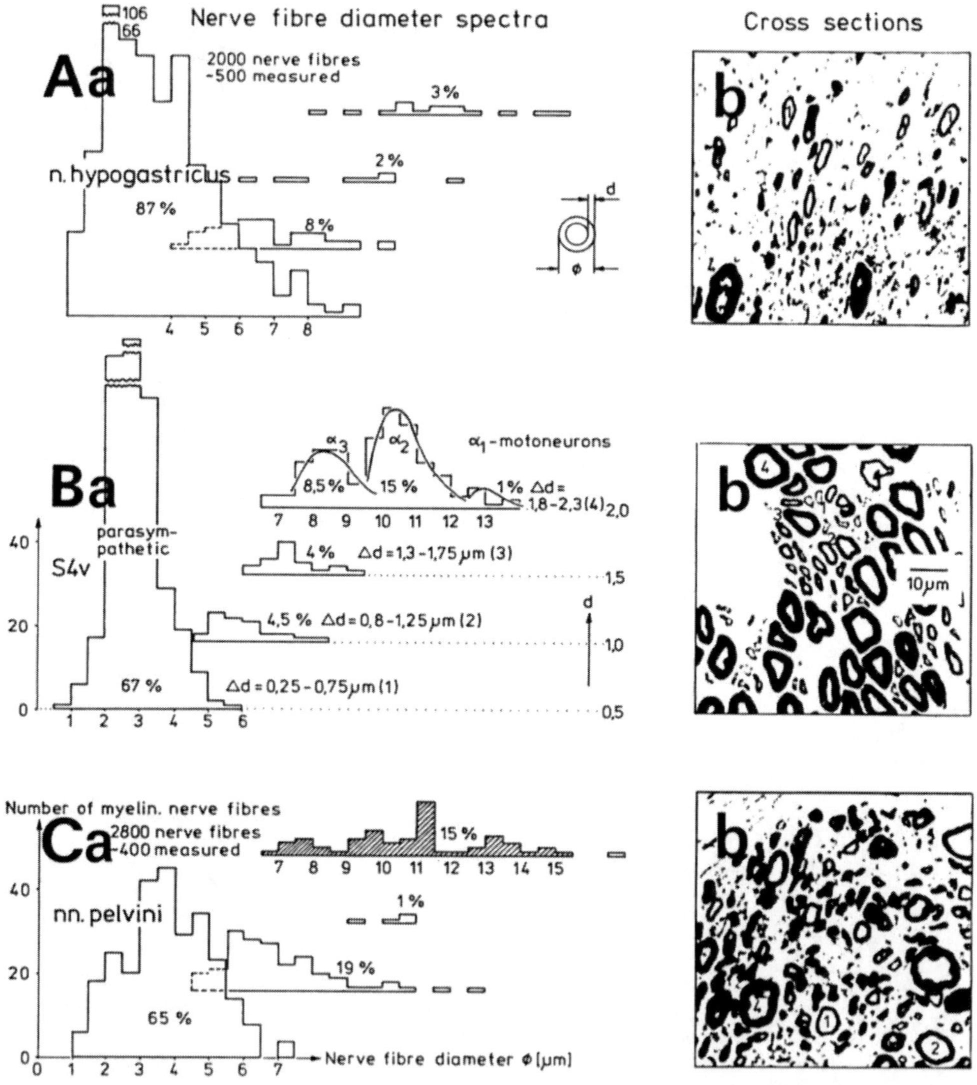

Figure 21. a. Nerve fibre diameter distribution histograms of the nervus hypogastricus, a S4 ventral root and the nervi pelvini classified by four ranges of myelin sheath thicknesses, indicated in Ba. % indicates percentages of fibres in classes or subgroups; b. Corresponding cross sections. A few fibres are numbered by the class of myelin sheath thickness to which they belong. Dimension scale is drawn in Bb. 8% shrinkage correction. – A. Nerve fibre diameter spectra of the nervus hypogastricus. Note, the nerve fibre diameter histogram class $0.3 \leq d < 0.8\mu m$ (87% of the fibres) shows a large proportion of thick fibres. B. Nerve fibre diameter spectra of a fourth ventral sacral root. In the distribution histogram of class $1.8 \leq d < 2.3\mu m$, the distribution curves of the three α-motoneuron classes are drawn. C. Diameter spectra of the nervi pelvini. Note, 15% are thick and have a very thick myelin sheath (class $1.8 \leq d < 2.3\mu m$, crosshatched). α_1, α_2, α_3, SP1, T1 = peak group diameters of α-motoneurons, primary spindle afferents and T1 skin afferents.

4.2.3.2 Nervus hypogastricus: Figure 21Aa shows the nerve fibre diameter spectra of the nervus hypogastricus. The majority of myelinated fibres are contained in the spectrum with myelin sheath thicknesses between 0.3μm and 0.8μm. Most sympathetic fibres efferent fibres will be unmyelinated since the hypogastric nerve lies caudal to the sympathetic chain (Figure 16). The two nerve fibre peaks at about 2.3 and 4.3μm lie at similar diameters as the two peaks of afferent fibres in the L4 dorsal root of Figure 20Aa and contain therefore most likely afferent fibres. Probably a few preganglionic parasympathetic fibres will be intermingled. Their diameters lie also in that range (Figure 21Ba). The nervus hypogastricus contains more thick thinly myelinated afferents in comparison with the afferents of the L4 dorsal root (Figure 20Aa; $0.3\mu m \leq d < 0.8\mu m$).

4.2.3.3 Nervi pelvini: The nerve pelvini of Figure 21Ca have nearly no fibres in the spectrum for myelin sheath thicknesses between 1.3 and 1.8μm, as can also be seen in Figure 16. The fibres in the myelin sheath thickness ranges $0.3\mu m \leq d < 0.8\mu m$ and $0.8\mu m \leq d < 1.3\mu m$ mainly represent parasympathetic afferent and efferent fibres. For the efferent parasympathetic fibres, this becomes clear by comparing Figure 21Ca with Figure 21Ba. For the afferent parasympathetic fibres, such direct comparison is not possible. The large number of thick fibres in the myelin sheath range $1.8 \leq d < 2.3\mu m$ (hatched in Figure 21Ca) is unexpected. It is likely that these fibres are all of parasympathetic afferent type. One would expect the thickest parasympathetic afferents innervate mechanoreceptors, flow receptors, Pacinian corpuscles (distributed over large areas in the tissue [65, 128]) and tension receptors in the detrusor wall. Their diameters probably range up to values of about 12μm (Figure 2). The largest at 11.3μm and the existence of nerve fibres with diameters larger than 12μm suggest that there are also motoneurons and may be a few spindle afferents in the spectrum with myelin sheath thicknesses between 1.8 and 2.3μm. It will be discussed below that these thick nerve fibres could innervate the external urethral sphincter.

4.2.4 Number of efferent parasympathetic fibres in the nervi pelvini: The parasympathetic peak of S4 ventral root fibres of Figure 21Ba contains about 380 fibres. The number of parasympathetic efferent fibres in the nervi pelvini on one side is therefore about 800, if one assumes that parasympathetic fibres are contained in three roots (S3, S4, S5 or S2, S3, S4) and that 20 to 30% of the fibres are ventral root afferents [172, 175, 180] ($(380 \times 3) \times 0.7 \approx 800$). Taking the value for myelinated fibres in the nervi pelvini (Figure 16) and disregarding the existence of possibly a few thick somatic motoneurons, then one finds that among myelinated fibres the nervi pelvini about 20% are efferents. In the monkey, it was found that the parasympathetic nucleus on one side contains about 410 motoneurons [145].

4.2.5 Innervation of the external bladder sphincter: In the nervi pelvini and in the nervus (plexus) hypogastricus are thick fibres with a thick myelin sheath ($1.8 \leq d < 2.3\mu m$) (Figure 21). At least for the nervi pelvini, it is unlikely that all these fibres are afferent fibres. In the nervi pelvini, these fibres make up 15% (hatched in Figure 21Ca). The possibility that a thin direct muscle branch to the pelvic floor [131, 156] was removed with the nervi pelvini is unlikely, since the direct branch to the pelvic floor tends to lie more laterally and the nervi pelvini more medially [19], and such thick fibres were found in all subnerves. It is possible that these thick fibres of the nervi pelvini innervate the external bladder sphincter. It has been reported that the external bladder sphincter could also be innervated through a direct branch from the S3 and S4 roots [156] or via the pelvic nerves [72]. In five male cadaver dissections, the author could not find the branch from the nervus pudendus or nervus dorsalis penis leading to the external bladder sphincter, situated caudally to the prostate gland. The uncertainty of the innervation pathway to the external bladder sphincter is indicated in Figure 16 by a question mark. To learn more about the innervation of the human urinary bladder by a comparison with animal data caution is necessary, since it was reported that similarly named muscles (for our purpose the external sphincter muscle) may not have homologous functions in humans and cats (58). The representation of the urinary bladder in the sacral nerve roots is different in dogs and humans [181-183].

4.2.6 Nerve anastomoses to repair to repair the urinary bladder [177-180]: Nerve anastomoses to repair the urinary bladder (for example, from the intercostal nerves to cauda equina nerve roots or bladder innervating nerves) and other functions in SCI are partly destructive operations (even if highly qualified) and no longer necessary because upon CDT, including structural repair, the bladder and other functions can partly repaired naturally (Chapter II).

4.2.7 Thickness of nerve fibres at different distances from the soma (tapering of nerve fibres): Nerve fibre diameter distributions of nerve roots and peripheral nerves (outside the spinal canal) were compared here, assuming that the nerve fibres have the same diameter along their length. If the peaks in Figure 21Ca at diameters of 7.8 and 9.8μm ($1.8 \leq d < 2.3$μm) are really the distribution peaks of α_2 and α_3-motoneurons, then these values would indicate that the motoneuron axons are distal slightly thinner.

In the comparisons made here, it was assumed that the reduction in nerve fibre diameters at distal parts is small in comparison to other errors or variations. The reduction of nerve fibre with increasing distance from the soma is small (0.2%/per 13cm).

5. Properties of Human Nerve Fibres

5.1. Tapering of Nerve Fibres

Summary of Tapering

To determine the tapering (203) of human nerve fibres (Figure 22), rostral and caudal root pieces of cauda equina nerve roots (Figure 23) were removed from four cadavers with no known neurological disease, and nerve fibre diameter distributions were constructed for four myelin sheath thickness ranges for the two sites and compared with each other (Figures 24,25). The reduction of the group diameter in the different α-motoneuron groups was 0.2 % per 13cm. Accounting for systematic errors, there may be even less tapering. An identified single-nerve fibre showed no tapering. Further, there is indication that γ-motoneurons, preganglionic sympathetic and parasympathetic fibres and skin afferents also reduce their fibre diameter by 0.2 % per 13 cm or less. Consequently, a nerve fibre with a diameter of 10 μm would be reduced to approximately 9.8 μm at 1m from the cell soma. Preganglionic parasympathetic fibres were found to be represented in roots S1 to S5. At similar distances from the spinal cord, the mean diameter of ventral root α_1-motoneuron (FF) axons increased from the thoracic towards the lumbo-sacral region before decreasing again in the lower sacral region. Usually no α_1-motoneuron axons were found in S5 roots. The diameter distribution of unmyelinated nerve fibres of a ventral S5 root showed three peaks at 0.25, 0.95 and 1.2 μm.

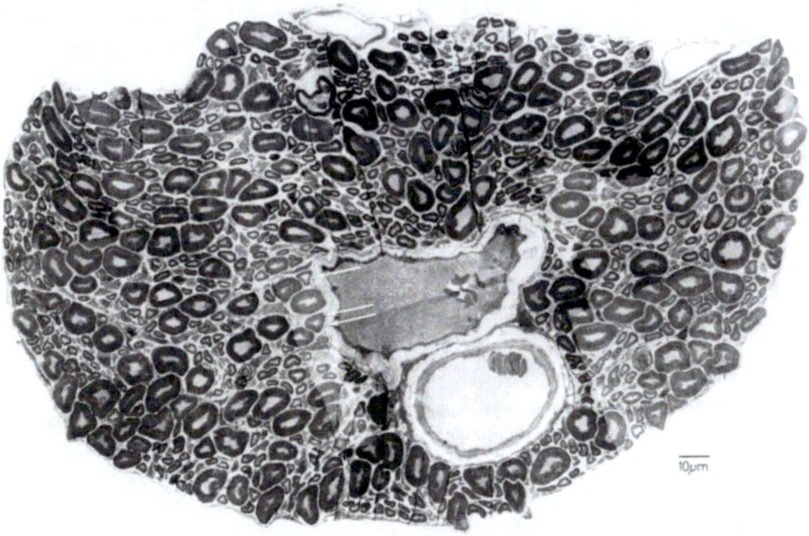

Figure 22. Montage of several electron microscope photographs of the S4 ventral root from HT1. Single arrow marks a nerve fibre with a comparably thin myelin sheath; double arrow marks a nerve fibre of same size with a comparable thick myelin sheath. Scale not corrected for shrinkage.

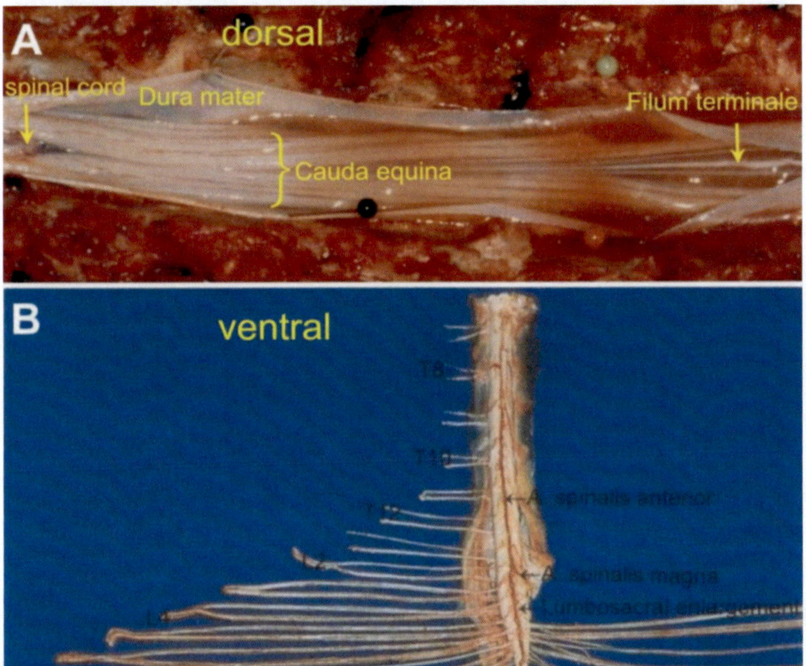

Figure 23. Schematic layout for the measurements of the tapering in human nerve fibres. After opening of the spinal canal (A) and removing of the lower spinal cord with the cauda equina nerve roots and the dura mater (B), rostral and caudal nerve root samples of 1 to 2cm length were removed from identified dissected and fixed nerve roots (mounting in some similarity to B).

The unmyelinated fibres with diameters around 0.25 μm may represent parasympathetic fibres. In six selected areas of the ventral S5 root, 6.6 times more unmyelinated nerve fibres than myelinated fibres were found on average.

5.1.1. Measuring Procedure

To measure the tapering of human nerve fibres, the spinal cords were removed together with the cauda equina nerve roots from four cadavers (Figure 23), fixated, and the nerve roots were then dissected to a maximum extent under the dissecting microscope in a chamber filled with the fixative.

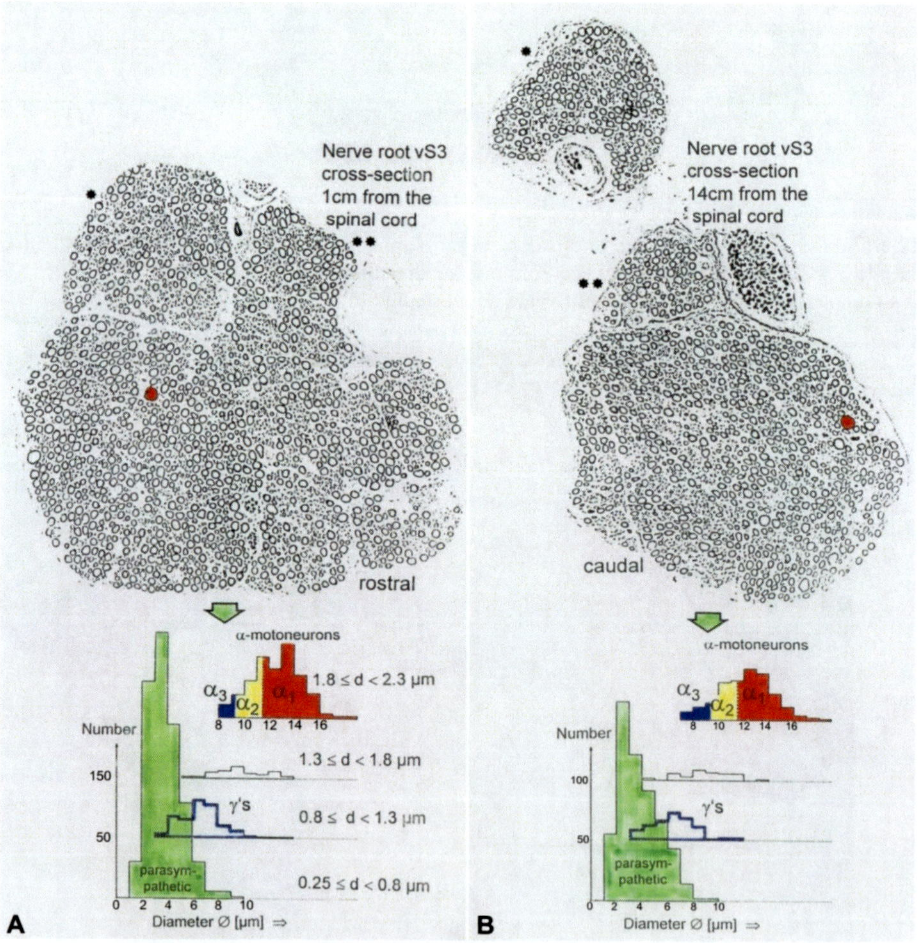

Figure 24. Comparison of nerve fibre diameter distributions of a vS3 root rostral (A) and caudal (B) to measure tapering. After embedding, cutting and staining (thionin acridine-orange), the cross sections were photographed (x1000) (D), and mean nerve fibre diameter and myelin sheath thickness were measured for the whole cross section or parts of it. Nerve fibre diameter distribution histograms (with nerve fibre group characterisations (α's and γ's) from the rostral and caudal cross sections for the four myelin sheath thickness ranges (d) and the corresponding rostral and caudal distributions were compared with respect to tapering. Probable fasciculation of the vS3 root is marked by one or two stars. An identified single fibre in the rostral and caudal cross sections is marked. If only a few ventral root afferents are present in the ventral S3 root, then the fibre diameter distribution of the myelin sheath thickness range 0.25 μm ≤ d < 0.8 μm represents mainly parasympathetic fibres.

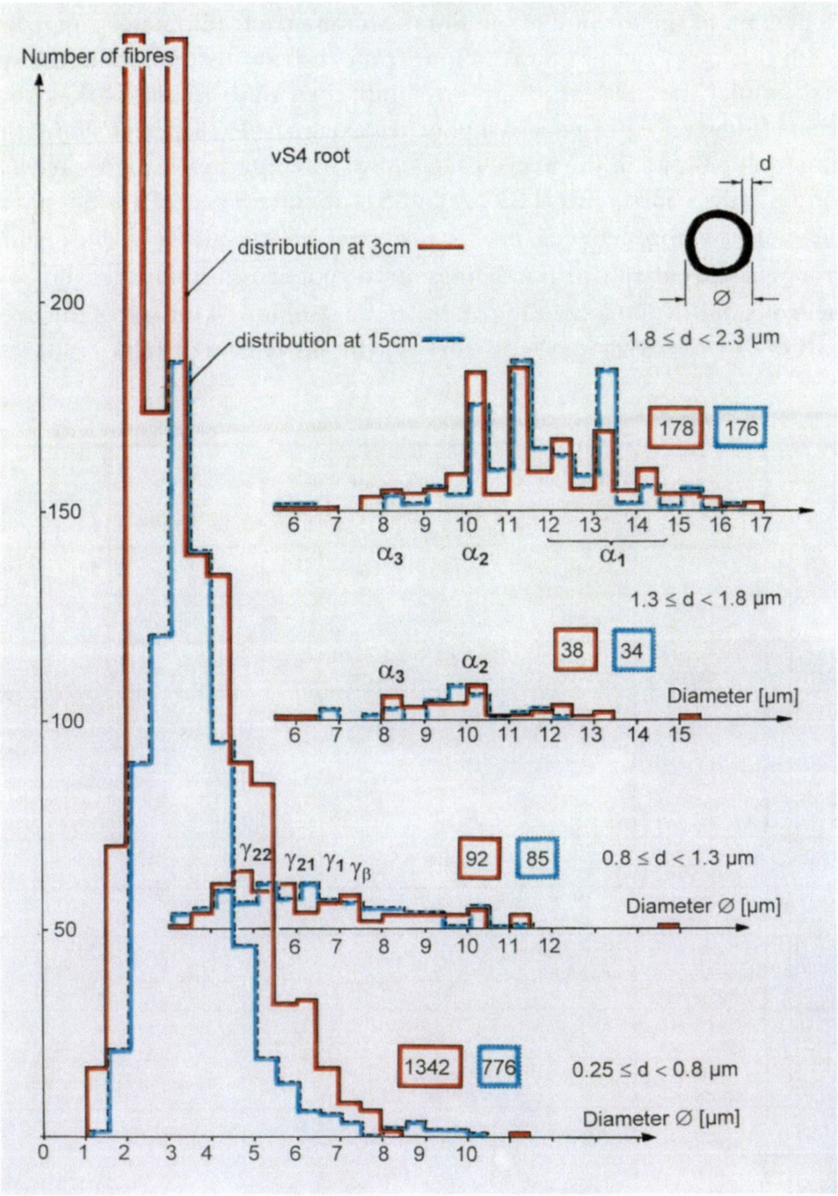

Figure 25. Nerve fibre diameter distributions for four myelin sheath thickness ranges "d" of the rostral (solid line) and the caudal (dashed line) cross section of a ventral S4 root. Nerve fibre groups are partly marked at the corresponding distribution ranges or peaks. Figures in the solid and dashed line rectangles give fibre numbers from the fibre distributions of certain myelin sheath thickness ranges rostrally and caudally. α_1, α_2 and α_3 = α_1 (FF), α_2 (FR) and α_3-motoneurons (S). γ_1 = dynamic γ-motoneuron. γ_{21}, γ_{22} = static γ-motoneurons.

The thin roots were of special interest because they are long and thin and can be expected to allow performing morphometry of the whole root cross sections over a long distance between the rostral and the caudal root parts. After freeing a root from the other roots, samples of rostral and caudal roots were taken, post-fixated, embedded, stained, and photographed, and the mean diameter and myelin sheath thickness were measured and nerve fibre diameter distribution histograms were constructed (Figures 24,25).

Table 1. Reduction of the mean diameter of α-motoneurons (diameter range 8 to 16 μm; myelin sheath thickness range 1.8 to 2.3 μm) from rostral (distance from the spinal cord 1 to 3 cm) to caudal site (distance from the spinal cord 12 to 16 cm) of the cauda equine nerve roots (Figure 1). Presence symp. or parasymp. = Presence of sympathetic or parasympathetic fibres in the nerve root, symp. = sympathetic, para = parasymp. = parasympathetic. vS3 = ventral S3 root, dS5 = dorsal S5 root, Para 5 = paraplegic patient 5, α-mot = α-motoneurons, prox = proximal, γ's = γ-motoneurons. Number of α-motoneurons (in the entirely or partly measured root cross sections) rostal → caudal = α-motoneurons in the diameter range 8 to 16 μm. Column 4 indicates the presence of sympathetic or parasymapathetic fibres in the nerve root or nerve root fascicle

Case	Root	Part measured	Symp. or para. fibres	Reduction of mean diameter: rostral → caudal distance from cord	Reduction of mean diameter: %	Number of α-mot rostral → caudal	Special notes
1	vT12	whole spectrum	symp.	3 cm –			
1	vS1	whole spectrum	para.	3 cm –			
1	vS3	whole spectrum	para.	12.0 μm → 11.7 μm 1 cm 14 cm	2.5	282 → 205	
1	vS3	single fibre		23.5 μm → 23.5 μm 1 cm 14 cm	0		identified by thin myelin sheath and 2 blood vessels
1	vS4	whole crosssection, only α-mot	para.	11.4 μm → 11.4 μm 1 cm 13 cm	0	37 → 30	more fibres prox., root connections should exist
2	vL1	whole spectrum	symp.	1 cm –			
2	vL5	whole spectrum	–	12.0 μm – 1 cm –			
2	vS4	whole spectrum of whole root	para.	11.5 μm → 11.3 μm 3 cm 15 cm	1.7	174 → 175	73 % more fibres rostrally for 0.25 ≤ d < 0.8 μm
3	vL1	whole spectrum	symp.	3 cm –			
3	vL5	whole spectrum	–	13.2 μm → 12.7 μm 1 cm 13 cm	3.8	129 → 119	
3	vS1	whole spectrum	para.	2 cm –			
3	vS4	whole spectrum of whole root	para.	11.7 μm → 11.7 μm 1 cm 16 cm	0	27 → 52	root fascicle of 20 μm measured
3	vS4	whole crosssection, only α-mot	para.	12.3 μm → 12.3 μm 1 cm 11 cm	0	23 → 15	
4	vL5	whole spectrum	–	12.8 μm → 13.1 μm 1 cm 12 cm	2.3	163 → 159	no tapering for γ's
4	vL5	α-mot spectrum	–	16 cm			
4	dS5	whole root, α-mot	/	13.4 μm → 13.5 μm 1 cm 16 cm	–0.7	26 → 22	
Para 5	S5	whole spectrum, whole fascicle S5	para.	8.3 μm ↔ 8.2 μm 3 cm 6 cm 7 ≤ d < 11 μm	1.2	70 → 70	

The nerve fibre diameter distributions for the rostral and the caudal part were compared quantitatively to find out whether human nerve fibres do or don't taper. The data of all analysed roots are summarised in Table 1.

For thicker roots, the fibres were only measured in a part of the cross sections, and only nerve fibre diameter distributions were compared for the different myelin sheath thickness (d) ranges. The different α-motoneuron groups were mainly compared for the myelin sheath thickness range $1.8 \leq d < 2.3$ μm. Since there was often a considerable reorganisation and fasciculation of the roots on their way caudally (Figure 24), the same root cross section area could not be identified safely rostrally and caudally. Only comparable areas were chosen for the analyses. In five cases, the root cross sections could be measured completely, and the rostral and caudal distributions were compared.

5.1.2. Reduction of the Mean Diameter of Completely and Partially Measured Roots

The reduction of the mean diameter of α-motoneurons was mainly measured in the diameter range from 8 μm to 16 μm. The mean diameter of the α-motoneurons in all measured cases dropped by -2.3 % to 3.8 % between the cord and, on average, 13 cm caudally on the roots (Table 1). A mean value of tapering of 0.6 % per 13 cm was obtained for the α-motoneurons (and afferent fibres in one case).

5.1.3. Reduction of the Mean Diameter of Completely Measured Roots

In the roots that could not be measured completely, there sometimes were slight differences in the composition of α_1, α_2, and α_3-motoneurons in the chosen areas of the cross sections. More reliable tapering values are therefore obtained from Table 1 by only selecting cases in which the whole root or the fascicle have been analysed rostrally and caudally. In Figure 25, the nerve fibre diameter distributions are shown for a ventral S4 root for the four myelin sheath thickness ranges, measured close to and at a distance from the spinal cord. The distributions for the different myelin sheath thickness ranges are very similar, apart from the distribution for the myelin sheath thickness range $0.25 \leq d < 0.8$ μm. It seems that at 3 cm from the spinal cord, there is a thin myelinated fibre population ($d \approx 2$ to 2.5 μm), which on its way to the caudal parts leaves the root via root interconnection (see below). From the five cases for which the whole cross sections could be measured (in Table 1 marked by the thick frame), a mean tapering of 0.2 % per 13 cm of α-motoneurons was obtained. If one considers that, on average, distal parts of the roots showed a stronger desiccation in places where the cauda equina consisted of fewer thinner roots, then the tapering of the nerve fibres may even be smaller than 0.2 %.

5.1.4. No Tapering of an Identified Single α-Motoneuron Axon

In one case (ventral S3 root), it was possible to identify a single α-motoneuron axon rostrally and caudally by its size (the thickest fibre), its thin myelin sheath thickness (1.8 μm) and its close location to two thin blood vessels (marked red in Figure 24A,B). The measured mean diameter was 23.5 μm both rostrally and caudally and showed, therefore, no tapering. The single-fibre measurement makes it likely that there may be very little or no tapering at all of nerve fibres if there is no branching of nerve fibres in between the sites where the measured root samples were taken.

5.1.5. Tapering of Afferents, γ-Motoneurons and Preganglionic Parasympathetic and Sympathetic Fibres

The rostral and caudal nerve-fibre diameter distributions of the same root (Figure 25) and other cases seem to indicate that the γ-motoneurons and the preganglionic parasympathetic fibres do taper only little or not at all.

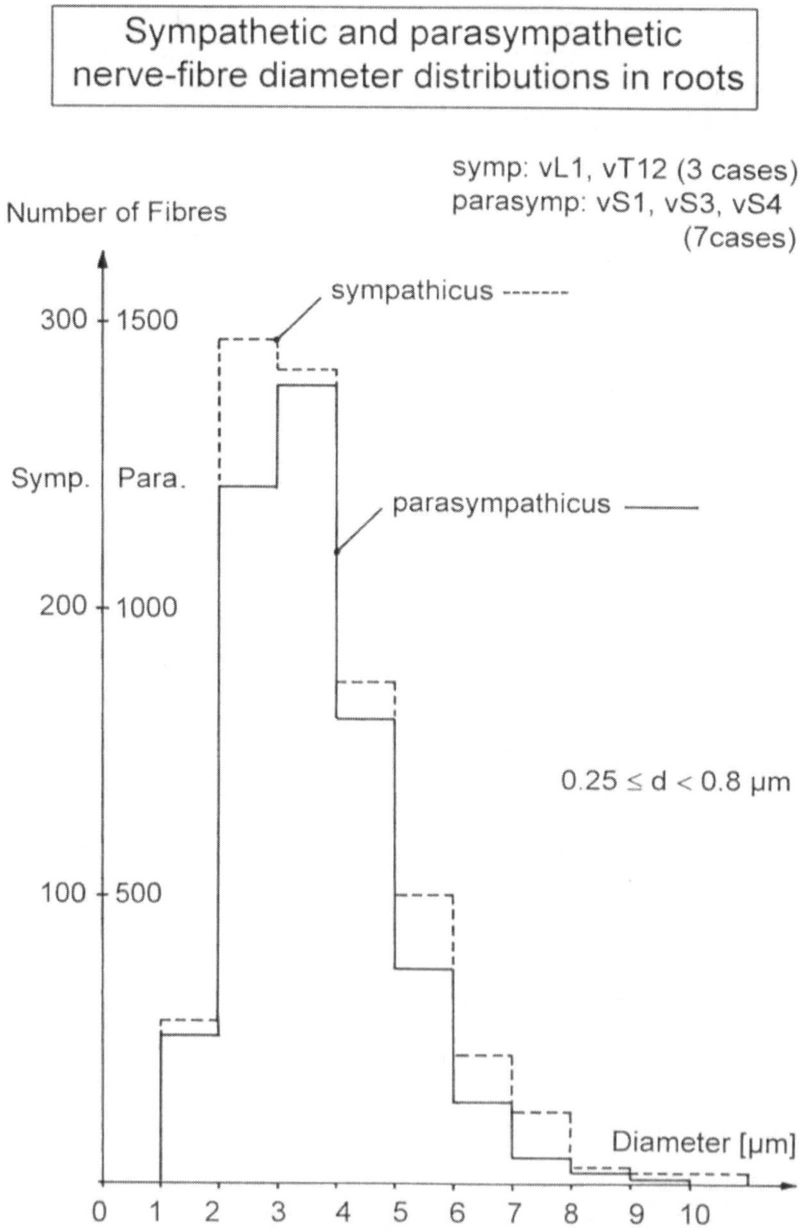

Figure 26. The sum of sympathetic (dashed line) and parasympathetic (solid line) efferent nerve-fibre diameter distributions of several roots indicated in the upper right Figure. Myelin sheath thickness between 0.25 μm and 0.8 μm. Note that the sympathetic and parasympathetic distributions are very similar. Sympathicus (symp): vL1, vL2 (three cases); parasympathicus (para): vS1, vS3, vS4 (seven cases).

The tapering of the preganglionic sympathetic fibres, probably leaving the spinal cord via the TH10 to L2 roots, is more difficult to measure since these roots are much thicker than the lower sacral roots in which the parasympathetic fibres are contained. Since, however, the preganglionic sympathetic and parasympathetic root fibres had very similar nerve-fibre diameter distributions rostrally and caudally (Figure 26), it seems that also the preganglionic sympathetic fibres (contained even more caudally in the ramus albeus) taper only very little or not at all. The much thinner and unmyelinated postganglionic sympathetic fibres (contained in the ramus griseus) have not been measured here.

5.1.6. Representation of Preganglionic Parasympathetic Fibres in Nerve Roots S1 to S5

The measurements summarized in Table 1 indicate that the S1 roots also contained a substantial number of preganglionic parasympathetic fibres. The S1 root was identified by counting from the 12th intercostal nerve (subcostalis) and by the fact that the ventral S1 root is the last thick root caudally (Figure 23B). Parasympathetic fibres were found earlier in thin S5 roots. Since S5 roots are very thin (much thinner than the S4 root (Figure 13)), their fibre numbers are not as important as those of the S1 root.

5.1.7. Root Inter-Connections

As shown in Figure 23B (right lower part), there are root inter-connections among the dorsal and among the ventral roots close to the spinal cord. More caudally at the nerve rootlets from the dura, there are also root inter-connections between the whole roots. Actually, the roots or root fascicles were dissected in between these two root connections. When dissecting the nerve roots, arachnoidea connections were cut, which kept the roots together. But since a root fascicle of 20 μm diameter (with 14 myelinated nerve fibres) has been found here and since sometimes the roots had different nerve fibre group populations rostrally and caudally (for different numbers of α-motoneurons rostrally and caudally, see Table 1), there is indication that some of the arachnoidea cross-connections were really nerve root connections with an interchange of nerve fibres, which gave rise to a regrouping of the nerve groups in the roots; this is known to take frequently place in peripheral nerves.

5.1.8. Representation of Differently Sized α_1-Motoneuron Axons at Different Spinal Cord Levels

Figure 27 shows that the mean of the α_1-motoneuron axon diameter distributions (indicated by an arrow) increase from the ventral (v) TH12 root towards the lumbar roots, before decreasing again in the lower sacral roots. In the ventral S5 root, there were no α_1-motoneuron axons at all; this agreed with the electrophysiologic identification, when recording single-nerve fibre action potentials from undissected nerve roots during surgery. In the ventral L1 root, there were only a few α_2 and α_3-motoneuron axons contained. Even though these measurements were obtained from an arbitrarily selected set of cross sectional areas, and the fibre distributions could vary in other sets, these measurements still reflect somehow the different distributions of differently sized α_1-motoneuron axons at different spinal cord levels. It will be discussed below that the different representation of α_1, α_2 and α_3-motoneuron axons in muscle nerves has consequences for the understanding of the size principle of motor unit recruitment.

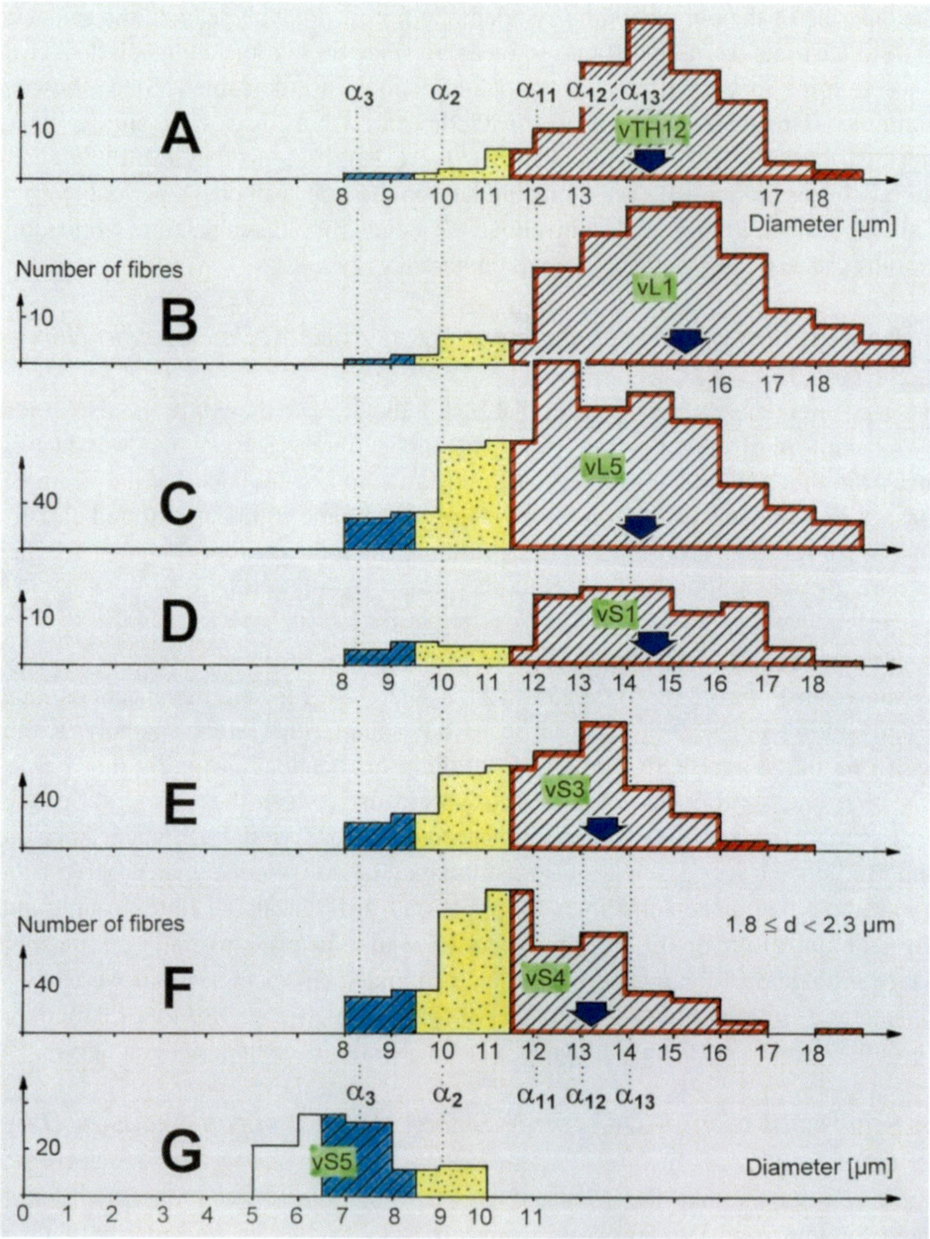

Figure 27. α-Motoneuron type representation in nerve roots (A-G). Composition of submotoneuron groups in a number of diameter distributions of ventral roots in the myelin sheath thickness range $1.8\ \mu m \leq d < 2.3\ \mu m$. α_1, α_2, α_3 = α_1 (FF), α_2 (FR), α_3-motoneurons (S). α_{11}, α_{12}, α_{13} = possible subgroups of α_1-motoneurons.

5.1.9. Unmyelinated Nerve Fibres in a Ventral Sacral S5 Root

In the cross sections of the ventral S5 root of the paraplegic patient 5 (24 years of age), unmyelinated nerve fibres were observed (Figure 28A). Because of the existence of ventral root afferents in the lower sacral nerve roots [175, 176], the finding of unmyelinated fibres was to be expected. Since in different cross sections there seemed to be sometimes mainly thick or mainly thin unmyelinated fibres, mean diameter of unmyelinated fibres were measured, and diameter distributions were constructed (Figure 28B).

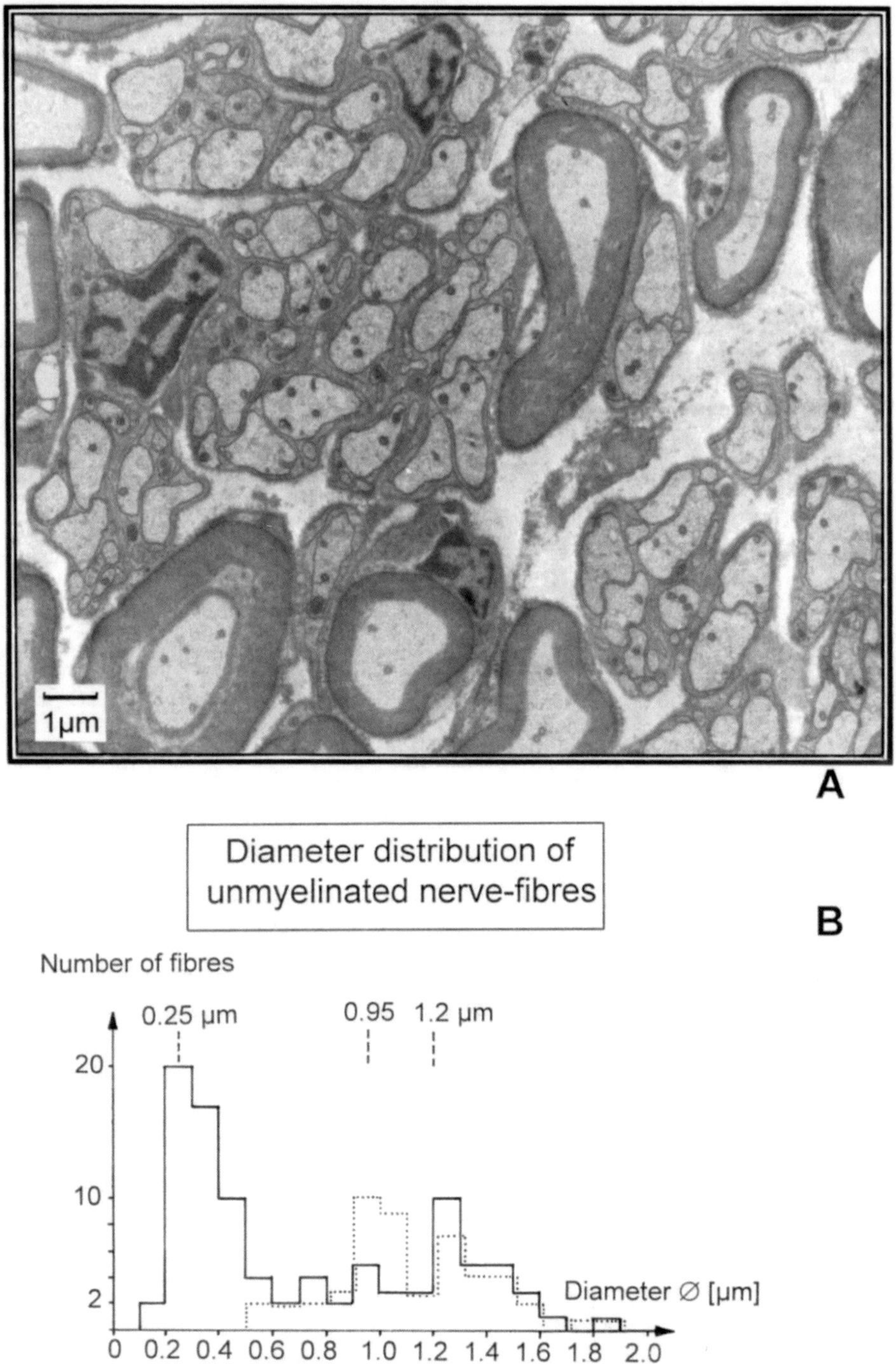

Figure 28. A. Section of a ventral sacral S5 root of a 24-year-old female paraplegic patient (para 5). Thin and thick unmyelinated nerve fibres can be seen besides myelinated nerve fibres. B. Nerve fibre diameter distributions of unmyelinated nerve fibres. The solid line distribution was taken from an area of which the section in A was a part. The dotted line distribution stems from an area in which mainly only thick myelinated fibres were present. Note that the peak at 0.25 µm is missing in the distribution marked by the dotted line.

The distribution in one area of unmyelinated fibres showed three peaks (solid line distribution) at 0.25, 0.95 and 1.2 μm. In another part of the cross section with mainly thick unmyelinated fibres (dotted line), the distribution showed only peaks at 0.95 and 1.2 μm. Because of the three distribution peaks at least three groups of unmyelinated nerve fibres seem to exist in the lower sacral nerve roots, which may belong to the somatic and parasympathetic nervous system divisions.

5.1.10. Numbers of Unmyelinated Nerve Fibres in Relation to Myelinated Fibres

In six areas of the ventral S5 root of the paraplegic patient 5, the number of unmyelinated fibres was measured. The fibre number varied quite much according to whether more thin fibres with a group diameter around 0.25 μm were contained or more thickly unmyelinated fibres were contained with probable group diameters of 0.95 or 1.2 μm. In the chosen areas (approx. 23 μm x 17 μm), the number of unmyelinated fibres ranged from 42 to 173, and the number of myelinated fibres from 10 to 16. The relation between unmyelinated and myelinated fibres ranged from 3.5 to 14. The mean ratio was 6.6. The ratio between unmyelinated and myelinated fibres depended therefore strongly on the number of thin myelinated fibres contributing to it.

5.1.11. Tapering of Nerve Fibres of Nerve Roots and Peripheral Nerves

A tapering of nerve fibres of nerve roots by 0.2% per 13 cm was identified by the present measurements when the whole cross sections were measured. Concerning the systematic error due to the mostly better fixation rostrally as a result of a slightly different osmotic pressure and different autolysis, and taking into account that an identified single fibre showed no tapering, it is safe to conclude that the tapering of nerve fibres is 0.2% per every 13 cm of length or smaller. First measurements show that the phrenic nerve fibres (0.8μm ≤ diameter) reduce their diameter by 0.2 % per every 10 cm of length. There is indication for human nerve fibres showing a slight tapering tendency. The conclusion that nerve fibres only show little tapering is in accordance with an earlier comparison of morphometric measurements on skin nerve branches of intercostals and a coccygeal nerve root [180] through which mainly skin nerve fibres run to the coccygeal dermatome. The nerve fibre diameter distributions of the nerves and the root were very similar, even though the coccygeal root piece was taken approximately 10 cm away from the ganglion and the spinal cord, whereas the skin nerve branch sample of the intercostal was taken approx. 25 cm away from the ganglion. In Table 1, the dS5 root of case 4 shows a negative tapering value of 0.7%. This negative value does not mean that afferent fibres get thicker the farther they are measured from the ganglion. Most likely, the systematic errors of fixation and staining gave rise to the negative value of tapering. By taking the rostral root sample at a distance of 1 to 3 cm from the spinal cord, the central-peripheral transition zone [37] was not touched. The analysed fascicle of the S5 root of paraplegic 5 was most likely a ventral one, since a dorsal root fascicle would have thicker fibres, at least from the T1 (PC) skin afferents. The distance of approx. 3 cm between the rostral and the caudal (it is unclear what is rostral and what is caudal) cross sections is too small in comparison to the variation of the morphometry due to the systematic error, even though as expected, the quality of the root cross sections from the patient was better than those from the cadavers (no autolysis). Long distances are needed to measure tapering.

5.1.12. Root Inter-Connections and Branching of the Cauda Equina Nerve Root Fibres

When these measurements were designed, it was thought that there only are root interconnections close to the spinal cord (very rostrally, Figure 23B) and distant to the spinal cord close to the ganglion. There should be no nerve fibre exchange between the different rather free roots in between the two boundaries: the dorsal and the ventral roots each stick together when leaving the spinal cord, whereas shortly before entering the ganglion, dorsal and ventral root fibres get combined. It was found here directly that some of the arachnoidea connections were actually really root connections with a transposition of nerve fibres. Indirectly, transposition of nerve fibres between the roots was found because sometimes more fibres were found rostrally (and caudally) in completely measured roots. More fibres caudally can be explained by branching, but more fibres rostrally can only be explained by a nerve fibre transposition from one root to another one (or blindly ending nerve fibres). Since often the number of α-motoneurons was very similar rostrally and caudally, it seems that there was no numerous branching of fibres in the cauda equina nerve roots. As could be expected, a mild peripheral neuropathy was observed in these cadavers of elderly individuals: axonal degeneration and regeneration was suggested by circularly arranged groups of thinly myelinated nerve fibres. Some fibres may be expected to end blindly in the roots. Also, some loops may be performed by a few nerve fibres. Even though there most likely were many minor variations (including mixing of afferents and efferents in dorsal and ventral roots), it still turns out that there is nearly no tapering of nerve fibres. The quality of the measurements might be improved, especially with respect to a better morphologic identification of the different nerve fibre group distributions (identification of groups), by measurements performed in the cauda equina nerve roots and certain nerves from brain-dead humans. However, such samples are extremely difficult to get, because of so-called ethical problems.

5.1.13. Unmyelinated Nerve Fibres in Lower Sacral Nerve Roots

In the S5 root of the paraplegic patient 5, peaks at 0.25, 0.95 and 1.2 μm were observed in the distributions of the unmyelinated fibres. In the sural nerve (with mainly only skin afferent fibres), the unmyelinated fibres had diameters from 0.2 to 2.8 μm, in some similarity to the values observed in this work. The peak value for the sural nerve ranged from 0.6 to 1.4 μm (page 72 of [51]). The pronounced peak at 0.25 μm measured in this work may therefore be correlated to afferent or efferent parasympathetic nerve fibres. The number of unmyelinated fibres in the selected areas ranged from 100,000 to 400,000 fibres per mm^2. In the sural nerve, values of 19,000 to 65,000 unmyelinated fibres per mm^2 were found (page 72 of [51]). The values found here are most likely higher because of the existence of additional thin parasympathetic unmyelinated fibres and because selected areas of unmyelinated fibres were taken (values taken from whole roots could be smaller). The number of myelinated fibres observed in the present work was approx. 6.6 times higher than that of the myelinated fibres. In the sural nerve, the number of myelinated fibres was approx. five times higher than the number of myelinated fibres [51]. If the unmyelinated fibres with a group diameter of 0.25 μm were specific for parasympathetic fibres, then the morphometry of unmyelinated fibres could be used to identify parasympathetic pathways.

5.1.14. Little or No Tapering Helps to Further Characterise and Identify Human Nerve Fibres

To obtain knowledge on the intricate functions and regulations of the somatic and autonomic nervous system divisions, a precise classification of the human peripheral nerve fibre groups is needed to identify safely what natural impulse patterns are conducted along what fibre types into the CNS for self-organisation. The fact that there is only very little or no tapering in human nerve fibres will help in the further characterisation and identification of nerve fibres.

It has been argued that the organisation of the human CNS is mainly determined by the integrated properties of the neurons when activated integratively in the neuronal networks [120] rather than by the properties of single isolated neurons. This may be so. But to understand specific CNS organisation in relation to behaviour, we also have to measure the specific coordinated afferent input patterns with respect to time and space in many identified neurons, giving rise specific CNS organisation.

5.1.15. Very Little Tapering of Nerve Fibres Allows the Comparison of Natural Impulse Patterns at Different Sites of the Human Body and Aids the Identification of Natural Impulse Patterns and the Understanding of the Self-Organisation Neural Networks of the CNS

The tapering of nerve fibres is especially of interest in human neurophysiology since afferent and efferent nerve fibres can exceed 1m in length. The morphometric measurements indicate that there is only very little tapering of human nerve fibres. Consequently, it seems justified to compare nerve fibre diameter distributions of peripheral nerves with those of the nerve roots. Secondly, the classification scheme of the human CNS holds close to the CNS as well as at a distance from it, and thirdly, conduction velocities measured in the cauda equina can be compared with those measured from peripheral nerves. A consequence of the very little tapering of nerve fibres is that natural impulse patterns of efferents in nerve roots can be compared with the natural impulse patterns of single motor unit action potentials (MUAPs) of muscles as the impulse patterns will not change essentially along the way to the periphery (Figure 70). An improvement of the electromyography (EMG) in general and an improvement of the identification of natural impulse patterns of single MUAPs in multi-unit EMG recordings seem possible by simultaneous recording of single-nerve fibre action potentials from nerve roots and single motor unit action potentials from muscles innervated by the motoneurons recorded from in the nerve roots, or simply by using the known natural impulse patterns obtained previously from motoneurons for the identification of single motor unit firings.

5.2. Saltatoric Conduction of Myelinated Nerve Fibre

5.2.1. Saltatoric Conduction and Nerve Fibre Diameters for Different Internodes (Teased Fibre Dissection)

Measurements of nerve fibre diameters in general and tapering especially depend on the assumption that the nerve fibre diameters are very similar along different internodes. Internodal lengths and their nerve fibre diameters have to be measured for clarification [172].

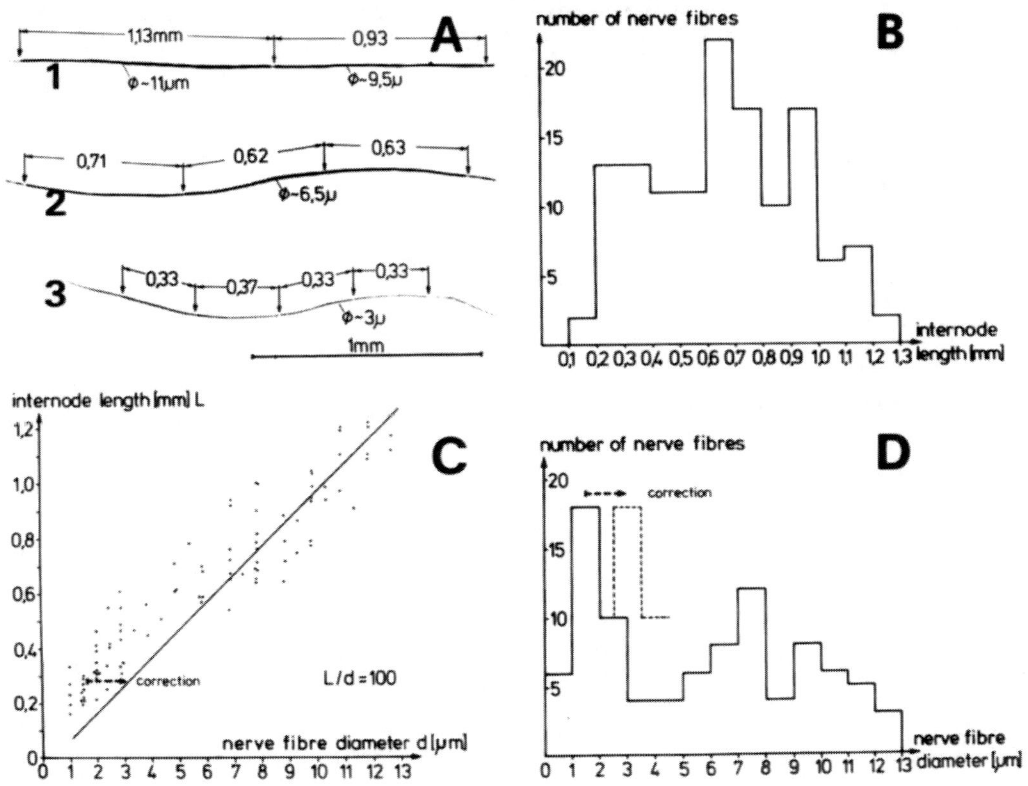

Figure 29. Quantified results of the teased fibre dissection of the whole S4 ventral root from HT1. No shrinkage correction. A. Three original nerve fibre dissections, several internodes are measured and marked. Approximate fibre diameters (Ø) are indicated. B. Frequency distribution histogram of internodes lengths. Numbers indicate peaks. Class borders are < and ≤. If more than one internodes length of a nerve fibre was measured (30% of the cases), then the mean value was used. C. Internodes length in relation to its nerve fibre diameter. A small correction for small nerve fibres was performed (see D). D. Very approximate nerve fibre diameter frequency distribution. Since the diameter of the nerve fibres were measured under the stereo-microscope with a 100 times magnification, the small nerve fibres were estimated more than measured. By comparing this diameter frequency distribution of the S4 ventral root with that of a S4 ventral root measured with a light microscope (Figure 19Ba), a correction for small nerve fibre diameters was performed and indicated in the Figure.

With normal effort, 130 (Figure 29B) of the 570 myelinated fibres of a S4 ventral root (Figure 21B) were teased and, additionally, the diameters of 95 of these were measured (Figure 29D). As the similar shape of the diameter histogram of the teased fibre dissection (Figure 29D) in comparison to that of Figure 21B indicates, a somewhat representative selection was teased. In order that several internodal lengths could be determined, as much of a length of a fibre as possible was teased. Figure 29A shows a montage of three photographed teased single fibres of different thicknesses. It can be seen that the internodal lengths are not always similar. It seemed that every third to fifth was markedly different from the other ones. On three thick fibres, it was found that the longer internode corresponded with a thicker nerve fibre as is shown in Figure 29A, fibre 1, so that the ratio of internodal length to nerve fibre diameter (Figure 29C) was not changed. Figure 29B shows the frequency distribution histogram of the internodal length. This distribution also shows four characteristic peaks, as the frequency distribution of the nerve fibre diameters does (Figure 21B). This similarity can be understood since the nerve fibre selection of the teased fibre dissection of Figure 29D was somewhat representative, and the relation between internode length L and fibre diameter d is proportional (Figure 29C). By dividing the internodal length in Figure 29B by L/d = 100, one gets a somewhat similar histogram as in Figure 21B (if the four histograms are added). By teased fibre dissections, it was therefore shown [172] (Figure 29C) that in human sacral nerve roots, the internodal length (L) is related to the nerve fibre diameter (d, including myelin sheath) by L/d = 100. Therefore, in frequency distribution histograms of internodal lengths, grouping of nerve fibres can also be observed (Figure 29B).

5.2.2. Action Potential Waveform Changes Due to Saltatoric Conduction

The action potential waveform may vary a bit along the nerve fibre, according to the different localization of the node of Ranvier in respect to the recording electrodes [105, 161]. Only three internodes lie between the electrodes of a pair of electrodes for the nerve fibres contained in peak 1 of Figure 21B (Mean internodal length = 1.2mm (Figure 29B); electrode pair distance = 4mm). Furthermore, the measurement of several internode lengths from one fibre indicated that every third to fifth internode length was quite different from the other ones (Figure 29A) and also the nerve fibre diameter was then different from the other ones. Therefore, one cannot exclude some changes in the action potential form from differences of the internodes, even though the nerve fibre could have partly compensated for such differences by a different internodal cleft or a different channel density.

5.2.3. Space Distribution of Single-Nerve Fibre Action Potentials

Figure 8 shows the action potential (AP) durations of an α_2-motoneuron, a γ_1 (dynamic), a γ_{21}-motoneuron (static), and an parasympathetic efferent fibre at a measuring temperature of 33°C. The main peaks of the triphasic waveform were approximately 0.15, 0.4, 0.5, and 0.6ms long, respectively. The space distribution (v = s/t; s = v times t) of the single-fibre APs along the fibres were approximately six (40m/s times 0.15ms = 6mm; α_2), 5.3, 5, and 3.2 (par) mm, respectively. With the approximate nerve-fibre diameters of 10.2, 6.8, 5.8, 3.5μm (Figure 2), we obtain internode lengths of (L/d = 100) 1.2, 0.68, 0.58, and 0.35mm, respectively. The main peak of the triphasic AP was approximately distributed over five (6mm : 1.2mm = 5; α_2), 7.8, 8.6, and 9.1(3.2 : 0.35 = 9.1; par) internodes, respectively. In a first approximation, one can disregard the saltatoric conduction, because the APs are distributed over several until many internodes. The values change a slightly according to the measuring temperature. The conduction velocity and the AP duration depend on the temperature.

5.3. Temperature Dependence of Group Conduction Velocities

In all measured cases, the AP durations decreased and the conduction velocities of the single-nerve fibres increased with increasing temperature. It seemed further as if the different nerve fibre groups had different temperature dependences, because for lower temperatures, it was much more difficult to identify the different nerve fibre groups by their peaks in the velocity distributions. The operational field was, therefore, often warmed with an infrared lamp to bring the root temperature as close as possible to 36°C.

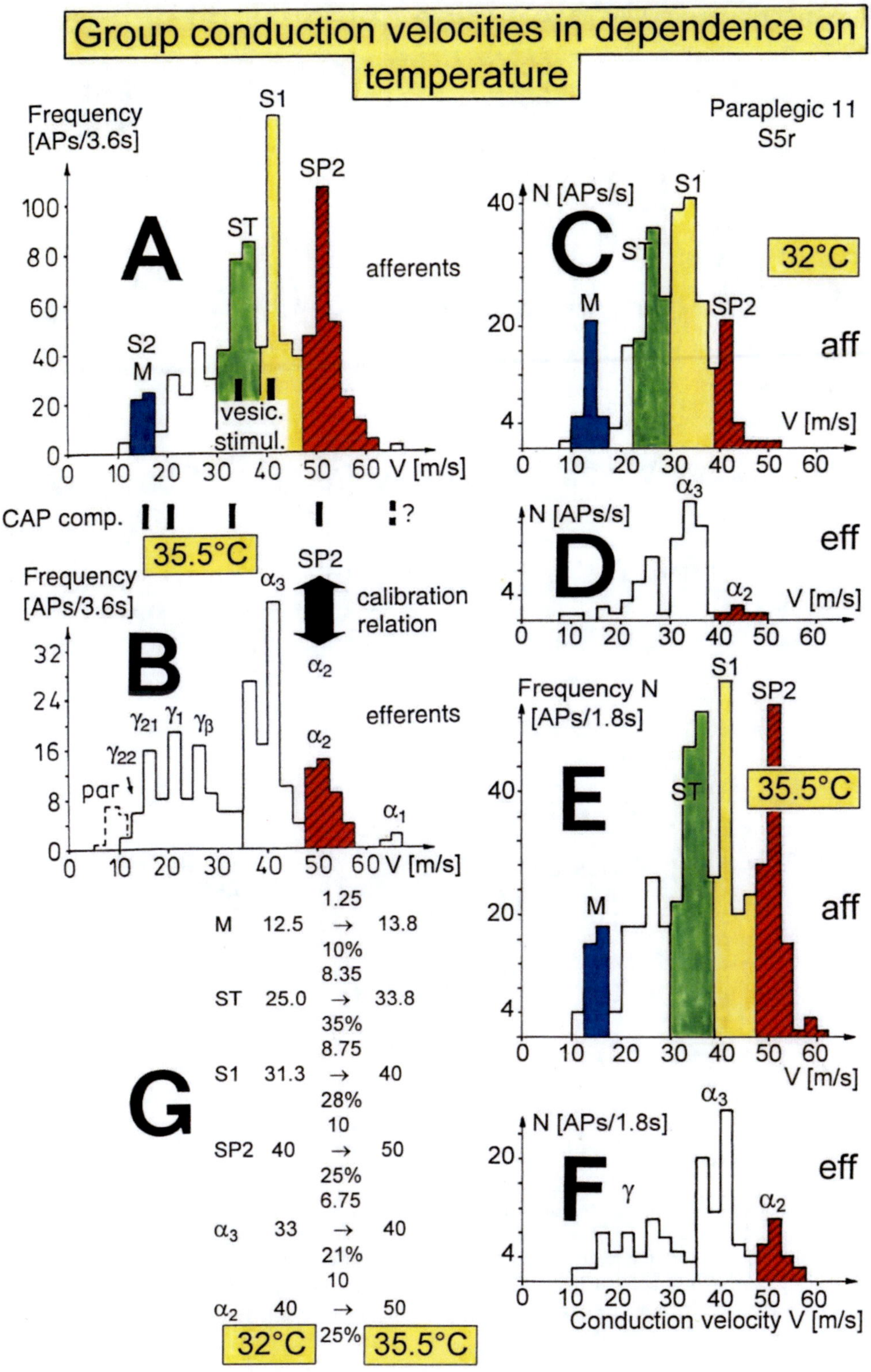

Figure 30. (Continued).

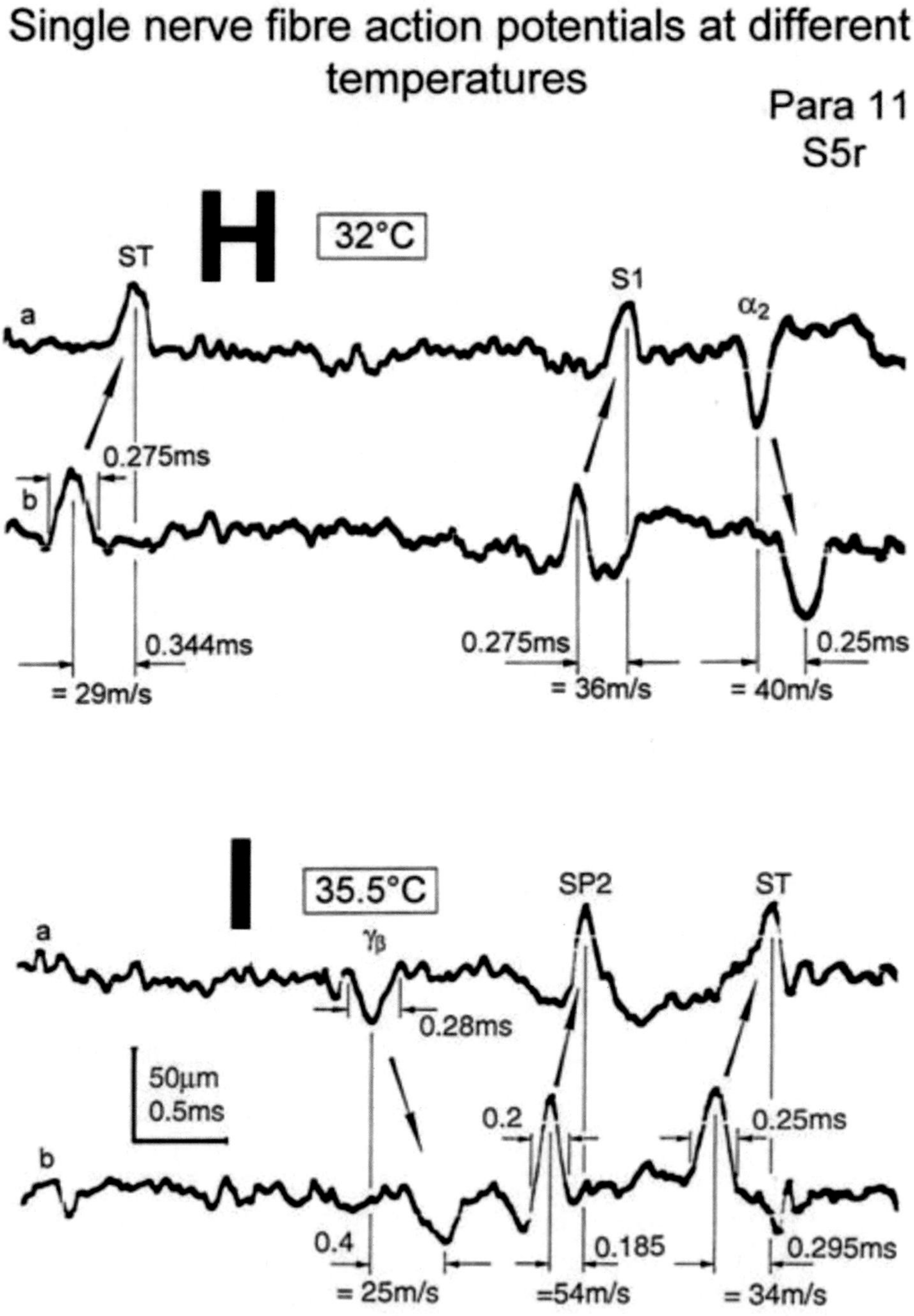

Figure 30. Conduction velocity distribution histograms with marked group distribution peaks in dependence on the nerve root temperature. For symbols of the nerve fibre groups, see legend to Figure 2. In G, the velocity values are given for both temperatures for the different afferent and efferent nerve fibre groups along with the percentages of change. Note that the secondary muscle spindle afferents (SP2) and the α_2-motoneurons (FR) conduct with the same velocity (calibration relation) for both temperatures (temperature-independence of the calibration relation).

In paraplegic 11, good conduction velocity distribution histograms from the same root under the same recording conditions were obtained for low (32°C) (Figure 30C,D) and for high (35.5°C) (Figure 30E,F) root temperatures. The secondary muscle spindle afferents (SP2) and the α_2-motoneurons increased their conduction velocity from 40 to 50 m/s (25% increase), the S1, ST and M afferents from 31.3 to 40 (28%), from 25 to 33.8 (35%) and from 12.5 to 13.8 m/s (10%) respectively with increasing temperature (Figure 30G). The α_3-motoneurons increased their conduction velocity from 33 to 40 m/s (21% increases). In the original recording piece of Figure 30H,I, an ST nerve fibre increased its velocity of conduction from 29m/s (32°C, H) to 34m/s (35.5°C, I). Even though only one set of good quantitative measurements could be obtained so far, the two sets of velocity values clearly show that different nerve fibre groups have different temperature dependences of their group conduction velocity. The SP2 afferents and the α_2-motoneurons showed the same temperature dependence of the group conduction velocity. The calibration relation of the velocity distributions of afferent and efferent fibres, namely that the SP2 fibres conduct with the same velocity as the α_2-motoneurons, is therefore temperature independent. To make the different temperature dependence of nerve fibre groups better visible, values of group conduction velocity values and group nerve fibre diameters will be plotted into the conduction velocity – nerve fibre diameter plane (Figure 32).

5.4. Correlation between Electrophysiology and Morphometry

5.4.1. Correlation between Group Nerve Fibre Conduction Velocities and Group Nerve Fibre Diameters

In Figures 1 and 2, the classification of human peripheral nerve fibres was obtained by measuring single-nerve fibre action potentials in brain-dead humans (HTs) and patients with cancer in the spinal canal during an operation. Nerve fibre diameters were measured in HTs and human cadavers.

By correlating the peak group values of distributions of velocities and diameters and taking the temperature dependence of the single-fibre velocities into account, this qualified classification scheme was obtained. A part of the measured values are summarized in Table 2. Further values in such table form can be found in Ref.172.

5.4.2. Unchanged Group Nerve Fibre Conduction Velocities and Group Nerve Fibre Diameters Following Spinal Cord Injury

But to find out the changes in CNS organisation following injury, we need to know that this classification scheme is also holding following CNS injury. Therefore, single-nerve fibre action potentials (APs) were recorded extracellularly from sacral nerve roots during surgery in nine paraplegic patients (Para 3 to Para 11) with spinal cord injuries for six months to six years. Again, conduction velocities of single fibres were calculated, and group conduction velocities were determined from peaks in the velocity distributions. The measured values are summarized in Table 3. It has been shown in an earlier paper for the same patients that the group conduction velocities did not change following spinal cord injury [192]. It was further shown that neither the group nerve fibre diameters did change following spinal cord injury [192].

Table 2. Peak conduction velocity values of conduction velocity distributions of single groups taken from conduction velocity histograms and peak nerve fibre diameter values of fibre diameter distributions of single identified groups taken from nerve fibre diameter histograms from two cadavers (Cad), four brain dead human cadavers (HTs) and one patient (Pat). Values without brackets are afferents, values in round brackets from efferents. T0, Tl, T2, T3, T4 = touch stimulated afferents from skin and anal canal, P = skin pain afferents, M = touch afferents from the mucosa of the urinary bladder and the anal canal, SI, ST = stretch afferents of the bladder, probably measuring mural tension, S2 = flow receptor bladder afferents, stimulated by high pressure and fluid movements (may be identical with M), SP1, SP2 = primary and secondary spindle afferents. (α_1), (α_{11}), (α_2), (α_3) = extrafusal motoneurons, (α_{11}) - subgroup of (α_1). (γ_β)?, (γ_1), (γ_2) = intrafusal motoneurons. (--), "-- " = no peak present, f = female, m = male, number behind "f" or "m " is age in years, numbers below "f" or "m" are the central temperature and the temperature of the fluid in the spinal canal in °C. v = ventral, d = dorsal, L = lumbal, S = sacral, Co = coccygeal, 19 = 9th intercostal nerve, musc = muscle branch, skin = skin branch, Ø = diameter

case	root	peak conduction velocity [m/sec]												peak nerve fibre diameter [μm]												root
		P (γ_2)	M	S2	T4 (γ_1)	T3 (γ_β)	ST	T2 (α_3)	S1	T1	T0	SP2 (α_2)	SP1 (α_1)	P (γ_2)	M (γ_1)	S2 (γ_β)	T4 (α_3)	T3	ST	T2 (α_2)	S1	T1	T0	SP2 (α_{11})	SP1 (α_1)	Ø [mm]
Cad f 47	dL4 vL4																8.2 (8.5)	9.1		10.3 (10.5)		11.1		12.2	13.3 (12.7)	
Cad3 m 64 I9	musc skin																 8.2	 9.1		 10.3		 11.1	 13.2	11.4	13.3	
HT3 f 21 33,5/36°C	vS4				21	31		39 (37)		44	49	51 (51)					(8.5)			(10.1)		11.1		(12.1)	13	0.15
	L?	(15)			(20)							51 (50)	59 (60)													0.1
HT4 f 56 32/32,5°C	dS5	4	7		7	14 (14)		21 (21)		26	29	(29)						9.4		10.3 (10.3)		11.3				0.25
HT5 f 58 38/32°C	dS3		6			14		19 (12)				(18)					8.2	9.3		10.1		11.2				0.4
HT6 f 37 33,5/33°C	dCo	5	--		6?	8		11 (--)		14	17	(--)	(--)	7.1 or 7.6			8.5	9.1		10.1		11.2		--	(--)	0.12
	dS4	9 (10)	12	12	14 (14)	21 (20)	28	29 (30)	34	36	41	43 (43)		5.5-8.2 (6.2)	 (6.7)	 (7.2)	8.4	9.1		10.1		11.1	12-12.7		(13)	0.2
	dS3				efferents present													9.1		10.1		11.2		12.1	13.1	0,3
	vS3																(8.4)			(10.2)				(12.3)	(13.1)	
	vL5																(--)			(10.2)				(12.2)	(13.2)	
Pat 24 m 10 37/33°C?	dS?		5					20		25		30 (28.5)	43 (40)													

Thus, the classification scheme of the human peripheral nervous system (Figures 1,2) can be used even in individuals with cord injuries. It is assumed that this classification scheme holds also in other CNS injuries, if the peripheral nerves are not directly injured. For a serious comparison of animal and human data (translational medicine), to repair the human nervous system, such data also need to be known for animals. For dogs and rats, some data will be given in this Chapter and Chapter IV.

Table 3. Group conduction velocities and group diameters (peak values of the distributions) obtained from the same sacral nerve roots. Para = paraplegic, centr. temp. = central temperature in the patient during surgery, f = female, m = male, age in years, temp measured = temperature measured close to the recording electrodes, real = approx. real temperature in the nerve roots. For designations of nerve fibre groups see legend to Table 2. Conduction velocity and fibre diameter values refer to afferents, those in round brackets to efferents. Edged brackets give conduction velocities of compound action potential components. Several peak diameters in the myelin sheath thickness class 0.25 ≤ d < 0.8 µm are indicated and separated by "+"- signs. Percentual proportions of myelinated fibres in classes 0.25 ≤ d < 0.8 µm to 1.8 ≤ d < 2.5 µm. Arter. = Arteries. One to two APs for the root S5 in para 11 indicate that the α_1-motoneuron group in this root consisted of approx. two fibres only. In para 5, recording was performed from whole root S5, but the diameter distribution was taken only from the ventral root S5

Case Centr. temp. [°C]	Sex/Age Lesion level/age of lesion	root temp measured /real	Group conduction velocity [m/s]												Group diameter [µm]					Root Ø [mm]	Number of myel. fibres % in classes 0.25 ≤ d < 0.8 µm/ 0.8 ≤ d < 1.3 µm/ 1.3 ≤ d < 1.8 µm/ 1.8 ≤ d < 2.5 µm	Contribution in % to the cross section area from	
			(par)	(γ_{22})	M/S2 (γ_{21})	T4 (γ_1)	T3	ST (γ_β) [CAP-components]	T2	S1 (α_3)	T1	SP2 (α_2)	GO	SP1 (α_1)	Peak diameters in myelin class 0.25 ≤ d < 0.8 µm diameter range µm	T2 (α_2)	T1	SP2 (α_{11})	SP1 (α_{12})			Arter.	Veins
Para 3	f 33	S5l 28/31.5				(16)				(31)		41 (41)	51	58 (-)	2.5 + 4.0 1.8 - 5.5			12.1		0.33	2100 81/6/3/10	9	12
34.7°C	C6/3	vS4r 28/31.5										53 (52) [53]		(59?)				-					
Para 4 35.7°C	f 26 TH4/6	vS3r 29/32.5				(16?)		(21?) [21]		33 (33)		36 (36) [36]		41 (41) [41]				-					
Para 5	f 24	S5r(vS5) 26.5/30	(5)			13 (14?)	20	(21)	25	(26)	33	35 (35)		40 (40)	2.1 + 3.1 + 4.0 1.5 - 6.0	10.2 (10.2)	-	-	-	0.39	1300 52/21/12/15	15	41
34.6°C	TH12/2	dS4r 29/32.5						35		41 (41)		50 (50)		(65)	2.0 + 3.0 1.5 - 5.5			12.2	13.2	0.60	5500 70/10/7/13	9	8
Para 7 35°C	f 37 TH5/2.5	S5l 29°/32.5		[15]		17 (25?) [26]	25	33 (33) [28]	36	41 (41)	44	50 (50)		(60)	3.1 2.0 - 5.5				12.8	0.48	2200 63/7/3/27	19	10
Para 8	f 20	S5l 28.5/32			(6)	(8.5?) [8.5]		[21]		(35) [34]		41 (41) [41]		(49)	3.0 + 4.0 + 5.0 1.5 - 6.0			12.0	(13.1)	0.49	2400 68/5/7/20	21	6.5
35.2°C	TH6/1.5	S5r 27.9/31.5	[7]	[14]	[16]	(21) [20]		26 (26) [26]		33 (33)		40 (40) [50?]		(50)	3.0 + 4.0 1.5 - 5.5	(10.5)	11.8	(12.5)		0.50	2500 67.5/7/7.5/18	23	23
Para 9	f 23	S5r 26/29.5				(20?)		(26?)		32 (32)	36	42 (45)			3.3 2.0 - 5.5			12.1		0.77	8300 75/6/5/14	10	16
34.5°C	TH1/0.5	vS4r 25/28.5				(5)				(8)		16.5 (14) [16.7]		20 - [18.2]				-					
		S5r ?								(26) [26]	28.5	33 (32) [30]											
Para 10 34.3°C	m 30 TH12/0.5	S5l 29.4/33				14 (15)	21 [21]		26	(26) [26]	31	33 (33) -		reflected ↓ [36]									
		dS4r ?						28.5 [28.5]		34 (34) -		41 (41) -		50 (50) -	3.1 2.0 - 5.1		11.2?	12.2?	13.1 (13.1)	0.79	9500 69/9/6/16	2.5	4
		S5r 32.0/35.5	(9)	(13)	16 (16) [16]	21 (21) [21]		36 (26/30) [33]		41 (41)		51 (51)		1-2 APs (65) [65]	3.0 + 3.7 2.0 - 5.2	(10.4)	-11.1	12.1		0.34	907 70/5/4/21	29	12.5
		vS3r 32.0/35.5						36 (28)		(41) (40)		50 (50)											
Para 11 35.8°C	f 25 TH5/2	vS4r						-							2.9 + 3.7? 2.0 - 6.0	(10.2)	11.1	12.1 (12.1)	(13.1)	0.31	898 77/4/1/18	6.5	20.5
		dS4r						-							3.0 2.0 - 5.0	10.1	11.1	12.1	13.1	0.46	4848 75/6/5/14	1	5.5
		dS2r						35 (intravesical		41 stimulation)					3.1 1.8 - 4.6	10.1	11.1	12.1	13.0	1.6	31600 59/3.5/3.5/34	1.4	4.7

5.4.3. Correlation between Group Nerve Fibre Conduction Velocities and Group Nerve Fibre Diameters in the Velocity-Diameter Plane [175, 181]

Peak values of group nerve fibre conduction velocities and group nerve fibre diameters are plotted in Figures 31 and 32 into the velocity-diameter plane.

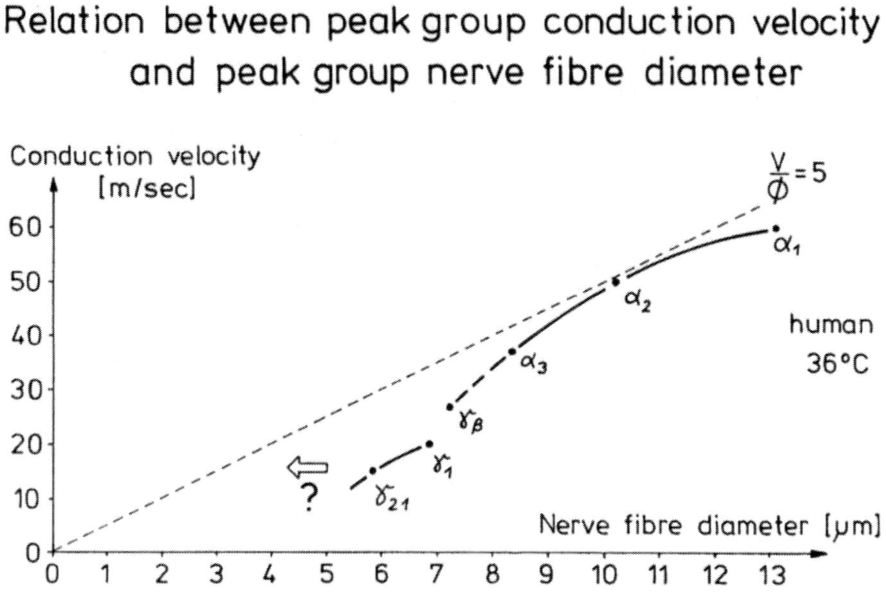

Figure 31. Relation between peak group conduction velocity and peak group nerve fibre diameter, 36°C, human.

It seems that the γ-motoneurons have different velocity-diameter dependence than the α-motoneurons, and the γ_β-motoneurons seem to belong more to the α-motoneurons than to the γ-motoneurons (Figure 31). By looking at the correlation curves of the skin touch afferents (T0 to T4) in the velocity-diameter plane, it is that the touch afferent fibres (Figure 32A) conduct slower than the α-motoneurons at a given fibre diameter (Figure 32B). One possible reason for the reduced conduction velocity could be a thinner myelin sheath. But all α-motoneurons and all skin touch afferents lie mainly in the myelin sheath thickness range d = 1.8-2.3μm. Therefore, a different myelin sheath thickness may not be the main factor. There are probably differences in the axon membrane properties of skin touch afferents and α-motoneurons.

5.4.4. Correlation between Group Nerve Fibre Conduction Velocities and Group Nerve Fibre Diameters in Dependence of Temperature

Exact temperatures of the measured roots were not known. Absolute conduction velocity dependency could, therefore, not be calculated. But relative dependency could be studied since the two temperatures in Figure 32, even though only approximately known, were still the same for the α-motoneurons and the touch afferent groups, because conduction times were measured simultaneously. In Figure 32A, the temperature dependence of the touch afferent groups changes from 16%/4°C to 34%/4°C when going from the T0 group to the T4 group. This clearly indicates that in thinner touch afferents, the conduction velocity depends more strongly on the temperature. α_2 and α_3-motoneurons (Figure 32B) show a similar behaviour, namely that the group with the thinner fibres and the lower conduction velocities are more temperature dependent. Another observation from Figure 32 is that the conduction velocities of the α-motoneurons are about only half as dependent on the temperature as the touch afferents.

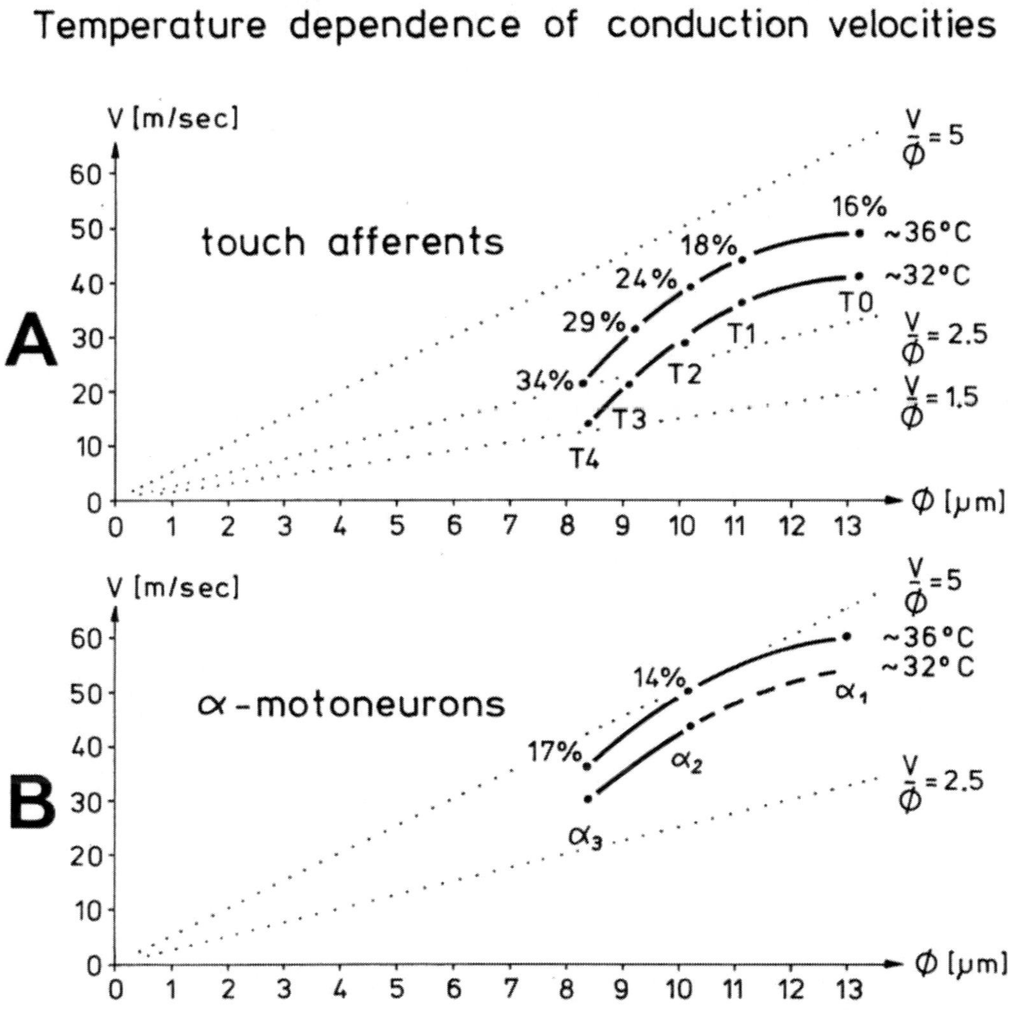

Figure 32. Temperature dependence of group conduction velocities (peak values) of skin afferents (A) and α_2 and α_3-motoneurons (B). 36°C-trace from HT3, 32°C-trace from HT6. Temperature values are only approximate. Percentages give the drop of conduction velocity values when changing from 36°C to 32°C.

5.4.5. Identification of Peaks of γ-Motoneurons and Parasympathetic Fibres in Conduction Velocity Distributions on Log Scale [189]

One of the drawbacks with the identification of nerve fibre groups is that it is difficult to identify the different intrafusal motoneuron peaks in distribution histograms of conduction velocity frequencies and to identify impulse patterns of single γ-motoneurons and parasympathetic efferents. On a linear scale, velocity peaks fused and can be separated by their different functions only partly. By constructing first histogram classes for conduction times, however, and plotting them on a log scale from the column values, the velocity peaks separated. Figure 33D,E shows that the fused γ-peaks of the linear plots (Da, Ea) split up into different peaks on the log scale (Db, Eabc). Since first histogram classes of conduction times were constructed using a linear scale, it was possible to study the dynamics of peaks or peak values also upon stimulation.

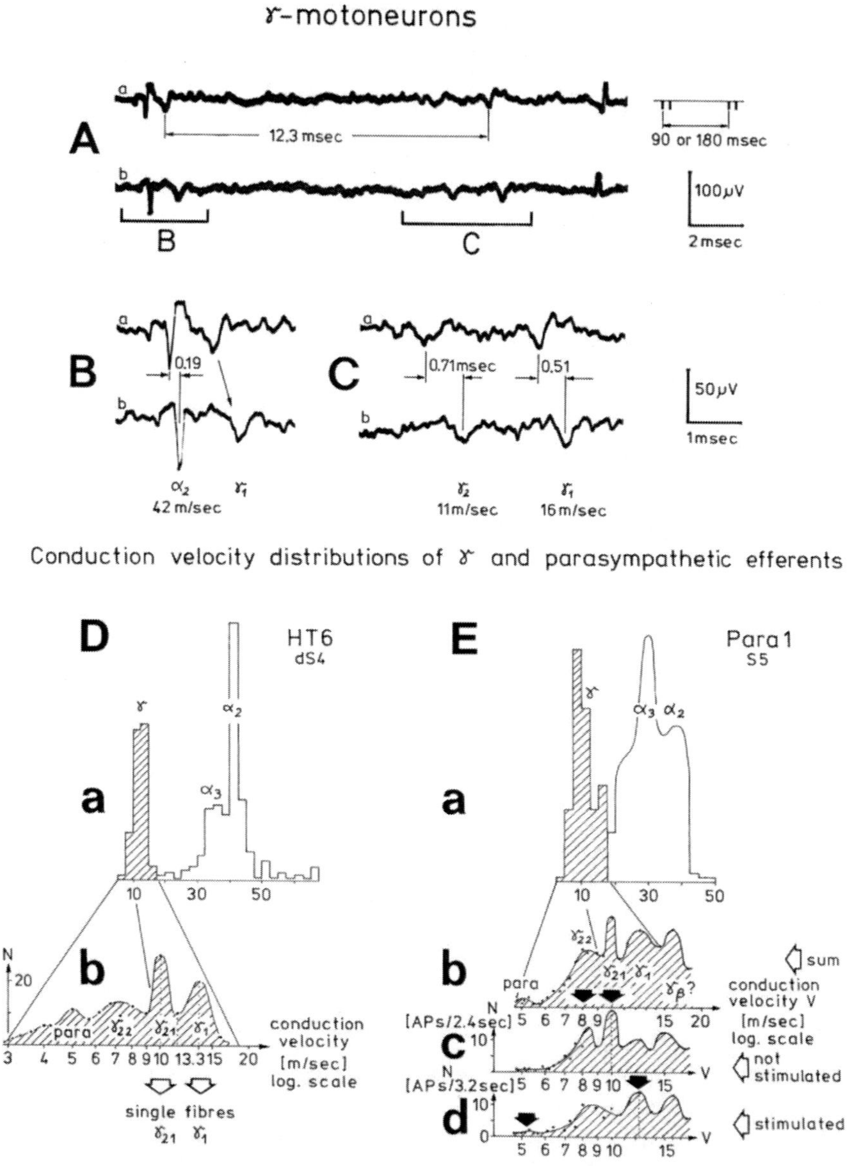

Figure 33. A. Activity from 2 γ-motoneurons and an α_2-motoneuron. The γ_1-motoneuron may have fired in the oscillatory mode for a few cycles with a period duration of 90 ms or 180 ms. Interspike intervals of the possible impulse train, consisting of two APs, are indicated (12.3 ms). The insert shows the possible oscillation cycle period. HT6; dS4. B,C. Time and amplitude expansions of A. Conduction velocities are indicated. Note that with the increasing conduction velocity, the AP amplitude increases and the duration decreases. D,E. Conduction velocity distributions of motoneurons for the brain-dead human HT6 (D) and paraplegic 1 (E). To make intrafusal motoneuron and parasympathetic peaks visible, the main γ-peak of Da and Ea was plotted on a log scale in Db and Eb. The distribution peaks are labelled with the groups that they most likely represent. In E, the distribution Eb is split into the distribution upon no additional stimulation (Ec) and upon additional stimulation (Ed). Note that in the non-stimulated distribution (Ec), the static γ-motoneuron peaks (γ_{22}, γ_{21}) are highest, whereas under stimulation (Ed), the parasympathetic (Para) and the dynamic γ-motoneuron peaks (γ_1) are highest. When plotting the velocities in Db and Eb logarithmically, the conduction times were first grouped by a conduction time histogram, and the column values were then used (conduction distance = 8 mm) to construct conduction velocity distribution curves.

Figures 33Db and Eb show several distribution peaks in which the γ_1 and γ_{21}-peaks could be identified from earlier functional considerations. By comparing (Figure 33E) the velocity distributions obtained upon stimulation with those without stimulation, and assuming reasonably that the dynamic intrafusal motoneuron peaks are higher upon stimulation (Figure 33Ed) and the static ones are higher with no stimulation (Figure 33Ec), a second static γ_1-peak (γ_{22}) can be identified and at least one further rather dynamic motoneuron peak (para) can be seen. This peak distribution observed in the paraplegic patient could also be found in the measurements in HT6 (Figure 33Db). That the "Para" peak contains activity from parasympathetic fibres will be shown later. The identification of parasympathetic efferents contributes to the understanding of urinary bladder functions under physiologic and pathophysiologic conditions and its repair by coordination dynamics therapy. By comparing Figure 33 with Figure 35, it can be seen that there is a similarity in the distributions of the motoneuron groups between the compound AP with its subcompound APs and the nerve fibre groups in conduction velocity distributions for single fibres. To get more safety for the identification of nerve fibres, nerve fibre groups are additionally identified by compound APs with their subcompound APs.

5.5. Nerve Compound Action Potentials Analysed with the Simultaneously Measured Single-Nerve Fibre Action Potentials

Summary

1. Compound action potentials (CAPs) and single-nerve fibre APs were recorded with two pairs of wire electrodes from an S5 root from a paraplegic patient during surgery. The compound APs were evoked by electrical nerve root stimulation.
2. Frequency distribution histograms of single-nerve fibre conduction velocities were constructed, and nerve fibre group conduction velocities were compared with the conduction velocities of the peaks of the CAPs. By increasing the strength of the stimulation, the peaks in the CAP could be identified, and the threshold order of efferent nerve fibre groups determined. Using additional literature data, it is likely that the primary spindle afferents have the lowest threshold due to electrical nerve root stimulation followed by the α_1-motoneurons (FF), the secondary muscle spindle afferents, the α_2-motoneurons (FR), the a α_3-motoneurons (S), the γ_β, γ_1 (dynamic), γ_{21} (static), γ_{22} (static), and the parasympathetic motoneurons.
3. By comparing the mean area of a single-nerve fibre AP with the areas of the peaks of the compound APs, it was found that 230 single-nerve fibre APs contributed to the compound AP. In the secondary spindle afferent fibre and α_2-motoneurons groups, 53 fibres were stimulated (23%). The α_3-motoneuron peak and the afferents in the same velocity range contained 101 fibres (44%), the γ_β peak contained nine fibres (4%), the γ_1 32 (14%), the γ_{21} 23 (10%) and the γ_{22} 12 (5%) fibres. Additionally, two primary spindle afferents and two α_1-motoneurons most likely contributed to the compound AP. Since the large peaks in the compound APs did not change their area with increasing stimulation, most likely all muscle spindle afferents and α-motoneurons were activated to contribute to the compound AP.

4. Electrical intravesical stimulation evoked single-nerve fibre action potentials (APs) and compound action potentials (CAPs) of stretch (S1) and tension receptor afferents (ST). From the latencies (7.4ms and 9.5ms) and the group conduction velocities (41 (S1) and 35m/s (ST)), it can be calculated that the afferent activity travelled from the urinary bladder approximately 320mm to reach the recording electrodes of an dS2 root.

5.5.1. Single-Nerve Fibre Action Potentials and Compound Action Potentials Measured Simultaneously from a Nerve Root [190]

By recording simultaneously single-nerve fibre APs following no or natural stimulation and compound nerve action potentials (CAPs) elicited by electro-stimulation of the nerve roots (Figure 34D), single-nerve fibre conduction velocity spectra can be constructed and nerve fibre groups identified and compared with the subpeaks of the CAPs.

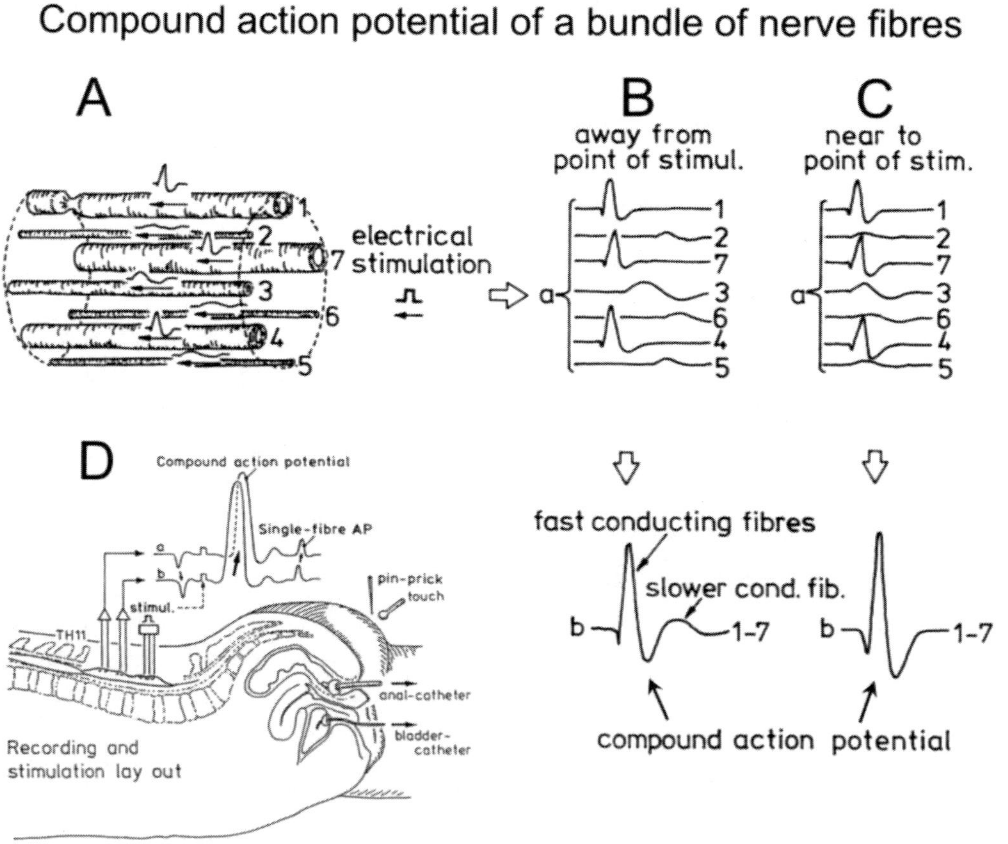

Figure 34. A,B,C. Schematically generated compound action potential. The real compound action potential (CAP) generated in a nerve root corresponds to the case “B.” Subcomponents of the CAP are not drawn in “B.” D. Layout for stimulation and recording of single-fibre action potentials (APs) and compound action potentials (CAPs). Traces a and b are records taken from the two sites; downward arrow = efferent, upward arrow = afferent, the natural stimulation used were touch, pinprick, and anal and bladder catheter pulling. Following electrical (artificial) stimulation with a “Brindley” stimulator (stimul.), the compound action potential passes the two electrode pairs as a large afferent-like AP.

Compound action potentials and single fibre action potentials

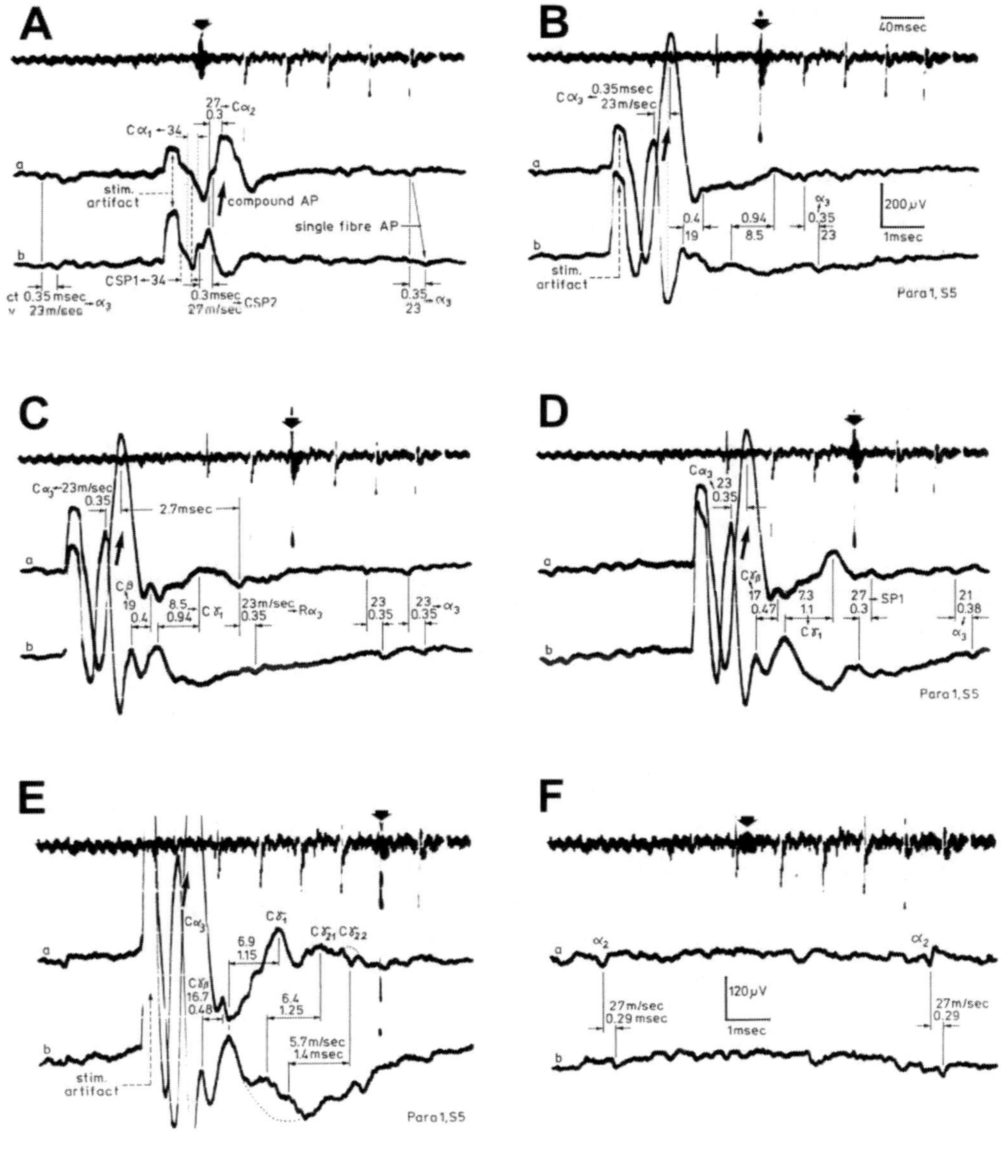

Figure 35. Size and form of the compound action potential (AP) in dependence on the strength of the stimulation and in relation to the single-fibre APs. Single-fibre APs are labelled according to the group to which they belong (e.g., SP2, α_2); components of the compound AP are labelled according to the nerve fibre group to which they belong, namely CSP2, Cα_3. Rα_3 designates possible reflected antidromically propagating α_3-motoneuron APs. Conduction time (ct) and conduction velocity (v) are indicated at single-fibre APs and peaks of compound APs. The large arrows in A, B, C, D indicate the main compound AP. Note that with increasing stimulation artefact (stim. artefact) from A to D, also the more slowly conducting nerve fibre groups get excited. The small thick arrows on the summary trace (trace "a"; time scale 40 ms) on top of A to F marks the first (A), second (B), third (C), fourth (D) and fifth (E) stimulation artefact. In E, the amplitudes are stretched (120 µV) to make compound AP components Cγ_{21} and Cγ_{22} visible. F shows single-fibre APs recorded between the first and the second stimulation. Para 1, S5 root, root temperature unknown.

Figure 34 shows schematically how a CAP is generated. The electro-stimulated elicited single-nerve fibre APs (Figure 34A) add up algebraically and give rise to a triphasic AP close to the site of stimulation (Figure 34C). Distant to the site of electro stimulation, the CAP changes its shape due to the different fast conduction fibres (Figure 34B). Under suitable recording conditions, sub CAPs can be identified (Figure 35) and correlated to the group conduction velocities obtained from single-nerve fibre conduction velocity distribution histograms. In a paraplegic patient, subpeaks of compound APs could be recorded from a S5 root and the subpeaks could be identified (Figure 35) by comparing them to the single-nerve fibre conduction velocity group values and to the available knowledge concerning electrical stimulation thresholds. Using additional literary data, it was found that the primary spindle afferents likely have the lowest threshold upon electrical nerve root stimulation, followed by α_1-motoneurons (FF), secondary muscle spindle afferents, α_2-motoneurons (FR), α_3-motoneurons (S), γ_β, γ_1 (dynamic), γ_{21} (static), γ_{22} (static), and parasympathetic motoneurons. In a first approximation, the AP duration increases in the same way as the AP amplitude decreases with the decreasing conduction velocity, and the area between the average single-fibre AP curve, and the baseline is the same for all single-fibre APs with the same distance from the recording electrodes in the root cross section. By comparing the mean area of a single AP with the areas of the peaks of the compound APs, 230 single-fibre APs were found to contribute to the compound AP in the measured case. In secondary spindle afferent fibres and α_2-motoneuron groups, 53 fibres were stimulated (23%). The α_3-motoneuron peak and the afferents in the same velocity range contained 101 fibres (44%), the γ_β peak contained nine fibres (4%), the γ_1 32 (14%), the γ_{21} 23 (10%), and the γ_{22} contained 12 (5%) fibres. Additionally, two primary spindle afferents and two α_1-motoneurons most likely contributed to the compound AP. Since the large peaks in the compound APs did not change their area with the increasing stimulation, all muscle spindle afferents and α-motoneurons were most likely activated to contribute to the compound AP [190]. The different subcompound APs comprised in the compound AP (CSP1, $C\alpha_1$, CSP2, $C\alpha_2$, $C\alpha_3$, $C\gamma_\beta$, $C\gamma_1$, $C\gamma_{21}$, $C\gamma_{22}$) (Figure 35) may provide more conclusive information with respect to splitting of nerve fibre groups. In unfavourable recording conditions (for example, low temperature), the compound AP can give additional information that can be used to identify nerve fibre groups [190].

5.5.2. Single-Nerve Fibre Action Potentials and Compound Action Potentials Recorded from Nerve Roots Elicited by Electrical Stimulation of Urinary Bladder Receptors or Afferents

Electrical intravesical stimulation of the nearly empty urinary bladder with up to 30V evoked single-nerve fibre action potentials (APs) and compound action potentials (CAPs) of stretch (S1) and tension receptor afferents (ST). In the recording from a dS2 root, mainly two CAPs followed the stimulation artefact, with latencies of 7.4ms and 9.5ms. The CAP conduction velocities were 41 and 35m/s. From the latency and the group conduction velocity, it can be calculated that the afferent activity travelled approximately 320mm to reach the recording electrodes. Further details of the electrical intravesical stimulation and the recording and stimulation layout will be given in Chapter V (Figure 10 of Chapter V, Part I), where the scientific knowledge is developed to repair the urinary bladder by coordination dynamics therapy following spinal cord injury. After having developed a powerful tool to explore the functions of the human nervous system, we will analyze the natural afferent input patterns that inform the CNS about the changes in the outside world, and then we will measure the efferent output and CNS neuronal network organization being only partly dependent on the afferent input. It will be started with the afferent input from the skin and the mucosa of the anal canal end urethra.

6. Properties of Mechanoreceptors of the Skin

Summary

1. Identified touch skin nerve afferents, characterized by their group peak values of conduction velocity and fibre diameter were at about 36° C (age 30 years) : TO (49 ms^{-1}/13.013.2μm), Tl (44/11.2), T2 (39/10.1), T3 (31/9.1), T4 (20/8.3); pain afferents : P (13 ms^{-1}/?).
2. Single-fibre action potentials (APs) were recorded with two pairs of wire electrodes from a coccygeal nerve root of a brain-dead human. Simultaneous impulse patterns of 3T0 + 6T1 + 3T2 +11T3 + 5T4 and 4P single skin afferent units were separated by their different AP waveforms and conduction velocities. For a rapid touch or pinprick, the afferent units fired on average with two APs.
3. The touch units had most likely the following receptor properties. Tl: rapidly adapting, medium-high threshold, sensitive to skin traction over a distance of more than 10 cm. T2: rapidly adapting, high threshold. T3: slowly adapting, lowest threshold among T afferents for a light touch with no movement along the skin. T4: slowly adapting, lowest threshold among T afferents for a light touch with a movement component parallel to the skin surface. The Tl to T4 afferent receptors are the PC, RA, SAI and SAII receptors, respectively. The correspondence is not safe.
4. With nerve fibre diameter distribution histograms (morph), myelinated fibre numbers, conduction velocity distribution histograms (electro) and AP occurrence patterns, innervations were determined. With estimated dermatome areas, innervation densities i of skin dermatomes were calculated for nerve fibre diameters larger than 5μm:

i (electro, morph; Co) = (0.6T0+ 1.2T1 + 2.8T2 + 2.4T3 +1.2T4 +1.6P)/cm^2,

i (morph; S4) = (0.4T0 + 1T1 + 0.8T2 + 1.1T3 + 0.5T4)/cm^2,

i (electro; S4) = 0.4T0 + 2.3T1 + 2.1T2 + 2T3 + 1.5T4) cm^2,

i (morph; T9) (lateral)) = (0.8 (TO or hair follicle) + 1.7T1+ 0.9T2 +0.7T3 + 0.4T4)/ cm^2.

5. The two point discrimination of a volunteer was for a light touch 30 mm in the S3 and S4 dermatomes and reduced to 25 mm with faster and stronger touch. In the direction of the anus and the os coccygeum, the two point discrimination reduced to

25 mm. In the axillary line of the T9 dermatome, the two point discriminations were 35 mm (transverse axis) and 45 mm (longitudinal axis).
6. By comparing the innervation densities with the two point discriminations, it was found that in the rather untrained situation, the discrimination sensitivity increased with the innervation density. Summing up the touch unit innervations from TO to T4, one obtains the following correlation between the two point discrimination and touch innervation density: coccygeal skin (25mm two-point discrimination / 7.2 innervations/cm^2), S4 dermatome (30mm/6 innervations), T9 dermatome (40/4.5 innervations). It is calculated that there are seven to eight touch unit innervations between the two points of the two point discrimination (fibre diameters of afferents larger than 5μm).
7. It is analysed that the Tl afferents can serve as an information transport across the level of injury if motor functions are restored in spinal cord injury patients. Twenty percent of the afferent fibres may only be needed for a re-innervation of bladder, anal canal and sacral dermatomes. Because of similar properties of the receptors of T4 afferents and flow receptors, the sense of bladder fullness may also be reconstructed.
8. Receptor density in the skin does not change following spinal cord injury, provided the skin has not been damaged by trauma or pressure ulcers.

6.1. Introduction: Identification of Skin Afferents

Johannson has nicely identified the impulse patterns of the different kinds of myelinated skin afferents of the hand with Tungsten electrodes [115], but the group conduction velocity identification was poor. In this research project, when recording from undamaged (not dissected) nerve roots, the group conduction velocities of the different skin afferents could be identified nicely. The impulse patterns of the different skin afferents can be simultaneously measured [175, 180], but the characterisation with respect to the patterns obtained by Johannson are not as accurate. Mucosal and urinary bladder have not been measured before in humans. The comparison with the skin afferents from the animal research is further complicated by the fact that the animals have different skin. It is of importance for the progress of research in general and the self-organisation of neuronal networks in particular that we can identify nerve fibre groups with different methods safely, to be sure we are speaking about the same nerve fibre groups. Since there is practically no tapering in human nerve fibres (see above), group conduction velocities obtained with the single-nerve fibre action potential recording method can be compared with the conduction velocities of the components of compound action potentials [190].

6.2. Calibration of Skin and Other Afferent and Efferent Conduction Velocities

Since in the lower sacral roots there are mostly a few afferents in the ventral (motor) roots and a few efferents in the dorsal root, the correlation between easily identifiable afferent and efferent conduction velocity distribution peaks offers the possibility of calibrating conduction velocity distributions. Conduction velocity peaks (of non-symmetrical distributions) can then be identified relatively by their position in relation to the calibrating peaks, which are measured simultaneously, and absolute peak values can be obtained from the known conduction velocity distributions at a useful temperature (36°C) (Figure 36).

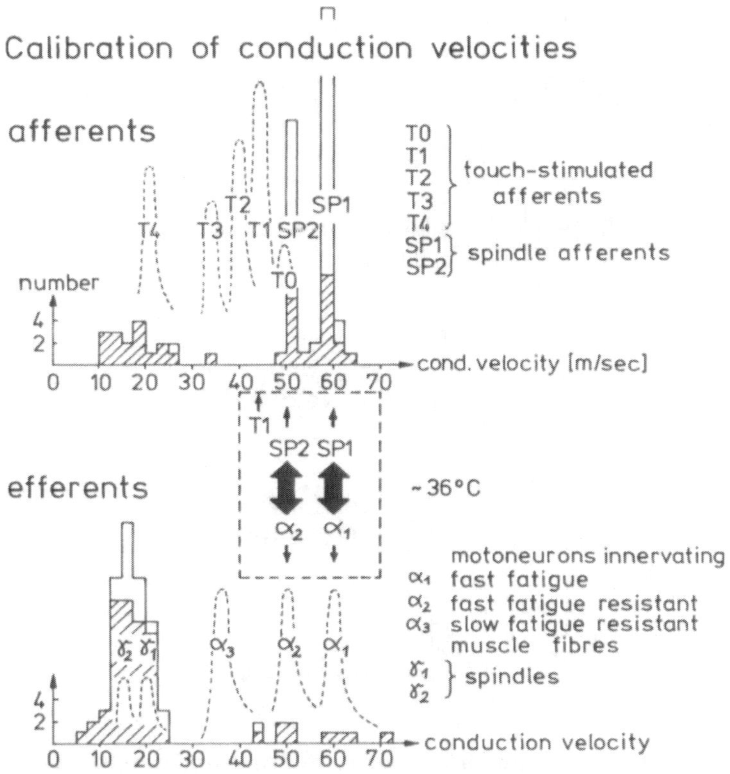

Figure 36. Calibration of conduction velocities of afferent and efferent nerve fibres. Black double arrows mark similar conduction velocities of afferents and efferents. Note, the α_2-motoneurons have about the same conduction velocity as the secondary muscle spindle afferents (SP2), the T1 touch afferents conduct about 10% slower. L4 plus S4 roots, HT3. Root-temperature ≈ 36°C.

6.3. Properties of Mechanoreceptors of the Skin

In Figure 37 the recorded activity is shown when sticking S5 or coccygeal dermatomes with a pin. The activity pattern shows a touching and a releasing part of activity, as can be seen from the action potential (AP) appearance of large amplitude (T1 and T2 units) and from the schematic activity pattern of the T1 units in the lower part of Figure 37A. All the four touch (T1 to T4) and one pain group, recorded here, seem to have their own touching and releasing part but, according to their latencies and conduction velocities, at different times. It can be seen from Figure 37 that the touching activity part of the slower conducting T3 and T4 fibres with their smaller AP amplitudes (Figure 37A and C) started about in the middle of the T1 and T2 activity part and lay mainly between the touching and releasing parts of the T1 and T2 units. The releasing activity of the T3 and T4 fibres appeared after that of the T1 fibres.

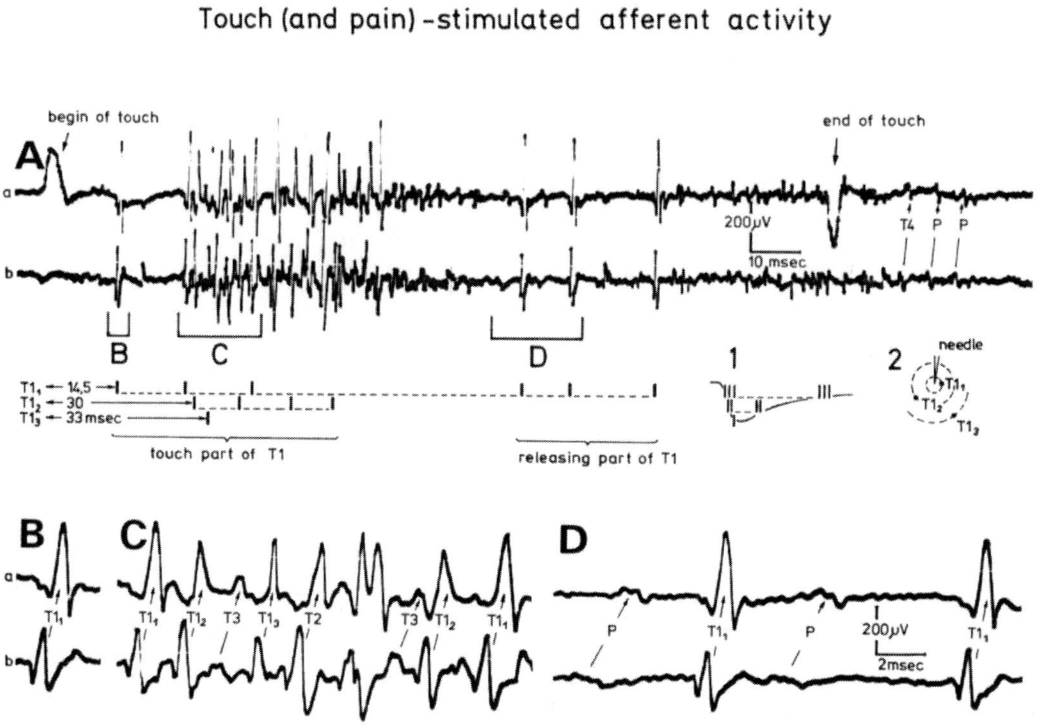

Figure 37. Touch (and pain)-stimulated afferent activity. Touch and pain activity stimulated by pinpricking with a pin S5 or Co dermatomes and recording extracellularly from a dorsal coccygeal root (brain-dead human HT6). T1, T2, T3, T4, P = mark action potentials (APs) from single touch and pain fibres. Subscripts 1, 2, 3 mark single fibres. A. Whole sweep following pinprick one shown at a slow time base. The large upward artefact on trace "a" marks electronically the beginning of the pinprick. The large downward artefact on trace "a" marks the end of the pinprick. Note that two intervals of high activity of large APs occur, one after the beginning of the pinprick with one AP in front, and a second before the end of the pinprick; potentials with small amplitude follow potentials of large amplitude. Time intervals B, C and D are shown in a time-expanded form in Figures B, C and D. The APs from three T1 units are schematically shown by dashed lines and upward deflections in the lower part of A. Latencies are indicated. Insertion "1" shows schematically the activity pattern (dashed lines with deflections) in relation to the indentation of the skin. Indentation "2" shows schematically the spatial distribution of the corresponding T1 receptors. The dashed circles indicate possible places of one receptor belonging to the $T1_1$, $T1_2$, and $T1_3$ units. B, C, D. Time expanded sweep pieces of A. Identified APs are indicated. Note that the APs from the $T1_1$ touch unit can be safely identified by the waveforms in B, C, D.

The activity of the pain fibres (here the slowest conducting fibres) appeared even later at the recording electrodes and stopped appearing after the end of the pinpricking (recorded at the proximal cauda equina). The very late pain after pinpricking, which one normally also feels, is probably conducted by very thin myelinated or non-myelinated fibres and cannot be recorded here.

In the T1 touch group, three units could be identified by waveform considerations. It will now be analysed in more detail in Figure 37, together with the activity pattern of these three units, why the touching part of activity normally had a higher and longer lasting activity than the releasing part, and it will be attempted to clarify why T1 units can have quite different delays, which cannot be explained by conduction velocity differences alone. In the lower part of Figure 37A, the activity pattern of each T1 unit is schematically redrawn. It can be seen that all three units had a different delay and that the releasing activity of the $T1_2$ unit nearly fell into the touch activity part of the $T1_1$ unit and that the $T1_3$ unit responded only with a touching part of one AP. With the additional information that the $T1_1$ unit responded with three APs, the $T1_2$ unit with two and the $T1_3$ only with one AP, these activity patterns are interpreted with the spatial distribution of the receptors of these three T1 units (Figure 37A, insertion "2"). The $T1_1$ unit had its receptor nearest to the needle (Figure 37A, insertion "2") and responded fastest and longest since the skin indentation first reached the $T1_1$ receptor, gave rise to the largest skin shift and turned away last (Figure 37A, insertion "1"). The $T1_2$ receptor, being further away from the centre of skin touch, was reached later by the skin indentation, the skin shift was not so large, and the skin turned back to the former position quicker. The APs in the $T1_2$ unit therefore appeared later, showed only two APs in the impulse train and the release activity part also appeared earlier. The $T1_3$ receptor, lying furthest away from the centre of skin indentation, was reached last, with just a threshold skin indentation, and responded with 1AP. From Figure 37A, it can be seen that the T1 units adapted rapidly. But the dependence on the time course of the skin indentation may be even stronger than the dependence on velocity. As can be seen from the lower part of Figure 37A, the $T1_1$ unit had the shortest latency, the $T1_2$ unit the second shortest and the $T1_3$ unit the longest latency. This was explained simply by the time the beginning of the skin indentation needed to propagate to the different T1 receptors, which means it was explained by the closeness at which a receptor is stimulated. If one receptor is stimulated very close and the other ones not, then one AP will be quite far in front of the other ones in Figure 37A due to the very short latency of the focally stimulated receptor. But if two receptors are stimulated only half focally, then no AP will lead clearly. The focally argument can explain the often observed phenomenon that one AP led ("B" in Figure 37A). But it is difficult to understand why that leading was sometimes so strong or why the latencies in the T1 group were so different. The $T1_2$ unit had a latency more than double as long as the $T1_1$ unit, even though its conduction velocity was a bit faster (for shorter conduction time, see Figures 37B and C), whereas the $T1_3$ unit had an unproportional, only slightly longer latency than the $T1_2$ unit.

If one splits the latency into a first part – the time needed from the beginning of the touching of the skin until the occurrence of the APs in the receptor (l_1) (l = latency), a second part – the time needed for the AP to travel from the receptor to the place, where the unit fibre starts, which will be neglected here and does not exist if a unit has only one receptor, and a third part – the time needed for the AP to travel from the beginning of the unit-afferent fibre to the measuring place (l_2), then one can calculate the latencies l_1 and l_2 from the temperature corrected T1 group conduction velocity of 44m/s from Figure 2, the distance of about 550mm from the receptors to the recording place and the three overall latencies of the 3 T1 units, namely 14.5ms ($T1_1$), 30ms ($T1_2$) and 33ms ($T1_3$) from Figure 37A: l_2 ($T1_1$, $T1_2$, $T1_3$) = 550mm/44ms^{-1} = 12.5ms; l_1 ($T1_1$) = l-l_2 = 2ms; l_1 ($T1_2$) = 30 – 12.5 = 17.5ms and l_1 ($T1_3$) = 33-12.5 = 20.5ms. The differences of the l_1 latencies between the $T1_1$ and $T1_2$ units is l_1 ($T1_2$) - l_1 ($T1_1$) = 17.5-2 = 15.5ms and the one between the $T1_2$ and the $T1_3$ unit is 3ms (20.5-17.5). It does seem that not only spatial differences (Figure 37A, insertion "2") can account for a l_1 latency difference of a factor of 5. It is likely that the T1 receptors also depend on the acceleration with which the skin around the receptor is shifted. The acceleration of skin shift will be highest at the most focally stimulated $T1_1$ receptor. It is, therefore, concluded that the T1 receptors are also acceleration dependent receptors.

As can be seen from Figure 37C, the first AP from a T2 touch unit arrives later than those from the T3 unit. It could be that the T2 touch units have similar properties with respect to latencies than the T1 units. This would explain the relatively long latency of the $T2_1$ unit in Figure 10. Properties of T1 and T2 units will now be further compared with the ones of the T3 and T4 units. A touch with a metal ball or a pin of medium strength, as analysed before in Figure 37, stimulated the T1 to T4 units to generate a response with the beginning and with the end of the touch. The order of the activity appearance of the different groups is roughly given by the conduction velocity of each group, which means the activity parts of the T1 and T2 touch units are recorded before the activity parts of the T3 and T4 (Figure 10). This order of the occurrence of APs from the different touch groups changed when a light or a very light touch was applied. In the case of a light touch, the T3 and T4 units occurred before the T1 and T2 units. For a very light touch, one T1 potential, but a few T3 and T4 APs occurred. It is concluded that the T3 and T4 touch units are more sensitive to light touch and/or have a higher receptor density than the T1 and T2 touch units. From six touch stimulations, like the one in Figure 37A, the latencies, the strength of the touch and the releasing activity parts were measured from T1 and T2 units and from the T3, T4 and P (pain) units. The mean latencies of the T1 and T2 (the first AP was taken as the measure) was 16ms (two latencies were left out because of enormous length due to a very light touch), and the mean latency of the T3, T4 and P units was 32 ms (see also Figure 37C). This latency length can be mainly understood with the differences in the conduction velocities. The duration of the touch and the releasing activity parts of the T1 and T2 units were 35 and 16ms, respectively, those of the T3, T4 and P units were 68 and 45ms. It is concluded that the T1 and T2 units adapt more rapidly than the T3, T4 and P units.

6.4. Types of Mechanosensitive Units of the Skin: T1, T2, T3, T4 Units Correspond to PC, RA, SAI and SAII Units

Disregarding the T0 afferents, four Types of touch unit afferents, T1, T2, T3 and T4 (Figure 36), have been identified from the nearly hairless skin on top of the gluteus maximus (S3 to coccygeal dermatomes). Four different types of mechanosensitive units with low threshold have been identified from the glabrous skin [231, 115, 123]. The interpretation is [231] that the end organs of the four types of unite are paciniform nerve endings for the PC units, Meissner's corpuscles for the RA units, Merkel's cell neurite complexes for the SAI units and Ruffini's nerve endings for the SAII units. It is quite clear that the PC, RA, SAI and SAII units are identical to the T1, T2, T3, and T4 touch units; it is not clear which is which. There is indication that the T1 and T2 units adapt more rapidly than the T3 and T4 touch units. The T1 and T2 units therefore correspond to the PC and RA units (rapidly adapting) and the T3 and T4 units to the SAI and SAII units (slowly adapting). The lower part of Figure 37A, including insertion "1" and "2," indicates further that the three T1 units come from three different points and have no distinct receptive field borders and also signal the acceleration of skin indentation (see the calculations in connection with the delays). The T1 units are therefore most likely the PC units with the single Pacinian corpuscles. The Pacinian corpuscles are innervated by thick myelinated nerve fibres and are also quite often present, similar to the T1 units. Maybe SAII units have a slower conduction velocity than the SAI units [123] and correspond to the T4 units. A safe correlation of the different units on the basis of their conduction velocities (0054,0055) is not possible, since Figure10 and Figure 37 demonstrate that the delays from the moment of touching the skin to the appearance of the

APs, used by Knibestöl [122, 123], are not always in relation to the conduction velocity of the fibre. PC, RA, SAI and SAII units are probably the T1, T2, T3 and T4 units, respectively.

6.5. Afferents from Hair Follicles

The function of the T0 units is unclear. But the small distribution peak (Figure 36) seems to be more than just misjudged secondary spindle or Golgi afferents. Since in the hairy skin only one additional hair follicle unit could be identified [231], it seems that the T0 afferents innervate hair follicles. This interpretation is supported by the finding that in the cat, the fastest conducting mechano-receptive units come from the hair follicle type T [32] units, but hair follicle afferents are much more important and different in the cat than in the human. The afferents from the hair follicles are still unclear, since in the measured areas are also down hair (thin blond short hair, directed more downwards), which may have been stimulated in addition to pubic hair.

6.6. Comparison to Animal Values

The conduction velocity distributions of the touch afferent fibre groups in this Chapter have similarities with those of the cat [32, 101], only the peak velocity values have to be reduced by about 30 to 35%. The peak conduction velocity of myelinated pain afferents (high-threshold mechanoreceptor afferents of A_δ type in the Erlanger and Gasser classification) is 13m/s. The peak diameter is in the range between 5.5 and 8.2μm. A conduction velocity distribution histogram from nociceptive A_δ afferents of the cat [33] has its peak at about 18m/s (range from 6 to37m/s) and shows nicely the typical skewed distribution with the “long foot” to higher values. The peak conduction velocity value reduced by 30% fits approximately the human value of 13m/s.

6.7. Need for a Deeper Understanding of the Population Response upon Touching the Skin

For a repair of CNS injury, receptor properties of skin, urinary bladder and anal canal afferents (see below) have to be further clarified and innervation densities calculated to understand the population response, which give rise to pattern change in the human CNS. Even though there is the general belief that everything is known about skin afferents in humans [115, 122, 123, 231], the knowledge about group conduction velocities and group nerve fibre diameters is poor. Very little is known about innervation densities. Receptor thresholds of different skin afferents were not measured simultaneously. Receptor properties of skin afferents have not been compared with those of bladder afferents. To get more safety and exactness upon the natural impulse patterns from the skin, giving rise to pattern organization, the natural activity in a coccygeal root will be analysed in detail. It will also show how powerful and exact the single-nerve fibre action potential recording method is under suitable recording conditions, which can be achieved in recordings from brain-dead humans (HTs).

6.8. Nerve Fibre Diameter Distribution of the Skin-like Nerve Root dCo

Figure 38 shows the cross section of a dorsal coccygeal root, which will be analysed electrophysiologically (and morphologically) in detail in the following. The cross section of Figure 38A shows many thick, thickly myelinated fibres and thin, thinly myelinated fibres. The corresponding nerve fibre diameter distribution histograms (Figure 38A supports this overall view). A large number of fibres is contained in the myelin sheath thickness range d between 1.8 and 2.3μm. Only few fibres have myelin sheath thicknesses between 0.8 and 1.8μm. The large population of thin (Ø ~ 3μm) thinly myelinated (d < 0.8μm) nerve fibres are not measured here; because of their smallness, their APs are too small to be recorded and the fibre diameters can therefore not be compared with the corresponding conduction velocities. By comparing the nerve fibre diameter distribution of Figure 38 with the ones Figures 7,9 of Ref. 177, it is concluded that this coccygeal root has similarity with a skin nerve. Thick proprioceptive afferents are mainly missing. Electrophysiologically, no dorsal root efferents could be found, probably because this coccygeal root was caudal to the motor outflow of the conus medullaris. Therefore, also motoneurons with diameters larger than 5μm could not have contributed to the spectra of Figure 38B. With the help of the known group nerve fibre diameters (Figure 2), some peaks of skin afferents were identified and designated in Figure 38B.

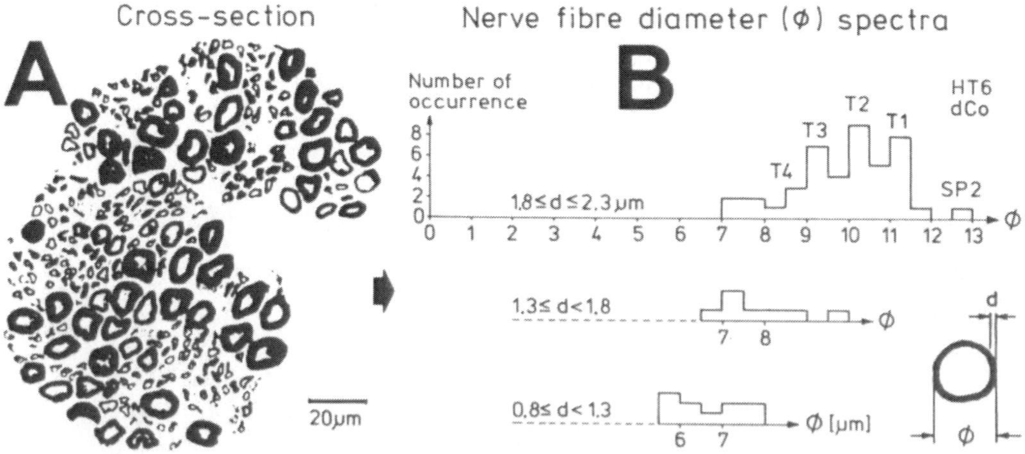

Figure 38. A. Cross section of a dorsal coccygeal nerve root (dCo), HT6. vCo did not exist. B. Corresponding nerve fibre diameter frequency distribution histogram, classified by the myelin sheath thickness range "d." Fibre diameter class borders ≤ and <. The very thin fibres d < 0.8μm are not measured.

6.9. Conduction Velocities of Thick Skin Afferents in the dCo Root

Conduction velocity distributions of the coccygeal root following touch and pinprick are shown in Figure 39A and B.

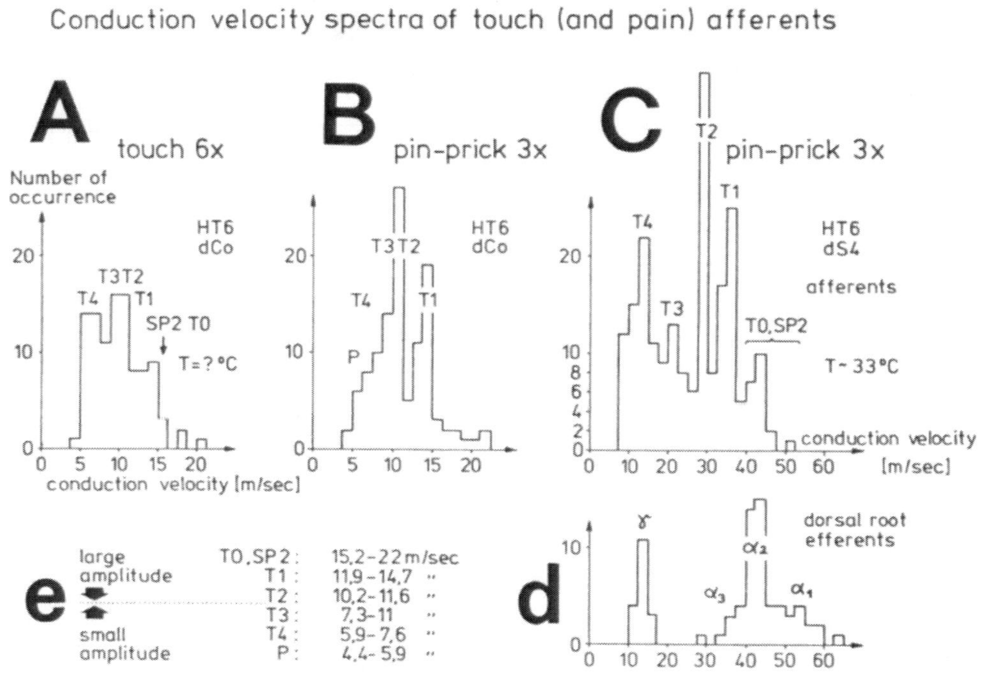

Figure 39. A, B. Conduction velocity frequency distribution histograms of the dCo root of Figure 38 following touch (A) and pinprick (B). The one active secondary muscle spindle afferent fibre is marked in “A” with SP2. Note, the touch with a ball (Ø = 5mm) activated a different selection of touch afferents (different population response) than the pinprick. Touch afferent peaks are designated, HT6. C, d. Conduction velocity distribution histograms of afferent (C) and efferent nerve fibres (d) of the same HT as in “A” and “B,” but from a different root (dS4). e. Conduction velocity ranges of touch afferent groups, indicated in “A” and “B,” of the coccygeal root used in the following Figures.

Because of the unknown low temperature, the peaks of the different skin afferents fused and the typical peak distribution did not occur. To obtain the velocity ranges of the different skin afferents, the velocity ranges of a S4 dorsal root of the same HT (HT6) (Figure 39C), calibrated by the simultaneously recorded motoneurons (Figure 39d) were used for comparison. Further, the conduction velocity of a continuously firing secondary spindle afferent fibre (SP2) and the different AP amplitudes of the skin afferents were used to establish the velocity ranges of Figure 39e. A few uncertain cases remained and are marked in the following impulse patterns. With the three criteria “AP waveform on trace a,” “AP waveform on race b” and the “conduction velocity of a single AP,” the simultaneously occurring impulse patterns of about 32 active skin afferent units will be extracted from the population response. Some patterns will be given in following Figures to understand what natural impulse patterns from the skin, upon natural stimulation, give rise to physiologic pattern organization in the human CNS.

6.10. Comparison between Fibre Numbers Obtained from Diameter Distributions and AP Occurrence Patterns

Thirty-two patterns of single fibres were extracted from the following population responses (Figure 40).

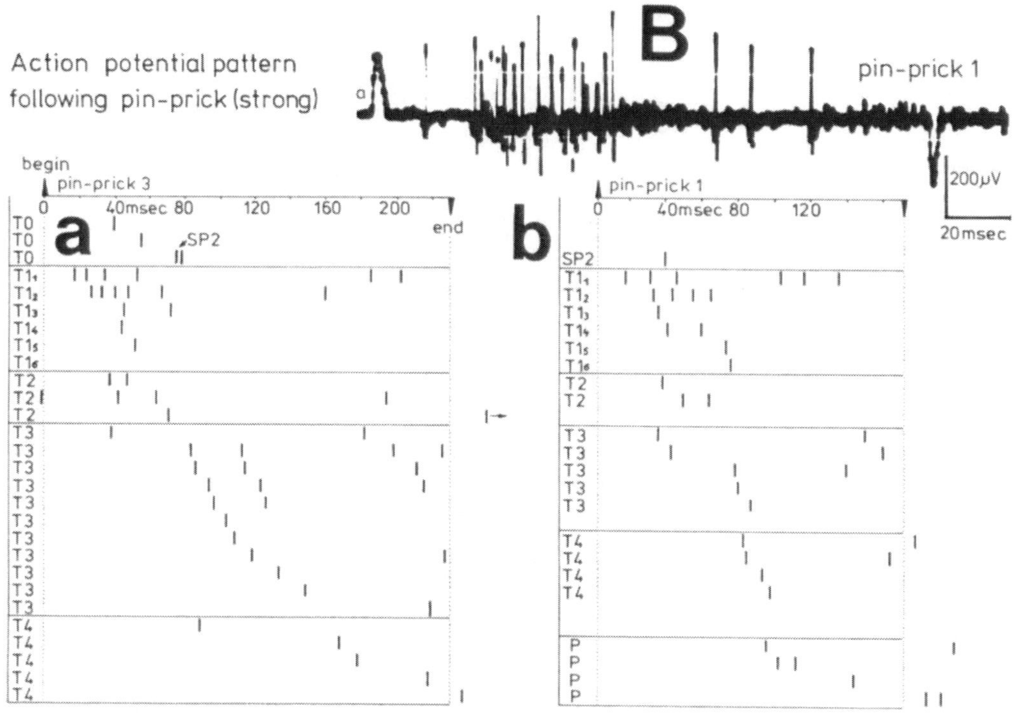

Figure 40. a. Action potential (AP) occurrence patterns of single touch fibres following strong long pinprick 3 (no pain receptors stimulated). The single-fibre AP activity of the different touch groups are identified by the AP forms on the traces "a" and "b" and the conduction velocities. The single touch afferents of the T1 group are marked with subscripts and can be followed up in all the stimulations of the following Figures. The one active secondary muscle spindle afferent fibre (SP2) could always safely be identified and is marked in this and the following Figures. B. Original recording of pinprick 1 (only trace "a" shown) with single-fibre analysis in "b." b. Single-fibre AP occurrence patterns of the short pinprick 1 shown in B. Note, the $T1_1$-unit adapts in "a" following the strong and slow pinprick (release activity is reduced), but not in "b" following the strong and short pinprick (release activity is not reduced).

In the diameter distribution for myelin sheath thickness $1.8 \leq d < 2.3\mu m$ (Figure 38B), 37 touch and pain fibres are contained. The numbers of fibres measured morphometrically and identified electrophysiologically are in good agreement. Probably not all afferents from the dCo dermatome leading through the dCo root were stimulated. The correlation between nerve fibre numbers obtained from impulse patterns and nerve fibre diameter distributions also hold within the skin afferent groups. It can be seen from Figures 40a and b that with strong pinpricks, four fast conducting pain fibres and five T4 afferents were activated. In the fibre diameter spectrum of Figure 38B for $1.8 \leq d < 2.3\mu m$ and $7 \leq Ø < 9\mu m$ four pain fibres and four T4 afferents are contained. The T3 impulse patterns of Figure 40a contain 11 fibres; in the fibre diameter peak of Figure 38B, 11 fibres are also contained. In the T2 group (Figure 40a), three fibres were stimulated; the morphometry group of Figure 38B contains about 14 fibres. In the T1 and T0 group, nine fibres were stimulated (Figure 40a); the corresponding group of Figure 38B contains nine fibres. Apart from the T2 group, fibre numbers from the diameter spectrum $1.8 \leq d < 2.3\mu m$ and the occurrence patterns are in good agreement. Taking the few fibres of the diameter spectrum $1.3 \leq d < 1.8\mu m$ into account does not change the good agreement very much. The discrepancy in the T2 group is a bit unexpected, since one would expect uncertainties more for the thin, slower conducting fibres. That the T2 fibres accidentally innervated another, but did not stimulate skin area is possible, but such situation would probably not account for the whole discrepancy. Further analysis will show that the T3 and T4 skin afferents are very sensitive to light skin touch. They were therefore easily stimulated and all recruited. The T1 afferents with much higher thresholds were also sensitive to skin traction. They were therefore easily accessible to stimulation. One is left with the likely explanation that many T2 afferents were not stimulated because of their high thresholds and their short range of receptive fields.

6.11. Adaptation and Thresholds of Skin Afferents

For the identification of certain APs and their reoccurrence, the recordings were analysed on an expanded time scale as in Figure 37B,C,D. In the following, the adaptation and the thresholds will be compared between the different skin afferent groups. The adaptation will be judged by comparing the number of APs per impulse train at the beginning and end of the pinprick or touch and the overall activity. Figure 40a shows that the receptors of the $T1_1$ and $T1_2$ afferents adapted fast following the rather long lasting pinprick 3, since the release activity consisted of fewer APs. Following short strong pinprick (Figure 40b), the $T1_1$ receptor showed no adaptation since pinprick and pinprick release impulse trains consisted both of three APs. In Figure 40a, it seems as if the T2 afferent receptors also adapted fast. From the long-lasting pinprick of Figure 40a, it can be seen that the receptors of the T3 afferents did not adapt fast. That the T3 receptors adapted much more slowly than the T1 receptors can also be seen by comparing the overall activity for long (Figure 40a) and short (Figure 40b) pinprick. The T1 afferents fired with 16 APs for the long and with 15 APs for the short pinprick. The T3 afferents fired 21 APs for the long and with eight APs for the short pinprick. This shows that the activity of the T1 afferents did not depend on the duration of the stimulation, whereas the activity of the T3 afferents did very much. Since the T3 receptors have a lower threshold than the T1 receptors, as will be shown below, the stimulation threshold will not disturb the adaptation difference between the T1 and T3 receptors. For the T4 receptors, the situation is more complicated. It seems as if they are slowly adapting (see below), but for the short pinprick, the activity consisted of six APs and for the long pinprick of five APs. This would point towards a fast adaptation. An explanation could be that the T4 receptors are adapting slowly, but their threshold could be velocity dependent (see below). The impulse patterns of the pain afferents (P) in Figure 40b give no clear answer about their adaptation speed. Figure 41 shows the impulse patterns with light and rather long-lasting pinprick. Mainly, only T3 afferents were stimulated. This indicates that the T3 afferents have the lowest threshold among the skin afferent receptors measured in this experiment. No pain afferents were stimulated. The pain activity property found in Figures 40a and 41 are in accordance with normal feeling. With light pinprick, no pain is felt, probably no receptors are activated (high threshold). With rather strong pinprick, it is sometimes possible to miss pain receptors sensing fast pain (Figure 40a).

Touching the skin at the former place of the pinprick (Figure 42) with a ball gives similar skin afferent impulse patterns as by pinpricking (Figure 40), apart from the pain afferent activity and the overall population responses (compare Figure 39A with Figure 39B). For rather long touch $T1_1$ and $T1_2$ receptors adapted (Figure 42a) (release impulse part is much shorter than the touch part) but for the short touch, they did not (Figure 42b).

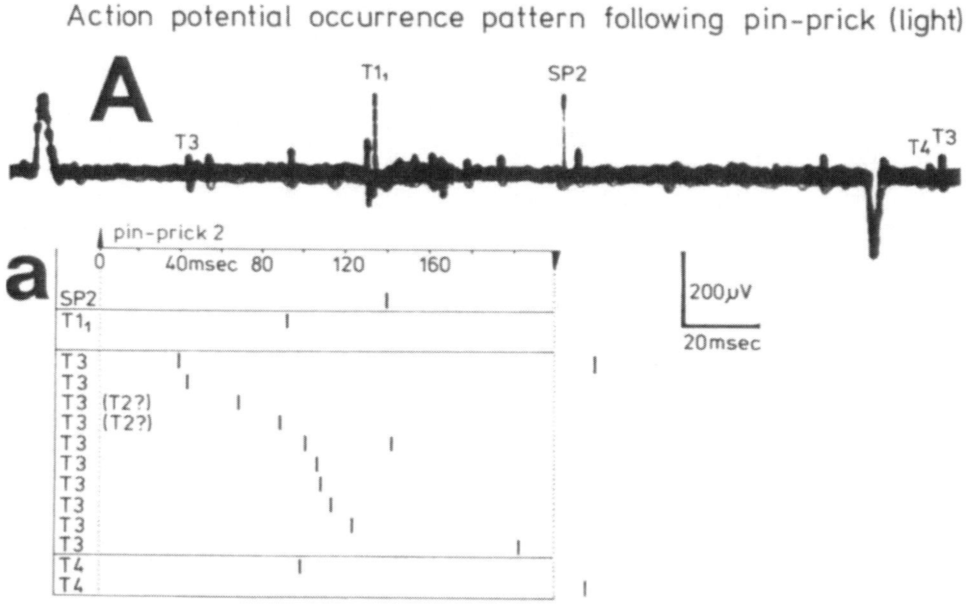

Figure 41. AP occurrence pattern following light and long pinprick 2. Original recording in A; single-fibre analysis in "a." Note, mainly only T3 touch afferents were stimulated.

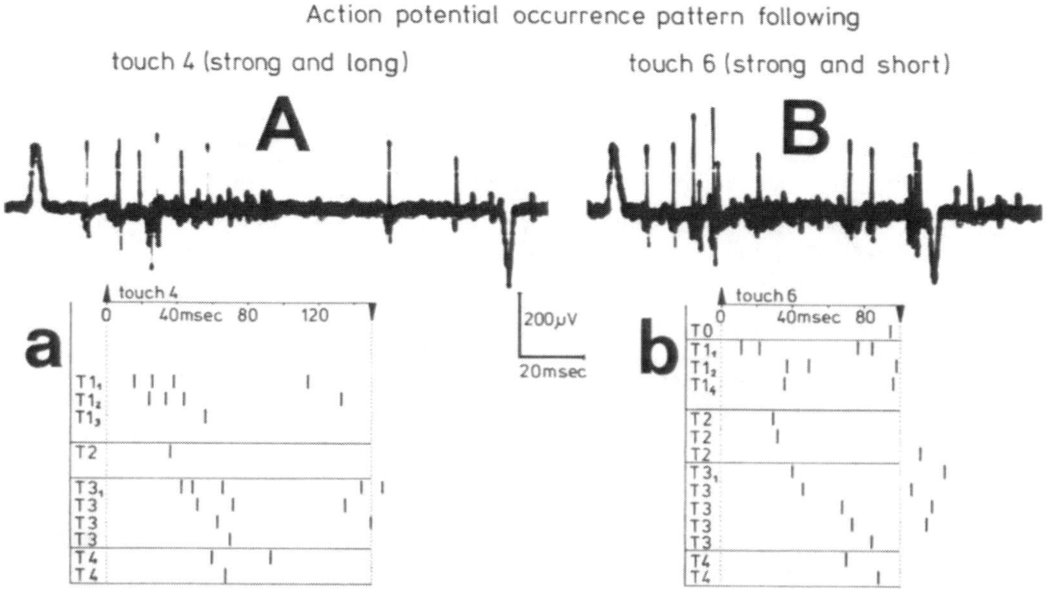

Figure 42. AP occurrence pattern following strong and long touch 4 with a ball of 5mm in diameter and strong and short touch 6 (B, "b"). Note, the stimulated $T1_1$-unit adapted in the case of the strong and long touch (A, "a") (release activity strongly reduced) and not for the strong and short touch (B, "b"). Note further, the $T3_1$-unit adapted more slowly than the $T1_1$-unit ("a").

The $T3_1$ receptor showed less adaptation with the longer touch (Figure 42a) than the $T1_1$ and $T1_2$ units. With the shorter touch, the T3 afferents fired mainly only with threshold APs. It seems that the slowly adapting T3 receptors needed also more time to build up an impulse train than the T1 receptors. The slower adapting receptors are also slower in responding. The overall touch stimulated activity of the T1 receptors was nine for the long and nine for the short touch. The T1 afferent response activity was therefore, as for the pinprick, independent of the length of the touch. The T3 afferents fired with a slightly higher activity for longer lasting touch (11 against 9). With the skin touch, not as many afferents were stimulated as with the pinprick. A likely explanation is that the pinprick was normally stronger than the touch and the stimulated skin area increased with a stronger touch or pinprick, since a larger area will be intended with a stronger touch or pinprick (see below). This interpretation is supported by the choice of T1 receptors stimulated. With touch (Figure 42), only those four T1 receptors were stimulated, which showed the shortest delay with pinprick (Figure 40), which had therefore on average the lowest threshold with respect to the point of stimulation and which were, therefore, on average, nearest to the point of stimulation (see also Figure 45b). Figure 43 shows for touch, as Figure 41 showed for pinprick that with light and long touch, mainly only T3 receptors were stimulated, which means that the T3 receptors have the lowest threshold for slowly applied rather long lasting touch.

So far, no properties could be obtained of the T4 afferents. With the stimulation shown in Figure 44, the T4 afferents were more activated than the T3 afferents. Since four T1 receptors were also stimulated, the strength of touch was medium and the T3 afferents should also have been activated strongly. But they were not. The impulse patterns of the T4 afferents in Figure 44a show a more slowly adapting nature, whereas the slowly adapting T3 afferents in Figures 44a,b show only threshold activation. The slowly adapting nature of the T4 afferents can also be seen, if one compares their overall activity of the rather long touch 2 with the shorter touch 3. For the longer touch, the T4 afferents fired with 11 APs and for the shorter touch with eight APs. The fast adapting T1 afferents fired with six and five APs roughly with the same activity. The velocity of touch is probably also not the reason for the activation of the T4 afferents, since the activity with shorter and probably higher speed of touch was shorter.

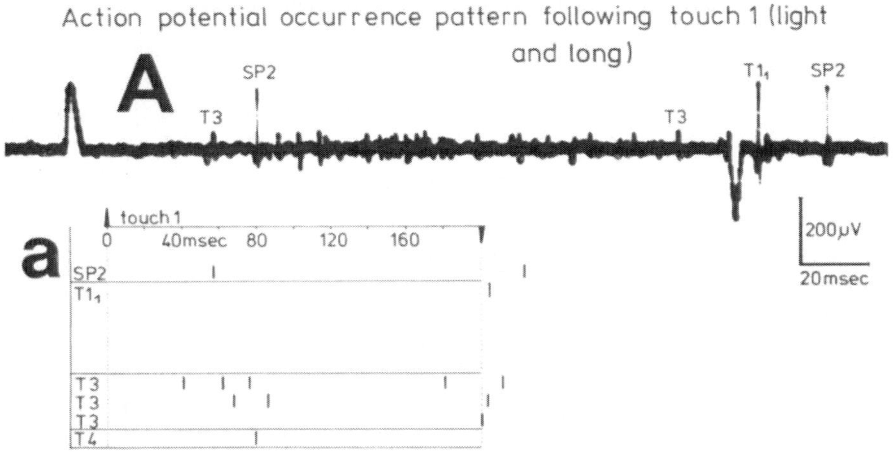

Figure 43. AP occurrence patterns following light and long touch 1. Note, mainly only T3 units were activated.

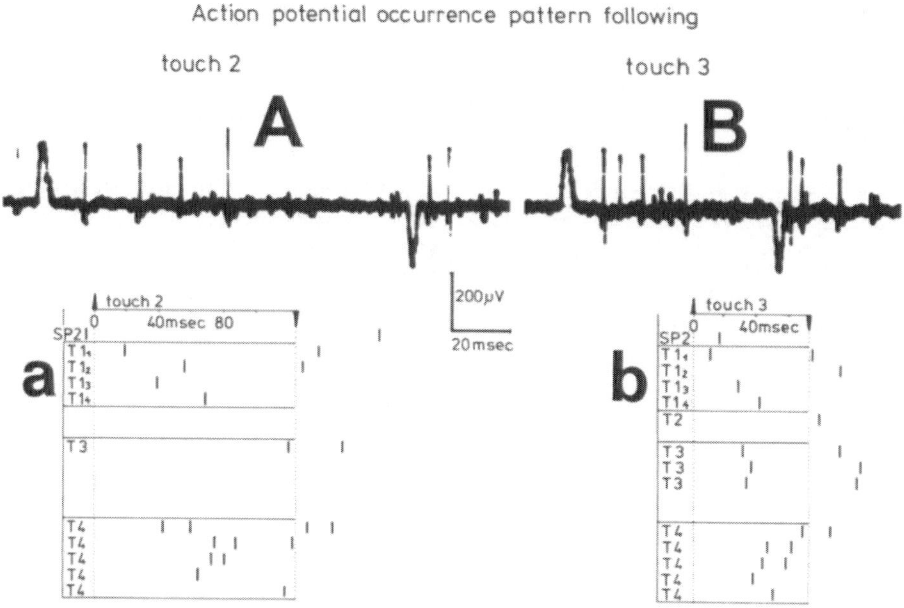

Figure 44. AP occurrence patterns following different touch stimulations. Note, the T4-units are more activated than the T3-units, especially in "a," in difference to the stimulations of the Figures 40 to 43.

It is concluded that the T3 and T4 receptors are both slowly adapting, both have a low threshold, but sometimes the T3 receptors are more activated and sometimes the T4 receptors are more activated.

Therefore, there must be a difference in the adequate stimulus of the T3 and T4 receptors. Looking once more over all stimulations, it can be seen from Figures 40 to 44 that with pinpricking, the T3 afferents were always activated the most and with the touch of the skin with a ball sometimes the T3 and sometimes the T4 afferents were activated more. Since touching with a ball, a movement component parallel to the skin is also possible and for the pinprick not (only skin traction), it is concluded that the T4 receptors are very sensitive to touch movements along the skin, whereas the T3 receptors are more sensitive to a touch without a movement along the skin. The T4 receptors are therefore the "Streichel receptors" (caress receptors). The analysis of the adaptation of touch receptors by comparing the "on" impulse train with the "off" parts is questionable. Normally, adaptation is judged by applying long-lasting skin indentation and measuring whether a receptor stops firing rapidly or slowly.

The T0 afferents could innervate hair receptors, because of their high conduction velocity, or they could be developmental rudiments. Since the T2 receptors were activated most with strong pinprick, they will have a higher threshold than the T1 receptors and probably do not respond so much to skin traction as the T1 receptors do, as will be analysed below.

Since further only three T2 afferents were activated (Figure 40) of the about 14 present (Figure 38B), whereas nearly all of the T1, T2 and T3 afferents were activated, it is likely that the T2 receptors obtained only little adequate stimulus, namely pressure. It will be difficult to apply pressure to a small defined skin area on top of the gluteus maximus.

6.12. Receptors of T1 Afferents Are Sensitive to Skin Traction

Following bladder and anal catheter pulling, no clear activity could be found of mucosal afferents. The coccygeal root seems to be more like a skin nerve and originates more caudally than the nervous outflow to the bladder and the anal canal.

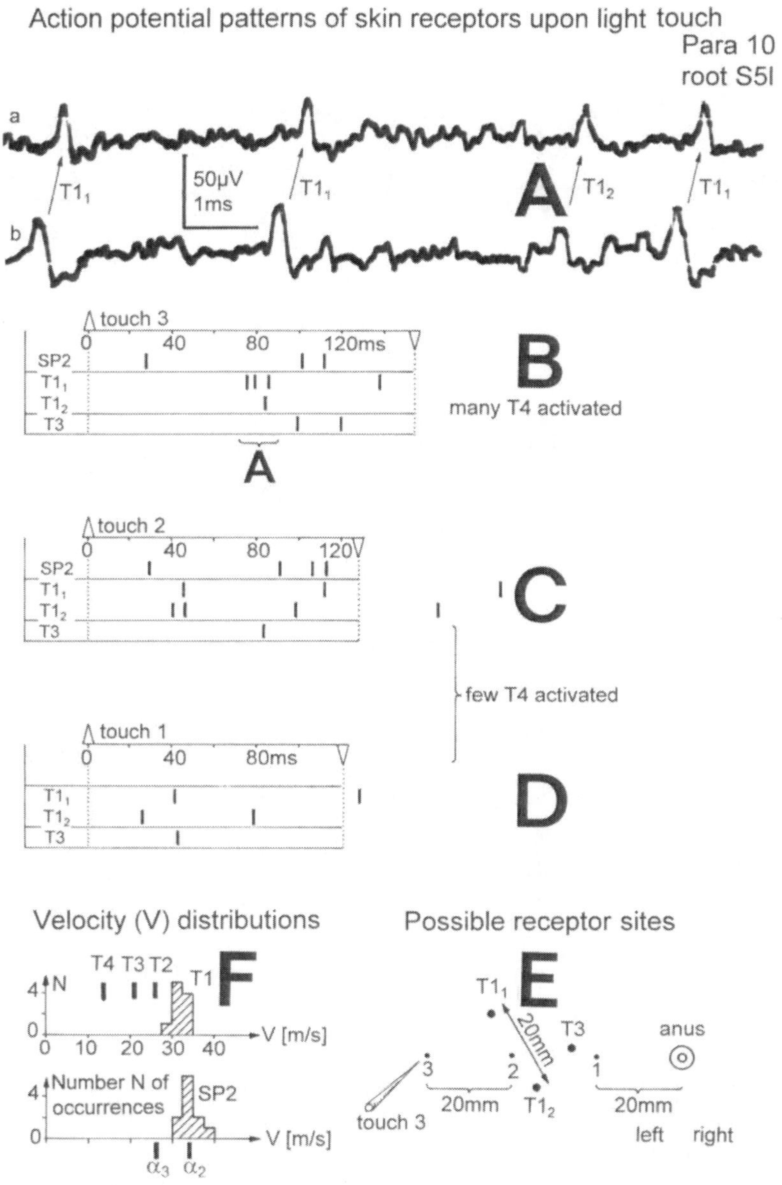

Figure 45. Recording and stimulation arrangement for measuring simultaneously many single touch and pain units. A = by skin folding stimulated area, which is drawn more in detail in "b." $T1_1$, $T1_6$ = suggested touch points of the $T1_1$ and $T1_6$-units. b. Drawing of the very approximate by skin folding stimulated skin area. $T1_{1-6}$ = suggested focal T1 touch points. Two-point discrimination indicated for comparison. N_A = number of stimulated units in the dorsal coccygeal root of Figure 38. Skin tractions, evoked by anal and bladder-catheter pulling, are indicated by large open arrows.

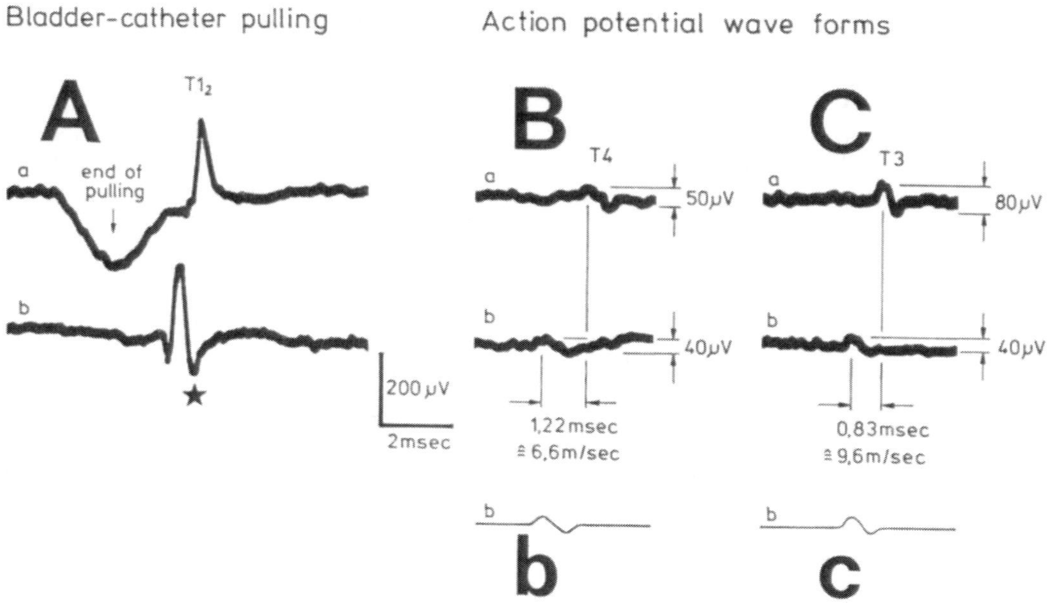

Figure 46. A. AP occurrence pattern following bladder-catheter pulling 1. Two potentials of T4 or M-type occurred (not shown) and this $T1_2$ touch afferent fibre AP at the end of the pulling, which was also mostly stimulated following pinprick and touch. B, C. Action potential waveform analysis and comparison of T3 and T4 touch afferents. Rather focal recorded T4-Aps are shown in B and idealised in "b." The T4 identification is done by the conduction velocity (see Figure 39e). The T3-AP (compare with the velocity ranges of Figure 39e) is rather focally recorded with the proximal electrode pair (trace "a," amplitude = 80μV) and non-focally recorded with the distal electrode pair (trace "b," amplitude = 40μV). The non-focal registration is schematically redrawn in "c." Note, a focal to a non-focal registration reduces the amplitude from 80μV to 40μV. Root diameter 0.13mm; electrode pair distance = 8mm.

With anal-catheter pulling, several kinds of above skin afferents including T1 units fired. Because it is not clear from the anatomy (Figure 45a) which activity was due to skin traction and which to skin touch from right and left body parts, the bladder-catheter pulling will give more clear answers than anal-catheter pulling about skin traction.

Figure 46A shows on a fast time scale the end of a sweep piece of a bladder-catheter pulling. By comparing the waveform of the AP of Figure 46A with those of the skin stimulation of Figure 37C, it can be seen, that the AP of Figure 46A was conducted by the $T1_2$ afferent fibre.

With this bladder catheter pulling, no low threshold T3 afferents were activated. A stronger touch of skin by the touching of adjacent skin parts is therefore unlikely. The $T1_2$ receptor was therefore stimulated from far away (at least 10cm, see Figure 45bA) by skin traction, which was caused by the pulling of the bladder-catheter.

Since the $T1_6$ receptor was not stimulated, which seemed to be most far away from the place of pinprick and nearest to the meatus of the urethra (activated by anal-catheter pulling) (Figure 45bA), it seems that not all T1 receptors had the same sensitivity to skin traction. The sensitivity to skin traction probably also depends on the receptor arrangement itself and on the direction of skin traction and how the skin is folded or bended.

6.13. Summarized Receptor Properties of the T1, T2, T3 and T4 Skin Afferents

The T1 and T2 skin afferent units were rapidly adapting and the T3 and T4 afferents were slowly adapting according to the measure for adaptation used. With skin touch and no movement component along the skin, the T3 receptors had the lowest threshold among the four receptor kinds. For a touch with a movement component parallel to the skin, the T4 afferent receptors had the lowest threshold. T1 and T2 receptors had higher thresholds. The T2 receptors probably had the highest threshold. T1 receptors could be stimulated by skin traction from more than 10cm away. Since the APs from one T1 receptor were often clearly leading (Figure 37A,B), the T1 units seem to have a very sensitive single focus and receptive fields, which arc not sharply delineated. T2 afferent receptors probably have small receptive fields. High thresholds and small receptive fields could explain that only few T2 afferents were activated with pinprick and touch with a ball of 5mm in diameter. The T3 receptors had very low thresholds to skin indentation and did not respond very much to skin traction or touch movement along the skin, but many of them were activated simultaneously with a single touch or pinprick. May be their receptive fields are large and there is a large overlap of the receptive fields or T3 receptors are very sensitive to skin folding or bending.

6.14. Innervation and Innervation Densities of Skin Afferents for Co, S4 and T9 Dermatomes

To calculate different innervation densities of sacral and thoracic skin segments, first the innervations “I” were determined from nerve fibre diameter distribution histograms, myelinated fibre numbers, conduction velocity distribution histograms and AP occurrence patterns. From the diameter spectra of Figure 38B for myelin sheath thicknesses $1.3\mu m \leq d$, the innervation on the coccygeal dermatome on one side was about I (morph; Co) = 9 (T0 + T1) + 14T2 + 12T3 + 6T4 + 8P. From the AP occurrence pattern of Figure 40, an innervation I (electro; Co) = 3T0 + 6T1 + 3T2 + 11T3 + 4T4 +4P could be obtained. As analysed before, the agreement is good between the values obtained from the morphometry and the electrophysiology. Only the values of the T2 afferents show a big discrepancy, which were explained by the high threshold of the receptors and their small receptive fields. To calculate innervation densities “i”, dermatome areas and stimulated areas have to be known. Lower sacral dermatomes were measured by Bohm (17) by cutting dorsal roots. In certain areas, the sensibility was reduced and in small areas absent. The remained sensibility can be explained by the existence of ventral root afferents. It is likely that the ventral root afferents contributed to the remained sensibility, for example, of the S4 dermatome following cutting the dorsal S4 root. But there is also a large overlap and variation in the lower and coccygeal dermatomes. Therefore, only very rough estimations can be obtained for sacral dermatomes and stimulated skin areas. The coccygeal dermatome area is assumed to be $5cm^2$ large on one side. The stimulated skin area was with $40cm^2$ (Figure 45) much larger than the coccygeal dermatome, and it seems as if nearly the whole coccygeal dermatome on one side was stimulated. The calculated innervation density from the electrophysiological and morphometric measurements for the coccygeal dermatomes is about i (electro, morph; Co) = (0.6T0 + 1.2T1 + 2.8T2 +

2.4T3 + 1.2T4 + 1.6P)/cm^2. From the nerve fibre diameter distribution histogram of a S4 dorsal root of Figure 47 [175] with identified skin afferent peaks, the following innervation on one side was obtained for myelin sheath thicknesses 1.3μm ≤ d I (morph; S4 = 8T0 + 20T1 + 22T2 (-6α_2) + 22T3 + 14T4 (-3α_3). "-6α_2" and "-3α_3" indicate that some α_2 and α_3-motoneurons of that dorsal root were subtracted. A few misjudged other afferent fibres were probably also intermingled. If one assumes a S4 dermatome area of 20cm^2 on one side, one obtains an innervation density of i (morph; S4) = (0.4T0 + 1T1 + 0.8T2 + 1.1T3 + 0.5T4)/cm^2 (ventral root afferents are not taken into account). In these calculations, the main uncertainty comes from the assumed dermatome area. More difficult is to calculate the innervation density from conduction velocity frequency distribution histograms. Since the activity of skin touch was very large in dorsal roots, many fibre APs merged, and the obtained velocity histograms (Figure 2B of Ref.175) gave no representative selections and can therefore not be used for calculations of innervations. The innervation density will therefore be calculated from the skin afferents of a ventral S4 root (Figure 2A of Ref.175). Taking five touch stimulations and 20% ventral root afferents into account, one obtains the following number of nerve fibres I (electro; S4) = 8T0 + 46T1 + 42T2 + 40 T3 + 30T4. With the assumed dermatome area on one side of 20cm^2, one obtains the innervation density i (electro; S4) = (0.4T0 + 2.3T1 + 2.1T2 + 2T3 + 1.5T4)/cm^2. Because ventral roots are thinner than dorsal roots (Figures 12,13) and dorsal and ventral root afferents have to be taken into consideration, i (electro; S4) is probably underestimated.

From the myelinated nerve fibre number of the skin branches of the intercostal nerve (3,000) to the lateral skin (Figure 17) and the corresponding nerve fibre diameter distribution histogram (Figure 7 of Ref.177) for 1.3μm ≤ d and 8μm ≤ Ø (36% of 3,000), the following innervation is obtained I (morph; T9) = 19 (T0 + hair follicle) + 40 T1 + 21T2 + 17T3 + 10T4) (3000 x 0.36)/310 x 80 cm^2 (20cm x 4cm (assumed dermatome broadness)) the innervation density results to i (morph; T9 (lateral = (0.8 (T0 + hair follicle) + 1.7T1 + 0.9T2 + 0.7 T3 + 0.4T4)/cm^2.

6.15. Two-point Discrimination

A 25-year-old female volunteer was analysed for the shape of skin indentation and two point discrimination. Two, not to sharp pencils were used for the touching. Stimulation was attempted in the same way and at similar places as was done during the HT measurements. The skin touch with one pencil resulted more laterally in a round-shaped skin indentation and more medially in an elliptic-shaped indentation with the long axis in direction of the anus, as in shown rather schematically in Figure 45. The area of the skin indentation (indentation amplitude 20 to 40mm) due to rather strong touch was measured and resulted in about 40cm^2, as indicated in Figure 45b. Then the two-point discrimination was measured, which was for s slow rather light touch 30mm in the S3 and S4 dermatomes. Figure 45b shows the two-point discrimination with the two circles of skin indentation according to scale. With faster or stronger touching, the two-point discrimination reduced to about 25mm. The two-point discrimination reduced when the skin was stimulated more close to the anus and was 25mm on top of the os coccygeum. The two-point discriminations in the axillary line of the T9 dermatome were about 35mm (transverse axis) and 45mm (longitudinal axis). The measured values are in accordance with the literature [206]. A further detailed analysis in the direction of the anus was not possible since firstly it was difficult to measure at the anal region without touching other skin parts and secondly the volunteer refused cooperation after some time. A measurement of the two-point discrimination in the anal canal and bladder was not possible with such simple means.

6.16. Comparison between Innervation Density and Two-point Discrimination

The two-point discrimination increased from the lateral T9 dermatome (40mm) to the S4 dermatome (30mm) and the coccygeal dermatome (25mm). With the used innervation density:

$$i = 1/cm^2\ (n_0T0 + n_1T1 + n_2T2 + n_3T3 + n_4T4)$$

T_v = kind of skin afferent fibre, n_v = number of innervation of a certain kind, the summed up innervation density N equals then to:

$$N = (n_0 + n_1 + n_2 + n_3 + n_4) / cm^2$$

N increases from the T9 dermatome (N = 4.5 touch points/cm^2) to the sacral (8.3; 3.8; mean = 6) and coccygeal dermatome (7.2). the increase of the two-point discrimination (shorter distance) with a stronger or shorter touch is in accordance with the number of recruited afferents in Figure 40 to 44, since with slow and light touches mainly only the receptors of the T3 (may be T4) afferents were activated, whereas with stronger and faster touch also the receptors of the T1 and T2 afferents were activated. If one takes the root from the densities at the point of touch and multiplies it with the distance of the two-point discrimination, then one finds that about seven to eight touch units are innervated between the two points (T9: $(4.5/cm^2)^{1/2}$ x 4cm = 8.4). On the basis of seven to eight touch units per two-point discrimination, it should be possible to reduce the distance of the two-point discrimination by learning.

6.17. Innervation Densities of the Hand

Johannson and Vallbo estimated innervation densities per cm^2 of 212 on the distal part of the fingers and 43 in the palm of the hand [114]. Comparable innervation densities are 4.5, 6 and 7.2 for the T9, S4 and coccygeal dermatomes, respectively. These innervation densities are rather compatible if one takes into account the differences in the two-point discrimination distances. It was calculated here that seven to eight touch point innervations were present between the two points of the two-point discrimination. If on the palm of the hand the two-point discrimination is 1cm [206], then the innervation density would be i = 7x7 = 49 touch innervation points/cm^2, which is about the value of above (43 innervations). For the distal fingers, one probably has to take also training into account, which reduces the two-point discrimination [206].

6.18 T1 Afferents and the Communication Across the Level of Spinal Cord Injury

Restoration of motor functions in patients with complete transaction of the thoracic spinal cord is possible in some cases by a training to activate paralysed muscles that are innervated by distal segment of the spinal cord [146]. The spinal reflexes of the isolated spinal cord are activated by muscles of the shoulder girdle, which ordinarily do not participate in lifting the leg. One hypothesis of the mechanism for the activation of the spinal reflexes is that the skin overlying the attachment area of the lower portions of the muscles becomes displaced by contraction of the upper portions of the long muscles of the trunk. This skin displacement then acts as a stimulus for the contraction of the lower portions of the muscles. As these lower portions of the long muscles of the trunk progressively regain their ability to contract, and with the aid of supplementary exercises, EMG activity begins to appear in the gluteus medius muscle, and this muscle begins to contract [146]. PC receptors can be activated by skin traction over a distance of more than 10cm. It seems therefore that information across the level of injury is made possible by skin displacement, since T1 afferents can be activated from far away and there is a large overlap in the innervation of adjacent dermatomes. The loss of the skin afferent input of one or two dermatomes at the level of injury, where the spinal cord is destroyed, may not stop the communication across the injury area. Injury-induced reorganisation of the CNS following dorsal rhizotomies has been demonstrated (52). Other kinds of afferents will contribute, but the major candidates for the information transport across the injury are the T1 afferents.

6.19. Skin Displacement and Movement-Based Learning

Generally, in movement-based learning (like coordination dynamics therapy), the activated receptors of the skin will substantially contribute to learning and learning transfer in CNS injury (besides muscle spindles and Golgi tendon organs). Integrative movements with large movement amplitudes of arms, legs and trunk are activating the receptors of the skin most integrative and strongest. Orthopaedic operations in cerebral palsy, on the other hand, in which tendons are transposed and changed in length, result in false information of the CNS about the movements in the periphery and hampers CNS repair by learning. After tackling the skin afferent input to inform the CNS about the periphery, the receptors of continence organs (urinary bladder and rectum) will be tackled now. The repair of the urinary bladder is one of the biggest problems in CNS repair.

6.20. Unchanged Skin Receptor Density Following Spinal Cord Injury

For the repair of the nervous system, it has to be known what changes in the nervous system took place following injury and how these damages can be repaired. Therefore, in this Chapter and in Chapter V, it will be tried to compare the rather physiologic physiologic nervous system functioning (in HTs) with that after injury (in patients with SCI). In a brain-dead human (HT6), it was found on the basis of AP occurrence patterns of certain touch afferents stimulated at different sites that T1 skin receptors (PC) in the skin adjacent to the gluteus maximus of a brain-dead individual were positioned approximately 20 mm apart (Figure 45b). In a patient with spinal cord injury, it was found during an operation (for the implantation of a urinary bladder stimulator) that the receptor density of T1 skin afferents was very similar to that of the brain-dead human HT6. As can be seen in Figure 47a, the distance between two T1 (PC) receptors was also approximately 20 mm at similar sites of sacral dermatomes. The receptor sites were obtained from the impulse patterns (Figure 47aA-D) recorded upon touching the sites 1, 2 and 3 (Figures 2,3 of Ref.197). It is concluded that neither the receptor density in the skin had changed following spinal cord injury, provided the skin has not been damaged by trauma or pressure ulcers.

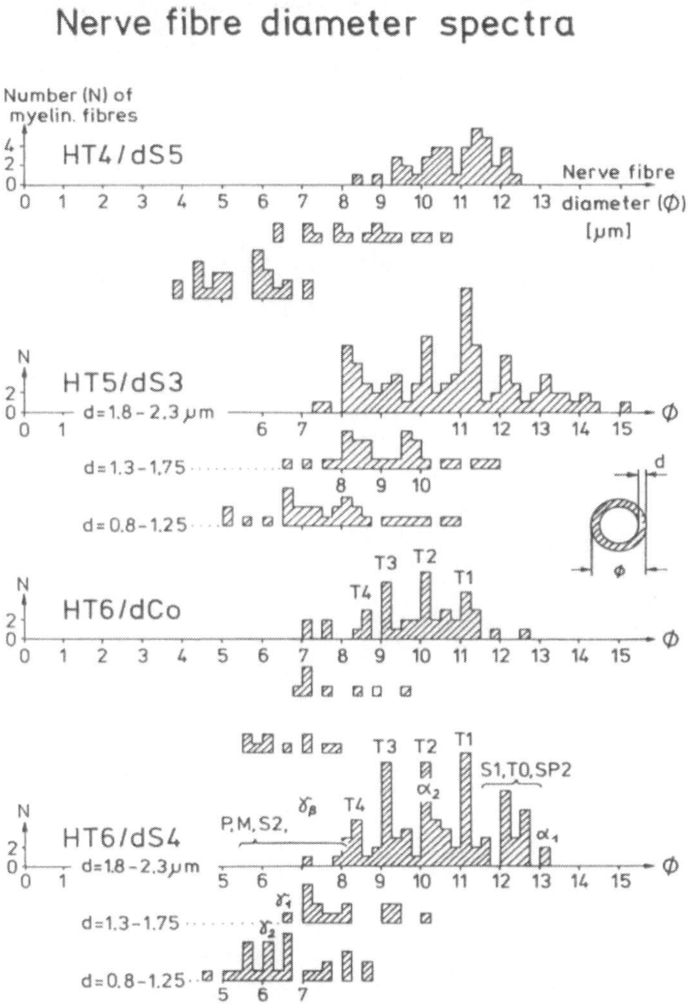

Figure 47. Nerve fibre diameter frequency distribution histograms from light microscope cross sections. Each set, containing three histograms, is designated by the case (number of HT) and the root measured (d = dorsal, S = sacral, Co = coccygeal). The three histograms in each set are partly marked with the myelin sheath thickness range (d) of the nerve fibres, which are contained in it. T0, T1, T2, T3, T4 = touch afferent peaks, M = mucosal afferent peak, P = pain fibres, α_1, α_2 = extrafusal motoneurons, γ_β, γ_1, γ_2 = intrafusal motoneurons.

7. Properties of Mechanoreceptors of the Urinary Bladder and Anal Canal

Summary

1. Identified afferents from bladder and anal canal characterized by their group peak values of conduction velocity and fibre diameter were at about 36° C (age 30 years): mucosa touch afferents: M (12.5 ms^{-1}/?); bladder afferents from stretch receptors measuring probably mural tension: SI (42.5 ms^{-1}/?), ST (38 ms^{-1}/?) and from flow receptors: S2(12.5 ms^{-1}/?).
2. Innervation densities of mucosal afferent units M of urethra, trigonum vesicae and anal canal through S3 and S4 roots had the following lower limits:

i (electro; urethra + trigonum) = 2 M/ cm^2,

i (electro; anal canal) = 3 M/ cm^2.

3. The innervation density of skin like receptors of the anal canal through an S4 ventral root was with i (electro; anal canal) = (0.5T0 +2.4T1 + 1.6T2+ 1.2T3+ 0.8T4)/ cm^2 at least as high as the innervation density of the S4 dermatome. The urethra possesses no skin-like receptors in the mucosa. Mechanoreceptor afferents of the mucosa (M) of urethra and anal canal had thresholds similar to those of T3 and T4 afferents.

7.1. Afferents from Mechanoreceptors of the Mucosa of the Bladder

Due to later use for diagnosis and repair, mechanoreceptors of the bladder mucosa were stimulated with a bladder catheter (Figure 45a). In this way, the trigonum vesicae, the urethra, the ostium urethrae externus and maybe the pubic region are stimulated. Figure 48 shows conduction velocity distribution histograms, constructed from the recordings of the activity increase due to pulling of the bladder catheter. By recording from an S3 dorsal root, one obtains a histogram with only one peak (Figure 48A). This peak (M) is interpreted as the conduction velocity distribution of afferents from one kind of mechanoreceptors from the mucosa of the bladder. If one records from a more distal S4 dorsal root (Figure 48B), one gets a histogram with many peaks. But all the additional peaks (T0, T1, T2, T3, T4) are the known skin afferent peaks.

This histogram is therefore interpreted as originating from the stimulation of mucosal mechanoreceptors and skin mechanoreceptors of the pubic region and the ostium urethrae externus.

The S4 root recording from a light stimulation (Figure 48C) supports this view, since that histogram really only shows the M peak and is therefore similar to the recording of Figure 48A, probably because only a few skin receptors of the pubic region had been stimulated. A strong pull (Figure 48D) probably gave a histogram with afferent peaks of the skin and mucosa because most possible regions had been stimulated.

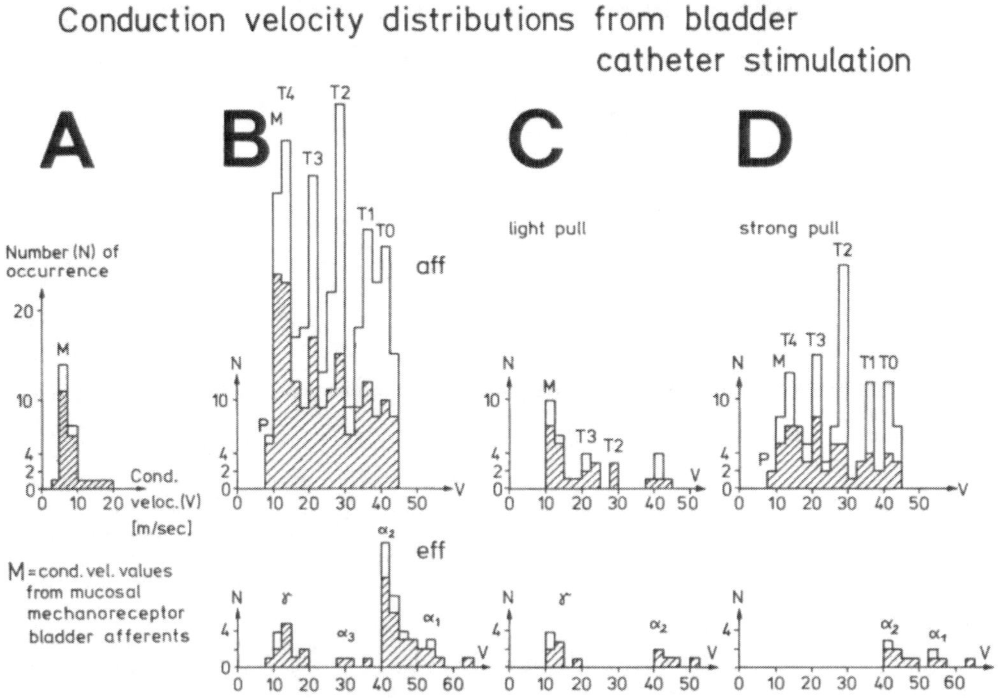

Figure 48. Conduction velocity distribution histograms from mechanoreceptors of the bladder (M) and skin (T0, T1, T2, T3, T4, P) stimulated by pulling a bladder catheter (upper Figure part). Lower part, simultaneously measured efferents (α_1, α_2, α_3, γ_1). A, recording from a dorsal S3 root, HT5. B, C, D, recording from a dorsal S4 root, HT6. A. Histogram from two recordings of single stimulations. Note the typical shape of a distribution peak from a single class of nerve fibres. B. Sum of four histograms, each being the response to a single stimulation. C. Histogram from a single sweep of one light stimulation. Note, the "M" peak has similar shape to the one in "A." D. Histogram of a single sweep of one strong stimulation. In the efferent histogram, the γ- motoneurons are missing, because high afferent activity masked them.

Only afferent activity from the mucosa was obtained from the S3 recording, and this can be interpreted in such a way that the S3 root innervated the mucosa of the bladder and the urethra, but not the pubic skin region and the ostium urethrae externus, whereas the S4 root did at least partly. Indirect stimulation of other regions of the body, especially the peri-vesicular parts, cannot be excluded in the case of the strong catheter pull. It also seems as if the skin pain peak (P) occurred (Figure 48B,D), but the separation from the M peak is not safe.

7.2. Afferents from Mechanoreceptors of the Mucosa of the Anal Canal and the Ampulla Recti

For the physiologic stimulation of the mechanoreceptors of the anal canal and the ampulla recti, an anal catheter was used with the balloon in the ampulla recti (Figure 45a). The stimulated areas were the ampulla recti, the anal canal and the perianal skin. Figure 49 shows the conduction velocity distribution histograms constructed from the recordings of the activity increase due to the pulling of the anal catheter.

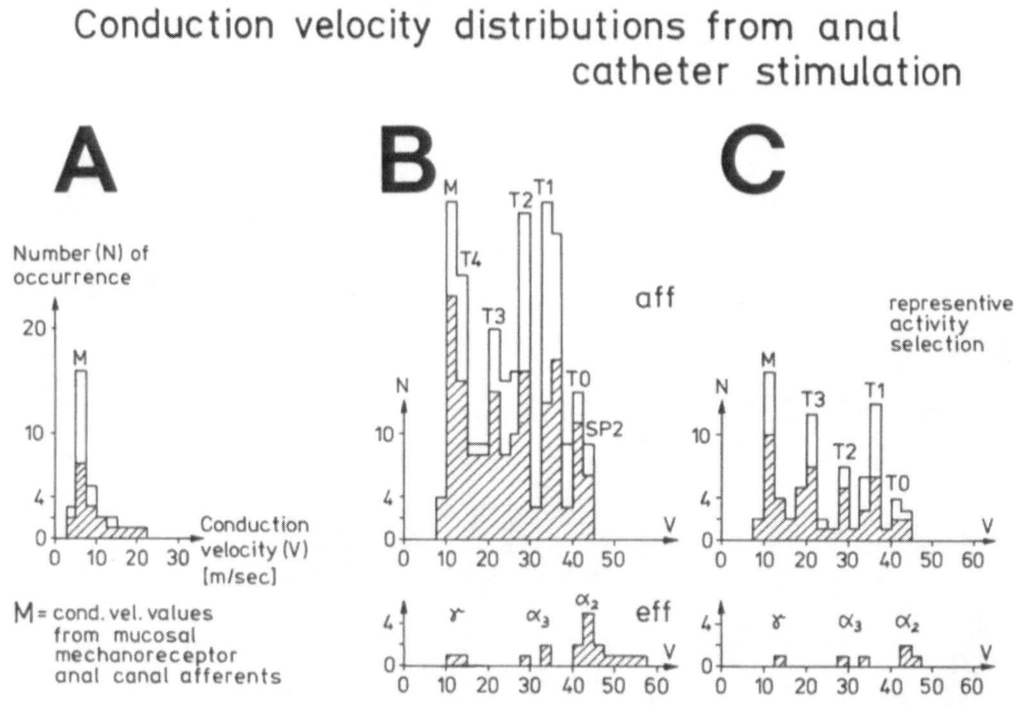

Figure 49. Conduction velocity distribution histogram of afferents from mechanoreceptors of the anal canal and peri-anal skin (M, T0, T1, T2, T3, T4) stimulated by pulling the anal catheter (upper part). Lower part, simultaneously measured efferents (α_2, α_3, γ_1). A, recording from a dorsal S3 root, HT5. B, C. recording from a dorsal S4 root, HT6. A. Histogram of two stimulations. Note, same distribution shape as in Figure 48A. B. Sum of four histograms, each of a single stimulation. C. Histogram from a single stimulation.

The histograms of the Figures 49A, B and C can be directly compared with the ones of Figures 48A, B and C. They are very similar, only the afferent peaks do not disappear in the case of the slight catheter pull (Figure 49C), indicating that there are also skin mechanoreceptors in the anal canal. Figure 49A shows the single peak of M afferents from the mucosal mechanoreceptors (S3 recording) and Figures 49B, C the histogram of the mucosal afferents in connection with the skin afferents (S4 recording). The segment interpretation is similar to the one of the bladder catheter stimulation.

7.3. Mean Innervation Density of Urethra and Trigonum Vesicae and Anal Canal

From Figures 40, 42 and 44, it can be calculated That T1, T2, T3, T4 and P afferent units fired on average two times following rather short touch or pinprick. From Figure 48C, it can be calculated that 20APs from mechanoreceptors of the mucosa (M) were recorded from a S3 dorsal root following bladder-catheter pulling. A similar number of APs was recorded from a S3 dorsal root. A lower limit for the innervation of the mechanoreceptors of the mucosa of urethra and trigonum vesicae through S3 and S4 dorsal nerve roots on one side, assuming an activation number of two APs as above to hold also for the mucosa, is about I (electro; S3+S4) = 20M ((20+20)/2). These values are compatible with the nerve fibre diameter histogram of Figure 47, even though the peak group nerve fibre diameter of mucosal afferents is still unknown. The stimulated area of urethra and trigonum vesicae is in the range of 10cm^2 on one side (urethra: Ø = 0.4cm, L = 4cm; trigonum side length = 3 cm). The lower limit for the mean innervation density would then be I (electro; urethra + trigonum) = 2M/cm^2. The estimated innervation densities of the T4 afferents varied between 0.4 and 1.5 per cm^2. The lower limit of the innervation density of mucosal afferents of urethra and trigonum vesicae is therefore higher than the one of the T4 afferents of the skin.

From Figure 49, it can be further estimated in the same way that the innervation of M units of the anal canal is about I (electro; anal canal) = 25M. Since the mucosal receptors are distributed only until to the linea arcuata, the area will be smaller. The lower limit of the innervation density of the anal canal (diameter = 1.2cm, L = 4cm, A = 15/2cm^2) is with I (electro; anal canal) = 3M/cm^2 in the range than the one of urethra and trigonum. From Figure 49, it can be further estimated that the innervation of skin receptors of the anal canal through an S4 dorsal root (AP occurrences of Figure 49B) divided by 4 (four stimulations) is about I (electro; anal canal) = 4T0 + 18T1 + 12T2 + 9T3 + 6T4. With a stimulated mucosal area of 15/cm^2 on one side, an innervation density of about I (electro; anal canal) = (0.5T0 + 2.4T1 + 1.6T2 + 1.2T3 + 0.8T4)/cm^2 is obtained. The skin mechanoreceptors innervation density of the mucosa of the anal canal is at least as large as the innervation density of the S4 dermatome, since afferents from the anal canal mucosa could also lead through S3 and S5 roots.

7.4. Comparison to Mucosal Afferents in Cat

Todd found in the cat [228] a variety of lamelled end-organs in and around the wall of the urethra. They varied from large Pacinian corpuscles, as much as 500µm in diameter, in the peri-urethral connective tissue, down to small endings in the urethral mucosa about 30µm in diameter. The small endings were found only in the most superficial layers of the urethral mucosa in which they lay longitudinally under the epithelium: their long axis was thus parallel to the direction of flow of the urine. They were usually situated at the summit of the longitudinal folds, which develop in the urethral mucosa when the urethra is empty. They were found in groups of two to five endings [228].

For the light and strong anal and bladder-catheter pulling, it was shown in Figures 48 and 49 that for a light pulling, the skin like afferent activity (at least from the T1 and T2 afferents) reduced more than the mucosal activity (the skin afferent activity occurring with bladder-catheter pulling stems from the stimulation of the skin around the ostium urethrae). The threshold of the mucosal mechanoreceptors is therefore similar to those of the T3 and T4 afferents or lower. A closer comparison is difficult, since a strong pulling will result in movement and will stimulate more the T4 receptors, whereas light pulling results in no or less movement along the mucosal surface and will stimulate comparably more the T3 receptors. The mucosa of the urethra and the anal canal (apart from skin afferents) is innervated by mechanoreceptors (M), flow receptors (S2) (see below) and pain afferents (P) for nerve fibre diameters larger than about 5µm. The high sensitivity of the urethra to touch and pain also results from afferents with diameters smaller than 5µm. It seems that the extreme sensitivity of the urethra to stimulation originates in small end-organs innervated by thin afferents. In male SCI patients with substantial remained sensitivity, the urethra has to be anesthetized during intermittent catheterization. Still unclear is whether the M and S2 (see below) afferents are different kinds of afferents or are only one kind of afferent fibres that are stimulated by pulling the catheter (M) or by changing the flow of fluid during retrograde bladder filling (S2).

7.5. Afferents Activated by the Retrograde Filling of the Bladder from a Reservoir

To record from afferents that inform the CNS about the filling stage of the urinary bladder, the bladder was continuously filled in a retrograde manner by gravity or light pressure with 0.9% saline solution of about 37°C through the bladder catheter from a bag. The recorded activity was continuously stored on a tape, and the filling stage was marked every 50ml. Single sweep pieces are not very instructive and are therefore not demonstrated here (for anal canal stimulated afferents, see Figure 11). Since the activity varied quite a lot with time, the activity from three sweeps of 400ms was collected for the histograms shown in Figure 50. The histograms at different stages in the filling of the bladder seem to show three peaks, and they were marked with S1, ST and S2. The ST peak is less safe. At high "bladder-filling" stages (here 800ml), it could be that other receptors outside the bladder contributed due to the pressure the bladder may have exerted on the surrounding tissue.

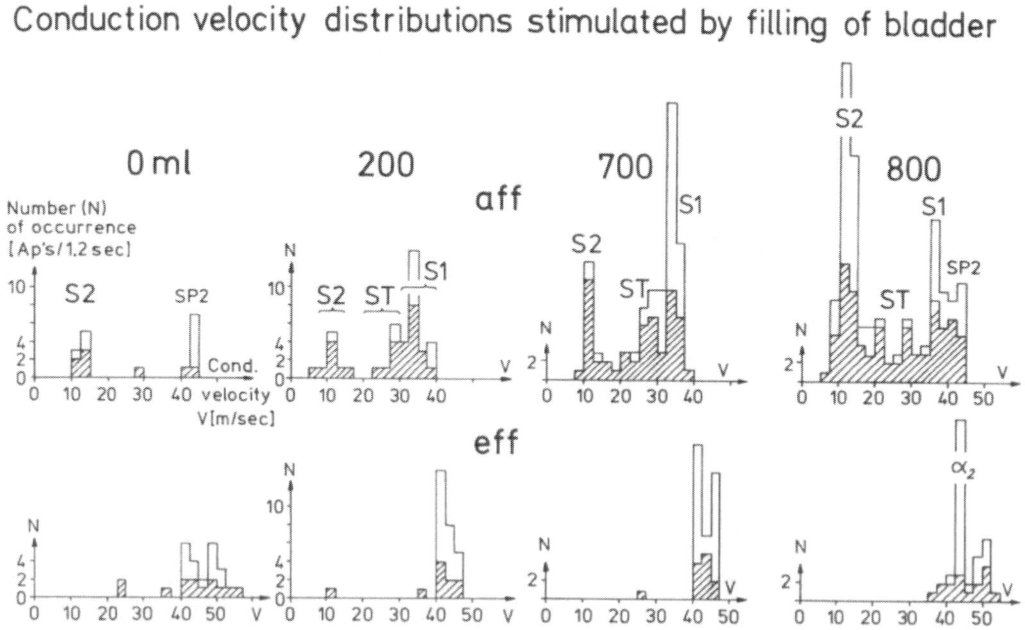

Figure 50. Conduction velocity distribution histograms of afferents from urinary bladder receptors responding to the filling of the bladder (upper part). Lower part, simultaneously measured efferents. Afferent conduction velocity distribution peaks are marked with S1 (30m/s $\leq v < 40$), ST (20m/s $\leq v < 30$) and S2 (7.5m/s $\leq v < 15$); v = conduction velocity. Dorsal S4 root recording, HT6. Speed of filling = 100ml/min, one stop in between for root wetting. The histograms at 0, 200, 700 and 800ml filling are the sum of three histograms, each constructed from a 400ms sweep. Note, the S2 peak is similar for 0 and 200ml and increases with 700 and 800ml, the S1 peak increases from 0ml to 700ml and is smaller for 800ml.

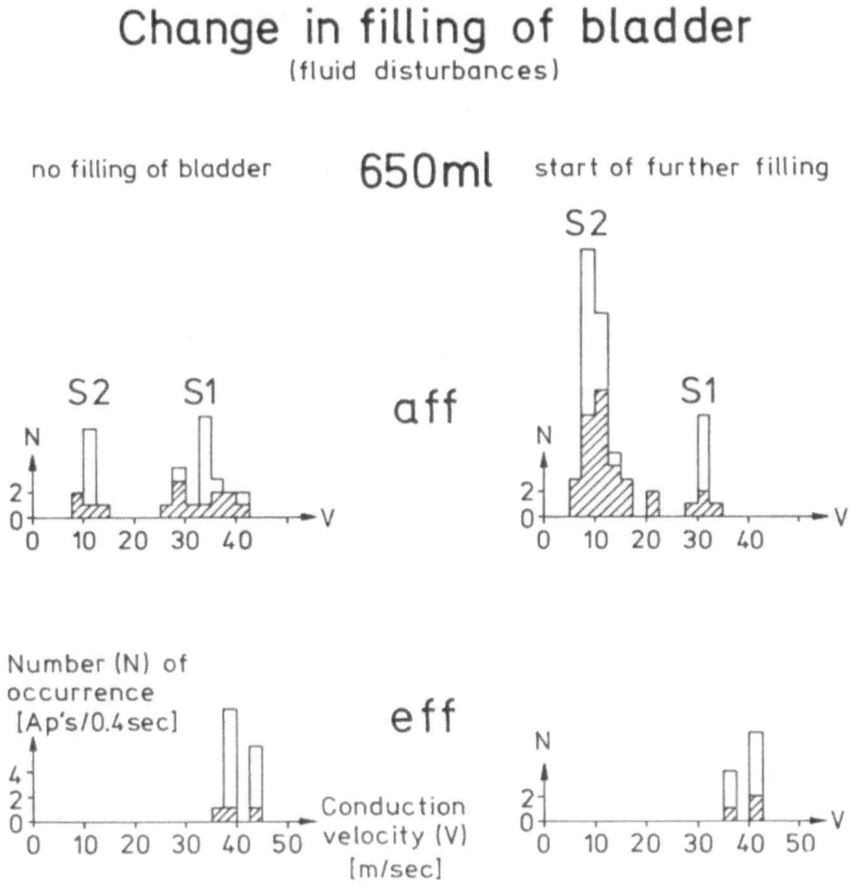

Figure 51. Conduction velocity distribution histograms for change of bladder filling between 620 and 650ml. Left, no filling; right further filling started. S1, S2 afferent peaks are the same as in Figure 50. HT6. Each histogram from a 400ms sweep piece (1/3 of the time as in Figure 50). Note, the S2 peak changed very much with further filling; the S1 peak did not.

It can be seen from Figure 50 that the amplitude of the S1 peak increased from 0ml to 700ml, and the afferent activity of that peak is interpreted as coming from stretch receptors of the bladder. The ST peak also increased with the filling of the bladder, and its activity probably originated from a second group of stretch receptors. The S2 peak showed different behaviour. This peak was already present when the bladder was empty and did not increase with the filling of the bladder up to about 600ml. From 600ml on, this peak increased strongly, as can be seen from Figure 50 (800ml). To get more information about the receptors giving rise to that peak, the part of the filling of the bladder where the filling procedure was stopped was looked for. Figure 51 shows the activity peak just before further filling of the bladder and just after. It can be seen that the S2 peak increased transiently very strongly, whereas the S1 peak was practically unchanged. The S2 receptors are therefore sensitive to fluid movements and high pressure in the bladder. With the interpretation that the S2 receptors are mechanoreceptors (see below), the ever-present S2 activity can be explained by the mechanical stimulation that the bladder catheter exerted onto the urethra.

7.6. Afferents from the Urinary Bladder and the Anal Canal

In the cat, afferents have been identified in the pelvic nerve and in the hypogastric nerve responding to tension changes in the detrusor musculature and mechanical distension of the mucosal-submucosal surface [244]. The tension afferents conducted below 16m/s and the mucosal mechanical afferents between 18 and 22m/s. It is possible that these afferents correspond to the S2 and M afferents, respectively, which both had conduction velocities of 12m/s at ≈ 32°C and 17.5m/s at 36°C (corrected). Todd [228] measured indirectly from cat pudendal nerves by recording from fine strands of sacral dorsal roots and cutting all branches of the sacral plexus, except those giving rise to the pudendal nerves. Distending the urethra with fluid, he recorded discharges of large amplitude that means from rather thick and fast conducting afferents. The receptors of these afferents were not sensitive to the flow of saline solution along the urethra and are therefore probably similar to the stretch receptors S1 and ST (Figures 50,51). When the urethra was obstructed by a ligature, a steady stream of fluid along the urethra was accompanied by action potentials of low amplitude (from thinner, more slowly conducting fibres), which increased in frequency especially when the flow started or its rate was changed (compare with Figure 51). The afferents from these flow receptors are most likely the S2 afferents (Figures 50,51). Todd [228] argued that the flow receptors are probably rapidly adapting mechanoreceptors to higher fluid velocities, when the laminar flow breaks down into turbulent flow. It seemed to him that the turbulence on the flow represents the normal mode of stimulation of the flow receptors. Irregularity of the wall or narrowing of the lumen ought to increase the possibility of local turbulence. Small lamelled end-organs were found in the most superficial layers of the urethral mucosa at the summits of mucosal folds with the long axis parallel to the direction of flow [228].

The S1 and ST afferents are most likely stretch receptors, which probably monitor tension, since their activity increased with the filling of the bladder (Figure 50) and decreased when the filling of the bladder was stopped (Figure 51), a stage when the bladder adapts and the pressure reduces [46]. The conduction velocity distributions suggest that there are two stretch receptor populations, even though a subgrouping of one receptor kind cannot be excluded. But since the ST afferents seem to start to respond later on with increasing filling of the bladder, it is conceivable that the ST (stretch-tension) afferents have to secure overstretch of the bladder in similarity to the function of the Golgi tendon organ afferents in skeletal muscle. The S2 afferents responded to fluid movement and high pressure (Figures 50,51). It is possible that Todd's specialized mechanoreceptors, which responded to fluid disturbances, also respond to high pressure. But it is likely that the S2 conduction velocity distribution population originated from two afferent populations, one responding to flow and the other one to high pressure in the bladder. Also the mechanoreceptor afferents M of the bladder mucosa, responding to pulling of the bladder catheter, could be identical to all the S2 afferents or the part of them, which responded to high pressure, since they have the same peak conduction velocity.

The mechanoreceptor interpretation of the S2 flow receptors is in accordance with human feeling. If one lets a drop of water run along the skin or lets water stream alongside the skin while sitting in the bathtub, one can experience a feeling similar to the beginning of micturition when the urine enters the proximal urethra. The flow of fluid at the skin is probably registered by the very low-threshold mechanoreceptors T3 and T4 (sensitive to very light touch). The S2 afferents responding to flow in the urethra and trigonum vesicae will not give rise to exactly the same feeling as the stimulated T3 and T4 afferents of the skin, since the S2 afferents probably have different receptors, may have a different firing pattern and probably are connected differently in the CNS. With receptors for stretch (S1) and maybe overstretch (ST) and one or two mechanoreceptors (S2, M) for urine flow, touch and high pressure, the bladder has an elaborate sensory system for monitoring its functional stage. The impulse pattern of the S1, ST and S2 afferents were more burst like than continuous. With the turbulence interpretation, this can be expected for the flow receptors. But also the stretch afferents, probably monitoring tension, could respond more burst-like, since Iggo [106] suggested that bladder tension-receptors are "in series" with the muscle fibres. Also in a continuously stretched bladder, due to the continuously retrograde filling of the bladder, the stretch receptor firing patterns could be irregular. It seems as if there was also some pain activity recorded from the bladder wall and the urethral wall, since the skin pain peak seemed to occur with strong pulling of the bladder catheter (Figure 48B). Patients report pain with strong bladder catheter pulling. Touch and pinpricks during cystoscopy are interpreted as pain in the bladder and also in the prostatic urethra (19, page 154).

By pulling the anal catheter, the conduction velocity peaks of the five skin touch afferent classes and one additional M peak were obtained (Figures 49B,C). This was to be expected, since the lining membrane of the anal canal in man contains all skin modalities [49]. The sensitivity seemed to be more acute than in the majority of cutaneous surfaces [49]. Duthie and Gairns [49] found additionally a profuse innervation with a highly specialised receptor in the lining membrane of the anal canal, which probably gave rise to the activity of the M afferents. But they did not find these receptors in the mucosa of the rectum. It still seems that there are M mechanoreceptors in the mucosa or submucosa in the caudal rectum or, at least, at the transition zone from the anal canal to the rectum (transitional-columnar epithel junction), since Figure 49A shows a velocity distribution from afferents only from the additional mucosa receptors. This sensory innervation of the anal canal with its important role in preserving continence can be understood. When the rectum is distended by contents, the fullness is probably sensed by the receptors of the M afferents, not allowing discrimination between flatus and faeces. Reflex relaxation of the internal sphincter and contraction of the external sphincter occur with further rectal distension and permit momentary contact of the rectal contents with the sensitive epithelial lining of the anal canal; this provides warning of the presence of the material and permits discrimination of its nature since the additional mechanoreceptors T1, T2, T3 and T4 have different thresholds so that probably the relative high threshold Pacinian corpuscles of the T1 afferents, lying deep in the region of the muscle bundles [214] (not shown in Figure 52), are only stimulated by faeces but not by flatus. Voluntary control or differential passage of flatus or faeces is therefore possible.

By comparing Figure 48 with Figure 49, one can see that with a light pull of the bladder catheter, there are not so many kinds of skin afferents that are stimulated simultaneously with the mucosal afferents with the pull of the anal catheter (S4 root recording). It indicates that in the urethra, there are no or not so many skin-like receptors as in the anal canal. This can be understood with the above interpretation. The receptors of the urethra need not to distinguish between different contents; only urine normally passes through. Even though they are of the same cloacae origin (18, page 7), the urethra and the anal canal probably differentiated a bit a differently according to the contents they transport.

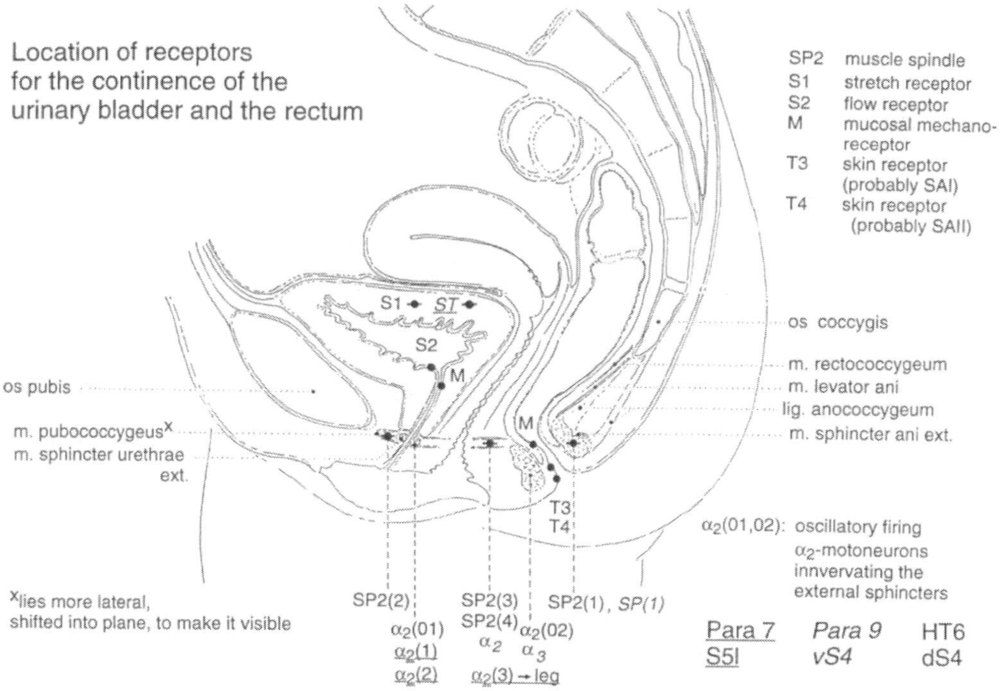

Figure 52. Location of receptors and muscles for the continence of the urinary bladder and the rectum, innervated by motoneurons the activities of which were recorded, in the brain-dead human HT6 (dS4 root), paraplegic 9 (vS4 root) and paraplegic 7 (nerve root S5).

The S3 dorsal root recording of the HT5 in Figure 48A and Figure 49A shows only mucosa mechanoreceptors afferent activity from the bladder, anal canal and rectum indicating little overlap of the S3 segments with the ones of S4 and S5. But Bohm's [17] large variation of the area and overlap of the S2 to S5 segments, therefore large variations of the lower sacral segments have to be expected. This is similar to the very large dermatome overlap from the S3 to S5 experienced in these measurements. The conduction velocity distributions from the afferents of low-threshold mechanoreceptors of the bladder (M) and the anal canal (M) are very similar. It seems possible that these afferents have the same receptor type. More functional aspects of afferents and efferents of the bladder and anal canal will be measured and analysed in the following and will lead to their repair following CNS injury.

7.7. Comparison of Receptor Properties of T4 and Flow Receptor Afferents of Urethra and Trigonum Vesicae

It was shown (Figure 51) that there are flow receptors in the urethra (S2) that responded to changes in the velocity of the flow of fluid. The interpretation for the adequate stimulus of the flow receptors was that turbulences, occurring with fluid flow, stimulated the receptors. These turbulences produced pressure changes and stimulated the flow receptors, which are mechanorecetors. The mechanoreceptors of the T4 afferents, which are most sensitive to movement along the skin, seem to have similar adequate stimulus than the flow receptors. This similarity may be important for a repair by learning of sensory function in the urethra, bladder and anal canal. The movement along the skin with very light pressure (caress, "streicheln") seems to be a certain quality of feeling. In the population response, the T4 receptors probably contribute the most to that feeling. It is unlikely that such feeling arises from the processing in the CNS from the changing afferent input while moving a finger along the skin. A series of light touches along the skin gives rise to a different feeling, even if the touch points are set very close to mimic a "streicheln." Movement along a surface seems to be an important perception outside and inside the human (and animal) body and is probably monitored at the skin, genitals, urethra and anal canal. In the walls of the intestine and the blood vessels, such receptors could also be positioned. Urinary bladder and anal canal functions (continence and defecation) will be picked up again, including the identification of the S1, S2, and ST groups by compound action potentials. But first, more knowledge is needed about sphincter muscles, their muscle spindles and their innervation to clarify physiologic and pathophysiologic bladder functions for bladder repair by learning (Chapter V).

7.8. Location of Identified Receptors of Urinary Bladder and Rectum for the Control of Storing, Continence, Voiding and Defecation

The receptors of skin, urinary bladder and anal mucosa, the muscle spindles that were innervated by secondary spindle afferents, and the muscles that most likely were innervated by α-motoneurons in paraplegics 7 and 9 and in the brain-dead human HT6 are marked in Figure 52 in the lower pelvis. The (rhythmic) activation of these receptors by anal and bladder catheter pulling and skin stimulation of sacral dermatomes, to stimulate simultaneously the somatic and parasympathetic divisions to induce learning transfer for repair, can be simulated and achieved during therapy, if a patient with severe cervical spinal cord injury is jumping rhythmically on springboard upon no weight support (Chapter V).

8. Identification of Muscle Spindle Afferents and α, β, and γ-Motoneurons in Animal and Human

8.1. Muscle Spindle Afferents

Most receptors of the periphery, e.g., skin receptors, are defined by the receptor properties alone. However, muscle spindle afferent activity also depends on the γ (and probably β) -motoneuron drive of the CNS and, depending on the site of location, sometimes on the drive of sympathetic or parasympathetic efferents. The receptor properties are therefore not uniquely determined by the peripheral receptor properties alone. A false efferent innervation during the development may further be compensated for by a different neuronal network organisation in the CNS. With respect to neuronal network functions, it has been argued that the network connections may be even more important than its buildings blocks, namely the different nerve cells [120].

Primary and secondary muscle spindle afferents [70, 71] have been distinguished from each other in animal research by their responses to a ramp stretch. Primaries respond stronger to stretch and show an instantaneous frequency rise. The secondaries show only an initial burst and frequency changes as if they were smoothed by a condenser [24]. The conduction velocities of the primary and secondary spindle afferents are clearly different. If spindle afferents from several muscles are pooled, the conduction velocity ranges start overlapping, and if velocities are even pooled from different animals, the velocities from primaries and secondaries overlap more strongly. In the recordings from the human lower sacral roots, most spindle afferents were mostly secondaries, because they conduct in the secondary range, they are projecting poly-synaptically onto α_2 and α_3-motoneurons (Figure 27F,G), innervating fast fatigue resistant (FR) and slow muscle fibres (S) (Figure 3), and they drive premotor spinal oscillators with phase relations of the broad peak type [196]. The possibility cannot be excluded that the primaries in continence muscles have different properties. Probably, a few primaries do innervate the spindles of the pelvic floor and the musculus flexor hallucis brevis (via S4 (!) and more rostral roots). The secondary muscle spindle afferents conduct with a group conduction velocity of 50 m/s at 36 °C (group nerve fibre diameter 12.1 μm) and the primaries with a group conduction velocity of 60 m/s (group fibre diameter = 13.2 μm), as identified by the group conduction velocities [192] and the conduction velocities of components of compound action potentials (Figures 34,35) [190]. Since the conduction velocity ranges of primaries and secondaries overlap, further criteria are needed for a safe identification and will be given below. In one measurement, a primary spindle afferent fibre could safely functionally be identified by its monosynaptic drive of an oscillatory firing α_1-motoneuron innervating fast fatigable muscle fibres (FF) (Figure 3) [195].

8.2. Identification of Dynamic and Static γ-Motoneurons and α-Motoneurons

Since nerve fibres taper only very little , the conduction velocities also vary only little if there is no branching and change of membrane properties along the fibres. The characterisation of nerve fibre groups by a group conduction velocity and a group nerve fibre diameter can therefore be used to identify nerve fibre groups and neuron types. In animal research, it has been found that the velocity ranges of static and dynamic γ-motoneurons overlap [23, 226]. In human research, measurements indicate that there are different peaks of γ-motoneurons according to the diameter and velocity distributions, even though the distributions strongly overlap. A better splitting of the different γ-peaks and preganglionic parasympathetic fibre peaks in conduction velocity distributions can be achieved if the velocities are plotted on a log-scale (Figure 33). For further comparisons of human and animal velocity and fibre diameter distributions, we have to consider the different measuring conditions.

In the human cauda equina, where a nice alignment of nerve fibres was expected (Figure 12), still quite a lot of variability was encountered. In peripheral nerves, more changes in the regrouping of nerve fibres and branching have to be expected. In CNS research, there is much discussion whether the branching of the reticular-spinal pathways already occurs in the reticular formation [204, 235] or during their course in the spinal cord [138, 153]. Further, if nerves are bathed in paraffin oil, as was frequently done in animal research, the conduction velocity will change at the bathed site [90]. It has to be measured how large the change in the conduction velocity is if parts of the nerve fibres are bathed in paraffin oil for temperature regulation. There may well be, therefore, variabilities in the conduction velocity measurements in animal experiments, due to the fact that a sufficient number of values (more than 100) are not measured simultaneously to obtain representative distributions and due to other reasons (see above), so that group conduction velocities of the static and dynamic γ-motoneurons overlap more strongly than in reality. Simultaneous velocity distribution measurements in animals indicate that there is not so much overlap in the velocity distributions of static and dynamic γ-motoneurons (Figure 53). A velocity plot on a logarithmic scale enables further separation of the peaks.

In man, on the other hand, α and γ-motoneuron groups were identified by the group conduction velocities and partly by group nerve fibre diameters and by the measured impulse patterns conducted along the fibres. Even though the conduction velocities were measured over small distances (10 mm), the relative group conduction velocities were very reliable, because the conduction velocities of all nerve fibre groups were measured simultaneously and were calibrated by the calibration relation: secondary muscle spindle afferents conduct with the same velocity as the α_2-motoneurons, and the calibration relation is temperature independent, even though the conduction velocities themselves are strongly temperature dependent (Figure 30). Sufficient characterisation of nerve fibre types is a principal problem when nerve fibres have to be safely identified for the understanding of neuronal network self-organization or the clarification of peripheral receptor properties of muscle spindles.

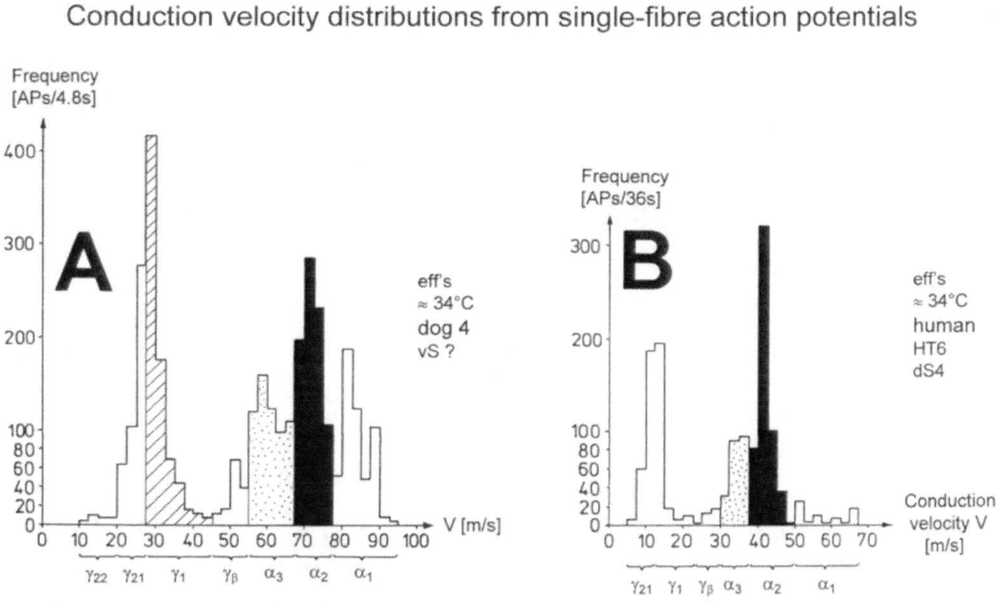

Figure 53. Distribution histograms of conduction velocity frequencies: efferent APs from a dog (A) and human (HT6; dS4) (B) lower sacral nerve roots. The distribution peaks are labelled according to the respective groups they represent. Motoneuron velocity ranges are indicated. In A, 24 sweeps of 0.2s, and in B, 30 sweeps (stimulated and non-stimulated) of 1.2s duration were used.

Since group conduction velocity distributions overlap, γ (and α) -motoneurons have to be identified additionally by their functions (by their natural impulse patterns).

In measurements in man, the groups of α and γ-motoneurons were identified by the speed and the dynamics to respond to natural stimulations such as touch, pinprick and catheter pulling, and the conduction velocity [189]. The characteristics of the static (γ_2) and the dynamic (γ_1) γ-motoneurons are not the same as in animal research. In humans, the γ-motoneurons were therefore characterised by the central nervous system properties, namely by the cell soma and in what neuronal networks the motoneuron cells are integrated, whereas in animal research, they were characterised by the periphery, namely how do primary muscle spindle afferents respond to repetitive 75 Hz stimulation (not always 75 Hz) of the static and dynamic γ-motoneurons in parallel with a ramp stretch [24]. To fulfil the needs of the human research and the clinical demands, we need a multiple group characterisation of γ-motoneurons, which can be used in animal research and in the clinical setting. For α-motoneurons, this has partly been achieved. α_2-Motoneurons, for example, fire for continuous high activation at a frequency of about 6.7 Hz, have a group conduction velocity of 50 m/s, a group nerve fibre diameter of 10.2 μm and innervate FR-type muscle fibres, which are fast oxidative glycolytic fibres (Figure 3). For dynamic and static γ-motoneurons, such a multiple characterisation is not known in humans. In human measurements, three conduction velocity peaks of γ-motoneurons (γ_1, γ_{21}, γ_{22}) seem to exist. The γ_1-motoneurons respond faster to stimulation than the γ_2-motoneurons [189]. It has to be shown whether the human γ_1-motoneurons (identified centrally) actually correspond to the dynamic γ-motoneurons as identified in animal research (identified in the periphery) and whether the γ_2-motoneurons correspond to the static γ-motoneurons. The notion will be helpful that there is no tapering of nerve fibres. First electrophysiologic measurements from single human muscle spindles in the periphery indicate that the human muscle spindles may function similarly as do cat spindles, even though the human muscle spindles seem to be more complex [70], and human muscles may be quite different in different parts of the body.

8.3. β-Motoneurons

In conduction velocity distributions and in nerve fibre diameter distributions, conduction velocities and fibre diameter values intermediate between α_3 and γ-motoneurons were found. But no specific impulse patterns of these intermediate fibres could be found so far. These fibres might thus be β-motoneurons or just α and γ-motoneurons lying rather outside of the normal diameter or velocity ranges for α_3 and γ-motoneurons. So far, the α_3-motoneuron distribution has not been clearly identified anyway.

In animals, it could be found that axon branches of α_2 (FR) and α_3-motoneurons (S) form the β-innervation of muscle spindles close to the spindles. This is a bit difficult to understand, since the dynamic β-motoneurons (β_d) are axon branches from α_3-motoneurons (S) (innervating bag 1 fibres), and the static β-motoneurons (β_s) are axon branches from α_2-motoneuron axons (FR) (innervating long chain fibres) as revealed by glycogen depletion and myosin ATPase staining [7]. Such β-innervation [78] contradicts the characterisation that the α_1, α_2 and α_3-motoneurons innervate their extrafusal muscle fibre type, namely FF, FR and S-type fibres, respectively.

9. Natural Firing Patterns of the Different α and γ-Motoneurons and Muscle Spindle Afferents Upon Natural Stimulations

9.1. Occasionally Firing Mode of α and γ-Motoneurons - Recruitment According to the Size Principle

9.1.1. Occasionally, Transient Oscillatory and Continuous Oscillatory Firing Mode

The method of single-nerve fibre action potential recording from whole nerve roots in humans was used to develop a precise classification scheme of nerve fibre groups in the PNS: the individual nerve fibre groups are identified by the group nerve fibre diameter and their group conduction velocity. Based on this classification scheme, receptor properties of skin, urinary bladder and anal canal afferents have been analysed (see above). Now the natural patterns of motoneuron firing, generated in the human lower spinal cord in response to natural stimulation, will be analysed. The natural impulse patterns of motoneurons are characterized by the occasional firing mode, the transient oscillatory and the continuous oscillatory firing mode. The recruitment of motoneurons in the occasional and oscillatory firing mode, which drive anal and external urinary bladder sphincters (and other skeletal muscles), will be analysed below. It is started with the recruitment of motoneurons in the occasional firing mode.

9.1.2. Comparison of Conduction Velocity Distributions of Humans and Animals

To analyse recruitment of α and γ-motoneurons in the occasional firing mode in animals and humans, first a comparison between the velocity distributions of a human and a dog is given (Figure 53). Figure 53 shows such conduction velocity distribution histograms of a human (B) and a dog (A) measured with the same method, the same equipment and at similar temperatures. It can be seen that the velocity distribution of the dog is very similar in their peaks, apart from the higher group conduction velocity values. The dog α_2-motoneurons, for example, conducted at 70 m/s whereas the human ones at 40 m/s at 34 °C. The difference between the human and the animal peripheral nervous system, however, may not just be that human nerve fibres conduct proportionally more slowly. Transmission frequencies of secondary muscle spindle afferents in man were measured to be as high as 5,000 Hz. It may be, therefore, that other human neuron membrane parameters differ from those of animals. The lower conduction velocity values in man may be the price for a higher transmission frequency. A higher transmission frequency may allow higher pattern variability and complexity.

9.1.3. Recruitment According to the Size Principle in Each Group of α and γ-Motoneurons in Dog and Man with Unspecific Low Level Stimulation – Muscle Tonus

Distribution changes of conduction velocities in each group of α_2, α_3 and γ-motoneurons were used for recruitment analysis. As will be shown below, the α-motoneurons will increase their firing rate upon the increase of their adequate afferent input from firing occasionally to firing oscillatory.

It has been shown in humans for low activation that the α and γ-motoneurons are recruited with the increasing conduction velocity, i.e., according to the thickness of the axon, thus according to the motoneuron size.

The slowly conducting fibres are recruited before the faster conducting ones in each motoneuron group (182, 10).

In man, with no additional stimulation, the slowly and fast conducting α_2 and α_3-motoneurons are recruited repeatedly every 2s (Figure 54A).

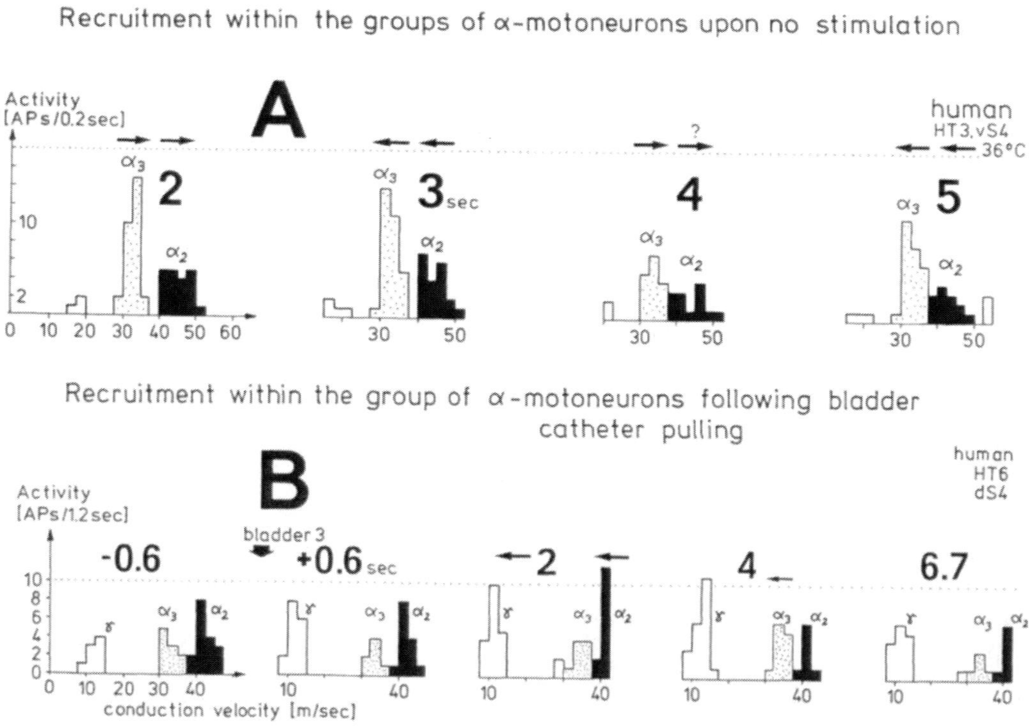

Figure 54. Frequency of the occurrence changes of single-fibre conduction velocities (recruitment changes) within the groups of α- and γ-motoneurons in man upon no stimulation (no catheter positioned, A) and following strong bladder catheter pulling (B). The velocity ranges in B are taken from Figure 53. Time in seconds following start of measurement or stimulation is indicated (e.g., 2: 2.0 to 2.2s). The arrow directions on top of the histograms indicate stages, at which motoneurons in a certain group with low or high velocities are preferentially activated, as can be calculated from the histograms. Note in B, the co-recruitment of γ_1 and α_2-motoneurons, and that α_2-motoneurons are recruited before α_3-motoneurons. HT3; vS4 (A) and HT6; dS4 (B).

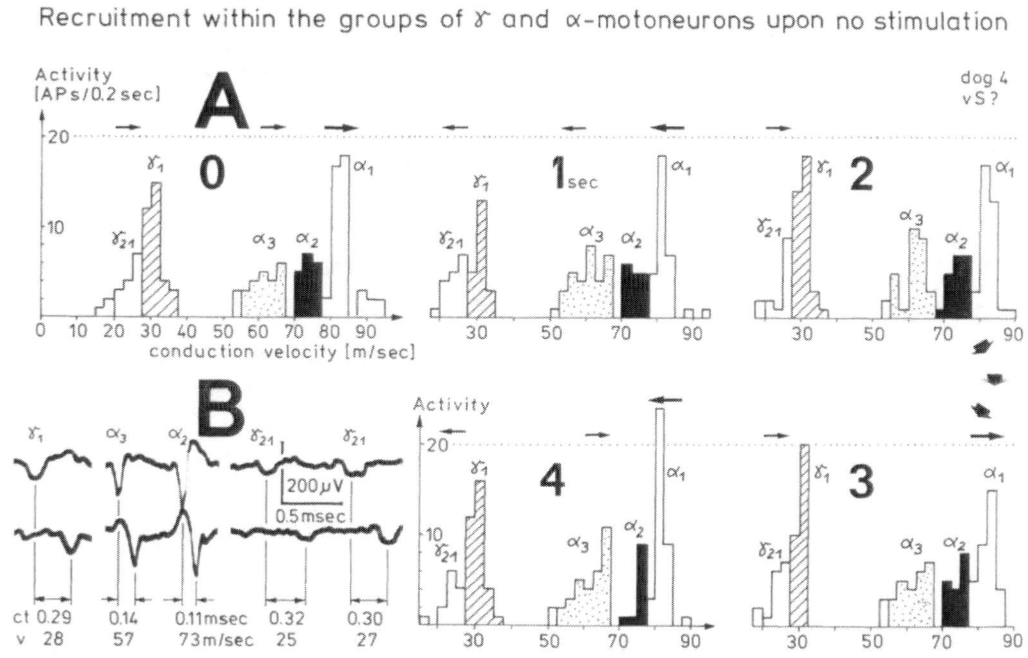

Figure 55. A. Frequency of occurrence changes of single-fibre conduction velocities within the groups of γ_1, α_3 and α_2-motoneurons (recruitment) in a dog sacral ventral root upon no stimulation. The group conduction velocity ranges are taken from Figure 53. Measurement times in seconds are indicated (e.g., one second: 1.0 to 1.2s). The arrow directions on top of the histograms indicate stages, at which motoneurons in a certain group with low or high velocities are preferentially activated, as can be calculated from the histograms. Columns for γ_1-motoneurons are crosshatched, those for α_3 are dotted, and those for α_2 are filled. Dog 4, bladder catheter positioned. B. A sweep piece from a recording from dog 4. Conduction times (ct) and conduction velocities (v) and α and γ-motoneurons are indicated.

In the dog (for identification of nerve fibre groups, see Figure 53A), with no additional stimulation, slowly (and fast) conducting γ_{21} and α_3-motoneurons showed repeated activation every 3 to 4s.

A subgroup of α_1-motoneurons showed repeated recruitment every 2s (Figure 55). In a slowly conducting rat subgroup of α_1-motoneurons, without specific stimulation, the motoneurons were recruited according to the size principle every 2s.

It seems, therefore, that in rats, dogs and man with unspecific low-level stimulation, α and γ-motoneurons are recruited repeatedly according to the size principle in each motoneuron group [182].

The period of repeated recruitment varies between 2 and 4s. The recruitment in the different groups may be synchronized or not.

The recruitment within the different α and γ-motoneuron groups for low level activation is important to understand muscle tone (apart from mechanical muscle properties); however, it is difficult to measure the recruitment because of low activity, which makes the sampling intervals large to get enough events.

9.1.4. Recruitment of α and γ-Motoneurons in Dog and Man Following Specific Stimulation

Following bladder catheter pulling, the human slowly conducting γ_1 and α_2-motoneurons were recruited 2s after the pulling, the slowly conducting α_3-motoneurons 4s after the pulling (Figure 54B).

In the dog, following bladder catheter pulling, the slowly conducting γ_1 and α_2-motoneurons were recruited directly following the pulling, and the slowly conducting α_3-motoneurons were recruited 1s later (Figure 56). In the rat, the slowly conducting α_1-motoneurons were recruited directly following pinpricking of the limb, the slowly conducting α_2-motoneurons 0.2s later, and the slowly conducting α_3-motoneurons 1s later. Therefore, following specific additional stimulation, the more dynamic motoneurons (α_1, α_2, γ_1) are recruited according to the size principle before the more static motoneurons (α_3, γ_2), depending on the function stimulated. Often, there was co-recruitment of α and γ-motoneuron groups.

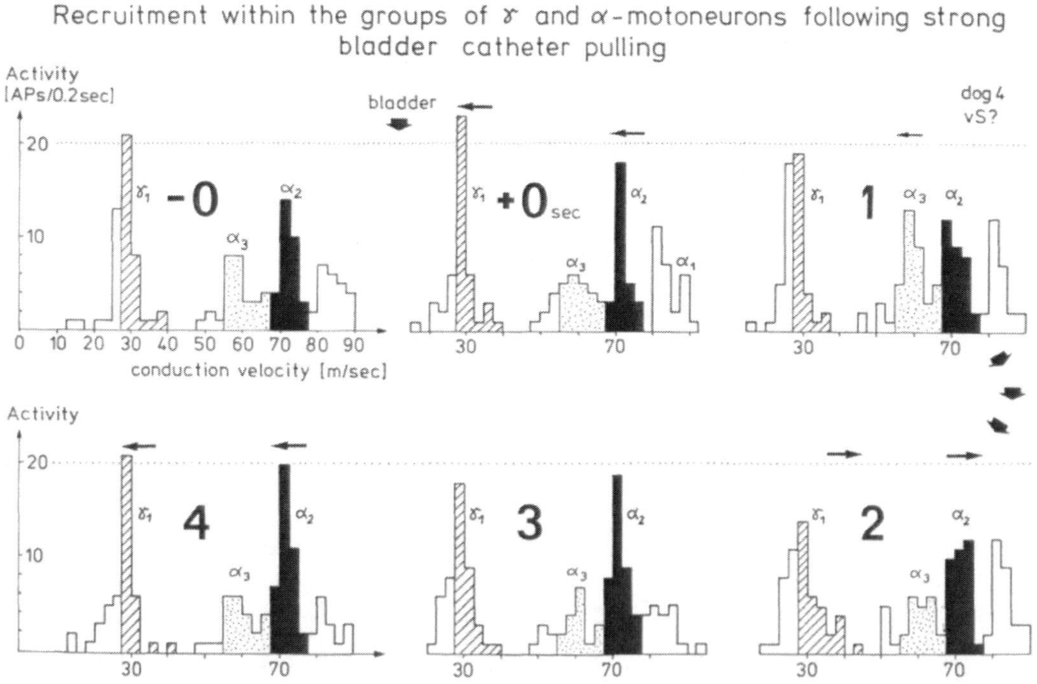

Figure 56. Recruitment of α and γ-motoneurons in a sacral ventral root of dog 4 following strong bladder catheter pulling. For further description, see legends to previous Figures. Note that α_2-motoneurons are recruited before α_3-motoneurons, and the co-recruitment of γ_1 and α_2-motoneurons.

9.1.5. Recruitment of Motoneurons in the Occasional Firing Mode Following Spinal Cord Injury - Pathologic Recruitment

Upon no additional stimulation in a paraplegic patient, slow and fast conducting motoneurons were recruited repeatedly within the groups of α_2 and α_3-motoneurons every 2.5s in similarity to the rather physiologic case (HT3, Figure 54). But the level of activation of the α and γ-motoneurons was higher in the paraplegic patient (Figure 57) than in the brain-dead human (Figure 54), even though the S4 roots (Figure 54) are expected to contain more fibres than an S5 root (Figure 57).

The higher activation of the neuronal networks of the spinal cord below the injury is in accordance with the pathologic over-activation of muscles (spasticity) in para- and tetraplegic SCI patients.

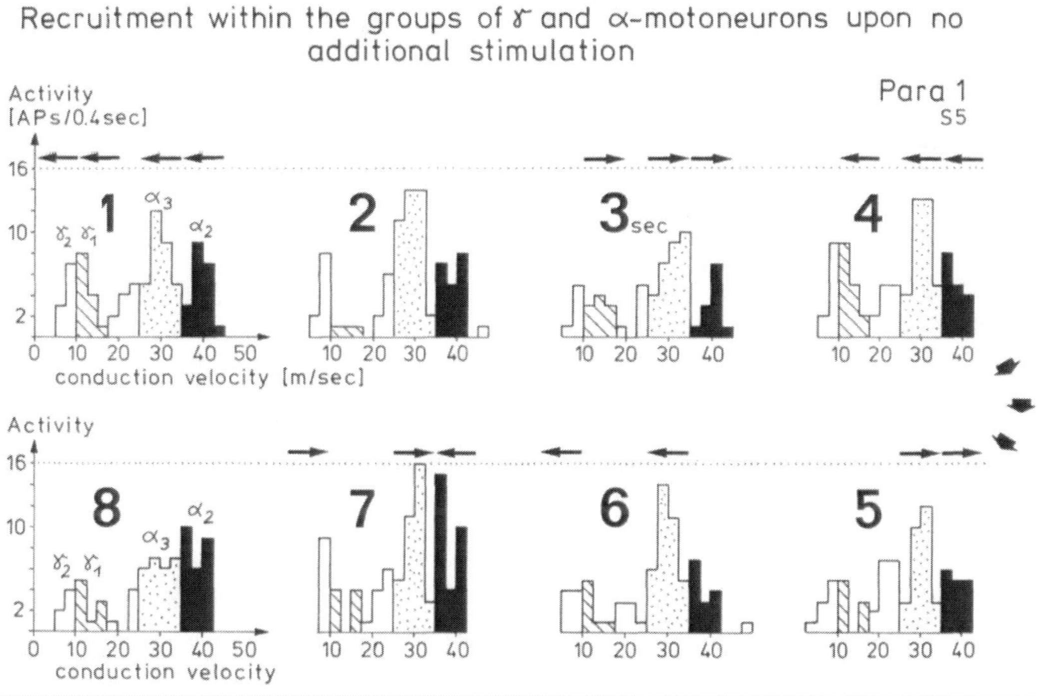

Figure 57. Frequency of occurrence changes of single-fibre conduction velocities in paraplegic 1 upon no stimulation, with the anal and bladder catheters positioned. The arrows on top of the histograms indicate stages at which α and γ-motoneurons in a certain group are preferentially activated with low or high conduction velocities. Note that there is mainly co-recruitment of γ_1, α_3 and α_2-motoneurons with rhythmicity.

Following anal catheter pulling, the slowly conducting α_3-motoneurons (S) were recruited directly following the stimulation, whereas the faster conducting α_2-motoneurons (FR) were recruited approx. 1s later (Figure 58).

The faster recruitment of α_3-motoneurons (more static) in comparison to the α_2-motoneurons (more dynamic) differed from the recruitment observed during measurements in brain-dead humans and in rats and dogs and was discussed with respect to the loss of interneurons in the spinal cord and is in accordance with the pathologic activation of muscles in paraplegia.

Bladder catheter pulling does not activate so many α_3-motoneurons (the external urethral sphincter mainly is mainly innervated by α_2-motoneurons) and is therefore not suitable for comparing the recruitment speed between α_2 and α_3-motoneurons.

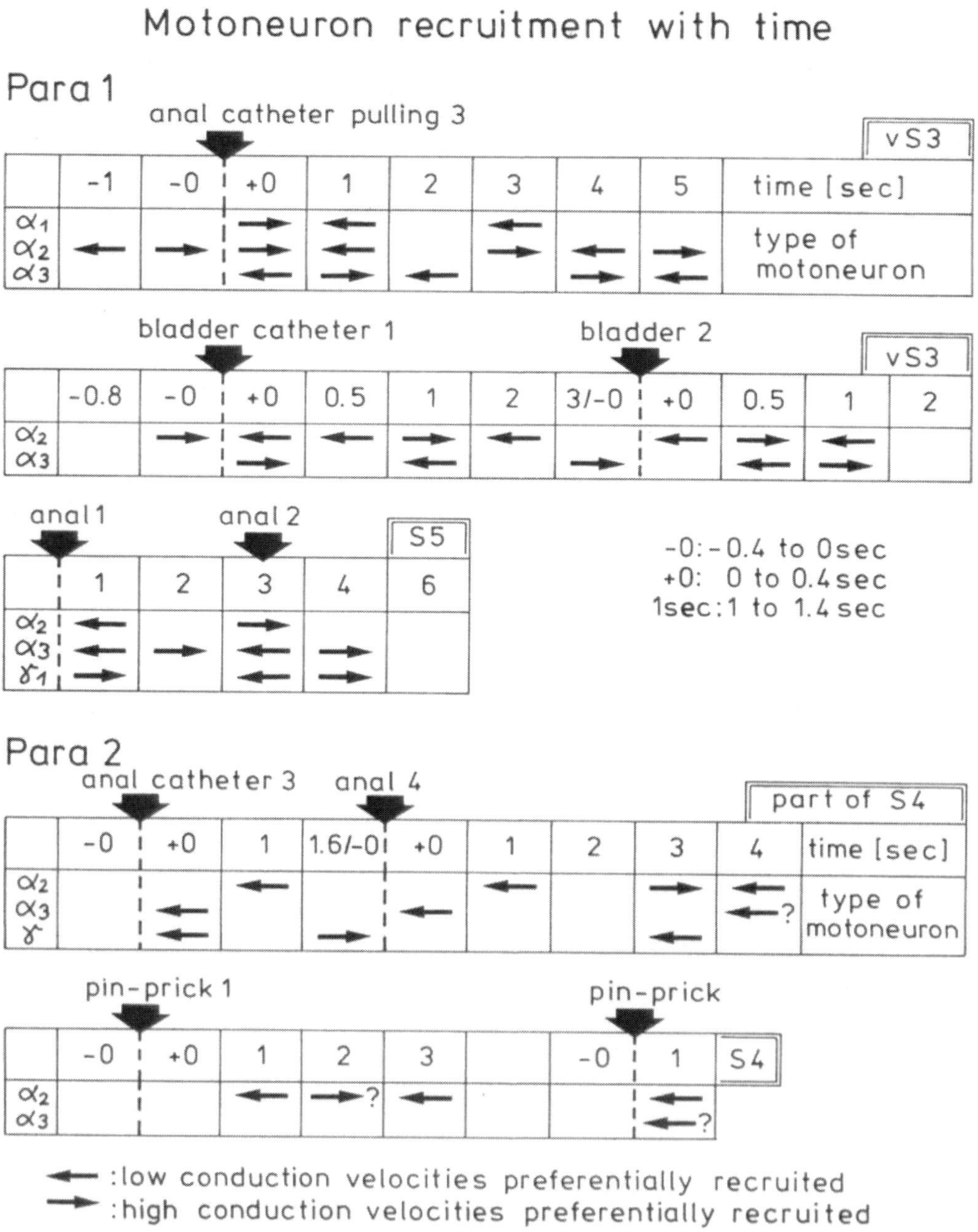

Figure 58. Summed recruitment changes in the occasional firing mode with ongoing time as measured in two paraplegics (1 and 2). The arrows direction indicates whether preferentially motoneurons with low or high axon conduction velocities are recruited (within the time interval of 0.4s) after a certain time. Para 1; Para 2.

9.1.6. Recruitment According to the Size Principle in Each Motoneuron Group

In an easy understanding, Hennemann's recruitment according to the size principle [9] can be understood as a special case of motoneuron recruitment (namely, recruitment among α

$_1$-motoneurons only) of the above-mentioned more general recruitment principle. Upon recording with EMG surface electrodes, e.g., from the musculus tibialis anterior of a healthy human individual, it can be seen that with low muscle activation, the single motor unit APs have a small amplitude, whereas with higher activation motor units with larger AP amplitudes are recruited. Since FF type motor units have much larger AP amplitudes than FR and S type motor units, one mainly recognizes the recruitment of FF type motor units. Taking the recruitment of motoneuron groups according to the size principle in each motoneuron group into account, the recruitment of motoneurons in the occasional firing mode is quite complex. In the occasional firing mode in man (low activity mode, when premotor spinal oscillators are not organised), the different α and γ-motoneuron groups have been shown to be recruited according to the conduction velocity (and axon diameter) in each nerve fibre group approx. every 2.5s. Directly following natural stimulation (touch, pinprick, anal and urinary bladder catheter pulling), the faster conducting nerve fibres (FF) are activated before slower conducting nerve fibres (S). But the activation order can change following spinal cord injury.

The recruitment of motoneuron axons according to the size principle in each motoneuron group seems to be in contradiction to the Hennemann's motor unit recruitment [9] according to the size, disregarding the motoneuron group.

When recording single motor unit action potentials (MUAPs) electromyographically from human leg muscles, it is obvious that with the increasing volitional muscle activation, on the average, the activated motor unit potential amplitudes increase in size. Since we can assume that the larger motor unit potentials will on the average be innervated by the larger sized motoneurons, this motor unit recruitment seems to be in contradiction to the motoneuron axon recruitment while in accordance with the Hennemann's size principle. Figure 27 shows that the motoneurons emerging from the intumescentia lumbosacralis are mostly of α_1-motoneuron type. Consequently, records from, e.g., most leg muscles are mainly records from α_1-motor units. In a first approximation, there is, therefore, no contradiction to the above motor axon recruitment if there is mainly only one motoneuron type innervating the different muscles. Further, since there is indication that the motor unit potential amplitudes decrease from the FF to the FR and S-type motor units, it is anyway more difficult to identify the FR and S-type motor unit potentials innervated by the α_2 and α_3-motoneurons. Potentials not fitting the size principle can be falsely excluded by claiming that the not-fitting small motor unit action potentials were probably generated too far from the recording electrodes to give a large potential (MUAP amplitude reduction due to volume conduction). Further, in the oscillatory firing mode, the very low amplitude α_3-motor units are difficult to identify, since close potentials start to fuse, because the duration of the interspike intervals of the impulse trains can be close to the range of those of the motor unit potentials ($\approx$ 5 ms).

The problem of the type identification is less difficult when recording single-nerve fibre action potentials, because the single-nerve fibre action potential durations are approximately shorter by a factor of 10 than the motor unit action potentials, and the action potential amplitudes may not decrease as much when changing from α_1 to α_3-motoneuron axons. For a comparison between single-nerve fibre action potentials and motor unit action potentials, see Figure 9 of Chapter II. The recording from a population of nerve fibres therefore provides more precise knowledge concerning the recruitment of motoneurons and in turn on the regulatory mechanisms of neuronal networks.

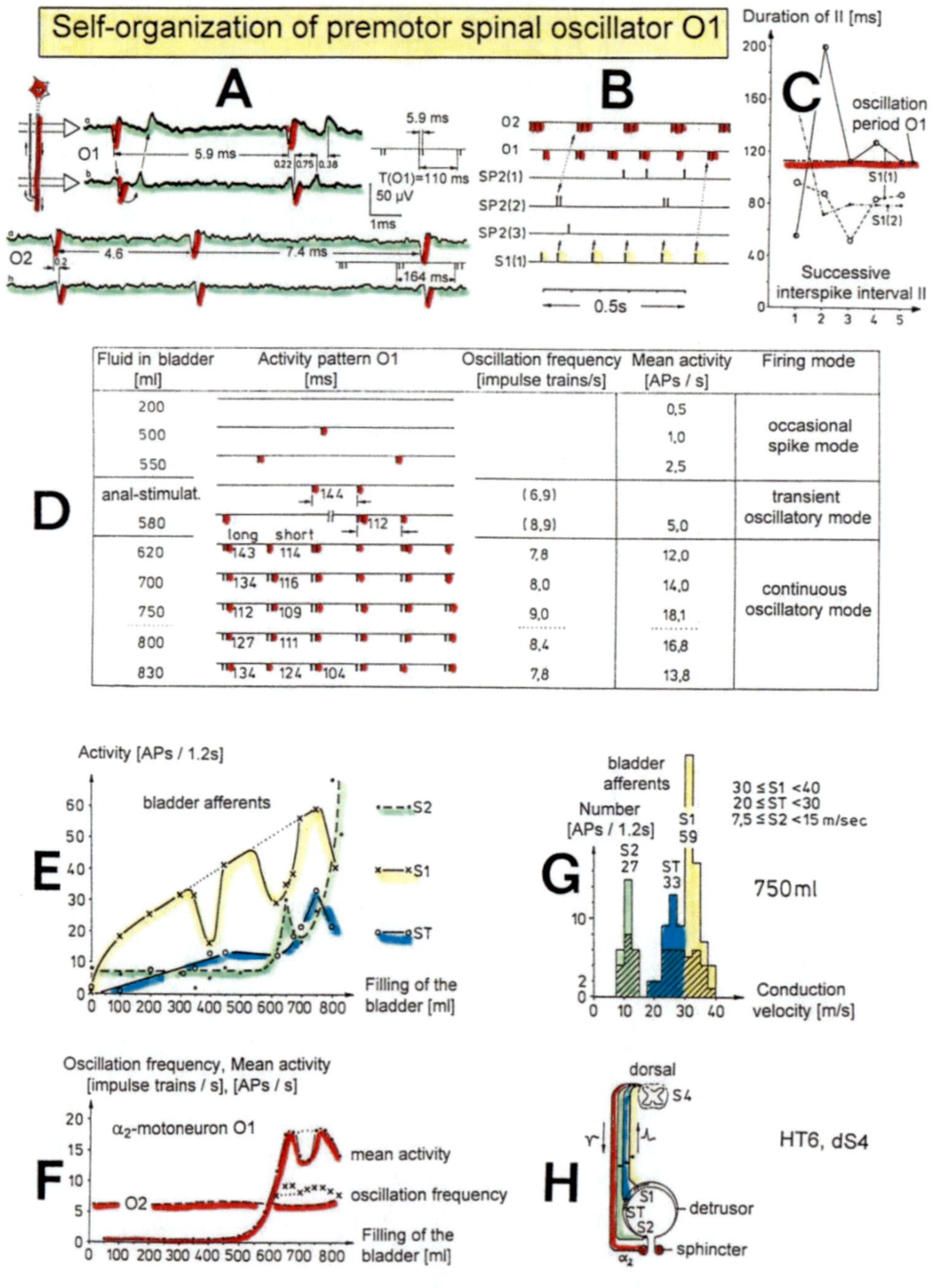

Figure 59. Self-organization of premotor spinal α_2-oscillator O1, which innervates the external urinary bladder sphincter (skeletal muscle). Brain-dead human HT6; recording from a dorsal S4 nerve root. A. Recordings from α_2-motoneurons O_1 and O_2, firing in the oscillatory mode with impulse trains of two (upper recording) and three (lower recording) action potentials (APs). The durations of the oscillation periods were 110 (O_1) and 164 ms (O_2). The interspike intervals of the impulse trains were 5.9 ms (O_1) and 4.6 and 7.4 ms (O_2). Motoneuron O_1 conducted at 36 m/s; its recurrent fibre conducted at 21 m/s. The measurement layout is shown schematically. The inserts show the oscillatory firing modes; they have not been drawn to scale. B. Impulse patterns of oscillatory firing α_2-motoneuron O_2 innervating the external anal sphincter, in relation to

the muscle spindle afferent activity SP2(1 to 3), activated by the stretch of the anal sphincter by the anal catheter, and impulse patterns of oscillatory firing α_2-motoneuron O1 innervating the external urethral sphincter, in relation to the stretch receptor afferent activity (S1(1)) of the urinary bladder, activated by 750 ml bladder filling. Phase relations between APs of SP2(2) and O_2 and between APs of S1(1) and O_1 are indicated by the small arrows. C. Three series of successive interspike intervals of the two stretch receptor afferent fibres S1(1) and S1(2) activated by retrograde bladder filling. The oscillation period of oscillatory firing motoneuron O1, activated only by bladder filling, is shown. D. The firing in the occasional spike mode, the transient and the constant oscillatory firing mode of α_2-motoneuron O_1 in response to filling of the bladder. In the "activity pattern" column, changing durations of oscillation periods are given. The oscillation frequencies in the brackets give the frequencies at the moment of oscillation for the transient oscillatory mode. Downward deflections are schematized APs. Interspike intervals of the close APs $\approx$ 6.0 ms (A). E. Activity levels of stretch (S1) and tension (ST) and flow receptor afferents (S2) (E) and of sphincter α_2-motoneuron O_1 (F) in response to retrograde filling of the bladder. The activity values of the S1, ST and S2 afferents are taken from histograms like the one in G. Filling of the bladder was stopped once between 600 and 650 ml. F. The small dotted lines represent mean activity (APs/s) and oscillation frequency (impulse trains/s) of α_2-motoneuron O_1 if bladder filling were not stopped in between. Note that the mean activity increases continuously with the filling of the bladder from 550 to 650 ml, even though motoneuron O_1 started to fire in the oscillatory mode from 620 ml on (D). Note further that the oscillatory firing motoneuron O_2 (frequency of firing with impulse trains is shown) is nearly not affected by the filling of the bladder and by the start of the oscillatory firing of motoneuron O_1. G. Conduction velocity frequency distribution histogram of stretch, tension and flow receptor afferent activity at 750 ml. The activities of afferents S1, ST and S2 are quantified by counting the afferent conduction velocities under the peaks (open plus hatched part), with the conduction velocity limits given in the insert. The counts (27, 33, 59) are given below the peak labelled S1, ST and S2 and plotted into E for the afferent activity at 750 ml. H. Schematic drawing of the anatomical arrangement of the afferents and the motoneuron O_1.

The main disadvantages of recording from nerve fibres include the invasive nature of the method and difficulties of getting an opportunity for recording. On the other hand, the electromyographic recording of motor unit action potentials with surface electrodes is simple to perform and presents practically no risks. It seems therefore worthwhile to try to improve the interpretation of motor unit action potentials recorded with surface electrodes.

9.2. Self-Organization of a Premotor Spinal Oscillator

The self-organization of a premotor spinal α_2-oscillator innervating the external striated urinary bladder sphincter is shown in Figure 59D.

Because the motoneuron axon of O1 had a recurrent fibre (A) at the recording site, each single AP of this motoneuron could be identified safely by the AP of the recurrent fibre. The function of the motoneuron was to secure bladder continence. The activity from urinary bladder receptors, i.e., the activity of bladder stretch (S1), tension (ST) and flow receptor afferents (S2) (Figure 59E), was an adequate afferent input to the motoneuron. For retrograde bladder filling up to 550 ml, motoneuron O1 only fired occasionally (D,F). This was the storage phase of the bladder, during which the intravesical bladder pressure increased only little. For higher bladder filling volumes, the motoneuron switched via the transient oscillatory firing mode to the continuous oscillatory firing mode (D) to generate a higher activity (F) for a stronger drive of the urethral sphincter to more strongly secure continence when the storage phase was nearly passed and the bladder pressure increased more strongly. For bladder filling volumes higher than 800 ml, the activity of the spinal oscillator decreased again; probably, the oscillator became inhibited (D). Pain fibres (not shown in Figure 59G) may have inhibited the oscillatory firing to protect the bladder from mechanical damage. The overflow mechanism was started. Probably, fluid entered the trigonum vesicae to activate flow receptors, so that the flow receptor activity (S2) increased strongly (E). The premotor spinal oscillator O1, of which motoneuron O1 is probably a part [186], was organized by the adequate afferent input (S1, ST, S2) induced by bladder filling. The oscillator to be formed in the premotor neuronal network in the spinal cord, consisting of motoneurons and interneurons, needs a certain preformation of the networks (connectivity, synapse efficacies, membrane properties of neurites) and adequate afferent input patterns. In modelling such networks, the organization of the premotor spinal oscillators in the spinal cord neuronal networks cannot be separated from its space-time distributed adequate afferent input patterns, giving rise to self-organization. From Figure 59B, it can be seen that the stretch receptor afferent APs (S1(1)) show a relative phase correlation to the impulse trains of the oscillatory firing motoneuron O1. The successive interspike intervals (IIs) of the S1(1) afferents have, on average, a value similar to that of the oscillation period of oscillator O1 (Figure 59C).

Other premotor spinal oscillators, activating the external striated bladder sphincter not shown here, will also have self-organized themselves by the same or similar afferent input. However, these oscillators subserving the same function will be correlated in their firing in the way that they do not fire in synchrony but distributed in their phases with respect to one another (probably by relative inhibition with respect to each other) to secure that the sphincter muscle does not show rhythmic movements (tremor). Such non-synchrony correlated oscillatory firing of several oscillators has in a small time window been measured (186) (Figure 71). In the case of a non-lesioned CNS, this premotor spinal oscillator O1 could also be activated volitionally from supraspinal centres. It would be very interesting to see what the afferent input patterns from supraspinal centres to the spinal premotor network would look like. By recording from the dorsal S4 root of the brain-dead human HT6, impulse trains from another oscillatory firing motoneuron and its driving afferents were measured, which served quite a different function (Figure 59F). The α_2-oscillator O2 (Figure 59A,B) innervating the striated external anal sphincter was activated by secondary muscle spindle afferent activity, induced by the anal catheter-stretched muscle spindles, probably located in the anal sphincter or functionally associated pelvic floor muscles. Also, mucosal and skin receptors within the anal reflex area will induce self-organization of premotor oscillators activating the external anal sphincter to secure anal continence. The self-organization and drive of this spinal oscillator upon adequate afferent input from muscle spindles and skin receptors will be tackled below. It is evident from the impulse patterns shown schematically in Figure 59B that the impulse trains of this oscillator O2 show a phase relation and an interspike interval relation to its driving spindle afferent APs.

The self-organization of the spinal oscillators in the premotor neuronal network cannot be separated from its adequate forming and driving afferent impulse patterns. It is not clear whether during phylogenies and ontogenesis, the CNS neuronal networks adapted to the impulse patterns of the receptors or whether the receptors adapted their firing to the network organization or both. But it would be far from reality to model self-organization of the human spinal cord (and more generally the CNS) neuronal networks without taking into account the space-time distributed adequate afferent impulse patterns from the periphery and the supraspinal centres, giving rise to self-organization. Before going into further details of the self-organization of premotor spinal oscillators, some data will be presented on the simultaneous firing of motoneurons in the occasional and oscillatory firing mode.

9.3. Simultaneous Firing in the Occasionally and Oscillatory Firing Mode

If motoneurons fire occasionally, transiently oscillatory or continuously oscillatory in dependence on their activation (Figure 59D), then in a population of motoneurons, one should observe all the three firing modes. Figure 60 (and Figures 62-64) schematically show how all the three firing modes of the motoneurons interact with each other. The data were obtained from a recording from a paraplegic patient. The α_2 and α_3-motoneurons are recruited repeatedly every ≈ 0.4 Hz according to the size principle in each motoneuron group. One α_3-motoneuron and one α_2-motoneuron were firing continuously oscillatory.

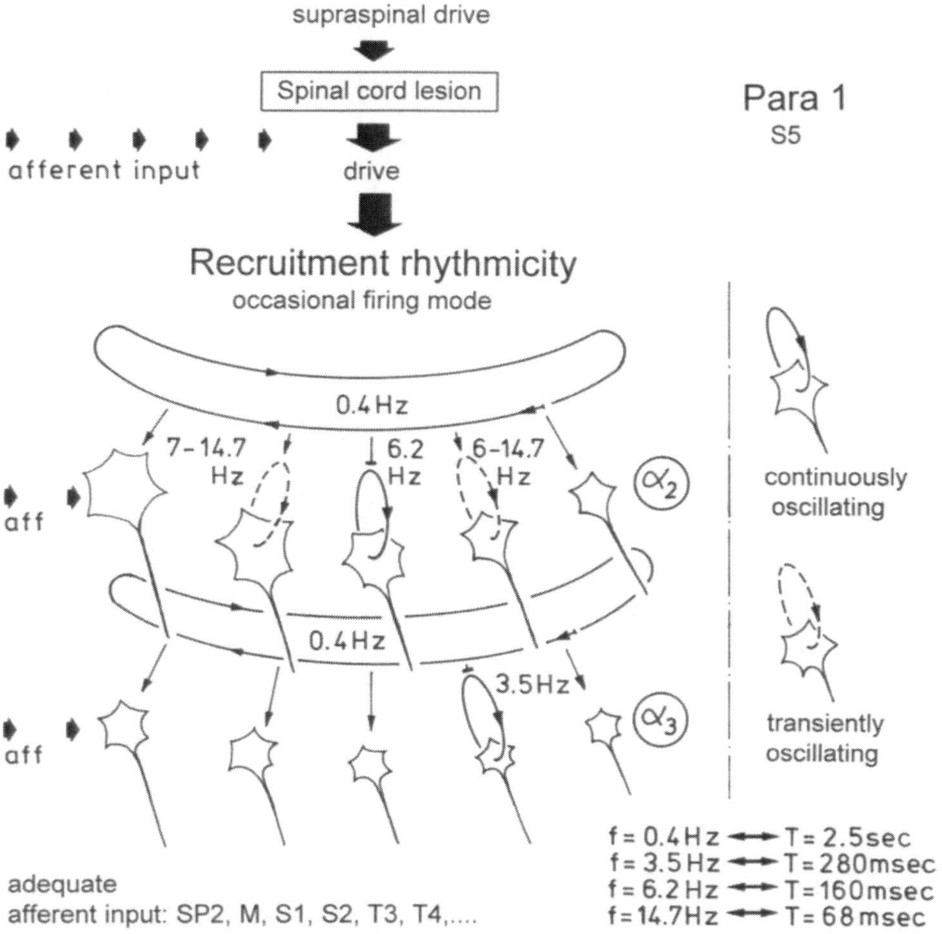

Figure 60. Recruitment rhythmicity of firing, drawn for α_2 and α_3-motoneuron axons, in an S5 nerve root in paraplegic 1. The changes in γ-motoneuron recruitment are not shown; neither the co-recruitment of γ_1 or α_2-motoneurons is indicated. The arrows pointing towards the motoneuron somas indicate recruitment; no arrow = no recruitment. SP2 = secondary muscle spindle afferents (affs), M = mucosal affs, P = pain affs, S1 = bladder stretch receptor affs, S2 = flow receptor affs, T3 = low threshold skin affs.

Two further α_2-motoneurons fired transiently oscillatory. It is unclear how the firings of the motoneurons in the occasional and oscillatory firing mode were correlated. By splitting the summed impulse traffic (Figure 61), it was found that very approx. 50% of the motoneurons in the occasional firing mode fired in relative coordination with the oscillatory firing motoneurons (they fired in relative coordination with the spinal oscillators), and the remaining 50% of motoneurons seemed to fire with no correlation to the spinal oscillators. This can be understood in the way that the motoneurons and the interneurons contributing to a network organization to generate a certain function fire relatively coordinately, and that the neurons not firing in relative coordination with the spinal oscillators are not or only contributing little to the organization of that particular function.

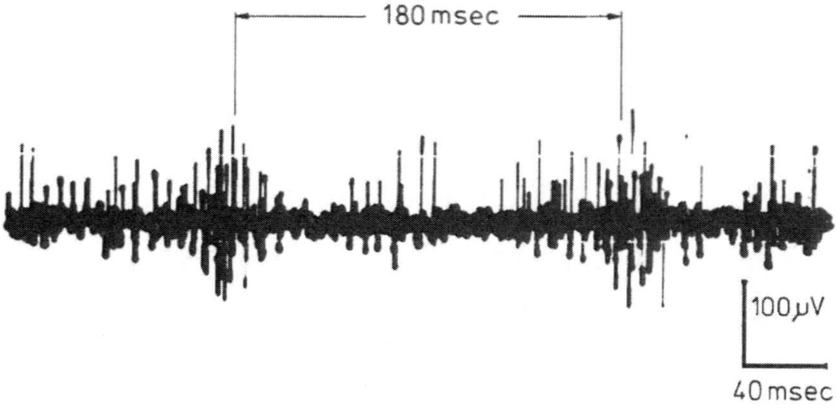

Figure 61. Overall efferent activity recorded from an S5 dorsal root 20s after pulling the bladder and anal catheters. The rhythmic activity interval is marked by a 180 ms distance. Brain-dead HT4.

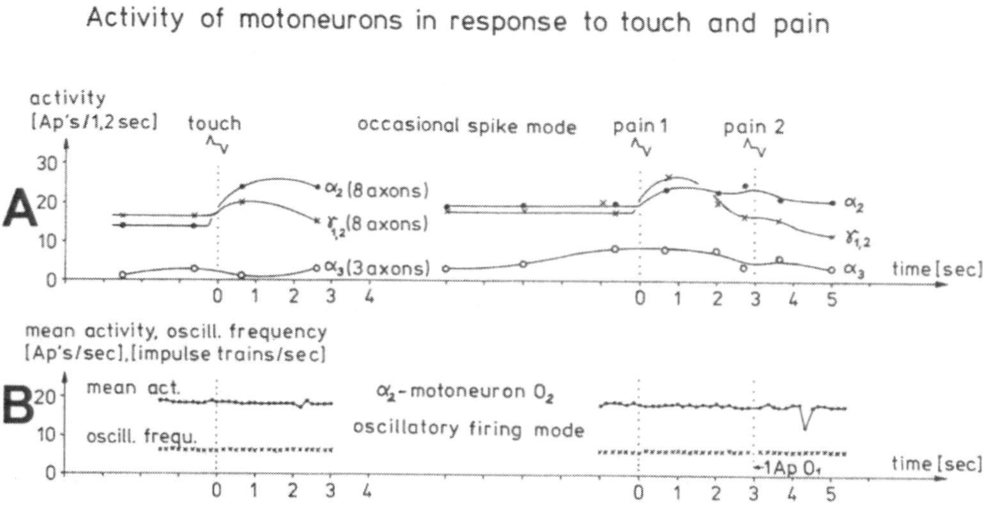

Figure 62. Activity of occasionally active α_2, α_3 and γ-motoneurons (A) and simultaneously measured mean activity and oscillation frequency of α_2-motoneuron O_2 (B) in response to touching the lower sacral dermatomes (outside the anal reflex area) with a ball or pricking them with a pin (pain). The activity values have been obtained from conduction velocity distribution histograms. HT6; dS4.

The self-organization of a certain function in the neuronal networks therefore occurs through the neurons in the subnetwork to generate this function becoming more activated and correlated with each other in response to the adequate afferent input. The motoneurons most strongly contributing to a certain function (contraction of the external anal or bladder sphincters to secure continence) fire oscillatory, motoneurons contributing less strongly to the generation of this function or pattern fire occasionally but correlated to the oscillators, and the motoneurons firing occasionally with seemingly no phase correlation to the firing of the oscillators contribute little if at all to that function (pattern). If the recruitment according to the size principle is given by the different thresholds of the motoneurons (the smaller motoneurons have a lower threshold), then the drive potential to activate the motoneurons in the occasional firing mode is changing periodically (in this case at 0.4 Hz). It is interesting to know how this rhythmicity is correlated to the rhythms of the self-organized premotor spinal oscillators. Figures 62 through 64 show some data on the simultaneous firing of motoneurons in the occasional and oscillatory firing mode (174).

Figure 62A shows the response of occasionally firing α and γ-motoneurons to touch and pain applied to sacral dermatomes outside the anal reflex area. While the α_2-motoneurons and γ-motoneurons showed a transient activity increase for about two seconds, the α_3-motoneurons with their polymodal afferent drive (see below) did not respond specifically. Also, oscillatory firing α_3-motoneurons (Figure 75) showed little dynamic response. The one or two α_1-motoneurons that may be running through the dorsal S4 root were not activated. Mostly α_1-motoneurons do not contribute to continence function. The very constantly oscillatory firing α_2-motoneuron O2 (Figure 62B), innervating the external anal sphincter and activated by the constant drive of muscle spindle afferents and anal mucosa afferents induced by the positioned anal catheter, showed nearly no response to the touch. Upon pinpricking, the α_2-motoneuron O2 fired further continuously oscillatory but switched once from firing with three AP impulse trains to two AP firing and backwards. The α_2-motoneuron O1, activated later on to fire oscillatory upon urinary bladder filling (Figure 59), was not firing because of no bladder filling. Upon the pain application, motoneuron O1 fired once with 1AP (Figure 62B). Upon pulling the bladder catheter moderately or strongly (Figure 63A,B), the occasionally firing α_2-motoneurons increased their activity transiently (for approx. 3s) (A). The α_3-motoneurons may have shown a bit of transient activity increase. The response of the γ-motoneurons seemed to be a mixture of excitation and inhibition. Following strong bladder catheter pulling (bladder 3), the prolonged response was due to the activation of also static γ-motoneurons ($\gamma_1 < \gamma_2$). The continuously oscillatory firing α_2-motoneuron O2, innervating the anal sphincter, responded more strongly to bladder catheter pulling (Figure 63B) than to touch and pinprick (Figure 62). Because in most (but not all) people the urinary bladder function and the anal canal function inhibit each other, the strong bladder catheter pulling probably induced inhibition of the oscillatory firing α_2-motoneuron O2, so that the motoneuron reduced its activity by firing alternating with two APs and three APs and then even stopped oscillatory firing for one second. The α_2-motoneuron O1 fired once with one AP when the motoneuron O2 stopped firing (reciprocal response). Figure 63C illustrates an efferent conduction velocity distribution histogram based on the above measurements. For a comparable low activity, it is important to distinguish between the activity of occasionally firing motoneurons and that of oscillatory firing motoneurons (hatched horizontally, Figure

63C). Best would be to count the conduction velocity of each motoneuron in the summed impulse traffic of a nerve root only once (crosshatched in C), which is sometimes roughly possible. For a high activation of nerve fibres, the separation of the oscillatory activity is not so important since the time interval within which the velocities are collected is short. Anyhow, for a detailed analysis of the afferent and efferent impulse traffic in a nerve root, including the identification of the nerve fibres also from conduction velocity distributions, one has to be careful when counting APs of the individual nerve fibres. Namely, if a fibre has a higher activity than another one and its APs are therefore counted several times, the velocity distribution histogram (used to identify the fibre types) can be strongly biased.

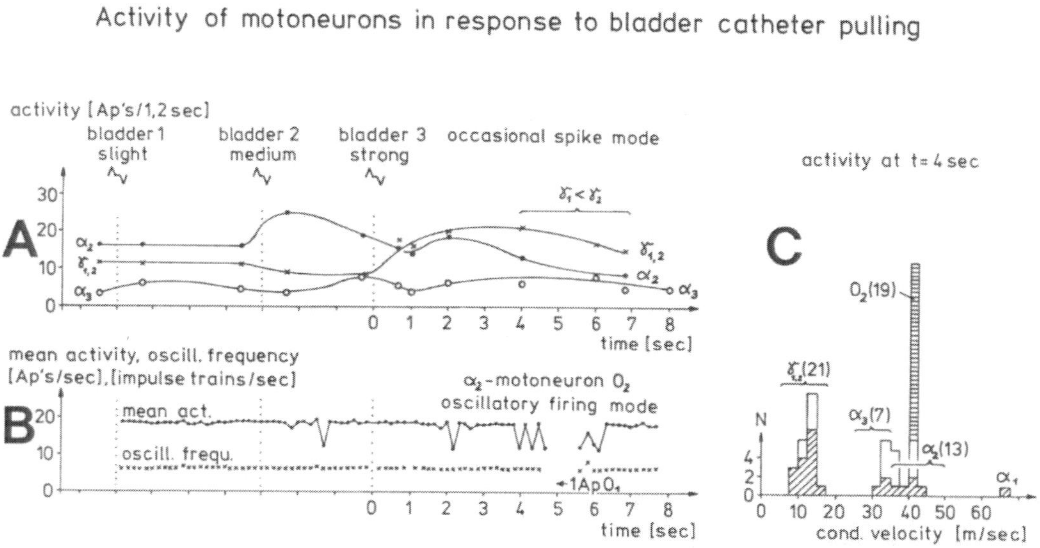

Figure 63. A, B. Activity of occasionally firing α_2, α_3, γ_1 and γ_2-motoneurons (A) and mean activity and oscillation frequency of oscillatory firing α_2-motoneuron O_2 (B) in response to pulling the bladder catheter slightly, moderately and strongly. Note that in the case of a strong bladder catheter pulling, the increased γ-activity lasted longer than the α_2-motoneuron activity, and this prolonged activity increase came from the γ_2-motoneurons (the higher activity of γ_2-motoneurons in comparison to the activity of γ_1 is indicated as $\gamma_2 > \gamma_1$). "1 AP O_1" in B marks the appearance of one AP of sphincter motoneuron O_1. The α_2-motoneuron O_2 stopped transiently firing oscillatory 5s after strong bladder catheter pulling (B). C. Conduction velocity distribution histogram of motoneuron activities at t = 4s. Distribution peaks and ranges of single motoneuron types are indicated. The Figures in the round brackets give the activities counted from open plus crosshatched histogram parts (all conduction velocity values are used). The horizontally hatched part of the α_2-motoneuron peak indicates the activity from oscillatory firing α_2-motoneuron O_2. The approximate number of activated motoneurons ($\alpha_2 = 8$, $\alpha_3 = 3$, $\gamma_{1,2} = 8$) was calculated from means of the crosshatched parts of the histograms (each conduction velocity value only taken once). HT6; dS4.

A comparison with the components of compound APs may help (see Figure 35) to identify group conduction velocities in critical cases, because with electrical stimulation, each fibre will be excited only once (apart from rebound APs). Upon anal catheter pulling (Figure 64), the occasionally firing α_2-motoneurons increased transiently their firing, the γ-motoneurons decreased their firing, and the α_3-motoneurons showed no specific response. The oscillatory firing α_2-motoneuron O2 was strongly affected; it changed the impulse train length, the oscillation period and stopped oscillating several times. Obviously, the α_2-

motoneuron O2 was activated, and it may have been over activated (inhibited) by its adequate afferent input from anal canal receptors. The α_2-motoneuron O1 fired once transiently oscillatory. It is unclear whether all motoneurons can fire oscillatory. In records from ventral sacral roots, often many motoneuron axons fired oscillatory. Then, the problem arose that it was difficult to extract the firing patterns of single fibres because the template recognition was not safe and the summed rhythmic activity became too complicated for a separation on the basis of rhythmic firing patterns of single motor axons.

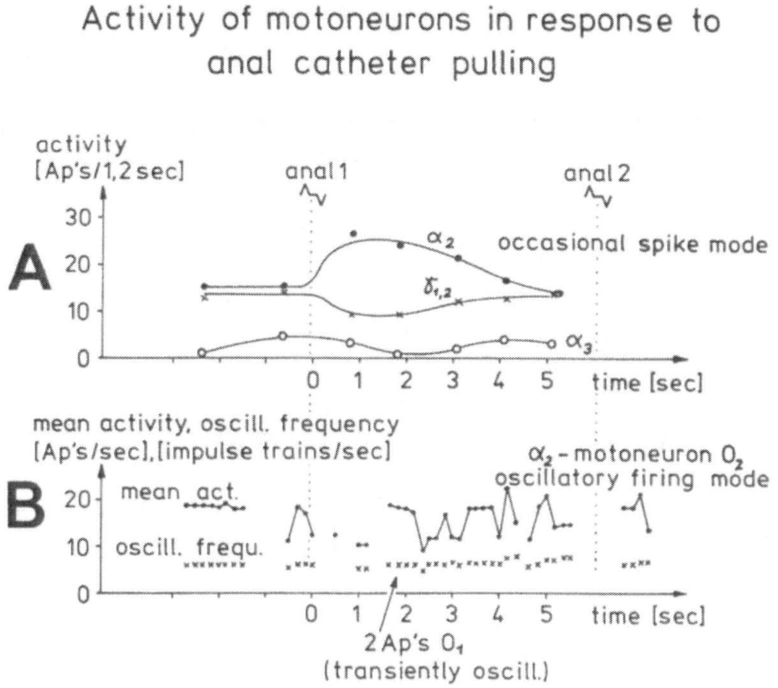

Figure 64. Activity of occasionally firing α_2, α_3 and γ-motoneurons (A) and mean activity and oscillation frequency of α_2-motoneuron O_2 (B) in response to pulling the anal catheter, marked "anal 1." "2 APs O_1" mark transient oscillatory firing of α_2-motoneuron O_1, innervating the external bladder sphincter. The α_2-motoneuron O_2 stopped firing oscillatory several times (B). HT6; dS4.

9.4. Mixed Functions in Dog Sacral Nerve Roots

It has been shown that the conduction velocity distribution histogram for dog efferents is very similar to those of humans, apart from the higher velocity values (Figure 53). It should therefore also be possible to set up a precise classification scheme for the dog peripheral nervous system in analogy to the one in humans. As has further been shown (Figures 55,56), the recruitment according to the size principle in each motoneuron group can also nicely be measured. The drawback with respect to rat, cat and dog recordings from nerve roots is that the anatomy of the spinal canal of those animals is different from that of man. Because of the ascensus of the human spinal cord, and as man has no tail, the continence functions are mainly represented in the lower sacral nerve roots only, and they are not mixed with tail functions. In the dog (and other animals), functions are mixed in the sacral nerve roots. Even though recruitment can nicely be measured in the dog, functions could not be identified so far in the measurements performed in the same way as in man (Figure 2 of Ref.182). Animal research certainly allows the use of additional invasive methods to split the functions conducted in the roots.

9.5. Need for Improved Classification Schemes of the Peripheral Nervous System of Animals

An improved classification and identification of peripheral nerve fibres in the frog, rat, cat, dog, pig and monkey would be of great value, so as to enable a direct comparison of the human data with those of animals, to make use of the animal data in human research, and thus also make patients benefit from animal research. In Chapter IV (Translational Medicine), single-nerve fibre action potentials will be recorded in frog and rat peripheral nerves to enable a comparison to humans, since the rat is often used for regeneration experiments, and the frog is a very good animal to demonstrate to medical students nervous system functioning. The classification schemes for man of Grundfest, Erlanger and Gasser, and Lloyd and Hunt [66, 100, 103, 132], and the measurements of Boyd and Davey [21] on peripheral nerve fibres were brilliant for their times but are insufficient for the present needs of measuring natural impulse patterns to get information concerning the self-organization of neuronal networks. In physics and mathematics, the basis for the resolution of new problems keeps improving. Often, the coordinate system is adjusted to the specific problem to be solved. It is important to have a precise classification of peripheral nerve fibres by group conduction velocities, group nerve fibre diameters and impulse patterns as it would enable extraction of natural impulse patterns of several identified afferent and efferent fibres from the summed impulse traffic in nerve roots or nerves [179], so that the self-organization of CNS neuronal networks (at least in the spinal cord) can be analyzed under physiologic and pathophysiologic conditions.

9.6. Correlation between the Activity of γ-Motoneurons and Secondary Muscle Spindle Afferents

So far, the oscillatory firing of γ-motoneurons has not been observed under physiologic conditions. Following spinal cord injury, the firing of the γ-motoneurons can become transiently rhythmic upon pinprick stimulation (see Chapter V). Since α and γ-motoneurons are partly co-recruited in the occasional firing mode, it could be that the firing of some γ-motoneurons becomes more rhythmic if the α-motoneurons fire oscillatory, and the α and γ-motoneurons are co-recruited for a certain function. The γ-motoneurons drive muscle spindle afferents via the muscle spindles, and a time lag of 400 ms has been measured upon natural stimulation (Figure 65), [174] in accordance with the γ-loop scheme. It will be shown below by the impulse patterns of single fibres and cross-correlations that α and γ-motoneurons and secondary muscle spindle afferents can have similar interspike intervals and similar phases in their firing (if harmonics are included), so that premotor spinal oscillators can build up an external oscillator loop (via the γ-loop) to the periphery. But we have to remember that the muscle spindles are included in a network of γ-motoneurons and spindle afferent connections: one γ-fibre may innervate several muscle spindles, and one spindle afferent fibre will project onto many α-motoneurons.

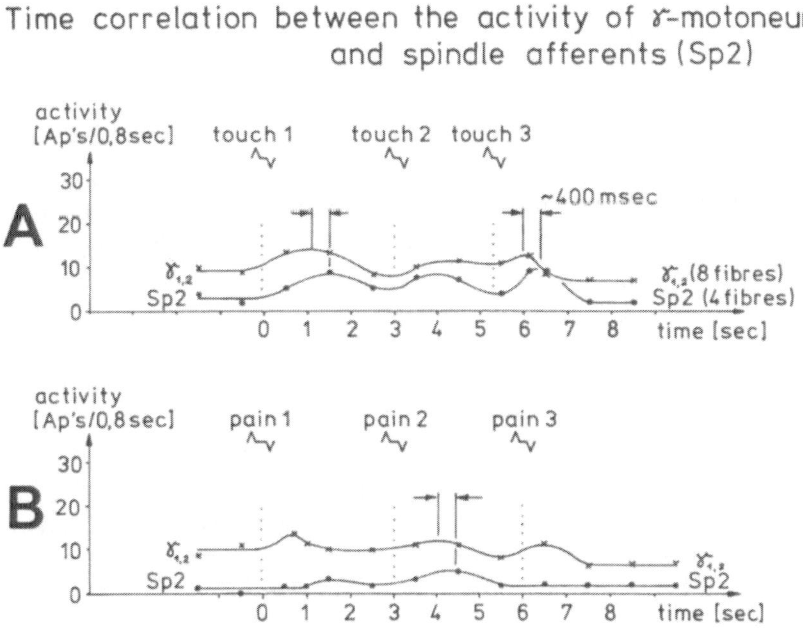

Figure 65. Time correlation between the activity of γ-motoneurons and secondary muscle spindle afferents in response to touch (A), marked "touch 1, 2, 3," and pricking with a pin (B), marked "pain 1, 2, 3," lower sacral dermatomes. The delay of the spindle afferent activity is indicated as 400 ms. HT6; dS4.

A single muscle spindle afferent fibre may even have several encoding sites [187] (see below). It is therefore likely that the firings of many α and γ-motoneurons and spindle afferent fibres are relatively coordinated, as can nicely be seen when α-motoneurons fire oscillatory.

10. Premotor Spinal α_1, α_2 and α_3-Oscillators - Oscillatory Firing of Single Human Sphincteric α_2 and α_3-Motoneurons for the Continence of Urinary Bladder and Rectum

Summary

1. By recording with two pairs of wire electrodes from human sacral nerve roots (S3-S5), rhythmic as well as occasional firing was observed in α_1, α_2 and α_3-motoneurons in response to physiologic stimulation of the urinary bladder and the anal canal. The

rhythmic firing consisted of periodically occurring impulse trains, most likely produced by true spinal oscillators which drove the motoneurons.

2. α_2-Motoneurons, innervating fast fatigue-resistant muscle fibres, were observed to fire with impulse trains of about two to four action potentials (APs). These impulse trains occurred every 110 to 170ms (5-9 Hz). α_3-Motoneurons, innervating slow fatigue-resistant muscle fibres, fired approximately every 1400ms (~ 0,7 Hz), with impulse trains of about 11 to 60 APs. α_1-Motoneurons, innervating fast fatigue muscle fibres, fired with one AP approximately every 0.1s (10Hz). γ-Motoneurons were not observed to fire continuously in the oscillatory firing mode.
3. Sphincter motoneurons were observed most likely in the oscillatory firing mode in response to the sustained stretch (reflex) of the external anal sphincter or to retrograde filling of the bladder (urethro-sphincter guarding reflex), in order to preserve continence. A urethral sphincter α_2-motoneuron increased its mean activity from 0.5 to 18 APs/s during retrograde filling by changing its firing pattern from the occasional spike mode via the transient oscillatory firing mode to the continuous oscillatory mode. Up to a filling of the bladder of 500 ml, the mean activity of the stretch receptors, measuring probably mural tension, increased roughly proportionally, and the sphincter motoneuron increased its activity to about one AP/s in the occasional spike mode. Up to 600 ml, the motoneuron responded in the transient oscillatory mode with mean activities of up to five APs/s. With higher bladder fillings, the flow receptors afferents fired additionally, probably according to pressure symptoms, and the motoneuron switched into the continuous oscillatory firing mode and increased its activity up to 18 APs/s at 700 ml. When the bladder was about 800ml full, the stretch afferent activity decreased, the flow receptor activity increased strongly and the α_2-motoneuron activity decreased; the overflow incontinence had probably started. Micturition was not observed, probably because of brain death.
4. It is suggested that one adequate stimulus for an α_2-motoneuron of the external anal sphincter to jump into the oscillatory firing mode was the activity from secondary spindle afferent (SP2) fibres from external anal sphincter muscle spindles. The interspike intervals (IIs) of the SP2-fibres were often similar to the length of the oscillation period or to the half of it. Two adequate stimuli of an α_2-motoneuron of the external urethral sphincter for switching into the high activity mode of oscillatory firing was the increased activity from stretch and flow receptors. The drive was in relative coordination. Similarities in the IIs of the afferents and the oscillation period of the motoneuron were not found until now. A clear adequate stimulus of a sphincter α_3-motoneuron to switch into the oscillatory firing mode was not found. α_1-Motoneurons do not substantially contribute to continence functions. They innervate fast fatigue muscle fibres of skeletal muscles and are activated volitionally and by primary muscle spindle afferents (SP1). The drive of the SP1-fibres is in absolute coordination. The α_1 oscillators have little oscillatory property in comparison to α_2 and α_3-oscillators. By surface EMG, coordinated oscillatory firing was found among several α_1-motor units in the same skeletal muscle and among different muscles.
5. The neuronal networks that drive sacral motoneurons in the oscillatory firing mode lie in the spinal cord. These spinal oscillators oscillated independently of each other but were coordinated. α_2 And α_3-motoneuron showed quite a large dynamic range in

response to different afferent inputs, with changes in the oscillation frequency and the number of APs per impulse train in order to decrease or increase the mean activity. α_1-Motoneurons changed the firing activity only by changing the frequency of firing. The oscillatory firing of α_2-motoneurons was quite stable in brain-dead humans with respect to oscillation frequency, APs per impulse train and length of the IIs in the impulse train.

6. In the impulse trains, the IIs increased in the range from 3.5 ms up to about 15 ms for α_2-motoneuron and α_3-motoneuron; 3.5ms is the known shortest soma-dendrite II in rats and cats. Longer impulse trains had shorter first IIs and longer last IIs. Typical IIs of impulse trains of three APs of α_2-motoneuron were 4.6 and 7.4 ms long. In comparison with animal data, it was found that the length of the impulse train mainly depended on the amplitude of the depolarization produced by long current pulses and not on the length of the depolarization. The strength of the depolarization is already manifested in the length of the first II. With increasing depolarization, the length of the impulse train increases and the length of the first II decreases. From further comparison with animal data, it is suggested that the oscillation cycle of a motoneuron, consisting of the impulse train part and followed by an inactive part, is produced by a drive potential, consisting of a short depolarizing part from excitatory postsynaptic potentials and followed by a longer hyperpolarizing part from inhibitory postsynaptic potentials.
7. It is proposed that the drive potential in a motoneuron is produced by exciting and inhibiting impulse trains from different oscillatory interneurons of a true spinal oscillator.

The external urinary bladder and anal sphincters are mainly innervated by α_2-motoneuron and probably by a few α_3-motoneurons.

10.1. Introduction to Oscillatory Firing of Motoneurons and Rhythmic Firing of Neurons in Cats

So far, the firing of motoneurons in the occasional firing mode was mainly considered. The demonstration that neurons of the CNS (here motoneurons) can fire oscillatory and non-oscillatory (occasionally) (Figure 59) is very important for the understanding of the functioning of the human CNS. To describe the functioning of the CNS merely by reflex pathways and loops or coupling of rigid oscillators (of cellular or network origin) is in contradiction to hard human data, namely that premotor spinal oscillators self-organize as was concluded from measurements of simultaneous natural impulse patterns of afferent and efferent fibres. In what follows, we shall concentrate on the oscillatory firing of motoneurons, which takes place for high activation. The analysis of this oscillatory firing mode will give more understanding about the range of activity of motoneurons and will give the membrane property of firing repetitively [9, 90] in human a physiologic meaning. But first, animal data will be presented, which makes oscillatory firing human motoneurons more understandable.

Hodgkin [90] classified the motoneurons in liligo into three groups with respect to the repetitive activity of membranes following intracellular applied current pulses. In the first group, the motoneurons showed strong repetitive activity. With long near threshold rectangular current pulses, the motoneurons responded with long impulse trains, whose interspike intervals and impulse train lengths, depending on the depolarization strength, are strikingly similar to the properties of impulse trains obtained from α_3-motoneurons when the anal canal and the urethra are constantly stretched. The α_3-motoneurons innervate muscle fibres that are responsible for posture, weight carrying of the lower abdomen and perhaps the continuous low level activity of the external anal sphincter. Hodgkin's second group contained motoneurons whose membranes showed reduced repetitive activity. These motoneurons probably correspond to the α_2-motoneurons, who fire with short impulse trains of about three action potentials (APs). As will be shown below, these α_2-motoneurons respond more specifically than the α_3-motoneurons, and guard in external (striated) sphincters the continence of rectum and urinary bladder. Hodgkin's third group contained motoneurons, which showed no or only little repetitive activity, which probably correspond to the α_1-motoneurons (FF-type) from which no repetitive activity has been measured so far. But there were only a few α_1-motoneurons in the lower sacral nerve roots. It may be that α_1-motoneurons also show some repetitive activity, especially the thinner, more slowly conducting ones, belonging to the F (int) subgroup [121], called α_{11} in Figure 2. But thicker, faster conducting α_1-motoneurons may not fire repetitively [144]. The correlation between the strength of the repetitive activity, the oscillation frequency and the mean activity for the different α-motoneurons shown in human here indicates that α_1, α_2 and α_3-motoneurons do not only drive their own muscle fibres (FF, FR, S-type), but that they are also rostrally driven by their own central neuronal network (Figure 3) to obtain their own activity pattern and level.

10.2. Oscillatory Firing of Spinal Interneurons

McMahon and Morrison [136] recorded with tungsten electrodes from sacral spinal interneurons of cats that responded to pelvic and pudendal nerve stimulation. The interneurons had receptive fields in bladder, colon, perineum and adjacent structures and often showed duration of the inhibitory response of about 500ms. These interneurons showed various patterns of responses following nerve stimulation, namely excitation only, excitation followed by inhibition followed by excitation and excitation followed by secondary excitation after a time interval. The time courses and the properties of these interneurons are compatible with oscillations. If, for example, the group of interneurons, responding to a single stimulation with excitation, inhibition and excitation, were to get a new stimulus at an appropriate time, the alternation of excitation and inhibition would continue. But these interneurons must not be the principle components of the spinal oscillators. Functionally, nearby interneurons may resemble the oscillator cells in some properties. In the medicinal leech, it was found that the basic swimming rhythm was generated mainly by rings of a few small oscillatory interneurons [61], which oscillated with impulse trains. The fundamental principle to which the network owed its oscillatory character was the recurrent cyclic inhibition of the neuronal loop [198]. The cycle period of the basic rhythm had durations in the range of 400 to 2000ms, and the time required for each interneuron to recover from inhibition was between about 30 and 150ms [61]. Even though rhythms can be generated by several kinds of mechanisms [216], it is assumed as a working hypothesis that the true spinal oscillators consist of an ensemble of spinal interneurons, which are activated or inhibited by peripheral and/or central input. In the HT measurements, the afferent input stimulated the oscillators; the central inhibition was maybe reduced. A network of synaptic connections links the sets of oscillatory interneurons to the motoneurons commanding the sphincters and functionally associated muscle groups (Figure 69).

10.3. Three Types of Human Premotor Spinal Oscillators

For high and rather constant afferent input, it was found that α-motoneurons fire repeatedly with impulse trains according to their type (Figure 3). The α_1-motoneurons (FF) fire rhythmically at around 10 Hz (range 8 to 12 and higher) with an impulse train consisting of one AP; α_2-motoneurons (FR) fire at approx. 6 to 9 Hz with two to five APs per impulse train, and α_3-motoneurons (S) fire with a frequency in the range of 1 Hz and with long impulse trains consisting of up to 40 APs (and more) (Figure 3) [176, 186]. The rhythmic firing patterns of α-motoneurons are generated by local neuronal networks of the spinal cord since oscillatory firing can be recorded from motoneurons of the disconnected spinal cord.

10.4. The Motoneuron is a Part of the Spinal Oscillator

There is indication that the motoneuron is a part of the spinal oscillator [186]. It seems that the information on the oscillation of stable oscillators is contained in the length of the first interspike interval (II) of the impulse train and also in the oscillation period.

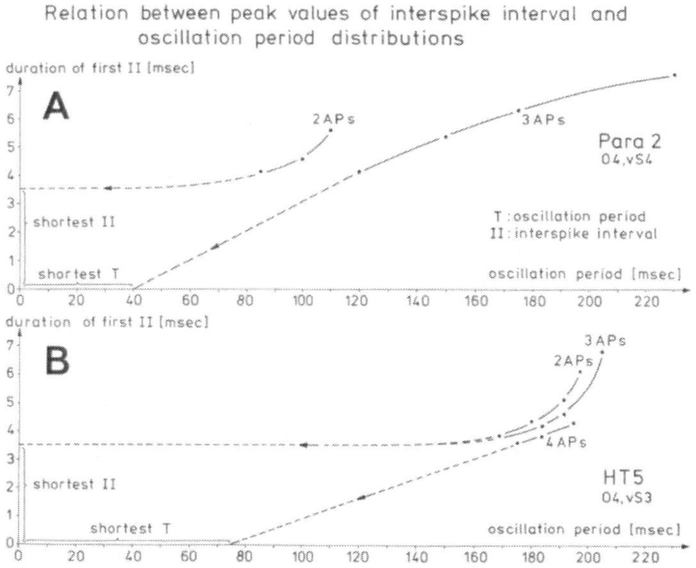

Figure 66. Relations between corresponding peak values of interspike interval and oscillation period frequency distributions for the α_2-motoneuron O4 in paraplegic 2 (A) and O4 in HT5 (B). The relation points are obtained from Figures 89, 90. Lines connect the dots to show trends. The dashed lines are extrapolations into the unphysiologic range to obtain characteristics constants of the oscillation. The constants on the ordinate have the same value as the shortest soma-dendritic spike interval and those on the abscissa are values of the oscillation period for impulse trains of zero APs (Figure 67).

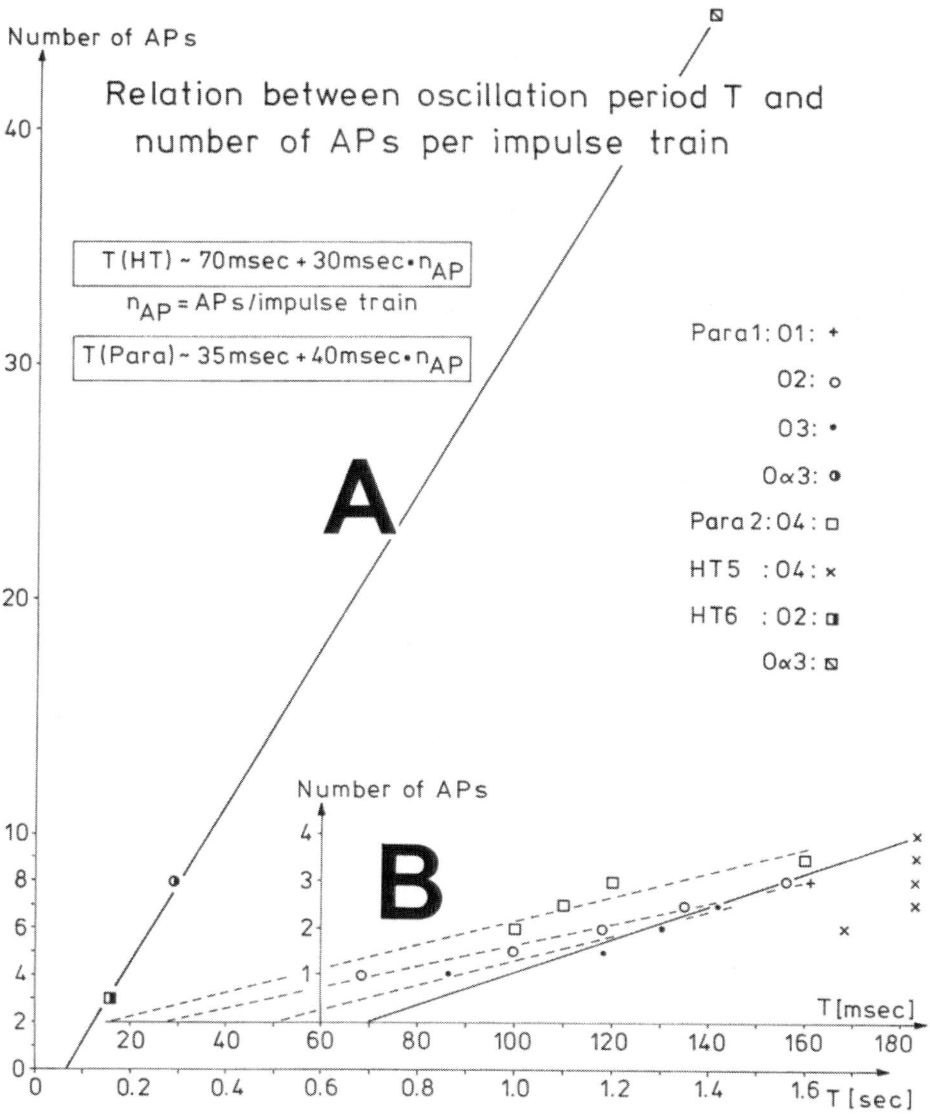

Figure 67. A. Relation between the oscillation period (T) and the number of action potentials (APs) per impulse train of stable continuously oscillatory firing α_2 and α_3-motoneurons. The straight line can be described by the relation T(HT) = 70ms + 30ms • n_{AP}. B. Expanded scales. The solid line is the straight line from A. The dashed lines characterize the relationship between the number of APs and the oscillation period for the oscillatory firing systems of the paraplegics where the motoneuron fired with different impulse train length. T(Para) describes the different linear relations. Note that the unstable oscillatory firing α_2-motoneuron in HT5 behaves differently.

With the information on the oscillation in both distributions, it is tried to estimate the constants of oscillation by relating the peaks of the distributions from the first II to those of the oscillation period. Figure 66 shows the correlation between the peaks of the distributions and the oscillation period for the oscillator in Para 2 (Figure 66A) and brain-dead human HT5 (Figure 66B). By extrapolating the curves for different AP impulse train lengths to the coordinate axes, one obtains the shortest II and the shortest oscillation period. Approximately 3.5ms are obtained for the shortest II from the oscillation itself in Para 2 and HT5 (Figure 66).

The shortest oscillation period for Para 2 is about 40ms, and for HT5 very approximately 70ms. The shortest II of 3.5 ms, which is approximately the expected shortest soma-dendritic spike interval of motoneurons (no information is available about differences between different α-motoneurons) is taken as an indication for the motoneuron itself being an important part of the spinal oscillator. The values of shortest oscillation period of 40ms and 70ms obtained from the oscillation itself for Para 2 and HT5, respectively, are approximately the same as those obtained from different oscillators (Figure 67).

10.5. Relation between Oscillation Period and Number of APs Per Impulse Train in HTs (Rather Physiologic)

Since more often in HTs than in paraplegics the motoneuron fired in the high activity mode with a certain number of APs per impulse train and a more fixed oscillation period, stability seems to be closer to physiologic conditions. The most stable oscillators were taken to plot the relation between the number of APs and the oscillation period T (Figure 67A). As can be seen, the relationship is a straight line not crossing the origin. The approximate equation derived for this linear relation is: Oscillation period = 70ms + 30ms • (number of APs per impulse train). For rather normally firing α_2 and α_3-motoneurons, this means that the oscillation period increases by 30ms if the number of APs increases by one AP per impulse train. Since in addition another interval of 70ms is always needed for a complete oscillation period, the linear relationship is interpreted by the oscillator consisting of at least two kinds of elements: one element always present and contributing 70ms to the oscillation period and a set of other elements that contribute according to their numbers, a multiple of 30ms to the oscillation period. Assuming that the oscillator consists of the motoneuron itself and of time-determining interneurons, each one responsible for a time consumption of 30ms, the motoneuron would yield an oscillation period of 70ms + 30ms (10kHz). In conclusion, the oscillation period (T) of a rather physiologic premotor spinal network oscillator is roughly related to the number of action potentials (APs) per impulse train (n_{AP}), and this can be expressed by the formula: $T = 70ms + 30ms \bullet n_{AP}$. A typical premotor α_2-oscillator fires with three APs every 160 ms ($T = 70$ ms + 30 ms • 3 = 160 ms), and can change its firing pattern to two APs every 130 ms for less activation or to four APs every 190 ms for higher activation (Figure 67).

10.6. Relation between Oscillation Period and Number of APs Per Impulse Train in Paraplegics (Pathologic)

Since there was a tendency for stable α_2-oscillator to oscillate with a certain number of APs per impulse train (e.g., 3) and a rather fixed oscillation period (e.g., 160ms), it might be interesting to plot the oscillation period against the number of APs per impulse train for the situation where the oscillator changes the oscillation period and the number of APs per impulse train for different stimulations. Variations of first IIs of the impulse trains and of the oscillation period (frequency) occurred on average more often in paraplegics than in HTs. It is followed that changing first IIs (for a fixed number of APs per impulse train) and frequencies is an indication for more extensive pathologic oscillatory firing. Figure 67B (insert) shows (the dashed lines) the relationship between the oscillation period and the number of APs per impulse train for the unstable, more pathologic, oscillators. Averaging the three dashed lines gives, for these more pathologic cases, an oscillation period of approximately 35ms plus 40ms times the number of APs per impulse train. On the background of the previously drawn oscillation picture, this would mean that the contribution to the oscillation period of the motoneuron and the time consuming interneurons have changed. As can be seen from Figure 67B, the most unstable oscillator measured in HTs (crosses) showed variability different from those of unstable oscillators in the paraplegics. Generally, the variability in the duration of the IIs was larger in the patients with spinal cord injury than in HTs. Further, it seems as if the shortest II (of about 3.5ms) of α_2 and α_3-motoneurns were slightly reduced in paraplegics. In rats [77] and cats [211], the shortest soma-dendrite spike interval was between 3.4 and 3.5ms. Judged by the shortest II of the impulse train humans, the shortest soma-dendrite spike interval also seems to be 3.5. The on–the-average slightly reduced shortest soma-dendrite spike intervals could mean that the properties of the soma and the dendrite trees of the α-motoneurons have slightly changed following spinal cord injury.

10.7. Hypothetical Structures of Premotor Spinal Oscillators

It is assumed as a simplified working hypothesis that these spinal oscillators are organized by excitatory reverberatory loops of interconnected interneurons in the form of synfire chains [186] to which inhibitory interneurons essentially contribute because their inhibition time, 30 ms (rat) and 150 ms (cat) [151], falls within the range of the 30 ms building time block of the oscillation period and the oscillation period itself (160 ms) (Figure 68). Premotor α_1-oscillators show little rhythmicity generated by small networks, whereas α_3-oscillators show much rhythmicity generated by a larger local subnetwork [186]. The oscillator loops consisting of chains of neurons (Figure 68) are therefore only schematized network loops. This network loop seems to be rather continuous for α_3-oscillators, whereas the α_2-oscillator loop may show rather discrete subloops (see below) accounting for the firing with different impulse trains length (Figure 68) and therefore different loop periods according to $T = 70ms + 30ms \bullet n_{AP}$. The network loop of an α_1-oscillator is more in the direction of a single neuron chain loop. The measurements give no direct information about the number of neurons contributing substantially to the self-organization of the spinal oscillators. But since α_2-motoneurons show multi-frequency oscillation and α_3-motoneurons show a rather continuous change of the frequency (indicating rather complex local neuronal networks), one can speculate that the number of neurons contributing substantially to the oscillation may vary between a few (α_1-oscillators) and a few hundred (α_3-oscillators). The characteristic of α_2-oscillators to be able to change their organization when activated additionally by the parasympathetic nervous system division (the number of driving phases changes from two to three) (Chapter V) points towards complex local oscillators consisting of many interneurons.

Another not so likely premotor spinal oscillator consists of a ring of oscillatory firing interneurons (see oscillatory firing interneurons, above) driving the motoneuron, as in Figure 69. But there are also some animal measurements supporting this model, especially the oscillatory drive potential gives some understanding.

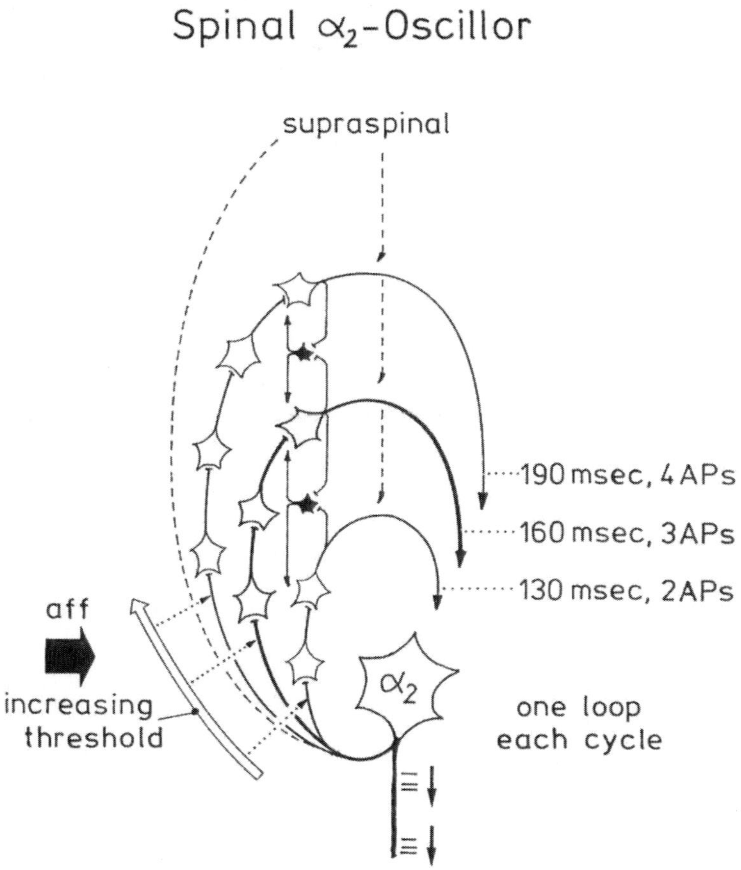

Figure 68. Working hypothesis of principle circuitries of the spinal α_2-oscillator. $\alpha_2 = \alpha_2$-motoneuron soma; open cell somas = interneurons; filled somas = inhibitory interneurons (two are indicated) for lateral field inhibition (unclear whether they work pre- or post-synaptically). Three loops with sets of interneurons are indicated, up to six are measured. Threshold arrow indicates higher thresholds for longer loops with respect to the adequate afferent input. I all thresholds are too high, so that no loop is opened (self-organized), the α_2-motoneuron fires in the occasional firing mode. If working hypothesis would be right, still many more interneurons are necessary to realise the measured interspike interval and oscillation period distributions. For example, the interloop interaction is not indicated, which becomes important if the afferent input changes quickly; more inhibiting interneurons are necessary; the preference for the three AP loop is not explained by the model; the function of the oscillator is not known: Oscillates the α_2-motoneuron in itself (impulse train) and is transiently inhibited by the interneurons (time between the impulse trains) or oscillates the α_2-motoneuron together with the interneurons. One loop each cycle = probably the excitation uses in a first approximation one pathway for an oscillation cycle, but which pathway is taken depends on the probability distribution for a certain afferent input.

It is possible that the motoneurons oscillate with a few interneurons among themselves in a pool of α_2 and α_3-motoneurns with specific functions via recurrent facilitation and inhibition [85, 162, 242], as has been observed in gastric and pyloric functions of the lobster [216]. Recurrent collaterals are likely to be of greater importance for the tonic than for the phasic activity [74, 75]. The frequency range of the operation of recurrent inhibition (5 to 40 shocks/s [76], cats) is similar to the mean activity range of α_2 and α_3-motoneurns (5 to 40 APs/s, see below). For an α_2-motoneuron, a recurrent collateral could even be found in the ventral root (Figure 105). However, it has been reported that the cat sphincteric motoneurons lack Renshaw inhibition [134] and crossed disynaptic inhibition [111].

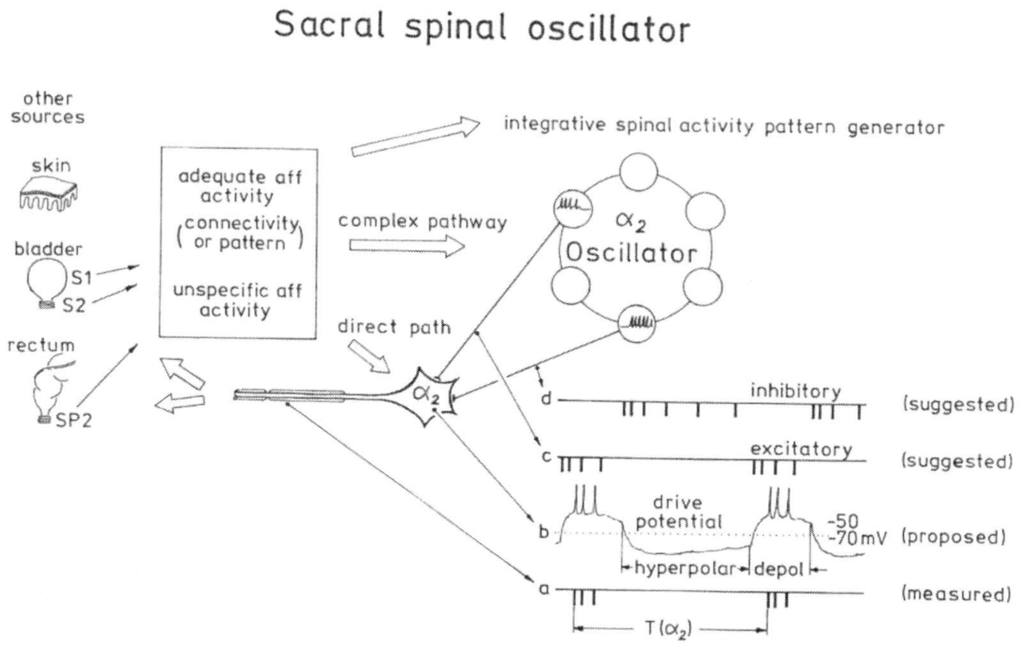

Figure 69. Schematic drawing of a proposed spinal α_2-oscillator with its afferent input and efferent output, not drawn to scale. On the left side are the possible sources from which the S1, S2 and SP2 afferents are found as possible sources for contributing specifically to the stimulation of the oscillation. The afferent input (adequate and unspecific afferent activity) can go three ways to the motoneuron: direct pathway (mono- or disynaptically), complex pathways (multisynaptically) and even more complex pathways. The α_2-oscillator consists of an ensemble of interneurons, each oscillating at a different phase (indicated in two interneurons by different bursts), and drives the α_2-motoneuron via excitatory ("c") and inhibitory ("d") pathways of unknown number. The suggested extracellular derived oscillatory firing patterns are shown in traces "c" and "d." Trace "b" shows the proposed intracellularly derived oscillatory drive potential with three APs in each depolarizing part, marked with depol, and APs in hyperpolarizing part marked with hyperpolar. Note, the AP amplitudes are too small; they should reach +30mV! Trace "a" shows the actual extracellular measured oscillatory firing mode at one site; $T(\alpha_2)$ = oscillation cycle period of the α_2-motoneuron. The axon of the α_2-motoneuron innervates one of the striated sphincters as indicated by the arrows.

From this point of view, it is likely that mainly spinal interneurons oscillate among themselves and drive with their oscillation in the high activity mode α_2 and α_3-motoneurns (Figure 69), innervating striated sphincters and functionally associated perineal and pelvic floor muscles.

The oscillation in these measurements has always been observed in connection with the repetitive activity. Oscillation with repetitive activity has been recorded extracellularly from inspiratory and expiratory intercostal motoneurons [42, 209] (cat). Intracellular "central respiratory drive potential" could be recorded, which gave rise to the repetitively occurring impulse trains [210, 212].

The drive potential had a depolarizing and a hyperpolarizing part, of which the depolarization stimulated the repetitive discharges. The impulse trains are probably produced by depolarizing drive potentials, which have similarity to the ones produced by rectangular current pulses (see Chapter I).

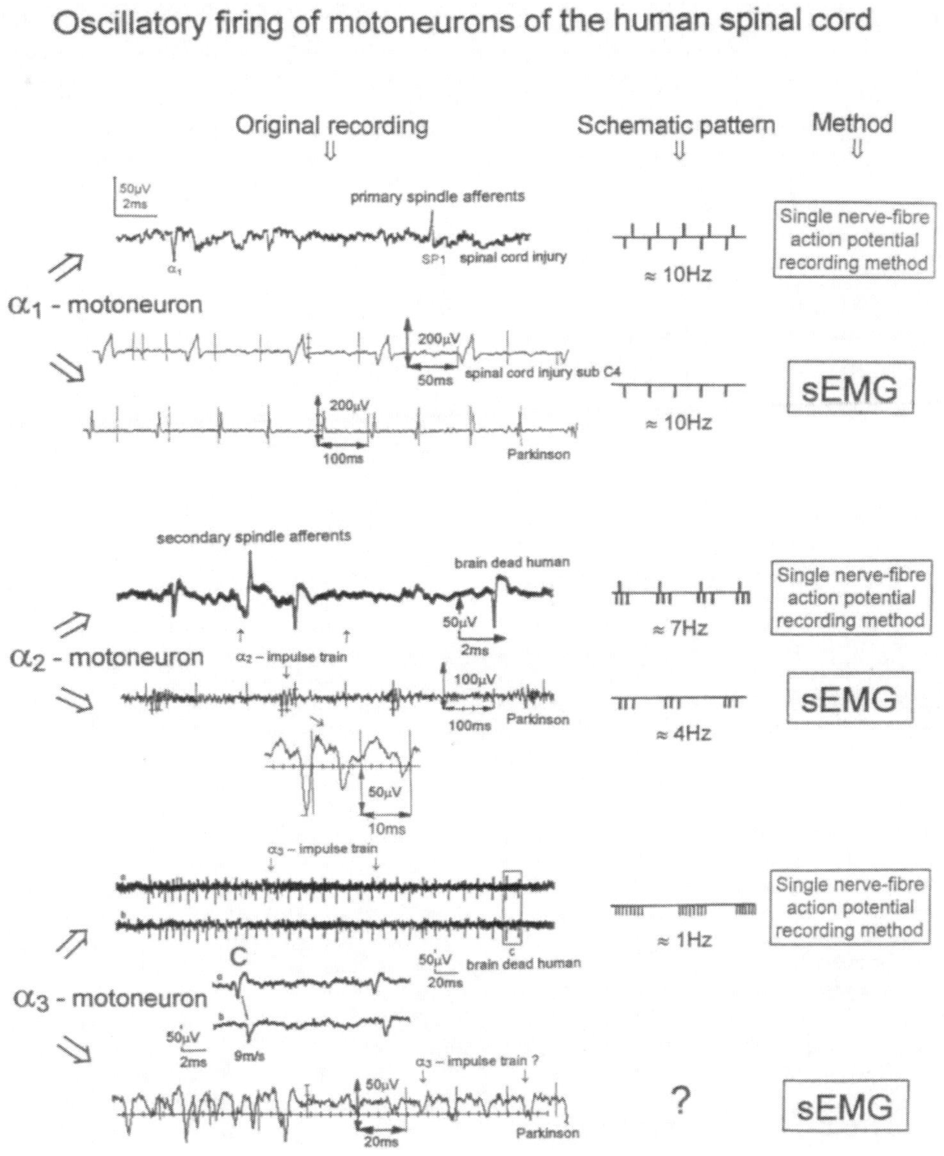

Figure 70. Oscillatory firing patterns of α_1, α_2, and α_3-motoneurons recorded from motoneuron axons with the single-nerve fibre action potential recording method and by surface electromyography (sEMG) from FF, FR, and S-type motor units. The left panel shows original recordings, the middle panel the schematic patterns; the recording methods are indicated on the right side. The recordings were taken from patients with spinal cord injury and Parkinson's disease and from brain-dead humans.

This depolarizing potential will most likely be followed by a hyperpolarizing potential to fully activate the membrane again after the long depolarization (Figure 69). By recording from the pudendal motor nerves of cats, when the afferent fibres of the pelvic or contralateral pudendal nerves were stimulated, it was found that the initial excitatory phase of pudendal nerve activity was followed by a second inhibitory phase of between 150 and 2500ms duration [137]. Intracellular drive potentials could be recorded from pudendal α_2-motoneurns [30], which consisted of a short depolarizing part from excitatory post-synaptic potentials followed by a hyperpolarizing part from inhibitory post-synaptic potentials.

The time course of this drive potential is similar to the time course of the oscillation cycle period, consisting of the repetitive activity part and the silent part of α_2-motoneurns. It is concluded that the oscillatory drive potential of one cycle giving rise to the impulse train followed by an inactive part consists of a depolarizing part from excitatory, post-synaptic potentials and hyperpolarizing part from inhibitory post-synaptic potentials (Figure 69, trace "b"). In humans, the motoneurons fire oscillatory with impulse trains for high activation as has been measured with the single-nerve fibre action potential recording method (see below) and with surface Electromyography (sEMG) (Figure 70). Even though the ring oscillator (Figure 69) seems to be supported by many measurements in animals, the oscillator with the reverberatory loops seem to be more likely for humans, especially because the motoneurons are a part of the oscillator. The three different types of premotor spinal oscillators will be analyzed now in detail.

10.8. Coordinated Firing of Premotor Spinal Oscillators

The rhythmically firing of the network oscillators is mainly coordinated (Figure 71), similarly as the whole neuronal network of the CNS, by the organization tendencies of the network, the descending impulse patterns and the spatiotemporal afferent impulse patterns (see below).

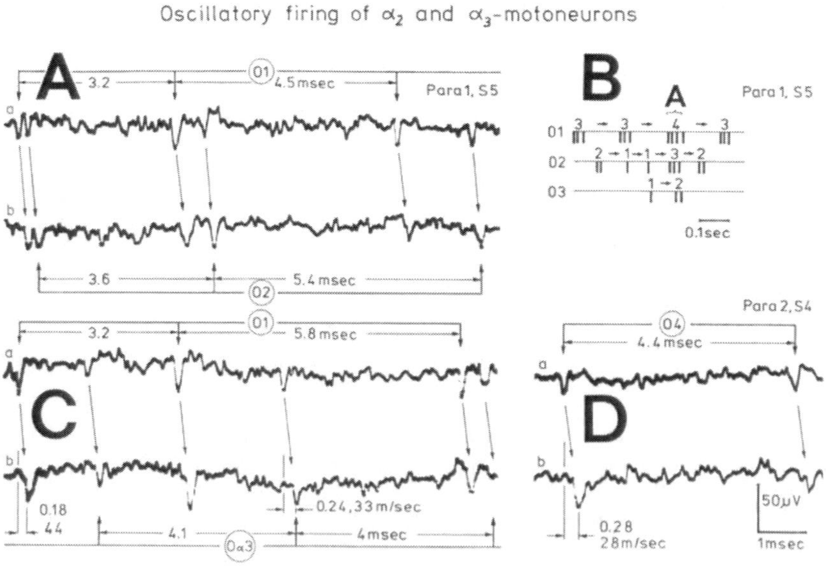

Figure 71. Recordings of impulse trains of oscillatory firing motoneurons in paraplegic 1 and 2. A. Impulse train of the continuously oscillatory firing α_2-motoneuron O1 (three of the four APs are shown) together with the impulse train of the transiently oscillatory firing α_2-motoneuron O2. Interspike intervals are indicated. B. Impulse patterns of the three oscillatory firing α_2-motoneurons O1, O2 and O3: O1 continuously oscillatory firing, O2 and O3 transiently oscillatory firing. "A" marks the sweep piece shown in A. Paraplegic 1. C. Impulse train of the α_2-motoneuron O1 together with a part of the impulse train of the oscillatory firing α_3-motoneuron Oα3. Interspike intervals, conduction times and conduction velocities are indicated. Paraplegic 1, S5 root recording. D. Impulse train (consisting of two APs) with the corresponding interspike interval, conduction time and conduction velocity of the continuously oscillatory firing α_2-motoneuron O4. Paraplegic 2, S4 root recording.

If the premotor spinal oscillators would not coordinate their firing and partly synchronize their firing, tremor would occur. Pathologic synchronization can be observed in Parkinson patients [202]. The differences between the three premotor oscillator types will be considered now in more detail.

10.9. α_3-Oscillators

In Figure 72, an original recording from an α_3-oscillator is shown. The insert "C" shows a schematic layout of the firing. The duration of the interspike intervals of the impulse train increases regularly at the beginning of the impulse train and does alternate between long and short values at the end of the impulse train. The firing with alternating long and short interspike intervals from the 30th AP on can be better seen in Figure 73. This long-short alternation of interspike intervals may be due to a coupling phenomenon between oscillators. The alternation becomes only visible at the end of the impulse train close to the threshold of excitability of the motoneuron. Alternating firing with short and long oscillation periods has been observed for α_2-oscillators (Figure 88) and may be connected with the existence of antagonistic inhibition of network structures similar to half-centre oscillators. If one measures the mean interspike interval of different parts of an α_3-impulse train (Figure 74), then the feature of alternating interspike intervals is lost.

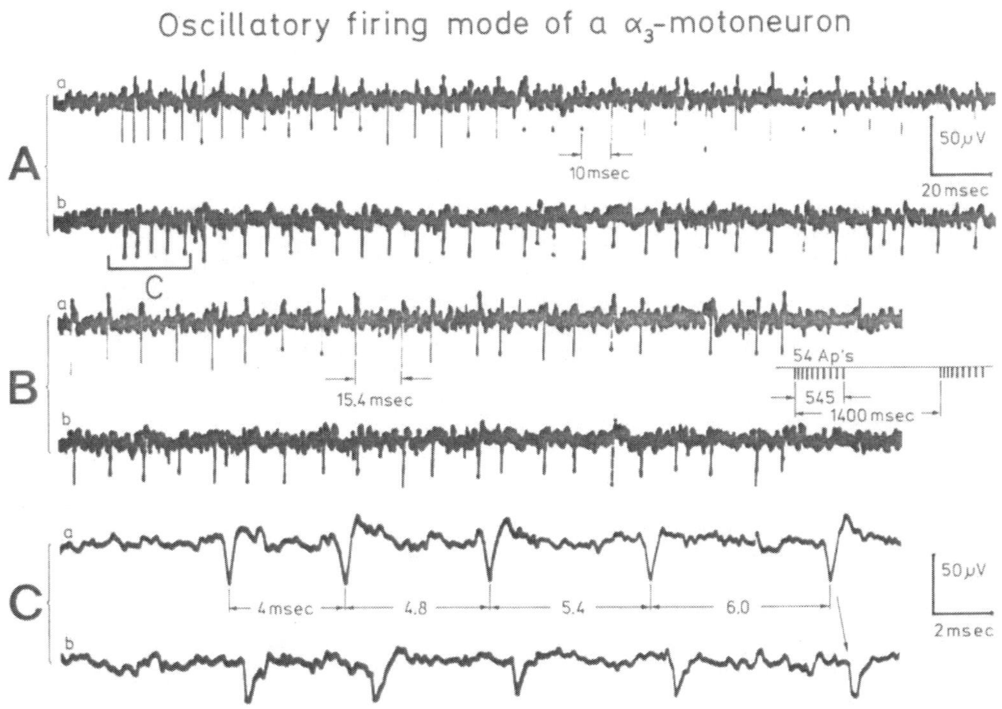

Figure 72. An impulse train (54 APs) of oscillatory firing α_3-motoneuron O_3 of brain-dead HT5. "B" is the continuation of "A." The duration of the oscillation period was approximately 1400 ms (insert "B"). The start of the impulse train is shown time-expanded in C.

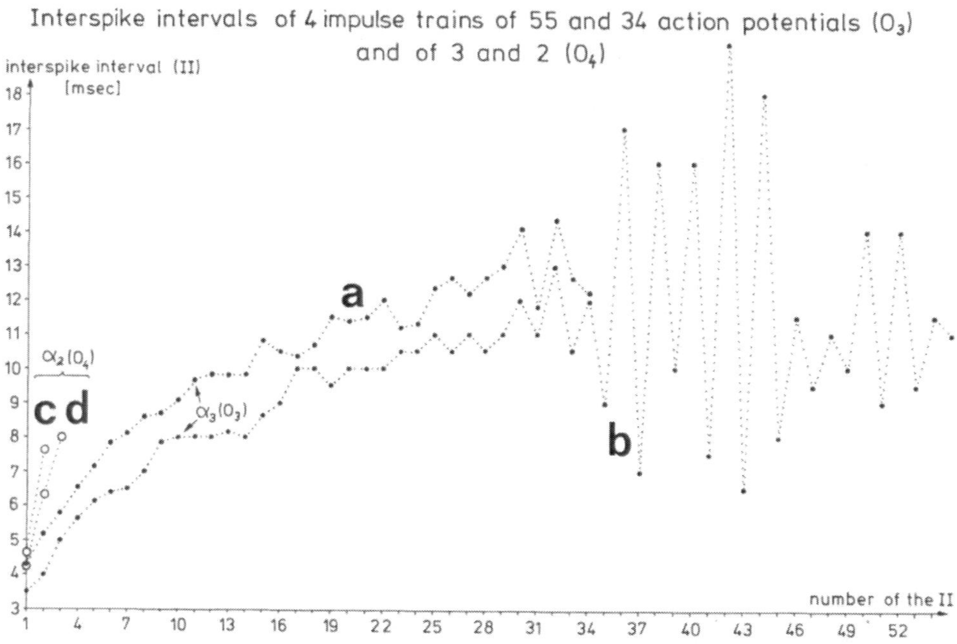

Figure 73. Interspike intervals (IIs) of successive APs of a short (a) and a long impulse train (b) of α_3-motoneuron O_3. The large dots give the values of the IIs, and the dotted lines connect successive IIs. Note that the IIs from impulse train "b" are very unequal from II 34 on. HT5; dS3 root.

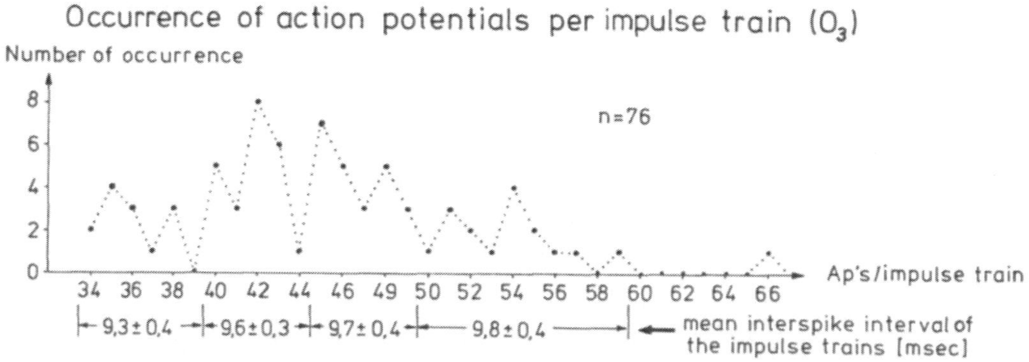

Figure 74. Occurrence of APs per impulse train. The large dots give the numbers of how often an impulse train of a certain number of APs occurred; the dotted lines connect the values. Note that the AP numbers per impulse train do not follow a Gaussian distribution. The lower part of the Figure gives mean IIs for ranges of impulse train lengths.

The static behaviour of α_3-motoneurons can also be seen when they fire oscillatory. In Figure 75, the oscillatory firing α_3-motoneuron shows only little or no specific response to natural stimulation like bladder or anal catheter pulling. The dynamics of the response is weaker in comparison to that of the oscillatory firing α_2-motoneuron. In Figure 76 an α_2-oscillator responded still less dynamically than an α_1-oscillator.

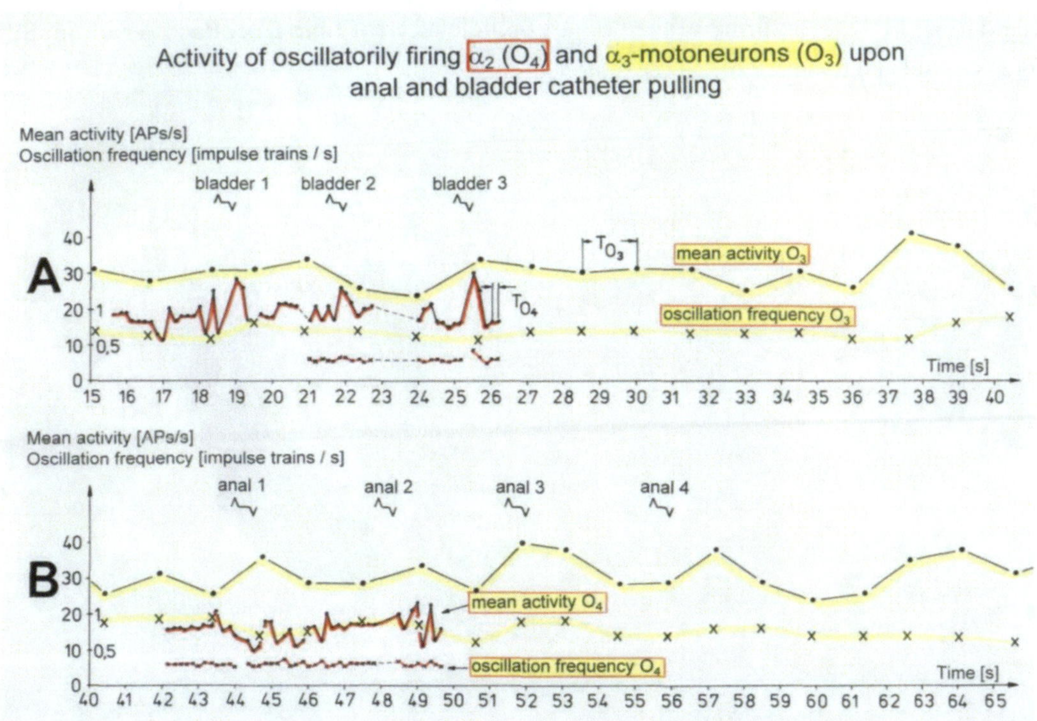

Figure 75. Mean activity (APs/s) and oscillation frequency (impulse trains/s) in response to urinary bladder (A) and anal catheter pulling (B) from oscillatory firing α_2-motoneuron O_4 and α_3-motoneuron O_3, recorded simultaneously. Dots = mean activity, crosses = oscillation frequency, large dots and crosses refer to α_3-motoneuron O_3, the small ones to α_2-motoneuron O_4. Values from α_2-motoneuron O_4 only calculated for certain periods; the dashed lines indicate that the oscillation was not interrupted. The scale 0 to 1 relates to the oscillation frequency of α_3-motoneuron O_3; scale 0 to 40 relates to the mean activity of α_3-motoneuron O_3, and the mean activity and oscillation frequency of α_2-motoneuron O_4. $T(O_3)$ and $T(O_4)$ mark oscillation periods. Dots and crosses mark the mean activity and oscillation frequency at the end of the oscillation cycle period. Lines connect discrete values of mean activity to make trends more obvious. The time and duration of the pulling of bladder and anal catheters are marked; repeated stimulation is marked with numbers. Time, in seconds, continuous. HT5, dS3 root.

The dynamics, therefore, increase from α_3 to α_2 to α_1-oscillator, in accordance with the dynamics of responses of the α-motoneuron firing in the occasional firing mode and in accordance with the muscle fibre types the three types of α-motoneurons innervate. The slow (S), medium fast (FR) (fast fatigue-resistant) and fast contracting muscle fibres (FF) (fast fatigable) have their own corresponding premotor networks in the spinal cord, namely the α_1, α_2 and α_3-networks (Figure 3).

10.10. α_1-Oscillators

With respect to dynamics, α_1-oscillators are the opposite of α_3-oscillators. The α_1-oscillators respond very dynamically but have only little oscillator network properties. In Figures 77 and 78, it can be seen that the firing of the α_1-motoneuron is absolutely correlated (phase $\alpha_1 \rightarrow$ SP1) to the firing of the primary spindle afferent fibre SP1. Each AP of the SP1-fibre is correlated to an AP of the α_1-motoneuron.

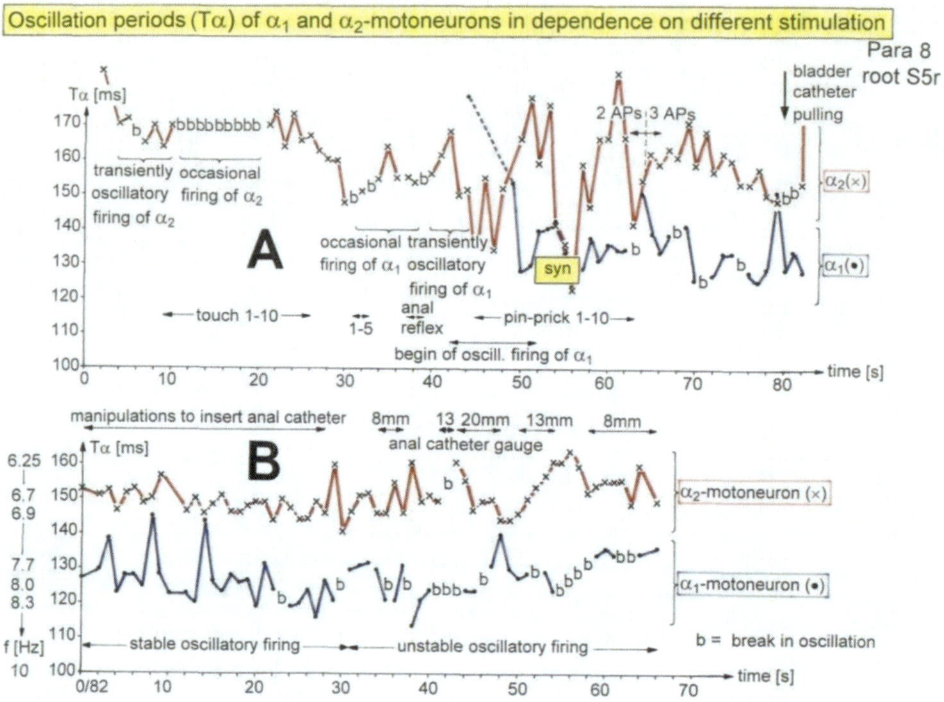

Figure 76. Oscillation periods (Tα) of an α_1 (FF) and an α_2-motoneuron (FR) in dependence on different stimulations; "B" is a continuation of "A." Touch 1 - 10 = Touching with a needle at points 1, 2,3,4,5 (upper row) and 6,7,8,9,10 (lower row). 1-5 = Touching from point 1 to point 5 (perpendicular to the direction of anal stimulation). 8, 13, 20, 13, 8mm = changing the diameter of the anal catheter from 8 to 13 to 20 to 13 and back to 8 mm (three gauges of anal catheter were used). syn = partial synchronization of the oscillatory firing patterns of the α_1 and the α_2-motoneurons. Two APs / three APs = impulse train of the α_2-motoneuron mainly consisted of two or three APs respectively. Note that α_1 and α_2-motoneurons first fired transiently oscillatory before firing continuously oscillatory. Note further that the oscillatory firing was most regular upon manipulation at the anus (0-30s in "B," anal reflex stimulation to secure continence).

The SP1-fibre and the α_1-motoneuron fire rhythmically in absolute correlation (Figure 77D,F,H; Figure 78C,E,F), but the firing of α_1-motoneurons has also some correlation to secondary muscle spindle afferent activity: When the SP1-fibre stopped firing, the oscillatory firing α_1-motoneuron continued firing for one more oscillation period; in other words, it fired one further AP (Figure 78F).

The α_1-motoneuron could also start to fire before the SP1-fibre (Figure 78E), which means that the α_1-motoneuron was getting also drive from other afferents, probably secondary muscle spindle afferents as the α_1-motoneuron fired also in relative correlation with a secondary spindle afferent fibre (Figure 78C, phase distribution $\alpha_1 \rightarrow$ SP2(1)). A simultaneously recorded oscillatory firing α_2-motoneuron fired oscillatory in relative coordination with a secondary muscle spindle afferent fibre (SP2) (Figure 78D).

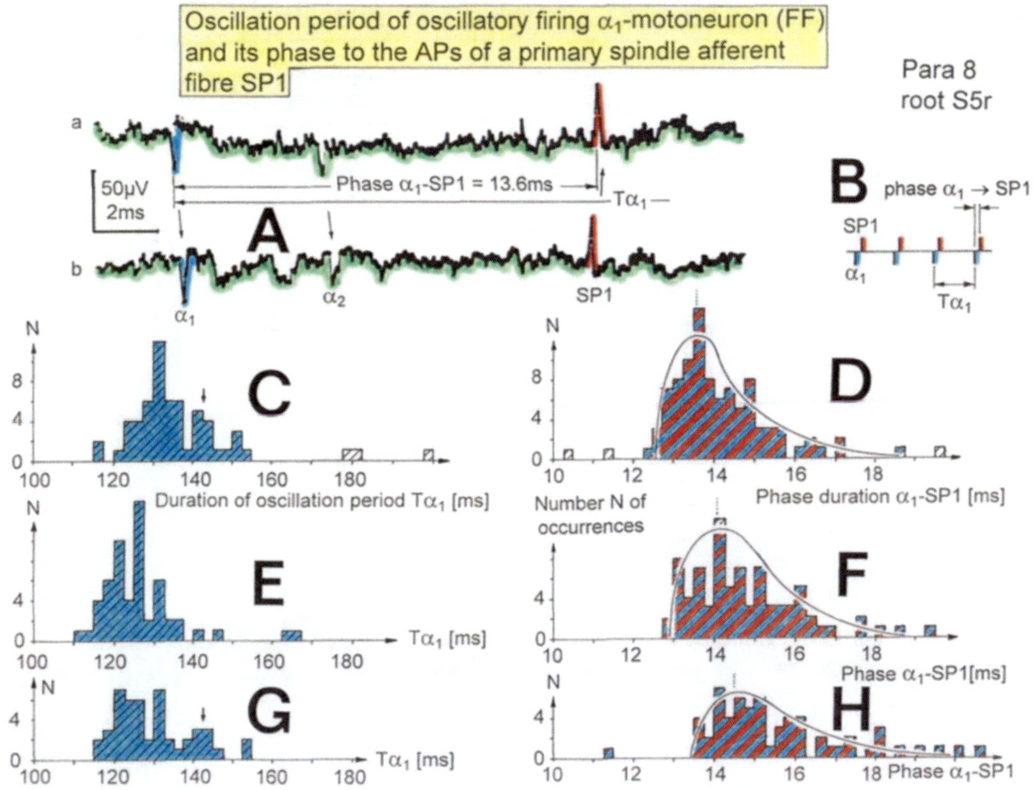

Figure 77. Distributions of the oscillation period of an oscillatory firing α_1-motoneuron (FF) and the phase to its driving primary spindle afferent fibre SP1. A. Original recording of APs of the α_1-motoneuron and the SP1 fibre. B. Definition of oscillation period $T\alpha_1$ and phase $\alpha_1 \rightarrow$ SP1. C.-H. Oscillation periods and the corresponding phases following different stimulations (C;D, oscillation period and phase distributions for the time interval 65-82s (strongest afferent drive). E, F. Distribution for the time interval 1-30s. G, H. For 31-66s. In "C" and "G," small arrows mark subpeaks in the oscillation period distributions. Para 8, right root S5.

The weak oscillatory characteristic of the α_1-motoneuron (Figures 77,78) can also be judged from its sharp (absolute) firing correlation with the SP1-fibre ($\approx$ 5ms) (Figure 77D, 78C) in comparison to the variation of the duration of oscillation periods of α_1-oscillators (40 (Figure 77C) until 120ms (very similar to Figure 78G)).

The firing patterns of α_1, α_2 and α_3-motoneurons can easily recorded with the single-nerve fibre action potential recording method but not with sEMG (Figure 70). From spinal cord injury patients, on the other hand, single-motor unit APs can be easily recorded from α_1 motor units but not from α_2 and α_3 motor units (Figure 70), because their AP amplitude seems to be very small. Clinical EMG recordings therefore show mainly the activity of α_1 motor units. Since only few α_1 motoneuron axons are contained in the lower sacral roots (not needed for continence functions), α_1 motor activity has to be obtained from suitable sEMG recordings. The generation of motor patterns of α_1 motoneurons with increasing load and the phase and frequency coordination among the motoneurons has to be taken from sEMG recordings (Figures 79, 80).

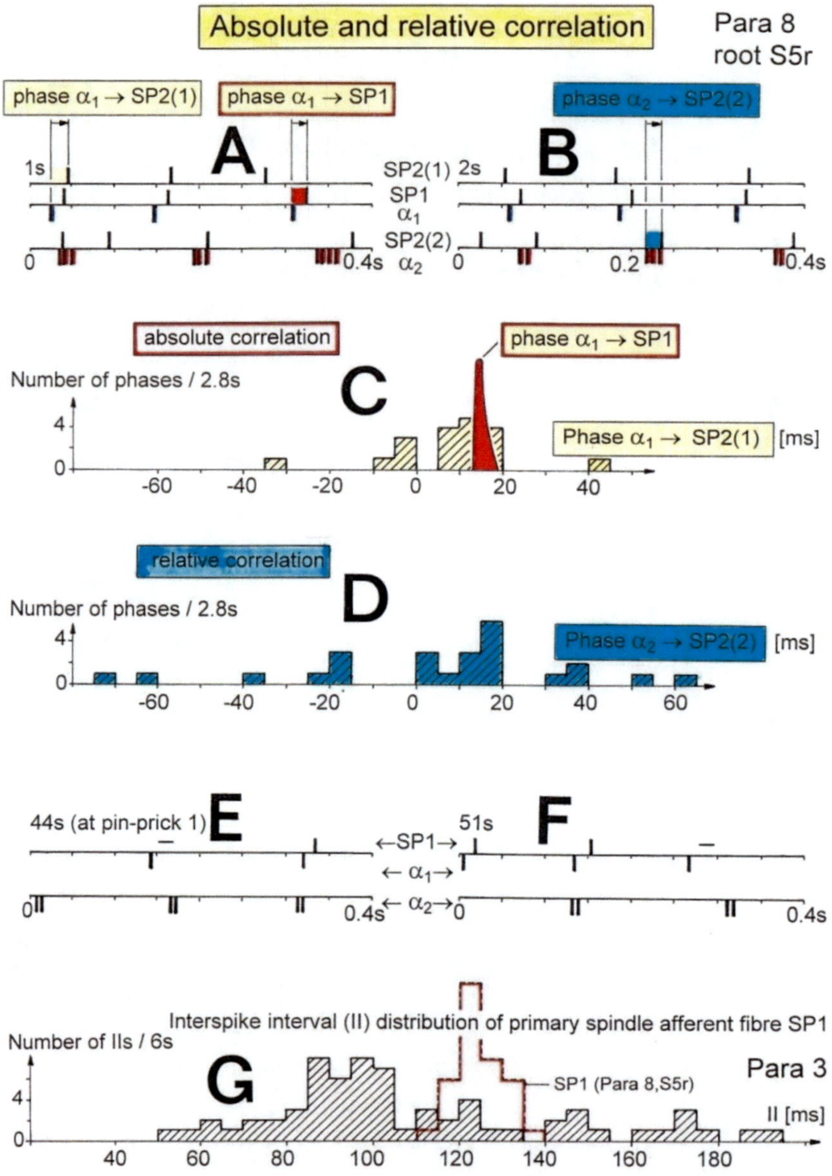

Figure 78. Absolute and relative correlation quantified by phase relations between the α_1 (FF) and α_2-motoneurons (FR) and their driving primary (SP1) and secondary (SP2) muscle spindle afferent fibres. A., B. Definition of the different phases. C. Distribution of the phases between α_1-motoneuron and the secondary muscle spindle afferent fibre SP2(1). Note that the phase distribution $\alpha_1 \rightarrow$ SP2(1) is approx. 40 times wider than that of the $\alpha_1 \rightarrow$ SP1 distribution (phase $\alpha_1 \rightarrow$ SP1 taken from [128]). D. Distribution of the phases between α_2-motoneuron and the secondary muscle spindle afferent fibre SP2(2). Note that the phase distribution $\alpha_2 \rightarrow$ SP2(2) is similar to that of $\alpha_1 \rightarrow$ SP2(1) (approx. four times wider). E., F. Note that every α_1-AP of the oscillatory firing α_1-motoneuron is accompanied by a time-locked SP1-AP. "A"- "F" = Para 8. G. Interspike interval (II) distribution for a primary spindle afferent fibre of paraplegic 3. Note that the distribution peak is at approx. 95 ms; probably, this fibre was driving an α_1-motoneuron with an average oscillation period of 95 ms. The SP1 fibre distribution for paraplegic 8 is included for comparison. Para 8; Para 3.

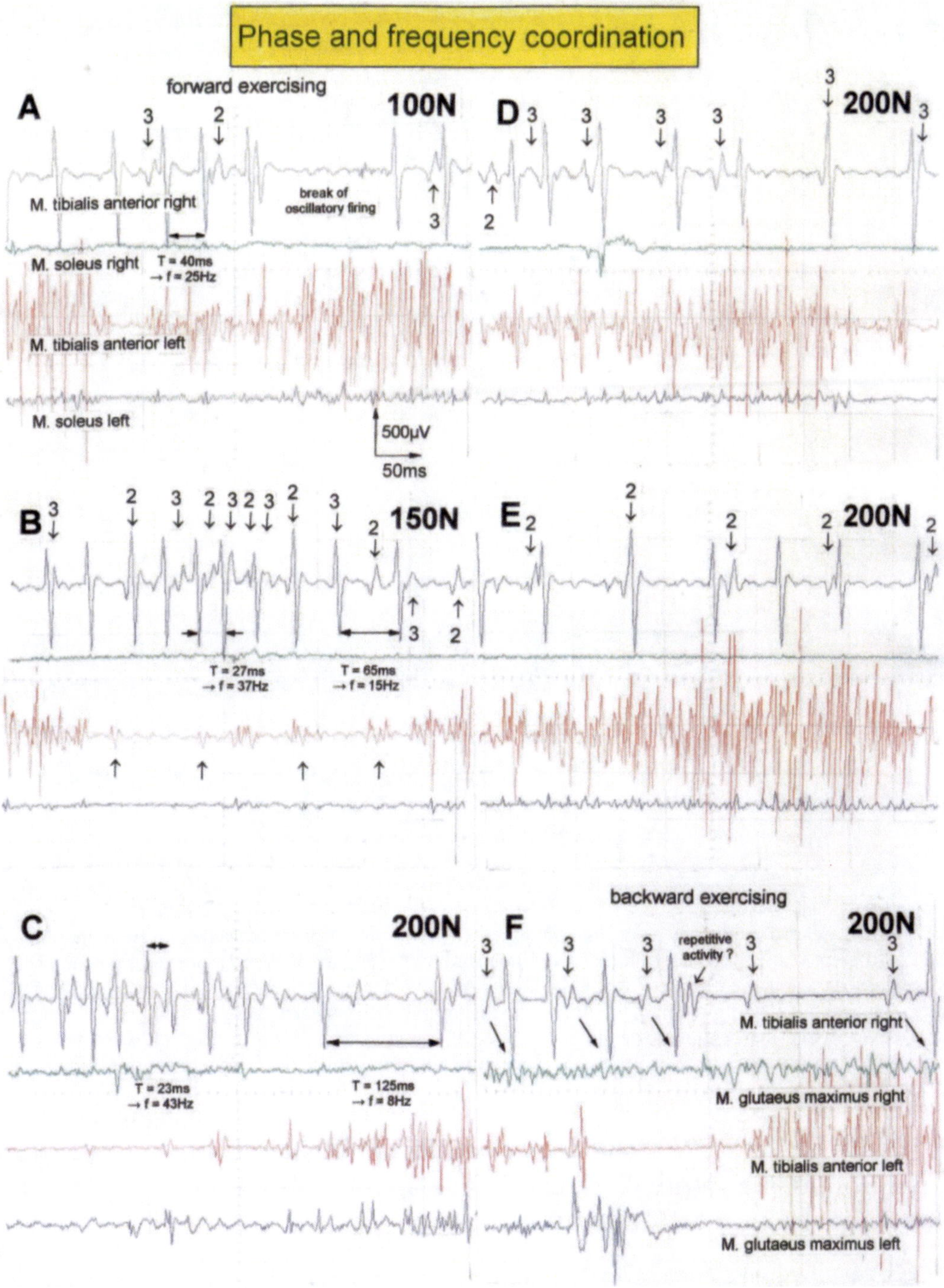

Figure 79. Phase and frequency coordination between oscillatory firing motor units (FF-type) during the generation of a motor program during exercise on the special coordination dynamics therapy device at loads increasing from 100 to 200Newton. Oscillation periods (T) and oscillation frequencies (f [Hz]) of oscillatory firing motor unit 1 are partly indicated. "C,F" soleus electrodes shifted to gluteus muscles to check early reinnervation upon therapy. The waveforms of the three identified FF-type motor unit potentials "1," "2," and "3" are indicated; motor units "2" and "3" are partly marked. In "F," some coordination's between motor unit "3" and "1" are marked.

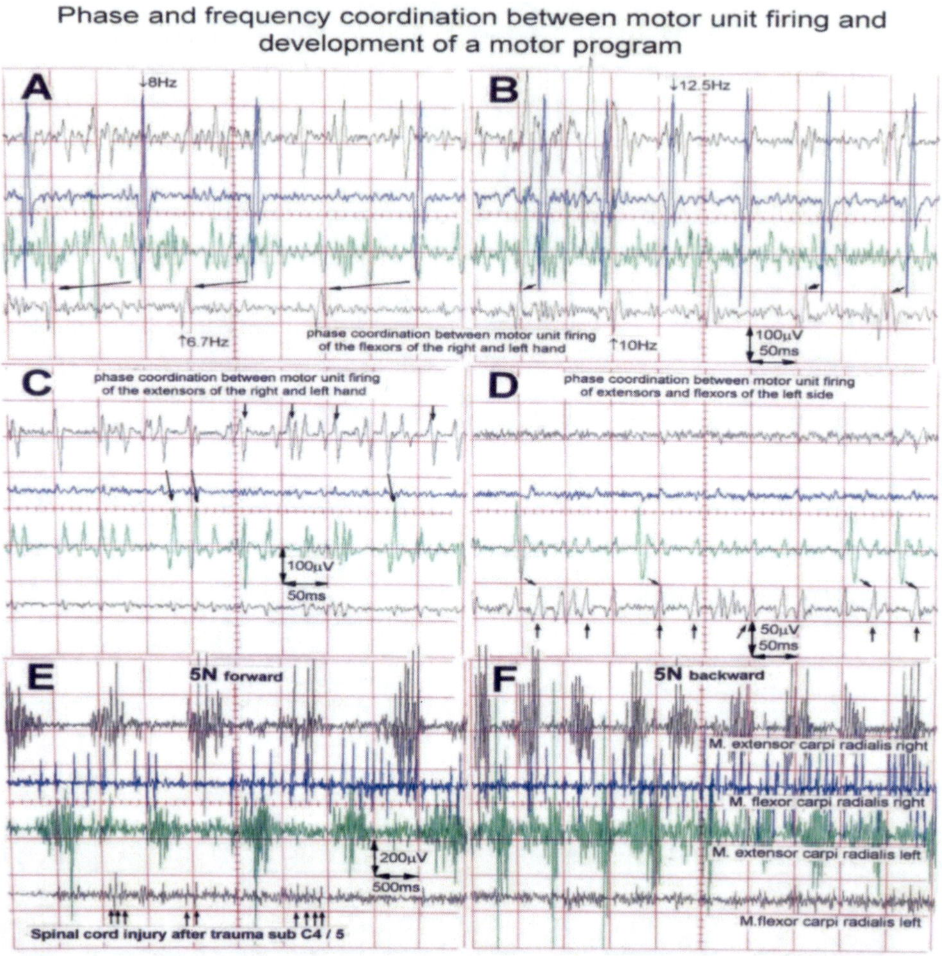

Figure 80. Phase and frequency coordination of single-motor unit firing between different muscles and between different arms. Phase coordinations are indicated by the arrows between motor unit potentials. A-D. The patient with a spinal cord injury was in position at the special coordination dynamics therapy device but activated muscles on volition upon looking onto the oscilloscope screen. E,F. Patient exercising on the special device. Note the beginning of the appearance of a motor program in the right flexor carpi radialis muscle for 5N forward turning. Recordings for a patient with a complete spinal cord injury sub C4/5.

10.11. α_2-Oscillators in Comparison with α_1 and α_3 Oscillators: Oscillation Period, Interspike Intervals of the Impulse Train, and Afferent Drive of Oscillatory Firing α-Motoneurons

The characteristics of α_2-oscillators are in between those of the α_1 and the α_3-oscillators. They are self-organized by the adequate afferent input patterns from several kinds of receptors. The α_2-oscillator O1 (Figures 59F,81) was, e.g., organized from receptors sensing bladder filling and emptying (stretch (S1), tension (ST) and mucosal (S2) afferents) (Figures 81, 82,83).

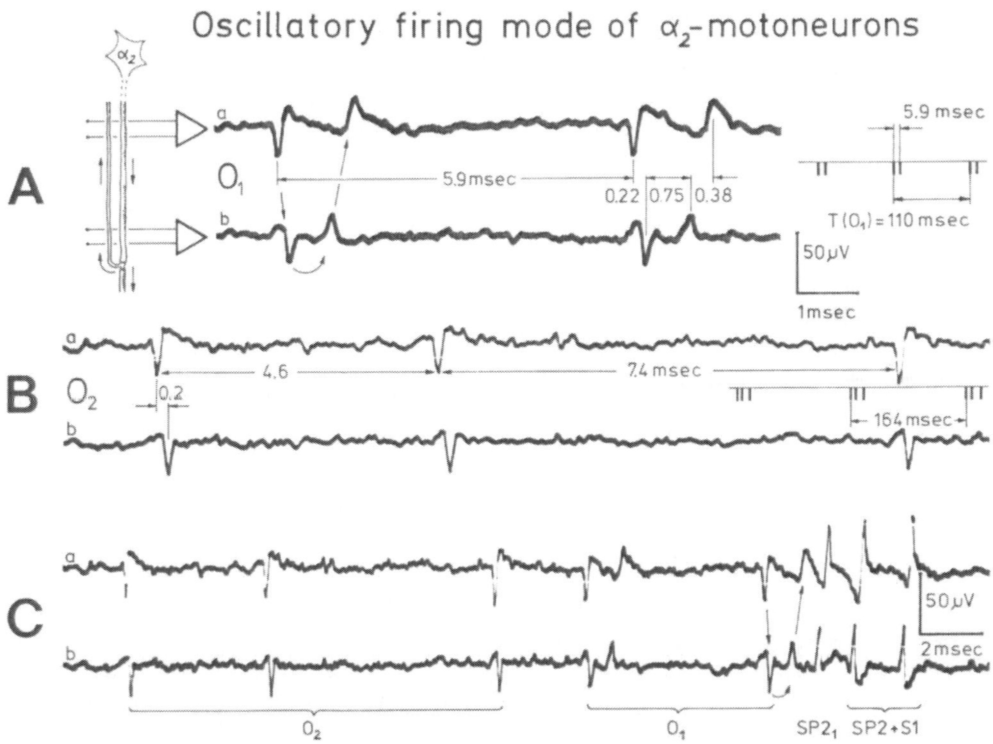

Figure 81. A. Recordings from α_2-motoneurons O_1 and O_2, firing in the oscillatory mode with impulse trains of two (A) and three (B) action potentials (APs). The durations of the oscillation periods were 110 (O_1) and 164 ms (O_2) (B). The interspike intervals of the impulse trains were 5.9 ms (O_1) and 4.6 and 7.4 ms (O_2) (B). Motoneuron O_1 conducted at 36 m/s; its recurrent fibre conducted at 21 m/s. The measurement layout is shown schematically in A. The inserts in A and B show the oscillatory firing modes; they have not been drawn to scale. The impulse trains of the α_2-motoneurons O_1 and O_2 are shown in C together with the APs of the identified secondary muscle spindle afferent fibre $SP2_1$ and two other spindle or bladder stretch (S1) afferents. S4 dorsal root. HT6. See also Figure 59.

fluid in bladder [ml]	activity pattern [msec]	oscillat. frequency [impulse trains/sec]	mean activity [Ap's/sec]	firing mode
200			0,5	occasional spike mode
500			1,0	
550			2,5	
anal-stimulat.	144	(6,9)		transient oscillatory mode
580	112	(8,9)	5,0	
620	long 143 short 114	7,8	12,0	continuous oscillatory mode
700	134 116	8,0	14,0	
750	112 109	9,0	18,1	
800	127 111	8,4	16,8	
830	134 124 104	7,8	13,8	

Figure 82. The firing in the occasional spike mode, the transient and the constant oscillatory firing mode of α_2-motoneuron O_1 in response to filling of the bladder. In the "activity pattern" column changing durations of oscillation periods are given. The oscillation frequencies in the brackets give the frequencies at the moment of oscillation for the transient oscillatory mode. Downward deflections are schematized APs. Interspike intervals of the close APs ≈ 6.0 ms. HT6. See also Figure 59.

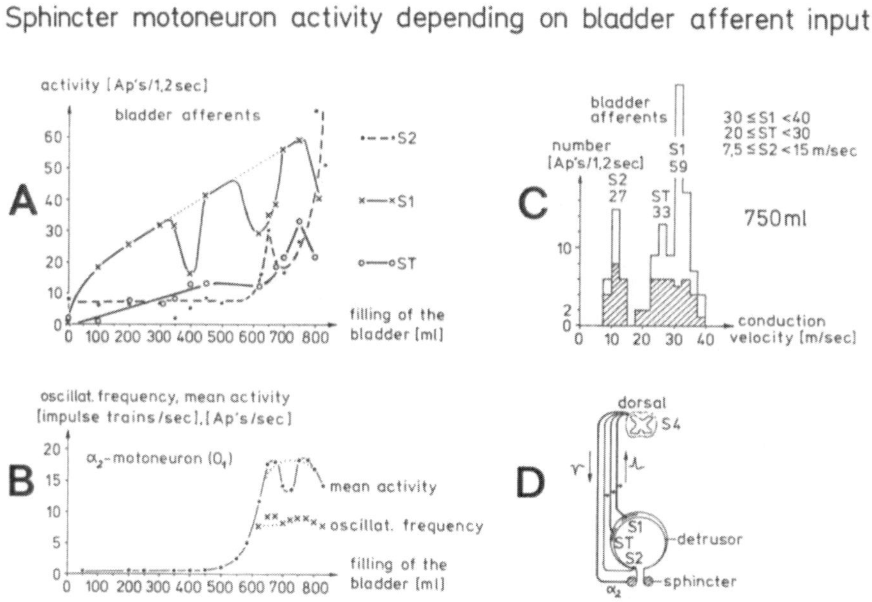

Figure 83. A. Activity levels of stretch (S1) and tension (ST) and flow receptor afferents (S2) (E) and of sphincter α_2-motoneuron O_1 (F) in response to retrograde filling of the bladder. The activity values of the S1, ST and S2 afferents are taken from histograms like the one in C. Filling of the bladder was stopped once between 600 and 650 ml. B. The small dotted lines represent mean activity (APs/s) and oscillation frequency (impulse trains/s) of α_2-motoneuron O_1 if bladder filling were not stopped in between. Note that the mean activity increases continuously with the filling of the bladder from 550 to 650 ml, even though motoneuron O_1 started to fire in the oscillatory mode from 620 ml on (D). Note further that the oscillatory firing motoneuron O_2 (frequency of firing with impulse trains is shown) is nearly not affected by the filling of the bladder and by the start of the oscillatory firing of motoneuron O_1. C. Conduction velocity frequency distribution histogram of stretch, tension and flow receptor afferent activity at 750 ml. The activities of afferents S1, ST and S2 are quantified by counting the afferent conduction velocities under the peaks (open plus hatched part), with the conduction velocity limits given in the insert. The counts (27, 33, 59) are given below the peak labelled S1, ST and S2 and plotted into E for the afferent activity at 750 ml. D. Schematic drawing of the anatomical arrangement of the afferents and the motoneuron O_1. HT6. See also Figure 59.

The change from the occasional via the transient oscillatory to the continuous oscillatory firing mode to secure continence upon bladder filling and increase of S1, S2 and ST receptor activity (Figure 83) can nicely be seen. The α_2-oscillator O2 (Figures 59F,81) was stimulated to organize itself in neural networks completely differently, namely by the activity from secondary muscle spindle afferents (SP2) and skin and mucosal afferents from the anal reflex area.

It will be shown that such self-organized oscillators fire in relative coordination with muscle spindle afferents and efferents and other spinal oscillators (Figure 71). The α_2-oscillators may have several rather discrete network loops (see below) and oscillation frequencies, and they do respond less dynamically than α_1-oscillators (Figure 76) and more dynamically than α_3-oscillators (Figure 75).

It is unexpected that only one oscillatory firing α_3-motoneuron could be found in patients with spinal cord injury (SCI) (Table 5). This may indicate that α_3-motoneuronal networks are no more able to organize themselves to fire oscillatory in a physiological manner following SCI. α_2-Motoneurons fired frequently oscillatory in SCI patients, but their oscillatory firing was much less regular than observed in brain-dead and normal individuals [186, 194]. The data in Table 3 indicate therefore that whereas the α_1-neuronal network underwent little changes, the α_2-networks showed much more changes and the α_3-networks even considerably more changes. The α_3-neuronal networks seem no more able to organize themselves to fire oscillatory. As most interneurons are integrated in the oscillatory firing of α_3-motoneurons and very few in α_1-motoneuron firing, the integrative network functions of the spinal cord seem to have changed or deteriorated following spinal cord lesion. The jumping on spring-board (Chapter II) is especially designed to functionally repair the promoter spinal oscillators. The jumping frequency is in the range of the oscillator frequency of α_3-oscillators ($\approx$ 1 Hz).

Table 3 shows that the α_1-motoneurons are mainly driven by primary spindle afferents (SP1), and the α_2-motoneurons are driven by secondary spindle afferents (SP2). The distributions of the occurrence of interspike intervals within the impulse trains and of oscillation periods provide information concerning the organization of the neuronal networks of the medulla. The shape of the distributions of the interspike intervals shows similarities to coupling properties of mechanical oscillators (coupled pendulums). Also, similar subpeaks in oscillation period distributions of different motoneurons indicate rhythm coupling between oscillators. The shift of the corresponding distribution peaks in oscillation period distributions with the increasing oscillator network activation (increasing impulse train length) suggests the existence of several rather discrete network loops of α_2-oscillators. Interpolation of the oscillation period distributions shows how the excited network loops shift with increasing excitation (see below). The network loop of the α_2-oscillator of the neuronal network caudal to a spinal cord lesion is more distributed than that of a brain-dead human, indicating a widening of the network loops (further spread of the excitation) and is in connection with the loss of phase and frequency coordination (see below), an indication for the loss of specificity of oscillator properties due to the loss of supraspinal control including descending inhibition.

10.12. Regular Oscillatory Firing of α_2-Motoneurons in Brain-Dead Individuals

Figure 84 shows an impulse train triplet of an oscillatory firing α_2-motoneuron (O2) of the brain-dead individual HT6 [186] to secure bowel continence. The firing pattern is schematically drawn in Figure 84B. Distributions of the oscillation period and of the first and the second interspike intervals are plotted in Figure 84C,D,E.

The distributions of the oscillation period and of the first interspike interval are smooth and show no obvious subpeaks. The distribution of the second II shows subpeaks suggesting interactions with other oscillatory firing circuitries. The narrow distribution of the oscillation period, and the interspike intervals of the impulse train show that the oscillatory firing network fired very regularly.

Other spinal oscillators in brain-dead individuals [186, 191] fired just not as regularly as the one shown in Figure 84. It is concluded that in brain-dead and normal individuals, the neuronal networks, driving the motoneurons, fire regularly with some variation. It will be shown below that this regularity is lost following spinal cord injury.

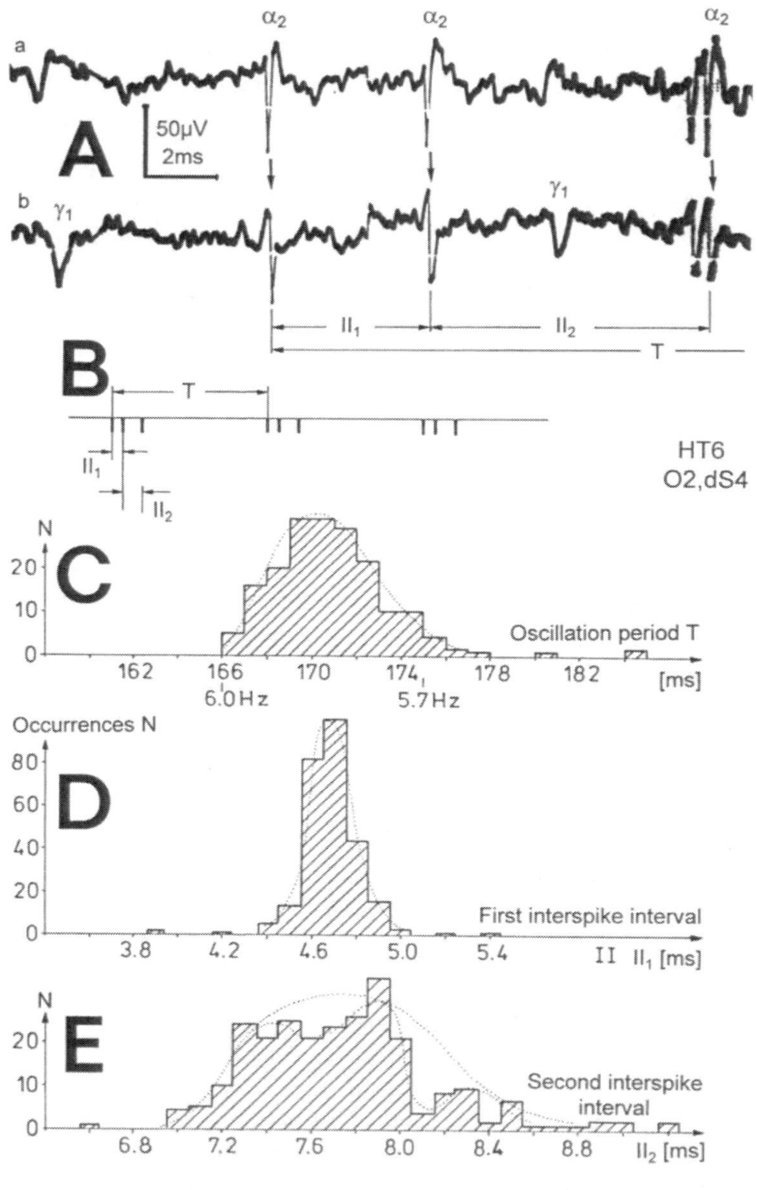

Figure 84. Distribution of oscillation period (T) (C), first (II_1) (D) and second interspike intervals (II_2) (E) of a α_2-motoneuron, firing repeatedly with three AP impulse trains. A. Original recording of the three AP impulse train; II_1, II_2 and T are indicated. γ_1 marks APs from a γ-motoneuron. B. Schematic drawing of the impulse pattern of the α_2-motoneuron. Note that, the distribution of II_2 (E) may consist of three subpeaks. Brain-dead human HT6.

10.13. Irregular Oscillatory Firing of α_2-Motoneurons Following Spinal Cord Injury

Following spinal cord injury, mean values of oscillation period and interspike intervals for oscillatory firing motoneurons show much more variation (Tables 5,6) than do normal values (HTs, Table 4) [176, 186].

Table 4. Oscillatory firing modes of α_2, α_3 and γ-motoneurons measured from three HTs and one patient (Pat), m = male, f = female, age in years, temp = central temperature, d = dorsal, v = ventral, S = sacral, fosc = oscillation frequency, mean II = mean inter spike interval, activ. = activity (APs/s), s = sec, ms = ms. Downward deflections in the schematically drawn activity modes indicate the APs of the repetitive activity. Time from impulse train to impulse train indicates the duration of the cycle period in ms. II = interspike interval in ms, ± = absolute error (standard deviation, number of measurements between 50 and 100). O1, O2, O3, O4 = designation of motoneurons, which will be referred to in all Figures, non- = occasional spike mode, transiently oscill. = transient oscillatory firing mode, constantly oscill. = constant oscillatory firing mode

Case root sex age temp.	γ_1-motoneurons	α_3-motoneurons firing mode	$f_{osc.}$ [1/s]	$\overline{II}$ [ms]	activ. [1/s]	α_2-motoneurons firing mode	$f_{osc.}$ [1/s]	$\overline{II}$ [ms]	activ. [1/s]
HT4 dS5 f 56 32°C		1.constantly oscill. 11Ap's 1400ms II=2,4···8,6 2.constantly oscill. 4Ap's 410 II=5/5,6/6,3	0,7 2,4		8 10	transiently oscill. 3Ap's 170ms II=4,7/6,4ms	6		17
HT5 dS3 f 58 38°C	γ's present	O_3constantly oscill. 45Ap's 1400±110 430	0,6- 0,83	9,6	24- 50	O_4 constantly oscill. 2Ap's a II=5,4 3Ap's b 170±13 II=4,6/7,7±2 4Ap's c II=4,2/6,3/8	4,8- 7,1	6,15	9,6- 28,6
HT6 dS4 f 37 35,5°C	transiently oscill. ? 1. 2Ap's II=12,4 90 or 180 2. 3Ap's II=4/8 370					O_1 non-transiently and constantly 1-2Ap's oscill. 110 II=6,0 O_2constantly and transiently oscill. 3Ap's 164 ±4 II=4,5/7,45±0,35	7,6- 9,0 6,1	6,0 6,0	12- 18 18,3
Pat 24 vS3 m 10 37°C	γ's present					transiently oscill. 3Ap's 144 II=3,5/5,4	6,9		21

The most regular oscillatory firing α_2-motoneuron, shown in Figure 85, fired nearly as regularly as the α_2-motoneuron of the brain-dead individual HT6 (Figure 84). On the other hand, some oscillatory firing α_2-motoneurons (Tables 5,6) fired extremely irregularly so that it was difficult or even impossible to identify the kind of the motoneuron from the impulse patterns. The distributions of the oscillation period and of the interspike intervals of the impulse train from an α_2-motoneuron in paraplegic 8 (Figure 85) show more scatter than those obtained for HT6 (Figure 84).

Further, it can be seen from Figure 85D that the α_2-motoneuron reduced its mean interspike interval from approximately 9 ms to 7.5 ms in the two AP-firing mode before switching into the three AP-firing mode with interspike intervals of 7 ms and 11.5 ms (Figure 85F).

Table 5. Oscillatory firing modes of α_2 and α_3-motoneurons from two paraplegics (Para 1,2) and two HTs (brain-dead humans). Centr. Temp. = central temperature; cont. osc. = continuously oscillating; trans. osc. = transiently oscillating. Mean T_{osc} = mean oscillation period; f_{osc} = mean oscillation frequency; II_1, II_2, II_3 = first, second and third interspike interval of the impulse train; activity measured in action potentials (APs) per second. Downward bars in the schematically drawn activity modes indicate APs of the repetitive activity. ± = error (standard deviation); n = number of observations. O1, O2, O3, O4, Oα3 = designations of oscillatory firing motoneurons, which will be referred to in all Figures. The central temperature of 38°C in HT5 was due to an infection

Case Root Sex Age	No	Kind of oscill.	$\overline{T}_{osc}$ [msec]	$\overline{f}_{osc}$ [Hz]	Length of impulse train	II_1 [msec]	II_2 [msec]	II_3 [msec]	T_{osc} [msec]	f_{osc} [Hz]	Activity [sec^{-1}]
Para 1	01 cont. osc.	α_2	161	6.2	←T→	3.26±.2 n=119	5.5±.9 n=118		161	6.2	18.6
S5 root					←II_1	3.2 n=1	4.4	6.6	166	6.0	21
female	02 trans. osc.	α_2				n=1			68	14.7	14.7
37 years						5.7 n=3			100	10	15
leason:						5.8 n=2			118	8.5	17
13 years ago						4.6 3.5 n=2	5.8		135	7.4	18.5
leason level: T8 incomplete						3.7 n=1	6.0		156	6.4	19.2
T12 complete	03 trans. osc.	α_2	86	11.6					68/82/112	14.7/12.1/8.9	14.7/12.1/8.9
centr. Temp. 35°C			118	8.5		5.7 n=7			82/112/142	12.1/8.9/7.0	18.1/13.4/10.5
root Temp. 25°C?			130	7.7		6.5 n=5			112/127/142	8.9/7.9/7.0	17.8/15.7/14
bladder dyssynergia			139	7.2		4.5 n=2	7.5 n=2		127/142	7.9/7.0	15.7/14
from 200 ml on			142	7.9		6.2 5.2	11.2		142	7.0	17.5
	Oα3 cont. osc.	α_3	285± 18.5 n=7 (256 to 302)	3.5 (3.9 to 3.3)	6 to 15 APs mean=8.4	see Fig.3B					29.4
Para 2			100	10		5.3±.9 n=71 (4.2/4.6/5.6)			85/100/110	11.8/10 9.1	23.6/20/18.2
S4 root male			110	9.1					100/110/150	10/9.1/6.7	25/22.8/16.7
23 years leason: 2 years ago	04	α_2	120	8.7		5.3±1 n=58 (4.2/5.6)	8.7±2.3 n=59 (4.4/8.8/12.8)		120/150	8.3/6.7	25/20
level: T8 centr. Temp. 36°C									160/200	6.25/5	22/17.5
root Temp.? bl.dyssynergia:						5.7±2 n=8	7.2±3 n=8	10.2±3 n=8	115?		
> 200 ml									100/170	10/5.9	30/17.4
HT5						4.4			168	6.0	12
dS3 root									184	5.4	10.9
female 58 years	04	α_2	184	5.4		3.8 (6.8)	7.0		168/184	6.0/5.4	18/16.2
									172?/184	5.4	18.9
centr. Temp. 38°C						3.8/4.3	4.5/6.8	7/8/9/10	184	5.4	21.6
									176/184	5.7/5.4	17/16.3
HT6	02	α_2	125 (160)	8 (6.25)		7.4±.5 n=6			125/160	8/6.25	16/12.5
dS4 root female			160	6.25		4.5/7.45					19
37 years c.Temp. 35.5°C	01	α_2	110.5	9.0		6.0					18.1

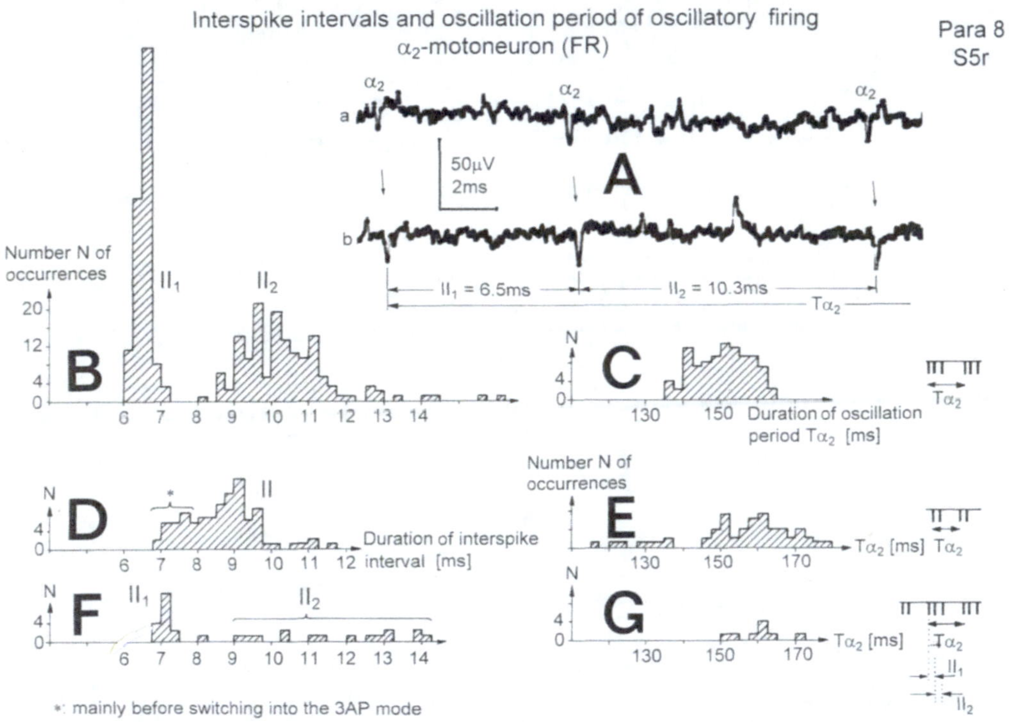

Figure 85. Distribution of oscillation period $T\alpha_2$, First (II, II_1) and second interspike intervals (II_2) of a α_2-motoneuron, innervating the external anal sphincter, firing repeatedly with impulse trains consisting of two and three APs. A. Original recording of a three AP impulse train; oscillation period and interspike intervals are indicated. B. Distributions of the first and second II of the three AP impulse train. C. Oscillation period distribution for the three AP-firing. D, E. Distributions of interspike intervals and oscillation period $T\alpha_2$, when the α_2-motoneuron fires with impulse trains consisting of two APs. F, G. Distribution of first and second interspike intervals and of the oscillation period of the three AP impulse train, when the α_2-motoneuron changed from the two AP- to the three AP-firing mode. Paraplegic 8, left root S5.

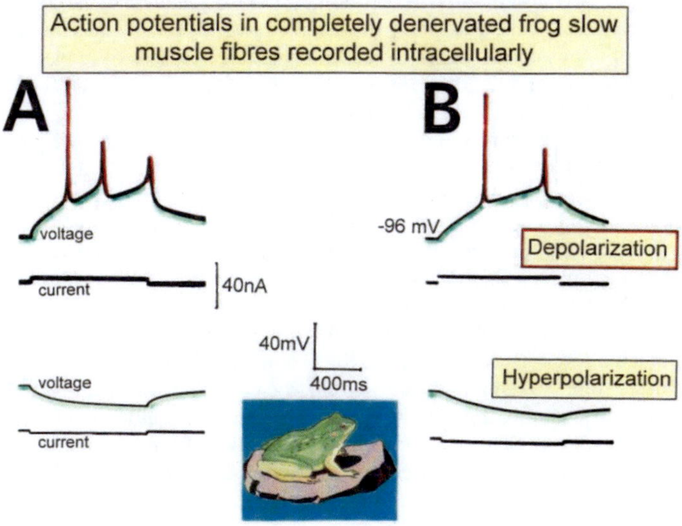

Figure 86. (Continued).

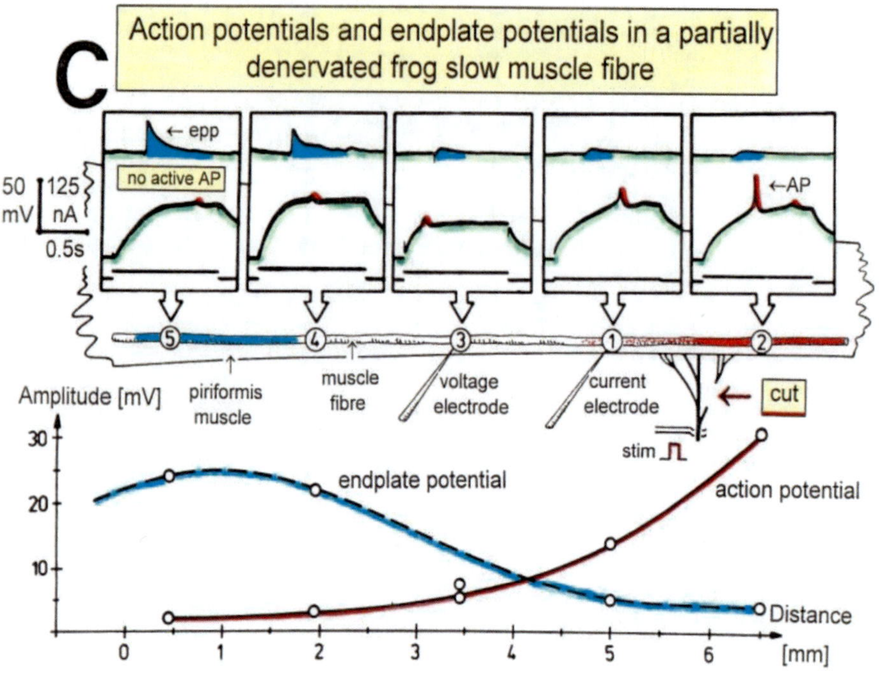

Figure 86. Simultaneous occurrence of action potentials (APs) and endplate potentials in slow muscle fibres of the frog rana temporaria. Slow muscle fibres of frogs normally do not generate APs to depolarize the muscle fibre to increase Ca^{++} concentration for contraction; the slow muscle fibres are depolarized by the endplate potentials of several distributed endplates connected to one or two thin motor axons. When impaling a voltage and a current microelectrode into the slow muscle fibre (C) and applying constant current, the membrane resting potential can be held at -90 or to -100 mV. Upon applying an additional current pulse for 100 ms, the muscle fibre can be transiently depolarized or hyperpolarized. When hyperpolarizing the muscle fibre, the slow membrane potential deflection indicates high membrane resistance (≈ 10 MΩ) typical for slow muscle fibres. Twitch muscle fibres (not shown here) show a faster time course and have a lower membrane resistance (≈ 1 MΩ). When depolarizing a normal slow muscle fibre by a depolarizing current pulse, no AP is generated (similarly to voltage electrode impaling site 5 in C). When depolarizing a denervated slow muscle fibre (cut nerve supply to the muscle), after 10 to 15 days, the frog (held at ≈ 18°C) slow muscle fibres generating APs similarly as in A and B. Upon partial denervation of a slow muscle fibre (C), which is innervated by two motor axons one of which is cut, the innervated muscle fibre part responds with a motor endplate potential (and no AP) and the denervated muscle fibre part with an AP (and no endplate potential). The AP and the endplate potential will spread electrotonically (passively) to other parts of the muscle fibre (lower part of C). It seems therefore that the different parts of the muscle fibre are genetically differently controlled by the different muscle fibre nuclei. A,B. APs generated in denervated slow muscle fibres of frog piriformis muscles upon depolarization by transient current pulses of 100 ms duration. The slow time course of the voltage upon hyperpolarizing the fibre shows that the microelectrodes were impaled into slow muscle fibres. With increasing depolarizing current pulses (and fully developed AP mechanism), first one AP is mostly generated (C, impaling site 2), then two APs (B) and then three APs (A) (in this case only shown for three different fibres). Note with respect to the relative length of the interspike intervals that the two AP impulse train in B and the three AP impulse train in A bears, similarity to the impulse trains recorded from oscillatory firing human α_2-motoneuron axons. C. Recording of APs and endplate potentials along a partially denervated slow muscle fibre by keeping the current microelectrode at site 1 (close to the nerve entrance to the piriformis muscle) and impaling the voltage electrode successively along the muscle fibre from site 1 to site 5. At each site of the voltage electrode impalement, the muscle fibre is directly stimulated by the current pulse from the current electrode (the reference bath-electrode is not shown, muscle in frog Ringer solution) to elicit an AP, and indirectly stimulated, by applying to the nerve supplying the muscle a voltage pulse of 0.5 ms duration and 1 to 10 V amplitude, to elicit an endplate potential. The results of both stimulations are shown in the pictures related to the sites of successive impalements of the voltage electrode. In the partially denervated slow muscle fibre (C), the endplate potential did not elicit an AP generated in the other

denervated part of the muscle fibre. But often, the endplate potential evoked in the still innervated muscle fibre part (here left side) is still high enough in the denervated muscle fibre part (here right side) to elicit an AP there, which is then actively conducted in the denervated area of the muscle fibre and spreads electrotonically (passively) into the innervated muscle fibre part. - This animal example shows how delicately membrane properties can change on the cellular level with functional consequences, which cannot be understood with molecular, genetical or pharmacological methods alone. Avoiding regeneration, it seems possible to increase by training the size of one or two left synapses [120] in the innervated area of the muscle fibre, so that the endplate potentials can generally reach the threshold for AP generation in the denervated muscle fibre segment. It is conceivable that similar mechanisms are working in the human CNS, e.g., with respect to the integrated functions of dendrites (part for electrotonic potential spread), cell soma (summing point) and axon hillock (AP generation for active conduction along the axon), when changing neuronal network properties by repreformation of the neuronal network structures due to changes in efficacy and properties of synapses and changes of the excitability of membranes.

Table 6. Oscillatory firing of α_1 (FF), α_2 (FR) and α_3-motoneurons (S) in humans 0.5 to six years following spinal lesion (paraplegic = Para). Centr. temp. = central temperature, f = female, m = male, vS4r = right ventral sacral root 4, O1; O2; O3 = firing modes of α_1, α_2 and α_3-motoneurons (see Figure 3). SP1; SP2 = primary, secondary muscle spindle afferents. Anal, bladder = motoneuron innervated external anal or urinary bladder sphincters

Case	Sex/Age	Root	α_1-Motoneurons					α_2-Motoneurons					α_3-Motoneurons		
Centr. temp. [°C]	Lesion level/Age		Oscillation period	Number of APs per impulse train	Interspike intervals of impulse train	Firing mode	Afferent drive interspike intervals (II)	Oscillation period T [ms]	Number of APs per impulse train	Interspike intervals of impulse train [ms]	Firing mode	Afferent drive	Period	Number of APs per impulse train	Firing mode
Para 3 34.6 °C	f 33 C6/3	S5l					SP1 II=95ms	1. 130 111 2. 134 118 3. 142 117	2 1 2 1 2 1	13 10 9.8	O2 O2				
Para 5 34.6°C	f 24 TH12/2	S5r						1. 167 2. 111	3 (anal) 2 (bladder)	4.3/8.5 7.4	O2 O2	SP2 parasymp. activated			
Para 7 35°C	f 37 TH5/2.5	S5r						1. 77 72 57	3 2 1	8.15/8.95 8.9	dys irregular				
		S5l	3. (134	2/1	7.4ms	O2)=α_1?		1. 107 2. 130	2 1 2	4.6 6.2	O2 O2				
		vS4						1. 14	2 1		dys very irregular				
Para 8 35.2°C	f 20 TH6/1.5	S5r	1. 122	1		O1	SP1 SP2	1. 150 172	3 2	6.4/10.4 7.9	O2				
Para 9 34.5°C	f 23 TH1/0.5	vS4r						1. 300	1 Synchronization following pin-prick and anal stimulation	-	O1/O3	SP2	1. 100 γ_1-fusimotor fired partly oscillatory	1	O1
Para 10 34.3°C	m 30 TH12/0.5	dS4r						1.150	4 3 2	4.2/7.1/9.1; 4.1/7.1 8.6	O2	SP1 or SP2 phase lag ~6.3ms later			
Para 11 35.8°C	f 25 Th5/2	vS3r										SP2 fibres parasymp. activated			

Firing regularly in the three-AP mode, the first and the second interspike intervals were approx. 6.5 ms and 10 ms (Figure 85B). Such a change from two to three AP impulse train firing can be observed if one records from twitch or denervated slow muscle fibres of the frog intracellularly and when the depolarizing pulse amplitude is increased to mimic depolarization by excitatory postsynaptic potentials (Figure 86). Upon increasing the repeated transient depolarization, the interspike interval of the two AP impulse train reduces (Figure 86B) before a three AP firing occurs (Figure 86A).

10.14. Alternating Long and Short Oscillation Periods in Some Oscillatory Firing α-Motoneurons

It was shown previously that an α_2-motoneuron, innervating the external bladder sphincter in a brain-dead individual, fired oscillatory with long and short oscillation periods and long (two APs) and short (one AP) impulse trains [176].

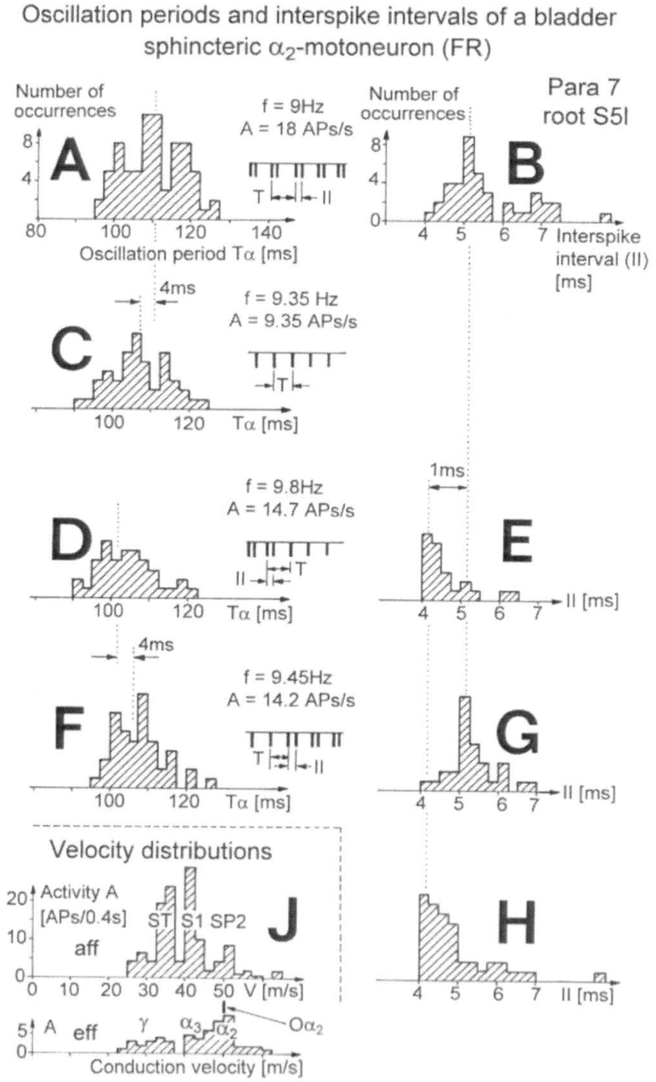

Figure 87. Oscillation period (T) and interspike interval (II) of an α_2-motoneuron innervating the external bladder sphincter, in dependence on the length of the impulse train (impulse train is one or two AP long). A, B. T and II for regular firing with two APs. C. T for regular firing with two AP. D, E. T and II when the pattern changes from regular two AP firing to regular one AP firing and vice versa (F,G). H. II for other changes of the firing pattern. J. Conduction velocity distributions of afferents and efferents from which recordings were performed; the velocity of the oscillatory firing motoneuron $O\alpha_2$ is indicated. S1, ST = stretch and tension receptor bladder afferents, SP2 = secondary muscle spindle afferent fibre. f = frequency, A = activity. Note, that in different firing modes, the oscillation period differs by 4 ms and the interspike interval by 1ms.

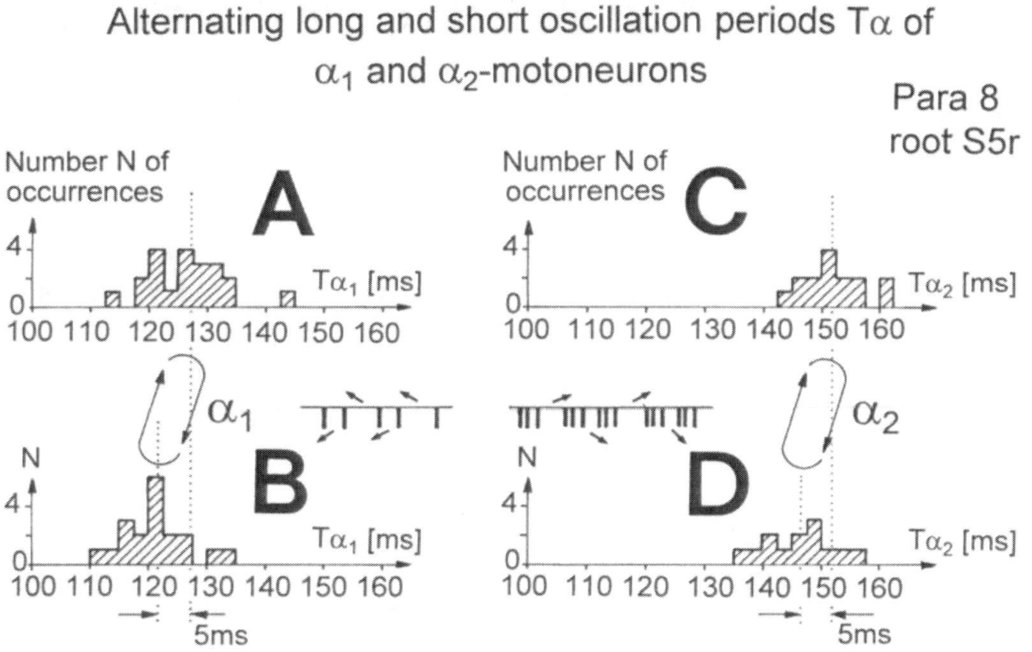

Figure 88. Alternating long and short oscillation periods of oscillatory firing α_1 (A,B) and α_2-motoneurons (C,D). Note, that on the average, long and short oscillation periods for both motoneurons differ by 5 ms.

In Figure 87 the oscillatory firing of an α_2-motoneuron is shown, which innervated the external bladder sphincter (paraplegic 7). No alternating firing with long and short oscillation periods and one AP and two AP impulse trains was observed. The probably pathologic firing (less regular oscillatory firing than in brain dead individuals) consisted of two AP firing (Figure 87A,B), one AP firing (Figure 87C), changing from two AP to one AP firing and vice versa (Figure 87D,F), and irregular oscillatory firing (Figure 87H). The one AP firing (C) showed a 4ms shorter oscillation period than the two AP firing (A). The interspike interval was 1ms shorter when changing from the two AP firing to the one AP firing (E) in comparison to the change from one AP firing to the two AP firing (G). According to the formula $T = 70\text{ ms} + 30\text{ ms} \bullet n_{AP}$, a change of approx. 30 ms would be expected in the duration of the oscillation period for the change from a longer to a shorter loop (see Figure 84). The measured oscillation period difference was only 4ms. One explanation for this large difference is pathologic firing. Another is that spinal oscillators of the trunk, including sphincter muscles for continence, may consist of two more or less coupled oscillators, which partly inhibit each other reciprocally. Also, afferent firing patterns may cause more or less directly alternating oscillation periods.

To allow differentiation between alternating oscillatory firing and the change of the oscillation loop, impulse patterns were looked for in which oscillators probably may not have changed their loop as indicated by a constant firing with a certain impulse train length (same number of APs per impulse train). In paraplegic 8, an α_1 (FF) and an α_2-motoneuron (FR) could be found, which showed alternating long and short durations of the oscillation period (Figure 88). The difference between the long and the short oscillation period was 5 ms for the same impulse train length. It may therefore be that some trunk muscles in humans are driven by spinal oscillators that are coupled in an antagonistic or alternating fashion.

11. Impaired Organisation of Human Spinal Oscillator Following Injury

Summary

1. Single-fibre action potentials (APs) were recorded from lower sacral nerve roots of brain-dead humans (HTs) and paraplegics with thoracic spinal cord injury (SCI). Impulse patterns of single oscillatory firing α_2 and α_3-motoneurons were identified and analysed. An α_2-motoneuron fired in HTs typically with three AP impulse trains every 160ms. Successive interspike intervals (IIs) of the impulse trains increased from about 3.5ms to over 10ms in paraplegics similarly as was the case in HTs.
2. In stable spinal oscillators of HTs, the oscillation period was 70ms plus 30ms times the number of APs per impulse train. The respective values for paraplegics were 35ms plus 40ms times the number of APs per impulse train, indicating an impairment of oscillatory firing following SCI.
3. By relating the peak values of the first II of an impulse train and the oscillation period, a shortest II of 3.5ms was obtained in a HT and a paraplegic, and a shortest oscillation period of 70ms was measured in a HT and 40m in a paraplegic. The derivation of the shortest II of 3.5ms from the oscillation itself is taken as an indication for α_2-motoneurons being included in the spinal oscillator function, since the shortest soma-dendritic spike intervals of motoneurons is expected to be 3.5ms.
4. The spinal oscillators of paraplegics showed more and higher activity changes than that in a HT as if spinal oscillators following SCI were insufficiently damped.
5. By interpreting the peaks in oscillation period distributions of unstable oscillators as being different oscillation loop pathways, it is speculated that the oscillators consist of the motoneuron, to which different interneuron pathways are connected. A contrasting of pathways with increasing excitation of the oscillator was observed.
6. The comparison of possible loop pathways of oscillation between paraplegics and HTs (closer to normal) indicates that the loss of descending tracts following SCI and the adaptation to it result in an increase of the oscillation pathways for α_2-oscillators from 1 to 3 in HTs to up to six in paraplegics. The pathway increase of premotor spinal α_2-oscillators in patients with spinal cord injury indicates a loss of specificity of neural subnetwork properties following SCI.
7. In HTs, measured and theoretically predictable spinal α_2-oscillators can cover altogether a frequency range between 5 and 10 Hz. A single unbalanced spinal α_2-oscillator following SCI can already cover a frequency range between 4 and 13.5 Hz, which is the full frequency range of the ankle clonus, postural and physiological tremor.
8. Promoter spinal oscillators are important because 1) they generate the high activity mode of motor units of muscles, 2) they consist, in addition to the motoneuron, of many interneurons and thus allow the study of interneuron connectivity under physiologic and pathophysiologic conditions, 3) the oscillation, namely the repeated firing with impulse trains, can be measured invasively from lower human sacral nerve roots and non-invasively with surface EMG, 4) the regular firing of spinal oscillators can be used as a reference basis to measure phase and frequency

coordination during the dynamic changes of CNS self-organization, 5) they can also be used to measure deterioration of CNS neural networks following injury or degeneration (tremor).

9. The frequencies of field potential oscillations in electroencephalography are similar to those of the premotor spinal oscillators. The frequencies within the delta range (0.4 to 4Hz) are similar to those of α_3-oscillators (0.4 to 4Hz), those within the theta range (6 to 7Hz) are similar to those of α_2-oscillators (6 to 9Hz), and the α-activity (8 to 12Hz) is similar to the frequency of α_1-oscillators (8 to 12Hz).

11.1. Unchanged Skin Afferent Activity Following SCI

It could be shown in Figure 45c that the T1 (PC) receptor density had not changed as compared to physiologic conditions (Figure 45b, [180]). It may therefore be concluded that the receptor densities of other skin receptors had not changed either. Since afferents from normal skin (no trauma-induced scar and no ulcer that may change the anchorage (for References see [115]) of the receptors) showed impulse patterns (Figure 45c) similar to physiological (Figure 45b), it is concluded that the afferent input from the skin had not changed following spinal cord injury. In regions with trauma-induced scar tissue or where pressure ulcer developed, the afferent pattern may differ, due to, e.g., changed anchorage of the receptors.

It was found earlier that neither muscle spindle undergo changes following spinal cord injuries [193]. On the other hand, a changed afferent activity from the urinary bladder was observed [192] because of bladder infections and a hypertrophy of the detrusor (reduced compliance). Activity from joint receptors and Golgi tendon organ afferents could not be measured so far. It seems therefore that if there are no pathologic changes in the periphery and any inflammation or irritation, then the afferent input from the periphery (skin, joint, muscle spindles) is physiologic. A dysfunction of the distal injury-isolated spinal cord will then be due to a false or unbalanced input from supraspinal centres including missing inhibition and the pathologic functioning of the spinal cord neural networks resulting from the isolation of the cord, and the adaptation to lost tract fibres and the loss of mobility (non-use) of the patients.

11.2. Physiologic Firing of Motoneurons in Man

The firing modes of α-motoneurons for constant afferent input are the occasional firing, the transiently oscillatory firing and the continuously oscillatory firing [176, 196] (Figure 59D). For low-level afferent input, the motoneurons are recruited in each of the nerve fibre groups (α_1, α_2, α_3) approx. every 3s according to the size principle [182, 184, 185] (Figures 54-58). With the increasing afferent input, the motoneurons switch transiently into the oscillatory firing mode (Figure 59D). For a high rather constant afferent input, the motoneurons fire continuously oscillatory. If a transient inhibition occurs, the α-motoneurons transiently break their oscillatory firing (Figures 75,76). The α_1-motoneurons (FF) fire oscillatory with a frequency around 10 Hz (range 8-12 Hz and higher), the α_2-motoneurons

(FR) fire oscillatory in the 6-8 Hz range, and the α_3-motoneurons (S) fire with frequencies below 1 Hz (Figure 3) [186, 194]. The time-related recruitment of the motoneurons in the oscillatory firing mode is distributed so that the motoneurons do not fire simultaneously (Figure 71). The firing is nicely time-distributed to smoothly activate muscles. Individual muscles show different proportions of the three muscle fibre types and are innervated by the corresponding motoneurons (Figure 3). Weight carrying muscles and probably muscles subserving posture are predominantly innervated by α_2 and α_3-motoneurons, and muscles that have to react quickly are prevailingly innervated by α_1-motoneurons. The muscle fibre composition of human muscles is only partly known [113, 133, 208]. Besides neuronal networks that activate the motoneurons to fire oscillatory, there may exist monosynaptic pathways from certain peripheral afferents onto α-motoneurons. Famous and clinically important, but overstressed with respect to the function of the CNS, is the so-called monosynaptic stretch reflex discovered by Hoffmann [92], in which the primary spindle afferents directly synapse onto α_1-motoneurons [196].

11.3. Pathophysiologic Firing of Motoneurons and Pathologic Coupling of Oscillatory Firing Subnetworks in Man Following SCI – Comparison to Oscillators in Animals

As is shown in Tables 5 and 6 and Figures 66 and 67, and was shown earlier [186], the oscillatory firing of motoneurons becomes irregular following spinal cord injury in comparison to brain-dead and normal individuals [176]. Similarities are found in animals. Sensory or descending control systems are responsible for inducing changes in gait. The coordinating system, especially in the cat, seems to be somewhat unstable in isolation, exquisitely sensitive to both descending and sensory control, and is easily acted upon to produce a wide variety of other patterns. The major phylogenetic trend has been towards increased separation and independence of the limb "oscillators" combined with an increased role to be played by descending and sensory systems, which lend a considerably increased flexibility. A possible conclusion could be that during evolution, what is changing in the central pattern generator control of the limbs is the access to and the need for extrinsic control and the general labiality of the coordinating systems [41].

Grillner and Zangger [82] have shown quite clearly in the cat mesencephalic preparation that in the absence of sensory input, the flexors and the extensors do not simply alternate. They also showed that no single pattern emerges in the simplified preparation. There was some evidence for alternation between some pairs of dorsal and ventral muscles, but it was inconsistent. Generally, in the mesencephalic as well as the decorticated curarized preparations [158], the removal of sensory feedback destabilizes the coupling between the burst generators and leads to a deteriorated pattern. The local oscillators of the lampreys, when separated from each other, burst at different frequencies. This has not been reported in dogfish. On the basis of this difference, Kopell [50] has suggested that dogfish have greater descending control of the coordinating system and perhaps less control intrinsic to the cord than do lampreys. It remains uncertain what role is played by sensory input with respect to intersegmental coordination [41]. The motor pattern observed in the intact swimming fish and those recorded from the isolated spinal cord are essentially identical [40, 234]. In dogfish [79,

80] and young chicks [108, 109], the intact and the fictive patterns are very similar. In cats, the motor pattern is more complex in intact animals. It also reflects a more complex modulation by sensory and descending control systems. Because of this, there is a great deal of variability among individual preparations, and there are some differences between the motor patterns of intact, chronic spinal [60], and functionally "isolated" cords [81].

In complete thoraco-lumbar spinal cord injury in patients, the functionally "isolated" caudal spinal cord has lost the descending control system and may have an impaired sensory control. Deteriorated organization of spinal oscillators has to be expected. In the chronic case with partial immobility, even more pathologic organization of the isolated cord has to be expected.

In the case of an incomplete cervical spinal cord injury sub C5, the sensory and the supraspinal control systems are impaired but not completely. The pattern generator for leg movements is not expected to be impaired in the lumbo-sacral enlargement of the cord, since in animals each unit oscillator spans no more than one to four segments [40]. If the lumbo-sacral enlargement for locomotion starts with L1, then the most rostral oscillators for locomotion will probably not reach Th9 segments. If we assume that the injury mainly spans three segments, then a lesion sub Th6 should only little affect the spinal locomotors centre. It has however been measured in animals that rhythm genesis was distributed along the rostro-caudal axis of the spinal cord with a stronger capability of the rostral segments of the lumbo-sacral cord compared with the more caudal segments (see below).

On the other hand, in cats, the walking seems mainly to be induced by the hip and in humans by the ankle (see [199]). Clinically, it is found that patients with incomplete spinal cord injury sub Th6 can relearn walking. In more caudal injuries, it becomes more difficult for patients with incomplete injury to relearn walking. In the future, it has to be explored with the functional MRI [34] for prognostic reasons where the spinal locomotors centre for different movements is located and what spinal segments and to what degree they are impaired. In the above-mentioned patient with an incomplete injury sub C5, the spinal locomotors centre was most likely not impaired.

However, the afferent input patterns are impaired because of pathologic movement due to the injury. With support, the patient moves rather physiologically, but the motor program remains more or less pathologic [199] (Chapter II). Therefore, the injury-induced unbalanced descending control system, partially caused by the injured ascending afferent input to supraspinal centres (see also [199]), gives rise to false activation times of some muscles. Since tetraparetics (incomplete tetraplegics) can improve their walking when walking on beach sand (deformable ground), it mainly is a stronger afferent input to the spinal cord and the supraspinal control system that improves walking.

It was shown in Chapter II that an incomplete tetraplegic could relearn running, besides other movements, by training rhythmic, stereotyped, dynamic, natural movements. Besides movement-based learning induced reorganization or changes of spinal cord neuronal networks, supraspinal reorganization of the descending control system took place to compensate for unbalanced injury of ascending afferent and control fibres. In Chapter I, it was shown within a frog model that localized membrane property changes in combination with synapse migration could be one possibility to achieve changes in the functional organization of neuronal networks.

11.4. Partial Loss of the Coordinating System and the Stabilizing Sensory Feedback May Change Spinal Oscillator Coupling and May Result in Pathologic Activation of Two-Joint Muscles

In a patient who relearned running [199], the main cause for the limitation of further improvements of running and other kinds of locomotion was false activation of the left rectus femoris [199]. The rectus femoris muscle extends over two joints, namely hip and knee. Also in cerebral palsy [16] and hemiplegia [15], the two-joint muscles are often activated at the wrong phase of the locomotors cycle or show rather permanent activity. In such cases, orthopaedics often transpose tendons to change partially a flexor into an extensor muscle or vice versa [63] to overcome the problem of false recruitment phases of muscles during the motor cycle. If the phases of activation of muscles during the motor cycle do change following spinal cord injury [199], cerebral palsy or hemiplegia and if coupled spinal oscillators have two possible phases for oscillator coupling per oscillation cycle [194, 198, 199] (per somatic nervous system), and if coupled premotor spinal oscillators can change from alternating to symmetrical oscillatory firing [199] (phase coupling changes), then it is conceivable that with the impairment of the control system supraspinally (cerebral palsy, hemiplegia) or spinally (spinal cord injury (tract fibres)), the critical two-joint muscles may show most dramatic recruitment changes per motor cycle. If there is with ongoing phylogenesis more need for extrinsic control because of the increase of the labiality of the coordinating system for motor control (and continence functions [194]), then with the partial loss of the coordinating system and the stabilizing sensory feedback (see above), a loss of specifics and a loss of functional splitting of muscle will follow. The lost coordination and stability can be explained in the oscillator concept with the deterioration of oscillator properties. Spinal oscillators with broadened frequency band will couple more easily among functionally near oscillators (loss of functional splitting) and may not couple at all with functionally far away oscillators (loss of specify) [199]. Especially the existence of two coupling phases per oscillation cycle of premotor spinal oscillators [198] gives rise to two possible pathologic muscle activations. First, the premotor spinal oscillators are activated at the wrong coupling phase of the oscillation cycle of two possible coupling phases per somatic nervous system. Motor units of an extensor muscle may now be recruited in the flexor phase. Second, the premotor spinal oscillators are coupling with other spinal oscillators at both phases of its oscillation cycle. Motor units are now recruited in the extensor and the flexor phase. This pathologic double-coupling in combination with broadened coupling phases may result in a rather permanent muscle activity. The loss of specifics and the loss of functional splitting of muscle will most seriously affect the critical two-joint muscles. There are at least two possible treatments for the false muscle recruitment during the motor cycle. One can think mechanically and tries to improve locomotion in the patient by transposing tendons, or one believes in the plasticity of the human nervous system and tries to improve locomotion by training-induced network plasticity (Chapter II). This research project follows the second line [199] in accordance with W.R. Hess, who argued that one should not think too mechanically with respect to the functioning of the human nervous system. Experience shows that transposed tendons change the movement-induced afferent input so much that patients with CNS injury have big problems relearning physiologic movements. However, patient with a not-injured CNS can quickly relearn physiologic movements following tendon transposition.

11.5. Tremor and Repair of Premotor Spinal Oscillators

It is unlikely that the most important firing mode of the motoneurons, the oscillatory firing, would not be observable macroscopically in certain situations. Indeed, it was already Descartes [47] who wrote about tremor. During the last 100 years, many authors have suggested rhythmic firing of neuronal networks that are responsible for rhythmic movements (trembling, tremor) of the body (trunk) and legs, arms and fingers (for Refs. see [117, 243]). The conclusions drawn by R. Jung from measurements on tremor and clonus in 1941 [117], are nearly identical with the findings concerning the self-organization of spinal oscillators in this human research. Only, R. Jung did not differentiate between different motoneuron types and analysed the tremor by using mechanical and electromyography recordings. The similarity between the tremor frequencies and oscillator frequencies of a brain-dead human individual (physiologic case) and two paraplegic patients (pathophysiologic cases) has been described elsewhere [185, 198] (see also below). Tremor in Parkinson's disease patients can be induced by uncontrolled activation and uninhibited synchronization of α_2-motoneuron firing to which α_1-motoneuron firing synchronizes [202]. In a very important measurement of Chapter V, it will be shown that spinal oscillators can build up external loops to muscle spindles, and the oscillation frequency of spinal oscillators can be modulated, via these external loops, in a certain frequency range. The possibility that the oscillatory firing can be controlled in a certain frequency range by this external loop gives rise to the possibility to formate the pathologically oscillatory firing networks to fire more physiologically again and to reduce spasticity (false activation of muscles with respect to phase and activity level) with the gain of useful locomotion. The jumping on spring-board repairs premotor spinal oscillators by the rhythmic movement-induced afferent input. The rhythm frequency lies in the range of those of α_3-oscillators (see Chapter V). From the changed properties of oscillators, it is likely that the tremor will change following incomplete spinal cord injury. From electromyography recordings, it can be seen that specific properties of motor unit firing are lost [202, 199]. The physiologic activation times of the muscles are partly lost [48, 199, 239]. In Parkinson's disease patients, CNS functioning can be improved upon administering coordination (pattern) dynamics therapy (Chapter II) [201]. This improvement can be quantified by measuring the coordination (pattern) dynamics (Chapter II) [200]. The problem in older people is that the patients cannot be pushed to the limits of exercising. Young patients, on the other hand, can be pushed to their limits. But most likely, more nerve cells are dying due to the progress of the disease than new ones can be built by movement-based learning, including the building of new nerve cells.

11.6. Coupled Half-Oscillators for Activating Human Trunk, Continence and Leg Muscles?

Since there is little doubt that spinal oscillators (premotor neuronal network, motoneuron included) and other rhythmically firing neuronal subnetworks (propriospinal oscillators) are the original reason for tremor, the oscillator theory for the understanding of parts of the functions of the CNS must be able to incorporate or explain the different kinds of tremor. In tremor, it is differentiated between tremor of the extremities and of the trunk. Tremor of the limbs may be independent or changing with coordinated bilateral asymmetry. In trunk muscles, there is absolute coordination in tremor. In head and trunk muscles of both sides, it shows a symmetric alternating pattern (phase difference 180°). In rostral caudal direction, tremor shows certain constant phase relations [117]. Trunk tremor has therefore very similar occurrence patterns as do phylogenetically old movements of the fins of certain fish [93] and the spinal locomotors organization in the lamprey [83, 84].

In the lamprey, spinal cord segments produce the motor output underlying locomotion throughout a frequency range of 0.25 - 10 Hz (the frequency range measured in humans is 0.3 -10 Hz (Figure 3)). The motor output can be subdivided into small pieces down to a few segments, each of which can produce rhythmic bursts of activity and a characteristic intersegmental coordination with a constant phase lag between consecutive segments. Under intact conditions, this results in an undulatory wave travelling down the body and pushing the animal through the water. Each segment contains about 100 motoneurons that supply different parts of the myotome. The motoneurons of the left and right side provide alternating burst activity. As the motoneurons of one side get excited, the motoneurons on the contralateral side become inhibited in a reciprocal fashion. During a swim cycle, each motoneuron thus goes through a phase of excitation followed by inhibition [83, 84].

It was shown in Figure 88 that an α_1 and an α_2-motoneuron fired continuously oscillatory. However, every second cycle the oscillation period was prolonged. This kind of firing can possibly be interpreted as every second oscillation cycle being inhibited. It is therefore conceivable that an oscillator circuit on one side was inhibited by an oscillator circuit from the other side and vice versa, in some similarity to the lamprey model. Whereas motor pools to antagonistic muscle pairs in the lamprey lie on the opposite side of the cord, in quadrupeds motor pools to antagonists within a single limb are on the same side of the cord. Further, alternation is the only coordination pattern observed in all intact, reduced, or spinal animals studied (page 159 of [41]). In many quadrupeds, limb muscles at a single cord level can exhibit a range of both alternating or symmetrical (walk and trot) and non-alternating or asymmetrical (canter, gallop, etc.) phase relations.

In the lamprey locomotors system, the motoneurons in each segment show alternating left-right burst activity [83, 84]. It is therefore conceivable that also in humans, some oscillatory firing networks consist of two more or less coupled halves that may or may not show alternating long and short oscillation periods. Motoneurons innervating antagonistic limb muscles (in humans, a few motoneurons, innervating leg muscles, run through the lower sacral roots) run through the same root, whereas motoneurons innervating antagonistic trunk muscles (and may be continence muscles) will run through the opposite roots. To provide evidence for half-centre networks of continence muscles, one would have to record from the two halves of the left and right side of the medulla. So far, this has not been possible with the single-nerve fibre AP recording method because this would require recording with four electrode pairs instead of two. If coupled antagonistic spinal oscillators fire pathologically, it may be that one part of the oscillator fires more pathologically than the other half. In the unbalanced case, it may then be that every second oscillation shows a different oscillation period even for coupled oscillators that normally show no alternating long and short oscillation periods. Most likely, the mutual inhibition (if existing) was not as strong as in the lamprey model, since the prolongation of the oscillation period was on average only 4 ms (4 and 3.3 %). Nevertheless, these measurements indicate that there may be some mutual inhibition between spinal oscillators. Further, this (reciprocal) inhibition may work differently for α_1 and α_2-motoneurons in that the projection onto the (antagonistic) motoneuron network for inhibition is more direct (less interneurons) for the α_1-motoneuron.

It was shown in Figure 3 that primary and secondary muscle spindle afferents project differently onto oscillatory firing α_1-motoneuronal circuits (the primary afferents project monosynaptically; the secondary ones project oligosynaptically). Therefore, also self-organizing projections can be expected from α_1 to α_2 oscillatory firing circuitry and vice versa. Further research has to show how directly the reciprocal inhibition acts. Simultaneous recordings from right and left nerve roots will provide further information. On the other hand, a change from alternating to symmetrical oscillatory firing has been observed electromyographically in humans, which was correlated to a simultaneous phase change in synchronization of two other oscillatory firing motor units [199]. This measurement indicates that alternating and symmetrical oscillatory firing are related to coupling changes among populations of interacting spinal oscillators in the human CNS. Alternating oscillatory firing may therefore be due to reciprocal inhibition of half-centre oscillators, changes in the couplings of non-equal oscillators or mutual inhibition of oscillators driving antagonistic muscles. So far, these measurements have mediated no information on how each oscillation cycle of each oscillatory firing circuit is terminated. Graham Brown, who proposed the half-centre model, hypothesized that a limiting factor could be fatigue of the inhibiting synapses, which causes the strength of synaptic transmission to decline with time. Other possible intrinsic processes, now considered more plausible, are adaptation, in which a neuron responds to constant excitatory input with a declining rate of output, and post-inhibitory rebound, in which the threshold for excitation of a neuron decreases transiently as a result of past inhibition [119].

Intrinsic processes cannot be the only or main cause for the termination of the oscillation cycle. The extreme changes in the duration of consecutive oscillation periods and the reduction of the duration of the oscillation period following spinal cord injury cannot be explained by changes of membrane properties alone. Changes of membrane properties on a large scale following spinal cord injuries are probably negligible since most projections onto motoneurons come from the interneurons, and interneuron cell death following spinal cord lesions has not been observed so far. On the other hand, this reasoning gives no explanation for the spinal shock. But localized changes of membrane properties can have substantial functional changes as is indicated in a frog model (Chapter I). In a partially denervated slow muscle fibre, the denervated part changes its membrane properties in the way to generate APs, which the innervated part cannot do (Figure 86C).

11.7. Modelling of Human Neuronal Subnetwork Functions

It has been argued that the oscillatory firing with alternating long and short periods can simply be generated by a propriospinal oscillator, which drives two interneurons, which in turn drive with different delays the motoneuron to fire alternating oscillatory. Such simple neuronal wirings are supported with the argument that most important constructions of living organisms (genetic mutation and natural selection) like elegant and simple solutions. Since nearly nothing is known concerning the integrative functions of the human nervous system, many speculations are possible. We therefore have to stick as much as possible to measurements. Indeed, the schematic model of Figure 16 of Ref. 195 can explain the alternating oscillatory firing of α_1-motoneurons, firing rhythmically with one AP. The alternating oscillatory firing of the α_2-motoneurons, firing with impulse trains consisting of several APs (Figure 88), cannot be explained with that model. Additional or different interneurons are necessary. Further, there is some indication that the motoneuron itself is a part of the premotor spinal oscillator (Figure 68). If this is so, then the model is not valid anymore. It has been measured further that when the motoneuron is not firing continuously or transiently oscillatory, then the motoneurons are activated approximately every 3s according to the size principle in each motoneuron group [182, 184, 185] (Figures 54-58). This important different functional state, generating a certain low tonus in the muscles, cannot be explained by the model. The change from alternating to symmetrical [199] cannot be explained by the model and neither can the changing couplings between different oscillators. The model can mimic only one special network function but not all the others. For modelling of neuronal functions of animals see [110].

Modelling of human subneural networks only makes sense if all the known human data are incorporated. Further, the problem for the time being is that nearly no human data are available. Animal data, as, for example, synaptic transmission time (~ 0.5 ms), time for the ascending phase of the exciting postsynaptic potential above threshold for spike generation (~ 2 ms) [110] or membrane properties, have to be measured, especially as there is indication that human membrane properties of neurons are different from those of animals. The conduction velocities, for example, of peripheral nerve fibres is much lower than those in the rat, cat and dog (Figure 53), and transmission frequencies of up to 5 kHz have been measured in paraplegics [183], whereas textbooks give a highest value of 1 kHz. There are more neurons in the human body than there are humans on earth, which does not point towards subnetworks that consist only of a few neurons. It is possible that hundreds of interneurons contribute to the oscillatory firing of neural subnetworks. The quality of the human nervous system lies in its complexity to be able to adapt to the surrounding and its plasticity. There are perhaps simple and elegant strategies for the self-organization of neuronal networks of the spinal cord according to afferent and supraspinal inputs.

As long as the modelling of human subneuronal networks does not incorporate human data, the models may be useful in robotics and computer sciences, but they have nothing or nearly nothing to do with the human reality. In the clinic, we need to understand the functioning of the human nervous system under physiologic and pathophysiologic conditions and its capacity for reorganization to adapt treatment to it. Actually, the enormous functional possibilities that are contained in the human nervous system seem to make it possible that may be only a small part of the system is able to run the basic functions [199].

11.8. Origin of Oscillatory Firing

What is necessary for a neuronal system to produce self-sustaining oscillations but not require phasic external input? The measuring situation, e.g., is if in a human with a complete spinal cord injury an α_2-motoneuron (FR) innervating the external anal sphincter fires oscillatory due to the isometric stretch of the anal sphincter by an anal catheter. Two features are recognizable. First, one action must inevitably be followed by its antagonistic action (sign reversal). Thus, excitation (or firing) of the motoneuron must be followed by inhibition. Likewise, inhibition must lead to excitation (or firing). For example, excitation to initiate the impulse train (burst) must ultimately be followed by inhibition to terminate the burst. Inhibition could include (i) activation of an intrinsic membrane current that tends to hyperpolarize the motoneuron, (ii) synaptic inhibition, or (iii) its functional equivalent, loss of excitation. The second major feature of an oscillatory firing motoneuron is that there must be a delay between the initial action and its antagonist [62]. A delay is necessary to allow one process (excitation or inhibition) to be expressed before the onset of its inevitable opposite action. The delays are relatively long and close in duration to the cycle period. Oscillatory firing can originate in properties intrinsic to single cells (cellular), from the action of single synapses (synaptic) and from the self-organization of an assembly of cells due to the input (network) and synapsing into complex circuits [67, 217]. It is indicated below that the main functional contribution of human spinal oscillators comes from the network properties.

11.9. Definition of (Biological) Spinal Oscillators

A premotor spinal oscillator is viewed as an assembly of interneurons (changing number of contributing interneurons) and a motoneuron, which, by virtue of their intrinsic properties (preformatted neuronal network = neuronal connectivity with a selection of synaptic strengths), is capable of self-organizing by pattern messages from supraspinal centres or the periphery in the form of spatio-temporal vectors. The premotor spinal oscillator fires rhythmically with impulse trains according to the motoneuron type and is capable of undergoing relative transient coupling with other oscillators of spinal (motoneuronal (premotor) or purely interneuronal (propriospinal)) or supraspinal origin (especially brain stem) to generate systematic macroscopic pattern, for example, locomotion or homeostatic regulation. In these self-organized circuits, the activation is most probably maintained by circulating synchronized firing (closed "synfire chains"). Stabilisation of the reverberatory activity in positive (re-excitatory) loops could be supported by processes similar to synaptic modulation.

Since there is indication that motoneurons are a part of the rhythmically firing premotor neuronal network, the spinal oscillators are divided here into the premotor spinal oscillators and propriospinal oscillators. For high activation, the premotor spinal oscillators are the output network of neuronal networks, and the propriospinal oscillators are oscillatory firing subsystems, which generate movement patterns (Figure 20 of [199]) or homeostatic regulation.

Transient coupling is generally necessary for relative coordination of the different rhythmic firing systems to generate an integrative function. Typical coupling response phenomena have been observed in firing occurrence patterns of spinal oscillators [186]. Rhythm coordination seems to be a basic principle of physiological self-organization of many rhythmic subunits. There are many kinds of possible transitions between full independence and strict synchronization. Complex physiological rhythmicity appears as a general principle of organization, allowing a compromise between stability and flexibility of adaptation. If α-motoneuronal spinal oscillators become synchronized over longer time periods, tremor will be created, which is disadvantageous. Prolonged coupling of many premotor and propriospinal oscillators most likely hinders differentiated organization and results in pathologic neuronal network organization like spasticity. Similarly, when in the cerebral cortex, the synchronized discharge of large masses of cells becomes excessive, and spontaneous movements appear to a pathologic degree as in the epileptic attack; the smaller cell groups losing their individuality of differentiated function [112].

11.10. Cellular Oscillator Properties

Central neurons contain more numerous and complex inonic currents, as originally described by Hodgkin and Huxley [91] for the squid axon. Different combinations of these currents can impart radically divergent response characteristics to cells (for review, see [2, 13, 142, 147]). Often, the underlying ionic currents are not as important as the nature of the response characteristics to which they give rise. The important cellular properties that contribute to pattern generation include (in similarity to statements of Getting [68]):

1. *Threshold:* This property determines the level of excitation needed for an oscillator to start firing. The threshold for self-organization of the spinal oscillator is given by adequate input (afferent or supraspinal) patterns in the nerve fibre group-spatio-temporal system. Adequate input patterns probably trigger as much self-organization among interneurons that excitation just travels from the motoneuron (premotor oscillators) to the interneurons and back to the motoneuron via the afferent input facilitated recurrent excitation pathways may be in the form of "synfire chains."
2. *Spike-frequency adaptation:* This property is manifested as a decrease in spike frequency (increase of interspike intervals of successive APs of an impulse train) with time during sustained depolarization of the motoneuron. - Spike frequency adaptation is associated with the activation of slow outward potassium currents. Adaptation is a self-inhibitory process, as it depends upon the recent firing history of the cell and tends to decrease the responsiveness of the cell for further firing. Spike frequency adaptation can serve to shape burst structure (of the impulse train) [104], and it contributes to the regulation of cycle period length [67] approximately following the formula (for the human premotor spinal oscillators) $T = 70$ ms $+ n_{AP} \bullet 30$ ms (T = oscillation period; n_{AP} = number of APs per impulse train) under physiologic conditions.
3. *Post-burst hyperpolarisation:* Following an impulse train, APs were not observed in humans. Most likely, the motoneurons became hyperpolarized. - Following a burst of spikes, many neurons become hyperpolarized for a period of time ranging from milliseconds to seconds (depending on the strength and duration of the preceding burst). The post-burst hyperpolarisation appears to be a manifestation of a mechanism similar to that responsible for spike-frequency adaptation [104, 141]. Both characteristics are usually present together. Post-burst hyperpolarisation also is a self-inhibitory process producing an inhibition delayed until after a burst. According to Ekeberg [54], temporal recruitment (spike synchronization) is an important mechanism for determining the transient response of a pool of neurons, and it may be related to the after-hyperpolarisation of the neurons.

4. In an α_3-motoneuron (S), firing with impulse trains consisting of approx. 40 APs [176] every 0.7 seconds, sometimes a short impulse train was observed (≈ five APs) in between the impulse trains. It could therefore be that post-inhibitory rebound and delayed excitation [68] contributes to the oscillatory firing of the different types of motoneurons in man. For endogenous and conditional bursters and synaptic properties, see [68].

11.11. Recurrent Loop Excitation and "Inhibition of Recurrent Inhibition" for the Oscillatory Firing of Motoneurons

Systems that are used infrequently and produce only a few cycles rely heavily upon a balance of synaptic excitation and inhibition for pattern generation [68]. It is likely that cellular properties (points 1-4 above) and excitatory and inhibitory connexions contribute to the rhythmic firing of spinal oscillators. The question is whether there is a need for recurrent excitation or whether the inhibition of the recurrent inhibition can explain rhythmic firing. Hultborn and Pierrot-Deseilligny [99] maintain that during voluntary tonic soleus contraction of increasing strength or during ramp contraction in man, the Renshaw cell recurrent inhibition is progressively diminished from an initially facilitated level, probably by supraspinal descending influences bypassing the motoneurons. The inhibitory influence on Renshaw cells from supraspinal sources appears to dominate and to offset the excitatory input derived from motoneuron excitation. In the vicinity of the inhibited Renshaw cells, motoneurons were found that were excited by the supraspinal stimulus with the appropriate discharge pattern to make them candidates for mediating the inhibition on the Renshaw cells. Ross and Wittstock [166] reported changes in lumbar Renshaw cell activity correlated with vertical whole-body movements, indicating that macular afferents exert an inhibitory influence on Renshaw cells.

Henatsch et al. [89], upon conditioning stimulation of the contralateral nucleus ruber, found predominantly facilitation of various monosynaptic reflexes associated with depression of antidromic and orthodromic Renshaw cell discharges, again indicating a flexible coupling of the latter to their input-giving α-motoneurons. It is not clear in many cases how the supraspinal effects on Renshaw cells are related to those on the α- (and/or γ-) motoneurons to which the Renshaw cells may be predominantly related. Both the Renshaw cells and Ia inhibitory interneurons exhibit rhythmical modulation of their discharge during fictive scratching and stepping [44, 45, 59, 116]. However, they do not belong to the rhythm generator since considerable changes of their activity upon stimulation of ventral roots do not influence the cycle duration [59, 116]. The Renshaw cells and Ia interneurons exert inhibitory action upon motoneurons, but they are not the neurons responsible for the basic pattern of rhythmic changes of the motoneuron membrane potential. Depolarization appears in motoneurons just at the moment of the maximum drive from Renshaw cells and Ia interneurons [44, 59]. Since there probably is similarity between the network organization of premotor spinal oscillators and the rhythmically firing subnetworks, constructing the pattern generators (propriospinal oscillators), Renshaw cells and Ia inhibitory interneurons do not essentially contribute to the rhythm generation.

In connection with the excitatory and inhibitory interneurons of the lamprey, it is likely that the alternating long and short durations of oscillation period (Figure 88) can be explained by mutual inhibition. An inhibition of recurrent inhibition, however, in humans, can at best cancel the recurrent inhibition but cannot explain the oscillatory firing of motoneurons. If, for example, the recurrent inhibition is completely inhibited by supraspinal activation of spinal neuronal networks, then the oscillatory firing of the α-motoneurons must originate in the motoneuron cell itself. Probably, the α-motoneurons contribute to the oscillatory firing (see above) but cannot explain the oscillatory firing for at least two reasons. Firstly, following spinal cord injury, the rhythmically firing motoneurons change dramatically their oscillation period duration from one cycle to the next. Such oscillation period changes cannot be explained by an endogenous oscillatory property of individual motoneurons. Membrane properties are known to change following denervation [143]. However, following spinal cord injury, no interneuron cell death has been observed so far (no published information). Since further the motoneurons get the majority of their input from interneurons, the denervation effect by the spinal cord lesion (destroyed spinal tracts) is only weak [12]. It has been shown that following partial denervation, membrane properties change in the denervated areas only ([168], Figure 21 of [199]). Since the synaptic contribution from tract fibres is but small, the overall membrane properties of the α-motoneurons probably change only little. Similar arguments may hold for the interneurons. On the other hand, it will be difficult with the above to explain the spinal shock. The large changes of the oscillation period duration seem to be explainable by membrane property changes alone. An endogenous burster cannot explain the changes in oscillation period. We have to think of a network producing the oscillatory firing. The spinal oscillators, which are under consideration herein, are those networks of which the motoneuron is a part (premotor neural network). It is most likely that oscillatory firing is generated by recurrent loop excitation. Secondly, in normal humans, endogenous bursters cannot explain transient oscillatory firing and resetting of the cycle [197]. Endogenous bursters are thought to generate chronic rather than episodic rhythms. If, as a working hypothesis, the motoneurons fire rhythmically with impulse trains, via recurrent excitatory loops (because the motoneuron seems to be a part of the premotor spinal oscillators [186]), why has no recurrent excitation has been measured? Upon stimulating axon collaterals through electrical stimulation of ventral roots, only inhibitory potentials are recorded in α-motoneurons. Sophisticated electrical stimulation of CNS neuronal networks are only a good research tool if the neuronal networks are "fixed wired." If the neuronal networks organize themselves according to the natural input patterns of afferent fibres and descending tract fibres, artificial electrical stimulation represent artificial experimental conditions and will give rise to artificial network organization.

It would be of importance to have intracellular recordings from oscillatory firing motoneurons. Most likely, one would record excitatory potentials similarly as for respiratory motoneurons [57]. Then, additional electrical stimulation of the ventral or dorsal root efferents may show competition between inhibitory and excitatory synaptic potentials. To explore natural functions and organizations, we have to stimulate naturally to generate natural organizations of the CNS. Probably, recurrent excitatory loops (maybe in the form of synfire chains), are only opened (facilitated) by adequate afferent input patterns if the necessary task groups of interneurons are activated. It is proposed that in the human spinal cord, there does exist recurrent loop excitation besides recurrent inhibition, since recurrent inhibition cannot explain oscillatory firing of a subneural network of which the motoneuron is a part.

In two sets of animal experiments, it has been shown that Ib effects change dramatically during locomotory compared with static conditions. Before locomotion, Ib inputs produced inhibitory postsynaptic potentials, and during locomotion, they produced excitatory postsynaptic potentials [73, 154]. It has been measured in animals that rhythmogenesis was distributed along the rostro-caudal axis of the spinal cord with a stronger capability of the rostral segments of the lumbo-sacral cord compared with more caudal segments. The synchronization of motoneuron activity, as well as synaptic drive to caudal motoneurons, was mediated, at least in part, by axons running within the ventro-lateral tract. Also, optical imaging showed that each successive rhythmic burst starts in the ventro-lateral region (where the motoneurons are located), and then a Ca^{2+} wave invades the lateral part of the intermediate area (Rexed lamina VII and others) of the grey matter (where spinal interneurons are located) [11, 96, 148].

11.12. Synergetic

It is most important to get more data concerning the natural functions of the human CNS. On the other hand, we also need theoretical concepts or approaches that would allow us to understand such a complex system as the human CNS and give us ideas for developing adequate research strategies. The most advanced theoretical tool currently available is possibly Synergetic. The principles of self-organization are the central point of its study [88]. In synergetic, instead of considering an extremely high-dimensional space of all the variables, it suffices to consider the low-dimensional space of order parameters and to discuss the kinds of attractors available in such a low-dimensional space. When a system relaxes towards an equilibrium state, one may say that it relaxes to a so-called stable fixed point. When it oscillates coherently, one speaks of a limit cycle attractor, which is described in particular by a specific frequency. One of the main achievements of synergetic is the insight into the mechanism that produces macroscopic structures, be it temporal, spatial, or functional. The system theory of pattern formation rests on synergetic and is used as one basis to repair the human CNS (Chapter II).

11.13. Mathematical Predictions from Populations of Interacting Biological Oscillators Compared to Measurements on Neuronal Network Data

Since in certain functional states of organization, at least parts of the neuronal networks of the human CNS do self-organize to oscillators that interact with each other to generate integrative functional structures, a biological oscillator approach of spinal cord functions seems (even though not most basic) of much practical value in explaining rhythmic network data with the properties of populations of interacting oscillators. The premotor spinal oscillators were functioning pathologically due to non-use and missing supraspinal drive, caused by the complete spinal cord injury.

The physiologically functioning α_1-motoneuron oscillators can mathematically be approximated by a single oscillator or better by a few interacting oscillators with mono-frequency oscillation. At least a system of several coupled oscillators is required to describe multi-frequency oscillations and the occasional firing mode of α_2-oscillators (page 151 of [152]). Since there are easy changes of the oscillation periods from 130 ms to 160 and to 190 ms, a strong coupling has to be assumed between the suboscillators (page 156 of [152]). The premotor α_3-oscillators, consisting of many APs per impulse train and therefore many units of 30 ms and rather continuously changing oscillation frequency within a certain range, probably consist of a large population of strongly interacting oscillators (page 156 of [152]).

To a certain extent, the assumed approximated network structure of oscillators by synfire chains is compatible with the mathematically obtained results on populations of oscillators giving rise to oscillation periods of 130, 160 and 190 ms. Several synfire chains with recurrent loop excitation, converging onto one motoneuron, can account for strong coupling of several oscillators with a few frequencies (α_2-oscillators). The repeated firing with impulse trains, probably generated by a transient long depolarization of the motoneuron, can also be generated electronically by a few coupled oscillators. The premotor α_3-oscillators can be set up of many synfire chains. The generation of numerous APs per impulse train can simply be achieved by coupling of a few oscillators. The rather continuous change in oscillation frequency of α_3-oscillators [176] can be simulated mathematically and explained structurally by a number of similar long synfire chains with little length change of neighbouring chains. The coupling of oscillators is achieved by an overlapping of synfire chains. From a mathematical point of view, most premotor oscillators are a small population of strongly interacting suboscillators.

With the deterioration of the neuronal networks because of non-use, and missing supraspinal drive or missing synchronization with supraspinal oscillators, most likely the properties of the oscillators and the strength of mutual coupling changed. Because of the broadening of the frequency band, masses of oscillators will couple and may give rise to mass contractions of muscles (loss of specify and loss of functional splitting of muscle functions). Because of a reduced strength of coupling, certain stable functional states of coupled oscillators will not be reached anymore (page 156 of [152]). Specific functions are lost and/or altered.

How the neuronal networks self-organize and the change from one functional state of organization to another (transition of functional states) cannot be solved mathematically or physiologically alone. Network changes by plasticity or deterioration will anyway change the system under consideration. Because of these and other difficulties in a mathematical approach, it seems important in the long term to establish a mixed strategy involving analysis, computer simulations and measurements with different methods, to understand certain ranges of functioning of the networks especially for stable or rather stable network states.

However most important is the need for more data on human spinal oscillators, i.e., data on the physiologic regulation of coupling and recoupling of basic oscillatory firing subsystems, and also, it is important to know what kinds of different basic oscillators do exist. At least for premotor oscillators, the assumption that neuronal network structures can be simulated by the interaction of one or two populations of very similar oscillators is not in accordance with the human measurements.

Two groups of oscillators were defined here, the premotor and the propriospinal oscillators. The premotor spinal oscillators split into α_1, α_2 and α_3-oscillators according to the type of the motoneuron integrated. In this work, no measurements were done on propriospinal oscillators. It was suggested earlier that preganglionic parasympathetic motoneurons may also fire oscillatory [194] or almost oscillatory.

Thus, premotor parasympathetic oscillators may also exist. Since false interaction between different oscillator populations may occur following spinal cord injury giving rise to important clinical problems (detrusor (driven by parasympathetic oscillators) - sphincter (driven by α_2-oscillators (Figure 3)) dyssynergia), the interaction between different oscillator populations should be analysed mathematically (for further details, see [199]). Rhythmic firing and coupling has been observed in many biological systems [124]. Rhythmic coupling in the common brain stem system gives rise to respiratory sinus arrhythmia [129, 164, 213]. Oscillators are not only a principle network structure of the human CNS since they keep organized at least sometimes during the transition from one functional state of organization to the next [197], but their functioning can comparably easily be measured and may provide for understanding of the functioning of the human CNS.

11.14. Natural Impulse Patterns and Electro Stimulation Patterns in the Treatment of Spinal Cord Injury

Figures 40 to 45 show simultaneous impulse patterns secondary to a pinprick or touch stimulation of sacral dermatomes with a needle. The impulse patterns of the more than 30 myelinated afferents (Figure 40) were conducted in the fibres of a coccygeal root. Many further afferents in the adjacent S4 root will also have been activated upon that pinprick since sacral dermatomes strongly overlap. One hundred to 200 myelinated afferent fibres will have probably been activated upon touching the skin inside the anal reflex area [196] and will have given rise with their impulse patterns to a contraction of the anal sphincter to secure continence. Such a collection of spatiotemporally related impulse patterns is the input to the sacral neuronal networks to organize them in a way that the premotor spinal α_2 and α_3-oscillators are activated to fire transiently oscillatory for the contraction time of the external anal sphincter.

One can imagine what an enormous number of specific spatiotemporally related input patterns enter the spinal cord with the movement-induced afferent input from the skin, muscle and joints to organize the neuronal networks of the spinal cord to induce and maintain the primary automatic stepping in newborn infants [199]. The sural nerve contains between 5,000 and 10,000 myelinated afferents (page 69 of [51]), which innervate a skin area of approx. 30 cm^2 of the foot sole.

The message from the receptors of the periphery for the self-organization of the CNS is enormous and specific. It is difficult to see how such a collection of spatiotemporally related afferent input patterns can be simulated by electro stimulation. Even though the natural impulse patterns in the motor system are less complex, one still has to consider that firstly the F, FR and S-type muscle fibres get their specific activation patterns (Figure 3) and secondly that they are not activated simultaneously but nicely distributed over time (Figure 3 of [194]). The action potentials retrogradely conducted upon the stimulation in the axons will enter the neuronal network via recurrent fibres and will supply at least the premotor spinal oscillators with artificial impulse patterns for pathologic network organization.

It further has been shown that the nerve-muscle junctions do sprout and remodel with training [240, 241], whereas a direct stimulation of muscle fibres will not increase the area covered by the motor endplate. Similar arguments will hold for stimulation of the CNS, especially the spinal cord. Further, one has to remember that the pattern of electrical stimulation changes the functional properties of muscle fibres [207], as is experienced in electrical stimulations of the nervous system in very high spinal cord injuries when the diaphragm changes its speed of contraction with electrical stimulation. It seems therefore that the electro stimulation could be a substitute for a missing natural therapy in a very few cases only.

11.15. Similar Frequencies of Premotor Spinal Oscillators and Field Potential Oscillations in the Cortex

The frequencies of field potential oscillations in electroencephalography (EEG) (220) are very similar to those of the premotor spinal oscillators. The frequencies within the delta range (0.4 to 4Hz) are similar to those of α_3-oscillators (0.4 to 4Hz), those within the theta range (6 to 7Hz) are similar to those of α_2-oscillators (6 to 9Hz), and the α-activity (8 to 12Hz) is similar to the frequency of premotor spinal α_1-oscillators (8 to 12Hz). The high-frequency oscillations in the EEG recordings in the β (15 to 30Hz) and γ range (30 to 60Hz) have so far no counterpart in the premotor neuronal network of the spinal cord. But the oscillations measured in EEG recordings are changes in field potentials, whereas the measured oscillations of the premotor spinal oscillators are rhythmic activity of individual nerve cells (motoneurons) of the CNS. Moreover, the natural impulse patterns of single neurons provide more information on the functioning of the human CNS than do frequency ranges in a Fourier spectrum.

A typical α_2-premotor spinal oscillator fires every 160 ms (= oscillation period T) with an impulse train of three APs. The oscillation period of 160 ms is most likely generated by the local neuronal network (the loop (as one possibility) time is 160 ms) whereas the three-AP impulse train with interspike interval durations of 4.6 and 7.4 ms is most likely generated in the motoneuron itself (a longer depolarizing pulse excites the motoneuron to fire repetitively (with an impulse train) [176]).

Therefore, the directly measured natural impulse pattern provides structural information; the interspike intervals of the impulse train of 4.6 and 7.4ms duration provide primarily information concerning the membrane properties of the activated motoneuron, and the oscillation period of 160ms provides primarily information concerning the network properties of the oscillatory firing cell assembly.

And of course, the natural impulse patterns of motoneurons can be compared with the natural impulse patterns of afferents, and the changing phase relations between motoneurons and afferents can be analyzed to provide clues for an analysis of integrative CNS functions. Fourier analysis provides mixed information concerning rhythmicity of network properties and membrane properties of neurons.

When analyzed using Fourier analysis, the membrane properties of the motoneuron, giving rise to the interspike intervals of 4.6 and 7.4 ms of the impulse train, suggest rhythmic network or single neuron firing in the range of 22 and 14 Hz, even though these frequencies do probably reflect the rhythmicity of repeated firing of the depolarized motoneuron membrane. Therefore, direct analysis of natural impulse patterns provides more information concerning the structure of the neuronal network of the human CNS than does Fourier analysis.

Spontaneously occurring field potentials appear to follow the rule that the amplitude of the fluctuations decreases with the increasing frequency of oscillations. This suggests that rapidly oscillating cell assemblies comprise fewer neurons than slowly oscillating assemblies [220].

The same conclusion was arrived at with the formula for the oscillation period of premotor spinal oscillators $T = 70ms + 30ms \bullet n_{AP}$ (n_{AP} = number of APs per impulse train), namely that the premotor spinal α_1-oscillator (10Hz) comprises fewer building blocks of T, and therefore interneurons [176] than α_2 (6-9Hz) and α_3-oscillators (0.4-4Hz) if one assumes that the oscillation period in the local neuronal network is generated by a series of neurons connected to a reverberatory loop, of which the motoneuron is a part (Figure 3). Since there is similarity in the frequencies of oscillatory firing cell assemblies of the most caudal CNS (premotor spinal oscillators of the conus medullaris) and the rostral CNS (cortex), and the oscillatory firing cell assemblies are most likely quite different (there are no motoneurons in the cortex), the similarity of the frequencies of the spinal cord and the cortex may indicate rhythm coupling as a coupling or activation mechanism of the spinal cord by the cortex. Relative coordination, with synchronization as the simplest type of coordination, may be used throughout the CNS for self-organization. It should be simultaneously recorded from the cortex (EEG) and from muscles (EMG) during coordinated movements to see whether there are phase relations (relative coordination) between the firings of single motor units and the changes of the field potential.

11.16. Space-Time Distributions of Oscillatory Firing Neuronal Networks and Network Oscillator Coupling

In stable oscillators, the duration of the first interspike interval (II) already provides information on how long the impulse train will be and how long the oscillation period will be. This means the complete information about the ongoing oscillation is already contained in the length of the first II. In other words, the oscillation period in stable oscillators "tells" how long the impulse the impulse train and the duration of the first II were. Distributions of the occurrence of interspike intervals within the impulse trains (Figures 89A, 90A) and of oscillation periods (Figures 89B, 90B) therefore provide information concerning the organization of the neuronal networks of the medulla. The shift of the corresponding distribution peaks in oscillation period distributions with increasing oscillator network activation (increasing impulse train length) suggests the existence of several rather discrete network loops of α_2-oscillators (Figures 89B, 90B).

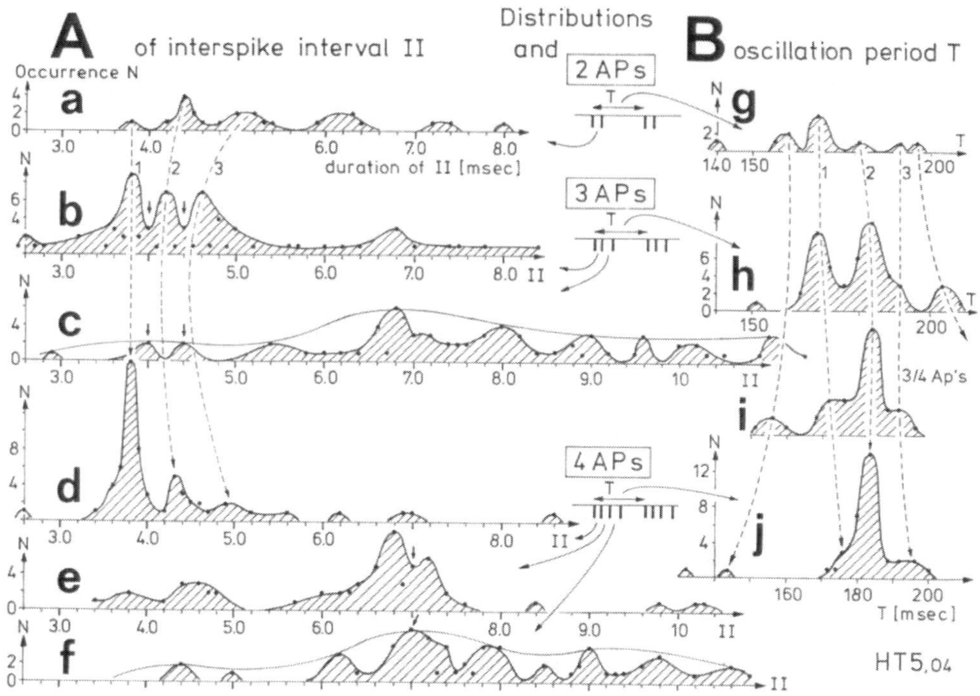

Figure 89. Interspike interval frequency distribution (A) and the corresponding oscillation period frequency distribution (B) for different lengths of impulse trains. The corresponding distribution peaks of the first interspike intervals and the oscillation period are connected by the dashed lines. The corresponding interspike interval peaks and oscillation period peaks are labelled in the same manner (1, 2 and 3) (b and g). Distributions of the second (c, e) and third interspike intervals (f) are also shown. The small arrows indicate opposite trends of occurrence between the first (b) and the second (c), and between the second (e) and the third (f) interspike intervals. HT5; dS3.

Such oscillator loop distributions are pictured on the right side of Figure 91. Interpolation of the oscillation period distributions of Figures 89, 90 shows how the excited network loops shift with increasing excitation (Figure 91).

The network loop of an α_2-oscillator of the neuronal network caudal to a spinal cord injury is more distributed than that of a brain-dead human (Figure 91), indicating a widening of the network loops (further spread of the excitation) and is in connection with the loss of phase and frequency coordination (see below), an indication for the loss of specificity of oscillator properties due to the loss of supraspinal control including descending inhibition. Further, the shape of the distributions of the interspike intervals shows similarities to coupling properties of mechanical oscillators (coupled pendulums).

Similar subpeaks in oscillation period distributions of different motoneurons indicate rhythm coupling between oscillators (compare Figure 85C (α_2-oscillator) with Figure 78C,G (α_1-oscillator)).

The coupling of oscillatory firing networks can be seen in subpeaks of oscillation period distributions. The activation of premotor neural networks via the external loop via skin and muscle spindle afferent input when jumping for example on the springboard (see Chapter V) cannot only formate the different oscillator families but also those ones to which they are coupled.

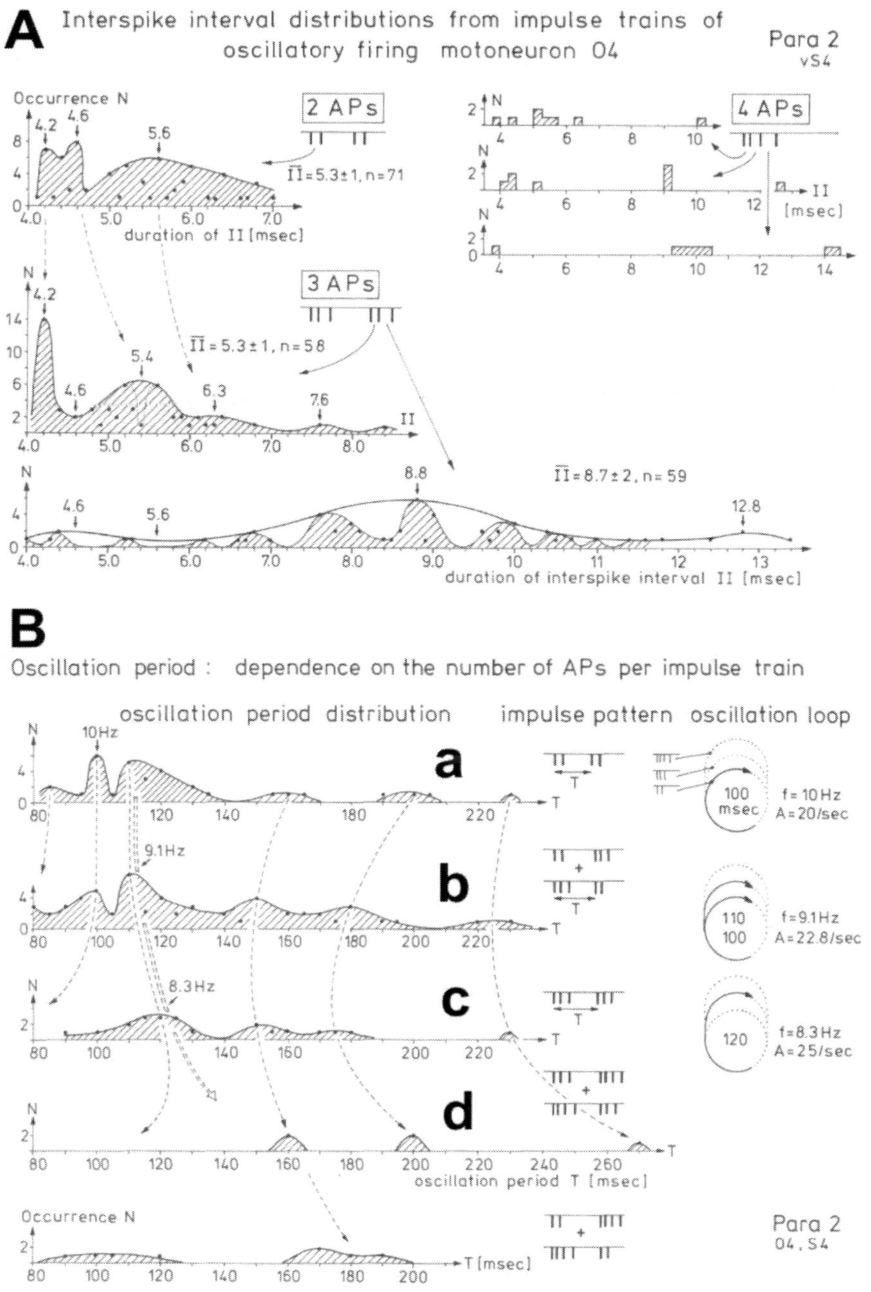

Figure 90. A. Interspike interval frequency distributions from impulse trains of oscillatory firing α_2-motoneuron O_4. Owing to systematic errors, e.g., between even and odd values, enveloping curves were drawn. For impulse trains of four APs, only histograms were plotted because of rare occurrence. The dotted lines mark the corresponding peaks in the occurrence patterns of the first interspike interval (II). Note that mean interspike intervals ($\overline{II}$) provide only little information about the distribution of IIs. Two enveloping curves are drawn for the distribution of the second II. B. Distributions of oscillation period frequencies: impulse train lengths of two APs (a), 2/3 APs (b), three APs (c), 3/4 APs (d) and 2/4 APs (e). The corresponding peaks are connected by the dashed lines. The double dashed line indicates mean shift of the distribution. A very simple loop interpretation is given for the increasing oscillation period at the right side of the Figure. Para 2; vS4.

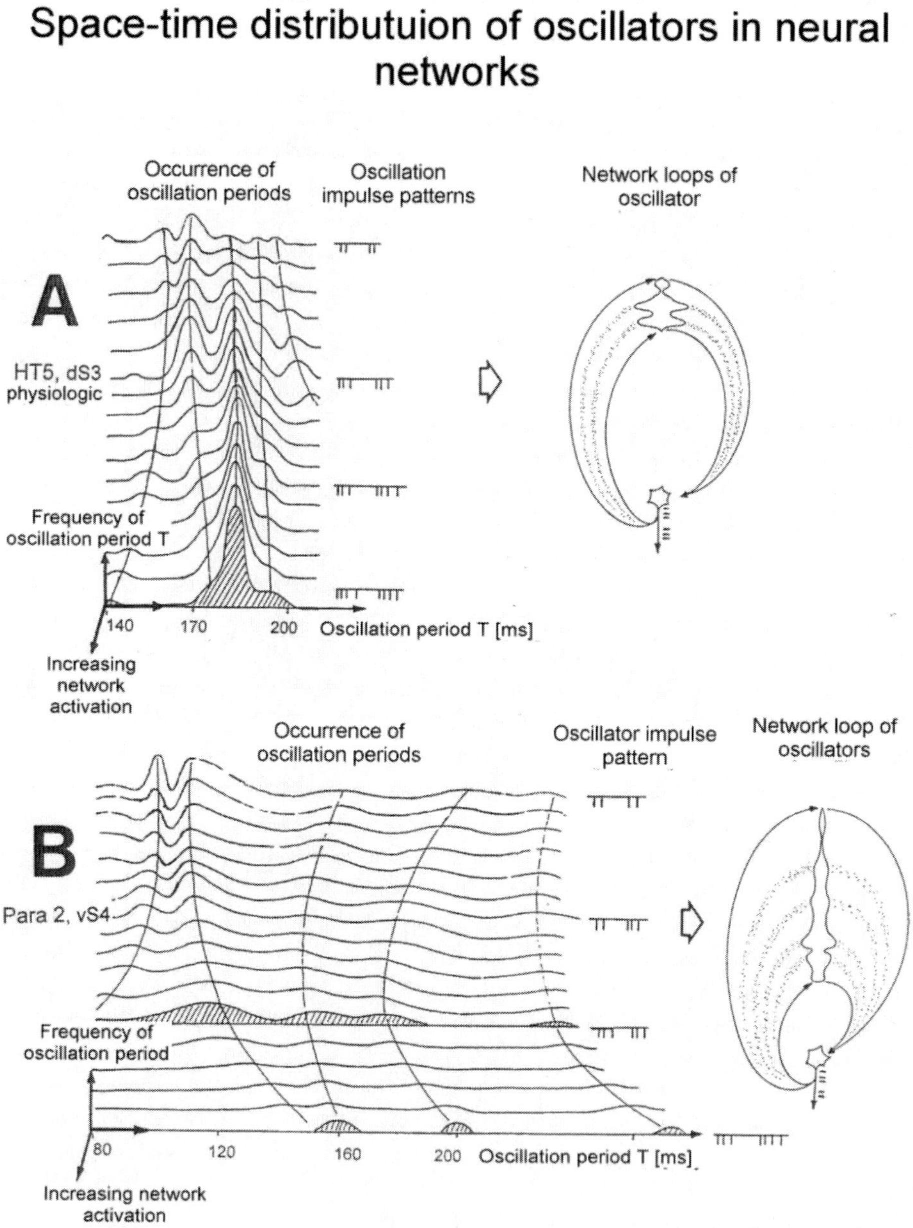

Figure 91. Space-time distribution of activated neuronal networks of α_2-oscillators from a brain-dead human (HT5; dS3) (A) and following spinal cord lesion (Para 2; vS4) (B). The oscillation period distributions for different impulse lengths were taken from Figures 26 and 27 and interpolated so that the frequency of occurrence of the oscillation period be obtained in dependence on increasing network activation (the activity of an oscillator increases with the increasing impulse train length, as can easily be calculated). The frequency of the occurrence for a medium impulse train length was used to draw schematically the network loops of oscillators in brain-death and in paraplegia (right side of the Figure). The network loops give the impression of probability distributions of excitation in neuronal networks. Note that in paraplegia, the activated network loop is much more space-time distributed, as if lateral field inhibition were missing.

The coupling of premotor spinal oscillators gives, therefore, an example how the integrative functions of networks are changed upon training. The system theory of pattern formation (Chapter II) gets with the human neurophysiology a physiologic basis.

11.17. Oscillator Pathway Interpretation

Figure 90B (A through E) shows the frequency distributions of the oscillation periods for increasing impulse train length of Para 2. Peaks can be recognized in each distribution, and the corresponding peaks of different distributions, for different impulse train lengths, are indicated by dashed line arrows. With the increasing impulse train length, the largest peak of the oscillation period shifts towards longer oscillation periods and, as can be calculated, the activity produced by the oscillator increases (10Hz: A = 20APs/s; 9.1Hz: A = 22.8; 8.3Hz: A = 25). Considering the increased activity of the oscillator and interpreting the oscillation period as a loop spread into the space of interneurons in the spinal cord, Figure 90B would suggest that with the increasing oscillator activity, the oscillation loop is extended, and it probably spreads away from the motoneurons. But at the same time, the dashed lines draw near (closest in Figure 90C). This seems to be indication of some kind of contrasting. When attributing the peaks to certain interneuron pathways, Figure 90A through E show how the interneuron pathways change with the increasing activity of the oscillator. In A, the oscillator uses mainly two out of six pathways. In B, the oscillator still uses six pathways but prefers the one with 110ms. In C, the pathway with 120ms is mainly used. A simplified interpretation of the oscillation loop is shown at the right side of Figure 90. The double dashed line arrow (left part of Figure 90) shows how the mean loop pathway increased. This loop pathway interpretation suggests that the oscillator can use different neural pathways and that certain pathways are preferentially used according to the activation; a kind of contrasting occurs with increased spreading. The major contrasting in these oscillator pathways occurred for three APs per impulse train (Figure 90C). The oscillator loop pathway interpretation suggests the spinal oscillator structure of Figure 68, if the motoneuron is a part of the oscillators. There is no clear-cut information that can be derived from the distributions of the second and third II. Comparing of the first II (Figure 89b) with those of second (Figure 89c), peaks in the first II distribution partly correspond to valleys in the second II distribution and vice versa (small arrows). A similar case is seen in Figure 89e and f. If the α_2-motoneuron is a part of the oscillator, and this has been shown above, then this partly opposite distribution behaviour may reflect inactivation of parts of dendrites when the motoneuron oscillates itself to generate the impulse train or when depolarization spreads into the dendrites tree endings to produce the third and fourth soma-dendritic spike (see Discussion of Ref. 186).

11.18. Importance of Spinal Oscillators

The spinal oscillators are of interest for at least of four reasons. 1) They are important neural networks of the CNS, which generate the high-activity mode for sphincters and other muscles. 2) They consist, in addition to the motoneuron, of many interneurons. The spinal oscillators allow therefore the study of interneuron connectivity under physiologic and pathophysiologic conditions. 3) The oscillation, namely the repeated firing with impulse trains, is easy to measure invasively from lower human sacral nerve roots, and tremor, a result of the oscillation, is nearly as easily measurable non-invasively as the reflexes. The oscillators are somehow the CNS interneuron counterpart to the monosynaptic reflexes, which include only the motoneurons. 4) The regular firing of premotor spinal oscillators can be used as a reference basis to measure phase and frequency coordination during CNS changing self-organization.

11.19. Frequency Ranges of Premotor Spinal Oscillators

The oscillation periods of the oscillators in Para 2 and HT5 were transposed into frequency distributions and plotted in Figure 92. The possible frequency ranges of normal humans are not known. To have an approximate comparison to the physiologic case, somehow normal oscillation period distributions were constructed in the following way. With the linear relation between the oscillation period and the number of APs per impulse train (Figure 67), the frequency values of stable oscillators were plotted into Figure 92. The small changes in oscillation frequency with different stimulations have been accounted for by small crosshatched areas.

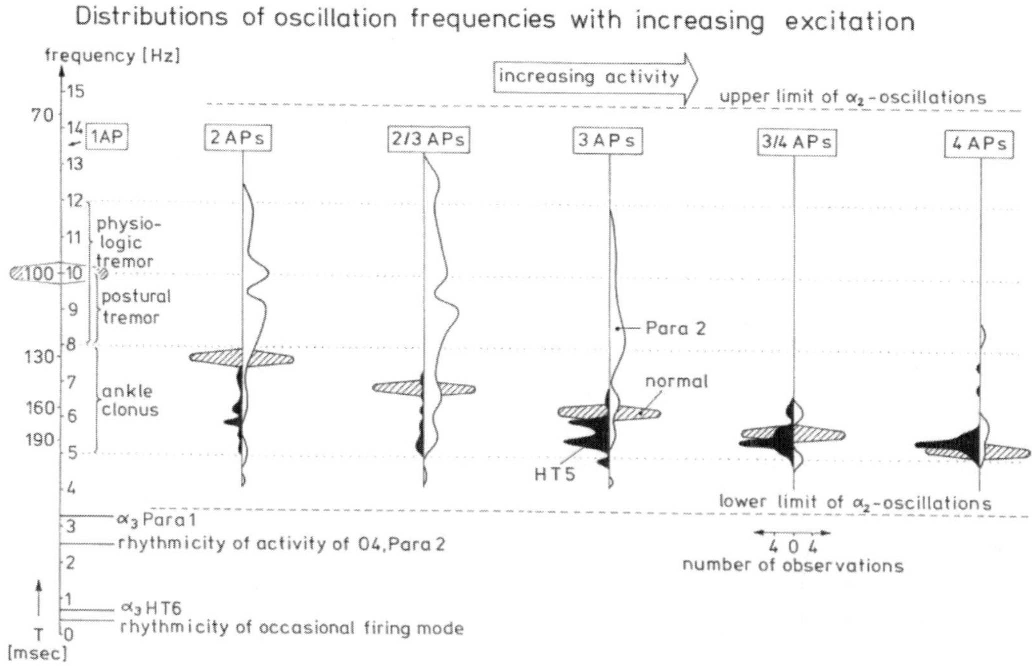

Figure 92. Frequency distributions of oscillation frequencies of continuously oscillatory firing α_2-motoneurons with increasing number of APs per impulse train (increased activity) in paraplegic 2 (open), in brain-dead HT5 (filled), and probably normal human (crosshatched). Frequencies and rhythmic activity changes in the occasional and oscillatory firing mode are indicated. Ranges of physiologic tremor, postural tremor and ankle clonus are also drawn. Note that frequencies for the brain-dead HT5 are too low, and the oscillation frequencies of the spinal cord isolated for a long time (Para 2) are too high and too spread as compared to the theoretically predicted frequency ranges (crosshatched). T = oscillation frequency.

The frequency distributions of stable oscillators also have to be measured in future. The normal oscillators have to be understood in the following way. Let a stable oscillator oscillate with 6.25Hz (T = 160ms) with three AP impulse trains. If a higher activity is needed, the frequency is changing only little to about 6.5Hz, but the impulse train increases by one AP.

Therefore, the crosshatched areas represent six oscillators, each one firing with its own frequency. Only once has it been observed that a stable oscillator changed strongly its frequency for one oscillation. Such quick changes of the oscillation frequency are not possible with mechanical oscillators. Spinal oscillators have no analogy of the kinetic energy of mechanical oscillators. Also, magnetic field may not occur with the loop excitation. They are run by relative timing of activity with pre- and postsynaptic potentials. In unstable oscillators, the frequency and the number of APs per impulse train change strongly with increasing or decreasing stimulation (afferent input). Unstable oscillators cover, for a certain AP number, a large range of possible frequencies.

From Figure 92, it can be seen that for low activation, the oscillator in HT5 (2APs) covers a larger frequency range than does the theoretically normal oscillator, and the frequencies are lower than normal. For high activation (four APs), the oscillator frequency in HT5 is rather normal. That means that for low activation, the oscillator in HT5 is under activated and for high activation is rather normal. A partial spinal shock may be responsible for this under activation for low activation. The situation is opposite in Para 2. For low activity demands (two APs), the frequency distribution range is spread strongly, and the oscillator is strongly over activated. For high activation (four APs), the frequency is closer to normal, even though increased. Most dramatic over activation of spinal oscillators in paraplegics is to be expected in the lower and middle activation range. As Figure 92 shows, for activations with two to three APs, the oscillator in paraplegic 2 covered the whole range of frequencies for ankle clonus, postural and physiologic tremor, even though certain frequencies occurred preferentially. Since the oscillatory firing mode represents high motoneuron activation, there was over activation in the in the α_2-motoneuron system in paraplegics (Figure 92). Further, in Para 1 in a lower sacral root (S5), three α_2-oscillators and one α_3-oscillator could be identified (Table 5), double the number found in HTs with no more than two oscillators detected in a root. This points towards a higher recruitment of the oscillatory firing mode in paraplegics than in HTs. It was reported previously that in the occasional firing mode, α_2-motoneurons (and may be γ-motoneurons) were stronger activated [185]. All these measurements point toward a situation in which the isolated spinal cord is over activated as compared with HTs, and maybe also in comparison to the normal case (Figure 92). However, over activation is not only a general increase, but acts specifically on different structures in the CNS. The summing of different effects of this kind may result, in certain cases, in complete pathologic function.

An oscillatory system only oscillates if there is energy input in phase. The system under consideration is a constant stretch reflex of the anal sphincter (and the bladder function), which is mainly activated through stretching muscle spindles. The following will therefore be concerned with the function of the stretched muscle spindles driving the oscillators. For the sake of a later comparison to the efferent innervation of the muscle spindles [189], the impulse patterns of single encoding sites of secondary muscle spindle afferents will be analysed first before comparing muscle spindle and oscillator activities.

By comparing premotor spinal oscillators in HTs (rather physiologic case) with patients with SCI (pathologic case), it was shown that the organisation of the neural networks of the spinal cord changed specifically. It is difficult to understand how by simple stem cell therapy, and an assumed integration of these cells, the CNS should function physiologically again after the administration of these cells.

12. Impulse Patterns of Single Encoding Sites of Human Secondary Muscle Spindle Afferents

Summary

1. By recording single-nerve fibre action potentials (APs) from human sacral nerve roots, the impulse patterns of six secondary spindle afferents could be identified, analysed and compared with the impulse patterns of oscillatory firing α_2-motoneurons.
2. Secondary spindle afferents fired with single APs, doublets, bursts and equidistant APs. On the assumption that the most simple firing pattern of a single fibre with a shortest interspike interval (II) of 80ms was generated by a single encoding site, the firing patterns of the other afferents could be split into firing patterns of single encoding sites with similar II distributions and also with a shortest II of about 80ms.
3. Successive IIs also reflected the different ways of firing of the secondary muscle spindle afferents. A fibre from a paraplegic with urinary bladder dyssynergy seemed to show a higher activity than a comparable probably normal fibre from a brain-dead human. The doublet IIs of two secondary spindle afferents ranged from about 9 to 14ms with peaks at 10.2, 11.2, and 13ms in the distributions. Multiple encoder sites increased discharge rates of parent fibres but did not always regularize the output discharge.
4. The simultaneously plotted impulse patterns of four secondary spindle afferents showed coordinated firing: they did not fire simultaneously but regularized the summed activity. A phase of 50ms (range 30 to 70ms) often occurred between the spindle afferent APs and the impulse trains of an oscillatory firing α_2-motoneuron. An afferent drive of the oscillator with a certain phase is likely.

12.1. Introduction to Human Secondary Muscle Spindle Afferents

Above, the oscillatory firing of motoneurons was analysed, which activate the sphincters for the continence of the rectum [176, 186] and the urinary bladder [176, 188]. However, an oscillatory system only oscillates if supplied with energy in phase. A pendulum needs a certain push in phase to swing because of the existing damping. An electronic oscillator needs recoupling with a certain phase to oscillate. The oscillatory systems under consideration here are the sphincter motoneurons with sets of interneurons (Figures 68,69), which start to oscillate in brain-dead human (HTs) or paraplegics (Para), if certain afferent input in phase activates and drives the oscillatory systems. Since at least some motoneurons innervating the external anal sphincter fire in the oscillatory firing mode because of the constant stretch reflex of the anal sphincter, the afferent drive comes mainly from the stretched muscle spindles. The secondary muscle spindle afferents (only few primaries were found until now [175]) should have a suitable pattern to drive the oscillatory systems (176). The afferent impulses contributing to the oscillation must also have a rather constant phase to the motoneuron impulse trains. The afferent impulsation pattern can originate from the spindle itself (stretch) or be formed in cooperation with the activity of the efferent spindle innervation. Synchronicity of the spindle afferent action potentials (APs) with the motoneuron impulse trains, however, can only come from the activity of the efferent spindle innervation. A further step to be clarified is from where the efferent spindle innervation, consisting mainly of γ-motoneurons (intrafusal), gets their fixed phase drive. Since secondary spindle afferents possess several myelinated endings [6], multiple sources of APs can be expected [6, 53, 97, 159]. A thorough analysis of possible phase relations between the afferent and efferent spindle APs require the knowledge of the impulse patterns of single afferent encoder sites. Here, mainly impulse patterns of secondary spindle afferents are analyzed, which most likely drive the oscillatory motoneurons. On the assumption that one afferent fibre had only one encoding site, the more complex impulse patterns of the afferents could be split into the impulse patterns of single encoders, all having roughly identical interspike interval distributions. The known impulse patterns of single encoders will be correlated with the impulse patterns of the intrafusal motoneurons in the following section (189). Since one objective of this research is the understanding of the drive of the oscillatory firing extrafusal motoneurons, the afferent impulse patterns will be compared with the impulse trains of the oscillatory firing motoneurons in this [187] and the following section.

Secondary muscle spindle afferents branch to innervate the equatorial regions on one or both sides of the primary nuclear bag 1, nuclear bag 2, but mainly nuclear chain fibres [6, 20, 127]. Since the branches become only unmyelinated 25-30μm before their sensory terminals, all main branches are myelinated [8, 36, 127]. Ferric and ferro-cyanide ions were used to stain specifically those sites where APs are initiated or propagated [159]. Probably sites of APs' origin were identified in the distal portions of Ia axons in neuromuscular spindles of cat and frog [159]. Terminal hemi-nodes and some of the penultimate nodes are potential sites of spike generation [159]. When muscle fibres in a spindle are stretched, the final branches of spindle afferents are deformed and a generator potential (or algebraic sum of several receptor potentials (6)) is produced across the membranes of the terminals. This graded potential spreads electronically and excites one or more encoder sites to produce APs. The summed AP activity is then transmitted to the CNS via a single myelinated fibre. There has been quite a discussion about how single encoders contribute to the summed impulse activity of the parent axon, especially since in mammals it has not been positive to precisely locate the encoders, because the distal afferent branches are bound together inside the capsule and are not accessible or separable with current physiological methods. When an impulse, propagating centrally in a first order branch, reaches the branching node, it will subsequently propagate orthodromically into the parent axon and, provided that it is not refractory, antidromically into the first-order branch. During the subsequent refractory period, this branch would then be less likely to transmit another impulse, thus suppressing its pacemaker. However, the antidromic impulse itself may fail to propagate into some of the pacemaker branches or be eliminated by collisions with abortive orthodromic spikes within them. Since these effects would be cumulative, a profusely branching first-order system would recover from an antidromic impulse more rapidly than with fewer branches.

The single-nerve fibre action potential recording from nerve roots or nerves is another approach to clarify the problem of the contribution of single encoder sites to the summed impulse traffic in the parent secondary muscle spindle afferents. By recording simultaneously from many fibres and comparing their impulse traffic, the simplest possible source of AP production is extracted. Probably the impulse patterns of such sources are the impulse patterns of single encoders. Since muscle spindles can be specialised according to their function in different parts of the body [163] and encoder sites may influence each other, it is still possible that encoder fire with different impulse patterns. Since impulse patterns of secondary spindle afferents of the accuracy presented herein are not available even from animals, in particular for direct comparison to physiologic α (extrafusal) and γ-motoneuron (intrafusal) impulse patterns [189], it will be shown that a recording from many fibres can be superior to single-fibre recordings.

12.2. Identification of Single Muscle Spindle Afferents by the Action Potential Waveform

From the muscle spindle afferent activity in a dorsal root of HT6, the activities of five single fibres could be separated by their waveform and partly also by their firing pattern. The five fibres were the secondary muscle spindle afferents SP2(1), SP2(2), SP2(3) and SP2(5), and probably the tertiary muscle spindle afferent fibre SP2(4). Approximately three other muscle spindle afferents were contained in the root, but they were seldom active during the different stimulations. The activity of the five identified muscle spindle afferents, which contributed more than 95% to the muscle spindle afferent activity of the HT6 S4 dorsal root, and the identified activity of two secondary muscle spindle afferents from a S4 root of paraplegic 1 will be used to analyse the impulse patterns of the single encoding sites of secondary muscle spindle afferents.

Figure 93 shows that the five spindle afferent action potentials (APs) of HT6 could be safely distinguished from each other. Even though the APs of the SP2(2)-fibre could be safely distinguished from the APs of the other four fibres shown in Figure 93, its APs looked rather similar to those of the three other fibres not under consideration. The activity of the SP2(2)-fibre was mainly distinguished from them by the doublet firing shown in Figure 93D. The APs of the fibres SP2(1), SP2(2), SP2(3) and SP2(5) were conducting with velocities between 38 and 40m/s and are considered as secondary muscle spindle afferents based on the conduction velocity frequency distribution histogram constructed for HT6 (Figure 53B). In SP2(5)-fibre, the conduction from the distal electrode pair (trace b) to the proximal pair (trace a) was partly blocked (Figure 93D), probably by some mechanical alterations. But occasionally also in this fibre, AP was conducted across the two electrode pairs quite normally, so that the conduction velocity could be measured (38m/s).

The APs of the SP2(4)-fibre were conducted at 13.3m/s (Figure 93E), much more slowly than in secondary spindle afferents. The velocity of this fibre may fall outside of the normal velocity range. However, since the activity of the SP2(4)-fibre was very constant (see later), this fibre probably was a tertiary muscle spindle afferent fibre. Since such a fibre has been detected here for the first time, its data were pooled with those of the secondary spindle afferents.

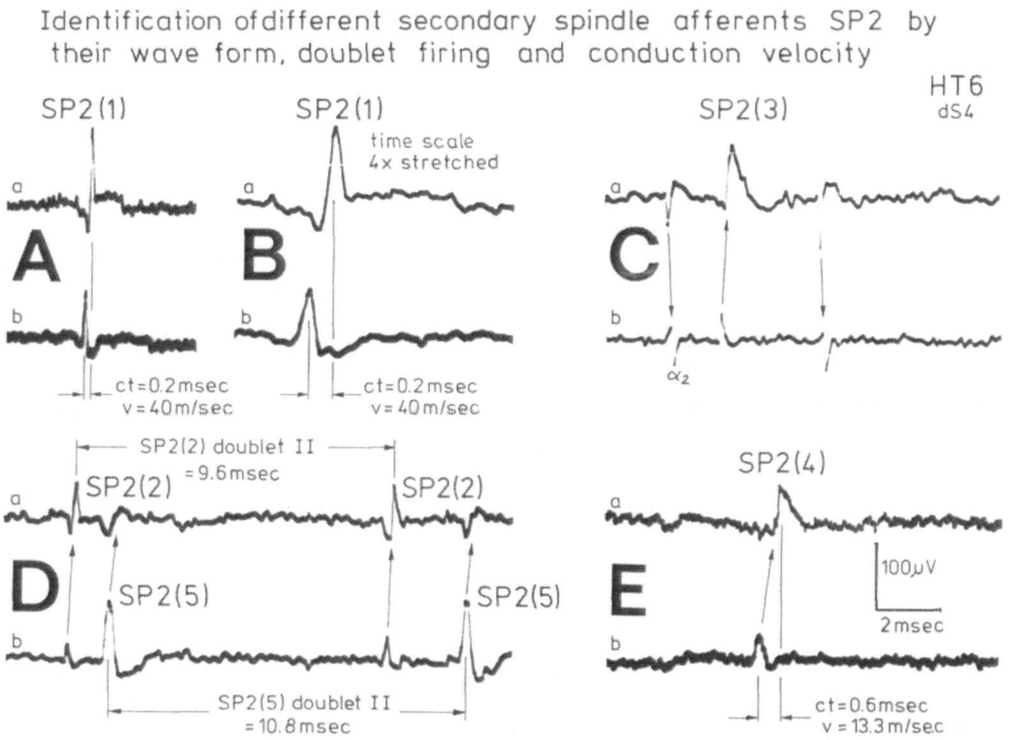

Figure 93. Extracellular action potentials (APs) of five secondary muscle spindle afferents SP2 (1 to 5), which are simultaneously identified by their waveforms. The SP2(4)-fibre (E) is likely a tertiary muscle spindle afferent because of the low conduction velocity and the very little dynamic behaviour. The APs from the SP2(2)-fibre were sometimes identified by the doublet firing. II = interspike interval, ct = conduction time, v = conduction velocity, $\alpha_2 = \alpha_2$-motoneuron AP.

Golgi tendon organ afferents are not at all or only a little present in the lower sacral range, and they conduct with velocities between those of the primary and the secondary muscle spindle afferents (Figure 2). Based on the different AP waveforms (Figure 93), the APs of the five muscle spindle afferents could be identified in various stimulation situations, and the activity patterns of these five single fibres could therefore be simultaneously measured. The doublet firing of the SP2(2) and the SP2(5)-fibres (sometimes also the SP2(3)-fibre) is a new observation and will be analysed later. Original recordings of the two secondary muscle spindle afferents of paraplegic 1, identified by the waveform, have been presented elsewhere [186, 188].

12.3. Single AP, Doublet, Burst and Multi-Ending Regular Firing of Secondary Muscle Spindle Afferents

Being able to identify the different secondary muscle spindle afferent APs, it is now time to try to analyse the impulse patterns from single encoding sites. Figure 94A shows the distribution of interspike intervals (IIs) of the SP2(1)-fibre for anal and bladder catheter pulling, touch and pinprick. It can be seen that the duration of the shortest II measured is 80ms. Since Figure 94A includes numerous IIs and responses to various kinds of stimulations, it is concluded that the duration of the shortest possible II is 80ms for the constantly stretched anal sphincter (catheter Ø = 12mm). Since in addition, the distribution of the other spindle afferents (Figure 94B,C,D) showed less specific properties and more intricate structure, the distribution in Figure 94A is interpreted as originating from a single encoding site; this means that the SP2(1)-fibre possessed only one active encoding site. Based on this assumption or interpretation, it will be tried below and later to understand the more complex impulse patterns of single secondary spindle afferent fibres with several encoding sites. The frequency distribution shown in Figure 94A suggests that the single encoding site preferred certain IIs, in particular 80, 140 and 200ms. The peak between 140 and 170ms is split. It could be that this splitting is due to a hysteresis-effect following the increase and decrease of the contracture of the intrafusal muscle fibre due to different intrafusal (γ) activation. The oscillation periods of possible spinal oscillators, taken from the relationship between oscillation period and the number of APs per impulse train [186], are indicated in Figure 94A with one AP to five APs. The actually measured α_2-oscillator O2 is labelled three AP. It could be that due to the hysteresis-effect induced splitting of the II peak between 140 and 170ms, the α_2-oscillator O2 is driven by IIs shorter than the oscillation period and is inhibited by IIs longer than the oscillation period.

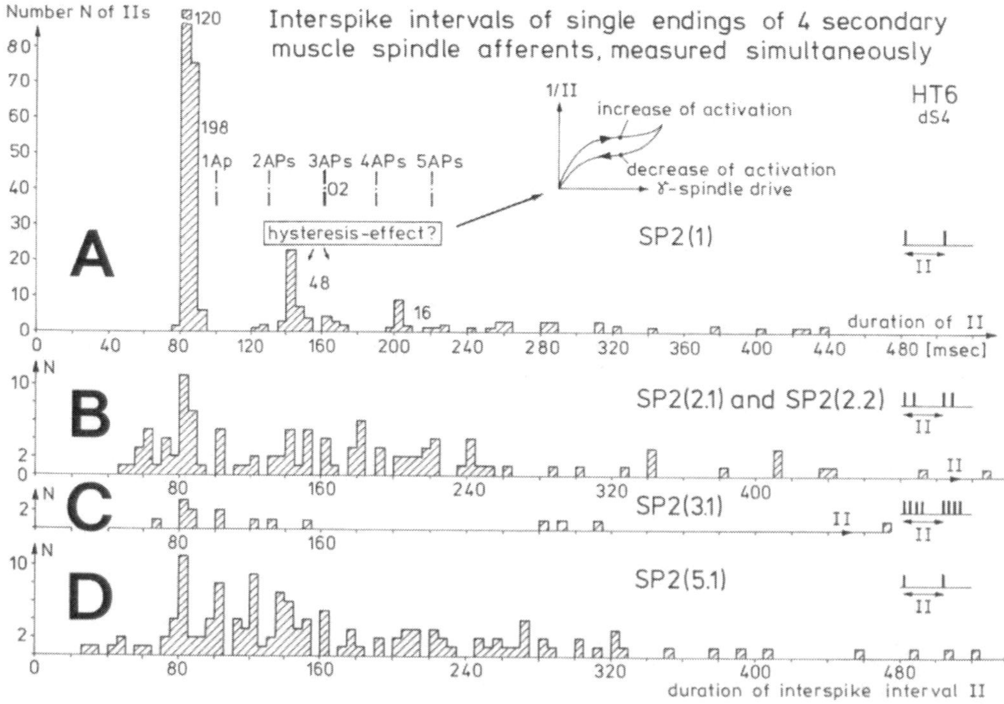

Figure 94. Interspike interval frequency distributions of single endings of four secondary muscle spindle afferents, measured simultaneously. In A, the theoretical oscillation periods are drawn for comparison. The oscillation period of the oscillatory firing α_2-motoneuron O2 of HT6 is also indicated. Note that the discretely distributed SP2(1)-IIs (A) are similar to the oscillation periods of the O2-oscillator for different impulse train length. The doublet peak at about 160ms may be due to a hysteresis-effect of the intrafusal muscle fibre (inset in A). The Figure after the point in the parentheses (SP2(5.1)) indicates the single encoding site of a single secondary muscle spindle afferent.

Probably, certain combinations of shorter and longer IIs could or could not drive the oscillator. Below comparing the afferent drive with the impulse activity of spinal oscillators, the hysteresis-effect will be picked up again. However, the impulse patterns of the other spindle afferents (Figure 94B,C,D) also have to be considered for the drive of the oscillator.

Assuming that an II of about 80ms is the shortest II of a single encoding site, the impulse patterns of the single afferents SP2(2), SP2(3) and SP2(5) could be split into probable patterns of single encoding sites (Figures 99,100) and their II distributions were plotted in Figure 94B,C,D. The very static fibre SP2(4) had such a low activity that no short IIs occurred.

The secondary muscle spindle afferents SP2(1) through SP2(5) showed four ways of firing: single-AP firing (SP2(1) and SP2(5)), doublet firing (SP2(2) and SP2(5)), burst firing (SP2(3)), and multi-ending regular firing (SP2(2)). The muscle spindle afferents partly changed their way of firing according to the kind and strength of stimulation and maybe according to the activity of the efferent parasympathetic innervation of the spindle (see next section).

It was assumed that the doublet firing originated in a coordinated activation of two encoding sites, and the II distribution for the doublet firing of the SP2(2)-fibre is shown in Figure 94B. The distribution is similar to that shown in Figure 94A. The II distribution of the SP2(2)-fibre firing is not as specific and has values below 80ms. Since in this case identification of the AP waveform was not as safe as for the SP2(1)-fibre, these shorter IIs could result from misjudged AP waveforms. This can be the reason only partly, since IIs shorter than 80ms were also observed for the SP2(5)-fibre (Figure 94D), and their APs could be identified safely. An explanation for IIs shorter than 80ms is that in spindle afferents with more than one encoding site, there is some interaction between the encoding sites, so that the shortest II of a single encoding site is reduced. The burst firing of the secondary spindle afferent fibre SP2(3) is analysed in a similar way. It is assumed that the burst firing originates from the activity of at least four encoding sites. The resulting distribution is shown in Figure 94C. Figures 99 and 100 show the splitting of the burst firing, which resulted in the distribution shown in Figure 2C. The secondary spindle afferent fibre SP2(5) fired with single APs and with doublets. During the doublet firing, the activity was very low and the IIs were too long to provide information about shortest II. During the single-AP firing mode, however, the activity was sometimes high, and the II distribution could be determined as presented in Figure 2D. Following the splitting into different encoding sites, the interspike interval (II) distributions were very similar, and the fibre with the expected single encoding site (SP2(1)) showed most specifically. The IIs of the secondary spindle afferents of paraplegic 1 showed similar distributions. Nevertheless, since the AP waveform identification was not as safe as in HT6, a splitting into single endings was not tried: just a few misjudged APs can change the distribution strongly. The SP2(2)-fibre did not fire with doublets only. Following very strong bladder catheter pulling, it fired with increasing high activity, with rather similar spaced APs. Figure 97C shows the II distributions with the activity increasing from a through d (note the different scale of "N" in d). Assuming a shortest II of 80ms for each encoding site, it is found that up to six encoding sites were activated with increasing activity. The firing was mostly regular with respect to the IIs. When a new encoding site became active, there occurred some irregularity in successive IIs, but quickly the IIs became regular again. This multi-ending regular firing is probably due to the parasympathetic innervation of the spindle and will be analysed further in the next section. The recruitment of new encoding sites and the regularity of the firing pattern can be directly seen in Figure 110A.

12.4. Successive Interspike Intervals of Secondary Muscle Spindle Afferents

Interspike interval (II) distributions provide information about the average occurrence of certain II durations, but no information about the decrease and increase of the IIs. Successive IIs of secondary muscle spindle afferents show directly how quickly changes occurred. To turn out the dynamic properties of muscle spindle afferents is of interest for the understanding of the physiologic function of spindles.

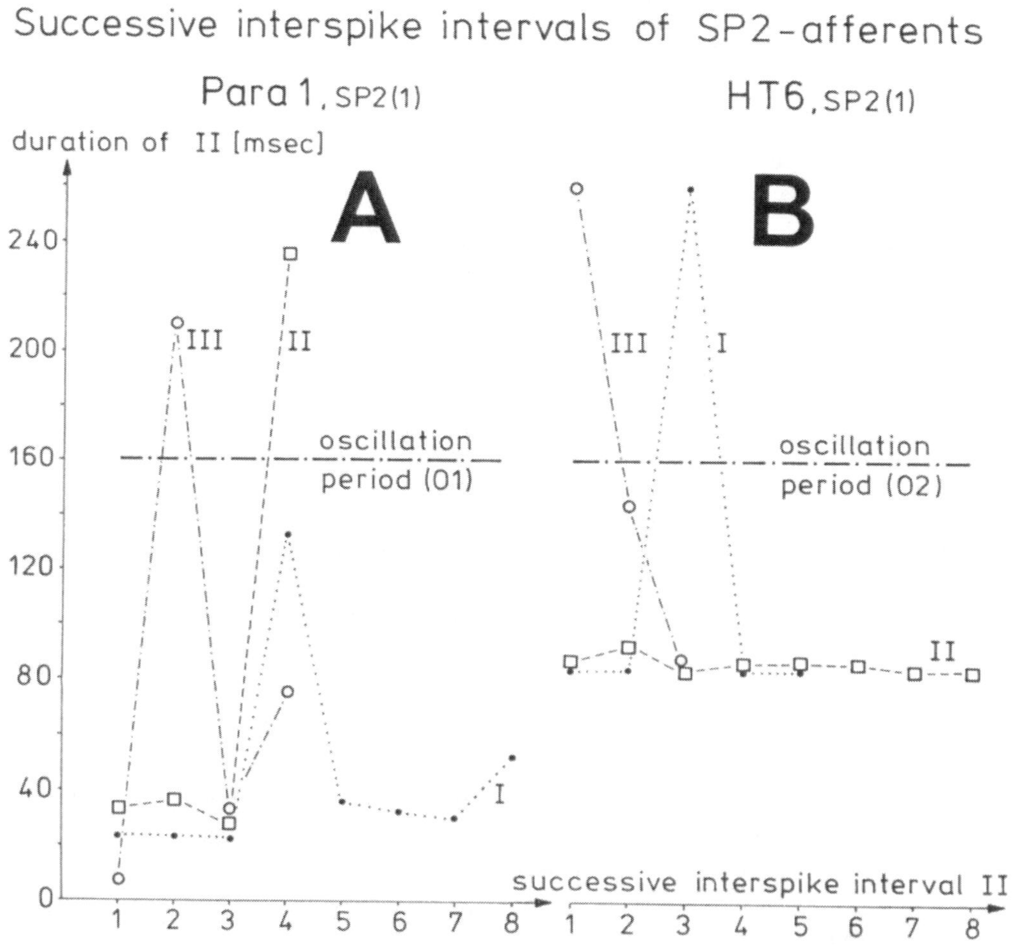

Figure 95. Successive interspike intervals (IIs) of single muscle spindle afferents of paraplegic 1 (SP2(1)) and the brain-dead human HT6 (SP2(1)). The Figures I, II, and III designate three series of successive IIs. The dotted and dashed lines link the points representing interspike intervals values. The oscillation periods of the oscillatory firing α_2-motoneuron O1 and O2 are indicated.

But since in the anal sphincter, and associated functions of the pelvic floor, the extrafusal muscle fibres change their length only little, changes of spindle afferent activity will be mainly due to contraction-associated changes of intrafusal muscle fibres. Changes of the muscle spindle afferent activity are therefore mainly due to changes of the activity of the efferent innervation of the spindles, which originates in the circuitry of the CNS. The aim of this part of research is to find out what is pathologic in the function of the spinal cord (following SCI), in addition to exploring normal functions, hopefully present in HTs. Comparisons between spindle functions of in subjects with paraplegia and in HT are expected disclose pathologic behaviour, if present.

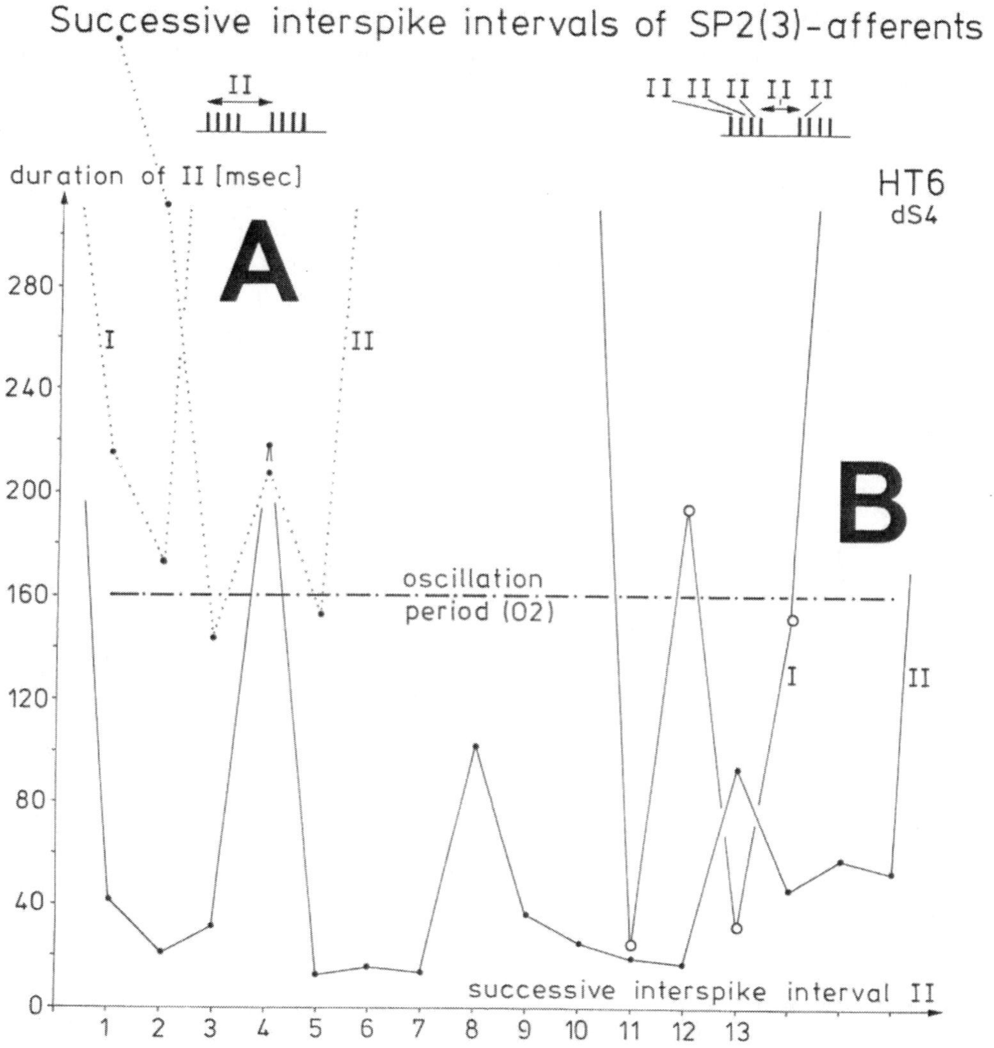

Figure 96. Successive interspike intervals (IIs) of the secondary muscle spindle afferent fibre SP2(3). The way the interspike intervals of burst-like firing APs were measured is indicated. A. Two series (I, II) of single-ending IIs (dotted lines). B. Two series (I, II) of single afferents but not of single endings (solid lines). Note that the single-afferent IIs are similar to those shown in Figure 3A, and the single-ending are similar to those shown in Figure 95B. The oscillation period of the oscillatory firing α_2-motoneuron O2 is indicated.

Figure 95B shows three characteristic sets of successive IIs of the secondary spindle afferent fibre SP2(1) measured in HT. Set II (B) shows the highest activation possible, namely firing with the always shortest II of 80ms. In sets I and III, IIs change from 140 and 260ms to 80ms and vice versa. The changes in the duration of successive IIs in paraplegic 1 (Figure 95A) were much more pronounced. In set II, there is a change from 240ms to 25ms. In general, the short IIs in the paraplegic subject were much shorter than in the SP2(1)-fibre measured in HT. The spindle afferents of paraplegic 1 seemed to be less stable and over-activated.

From the above section, it is known that the SP2(1)-afferent fibre in the HT possessed only one encoding site with a rather simple firing pattern, whereas the SP2(1)-fibre of paraplegic 1 could have possessed several encoding sites with higher and more structured activity of the parent fibre. The search for different encoding sites and firing pattern will provide more information on whether the spindle afferent fibre 1 was really over-activated. The oscillation periods of the continuously oscillatory firing α_2-motoneurons are indicated in Figures 95A and B. The IIs of 130ms and 140ms duration could contribute to the drive of the oscillator.

Figure 96B (solid lines) shows successive IIs of the spindle afferent fibre SP2(3) which sometimes fired in a burst-like way. No regard was paid to different encoding sites. Sets I and II show similarity to the successive IIs of paraplegic I (Figure 95A). Based on this comparison, it is concluded that the secondary muscle spindle afferent fibre SP2(1) of paraplegic 1 probably fired in a burst-like way. Pathologic function cannot be concluded from these sets of successive IIs.

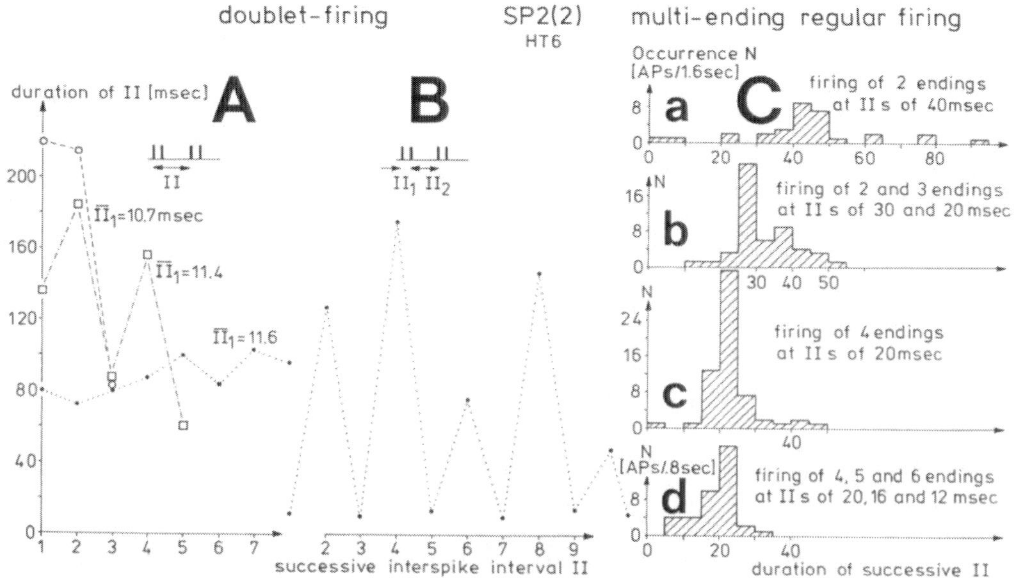

Figure 97. A., B. Successive interspike intervals (IIs) of the secondary muscle spindle afferent fibre SP2(2). The way of measuring is indicated. In A, three series of successive single-ending IIs are shown. $\overline{II}$ = mean doublet II. In B, saw-tooth like successive IIs are shown. C. Interspike interval frequency distribution histograms (a, b, c, d); increasing activity following bladder and anal catheter pulling. Peak suggests the number of activated endings, when the shortest single-ending II (80ms) is divided by the value of the peak II (e.g., 80ms : 40ms = 2 endings).

However, since in HT the fibre only showed burst-like firing following painful stimulations and the fibre of paraplegic 1 seemed to show always that kind of firing, it is probable that the afferent fibre of the paraplegic subject fired with a higher than normal activity. Considering now different encoding sites for the SP2(3)-fibre of HT6, successive IIs are obtained from the form shown in Figure 96A (dotted lines). Figure 97B shows successive IIs of the doublet firing of the secondary spindle afferent fibre SP2(2). The shape of the IIs makes up a saw-tooth pattern. Saw-tooth successive IIs are therefore an indication for doublet firing.

Assuming that the doublet firing stems from the activity of two encoding sites, the successive IIs of the supposed single encoders are plotted in Figure 97A. Mean IIs of the doublets are indicated at each set. The mean doublet IIs of 10.7, 11.4 and 11.6ms seem rather similar.

They will be further analysed below. As stated earlier, the SP2 (2)-fibre was able to fire with much higher activity keeping all the time distances between the APs similar, as can be judged from the distributions in Figure 97C. Most likely, this coordinated activation of different encoding sites is due to the parasympathetic innervation of the intrafusal muscle fibres innervated by the SP2(2)-fibre, The parasympathetic innervation of muscle spindles will be analysed in connection with the somatic efferent innervation of muscle spindles in a following section.

12.5. Doublet Firing

The single-AP firing (SP2(1)-fibre) was the expected firing pattern. Burst-like firing was activated by pinprick. Several encoding sites were probably activated for the response to pain to serve the escape reaction. Multi-encoding site regular firing will be shown to be activated by parasympathetic fibres.

A reason for the doublet firing is difficult to find, especially since the average activity level is rather low. Figure 98 shows the distributions of the doublet IIs for the two secondary spindle afferents SP2(2) and SP2(5) during two series of measurements. At the end of the first series of measurements, a very strong bladder catheter pulling most likely activated the parasympathetic system (Figure 111).

Some doublet firing was recorded in the first series of measurements from both afferents. More doublet firing was measured in the second series after the activation of the parasympathetic nervous system.

One possibility is that the doublet firing is connected to the activity level of parasympathetic intrafusal motoneurons. A second possibility, as will be shown later, is that the activation pattern is caused by γ-motoneurons. The doublet firing distribution of Figure 98C shows three peaks at 10.2, 11.2 and 13ms, represented by the solid line. Also, the three peaks are suggested on the distribution shown in Figure 98A, which is based only on a few measurements.

The doublet firing of these two afferences seem to have three sources. The 13ms peak doublet is labelled γ_1 to indicate that a single γ_1-motoneuron (intrafusal) was partly firing with doublets, which had peak II duration of 13ms.

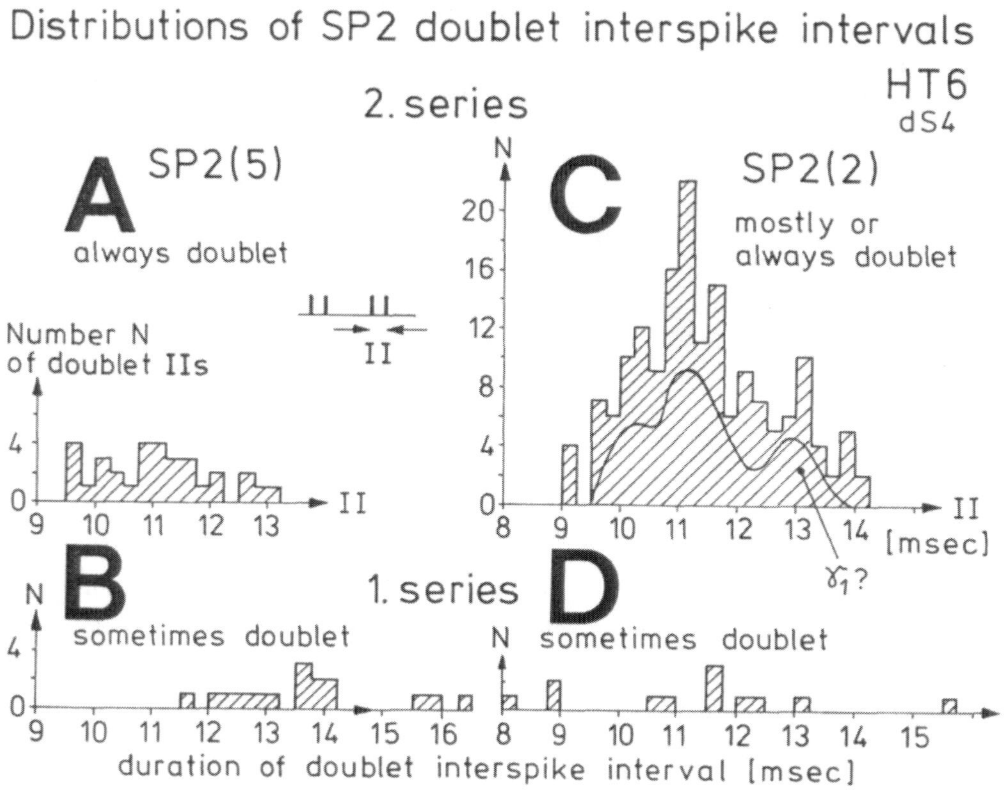

Figure 98. Doublet interspike interval frequency distributions of the secondary muscle spindle afferent fibre SP2(5) (A, B) and SP2(2) (C, D), HT6. The distributions in A and C may contain three peaks, one of which is labelled γ_1.

12.6. Impulse Pattern of Secondary Muscle Spindle Afferents in Relation to the Continuously Oscillatory Firing of α_2-Motoneuron O2

In addition to interspike interval (II) distributions, durations of successive IIs and doublet II distributions, even more information can be obtained from impulse patterns by directly plotting the action potentials (APs) of the five spindle afferents vs. the impulse patterns of the oscillatory firing motoneuron. In Figure 99A,B,C the impulse patterns of the different fibres can be seen.

Figure 99D shows the IIs between the first encoding site of the SP2(3)-fibre (SP2(3.1)) and the SP2(4)-fibre, which is probably a tertiary muscle spindle afferent fibre. As the distribution in D shows, there is a correlation between these two impulse patterns. Thus, the afferent fibre SP2(4) belongs to the same spindle as the SP2(3)-fibre (in A,B,C indicated by the dashed box), or the SP2(4)-fibre belongs to a spindle, which is in tandem [163] with the spindle of the SP2(3)-fibre. Figure 99A shows the activity of four spindle afferent fibres following light pinprick 1. The activity of the SP2(1)-fibre is high; those of the other fibres are low. The open arrow marks a quick change of the oscillation period and a change of the impulse train.

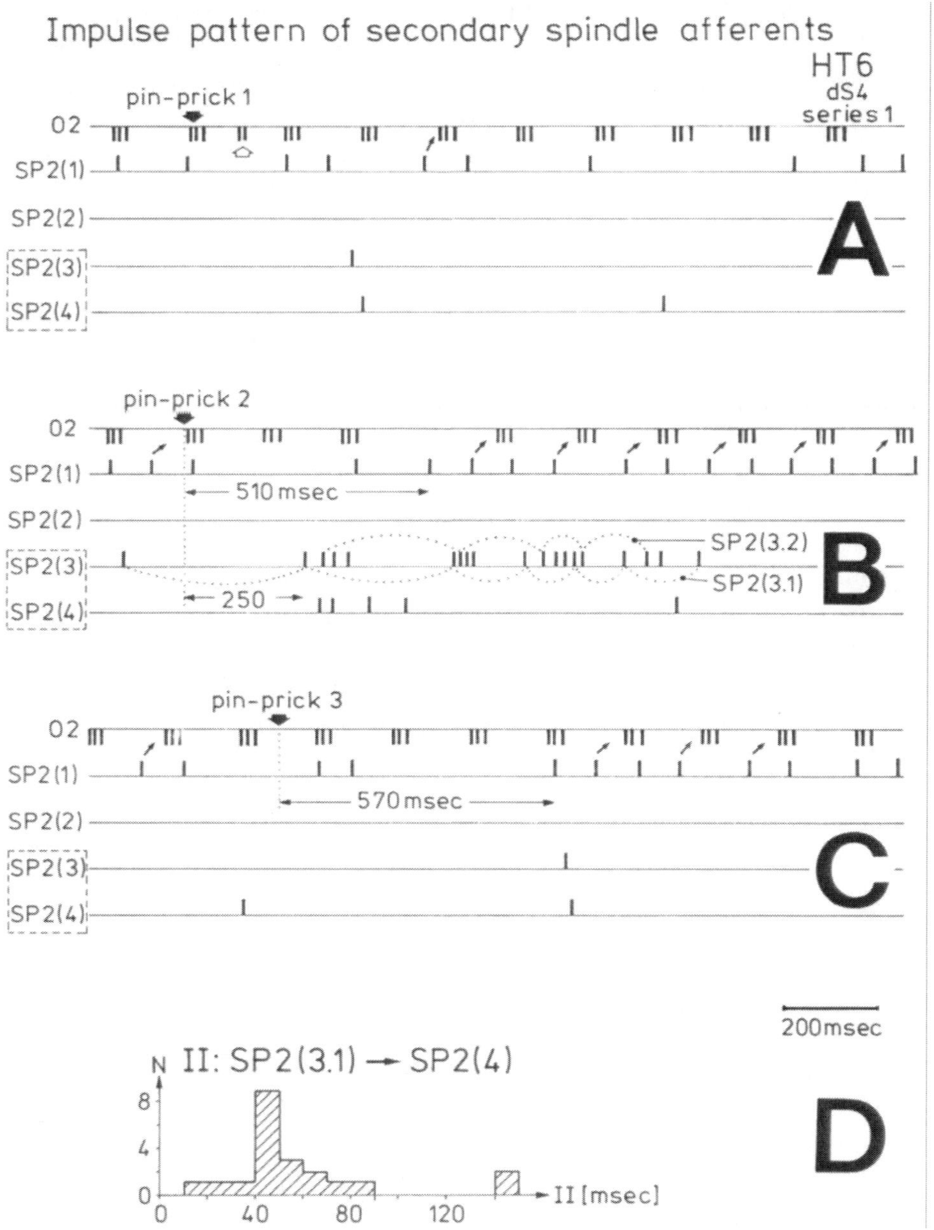

Figure 99. Impulse patterns of simultaneously recorded secondary muscle spindle afferent fibres (A, B, C) following pinprick. Each bar represents an AP; upward = afferent, downward = efferent. In B, the activity from two single endings (SP2(3.1) and SP2(3.2)) is marked by dotted lines. The small arrows indicate similar time intervals between muscle spindle afferent APs and motoneuron O2 APs. D. Interspike interval distribution between SP2(3.1) and SP2(4)-fibres (cross-correlation). Since there is relative phase coordination between fibres SP2(3.1) and SP2(4) fibres, the spindle afferents may belong to the same spindle (indicated in A,B,C by the dashed box), or fibre SP2(4) belongs to a spindle, which is in tandem with the spindle of fibre SP2(3) or the spindles of the two fibres are innervated partly by the same γ-motoneurons. HT6; dS4.

Such a quick change of the oscillatory firing was only observed once. Following strong pinprick 2 (Figure 99B), the SP2(1)-fibre increased its activity with a delay of about 510ms, and the probably pain-sensitivity SP2(3)-fibre fired transiently with bursts 250ms after the pinprick (Figure 99B). Splitting up the SP2(3) activity into the activities of the probable single encoding sites, two of them indicated by the dotted lines (labelled SP2(3.1) and SP2(3.2)), it can be seen that the burst-like firing can be explained by the recruitment of four to five encoding sites, in which the II decreased and increased in a similar way as for the SP2(1)-fibre. The SP2(4)-fibre was activated slightly later than the SP2(3.1)-encoding site. With a shortest II of about 80ms of a single encoding site, also the SP2(4)-fibre had more than one encoding site. Following pinprick 3, only the activity of the SP2(1)-fibre increased with a delay of 570ms (Figure 99C). The light and strong pinprick was quantified above (Figure 40) by measuring only pain afferent activity following the strong pinprick.

Since it can be expected that during the constant stretch reflex (214) of the anal sphincter the spindle afferent activity drives the oscillatory firing α_2-motoneurons innervating the anal sphincter [176], similarities have to be looked for the IIs of the afferents and the oscillation periods, as was done [176] (see above), and for a rather constant phase relation between the afferent APs and the oscillation cycle, which is marked by the first AP of the impulse train. Assuming the existence of a phase relation for the drive of the oscillator, if the afferent AP leads 30 to 70ms (Means ± s.d. = 50 ± 10ms, n = 23 of SP2(1) and SP2(2) to the first AP of the impulse train, then the small arrows in Figure 99 (and 100) from the SP2(1)-AP to the first AP of the motoneuron impulse train indicate phase relation. Driving phases of about 50ms occurred 17 times between the SP2(1)-AP and the motoneuron impulse train, and six times between the SP2(2)-AP and the impulse train (Figures 99,100). As can be seen, most phase relations are present in Figures 99 and 100, following the painful pinprick 2. It is reasonable that the anal sphincter enhanced its contraction following pinprick into the gluteus maximus in addition to the constant stretch reflex induced by the anal catheter. In Figure 99, it can be seen that the SP2(1)-fibre is almost the only one to contribute to the driving phase relation. Since in Figure 99A there is only one afferent AP with the right phase, probably also other afferents, not contained in that S4 root, contributed to the drive of the oscillatory firing α_2-motoneuron.

In a second series of measurements, after the activation or increased activation of the parasympathetic fibres, the impulse patterns of the five muscle spindle afferents were quite different, whereas the oscillator was still continuously oscillating. The activity of the SP2(1)-fibre is low (Figure 100A,B), and nearly no phase relations occur for the motoneuron drive. Only following bladder catheter pulling 4, the SP2(1)-activity increased and more phase relations occurred. The activity of the SP2(2)-fibre was different from the first series. The SP2(2)-fibre fired now with doublets (Figure 100A), and this doublet firing increased with the bladder catheter pulling (Figure 100B,C). Also, phase relations (indicated by the small arrows) occurred now between the SP2(2)-APs and the impulse train of the motoneuron.

There seemed to also be some coordination between the SP2(1)-fibre activity and that of the SP2(2)-fibre, since only rarely their APs occurred at similar times. They shared the time for firing. The activity of the SP2(3)-fibre strongly increased following bladder catheter pulling 4. Patients report the wish to micturate upon light to medium catheter pulling and report pain upon strong bladder catheter pulling. In Figure 100A, the activity of the most sensitive encoding site (SP2(3.1)) is indicated by the dotted line. The firing pattern was something in between repeated burst-like firing (Figure 100A) and doublet firing (Figure 100C). The SP2(3)-fibre preferred to fire when the SP2(1) and SP2(2)-fibres were silent. The activation of the SP2(3)-fibre was therefore included in the coordinated activation of the

SP2(1) and SP2(2)-fibres. Possible phase relations are not indicated, since constant drive for the continuously oscillatory firing was searched for.

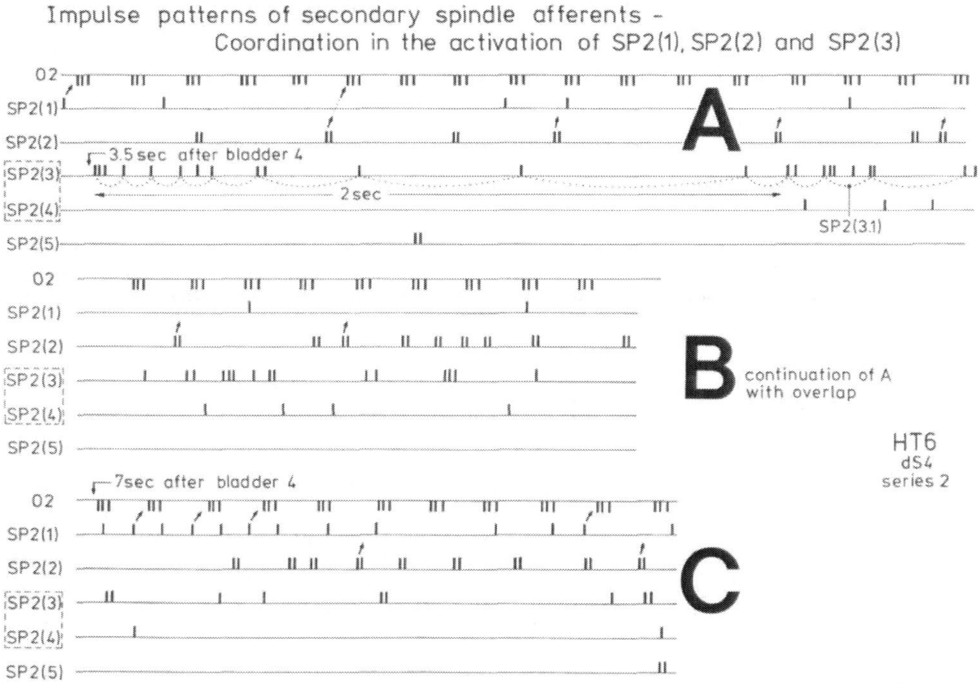

Figure 100. Impulse patterns of the five secondary muscle spindle afferent fibres SP2(1) through SP2(5) in relation to the impulse pattern of oscillatory firing α_2-motoneuron O2 following bladder catheter pulling. Bars represent APs. Very short IIs (as in doublets) are not drawn to scale; only the first APs are in their exact position. The dashed rectangle around SP2(3) and SP2(4) indicates that both afferents probably innervate the same spindle. The dotted curves, connecting certain APs of the SP2(3) activity, most likely represent the activity from single endings. Note that doublet firing occurs in spindle afferents SP2(2), SP2(3) and SP2(5); the doublet firing builds up in fibre SP2(3) from A (no doublet firing) to C (rather doublet firing). Note further the more static behaviour of fibre SP2(4) in comparison to fibre SP2(3) (delayed response, lower activity changes); fibre SP2(4) may be a tertiary spindle afferent fibre. The small arrows indicate similar time intervals from the afferent APs to the motoneuron APs (phase coordination). HT6; dS4.

The activity of the SP2(4)-fibre also increased, and it seemed to also be coordinated with those of the former spindle afferents. The SP2(4)-fibre was activated by painful pinprick (Figure 99B) and by painful bladder catheter pulling (Figure 100A), similar to the SP2(3)-fibre. The activity increase was delayed to that of the SP2(3)-fibre and was pronounced. The activation of the SP2(4)-fibre was thus more static as compared to the SP2(3)-fibre. The SP2(5)-fibre fired seldom with doublets. Further comparisons between the activities of the secondary muscle spindle afferents and that of the oscillatory firing motoneuron will be presented in the following section [188].

12.7. Splitting of the Summed Activity into Those of Single Encoding Sites

It was shown above that human secondary muscle spindle afferent fibres in the sacral range can fire with single action potentials (APs), with doublet APs, with AP bursts and with equidistant APs. Single fibres changed their mode of firing upon different stimulations. Afferent spindle activity increases occurred several seconds following stimulation. On the assumption that the secondary spindle afferent with the simplest impulse pattern (SP2(1)) possessed only one source of AP generation, the impulse patterns of the other spindle afferents were split into the patterns of single encoders. The justification of this splitting is given by the similarity of interspike interval (II) distributions (Figure 94) of the constructed AP generation sources with those of the SP2(1)-fibre. Also, the decreasing and increasing of the encoding site IIs were similar, as can be seen from successive IIs (Figures 95B, 96A, 97A) and the direct impulse patterns (Figures 99,100). Further, it will be shown below (85) that single γ-motoneuron II distributions are very similar to those of single encoders of secondary spindle afferents. In the multi-ending regular firing mode in the parent fibre, the recruitment of a new encoder to further enhance activity was reflected by a transient irregularity (Figure 110A). In non-contracting muscles, Vallbo measured IIs larger than 50ms [230] and Burke et al. measured mean IIs of 90ms [35].

Up to six encoder sites could be constructed from the impulse patterns of the SP2(2)-fibre (Figure 97C) and up to four or five for the SP2(3)-fibre (Figure 99); in a human muscle spindle, eight myelinated branches of primary afferents were found [127]. In cat hind limb spindles, six secondary afferent axons showed between two and seven myelinated branches [6]. With the finding that the APs are most likely generated in the myelinated branches at the hemi-nodes or penultimate nodes [159], and assuming that the human secondary muscle spindle afferents in the sacral range are similar to those of the cat hind limb, it follows that the number of encoder sites constructed from the impulse patterns is in accordance with the number (and range) of myelinated branches. It is therefore likely that the encoder sites constructed in these measurements were in the myelinated branches of the parent fibres.

12.8. Coordination of Different Encoding Sites

It is difficult to understand how the different encoder sites coordinated their firing. The burst-like firing of the SP2(3)-fibre (Figure 99B) occurred because the four single encoders (SP2(3.1) through SP2(3.4)) fired in a somehow coordinated manner. An uncoordinated firing of the four encoder sites would have looked differently. The doublets of the doubler firing of the SP2(2)-fibre (Figure 100) showed a striking coordination. An easy explanation would be that the APs are generated in two myelinated branches. But how are the encoders coupled to give certain doubler II? Figure 98C shows doublets between 9 and 14ms. The value of 9 to 14ms cannot be explained easily by one encoder firing, antidromically invading the other side, and resetting the second encoder, since the value for the reset time is in the range of 1 to 2ms [53]. The doublet II distribution in Figure 98C shows three peaks. The peak at 13ms is labelled γ_1. This is to indicate that a simultaneously firing γ_1-motoneuron (intrafusal) in that nerve root fired with doublets, with a peak in the doublet II distribution also of about 13ms (Figure 108C). Considering the lack of viscosity of nuclear chain fibres and that secondary afferents innervate bag 1, bag 2 and chain fibres, but mainly chain fibres [6, 20], small fluctuations up to 60Hz in polar length are readily transmitted to the equatorial region [6, 25]; it could then be that doublet firing is due to the doublet firing of γ_1-motoneurons innervating the chain fibres. γ_1-doublets between 12 and 16ms [189] correspond to frequencies between approximately 83 and 62Hz. γ_1-doublets between 16 and 21ms [189] correspond to frequencies between 62 and 48Hz. Note before the impulse patterns of the efferent input (γ-motoneurons) to the muscle spindle fibres has been compared to the efferent output (spindle afferents), one should try to understand the impulse patterns of muscle spindle afferents, based on electronic generator potential spreading and concurrence of different encoder sites by antidromic invasion of one encoder site by the other and vice versa. First, it has to be known what part of the structure of the afferent impulse patterns is due to the activity patterns of the efferent input. The remaining structure of the afferent impulse patterns, which is only due to the muscle spindle properties, should be analysed with the competing encoder sites and electronic generator potential spread on the background of the detailed morphologic structure of the electrophysiologically measured spindles or similar units.

The doublet II distribution in Figure 98C shows three peaks. If doublets were generated by two encoder sites, this could mean that six encoders contributed to the firing of the parent fibre. The multi-ending regular firing suggested that the SP2(2)-fibre possessed at least six branches, possibly myelinated. The doublet II distribution peaks at 10.2, 11.2 and 13ms may have occurred because certain anatomical structures: e.g., the two branches producing the doublets may innervate different combinations of intrafusal muscle fibres, such as chain-bag 1, chain-bag 2 and bag 1-bag 2, or chain-chain, bag 1- bag 1 and bag 2-bag 2, or other combinations. Nevertheless, three sources of doublet AP generation were coordinated, since no IIs shorter than about 80ms occurred. If the doublet firing of the secondary muscle spindle afferents was generated by the doublet firing of the innervating γ_1-motoneurons, then three γ-motoneurons contributed each one with its own doublet II distribution. It could also be that the γ-doublets generated different afferent doublets at different intrafusal muscle fibres.

But since doublets occurred in γ-motoneuron and secondary spindle afferents, it could also be that not only the γ-motoneurons drove the spindle afferents, but also, at least in the pathologic case, the spindle afferents drove the γ-motoneurons, and they in turn drove the spindle afferents. In the pathologic case, a loop excitation may exist from the γ-motoneuron circuitry of the CNS to the muscle spindles. Oscillations between the γ-system and the muscle spindles have been suggested for some cases of pathologic tremor in humans [224]. It is reasonable to attribute some coordinated firing of different encoder sites of the secondary spindle afferents to the γ-systems. Figure 100 shows that the four parent afferents SP2(1) through SP2(4) fired mainly in a coordinated manner [14]. Only seldom, these fibres fired simultaneously. A coordinated firing of the different efferent endings is therefore likely. Progress in understanding of the function of muscle spindles will come from further simultaneous measurements of the impulse patterns of the muscle spindle afferents and the intrafusal motoneurons, and from further morphologic details of human spindles in the body part of interest. The identification of impulse patterns of single γ-motoneurons measured simultaneously with the afferent impulse patterns was a key finding [189] for the understanding of the function of human muscle spindles; in addition, the parasympathetic activity was detected, since the γ-motoneuron impulse activity alone cannot explain the spindle afferent activity increase (Figures 111,112). As will be shown below, [189], the multi-ending regular firing of the SP2(2)-fibre (Figure 97) is generated by the activity of parasympathetic muscle spindle innervation. Because of the still low level of parasympathetic activity, it has not been possible so far to safely pick up the pattern of the parasympathetic activity. A correlation between the activity pattern of the SP2(2)-fibre (Figure 97C) and that of the parasympathetic axons to find a reason for the multi-ending regular firing was therefore not possible. One possibility is that the recruited endings were unmyelinated, and the heminode generated equidistant APs according to the addition of evoked generator potentials from the different activated endings.

The high complexity of the possible impulse patterns of parent muscle spindle afferents and the most likely high complexity of recruitment possibilities of the different encoding sites, including their mutual interactions, are in accordance with the high complexity of the afferent innervation of chain, bag 2 and bag 1 fibres and the probably complex firing patterns and innervations of intrafusal motoneurons.

13. Phase Correlated Adequate Afferent Action Potentials as a Drive of Human Spinal Oscillators

Summary

1. By recording, with two pairs of wire electrodes, single-nerve fibre action potentials (APs) from lower sacral nerve roots of a brain-dead human and a patient with spinal cord injury, impulse patterns of afferent APs and impulse trains of oscillatory firing motoneurons could be identified and correlated.
2. Two highly activated secondary muscle spindle afferents increased and decreased their activity at about 0.3Hz. The duration of the doublet interspike interval of a secondary spindle afferent fibre showed no correlation to the oscillation period of the motoneuron.
3. A continuously oscillatory firing motoneuron innervating the external anal sphincter showed more transient breaks with the reduction of the number of phase correlated APs from two spindle afferents, indicating a looser oscillation. A transient brake of a 157ms period oscillation could be correlated to the shift of an interspike interval distribution peak from 150 to 180ms of the adequate afferent input, which suggests a transient loss of the necessary phase relation.
4. Oscillatory firing α_2-motoneurons innervating the external bladder and anal sphincters fired independently according to their phase-correlated APs from the urinary bladder stretch receptor and muscle spindle afferents respectively; the bladder motoneuron slightly inhibited the anal motoneuron.
5. Receptors of the afferents and innervation sites of oscillatory firing motoneurons could be located within the urinary tract and anal canal.

13.1. Introduction

After having analysed above the impulse patterns of oscillatory firing motoneurons [186] and the impulse patterns of parent secondary spindle afferent fibres and their probable encoder sites at the hemi-nodes or penultimate nodes [7] of the myelinated branches [6, 187], in this section the afferent activity will be compared, most likely driving the oscillatory system, with the motoneuron activity; this will continue the analysis of the previous section [187].

Only the oscillatory system with its secondary muscle spindle afferent input, driving the anal sphincter, has been analysed so far. In this section, also an oscillator driving external bladder sphincter muscle fibres is considered with its stretch receptor associated afferent input from the bladder. It will be shown that both oscillatory systems are firing simultaneously, but independent of each other, even though small interactions exist.

13.2. Simultaneous Recordings from Oscillatory Firing Motoneurons and the Afferents Driving Them

Figure 101 shows impulse trains of oscillatory firing α_2-motoneurons innervating the anal sphincter of brain-dead human HT6 and paraplegic patient 1 (Para 1). Sweep pieces were selected to present afferent (upwards) and efferent (downwards) action potentials under consideration.

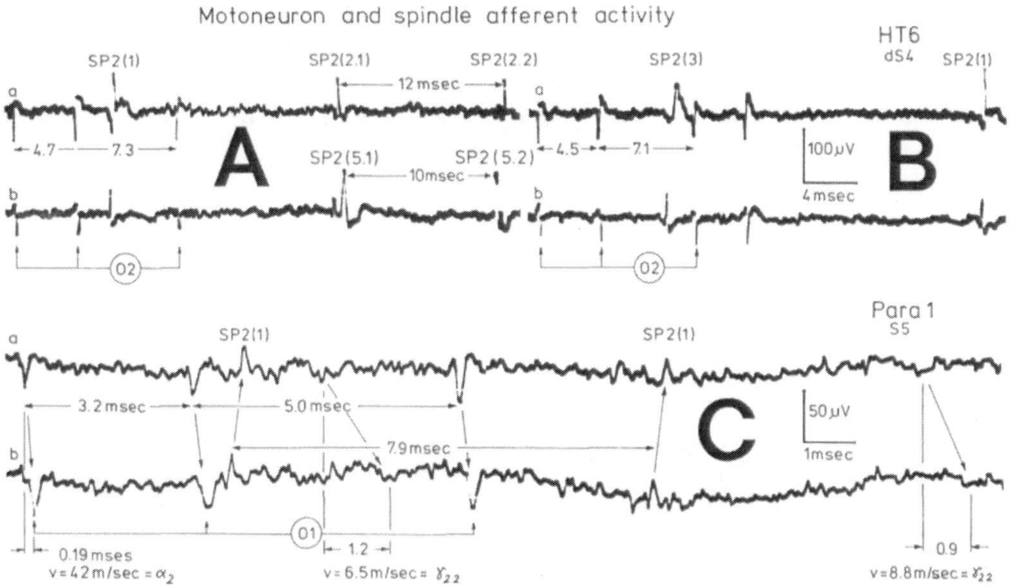

Figure 101. Recordings of the activity of α_2-motoneuron O2 and secondary muscle spindle afferents SP2(1), SP2(2), SP2(3) and SP2(5) (A, B) in HT6, and of α_2-motoneuron O1 and secondary muscle spindle afferent fibre SP2(1) in paraplegic 1. In A, trace b, the AP of SP2(2.2) is merged with that of SP2(5.2), trace b, does not appear on trace a, being partly blocked.

Traces a and b are from the proximal and distal electrode pairs respectively. Usually, there was more spacing between the impulse trains of the motoneurons and the afferent action potentials (APs). In Figure 101A, an impulse train of α_2-motoneuron O2 is shown together with those of the secondary muscle spindle afferent fibres SP2(1), SP2(2) and SP2(5). Afferents SP2(2) and SP2(5) fired with doublets of 12 and 10ms. SP2(2) AP is merged on trace b with SP2(5.2) AP. APs of SP2(5.1) and SP2(5.2) (trace b) were not conducted across the second electrode pair (trace a) or were strongly weakened. In Figure 101B, another impulse train of oscillatory firing α_2-motoneuron O2 is shown together with APs of afferents SP2(1) and SP2(3). It can be seen in Figure 101A,B that the impulse trains could be identified by the interspike intervals (4.7 and 7.3ms in A) and by the oscillation period (not shown); identification of the secondary muscle spindle afferent APs by the waveforms. Time expanded waveforms of the afferents under consideration were shown in the previous section. Approximately three other activated but unidentified spindle afferents of the same dS4 root had low activity and therefore did not contribute to the strong excitation of the motoneuron. Figure 101C shows an impulse train of oscillatory firing α_2-motoneuron O1 of para 1 together with APs of the secondary muscle spindle afferent fibre SP2(1). Some of the interspike intervals, conduction times and conduction velocities are indicated. Even though the quality of the recordings from Para 1 was not as good as that from the HT6, it was also possible to identify the motoneuron and the SP2(1) APs after some training. Even the marked γ_{22} AP could be identified.

13.3. Stability of Oscillation in Dependence on the Number of Adequate Afferent APs within a Certain Phase Range

Figure 102A shows the activity levels of the afferents at least partly driving the oscillatory firing α_2-motoneuron O2 in HT6. Low activity of afferent fibres is not always plotted. It can be seen that throughout the activities, the different afferents are increasing and decreasing at a mean frequency of about 0.3Hz.

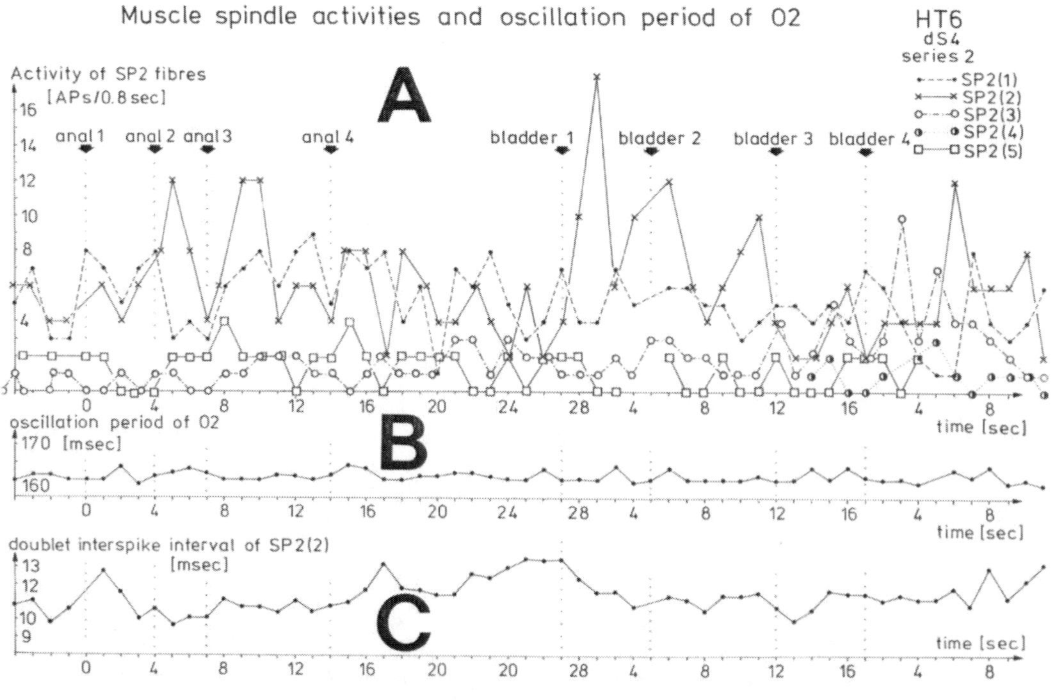

Figure 102. Activity of five secondary muscle spindle afferents (SP2(1 through 5)) (A) in relation to oscillation period of continuously oscillatory firing α_2-motoneuron O2 (B), and in relation to the doublet interspike interval of afferent SP2(2) (C) following anal and bladder catheter pulling. All values measured simultaneously. HT6.

Even during no stimulation, between anal catheter pulling 4 and bladder catheter pulling 1, the increasing and decreasing continued on a reduced activity level. This finding is in accordance with the coordination of the impulse patterns of several spindle afferents (sharing activity) discussed above. The summed activity of all afferents is smooth, as was shown previously for the same case. Following stimulation, the single spindle afferent fibres responded differently. Fibre SP2(2) responded strongly to bladder catheter pulling (bladder 1) but not always in the way shown in Figure 2. Thus, the firing of the spindle afferents was at least partly regulated by the efferent innervation, since the length of the anal sphincter muscle and of the functionally associated muscles of the pelvic floor (Figure 52) probably remained rather constant.

In the second series of measurements (Figure 102), the activity levels of the spindle afferents were higher in comparison to those of the first series (Figure 103). There was an activation of the parasympathetic nervous system in between, which innervated the SP2(2)-fibre spindle and maybe also some other spindles [189] (see below). With these high afferent activity levels (Figure 102A), α_2-motoneuron O2 (Figure 102B) fired continuously in the oscillatory mode with no breaks.

It was shown in the previous section that many afferent APs were contributing, with a mean phase of 50ms between the afferent APs and the first AP of the impulse train of the oscillatory firing α_2-motoneuron (50ms/160ms = 0.3, cycle phase 112.5°). In the first series of measurements (before the activation of the parasympathetic fibres), the situation was different (Figure 103).

The activity levels of the different spindle afferents were lower (Figure 103A), and fewer APs were contributing to the oscillation, with a mean phase of 50ms prior to the impulse train (previous section).

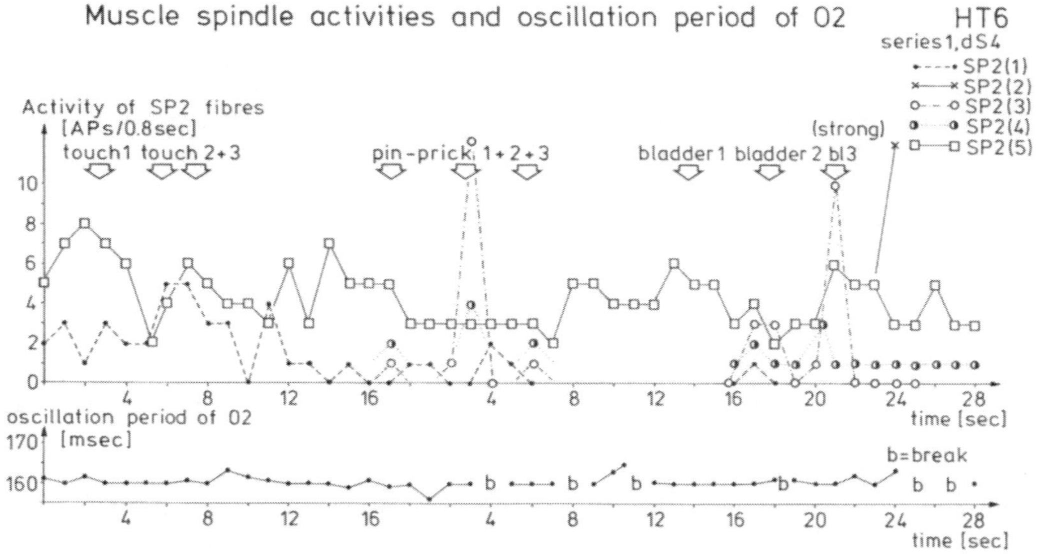

Figure 103. Activity changes of five secondary muscle spindle afferents SP2(1 to 5) in relation to oscillation period of continuous oscillatory firing α_2-motoneuron O2. Note that oscillation breaks more easily than in the case illustrated in Figure 102. The activity of fibre SP2(2) is only suggested. Low activity is often omitted. HT6.

Several times, the oscillation broke transiently (Figure 103B). It is concluded that the lower activity levels with less contributing spindle afferents APs in phase deove the oscillator O2 less strongly, so that more breaks could easily occur.

Figure 102C shows the development of the mean doublet interspike interval of the SP2(2)-fibre with time and in dependence on its own activity level (Figure 102A). No direct dynamic correlation between the activity of the SP2(2)-fibre and the duration of the doublet II can be seen.

Following bladder catheter pulling 1, e.g., the activity first increased (Figure 102A), and the doublet II increased; however, when the activity quickly decreased, the doublet II did not increase as quickly.

Doublet II decreased slowly with the increasing activity and increased slowly with the decreasing activity. This slow change could come from the more static behaviour of the second encoder site generating the second AP (of intrafusal muscle fibre origin); it could be due to slow and fast activity changes of the somatic intrafusal innervation; or it could be due to the slowly changes of the parasympathetic nervous system (189). The doublet II duration changes (Figure 102C) showed no direct correlation to the duration of the oscillation period of the oscillatory firing α_2-motoneuron O2 (Figure 102B).

13.4. Break of α_2-Oscillation in Paraplegic 1 Secondary to SP2(1)-Fibre Associated Changes Following Anal Catheter Pulling

Probably the transient break of continuous oscillation is functionally unimportant. Nevertheless, it is of high interest here, because it may throw light on the mechanism that drives the oscillator.

The drive of the oscillator includes adequate afferent input, certain activity and certain phases between the afferent impulse patterns and the oscillation cycle. The phases between the afferent APs and the first AP of the motoneuron impulse trains can be directly measured as the time difference between them. An indication for the existence of a necessary phase relation is a rather fixed relation of the respective afferent II distributions to the oscillation period. In Para 1, it could be shown that a break in continuous oscillation correlated with a strong shift of the afferent II distribution peak at about 150ms in relation to the oscillation period.

Figure 104A shows the activity of the secondary muscle spindle afferent fibre SP2(1) in Para 1 before stimulation (a), upon anal catheter pulling (b), shortly thereafter (c), at a later interval (d) and after a longer interval following catheter pulling (e). After stimulation, only the activity actually increased in response to catheter pulling (c). The light activity increase before catheter pulling, between 4 and 14s, was due to the light stimulation caused by the handling of the connection to the catheter. Long after stimulation (38s, d), the activity only returned to the pre-stimulation level. A transient break of the continuously oscillatory firing α_2-motoneuron O1 occurred only long after stimulation and afferent activity increase. In Figure 104C, the break in oscillation is correlated to the distribution of interspike intervals (IIs) at different time periods. Figure 104B shows the approximate intervals when oscillation occurred.

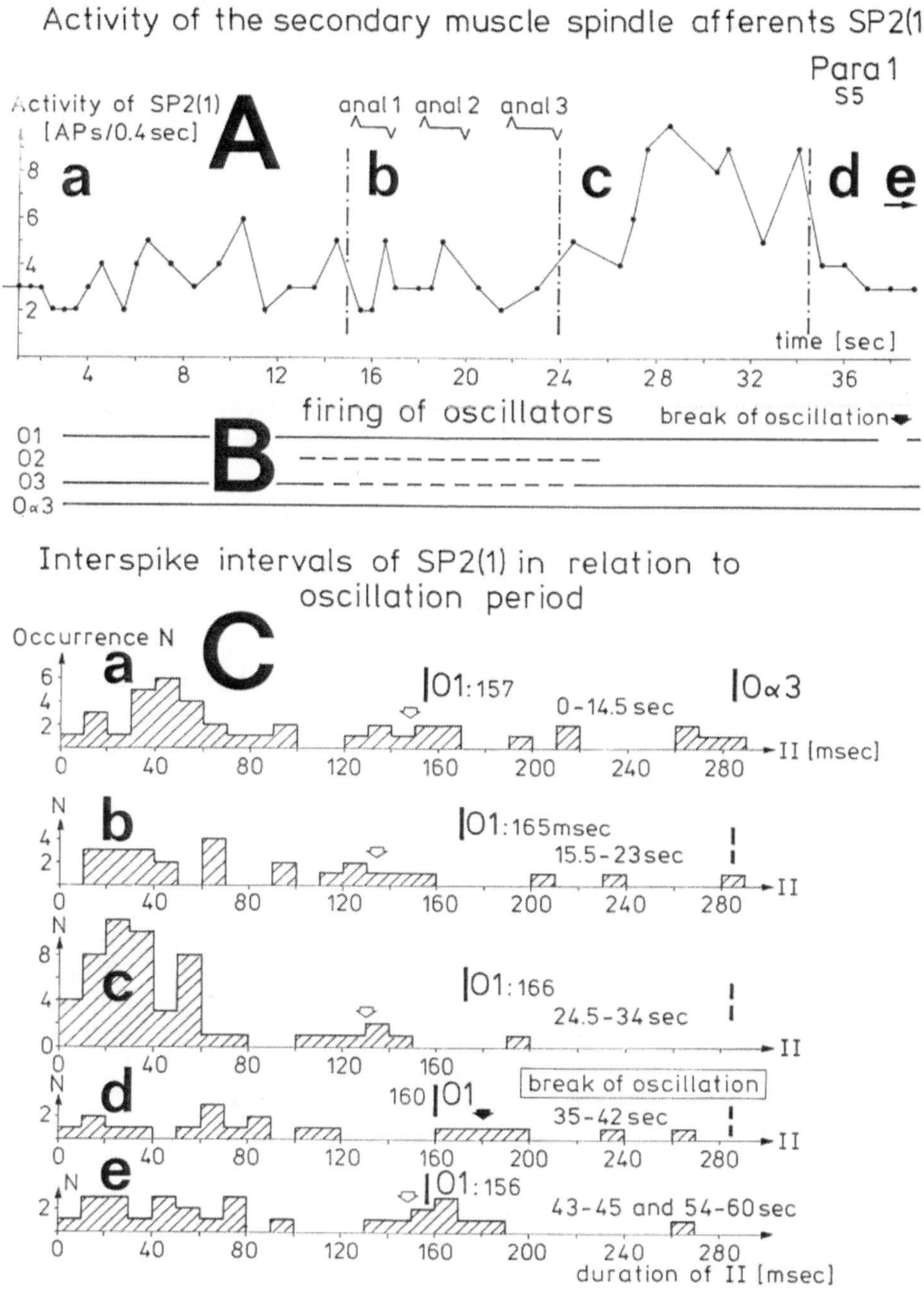

Figure 104. Activity and interspike interval changes of the secondary muscle spindle afferent fibre SP2(1) in paraplegic 1 before and following stimulation. A.a. Activity changes before anal-catheter pulling. Some stimulation appears between 4 and 15 sec due to catheter handling, b. Anal catheter pulling period, c. Immediate post-stimulation interval, d. and e. Later post-stimulation intervals. Intervals a to e same as in C. B. Oscillatory firing α_2-motoneurons Ol, O2 and O3 and α_2-motoneuron Oα3. Full line = continuous oscillatory firing, dashed line = transient oscillatory, no line = no oscillatory firing. C. Interspike interval frequency distribution histogram of fibre SP2(1) before (a), upon (b), and after stimulation (c, d, e). The time intervals used are indicated at each histogram. Exact mean oscillation periods of Ol at the same intervals are shown. Note that in d, the 150 ms peak is shifted to 180 ms (filled arrow) (hysteresis-effect), at a time when oscillation broke. The oscillation period of the oscillatory firing α_3-motoneuron Oα3 is given.

The dashed line indicates transient oscillatory firing and the solid line continuous oscillatory firing. Numerous motoneurons were identified in the same S5 root of Para 1, which fired in the occasional firing mode [185] (Figure 57,58), and four motoneurons were identified, which fired in the oscillatory firing mode [186] (Figure 71). Three of the four motoneurons were mainly continuously oscillatory firing, and one motoneuron was transiently oscillatory firing (O2). Most likely, all the four oscillatory firing motoneurons contributed to the continence of the rectum.

The II distributions shown in Figure 104C are those of the parent fibre SP2(1). It was not attempted to split the activity into those of single encoder sites. As shown above, however [187] (Figure 95A), fibre SP2(1) in Para 1 fired with bursts. It seemed therefore that the fibre was over-activated in relation to the activity of a comparable fibre in HT6 (SP2(3)) [187] (Figure 96). Before anal catheter pulling, the distribution peak of fibre SP2(1) at 150ms (Figure 104Ca) had a similar duration as the oscillation period of the oscillatory firing α_2-motoneuron O1 (157ms). Upon anal catheter pulling (b), this peak slightly shifted towards shorter II durations, and the oscillation period was prolonged; this may be an indication for a less effective drive; nevertheless, oscillation continued. With the activity increase in c, the driving peak shifted to even shorter durations (130ms), the oscillation period showed a further increase to 166ms, but still, the oscillation held. Then with a drop of the activity of fibre SP2(1) (d), this peak shifted drastically to longer duration (180ms) (lower frequency), and the oscillation broke (filled arrow).With a backward shift of the SP2(1) fibre distribution peak, the oscillation started again with a period of 160ms. Later on (e), the former distribution peak of fibre SP2(1) shifted further backwards to shorter values closer to those of before stimulation (150ms), and the oscillation period returned to the pre-pulling value (156ms). This series of correlations indicate that the transient break in oscillation was related to the shift of the SP2(1)-fibre II distribution peak at about 150ms (arrow). However, as can be seen from II distribution shown in Figure 104C, not only the peak at 150ms shifted with the break in oscillation (d): also, the peak between 40 and 80ms moved to longer times, and the whole activity was reduced. Thus, the break in oscillation was probably due to the change of the entire impulse pattern of fibre SP2(1), and this change could be made visible by the shift of the 150ms peak. Several spindles must have contributed to the afferent drive of oscillation. Probably, the other spindle afferents showed a similar behaviour. Fibre SP2(2) in paraplegic 1 showed a similar distribution as that shown in Figure 104Ca. However, the waveform identification was not safe enough to allow considerations concerning longer intervals. That not only the afferent distribution peak at about 150ms contributed to oscillation can be understood owing to a rather constant phase that must occur quite often between afferent APs and oscillation cycle, as determined by the first AP of an impulse train. If the oscillation period is of duration of 157ms and the driving peak of 150ms, then after a couple of 150ms intervals, the necessary phase range for the drive of oscillation will be lost. Therefore, some IIs of other durations have to be interposed to regulate the phase. This can be best seen with the IIs of fibre SP2(1) in HT6 (Figure 95A). A combination only of at least 145 and 85ms IIs ($2 \bullet 85 = 170$ms) can yield a mean II of about 160ms, which is the interval for the oscillation period to keep a constant phase.

Figure 94A suggested that the splitting of the II distribution peak of fibre SP2(1) in HT6 had been due to a hysteresis effect of the muscle spindle; namely, with further contraction, the afferent IIs slightly differed from those for a decreased contraction of the same intrafusal muscle length (after effects and creep). Figure 104C supports this hypothesis, since the states shown in Figures 104Ca and 104Cd had about the same activity levels (Figure 104A, note the different time interval), but the II distributions were quite different. Figure 104C also gives the oscillation period of the oscillatory firing α_3-motoneuron Oα3 (about 280ms). Little is known about how α_3-motoneuron oscillators are driven (most likely polymodal). The distribution peaks at 280ms and 140ms, and also shorter ones, of fibre SP2(1) could also drive the oscillatory firing α_3-motoneuron. Phase and frequency coordination, and their impairment following CNS injury, will be picked up again in Chapter V.

13.5. Simultaneous Oscillatory Firing of α_2-Motoneurons Innervating the External Sphincters of the Bladder and the Anal Canal with Their Different Adequate Afferent Driving Inputs

Most measurements in the present work so far were devoted to study the function of the external anal sphincter, even though a good function of the external bladder sphincter is of considerable more importance for patients with spinal cord injury (bladder infections, dyssynergy of the urinary bladder). However, it is easier to measure anal sphincter functions and its coordinated drive. Nevertheless, some progress in the understanding of the external bladder sphincter function could also be achieved and will partly be reported here. Substantial further knowledge will be achieved by comparing neural firing patterns with bladder pressure measurements (urodynamics) and will be presented in Chapter V. It was shown above (Figures 50,51,59) that with the retrograde filling of the urinary bladder, stretch and tension receptor activity activated an α_2-motoneuron to switch into the oscillatory firing mode for fillings larger than 600ms. More details will be given here.

Figure 105A shows three sets of successive interspike intervals (IIs) of stretch receptors [175]. The indicated oscillation period of about 110ms at the given stage of filling was similar to some IIs of the stretch receptors (Figure 2), at least those of fibre S1(1) (Figure 105A). Since the number of APs per impulse train of this oscillatory firing α_2-motoneuron O1 in HT6 changed from 1.5 to 2 in different combinations with increasing filling of the urinary bladder, the oscillation period changed [175]. For stable oscillations, the theoretical value of the oscillation period for a one AP impulse train is 100ms, and 130ms for a two AP impulse train [186] (Figure 67) (oscillation period $T = 70\text{ms} + 30\text{ms} \bullet n_{AP}$; n_{AP} = number of APs per impulse train). Figure 105B shows an impulse train consisting of two APs. Since with this α_2-motoneuron O1 a recurrent fibre was identified, it was possible to identify each single AP of this motoneuron during all recordings. In Figure 105B, the conduction times and the conduction velocities of the oscillatory firing α_2-motoneuron O1 and of a mucosal afferent fibre (M) [175] are indicated.

With no bladder filling α_2-motoneuron O2 (HT6), innervating the external anal sphincter was oscillatory firing. With the filling of the bladder α_2-motoneuron O1, innervating the external bladder sphincter started additionally firing in the oscillatory firing mode for bladder fillings larger than α_2-motoneuron 600ml (Figures 106B, 59D). With more than 600ml in the bladder, oscillator O1 and O2 oscillated simultaneously, but they were excited by their own different adequate afferent inputs. The oscillatory firing motoneuron O2 (Figure 105C) received its adequate afferent input from APs in phase of the secondary muscle spindle afferents SP2(1) and SP2(2) and partly SP2(3).

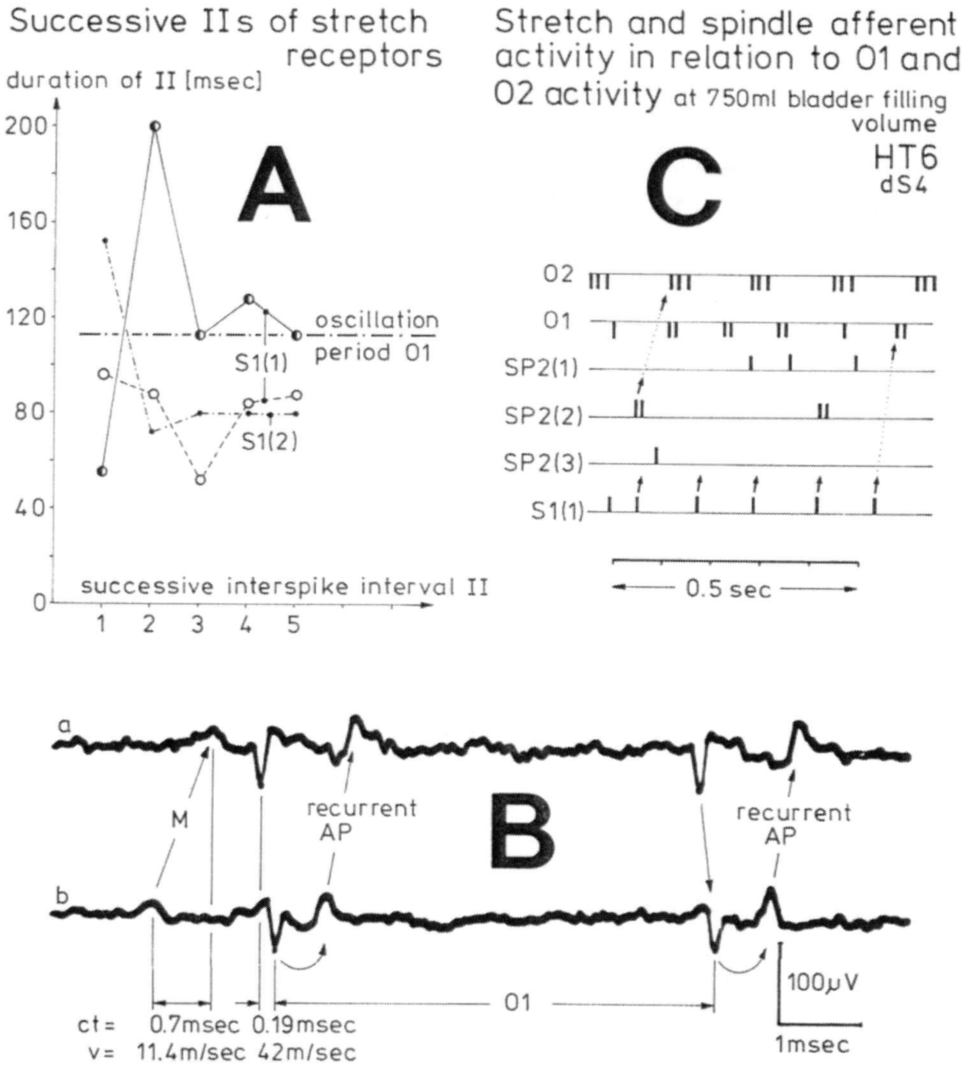

Figure 105. A. Three series of successive interspike intervals of two stretch receptor afferent fibres (S1(l), S1(2)), activated by retrograde urinary bladder filling. Oscillation period is given of oscillatory firing α_2-motoneuron Ol, activated only by bladder filling. B. Recording of (α_2-motoneuron Ol), which most likely innervated the external urethral sphincter. The motoneuron possessed a recurrent fibre (time-locked afferent-like action potential), ct = conduction time, v = conduction velocity, M = mucosal mechanoreceptor afferent AP. C. Impulse patterns of oscillatory firing α_2-motoneuron O2 innervating the external anal sphincter, in relation to the muscle spindle afferent activity (SP2(1 to 3)), activated by the stretch of the anal sphincter by the anal catheter, and impulse patterns of oscillatory firing α_2-motoneuron Ol innervating the external urethral sphincter, in relation to the stretch receptor afferent activity (Sl(l)) of the urinary bladder, activated by 750 ml bladder filling. Phase relations between APs of SP2(2) and O2 and between APs of Sl(l) and Ol are indicated by small arrows.

An AP in phase from fibre SP2(2) is marked with a small arrow in Figure 105C. The simultaneously oscillatory firing α_2-motoneuron O1 (Figure 105C) received its adequate afferent input in phase from APs of the stretch receptor afferent fibre S1(1) and maybe other fibres.

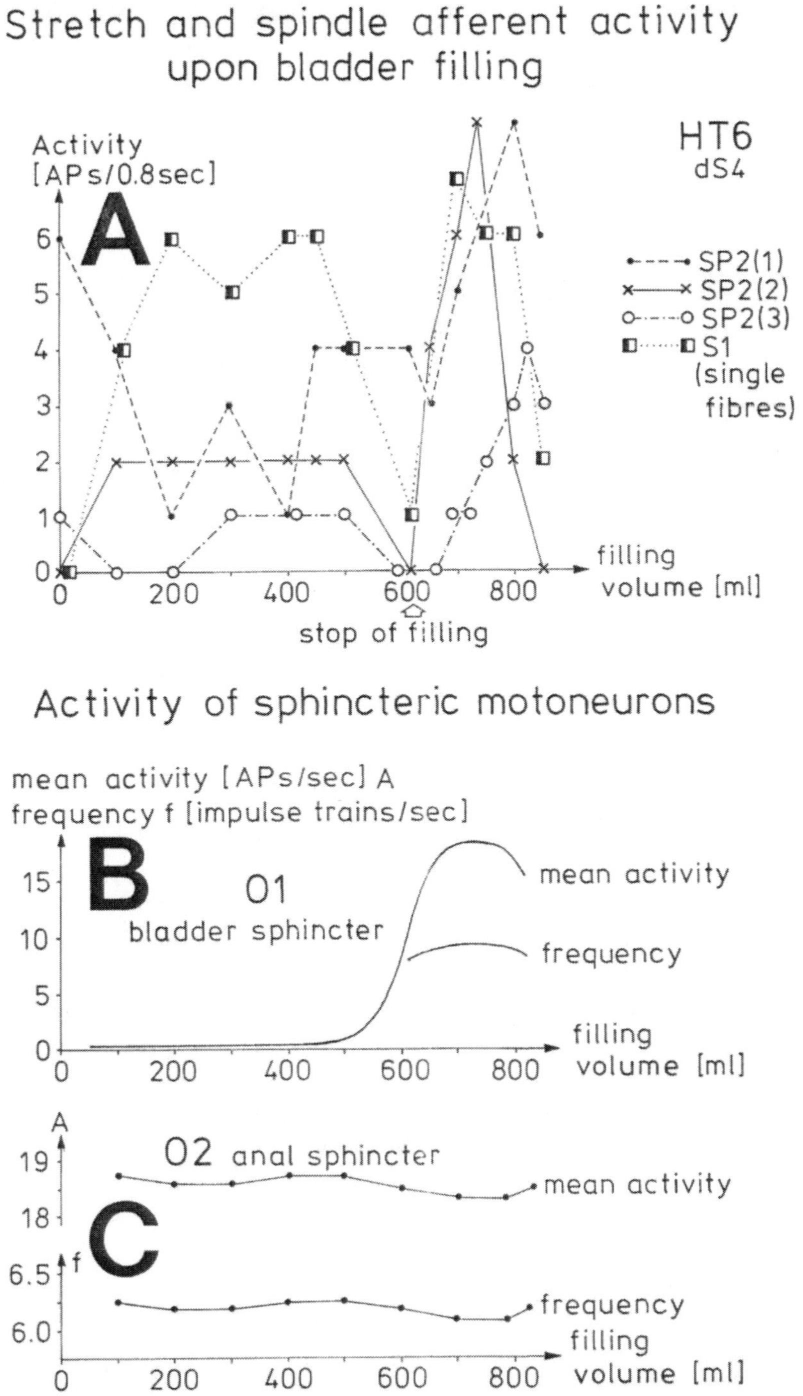

Figure 106. Activity changes of secondary muscle spindle (SP2) and urinary bladder stretch receptor (S1) afferents (A) in relation to activity changes of sphincter α_2-motoneuron Ol of the external urethral sphincter (B), and in relation to the activity changes of sphincter α_2-motoneuron O2 innervating the external anal sphincter; increasing filling volume of the bladder. Motoneuron O2 fired in the constant oscillatory firing mode, slightly affected by the filling of the bladder. Motoneuron Ol fired until approximately 550 ml filling volume in the occasional firing mode (very low activity), until 620 ml in the transient oscillatory firing mode (medium activity level), and until 820 ml in the constant oscillatory firing mode (high activity) (9). HT6.

The phase between these five APs and the AP of the O1 motoneuron impulse train was about 50ms ± 6ms. As shown above [187], the phase between APs of the SP2(1) and SP2(2)-fibres and the first AP of the impulse train of α_2-motoneuron O2 was also 50ms. The similarity of the two phases is probably accidental. Each oscillatory firing α_2-motoneuron fired according to its own adequate afferent APs in phase. For longer periods of time, it was difficult to identify single stretch receptor afferents by the waveform.

Since it is known that there are some interactions between the external anal and the external bladder sphincters in humans [214], the frequency and the activity of both oscillatory systems were plotted (Figure 106B,C). The mean activity and the frequency of the bladder sphincter motoneuron O1 were taken from a previous work [176]. As can be seen from Figure 106C, the anal sphincter α_2-motoneuron O2 reduced its activity and mean frequency with the increasing activity and frequency of the bladder sphincter α_2-motoneuron O1. Partly antagonistic function of external anal and bladder sphincters is known for humans [214]. With the beginning of the overflow mechanism of the bladder for bladder fillings larger than 800ml, the mean activity and the frequency of motoneuron O1 decreased, and the mean activity and frequency of motoneuron O2 increased again.

13.6. Function and Location of Receptors Contributing to Micturition and Defecation

Further knowledge about the function and the anatomical location of the receptors of the secondary muscle spindle afferents SP2(1) through SP2(3) and the stretch receptor afferent fibre S1 could be obtained from their activity changes associated with the filling of the bladder (Figure 106A). The single afferent fibre S1(1) of a urinary bladder stretch receptor increased its activity with the filling of the bladder. The other stretch receptor afferents in the same root showed similar activity changes [176] (Figures 50,51). The single fibre S1(1) reduced its activity during a 10 to 20min break in filling (Figure 106A), most likely the bladder adapted to the filling [46] (smooth muscle fibre shift). With further filling, the activity of fibre S1(1) increased again and decreased upon onset of the overflow mechanism of the bladder at filling volumes exceeding 800ml. Fibre S1 indicated pressure in the bladder and most likely monitored tension in the bladder wall. It is likely that its receptor was located in the bladder wall, as illustrated in Figure 52 on the lower female pelvis. The secondary muscle spindle afferent fibre SP2(2), which was very sensitive to strong bladder catheter pulling (Figure 102A), increased its activity with the bladder filling and reduced it upon the filling stoppage (Figure 106A).

A strong increase in activity was again recorded upon further bladder filling, and a sudden decrease just before the onset of the overflow mechanism at 800ml. Thus, the muscle spindle of this secondary afferent fibre must have been located in a muscle involved in performing of the continence function of the urinary bladder. Since most likely there are no muscle spindles in the external urinary bladder sphincter [176], the SP2(2) muscle spindle was probably located in the musculus pubococcygeus (Figure 52), which contributes to the continence of the urinary bladder [72].

Therefore, in Figure 52, the SP2(2) spindle is sited into this muscle. The secondary muscle spindle afferent fibre SP2(3) responded to painful pinprick and painful (strong)bladder catheter pulling (Figures 103,99,100). The activity of fibre SP2(3) increased slightly with the bladder filling but increased most strongly upon reaching the highest pressure in the bladder or immediately after (Figure 106A), as measured by the S1(1) fibre activity. Thus, fibre SP2(3) somehow monitored painful pressure in the bladder, which is the highest with the highest pressure in the bladder.

Upon the painful application of painful stimuli, the sphincters and the pelvic floor will contract. Since fibre SP2(3) did not respond strongly to bladder and anal catheter pulling of medium intensity, the spindle belonging to the SP2(3) afferent fibre was probably located in the pelvic floor (Figure 52). The secondary muscle spindle afferent fibre SP2(1) showed the highest activity before bladder filling and the highest filling and reduced its activity for medium bladder fillings (Figure 106A). Stoppage of the bladder filling had little effect on this activity. Thus, the spindle of fibre SP2(1) is not directly involved in the continence of the urinary bladder. The reduced activity upon medium-sized bladder fillings, and thus slightly antagonistic bladder function, indicates association with the continence of the rectum. Since fibre SP2(1) responded to anal catheter pulling (Figure 102) and to touch (Figure 103), and not to bladder catheter pulling (Figure 103), spindle SP2(1) was probably situated in the superficial parts of the external anal sphincter (Figure 52). The pain sensitive tertiary muscle spindle afferent fibre SP2(4) innervated the same muscle spindle as did fibre SP2(3) (Figures 99,100; indicated by a dashed box). It was not possible to locate spindle SP2(5), as it did not responded specifically.

Even with recording from a S4 dorsal root and having no anatomical knowledge about which are the body parts certain fibres run to, the changes of impulse patterns and activity levels provide quite a lot of information about which parts of the pelvis are innervated by the afferents and what functions they transmit. Figure 52 shows the female pelvis with the bladder and rectum and the muscles contributing to their continence. It was reasoned above to which muscles the muscle spindles and the stretch receptors have to be attributed. Moreover, Figure 52 shows the two sphincters that are innervated by the oscillatory firing α_2-motoneuron O1 and O2. Flow receptors (S2), which were strongly activated with retrograde bladder filling [175] (Figures 50,51) are sited to the trigonum vesicae (Figure 52), where the density of the flow receptors is probably the highest. Mucosal afferents (M) were stimulated upon bladder and anal catheter pulling [175] (Figures 48,49) and are thus sited into the urethra and the anal canal (Figure 52). Skin receptor afferents were stimulated upon anal catheter pulling and only little upon light bladder catheter pulling (Figures 48,49). The low threshold T3 and T4 skin afferent receptors were sited into the anal canal. Their function is important, similar to that of afferent skin receptors T1 and T2 [175] for the discrimination between gas, fluid and solid substances [157, 175, 214]. Quite a lot of specific afferent and efferent nerve fibre functions can be measured with different stimulations necessary for micturition and defecation by recording with two pairs of wire electrodes from lower sacral dorsal nerve roots.

13.7. Phase Relation between Oscillation Cycle and APs of the Adequate Afferent Input

It was shown that α_2-motoneuron O2 fired in the oscillatory firing mode with no breaks (Figure 102) when the activity of the secondary spindle afferents SP2(1) and SP2(2) was high.

Breaks occurred (Figure 103) when the activity of fibres SP2(1) and SP2(2) was low. Thus, the activity levels of the adequate spindle afferents were one factor for the excitation of the oscillators. The oscillatory systems consisted mainly of the motoneuron itself and of sets of coupled interneurons. Two possibilities of oscillator structure are given in Figures 68 and 69. Mechanical and electronic oscillators need rather continuous energy supply in phase for the oscillation to overcome damping. As the phases in Figure 105C indicate, not only the number of APs of the adequate afferent type is sufficient for continuous oscillation to occur, but also the impulse patterns must be in a way that the APs have a certain time correlation to oscillation cycle. From Figure 4, it can be seen that oscillation of α_2-motoneuron O1 broke upon the anal catheter pulling, even though the activity level was the same as before (Figure 104Aa (left part)). During the break of oscillation, the interspike interval distribution had completely changed (Figure 104Cd) in comparison to the pre-stimulation pattern (Figure 104C,a-c). In particular, the peak at about 150ms changed its position. Figures 99 and 100 showed (small arrows) moments when SP2(1) and SP2(2) fibre APs were in phase, within a certain range (50 ± 20ms), to the first AP of the impulse train of the oscillatory firing motoneuron. Also in this recording, there was a phase relation of 50 ± 6ms between S1(1) APs and motoneuron O1 impulse trains in a completely different oscillatory system. The identity of phases is most likely accidental. The actual value of the phase (range) is of less importance, as there could be time consuming interneurons between the afferents and the oscillation loop to adjust the phase. In mechanical and electronic oscillators, only energy in phase contributes to oscillation. In further studies, the distribution of phases of the afferent APs will have to be correlated to oscillation period and the stability of oscillation as will be done in Chapter V.

The duration of the interspike intervals of the spindle afferents, which in different combinations were similar to that of the oscillation period, could have been generated by the muscle spindle itself or together with the impulse patterns of the efferent muscle spindle innervation. At least to keep a constant phase, however, the spindle has to be under a control system, which guards that the spindle activity is not drifting. In electronic oscillators, the output is coupled back to the input to establish a constant phase. Since only efferent innervation can precisely control the muscle spindle, the intrafusal motoneurons (and maybe the autonomic system) need information on the phases of oscillation. It will be shown in the following section that there was a rather constant phase relation between the occurrences of γ-APs and the first AP of the oscillatory firing α_2-motoneuron. The constant phase from the activity of the efferent spindle innervation, via that of the muscle spindle afferent fibre, to the oscillatory system, can be broken more directly by the γ-efferent activity, not under consideration here, and by intrafusal muscle fibres, coupling the activity of the afferent and efferent spindle innervation. The different properties of the bag 1, bag 2 and chain fibres (6) can change the phase relation. Following stimulation, the summed or single (chain fibre) contributions from intrafusal muscle fibres can result in a hysteresis effect: too few APs of the adequate afferents in the necessary phase range will give rise to the break of the oscillation. When the steady state is nearly achieved again, sufficient APs with the right phase occur again and the oscillator starts oscillating again. The steady state is given by the constant stretch of the anal sphincter maintained by the catheter. Since each spindle afferent fibre innervates with its branches several intrafusal muscle fibres [6], so that several encoding sites exist [187], and since also each γ-motoneuron innervates different intrafusal fibres [127], there is a considerable complexity in the muscle spindle receptor.

A single encoding site can send unmyelinated branches to different intrafusal muscle fibres, so that a single encoder can obtain information from different intrafusal muscle fibres, and can therefore receive drive from different intrafusal motoneurons. Generator potentials may spread from one branch to another, giving rise to interactions of different encoders belonging to the same parent fibre.

However, using this method to determine the impulse patterns of the muscle spindle afferents, and being able to split them into impulse patterns of single encoders (probably, only one or two encoders are sometimes active of the several existing), and recording simultaneously the impulse patterns of single intrafusal motoneurons [189] (see below), the impulse pattern changes through the spindle can be measured. Concerning the continence of the bladder, the APs of one stretch receptor afferent fibre also very often were in phase with the impulse train of the oscillatory firing α_2-motoneuron, innervating the external bladder sphincter (Figure 105C). The parasympathetic division might be a candidate for supplying certain phase or for stopping the receptor activity from drifting.

13.8. Different Oscillatory Systems

One of the kinds of oscillators are here the α_2-motoneurons innervating the external anal sphincter, with its interneurons, excited by the afferents of muscle spindles of the anal sphincter or functionally associated muscles. Activity from mucosal and skin afferents (Figure 52) may contribute to drive the oscillator. The other kinds of oscillators are α_2-motoneurons innervating the external urinary bladder sphincter with their interneurons, which are excited by the activity of the stretch receptor afferents of the bladder wall. Activity from mucosal and flow receptors (Figures 48-52) of the urethra and trigonum vesicae contribute to the excitation or inhibition [176] of the oscillator.

The interactions between the two kinds of oscillatory systems contribute to the agonistic and antagonistic functions of micturition and defecation. α_3-Motoneuron oscillators have also been detected (see above). The irregularity in the interspike intervals of long α_3-motoneuron impulse trains [176, 186] may indicate coupling of the oscillators. In the occasional firing mode, extrafusal and intrafusal motoneurons are recruited rhythmically (Figures 54-58). Rhythmicity was observed in the activity changes of the oscillatory firing α_2-motoneuron O4 in Para 2 [186].

Secondary spindle afferents increased and decreased their activity with a frequency of 0.3Hz (Figure 102). α_1-Motoneurons (Figures 76,77,78) do not contribute substantially to continence functions, and they were only sometimes found in these recordings. They are substantially involved in movements and can be easily measured by electromyography (EMG) by detecting single-motor unit potentials (Figures 79,80). Recordings (with the same equipment) of single-nerve fibre APs and single-motor unit APs of the three kinds of motoneuron oscillations are given in Figure 70. In the brain stem, oscillators have been detected for breathing and the control of the heartbeat [130]. In Chapter V, the coordination between the somatic and the parasympathetic nervous system divisions will be analysed to clarify the synergy and dyssynergy of the urinary bladder for repair by movement-based learning.

14. Natural Action Potential Patterns of Intrafusal γ and Parasympathetic Motoneurons, Secondary Muscle Spindle Afferents and an Oscillatory Firing α_2-Motoneuron, and the Phase Relations among Them in Humans

Summary

1. Single-fibre action potentials (APs) were recorded from an S4 root of a brain-dead human (HT) and a patient with a thoracic spinal cord injury (Para). Apart from the doublet firing, a single γ_1 (dynamics) and a single γ_{21}-motoneuron (static) fired with shortest interspike intervals of 80ms, similar to a simultaneously firing secondary muscle spindle afferent fibre; with 13ms, the doublet interspike interval of the γ_1-motoneuron was very similar to that of the spindle afferent fibre. The intrafusal motoneurons fired with patterns very similar to those of secondary spindle afferents, and they did not fire in the oscillatory mode for high activations as α_2 and α_3-motoneurons do.
2. The stability of the oscillatory firing of a sphincter α_2-motoneuron increased with the increasing phases of approximately 50ms between the APs of the γ-motoneurons and the impulse train of the motoneuron following bladder catheter pulling. Moreover, in a certain phase range, the number of phases between the APs of a γ_1-motoneuron and a secondary muscle spindle afferent fibre increased upon stimulation and, as shown earlier, also the number of phase correlations between the APs of some secondary spindle afferents and the impulse train of the oscillatory firing α_2-motoneuron increased. Phase relations between the APs of two spindle afferents also increased upon bladder catheter pulling. During the oscillatory firing of the α_2-motoneuron, a phase loop seemed to exist from the APs of the α_2-motoneuron to the APs of certain γ-motoneurons, to the APs of certain secondary spindle afferent fibres and back to the APs of the α_2-motoneuron, in similarity to the phase recoupling of electronic oscillators.
3. Single γ-motoneurons changed their activity levels rhythmically, similar to the secondary spindle afferents, often with a frequency of 0.3Hz. Parasympathetic intrafusal motoneurons fired at a constant activity level for about 10s and activated up to six single encoding sites of a parent secondary spindle afferent fibre. Shortly after the recruitment of a new encoding site, the interspike intervals of the parent fibre were similar again. The parent spindle afferent fibres raised its activity in response to the parasympathetic activation with a time constant of 3s (e-fit) and decreased it with a time constant of 40s.
4. The clinical implication of the measurements of the coordinated firing of α and γ-motoneurons and muscle spindle and urinary bladder stretch receptor afferents is that not only a structural repair is needed in CNS injury but also the coordinated firing of the neurons has to be repaired by learning. Since during development the healthy human CNS is using movements for optimizing and repairing developmental errors of CNS functioning, movement-based learning is a CNS repair strategy to relearn the

coordinated firing of neurons via performing coordinated movements. Since learning also most likely includes the building of new nerve cells in different parts of the CNS, a structural repair seems (and is (Chapter II)) also possible by movement-based learning.

14.1. Introduction

In the previous section, it was shown that the constant stretch reflex of the external anal sphincter is driven by constantly firing oscillators, which consist of the motoneuron itself and many interneurons (186) (Figures 69,69). Probably, only a part of the sphincter motoneurons will fire in the oscillatory firing mode. Others will still fire in the occasionally firing mode (Figures 54-58). It was further shown that a spinal oscillator is excited by afferent action potentials (APs) of secondary muscle spindle afferents, which have a certain phase to the APs of the oscillatory firing α_2-motoneuron. Constant stretch reflex was induced by rectal distension with an anal catheter. The term constant stretch reflex is not accurate, since this tonic reflex will probably decrease within minutes to hours [157]. With additional stimulations (bladder and anal catheter pulling), the phase relations between the secondary spindle afferent and oscillatory motoneuron APs changed transiently, and the spinal oscillator stopped firing transiently (Figure 104). A transient brake of oscillation occurred probably due to reduced occurrence of phases in a certain range between the secondary spindle afferent APs and the motoneuron APs. The reduced number of occurring phase relations was in turn probably due to a hysteresis effect of the secondary spindle afferents caused by creep and other after effects of the intrafusal muscle fibres [55]. The three secondary muscle spindle afferents SP2(1), SP2(2) and SP2(3), contributing to the drive of the oscillator in brain-dead human HT6, showed pronounced sensitivity to anal catheter pulling, bladder catheter pulling and painful pinprick, respectively (previous two sections). Since the muscle spindles are not innervated by skin, mucosal and pain receptor afferents, the input to these receptors has to run via the intrafusal motoneuron systems. The identification of natural impulse patterns of single intrafusal motoneurons is therefore of high importance for the understanding of the drive of the oscillatory firing α_2-motoneuron by the intrafusal motoneurons via the secondary muscle spindle afferents (final common path). With the identification of the impulse pattern of two single intrafusal motoneurons here, it will be shown that the intrafusal motoneurons respond differently to changing afferent input from skin and mucosal receptors and drive differently the secondary spindle afferents. It will further be shown that there are certain phase relations between the APs of the γ and the oscillatory firing α_2-motoneurons, the γ-motoneurons and the secondary muscle spindle afferents and, as shown above, between the spindle afferents and the α_2-motoneuron. A phase loop seems to exist for excitation of the spinal oscillator in phase (Figure 113).

Even though exact measurements of human spinal oscillators with their mainly secondary muscle spindle afferent drive and the phase loop is a completely new development for the understanding of the constant stretch reflex, these findings are partly in accordance with the current opinions concerning the function of muscle spindles. It is the Matthews' standpoint that secondary muscle spindle sensory endings induce a powerful excitation in human stretch reflexes [140], which in the case here is the oscillatory firing α_2-motoneuron in the constant stretch reflex. Since secondary spindle afferents predominantly innervate nuclear chain muscle fibres [22], and chain fibres can contract locally [29] and can oscillate up to 60Hz [23, 25]; the chain fibre activity provides a mechanism for maintaining a high-level input to α_2-motoneurons that is independent of changes in muscle length [28]. Hulliger and Prochazka proposed a concept of fusimotor set [98], claiming that fusimotor firing levels, and the balance between γ_1 (dynamic) and γ_2-motoneurons (static), is set in a way characteristic of, and specific to, a particular movement. Through the α-motoneuron output must also have a characteristic pattern in any movement, the α and γ-motoneuron patterns are usually independent unless the movement is powerful [69]. With the identification of impulse patterns of γ-motoneurons, it will be shown that somatic γ-motoneuron activities alone cannot explain a strong long-lasting activity increase of secondary spindle afferent fibre SP2(2), caused by strong bladder catheter pulling. Parasympathetic muscle spindle activation was therefore introduced and has to be included into the concept of fusimotor set. This autonomic spindle innervation was in accordance with certain activity changes of fibres whose conduction velocities lay in the expected parasympathetic peak in the conduction velocity distribution. Both histological [5, 8, 225] and functional [150] indication for autonomic muscle spindle innervation have been found in animals.

14.2. Identification of Peaks of γ-Motoneurons and Parasympathetic Fibres in Conduction Velocity Distributions

The limits of this classification scheme of peripheral nerve fibres are the small fibres, because the extracellular recorded action potentials (APs) become too small to be easily separated from the noise and artefact level. Further, in conduction velocity distribution histograms, it becomes difficult to identify the peaks of thin nerve fibre groups, because the peaks fuse on the linear velocity scale. By constructing first histogram classes for conduction times on a linear scale, however, and plotting them on a log scale from the column values, the velocity peaks separated. As shown in Figure 33D,E, the fused γ-peaks of the linear plots (Figure 33Dab) split up into different peaks on the log scale (Figure 33Ea-d), and the different γ-motoneurons and parasympathetic fibres could be identified. Further, by comparing Figure 33 with Figure 35, it can be seen that there is a similarity in the distributions of the motoneuron groups between the compound AP with its subcompound APs and the nerve fibre groups in conduction velocity distributions for single fibres. This improvement of the classification scheme made it possible to further analyse the self-organization of neuronal networks of the human CNS with respect to the involvement of muscle spindles and the parasympathetic nervous system division.

14.3. Impulse Patterns of Dynamic and Static Intrafusal Motoneurons in Relation to Those of Secondary Spindle Afferents

By looking repeatedly through the high-quality HT recordings at high γ-motoneuron activation, it was possible to identify the impulse patterns of a single γ-motoneuron by the AP waveform, duration, conduction velocity and doublet firing. Because of the importance of the finding, five successive doublet interspike intervals (IIs) are shown in Figure 107.

The approximate IIs from doublet firing are shown in the upper part of Figure 107 (see also Figure 111A). The γ-motoneuron doublet II had very similar durations of about 13 to 14ms. Before plotting the doublet IIs and the IIs from doublet to doublet, let us look at the original registrations of other nerve fibres (Figure 108).

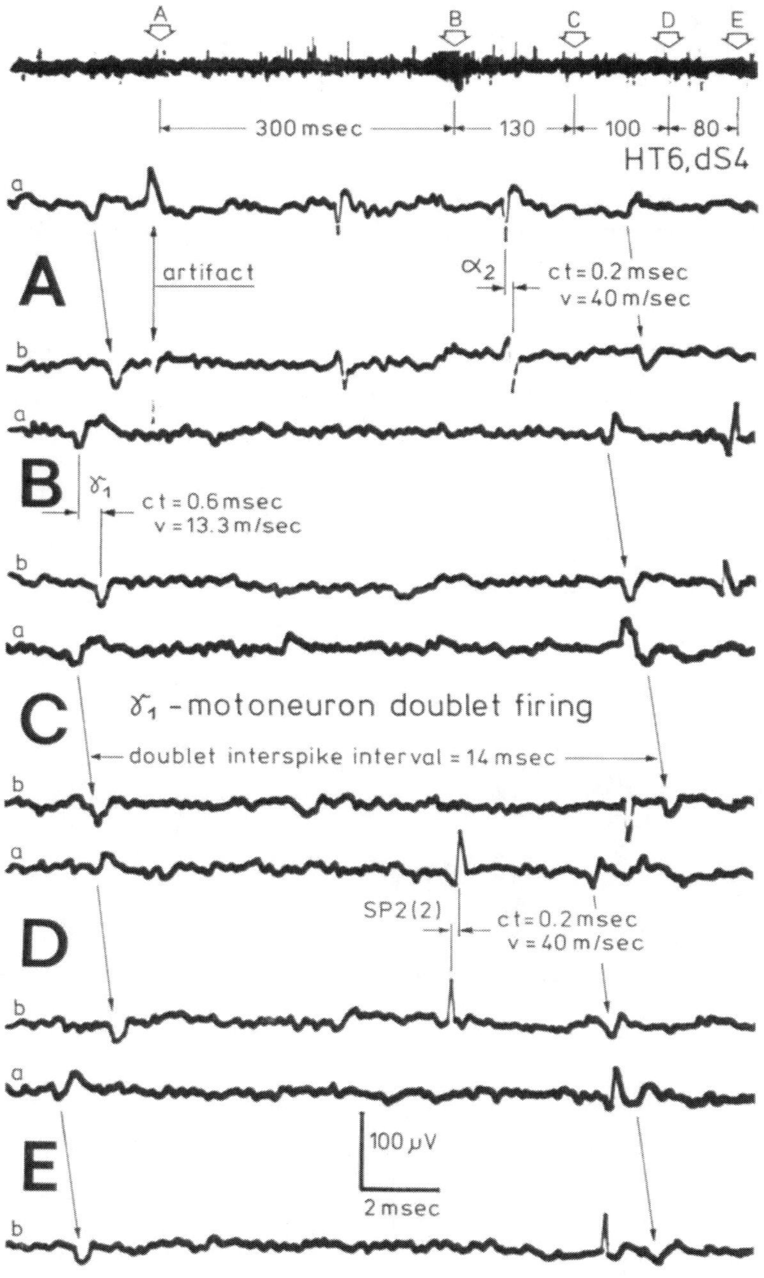

Figure 107. Doublet firing from a single γ_1-motoneuron (intrafusal, dynamic). From the sweep on the top of the Figure, successive AP doublets are shown on an expanded time scale in A to E. Conduction times (ct), conduction velocities (v), a doublet interspike interval and interspike intervals from doublet to doublet are indicated. An artefact (A) can clearly be distinguished from real action potentials by the mirror picture potentials and the missing of the conduction time. HT6, dS4.

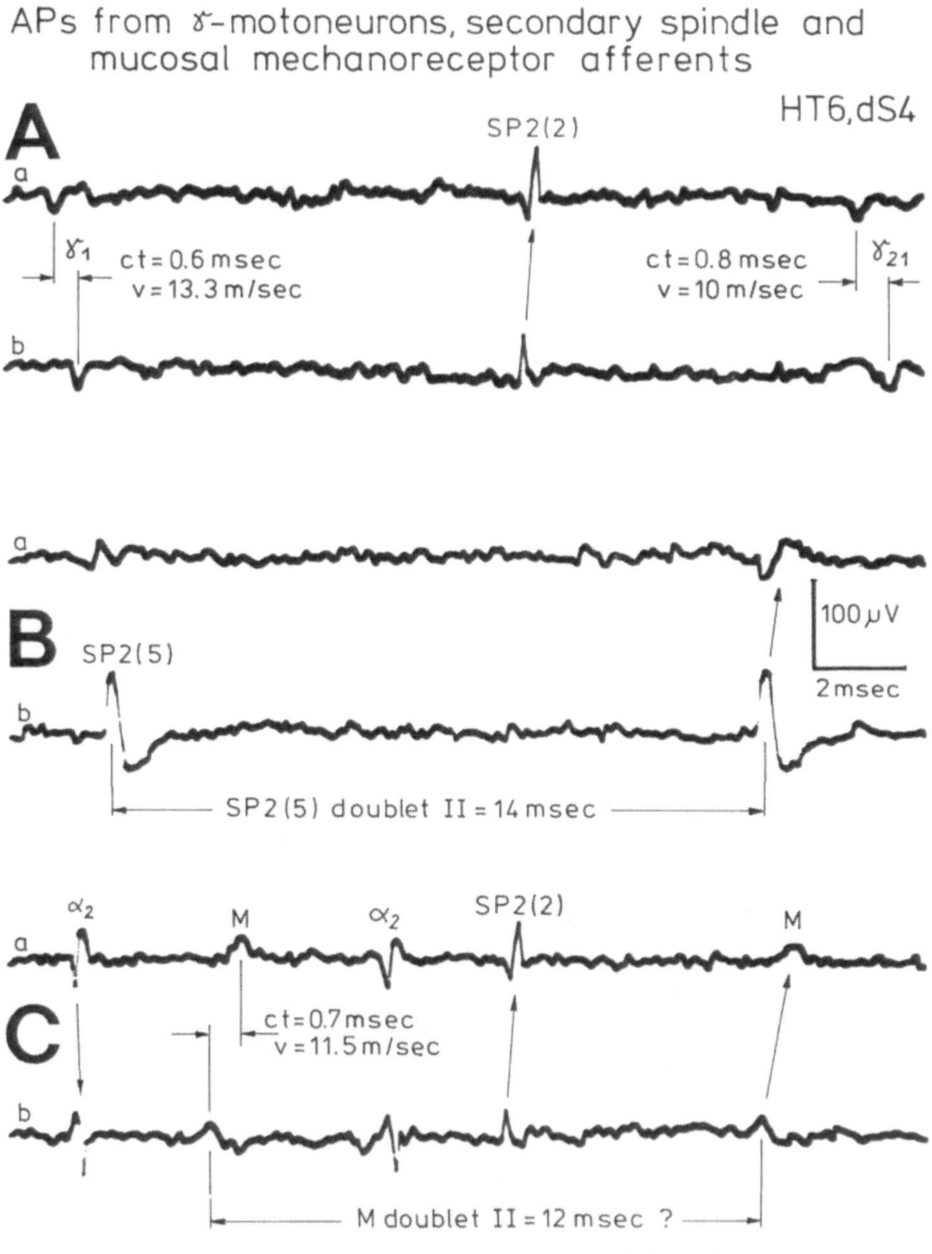

Figure 108. A. Recording of a single γ_{21}-motoneuron action potential (AP) (intrafusal, static) in comparison to a single γ_1-motoneuron AP. B. AP doublet of the secondary muscle spindle afferent fibre SP2(5), recorded at the same time period as the APs of the single γ_1 and γ_{21}-motoneurons. Note, that the SP2(5) doublet has the same duration as the γ_1-doublet in Figure 107 (14ms). C. APs recorded from a mucosal afferent fibre (M). Conduction time (ct) and conduction velocity (v) are indicated. APs from other fibres are intermingled.

It was further possible to identify the impulse pattern of a single γ_{21}-motoneuron. An AP of it is shown in Figure 108A, together with an AP of the γ_1-motoneuron.

Figure 108B shows a doublet II of the simultaneously firing secondary muscle spindle afferent fibre SP2(5).

Only the second AP of the doublet is partly conducted to the proximal electrode pair (trace a).

So far, only the muscle spindle afferent fibres were analysed as candidates for the afferent excitation of the oscillatory firing of motoneurons for the constant stretch reflex of the external anal sphincter. However, mucosal afferent APs could also contribute to the oscillatory firing of the motoneuron. Figure 108C shows two mucosal APs with II duration in the range of those of SP2 doublets. Probably, this M-doublet occurred only accidentally. It has not been possible so far to pick up the impulse patterns of single mucosal receptor afferents.

Figure 109 shows the impulse patterns of a γ_1 and a γ_{21}-motoneuron in the form of IIs between doublets (Figure 109A,B) and of IIs of the doublets (Figure 109C,D). The dynamic (γ_1) and the static (γ_{21}) intrafusal motoneurons fired with a shortest II of about 80ms. Also, the single encoding sites of APs of secondary muscle spindle afferents fired with a shortest II of about 80ms (Figure 94).

The doublet II distribution for the γ_1-motoneuron showed a peak between 13 and 13.5ms (Figure 109C). The γ_{21}-motoneuron fired only sometimes with doublets of durations of about 18ms (Figure 109D).

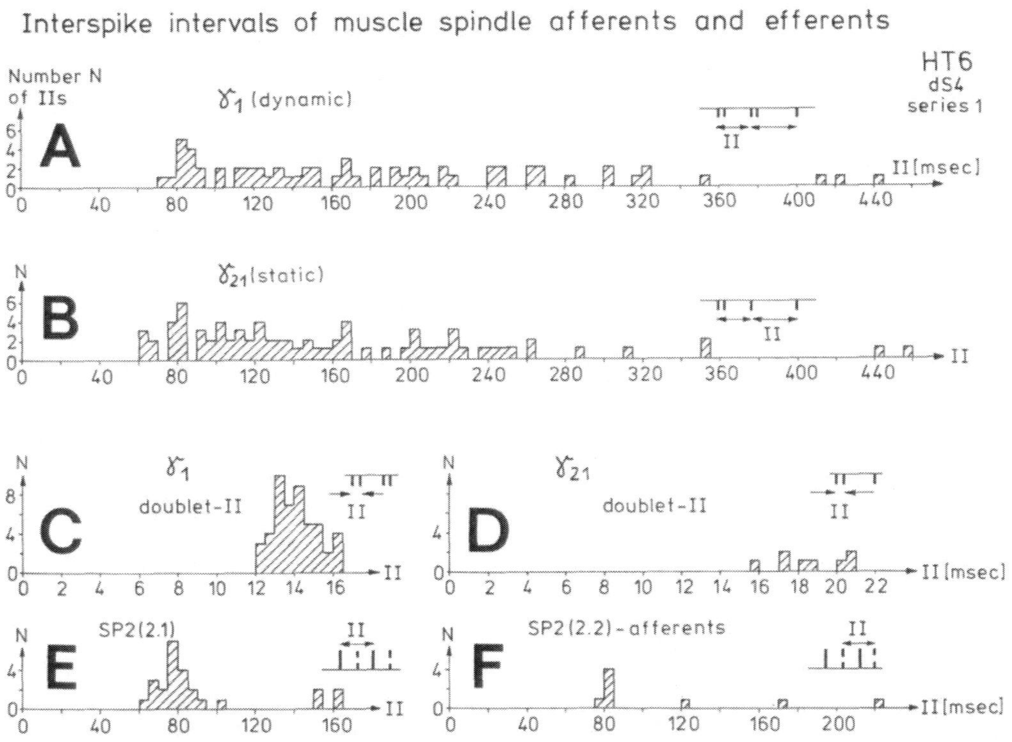

Figure 109. A, B. Interspike interval frequency distribution histogram of a single γ_1-motoneuron (A) and a single γ_{21}-motoneuron (B). The way the intervals were measured is shown. C, D. Doublet interspike interval distribution histogram of a single γ_1 (C) and a single γ_{21}-motoneuron (D). E, F. Interspike interval distribution histogram of the two single endings SP2(2.1) (E) and SP2(2.2) (F) of the single parent secondary muscle spindle afferent fibre SP2(2) measured at the same time interval as the γ interspike intervals in A to D.

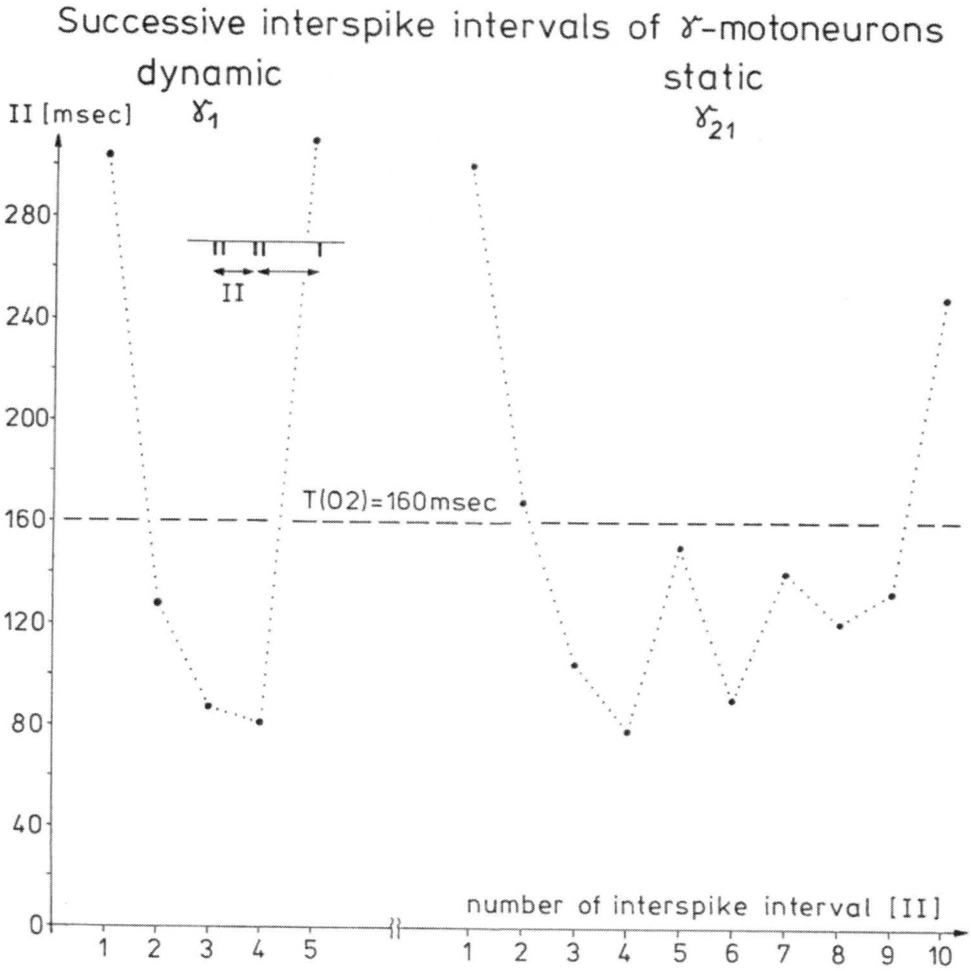

Figure 110. Successive interspike intervals (IIs) of single γ_1 and γ_{21}-motoneurons. The way IIs were measured is indicated. The oscillation period of the oscillatory firing α_2-motoneuron O2 is indicated.

Maybe the γ_{21}-motoneuron neural network was not as excited as that of the γ_1-motoneuron. By firing with shortest IIs of 80ms and doublets of 12 to 21ms, single γ_1 and γ_{21}-motoneurons fired with the pattern of the secondary spindle afferents rather than with that of α_2-motoneurons in the high activity mode of oscillatory firing. A α_2-motoneuron, firing with an impulse train of two APs, about 6ms apart (Figure 66) would fire with an oscillation period of 130ms (Figure 66). The two γ-motoneurons showed, at the moment of firing with doublets and a shortest II of 80ms between the doublets, a higher activity than did the α-motoneuron, firing in the oscillatory mode with impulse trains consisting of two APs.

The two γ-motoneurons fired in the way the secondary spindle afferents did. The distributions of two encoding sites of the simultaneous firing SP2(2) fibre are shown in Figure 109E,F. It will be discussed below whether the intrafusal motoneurons drove the secondary spindle afferents via the intrafusal fibres or whether the secondary spindle afferents drove the γ-motoneurons. The single encoding site SP2(2.1) of the secondary spindle afferent fibre SP2(2) showed more IIs of 80ms duration (Figure 109E) than did single intrafusal motoneurons γ_1 and γ_{21} (Figure 109A,B); its degree of regularity was therefore higher at the time of these measurements.

Successive interspike intervals (IIs) of the dynamic (γ_1) and the static (γ_{21}) intrafusal motoneurons are shown in Figure 110. The changes of the γ_{21}-motoneurons were slower and longer lasting that that of the γ_1-motoneuron, supporting the identification of static and dynamic patterns by the velocity distributions (Figure 33). The successive IIs of these two γ-motoneurons (Figure 110) were very similar to those of certain spindle afferents (Figure 96).

Figure 111 shows the impulse patterns of γ_1 (dynamic) and γ_{21}-motoneurons (static) simultaneously with the impulse patterns of secondary muscle spindle afferent fibres SP2(2) and SP2(5) (and SP2(4)) and the continuously oscillatory firing α_2-motoneuron O2 in HT6. It can be seen from these direct patterns that the γ-motoneurons did not show oscillatory firing as did α_2-motoneuron O2 (no co-activation). On recordings shown in Figure 111A, the γ_1-motoneuron is mainly firing with doublets following bladder catheter pulling. The marked part (marked with Figure 2) of it is shown in Figure 107. Before bladder catheter pulling (Figure 111B), the γ_1-motoneuron was only sometimes firing with doublets.

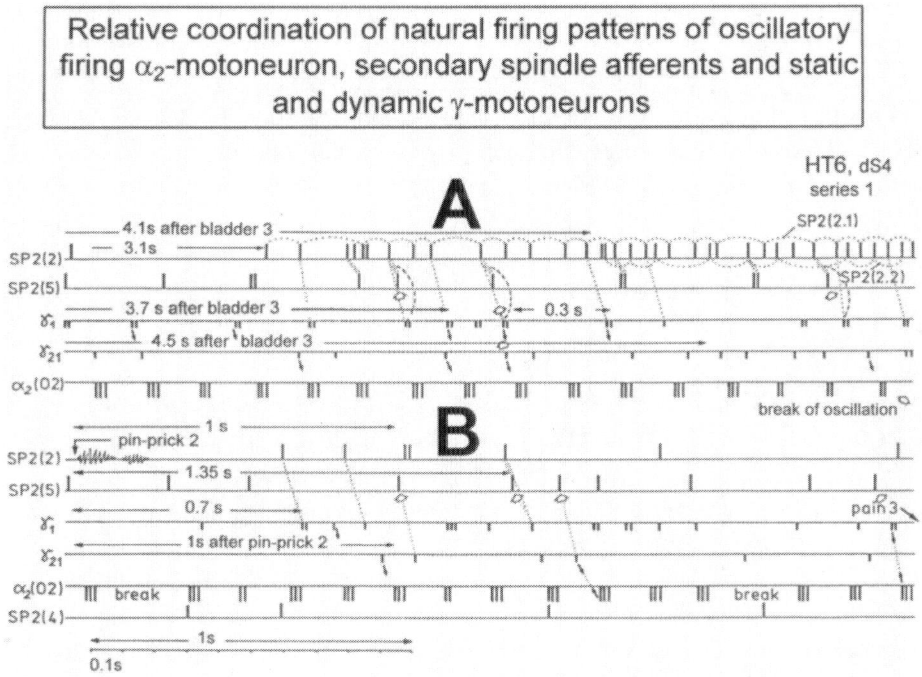

Figure 111. Impulse patterns of simultaneously recorded γ-motoneurons (γ_1 and γ_{21}), secondary spindle afferent fibres (SP2(2), SP2(4), SP2(5)) and oscillatory firing α_2-motoneuron O2 following bladder catheter pulling (bladder 3) (A) and pinprick 2 (B). B was recorded before A. In A the impulse patterns of the two encoding sites SP2(2.1) and SP2(2.2) of the single parent fibre SP2(2) are indicated by the dotted curves. Times to the activity increases of γ-motoneurons and secondary spindle afferents following stimulation are indicated. Similar time intervals of the occurrence of γ-motoneuron APs and SP2(5) fibre APs (phase coordination) are indicated by the open arrows, and the similar time intervals of γ-motoneuron APs and α-motoneuron APs are indicated by small arrows. Similar time intervals of the APs of fibres SP2(2) and SP2(5) are indicated by the double dotted lines, those of γ_1-APs and the SP2(2) fibre APs by a dotted line, and those of γ_1-APs and the SP2(2)-SP2(5) correlation by a curved dashed line. HT6; dS4.

Thus, repeated bladder catheter pulling may have activated the doublet firing. The increase in activity following bladder catheter pulling (Figure 111A, 25APs) as compared to that following pinprick (Figure 111B, 20APs) is only slight. Since this repeated bladder catheter pulling activated parasympathetic spindle innervation (see below), the increase in the doublet firing maybe due to an increase of the low level parasympathetic activity. Also, the spindle afferent fibre SP2(5) raised its doublet firing from no doublets (Figure 111B) to occasional doublets (Figure 111A). Upon strong activation of the parasympathetic nervous system (Figures 112,113) this SP2(5) fibre showed about 10min later low activity, consisting only of doublets. Also, fibre SP2(2) fired about 10min following this parasympathetic activation with doublets (Figure 100).

The γ_{21}-motoneuron fired with doublets but occasionally. As can be seen from Figure 111, the dynamic γ_1-motoneuron raised its activity more rapidly following pinprick (0.7s) and bladder catheter pulling (3.7s) than did the static γ_{21}-motoneuron (1s and 4.5s, respectively). The spindle afferents SP2(2) and SP2(5) raised their activity following pinprick later than did γ_1 and γ_{21}-motoneurons (Figure 111B, 1s and 1.35s). However, the spindles were most likely innervated by more intrafusal motoneurons (β-motoneurons were nearly not activated (Figure 33D)). Some of them may have raised their activity earlier than γ_1 and γ_{21}-motoneurons, so that the time correlation of the two single γ-motoneurons with their spindle afferents is conclusive but partly. As can be see from Figure 112B, the remaining γ-motoneurons (γ-rest) raised their activity slightly earlier than did γ_1 and γ_{21}-motoneurons, indicating that there may have been a γ-motoneuron (or parasympathetic fibre), which had raised its activity slightly earlier than did γ_1 and γ_{21}-fibres. The strong activity increase of fibre SP2(2) (Figure 111A) was not induced by γ_1 and γ_{21}-motoneuron activity or by that of the remaining γ-motoneurons (Figure 112B). It was due to the parasympathetic spindle innervation, as will be analysed below. The dotted lines link the impulses of single encoding sites of parent fibre SP2(2).

14.4. Phase Relations between Extra- and Intrafusal Motoneurons and Secondary Spindle Afferents: Phase and Frequency Coordination

Following bladder catheter pulling, the oscillatory firing α_2-motoneuron O2 was more stable (Figure 111A) than before pulling (Figure 111B) following pinprick. Two breaks and two impulse trains of two APs each occur in Figure 111B, and there is no break in Figure 111A. The activity of γ_1 and γ_{21}-motoneurons is slightly higher following bladder catheter pulling. Figures 102 and 103 show that there were fewer breaks in oscillation with higher spindle afferent activity. With a phase of 50 ± 10ms (n = 5) (mean ± s.d.) for γ_1 and 42 ± 15 for γ_{21}-motoneuron between γ-motoneuron O2 and α_2-motoneuron O2, nine phase-correlated APs occurred in Figure 111A and 4 (phase = 44 ± 20ms) are seen in Figure 111B. The number of phase correlations between γ-motoneurons and α_2-motoneuron O2 increased more rapidly than did γ APs occurrence. The strength of the drive of oscillation therefore can be better measured by the number of phases occurring in a certain range between γ and the α-motoneurons than by the number of γ or SP2 fibre APs. In previous work, it seemed that co-recruitment of γ_1 and α_2-motoneurons in the occasional firing mode increased with the strength of stimulation.

In Figure 111B, there were three phase correlations between γ_1 and γ_{21}-motoneurons and the ten SP2(5) fibre APs. In a higher excited stage of Figure 111A, four phase correlations occurred between the APs of the two γ-motoneurons and the 12 SP2(5) fibre APs. The fibre SP2(5) thus received only little drive from encoder sites by single γ_1 and γ_{21}-motoneurons. Nevertheless, the activity of fibre SP2(5) was strongly correlated to that of the single γ_1-motoneuron. Already three phase correlations occurred for seven afferent APs (67 ± 11ms) following pinprick 2 (Figure 111B). The second γ APs of the doublets contributed further phase correlations. There is indication for the γ_1-motoneuron driving the single encoding site of fibre SP2(2), active at the time shown in Figure 111B. About six phase correlations occurred (Figure 111A) between the APs of γ_1-motoneuron and fibre SP2(2) (phase = 52 ± 13ms). However, additionally, there were at least parasympathetic intrafusal motoneurons (see below), which additionally drove encoding sites of the SP2(2) parent fibre (right side of Figure 111). With stronger oscillation of α_2-motoneuron O2 (Figure 111A), there were more phase correlations occurring between γ_1-motoneuron and fibre SP2(2) and between γ_1-motoneuron and oscillatory firing α_2-motoneuron O2. With stronger afferent input, the number of phase correlations between APs of γ_1-motoneuron and α_2-motoneuron O2 increased, as did those between γ_1-motoneuron and the secondary afferent fibre SP2(2) (Figure 111); as shown previously [84] (above), there were also more correlations between fibre SP2(2) and α_2-motoneuron O2. As the afferent inputs from the bladder, anal canal and skin were not constant [72], and in response to it, γ-motoneurons changed their activity levels (Figures 111,112B), synchronization with the activity of the extrafusal motoneurons was often transiently lost. This loss of synchronization will probably also depend on the kind of intrafusal muscle fibre (bag or chain fibre) and on whether there is increasing or decreasing activation (hysteresis effect). Even with many APs from muscle spindle afferents, the oscillation can transiently break (Figure 104). Apart from these dynamic changes, with the increasing drive from the bladder, anal and skin afferents, the APs of γ-motoneurons and the oscillatory firing α_2-motoneuron O2 were closer phase-correlated, and thus the single encoder sites of the secondary spindle afferent fibre SP2(2) and the α_2-motoneuron O2 were also correlated, as shown for fibre SP2(2) (Figures 99,100).

In figure 111A, there were many SP2(5) APs correlated to SP2(2) APs (indicated by the double-dotted line; phase = 35 ± 9ms, n = 6). Most phase relations occurred between SP2(5) APs and the encoder site SP2(2.1). Since phase correlations also occurred to other SP2(2) fibre APs, it is likely that the splitting of the SP2(2) fibre activity into APs of single encoding sites was only approximately correct. A correction is possible (Figure 111A) but was not performed. The increase of phase correlations between APs of fibre SP2(2) and those of SP2(5) indicates that partly the fibres were activated by the same intrafusal motoneurons. Since three phase relations occurred between the γ_1-doublets and the SP2(2)-SP2(5) phase relation (indicated by the bended dashed line in Figure 111A), it is likely that the γ_1-activity contributed essentially to the SP2(2.1) encoder site and slightly to the SP2(5) afferent fibre, so that SP2(2)-SP2(5) phase relations occurred. Thus, the γ_1-motoneuron probably innervated intrafusal muscle fibres of the muscle spindles, afferently innervated by SP2(2) and SP2(5) afferents. Since fibre SP2(2) responded mainly to bladder catheter stimulation, and fibre SP2(5) responded non-specifically [188], it is unlikely that afferents SP2(2) and SP2(5) innervated the same muscle spindle. These analyses, based on measurements, indicate already on the rather simple level of organization that the organization of the human CNS is extremely complex and exact – no sign of chaos.

14.5. Parasympathetic Muscle Spindle Innervation

In previous work, it was argued that the strongly increased spindle afferent activity (lasting for more than 10min) following strong bladder catheter pulling was due to the cumulative response of static γ-motoneuron activity [174, 183]. With the identification of impulse patterns of single dynamic and static γ-motoneurons, it will turn out that this long-lasting increased summed SP2 fibre activity was due to the activity increase of fibre SP2(2) alone, secondary to parasympathetic activation of the secondary spindle afferent fibre.

Figure 112A shows that fibre SP2(5) changed only very little its activity level, whereas the activity of fibre SP2(2) strongly increased following strong bladder catheter pulling. About 10min later, the activity still did not fully return to pre-stimulation level. This long-lasting activity increase cannot be due to the activity increase of the single dynamic γ_1-motoneuron, since there was a similar increase in γ_1-motoneuron activity following pinprick (left part of Figure 112B) with no increase in that of fibre SP2(2). Either the single static γ_{21}-motoneuron cannot be responsible for this long-lasting activity increase of fibre SP2(2), since after bladder catheter pulling the γ_{21}-activity was nearly as high as with anal catheter pulling, but fibre SP2(2) activity decreased slowly instead of strongly increasing. It is also very unlikely that another γ-motoneuron is responsible for this strong activity increase of fibre SP2(2), since activation of the remaining fibres of the same root with low and very low conduction velocities (γ-rest) was nearly identical upon both anal catheter pulling (Figure 112B, right) and bladder catheter pulling (Figure 112B, middle), but SP2(2) fibre activity did not increase. Still another possibility is that there were slowly conducting fibres that contributed to the activity of "γ-rest" fibres, which had a different effect on the intrafusal muscle fibres innervated by the myelinated afferent endings of SP2(2). In Figure 33A,B, there is a conduction velocity peak at about 5m/s, which probably contained activity of parasympathetic fibres. Since these very slowly conducting fibres responded more dynamically (Figure 33Bd) and it can be expected that the parasympathetic fibres (first motoneuron) being contained in a S4 root [172] conduct only slightly than the static γ-motoneurons (the fibre diameter of the static γ-motoneurons is about 5 to 6μm and that of the parasympathetic fibres is about 3 to 5μm [171, 175, 181]), it is likely that SP2(2) was also activated by parasympathetic motoneurons.

This analysis is supported in Chapter V by measurements of urinary bladder functions including electrical intravesical stimulation and clinical urinary bladder function tests (urodynamic (pressure-flow studies)). The measurements of compound action potentials and single-nerve fibre action potentials of bladder stretch (S1) and tension receptor afferents (ST) caused by electrical intravesical stimulation will give more safety concerning the this human classification scheme of human peripheral nerve fibres. Muscle spindle activation, due to parasympathetic spindle activation, will be compared to urinary bladder pressure changes upon retrograde bladder filling.

Parasympathetic muscle spindle activation of the brain-dead human HT6 (rather physiologic) will be compared with those following spinal cord injury to find out what is impaired in the self-organization of parasympathetic nervous system division following spinal cord injury. Phase correlations between the firings of the different neurons will be presented in the form of distributions. The reasoning about parasympathetic muscle spindle activation will get therefore strong support with the measurements of Chapter V.

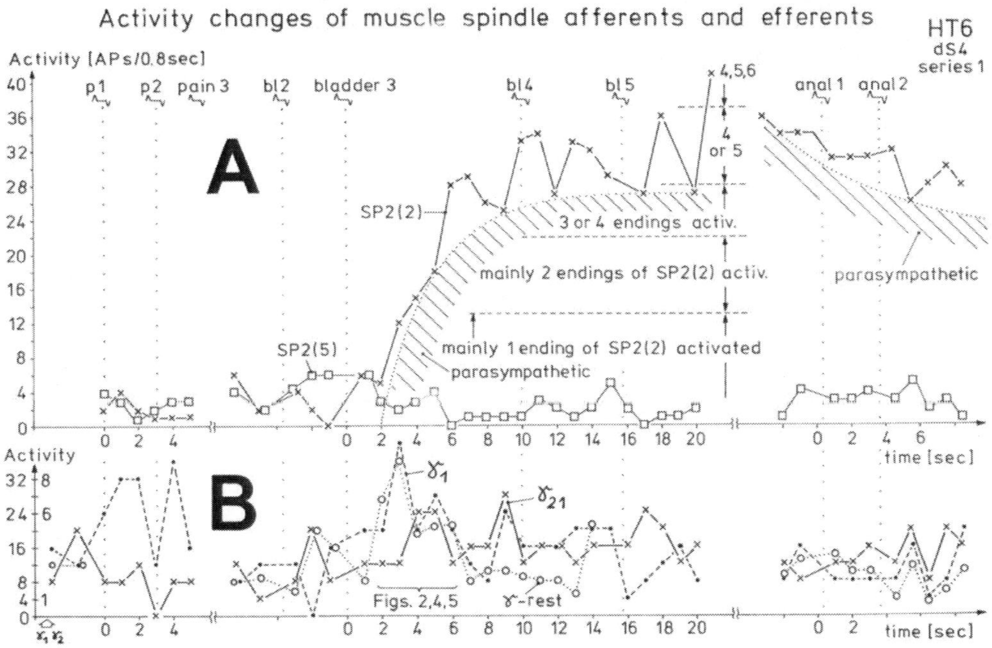

Figure 112. Activity changes of the secondary muscle spindle afferent fibre SP2(2) (and SP2(5)) (A) in relation to the simultaneously measured activity changes of the single intrafusal motoneurons γ_1 and γ_{21} and the remaining intrafusal motoneurons γ-rest (including parasympathetic efferents) (B). The very strong long-lasting baseline activity increase of fibre SP2(2) following very strong bladder catheter pulling (bladder 3) is indicated by the dotted, and the crosshatched lines and represents, most likely, the activity increase due to parasympathetic muscle spindle activation. It is indicated in A how many single endings of fibre SP2(2) were activated, p1, p2, pain 3 = pinpricks; bl 2, bladder 3, bl 4, bl 5 = bladder catheter pullings; anal 1, anal 2 = anal catheter pulling. In B, the activity 0 to 8 holds for the γ_1 and γ_{21}-motoneurons and the scale 0 to 32 for the rest of efferent fibres (γ-rest).

It can be seen from Figure 112B (middle part) that the single γ_1 and γ_{21}-motoneurons increased and decreased their activity following stimulation with a rhythm of about 3s. It could well be that parasympathetic fibres, belonging to another division of the nervous system, are activated with different impulse patterns. This possibility was tested by plotting the activity changes of motoneurons conducting with different velocities (low ranges) in relation to the activity increase of fibre SP2(2) suspectedly caused by parasympathetic fibres (Figure 113). To obtain the proportion of the activity increase of fibre SP2(2) mainly due to parasympathetic fibre activity (Figure 112A), fibre SP2(2) activity following bladder and anal catheter pulling was compared with the activity level and its changes before the catheter pulling. The strong baseline activity increase, most likely due to parasympathetic activity, is indicated in Figure 112A by the dotted line and the partly crosshatched area. Based on shortest interspike intervals of 80ms of single encoding sites of human secondary spindle afferents (Figure 109E,F; Figure 97), the parasympathetic fibres could be assumed to activate up to six single encoding sites (Figure 112A). Figure 113 shows the activity increase of fibre SP2(2) due to parasympathetic activation, replotted from Figure 112A into a different scale. Figure 113L shows the velocity distribution of γ-motoneurons and probably parasympathetic fibres from Figure 33Db. Figure 113M shows the conduction velocity distribution histogram at the beginning (or increasing) of parasympathetic activation measured by the afferent response (Figure 113A).

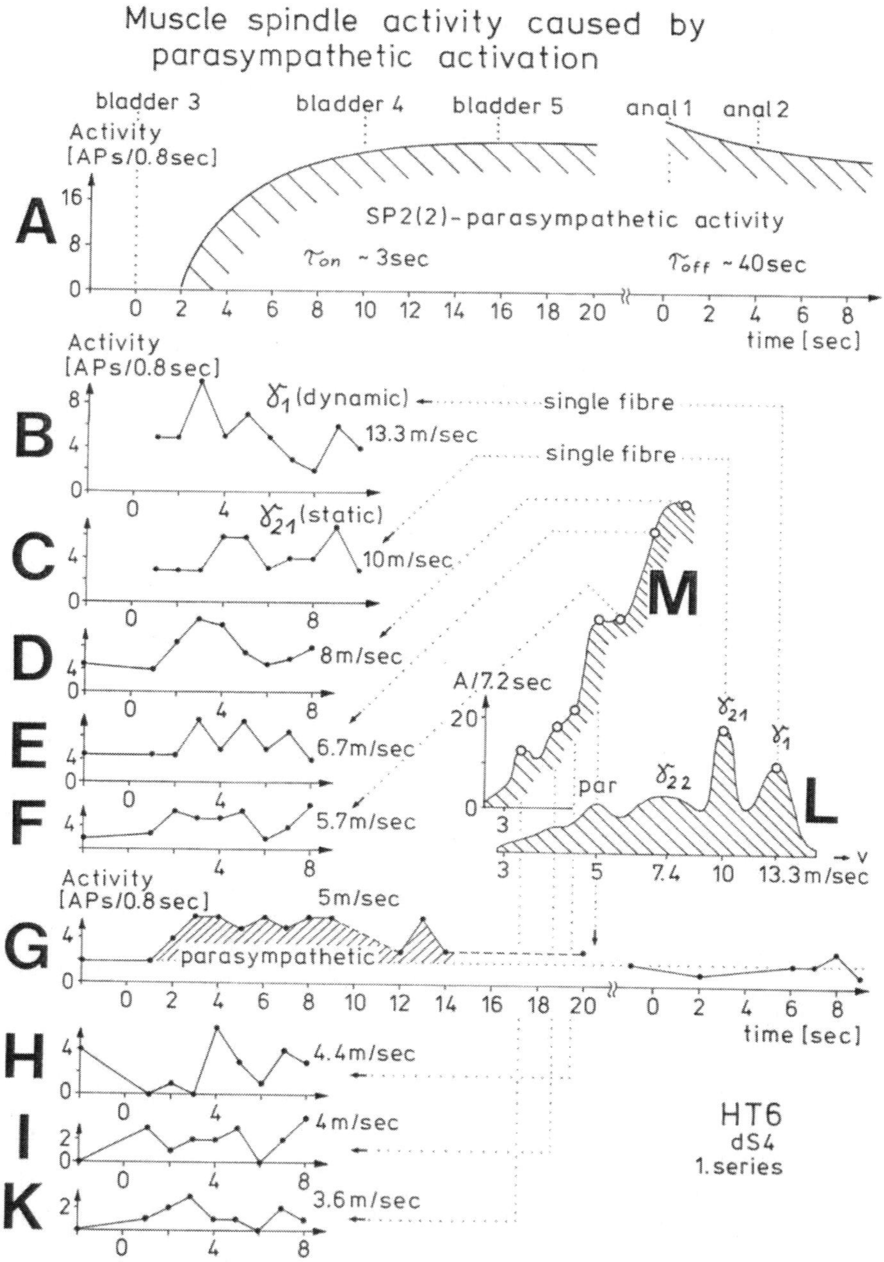

Figure 113. Long-lasting activity changes of the single secondary muscle spindle afferent fibre SP2(2) in relation to the activity changes of the efferents, most likely including the efferent innervation of the spindle, which is innervated by SP2(2) fibre. A. Changes of the baseline activity, taken from Figure 112A (different scale). B. to K. Activity changes of the efferents in the dorsal S4 root of HT6, conducting at different group peak conduction velocities (or certain intermediate values) indicated by circles on the conduction velocity frequency distribution curves M and L. L. Velocity curves of Figure 33A, taken from measurements at different times. M. Velocity curve from the measurements used for the construction of Figures B to K. B and C are single-fibre activity changes, taken from Figure 112B. D to K. Activity changes from probably several fibres. Note that the activity changes in G are long lasting, the other ones in B to K are not; only the activity in F is similar, with probably the activities of one or two parasympathetic fibres intermingled. Note further that at the time of anal-catheter pulling, where fibre SP2(2) shows still parasympathetic induced activation, there is no or no increased parasympathetic efferent activity, v = conduction velocity. HT6,dS4.

Figures 113B through K show the activity changes of fibres with conduction velocities indicated in Figure 113M. The activity changes of the single γ-motoneurons of Figures 113B and C are partly those seen in Figure 112B. Time-related changes of fibre groups with selected conduction velocities are shown in Figures 113B through K. It can be seen from Figures 8B to K that following bladder catheter pulling the activities of all fibre groups increased and decreased similarly as that of single γ-motoneurons (Figure 113B,C) with the exception of the (assumed) parasympathetic fibre group shown in Figure 113G. The parasympathetic fibre activity remained a certain level up to 9s following bladder catheter pulling. Irregularities occurred because of further bladder catheter pulling. The only other fibre group that showed an increased activity were fibres represented in Figure 113F. However, the fibres giving rise to this increased activity (Figure 113F) are a mixture of parasympathetic fibres and γ_{22}-motoneurons. This measurement allowed identifying, by the long-lasting activity increase, the parasympathetic motoneurons generating the strong long-lasting activity increase of muscle spindle afferent fibre SP2(2). The γ-motoneurons showed no constant activity increase; they increased and decreased their activity with a rhythm of 0.3Hz. The actual activity patterns of single parasympathetic fibres could not be identified so far. As Figures 113M and L further show, there seem to be further groups, conducting at velocities between 3 and 5m/s at 33.5°C. Nothing can be said so far about whether parasympathetic motoneurons split or not into extrafusal and intrafusal fibres. By comparing Figure 113A with Figure 113G, it can be seen that the parasympathetic fibres were only active at the time of the activity increase of afferent fibre SP2(2) and not during the slow decrease of fibre SP2(2). The parasympathetic fibres seem to induce a very strong cumulative response on fibre SP2(2). Using the well-known fits of the natural exponential function ($A_0(1-e^{-t/\tau on}$) and ($A_0(1-e^{-t/\tau off}$), the time constant τ for the activity increase and decrease of fibre SP2(2) can be calculated:

τ_{on}: if $t = \tau_{on}$, the $A_0(1\text{-}1/e) = A_0(1\text{-}0.37) = 27$APs/0.8s •0.63 = 17APs/0.8s.

The value of A = 17APs/s was reached after 3s (Figures 112A,113A), $\tau_{on} = 3$s.

τ_{off}: if $t = \tau_{off}$, then A_0e^{-1} = 34APs/0.8s•0.37 = 12APs/0.8s. The reduction from A = 34APs/0.8s to A = 12 APs/0.8s was reached 40s later (Figures 112A,113A), $\tau_{off} = 40$s.

The time constant for the reduction of fibre SP2(2) activity was with $\tau_{off} = 40$s, more than ten times longer than the corresponding time constant for the activity increase (τ_{on} = 3s). Since fibre SP2(2) decreased and increased its activity quickly in parallel with the activity changes of γ-motoneurons (Figure 112A), the extremely slow reduction of the SP2(2) activity following parasympathetic activation is most likely due to the very slow relaxation of the intrafusal muscle fibre area innervated by parasympathetic motoneurons (see below).

14.6. Relationship between Conduction Velocity, Action Potential Amplitude and Action Potential Duration

With increasing conduction velocity (and the increasing diameter), AP amplitude increases and AP duration decreases (Figures 7-9). The scatter in these relations is large because different fibres in a root are at different distances from the recording electrodes. Also, nerve fibres change their diameters slightly from internode to internode (Figure 29A). And, as a matter of fact, damaged nerve fibres have strongly reduced conduction velocity and increased AP duration. Poor digitalisation and too strong filtering will reduce AP amplitudes. Nevertheless, good recordings clearly demonstrate that average AP amplitude decreases and AP duration increases with the reduction of conduction velocity when changing from one nerve fibre group to another. These relations are of interest here, since with known expected conduction times, AP amplitudes and AP durations, it is much easier to identify or to pick up certain APs on recording traces a and b, in particular in the presence of small amplitudes and high levels of noise and artefacts. From Figures 8B and 107, it can be calculated that the AP duration increased from α_2 through γ_1 to parasympathetic motoneurons (1: 3 : 6), the AP amplitude decreased (1 : 1/3 ; 1/8) and the conduction velocity decreased in the same sequence (6 : 2 : 1.5 : 1) (Figure 8C).

14.7. Natural Stimulation of the Stretch Reflex

The constant stretch reflex of the anal sphincter is considered. Upon constant stretch of the anal sphincter by an anal catheter of 12mm diameter, sphincteric α_2-motoneurons (FR) and α_3-motoneurons (S) responded partly by firing continuously in the oscillatory firing mode. This oscillatory firing was stimulated, when the catheter was not pulled, mainly by the activity of secondary muscle spindle afferents, activated by the sphincter distension. Apart from mucosal afferents, other afferents in the same dorsal S4 root with diameters larger than 5μm did not fire (limit of the recording method). This finding is in accordance with the Matthews' opinion that secondary spindle afferents cause a powerful autogenic excitation in human stretch reflexes [140]. In a first approximation, the anal sphincter contracted isometrically, since the catheter determined approximately constant length.

To explore this rather constantly acting recto-sphincteric-reflex [214], stimulations such as pinprick, anal and bladder catheter pulling were used to increase or decrease the reflex to study the reflex function. Of interest were particular breaks in continuous oscillation, i.e., transient breaks of constant stretch reflex, since the oscillator, of which the sphincteric motoneuron is a part, must have transiently lost adequate afferent input from the secondary spindle afferents. It was shown above that five single secondary spindle afferents could be identified, three of them responding preferentially to certain natural combinations of afferent input, namely painful pinprick, anal and bladder catheter pulling. Pinprick was shown (Figures 37-45c) to stimulate, within the limits of the recording method T1 ($v = 44ms^{-1}/Ø = 11.2μm$), T2 (39/10.1), T3 (31/9.1) and T4 (20/8.3) skin touch afferents (maybe corresponding to PC, RA, SAI, SAII skin receptors) and fastest pain afferents P ($13ms^{-1}/Ø$ = ?). Anal catheter pulling stimulated skin afferents T1 through T4 of the anal canal and mucosal afferents M ($17.5ms^{-1}/Ø$ = ?) (Figure 49). Bladder catheter pulling (Figure 48) stimulated mucosal afferents innervating the trigonum vesicae and the urethra (and may be flow receptors S2 ($17.5ms^{-1}/Ø$ = ?)) (and skin afferents), and strong pulling additionally stimulated pain afferents, the fastest component of which could be measured (P). The T4 afferent receptors seem to be especially sensitive to touch movements along the skin. As shown in Figures 99 and 100, the strength of stimulation changed the firing patterns of the secondary spindle afferents. But also the stimulated areas will change the response of the secondary spindle afferents. Spindle afferent fibre SP2(2) was strongly activated by pulling of the bladder catheter, i.e., mainly by the activity of the mucosal and pain afferents innervating the urethra and the trigonum vesicae. Fibre SP2(1) was strongly activated by the anal catheter pulling (Figure 102), i.e., by the activity of the mucosal and pain afferents of the anal canal and the rectal ampulla, and the skin afferents of the anal canal. It is obvious that bladder catheter pulling, by which the mucosal and pain afferents are stimulated, and a simultaneous touch of the S3 and S4 dermatomes by which the skin afferents are stimulated, will not activate strongly fibre SP2(1), which was strongly activated by the afferent input from skin, mucosal and pain receptors following anal catheter pulling. Electrical stimulation of groups of afferents is therefore far from reality to mimic natural spindle activation. Already, Sherrington advocated the need for a well-balanced peripheral input [219].

14.8. Impulse Patterns of Intrafusal Axons and Secondary Spindle Afferents

Since the skin, mucosal and pain afferents are not innervating the muscle spindles themselves, the different drives of the secondary muscle spindle afferents are channelled through the efferent muscle spindle innervation, consisting of the somatic dynamic and static γ-motoneurons and, as measured in the present work, of parasympathetic motoneurons. According to the different afferent inputs from mechanoreceptors of the skin, the urethra, the trigonum vesicae and the anal canal, the γ and parasympathetic motoneurons changed their activities, with secondary spindle afferents showing AP patterns as shown in Figures 99, 100 and 111. In addition to activities of afferent fibres SP2(2) and SP2(5), Figure 111 shows the natural impulse patterns of a single dynamic (γ_1) and a single static γ-motoneuron (γ_{21}). In the second series of measurements, where all five secondary afferents were strongly activated, it was difficult to identify the γ-motoneurons, probably because of an insufficient blood supply, following recording for more than 20min from the root. As Figure 109 shows, the interspike intervals of the γ-motoneurons and the secondary muscle spindle afferents are similar (see also Figure 94). It is therefore not clear whether the γ-motoneurons drove single encoding sites of the secondary spindle afferents or whether the parent secondary spindle afferent fibre drove partly neuronal networks, which produced the impulse patterns of the dynamic and the two static γ-motoneuron groups. A second static γ-motoneuron group was proposed by Boyd [26]. Since secondary spindle afferents predominantly innervate nuclear chain fibres in cats [22], and contractions of chain fibres can follow frequencies up to 60Hz [23, 25], it may well be that the muscle spindle capsule is rather constant, whereas the chain fibres show local movements according to the activity patterns of the γ-motoneurons [29]. These local contractions will increase or decrease the stretch of the equatorial region, innervated by the secondary spindle afferents. Single encoding sites will send impulse patterns in competition with, or in addition to, other encoding sites via the parent afferents, to γ-neural networked to form a loop (Figure 114) and the spinal oscillators. Owing to the time needed, the existence of a loop from the γ-neural networks to the spindles and backwards is possible, since the activation of micturition and defecation is in the time range of seconds to minutes. The proposed loop is too complex to follow up the AP patterns. Many fusimotor axons will activate nuclear chain and nuclear bag fibres, and different encoding sites of chain and bag fibres will compete with one another.

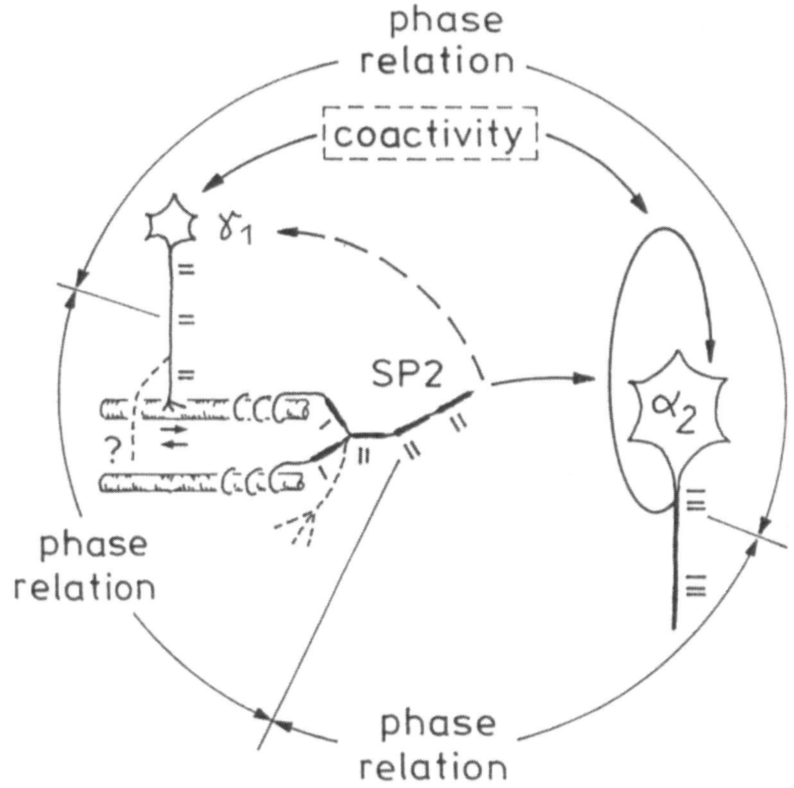

Figure 114. Schematized existing phase relation between α_2 and γ_1-motoneurons and a secondary muscle spindle afferent fibre (SP2). Parallel existing phase relations between other parent afferents and the α_2-motoneuron and between parent secondary spindle afferents are not shown. Phase relation means the increased occurrence of phases in ms in a certain phase range between the action potentials (APs) of the two compared nerve fibres. The complex afferent and efferent muscle spindle innervation was not attempted to be shown. Small arrows at intrafusal muscle fibre indicate local contraction, which is in nuclear chain fibres, readily transmitted to the place of afferent innervation. A possible reason of the doublet firing of the SP2 fibre is pictured to occur from single APs (schematized by bars) of two myelinated endings, not necessarily from pacemaker switching. More endings of the parent SP2 fibre and γ_1-motoneurons are indicated by dashed line branches. "Coactivity" indicates a correlation between γ and α-motoneuron spinal cord circuitries for higher activations.

As Figure 112B shows, the overall activity of the single γ_1, single γ_{21} and γ-rest-motoneurons are similar, but the actual AP patterns of the γ_1 and γ_{21}-motoneurons were different (Figure 111).

The single γ_1-motoneuron seem to innervate a single encoding site of the secondary spindle afferent fibre SP2(2), since i) the γ_1-motoneuron had a doublet interspike interval distribution peak at 13ms (Figure 109C) similar to one subpeak of fibre SP2(2) (Figure 98, labelled γ_1) and ii) many phase relations occurred between γ_1 APs and SP2(2) APs (Figure 111). However, the impulse patterns of γ_1-motoneurons and fibre SP2(2) were not identical and time-locked either. Since identical peaks of doublet interspike interval distributions existed at 13ms, the doublet firing may have been generated in SP2(2) fibre endings by the pacemaker switching of the two encoding sites; and this doublet firing projected into the γ-neural networks. The shortest interspike interval of about 80ms, with the exception of the doublet IIs, may be generated in the neural network generating γ-motoneuron activity. Also, the doublet and the 80ms IIs may be generated in γ-neural networks as a preform of self-organisation similar to the self-organisation of α_1, α_2 and α_3-oscillators. The activity level of a γ_1-motoneuron firing with a doublet every 80ms is higher (25APs/s) than that of an oscillatory firing α_2-motoneuron, which fires with two APS every 130ms (15.4APs/s). If α_2-motoneurons fires in the oscillatory firing mode with three APs every 160ms, which often occurs, is still lower with 19 APs/s.

14.9. Population of Intrafusal Motoneurons

The question was raised by Boyd whether there existed two types of static γ-axons (26). The distributions of conduction velocity frequencies obtained for HT6 and paraplegic 1 (Figure 33D,E) show three γ-peaks, two of which seem static. Towards lower conduction velocities, a peak of parasympathetic fibres appeared at about 5m/s (T = 33.5°C). However, as Figures 33E and 113M indicate, there were one or two other conduction velocity peaks, suggesting further types of motoneurons. In compound action potentials of a S5 nerve root of paraplegic 1, two static γ-motoneurons seem to occur (Figure 35E). It has to be clarified further whether there is but a single type of parasympathetic motoneurons or several ones and whether there are different parasympathetic motoneurons for extrafusal and intrafusal innervation.

Here, the identification of the parasympathetic fibres was mainly based on the identification of AP patterns of a single γ_1 and a single γ_{21}-motoneuron. The impulse patterns of single γ_{22} and parasympathetic motoneurons need to be separated from the summed impulse traffic of dorsal or ventral roots.

14.10. Phase Relations and Spinal Oscillators

As a comparison of Figures 111B and 111A indicate, the number of phase relations (phases in a certain range) between single-fibre APs of γ-motoneurons and certain secondary spindle afferents increased with bladder catheter pulling. Also, the number of phase related γ and oscillatory firing α_2-motoneuron O2 APs increased, which suggests a certain kind of α-γ coactivity.

Above, it was shown that the number of phase-related APs between secondary spindle afferents and oscillatory firing motoneuron O2 increased. A phase loop seemed to exist from the oscillatory firing α_2-motoneuron O2 through γ-motoneurons to the secondary spindle afferents and back to the oscillatory firing motoneuron (Figure 114). This phase loop is similar to recoupling in phase from output back to input in electronic oscillators. The analogy is only partly correct, since γ-motoneuron systems probably do not get the phase from the α-systems. Probably, the phase relation occurs because α-γ coactivity (or more exactly, α_2-γ_1 coactivity) for high activations. The absolute phases may not be so important for the drive of the spinal oscillators, since interneurons could serve as delay components to adapt to the right phase. These rather qualitative considerations concerning phase relation will be quantified in Chapter V.

14.11. Fusimotor and Skeletomotor Activity

During voluntary movements in man, Hagbarth and Vallbo measured the impulse rate in spindle afferents to run reasonable in parallel with the skeletomotor activity as estimated from EMG activity [87, 232]. During the constant stretch reflex of the external anal sphincter, some α_2 and α_3-motoneurons fired in the high activity oscillatory firing mode. Transiently, the identified impulse patterns of single γ_1 and γ_{21}-motoneurons showed here higher activity levels when firing with doublets every 80ms than the oscillatory firing α_2-motoneuron O2 (HT6) firing with impulse trains of three APs every 160ms. On a reasonable assumption that the highly activated intrafusal motoneurons were functionally connected to the oscillatory firing α_2-motoneuron, as supported by an increase of phase relations between their APs with increasing activation, the activity level changes of these extra- and intrafusal motoneurons run in parallel only sometimes. In the occasional firing mode following stimulation, α_2 and γ_1-motoneurons were activated in parallel also only sometimes [174]. α_3-motoneurons changed their activity level differently from those of α_2-motoneurons [174, 182].

As the actual impulse patterns show (Figure 111), there was no or no strongly linked α-γ co-activation. α_2-Motoneuron O2 fired in the oscillatory firing mode and γ-motoneurons did not. The fusimotors had impulse patterns very to the secondary spindle afferents. Since APs of the highly activated fusimotor are easier to pick up from the summed impulse traffic than APs of only little activated neurons, highly activated fusimotors of the dS4 root therefore most likely did not escape detection.

14.12. Dissociation of Function and Repair by Learning

From the plasticity of the CNS of animals and humans, one would expect more variability in the dissociation of functions [70] in the human CNS. Sperry transposed the nerve supplies of flexor and extensor muscles in a rat [221] and a monkey [222]: the monkey relearned the task after some time; the rat did not. Surprisingly, few trials were required for poliomyelitis patients to use transposed tendons successfully [236]. The voluntary movement probably activates the human CNS differently from the constant stretch reflex. Rigid integrative CNS organisations like automatisms may be more successful in repair of CNS injury than voluntary movements because pathologic subnetwork organisation cannot that easily escape repair by learning by dissociation of function.

14.13. Time Course of Activation of a Secondary Spindle Afferent Fibre Activated by Parasympathetic Fibres

From conduction velocity frequency distributions (Figure 33) and the time course of activation of fibre SP2(2) following strong bladder catheter pulling 3 (Figures 112,113), intrafusal parasympathetic action could be identified. This autonomic muscle spindle activation is of parasympathetic type: parasympathetic fibres emerge from the spinal cord via the lower sacral roots (where measurements were performed) to partly form the pelvic nerves (Figure 16), whereas sympathetic fibres leave the spinal cord through levels between TH12 and L2 and innervate the urinary bladder via the hypogastric plexus. Information about the autonomic innervation of animal muscle spindles has been reported elsewhere [5, 8, 150, 225]. A time constant τ of 3s has been calculated here for the activity increase of fibre SP2(2). The plateau of activation was reached within 10 to 15s (Figure 112). This value is in accordance with animal measurements, where sympathetic stimulation increased muscle afferent activity to a plateau value within 10 to 15s (150) as well. Following parasympathetic activation (Figure 113G), fibre SP2(2) decreased its activity very slowly with a time constant of 40s. The time constants for the increase and decrease of the SP2(2) fibre activity differed by more than a factor of 10, which indicates that the parts of the intrafusal muscle fibre that contracts with the parasympathetic activity, relaxes very slowly. Frog skeletal slow muscle fibres, used as a reference for contraction and relaxation speeds of intrafusal muscle fibres, do not relax following nerve stimulation as slowly (not a K^+-contracture). The time constant for the increase of the isometric tension in the frog rectus abdominis following selective small axon stimulation of motor branches of spinal nerves, which innervate extrafusal slow muscle fibres, was about 1.3s. A plateau was reached 10 to 15s following stimulation [50]. Upon switching off the 60Hz tetanus, the isometric tension decreased with a time constant of approximately 3.5s. In kittens [38], the slow muscle extensor digitorum longus (ELD) reduced its isometric tension following titanic stimulation with a time constant much shorter than 1s. The area of the human intrafusal muscle fibre activated by parasympathetic fibres has very slow properties. Creep and other after effects [27, 55] cannot account for such low time course. It could be that parts of the intrafusal muscle fibres have properties similar to those of smooth muscle. Human muscle spindles of the external anal sphincter and the musculus pubococcygeus (Figure 52) need to be examined histologically and histochemically. Still, intrafusal muscle fibres may have normal morphology and contract more slowly because of the parasympathetic fibres exerting a neural control and changing the excitation-contraction coupling of that part of the intrafusal muscle fibre. For frog slow muscle fibres, it has been shown that the action potential mechanism of the membrane and the contraction and relaxation speeds are under neural control [143]. The excitability of the membrane is under neural control even in parts of the slow muscle fibre [168]. The change of relaxation time following K^+-contracture during the development of the frog upon neural control changes can be seen in Figure 5 of Chapter I. It is conceivable that several intrafusal axons innervate the intrafusal muscle fibres and exert their neural control. A single intrafusal muscle fibre could even be innervated by different kinds of axons, each exerting its specific neural control. Different nuclei may show different gene expression. An injury of the CNS is not only changing the neuronal networks structurally and functionally but may even change the neural control of the muscle spindles.

The secondary spindle afferent fibre SP2(2) increased its activity following parasympathetic activation by recruiting new encoding sites (Figures 111,112A,97C). The activity of the parent fibre SP2(2) became regular again (similar interspike intervals) shortly after the recruitment of a new encoding site (Figure 111). Such a smoothening of the interspike intervals can be explained by a common pacemaker [53] for the six encoding sites. More likely, potentials generated in up to six unmyelinated endings added up at the first heminode or penultimate node, where the action potential was produced. A pacemaker switching of myelinated endings seems to be more unlikely to produce a coordination of up to six encoding sites.

14.14. Pathologic Spinal Cord Function Following Spinal Cord Injury: Impairment of Phase and Frequency Coordination

It was shown [185] that in spinal cord injury, in the presence of dyssynergy of the urinary bladder, α_3-motoneurons (S) changed their excitation properties being recruited too fast. This may indicate a specific loss of central inhibition. In another paper [186], an α_2-oscillator was reported to fire physiologically at a frequency of 6.25Hz, while firing at frequencies up to 13.5Hz under pathologic conditions (spinal cord injury). Such broadening of frequency range (Figure 92) can have consequences. It was shown above that oscillatory firing of an external anal sphincteric α_2-motoneuron was mainly driven by certain secondary muscle spindle afferents (SP2(1) through SP2(3)), and oscillatory firing of an external bladder sphincteric α_2-motoneuron was mainly driven by the stretch receptors most likely located in the urinary bladder wall (Figure 52). If these oscillators fire at broad ranges of frequencies according to certain probability distributions instead of firing at certain resonance frequencies (Figure 92), these oscillators may also be excited at different frequencies. Thus, following spinal cord injury, spinal oscillators driving the external bladder and anal sphincters (and other skeletal muscles) in the high activity mode may be excited not only by their respective afferent input but also by impulse patterns of other receptors, which normally do little or not at all contribute to drive. In dyssynergetic bladders, with parasympathetic fibres activating the detrusor for emptying the bladder, also the external or internal sphincters are transiently activated for bladder fillings exceeding 200ml. Parasympathetic impulse patterns may have hit pathologic frequencies of the spinal oscillators. It was shown in here that parasympathetic fibres also innervate muscle spindles. With the destruction of the descending tracts following spinal cord injury, afferents serving mainly parasympathetic activation may have sprouted to sites of degenerated tract synapses and could have enhanced parasympathetic actions. It is of importance to identify the impulse patterns of parasympathetic fibres to see whether they could activate spinal oscillators via spindle activation or directly under pathologic condition. Many functions of the neuronal networks of the spinal cord will have changed following injury.

Unbalanced damage is probably the worst one. Unbalanced injuries of the neuronal networks can already be induced by transient asphyxia of the spinal cord upon pressure from split vertebrae or destroyed blood supply. Anoxia is one of the pathologic processes occurring with spinal pressure following spinal cord injury, when edematized cord is trapped by the rigid pia mater. Patients with spinal cord injury have to be operated on as quickly as possible (best within seven hours after injury) to reduce pressure and improve blood supply. The mistakes done following injury until to the operation, including false transportation of the patient, can mostly not be compensated for even by the best therapy.

A spinal cord injury and also brain injury cannot just be repaired by administering stem/progenitor cells, even if integrated into the adult damaged neural networks. The whole CNS network has to be repaired in its functioning and a repair strategy is movement-based learning since the developing CNS is also using movement-based learning for eliminating developmental errors and optimizing and adapting to the development CNS functioning. If there is some similarity between development and repair of the human CNS, which may be 17 years, then we cannot expect a quick repair. Years are needed.

14.15. Continuation of the Research to Repair the Human CNS in Chapters IV and V

Physiologic and pathologic organisation of human CNS functioning will be measured and further analysed in Chapter V. The rather physiologic CNS functioning of brain-dead humans will be compared further with the pathologic CNS functioning in patients with spinal cord injury (SCI). Especially the pathology of the parasympathetic nervous system will be explored further following injury. After summarizing the human neurophysiology under physiologic and pathophysiologic conditions, the human knowledge will be used to develop therapy, and it will be shown that this coordination dynamics therapy (CDT) is really working in humans. The biggest problem in SCI, the repair of the urinary bladder, is shown to be solved by movement-based learning. This research shows clearly that human research leads to the repair of the human nervous system, whereas animal research in a first approximation leads only to the repair of the animal nervous system. When researchers of top universities worldwide think that they can repair the human nervous system by studying only the animal nervous system, then their knowledge is 20 years out-of-date, and the poor patients will have to suffer from that out-of-date thinking. If I understand the functioning of a radio, I still cannot understand the functioning of a TV. Or is there a reader who believes that one can study the repair of the speech (Chapter II) by animal experimentation?

For allowing a better comparison of animal data with human ones, the single-nerve fibre action potential recording method will be applied in Chapter IV to the rat. It is shown there that one can measure regeneration following SCI in rat (or mice) also in a qualified way and need not to judge improvement of CNS functioning by letting the rat run on a ladder or similar movements. For an animal comparison, actual dogs and pigs are more suitable because of size. Dog measurements have been used already here (conduction velocity frequency distributions) to improve human data (Figure 53). The dog measurements were obtained by sharing the animals with surgeons working on the dog urinary bladder. But the dog conus medullaris is functioning differently from the human one because of tail and the function also sited in the caudal spinal cord. In Chapter VI, it will be analysed whether stem/progenitor cells have a chance to be integrated in adult impaired neural networks for repair.

References

[1] Abeles, M.: *Local cortical circuits. An electrophysiological study*. Springer, Berlin, 1982.

[2] Adams, D.J., Smith, S.J. and Thompson, S.H.: Ionic currents in molluscan soma. *Ann. Rev. Neurosci.*, 3: 141-167, 1980.

[3] Adams, J.A.: Historical review and appraisal of research on the learning, retention, and transfer of human motor skills. *Physiol. Bulletin*, 101: 41-74, 1985.

[4] Arbib, M.A.: *The Handbook of Brain Theory and Neural Networks*, MIT Press, Cambridge, 1995.

[5] Arbuthnott, E.R., Gladden, M.H. and Sutherland, F.: Autonomic innervation of muscle spindles from the hindlimb of the cat. *Suppl. Proc. Roy. Soc.*, 17: 1, 1982.

[6] Banks, R.W., Barker, D. and Stacey, M.J.: Form and distribution of sensory terminals in cat hindlimb spindles. *Phil. Trans. R. Soc. Lond.*, B299: 329-364, 1982.
[7] Barker D., Scott J.J.A. and Stacey M.J.: A study of glycogen depletion and the fibre-type composition of cat skeleto-fusimotor units. *J. Physiol.* 450: 565-579, 1992.
[8] Barker, D. and Banks, R.W.: The muscle spindle. In: Engel, A.E. and Banker, B.Q. (Eds.), *Myology, Vol. 1, Part 1*, McGraw-Hill, New York, pp. 309-341, 1986.
[9] Barron, D.H. and Matthews, B.H.C.: The interpretation of potential changes in the spinal cord. *J. Physiol.*, 92: 276-321, 1938.
[10] Bawa, B., Binder, M.D., Ruenzel, P. and Henneman, E.: Recruitment of motoneurons in stretch reflexes is highly correlated with their axonal conduction velocity. *J. Neurophysiol.*: 52: 410-420, 1984.
[11] Bayev, K.V., Dekhtyarenko, A.M., Zavadskaya, T.V. and Kostyuk, P.G.: Activity of lumbar interneurons during fictitious locomotion of thalamic cats. *Neurophysiology* (Kiev), 11: 329-338, 1979.
[12] Bernstein, J.J., Wells, M.R. and Bernstein, M.E.: Spinal cord regeneration: synaptic renewal and neurochemistry. In: Cotman, C.W.(Ed.), *Neuronal Plasticity*. Raven Press, New York, pp. 49-71, 1978.
[13] Berridge, M.J. and Rapp, P.E.: A comparative survey of the function, mechanism and control of cellular oscillators. *J. Exp. Biol.*, 81: 217-279, 1979.
[14] Binder, M.D. and Stuart, D.G.: Motor unit muscle receptors interaction: Design features of the neuro-muscular control system. *Progr. Clin. Neurophysiol.*, 8: 72-98, 1980.
[15] Bobath, B.: Adult Hemiplegia, *Evaluation and Treatment*. Butterworth Heinemann, Oxford, 1994.
[16] Bobath, K.: *A Neurophysiological Basis for the Treatment of Cerebral Palsy*. William Heinemann Medical Books Ltd., London, 1980.
[17] Bohm, E.: Sacral rhizopathies and sacral syndromes (SII-SV). *Acta Chirurgica Scandinavia*, 216 (Suppl.): 5-28, 1956.
[18] Bors, E. and Comarr, A.E.: *Neurological Urology*. Karger, Basel, 1971.
[19] Bors, E. and Porter, R.W.: Neurosurgical consideration in bladder dysfunction. *Urol. Int.* 25: 114-133, 1970.
[20] Boyd, I.A.: The structure and innervation of the nuclear bag muscle fibre system and nuclear chain muscle fibre system in mammalian spindles. *Phil. Trans.*, B245: 81-136, 1962.
[21] Boyd, I.A. and Davey, M.R.: *Composition of peripheral nerves*. Livingston, Edinburgh, 1968.
[22] Boyd, I.A.: The structure and innervation of the nuclear muscle fibre system and the chain muscle fibre system in mammalian spindles. *Phil. Trans.*, B245: 81-136, 1962.
[23] Boyd, I.A., Gladden, M.H., McWilliam, P.N. and Ward, J.: Control of dynamic and static nuclear bag fibres and nuclear chain fibres by gamma and beta axons in isolated cat muscle spindles. *J. Physiol.*, 265: 133-162, 1977.
[24] Boyd I.A.: The isolated muscle spindle. *Trends in Neurosciences*, 3: 258-265, 1980.
[25] Boyd, I.A.: The action of three types of intrafusal fibre in isolated cat muscle spindles and length sensitivity of primary and secondary sensory endings. In: A. Taylor and A. Prochazka (Eds.), *Muscle Receptors and Movement*, Macmillan, London, pp. 17-32, 1981.

[26] Boyd, I.A., Gladden, M.H. and Ward, J.: Two types of static γ axons having predominantly static bag 2 fibre action or predominantly chain fibre action in the several cat muscle spindle they supply. *J. Physiol.*, 343: 110-111P, 1983.

[27] Boyd, I.A.: Review. In: Boyd, I.A. and Gladden, M.H. (Eds.), *The Muscle Spindle*, Stockton Press, New York, pp. 129-150, 1985.

[28] Boyd, I.A., Murphy, P.R. and Mann, C.: The effect of chain fibre "driving" on the length sensitivity of primary sensory endings in the tenuissimus, peroneus tertius and soleus muscles. In: Boyd, I.A. and Gladden, M.H. (Eds.), *The Muscle Spindle*, Stockton Press, New York, pp. 195-199, 1985.

[29] Boyd, I.A., Sutherland, F.I. and Ward, J.: The origin of the increase in the sensitivity of secondary sensory endings produced by some fusimotor axons. In: Boyd, I.A. and Gladden, M.H. (Eds.), *The Muscle Spindle*, Stockton Press, New York, pp. 207-214, 1985.

[30] Bradley, W.E. and Teague, C.T.: Synaptic events in pudendal motoneurons of the cat. *Experimental Neurology*, 56: 237-240, 1977.

[31] Brindley, G.S., Polkey, C.E., Rushton, D.N. and Cardozzo, L.: Sacral anterior root stimulation for bladder control in paraplegia: the first 50 cases. *J. Neurol. Neurosurg. Psychiatr.*, 49: 1104-1114, 1986.

[32] Brown, A.G. and Iggo, A.: A quantitative study of cutaneous receptors and afferent fibres in the cat and rabbit. *J. Physiol.*: 193: 707-733, 1967.

[33] Buchthal, F. and Schmalbruch, H.: Motor unit of mammalian muscle. *Physiological Reviews*, 60: 91-142, 1980.

[34] Buonocore, M.H. and Hecht, S.T.: Functional magnetic resonance imaging depicts the brain in action. *Nature Medicine*, 1: 379-381, 1995.

[35] Burke, D., Skuse, N.F. and Stuart, D.G.: The regularity of muscle spindle discharge in man. *J. Physiol.*, 291: 277-290, 1979.

[36] Burke, D. and Gandevia, S.C.: Peripheral motor systems. In: G. Praximos (Ed.), *The human nervous system*, Academic Press, Inc., New York, pp.125-145, 1990.

[37] Carlstedt, T.: Observation on the morphology at the transition between the peripheral and the central nervous system in the cat. I. Preparative procedure useful for electron microscopy of the lumbosacral dorsal rootlets. *Acta Physiol. Scand.* (Suppl.), 446: 5, 1977.

[38] Close, R. and Hoh, J.F.Y.: Velocity properties of kitten muscles. *J. Physiol.*, 192: 815-822, 1967.

[39] Coggeshall, R.E., Applebaum, M.L., Fazen, M., Stubbs, T.B. and Sykes, M.T.: Unmyelinated axons in human ventral roots, a possible explanation for the failure of dorsal rhizotomy to relieve pain. *Brain,* 98: 157-166, 1975.

[40] Cohen, A.H. and Wallèn, P.: The neuronal correlate of locomotion in fish: "Fictive swimming" induced in an in vitro preparation of the lamprey spinal cord. *Exp. Brain Res.*, 41: 11-18, 1980.

[41] Cohen, A.V.: Evolution of the vertebrate central pattern generator for locomotion. In: Cohen, A.H., Rossignol, S. and Grillner, S. (Eds.), *Neural Control of Rhythmic Movements in Vertebrates*, pp.129-165, John Wiley and Sons, New York, 1988.

[42] Corda, M., Euler, C. and Lennerstrand, G.: Reflex and cerebellar influences on α and on "rhythmic" and "tonic" γ activity in the intercostal muscle. *J. Physiol.*, 184: 898-923, 1966.

[43] Creed, R.S., Denny-Brown, J.G., Eccles, J.G., Lidell, E.G.T. and Sherrington, C.S.: *Reflex Activity of the Spinal Cord*. Clarendon Press, Oxford, 1932.

[44] Deliagina, T.G. and Feldman, A.G.: Activity of Renshaw cells during fictive scratch in the cat. *Exp. Brain Res.*, 42: 108-115, 1981.

[45] Deliagina, T.G. and Orlovsky, G.N.: Activity of Ia inhibitory interneurons during fictitious stretch reflex in the cat. *Brain Res.*, 193: 439-447, 1980.

[46] Denny-Brown, D. and Robertson, E.G.: The state of the bladder and its sphincters in complete transverse lesions of the spinal cord and cauda equine. *Brain,* 56: 397-463, 1933.

[47] Descartes, R.: Les passion de l'âme. Article XI and CXVIII, Amsterdam 1649.

[48] Dietz, D., Colombo, G. and Jensen, L.: Locomotor activity in spinal man. *Lancet,* 344: 1260-1263, 1994.

[49] Duthie, H.L. and Gairns, F.W.: Sensory nerve endings and sensation in the anal region of man. *Brit.J. Surg.*, 47: 585, 1960.

[50] Forrester, T. and Schmidt, H.: An electrophysiological investigation of the slow fibre system in the frog rectus abdominis muscle. *J. Physiol.*, 207: 477-491, 1970.

[51] Dyck, P.J and Thomas, P.K.: *Peripheral Neuropathy, Volume I*, Saunders Company, Philadelphia, 1993.

[52] Eccles, J.C.: The plasticity of the mammalian central nervous system with special reference to new growths in response to lesions. *Naturwissenschaften*, 63: 8-15, 1976.

[53] Eagels, J.P. and Purple, R.L.: Afferent fibres with multiple encoding sites. *Brain Research*, 77: 187-193, 1974.

[54] Ekeberg, Ö.: Response properties of a population of neurons. *Int. J. Neur. Syst.*, 4:1-13, 1993.

[55] Emonet-Denand, F., Hunt, C.C. and Laporte, Y.: Fusimotor after-effects on bag 1 contraction. In: Boyd, I.A. and Gladden, M.H. (Eds.), *The Muscle Spindle*, Stockton Press, New York, pp. 187-188, 1985.

[56] Eriksson, P.S., Perfilieva, E., Björk-Eriksson, T., Alborn, A.M., Nordborg, C., Peterson, D.A., Gage F.H.: Neurogenesis in the adult human hippocampus. *Nature Medicine*, 4: 1313-1317, 1998.

[57] Euler, C. von: On the central pattern generator for the basic breathing rhythmicity. *J. Appl. Physiol. Respirat. Environ. Exercise Physiol.*, 55: 1647-1659, 1983.

[58] Evans, J.P.: Observation on the nerves of supply to the bladder and urethra of the cat, with a study of their action potentials. *J. Physiol.*, 86: 396-413, 1936.

[59] Feldman, A.G. and Orlovsky, G.N.: Activity of interneurons mediating reciprocal Ia inhibition during locomotion. *Brain Res.*, 84: 181-194, 1975.

[60] Forssber, H.: On integrative motor functions in the cat's spinal cord. *Acta Physiol. Scand., Suppl.* 474: 1-56, 1979.

[61] Friesen, W.O., Poon, M. and Stent, G.S.: Neuronal control of swimming in the medical leech. IV. Identification of a network of oscillatory interneurons. *J. Exp. Biol.*, 75: 25-43, 1978.

[62] Friesen, W.O. and Block, G.D.: What is a biological oscillator? *Am. J. Physiol.*, 262: 847-851, 1984.

[63] Gage, J.R.: Distal hamstring lengthening / release and rectus femoris transfer. In: Sussman M.D. (Ed.), *The diplegic child*, pp. 317-336, Rosement: AADS, 1992.

[64] Gage F.H. et al.: Regenerative responses in the spinal cord. *Rev. Neurol. Arg*. Suppl 1: 56, 1997.

[65] Garry, R.C. and Garven, H.S.D.: The ganglia, afferent nerve-endings and musculature of the urethra in the cat. *J. Physiol.*, 139: 1P-2P, 1957.

[66] Gasser, H.S., Grundfest, H.: Axon diameter in relation to the spike dimension and the conduction velocity in mammalian A fibres. *Amer. J. Physiol.* 127: 393-414, 1993.

[67] Getting, P.A.: Mechanism of pattern generation underlying swimming in Tritonia. II Network reconstruction. *J. Neurophysiol.*, 49: 1017-1035, 1983.

[68] Getting, P.A.: Comparative analysis of invertebrate central pattern generator. In: Cohen A.H., Rossignol, A.H. and Grillner, S. (Eds.), *Neural Control of Rhythmic movements in Vertebrates*, John Wiley and Sons, New York, pp. 101-127, 1988.

[69] Gladden, M.H. and Murphy, P.: Review. In: Boyd, I.A. and Gladden, M.H. (Eds.), *The Muscle Spindle*, Stockton Press, New York, pp. 243-254, 1985.

[70] Gladden M.H.: Efferent control of human muscle spindles. In: Boyd, I.A. and Gladden, M.H. (Eds.), *The Muscle Spindle*, Stockton Press, New York, pp. 243-254, 1985.

[71] Gladden M.H.: Muscle receptors in mammals. *Advances in Comparative and Environmental Physiology. Vol. 10*, Springer-Verlag, Berlin, pp. 281-302, 1992.

[72] Gosling, J.A., Dixon, J.S., Humpherson, J.R.: *Functional anatomy of the urinary tract.* Churchill Livingston distributed, Gower Medical Publishing, London, pp. 5-18, 1983.

[73] Gossard, J.-P., Brownstone, R.M., Barajon, I. and Hultborn, H.: Transmission in a locomotor-related group Ib pathway from hindlimb extensor muscles in the cat. *Exp. Brain Res.*, 98: 213-228, 1994.

[74] Granit, R., Pascoe, J.E. and Steg, G.: The behaviour of tonic α and γ motoneurons during stimulation of recurrent collaterals. *J. Physiol.*, 138: 381-400, 1957.

[75] Granit, R., Haase, J. and Rutledge, L.T.: Recurrent inhibition in relation to frequency of firing and limitation of discharge rate of extensor motoneurons. *J. Physiol.*, 154: 308-328, 1960.

[76] Granit, R. and Renkin, B.: Net depolarization and discharge rate of motoneurons, as measured by recurrent inhibition. *J. Physiol.*, 158: 461-475, 1961.

[77] Granit, R., Kernell, D. and Smith, R.S.: Delayed depolarization and the repetitive response to intracellular stimulation of mammalian motoneurons. *J. Physiol.*, 168: 890-910, 1963.

[78] Gril, S.E. and Rymer, W.Z.: Beta-contribution to fusimotor axons in triceps surae muscles of decerebrated cats. *J. Neurophysiol.*, 57: 574-595, 1987.

[79] Grillner, S., Perret, C. and Zangger, P.: Central generation of locomotion in spinal dogfish. *Brain Res.*, 109: 255-269, 1976.

[80] Grillner, S. and Kashin, S.: On the generation and performance of swimming in fish. In: Herman, R., Grillner, S., Stein, P. and Stuart, D. (Eds.), *Neural Control of Locomotion, Vol.18*, Plenum Press, New York, pp. 181-202, 1976.

[81] Grillner, S. and Zangger, P.: On the central generation of locomotion in the low spinal cat. *Exp. Brain Res.*, 34: 241-261, 1979.

[82] Grillner, S and Zangger, P.: The effect of dorsal root transection on the efferent motor pattern in the cat's hindlimb during locomotion. *Acta Physiol. Scand.*, 120: 393-405, 1984.

[83] Grillner, S., Buchanan, J.T., Wallen, P. and Brodin, L.: Neural control of locomotion in lower vertebrates: From behavior to ionic mechanisms. In: Cohen, A.H., Rossignol, S.

and Grillner, S: (Eds.), *Neural Control of Rhythmic Movements in Vertebrates*, Wiley, New York, pp. 1-40, 1988.

[84] Grillner, S., Wallen, P. and di Prisco, G.V.: The lamprey locomotor system-Basic locomotor synergy. In: Shimamura, M., Grillner, S. and Edgerton, V.R. (Eds.), *Neurobiological Basis of Human Locomotion*, Japan Scientific Societies Press, Tokyo, pp. 77-92, 1991.

[85] Haase, J., Cleveland, S. and Ross, H.-G.: Problems of postsynaptic autogenous and recurrent inhibition in the mammalian spinal cord. *Rev. Physiol. Biochem. Pharmacol.*, 73: 73-129, 1975.

[86] Hagbarth, K.E. and Finer, B.L.: The plasticity of the human withdrawal reflexes to noxious skin stimuli in lower limbs. *Progr. Brain Res.:* 1: 65-81, 1963.

[87] Hagbarth, K.-E.: A critique of the papers by Loeb and Hoffer and Prochazka and Wand. In: Taylor, A. and Prochazka, A. (Eds.), *Muscle Receptors and Movement*, McMillan, London, pp. 245-247, 1981.

[88] Haken, H.: Synergetics - Can it help physiology? In: Haken, H. and Koeppchen, H.P. (Eds.), *Rhythms in Physiological Systems*, Springer-Verlag, Berlin, pp.155-164, 1990.

[89] Henatsch, H.-D., Meyer - Lohmann, J., Windhorst, U. and Schmidt, J.: Differential effects of stimulation of the cat's red nucleus on lumbar alpha motoneurones and their Renshaw cells. *Exp. Brain Res.*, 62: 161-174, 1986.

[90] Hodgkin, A.L.: The local electric changes associated with repetitive action in a non-medullated axon. *J. Physiol.*, 107: 165-181, 1948.

[91] Hodkin, A.L. and Huxley, A.F.: A quantitative description of membrane current and its application to conduction and excitation in nerve. *J. Physiol.* (Lond.), 117: 500-544, 1952.

[92] Hoffmann, P.: Die physiologischen Eigenschaften der Eigenreflexe. *Ergebnisse der Physiologie*, 36: 15-108, 1934.

[93] Holst, E.v.: Die relative Koordination als Phänomen und als Methode zentralnervöser Funktionsanalyse. *Erg. Physiol.*, 42: 228-306, 1939.

[94] Holst, E.v. and Mittelstedt, H.: Das Reafferenzprinzip (Wechselwirkungen zwischen Zentralnervensystem und Peripherie). Naturwissenschaften, 20: 464-476, 1950.

[95] Holst E.v.: Relations between the central nervous system and the peripheral organs. *Brit. J. Anim. Behav.* 2: 89-94, 1954.

[96] Ho, S. and O'Donovan, J.: Regionalization and intersegmental coordination of rhythm-generating networks in the cord of the chick embryo. *J. Neurosci.*, 13: 1354-1371, 1993.

[97] Hullinger, M. and North, J.: Static and dynamic fusimotor interaction and the possibility of multiple pace-maker operating in the cat muscle spindle. *Brain Research*, 173: 21-28, 1979.

[98] Hulliger, M. Zangger, P., Prochazka, A. and Appenteng, K.: Simulations reveal large variations in fusimotor action in normal cats: "fusimotor set." In: Boyd, I.A. and Gladden, M.H. (Eds.), *The Muscle Spindle*, Stockton Press, New York, pp. 311-315, 1985.

[99] Hultborn, H. and Pierrot-Deseilligny, E.: Changes in recurrent inhibition during voluntary soleus contraction in man studied by an H-reflex technique. *J. Physiol.* (Lon.), 297: 229-251, 1979.

[100] Hunt C.C.: Relation of function to diameter in afferent fibres of muscle nerve. *J. Gen. Physiol.* 38: 117-131, 1954.

[101] Hunt, C.C. and McIntyre, A.K.: An analysis of fibre diameter and receptor characteristics of myelinated cutaneous afferent fibres in the cat. *J. Physiol.*, 153: 99-112, 1960.

[102] Hunt, C.C.: The physiology of muscle receptors. In: C.C. Hunt (Ed.), *Handbook of Sensory Physiology, Vol. III/2, Muscle Receptors*, Springer, New York, pp. 191-234, 1974.

[103] Hursh J.B.: Conduction velocity and diameter of nerve fibres. *Amer. J. Physiol.* 127: 131-139, 1939.

[104] Hume, R.I. and Getting, P.A.: Motor organization of Tritonia swimming. III. Contribution of intrinsic membrane properties to flexion neuron burst formation. *J. Neurophysiol.*, 47: 91-102, 1982.

[105] Huxley, A.F. and Stämpfli, R.: Evidence for saltatoric conduction in peripheral myelinated nerve fibres. *J. Physiol.*, 108: 315-339, 1949.

[106] Iggo, A.: Tension receptors in the stomach and the urinary bladder. *J. Physiol.*, 128: 593-607, 1955.

[107] Ingbert, C.: An enumeration of the medullated nerve fibres in the dorsal roots of the spinal nerves of man. *J.Comp. Neurol.*, 13: 53, 1903.

[108] Jacobson, R.D. and Hollyday, M.: A behavioral and electromyographic study of walking in the chick. *J. Neurophysiol.*, 48: 238-256, 1982.

[109] Jacobson, R.D. and Hollyday, M.: Electrically evoked walking and fictive locomotion in the chick. *J. Neurophysiol.*, 48: 257-270, 1982.

[110] Janco, J., Stavrovský, I and Pavlásek, J.: Modeling of neuronal functions: a neuron-like element with the graded response. *Computer and Artificial Intelligence*, 13: 603-620, 1994.

[111] Jankowska, E. Padel, Y. and Zarzecki, P.: Crossed disynaptic inhibition of sacral motoneurons. *J. Physiol.*, 285: 425- 444, 1978.

[112] Jaspers, H.H. and Andrews, H.L.: Brain potentials and voluntary muscle activity in man. *J. Neurophysiol.*, 1: 87-100, 1938.

[113] Jennekens, F.G.I., Tomlinson, B.E. and Walton, J.N.: Data on the distribution of fibre types in five human limb muscles. An autopsy study. *J. Neurol. Sci.*, 14: 245-257, 1971.

[114] Johannson, R.S. and Vallbo, A.B.: Skin mechanoreceptors in the human hand: An inference of some population properties. In: Zotterman, Y. (ed.), *Sensory functions of the skin in primates*. Pergamon Press, Oxford, pp.171-184, 1976.

[115] Johannson, R.S.: Tactile sensibility in the human hand: Receptive field characteristics of mechanoreceptive units in the glabrous skin area. *J. Physiol.*, 281: 101-123, 1978.

[116] Jordan, L.M., Pratt, C.A. and Menzies, J.E.: Intraspinal mechanisms for the control of locomotion. *Adv. Physiol. Sci.,* 1: 183-185, 1981.

[117] Jung, R.: Physiologische Untersuchungen über den Parkinsontremor und andere Zitterformen beim Menschen. *Zeitschrift für die gesamte Neurologie und Psychiatrie*, 173: 263-332, 1941.

[118] Kaufmann S.A.: *The Origin of Order: Self-Organization and Selection in Evolution.* Oxford University Press, New York, 1993.

[119] Kandel, E.R., Schwartz, J.H. and Jessel, T.M.: *Principles of Neural Science*, Prentice-Hall International Inc., London, pp. 581-584, 1991.

[120] Kelso J.A.S. (1995): Dynamic Patterns. *The Self-Organization of Brain and Behavior.* MIT Press, Cambridge.

[121] Kernell, D.: Organisation properties of spinal motoneurons and motor units. *Progress in Brain Research,* 64: 21-30, 1986.

[122] Knibestöl, M.: Stimulus-response functions of rapidly adapting mechanoreceptors in the human glabrous skin area. *J. Physiol.*, 232: 427-452, 1973.

[123] Knibestöl, M.: Stimulus-response functions of slowly adapting mechanoreceptors in the human glabrous skin area. *J. Physiol.*, 245: 63-80, 1975.

[124] Koepchen, H.P.: Physiology of rhythms and control systems: An integrative approach. In: Haken, H. and Koepchen, H.P. (Eds.), *Rhythms in Physiological Systems*, Springer-Verlag, Berlin, pp. 3-20, 1990.

[125] Kopell, N.: Modeling CPGs: A robust approach. In: Grillner, S., Stein, P.S.G., Stuart, D., Forssberg, H. and Herman, R. (Eds.), *Neurobiology of Vertebrate Locomotion*, Macmillan, London, pp. 383-385, 1986.

[126] Kopell, N.: Toward a theory of modelling central pattern generators. In: Cohen, A.H., Rossignol, S. and Grillner, S. (Eds.), *Neural Control of Rhythmic Movements in Vertebrates*, John Wiley and Sons, New York, pp. 369-413, 1988.

[127] Kucera, J.: Reconstruction of the nerve supply to a human muscle spindle. *Neurosci. Lett.*, 63: 180-184, 1986.

[128] Kugelberg, E. and Hagbarth, K.E.: Spinal mechanisms of the abdominal and erector spinae skin reflexes. *Brain*, 81: 290-304, 1958.

[129] Lambertz, M., Rittweger, J.E., Kluge, W. and Langhorst, P.: The common brain stem system (CBS): Dynamic properties of the functional organization and their influence on signal processing. Wiss. Z. Humboldt-Univ. *Berlin Richtung Med.*, 41: 83-84, 1992.

[130] Langhorst, P., Schultz, G. and Lambertz, M.: Integrative control mechanisms for cardiorespiratory and somatomotor functions in the reticular formation of the lower brain stem. In: Grossman, P., Janssen, K.H. and Vaitl, D. (Eds.), *Cardiorespiratory and Cardiosomatic Psychophysiology*, Plenum Publishing Corporation, pp. 9-39, 1986.

[131] Lawson, J.: Pelvic anatomy, I. Pelvic floor muscles. *Ann. R. Coll. Surg. Engl.*, 54: 244-252, 1974.

[132] Lloyd D.P.C., Chang H.T.: Afferent fibres in muscle nerves. *J. Neurophysiol.* 11: 199-207, 1948.

[133] Loughlin, M.: *Muscle Biopsy, A Laboratory Investigation*, Butterworth-Heinemann, 1993.

[134] Mackel, R.: Segmental and descending control of the external urethral and anal sphincters in the cat. *J. Physiol.*, 294: 105-122, 1979.

[135] Matthews, P.B.C.: Human long-latency stretch reflexes. A new role for the secondary ending of the muscle spindle. In: Barnes, W.J.P. and Gladden, M.H. (Eds.), *Feedback and Motor Control*, Croom-Helm, Kent, 1985.

[136] McMahon, S.B. and Morrison, J.F.B.: Two groups of spinal interneurons that respond to stimulation of the abdominal viscera of the cat. *J. Physiol.*, 322: 21-34, 1982.

[137] McMahon, S.B., Morrison, J.F.B. and Spillane, K.: An electrophysiological study of somatic and visceral convergence in the reflex control of the external sphincters. *J. Physiol.*, 328: 379-387, 1982.

[138] Martin G.F., Cabana T. and Humbertson A.O., Jr.: Evidence for collateral innervation of the cervical and lumbar enlargements of the spinal cord by single reticular and raphe

neurons. Studies using fluorescent markers in double-labelling experiments on the North American opossum. *Neurosci. Lett.*, 24: 1-6, 1981.

[139] Matthews, P.B.C.: Evidence from the use of vibration that the human long-latency stretch reflex depends upon spindle secondary afferents. *J. Physiol.*, 348: 383-415, 1984.

[140] Matthews, P.B.C.: Human long-latency stretch reflexes. A new role for the secondary ending of the muscle spindle. In: Barnes, W.J.P. and Gladden, M.H. (Eds.), *Feedback and Motor Control*, Croom-Helm, Kent, 1985.

[141] Meech, R.W.: Calcium influx induces a post-tetanic hyperpolarization in Aplysia neurons. *Comp. Biochem. Physiol. A.*, 48: 387-395, 1974.

[142] Meech, R.W.: Membrane potential oscillations in molluscan "burster" neurons. *J. Exp. Biol.*, 81: 93-112, 1979.

[143] Miledi, R., Parker, I. and Schalow, G.: Calcium transients in normal and denervated slow muscle fibres of the frog. *J. Physiol.* (Lond.), 318: 191-206, 1981.

[144] Mishelevich, D.J.: Repetitive firing to current in cat motoneurons as a function of muscle unit twitch type. *Experimental Neurology*, 25: 401-409, 1969.

[145] Nadelhaft, I., Ropollo, J., Morgan, C. and De Groat, W.C.: Parasympathetic preganglionic neurons and visceral primary afferents in monkey sacral spinal cord revealed following application of horseradish peroxydase to pelvic nerve. J. *Compar. Neurol.*, 216: 36-52, 1983.

[146] Nesmeyanova, T.N.: Physiological aspects in the restoration of motor functions of spinal cord injury patients. In: Kao, C.C., Bunge, R.P. and Reier, P.J. (Eds.), *Spinal cord reconstruction*. Raven Press, New York, pp. 475-480, 1983.

[147] Noble, D.: Ionic mechanisms of rhythmic firing. In: Roberts, A. and Roberts, B.L. (Eds.), *Neural Origin of Rhythmic Movements*, Cambridge University Press, Cambridge, pp. 1-28, 1983.

[148] Orlovsky, G.N. and Feldman, A.G.: Classification of lumbosacral neurons according to their discharge patterns during evoked locomotion. *Neurophysiology* (Kiev), 4: 410-417, 1972.

[149] Paintal, A.S.: The influence of diameter of medullated nerve fibres of cats on the rising and falling phases of the spike and its recovery. *J. Physiol.*, 184: 791-811, 1966.

[150] Passatore, G.M., Fillipi, G.M. and Grassi, G.: Cervical sympathetic nerve stimulation can induce an intrafusal muscle fibre contraction in the rabbit. In: Boyd, I.A. and Gladden, M.H. (Eds.), *The Muscle Spindle*, Stockton Press, New York, pp. 221-226, 1985.

[151] Pavlásek, J.: The binding problem in population neurodynamics: A network model for stimulus-specific coherent oscillations. *Gen. Physiol. Biophys*. 17: 1-18, 1998.

[152] Pavlidis, T.: *Biological Oscillators: Their Mathematical Analysis*, Academic Press, New York, 1973.

[153] Peacock, M.J. and Wolstencroft, J.H.: Projections of reticular neurons to dorsal regions of the spinal cord in the cat. *Neurosci. Lett.*, 2: 7-11, 1976.

[154] Pearson, K.G., Collins, D.F.: Reversal of the influence of group Ia afferents from plantaris on activity in medial gastrocnemius muscle during locomotor activity. *J. Neurophysiol.*, 70: 1009-1017, 1993.

[155] Peiper A.(1961): *Cerebral Functions in Infancy and Childhood.* Consultants Bureau, New York

[156] Percy, J.P., Swash, M., Neill, M.E. and Parks, A.G.: Electrophysiological study of motor nerve supply of pelvic floor. *Lancet*, 3:16-17, 1980.

[157] Percy, J.: The neural control of the bowel. In: Swash, M. and Kennard, C. (Eds.), *Scientific Basis of Clinical Neurology*, Edinburgh, pp. 503-510, 1985.

[158] Perret, C. and Cabelguen, J.M.: Main characteristics of the hindlimb locomotor cycle in the decorticate cat with special reference to bifunctional muscles. *Brain Res.*, 187: 333-352, 1980.

[159] Quick, D.C., Kennedy, W.R. and Poppele, R.E.: Anatomical evidence for multiple sources of action potentials in the afferent fibres of muscle spindles. *Neuroscience*, 5: 109-115, 1980.

[160] Rand, R.H., Cohen, A.H. and Holmes, P.J.: Systems of coupled oscillators as models of central pattern generators. In: Cohen, A.H., Rossignol, S. and Grillner, S. (Eds.), *Neural Control of Rhythmic Movements in Vertebrates*, John Wiley and Sons, New York, pp. 333-367, 1988.

[161] Rasminski, M., Kearney, R.E., Aguayo, A.J. and Bray, G.M.: Conduction of nerve impulses in spinal nerve roots and peripheral nerves of dystrophic mice. *Brain Research,* 143: 71-85, 1978.

[162] Renshaw, B.: Influence of discharge of motoneurons upon excitation of neighboring motoneurons. *J. Neurophysiol.*, 4: 167-183, 1941.

[163] Richmond, F.J.R., Stacey, M.J., Bakker, G.J. and Bakker, D.A.: Gaps in spindle physiology: Why the tandem spindle? In I.A. Boyd and M.H. Gladden (Eds.), *The Muscle Spindle*, Stockton Press, New York, pp. 75-81, 1985.

[164] Rittweger, J.E., Kluge, W., Langhorst, P. and Lambertz, M.: Rhythmical properties of the common brain stem system (CBS), Wiss. Z. Humboldt-Univ. *Berlin Richtung Med.*, 41: 91-94, 1992.

[165] Rosenfalck, P.: Intra- and extracellular potential fields of active nerve and muscle fibres. *Acta Physiol. Scand.*, 75 (Suppl. 321): 1-168, 1969.

[166] Ross, H.-G. and Wittstock, C.: Renshaw cell activity in falling cats. *J. Physiol.* (Lond.), 388: 83-99, 1987.

[167] Rushton, W.A.H.: Resistance artifact in action potential measurements. *J. Physiol.*, 104: 19P-20P, 1945.

[168] Schalow, G. and Schmidt, H.: Local development of action potentials in slow muscle fibres after complete or partial denervation, *Proc. R. Soc.*, B 203: 445-457, 1979.

[169] Schalow, G.: The problem of cauda equina nerve root identification, *Zentralblatt Neurochir.*, 46: 322-330, 1985.

[170] Schalow, G., Aho, A. and Lang, G.: Nerve fibre counts for an intercostal nerve to cauda equina nerve root anastomosis, *Zentralblatt Chir.*, 112: 457-461, 1987.

[171] Schalow, G. and Lang, G.: Electrodiagnosis of human dorsal sacral nerve roots by recording afferent and efferent extracellular action potentials, *Neurosurg. Rev.*, 12: 223-232, 1989.

[172] Schalow, G.: Efferent and afferent fibres in human sacral ventral nerve roots: basic research and clinical implications. Electromyogr. *Clin. Neurophysiol.*, 29: 33-53, 1989.

[173] Schalow, G.: Feeder arteries, longitudinal trunks and arterial anastomoses of the lower human spinal cord. *Zent. bl. Neurochir.*, 51: 181-184, 1990.

[174] Schalow, G.: Coactivity of secondary spindle afferents and α_2, α_3, γ_1 and γ_2-motoneurons innervating anal and urinary bladder sphincters in humans. *Electromyogr. Clin. Neurophysiol.* 31: 223-241, 1991.

[175] Schalow, G.: Conduction velocities and nerve fibre diameters of touch, pain, urinary bladder and anal canal afferents and α and γ-motoneurons in human dorsal sacral nerve roots. *Electromyogr. Clin. Neurophysiol.*, 31: 265-296, 1991.

[176] Schalow, G.: Oscillatory firing of single human sphincteric α_2 and α_3-motoneurons reflexly activated for the continence of urinary bladder and rectum. Restoration of bladder function in paraplegia. *Electromyogr. Clin. Neurophysiol.*, 31: 323-355, 1991.

[177] Schalow, G., Aho, A. and Lang, G.: Microanatomy and number of nerve fibres of the lower intercostal nerves with respect to a nerve anastomosis. Donor nerve analysis. I (IV), *Electromyogr. Clin. Neurophysiol.*, 32: 171-185, 1992.

[178] Schalow, G.: Number of fibres and fibre diameter distributions of nerves innervating the urinary bladder in humans. Acceptor nerve analysis. II (IV), *Electromyogr. Clin. Neurophysiol.*, 32: 187-196, 1992.

[179] Schalow, G. and Barth, H.: Single-fibre action potential recording from the nerve to the musculus obliquus externus abdominis following pinprick in humans. III (IV), *Electromyogr. Clin. Neurophysiol.*, 32: 197-205, 1992.

[180] Schalow, G.: Impulse pattern, innervation density and two-point discrimination of skin and mucosal afferents in humans. Consideration for a sensory reinnervation of urinary bladder and anal canal in spinal cord lesion. IV (IV), *Electromyogr. Clin. Neurophysiol.*, 32: 259-285, 1992.

[181] Schalow, G. and Barth, H.: Group conduction velocities and nerve fibre diameters of α and γ-motoneurons from lower sacral nerve roots of the dog and humans. *Gen. Physiol. Biophys.,* 11: 85-99, 1992.

[182] Schalow, G.: Recruitment within the groups of γ_1, α_2 and α_3-motoneurons in dogs and humans following bladder and anal catheter pulling. *Gen. Physiol. Biophys.*, 11: 101-121, 1992.

[183] Schalow, G.: Ventral root afferent and dorsal root efferent fibres in dog and human lower sacral nerve roots. *Gen. Physiol. Biophys.*, 11: 123-131, 1992.

[184] Schalow, G and Wattig, B: Recruitment of α and γ-motoneurons in rats, dogs and humans. *Electromyogr. Clin. Neurophysiol.*, 33: 387-400, 1993.

[185] Schalow, G.: Recruitment of motoneurons in the occasional firing mode in paraplegics. *Electromyogr. Clin. Neurophysiol.*, 33: 401-408, 1993.

[186] Schalow, G: Spinal oscillators in man under normal and pathologic conditions. *Electromyogr. Clin. Neurophysiol.*, 33: 409-426, 1993.

[187] Schalow, G.: Impulse patterns of single encoding sites of human secondary muscle spindle afferents. *Electromyogr. Clin. Neurophysiol.* 33: 451-464, 1993.

[188] Schalow, G.: Phase-correlated adequate afferent action potentials as a drive of human spinal oscillators, *Electromyogr. Clin. Neurophysiol.*, 33: 465-476, 1993.

[189] Schalow, G.: Action potential patterns of intrafusal γ and parasympathetic motoneurons, secondary muscle spindle afferents and an oscillatory firing α_2-motoneuron, and the phase relations among them in humans. *Electromyogr. Clin. Neurophysiol.*, 33: 477-503, 1993.

[190] Schalow, G. and Zäch, G.A.: Nerve compound action potentials analysed with the simultaneously measured single-fibre action potentials in humans. *Electromyogr. clin. Neurophysiol.*, 34: 451-465, 1994.

[191] Schalow, G., Bersch, U., Göcking, K. and Zäch, G.A.: Interspike intervals of secondary muscle spindle and urinary bladder afferents in relation to the oscillation periods of sacral spinal oscillators for continence in man. *Physiol. Res.*, 43: 7-18, 1994.

[192] Schalow, G., Zäch, G.A. and Warzok, R.: Classification of human peripheral nerve fibre groups by conduction velocity and nerve fibre diameter is preserved following spinal cord lesion. *J. Auton. Nerv. Syst.*, 52: 125-150, 1995.

[193] Schalow, G., Bersch, U., Göcking, K. and Zäch, G.A.: Detrusor-sphincteric dyssynergia in paraplegics compared with the synergia in a brain-dead human by using the single-fibre action potential recording method. *J. Auton. Nerv. Syst.*, 52: 151-180, 1995.

[194] Schalow, G., Bersch, U., Michel, D. and Koch, H.G.: Detrusor-sphincteric dyssynergia in humans with spinal cord lesions may be caused by a loss of stable phase relations between and within oscillatory firing neuronal networks. *J. Auton. Nerv. Syst.*, 52: 181-202, 1995.

[195] Schalow, G., Bersch, U., Zäch, G.A. and Warzock, R.: Classification, oscillatory and alternating oscillatory firing of α_1 (FF) and α_2-motoneurons (FR) in patients with spinal cord lesion. *Gen. Physiol. Biophys.* 15, Suppl.1: 5-56, 1996.

[196] Schalow, G. and Zäch, G.A.: Mono and polysynaptic drive of oscillatory firing α_1 (FF) and α_2-motoneurons (FR) in a patient with spinal cord lesion. *Gen. Physiol. Biophys.* 15, Suppl.1: 57-74, 1996.

[197] Schalow, G. and Zäch, G.A.: Reflex stimulation of continuously oscillatory firing α and γ-motoneurons in patients with spinal cord lesion. *Gen. Physiol. Biophys.* 15, Suppl.1: 75-93, 1996.

[198] Schalow, G. and Zäch, G.A.: External loops of human premotor spinal oscillators identified by simultaneous measurements of interspike intervals and phase relations. *Gen. Physiol. Biophys.* 15, Suppl.1: 95-119, 1996.

[199] Schalow, G., Blanc, Y., Jeltsch, W. and Zäch, G.A.: Electromyographic identification of spinal oscillator patterns and recouplings in a patient with incomplete spinal cord lesion: oscillator-formation training as a method to improve motor activities. *Gen. Physiol. Biophys.* 15, Suppl.1: 121-220, 1996.

[200] Schalow, G., Pääsuke, M., Ereline, J. and Gapeyeva, H.: Coordination dynamics in Parkinson's disease patients and healthy subjects quantified by the coordination dynamics recording method and sEMG. *Electromyogr. Clin. Neurophysiol.* 43: 473-485, 2003.

[201] Schalow, G., Pääsuke, M., Ereline, J. and Gapeyeva, H.: Improvement in Parkinson's disease patients achieved by coordination dynamics therapy. *Electromyogr. Clin. Neurophysiol.* 44: 67-73, 2004.

[202] Schalow, G.: Tremor in Parkinson's disease patients can be induced by uncontrolled activation and uninhibited synchronization of α_2-motoneuron firing to which α_1-motoneuron firing synchronizes. *Electromyogr. Clin. Neurophysiol.*, 45: 393-406, 2005.

[203] Schalow, G.: Tapering of human nerve fibres. *Gen. Physiol. Biophys.* 24: 427-448, 2005.

[204] Scheibel, M.E. and Scheibel, A.B.: Structural substrates for integrative patterns in the brain stem reticular core. In Jaspers, H.H. et al. (Eds.), *Formation of the Brain Stem Reticular Core*, Little and Brown, Boston, pp. 31-55, 1958.

[205] Schmidt, R.A., Bruschini, H. and Tanagho, E.A.: Urinary bladder and sphincter responses to stimulation of dorsal and ventral sacral roots. *Investigative Urology*, 16: 300-304, 1979.

[206] Schmidt, R.F. and Thews, G.: *Physiologie des Menschen*. Springer-Verlag, Berlin, 1985 (English edition, Human Physiology, Springer-Verlag, Berlin, 1983).

[207] Schmidt, H. and Schalow, G.: In-vitro and in-vivo experiments on the incorporation of Na- channels into the membrane of frog slow muscle fibres. In: *Membrane Permeability: Experiments and models*, A.H. Bretaa (Ed.), Techsearch Adelaide, 1983.

[208] Schröder, J.M.: *Pathologie der Muskulatur*, Springer-Vcrlag, Berlin, 1982.

[209] Sears, T.A.: Efferent discharges in alpha and fusimotor fibres of intercostal nerves of the cat. *J. Physiol.*, 174: 295-315, 1964.

[210] Sears, T. A.: The slow potentials of thoracic respiratory motoneurons and their relation to breathing. *J. Physiol.*, 175: 404-424, 1964.

[211] Sears, T.A.: Some properties and reflex connection of respiratory motoneurons of the cat's thoracic spinal cord. *J. Physiol.*, 175: 386-403, 1964.

[212] Sears, T.A.: The respiratory motoneuron: Integration at spinal level. In: Howell and Cambell (eds.). Breathlessness, International Symposium, Manchaster 1965. Proceedings Oxford, pp. 33-47, 1966.

[213] Schulz, G., Lambertz, M., Schulz, B., Langhorst, P. and Krienke, B.: Reticular formation of the lower brain stem. A common system for cardio-respiratory and somatomotor function: discharge pattern of neighboring neurons influenced by somatosensory afferents. *J. Auton. Nerv. Syst.*, 9: 433-449, 1985.

[214] Schuster, M.M: Motor action of rectum and anal sphincters in continence and defecation. In: Heidel, W. (Ed.), *Handbook of Physiology, Section 6, Alimentary canal, Vol. IV, Mobility*. American Physiol. Society, Washington D.C., pp. 2121-2146, 1968.

[215] Seitz R.J., Huang Y., Knorr U., Tellmann L., Herzog H., Freund H.-J.: Large-scale plasticity of the human motor cortex. *NeuroReport*, 6: 742-744, 1995.

[216] Selverston, A.I.: Neuronal mechanisms for rhythmic motor pattern generation in a simple system. In: Herman, R., Grillner, S., Stein, P. and Stuart, D. (Eds.). *Neural control of locomotion*. Plenum Press, New York, pp. 377-400, 1976.

[217] Selverston, A.I., Miller, J.P. and Wahdepuhl, M.: Cooperative mechanisms for the production of rhythmic movements. In: Roberts, A. and Roberts, B.L. (Eds.), *Neural Origin of Rhythmic Movements*, Cambridge University Press, Cambridge, pp. 55-88, 1983.

[218] Sherrington, C.S.: On the question of whether any fibres of the mammalian dorsal (afferent) spinal roots are of intraspinal origin. *J. Physiol.* 21: 209-212, 1897.

[219] Sherrington, C.S.: Nervous rhythm arising from rivalry of antagonistic reflexes: Reflex stepping as outcome of double reciprocal innervation. *Proc. R. Soc. Lond.* (Biol.), 86: 233-261, 1913.

[220] Singer, W.: Synchronization of cortical activity and its putative role in information processing and learning. *Annu. Rev. Physiol.*, 55: 349-374, 1993.

[221] Sperry, R.W.: The problem of central nervous system reorganisation and muscle transposition. *Quart. Rev. Biol.,* 20: 311-369, 1945.

[222] Sperry, R.W.: Effect of crossing nerves to antagonistic limb muscles in the monkey. *Arch. Neurol. Psychiat.* (Chicago), 58: 452-473, 1947.

[223] Stalberg, E.: Propagation velocity in human muscle fibres in situ. *Acta Physiol. Scand.* 70, Suppl. 287: 1-112, 1966.

[224] Stein, R.B. and Lee, R.G.: Tremor and clonus. In: V.B. Brooks (Ed.), *Handbook of Physiology, The Nervous System, Vol. II*, Washington DC, American Physiol. Society, pp. 325-343, 1981.

[225] Swash, M. and Fox, K.P.: Adrenergic innervation of baboon and human muscle spindles. In: Boyd, I.A. and Gladden, M.H. (Eds.), *The Muscle Spindle*, Stockton Press, New York, pp. 121-126, 1985.

[226] Taylor, A., Durbaba, R. and Rogers, J.F.: The classification of afferents from muscle spindles of the jaw-closing muscles of the cat. *J. Physiol.*, 456: 609-828, 1992.

[227] Thiermann, E.: Der sacrale Zugang in der Urologie. *Z. Urol.*: 46: 808-816, 1986.

[228] Todd, J.K.: Afferent impulses in the pudendal nerves of the cat. *Q. Il. Exp. Physiol.*, 49: 258-261, 1964.

[229] Tsukahara, N.: Synaptic plasticity in the mammalian central nervous system. *Ann.Rev. Neurosci.*, 4: 351-379, 1981.

[230] Vallbo, A.B.: Afferent discharge from human muscle spindles in non-contrating muscles. Steady state impulse frequency as a function of joint angle. *Acta Physiol. Scand.*, 90: 303-318, 1974.

[231] Vallbo, A.B., Hagbarth, K.-E., Torebjörk, H.E. and Wallin, B.G.: Somatosensory propioceptive, and sympathetic activity in human peripheral nerves. *Physiol. Rev.*, 59: 919-957, 1979.

[232] Vallbo, A.B.: A critique of the papers by Loeb and Hoffer and Prochazka and Wand. In: Taylor, A. and Prochazka, A. (Eds.), *Muscle Receptors and Movement*, McMillan, London, pp. 249-255, 1981.

[233] von der Malsburg, C.: Nervous structures with dynamical links. *Ber. Bunsen-Ges. Phys. Chem.*, 89: 703-710, 1985.

[234] Wallèn, P. and William, T.L.: Fictive locomotion in the lamprey spinal cord in vitro compared with swimming in the intact and spinal animal. *J. Physiol.*, 347: 225-239, 1984.

[235] Waltzer, R. and Martin, G.F.: Collateralization of reticulospinal axons from the nucleus reticularis gigantocellularis to the cerebellum and diencephalon. A double-labelling study in the rat. *Brain Res.*, 293: 153-158, 1984.

[236] Weiss, P. and Brown, P.F.: Electromyographic study on coordination of leg movements in poliomyelitis patients with transposed tendons. *Proc. Soc. Exper. Biol. Med.*, 48: 384-387, 1941.

[237] Weiss, S. et al.: Multipotent CNS stem cells are present in the adult mammalian spinal cord and ventricular neuroaxis. *J. Neurosci.* 16 (23): 7599-7609, 1996.

[238] Weiss, S.: Adult forebrain and spinal cord stem cells - Basic biology and implications for cell replacement and regeneration. *Neurol. Arg. Suppl* 1: 56, 1997.

[239] Wernig, A., Müller, S.: Laufband locomotion with body weight support improved walking in persons with severe spinal cord injuries. *Paraplegia*, 30: 229-238, 1992.

[240] Wernig, A. and Herrera, A.A.: Sprouting and remodelling at the nerve-muscle junction. *Prog. Neurobiol.*, 27: 251-291, 1986.

[241] Wernig, A., Anzil, A.P. and Bieser, A.: Formation and regression of synaptic contacts in the adult muscle. In: H. Flohr and W. Precht (Eds.), *Lesion-Induced Neuronal Plasticity in Sensorymotor Systems*, Springer, Berlin, pp. 38-50, 1981.

[242] Wilson, V.J.: Distribution of recurrent facilitation and inhibition in cat spinal cord. *J. Neurophysiol.,* 23: 144-183, 1960.

[243] Windhorst, U.: *How Brain-like is the Spinal Cord*. Springer-Verlag, Berlin, 1988.

[244] Winter, D.L.: Receptor characteristics and conduction velocities in bladder afferents. *J. Psychiat. Res.*, 8: 225-235, 1971.

Chapter IV

Translational Medicine

1. Rat Model to Study Neural Network Organization

1.1. Recruitment in the Occasional Firing Mode of α_1-Motoneurons in Rat in Relation to the Recruitment of α_2 and α_3-Motoneurons in Human and Dog

Since there are no or only a few α_1-motoneurons in the lower human sacral roots, and α_1-motoneurons in the comparable roots of the dog are only activated a little with anal and bladder catheter pulling (Figure 56 of Chapter III), activation changes of rat α_1-motoneurons innervating hind limb muscles will be added here to show the recruitment of α_1-motoneurons in relation to the recruitment of α_2 and α_3-motoneurons.

1.2. Recording of Single-Nerve Fibre Action Potentials from the Rat Sural Nerve

The rat sural nerve is long enough to allow recording of single-nerve fibre action potentials (APs) with the same set of recording electrodes as was used for the measurements in man and dogs. The rat sural nerve also contains motoneurons that innervate foot muscles. Pinpricking of the foot under light ether anaesthesia will activate the escape reaction of the rat, and α_1-motoneurons can thus be expected to be strongly activated. However, peripheral nerves outside the spinal canal have an epineurium envelope. The epineurium soaked with saline solution will shunt extracellular APs and should therefore be removed. Since this is not completely possible, peripheral nerve recordings will not achieve the quality of the nerve root recordings. In addition, the rat sural nerve shows variable fasciculation, which makes the dissection more complicated if one tries to choose a single fascicle. But, as will be shown below, the recording quality from the rat nervus suralis was still sufficient to yield reasonably good distribution histograms of conduction velocity frequencies for afferents and efferents to calibrate the motoneuron velocity distributions and to identify the different motoneuron groups with their velocity ranges in a representative velocity histogram; measurement of recruitment changes in α_1, α_2 and α_3-motoneurons classes is thus possible.

1.3. Classification Scheme of Peripheral Nerve Fibres in Rat

The layout for the recording from the rat nervus suralis is shown in Figure 1A.

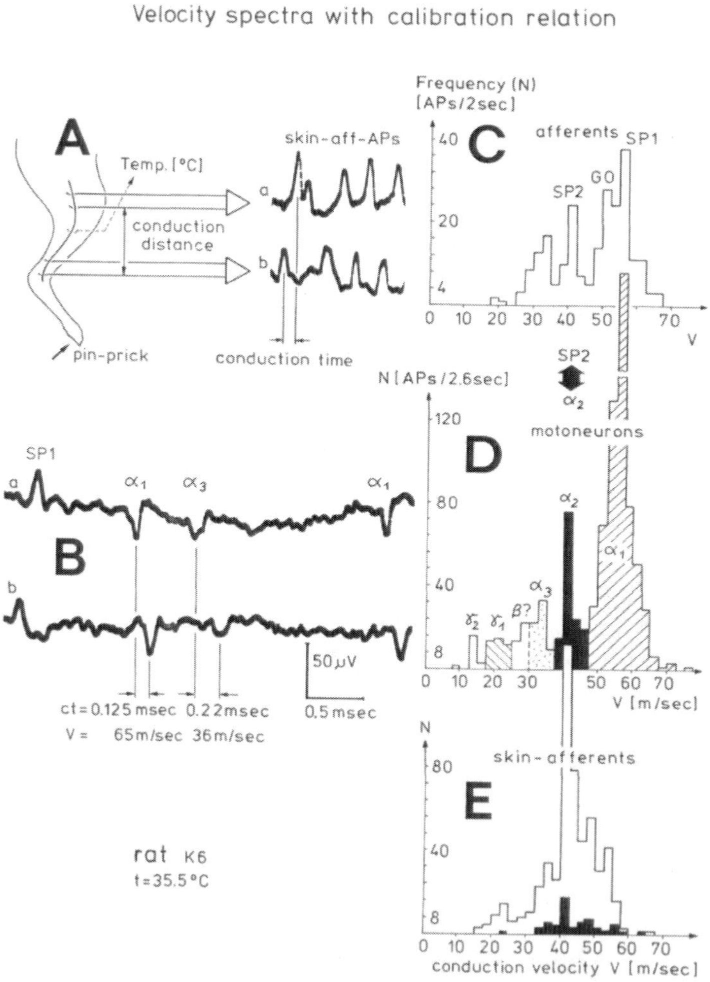

Figure 1. Recording of single-nerve fibre action potentials (APs) from the rat nervus suralis, which contains also nerve fibres innervating muscle fibres (nervus plantaris lateralis). A. Stimulation and recording layout. B. Sweep piece of rat K6, 110 days old; conduction times (ct) and conduction velocities (v) are indicated. C, D, E. Distribution histograms of conduction velocities for muscle nerve afferents (C), extra- and intrafusal motoneurons (D), and skin afferents (E). Identified distribution peaks are labelled according to the respective group they represent (C, D). Note that α_2-motoneurons and secondary spindle afferents (SP2) conduct at the same velocity (calibration relation). Velocity ranges of α_3, α_2, and α_1-motoneurons are represented by dotted, filled or crosshatched columns. In D, APs from 13 sweeps of 0.2 sec duration are summarized. The frequency ratios of AP occurrence with pinprick are: $\alpha_1 : \alpha_2 : \alpha_3 = 637{:}133{:}65 \approx 10{:}2{:}1$. E. Histogram of added velocities of skin afferents from 12 rats, each stimulated two times (the open column). The velocity histogram for rat K6 following one pinprick is represented by the filled columns.

The temperature of the nervus suralis was directly measured with a thermo-element and kept constant by heating with infrared light. A recording from rat K6 afferents and efferents is shown in Figure 1B. Two conduction times (ct) and conduction velocities (v) are indicated. In Figure 1B, the motoneuron classes of origin of the APs were identified based on the velocity ranges of the individual motoneuron classes (Figures 1C,D) and the respective conduction velocities.

Figure 1C shows the conduction velocity distribution histogram for the afferents with no stimulation, and Figure 1D shows the representative conduction velocity frequency distribution histogram for the motoneurons. In accordance with earlier measurements in humans [8], the nerve fibre classes that the peaks most likely belong to were identified. The calibration relation of the conduction velocity frequency distribution histograms, namely that the secondary muscle spindle afferents conduct at about the same velocity as α_2-motoneurons [8], seems to hold also for the rat nervus suralis (Figure 1D). The limits of the velocity ranges for the different motoneuron classes were drawn into Figure 1D. Only the velocity ranges of α_1 and α_2-motoneurons are safe. Since the large α_1-motoneuron peak does not show the typical distribution form very well, it might be that the rat α_1-motoneurons split into different subgroups similarly as α_1-motoneurons in the dog [10]. By comparing Figure 1D with Figure 53A of Chapter III, it is obvious that the rat α_1-motoneurons were much more activated than those of the dog.

1.4. Difference between the Hairless Skin Afferents between Rats and Humans

The distribution histograms of conduction velocity frequencies for afferents activated by skin stimulation (light pinprick) are shown in Figure 1E. The filled columns refer to rat K6 and were obtained with 1 pinprick. The open columns refer to 12 comparable rat measurements and two pinpricks each. It can be seen from Figure 1E that a single stimulation in rat K6 already yielded the typical velocity distribution. By comparing Figure 1E with Figure 1D, it can be seen that the largest skin-afferent velocity peak falls together with the α_2-motoneuron peak.

This is different from the distributions obtained from man. There, the skin-afferents in the largest peak (probably innervating Pacinian corpuscles) conducted about 10% slower than α_2-motoneuron [8]. Therefore, the afferents innervating the hairless skin of the rat hind limb are probably not directly comparable to those innervating the hairless skin in humans or the peripheral, and the central branch of the afferent cells are conducting with different velocities. Also the α_2-motoneuron may have a reduced diameter in the periphery.

1.5. Nerve Fibre Diameters of the Normal Rat Sural Nerve

Single-nerve fibre action potential recordings from whole nerves or roots give detailed information about the afferent and efferent impulse traffic of the activated nerve fibres; but they provide no information about fibres that are not activated.

A morphometric analysis of the nerve fibre diameters or similar values of the nerve cross section, on the other hand, provide information about all fibres, even though sometimes only little information can be extracted from it. Nevertheless, it says exactly what nerve fibre population was recorded from.

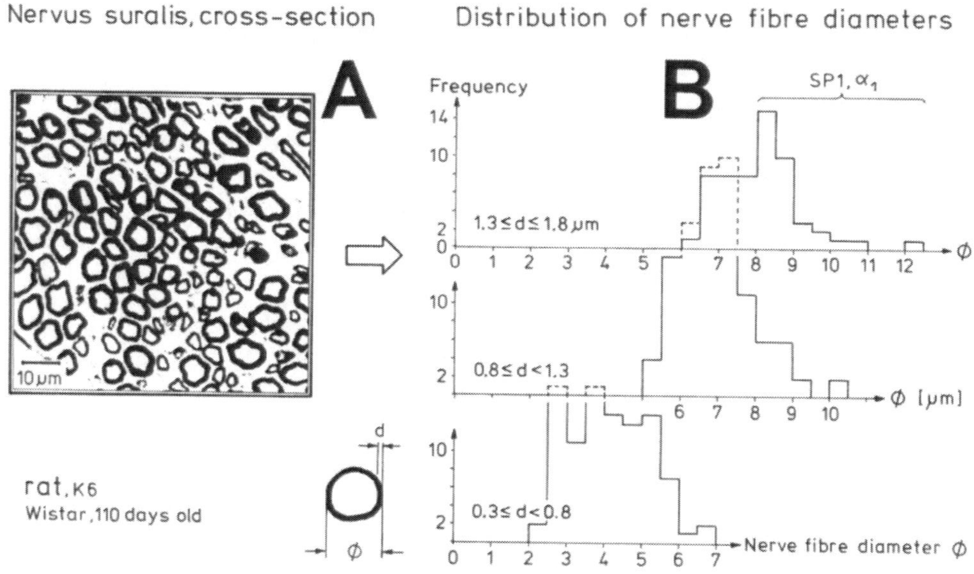

Figure 2. A. Portion of the cross section of the nervus suralis measured electro-physiologically (Figure 1). Thionin acridine-orange staining, contrasted with a copying-machine. B. Distribution histograms of nerve fibre diameter frequencies from the light microscope cross section, partly shown in A. The ranges of the myelin sheath thicknesses, d, of fibres are given at each column. The distribution peak of primary spindle afferents (SP1) and α_1-motoneurons is indicated.

The nerve fibre diameter spectra of the dog sacral root were discussed in Ref.10; those of the human dorsal root were analysed in Figures 19 through 29 of Chapter III. The nerve fibre diameter spectra with respect to different myelin sheath thicknesses of the rat K6 nervus suralis are shown in Figure 2.

A representative part of the cross section is shown in Figure 2A. The corresponding fibre diameter distribution histograms are shown in Figure 2B. Most likely, the marked peak in the class of thick myelin sheaths contains primary spindle afferents (SP1) and α_1-motoneurons. No further peaks could be identified. This was to be expected, since besides skin afferents, the rat nervus suralis contains also nerve fibres (afferents plus efferents) to foot muscles, so that the distribution histograms of nerve fibre diameter frequencies will consist of too many nerve fibre groups to allow the identification of single group peaks, with the exception of groups with the thickest fibres.

Taking 8.5μm as the peak value of the group diameters for α_1-motoneurons and primary muscle spindle afferents (SP1) (Figure 2B) and using 56m/s as the peak value of the group conduction velocity for α_1-motoneurons and SP1-fibres (Figure 1), a value of 6.6 is obtained for the ratio of conduction velocity and fibre diameter for α_1-motoneurons and SP1-fibres.

1.6. Recruitment of α_1-Motoneurons before α_2 and α_3-Motoneurons According to the Size Principle in Each Motoneuron Group upon Stimulation

Figure 3A shows that α_1-motoneurons in the nervus suralis were recruited during the escape reaction, when the sole of the hind limb was pinpricked.

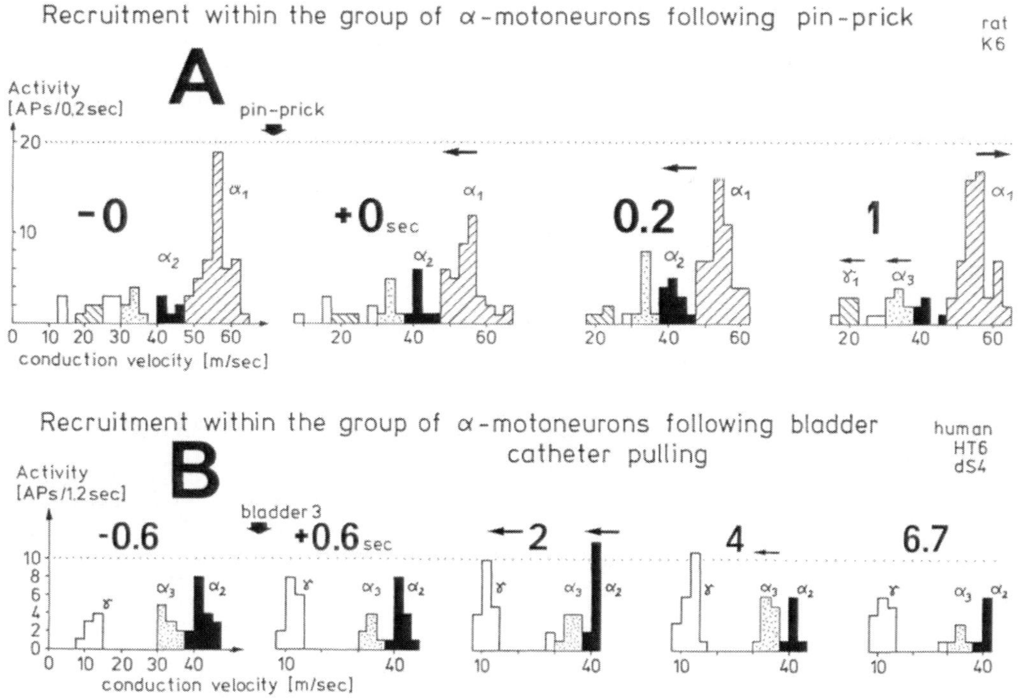

Figure 3. A. Frequency changes of single-nerve fibre conduction velocities within the groups of α_2 and α_3-motoneurons in the rat nervus suralis following pinprick. The velocity ranges are taken from Figure 1D. +0 indicates interval of 0.2 sec following pinprick. Note that α_1-motoneurons are recruited before α_2-motoneurons. B. Frequency changes of single-nerve fibre conduction velocities within the groups of α_2 and γ_1-motoneurons in a human (HT6) sacral nerve root following strong bladder catheter pulling. The group conduction velocity ranges are taken from Figure 53B. -0.6, +0.6, 2, 4 and 6.7sec are the times (half time of the 1.2 sec interval) in seconds before are after stimulation. Note the co-recruitment of γ_1 and α_2-motoneurons, and that α_2-motoneurons are recruited before α_3-motoneurons.

It can be seen that the slowly conducting α_1-motoneurons were recruited directly following stimulation (+0), followed by α_2-motoneuron (0.2sec) and α_3-motoneurons (1sec). This measurement in the rat extends the recruitment order with respect to the different motoneuron groups to α_1-motoneurons; the slowly conducting α_1-motoneurons are activated first, followed by the slowly conducting α_2-motoneuron and then the slowly conducting α_3-motoneurons. The identification of the recruitment of γ_1-motoneurons is not safe. It seems as if they were not co-recruited with α_2-motoneurons. Figure 3B shows that in man, the slowly conducting α_2-motoneuron were recruited (2 sec) before the slowly conducting α_3-motoneurons (4 sec). Maybe, γ_1-motoneurons were in this case co-recruited with α_2-motoneurons. For further data, see Ref. 11. As suggested by the measurements in the dog (Figure 56 of Chapter III) and by earlier results [11], the motoneurons in each group are re-

recruited rhythmically every 3 to 4s according to the conduction velocity (size principle) if the duration of stimulation, i.e., the afferent activity increase, is sufficiently long. The question arises: how is the recruitment of motoneurons regulated if there is no additional afferent input? How is the firing of motoneurons for the so-called muscle tone when the afferent input is non-specific and originates in different parts of the body from different receptors? The problem is rather intricate, since the activity levels of the motoneurons are low, and there is no fixed starting point for the recruitment analysis, so that often the analysed sweeps hit intermediate recruitment stages, if periodic recruitment exists at all.

1.7. Ongoing Recruitment Changes upon No Specific Stimulation (Muscle Tonus)

In the measurement in rats (Figure 4A), α_1-motoneurons were recruited approx. every two sec. Again, a more slowly conducting subgroup seems to be activated repeatedly. Of special interest here is that this measurement was taken before the first pinprick. Further, the depth of the ether anaesthesia was not constant: it strongly weakened from 0s to 3.3s, as reflected in an overall activity increase. During deep ether anaesthesia, no or nearly no efferent activity was recorded following pinprick. Therefore with the weakening of anaesthesia (Figure 4A), efferent activity appeared (upon no stimulation): α_1-motoneurons were recruited repeatedly, according to the size principle. Even though the rat was not stimulated (no catheter was positioned either), there probably was afferent input to the spinal cord from pain receptors from the wound area and with the weakening of anaesthesia; this pain (and other) afferent input was processed in the CNS.

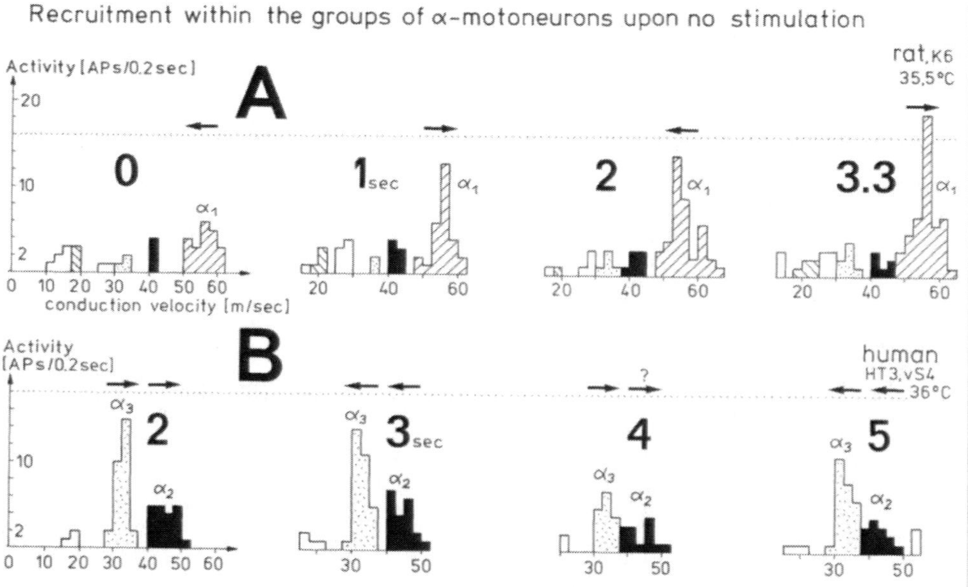

Figure 4. Recruitment changes of α and γ-motoneurons in the rat (A) and in man (B) (HT3) upon no stimulation (no catheter positioned). The velocity ranges in A are taken from Figure 1D. The activity increase in A with time (from 0 to 3.3s) is due to the weakening of ether anesthesia. In B, no histogram is shown to enable identification of the velocity ranges. Other symbols as in previous Figures.

Since the efferent activity of the dorsal S4 root of HT6 was too low in the non-stimulated case, recruitment with no stimulation was measured in another HT (the ventral S4 root of HT3). Figure 4B shows that α_2 and α_3-motoneurons are recruited rhythmically every two sec. The co-recruitment of α_2 and α_3-motoneurons is probably accidental. Since there was no catheter used with HT3, it is concluded that there are also ongoing recruitment changes with no specific stimulation in humans.

1.8. Single-Nerve Fibre Recording from the Sural Nerve Is a Comparable Powerful Research Tool

It has been shown that single-nerve fibre action potentials can also be recorded from the nervus suralis of the rat. Since the rat nervus suralis contains also nerve fibres to foot muscles, distribution histograms of conduction velocity frequencies can separately be constructed from skin afferents, motoneurons and non-skin afferents measured simultaneously. Also, the calibration relation of the velocity distributions is valid: α_2-motoneurons and secondary spindle afferents conduct with the same velocity [8]. Even though the single-nerve fibre action potential recording from the sural nerve provides less information than human root recordings (where no epineurium and nearly no perineurium in ensheathing the root) and in contrast to human ventral roots, the suralis morphometry supplies nearly no information about single peaks of nerve fibre groups, simultaneous analysis of single-nerve fibre action potentials and nerve fibre diameters of the rat sural nerve is a comparable powerful tool: it is cheap and easy to handle. With this animal model, one can study, e.g., regeneration (see below), artificially induced diabetic neuropathies or effects of drug therapy [13].

1.9. Conduction Velocity-Fibre Diameter Conversion Factor for Different Species

By correlating the group conduction velocity for α_1-motoneurons of a 110-day-old Wistar rat (Figure 1) with the group nerve fibre diameter (Figure 2), a conversion factor of 6.6 was calculated for conduction velocity to nerve fibre diameter. This value is roughly in accordance with classical measurements [4,5]. It has been shown in humans [8] and dogs [10], however, that the conversion factor of myelinated fibres is different for different nerve fibre groups [8]. For α_1-motoneurons of 110–day-old Wistar rat at about 36 °C, the conversion factor was 6.6, while being 5.1 in an Alsatian dog (α_{12}-motoneurons) [10] and 4.6 [8] in man. Boyd et al. [2] found a mean ratio of 5.7 for the largest fibres and 4.5 for the slower ones in the cat. In the baboon, the conduction velocity to fibre diameter ratio is about 5.2 for the fastest fibres [6].

1.10. Recruitment Phase

It has been shown that motoneurons were recruited in their group according to the size principle [11]; this means that slowly conducting motoneurons in each group were activated before the faster conducting motoneurons. But as shown previously (see Ref. 11), the motoneurons are recruited again three to four sec later if the afferent input was sufficiently long. As shown in Table 1 of Ref. 11, γ_1-motoneurons (intrafusal) were sometimes also repeatedly recruited every two sec. Upon no (specific) stimulation, α_3-motoneurons (Figure 4B) showed repeated recruitment every two sec. It has to be further clarified what the time periods are for repeated recruitment and whether the repeated recruitment is different with and without stimulation. It can be seen from Figure 4B (no specific stimulation) that α_3-motoneurons show more pronounced recruitment than α_2-motoneurons. This indicates that static motoneurons of the pelvic floor are activated quite strongly, associated with the weight carrying of the pelvic floor. It has been shown in here and previously that the motoneurons in each group are recruited at certain times following stimulation. On the other hand, there is ongoing recruitment in the occasional firing mode with no specific stimulation. Therefore, with strong and perhaps specific stimulation the motoneurons in each group are activated according to the size principle. Mostly, new activation is not in phase with the ongoing one, but the recruitment followed the phase of new activation. If the spinal circuitries responsible for the periodic recruitment of the different motoneuron groups are the same for strong specific activation and low-level non-specific activation, then with strong stimulation, the recruitment phase is changed, and the amplitude of activation declines with time to the low level of the non-specific afferent input. If there exists a separate circuitry responsible for the high activation and another one for the low-level activation (in the occasional firing mode), then there will be a break when the recruitment changes from high- to low-level activation. This speculation may not be hypothetical, because if α_2-motoneurons are strongly activated over a longer interval, they start to fire in the oscillatory mode with impulse trains at about every 160ms (9). But when the specific afferent input is terminated, another circuitry arrangement takes over, which activates α_2-motoneurons in the occasional firing mode at a periodicity of two to four sec. It remains unclear whether, with the recruitment in the occasional firing mode, all motoneurons of a certain motoneuron group are activated or not. Motoneurons can simultaneously belong to different motoneuron pools. Are both motoneuron pools recruited in phase? The observation of repeated recruitment in the sacral root of the dog (Chapter III), where leg, tail and sphincter functions are contained, point towards the same recruitment phase. The circuitry in the CNS may serve several motoneuron pools with the same recruitment speed, according to different afferent inputs to the different motoneuron pools; however, their activation patterns are quite different.

2. Rat Model to Study Regeneration

2.1. Steps in Regeneration

Enhancement of the regenerative capacity of the nervous system has to be achieved step by step with the final goal in mind to treat the patient. Efforts to develop a drug for the regeneration of the spinal cord have been underway since World War II with no benefit for the patient so far. Stem cell therapy has also not been successful until now. Coordination dynamics therapy, on the other hand, can partly repair the human nervous system (Chapter II). The rate of repair is very low. The rate of development of the human nervous system is also very low (17 years) in comparison to that of the rat (one year). The repair of the human nervous system may, therefore, also be much slower than that of the rat. To study the enhancement of regeneration capacity in animals, one needs a powerful animal model to measure the grade of regeneration functionally and morphologically. Such a model has been developed above for the peripheral nervous system and will be introduced here.

2.2. Regeneration in Rat Sural Nerve Quantified by the Nerve Fibre Diameter (Structure) and the Conduction Velocity (Function)

Single-nerve fibre action potentials were recorded from the sural nerve of a normal 110-day-old rat, with two pairs of wire electrodes, and conduction velocity frequency distribution histograms were constructed (Figure 1). The same recording equipment was used as for the recordings of single-nerve fibre action potentials in humans, dogs and frogs (see below). After recording, the sural nerve was removed, fixated, embedded, stained, and morphologically analysed, and histograms of nerve fibre diameter distribution were constructed (Figure 2). The same procedure was used with another rat (110 days old, Wistar) with the sural nerve crushed together with the nervus ischiadicus 60 days previously. A sweep piece of the recording from the regenerated nerve is shown in Figure 5A.

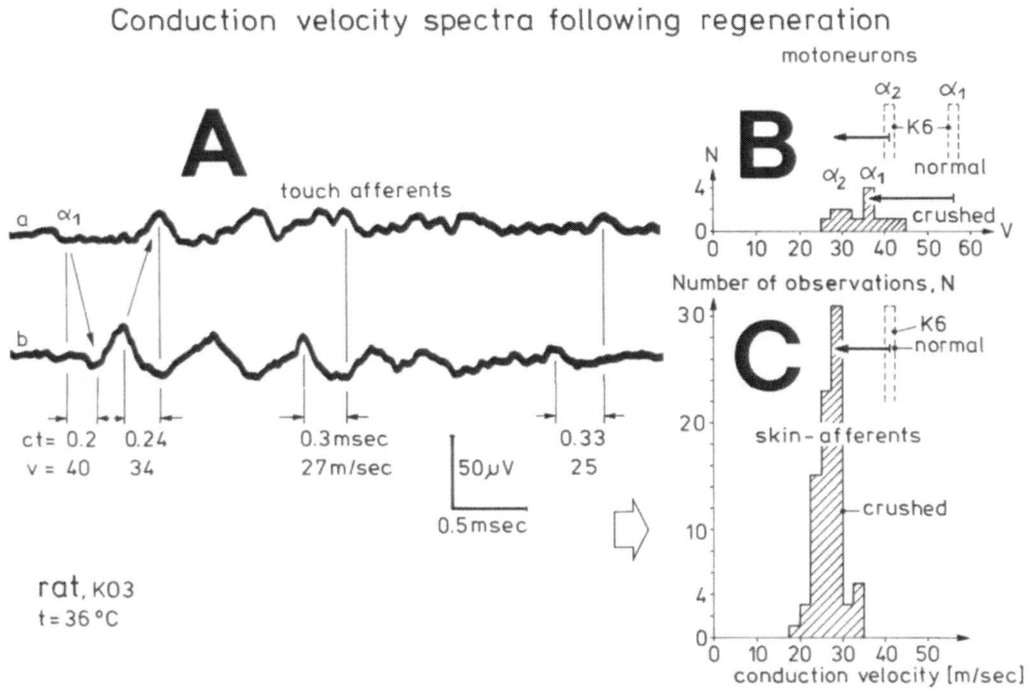

Figure 5. Conduction velocities of the regenerating sural nerve fibres of a Wistar rat, 110days old; nerve crushing 60 days previously. A. Recording of single-nerve fibre action potentials (APs). $\alpha_1 = \alpha_1$-motoneuron AP, ct = conduction time, v = conduction velocity. B, C. Conduction velocity distribution histograms of skin afferents (C) and α-motoneurons (B). Crosshatched histograms from rat KO3 from the regenerating nerve fibres. The dashed histograms were taken for comparison from normal rat K6 (Figure 1). The arrows indicate the shift of the velocity peaks due to crush and regeneration. Ether anesthesia.

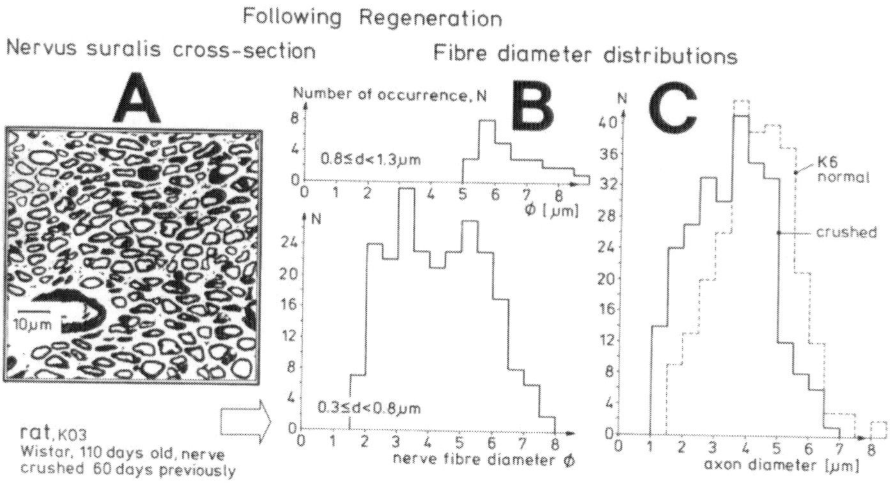

Figure 6. Nerve fibre diameters of regenerating fibres of the sural nerve of a rat, same as in Figure 5. A. Part of the cross section. B. nerve fibre diameter frequency distribution histograms with the myelin sheath thickness range d. C. Axon diameter frequency distribution histogram; solid line = regenerated rat, dashed line = normal rat for comparison.

In comparison with Figure 1A, APs from the regenerated fibres (Figure 5A) were of longer duration and were conducted more slowly. The regenerated fibres were not as matured as the normal ones. A comparison of the conduction velocity frequency distribution histograms of motoneurons and skin afferents of normal and regenerated fibres are shown in Figure 5B,C. The regenerated fibres conducted by 13 to 19m/s more slowly. The morphometry of the regenerated sural nerve is shown in Figure 6A,B. In comparison to Figure 2, it can be seen that both the myelin sheaths and the axons are thinner. A comparison of the axon diameters of the normal and the crushed fibres of the sural nerve is shown in Figure 6C. The regenerated α_1-motoneurons (FF) had a peak group nerve fibre diameter of 5.75μm (Figure 6B) and conducted with a peak group conduction velocity of 36.5m/s. The velocity-diameter relation of the regenerated α_1-motoneurons equals $V/Ø = 36.5/5.75 = 6.3$. In the normal rat, the conversion factor was 6.6 at 36°C. The structure-function coupling for α_1-motoneurons is roughly preserved in the rat during regeneration. If a regenerating nerve fibre is more matured, the nerve diameters and the conduction velocities increase. Regeneration can, therefore, be measured simultaneously structurally and functionally.

2.3. Enhancement of Regeneration Speed in Peripheral Nerve upon Drug Administration

Administering to a third group of rats, which had their nerves crushed previously, a drug that speeds up the process of regeneration, the regenerating fibres will have larger nerve fibre diameters and show higher conduction velocities than fibres in the group receiving no drug, but still smaller diameters and lower velocities than the fibres in the normal group. It has been shown in rats that the daily administration of 3.0mg/kg body weight uridine monophosphate and 2.5mg/kg cytidine monophosphate (Trommsdorf, Alsdorf, Germany) enhanced regeneration by about 20% as estimated based on nerve fibre diameters and conduction velocities [13] (clinical application and side effects were not tested). The enhancement of the regeneration speed in the peripheral nervous system is of importance, since nerve fibres in human regenerate at a rate of 1 to 2mm/day, so that often distal muscles are already atrophied before regenerating fibres reach the muscle. Even though urinary bladder muscles do not undergo atrophy following spinal cord injury because of the inner plexus and probably reflex arcs, a speeding up of the regeneration is useful, since the patient has not to wait as long for the first signs of regeneration, which can be expected to appear in the order of one to three years after the surgery.

2.4. Measurements of Conduction Velocity and Nerve Fibre Diameter in Nerve Roots Can Be Used for CNS Regeneration Quantification

This animal model can also be used for regeneration experiments in the CNS. Measurements of conduction velocity and nerve fibre diameter can also be performed on nerve roots, if the roots are sufficiently long. Electrophysiological measurements can even be performed much better because of the missing of the epineurium and the perineurium, and the structure-function relationship can be better estimated, since ventral root diameter distributions with the different α and γ-peaks can be used to calibrate diameter distributions. By crushing the root directly at the spinal cord, one can crush the PNS-CNS transition zone, since in some animals, this zone protrudes into the root [1, 3]. Response measurements can be used to demonstrate whether afferent nerve fibres have grown into the CNS. With small parts of the spinal cord in the conus medullaris range crushed, at first steps close to the root exit, regeneration in the CNS can be measured in the roots by measuring conduction velocities, nerve fibre diameters and stimulation responses. By progressive crushing from the root towards the PNS-CNS transition zone to the spinal cord and by measuring regeneration in the root with and without the application of drugs [12], it may be possible to find drugs that improve regeneration at the different levels (root, transition zone, CNS). For quick screenings of drugs, the quickly performable electrophysiological measurements can be used only. For large and expensive animals like dogs, pigs and monkeys, it is possible to measure regeneration stages several times on the same animal if one is restricted to electrophysiological measurements. Apart from the laminectomy, the nervous system is only little damaged with each measurement. Humans tolerate repeated laminectomies quite well.

2.5. Indication for High Structural and Functional Accuracy in the Nervous System

The new classification scheme in which group conduction velocities and group nerve fibre diameters (peak values) are correlated gives a more accurate basis for the exploration of the circuitries of the human nervous system and its functions. The power of this peripheral nervous system scheme rests upon the accuracy of nerve fibre group characterization based on conduction velocities and nerve fibre diameters. So far, it seems that the internal consistency of the nerve fibre groups and their activities, measured in the peripheral nervous system, is still higher than the accuracy of measuring it. Since this classification scheme seems partly to be preserved following regeneration, a rather exact measuring of the stage of regeneration is possible, which is a precondition when regeneration enhancement is studied.

2.6. The Human Sacral Spinal Cord and Cauda Equina Offer a Unique Anatomy for Studying the Human CNS

In humans, there is a unique anatomical situation in the lower sacral range. First, bladder functions are not intermingled with tail functions in the conus medullaris and in the roots leading to it, as, e.g., in rats and dogs. Second, lower sacral nerve roots are thin, long (Figure 12 of Chapter III), have a good blood supply (Figures 12,14,15 of Chapter III), and afferents and efferents are partly mixed in dorsal and ventral roots, which allows detailed measurements of nervous system functioning for diagnostic and research reasons. Third, there is plenty of space for operating procedures in the lower spinal canal because of the ascensus of the human spinal cord.

2.7. CNS Repair Strategy Difference between Animal and Human

Since there are no reports about specific regeneration upon drug application in the human CNS, most likely one has to consider necessary tract fibre numbers, mismatch and functional aspects for regeneration in the CNS. To enhance regeneration in rats by movement-based learning seems difficult. How can a rat turn on the special coordination dynamics therapy device, jump on springboard, follow instructions or activate certain neural subnetworks by intention? There seems to be a principle difference between CNS regeneration in rats and humans. The rat CNS is in tendency more hardwired, whereas the human CNS is more variable. If this is so, then the regeneration in rats has to be improved primarily by enhancing nerve fibre growing, whereas in the human CNS, the repair has to be enhanced by learning. The learning capacity of the human CNS is outstanding among other species. Through learning, nearly every neural subnetwork can be reached for repair, which means that every human nervous system (healthy or damaged) can be improved in its functioning by learning, and since the human nervous system is involved in many functions of the body, movement-based learning can be used for the improvement of many diseases. The dream of the repair of the human CNS is not to get stem cell therapy working but to influence the epigenetic mechanisms by (movement-based) learning to enhance the rate of repair.

3. Frog Model for Teaching and Research: The Power of Electrophysiology

3.1. Intracellular Recording of Action Potentials from Frog Muscle Fibres

It has been shown in Chapter I that by recording intracellular (action) potentials from muscle fibres upon direct muscle fibre and nerve stimulation, details about innervation changes of target muscle cells during development and regeneration were obtained in the frog. Muscle target cells changed their cell properties according to the innervating nerve cells (motoneurons).

Since following partial denervation parts of the muscle cell change their properties (Figure 86 of Chapter III), it seems possible that several different kinds of neurons can control different parts of a single muscle or nerve cells. A tremendous complexity of structural neural network organisation could, therefore, exist and could contribute to the functional complexity of neural network pattern organisation stimulated by intention and receptor input from the periphery. This structural and functional complexity could change following CNS damage and repair.

3.2. The Frog Slow Muscle Cell Function May Help to Understand the Functional Complexity of the Human Muscle Spindle

In Chapter III, it was analysed that the human muscle spindle is a receptor that is stimulated by changes of muscle length (changes of the periphery) and by the CNS via the efferent innervation. Since the efferent innervation of nuclear bag and nuclear chain fibres of the muscle spindle includes different kinds of γ-motoneurons and autonomic neurons (sympathicus and/or parasympathicus), CNS neuronal network organisation can specifically activate certain muscle spindles.

Also, since the primary and secondary muscle spindle afferents (SP1, SP2) may project partly into the networks they are driven by, the muscle spindle is a very important receptor for the control of movements, continence and other functions. The frog's slow muscle fibre with its complex neural control has been used to try to understand the functioning of the intrafusal muscle fibres.

As has been shown in Chapters I and III and will be shown in Chapter V, electrophysiology is a powerful tool to explore nervous system functioning of animals and humans. In Chapter II, it was shown that the combination of the System Theory of Pattern Formation and Electrophysiology lead to repair treatment of the human CNS.

3.3. Extracellular Recording of Single-Nerve Fibre Action Potentials from Thin Peripheral Nerves of the Frog

Since human nerve roots have no epineurium, single-nerve fibre action potentials (APs) could be recorded from un-dissected nerve roots and systems functions explored. Since the frog's peripheral nerve has nearly no epineurium, single-nerve fibre APs can be easily recorded from thin un-dissected nerves [7].

Such recording from the frog is shown in Figure 7A,B. The relation between conduction velocity and AP amplitude can be seen in Figure 7A and seems to be similar as in human (Figure 7D). Conduction velocity distributions can also be constructed (Figure 7C). Skin afferent activity can nicely be recorded upon the stimulation of the skin (Figure 7B). Since the AP recording quality is high, a classification scheme for the frog, similar to that in humans, can be constructed and nervous system functions explored. But since the understanding of the functioning of the human nervous system is more important for the repair of the human nervous system than that of the frog, the functioning of the human nervous system will be the focus.

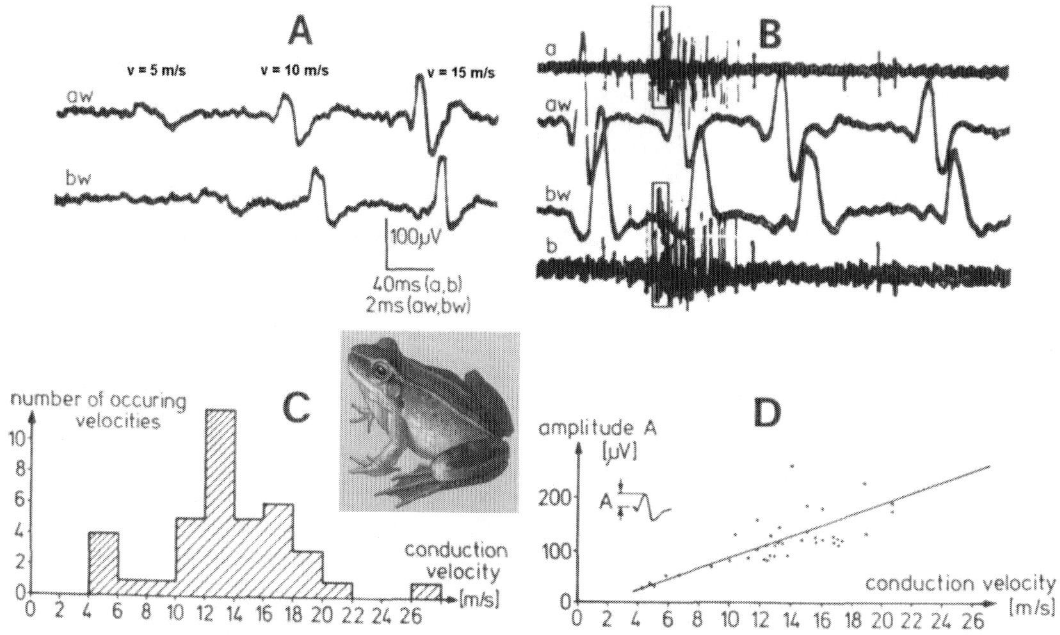

Figure 7. Single-nerve fibre action potential (AP) amplitude in relation to conduction velocity in the frog. A. Recording of three single APs of different conduction velocities and different amplitudes. Note, the AP with the highest conduction velocity (shortest conduction time from one electrode pair (trace a) to the other (trace b)) has the largest amplitude. B. Overall view of the activity increase on the traces a and b upon touching the skin of the frog. On the window traces aw and bw (stretched traces) the waveform of the APs of the touch afferents can nicely be seen. C. Conduction velocity frequency distribution histogram at room temperature of spontaneous and stimulated activity. D. AP amplitude in relation to conduction velocity.

3.4. Human Nervous System Research Starts with Animal Research

To learn about the recording technique of measuring single-nerve fibre APs and the practical problems of recording, I started with frog measurements, went then to other animals and came only then to measurements in human. In a human operation, one has only then a chance to successfully record single-nerve fibre APs, if one is familiar with all the practical problems of recording and has the necessary human anatomy in mind.

3.5. Demonstration of Frog Nervous System Functioning in Peripheral Nerves for Teaching

Since the recording of single-nerve fibre APs from frogs' peripheral nerves is comparably easy, such electrophysiology could be used for teaching medical students. Since 40% of the oxygen consumption in frogs is supplied by oxygen diffusion through the skin, not much care has to be taken with frog preparations, for sufficient oxygen to be supplied, allowing afferent and efferent single-nerve fibre AP conduction in peripheral nerves to be demonstrated to medical students or physicians.

When I demonstrated single-nerve fibre AP recording from the frog at a small conference, the physicians were amazed. Further, when I successfully recorded in the operational theatre the first times single-nerve fibre AP from human nerve roots at Ernst-Moritz-Arndt-University Greifswald (Germany), the surgeons were coming from other operational theatres to see human nervous system functioning in reality. They had never seen before afferent and efferent impulse traffic in human nerves or nerve roots. Some professors were so impressed that they offered me a professorship in Dresden for that new recording technique. It was the time before the fall of the Berlin Wall. After the fall of the Berlin Wall, professors from the former West-Germany took over, and the interest in this powerful recording method was lost.

References

[1] Berthold, C.-H., Carlstedt, T. and Corneliusson, O.: Anatomy of the root at the central-peripheral transition region. In: Thomas, P.J., Lampert, E.H. and Bunge, R.P. (Eds.), *Peripheral Neuropathy*. Vol. 1, pp. 156-170, 1984.

[2] Boyd, L.A. and Davey, M.R.: *Composition of Peripheral Nerve*, pp. 1-57, Livingston, Edinburgh, 1968.

[3] Carlstedt, T.: Regenerating axons from nerve terminals at astrocytes. *Brain Res.,* 347: 188-191, 1985.

[4] Gasser, H.S. and Grundfest, H.: Axon diameter in relation to the spike dimensions and the conduction velocity in mammalian A fibres. *Am. J. Physiol.*, 127: 393-414, 1939.

[5] Hunt, C.C.: Relation of function to diameter in afferent fibres of muscle nerves, *J. Gen. Physiol,* 38: 117-131, 1954.

[6] McLeod, J. G. and Wray, S. R.: Conduction velocity and fibre diameter of the median and ulnar nerves of the baboon. *J. Neurol. Neurosurg. Psychiat.*, 30: 240-247, 1967.

[7] Schalow, G.: Single unit potential amplitude in relation to the conduction velocity if frog and human. *Zent. bl. Neurochir.*, 48: 109-113, 1987.

[8] Schalow, G.: Conduction velocities and nerve fibre diameters of touch, pain, urinary bladder and anal canal afferents and α and γ-motoneurons in human dorsal sacral roots. *Electromyogr. Clin. Neurophysiol*, 31: 265-296, 1991.

[9] Schalow, G.: Oscillatory firing of single human sphincteric α_2 and α_3-motoneurons reflexly activated for the continence of urinary bladder and rectum. Restoration of bladder function in paraplegia. Electromyogr. *Clin. Neurophysiol.*, 31: 323-355, 1991.

[10] Schalow, G. and Barth, H.: Group conduction velocities and nerve fibre diameters of α and γ-motoneurons from lower sacral nerve roots of the dog and humans, *Gen. Physiol. Biophys.*, 11: 85-99, 1992.

[11] Schalow, G.: Recruitment within the groups of γ_1, α_2 and α_3-motoneurons in dogs and humans following bladder and anal catheter pulling. *Gen. Physiol. Biophys.*, 11: 101-121, 1992.

[12] Schnell, L. and Schwab, M.E.: Axonal regeneration in the rat spinal cord produced by an antibody against myelin-associated neurite grows inhibitors. *Nature*, 343: 269-272, 1990.

[13] Wattig, B., Schalow, G., Heydenreich, F., Warzok, R. and Cervos-Navarro, J.: Nucleotides enhance nerve fibre regeneration after peripheral nerve crush damage - Electrophysiologic and morphometric investigation. *Drug Res.*, 42 (II), 9: 1075-1078, 1992.

Chapter V

From the Measured Impaired Phase and Frequency Coordination in and between the Somatic and Parasympathetic Nervous System Divisions in Severe Cervical Spinal Cord Injury to the Repair of the Urinary Bladder Upon Three Years of Movement-Based Learning

1. The Classification and Identification of Human Somatic and Parasympathetic Nerve Fibres Including Urinary Bladder Afferents and Efferents is Preserved Following Spinal Cord Injury

Single-fibre extracellular action potentials were recorded with two pairs of wire electrodes from lower human sacral nerve roots during surgery. The roots from which recordings were taken were used for morphometry. Nerve fibre groups were identified by conduction velocity distribution histograms of single afferent and efferent fibres and partly by nerve fibre diameter distribution histograms. The values of group conduction velocity and group nerve fibre diameter measured in the paraplegics were very similar to those obtained from brain-dead humans and patients with no spinal cord injury. Thus, the classification and identification of nerve fibre groups remained preserved following spinal cord injury. Upon retrograde bladder filling, the urinary bladder stretch and tension receptor afferent activities were increased; on two occasions, they even fired when the bladder was empty. Two reasons are brought forward for insufficient storage volume of the urinary bladder in paraplegics: too high afferent activity of the bladder due to changed receptor field transduction mechanisms and insufficient compliance.

1.1. Summary

1. Single-nerve fibre action potentials (APs) of lower sacral nerve roots were recorded extracellularly with two pairs of wire electrodes during an operation for implanting an anterior root stimulator for bladder control in nine humans with a spinal cord injury and a dyssynergia of the urinary bladder. Roots that were not saved and that were used to record from were later used for morphometry.
2. Nerve fibre groups were identified by conduction velocity distribution histograms of single afferent and efferent fibres and partly by nerve fibre diameter distribution histograms, and correlation analysis was performed. Group conduction velocity values were obtained additionally from compound action potentials (CAPs) evoked by electrical stimulation of nerve roots and the urinary bladder.
3. The group conduction velocities and group nerve fibre diameters had the following pair-values at 35.5 °C:
4. Spindle afferents: SP1(65ms^{-1}/13.1μm), SP2(51/12.1); touch afferents: T1(47/11.1), T2(39/10.1), T3(27/9.1), T4(19/8.1); urinary bladder afferents: S1(41ms^{-1}/-), ST (35/-); α-motoneurons: α_{13}(-/14.4), α_{12}(65ms^{-1}/13.1μm), α_{11}(60?/12.1)[FF], α_2(51/10.3)[FR], α_3(41/8.2)[S]; γ-motoneurons: γ_β(27/7.1), γ_1(21/6.6), γ_{21}(16/5.8), γ_{22}(14/5.1); preganglionic parasympathetic motoneurons: (10ms^{-1}/3.7μm).
5. The values of group conduction velocity and group nerve fibre diameter measured in paraplegics were very similar to those obtained earlier from brain-dead humans and patients with no spinal cord injury. Also, the axon number and the axon density of myelinated fibres of lower sacral nerve roots remain unchanged following spinal cord injury. Thus the classification and identification of nerve fibre groups remained preserved following spinal cord injury. A direct comparison can thus be made of natural impulse patterns of afferent and efferent nerve fibres between paraplegics (pathologic) and brain-dead humans (supraspinal destroyed CNS, in many respects physiologic).
6. When changing the root temperature from 32°C to 35.5°C, the group conduction velocities changed in the following way in one case: SP2: 40m/s (32°C) to 50m/s (35.5%), S1: 31.3 to 40, ST: 25 to 33.8, M: 12.5 to 13.8; α_2: 40 to 50, α_3: 33 to 40. The group conduction velocities showed different temperature dependence apart from SP2 fibres and α_2-motoneurons.
7. Upon retrograde bladder filling the urinary bladder stretch (S1) and tension receptor afferent (ST) activity levels were undulating and increased. As compared to activity levels detected in a brain-dead human, S1 (designates afferents, not cord segment) and ST afferents fired even when the bladder was empty, with an activity level similar to those observed in a brain-dead human with the bladder half filled. Two reasons are brought forward for an insufficient storage volume of the urinary in paraplegics: too high afferent activity of the bladder due to changed receptor field signal transduction mechanisms and insufficient compliance.
8. With the newly developed "coordination dynamics therapy," applied early after spinal cord injury, such complications of bladder functioning can be avoided; the bladder can causally be cured in severe spinal cord injury.

1.2. Introduction

A movement and learning therapy, called coordination dynamics therapy, has been shown to be able to improve central nervous system (CNS) functioning after stroke [65], traumatic brain injury [66,74], hypoxic brain injury [73], cerebellar injury [75,76], spinal cord injury [67,68,84-87], in cerebral palsy [72], and in Parkinson's disease [69,71].

The research in this and the following four parts in basic human neurophysiology and clinical research [88-91,78] is aimed at developing cure for patients with spinal cord injury [35,36,41-44,49] in general and to improve micturition, defecation and sexual function in particular. Recent review articles on the repair of the spinal cord injury concentrated on the regeneration of the spinal cord in animals, therapeutic potential of neural stem cells, and molecular mechanisms [79-82]. These review articles did not include human research. The biggest problem among patients with spinal cord injury, namely how to cure urinary bladder functions, was also not reviewed. This current research project, on the other hand, approaches the problem, to cure spinal cord injury in human, differently. It concentrates on human research on the single neuron level, the integrative level of CNS organization, and the clinical level. It includes data from patients with spinal cord injury and brain-dead humans to compare the pathophysiologic with the physiologic CNS functioning to develop treatment. As will be shown, most of the dysfunctions of the urinary bladder can be understood by using the measured natural impulse patterns of humans and functional anatomy. But to understand how to cure detrusor-sphincteric dyssynergia by changing the stability of the attractor states "detrusor-sphincteric synergia" and "dyssynergia," the "System Theory of Pattern Formation [83] for Repair" [75,86,87], detailed anamneses of patients, and knowledge of movement learning, including learning transfer, has to be used. To improve the coordination between the activations of the somatic and parasympathetic nervous system divisions by stability changes, achieved by coordination dynamics therapy, integrative functions of the CNS have to be understood. Still, the essential step to cure partly spinal cord injury came from the measured natural impulse patterns in brain-dead humans by use of the single-nerve fibre action potential recording method, for which the author has been attacked ethically in the past. The goal of curing urinary bladder functions, following spinal cord injury, is achieved and will be presented in the last part [78]. Destructive operations are no longer needed to improve urinary bladder functions. Bladder problems are not only faced by patients with spinal cord injury, but also by elderly people and young women after the birth of the first child (stress incontinence).

Upon advancing with this basic human research, clinical consequences will be included and discussed. The general clinical consequences will be discussed in the treatment part [78].

This basic human research started from scratch by first clarifying the necessary anatomy [35,36,41,42] and developing a new basic human electrophysiological recording technique, the recording of single-nerve fibre extracellular action potentials (APs) with two pairs of wire electrodes from undissected nerve roots, a technique that can be used for intraoperative diagnosis and research [37,39,44].

The further development of the morphometry of nerves (classification of fibre diameters according to four myelin sheath thickness ranges) made it possible to simultaneously characterize nerve fibre groups by group conduction velocities and group nerve fibre diameters (Figure 1) [39,45,46].

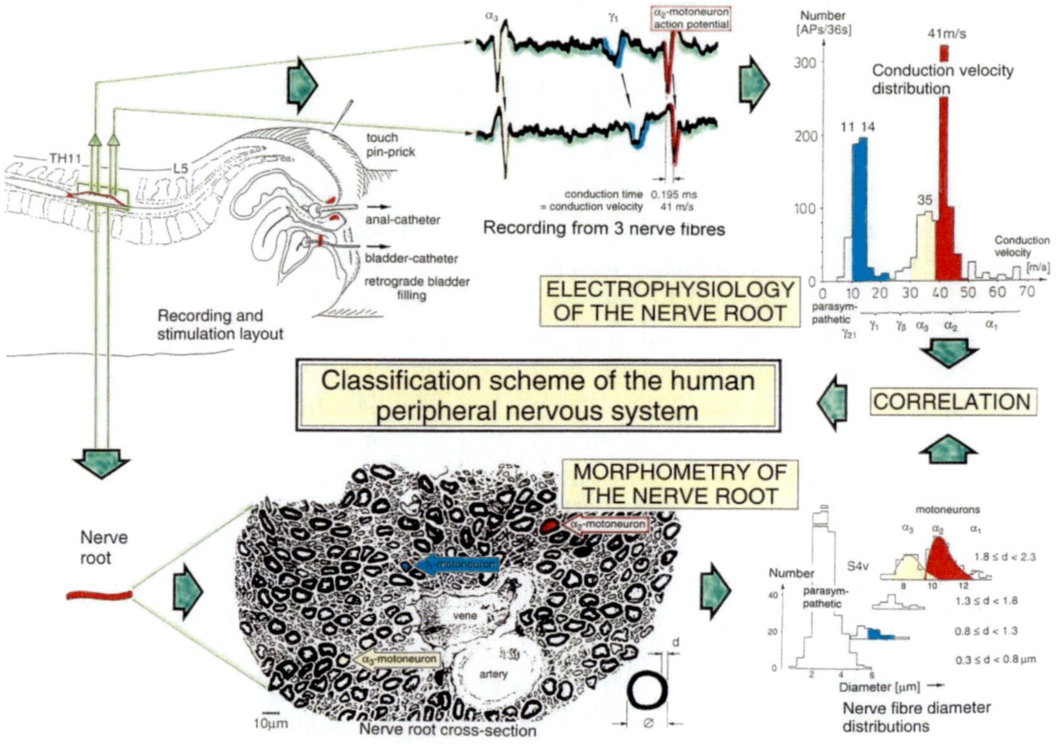

Figure 1. Schematic layout of the classification scheme for the human peripheral nervous system. By recording with two pairs of platinum wire electrodes from a nerve root containing approx. 500 myelinated nerve fibres, a recording is obtained in which three action potentials (APs) from three motoneurons (main AP phase downwards) can be seen. By measuring the conduction times and knowing the electrode pair distance (10 mm), a conduction velocity distribution histogram is constructed in which the nerve fibre groups are characterized by ranges of conduction velocity values and peaks in asymmetrical distributions. After recording, the root was removed, fixated, embedded and stained, light microscope cross sections were prepared and used to measure the mean diameter and the myelin sheath thickness (d). Distributions of nerve fibre diameters were constructed for four different ranges of myelin sheath thicknesses. Nerve fibre groups were characterized by the peak values of asymmetrical distributions. By correlating the peak values of the velocity distributions with those of the diameter distributions obtained from the same root, a classification scheme was constructed of human peripheral nervous system. Brain-dead human HT6.

In this way, an exact classification scheme could be constructed for the human peripheral nervous system. The scheme is a solid basis for human neurophysiology and pathophysiology, even though it is incomplete and holds so far only for nerve fibre diameters larger than approximately 3.5μm. Another basic electrophysiological method for use on humans, the recording of single-nerve fibre extracellular APs with tungsten electrodes [57], made it possible to record impulse patterns of single-nerve fibres; the method however rests on the peripheral nerve fibre classification schemes for animals, which do not apply to humans.

The following up of time-dependent conduction velocity distributions of afferent and efferent fibres allows an analysis of activity level changes of nerve fibre groups. Urinary bladder afferent activity changes upon retrograde bladder filling [40] or motoneuron recruitment in nerve fibre groups are examples of this approach [47,48]. Since it is possible to further distinguish the APs from afferent and efferent nerve fibres and to extract from the summed activity of a fibre bunch the discharge patterns of single fibres, it is possible to analyse receptor properties [44] and functions of the human central nervous system (CNS). An insight into the CNS functions can be obtained from simultaneous impulse patterns of single afferent and efferent fibres and from the phase relations between the impulses of afferents and efferents [49], following natural stimulation. With the discovery of the human spinal oscillators [40], CNS functions can be investigated based on the behaviour of spinal oscillators for long-lasting, rather constant, afferent input.

Since it is possible to record APs from preganglionic parasympathetic axons, to extract parasympathetic activity from secondary muscle spindle afferent fibre activity [49], and to partly correlate urodynamic parameters (bladder pressure) with the activity patterns of secondary muscle spindle afferent fibres [51], a simultaneous analysis of functions of the somatic and the parasympathetic nervous systems is possible. Such a simultaneous analysis allows the study of correlations between the somatic and the parasympathetic divisions occurring in micturition.

1.3. Materials and Methods

Measurements were collected from nine (mean age of the patients = 27 years) humans with spinal cord lesions (Table 1 of 64), a dyssynergia of the urinary bladder, spastic pelvic floor, and spasticity in general. The electrophysiologic measurements were performed during surgery with the new method of recording single-fibre action potentials (APs) extracellularly from undamaged nerve roots or nerve root fascicles. The surgery was performed in order to implant a sacral anterior root stimulator (according to Brindley [7]) to improve bladder control and to save the kidneys. This destructive operation is no longer justified. The strategy of the surgery is to deafferentiate the urinary bladder to increase its storage volume and to stimulate the motor roots afterwards (mainly S3 and S4) to empty the bladder. To deafferentiate the bladder, afferents and efferents in the nerve roots were identified by electrical stimulation and partly by using the single-fibre AP recording method to recognize afferents in certain roots. For the identification and the electrical stimulation after the operation, dorsal and ventral roots were separated or dissected when sticking strongly together. The extirpation of parts of dorsal roots is a part of the surgical procedure. Dorsal root fibres do further conduct 10 to 14 days following cutting. Their functioning is unsuitable for the electrical stimulation afterwards. Root parts were removed mostly between the spine segments L4 and S2; that is about the middle part of the S5 root and a more caudal part for the S2 root. Slight damage may have occurred with the handling of the roots, and slight desiccation may have occurred, when removing the roots. The remaining ventral roots were functioning well after the operation. Cut dorsal roots, from which recording was performed, were fixated and prepared for morphometry (Figure 1). One ventral S4 root was obtained because of an accidental cut. Light anaesthesia (paraplegics feel no pain) was administered with Propofol, not to reduce the AP amplitude too much. Intraoperative recording times were less than 0.5 hours. The heart rate during the operation was around 55/min (para 4, 85), the blood pressure was in the range 95/60 (para 5, 130/85). When with the electrical stimulation of the dorsal roots (for clarifying the representation of bladder afferents and efferents in the sacral roots), the systolic blood pressure increased over 140, further electrical stimulation was halted until the blood pressure decreased again. The respiratory rate was between 10 and

13/min, the O_2 saturation was mainly between 97 and 100%, and the CO_2 expiration was mainly between 4.0 and 4.5%.

1.3.1. Electrophysiology

Single-fibre APs were recorded extracellularly (Figure 1) from nerve roots with two platinum wire electrode pairs (electrode pair distance = 10mm; electrode distance in each pair = 4mm) at two sites, pre-amplified (x1000), filtered (RC-filter, passing frequency range 100Hz-10kHz) and displayed on a digital storage oscilloscope (Vuko Vks 22-16) and also stored using a PCM-processor (Digital Audio Processor PCM-501ES) and a video recorder. The beginning of a touch or a pinpricking was marked with an upward pulse; and the end with a downward pulse on trace "a." These pulses were generated by a markation pulse generator connected to the digital scope, which was switched on and off with a touch sensor working on the basis of resistance changes. Also, the pulling and releasing of anal and bladder catheters were mostly marked with the help of pull-switch connected to the catheters and working in connection with the same markation pulse generator. Trace "a" was the recording from the proximal electrode pair and trace "b" from the distal pair. Conduction velocities of single fibres were calculated from the conduction distance (electrode pair distance) and the respective conduction times, the time needed for an AP to cover the conduction distance (time difference between traces "a" and "b" for a particular AP). APs from afferents and efferents could clearly be distinguished since for the used electrode arrangements the main phase (second phase) from afferent fibres is upwards and that of efferents downwards (Figure 2A). E.g., the AP of a skin afferent fibre reaches a pair of electrodes first as negative and then as positive. According to the electrode setting used, the main phase is upwards. An AP of a motoneuron, coming from the opposite direction, would reach the electrodes in the order positive-negative. The potential changes are therefore opposite, and the main triphasic AP will point downwards. An AP in an afferent fibre reaches first the caudal electrode pair and then the rostral pair, whereas an AP of the efferent fibre reaches first the rostral electrode pair and then the caudal one. The conduction times are therefore also opposite (Figure 2A). If APs of two efferent fibres meet at one electrode pair at the same moment, the APs add up algebraically. The adding up of many fibres' impulses, following electrical stimulation, will give rise to a compound AP [50]. If the APs of an afferent and an efferent fibre meet at the same point of time at one electrode pair, they partly or fully abolish each other (subtraction of the AP amplitudes). However since these APs will not meet each other at the other electrode pair, the afferent and the efferent APs are clearly distinguishable on the trace of the other electrode pair. The conduction velocities of afferent APs were plotted in an afferent conduction velocity distribution histogram (Figure 2B) and those from efferent APs in an efferent velocity histogram (Figure 2C). Histogram classes were half-closed intervals (≤ and <); the left border belongs to a particular class; the right one does not. Group conduction velocity and nerve fibre diameter values were the peak values of asymmetrical distributions. Single-fibre APs were always recorded from whole roots or fascicles. I did not try to tease single fibres. In nerve roots thinner than 0.6mm in diameter (radial decline of AP amplitude due to volume conductance is approx. by a factor of 1/10 per 0.3mm; the root flattens when positioned on the wire electrodes), it is possible to record single-fibre APs from all fibres with a diameter larger than approx. 4μm.

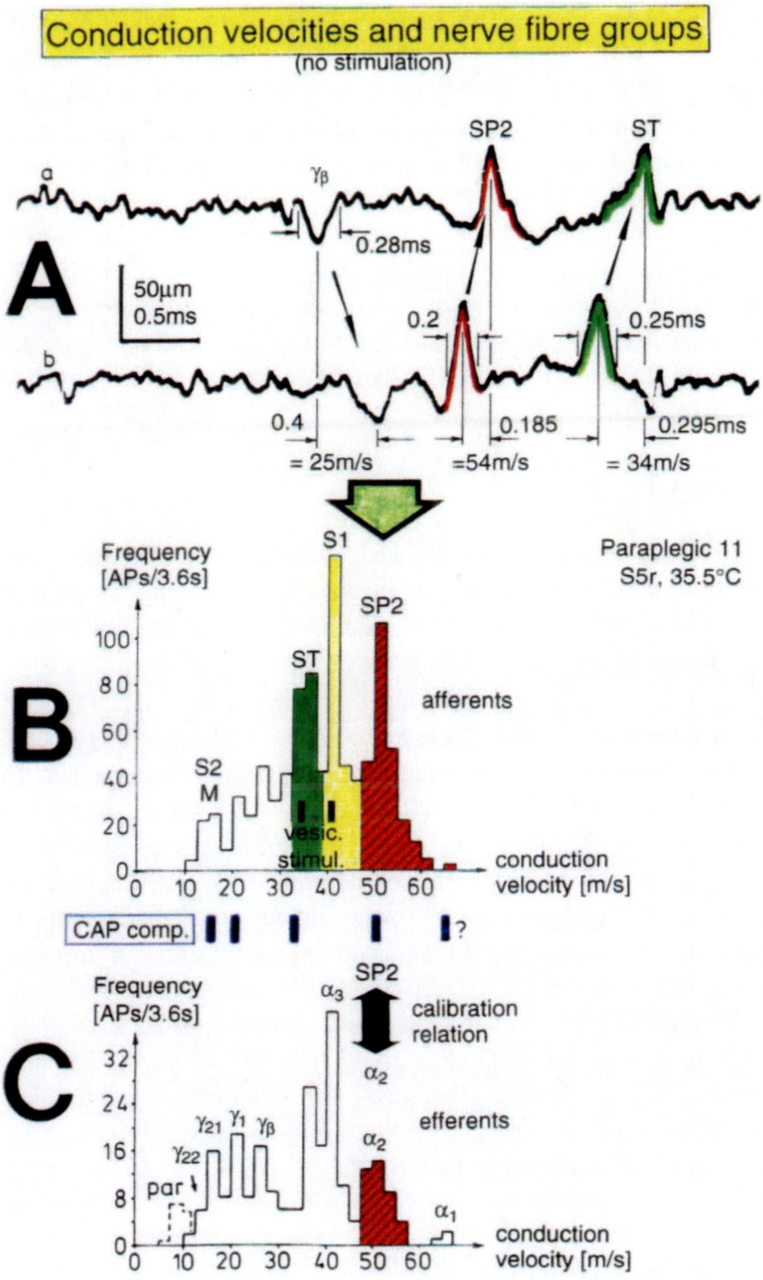

Figure 2. A. Sweep piece of recording; conduction times and corresponding conduction velocities are indicated. Root temperature at recording, 35.5°C. B, C. Conduction velocity distributions of afferents (B) and efferents (C) obtained for a time interval of 3.6s with no additional stimulation. SP2 = secondary muscle spindle afferents, S1 = stretch receptor afferents of bladder, ST = tension receptor afferents, M = mucosal afferents, S2 = afferents responding to fluid movement; $\alpha_1 = \alpha_1$-motoneurons (FF), $\alpha_2 = \alpha_2$-motoneurons (FR), $\alpha_3 = \alpha_3$-motoneurons (S), $\gamma_B = \gamma_B$-motoneurons, $\gamma_1 = \gamma_1$-fusimotors (dynamic), $\gamma_{21} = \gamma_{21}$-fusimotors (static), $\gamma_{22} = \gamma_{22}$-fusimotors (static), par = preganglionic parasympathetic motoneurons. CAP comp. = group conduction velocities obtained from the components of compound action potentials (CAPs). Vesic. stimul. = group conduction velocities of bladder afferents obtained upon electrical intravesical stimulation (see Figures 8, 9). Calibration relation indicates the same peak group conduction velocity of secondary spindle afferents and α_2-motoneurons (crosshatched). Velocity histogram classes ≤ and < (half closed (left) interval).

The radial decline of low-amplitude long-lasting extracellular APs, due to volume conductor embedding, of thin fibres, including preganglionic parasympathetic fibres, is less pronounced than it is for high-amplitude short-lasting APs since the low-frequency sinusoidal coefficients will have large amplitudes in a Fourier-expansion [17,32]. Since nerve roots have no epineurium and nearly no perineurium, the nerve fibres in the roots can easily be damaged when recording with wire electrodes. Pressure and stretch will change the AP waveform or even block conduction (most easily at the node of Ranvier) so that an AP can be recorded from one electrode pair only. Double-peaked APs can occur, probably when a node of Ranvier is blocked. Mechanical stimulation, nerve fibre compression, resistance artefacts and trigger zones may change impulse shape and activity [23,26,33,34,55,58,60,61]. For further literature, see Ref. [38]. Heavily distorted APs from damaged fibres were omitted from the analysis.

1.3.2. Morphometry

Pieces a few cm long were removed from dorsal roots, which were used to record from, fixated for four hours in 4% glutaraldehyde in cacodylate buffer, after fixated in 1% OsO_4 for two hours, and dehydrated and embedded in Araldite according to standard techniques. Pictures of semi-thin sections (approximate depth = 1μm) stained with thionin acridine-orange were taken with a light microscope (x1000). Nerve fibre diameters $\varnothing = 1/2(\varnothing_1 + \varnothing_2)$ ($\varnothing_1$ and $\varnothing_2$ are the larger and the smaller diameter of non-round shaped fibres) and the mean myelin sheath thickness "d" were measured by hand. A shrinkage correction of 8% was allowed. The nerve fibre diameters measured were divided into four ranges of myelin sheath thickness: $0.25\mu m \leq d < 0.8$; $0.8 \leq d < 1.3$; $1.3 \leq d < 1.8$; $1.8 \leq d < 2.5\mu m$. For each class of myelin sheath thickness, a diameter distribution histogram was constructed. In the case of damaged fibres (split myelin sheath), the myelin sheath thickness was measured at the most preserved part. Severely damaged fibres were not taken into consideration. Because of the preference of the authors of even to odd values, neighbouring histogram classes show large variations. This systematic error can be abolished by increasing the width of the histogram classes from 0.25 μm to 0.5 μm. For the diameter distribution $0.25 \leq d < 0.8\mu m$, the higher histogram class was used anyway. Computer-assisted morphometry would eliminate such systematic error but would have less accuracy, since computer programs cannot handle artefacts and altered nerve fibres as if they were normal fibre. Electron micrographs were only used to check the light microscope pictures.

1.3.3. Ethics

Informed consent was obtained from the patients for the recordings to be made. The method of recording single-fibre APs was used for intraoperative diagnosis to identify more safely (e.g., in the case of an anatomical variation) what nerve fibres are contained in what nerve root fascicles. An improvement of the intraoperative diagnosis may make it possible in the future to deafferentiate the bladder more specifically. It would be of great benefit if one could save, e.g., the afferents signalling sexual sensation. This qualified but destructive operation is not justified anymore, because urinary bladder function can be repaired by functional reorganization and regeneration of the human spinal cord [78,86,87].

1.4. Results

1.4.1. Conduction Velocity Distributions for Afferent and Efferent Nerve Fibres

From nine paraplegics (Para 3 to Para 11) with spinal cord lesions for 0.5 to 6 years, single-nerve fibre action potentials (APs) were recorded extracellularly from sacral nerve roots during the surgery, conduction velocities were calculated and group conduction velocities were determined from the peaks in the velocity distribution histograms. The measured peak values of the group conduction velocities for all paraplegics are summarized in Table 1 of 64 (Table 3 of Chapter III). The patients, aged 20 - 37 years, suffered from dyssynergia of the urinary bladder, had a spastic pelvic floor and showed spasticity of different degree in general. They underwent surgery for the implantation of a sacral anterior root stimulator (destructive operation) to enable urinary bladder control (Table 3 of Chapter III). Upon electrical stimulation of the ventral roots (often S3 plus S4), the external bladder sphincter and the detrusor contracted. Since the external bladder sphincter (striated muscle) relaxes earlier than the detrusor (smooth muscle) following the impulse train, urine emerges from the bladder. Repeated stimulation empties the bladder.

From single-nerve fibre AP recordings, like those in Figures 2A and 3D, conduction velocity frequency distribution histograms were constructed for single fibres with no additional stimulation (Figure 2B,C) and measurements obtained during changing of a thin for a thick anal catheter (Figure 3E,F). Following anal catheter stimulation, the peak of preganglionic parasympathetic fibres appeared (Figure 3F), indicating that the parasympathetic division was activated. The activation of the parasympathetic division will be tackled in a following part [51]. The identification of nerve fibre groups from conduction velocity distributions rests upon the assumption that nerve fibres of a certain group conduct at a velocity similar to those obtained from measurements in brain-dead humans (HTs) and in patients with no spinal cord lesion [39]. Further support for the use of these single-fibre conduction velocity distributions with group identification was obtained from electrical nerve root stimulation [50] and electrical intravesical stimulation (Figures 9,10).

Electrical nerve root stimulation with the Brindley stimulator [7] evoked compound action potentials (CAPs), the components of which were conducted at velocities similar to the group conduction velocities obtained from single-fibre AP recordings showing natural impulse patterns. Because of a large number of afferent and efferent nerve fibre groups contained in the roots, only few CAP components were obtained. The fast-conducting groups of secondary muscle spindle afferents (SP2), conducting orthodromically, and the α_2-motoneurons (FR), conducting antidromically if stimulated caudal to the recording electrodes (or vice versa if stimulated rostrally), could nevertheless be extracted from the CAP components. Because of time restrictions during the surgery, electrical stimulation at the thresholds of the different groups was only performed occasionally so that the CAPs often did not split into different components. Also, due to insufficient space, the stimulation electrodes were positioned approx. 2 cm rostral or caudal to the recording electrodes; for strong stimulations, this resulted in a partial fusion of the stimulation artefact and the CAP. Anyhow, the group conduction velocity values obtained from the components of the CAPs have confirmed the group conduction velocity values obtained from single afferent and efferent nerve fibres (Table 3 of Chapter III).

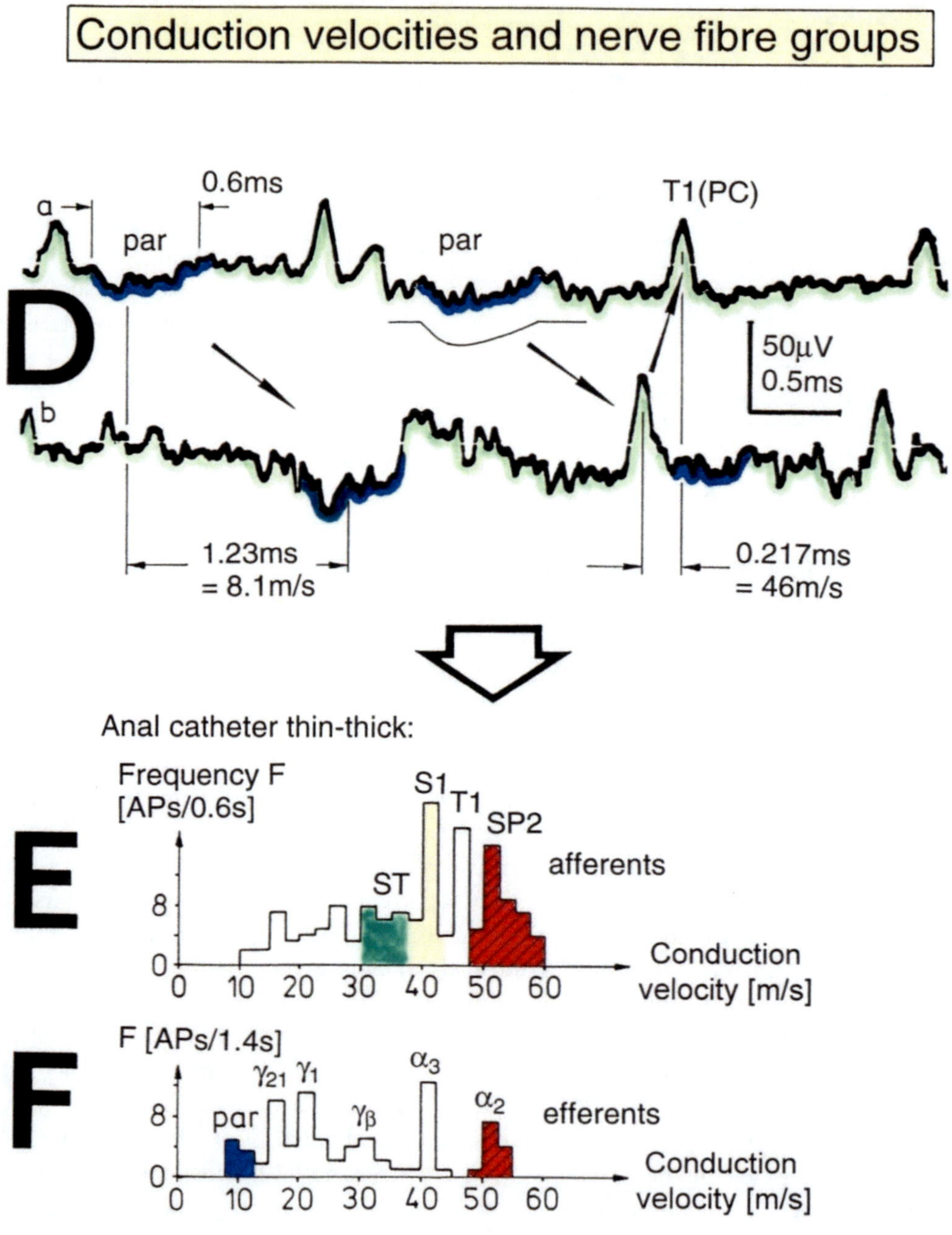

Figure 3. Sweep piece of recording (D) and conduction velocity distributions (E,F) taken from time intervals following a change of a thin anal catheter (∅ = 12 mm) for a thick one (∅ = 20 mm). Note the manifestation of the parasympathetic peak. For symbols, see legend Figure 2.

CAPs evoked by electrical intravesical stimulation (Figure 9A), also performed with the Brindley stimulator (pulse duration 0.3ms), were conducted at group conduction velocities very similar to those obtained for single stretch (S1) and tension receptor afferents (ST) of the urinary bladder (Figure 2B).

1.4.2. Fibre Diameter Distributions for Afferent and Efferent Nerve Fibres

After the recording of single-fibre APs, the dorsal nerve roots were removed, fixated and prepared for morphometry. Figures 4A,B,6B show examples of nerve root cross sections obtained. Manually determined values of mean nerve fibre diameters ($1/2(\varnothing_1 + \varnothing_2)$) and mean myelin sheath thicknesses d (and of other parameters) were used to construct nerve fibre diameter distribution histograms (Figures 5A,B,6C). Nerve fibre groups were identified with the help of conduction velocity distribution histograms and the assumption that nerve fibre group diameters had not changed too much from physiologic values. The identification was started with the thickest fibres. Identified peaks were assigned to the group to which they belonged (see Figure 5).

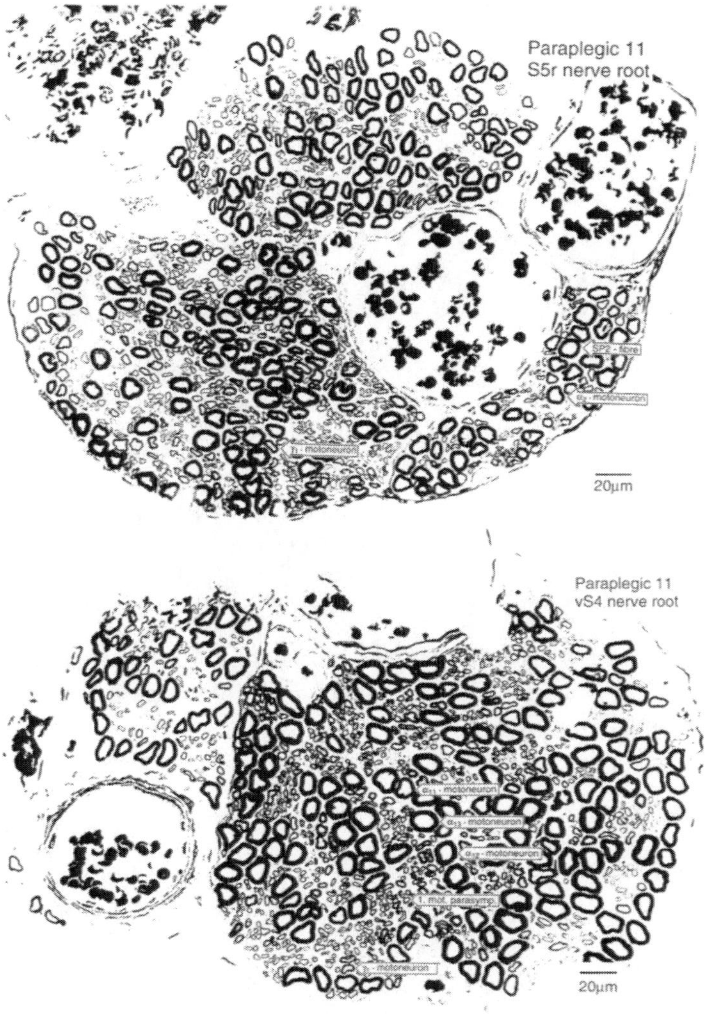

Figure 4. A. Light microscopic picture of a cross section of the right nerve root S5. Nerve fibres are indicated that, according to their mean fibre diameter and myelin sheath thickness, most likely are a secondary muscle spindle afferent fibre (SP2), a γ_1 or an α_2-motoneuron. B. Light microscopic picture of a cross section of a vS4 nerve root. Nerve fibres are indicated that, according to their mean diameter and myelin sheath thickness, are a preganglionic parasympathetic motoneuron, γ_1, α_{11}, α_{12}, or α_{13}-motoneuron. Thionin acridine-orange staining of the myelin sheath. Paraplegic 11.

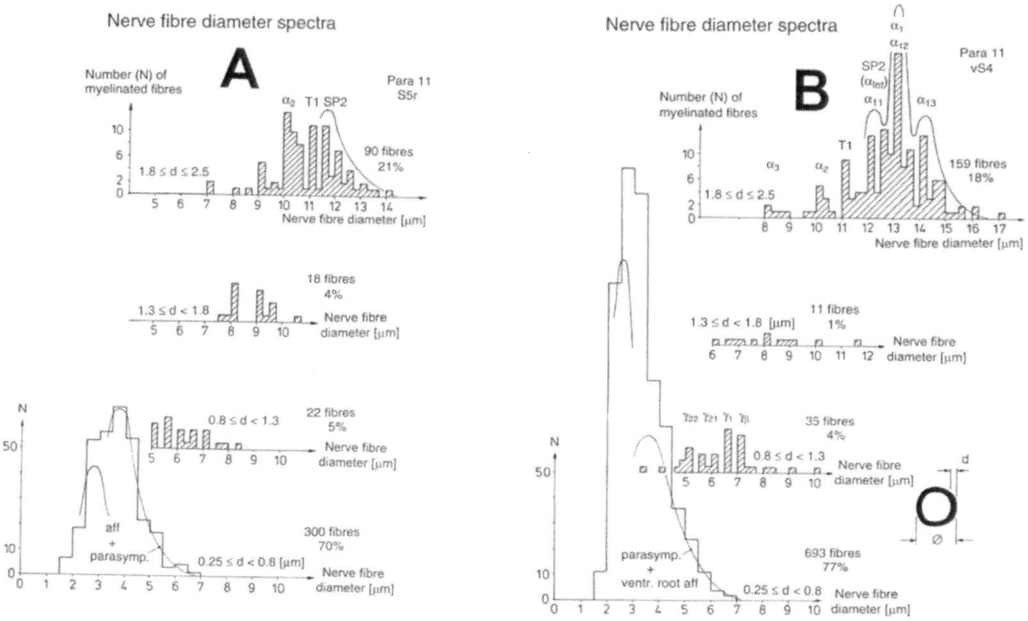

Figure 5. A. Nerve fibre diameter distributions of the right nerve root S5 (from Figure 4A); classification in four ranges of myelin sheath thicknesses (d). Fibre diameter peaks are indicated, which represent α_2-motoneurons, T1 skin afferents (PC), secondary spindle (SP2) afferents (aff) and preganglionic parasympathetic (parasymp.) nerve fibres. A distribution curve for SP2 fibres is indicated. In the range of myelin sheath thicknesses $0.25 \le d < 0.8$ µm, possible fibre distributions are drawn into the histogram, which represent parasympathetic fibres and afferent fibres. B. Nerve fibre diameter spectra for nerve root vS4 (from Figure 4B), according to four classes of myelin sheath thicknesses (d). Fibre diameter peaks are indicated, which represent γ_{22}, γ_{21}, γ_1, γ_β, α_3, α_2, α_{11}, α_{12}, and α_{13}-motoneurons, parasymp. = preganglionic parasympathetic fibres, ventr. root aff = ventral root afferents. Paraplegic 11. The high number of α_1-motoneurons (FF) in root S4 is unusual. Distributions of subgroups α_{11}, α_{12} and α_{13} are qualitatively drawn for α_1-motoneurons. In the class of myelin sheath thicknesses $0.25 \le d < 0.8$ µm, distribution curves are qualitatively drawn for parasympathetic fibres (parasymp.) and ventral root afferents.

The presence in the lower sacral nerve roots of ventral root afferents and dorsal root efferents complicated the group identification from nerve fibre diameter peaks. This problem did not arise with conduction velocity distributions since APs from afferents and efferents have opposite triphasic amplitude and opposite conduction times. Also, velocity distribution peaks could partly be identified by different kinds of stimulation. With electrical intravesical stimulation, for example, bladder afferents will mainly be activated, whereas stimulation of the parasympathetic division will activate parasympathetic motoneurons. As can be seen from Figures 5A,B,6C, the muscle spindle afferents, the four thickest skin afferents and the α-motoneurons had myelin sheath thicknesses between 1.8µm and 2.5µm. The fusimotor fibres (γ_1 - γ_{22}) had myelin sheath thicknesses between 0.8µm and 1.3µm, and the preganglionic parasympathetic motoneurons between 0.25 and 0.8µm. Only few axons had myelin sheaths with a thickness between 1.3µm and 1.5µm. In dorsal roots, there were only few fibres with a myelin sheath thickness between 0.8 and 1.3µm. The classification of nerve fibre diameters into four different ranges of myelin sheath thickness has the advantage that certain nerve fibre groups fit mainly within certain myelin sheath thickness ranges and can therefore be better identified. E.g., the parasympathetic preganglionic motoneurons have similar diameters as do the static fusimotors (γ_{22}) (Figure 5B). They can be distinguished from each other by a thicker myelin sheath of the γ_{22}-motoneurons.

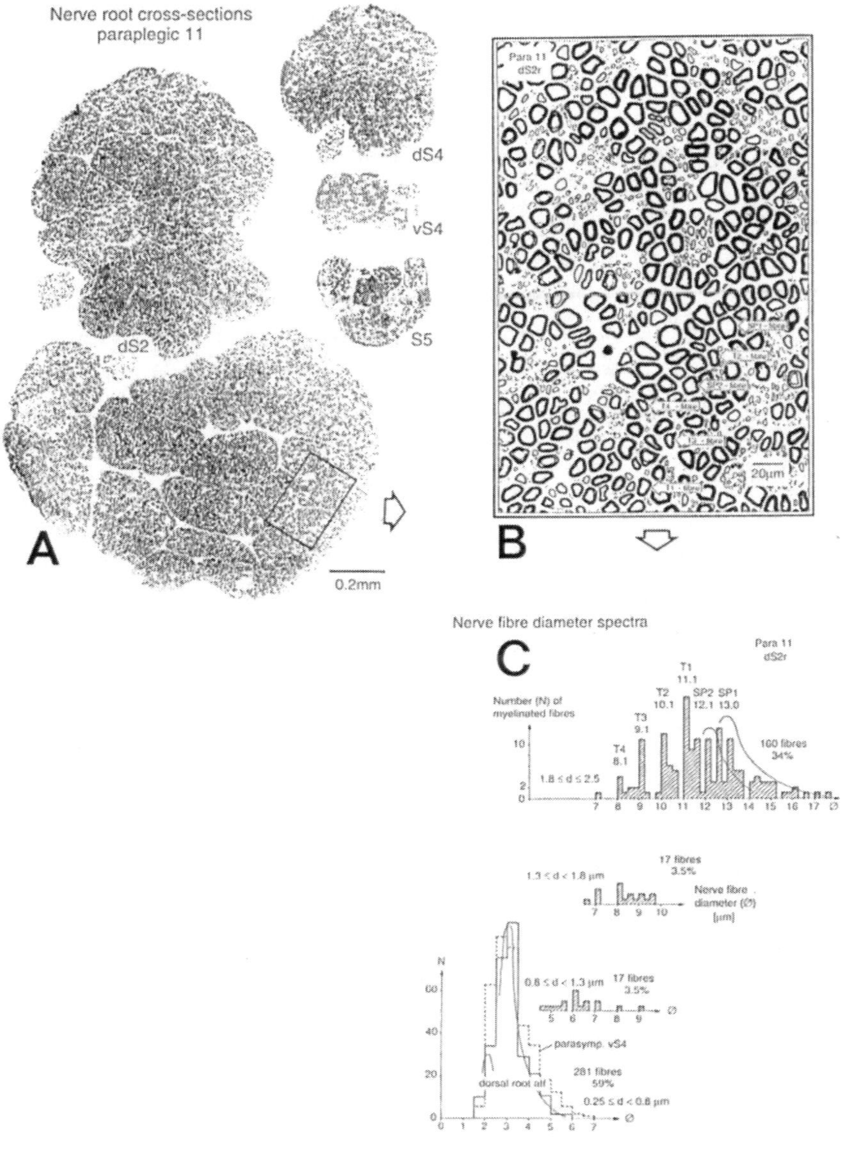

Figure 6. A. Cross sections of the right nerve roots S5, vS4, dS4, dS2. The dorsal root S2 consists of two main fascicles. The square area marked in the lower fascicle of the S2 cross section is shown in B at a higher magnification. B. Magnified area of the dS2 cross section from A. Nerve fibres are marked that, according to their diameter and myelin sheath thickness, most likely represent primary spindle (SP1), secondary spindle afferent fibres (SP2) and T1 (PC), T2, T3, and T4 skin afferent fibres. C. Nerve fibre diameter spectra of the dS2 cross sectional area in B according to four ranges of myelin sheath thicknesses (d). Fibre diameter peaks are indicated, which represent primary (SP1) and secondary muscle spindle afferents (SP2), and the myelinated skin afferent fibres T1 (Pacinian corpuscle), T2, T3 (probably SAI) and T4. Distribution curves for SP1 and SP2 fibres are indicated. The histogram for parasympathetic fibres (including ventral root afferents) from root vS4 in Figure 5B is represented by the dashed line (parasymp. vS4). The histogram for thin fibres ($0.25 \leq d < 0.8$ μm) includes possible afferent fibre distribution curves. Paraplegic 11. For further abbreviations and symbols, see legend to Figure 11.

In humans, there are approximately 20% ventral root afferents and dorsal root efferents in the S4 root, 1-2% in the S3 root and less than 1% in the S2 root [46]. Because of the myelinated ventral root afferents in the ventral S4 root, the fibres in the myelin sheath thickness class $0.25 \leq d < 0.8$ μm (Figure 5B) represent not only parasympathetic motoneurons but also afferent fibres, whereas the fibres in the dorsal S2 root represent mainly afferents (Figure 6B) in the myelin sheath thickness range $0.25 \leq d < 0.8$ μm. By plotting the distributions of thin myelin sheath fibres with reduced amplitude (Figure 5B) into Figure 6C (dashed line), one can compare the mixed parasympathetic-afferent fibre distribution with that of pure afferent fibre distribution.

It can be estimated that parasympathetic preganglionic fibres (peak group diameter approx. 3.5 μm) have a group diameter by approx. 0.6 μm thicker than the thin afferent fibres in the same myelin sheath thickness range, even though the distributions may not be identical. This numerical characteristic was already recognized during manual measurements of fibre diameters and myelin sheath thicknesses. Only seldom fibres were found in the dS2 root (Figure 6B), with medium diameters and very thin myelin sheaths. The qualitatively drawn distribution curves of parasympathetic preganglionic motoneurons and afferent fibres in the myelin sheath thickness class $0.25 \leq d < 0.8$ μm (Figures 5A,B,6C) slightly help to differentiate between afferents and parasympathetic efferents; as indicated in Figure 6C, however, the afferent fibre distribution itself most likely includes more than a single-fibre group with the different groups having similar peak group diameter values.

The fibre diameter distribution of the secondary muscle spindle afferents (SP2) in the myelin sheath thickness range $1.8 \leq d < 2.5$ μm (Figure 5A) was qualitatively constructed by broadening the histogram classes and subsequently drawing the distribution curve. The SP2-fibre distribution was then drawn into the thick fibre distribution of the right dS2 root for the same patient (Figure 6C). Now it is possible to draw qualitatively the distribution of the primary spindle afferents (SP1) again after broadening the histogram classes. The urinary bladder stretch (S1) and tension receptor afferents (ST) have group conduction velocities between those of touch 1 (T1) and touch 2 (T2), and between T2 and touch 3 (T3) skin afferents (Table 1 of 64) (for nomenclature, see also Figure 11). The group diameters of the bladder afferents may therefore be within the 9μm - 11μm range. But as Table 1 of Ref. 64 shows, the conduction velocities and the fibre diameters are not simply linked by a unique scaling factor in the range of 6. For the determination of more exact values, see the Discussion section.

The fibre class of α_1-motoneurons represented in Figure 5B seems to split up into three subgroups (α_{11} (= α_{Int}), α_{12}, α_{13}) in similarity to the results of measurements in dogs [45]. The subpeaks have been drawn qualitatively. Again ventral root afferents (T1, SP2) disturb the motor fibre distribution. As expected, the different γ-motoneuron classes can be found in the myelin sheath thickness class $0.8 \leq d < 1.3$ μm. Figure 6A shows a comparison of the sizes of different roots measured in paraplegic 11. It can be seen that the root dS4 has approximately twice the thickness of root S5 and twice the thickness of root vS4. Root dS2 is approximately four times as thick as root dS4. In Figure 4A,B,6B, some afferent and efferent fibres are marked with the groups to which they belong. This assignment rests purely on the nerve fibre diameters and the myelin sheath thicknesses. As can be seen from the diameter distribution histograms, there also is reasonable probability that these fibres belong to adjacent fibre classes, because of overlaps of adjacent group distributions; also, fibres of so far unidentified fibre groups could be represented by those fibres. Anyhow, these fibre designations clearly show that it is only detailed morphometry that can partially reveal the distinct nerve fibre groups.

1.4.3. Correlation between Group Conduction Velocities and Group Nerve Fibre Diameters

All the measured peak values of group conduction velocities and group nerve fibre diameters are summarized in Table 3 of Chapter III. The nerve roots used for electrophysiologic and morphometric determinations yielded a direct correlation between group conduction velocities and group nerve fibre diameters. As can be seen from Table 1, the group diameters vary only marginally from patient to patient, whereas the group conduction velocities vary significantly. The large variations of group conduction velocities are mainly due to different root temperatures, which are effected by the central temperature of a patient (34.3°C to 35.8°C). The temperature dependence of the conduction velocity is tackled below. The following velocity-diameter pairs of afferents and efferents were obtained from Table 3 of Chapter III at ~35.5°C:

α_{13}(-/14.1), α_{12}(65 m/s /13.1 μm), α_{11}(60?/12.1), α_2(51/10.3), α_3(41/8.2), γ_β(27/7.1), γ_1(21/6.6), γ_{21}(16/5.8), γ_{22}(14/5.1), par(10/3.7).

SP1(65/13.1), SP2(51m/s /12.1μm), T1(47/11.1), T2(39/10.1), T3(27/9.1), T4(19/8.1), S1(41/-), ST(35/-).

As will be discussed below, the group conduction velocities and the group nerve fibre diameters of paraplegics are very similar to those obtained from brain-dead humans (Table 1 of Ref. 39) (Table 2 of Chapter III) and patients with no spinal cord lesion (Figure 11).

1.4.4. Temperature Dependence of Group Conduction Velocities

In all measured cases, the AP durations decreased and the conduction velocities of the single-nerve fibres increased with increasing temperature. It seemed further as if the different nerve fibre groups had different temperature dependences, because for lower temperatures it was much more difficult to identify the different nerve fibre groups by their peaks in the velocity distributions. The operational field was therefore often warmed with an infrared lamp to bring the root temperature as close as possible to 36°C. In paraplegic 11, good conduction velocity distribution histograms from the same root under the same recording conditions were obtained for low (32°C) and for high (35.5°C) root temperatures. The secondary muscle spindle afferents (SP2) and the α_2-motoneurons increased their conduction velocity from 40 to 50 m/s (25% increase), the S1, ST and M afferents from 31.3 to 40 (28%), from 25 to 33.8 (35%) and from 12.5 to 13.8 m/s (10%), respectively, with increasing temperature. The α_3-motoneurons increased their conduction velocity from 33 to 40 m/s (a 21% increase). Even though only one set of good quantitative measurements could be obtained so far, the two sets of velocity values clearly show that different nerve fibre groups have different temperature dependences of their group conduction velocity. The SP2 afferents and the α_2-motoneurons showed the same temperature dependence of the group conduction velocity. The calibration relation of the velocity distributions of afferent and efferent fibres, namely that the SP2 fibres conduct with the same velocity as the α_2-motoneurons, is therefore temperature independent.

1.4.5. Nerve Root Cross section and Axon Numbers

Table 3 of Chapter III shows root diameter values. They vary between 0.31 and 1.6mm, depending on the root level. Root S5 has a mean diameter of 0.47 mm, and the dorsal root S4 has 0.62 mm. It cannot be excluded that in some cases, dorsal root fascicles of more rostral roots were measured together with the S5 root. The diameters were calculated from the cross sectional areas. Also, the percentual contributions of arteries and veins to the cross sectional area are listed in Table 3 of Chapter III. As can be seen from Table 3 of Chapter III, in thin roots, the blood vessels can cover more than 50% of the cross sectional area. For blood vessel diameters of nerve roots, further nerve root diameters and nerve root fibre counts, see Refs. [35,36,62,63]. Representative 350 to 500 fibres were measured in each root. The total number of nerve root fibres was obtained by comparing the area of the measured fibres with the area of all nerve root fibres in the root. The S4 ventral root was measured completely. The numbers of myelinated fibres increased towards more rostral roots from 900 to 31,600. Root S5 contained on average 2900 fibres. Most myelinated fibres were found in the myelin sheath thickness class $0.25 \leq d < 0.8$ μm followed by the class of thick myelin sheaths ($1.8 \leq d < 2.5$ μm). Few nerve fibres had a myelin sheath thickness between 0.8 and 1.8 μm. The small number of nerve fibres in the S5 root suggests that this root is mostly not important for pelvic organ innervation. This is actually the reason why the S5 root (dorsal plus ventral) was normally cut in patients and obtainable for research. The unimportance for urinary bladder function was verified in the operation by not having a bladder pressure increase due to electrical stimulation of the S5 root. Sometimes, a substantial urinary bladder pressure increase was obtained upon electrical stimulation of the S5 root. In such cases, the dorsal root was separated and removed and the ventral S5 root saved for electrical stimulation for bladder control.

1.4.6. Velocities of Urinary Bladder Afferents

Urinary bladder afferent activities for constructing conduction velocity distribution histograms were recorded at bladder fillings of approx. 200ml, depending on the compliances of the bladders measured preoperatively. The thin bladder afferents with small AP amplitudes and long AP durations (M, S2; Figure 11, Table 3 of Chapter III) (possibly also activated by the bladder catheter) could not reliably be detected so far because of not good enough recording conditions.

1.4.7. Activity of Urinary Bladder Afferents to Retrograde Bladder Filling

For measuring bladder afferent activity increase upon retrograde bladder filling, the bladder was emptied and then filled while recording (time-consuming procedure, successful so far only in two cases). With the groups of nerve fibres identified from conduction velocity distribution histograms, the increase in activity of the urinary bladder stretch (S1) and tension receptor afferents (ST) to retrograde bladder filling can be measured. Figure 7 shows the activity increases following filling of the bladder of Para 7, via a catheter, with up to 400 ml fluid.

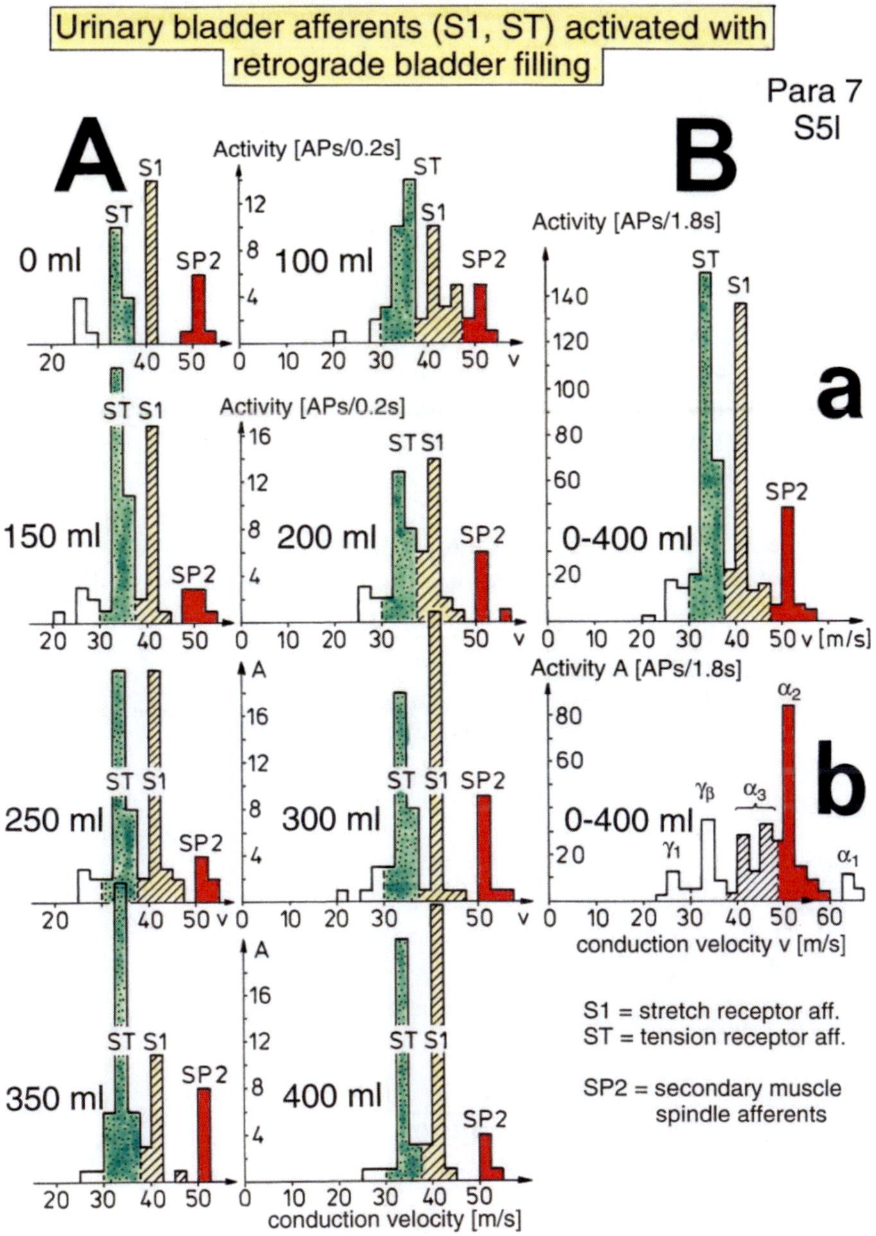

Figure 7. Conduction velocity distribution histograms for stretch (S1) and tension receptor afferents (ST) and secondary spindle afferents (SP2) for different retrograde filling stages of the urinary bladder (A). Summed histograms for afferents (a) and efferents (b) are plotted in B. α_1, α_2, γ_β, γ_1 represent velocity distributions of α_1, α_2, γ_β and γ_1-motoneurons. Para 7, left nerve root S5.

The bladder was filled to the maximal acceptable value as determined pre-operatively by compliance (bladder filling volume / bladder pressure) measurements. Figure 7B shows the sum of the individual velocity distributions for different filling stages for afferents and efferents. The peaks are identified by two calibration relations; α_2-motoneurons conduct with the same velocity as the secondary muscle spindle afferents (~50 m/s); and α_3-motoneurons conduct with the same group velocity as the stretch receptor afferents S1 (41 m/s) at about

35.5°C. Figure 7A gives velocity distributions for certain bladder filling stages. The most important obvious feature is that the bladder stretch and tension receptor afferents fire considerably even with no bladder filling.

By setting up conduction velocity borders of the S1 and ST groups according to the distribution curve (Figure 11, inset), the activity of each group could be measured for different filling volumes and plotted (Figure 8A). As can be seen in Figure 8A, the stretch (S1) and tension receptor afferents (ST) in the left root S5 fired with 15 APs during 0.2s at no bladder filling.

The bladder afferents increased their firing with the increasing bladder storage volume. At 160 ml filling, a transient very strong firing of ST afferents occurred. This was the filling volume at which before the surgery the detrusor was activated in an urodynamic measurement. So far, it has not been possible to activate the detrusor with retrograde bladder filling during the surgery, even despite the light anaesthesia, as paraplegics with a complete spinal cord lesion feel no pain.

From a ventral S3 root in paraplegic 11, a similar undulating activity increase of S1 and ST bladder afferents was obtained with retrograde bladder filling. When the bladder was empty, the afferents (S1 and ST) fired already with ten APs/0.2s. At 50 ml filling, they fired with 23, at 200ml with 6 at 250ml with 25 and at 500ml with 12 Aps/0.2s.

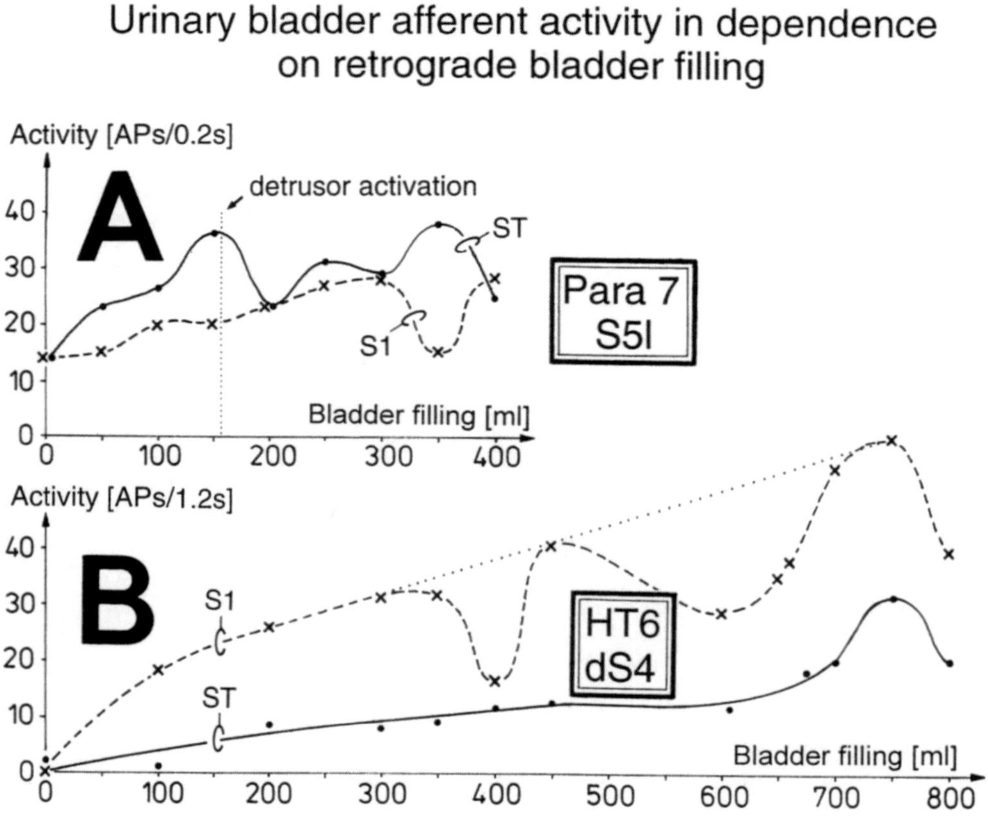

Figure 8. Afferent activity increase of stretch (S1) and tension receptor afferents (ST) to retrograde bladder filling in paraplegic 7 (left root S5) in relation to those of the brain-dead human HT6 (dorsal root S4). APs = action potentials. The speed of filling was 130 ml/min in para 7 and 100 ml/min in HT6.

The detrusor was activated pre-operatively at a filling volume of 80ml (continuously undulating contractions (0.08Hz) started). As in the para 7, the first high afferent input falls approximately together with the onset of the pre-operative detrusor contraction, when filling the bladder. The occurrence of bladder afferent APs in the ventral S3 root was higher (166 APs/2.2s) during the bladder filling than those of α-motoneuron APs (70 APs/2.2s); the secondary muscle spindle afferents fired with 64 APs/2.2s. The speed of filling in Para 11 was 300ml/min. In this ventral S3 root, there were at least a few percent of the fibres afferent.

1.4.8. Comparison of Bladder Afferent Activity Increase to Bladder Filling between a Paraplegic and a Brain-Dead Human

The pathology of the bladder afferent activity in the paraplegic 7 analyzed in Figure 8A, with dyssynergia of the bladder, is fully revealed by a direct comparison with the activity increase of the bladder afferents following retrograde bladder filling in brain-dead human HT6 (Figure 8B). The dependence of the bladder afferent activity on the bladder filling volume in paraplegic 7 shows four main differences in comparison to those in HT6. (1) The bladder afferents already fired with no bladder filling, which was not the case in HT6. (2) The activity increased more smoothly in HT than in the paraplegic, even though bladder filling was stopped twice in HT6 but was not stopped in Para 7. (3) In the Para 7, the tension receptor afferent activity was higher than the stretch receptor afferent activity, which was not the case in HT6. (4) The bladder afferent activity was higher in the paraplegic than in the HT6. By taking into account the different measuring times of the activity, the ST-activity in Para 7 was approx. 40 times higher than those in HT6 (210APs/1.2s in Para 7 against 5APs/1.2 in HT6). The recording conditions are not directly comparable. However, there are on average more bladder afferents in a root dS4 (HT6) than in a root S5. Still, the 40 times higher tension receptor afferent activity is probably overestimated. It is nevertheless safe to conclude that in some paraplegics, the bladder afferent activity, especially from tension receptor afferents, was several times higher than that in the brain-dead human, the latter probably representing the physiologic case in this respect. The finding of an increased afferent activity in two paraplegics is in accordance with the experience that in brain-dead humans, activity from bladder afferents was recorded only occasionally, whereas in paraplegics, the activity from S1 and ST-afferents was always obvious and prominent in the velocity distributions, rather independent of the extent the bladder was filled. Paraplegics with dyssynergia of the bladder suffer from high bladder pressure. As will be shown in the following papers [51,52], this increased afferent activity will have consequences on the excitation of the CNS activating the striated sphincters of the bladder and the rectum.

1.4.9. Electrical Intravesical Stimulation

In Figure 10A, the recording and stimulation layout for electrical intravesical stimulation is shown. The negative electrode was connected to the tip of the bladder catheter but was in no direct contact with the bladder mucosa. The indifferent positive electrode was attached to the leg of the patient. Following stimulation with 30V compound action potentials (CAPs) of stretch (S1) and tension receptor afferents (ST) could be evoked. In the recording from the root dS2 (Figure 9A), mainly two CAPs follow the stimulation artefact, with latencies of 7.4 ms and 9.5 ms.

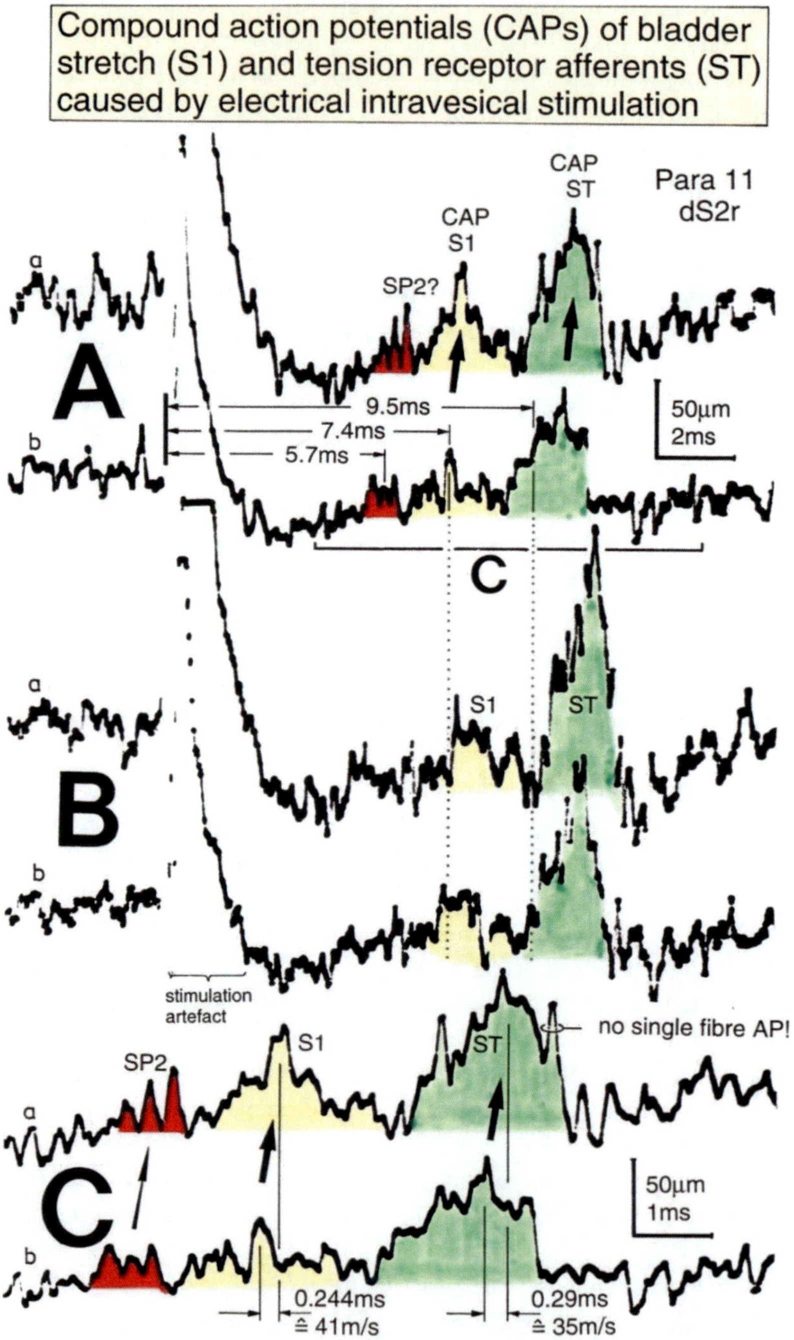

Figure 9. Compound action potentials (CAPs) of urinary bladder stretch (S1) and tension receptor afferents (ST) following electrical intravesical stimulation. Stimulation in B was slightly stronger than in A. The recordings in C represent a time-stretched portion of A. Mean conduction times and calculated mean conduction velocities of CAPs are indicated. In C, three single secondary muscle spindle afferent fibres are marked (SP2). The stimulation pulse (not rectangular) was 0.3ms long; no earthing between recording and stimulating electrodes. Earth electrodes positioned 5 to 10mm away from the recording electrodes reduced stimulation artefact and single-fibre AP amplitudes and were therefore not used. Horizontal line of stimulation artefact is due to the limited input range.

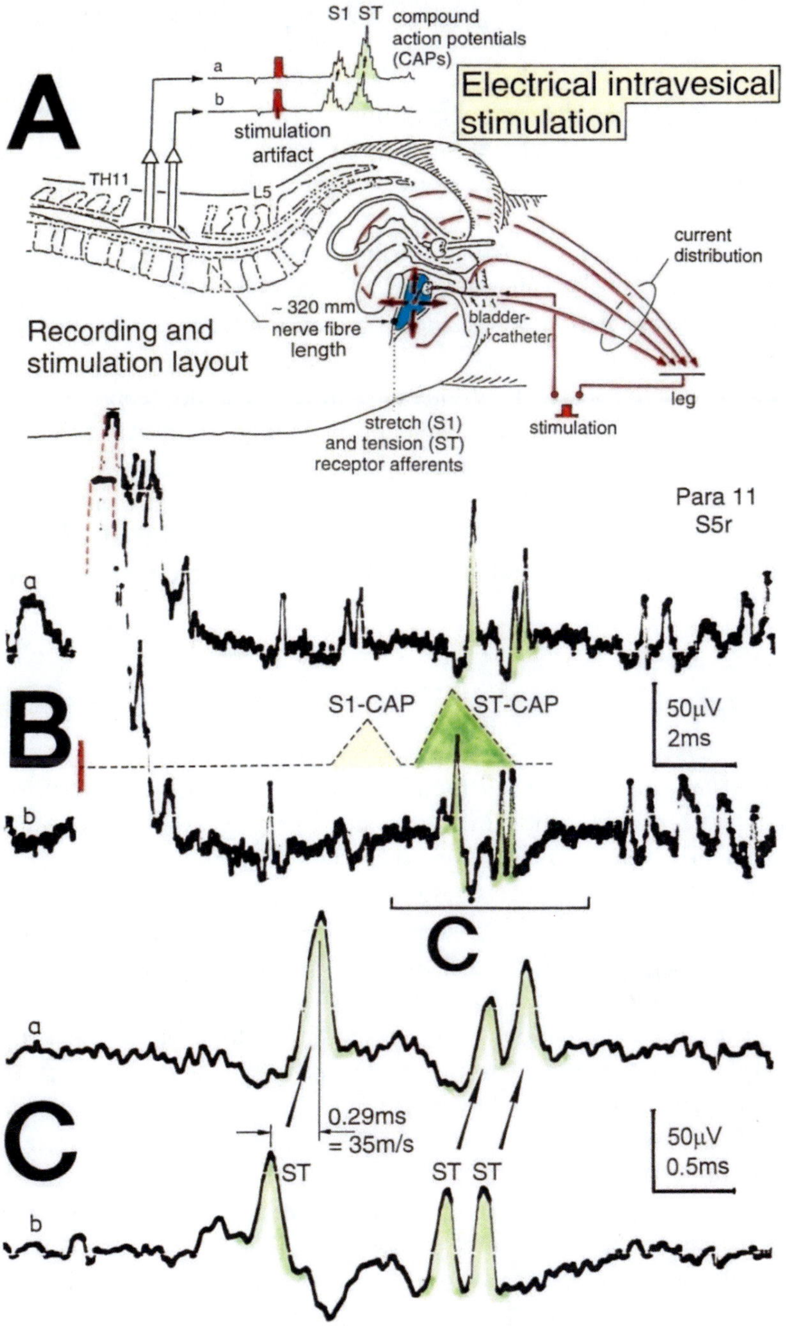

Figure 10. A. Recording and stimulation layout for electrical intravesical stimulation (electro negativity in bladder). B. Recording from an S5 root following electrical stimulation. Compound action potentials (CAPs) of S1 and ST afferents schematically redrawn from Figure 9A. C. Time-stretched sweep piece of the recording in B. Single tension receptor afferent APs (ST) are indicated.

The CAP conduction velocities in Figure 9C were 41 and 35 m/s. According to Figure 2B, the stretch receptor afferents contributed to the CAP with a latency of 7.4 ms and the tension receptor afferents to the CAP with a latency of 9.5 ms. The faster conducting S1 afferents give rise to the CAP with the shorter latency, as they would be expected to. From the latency and the group conduction velocity, it can be calculated that the afferent activity travelled approximately 320 mm to reach the recording electrodes, as indicated in Figure 10A. The surgeon estimated a distance of 300 mm.

With a slightly increased electrical stimulation, the S1 and ST CAPs increased slightly (Figure 9B), probably because more afferents were stimulated in the bladder wall. A recording from root S5 (same electrical stimulation) showed no pronounced CAPs. By schematically plotting S1 and ST CAPs from Figure 9A into Figure 10B and by calculating the conduction velocities of the single-fibre APs in the latency range of the ST CAP, it was found that the single-fibre APs were conducted by tension receptor afferents, as can be seen from the time-stretched Figure 10C. Thus, at least the ST CAP was composed of the single-fibre APs of the ST afferents. Normally, subpeaks in CAPs do not represent single-fibre APs, as is indicated in Figure 9C. The arithmetic addition of single-fibre APs of different triphasic AP waveforms often results in large amplitude potential changes, which do not represent single-fibre APs, as can be learned from recordings of high skin afferent activity. To obtain CAP recordings from stretch and tension receptor afferents a substantial number of bladder afferents have to be stimulated simultaneously, through the nerve root that is used to record from.

1.5. Discussion

1.5.1. Unchanged Conduction Velocities and Nerve Fibre Diameters

A classification scheme for the human peripheral nervous system has been developed by measuring conduction velocities and fibre diameters of single-nerve fibres in brain-dead humans and patients with no spinal cord lesion. This is based on a new recording technique in man, allowing the single-nerve fibre action potential (AP) recording from whole human sacral nerve roots and a further developed morphometry allowing for fibre diameters to be grouped into four ranges of myelin sheath thickness (Figure 1) [39,45]. Moreover, a new treatment has been developed (called coordination dynamics therapy), based on human neurophysiologic research based on earlier, this, and the following papers, which can cure urinary bladder function [78,86,87] and can partially cure motor functions in humans with spinal cord injuries [67,68,84-87]. The out-of-date bladder treatment, namely the electrical sacral anterior root stimulation, according to Brindley [7], is discussed here mainly for scientific reasons. The treatment involves de-afferentation of the bladder to increase bladder storage volume, and electrical stimulation of the motoneuron and preganglionic parasympathetic axons in the motor roots of the cauda equina. A precise electrodiagnosis during the surgery is crucial for the cutting of the bladder afferents and the saving of the motoneurons and is performed by nerve root stimulation and bladder pressure measurements. The recording of single-fibre APs and compound APs brings about an improvement of the intraoperative diagnosis. A more selective cutting of afferents and partial saving of afferents transmitting sexual sensations would improve the treatment. The hope is that the dorsal S2 root contains many afferents conducting sexual sensitivity and only few conducting bladder fullness. The detailed representation of the urinary bladder and sexual organs in the sacral roots is still unclear. The representation varies among patients [4]. Further, little is known which fibres conduct sexual feeling; the touch 4 (T4) (may be SAII) skin afferents ("Streichel receptor afferents") [44] probably contribute to it. A case is known in which a lady with a partial spinal cord lesion was operated for urinary bladder control by electrical anterior root stimulation. The dorsal S2, S3 and S4 roots and the whole S5 root were cut. The sexual sensitivity was partly preserved. Some sexual organ afferents may therefore also run through the plexus hypogastricus or the dorsal S1 root. Ventral root afferents are probably too few to account for the preservation of sexual feeling. Morphometric analyses may help to clarify the nerve fibre group composition of nerves innervating pelvic organs [42], since the tapering of nerve fibres is very small (0.2% per 13cm [70]). This new intraoperative electrodiagnosis may be of importance for refined surgical treatments.

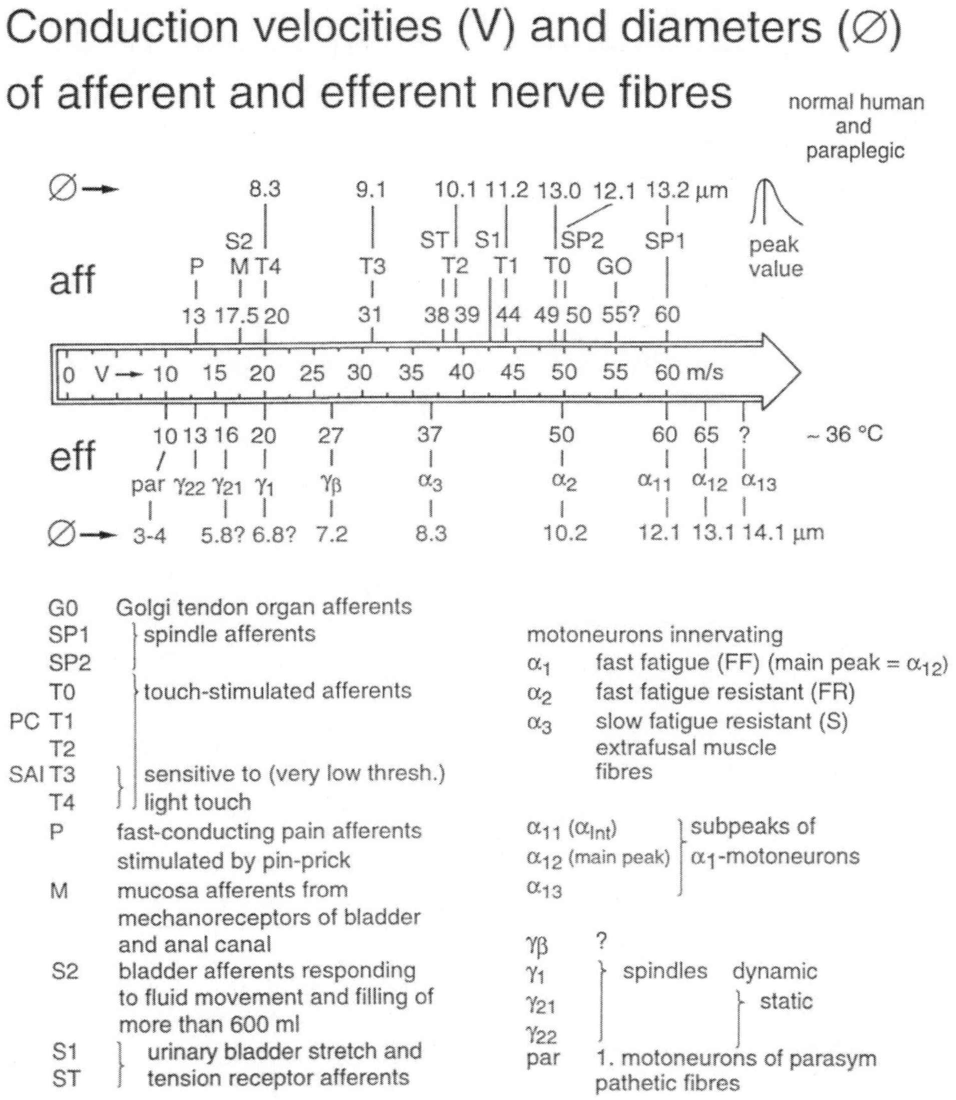

Figure 11. Conduction velocities (V) and nerve fibre diameters (∅) of afferent and efferent nerve fibre groups in normal humans and in patients with a traumatic spinal cord lesion present for 0.5 to 6 years. The splitting of the α_1-motoneurons into three subgroups, α_{11}, α_{12}, α_{13}, has not yet been confirmed.

The availability of single-fibre AP recordings and of dorsal roots, in which also dorsal root efferents are contained in the lower sacral nerve roots [39,46], allowed me to simultaneously measure conduction velocities and fibre diameters of single afferent and efferent nerve fibres. Group conduction velocities and group nerve fibre diameters could be obtained from distribution histograms, similar as in earlier measurements in brain-dead humans. The group conduction velocities and group nerve fibre diameters (peak values) obtained for paraplegics were found to be very similar to those estimated for brain-dead humans and patients with no spinal cord lesion. Thus, the classification scheme of the human peripheral nervous system is preserved at least 0.5 to 6 years following spinal cord lesion (Figure 11). The subdivision of the α_1-motoneurons (FF) in α_{11}, α_{12} and α_{13}-motoneurons needs further clarification. The preservation of the classification scheme is not a trivial conclusion since changes occur in the peripheral and central nervous system following spinal cord lesion. It has been shown in this paper that, at least for some patients, the tension receptor afferents of the bladder fired even when the urinary bladder was empty. In the following papers [51,52], it will be shown that functions of the spinal neuronal network change after injury, as judged by the impulse patterns of secondary muscle spindle afferents and spinal oscillators.

1.5.2. Classification Schemes

The classification schemes of Grundfest, Erlanger and Gasser and Lloyd and Hunt [1,3,6,11,16,20,21,27,29-31] do not apply to humans and are too inaccurate to be used for detailed analyses of human nervous system functions. The differentiation between α_1(FF), α_2(FR) and α_3(S)-motoneurons [9,12,18,24,39,45] is essential for the understanding of CNS functions, since the three α-motoneuron types are integrated in functionally different spinal circuitries and can be distinguished from each other by the different oscillation frequencies and impulse trains (high activity mode) [40,52], different recruitment in the occasional firing mode (low activity mode) [47,48] and different motoneuron drive [52]. The existence of slow and fast systems in the CNS has already been suggested by Sherrington [53]. There have been no other authors dealing seriously with the measurements of urinary bladder afferents in humans [39,40]. Animal data [13-15,22] are unsuitable for a direct comparison with human data, unless the classification schemes for animals are improved, which has been started [9,45,47,49]. The differentiation between stretch and tension receptor afferents [39,40] will contribute to the understanding of pathologic micturition. Application of classification schemes to myenteric neurons has been attempted [5,59].

1.5.3. Temperature Dependence of the Conduction Velocity

The conduction of APs in nerve fibres depends strongly on the temperature both in humans (Figures 30-32 of Chapter III) (Table 1 of Ref. 64, section "temperature dependence of group conduction velocities") and animals [29]. The increase of the conduction velocity of α_2-motoneurons and secondary muscle spindle afferents results to approximately 2.8 m/s / °C. In humans, a temperature dependence of 2.1 m/s / °C was measured; no differentiation between different groups was made [8]. In animals, a temperature dependence of 3.5%/°C was measured [30]; data for different groups were not given. In this paper, it was measured that in humans, the temperature dependence of nerve fibre groups is different for the different groups. The effect of a strong temperature reduction is that the conduction velocity distribution peaks fuse, and it becomes difficult to identify the groups (peaks) (see Figure 2 of [48]). In particular, the preganglionic parasympathetic motoneurons can be better differentiated from the static γ-motoneurons at a root temperature higher than 35°C. The temperature of the thin roots is not exactly measurable. Efforts should be spent to keep the temperature in the operational field close to the central temperature of the patient. In the future, conduction velocities obtained from the roots of the cauda equina have to be calibrated using conduction velocities obtained non-invasively from the same fibres of the leg. Such a comparison requires an analysis of CAPs with single-fibre APs [50]. Identification of nerve fibre groups however does not depend on the knowledge of the exact conduction velocity: conduction velocity distribution histograms can be calibrated using the same velocity of conduction by α_2-motoneurons and secondary muscle spindle afferents (SP2), α_3-motoneurons and stretch receptor afferents of the bladder (S1), and approximately by α_1-motoneurons and primary spindle afferents (SP1). The SP2 fibres and the α_2-motoneurons showed the same temperature dependence of the conduction velocity; this calibration relation is therefore temperature independent in the temperature range 32 to 36°C.

1.5.4. Recording of APs from Preganglionic Parasympathetic Motoneurons

In a first approximation, the AP amplitude depends on the diameter of the axon and declines due to volume conductance if the axons are not positioned close to the recording electrodes (see Method). Thinner axons have smaller AP amplitudes than thicker ones [37], even though the area between the AP curve and the baseline is, in a first approximation, independent of the AP amplitude [50], because the AP duration increases for thinner, slower conducting fibres [30,31,37]. The preganglionic parasympathetic motoneurons with a peak group diameter of approx. 3.5 μm have very small AP amplitudes (and very long duration) (Figure 3D), so that they are difficult to recognize in the noise and artefact level and are often overshadowed by large APs of thick fibres. Even though most likely, parasympathetic action can also be estimated from the impulse patterns of secondary muscle spindle afferents [49,51], sometimes it is very important to obtain impulse patterns of single preganglionic parasympathetic motoneurons. Only in this way is it possible to analyze neuronal networks of the sacral micturition centre. The asymmetrical distribution of preganglionic parasympathetic axons with a nerve fibre diameter of up to 6 μm will sometime make it possible to safely record parasympathetic APs and to extract natural impulse patterns of single fibres from the recordings, provided good recording conditions. Thin ventral S4 roots freed over long sections from other roots, with nearly no shunting blood vessels (see Method), are favourable for recording single-fibre APs in patients with no arachnoiditis.

1.5.5. Extension of the Classification Scheme to Thinner Fibres

As discussed above for the preganglionic parasympathetic motoneurons, the limit for recording single-fibre APs from undissected roots or fascicles are diameters between 3 and 4 μm. It was recorded therefore from myelinated fibres only. This does not mean that it is impossible to record from unmyelinated fibres. It should be possible to record extracellular APs from unmyelinated human axons thicker than 4 μm. To extend the classification scheme of the human peripheral nervous system to nerve fibres thinner than 3 to 4 μm, CAPs have to be analyzed with the single-fibre AP recording method to identify the nerve fibre group composition, as has been begun [50]. This seems possible since the radial decline of the AP amplitude due to volume conductance [17] is reasonably small for very thin roots or fascicles. The area between the AP curve and the baseline is in a first approximation, the same for all nerve fibres, so that CAPs can be analysed with the simultaneously recorded single-fibre APs [50]. In this way, it seems possible to measure the conduction velocities of thin nerve fibre groups, in which the single-fibre APs cannot be recorded anymore, using conduction velocities of CAPs (see Figure 9C). The identification of further peak group nerve fibre diameters is more difficult. Firstly, there are afferent nerve fibre groups with strongly overlapping diameter distributions, and afferent fibres cannot be distinguished from efferent fibres. For a further identification of nerve fibre group diameters, peripheral nerves of known group composition have to be analysed morphometrically, and this work has already been started [41-44]. A comparison of nerve fibre diameters of fibres of the spinal canal with those at more distance from the medulla in the periphery is possible (if there is no branching of fibres) because the tapering of nerve fibres is small (0.2% per 13cm [70]).

1.5.6. Recording Conditions

The optimal inter-electrode distance for obtaining the highest AP amplitudes is approximately 6mm; the distance of 4mm used was close to that value. Low-resistance electrical shunts reduce the AP amplitude and result from the saline solution around the root filaments, the cerebral fluid (plus 0.9% NaCl from washing the roots) of the spinal canal connected via the root filaments emerging from the fluid, and the blood vessels in the root. The first low-resistance shunt is reduced by removing the saline film around the root (soaking up with a wet tissue and drying the root). The second shunt is reduced by lifting the up to 200mm long root [35] 10 to 20mm out of the cerebral fluid, which is not always possible (anatomical variation, arachnoiditis). Since thicker sacral nerve roots have comparably less contributions from blood vessels to the root cross section than the thinner ones (Figure 6A, Table 1 of Ref. 64), single-fibre extracellular APs can be recorded up to a fascicle diameter of 1mm. S4 ventral roots are ideal for recording, because they are thin (large AP amplitude), do not contain too many afferents (~1%, low activity) and somatic and parasympathetic afferents and efferents are contained. Drying up of roots should be avoided because of axonal damage; normally, drying up results in a dramatic activity increase of mirror-picture potentials (an AP generated between the electrode pairs travels ortho- and antidromically and shows an afferent waveform on one electrode pair trace (upwards) and an efferent form (downwards) on the other trace), and in seemingly very fast AP conduction. Wetting of the roots with 0.9% NaCl stops the generation of artefact potentials. Incorrect plugging of the wire electrodes can change the AP waveform. With the recording of touch activity [39,44], it can easily be proved that the wiring is correct. One advantage of the recording with wire electrode pairs and differential input is that no Faraday cage (normally not available in operating theatres) is needed, which is an important point in this new recording technique. The unscreened length of the recording electrodes was about 15mm. Further screening was achieved by positioning the platinum wires close to the electrode holder, made of bronze. Covering the operational area with mull soaked with 0.9% NaCl at 37°C creates a Faraday cage and a moist chamber (no drying up), and the temperature is kept more constant because of reduced evaporation. The exact measurement of the root temperature is a problem; in the future, it might partially be solved by comparing conduction velocities measured from nerve roots of the opened spinal canal with those of the lower limbs where the temperature can easily be measured. For such comparisons, conduction velocities of nerve fibre groups in limb nerves have to be defined according to the classification scheme of the human peripheral nervous system and measured exactly. Paraffin oil cannot be used during the surgery as it changes the excitability of membranes [19].

1.5.7. Unchanged Axon Numbers and Axon Densities of the Lower Sacral Nerve Roots Following Spinal Cord Lesion

Figure 6A shows different nerve root cross sections. It can be seen that the right root dS4 is approximately twice as thick as either root vS4 or root S5. Root dS2 is approximately four times (2x2) as thick as root dS4. The findings are in accordance with the measurements on human cadavers, showing that dorsal roots in the sacral segments are approx. twice as thick as the ventral ones, and that ventral or dorsal sacral roots (~5 cm away from the medulla) are of approximately twice the diameter of the adjacent caudal root [36]. The dorsal S2 root (Figure 6A) has approx. 31,600 myelinated axons (patient aged 25 years). Humans aged 47-85 years without spinal cord lesions had 28,000 myelinated axons in the dorsal S2 root. Since the number of axons (and the nerve conduction velocity) reduces with the age [28], it can be stated that the fibre number in the dorsal S2 root did not change following spinal cord lesion. The axon density in the S2 root is $a = \text{number}/d^2\pi/4 = 31600/2 = 15{,}800$ myelinated axons / mm^2. For normal humans (age 47-85 years), the axon density per mm^2 is 12,000 [36]. Allowing for age, these values are very similar again. The three roots dS4 shown in Table 1 had mean myelinated axon numbers of 6,600 and a mean density of 22,600 axons per mm^2 (age 26 years). Comparable values for normal (older) humans are 4,800 axons and 13,800 axons / mm^2,respectively. With age taken into account, it can be concluded that the number of fibres in the sacral dorsal roots had not changed following spinal cord lesion. The ventral S4 root in Figure 6A contained 900 myelinated fibres and had a myelinated axon density of 12,000 per mm^2. The values for the normal (older) population are 800 myelinated axons and 9,000 myelinated axons / mm^2 [36]. Again, the values for the ventral root in the paraplegics are very similar to those for the normal population. Even though the evidence has been based on a few cases, it can be concluded that the numbers of myelinated fibres in dorsal and ventral lower sacral nerve roots did not change 0.5 to 6 years following a spinal cord lesion.

Further, hemorrhagic necrosis and oedema affect the cord segments adjacent to the injured cord segment. Later changes include gliosis along the degenerating tracts that normally connect the sacral cord and brain. Some patients with complete lesions well above the level of the sacral cord also have evidence of trauma or infarction of the lumbosacral cord. Such patients have a flaccid bladder and little sphincter tone. These pathological changes may partly explain the variation that can be seen in urodynamic findings of groups of paraplegic patients in some studies [54,56].

Since the number of myelinated fibres in the sacral nerve roots did not change (no evidence of motoneuron cell death), and the operated paraplegics had a spastic bladder, it is unlikely that the necrosis reached the sacral spinal cord. Because interneuron cell death has also not been found so far, bladder dysfunction probably results from changed afferent input to the spinal cord and a disorganization of the neuronal networks of the isolated cord (see below).

1.5.8. Increased Bladder Afferent Activity a Reason for Urinary Bladder Dysfunction

It has been shown above, by comparing with normal values, that the numbers of myelinated fibres and the density of myelinated nerve fibres in the lower sacral ventral and dorsal roots did not change following spinal cord lesion. Also, it has been shown, based on a comparison with values for brain-dead humans and patients with no spinal cord lesion, that the values of group conduction velocities and group nerve fibre diameters remained unchanged. Thus, the nerve fibres of the peripheral nervous system were not affected by any change following spinal cord lesion unless some additional degenerative process had occurred in the periphery (nerve roots emerging from the medulla close to the spinal cord lesion will have changed nerve fibre numbers because of a damage also to the gray matter). This does not mean, however, that the impulse traffic has not changed. A dramatic increase in the activity of the stretch (S1) and tension receptor afferents (ST) was shown with certainty in two paraplegics (Figure 8). Also, in several other patients, bladder afferent activity seemed increased, as the bladder afferent activity was prominent for even small bladder filling volumes. Nearly all patients who underwent surgery had a much smaller storage volume of the urinary bladder but not necessarily a reduced compliance (filling volume / pressure ratio). Following urinary bladder de-afferentiation, the bladder storage volume increased as did the compliance. However, there are also paraplegics with a storage volume of 50 ml and no increased compliance. They are said to have hyperreflexia of the bladder. Thus, there are patients with a stiff and infected bladders in whom the afferent activity of the bladder is much higher (see also [51]). The afferent activity is often even present with an empty bladder. The increased bladder afferent activity causes the detrusor being activated too early [52]. On the other hand, there are patients with no increased compliance and no infection and a detrusor that already contracts at filling volumes of 50 ml. It will be shown in the two following papers [51,52] that there are important changes in the somatic and parasympathetic neuronal network of the sacral micturition centre and in the coordination between them; this can account for an early activation of the detrusor and the external sphincter. It will further be shown in a following paper [78] that most of the complications of urinary bladder functions can be avoided and urinary bladder functions can be cured if the learning-based movement therapy, called "coordination dynamics therapy," is started early after the spinal cord injury (best one or two weeks after injury) and is administered for three years.

1.5.9. Electrical Intravesical Stimulation and Clinical Implications

Electrical stimulation of the nearly empty urinary bladder with 30 V (0.3 ms pulse duration, negative electrode in the bladder, contact gel used) evoked CAPs of bladder stretch and tension receptor afferents. Even though more experience is needed to reliably activate the bladder afferents by electrical stimulation, the group conduction velocities of bladder afferents obtained from the CAPs were in accordance with the values obtained from single-fibre AP recordings. This accordance is of importance at least for two reasons. Firstly, there are no other precise data on human urinary bladder afferents available. Secondly, if intravesical stimulation can be used as a reliable identification tool of urinary bladder afferents, the diagnosis of bladder innervation during a surgery can be more refined and provide for a more selective de-afferentiation and more selective electrical stimulation, thus improving this method of (invasive, destructive, non-causal) treatment. Also, the possible biological approach of improving urinary bladder function by a nerve anastomosis from the lower intercostal nerves to the cauda equina nerve roots with respect to functional aspects (invasive, partly destructive, only partly causal) [41-44] needs extremely detailed knowledge of the bladder innervation and function, and intraoperative verification of afferent and efferent fibres innervating the bladder, as the representation of the urinary bladder in roots S2 to S5 (mainly S3 and S4) varies.

1.5.10. Proper Treatment of Urinary Bladder Function

Since coordination dynamics therapy can improve CNS motor functions in incomplete spinal cord injury [67] to a near-total cure [68,86] and can cure urinary bladder function in severe cervical spinal cord injury [78], the sacral anterior root stimulation according to Brindley [7] and the intercostal nerve to cauda equina nerve root anastomosis to improve urinary bladder functions [41-44] are not justified anymore in (young) patients with spinal cord injury because they are destructive treatments. Coordination dynamics therapy, on the other hand, is a causal natural learning-based movement therapy, in which the patient is not loosing any function. He can only regain further functions. At the beginning of this research project, I developed myself a nerve anastomosis operation for urinary bladder control (most important function for patients to be reconstructed) on a high scientific level [41-44], but I did not want to apply the treatment at the end because it was still a destructive operation. Patients with spinal cord injury (especially in severe cervical spinal cord injury) have so few functions left that they should not loose any further functions with the treatment. For scientific reasons, however, it is still interesting to discuss these treatments. Also, there may be patients who do not want to train or cannot perform a movement therapy for other reasons. For them, there would then still be these treatments available.

References

[1] Arbuthnott, E.R., Boyd, I.A. and Kalu, K.U.: Ultrastructural dimensions of myelinated peripheral nerve fibre in the cat and their relation to conduction velocity. *J. Physiol.*, 308: 125-157, 1980.

[2] Behse, F. and Buchthal, F.: Normal sensory conduction in the nerves of leg in man. *J. Neurol. Neurosurg. Psychiat.*, 34: 404-414,1971.

[3] Boyd, I.A. and Davey, M.R.: *Composition of peripheral nerves*. Livingston, Edinburgh, 1968.

[4] Bohm, E.: Sacral rhizopathies and sacral root syndromes (SII-SV). *Acta Chir.* Scand., 216: 5-48, 1956.

[5] Bornstein, J.C., Furness, J.B. and Kunze, W.A.A.: Electrophysiological characterization of myenteric neurons - How do classification schemes relate (Review). *J. Auton. Nerv. Syst.*, 48: 1-15, 1994.

[6] Boyd, I.A. and Kalu, K.U.: Scaling factor relating conduction velocity and diameter for myelinated afferent nerve fibres in cat hindlimb. *J. Physiol.*, 289: 277-297, 1979.

[7] Brindley, G.S., Polkey, C.E., Rushton, D.N. and Cardozzo, L.: Sacral anterior root stimulation for bladder control in paraplegia: the first 50 cases. *J. Neurol. Neurosurg. Psychiat.*, 49: 1104-1114, 1986.

[8] Buchthal, F. and Rosenfalck, A.: Evoked potentials and conduction velocity in human sensory nerves. *Brain Research,* 3: 1-122, 1966.

[9] Burke, R.E., Levine, D.N., Tsairis, P. and Zajak III, F.E.: Physiological types and histochemical profiles in motor units of the cat gastrocnemius. *J. Physiol.*, 234: 723-748, 1973.

[10] Dorfman, L.J., Cummins, K.J. and Leifer, L.J.: *Conduction Velocity Distributions - A Population Approach to Electrophysiology of Nerve*, Alan R. Liss, Inc., New York, 1981.

[11] Erlanger, J. and Gasser, H.S.: *Electrical signs of nervous activity.* Philadelphia, University of Pennsylvania Press, 1937.

[12] Fleshman, J.W., Munson, J.B., Sypert, G.W. and Friedman, W.A.: Rheobase, input resistance, and motor unit type in medial gastrocnemius motoneurons in the cat. *J. Neurophysiol.,* 41: 1326-1338, 1981.

[13] Garry, R.C. and Garven, H.S.D.: The ganglia, afferent endings and musculature of the urethra in the cat. *J. Physiol.*, 139: 1P-2P, 1957.

[14] Garry, R.C., Roberts, T.D.M. and Todd, J.K.: Reflex responses of the external urethral sphincter of the cat to filling of the bladder. *J. Physiol.*, 139: 13P-14P, 1957.

[15] Garry, R.C., Roberts, T.D.M. and Todd, J.K.: Reflexes involving the external urethral sphincter in the cat. *J. Physiol.*, 149: 653-665, 1959.

[16] Gasser , H.S. and Grundfest, H.: Axon diameter in relation to the spike dimension and the conduction velocity in mammalian A fibres. *Am. J. Physiol.*, 127: 393-414, 1939.

[17] Gath, I. and Stahlberg, E.: On the volume conduction in human skeletal muscle: in situ measurements. *Electroencephalo. clin. Neurophysiol.*, 43: 106-110, 1977.

[18] Gustafson, B. and Pinter, M.J.: Relations among passive properties of lumbar α-motoneurons of the cat. *J. Physiol.*, 356: 401-431, 1984.

[19] Hodgkin, A.L.: The local electric changes associated with repetitive action in a non-medullated axon. *J. Physiol.*, 107: 165-181, 1948.

[20] Hunt, C.C.: Relation of function to diameter in afferent fibres of muscle nerve, *J. Gen. Physiol.,* 38: 117-131, 1954.

[21] Hursh, J.B.: Conduction velocity and diameter of nerve fibres. *Am. J. Physiol.*, 127: 131-139, 1939.

[22] Iggo, A.: Tension receptors in the stomach and the urinary bladder. *J. Physiol.*, 128: 593-607, 1955.

[23] Julian, F.J. and Goldmann, D.E.: The effect of mechanical stimulation on some electrical properties of axons. *J. Gen. Physiol.*, 46: 297-313, 1962.

[24] Kernell, D. and Zwagstra, B.: Input resistance, axon conduction velocity and cell size among hindlimb motoneurons of the cat. *Brain Research*, 204: 311-326, 1981.

[25] Koller, S., Reichertz, P.L. and Überla, K.: *Medizinische Informatik and Statistik.* Springer Verlag, Berlin, 1984.

[26] Kugelberg, E.: "Injury activity" and "trigger zones" in human nerves. *Brain*, 69: 310-324, 1946.

[27] Lloyd, D.P.C.: Neuron patterns controlling transmission of ipsilateral hind limb reflexes in cat. *J. Neurophysiol.*, 6: 293-315, 1943.

[28] Ludin, H.P.: *Praktische Elektromyographie*, Enke Verlag, Stuttgart, 1993.

[29] Paintal, A.S.: Effect of temperature on conduction in single vagal and saphenus myelinated nerve fibres of the cat. *J. Physiol.*, 180: 20-49, 1965.

[30] Paintal, A.S.: The influence of diameter of medullated nerve fibres of cats on the rising and falling phase of the spike and its recovery. *J. Physiol.*, 184: 791-811, 1966.
[31] Paintal, A.S.: Conduction in mammalian nerve fibres. In: J.E. Desmedt (Ed.). *New Developments in Electromyography and Clinical Neurophysiology*, Vol. 2, Karger, Basel, pp.19-41, 1973.
[32] Rosenfalck, P.: Intra- and extracellular potential fields of active nerve and muscle fibres. A physio-mathemathical analysis of different model. *Acta Physiol.* Scand., Suppl. 321: 1-168, 1969.
[33] Rudge, P., Ochoa, J. and Giliatt, R.W.: Acute peripheral nerve compression in the baboon, *J. Neurological Sciences*, 23: 403-420, 1974.
[34] Rushton, W.A.H.: Resistance artefact in action potential measurements. *J. Physiol.*, 104: 19P-20P, 1945.
[35] Schalow, G.: The problem of cauda equina nerve root identification. *Zbl. Neurochir.*, 46: 322-330, 1985.
[36] Schalow, G., Aho, A. and Lang, G.: Nerve fibre counts for an intercostal nerve to cauda equina nerve root anastomosis. *Zent. bl. Chir.*, 112: 457-461, 1987.
[37] Schalow, G. and Lang, G.: Electrodiagnosis of human dorsal sacral nerve roots by recording afferent and efferent extracellular action potentials. *Neurosurg. Rev.,* 12: 223-232, 1989.
[38] Schalow, G.: Efferent and afferent fibres in human sacral nerve roots: basic research and clinical implications. *Electromyogr. clin. Neurophysiol.*, 29 (1989) 33-53.
[39] Schalow, G.: Conduction velocities and nerve fibre diameters of touch, pain, urinary bladder and anal canal afferents and α and γ-motoneurons in human dorsal sacral nerve roots. *Electromyogr. clin. Neurophysiol.*, 31: 265-296, 1991.
[40] Schalow, G.: Oscillatory firing of single human sphincteric α_2 and α_3-motoneurons reflexly activated for the continence of urinary bladder and rectum. Restoration of bladder function in paraplegia. *Electromyogr. clin. Neurophysiol.*, 31: 323-355, 1991.
[41] Schalow, G., Aho, A. and Lang, G.: Microanatomy and number of nerve fibres of the lower intercostal nerves with respect to a nerve anastomosis. Donor nerve analysis. I (IV). *Electromyogr. clin. Neurophysiol.*, 32: 171-185, 1992.
[42] Schalow, G.: Number of fibres and fibre diameter distributions of nerves innervating the urinary bladder in humans. Acceptor nerve analysis. II (IV). *Electromyogr. clin. Neurophysiol.*, 32: 187-196, 1992.
[43] Schalow, G. and Barth, H.: Single-fibre action potential recording from the nerve to the musculus obliquus externus abdominis following pinprick in humans. III (IV). *Electromyogr. clin. Neurophysiol.*, 32: 197-205, 1992.
[44] Schalow, G.: Impulse pattern, innervation density and two point discrimination of skin and mucosal afferents in humans. Consideration for a sensory reinnervation of urinary bladder and anal canal in spinal cord lesion. IV (IV). *Electromyogr. clin. Neurophysiol.*, 32: 259-285, 1992.
[45] Schalow, G. and Barth, H.: Group conduction velocities and nerve fibre diameters of α and γ-motoneurons from lower sacral nerve roots of the dog and humans. *Gen. Physiol. Biophys.*, 11: 85-99, 1992.
[46] Schalow, G.: Ventral root afferent and dorsal root efferent fibres in dog and human lower sacral nerve roots. *Gen. Physiol. Biophys.*, 11: 123-131, 1992.

[47] Schalow, G. and Wattig, B.: Recruitment of α and γ-motoneurons in rats, dogs and humans. *Electromyogr. clin. Neurophysiol.*, 33: 387-400, 1993.
[48] Schalow, G.: Recruitment of motoneurons in the occasional firing mode in paraplegics. *Electromyogr. clin. Neurophysiol.*, 33: 401-408, 1993.
[49] Schalow, G.: Action potential patterns of intrafusal γ and parasympathetic motoneurons, secondary muscle spindle afferents and an oscillatory firing α_2-motoneuron, and the phase relations among them in humans. *Electromyogr. clin. Neurophysiol.*, 33: 477-503, 1993.
[50] Schalow, G. and Zäch, G.A.: Nerve compound action potentials analyzed with the simultaneously measured single-fibre action potentials in humans. *Electromyogr. clin. Neurophysiol.*, 34: 451-465, 1994.
[51] Schalow, G., Bersch, U., K. Göcking and Zäch, G.A.: Detrusor-sphincteric dyssynergia in paraplegics compared with the synergia in a brain-dead human by using the single-fibre action potential recording method. *J. Auton. Nerv. Syst.*, 52: 151-180, 1995.
[52] Schalow, G., Bersch, U. Michel, D. and Koch, H.G.: Detrusor-sphincteric dyssynergia in humans may be caused by a loss of stable phase relations between and within oscillatory firing neuronal networks of the sacral micturition center. *J. Auton. Nerv. Syst.*, 52: 181-202, 1995.
[53] Sherrington, C.S.: Nervous rhythm arising from rivalry of antagonistic reflexes: Reflex stepping as outcome of double reciprocal innervation. *Proc. R. Soc. Lond.* (Biol.), 86: 233-261, 1913.
[54] Thomas, D.G.: The urinary tract following spinal cord injury. In G.D. Chisholm and D.I. Williams (Eds.), *Scientific foundation of urology, 2nd edn.* Heinemann, London, pp. 421-433, 1982.
[55] Torebjörk, H.E., Hallin, R.G., Hongell, A. and Hagbarth, K.-E.: Single unit potentials with complex waveform seen in microelectrode recordings from the human median nerve. *Brain Research*, 24: 443-450, 1970.
[56] Torrens, M. and Morrison, J.F.B.: *The Physiology of the Lower Urinary Tract*, Springer Verlag, Berlin, p. 224, 1987.
[57] Vallbo, A.B., Hagbarth, K.-E., Torebjörk, H.E. and Wallin, B.G.: Somatosensory, proprioceptive, and sympathetic activity in human peripheral nerves. *Physiological Reviews,* 59: 919-957, 1979.
[58] Vallbo, A.B.: Prediction of propagation block on the basis of impulse shape in single unit recordings from human nerves. *Acta Physiol. Scand.*, 97: 66-74, 1976.
[59] Wood, J.D.: Application of classification schemes to the enteric nervous system (Review), *J. Auton. Nerv. Syst.,* 48: 17-29, 1994.
[60] Woodbury, J.W.: Direct membrane resting and action potentials from single myelinated nerve fibres. *J. Cellular and Comparativ. Physiol.*, 39: 323-339, 1952.
[61] Zotterman, Y.: Touch, pain and tickling: An electrophysiological investigation on cutaneous sensory nerves. *J. Physiol.*, 95: 1-28, 1939.
[62] Schalow, G., Aho, A. and Lang, G.: Nerve fibre counts for an intercostal nerve to cauda equina nerve root anastomosis. *Zent.bl. Chir.*, 112: 457-461, 1987.
[63] Schalow, G.: Feeder arteries, longitudinal arterial trunks and arterial anastomoses of the lower human spinal cord. *Zbl. Neurochir.*, 51: 181-184, 1990.

[64] Schalow. G., Zäch, G.A. and Warzock, R.: Classification of human peripheral nerve fibre groups by conduction velocity and nerve fibre diameter is preserved following spinal cord lesion. *J. Auton. Nerv. Syst.*, 52: 125-150, 1995.
[65] Schalow, G.: Stroke recovery induced by coordination dynamic therapy and quantified by the coordination dynamic recording method. *Electromyogr. Clin. Neurophysiol.*, 42: 85-104, 2002.
[66] Schalow, G.: Improvement after traumatic brain injury achieved by coordination dynamic therapy. *Electromyogr. Clin. Neurophysiol.*, 42: 195-203, 2002.
[67] Schalow, G.: Recovery from spinal cord injury achieved by three months of coordination dynamic therapy. *Electromyogr. Clin. Neurophysiol.*, 42: 367-376, 2002.
[68] Schalow, G.: Partial cure of spinal cord injury achieved by 6 to 13 months of coordination dynamic therapy. *Electromyogr. Clin. Neurophysiol.*, 43: 281-292, 2003.
[69] Schalow, G., Pääsuke, M., Ereline, J. and Gapeyeva, H.: Improvement in Parkinson's disease patients achieved by coordination dynamics therapy. *Electromyogr. Clin. Neurophysiol.*, 44: 67-73, 2004.
[70] Schalow, G.: Tapering of human nerve fibres. *Gen. Physiol. Biophys.*, 24: 427-448, 2005.
[71] Schalow, G., Pääsuke, M., Jaigma, P.: Integrative reorganization mechanism for reducing tremor in Parkinson's disease patients. *Electromyogr. Clin. Neurophysiol.*, 45: 407-415, 2005.
[72] Schalow, G., Jaigma, P.: Cerebral palsy improvement achieved by coordination dynamics therapy. *Electromyogr. Clin. Neurophysiol.*, 45: 433-445, 2005.
[73] Schalow, G.: Hypoxic brain injury improvement induced by coordination dynamics therapy in comparison to CNS development. *Electromyogr. Clin. Neurophysiol.*, 46: 171-183, 2006.
[74] Schalow, G. and Jaigma, P.: Improvement in severe traumatic brain injury induced by coordination dynamics therapy in comparison to physiologic CNS development. *Electromyogr. Clin. Neurophysiol.*, 46: 195-209, 2006.
[75] Schalow, G.: Symmetry diagnosis and treatment in coordination dynamics therapy. *Electromyogr. Clin. Neurophysiol.*, 46: 421-431, 2006.
[76] Schalow, G.: Cerebellar injury improvement achieved by coordination dynamics therapy. *Electromyogr. Clin. Neurophysiol.*, 46: 433-439, 2006.
[77] Schalow, G.: Cure of urinary bladder functions in severe (95%) motoric complete cervical spinal cord injury in human. *Electromyogr. Clin. Neurophysiol.*, 50: 155-179, 2010.
[78] Martino, G. and Pluchino, S.: The therapeutic potential of neural stem cells. *Nat. Rev. Neurosci.*, 7: 395-406, 2006.
[79] Thuret, S., Moon, L.D.F. and Gage, Gage, F.H.: Therapeutic interventions after spinal cord injury. *Nat. Rev. Neurosci.*, 7: 628-643, 2006.
[80] Bradbury, E.J. and McMahon, S.B.: Spinal cord repair strategies: why do they work? *Nat. Rev. Neurosci.*, 7: 644-653, 2006.
[81] Harel, N.Y. and Strittmatter, S.M.: Can regenerating axons recapitulate developmental guidance during recovery from spinal cord injury. *Nat. Rev. Neurosci.*, 7: 603-616, 2006.
[82] Kelso J.A.S.: Dynamic Patterns. *The Self-Organization of Brain and Behavior.* MIT Press, Cambridge, 1995.

[83] Schalow, G., Vaher, I. and Jaigma, P.: Overreaching in coordination dynamics therapy in an athlete with a spinal cord injury. *Electromyogr. Clin. Neurophysiol.*, 48: 83-95, 2008.

[84] Schalow, G.: Stem cell therapy and Coordination dynamics therapy to improve Spinal cord injury. *Electromyogr. Clin. Neurophysiol.*, 48: 233-253, 2008.

[85] Schalow, G., Jaigma, P. and Belle, V.K. : Near-total functional recovery achieved in partial spinal cord injury (50% injury) after three years of coordination dynamics therapy. *Electromyogr. Clin. Neurophysiol.*, 49: 67-91, 2009.

[86] Schalow, G.: Partial cure achieved in a patient with near-complete cervical spinal cord injury (95% injury) after three years of coordination dynamics therapy. *Electromyogr. Clin. Neurophysiol.*, 49: 199-221, 2009.

[87] Schalow, G.: The classification and identification of human somatic and parasympathetic nerve fibres including urinary bladder afferents and efferents is preserved following spinal cord injury. *Electromyogr. Clin. Neurophysiol.*, 49: 263-286, 2009.

[88] Schalow, G.: Coordination impairment between the somatic and parasympathetic nervous system divisions in the human sacral micturition centre following spinal cord injury. *Electromyogr. Clin. Neurophysiol.*, 49: 337-367, 2009.

[89] Schalow, G.: Phase relation changes between the firings of α and γ-motoneurons and muscle spindle afferents in the sacral micturition centre during continence functions in brain-dead human and patients with spinal cord injury. *Electromyogr. Clin. Neurophysiol.*, 50: 3-27, 2010.

[90] Schalow, G.: Scientific basis for learning transfer from movements to urinary bladder functions for bladder repair in patients with spinal cord injury. *Electromyogr. Clin. Neurophysiol.*, 50: 339-395, 2010.

2. Coordination Impairment between the Somatic and Parasympathetic Nervous System Divisions in the Human Sacral Micturition Centre Following Spinal Cord Injury

The detrusor-sphincteric dyssynergia is analyzed by comparing the natural impulse patterns of secondary muscle spindle afferents (SP2) contributing to continence (SP2 fibre activity changes are similar to detrusor pressure changes) and sphincteric motoneurons in a brain-dead human with those in patients with spinal cord injuries. In the brain-dead, the sphincteric motoneurons, subserving continence, were inhibited at a time when preganglionic parasympathetic efferents and a SP2 fibre increased their activity (physiologic). In paraplegics, the sphincteric motoneurons were not inhibited (pathophysiologic). In the brain-dead, an SP2 fibre showed doublet firing (interspike interval (II) 10 to 14ms) for low-level parasympathetic activation and multi-ending regular firing for high parasympathetic activation. In one paraplegic with strong bladder dysfunction, the multi-ending regular firing was replaced by a repeated burst firing with a shortest II of 0.2ms (transmission frequency = 5000Hz). The pathologic firing patterns of the SP2 fibres, the detrusor-sphincteric dyscoordination, and hyperreflexia in paraplegics are most likely a result of neuronal network changes in the parasympathetic and somatic nervous system divisions of the sacral micturition center after spinal cord injury.

2.1. Data Summary

1. Humans with spinal cord injury often show detrusor-sphincteric dyssynergia of the urinary bladder. The somatic external bladder sphincter is activated at the same time as the detrusor (smooth muscle) with the consequence that the urinary bladder cannot be emptied. This dyscoordination between the somatic and parasympathetic nervous system divisions in the human sacral micturition centre is reflected urodynamically in the simultaneous increase of the detrusor pressure and the electromyographic activity of the pelvic floor.
2. The time course of the increase in the secondary muscle spindle afferent activity, induced by the parasympathetic nervous system in muscle spindles contributing to continence, is very similar to that of detrusor pressure. The detrusor-sphincteric dyssynergia is therefore analyzed by comparing the natural impulse patterns of single secondary muscle spindle afferents (SP2) and sphincteric motoneurons in a brain-dead human (physiologic) with those in patients with spinal cord injury (pathologic). The parasympathetic nervous system was activated by painful bladder catheter pulling.
3. In a brain-dead human, the sphincteric motoneurons subserving continence were inhibited at a time when preganglionic parasympathetic efferents increased their activity (physiologic) for 10s and an SP2 fibre increased its activity for several minutes. In a paraplegic with a strong bladder dysfunction, the SP2 fibre activity increased, due to parasympathetic activation, lasted for approx. one minute, showed undulations, and its amplitude was smaller than that measured in a brain-dead human. The sphincteric motoneurons were not inhibited (pathologic).
4. In the brain-dead human, an SP2 fibre showed doublet firing (probably physiologic) with interspike intervals (IIs) of duration between 10 and 14 ms for low-level parasympathetic activation. For high-level parasympathetic activation, this single parent spindle afferent fibre showed multi-ending regular firing of up to six endings (probably physiologic) with IIs of a duration of predominantly 15 to 25 ms. In one paraplegic with a strong bladder dysfunction, the doublet firing was less regular, even though two II peaks at 10.2 and 11.2 ms occurred in a II distribution similar to in the brain-dead human. The multi-ending regular firing was replaced by a repeated burst firing (probably pathologic). In a second paraplegic with strong detrusor-sphincteric dyssynergia, the burst firing consisted of up to six impulses with increasing IIs and a first II of approx. 0.2 ms (transmission frequency = 5000 Hz,). In a third paraplegic with a less strong dysfunction of the bladder, a highly activated SP2 fibre showed an activity pattern intermediate to those of multi-ending regular firing and burst firing. The doublet and multi-ending regular firing of parasympathetically innervated muscle spindles may represent messenger function, which becomes impaired following spinal cord injury.

5. The time constant for the activity decrease of a spindle afferent fibre following parasympathetic activation was 31s in a paraplegic and approx. 40s in a brain-dead human. It is concluded that the muscle spindles are unchanged following spinal cord injury. The pathologic firing patterns of the SP2 fibres are thus probably a result of neuronal network changes in the parasympathetic and somatic nervous system divisions of the sacral micturition centre.

2.2. Introduction

The goal of this current research is to cure spinal cord injury in general and urinary bladder functions in particular. After having shown in the previous part that the identification of somatic and parasympathetic nerve fibres by the group conduction velocity and the group nerve fibre diameter had not changed following spinal cord injury, it will be analyzed here, on the basis of natural impulse patterns, travelling into and out of the sacral micturition centre, what became pathologic in the organization of the central nervous system (CNS) following spinal cord injury, with the goal in mind to develop cure.

Normal voiding is a voluntary act that results in sustained contraction of the bladder (detrusor) and relaxation of the urethra until the bladder is empty. To enable fluid flow along the urethra, it is necessary that the pressure in the urinary bladder exceeds that within the urethra lumen. Under normal circumstances, in order to initiate micturition, a fall in urethral pressure immediately precedes a rise in pressure within the lumen of the bladder (Figure 11.13 in [70]). Usually, this pressure rise is produced by active contraction of detrusor smooth muscle at the onset of micturition. The extensive vesical part of the pelvic plexus and the profuse distribution of autonomic motor nerves emphasize the importance of the autonomic nervous system in initiating and sustaining bladder contraction during micturition. Immediately prior to the onset of micturition, the tonus of the striated external bladder sphincter (rhabdosphincter) is reduced by central inhibition of its somatic motor neurons located in the third, fourth and fifth sacral spinal segments. This inhibition is mediated by descending spinal pathways originating in higher centres of the central nervous system (CNS) [70]. The central integration of the nervous control of the bladder and urethra is essential for normal micturition.

Immediately following complete spinal cord injury, there is a period of spinal shock that lasts about three weeks or longer. This period is characterized by muscular flaccidity and a loss of spinal reflexes. The recovery of reflex activities below the level of the injury occurs at different times. If bladder reflexes reappear, they differ in some important respects from those in a normal individual. The "autonomic bladder" contracts in response to distension, but the power is rarely that of the normal bladder, and the residual volume increases. Lesions above the level of the brain stem centres lead to involuntary voiding that is coordinated with sphincter relaxation (detrusor-sphincteric synergia). Lesions below the level of the brain stem centres but above the lumbosacral spinal cord lead, after a period of bladder paralysis associated with spinal shock, to involuntary reflex voiding that is not coordinated with sphincter relaxation (detrusor-sphincteric dyssynergia) (Figure 11.2 of [70]). In this paper, the coordination between the detrusor and the external bladder sphincter will be compared between paraplegic patients with lesions below the brain stem and detrusor-sphincteric dyssynergia (pathologic) and a brain-dead human with an at least partly destroyed brain stem and an at least partial detrusor-sphincteric synergia (partial physiologic).

If the spinal cord injury is complete, it is believed that paraplegics develop vesicosphincteric dyssynergia because of the disconnection from the pontine micturition center, which is responsible for the coordination of the detrusor and the external bladder sphincter (see above). In cases in which no detrusor-sphincteric dysynergia has developed, the spinal cord injury is believed to be incomplete, so that supraspinal centers can still coordinate bladder functions. The external bladder sphincter will not be "coactivated" with the detrusor in response to distension. The view that the important function of coordination is performed by spino-bulbo-spinal pathways has incorporated anatomical ideas of a hierarchy of functions of increasing complexity, the further rostrally one goes in the CNS. Still, the coordination for the mutual inhibition of detrusor and external bladder sphincter may be located in the sacral micturition center. In this and the following parts of this Chapter [84-86], I shall try to shed further light onto this dyscoordination problem, using the single-nerve fibre action potential (AP) recording method.

On the basis of such better electrophysiological understanding of human urinary bladder functions, a non-destructive learning-based movement therapy is developed to cure urinary bladder functions by using additional knowledge of the integrative functions of the human CNS [78,81] and learning transfer from special coordinated movements to urinary bladder functions [85]. It will be shown that such learning transfer is successful in repairing urinary bladder function [86], since the somatic and the autonomic divisions dovetail with one another with respect to those special movements and urinary bladder functions [84-86]. The necessary connection to the pontine micturition centre for the coordination control of the detrusor and the external bladder sphincter is achieved by limited regeneration of the human spinal cord [82,86] upon coordination dynamics therapy [81,82].

Functions of the CNS can be understood based on two important concepts, one old, another one new. The first is the evolutionary principle of levels of function, which implies that rostral segments of the brain have become dominant over caudal, and that when higher parts are removed many activities of lower segments are, after a time, "released" and can then be more readily analyzed [19].

The second, and a newer, concept relates to the extensive interaction normally occurring between somatic and autonomic reflexes. The autonomic division of the nervous system can no longer be regarded as a purely peripheral system but rather as an elaborately organized division of the CNS with representations at all levels. At each level, moreover, the somatic and the autonomic systems dovetail with one another [19].

For animals, reviews by de Groat [13] and de Groat et al. [15,16] provide a basis for understanding how the filling and voiding functions of the bladder work. de Groat [13] states that the primary stimulus for micturition is bladder distension, which induces reflex activation of the parasympathetic excitatory outflow to the bladder, depression of the sympathetic inhibitory outflow, and depression of the somatic efferent output to the external sphincter; secondary reflexes elicited by the passage of urine through the urethra may reinforce these primary reflexes and facilitate the complex emptying of the bladder. When the micturition reflex is exceeded, the parasympathetic reflex pathway through the brain stem is active. McMahon and Morrison [30-33,36] used a similar approach but attempted to separate functional pathways concerned with the reflex pathways and with the control of micturition threshold. For an introduction to the physiology of the lower urinary tract, see [70].

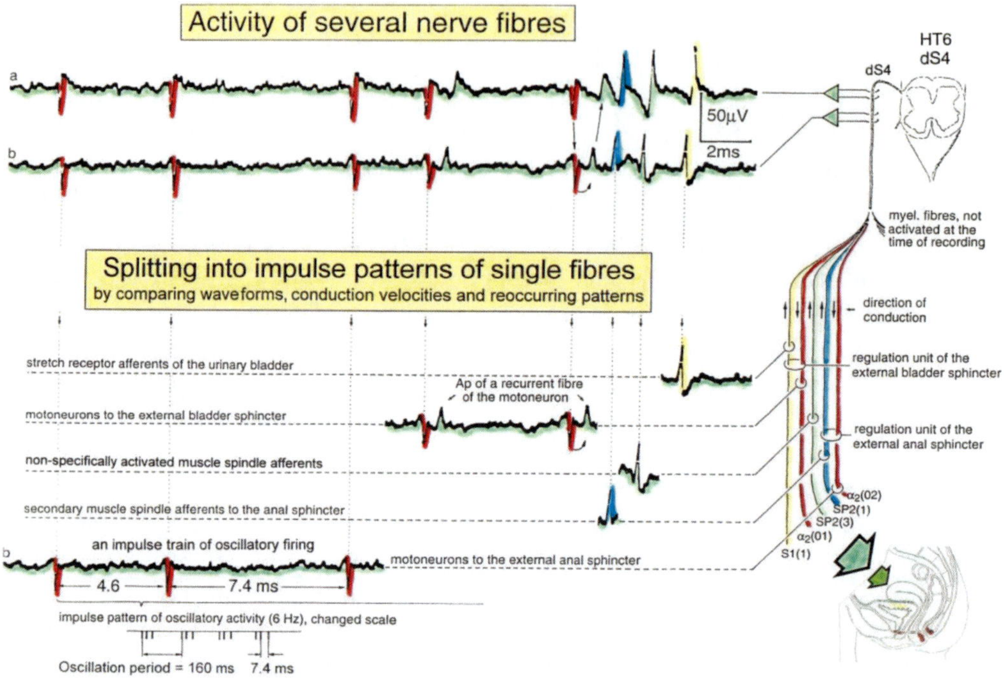

Figure 1. Schematic splitting of the activity of several nerve fibres into simultaneous impulse patterns of single fibres by comparing waveforms, conduction velocities and reoccurring characteristic impulse patterns (rhythmic firing of sphincteric motoneurons). The different conduction times and waveforms were recognized on an expanded time scale. "*" see Figure 10. of [59] for receptor sites and the location of the muscles that are innervated by the sphincteric motoneurons. Stretch receptor and secondary muscle spindle afferents contribute to the drive of sphincteric motoneurons and form, together with other afferents, regulation units.

Yates [74] has constructively criticized the quality of research in the area of human urinary bladder physiology. In particular, he has recommended that investigators should look at the methods being used in parallel disciplines.

On the basis of a newly developed basic human electro-neurophysiologic method, the recording of single-nerve fibre action potentials (APs), a classification scheme of the human peripheral nervous system has been developed [44,46,48], which still holds in paraplegics [58,83]. Since it is possible to simultaneously identify impulse patterns of single afferent and efferent nerve fibres (Figure 1), functions of the human CNS can be analyzed. The power of the method lies in the existence of a unique anatomical situation in humans: because of the ascensus of the human spinal cord, and as humans have no tails, urinary bladder functions are nearly purely represented in thin long, mainly S3 and S4, nerve roots. Because of the missing of epineurium and perineurium (apart from a thin layer of cells), single-nerve fibre APs can be ideally recorded.

Human lower urinary tract physiology and pathophysiology will be investigated with this electrophysiological method by comparing the measurements obtained from paraplegics with detrusor-sphincteric dyssynergia of the urinary bladder with those of a brain-dead human (HT6). The starting point will be routinely performed clinical urodynamic investigations.

2.3. Method

Measurements were performed in nine patients with spinal cord injury (paraplegics), dyssynergia of the urinary bladder, spastic pelvic floor and spasticity in general. Only those patients have been selected for demonstration who showed the properties (impulse patterns) of interest for the present investigations. The electrophysiological data obtained from a brain-dead human (HT6) were taken from previous analyses [54-56], in a few cases the original data were reevaluated. The ethical aspect of performing electrophysiologic measurements in patients is that the method of recording single-nerve fibre action potentials (APs) was primarily used for diagnosis during the surgery, and research was performed simultaneously. Actually, a close correlation between diagnosis and research measurements can be expected to bring progress into human physiology and pathophysiology. A close cooperation between clinic and research will be in the best interests of the patients in the long-term prospect.

2.3.1. Electrophysiology

Single-nerve fibre APs were recorded extracellularly from nerve roots with two pairs of platinum wire electrodes (electrode pair distance = 10 mm; electrode distance in each pair = 4 mm) at two sites, preamplified (x1000), filtered (passing frequency range 100Hz-10kHz) and displayed on a digital storage oscilloscope and also stored on a videotape using a PCM-processor and a video recorder (see also [58,83]). Trace “a” was the recording from the proximal electrode pair and trace “b” from the distal pair. Conduction velocities of single fibres were calculated from the conduction distance (electrode pair distance = 10mm) and the respective conduction times, the time needed for an AP to cover the conduction distance (time difference between traces “a” and “b” for a particular AP). APs from afferent and efferent fibres could clearly be distinguished since for the used electrode arrangements (differential electrodes), the main phase (second phase) from afferent fibres is upwards and that of efferent fibres downwards. Since the afferent and efferent APs reach the electrode pair from opposite directions, also the conduction times of afferent and efferent APs are opposite. The conduction velocities of afferent and efferent nerve fibres were plotted in velocity distribution histograms for afferent and efferent fibres. The nerve groups of afferent and efferent fibres were identified by the peaks and the ranges in the distributions and by the calibration relation (e.g., α_2-motoneurons conduct with the same velocity as do the secondary muscle spindle afferent fibres). The single-fibre APs of the multi-fibre recordings were then identified by the conduction velocity and assigned to their respective nerve fibre group (Figure 4). The extraction of single-fibre impulse patterns from a multi-fibre recording from a whole nerve root is performed using conduction time and AP waveform comparisons and reoccurring AP patterns (Figure 1). Different temperatures of the nerve roots do not change the patterns. A human ventral S4 root contains approximately 300 fibres with a diameter range down to 3 to 4 μm (Figures 4,5 of the previous part and Table 3 of Chapter III). In a certain recording, the situation may only be 5 to 50 fibres active. In general, the recording conditions have to be arranged so that the activity is not too high. The best approach under physiologic conditions is mostly to analyse motoneurons in the dorsal roots and afferent fibres in the ventral (motor) roots. Actually, the success of the method lies in the violation of the Dales principle. E.g., touching with a finger the S4 dermatome evokes a high transient skin afferent activity in the ipsilateral dorsal S4 root, so that single-nerve fibre APs can no longer be recognized. Upon pinpricking the coccygeal dermatome and recording from the coccygeal root on that side, excellent recordings can be obtained [46,52]. The members of the operating theatre were advised not to touch the patient or the covering towels to avoid response from the afferent and efferent fibres. Under pathophysiologic conditions, it may be difficult to obtain suitable low activity level recordings from lower sacral nerve roots. The afferent fibres of a ventral S3 or S4 root may fire with such a high activity that the APs of the afferent fibres cover the APs of the efferent fibres (cf. Figure 11e and Figure 11f). When retrieving signals from the tape after the surgery, good pieces of recordings were looked for: high AP amplitude, low activity, few artefacts, not too low a temperature. Sweep pieces of 0.4 and 0.8 s were fed from the tape to the oscilloscope and analyzed there. The APs from several single fibres are recognized by their shapes on trace "a" and trace "b" and the conduction times. By passing the window of the scope (0.5 to 2ms/cm) through a 0.8s sweep, the times of the occurrence of single-nerve fibre APs (Figure 4) are written down from the memory address of the scope. Simultaneous impulse patterns of afferent and efferent fibres are then drawn (e.g., Figure 8A,D). The calculation of interspike intervals (IIs) and II distribution histograms from the occurrence time data is easy. Also, activity levels can be easily constructed. The problem lies in the recognition of all APs from a particular fibre to obtain their times of occurrence. Often, the AP waveforms of several fibres are unique, and then comparably easy simultaneous impulse patterns can be constructed if the visual memory of the researcher is good. However, often the AP waveforms of several fibres are also rather similar on both traces as are the conduction times, especially because the digitization is mostly too poor (at least five digitization points per AP) to get a long sweep into the scope. The knowledge of impulse patterns will help because it can partly be predicted when a particular fibre will fire again. For example, a α_2-motoneuron firing with an impulse train of three APs in the oscillatory firing mode will on average fire again with a three AP impulse train after 160ms; thus, there is a high probability that the next firing occurs at that point. The procedure used for the analysis is the following if one has, e.g., three fibres from which APs can be recognized. By passing the window through a 0.8s sweep, the times of occurrence (memory address, cursor) are written down. If an AP waveform can only be identified with doubt or can be mixed up with those of another fibre, the situation is noted. Then, the impulse patterns of the three fibres of the 0.8s sweep are plotted. With the additional information of impulse patterns, and taking into account algebraic adding of potential changes of different APs and of noise and artifact potentials, the majority of doubtful AP waves can successfully be identified. If an important point cannot be identified, that particular sweep has to be omitted and another one taken from the tape. It is always good to have several stimulations of the same kind on the tape. Immediate plotting with a following reinspection of the doubtful cases is very efficient. It should not be concluded that the construction of impulse patterns is a speculation. Sometimes, there are only two motoneurons in a dorsal root (two thick dorsal root efferents) or one or two thick afferent fibres in a ventral root, so that their impulse patterns are obvious. In detail, thick fibres (fast conducting) generate APs of large amplitude, thin fibres (slowly conducting) of small amplitude [43,58,83]. The APs of unmyelinated fibres are normally not visible. The APs of a ventral root afferent fibre can be distinguished from the APs of the ventral root efferents by the opposite AP amplitude and conduction time. Often, the activity of the motoneurons is low, so that the impulse pattern of a single thick ventral root afferent fibre (multi-unit impulse pattern) is close to a single-unit impulse pattern. Because of the importance, it was always tried to get as much out of the recordings as possible. Each good recording is a small treasure. Since the brain of the researcher has to "store" the AP contours of several fibres (besides drawing them down) including their variations, the time basis should not be changed during the analysis. If fast (short AP duration) and slowly conducting fibres (long AP duration) are included in the identification, a sweep has to be inspected two times. The first time with the normal time base to recognize the fast potentials and a second time with a compressed time base to better recognize the slowly conducting fibres with a long AP duration, such as preganglionic parasympathetic neurons.

2.3.2. Urodynamic Measurements

The electromyographic recordings with surface electrodes from the pelvic floor and the bladder and rectal pressure measurements were performed with the cystometry apparatus Dantec UD5500.

2.4. Results

Single-nerve fibre action potentials (APs) were recorded extracellularly in three paraplegics during the implantation of an electrical anterior root stimulator for bladder control. Interspike intervals were calculated, impulse patterns constructed and activity levels compared with urodynamic parameters measured pre-operatively.

2.4.1. Dyssynergia of the Urinary Bladder

In Figure 2, typical detrusor-sphincteric dyssynergia of the urinary bladder is demonstrated, in a patient with a spinal cord injury sub TH12 present for two years. Because of a malformation of the ureters and ureteric reflux, this patient underwent surgery for the implantation of a sacral anterior root stimulator [9]. Without the malformation, the bladder dysfunction would have still been tolerable (see later). Upon retrograde bladder filling, the intravesical pressure increased. At a filling volume of 350 ml, the detrusor contracted automatically in response to bladder distension. The electromyographic (EMG) activity of the pelvic floor increased at a time when the detrusor pressure increased. If the external anal and the external (striated) bladder sphincters were activated at the same time, and this probably was the case, then the external bladder was activated at approximately the same time as the detrusor. This coactivation is called detrusor-sphincteric dyssynergia of the bladder, since in normal humans, the external bladder sphincter is relaxed (inhibited) upon the activation of the detrusor so that urine can pass through the urethra under low pressure.

The loss of sphincteric relaxation during detrusor contraction results in a high bladder pressure during voiding. A high bladder pressure will result in a detrusor hypertrophy with an increased bladder stiffness, as estimated from a reduced compliance (and, possibly, an increased bladder infection rate [65]); in addition, there probably will be an increase of stretch (S1) and tension (ST) [83] and pain receptor afferent activity of the bladder wall [70] resulting in a too early activation of the detrusor in response to it [59,84]. The storage capacity of the bladder may thus become reduced to 50 ml or even less. On the other hand, the intravesical pressure may cause ureteric reflux, and due to infections, progressive renal impairment.

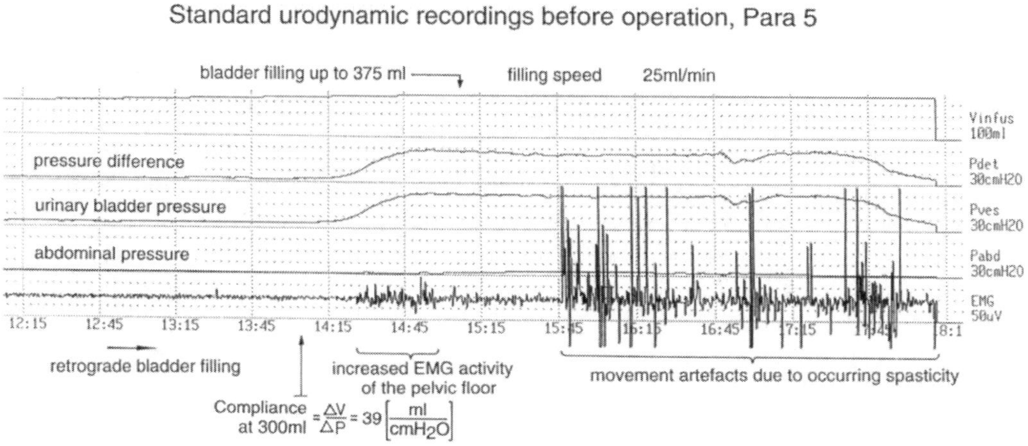

Figure 2. Standard urodynamic recording including intravesical pressure (urinary bladder pressure), abdominal pressure, detrusor pressure (pressure difference) and pelvic floor electromyography (EMG). Detrusor pressure is that proportion of the intravesical pressure that is produced by active and passive bladder wall properties. Detrusor pressure is calculated by electronically subtracting the value of abdominal pressure from that of intravesical pressure. This derivation is extremely important in order to be able to distinguish intrinsic and extrinsic contributions to intravesical pressure. Upon retrograde bladder filling, the detrusor started to contract at approx. 350 ml. Abdominal pressure is the pressure around the bladder, and this is measured as rectal pressure.

Upon further filling of the bladder (Figure 2) and a stoppage of the filling, movement artefacts indicate the occurrence of spasticity of the lower body. The avoidance of the seeming coactivation of the detrusor, the pelvic floor and the striated bladder sphincter is an important clinical problem (continence) as well as an interesting scientific problem. Since it is highly unlikely that parasympathetic motoneurons activate besides the detrusor (smooth muscle) also the external bladder sphincter (striated muscle), the problem lies in the pathologic coordination between the parasympathetic and the somatic nervous systems.

2.4.2. Doublet and Burst Firing of Secondary Muscle Spindle Afferents (SP2) in Paraplegics

I shall now try to show with the use of electrophysiological data that some muscle spindles in the lower sacral range, in the domain of the sacral parasympathetic nervous system, are innervated by parasympathetic efferent fibres. The response of SP2 fibres to the probable parasympathetic activation is a doublet and burst-like firing with short interspike intervals (IIs). Primary muscle spindle afferents are only seldom found in the lower sacral range.

Figure 3a,b shows the APs of an SP2 fibre with the second main phase of a triphasic AP upwards (according to the electrode arrangement). Since most likely this afferent fibre performed a loop in the root (see also Figure 12), the APs of this fibre were uniquely marked by the efferent-like AP of the backward running fibre. In Figure 3a, the II between two APs of that secondary fibre is 2.8ms. In Figure 3b, the II of 0.5 ms is even shorter (transmission frequency = 2000 Hz). In Figure 3c, the activity of another SP2 fibre in another paraplegic is shown. Those APs occur in a burst-like pattern with IIs increasing from 0.4 ms to 0.94 ms. The II of 0.4 ms represents a transmission frequency of 2500 Hz.

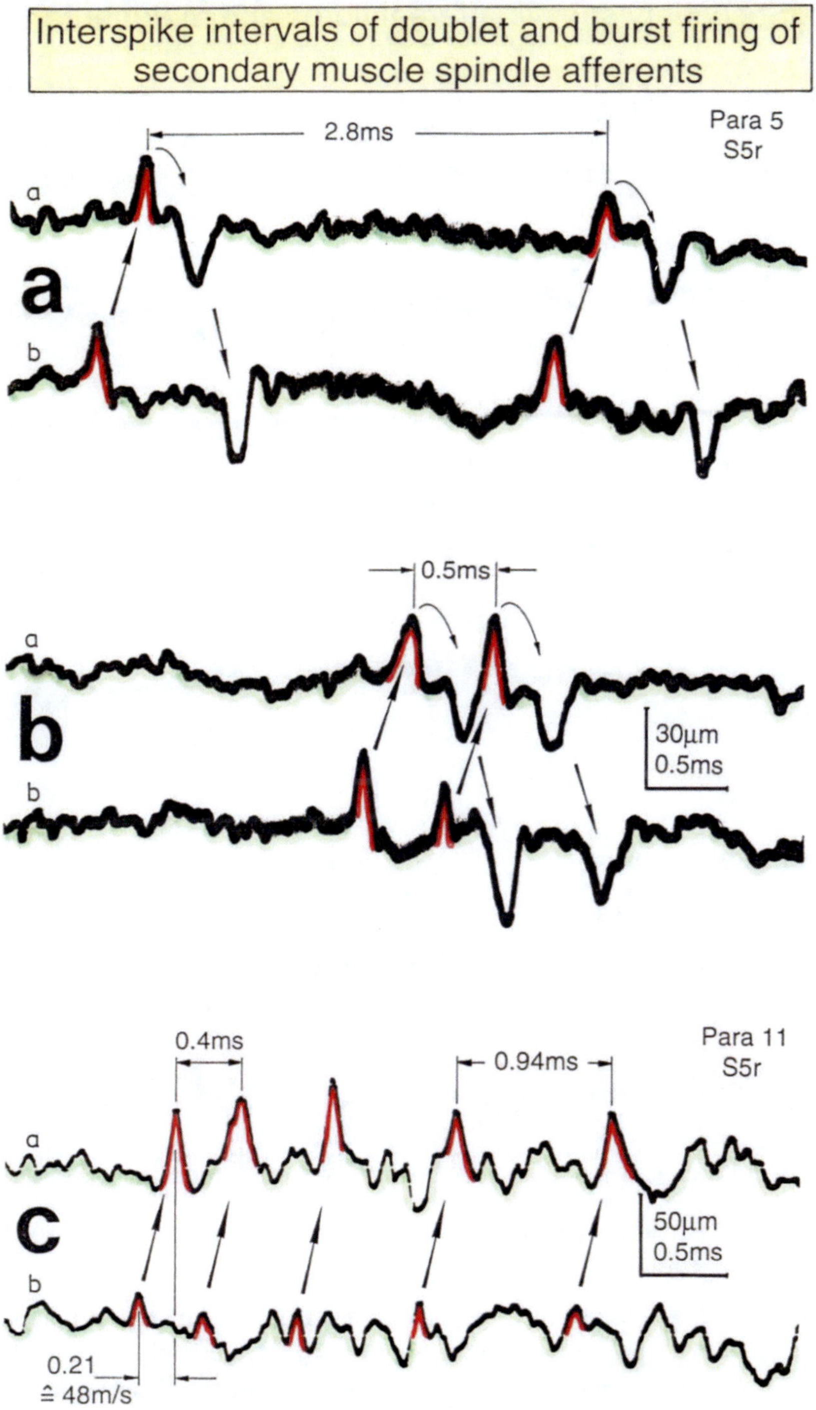

Figure 3. Doublet and burst firing of secondary muscle spindle afferent fibres. Recording a, b, paraplegic 5, and c, paraplegic 11. Interspike intervals (IIs) are indicated. In "c," the single afferent fibre conducted at 48 m/s, and II increased from 0.4 to 0.94 ms. The single-spindle afferent fibre in "a" and "b" is uniquely marked by performing a loop, which is demonstrated by an efferent-like AP time-locked to the afferent AP (see also Figure 12). The AP (action potential) amplitude calibration μm designates microvolt (μV).

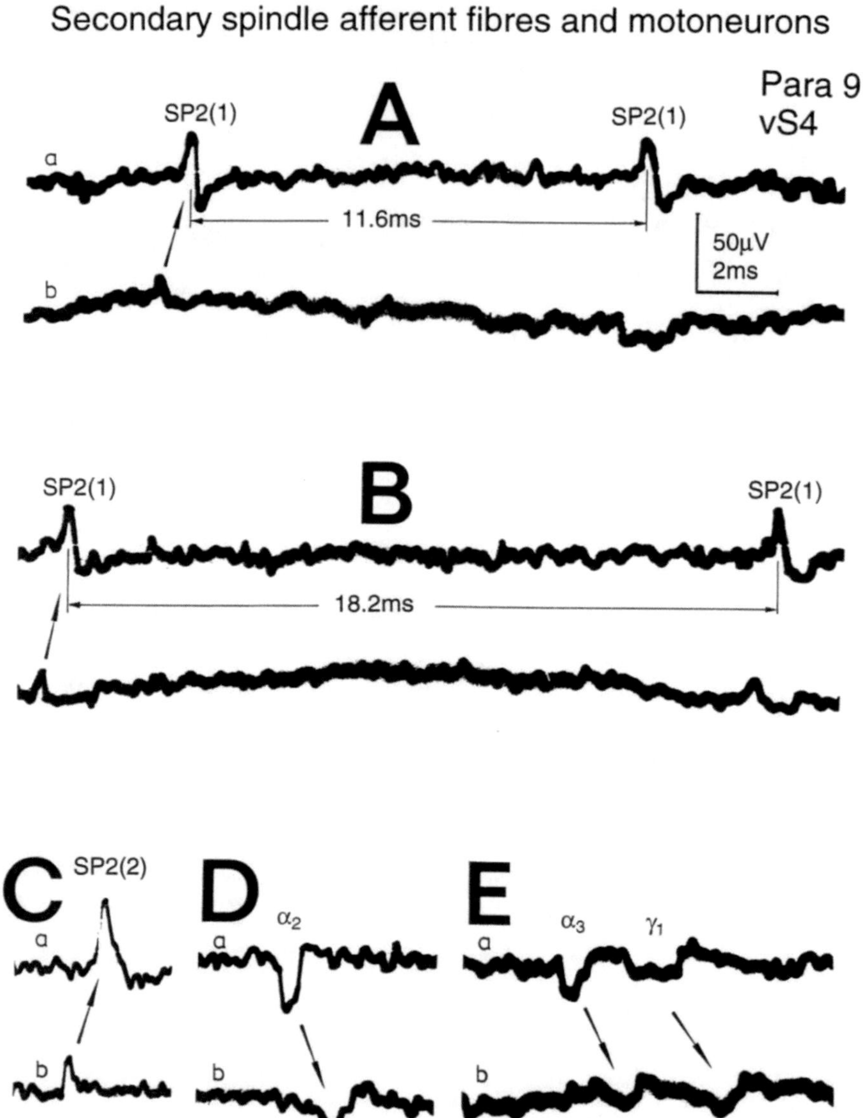

Figure 4. Characteristic waveforms of extracellular action potentials of secondary muscle spindle afferent fibres (SP2(1), SP2(2)), α-motoneurons (α_2, α_3) and a dynamic fusimotor (γ_1) used to construct single-nerve fibre impulse patterns. Two doublet interspike intervals (DIIs) of 11.6 ms (A) and 18.2 ms (B) are indicated. Paraplegic 9, nerve root vS4.

A systematic analysis of doublet (Figure 3a,b), burst (Figure 3c) and continuously short II firing of SP2 fibres will now be used to measure parasympathetic muscle spindle activation and to clarify, by direct comparison with the impulse patterns observed in a brain-dead human, whether such parasympathetic activation patterns are physiologic or pathophysiologic. Figure 4A,B shows two doublet APs of a secondary muscle spindle afferent fibre SP2(1) of paraplegic 9, with IIs of 11.6 and 18.2ms. In Figure 4C,D,E, typical AP waveforms (templates) of an SP2(2) fibre, α_2, α_3 and γ_1-motoneurons, are given; their impulse patterns will be followed up continuously over several seconds to analyse continence functions of the human nervous system.

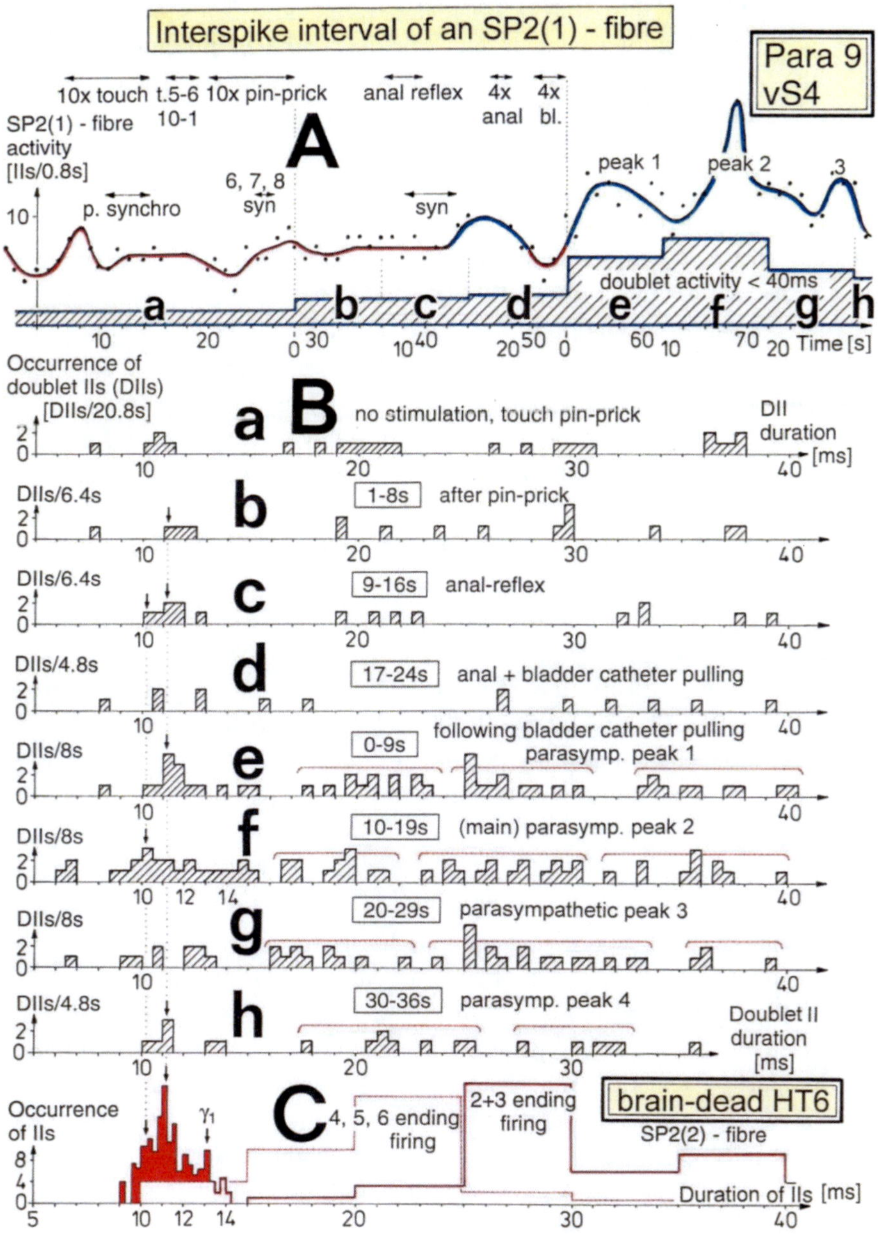

Figure 5. A. Activity of a secondary muscle spindle afferent fibre SP2(1) in dependence on time (interspike intervals = IIs = (APs – 1)/0.8s). 4x anal = Fourfold anal catheter pulling; 4xbl. = Fourfold bladder catheter pulling; t.5-6,10-1 = short anal reflex stimulation; syn = synchronization of oscillators, not under consideration here. Peak 1, peak 2, 3 = activity peaks of SP2(1) fibre (as will be shown, they are peaks of parasympathetic activity increase). The crosshatched histogram with histogram classes a-h is a histogram of doublet activity; doublet activity < 40 ms = activity of doublets with interspike intervals (doublet interspike intervals = DIIs) < 40 ms. Note the strong increase in doublet activity following bladder catheter pulling. B. Interspike interval (DII and burst interspike interval) distribution for the time intervals a-h. The arrows at 10.2 ms and 11.2 ms indicate peaks of DIIs similar to those measured in HT6 (C). The dotted parenthesis indicate possible grouping of IIs during burst firing, in similarity to the multi-ending regular firing observed in brain-dead human HT6. C. DIIs and multi-ending regular firing observed in brain-dead human HT6 [54] shown for comparison. Direct comparison is indicated by double framing of para 9 and HT6.

The activity level of the SP2(1) fibre in paraplegic 9 is shown in Figure 5A in dependence on time. Following anal and bladder catheter pulling and other natural stimulations, the SP2(1) fibre increased and/or decreased its activity. The trace of the activity changes shows several peaks. Especially the bladder catheter pulling induced large peaks in the activity level. It is known that in paraplegics, strong bladder catheter pulling activates the detrusor to contract. The activity peaks may therefore be due to parasympathetic muscle spindle activation, and they are marked with 1, 2, 3 and 4. A better justification for the probable parasympathetic activation will be given below. Since in an earlier publication it was suggested that the doublet firing can be attributed to low level parasympathetic muscle spindle activation [54,56], in Figure 5A also, the doublet activity level (doublet II < 40 ms) has been drawn. Also the doublet activity was highest following bladder catheter pulling in parallel with the occurrence of the activity peaks. To analyse the patterns of the doublet firing, doublet II frequency distribution histograms have been plotted for different time intervals (Figure 5Ba-h). To reveal signs of pathology in these patterns, the doublet firing histogram and the histogram for the multi-ending regular firing obtained from brain-dead human HT6 are given in Figure 5C (taken from Figures 97C and 98C of Chapter III).

2.4.3. Comparison of Doublet and Burst Firing of Paraplegics with Doublet and Multi-Ending Regular Firing of a Brain-Dead Human

In the brain-dead human HT6, the most likely physiologic firing pattern of the secondary spindle afferent fibre SP2(2) (Figure 5C) following parasympathetic activation included the doublet firing, the 2 plus 3 ending firing, and the 4, 5 or 6 multi-ending regular firing [54,56]. The doublet firing, of most likely two encoding sites of the parent SP2(2) fibre in HT6 [54], shows three peaks at 10.2 ms, 11.2, and 13.1 ms. The doublet duration peak at 13.1 ms is marked γ_1, indicating that a γ_1-fusimotor fibre in the HT6 also had a doublet histogram peak at that value [56], supposing that this peak occurred because of the γ_1-fusimotor activity or that the γ_1 neuronal network was controlled also by the parasympathetic division. Anyhow, this HT6 peak cannot be found in the doublet firing of the SP2(1) fibre in paraplegic 9 (Figure 5B). The other two doublet peaks firing at 10.2 and 11.2 ms can be seen in the SP2(1) fibre doublet histogram of Figure 5B. The general difference in the doublet firing of the SP2 fibres between the paraplegic and the brain-dead human is that the doublet firing histogram for paraplegic 9 is more broadened, in particular in the direction of short IIs. Since human secondary spindle afferents outside the sacral parasympathetic domain do not fire with IIs shorter than 50 ms [72], it was argued that the doublet firing stems from two endings (encoding sites) of a parent spindle afferent fibre, which are coordinated by the parasympathetic division in the way that a second ending fires with a delay of 10 to 14 ms [54,56]. In paraplegic 9, the doublet histograms show a more broadened distribution. Thus, the coordination of the firing of two encoding sites of the secondary spindle afferent fibre is partially lost. It remains unclear whether the coordination of the two encoding sites by the parasympathetic division occurs directly at the muscle spindle level or via the fusimotor system.

For high level parasympathetic activation, the secondary spindle afferent fibre in HT6 fired with equidistant impulses of up to six endings of the parent fibre. Again, the parasympathetic division probably coordinated up to six encoding sites to fire with equidistant APs. Only when a new encoding site was recruited with the increasing activity, the firing became transiently irregular for some time (see Figure 9B [55s]). This regular firing with equidistant APs was lost in paraplegic 9. Instead, the parasympathetically induced spindle afferent activity involved bursts with increasing II duration. The bursts consisted of up to six APs indicating a sudden activation of up to six encoding sites of the parent secondary spindle afferent fibre (see Figure 3C and Figure 11d,e). In Figure 5Be,f,g,h, the multi-ending firing histograms for paraplegic 9 show grouping (indicated by the dotted parenthesis). This grouping may suggest preforms or rudiments of multi-ending regular firing. Directly plotted impulse patterns will later show more clearly the differences between physiologic and pathophysiologic activation.

In any case, natural stimulation of the parasympathetic division in paraplegics, induced by anal and bladder catheter pulling, resulted in doublet firing of two endings and multi-ending firing of three and more endings. The regularity of firing (the specificity of firing), observed in HT6, was lost in the paraplegics. The physiologic specific doublet and equidistant AP firing (HT6, multi-ending regular firing, probably physiologic) changed into pathologic less specific doublet, burst, and non-equidistant AP firing (paraplegics, less regular firing, pathologic), which may indicate a loss of coordinated firing of parasympathetic neurons, innervating the muscle spindle. In the following section, it will be shown that parasympathetically induced muscle spindle activation is actually connected to the activation of the parasympathetic division and, most likely, does not originate from the somatic nervous system.

2.4.4. Time Course Comparison between the Proposed Parasympathetically Induced Muscle Spindle Afferent Activity and the Detrusor Pressure

Upon retrograde bladder filling in patients with complete spinal cord injuries, the detrusor was reflex activated at rather small storage volumes, as could be measured with urodynamics by the detrusor pressure. In paraplegic 5, the detrusor was automatically activated at a bladder filling volume of approx. 350 ml (Figure 2), and in paraplegic 9 at approx. 10 ml (!). Reasons for limited storage volumes were given in the previous paper with the increased afferent input from the bladder and strong disorganization of the neuronal networks in the spinal cord [83]. It was found there for the paraplegics 7 and 11 that the first high bladder afferent activity occurred at bladder filling volumes in the operation, at which the detrusor was activated pre-operatively. In Para 7, the first high afferent input occurred at a filling volume of 150ml; the detrusor was activated at 160ml (Figure 8 of [83]). In Para 11, the first comparatively high afferent input occurred at 50ml filling volume; the detrusor was activated at 80ml.

During surgery under light anaesthesia, the detrusor never responded to the filling of the bladder in 120 patients. Also, quick filling with 4°C saline solution did not activate the parasympathetic division. Only in one instance, during surgery on a tetraplegic patient with a lesion sub C3 (artificial ventilation), with no tubes placed in the trachea and extremely light Propofol-anaesthesia, there seemed to be a slight detrusor pressure increase upon retrograde bladder filling. On the other hand, strong (painful) bladder catheter pulling activated the detrusor to contract in paraplegics with no anaesthesia, most likely activated the parasympathetic nervous system in the brain-dead human HT6 (no anaesthesia, possibly partial spinal shock), and seemed to activate the parasympathetic division during very light anaesthesia. It is concluded therefore that painful stimulation of the bladder under light anaesthesia activated the detrusor.

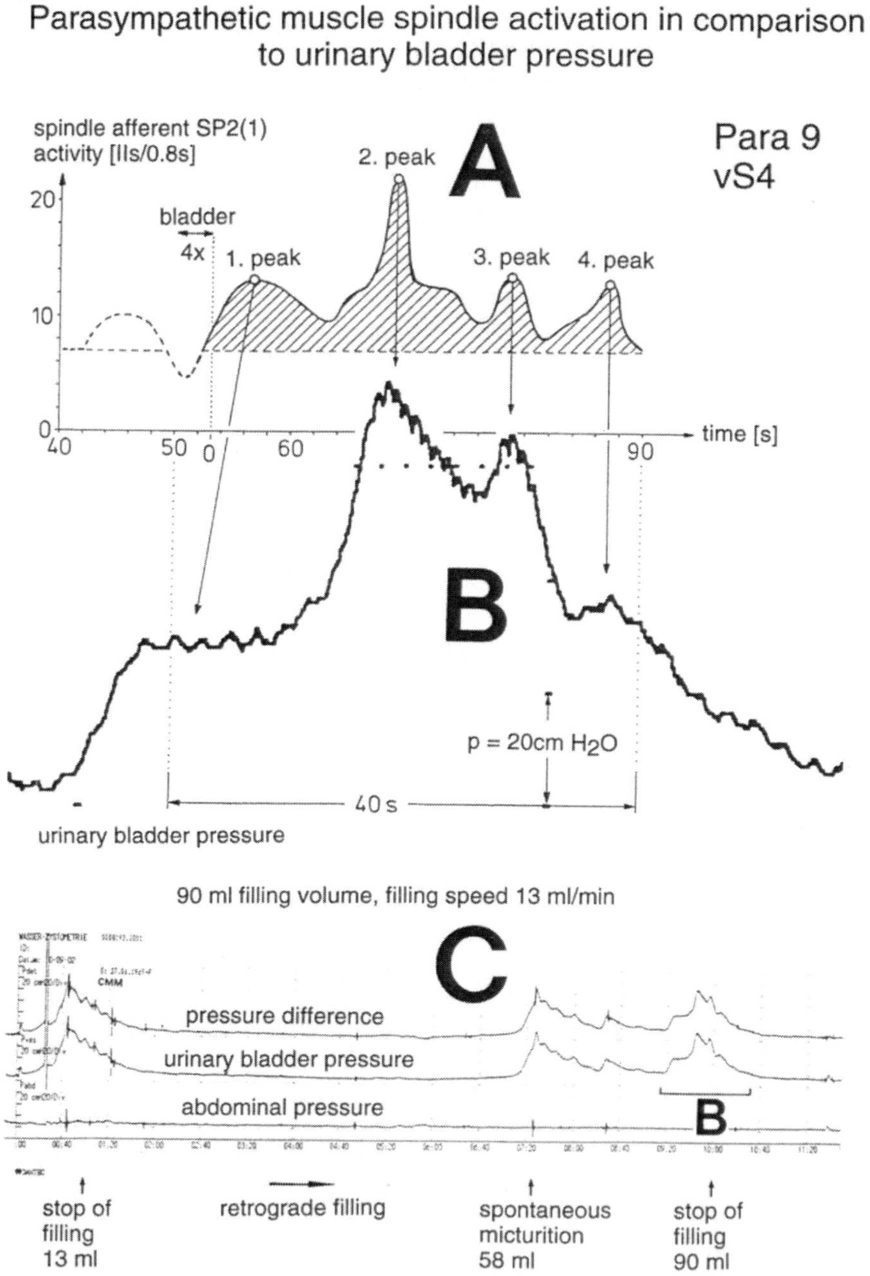

Figure 6. Comparison between changes in secondary muscle spindle afferent activity and detrusor pressure (measured before the surgery; for the implantation of a sacral anterior root stimulator). Paraplegic 9. A. Activity changes of the afferent fibre SP2(1) from Figure 5 following bladder catheter pulling. Approx. mean activity level is represented by a dashed line at 7.5 IIs/0.8s ((APs -1)/0.8s). The activity above the mean is crosshatched and is proposed to be due to parasympathetic activation. Root vS4. B. Detrusor pressure (pressure difference) changes taken from "C." Note that changes in the detrusor pressure show almost exactly the same time course as do the activated changes of the secondary muscle spindle afferent fibre SP2(1) of "A." Corresponding peaks are correlated by arrows. C. Abdominal pressure (measured as rectal pressure), intravesical pressure (urinary bladder pressure) and detrusor pressure (pressure difference) during retrograde filling before the surgery. One transient detrusor pressure increase, marked "B," is used, after enlargement in "B," to compare with the spindle afferent activity.

In Figure 6, the increase in detrusor pressure upon retrograde bladder filling before surgery is compared with the activity increase of the secondary muscle spindle afferent fibre SP2(1) following four times bladder catheter pulling. Figure 6A shows the undulating activity increase of the SP2(1) fibre from Figure 5A. In Figure 6C, the cystogram is shown. Upon bladder filling, spontaneous micturition occurred several times. If parasympathetic fibres really activated muscle spindles, then the activity increase of the secondary muscle spindle afferent fibres following bladder catheter pulling may have a similar time course as does the bladder pressure increase due to detrusor activation following retrograde bladder filling. To check this similarity of time course, one undulating increase of detrusor pressure (Figure 6C) has been brought to the same time scale as the measured changes in muscle spindle afferent activity (Figure 6A) and transferred into Figure 6B for a direct comparison with Figure 6A. By comparing Figure 6A with Figure 6B, it can be seen that the occurrence of activity peaks of secondary spindle afferents is very similar in its time course to that of the peaks of the detrusor pressure. From the similarity of changes of spindle afferent activity and detrusor pressure (4 peaks), it can be concluded that some muscle spindles in the domain of the sacral parasympathetic nucleus are partly controlled by the parasympathetic division and that the muscle spindle and the detrusor activation have similar time courses. The other retrograde bladder filling induced undulating transient increases in pressure shown in Figure 6C had a similar time course.

Another correlation could be found between detrusor pressure and muscle spindle activation. The bladder pressure in paraplegic 5 (Figure 2) increased steadily with no or only little undulations, probably because the functions of the parasympathetic nervous system were still relatively preserved. The activity increase of a secondary muscle spindle afferent fibre in paraplegic 5 was also sustained and only little undulating, as can be seen from Figure 10.

Thus there is indication for some muscle spindles being partly driven by the parasympathetic division. The drive can be by parasympathetic fusimotor activity or more indirectly by somatic fusimotors. Since in HT6 the somatic fusimotors did not change their activity levels strongly with the activation of preganglionic parasympathetic fibres (Figure 7 of [56]) and the activity increase of the secondary muscle spindle afferent fibre SP2(2) (Figure 8) followed the transient increase of parasympathetic activity, it is likely that some muscle spindles are directly controlled by parasympathetic fusimotors. The direct control of some muscle spindles by parasympathetic fusimotors is supported by the slow activity decrease following very high activation (Figure 12B,D) and the similarity with the time course of the detrusor pressure decrease following electrical stimulation of the preganglionic parasympathetic fibres during the surgery. The slow spindle afferent activity decrease would be difficult to explain by somatic fusimotor activation only.

2.4.5. Detrusor-Sphincteric Synergia of the Bladder in the Brain-Dead Human HT6, and Dyssynergia in Paraplegic 9

The measurement of parasympathetic activation of the detrusor by activity changes of secondary muscle spindle afferent fibres (the spindle is innervated by parasympathetic fusimotors) allows an analysis of detrusor-sphincteric dyssynergia using the natural simultaneous impulse patterns of secondary muscle spindle afferents and sphincteric α_2-motoneurons (and γ-motoneurons).

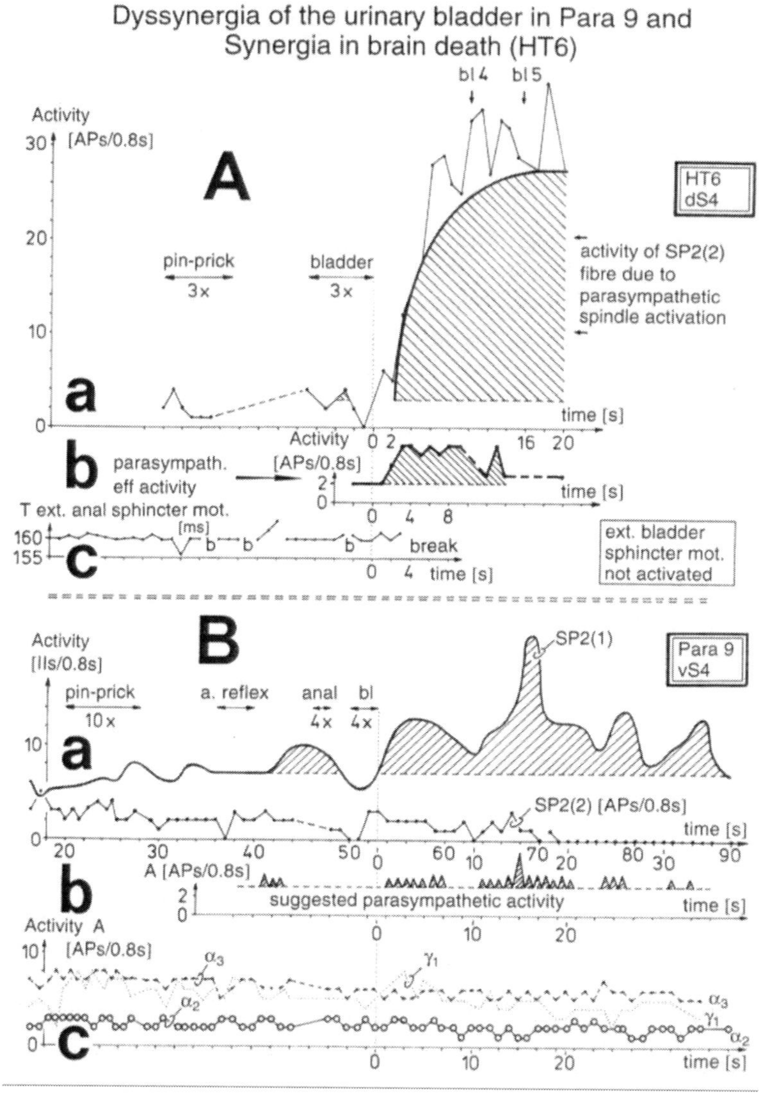

Figure 7. Direct comparison of secondary muscle spindle afferent and motoneuron activities between brain-dead human HT6 with no dyssynergia of the bladder (A) and paraplegic 9 with dyssynergia of the bladder (B). A. Simultaneous measurements of activities of secondary muscle spindle afferents (a), parasympathetic preganglionic motoneurons (b) and oscillatory firing (high activity mode) of a sphincteric motoneuron innervating the striated anal sphincter (c). Note that with the transient activity increase of the parasympathetic fibres (b), the secondary muscle spindle afferent fibre increased strongly its activity (a) for minutes, and the oscillatory firing sphincteric motoneuron discontinued its oscillation (c) to reduce strongly its activity. bladder 3x = 3 times bladder catheter pulling. T ext. anal sphincter mot. = oscillation period of the sphincteric α_2-motoneuron innervating the anal sphincter. For further details, see (56). B. Simultaneous measurements of activity of secondary muscle spindle afferent fibres SP2(1) and SP2(2) following anal (anal 4x) and bladder catheter pulling (bl 4x) (a), and the activity changes of an α_2-motoneuron (FR) and α_3-motoneuron (S) and a dynamic fusimotor fibre (γ_1) (c). Note that following bladder catheter pulling (and probably parasympathetic activity increase), the spindle afferent fibre SP2(1) (most likely contributing to continence) increased its activity in an undulating manner (a), whereas the SP2(2) fibre did not (probably not connected to continence) (a), and the α-motoneurons did not reduce their activity (c). The dynamic fusimotor γ_1 transiently increased its activity similarly as in HT6 measurements (see Figure 7 of [56]). In similarity to “Ab,” the suggested parasympathetic activity increase is pictured (b). a. reflex = anal reflex stimulation. IIs = interspike intervals; IIs/0.8s = (APS – 1)/0.8s (the activity measures IIs/0.8s and APs/0.8s differ by “1”).

Figure 7A shows that in the brain-dead human HT6, whose parasympathetic preganglionic neurons increased activity (Figure 7A,b) upon bladder catheter pulling, the SP2(2) fibre activity increased strongly, whereas the α_2-motoneuron innervating the external anal sphincter discontinued its oscillatory firing, which is a measure for a strong activity decrease. A α_2-motoneuron, innervating the external (striated) bladder sphincter, was not activated. This means that with the activation of the detrusor, the sphincteric motoneurons were relaxed by inhibition. Thus, the brain-dead human HT6 had a detrusor-sphincteric synergia of the bladder.

In paraplegic 9, who showed strong activity increase of the SP2(1) fibre, there was no sphincteric relaxation following bladder catheter pulling (Figure 7B). The secondary muscle spindle afferent fibre SP2(1) increased its activity in an undulating manner (Figure 7Ba). The parasympathetic fusimotors, driving the muscle spindle innervated by the SP2(1) fibre, probably were not continuously active as suggested by Figure 7Bb, in contrast to the parasympathetic activity observed in HT6 (Figure 7Ab). The other secondary muscle spindle afferent fibre in paraplegic 9 (SP2(2), Figure 7Ba) slowly reduced its activity upon bladder catheter pulling. This spindle afferent fibre was not connected to the continence of the bladder. Probably, its spindle was not parasympathetically innervated and was sited in leg muscles or parts of the pelvic floor muscles not contributing to continence. The α_2 and α_3-motoneurons (Figure 7Bc) showed a high activation, which is expressed in their oscillatory firing (see later) and probably contributed to the continence of the bladder and the rectum. These, probably sphincteric, motoneurons did not reduce their activity following parasympathetic activation, as can be seen from the SP2(1) fibre activity (monitoring parasympathetic activity). These motoneurons were not inhibited, and probably the external sphincters were not relaxed. The somatic fusimotor γ_1 (Figure 7Bc) increased transiently and slightly its activity upon painful bladder catheter pulling, in similarity to a γ_1 fibre in HT6 (Figure 7B of [56]). The measurements in paraplegic 9 indicate a loss of the inhibitory action of the detrusor onto the sphincteric motoneurons. However, there is no activity increase of the motoneurons with the increasing detrusor pressure, as suggested by Figure 2 (Figure 7Bc).

2.4.6. Natural Impulse Patterns of Afferents and Efferents of Para 9 and HT6

To reveal possible reasons for the loss of inhibition, simultaneous afferent and efferent impulse patterns have been analyzed. In Figure 8, the impulse patterns of HT6 and paraplegic 9 can be compared directly. In HT6, the secondary muscle spindle afferent fibre SP2(2) showed an impulse pattern, with the interspike intervals (IIs) rather similar (Figure 8Ab); irregularities in the pattern occurred only when the number of recruited encoding sites changed (Figure 8Aa,c). In paraplegic 9, the activity of the SP2(1) fibre was lower (Figure 8Da,c) and less regular (Figure 8D,b,d). Probably this reduced activity in the SP2(1) fibre was due to a reduced parasympathetic activation, since an only little parasympathetically activated spindle afferent fibre can be expected to show more doublet firing, somehow similar to the SP2(5) fibre in HT6 (Figure 8Aa). The impulse patterns of the parasympathetically driven spindle afferent fibres will be considered below. The other main difference in the impulse patterns between the brain-dead HT6 and the paraplegic was the different oscillatory firing of the α_2-motoneurons.

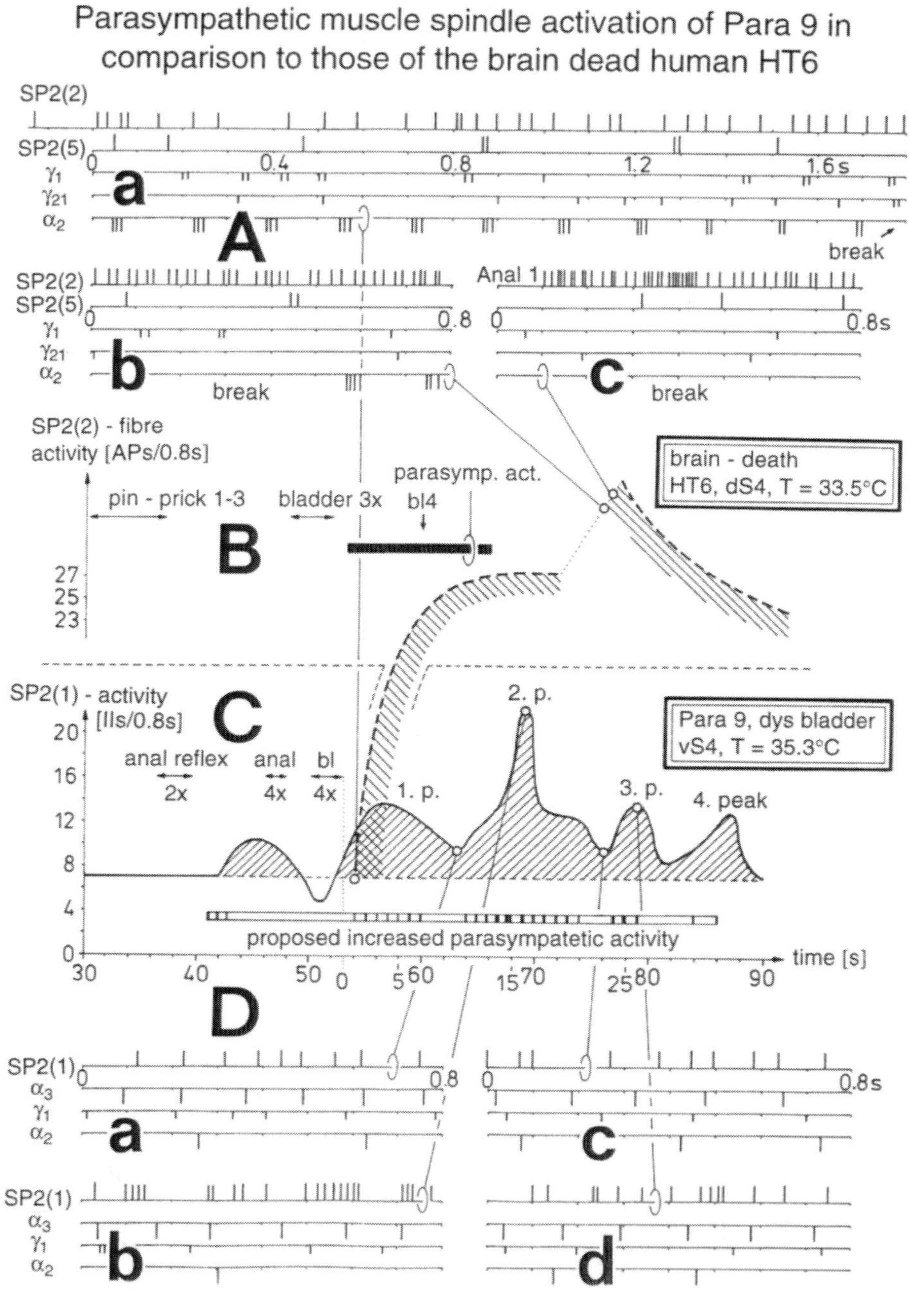

Figure 8. Comparison of muscle spindle afferent and efferent, and α-motoneuron activities between the brain-dead human HT6 (synergia of the bladder) (A,B) and paraplegic 9 with dyssynergia of the bladder (C,D). From the same recording as in Figure 7. Figures 8B,C are put close together to improve direct comparison. A. Natural impulse patterns of the secondary muscle spindle afferent fibres SP2(2) and SP2(5), the fusimotors γ_1 (dynamic) and γ_{21} (static) and the sphincteric α_2-motoneuron following bladder catheter pulling (bladder 3x). B. Activity increase of the SP2(2) fibre following bladder catheter pulling and parasympathetic activity (parasymp.act.) increase. C. Activity increase of the secondary spindle afferent fibre SP2(1) following anal (anal 4x) and bladder catheter (bl. 4x) pullings. Proposed parasympathetic activity increase indicated. D. Natural impulse patterns of the secondary muscle spindle afferent fibre SP2(1), the single α_2 and α_3-motoneurons and the fusimotor γ_1 following anal and bladder catheter pulling.

In HT6, the α_2-motoneuron fired regularly with impulse trains consisting of two or three APs every 160 ms (6.25Hz), whereas in paraplegic 9, the α-motoneurons fired irregularly in an oscillatory manner with impulse trains consisting of one AP every 130 ms (α_3, 7.7 Hz) or 320ms (α_2, 3Hz). The neuronal networks, driving the sphincters, had changed their properties completely. This indicates that the neuronal networks in paraplegics are damaged or malfunctioning. The function of the spinal oscillators in paraplegics will be analyzed in more detail in the following part [84], where the parasympathetic action onto the somatic division will be studied in detail. Here, the impulse patterns of the secondary spindle afferent fibres will be considered further, to find out what has changed in the parasympathetic nervous system of the micturition centre.

So far, there is indication for pathologic changes occurring in the functioning of the parasympathetic and the somatic divisions and in the interaction between both systems following spinal cord injury.

2.4.7. Differences in the Impulse Patterns of Secondary Muscle Spindle Afferent Fibres between Paraplegics and the Brain-Dead Human HT6

Figure 9 shows the impulse patterns of parasympathetically driven secondary muscle spindle afferents in paraplegic 9 and the brain-dead human HT6. In the brain dead individual (Figure 9B), the activity was low before strong bladder catheter pulling and increased strongly, continuously and regularly following stimulation. In paraplegic 9, the activity level was higher before bladder catheter pulling and increased more slowly, in an undulating and irregular manner following catheter pulling. The burst-like firing seen in paraplegic 9 did not occur in HT6. The doublet firing was less regular in Para 9 (Figure 5). Specific properties in the drive of the spindle seem to be lost in the paraplegic. Even though a burst-like firing also occurred once in HT6 (Figure 9Ba), the afferent fibre in HT6 could generate a sustained high regular activity level, whereas the fibre in paraplegic 9 could not.

The increase in the detrusor pressure in paraplegic 9 during retrograde bladder filling was very pathologic, since there was nearly no storage phase and the pressure increased transiently and in an undulating way (Figure 6C). In paraplegic 5, the detrusor pressure curve was more physiologic since there was a storage phase of 350 ml and a sustained and only little undulating pressure increase upon retrograde bladder filling (Figure 2). It is therefore of interest to see whether the impulse pattern of a secondary muscle spindle afferent fibre in paraplegic 5, driven by the parasympathetic division, was also closer to the impulse pattern obtained in HT6. Figure 10B,C shows the interspike interval (II) distribution histograms and impulse patterns in dependence on time for paraplegic 5. Indeed, the afferent fibre patterns in paraplegic 5 were quite similar to those observed in HT6 (Figure 9B). The impulse activity of the spindle afferent fibre was high, sustained, but still less regular than that in HT6. For low activity levels, impulse patterns similar to burst firing occurred (Figure 10C). The doublet firing was less regular (Figure 10C [32s]) than in HT6 (Figure 10A). Especially very short IIs occurred ($II < 1$ ms), which were not observed in HT6.

The bladder function in paraplegic 11 was also very pathologic. The storage volume before the surgery was 80 ml. The detrusor activity level was plateau-like and undulating with a frequency of 0.08Hz. After de-afferentiation, the storage volume increased to over 500 ml. A secondary muscle spindle afferent fibre fired with bursts of up to six APs with increasing II (Figure 11d,e).

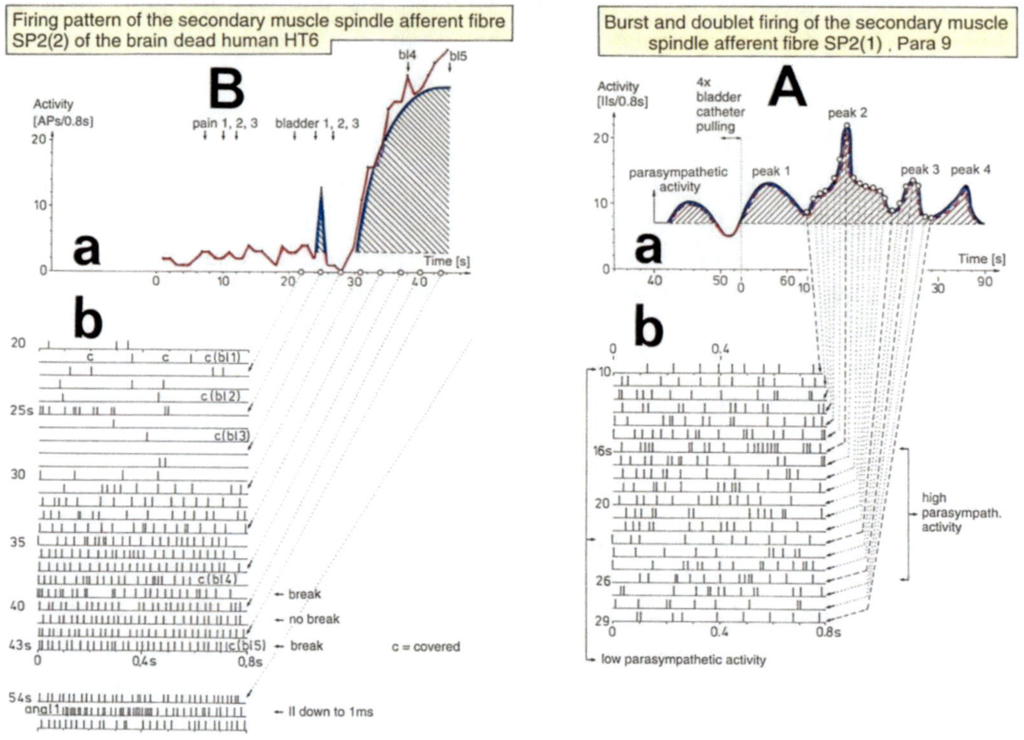

Figure 9. Comparison of the natural impulse patterns of single afferent fibres between paraplegic 9 with dyssynergia of the bladder (A) and the brain-dead human HT6 (B) with synergia of the bladder. Activity increase of the spindle afferent fibres following bladder catheter pulling (a) and the natural impulse patterns at different times following catheter pulling (b). A. Note that at activity peaks (a), there was burst firing in paraplegic 9 (marked by the arrows, 16s, 26s), and at low activity (a), there was no burst firing (10s, 23s). B. Note that even though burst firing appeared also in the brain-dead (25s), the activity increase sustained with similar interspike intervals (IIs). break = break of oscillation of sphincteric motoneuron. c = covered = skin afferent action potentials covered SP2 fibre impulses due to bladder catheter pulling (bl 1 - bl 5).

The shortest safely observed II had a duration of 0.19 ms (Figure 11c), which gives a transmission frequency of 5,260 Hz. The afferent activity in paraplegic 11 was very high (Figure 11d,e), as was the efferent activity (Figure 11f), as if the nervous system was strongly over activated and therefore malfunctioning. In this root S5, burst firing was observed in more than ten secondary spindle afferent fibres, consisting of up to six APs with increasing II. The shortest II observed had duration of 0.18 ms, and the II largest in a burst had duration of 18.5 ms. Since different spindle afferents showed different bursts, there seemed to be a different parasympathetic drive for each spindle. The parasympathetic fusimotor drive may be as complex as the somatic one.

From the S5 root was frequently recorded, because the root is thin [83] and normally not saved (de-afferentiation) when implanting an anterior root stimulator for bladder control. The S5 root was saved when, due to a variation, the root carried many preganglionic parasympathetic motoneuron fibres as verified by the bladder pressure increase upon electrical stimulation of the S5 root. In contradiction to anatomy books, feet muscles are frequently activated when lower sacral nerve roots (including S4) are electrically stimulated. Nerve fibres innervating leg muscles may even lead sometimes through the S5 root.

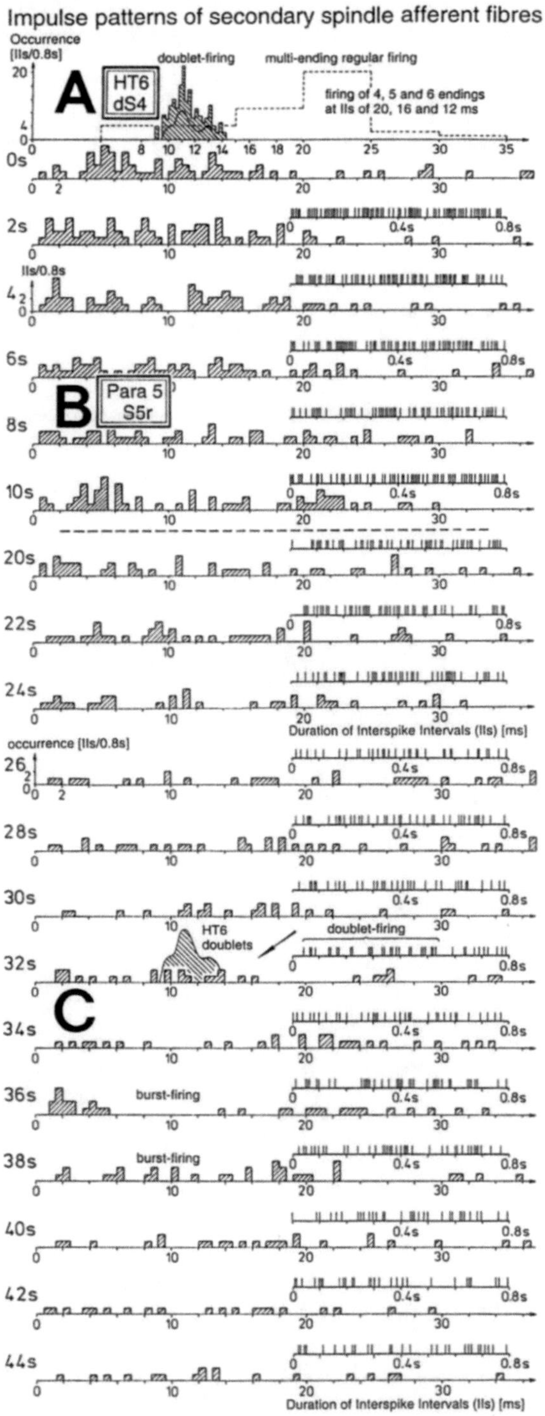

Figure 10. A. Histogram of doublet firing and multi-ending regular firing in the brain-dead individual HT6 (synergia of the bladder), same histograms as in Figure 5C. B,C (0s-44s). Interspike intervals and impulse patterns for a secondary muscle spindle afferent fibre of paraplegic 5 (with dyssynergia of the bladder) in dependence on time. The occurrence of doublet and burst firing indicated. Note that the impulse patterns are more regular than in paraplegic 9 (Figure 9Ab) with stronger dyssynergia but less regular than in the brain-dead individual (Figure 9Bb) with no dyssynergia.

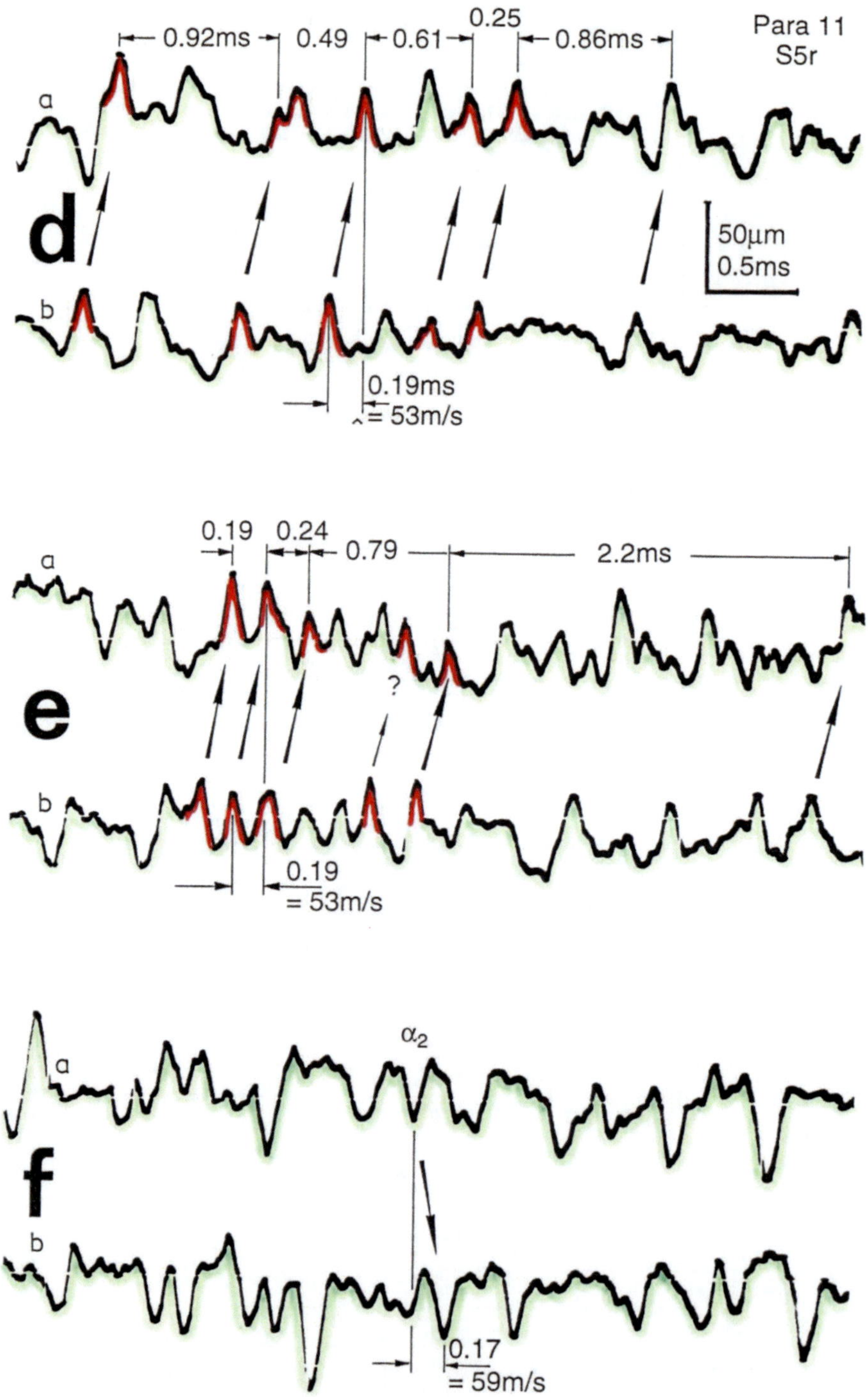

Figure 11. Afferent and efferent activity in an S5 root of paraplegic 11. d,e. Burst firing of secondary muscle spindle afferent fibres. Shortest II = 0.19 ms, which corresponds to a transmission frequency of 5260 Hz. f. Increased efferent activity. Note that if the afferent activity is high, the S5 root looks like a dorsal root (d,e), and if the efferent activity is high, the root looks like a ventral one. The AP (action potential) amplitude calibration 50 µm designates 50 microvolt (50µV).

In paraplegic 5, the bladder function was only slightly pathologic. It was therefore of interest to also consider the functional stage of the somatic division in this patient. Even though the oscillatory firing motoneurons showed quite a variation in their oscillation periods, the impulse train of each cycle was rather physiologic. The sphincteric α_2-motoneuron innervating the external anal sphincter fired with three APs every 167 ms on average and with a first and a second II of 4.3 and 8.5 ms very similar to the values observed physiologically (HT6, [47]). The α_2-motoneuron innervating the external bladder sphincter fired with impulse trains consisting of one to two APs (II = 7.4 ms) every 111 ms, also very similar to the firing pattern observed in HT6 [47]. Thus, the somatic division in the paraplegic 5 was, similar to the parasympathetic one, not too pathologic either. The main problem in paraplegic 5 was discoordination of the detrusor and the striated bladder sphincter.

The impulse patterns of γ-motoneurons and parasympathetic fibres could partly be identified in the S5 recording in paraplegic 5, but it was not possible to follow them up over longer time intervals. Besides the high parasympathetically activated SP2 fibre, a further identified secondary muscle spindle afferent fibre (SP2) fired with doublets at a low activity level, indicating a low parasympathetic activation. Simultaneous firing of secondary muscle spindle afferent fibres with low and high parasympathetic activation was also found in the brain-dead human HT6. In Figure 8Aa, fibre SP2(2) increased its activity strongly, whereas fibre SP2(5) fired with doublets at low activity levels.

It is satisfying that the results obtained with the single-fibre AP recording method from impulse patterns of afferent and efferent fibres support the results obtained urodynamically. The impulse patterns however will provide additional insight into the organization of the human nervous system under physiologic and pathophysiologic conditions.

2.4.8. Unaltered Muscle Spindles Following Spinal Cord Injury

It was shown in Figures 7,8,9 that the firing of the parasympathetically driven secondary muscle spindle afferents of paraplegic 9 had changed in comparison to those of the brain-dead human HT6. The impulse pattern changes could have been due to a pathologic self-organization of the neuronal networks mainly of the spinal cord, due to changes in the muscle spindles or both. Changes in peripheral ganglia cannot completely be excluded. If one could show that the muscle spindles themselves remained unchanged, then most likely the pathologic function would be due to organization changes in the CNS.

Figure 12B shows the activity decrease of a secondary spindle afferent fibre in paraplegic 5 following presumed parasympathetic activation. Original patterns are shown in Figure 10B,C. Since most likely the spindle afferent fibre turned backward in the root (see below), its APs were uniquely marked with the time-locked efferent-like AP, and the activity changes could therefore be measured safely.

The activity decrease of the secondary spindle afferent fibre (Figure 12B) was assumed to be of the e-functional type. The time constant τ_{off} (Figure 12D) was 25s. Since there was a slight additional activation between 30 and 50s, a successive calculation of the time constant was performed (τ_{off} = 36s) and a mean τ_{off} of 31s was obtained (Figure 12B,D).

The time constant for the activity decrease of the secondary muscle spindle afferent fibre in paraplegic 5 [31s] is similar to the value obtained for the brain-dead human HT6 ($\tau_{off} \approx$ 40s, [56]).

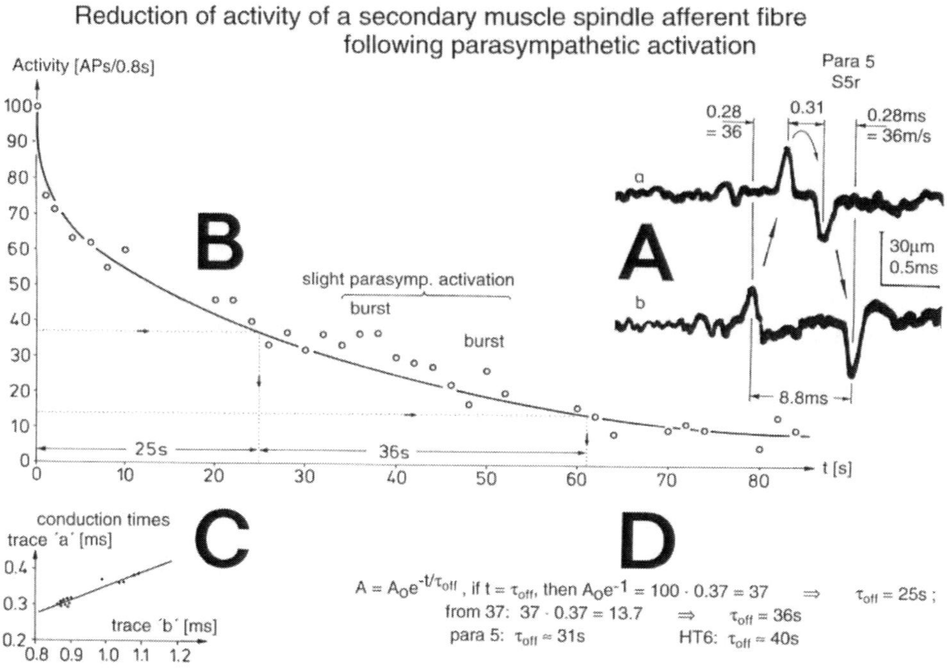

Figure 12. A. Recording from a secondary muscle spindle afferent fibre, which is uniquely marked by its efferent-like AP of the turning fibre. B. Activity reduction of the spindle afferent fibre following parasympathetic activation; impulse patterns are partly shown in Figure 10. e-functional decline is disturbed gently by further slight parasympathetic activation in the form of bursts. Time constants for e-functional decline τ_{off} are indicated. C. Relation between conduction times on trace "a" and trace "b" as indicated in "A." D. Calculation of the time constants for the e-functional decline. The AP (action potential) amplitude calibration 30μm designates 30 microvolt (30μV).

Following electrical stimulation of preganglionic parasympathetic fibres during the surgery, the detrusor pressure decreases slowly with an e-functional time constant of approx. 30 to 40ms. It is therefore concluded that the passive properties of the muscle spindle with respect to the parasympathetic innervation had not changed. It is further concluded that the changes in the natural impulse patterns of the parasympathetically activated secondary muscle spindle afferent fibre were caused by changes in the networks of the parasympathetic nucleus in the sacral micturition centre.

2.4.9. Turning of Nerve Fibres

Figure 12A shows an afferent AP, which is followed by a time-locked efferent-like AP. The conduction times on trace "a" and trace "b" are plotted in Figure 12C against each other, as indicated in "A." It can be seen that the correlation points lie on a straight line and are grouped in two areas. It seems therefore that sometimes the fibre conducted with a slower velocity. Probably the efferent like AP is not a reflected AP, since otherwise the correlation points would be more scattered. Also, a reflected and antidromically conducted AP would cancel out a closely following orthodromically conducted AP. This was not the case. An SP2 fibre AP reaching the recording electrodes 0.5 ms later (Figure 3b) would have a spatial delay of 25 mm ($v = s/t$; $s = vt = 50\text{mm/ms} \bullet 0.5\text{ms} = 25\text{mm}$). Previously [83], the distance between the urinary bladder and the recording electrodes in the spinal cord was calculated to exceed

300mm. A reflected SP2 fibre would therefore cancel out the AP arriving 0.5 ms later. It is concluded that the secondary spindle afferent fibre performed a loop or branched in the root. A cross-talk of nerve fibres (ephaptic transmission) is unlikely because of the regularity of firing.

The branching of nerve fibres in roots or the performing of loops is probably seldom and unimportant. Here, such situations are of special interest since they mark uniquely the activity of single fibres. For nerve fibre branching and nerve fibre loops in nerve roots, see [12].

2.5. Discussion

2.5.1. Membrane Properties

Figures 3,5,11 show interspike intervals of single secondary muscle spindle afferents down to 0.19ms. The corresponding naturally evoked transmission frequencies give 5,000 Hz and more. Transmission frequencies of less than 1,000 Hz were expected [24]; textbooks for medical students give values of transmission frequencies of between 500 and 1,000Hz [87]. Absolute refractory periods in the range of 0.2 ms therefore indicate that the membrane properties of secondary muscle spindle afferent axons in humans differ from those of animals. If axons of interneurons in the central nervous system (CNS) also had five times higher transmission frequencies, then CNS functions could be speeded up in humans in comparison to those of laboratory animals. It is difficult to understand why in humans, group conduction velocities range 40 % slower than in cats and dogs [46,48,83]. In comparing human and dog measurements, the higher conduction velocities in dogs seemed to be due to a larger nerve fibre diameter [48]. Human nerve fibres seemed to slightly enhance the conduction velocity through the presence of a thicker myelin sheath, which would increase the internode length [44] and in that way increase the conduction velocity.

In the squid, conduction velocities are increased phylogenetically by an axon diameter enlarged up to 1 mm [24]. Higher species developed the myelin sheath so that excitation jumps from one internode to the next one to increase conduction velocity. It is conceivable that in man, the nervous system tried to improve processing and complexity by increasing transmission frequency and reducing nerve fibre diameter at the price of reduced conduction velocities. In modeling human CNS functions, one probably has to consider membrane properties typical of human membranes.

Transmission frequencies, refractory periods and times of inactivation of nerve cells are important for the function of the CNS. To compare the trophic influence data for frogs (Chapter I) to those for humans, to look for similarities, when working on the trophic influence [40,60], the body temperature of living frogs was increased up to 37 °C. The frogs (Rana temporaria) died at temperatures of between 29 and 30 °C. Two kinds of frogs from Brazil also died at that temperature. Before the frogs died, they developed very strange behavior. Sexual reflexes or movements occurred. They resembled spasms in paraplegics. The CNS of the frogs probably developed pathologic functions due to the high temperature. Probably, too high impulse traffic occurred in the CNS. If one believes that the neuronal network of the CNS is not purely constructed of dedicated lines of excitation but also of synfire chains that cross each other (Figure 25 of [1]), then refractory periods of nerve cells may become of high importance. A reduction of refractory periods may allow a speeding up of CNS functions, since the crossing of excitation via crossing of synfire chains is possible at higher impulse traffic or gives more safety for physiologic functioning. One reason for the pathologic function of the disconnected spinal cord in paraplegics may be over excitation of the CNS. The exceedingly high activity in the disconnected spinal cord may contribute to spasticity and urinary bladder dysfunction. This problem will be picked up again in a following part [84], when discussing dysfunctions of spinal oscillators [53].

2.5.2. Muscle Spindle Afferents Activated by the Parasympathetic Division

In a previous paper, it has been reported that some muscle spindles in the lower sacral range were most likely activated by the parasympathetic division [56]. The physiologic parasympathetically induced activity patterns of the secondary muscle spindle afferents seemed to be the multi-ending regular firing for high activation and the doublet firing for low parasympathetic activation. Since a secondary spindle afferent fibre increased its activity following a transient activity increase of preganglionic parasympathetic motoneurons, even though the somatic fusimotors did not change their levels strongly, it was concluded that most likely the parasympathetic division directly activates the spindles; this means that muscle spindles are innervated by parasympathetic fusimotors [56]. The direct control of some muscle spindles by parasympathetic fusimotors is supported by the slow activity decrease following very high activation (Figure 12B,D) and the similarity with the time course of the detrusor pressure following stimulation of the preganglionic parasympathetic fibres during surgery. Such a slow spindle afferent activity decrease would be difficult to explain by somatic fusimotor activation only. More direct comparisons between activity levels of parasympathetic fibres and those of secondary muscle spindle afferents are needed. The recordings for such comparisons are difficult to obtain because the thin preganglionic parasympathetic motoneuron fibres generate only APs of small amplitude (Figure 8 of Chapter III) (Figure 3 of [83]). Further support for parasympathetic fusimotor drive comes from Figure 12. No slow e-functional decrease in the secondary muscle spindle afferent activity from very high levels is found following somatic fusimotor activation. Sometimes, intrafusal muscle fibres are compared with the slow muscle fibres of the frog. It has been shown that slow muscle fibres are under the trophic control of the nerve fibres they are innervated by, even before synapses start to function electrically [34,42] (Chapter I). The controlled functions include excitability of membranes and excitation-contraction coupling. It has further been shown that partially denervated slow muscle fibres of the frog partly change their functional properties in the denervated muscle fibre but remain unchanged in the still innervated area [39,41] (Figure 86 of Chapter III). Protein synthesis in the several nuclei of a muscle fibre is probably stimulated differently for corresponding membrane changes of the muscle fibre. It is therefore conceivable that parts of intrafusal muscle fibres are innervated by parasympathetic fusimotors. Morphologic support is needed. So far, few data are available concerning the morphology [28] and the electrophysiology [20,21] of the human muscle spindle outside the parasympathetic domain. In animals, some muscle spindles in the domain of the sympathetic division are innervated by sympathetic fibres [2,3,10,37]. It was further shown that parasympathetically driven muscle spindles are found in the domain of sacral parasympathetic division but not outside of it (Figures 3,4 of [57]). The high activity of secondary muscle spindle afferents (primaries are seldom found in the lower sacral range) can partly be split into the impulse patterns of the up to six encoding sites of parent fibres [54,56,57] with impulse patterns of the single endings similar to those obtained from spindle afferent fibres activated by somatic fusimotors only [10,72] (Figures 94,111 of Chapter III).

It remains unclear whether the supposed parasympathetic fusimotors have the same caliber as the preganglionic parasympathetic motoneurons synapsing on ganglion cells to innervate the detrusor. The classification scheme of the human peripheral nervous system [46,83] needs further improvement. A previous paper (Figures 1,8 of [56]) suggested a possible existence of several kinds of parasympathetic fibres (Figure 33D,E and Figure 113 of Chapter III). In paraplegic 11 (Figure 11), there were more than ten secondary spindle afferent fibres firing with bursts of different interspike intervals, indicating that the spindles may have obtained different fusimotor drives. In recordings of Para 5, it could be seen that at the same time some muscle spindles were highly (multi-ending regular firing or burst firing) while others only little parasympathetically activated (doublet firing). It seems therefore possible that parasympathetic efferent fibres also split up into different groups with different properties. The similarity of the time course between activity level changes of parasympathetically driven secondary muscle spindle afferents and detrusor pressure changes (Figure 6) point more towards simpler common parasympathetic activation, but since the time course is slow and includes smooth muscle fibre activation, changes are covered.

2.5.3. Similar Time Courses of Detrusor Pressure and Secondary Spindle Afferent Activity Changes

The similarity between detrusor pressure changes before the surgery, induced by retrograde bladder filling, and spindle afferent activity changes under light anaesthesia, following painful bladder catheter pulling, in the three paraplegics studied are of interest for several reasons. The clinically most important question is why does the detrusor get not activated during anaesthesia by retrograde bladder filling, but is activated by painful bladder catheter pulling? Somnolent patients who do not respond to speech and touch can be awakened with the application of somatic pain. Similar to somatic pain application, it is conceivable that pain application in the parasympathetic division is a stronger stimulus for the activation of the detrusor than the firing of stretch (S1) and tension receptor afferents (ST). If this is so, then it could further be that parasympathetic afferents, conducting pain, are activated by inflammation as probably do stretch and tension receptor afferents. A hyper-reflexia of the detrusor could then partly be due to sustained activation of parasympathetic pain afferents. A nearly complete de-afferentiation of the bladder stops the hyperreflexia of the detrusor, and large storage volumes of the bladder can be obtained after some time again. The de-afferentiation surgery not only cuts the thick afferents of stretch and tension receptors but also the thin fibre populations, including parasympathetic pain afferents. Also, the rather thin afferent fibres of the flow receptors are cut [46]. If these afferent fibres were stimulated by inflammation, the sacral micturition center would all the time get the message that urine is passing through the trigonum vesicae and the urethra, and bladder reflexes would probably be activated to sustain voiding (possible descending inhibitory control of the afferent input is destroyed by the spinal cord injury). Tetraplegics with a partial spinal cord injury report that during bladder infection, there was a complete dysfunction of the bladder. However, as soon as the infection wore off, the bladder started to work acceptably again. Therefore, patients with an inflammation must be free of infection for some time before prognosis concerning their bladder function can give meaningful answers.

Nearly all humans get bladder infections following spinal cord injury. The problem is how to improve bladder function to get rid of reoccurring infections. It was the benefit of Sir Guttmann [23] to introduce the "intermittent catheterization" so that the patients could avoid infections caused by the bladder catheter. But patients with severe cervical spinal cord injury have no hand functions left to perform this intermittent catheterization by themselves. It is the benefit of this current human research that it is possible to cure urinary bladder function in severe cervical spinal cord injury [86] (part 5, below), so that all urinary bladder complications including reoccurring infections, hyperreflexia and dyssynergia (with development of compliance changes, high bladder pressure and urinary tract damage) are not occurring anymore or cured if proper coordination dynamics therapy is started one or two weeks after the injury and is administered for at least three years.

The similar time courses of detrusor pressure and secondary muscle spindle afferent activity provides information about the time course of the bladder pressure from the activation of spindle afferent fibres and allows analyzing dysfunctions, even though it is not well understood what underlies these similar time courses. Probably, parasympathetic motoneurons activate the detrusor and intrafusal muscle fibres with approximately the same time course. Interesting is further that this parasympathetic motoneuron activation is rather stereotypic and does not change very much under light anaesthesia.

2.5.4. Replacement of Physiologic Multi-Ending Regular Firing by Pathophysiologic Burst Firing

As observed in the brain-dead individual, the low activity level doublet firing and the high activity level multi-ending regular firing (Figure 9B) probably represents physiologic parasympathetic activation of secondary muscle spindle afferent fibres with coordinated firing of several endings of the parent spindle afferent fibre. In paraplegics with detrusor-sphincteric dyssynergia and spasticity, these firing patterns were changed and mixed. In paraplegic 5 with slight bladder dysfunction, the multi-ending regular firing was somehow preserved (Figure 10B,C). In paraplegics 9 and 11 with strong bladder dysfunctions, the multi-ending regular firing was replaced by a burst-like firing (Figure 9A, Figure 3C, Figure 11d,e). If one assumes that on average the spindles are activated in parallel to the activity of the preganglionic parasympathetic neurons, as suggested in Figures 7Bb,8C, then the parasympathetic nucleus in the sacral micturition center seems to be unable to develop a sustained activation anymore. It is not known whether parasympathetic motoneurons also can fire in an oscillatory manner for high activation levels (see [84]). If it were so, then these oscillators could only fire in a transient oscillatory mode. In the brain-dead individual, a sphincteric α_2-motoneuron (FR) fired transiently oscillatory upon retrograde bladder filling, because the activity level from stretch and tension receptor afferents was low [47]. The motoneuron fired continuously oscillatory for high afferent input but was slightly inhibited for very high afferent input (Figure 9 of [55]). In the paraplegics, the stretch and tension receptors were highly activated or partly over-activated [83], but the preganglionic parasympathetic neurons could not fire continuously oscillatory. It is therefore conceivable that there was a dysfunction in the neuronal network activating the parasympathetic motoneurons, independent of whether the neuronal networks fired oscillatory or not. An over-activation of the neuronal network (see previous paragraph) may have contributed to the dysfunction.

The recorded bursts (Figures 3c,11d,e) with the extremely short interspike intervals (shorter than 0.2ms) suggest that parasympathetic neuronal networks tried strongly to generate a high activity level, but they failed to sustain the activity, so they tried again and again. One could even imagine that the loss of inhibition onto the somatic sphincteric motoneurons exerted by the parasympathetic division (see below) is due to the loss of sustained activation. In paraplegic 5 (slight bladder dysfunction), however, the developed detrusor pressure was sustained (Figure 2) and also the impulse activity level of a secondary spindle afferent fibre was rather sustained (Figure 10B(2s)), but still the pelvic floor was activated rather than inhibited with the increasing detrusor pressure (Figure 2). Thus, there may be several reasons for the dysfunction of the urinary bladder in paraplegics. Probably, some of them are key changes in or damages to the neuronal network below the spinal cord injury level that cannot be compensated for by the network itself. Only supraspinal control could substitute for the necessary control, which can be established [86]. In the following part [84], failures of the neuronal networks of the sacral micturition centre will be analyzed by making use of the spinal oscillators.

2.5.5. Simultaneous Action of the Somatic and Parasympathetic Divisions

As outlined in the Introduction, physiologically there is hardly any somatic response to natural stimulation that is not accompanied by an autonomic reaction [19]. In the cat's spinal cord, there is anatomical evidence for direct brain stem projections to the somatic motoneuronal cell groups and autonomic preganglionic cell groups [25]. In paraplegics and tetraplegics (lesion sub C5/6 or more rostral), often somatic functions are destroyed because of the spinal cord injury, but autonomic responses are preserved or even enhanced. Such autonomic responses are often helpful for the patients to monitor functions of the body, but they also can be unpleasant. A tetraplegic patient had been reporting all day long sweating in one half of the face for years. When a stricture of the urethra was revised, the sweating stopped. Sympathetic responses are common. The sympathetic chain is often undamaged following spinal cord injury. Reflex voiding is often accompanied by sweating or dangerous blood pressure increases. Bladder distension by retrograde filling can activate somatic spasticity (Figure 2). Especially when the somatic and the autonomic nervous system have receptors in common, the stimulation of those receptors will simultaneously result in strong somatic and autonomic responses.

Parasympathetic activation of different strength can be seen in Figure 5. In Figure 5A, the doublet activity levels of a secondary spindle afferent fibre are drawn, and in Figure 5B, the durations of the doublet interspike intervals following different natural stimulations are plotted. It has been mentioned that doublet firing is a measure for low-level parasympathetic activation. As can be seen from Figure 5A,B, only few doublets occurred upon touch and pinprick stimulation. There was only little activation of the parasympathetic division. Upon anal reflex stimulation, the parasympathetic response increased. Anal and bladder catheter pulling strongly activated the parasympathetic division as measured by the activity levels of doublet and burst firing. This was to be expected since catheter pulling activates continence functions.

Somatic and parasympathetic functions can be activated by common receptors. The anal canal is innervated by skin and mucosa afferents [52]. Especially the T4 (possibly similar to SAII receptors in animals) skin afferent fibres are of interest, since their receptors are very sensitive to skin displacement. Probably, the T4 mechanoreceptors are strongly activated when a finger is moved along the skin with very light pressure ("streicheln"). They will contribute to the discrimination between wind, fluid and solid content, when material is passing through the anal canal. The skin afferents T1 to T4 will activate somatic (external sphincter contraction) and parasympathetic functions (sphincter inhibition, see following part [84]).

Somatic and parasympathetic functions can be activated simultaneously by simultaneous stimulation of specific parasympathetic receptors and unspecific receptors. With the pulling of the bladder catheter, stretch (S1) and tension receptor afferents (ST) were activated besides flow (S2), mucosal (M) and other receptor afferents including pain afferents. Even though stretch and tension receptors themselves probably activate parasympathetic responses in a rather specific manner, natural stimulation such as pulling of the bladder catheter or even distension of the bladder will simultaneously activate less specific receptors or even receptors that stimulate somatic functions. If the bladder catheter is accidentally pulled strongly in patients with remained sensibility in the bladder, the patients report pain instead of a wish to micturate. Patients with lost sensibility of only the bladder may still feel, to some extent, a filled bladder, because the distended bladder displaces adjacent structures or pushes against them. It has been reported that Pacinian corpuscles (PC = T1) can be activated over distances of up to 100 mm [52]. The anchorage of the mechanoreceptors in the skin or other tissues is one reason for stimulation over long distances [26,27].

If in humans with spinal cord injuries the specific receptors (S1,ST) monitoring bladder storage volume (increase) are disconnected from supraspinal centres and if there are still connections preserved from unspecific mechanoreceptors, the training of these unspecific receptors may be of benefit for bladder fullness control, since normally the afferent input from the receptors is not being used efficiently. It has been reported for the skin that the two-point discrimination can be increased, since in the rather untrained situation there are seven to eight receptor innervations in between the two points of the two-point discrimination [52] (Figure 45 of Chapter III).

Possibly, also the afferent fibres innervating the adjacent structures of the urinary bladder may converge onto the same nerve cells as do bladder afferents. Convergence of visceral and cutaneous (Head's zone) afferent pathways has been reported in animals, including primates [17,62].

In part 5, it will be shown that the urinary bladder function can be cured by a functional reorganization and limited regeneration in severe cervical spinal cord injury [86]. A key movement for bladder repair is the jumping on springboard when no body-weight support is provided! Upon jumping rhythmically on springboard, the intestine, resting on the pelvic floor, will induce rhythmic stretch and tension increases in the pelvic floor, including the external anal and bladder sphincters. Also the detrusor and the urethra will be stretched and pressed rhythmically. Many of the receptors of the urogenital domain, including stretch (S1), tension (ST), flow (S2), mucosal receptors (M), and mechanoreceptors T1 through T4, will be activated to give rise to an intermingled activation of the somatic and parasympathetic nervous systems. But especially the muscle spindles, involved in continence functions, innervated simultaneously by somatic and parasympathetic fusimotors with their spindle afferents projecting into somatic and parasympathetic networks, will give rise that the somatic and parasympathic divisions dovetail with one another. The doublet firing of the spindle afferents may even have a messenger function for communication between the systems. With the learning-based movement therapy "coordination dynamics therapy," especially the jumping on springboard, a learning transfer [85] from movements to urinary bladder function for repair is therefore conceivable. But since there are several reasons for the dysfunction of the urinary bladder in patients with severe spinal cord injury with key changes in or damages to the neuronal network below the spinal cord injury level that cannot be compensated for by the network itself, only supraspinal control can substitute for full necessary control and learning. But it is not only the functional and structural repair that is needed. Within the system theory of pattern formation for repair [78,81] (Chapter II), for describing integrative CNS functions, there are attractor states for the synergia pattern (for micturition) and the dyssynergia pattern (for stopping micturition) with different stabilities, which can be changed by learning and intention (Figure 5 of Chapter II). Therefore, some supraspinal control is needed for the repair of the networks and for choosing between different network patterns (see below).

2.5.6. Detrusor-Sphincteric Coordination

It is generally assumed that it is the level of the pons and the adjacent medulla at which coordination and integration of the bladder and the external sphincter activity occurs in the adult. Lesions above the brain stem lead to coordinated incontinence, and lesions below the brain stem centres but above the lumbosacral spinal cord lead, after a period of bladder paralysis associated with spinal shock, to involuntary reflex voiding that is not coordinated with sphincter relaxation (detrusor-sphincteric dyssynergia) [22,69]. The measurements in the three paraplegics studied with clinically complete lesions below the brain stem and above the lumbosacral spinal cord and with detrusor-sphincteric dyssynergia support this view. Upon the activation of the parasympathetic division (Figure 7B), the detrusor pressure increased (Figure 6), but sphincteric α-motoneurons were not inhibited. In discordance with present view is that the sphincteric motoneurons in the brain-dead human were inhibited upon the activation of the parasympathetic division (Figure 7A). One can escape the conclusion that in the brain-dead individual there was coordinated action between the detrusor and the sphincter, in arguing that the breathing centre and other parts of the brain stem were destroyed but the pontine micturition centre was still working. Such a situation may in principle be possible, since brain-death can be manifested in different ways. However, it is more likely that the whole brain stem was destroyed.

These measurements then seem to indicate that the detrusor-sphincteric coordination is not located in the pontine micturition centre and that detrusor-sphincteric dysynergia develops during the three to four weeks lasting spinal shock phase when the micturition reflex via supra-spinal centres is replaced by a reflex via the sacral micturition centre, in some similarity to that in infants. There is little doubt that important changes occur in the spinal shock phase. Nevertheless, putting the problem behind the term of the not yet understood spinal shock would not promote the solution to this problem. There seem to be important rearrangements during the spinal shock, so that a training of the sublesional CNS in paraplegics already has to start during the spinal shock phase. Of high importance for the understanding of micturition following spinal cord injury would be to obtain recordings, at least urodynamic, from paraplegics below the age of one year, to see whether micturition in infants is changed following spinal cord injury. Infants possess prominent micturition responses that can be initiated by light tactile stimulation of the perineum, the lower back, or by cutaneous contact with water [6,66].

Micturition in neonatal mammals of many species (rat, cat, guinea pig) [4,5,14,29,38] is regulated by the mother. She initiates excretion by licking the perineum, and then she consumes the excreted waste, which conserves her bodily fluids to meet the demands of lactation [18] and prevents the odors of the wastes from attracting predators to the nest. Thus, micturition in the neonate is mediated by a spinal somatovesical reflex that is initiated by exteroceptive stimulation. In kittens, the somatovesical reflex pathway disappears about the time of weaning and is replaced by a vesicovesical reflex pathway that produces micturition via a supraspinal reflex pathway that is activated by interoceptive stimulation, i.e., bladder distension and, in addition, may be under voluntary control. Stimulation of the perigenital region in adult cats actually inhibits the supraspinal vesicovesical micturition reflex. Presumably, this inhibition serves to suppress micturition during copulation. Following spinalization, in older kittens, the excitatory somatovesical reflex consistently returns in less than a day, while in adult cats, the reflex returns more slowly and with greater variability, i.e., two days to two weeks [66,68].

Since micturition is a phylogenetically old function, coordination of the detrusor and the sphincter may well get support from several levels. There may already be some detrusor-sphincteric coordination in the sacral micturition center. The malfunctioning of different components in the reflex arc may result in the overriding of the weak detrusor-sphincteric inhibition. An already slight inflammation could be one cause. In normal humans with hemorrhoids, the external anal sphincter may be sustained contracted following slight inflammation. Upon suppressing the inflammation (by eating garlic with some anti-infectious action), the external anal sphincter relaxes. With the inflammation intensifying, the external sphincter develops sustained contraction again, and it is nearly impossible for a normal human to voluntarily inhibit this contracture. It will be shown in the following part what a dramatic activity increase does occur in the motoneurons innervating the external bladder sphincter upon pathologically high activity of stretch (S1) and tension receptor afferents (ST) from the bladder wall [84]. As long as inflammation and infection of the lower urinary tract are not fully relieved, no information can be obtained about the possible functions of the sacral micturition centre following spinal cord injury.

It has been suggested that bladder outlet obstruction in rats is accompanied by some degree of neural plasticity resulting in a more prominent spinal reflex that could contribute to the development of the unstable bladder following obstructions in humans and un-anaesthetized animals [71].

In part 5 on the cure of the human urinary bladder function [86], there is support for the view that the coordination between the detrusor and the external bladder sphincter may get support from several levels, because before a cure was achieved, there seem to exist both coordination patterns, the synergia and the dyssynergia. Within the system theory of pattern formation [78,81], the coordination possibilities of synergia and dyssynergia can be understood as two micturition patterns with certain stabilities. With the improvement of bladder function, a repair stage is passed at which the synergia and the dyssynergia patterns co-existed [86]. The further repair consisted in the increase of the stability of the synergia pattern (Figure 5 of Chapter II). For this important repair step in that patient, some supraspinal control was necessary, which was established by limited regeneration of the spinal cord.

2.5.7. Control Group

A typical control for the humans with spinal cord injuries may be a healthy human, which is operated under very light Propofol-anesthesia (pain may still partly be present). Such controls are not available. A brain-dead human may perform leg movements upon touch following one to two weeks of brain-death.

It may therefore be that in fresh brain-dead humans, a partial spinal shock is present. On the other hand, patients that undergo supraspinal CNS operations, which are not in connection with micturition (for example, cancer operations), have no bladder function for one to two days following the operation (partial spinal shock?).

Therefore a fresh brain-dead human is probably a reasonable control, even though extremely difficult to have excess to. Also, the activity increase of the SP2 fibre of the HT6, due to parasympathetic spindle activation, has a similar time course as the bladder pressure increase of paraplegics following electrical ventral root stimulation for several seconds with 30 Hz (direct detrusor stimulation) in the operation.

2.5.8. Safety in Splitting Multi-Unit Recordings into Single-Unit Impulse Patterns

Single-nerve fibre APs should have same shape and size, if the baseline is straight and if the interspike intervals are longer than the refractory period. But firstly, APs may fall sometimes into the partial refractory period, so that the AP amplitude is reduced. Secondly, APs of small thin myelinated fibres (γ-motoneurons, preganglionic parasympathetic fibres, mucosal afferents, ...), which have a long duration and a small AP amplitude (see Figure 3 of [83]), may change the baseline (like large noise), even though the individual AP waveforms cannot be recognized. Activity of parasympathetic fibres and γ-motoneurons will have been present and will have added up with the spindle afferent activity. Therefore the single unit shape varies.

The APs of the bursts of the single spindle fibres in Figure 11d,e were recognized by similar waveforms on both traces and same conduction times and marked by arrows. One or two APs were unsafe. Because of baseline changes, the single-fibre APs are not identical on the same trace.

The APs of a single fibre on the two traces are anyway different because of the different recording conditions at the two electrode pairs. In principle, it is possible that the APs of several fibres have accidentally an occurrence on trace "a," which looks like a burst firing from a single unit.

Such an occurrence is extremely seldom, whereas the burst firing often occurred. Further, the conditions for the time matching are different on trace "b" from the 10mm distant electrode pair; the pattern will be different there. Actually, the second electrode pair is the power in the method of splitting multi-unit recordings into single-unit patterns. A third electrode pair could give even more safety in splitting.

2.5.9. Automatic Bladder, Hyperreflexia and Dyssynergia of the Bladder

The parasympathetic division activated by secondary muscle spindle afferents showed different firing patterns. The activity level changes of the spindle afferents, measured intra-operatively, were similar to the active detrusor pressure changes upon retrograde bladder filling, measured pre-operatively.

It seems therefore as if the detrusor pressure gives information about the functional stage of the highly activated parasympathetic division. Low level parasympathetic activation, which can probably be measured by the doublet firing of the secondary spindle afferents cannot be judged by the detrusor activation. Since there are several kinds of too early detrusor activation (hyperreflexia), there are probably several kinds of dysfunction of the parasympathetic division.

Even though a hyperreflexia of the bladder (to early detrusor activation, generalization of the adequate afferent input (see below)) and a dyssynergia of the bladder (simultaneous activation of detrusor and ext. sphincter) were present in the three examined paraplegics, they must not necessarily accompany each other.

The three paraplegics had a complete spinal cord injury well above the lumbosacral cord and below the brain stem; they all had an automatic bladder, which means the detrusor was activated involuntary with retrograde bladder filling and the activation could not be stopped on volition.

All patients had a hyperreflexia of the bladder, which means the detrusor was activated for too small filling volumes. There are mainly two kinds of hyperreflexia: a spot like too early activation of the detrusor (Figure 6, Para 9) and a plateau-like undulating-like (up to several minutes) activation (cystometry not shown, Para 11). In hyperreflexia, the detrusor cannot be activated by bladder afferent activity only. Touching, tapping or manipulating the skin of the perineum, of the Head's zone of the bladder (dermatomes TH12-S1) or of the penis (not accompanied by a penial erection) may activate the detrusor. Paraplegics with incomplete spinal cord injuries may have a hyperreflexia of the bladder but no dyssynergia. On the other hand, there are paraplegics with a dyssynergia of the bladder, but they have no hyperreflexia (similar to Para 5).

The three examined cases are therefore somehow characteristic cases. All three paraplegics had a clinical complete spinal cord injury (few functioning tract fibres were probably spared) and (therefore) a dyssynergia of the bladder. Para 9 and 11 additionally had a hyperreflexia of the bladder. In Para 9, the detrusor was spot-like activated and in Para 11, plateau-undulating-like.

How much the different detrusor activation correlates with the parasympathetically induced secondary muscle spindle activity changes has further research to show. Since all spinal cord injuries are different, the lesion-induced changes of the nervous system will show delicate equilibriums [65]. It seems possible to explore the human nervous system by detailed analyses of the lesion-induced changes of different paraplegics.

In part 1 [83], it was found that in patients with a hyperreflexia of the bladder, the urinary bladder stretch (S1) and tension receptor afferent (ST) activity levels were undulating and increased. It was proposed that receptor field signal transduction mechanisms of bladder afferents had changed.

The hyperreflexia of the bladder can therefore be explained by the changed afferent input from the bladder and other receptors of the urogenital domain and a probable changed processing of the afferent input in the spinal cord.

With respect to human electrophysiology, the detrusor-sphincteric dyssynergia is probably due to a disconnection from the pontine micturition centre and/or a pathologic interaction between the somatic and the parasympathetic divisions, which has in terms of system theory of pattern formation [81] the consequence that the stability of the dyssynergia pattern is too high (the potential well too deep) in comparison to that of the synergia pattern

[86]. Further, the attractors for the synergia and the dyssynergia patterns are probably of unphysiologic form in the attractor layout.

2.5.10. Clinical Implications: Some History

It has been reported that a neonatal somatovesical reflex in adult cats can be unmasked with the serotonin autoreceptor antagonist 5-MeODMT [75] and that delta opiod receptors regulate reflex pathways to the bladder, whereas kappa opiod receptors regulate reflex pathways to the sphincters [67]. This differential regulation of bladder and sphincter pathways suggests possible means for the pharmacological management of urine retention and/or incontinence.

Pharmacological intervention is of importance following spinal cord injury in humans, since disorders of the spinal cord can be of such strength that even joints may dislocate. If, on the other hand, paraplegics don't want to reduce their life expectation times by the side effects of drugs and want to regain a higher quality of life especially in severe cervical spinal cord injury, other treatments have to be found and are found [86]. Also, drugs may hinder the reorganization and regeneration of the spinal cord. The geographical landscape of neuroattractive gradients during endogenous stem cell therapy [81,82,86], induced by coordination dynamics therapy, may be disturbed with the consequence that the homing of cells and connections becomes impaired.

Spontaneous healing of human spinal cord injury has not been observed so far. Since neurogenesis is very limited in the adult, the CNS seems to recruit other nerve cells to take over for destroyed ones. The substitution of a function from other parts of the CNS is only possible if the necessary connections exist or can be built and if there is enough plasticity in the human CNS for relearning the functions. In spinal cord injuries, nerve cells and tracts are destroyed. Injuries of the tracts are worse since the caudal spinal cord is disconnected from supraspinal control.

Tract cells regenerate only over small distances, are impeded by scar tissue formation and form inappropriate synapses (for references, see [56]). The normal surgical strategy to cut unhealthy tissue away and substitute nervous tissue (embryonic) is not feasible. Firstly, only few spinal cord injuries are complete. Resection of the damaged spinal cord parts would destroy still existing tract fibres. Secondly, by cutting the damaged spinal cord, e.g., in the lumbal range, the caudal spinal cord will be disconnected from the blood supply and also die, because 70% of blood is supplied via the largest feeder artery (Adamkiewicz) and is distributed via the longitudinal trunks [45] (Figure 14 of Chapter III). Replacement of damaged spinal cord tissue by embryonic nervous tissue in animals is of interest for the study of the mechanism of regeneration but had no consequences so far for the reconstruction of functions in humans.

The injection of different kinds of stem cells had also so far no consequences for the repair of the injured spinal cord in human. Worldwide performed stem cell therapies are doubtful because of missing scientific publications and necessary diagnostics, doubtful theory [81] and missing knowledge on the functioning of the human CNS under physiologic and pathologic conditions.

The therapeutic potential of neural stem cells has recently been reviewed [76]. A qualified study by the author on stem cell therapy and coordination dynamics therapy on a patient with a thoracic spinal cord injury was not successful [79,80]. For discussions of stem cell therapies and literature, see Refs. [80,81]. Stem cell therapies are still in the experimental stage. Chapter VI is offering information why exogenous stem cell therapy does not work (new cells cannot be integrated into the existing adult networks).

Since the damaged spinal cord part cannot be removed, neurosurgeons used to bypass it by performing a nerve anastomosis from the lower intercostal nerves rostral to the level of injury to the cauda equina nerve roots caudal to the lesion. Regeneration occurs, but no useful functions are obtained. In the most advanced operation [11], the regeneration of the urinary bladder function could be obtained, but the detrusor and the sphincter contracted simultaneously. At the beginning of this research project, the author also tried to improve such nerve anastomosis, so that the patients get useful functions reconstructed [49-52]. The number of total nerve fibres, of afferents and efferents of the donor nerves has to be compared with the acceptor nerves. E.g., it is useless to try to reinnervate the lower human body on one side with one intercostal nerve, which contains 10,000 myelinated fibres, if 250,000 are needed. However, the number of myelinated fibres of one intercostal nerve is sufficient for regeneration of the urinary bladder. Mismatch and functional aspects have to be taken into consideration.

To make a separate function of the detrusor and the sphincter possible, the bladder has to be reinnervated by the two different motoneuron pools, present in the intercostal nerve distal to its branching into two skin nerves (nervus cutaneus lateralis), a pure muscle branch supplying the musculus obliquus externus abdominis and the mixed nerve running to the musculus rectus abdominis. Specific bladder afferents can be reconstructed by using the skin afferents from the Head's zone (ramus cutaneus anterior of the intercostals T12 to S1). For further details, see [49-52,56]. The author did not proceed further with this nerve anastomosis therapy, since firstly it is still a partial destructive operation, secondly the patients need more than a functioning bladder, and thirdly in cervical spinal cord injuries, this anastomosis cannot be performed.

It has been reported that in adult rats, axonal regeneration after partial cord transection is greatly increased by blocking a subset of inhibitory myelin-associated proteins with neutralizing antibodies [61]. It has then been discussed with respect to human applicability that it is "just" a question of making it work [7]. In a recent review article on the repair of spinal cord injury in the same journal, 12 years later [77], it is still "just" a question of making it work. Human research is not included in the review. It seems that many animal researchers seem to think that one can repair the human CNS without understanding its functioning. The enhanced regeneration of corticospinal tracts of the treated rats has been demonstrated by an enhancement of regenerating axons in the not lesioned cord and by the improved running. It has not been proven electrophysiologically that the corticospinal tract fibres and the formed synapses were functioning. In Chapter IV, human data are partly "translated" to rat data. Therefore, also in rat regeneration, it can be measured scientifically. It has also been reported that late-growing axons are able to take an aberrant route [8]. Still, the improved running of the treated rats is difficult to understand. Sperry transposed the nerve supplies of flexor and extensor muscles in rats [63] and monkeys [64]: the monkey relearned the task after some time; the rat did not. Surprisingly few trials were required for poliomyelitis patients to use transposed tendons successfully [73].

When cutting in experimental rats only 50% of the spinal cord, the repair results have no consequences for the regeneration of the human spinal cord, because firstly the rat has a higher capacity for regeneration and secondly it has been shown in a human patient with a 50% injured spinal cord (according to magnetic resonance imaging) that movement and bladder functions could be repaired by a mainly functional regeneration induced by coordination dynamics therapy [81] (see Chapter II). Further, with respect to the nerve anastomosis from the lower intercostal nerves to the cauda equina nerve roots, nerve fibres of the intercostal nerves reinnervated the urinary bladder, but useful functions were not obtained [11]. Something like 20 years were not enough time, "just" to make this nerve anastomosis work.

In the human nervous system, the regenerative capacity is poor, but the learning-induced plasticity seems to be extremely high. A near-total functional recovery was achieved in a patient with an incomplete spinal cord injury after three years of coordination dynamics therapy [81] (Chapter II). Even in severe cervical spinal cord injury with a 95% injured spinal cord (according to magnetic resonance imaging), a partial cure is possible if coordination dynamics therapy is administered [82,85,86]. The training and learning-induced plasticity of the human CNS has by far been underestimated.

References

[1] Abeles, M., *Local cortical circuits. An electrophysiological study.* Springer, Berlin, 1982.

[2] Arbuthnott, E.R., Gladden, M.H. and Sutherland, F.: Autonomic innervation of muscle spindles from the hind limb of the cat. *Suppl. Proc. Roy. Microsc. Soc.*, 17: 1, 1982.

[3] Barker, D. and Saito, M.: Autonomic innervation of receptors and muscle fibres in cat skeletal muscle. *Proc. R. Soc. Lond.* (Biol.), 212: 317-332, 1981.

[4] Baverstock, P. and Green, G.: Water recycling in lactation. *Science*, 1987: 657-658, 1974.

[5] Beach, F.A.: Ontogeny of "coitus-related" reflexes in the female guinea pig. *Proc. Natl. Acad. Sci. U.S.A.*, 56: 526-533, 1966.

[6] Boehm, J.J. and Haynes, J.L.: Bacteriology of "midstream catch" urines. *Am. J. Dis. Child.*, 111: 366-369, 1966.

[7] Björklund, A.: A question of making it work, spinal cord repair. *Nature*, 367: 112-113, 1994.

[8] Bregman, B.S.: Development of serotonin immunoreactivity in the rat spinal cord and its plasticity after neonatal spinal cord lesion. *Brain Res. Dev. Brain Res.*, 34: 245-263, 1987.

[9] Brindley, G.S., Polkey, C.E., Rushton, D.N. and Cardozo, L.: Sacral anterior root stimulation for bladder control in paraplegia: the first 50 cases. *J. Neurol. Neurosurg.* Psychiat., 49: 1104-1114, 1986.

[10] Burke, D., Skuse, N.F. and Stuart, D.G.: The regularity of muscle spindle discharge in man, *J. Physiol.* London., 291: 277-290, 1979.

[11] Carlsson, C.A. and Sundin, T.: Reconstruction of afferent and efferent nervous pathways to the urinary bladder in two paraplegic patients. *Spine*, 5: 37-41, 1980.

[12] Coggeshall, R.E.: Law of separation of functions of the spinal roots. *Physiological Reviews*, 60: 716-755, 1980.

[13] de Groat, W.C.: Nervous control of the urinary bladder of the cat. *Brain. Res.*, 87: 201-211, 1975.

[14] de Groat, W.C., Douglas, J.W., Glass, J., Simonds, W., Weimer, B. and Werner, P.: Changes in somato-vesical reflexes during postnatal development in the kitten. *Brain Res.*, 94: 150-154, 1975.

[15] de Groat, W.C., Nadelhaft, I., Milne R.J., Booth, A.M., Morgan, C. and Thor, K.: Organisation of the sacral parasympathetic pathways to the urinary bladder and large intestine. *J. Auton. Nerv. Syst.*, 3: 135-160, 1981.

[16] de Groat, W.C., Booth, A.M., Milne, R.J. and Roppolo, J.R.: Parasympathetic preganglionic neurons in the sacral spinal cord. *J. Auton. Nerv. Syst.*, 5: 23-43, 1982.

[17] Foreman, R.D., Hancock, M.B. and Willis, W.D.: Responses of spinothalamic tract cells in the thoracic spinal cord of the monkey to cutaneous and visceral inputs. *Pain,* 11: 149-162, 1981.

[18] Friedman, M.I., Bruno, J.P. and Alberts, J.R.: Physiological and behavioral consequences in rats of water recycling during lactation. *J. Comp. Physiol. Psychol.,* 95: 26-35, 1981.

[19] Fulton, J.F.: *Physiology of the Nervous System*, pp VI-VII, Oxford University Press, London, 1943.

[20] Gladden, M.H., Wallace, W. and Craigen, M.L.: Movement convergence and motor endplate location in a human muscle spindle. In: I.A. Boyd and M.H. Gladden (Ed.), *The Muscle Spindle*, pp. 115-120, Macmillan, London, 1985.

[21] Gladden, M.H.: Efferent control of human muscle spindles. In: I.A. Boyd and M.H. Gladden (Ed.), *The Muscle Spindle*, pp. 161-166, Macmillan, London, 1985.

[22] Graber, P.: Mechanismus der normalen und neurogen gestörten Blasenentleerung. In: Stöhrer, M., Palmtag, H., Madersbacher, H. (Eds.), *Blasenlähmung*, pp. 12-24, Thieme, New York.

[23] Guttmann, L.: *Spinal Cord Injuries, comprehensive management and research*, Blackwell Scientific Publications, Oxford, 1976.

[24] Hodgekin, A.L.: *The conduction of the nervous impulse*, Liverpool University Press, 1964.

[25] Holstege, G., Kuypers, H.G.J.M. and Boer, R.C.: Anatomical evidence for direct brain stem projections to the somatic motoneuronal cell groups and autonomic preganglionic cell groups in the cat spinal cord. *Brain Res.*, 170: 329-333, 1969.

[26] Johannson, R.S.: Tactile sensibility in the human hand: Receptive field characteristics of mechanoreceptive units in the glabrous skin area. *J. Physiol.*, 281: 101-123, 1978.

[27] Johannson, R. and Vallbo, A.B.: Skin mechanoreceptors in the human hand: An inference of some population properties. In: Zotterman, Y. (Ed.), *Sensory Function of the Skin in Primates*, pp. 171-184, Pergamon Press, Oxford, 1976.

[28] Kucera, J.: Reconstruction of the nerve supply to a human muscle spindle. *Neurosci. Lett.,* 63: 180-184, 1986.

[29] Maggi, C.A., Santicoli, P. and Meli, A.: Postnatal micturition reflex in rats. *Am. J. Physiol.*, 250: R926-R931, 1986.

[30] McMahon, S.B.: Sensory-motor integration in urinary bladder function. In: Cervero, F., Morrison, J.F.B. (Eds.),*Visceral sensation*, pp. 245-253, Elsevier, Amsterdam, 1986.

[31] McMahon, S.B., Morrison, J.F.B.: Spinal neurons with long projections activated from the abdominal viscera of the cat. *J. Physiol.*, 332: 1-20, 1982.

[32] McMahon, S.B., Morrison, J.F.B.: Two groups of spinal interneurones that respond to stimulation of the abdominal viscera of the cat. *J. Physiol.*, 332: 21-34, 1982.

[33] McMahon, S.B., Morrison, J.F.B.: Factors that determine the excitability of parasympathetic reflexes to the bladder. *J. Physiol.*, 332: 35-42, 1982.

[34] Miledi, R., Parker, I. and Schalow, G.: Calcium transients in normal and denervated slow muscle fibres of the frog. *J. Physiol.*, 318: 191-206, 1981.

[35] Milne, R.G., Foreman, R.D., Giesler, G.J. and Willis, W.D.: Convergence of cutaneous and pelvic visceral nociceptive inputs onto primate spinothalamic neurons. *Pain*, 11: 163-183, 1981.

[36] Morrison, J.F.B.: The neural control of the bladder. In: Bloom, S.R., Polak, J.M., Lindenlaub, E. (Eds.), *Systemic role of regulatory peptides*, pp. 381-396, Schattauer, Stuttgart, 1982.

[37] Passatore, G.M., Fillipi, G.M. and Grassi, G.: Cervical sympathetic nerve stimulation can induce an intrafusal muscle fibre contraction in the rabbit. In: Boyd, I.A. and Gladden, M.H. (Eds.), *The Muscle Spindle*, pp. 221-226, Stockton Press, New York, 1985.

[38] Rheingold, H.L.: *Maternal behavior in mammals*, pp. 169-202, Wiley, New York, 1963.

[39] Schalow, G. and Schmidt, H.: Action potential induced in slow muscle fibres by partial denervation. *Nature* (London), 253: 122-123, 1975.

[40] Schalow, G. and Schmidt, H.: Effect of length and temperature on the induction of action potentials in denervated slow muscle fibres of the frog. *Pflügers Arch.*, 372: 17-22, 1977.

[41] Schalow, G. and Schmidt, H.: Local development of action potentials in slow muscle fibres after complete or partial denervation. *Proc. R. Soc. Lond.*, B203: 445-457, 1979.

[42] Schalow, G.: Elektrophysiologische und morphometrische Untersuchungen zur Differenzierung der Tonusfasern des Frosches, Inauguraldissertation, Universität des Saarlandes, 1983.

[43] Schalow, G. and Lang, G.: Electrodiagnosis of human sacral nerve roots by recording afferent and efferent extracellular action potentials. *Neurosurg. Rev.,* 12: 223-232, 1989.

[44] Schalow, G.: Efferent and afferent fibres in human sacral ventral nerve roots: basic research and clinical implications. *Electromyogr. clin. Neurophysiol.*, 29: 33-53, 1989.

[45] Schalow, G.: Feeder arteries, longitudinal arterial trunks and arterial anastomosis of the lower human spinal cord. *Zbl. Neurochir.*, 51: 181-184, 1990.

[46] Schalow, G.: Conduction velocities and nerve fibre diameters of touch, pain, urinary bladder and anal canal afferents and α and γ-motoneurons in human dorsal sacral roots. *Electromyogr. clin. Neurophysiol.*, 31: 256-296, 1991.

[47] Schalow, G.: Oscillatory firing of single human sphincteric α_2 and α_3-motoneurons reflexly activated for the continence of urinary bladder and rectum. Restoration of bladder function in paraplegia. *Electromyogr. clin. Neurophysiol.*, 31: 323-355, 1991.

[48] Schalow, G. and Barth, H.: Group conduction velocities and nerve fibre diameters of α and γ-motoneurons from lower sacral nerve roots of the dog and humans. *Gen. Physiol. Biophys.*, 11: 85-99, 1992.

[49] Schalow, G., Aho, A. and Lang, G.: Microanatomy and number of nerve fibres of the lower intercostal nerves with respect to a nerve anastomosis. Donor nerve analysis I(IV). *Electromyogr. clin. Neurophysiol.*, 32: 171-185, 1992.

[50] Schalow, G.: Number of fibres and fibre diameter distributions of nerves innervating the urinary bladder in humans. Acceptor nerve analysis. II (IV). *Electromyogr. clin. Neurophysiol.*, 32: 187-196, 1992.
[51] Schalow, G. and Barth, H.: Single-fibre action potential recording from the nerve to the musculus obliquus externus abdominis following pinprick in humans. III(IV). *Electromyogr. clin. Neurophysiol.*, 32: 197-205, 1992.
[52] Schalow, G.: Impulse pattern, innervation density and two point discrimination of skin and mucosal afferents in humans. Consideration for a sensory reinnervation of urinary bladder and anal canal in spinal cord lesions. IV (IV). *Electromyogr. clin. Neurophysiol.,* 32: 259-285, 1992.
[53] Schalow, G.: Spinal oscillators in man under normal and pathologic conditions. *Electromyogr. clin. Neurophysiol.*, 33: 409-426, 1993.
[54] Schalow, G.: Impulse patterns of single encoding sites of human secondary muscle spindle afferents. *Electromyogr. clin. Neurophysiol.* 33: 451-464, 1993.
[55] Schalow, G.: Phase correlated adequate afferent action potentials as a drive of human spinal oscillators. *Electromyogr. clin. Neurophysiol.*, 33: 465-476, 1993.
[56] Schalow, G.: Action potential patterns of intrafusal γ and parasympathetic motoneurons, secondary muscle spindle afferents and an oscillatory firing α_2-motoneuron, and the phase relations among them in humans. *Electromyogr. clin. Neurophysiol.*, 33: 477-503, 1993.
[57] Schalow, G., Bersch, U. Göcking, K. and Zäch, G.A.: Interspike intervals of secondary muscle spindle and urinary bladder afferents in relation to the oscillation periods of sacral spinal oscillators for continence in man. *Physiol. Res.*, 43: 7-18, 1994.
[58] Schalow, G., Zäch, G.A. and Warzok, R.: Classification of human peripheral nerve fibre groups by conduction velocity and nerve fibre diameter is preserved following spinal cord lesion. *J. Auton. Nerv. Syst.* 52: 125-150, 1995.
[59] Schalow, G., Bersch, U. Michel, D. and Koch, H.G., Detrusor-sphincteric dyssynergia in humans with spinal cord lesions may be caused by a loss of stable phase relations between and within oscillatory firing somatic neuronal networks of the sacral micturition centre. *J. Auton. Nerv. Syst.* 52: 181-202, 1995.
[60] Schmidt, H., Schalow, G. and Stefani, E.: Trophic influence of small motor axons on the membrane of frog slow muscle fibres. *Exp. Brain Res.*, Suppl. 183, 1975.
[61] Schnell, L. and Schwab, M.E.: Sprouting and regeneration of lesioned corticospinal tract fibres in the adult rat spinal cord. *Eur. J. Neurosci.*, 5: 1156-1176, 1993.
[62] Selzer, M. and Spencer, W.A.: Convergence of visceral and cutaneous afferent pathways in the lumbar spinal cord. *Brain Res.,* 14: 331-438, 1969.
[63] Sperry, R.W.: The problem of central nervous reorganosation after nerve regeneration and muscle transposition. *Quart. Rev. Biol.,* 20: 311-369, 1945.
[64] Sperry, R.W.: Effect of crossing nerves to anatagonistic limb muscles in the monkey. *Arch. Neurol. Psychiat.* Chicago, 58: 452-473, 1947.
[65] Störer, M., Palmtag, H. and Madersbacher, H.: *Blasenlähmung*. Thieme Verlag, Stuttgart, 1984.
[66] Thor, K.B., Blais D.P. and deGroat, W.C.: Behavioral analysis of the postnatal development of micturition in kittens. *Brain Res. Dev. Brain Res.*, 46: 137-144, 1989.

[67] Thor, K.B., Hisamitsu, T., Roppolo, J.R., Tuttle, P., Nagel, J. and deGroat, W.C.: Selective inhibitory effects of Ethylketocyclazocine on reflex pathways to the external urethral sphincter of the cat. *J. Pharmacol. Exp. Ther.,* 248: 1018-1025, 1989.

[68] Thor, K.B., Hisamitsu, T. and deGroat, W.C.: Unmasking of a neonatal somatovesical reflex in adult cats by the serotonin autoreceptor agonist 5-methoxy-N,N-dimethyltryptamine. *Brain Res. Dev. Brain Res*., 54: 35-42, 1990.

[69] Torrens, M.: Human physiology. In: Torrens, M., Morrison, J.F.B. (Eds.). *The Physiology of the Lower Urinary Tract*, pp. 333-350, Springer Verlag, 1987.

[70] Torrens, M. and Morrison, J.F.B.: *The Physiology of the Lower Urinary Tract*, Springer Verlag, 1987.

[71] Steers, W.D., deGroat, W.C.: Effect of bladder outlet obstruction on micturition reflex pathways in the rat. *J. Urol.,* 140: 864-871, 1988.

[72] Vallbo, A.B.: Afferent discharges from human muscle spindles in non-contracting muscles. Steady state impulse frequency as a function of joint angle. *Acta Physiol. Scan.*, 90: 303-318, 1974.

[73] Weiss, P. and Brown, P.F.: Electromyographic study on recoordination of leg movements in poliomyolitis patients with transposed tendons. Proceedings of the Society for Exper. *Biology and Medicine*, 48: 384-387, 1941.

[74] Yates, F.E.: Urodynamics and the study of female incontinence: the state of art as perceived by an outsider, with a modest proposal. In: Zinner, N.R., Sterling, A.M. (Eds.). Female Incontinence, *Prog. Clin. Biol. Res.* 78, pp. 1-14, 1980.

[75] Yoshiyama, M., Roppolo, J.R., Thor. K.B. and deGroat, W.C.: Effects of LY274614, a competitive NMDA receptor antagonist, on the micturition reflex in the urethane-anaesthetized rat. *Br. J. Pharmacol*., 110: 77-86, 1993.

[76] Martino, G. and Pluchino, S.: The therapeutic potential of neural stem cells. *Nat. Rev. Neurosci*., 7: 395-406, 2006.

[77] Bradbury, E.J. and McMahon, S.B.: Spinal cord repair strategies: why do they work? *Nat. Rev. Neurosci*., 7: 644-653, 2006.

[78] Schalow, G.: Symmetry diagnosis and treatment in coordination dynamics therapy. *Electromyogr. Clin. Neurophysiol*., 46: 421-431, 2006.

[79] Schalow, G., Vaher, I. and Jaigma, P.: Overreaching in coordination dynamics therapy in an athlete with a spinal cord injury. *Electromyogr. Clin. Neurophysiol*., 48: 83-95, 2008.

[80] Schalow, G.: Stem cell therapy and Coordination dynamics therapy improve Spinal cord injury. *Electromyogr. Clin. Neurophysiol*., 48: 233-253, 2008.

[81] Schalow, G., Jaigma, P., Belle, V.K.: Near-total functional recovery achieved in partial cervical spinal cord injury (50% injury) after three years of coordination dynamics therapy. *Electromyogr. Clin. Neurophysiol*., 49: 67-91, 2009.

[82] Schalow, G.: Partial cure achieved in a patient with near-complete cervical spinal cord injury (95% injury) after three years of coordination dynamics therapy. *Electromyogr. Clin. Neurophysiol*., 49: 67-91, 2009.

[83] Schalow, G.: The classification and identification of human somatic and parasympathetic nerve fibres including urinary bladder afferents is preserved following spinal cord injury. *Clin. Neurophysiol*., 49: 263-286, 2009.

[84] Schalow: Phase relation changes between the firings of α and γ-motoneurons and muscle spindle afferents in the sacral micturition centre during continence functions in

brain-dead human and patients with spinal cord injury. *Electromyogr. Clin. Neurophysiol.*, 50: 3-27, 2010.

[85] Schalow, G.: Scientific basis for learning transfer from movements to urinary bladder functions in human. *Electromyogr. Clin. Neurophysiol.*, 50: 339-395, 2010.

[86] Schalow, G.: Cure or urinary bladder function in severe cervical spinal cord injury (95% injury, motoric complete). *Electromyogr. Clin. Neurophysiol.*, 50: 155-179, 2010.

[87] Schmidt, R.F. and Thews, G.: *Physiologie des Menschen*, page 16, Springer Verlag, Berlin, 1985.

3. Phase Relation Changes between the Firings of α and γ-Motoneurons and Muscle Spindle Afferents in the Sacral Micturition Centre during Continence Functions in Brain-Dead Human and Patients with Spinal Cord Injury

Summary

1. Single-nerve fibre action potentials (APs) were recorded with two pairs of wire electrodes from lower sacral nerve roots during surgery in patients with spinal cord injury and in a brain-dead human. Conduction velocity distribution histograms were constructed for afferent and efferent fibres, nerve fibre groups were identified and simultaneous impulse patterns of α and γ-motoneurons and secondary muscle spindle afferents (SP2) were constructed. Temporal relations between afferent and efferent APs were analyzed by interspike interval (II) and phase relation changes to explore the coordinated self-organization of somatic and parasympathetic neuronal networks in the sacral micturition centre during continence functions under physiologic (brain-dead) and pathophysiologic conditions (spinal cord injury).
2. In a paraplegic with hyperreflexia of the bladder, urinary bladder stretch (S1) and tension receptor afferents (ST) already fired when the bladder was empty and showed a bladder afferent activity increase several times higher upon retrograde bladder filling than observed in the brain-dead individual. Two α_2-motoneurons (FR) innervating the external bladder sphincter were already oscillatory firing to generate high activity levels when the bladder was empty. They showed activity levels with no bladder filling, comparable to those measured at a bladder filling of 600 ml in the brain-dead individual. A bladder storage volume of 600 ml was thus lost in the paraplegic, due to a too high bladder afferent input to the sacral micturition center, secondary to inflammation and hypertrophy of the detrusor.
3. In a brain-dead human, two phase relations existed per oscillation period of 160 ms between the APs of a sphincteric oscillatory firing α_2-motoneuron, a dynamic fusimotor and a secondary muscle spindle afferent fibre. Following stimulation of mainly somatic afferent fibres, the phase relations changed only little.

4. In a paraplegic with dyssynergia of the urinary bladder, also two phase relations (less stable) existed per oscillation period of 110 ms in a functional unit between the APs of a sphincteric α-motoneuron, a fusimotor and a secondary spindle afferent fibre. The phase relations changed with time following stimulation of mainly somatic afferents. A second functional unit organized by phase related interactions was phase related to the first functional unit.
5. Following painful bladder catheter pulling, the parasympathetic division was transiently activated several times in the paraplegic. At times of activation of the parasympathetic division, three broad phase relations occurred within and between the two functional units, indicating that the parasympathetic division in the sacral micturition and defecation center channeled an additional input to the somatic oscillatory firing neuronal networks driving motoneurons that innervate the external bladder and/or anal sphincters.
6. It is conceivable that the mutual inhibitory action of detrusor and external bladder sphincter has the capacity to recover, if the functional neuronal organization of the sacral micturition center is improved in the direction of more stable phase relations between the firings of neurons and neuronal ensembles by natural coordinated afferent inputs from continence organs, supraspinal neurons, and functionally connected neuronal networks. For supraspinal control and improvement of neuronal organization, some kinds of bulbo-spinal-bulbo pathways have to exist or to be reconstructed by regeneration.

3.1. Introduction

I have shown that the classification scheme for the human peripheral nervous system is preserved following spinal cord injury (part 1) [57] and have analyzed the detrusor-sphincteric dyssynergia of the urinary bladder (part 2) [58]. Parasympathetically evoked impulse patterns of secondary muscle spindle afferents and sphincteric motoneuron activity were compared between paraplegics and a brain-dead human. It was shown that inflammation, hypertrophy and over-activated bladder stretch (S1) and tension receptors (ST) contributed to bladder dysfunction. The main reason for a detrusor-sphincteric dyssynergia of the bladder, however, seemed to originate from the central nervous system (CNS) itself. Following spinal cord injury, with the reorganization of the isolated sacral micturition centre, the mutual inhibition of the detrusor and the external bladder sphincter is lost or impaired. The mechanism underlying the loss of the mutual inhibition of detrusor and external sphincter may be, besides other reasons, a false sacral reflex reorganization or disconnection from the pontine micturition center, which is believed to coordinate the activation of detrusor and external bladder sphincter. Intermediate stages of the coordination of detrusor and external sphincter are conceivable. Since in a brain-dead human the parasympathetic division inhibited sphincteric motoneurons with detrusor contraction, pathologic reorganization and adaptation of the CNS, caudal to the level of injury, probably takes place in the spinal shock phase and thereafter. Of interest now are the pathologic changes in the neuronal networks of the caudal spinal cord and whether those changes are reversible.

An electrophysiologic tool has been developed for analyzing CNS functions in humans with spinal cord injuries. Based on the classification scheme for the human peripheral nervous system [27,57], it is possible to construct with the single-nerve fibre action potential (AP) recording method simultaneously natural impulse patterns of single afferent and efferent nerve fibres and the phase relations among them. Functions of the CNS can be analyzed following natural stimulation by studying time correlations between the natural discharge patterns of secondary muscle spindle afferents and sphincteric motoneurons innervating the external anal sphincter, and those of urinary bladder afferents and sphincteric motoneurons innervating the striated bladder sphincter (Figure 1 of [58]) (Chapter III).

Even though the structure and the function of the neuronal networks of the human CNS are far from being even partially understood, progress in the understanding of spinal cord dysfunction is possible by making use of the discovered spinal oscillators [28].

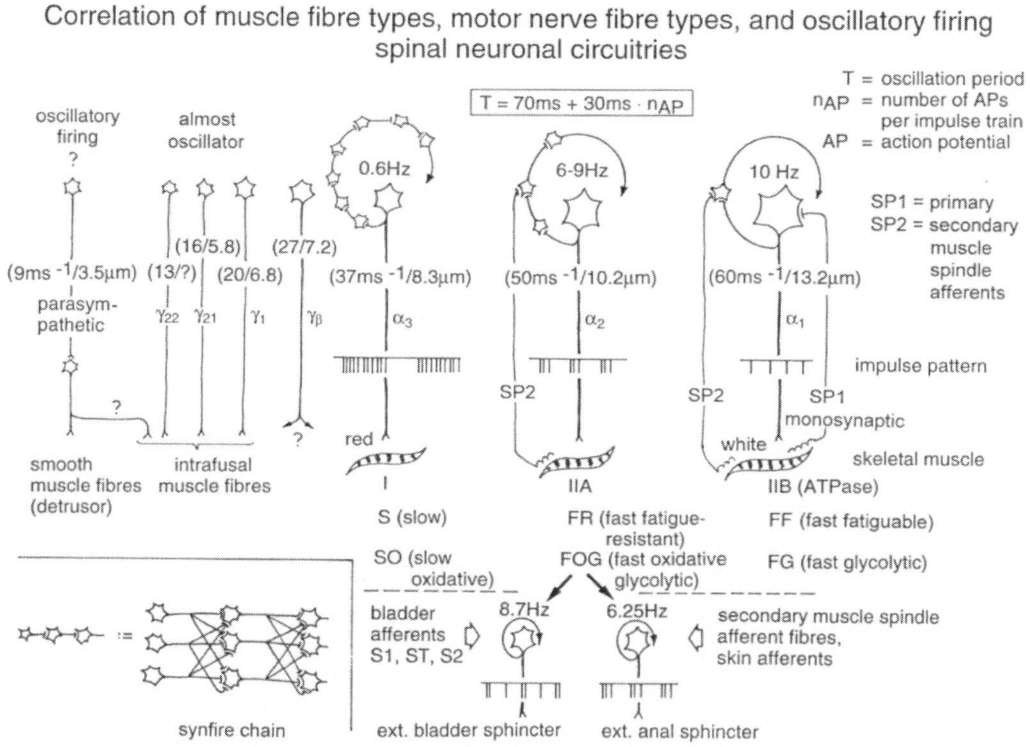

Figure 1. Correlation of muscle fibre types, motor nerve fibre types, and oscillatory firing spinal neuronal networks, based on histochemical, morphological and physiological properties. This Figure provides a simplified correlation between muscle fibre, motoneuron and sacral oscillator types. No additional subtypes have been included. The existence of α_1-motoneuron (FF) oscillators firing at 10 Hz has been predicted [34], and they have been identified in paraplegics (unpublished observation). α = motoneuron, γ_1, γ_2 = dynamic and static fusimotors, parasympathetic = parasympathetic preganglionic motoneuron. S1, ST, S2 = stretch, tension and flow receptor afferents.

These spinal oscillators are functional units that organize themselves in the neuronal networks of the CNS by natural impulse patterns of the adequate afferents. They are characterized by the rhythm of firing, namely by the frequency of repeated firing with impulse trains and the interspike intervals of the impulse train (Figure 1). If the adequate afferent input to the neuronal network is too low to fire oscillatory or if the neurons in the circuitry are inhibited, the sphincteric motoneurons are integrated in another organization form of network. Then, the motoneurons fire in the occasional firing mode [29,32,33] repeatedly every 3s approximately, or they fire transiently oscillatory [28] (Figure 60 of Chapter III).

Figure 1 summarizes what has been known so far about these spinal oscillators, which, under physiologic conditions, fire rhythmically with impulse trains [28,34]. α_1 (FF), α_2 (FR) and α_3-motoneurons (S), characterized by group conduction velocity and the group nerve fibre diameter, are integrated in their own neuronal network to fire oscillatory for high activation by their adequate afferent input and to innervate their own muscle fibre type. In humans, there are mainly three organization forms of somatic neuronal networks in the sacral micturition centre, which drive three kinds of striated muscle fibres (Figure 1). The α_2-motoneurons, for example, subserving continence functions, are integrated in neuronal network organization forms firing at 6 to 9 Hz; they innervate fast fatigue-resistant muscle fibres (FR) of histochemical type IIA, which are fast oxidative glycolytic (FOG) [3-9,11,16,22,25,27,28,34,63,64]. The α_2-motoneurons innervating the external bladder and anal sphincters are mainly specifically driven by oscillators, which fire with one to two APs every 110 ms (8.7 Hz) and with three APs every 160 ms (6.25 Hz), respectively (Figure 1). The neuronal network driving the external bladder sphincter is organized to fire oscillatory in response to the activity of bladder afferents from stretch (S1) and tension (ST), flow (S2) and other receptors. The neuronal network driving the external anal sphincter is channeled to fire oscillatory upon the activity of secondary muscle spindle afferent fibres (probably innervating the spindles of the anal sphincter) and skin and mucosa afferents innervating the anal canal. The oscillation period T of the spinal oscillators is related to the number of APs per impulse train n_{AP} by the relationship: $T = 70\text{ms} + 30\text{ms} \bullet n_{AP}$. Further properties of the spinal oscillators have been discussed elsewhere [28,34] (Chapter III). So far, preganglionic parasympathetic neurons and fusimotors could not be observed to fire oscillatory.

Since these spinal oscillators are essential organization forms of the neuronal network of the spinal cord, they can be expected to change their properties when the neuronal network of the spinal cord is damaged or when there is some pathology in its functioning. Early measurements of the dysfunction of spinal oscillators in paraplegics were used to study principal structures of their neuronal network [34]. In this part, mainly the interaction of the parasympathetic division with the somatic oscillators is under consideration, with the aim to tackle the problem of detrusor-sphincteric dyssynergia. This approach uses measurements of interspike intervals (II) and phase distributions (synchrony of discharge [65]) between the oscillatory firing of motoneurons and the firing of secondary muscle spindle afferents following stimulation of the parasympathetic division. It will be shown how powerful certain human CNS functions can be described by the use of these spinal oscillators. The rhythmic firing with impulse trains of the oscillators allows an easy recognition of the impulse patterns and opens possibilities for analyzing their circuitry [34]. I shall show in this part that the parasympathetic division of the sacral micturition centre interacts with the somatic neuronal circuitries, partially organized in oscillators, at points when the detrusor should inhibit the external sphincter.

In 1939 [17] and 1950 [18], E. von Holst criticized that the CNS was commonly regarded as a reflex apparatus, producing motor output. His "relative coordination" of different rhythms [17] of the CNS from different species including humans is very similar to recent findings on the correlation of human spinal oscillators [28,34,35]. According to Moshe Abeles, the nervous system is essentially a statistical machine [1]. According to von der Malsburg [65], the information in this statistical machine is encoded not only in the number of active nerve cells and their mean firing rates ("frequency code") but also in interactions between neuronal and neuromuscular elements of cell assemblies expressed in correlation of their firing patterns. This author argues that the coordination of discharges has a strong influence on nervous system dynamics, since neurons essentially are "coincidence detectors" (more generally coordination detectors [47]), because synchronous inputs excite a postsynaptic cell more effectively than do uncorrelated inputs. Temporal correlations, quantified by phase and frequency coordination, can therefore be processed by neuronal networks. Correlations with respect to time and space can be initiated in the network from an external source and then be propagated by "synfire chains," as suggested by Abeles [1]. The functional organization of the common brain stem system [21,26,61] and its dynamic changes is a consequence of afferent inputs and of the rhythmical properties of the common brain stem system network. The neuronal network reveals different types of functional organization. In the state of functional organization, in which the neurons discharge rhythmically, the phase relations between the different rhythms are essential in the transfer of information [21,26,39,58]. For references concerning possible functions of the spinal cord, see [68]. Decelerating burst and complex repetitive discharges were observed in the striated muscle of the urethral sphincter in women with urinary retention [14,15].

With the analysis of natural impulse patterns of urinary bladder functions, by use of the single-nerve fibre action potential recording method, the pathologic functioning of the sacral micturition centre is revealed. Possible clinical complications when the detrusor cannot inhibit the external bladder sphincter and the bladder cannot be emptied physiologically anymore are also known: inflammation, detrusor hypertrophy, lost storage volume, repeated bladder infections that can reach the kidneys, destruction of the whole urinary tract system including the kidneys, extirpation of the urinary tract system and leading the urine into the colon, which will ruin the electrolyte equilibriums (because the colon reabsorbs electrolytes). Before World War II, most patients with spinal cord injury died of these complications. Depending on the degree of bladder dysfunction, neuro-urologists suggested the implantation of an electrical anterior root stimulator (according to Brindly [69]). The anterior root stimulator empties the bladder and avoids the upper urinary tract complications. But this operation is destructive because the bladder is de-afferentiate (S2 through S5 dorsal sacral roots are cut and ventral roots are electrically stimulated later on), which means that the sexual function, running also through the lower sacral roots, is also destroyed or impaired. A natural causal repair of the spinal cord injury is no longer possible.

A substantial step further in this research project is that all these complication can be avoided by curing urinary bladder functions naturally by a movement-based learning therapy called coordination dynamics therapy [55,56,60]. By using the System Theory of Pattern Formation for Repair [55] (Chapter II), synergia and the dyssynergia of the bladder can be seen as two organizational patterns of the CNS with different stabilities. The causal therapy to cure urinary bladder function is to improve CNS functioning and to enhance the stability of the synergia pattern (the physiologic pattern) and to decrease the stability of the dyssynergia pattern (the pathophysiologic pattern) by learning transfer from certain movements [59]. To create and understand this learning transfer for cure (part 4), the author needed 20 years of human neurophysiologic and clinical research. The learning transfer includes a functional reorganization and some structural repair, which means some regeneration of the human spinal cord [56,60]. In four recent articles [72-75], possible treatments in spinal cord injury were reviewed. The biggest problems in spinal cord injury, the urinary bladder and sexual functions were not considered. Human research was not reviewed. More than 5% (up to 10%) of the society suffers from incontinence.

3.2. Materials and Methods

Measurements were performed in nine patients with spinal cord injuries (paraplegics), dyssynergia of the urinary bladder, spastic pelvic floor and spasticity in general. The group of paraplegics was the same as that studied and described in the two previous parts [57,58]. Only those patients have been selected (paraplegics 7 and 9) in whom impulse patterns of oscillatory firing motoneurons could be extracted from the summed impulse traffic. The other cases were not used for this kind of analysis, because the recording conditions were unsuitable (small AP amplitudes, limited uniqueness of AP waveforms of certain fibres) or the obtainable impulse patterns did not contribute further information to this paper. Electrophysiologic measurements from the brain-dead human HT6 (and patients) were partly taken from previous analyses [34-39,57,58]; in a few situations, the original recordings were reevaluated. In basic clinical research, one can only use available cases; and one reason why performing research in human neurophysiology is to develop treatment on the basis of those qualified measurements. It is extremely difficult to obtain suitable brain-dead humans for measurements, even though 5 to 10% of the society suffers from incontinence. After kidney explantation from brain-dead humans, nervous system functions are lost, because of lack of sufficient blood supply. This brain-dead human (HT6) is a reasonable control, since the results of the electrophysiologic measurements fit with the knowledge obtained from urodynamic measurements of healthy humans (for further discussion, see section control group in [58]). The practical details for curing the urinary bladder in the patient with the motoric complete cervical spinal cord injury is given in part 5 [60].

3.2.1. Electrophysiology

Single-nerve fibre action potentials (APs) were recorded extracellularly from nerve roots with two platinum wire electrode pairs and stored on a videotape. Trace “a” was the recording from the proximal electrode pair and trace “b” from the distal pair. APs from afferent and efferent fibres could clearly be distinguished since for the electrode arrangements used (differential electrodes), the main phase (second phase) from afferent fibres is upwards and that from efferent fibres downwards. Since the afferent and efferent APs reach the electrode pairs from opposite directions, also the conduction times for afferent and efferent APs are opposite. The conduction velocities of afferent and efferent nerve fibres were plotted in velocity distribution histograms for afferent and efferent fibres (Figure 2B). The afferent and efferent nerve fibre groups were identified by peaks and ranges in the distributions (Figure 2B), and by the calibration relations (e.g., α_2-motoneurons (FR) conduct with the same velocity as the secondary muscle spindle afferents (SP2)). The fitting of a single-nerve fibre AP to a certain group was recognized by the conduction velocity. The extraction of single-fibre impulse patterns from the multi-fibre recording of the whole nerve root was based on comparisons of conduction times and AP waveforms and on the recognition of reoccurring impulse patterns (for further details, see [57,58]). Under physiologic conditions, probably present in brain-dead humans (HTs), motoneurons in the oscillatory firing mode can be identified by the rhythmic firing (Figure 1).

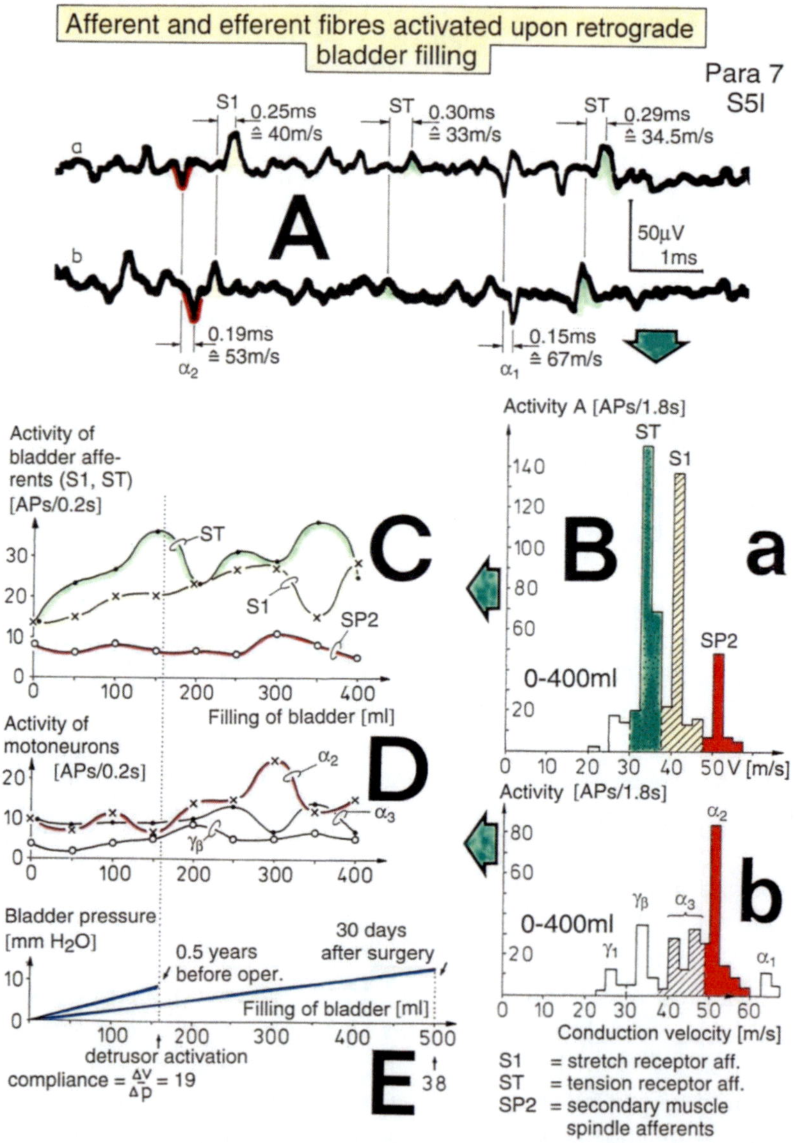

Figure 2. A. Action potentials (APs) recorded from α_1 and α_2-motoneurons and stretch (S1) and tension (ST) receptor afferent fibres. Conduction times and conduction velocities are indicated. B. Conduction velocity distribution histograms for afferent (a) and efferent fibres (b). C. Activity level changes of bladder afferent fibres (S1,ST) in dependence on bladder filling. D. Activity level changes of α_2(FR) and α_3-motoneurons (S) (no single fibres) in dependence on bladder filling, recorded simultaneously with the activity of bladder afferents. E. Detrusor pressure (bladder pressure) in dependence on bladder filling, before and after the surgery; redrawn from urodynamic recordings. Paraplegic 7.

Under pathologic conditions, the identification of the motoneuron type by the firing pattern is only possible if the pattern is not too much altered. Very pathologic oscillatory firing motoneurons can only be identified on the basis of group conduction velocities, since these velocities do not change following spinal cord injury [57]. As group conduction velocity ranges overlap, the identification is then only safe to approx. 80% for a nerve root temperature of 36°C. The identification error can in principle be estimated from the overlap of drawn conduction velocity distribution curves of neighbouring nerve fibre groups of a substantial number of active nerve fibres.

The conduction velocity distribution curves have a similar shape as the nerve fibre diameter distribution curves (Figures 1,5,6,11 of part 1 [57]). Exact distribution curves however have to be determined in future research. Since group conduction velocities of different nerve fibre groups have different temperature dependence [57], the overlap of neighbouring nerve fibre conduction velocity distribution curves depend also on the temperature. The overlap of the curves is smaller for higher temperatures (whereas the single-fibre AP amplitude is higher for lower temperatures (better recognition)).

3.3. Results

3.3.1. Premature Activation of Sphincteric Motoneurons in Paraplegics Following Too High Input from Tension and Stretch Receptor Afferents

In Figure 2A, a recording from an S5 root of paraplegic 7 is shown. Conduction velocity distributions of afferent and efferent fibres were constructed (Figure 2B), and nerve fibre groups were identified by the velocity ranges of the peaks. The identification of the nerve fibre groups rested upon the calibration relation, namely that secondary muscle spindle afferent fibres (SP2) conducted with the same velocity as the α_2-motoneurons (FR) and stretch receptor afferents (S1) conducted with the same velocity as α_3-motoneurons (S) [38,57].

The activities of bladder afferents and α_2 and α_3-motoneurons during retrograde bladder filling are shown in Figure 2C,D. The activity values of afferent and efferent fibre groups were obtained from histograms similar to those shown in Figure 2B. It can be seen that the motoneuron activity slightly increased during the bladder filling (Figure 2D) as did the stretch and tension receptor afferent activities (Figure 2C). The most important point, namely that bladder afferents fired already quite strongly when the bladder was empty, shows no characteristic consequences on the firing levels of the motoneurons. Most likely, the mixing of the activities from sphincteric motoneurons with those from other motoneurons, not associated with continence (and innervating, e.g., leg muscles), weaned off specific properties. A similar loss of specific properties of the activities of motoneurons with different functions was observed in dog sacral nerve roots (Figure 2 of [29], no specific properties in similarity to urodynamic recordings). In that case, motoneurons may have innervated leg, tail and continence muscles. The mixing of activities of the motoneuron firing in different modes, namely oscillatory and occasional, may have contributed to the loss of specific functions.

Specific properties of sphincteric motoneurons could be revealed separating impulse patterns of single motoneurons from the summed impulse traffic by waveform comparisons [39,58]. Figure 3A,B illustrates the activity increase of motoneurons, innervating the external bladder sphincter, upon retrograde bladder filling and illustrates simultaneously the activity increase of motoneurons with tension receptor afferent activity changes in a brain-dead individual and in the paraplegic 7.

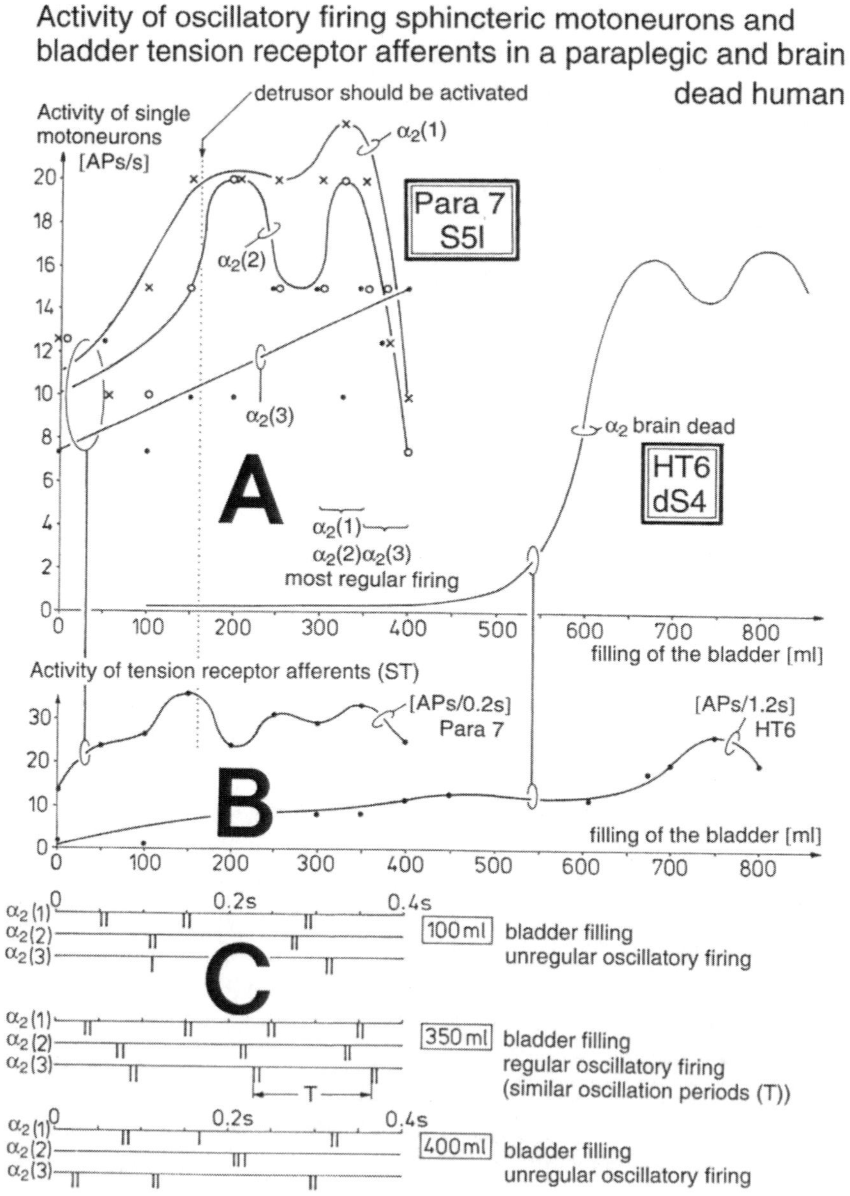

Figure 3. A. Activity of α_2-motoneuron (FR) in dependence on retrograde bladder filling, in paraplegic 7 and brain-dead human HT6 (taken from Figure 9 of Ref. 28). B. Activity of tension receptor afferent fibres (ST), recorded simultaneously with the motoneuron activity in "A" in paraplegic 7 (taken from Figure 2C) and the brain-dead human HT6 (taken from Figure 6 of (35)). Note that the α_2(1) and α_2(2)-motoneurons in paraplegic 7 show similar activity increases as the α_2-motoneuron in the brain-dead individual, with only the storage phase of the bladder being lost. Note further that with respect to the activity levels of the tension receptor afferent fibres (B), the α_2(1) and α_2(2)-motoneurons in paraplegic 7 are not activated earlier than the motoneuron in the brain-dead human HT6 (A). C. Discharge patterns of the α_2(1), α_2(2) and α_2(3)-motoneurons at 100, 350 and 400 ml bladder fillings. Paraplegic 7, nerve root S5.

In the brain-dead human, the sphincteric motoneurons show only a little activity in the occasional firing mode for a bladder filling smaller than 500 ml (storage phase). The activity of the bladder tension receptor afferents is nearly zero for no bladder filling. In paraplegic 7, the sphincteric motoneurons α_2 (1) and α_2 (2) fired already oscillatory to generate high activity levels when the bladder was still empty. Since in the paraplegic the activity of the tension receptor afferents was already very high for the empty bladder, the high excitement of the sphincteric motoneurons was most likely caused by a high bladder afferent activity (Figure 2C), especially from the tension receptors of the bladder wall.

With the increasing bladder filling, the activity of the sphincteric motoneurons increased in a manner similar to that observed in the brain-dead individual, if only at much smaller filling volumes. By comparing the activity increases of sphincteric motoneurons and tension receptor afferent fibres in paraplegic 7 with those in the brain-dead individual, it is concluded that the sphincteric motoneurons in the paraplegic behaved similar to the motoneuron in the brain-dead, the latter possibly representing the physiologic case in this respect. Only the sphincteric motoneurons in paraplegic 7 were activated too early because of a too high afferent input. The too high activity of the bladder afferents in the paraplegic mimics a rather full bladder, even though the bladder is still empty. The storage phase of the bladder is lost because of the too high afferent input. If the detrusor is also activated at certain bladder afferent inputs, similar to the sphincteric motoneurons, then the detrusor will also be activated at too small storage volumes (see below, Figure 2B). This means that the detrusor in this paraplegic was activated too early because of the too high bladder afferent activity at small bladder filling. The so-called hyperreflexia of the bladder, namely the automatic detrusor activation for too small bladder volumes in this case, can be explained by the too high bladder afferent activity, especially from the stretch receptors of the bladder wall.

In the brain-dead human the sphincteric α_2-motoneuron fired in the occasional firing mode at low bladder filling, in the transient oscillatory firing mode at bladder fillings between 500 and 600 ml, and in the continuous oscillatory firing mode at high bladder filling (Figure 82 of Chapter III). In the paraplegic, the α_2 (1) and α_2 (2)-motoneurons, innervating the external bladder sphincter, fired continuously oscillatory for any bladder filling.

The motoneurons innervating the external bladder sphincter are rather specifically driven by the bladder afferents (Figure 1). The secondary muscle spindle afferents (SP2), activating motoneurons that innervate the external anal sphincter (Figure 1), did not increase their activity with the increasing bladder filling (Figure 2C) and, for little bladder filling, did not contribute to the high activity of the α_2 (1) and α_2 (2)-motoneurons innervating the external bladder sphincter. The little understood γ_β-motoneurons (Figure 2D) showed no specific properties upon bladder filling.

During the surgery, the detrusor was not activated upon bladder filling (see discussion in [58]). Before the surgery, the detrusor was activated at approximately 160 ml bladder filling (Figure 2E), at a volume when the sphincteric motoneurons α_2 (1) and α_2 (2) showed their first activity peak. If the parasympathetic division had not been suppressed by anesthesia, it is likely that the detrusor would have been activated at that filling volume during the surgery. The motoneurons normally are activated strongly at high bladder filling to secure continence. At those bladder fillings, the urge to void is probably high. If in paraplegic 7 the bladder sensibility had been preserved, the urge to void would have been strong for small storage volumes. The relative extensive de-afferentation of the bladder (cutting of dorsal roots S2 to

S5 (S5 root was sometimes not cut, if many parasympathetic fibres were contained), ventral root afferent fibres and possible bladder afferents running through the plexus hypogastricus remain preserved) increased the storage volume from 160 to 500 ml and the compliance from 19 to 38 (Figure 2E).

In Figure 3A, there is another α_2-motoneuron ($\alpha_2(3)$) that increased its activity upon bladder filling. Since this α_2-motoneuron (FR) showed a different activity increase than the $\alpha_2(1)$ and $\alpha_2(2)$-motoneurons, it is likely that this motoneuron did not contribute to continence. It was activated by reflex or response generalization. In Figure 3C, the impulse patterns of the α_2 (1), α_2 (2) and α_3 (3)-motoneurons are shown for different bladder fillings. The motoneurons fired mainly rhythmically with impulse trains consisting of two action potentials (APs), in accordance with the expected pattern of bladder sphincteric motoneurons (Figure 1 and Table 2 of [28]). In accordance with the measurements in the brain-dead individual (Figure 82 of Chapter III), the oscillatory firing is most regular for the highest motoneuron activation and less regular for smaller bladder fillings and probably overfilled bladders. But the oscillatory firing neuronal network, driving the motoneurons, is dysfunctional, as the oscillation period was strongly changing. In paraplegic 9 (Figure 6), the disorder in the neuronal network of the functionally disconnected spinal cord was so grave that even the oscillation period and the number of APs per impulse train changed (Figure 1). In the next section, the interaction between the somatic and the parasympathetic neuronal networks will be considered at points where the detrusor activation is expected to inhibit the motoneurons innervating the striated (external) bladder sphincter. Again, recordings obtained from the brain-dead individual HT6 with bladder synergia will be compared with those from a paraplegic with detrusor-sphincteric dyssynergia.

3.3.2. Phase Relations between the APs of the α and γ-Motoneurons and Secondary Muscle Spindle Afferents in the Brain-Dead Individual

In Figure 4A, impulse patterns of fusimotors (γ_1, γ_{21}), secondary spindle afferent fibres (SP2(2), SP2(5)) and a α_2-motoneuron (α_2(O2)) innervating the anal sphincter after bladder catheter pulling at the beginning of a strong activation of the parasympathetic division are shown. Small arrows, open arrows, dotted lines, doublet lines and dashed lines mark similar phases (in ms) between the APs (represented by bars) from the different fibres. Phase relations seem to exist. Half quantitative analysis of phase relations has been performed earlier. Here, the phase analysis is extended to include a full analysis over long periods of time. In the impulse patterns of Figure 4Ba, phases (P) are defined between the APs of the motoneuron α_2 (O2) and the spindle afferent fibres and the fusimotors. Phase histograms (Figures 4Bb,4Cb) were drawn from the measured phases of the 0.8s long impulse pattern recordings (Figures 4Ba,4Ca). In the histogram of Figure 4Cb, two phase relations seem to exist between the APs of the γ_1-motoneurons, the SP2(2) fibre and the α_2-motoneuron (O2). To better quantify existing phase relations, phase histograms, such as those in Figure 4Bb,Cb, of selected time intervals were lumped together (see Figure 5).

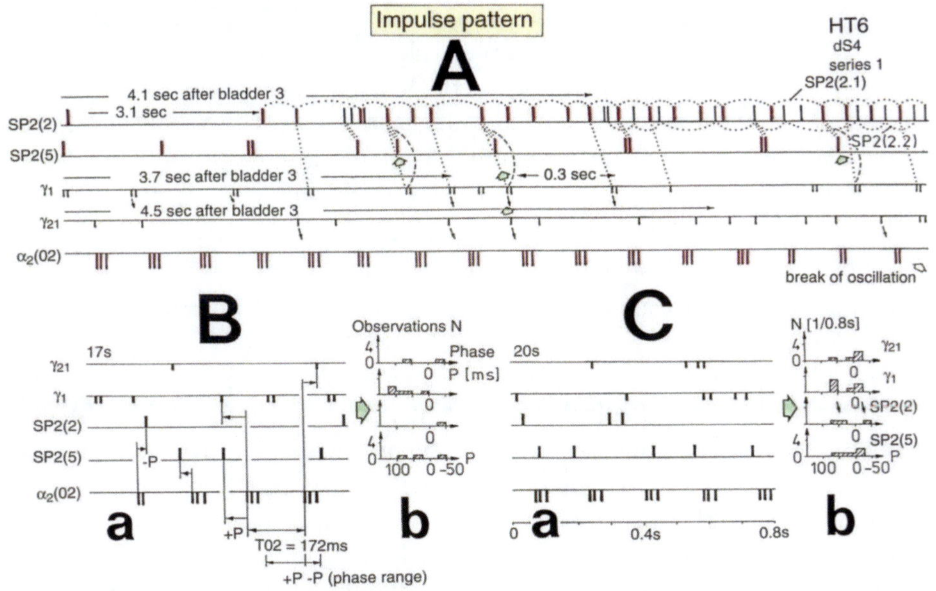

Figure 4. A. Impulse patterns recorded simultaneously from γ-motoneurons (γ_1 and γ_{21}), secondary spindle afferent fibres (SP2(2), SP2(5)) and the oscillatory firing α_2-motoneuron O2 upon bladder catheter pulling (bladder 3); afferent impulses = bars upwards, efferent impulses = bars downwards. Dotted half-circles in the SP2(2) fibre impulse pattern mark impulses from single encoding sites of the parent SP2(2) fibre. Dotted lines, double dotted lines, dashed lines and small arrows between the impulse patterns indicate similar phases. B. Definition of phases (P) between the APs of the afferent and efferent fibres and the α_2-motoneuron (O2). Phase durations given in ms. C. Another piece of impulse pattern (a) with corresponding phase histograms (b). Note that in "Cb," there seems to exist a phase relation between the γ_1-motoneuron and the secondary muscle spindle afferent fibre SP2(2), indicated by two small arrows; there is no similar relation in "Bb."

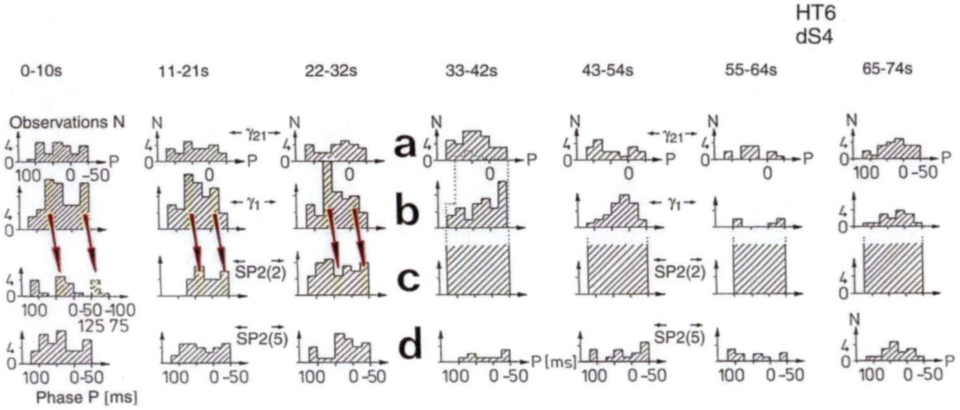

Figure 5. Phases (in ms) between the APs of the secondary muscle spindle afferent fibres SP2(2) (c) and SP2(5) (d), the dynamic fusimotor γ_1 (b), the static fusimotor γ_{21} (a) an the oscillatory firing α_2-motoneuron innervating the external anal sphincter, in dependence on time for selected time intervals (mostly including ten histograms of 0.8s long impulse patterns). Note that for the time interval 0-32s, the APs of γ_1 and SP2(2) fibres occur with two time correlations to the oscillation cycle of the α_2-motoneuron (see Figure 4Ba). Two rather constant phase relations per oscillation cycle (the long arrows) occurred between the impulses of the static fusimotor γ_1 and the secondary spindle afferent fibre SP2(2). With the activation of the parasympathetic division for times later than 33s (see Figures 7A,9B of (38)), the SP2(2) fibre activity increased, and it became difficult to further follow up the phase relations. HT6, dS4.

With respect to the excitation output the neuronal networks producing the γ_1 and α_2-motoneuron, activities had on average a constant phase against each other. The activities in the organized neuronal network were time-related. On the average, there was certain timing in the excitation of interneurons between these subneuronal networks. Since the secondary muscle spindle afferent fibre will get additional fusimotor drive different from that of the γ_1 fibre, some other fusimotor neuronal network organizations will, in general, also have fired in a phase-related manner to the oscillatory firing α_2-motoneuron circuitry.

The firing of the static fusimotor γ_{21} and the secondary muscle spindle afferent fibre SP2(5) was not tightly correlated to the firing of the α_2-motoneuron, since their phase relations were less pronounced.

With the activation of the secondary muscle spindle afferent fibre SP2(2) (see Figure 9B of part 2 [58]) by the parasympathetic division, the phase relations between the different fibres were lost. Most likely, a rearrangement in the organization of the neuronal network of the sacral micturition occurred. The important conclusion is that in the brain-dead human, some motoneuron circuitries had synchronous output firing. Probably, the firings of the interneurons in the subneural networks were, on the average, time related to each other. With the painful bladder catheter pulling, the strong afferent input rearranged the functional units.

3.3.3. Interspike Intervals of, and Phase Relations between the APs of the α and γ-Motoneurons and Secondary Muscle Spindle Afferents in Paraplegic 9 with a Dysfunction of the Bladder

Impulse patterns of γ (γ_1), α-motoneurons (α_2, α_3) and secondary muscle spindle afferent fibres (SP2(1), SP2(2)) in paraplegic 9 are shown in Figure 6b-g. To show the activation of the parasympathetic division, the activity (IIs/0.8s) level changes of the SP2(1) fibre are given in Figure 6a. The times are indicated of selected impulse pattern sampling and the time intervals are marked for which interspike intervals and phase relations were lumped together. The α_2 and α_3-motoneurons fired rhythmically with impulse trains consisting of one AP, in contrast to the physiologic firing patterns (Figure 1), in which α_2 and α_3-motoneurons fired with impulse trains consisting of more than one AP. The identification of motoneurons by the conduction velocity is not absolutely safe, since group conduction velocity ranges overlap [57]. It is very unlikely nevertheless that one of the motoneurons was an α_1-motoneuron (FF), even though they fire physiologically rhythmically with impulse trains consisting of one AP. Oscillatory firing α_1-motoneurons are driven by time-locked primary spindle afferent fibres (Figure 1, Figures 77,78 of Chapter III). The firing patterns of the α_2 and α_3-motoneurons are strongly pathologic with respect to the length of the oscillation period and the impulse train length so that it is impossible in this paraplegic 9 to identify the motoneurons by their discharge patterns of oscillatory firing; this is possible if the neuronal network driving the motoneurons were to fire in a physiologic manner. The dysfunction of the neuronal networks could originate in interneuron cell death, false functional organization due to loss of natural afferent input or other reasons (see below).

The time-correlation of afferent and efferent impulse patterns was easy to detect in the brain-dead individual, as the α_2-motoneuron O2 fired regularly like an inner clock. The phases of fusimotor and spindle afferent APs could be defined with respect to the impulses of that inner clock. In paraplegic 9, the rhythmic firing was very irregular. The motoneuron firing could therefore not be used as a time reference basis. More phases between the impulses of the different fibres are necessary to fully describe the correlation between the simultaneous impulse patterns. In Figure 6d,g, the mutual phases between the APs of the different fibres are defined. Figure 6h,i shows the corresponding phase distribution histograms. Since too few phases occurred in a sweep piece of 0.8s duration, phases occurred at certain time intervals were lumped together and plotted in Figure 7.

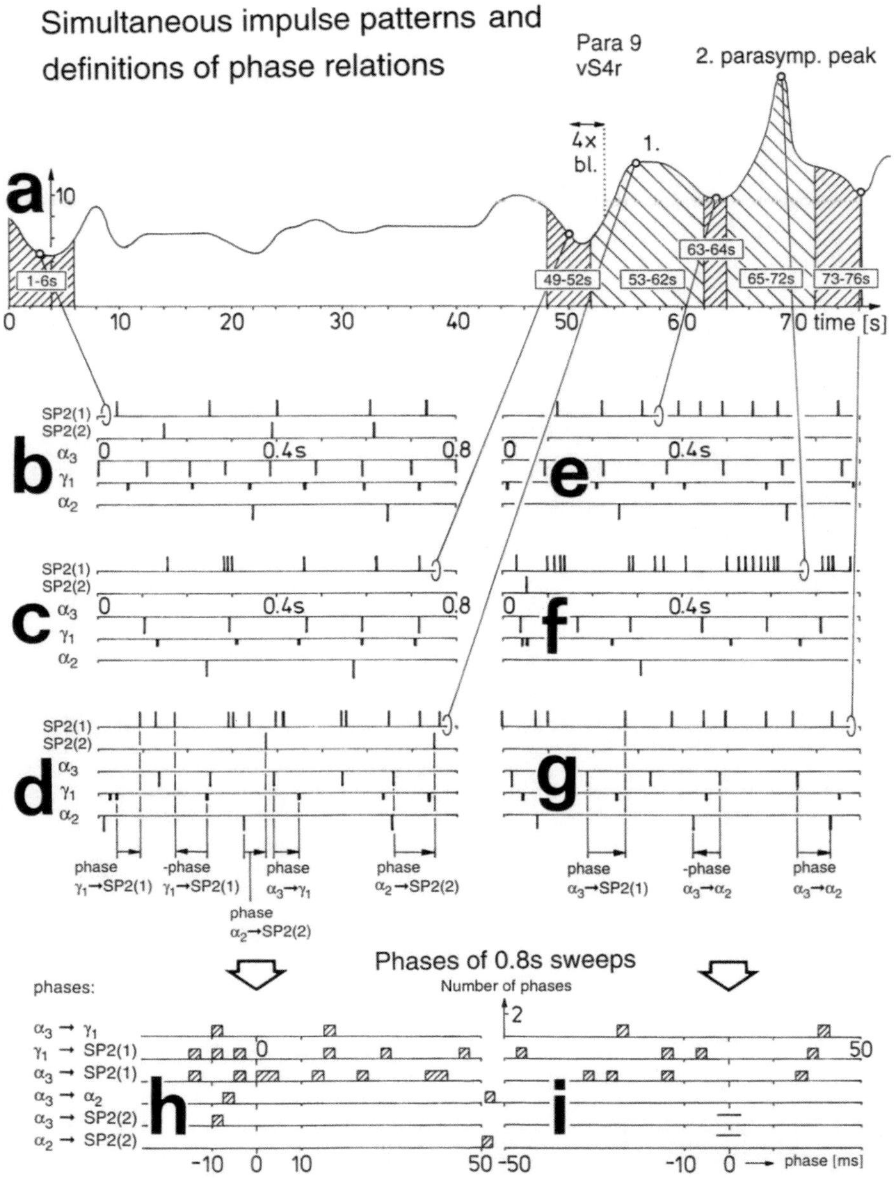

Figure 6. a. Activity level of secondary muscle spindle afferent fibre SP2(1) (IIs/0.8s) in dependence on time (taken from Figure 5A of (58)). 4x bl = fourfold bladder catheter pulling. 1, parasymp. peak 2 = the first and the second peak of spindle afferent activity due to parasympathetic activation. b.-g. Natural impulse patterns of secondary muscle spindle afferent fibres SP2(1) and SP2(2), α_2 (FR) and α_3 (S)-motoneurons and static fusimotor γ_1. In "d" and "g," the phases (in ms) between the impulses from the different fibres are defined. h.,i. Histograms of the phases between the APs of the afferent and efferent fibres obtained from the impulse patterns "d" and "g."

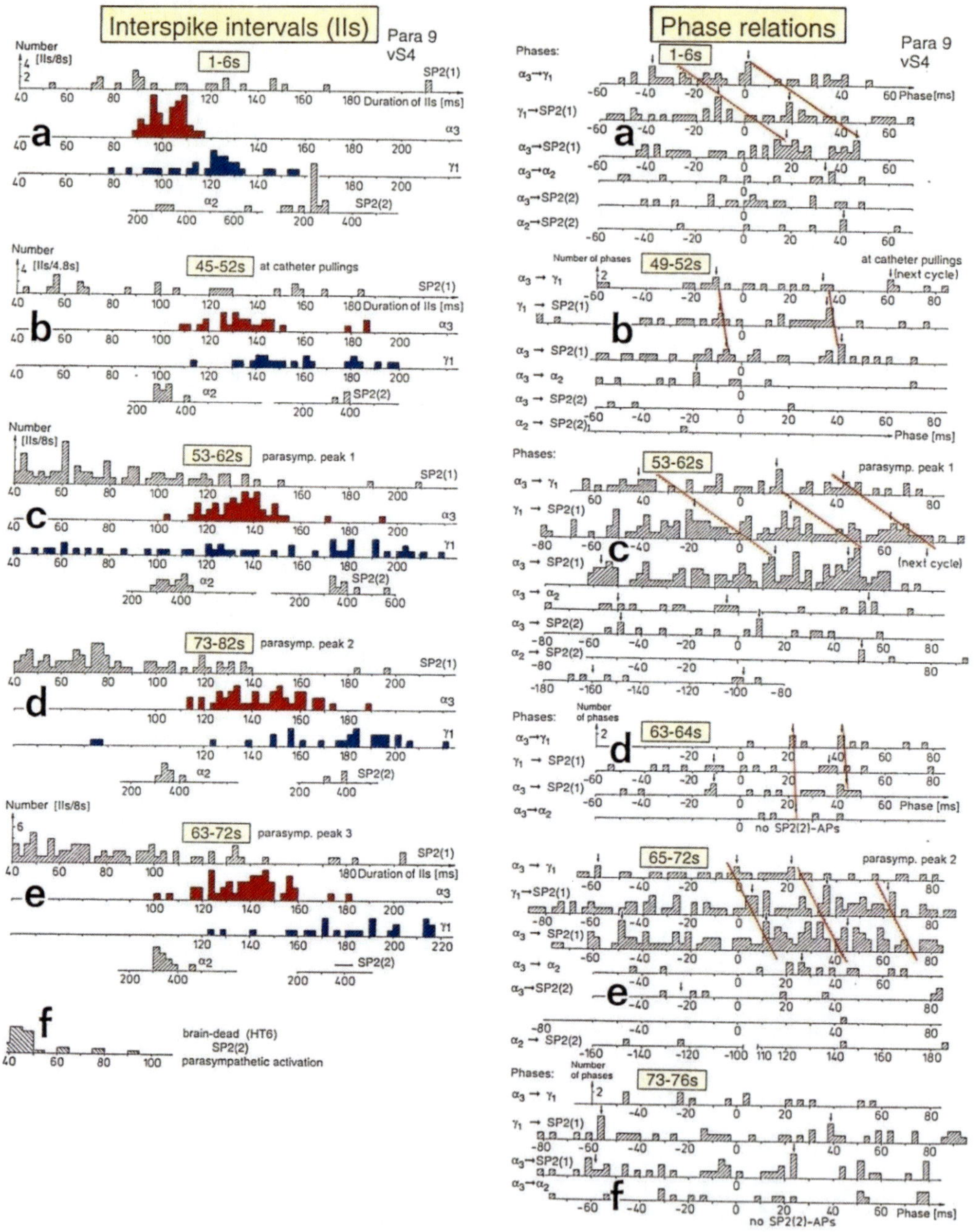

Figure 7. A. Interspike interval (II) distributions of secondary muscle spindle afferent fibres SP2(1) and SP2(2), α_2 (FR) and α_3-motoneurons (S) and the dynamic fusimotor γ_1 for different time intervals, as indicated in Figure 6, before and after painful bladder catheter pulling. IIs were collected from sweeps of 0.8s duration per second. Paraplegic 9, root vS4. In "f," the interspike interval distribution of the secondary muscle spindle afferent fibre SP2(2) following parasympathetic activation in the brain-dead human HT6 is drawn for comparison. B. Histograms of the phases between α_2 and γ_1-motoneurons and a secondary muscle spindle afferent fibres (SP2) (defined in Figure 6d,g) for time intervals (as indicated in Figure 6a) before and following painful bladder catheter pulling. The phases were collected from several histograms (defined in Figure 6hi) of sweeps of 0.8s duration per second. Paraplegic 9, nerve root vS4.

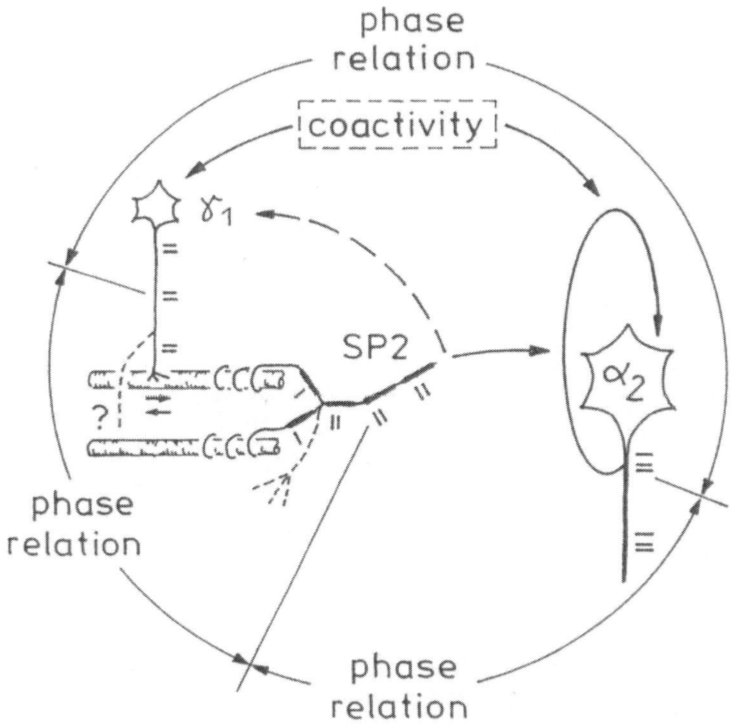

Figure 8. Schematized existing phase relation between α_2 and γ_1-motoneurons and a secondary muscle spindle afferent fibre (SP2). Parallel existing phase relations between other parent afferents and the α_2-motoneuron and between parent secondary spindle afferents are not shown. Phase relation means the increased occurrence of phases in ms in a certain phase range between the action potentials (APs) of the two compared nerve fibres. The complex afferent and efferent muscle spindle innervation was not tried to show. Small arrows at intrafusal muscle fibre indicate local contraction, which is in nuclear chain fibres readily transmitted to the place of afferent innervation. A possible reason of the doublet firing of the SP2 fibre is pictured to occur from single APs (schematized by bars) of two myelinated endings, not necessarily from pacemaker switching. More endings of the parent SP2 fibre and γ_1-motoneurons are indicated by dashed line branches. "Coactivity" indicates a correlation between γ and α-motoneuron spinal cord circuitries for higher activations.

In Figure 7, the interspike intervals (IIs) and the phases are shown for similar time intervals. Before stimulation in the time interval 1-6s, the α_3-motoneuron fired every 100ms, the γ_1-motoneuron every 100 to 130 ms, and the SP2(1) fibre every 80 to 150 ms (Figure 7Aa). The α_2-motoneuron mostly fired every 300 ms and the SP2(2) fibre every 250 ms. At that particular time interval, increased similar phases (phase relation) occurred two times per an α_3 oscillation period between the APs of the α_3 and γ_1 fibres, between the γ_1 and the SP2(1) fibres and between the α_3 and the SP2(1) fibres (Figure 7Ba). One phase relation occurred between the impulses of the α_3 and α_2-motoneurons and one between the α_3 and the SP2(2) fibres. Existing broad phase relations between motoneuron discharge patterns are interpreted as interactions between populations of neurons.

Since α and γ-motoneurons are partly correlated [58] (Figure 8) and muscle spindle receptors integrate and may correlate somatic and vegetative drive (part 2), motoneuron driving circuitries are both a source and a target of rhythmical events; thus, the rhythmical events are both means for, and consequences of, functional organization.

3.3.4. Differences in the Phase Relations of Motoneuronal and Spindle Afferent Discharges between Paraplegic 9 and Brain-Dead HT6

In the brain-dead individual, the phase relations between the impulses of the dynamic fusimotor γ_1, the secondary spindle afferent fibre SP2(2) and the α_2-motoneuron O2 changed only little upon different stimulations. Two phase relations existed per α-motoneuron oscillation period (160 ms) (Figure 5). With the activation of the parasympathetic division (time interval 33-74s, Figure 5), the stable relation between the γ_1 and the α_2-motoneuron was lost. A detailed analysis of the phases between the APs of the SP2(2) fibre and the α_2-motoneuron O2 was not performed.

Also in paraplegic 9, two phase relations existed between the activity of the α_3, γ_1 and SP2(1) fibres per oscillation period (100-140 ms) of the α_3-motoneuron (Figure 7Ba,b), but the phase relations changed with ongoing time (Figure 7Ba,b). With the activation of the parasympathetic division, three phase relations occurred per α_3-motoneuron oscillation period (Figure 7Bc [53-62s]). At the peak "1" of parasympathetic activation (Figure 6a [53-62s]), three phase relations occurred, and only two phase relations were present with little parasympathetic activation (Figure 7Bd [63-64s]) (times between the peaks (Figure 6a)) (Figure 9).

Upon strong parasympathetic activation (parasymp. peak "1" [53-62] and parasymp. peak "2" [65-72]), the somatic and parasympathetic nervous systems became functionally fully intertwined in the way that in the subnetwork, consisting of the coordinated activity of the α_3 and γ_1-motoneuron neuronal networks and SP2(1) spindle afferent fibre, the number of phase relations changed from two to three per α_3-motoneuron oscillation period. At least, the rhythmically firing α_3-motoneuron is in itself already an ensemble, namely a network oscillator of which the α_3-motoneuron is a part.

To make the changing phase relations per oscillation period of the α_3-motoneuron between the activity of the different α and γ-motoneurons and secondary spindle afferent fibres, occurring in the phase histogram of Figure 7B, easier to recognize, the functional units, consisting of the coordinated activity of fusimotor and α-motoneuron neuronal networks and spindle afferents fibres, were schematized drawn in Figure 8. A possible parasympathetic spindle innervation is not pictured. With further schematization, the change of the number of phase relations in and between two functional units upon different stimulations, plotted in Figure 7B, are made easily visible in Figure 9. In Figure 9, it can be easily seen now how the number of phase relations changed for the different time intervals and stimulations, defined in Figures 6,7B.

The difference of network organization between the brain-dead human and paraplegic 9 is that the functional units, consisting of fusimotor and α-motoneuron neuronal networks and spindle afferent fibres, were rather unstable in the paraplegic. Further, another (third) phase relation per α_3-motoneuron oscillation period occurred with the activation of the parasympathetic division (Figure 9).

The activated parasympathetic neuronal network of the sacral micturition and defecation centre seems to have channeled input to the oscillatory firing somatic neuronal network. An analysis of the change of the number of phase relations in the brain-dead was not performed.

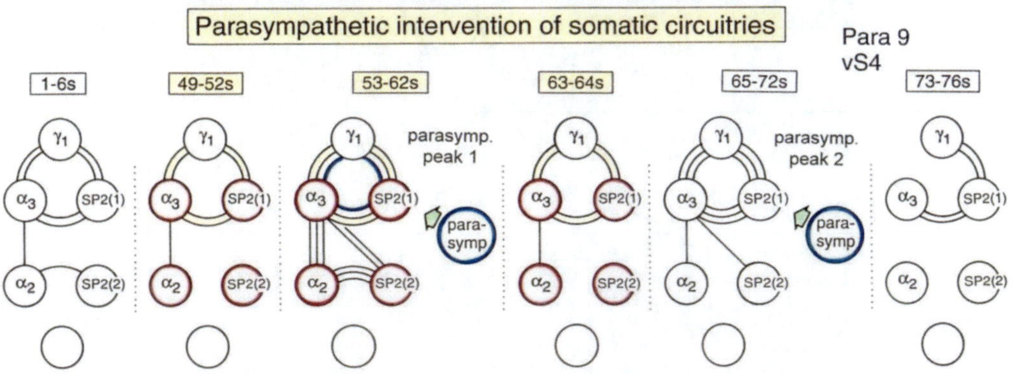

Figure 9. Schematic drawing of the occurrence of phase relations (of and between two units, one pictured in Figure 8) following activation of the sacral parasympathetic division (parasymp) by bladder catheter pulling (Figure 6a) between the APs of the α_2, α_3 and γ_1-motoneurons and the secondary muscle spindle afferent fibres SP2(2) and SP2(5). Note that an additional phase relation occurred per oscillation cycle with the activation of the parasymp. division. The time periods are same as in Figure 8.

3.3.5. Location and Stimulation of Receptors for Continence

The receptors of skin, urinary bladder and anal mucosa, the muscle spindles that were innervated by secondary spindle afferents, and the muscles that most likely were innervated by α-motoneurons in paraplegics 7 and 9 and in the brain-dead human HT6 are marked in Figure 10 in the lower pelvis.

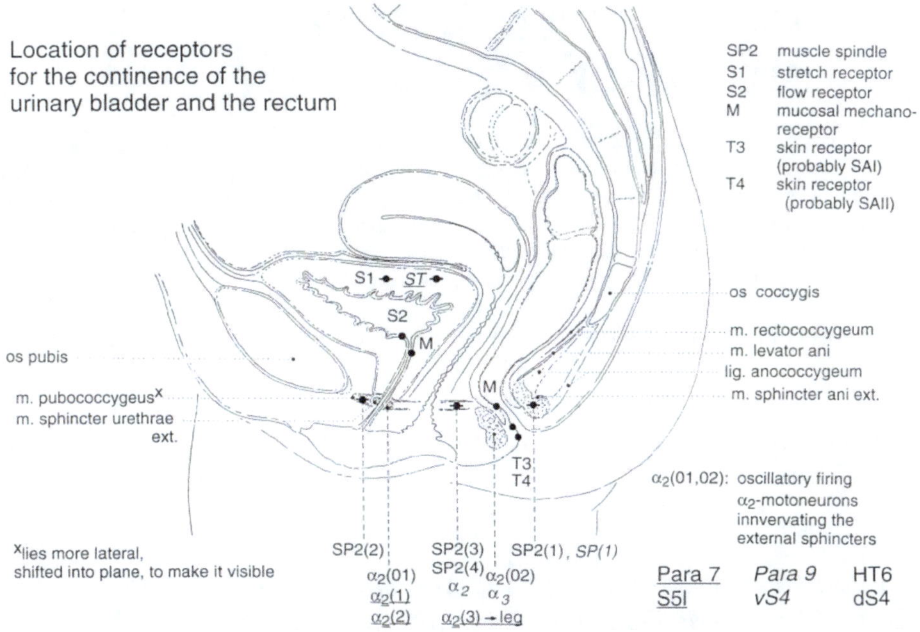

Figure 10. Location of receptors and muscles for the continence of the urinary bladder and the rectum, innervated by motoneurons the activities of which were recorded, in the brain-dead human HT6 (dS4 root), paraplegic 9 (vS4 root) and paraplegic 7 (nerve root S5).

Figure 11A. A. Patient with a severe cervical spinal cord injury (C5/6) during rhythmic jumping on springboard. The author is supporting the legs. Upon jumping, the different receptors of the intestine and the pelvic floor, including the external urinary bladder and anal sphincters, are rhythmically activated (including those receptors that are marked in Figure 10). Such repetitive mechanical stimulation has some similarity with the repetitive bladder and anal catheter pulling and activates simultaneously the somatic and parasympathetic division and increases learning transfer (Figures 6,7,9).

The locations for the brain-dead human were taken from Figure 6 of [35]. The (rhythmic) activation of these receptors by anal and bladder catheter pulling and skin stimulation of sacral dermatomes to stimulate simultaneously the somatic and parasympathetic divisions to induce learning transfer can be simulated and achieved during therapy, if a patient with severe cervical spinal cord injury jumps rhythmically on springboard with no weight support (Figure 11A).

3.3.6. Stimulation of the Parasympathetic and Somatic Division via Their Receptors of the Pelvic Floor and Intestine to Induce Learning Transfer from Movements to Urinary Bladder Functions for Cure

It was shown that the parasympathetic division can be activated besides the somatic one by anal and bladder catheter pulling (Figures 6,7).

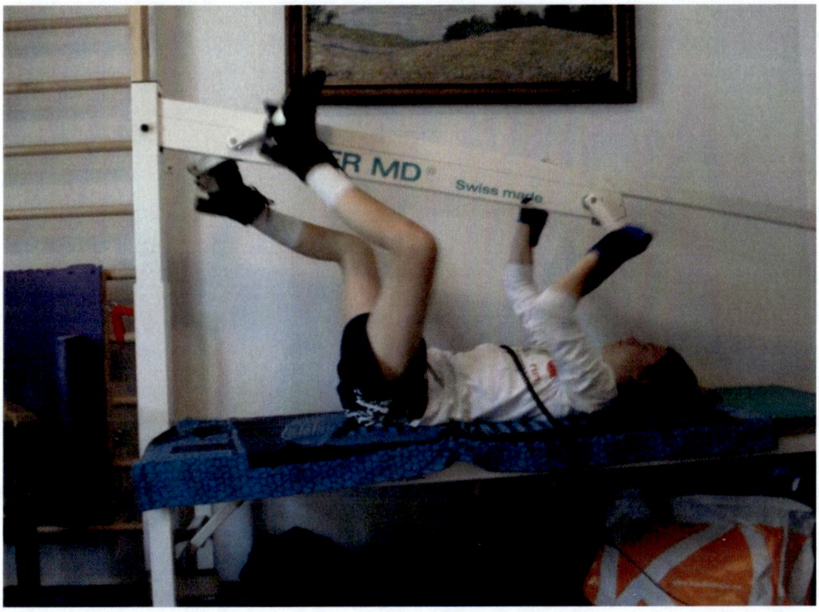

Figure 11B. Same patient as in Figure 11A but exercising now on the special coordination dynamics therapy device in the lying position. Such very coordinated turning with arms and legs is stimulating autonomic functions like bladder, intestine (peristaltic), and blood circulation functions, probably by improving the phase and frequency coordination of the networks driving those autonomic functions.

The activated parasympathetic neuronal network channeled input to the rhythmically firing somatic neuronal network via an additional phase relation (Figure 9). This joint stimulation of the somatic and parasympathetic division is used in "coordination dynamics therapy" to use movements that substantially activate the somatic and the vegetative nervous system to induce learning transfer from movement to bladder patterns to cure urinary bladder function. Upon jumping on springboard, without weight support in severe cervical spinal cord injury (Figure 11A), the pelvic floor (including the sphincters) is mechanically stimulated (stretch of muscle spindles) under the weight of the intestine.

Also, the stretch and tension receptors of bladder and rectum are activated upon the rhythmic up and down movements. The desire or urge to void is also strongly stimulated when the patient is exercising on the special coordination dynamics therapy device, especially in the lying position (Figure 11B). But this stimulation of the parasympathetic and somatic divisions is related to the improvements in the short-term memory of the phase and frequency coordination of the neuronal networks of the sacral micturition centre including the functional units pictured in Figure 9. This analysis will be picked up in a following part 4 [59]. Both the rhythmic mechanical stimulation of the continence organs (Figure 11A) and the improved CNS organization in the short-term memory upon exercising on the special device for turning movements (Figure 11B) lead to a strong activation of the sacral micturition centre, which may lead in turn to strong increases of the heart rate and blood pressure, which are vegetative symptoms to inform supraspinal centres about bladder fullness in complete spinal cord injury, if the urinary bladder is filled and if there is plenty of material in colon and rectum. Before exercising, the bladder should be emptied.

3.4. Discussion

3.4.1. Premature Activation of Sphincteric Motoneurons

As can be seen in Figure 3A, the sphincteric motoneurons $\alpha_2(1)$ and $\alpha_2(2)$ in paraplegic 7 were activated too early with respect to the extent of bladder filling (double-peaked activation shape) in comparison to the sphincteric motoneuron in the brain-dead individual HT6. With respect to the input of the tension receptor afferents, the sphincteric motoneurons in paraplegic 7 were not activated too early (Figure 3B). Thus the activation in the paraplegic was rather normal, with only the too high bladder afferent input mimicking a filled bladder. The third α_2-motoneuron ($\alpha_2(3)$) in paraplegic 7 was also highly activated but showed a different activation shape (Figure 3A). It is likely that normally this motoneuron would not directly contribute to continence. Probably this motoneuron was activated by reflex generalization and most likely innervated the pelvic floor. Since few motoneurons in S4 and S5 roots innervate also leg muscles [36], it cannot be excluded that this motoneuron innervated a leg muscle.

Even though in paraplegic 7 the too early activation of sphincteric motoneurons was caused by a too high afferent input from the bladder afferent fibres, there were also changes in the CNS as suggested at least by the detrusor-sphincteric dyssynergia. An early activation of the striated sphincters is not in contradiction with detrusor-sphincteric synergia. In normal women with a urinary bladder infection, the detrusor is activated too early, most likely because the infection activates bladder afferent fibres, but the external bladder sphincter still relaxes when the detrusor is activated (detrusor-sphincteric synergia). Interspike intervals and phase relations will now be used to analyse pathologic functions of the spinal cord. It will be shown that the oscillatory firing of motoneurons [34] and the coordination of firing between motoneurons and secondary muscle spindles changed following spinal cord injury, which could be one reason for the development of detrusor-sphincteric dyssynergia.

3.4.2. Interspike Intervals of, and Phase Relations between, the APs of the α and γ-Motoneurons and Secondary Muscle Spindle Afferents in Paraplegic 9 upon Activation of the Sacral Micturition Centre

Since two phase relations occurred per oscillation cycle between the α_3 and γ_1-motoneurons and the SP2(1) fibre (Figure 7Ba) in paraplegic 9, and also their IIs were rather similar, it is concluded that the neuronal networks of the α_3 and γ_1-motoneurons formed together with the spindle afferent fibre SP2(1) a part of a functional unit. The functional unit is built by efficiencies of synapses and projections between the convergence of several fusimotors on one muscle spindle and by the divergence of muscle spindle projections onto several rhythmically firing populations of neurons driving α and γ-motoneurons. Such a functional unit is partly pictured in Figure 8 and schematized drawn by three circles in Figure 9. The α_2-motoneuron and the SP2(2) fibre belonged to another functional unit (longer IIs and the existence of only one phase relation). The two functional units are characterized in Figure 9 by two sets of three circles each. The two functional units interacted with each other as there existed a phase relation between the α_2 and α_3-motoneurons.

Following touch, pinprick and anal reflex stimulation, but not painful catheter pulling, the functional organization in that subneural network of the spinal cord only changed slightly between the time intervals 1-6s and 49-52s (Figures 6a,7Ab,7Bb), since the durations of IIs and the existing phase relations changed only little. Functional unit one, consisting of the α_3 and γ_1-motoneurons and the SP2(1) fibre, only changed the values of their phases. Functional unit two (α_2/-/SP2(2)) became disorganized as the phase relation between the APs of the α_2-motoneuron and the SP2(2) fibre was lost (Figure 9 [49-62s]). The neuronal network driving the α_3-motoneuron was still interacting with the neuronal network driving the α_2-motoneuron, since a phase relation existed between the impulses of the α_2 and α_3-motoneurons. Following strong (painful) bladder catheter pulling with a first strong activation of the parasympathetic division (time interval 53-62s), measured by the burst firing (see [58]) of the SP2(1) fibre, the functional organization of the sacral micturition centre of the disconnected spinal cord changed completely. Functional unit α_3/γ_1/SP2(1) was now correlated by three phase relations per an α_3-oscillation cycle. The functional unit α_2/-/SP2(2) also showed three phase relations per an α_2-oscillation cycle and interacted with functional unit α_3/γ_1/SP2(1) by three phase relations as well (between the α_3 and α_2-motoneurons; Figure 9 [53-62s]).

Between the first and second parasympathetic peak at the time interval 63-64s (Figures 6a,9), the organization form of the two functional units was similar to that before the first parasympathetic activation [49-52s] (Figure 9), only the values of the phase relations changed (Figure 7Bb,d). With the second strong activation of the parasympathetic division (parasymp. peak 2, time interval 65-72s) the functional unit α_3/γ_1/SP2(1) was bound together again by three phase relations (Figure 9), in similarity to the first strong activation of the parasympathetic division, measured by the activity of the secondary muscle spindle afferent fibre SP2(1). The functional unit α_2/-/SP2(2) was disorganized, but phase relations still occurred between the α_3 and the α_2-motoneurons and the SP2(2) fibre (Figures 7Be,9). The α_2-neuronal network and the fusimotor networks, driving the SP2(2) spindle afferent fibre, were integrated differently.

After the second strong parasympathetic activation, in the time interval 73-76s (Figure 9), the functional organization of the two functional units in the spinal cord was similar to that before the activation of the parasympathetic division. Functional unit α_2/-/SP2(2) was slightly disorganized as the SP2(2) fibre strongly reduced its firing (Figure 6g).

As analyzed in a previous paper [58], the secondary spindle afferent fibre SP2(2) in the brain-dead HT6 also showed short IIs but no IIs of long duration (Figure 7Af) when activated by the parasympathetic division, as compared to the occurrence of also long IIs in the SP2(1) fibre in paraplegic 9.

3.4.3. Neuronal Network, Synfire Chains and Reverberating Circuits

It was shown in the brain-dead human that the phases in the functional unit, consisting of the fusimotor γ_1, the secondary spindle afferent fibre SP2(2), and the sphincteric α_2-motoneuron O2 and, probably, other fusimotors and secondary muscle spindle afferent fibres, changed only little in the time interval 0-32s (Figures 4,5). Since there was no strong phase relation between the static fusimotor γ_{21}, the secondary muscle spindle afferent fibre SP2(5), and the oscillatory firing α_2-motoneuron O2, the subpopulations of the neuronal network driving those fusimotors contributed only little to the information transfer to the network driving the α_2-motoneuron O2. The interspike interval and oscillation period distributions of the α_2-motoneuron APs, suggest that the α_2-motoneuron O2 was integrated in the organized neuronal network [34]. As a second working hypothesis, based on the relation between the oscillation period T and the number of APs per impulse train n_{AP}, the periodic firing is generated by reverberations in neuronal circuits in which the α_2-motoneuron O2 is integrated. Such reverberations travel in dynamically changing closed loop pathways, in which the excitation follows multiple inter-neural routes, which are changing from time to time. In each cycle, the reverberation traverses a certain chain of interneurons for certain duration of the oscillation period. If the α_2-motoneuron fires with impulse trains consisting of three APs every 160 ms, the inter-neural chain is three neurons, or neuron units (including inhibition), long. If the motoneuron fires with two APs every 130 ms, the inter-neural chain is only two neuron units long [34]. This simple working hypothesis that such reverberations travel well-defined anatomical circuits has to be given up, since single dedicated routes cannot explain the pathological firing of the oscillators as reported herein. Also, the excitation of inhibitory interneurons is not included (see page 81 of [40]). Without trying to improve the details of the oscillator model, some changes have to be discussed to understand the pathologic firing of sphincteric motoneurons in paraplegics with detrusor-sphincteric dyssynergia.

The dedicated lines of the working hypothesis have to be replaced by synfire chains, which consist of chains of neuronal sets. Each neuron in one stage excites every neuron in the next stage. The excitation transmission in a synfire chain from stage to stage is secured by the synchronous firing of all the cells in one stage [1]. The important point in the synfire chains is that excitation streams can cross. Two dedicated lines cannot cross over because once the cross-over point is excited, there is no way to tell which direction the activity will proceed. In the synfire chain, the same neuron may participate in many different chains according to whether it fires synchronously with one set of cells or with another. From Figure 6, it can be seen that α_2 and α_3-motoneurons in paraplegic 9 fired with one AP per impulse train every 100 to 160 ms (α_3) or every 300 to 400 ms (α_2) (Figure 7A), in complete contradiction to the physiologic firing mode (Figure 1). In paraplegic 7, the sphincteric motoneurons fired often with two APs every 200 to 300 ms (Figure 3C). The number of APs per impulse train of the motoneurons innervating the external bladder sphincter is in accordance with Figure 1, but the duration of the oscillation period is not. Essential changes must have taken place in the oscillatory firing neuronal networks, in which the motoneurons are integrated. The functional organization into different subpopulations should be different, because the oscillation period became much more variable for a certain impulse train length and the necessary phase relations became instable (Figure 7B). On the basis of the dedicated line hypothesis, the changing oscillation period can be explained by many different pathways. But these changes cannot be understood on the basis of dedicated lines and interneuron cell death. The dedicated lines are very sensitive to failure of individual neurons. If one neuron in a chain dies, the chain becomes useless as a whole. The large loops with long dedicated lines and long oscillation periods will become useless first, because they consist of many interneurons. Long oscillation periods still occurred in the very pathologically functioning spinal cord network, even though they were quite irregular. Dedicated line neuronal network and interneuron cell death cannot explain the pathologic firing of α-motoneurons. The hypothesis of dedicated lines has to be abandoned. Even if there were no interneuron cell death, one could not understand where the pathologic changes should come from, if networks are constructed of dedicated lines. If one thinks of a neuronal network consisting of synfire chains that cross each other and changes of synaptic efficacies (see later), then the measured pathologic impulse patterns are conceivable. If, for example, synfire chain loops cross each other, integrated in a network or matrix of synfire chains, then the reverberating excitation may change several times the loop. An over-excitation of the spinal cord neuronal network, often present in paraplegics, can result in a block of transmission at synfire chain crossings. When two synfire chains become active together and they share neurons in the same network, they may destructively interfere with each other (1). Therefore the synfire processing may be limited to one process at a time. The dying of frogs at temperatures between 29 and 30 °C (see previous paper [58]) when their nervous system gets overexcited can be understood on the basis of synfire chain networks. The synfire chain is a good candidate for the carrier of high-level processing in the central nervous system. The coupling of oscillatory firing networks to a different degree, called "relative coordination" by E. v. Holst [17], could be understood on the basis of synfire chains. The size of each set in a synfire chain is small, but there is a large subsynchronous fringe of neurons affected by the activity in the chain. A single neuron may be affected by many synfire chains and may even be a part of several synfire chains [1].

3.4.4. Different Neuronal Network Systems May Show Different Degrees of Pathology: More Integrative Network Organizations Get More Impaired

The common brain stem system integrates and regulates the activity level of the brain and the nervous activity for the cardiovascular and respiratory regulation and the regulation of the motor tone. With respect to the functional organization of the reticular formation of the lower brain stem, it is argued that exact frequencies or sharply defined frequency ranges are not very helpful because of the variability of the rhythmical events, as described above; their amplitudes and period durations depend on the functional organization of the system, the character of afferent inputs and phase relations between different, simultaneous rhythms [26]. Also, in the sacral micturition centre of humans, the frequency of the oscillatory firing motoneurons and the number of APs per impulse train depends on the afferent input [35] and on the coupling with other oscillatory firing networks [35]. Under physiologic conditions, however, the α_2-motoneurons (FR) fire rather stably in the oscillatory mode, the α_1-motoneurons (FF) probably very stably and the α_3-motoneurons (S) unstably [28]. Probably the stability under physiologic conditions depends on the size of the functionally organized subneuronal network, on the kind of the respective afferent inputs, and on whether the network serves faster or slower motor functions (activation against gravity). The neuronal networks in which the α_3-motoneurons are integrated probably consist of many hundreds of interneurons, and different kinds of afferent inputs will contribute to the drive. The α_3-motoneuron networks probably are most similar to brain stem networks with respect to the organization, because of the low and changing frequencies and multi-afferent drive. But even in α_3-motoneuron firing, there seems to exist a correlation between the number of APs per impulse train and the oscillation period [28].

Interestingly, rather physiologically firing α_3-motoneurons could not be detected at all in paraplegics, even though they are easy to detect because of their firing with long impulse trains. The α_3-motoneuron of Figures 6,7 fired so atypically (Figure 1) that one has doubts that it really was an α_3-motoneuron rather than an α_2-motoneuron (the identification rests only on the conduction velocity and not on the firing pattern). The slow neuronal network systems (α_3) seem to have changed more than the fast systems (α_1, α_2) following spinal cord injury. Since most likely the α_3-motoneuron neuronal networks consist of many more interneurons than the faster motor systems, it seems that it were the more integrative functions of the neuronal networks in the spinal cord that underwent changes. These integrative functions can be understood within the framework of System Theory of Pattern Formation, measured non-invasively by the coordination (pattern) dynamics and can be repaired by applying integrative coordinated input and reconstructing supraspinal control in motoric complete spinal cord injury (see below). During regeneration of the cord, following cervical spinal cord injury, the integrative functions (quantified by the coordination dynamics) get transiently impaired (increase of coordination dynamics values), very much like following injury (Figure 3 of 56).

Since the detrusor is activated more slowly than the α_3-motoneurons, the parasympathetic neuronal network in the sacral micturition centre is probably slower and more integrative in processing than the α_3-motoneuron neuronal networks. Probably, numerous interneurons contribute to the activation of the preganglionic parasympathetic neurons. One can expect therefore quite a bit of dysfunction of the parasympathetic neuronal network system of the sacral micturition centre. The parasympathetically activated secondary muscle spindle afferent fibres in paraplegics fired quite differently from those of the brain-dead human HT6 [58]. The parasympathetic neuronal network can be repaired by specific input to the parasympathetic division in coordination with the input to the somatic division. The jumping on springboard (Figure 11A) activates rhythmically the pelvic floor (with the external anal and bladder sphincters), the bladder (detrusor) and the colon and rectum via dynamic movements (Figure 10). The movement induced afferent input from the continence organs entrains parasympathetic (and somatic) neuronal networks.

It was shown in this paper (Figure 6) and elsewhere [34] that the oscillatory firing mode of α_2 and α_3-motoneurons changed following spinal cord injury. It was further shown that following spinal cord injury in the occasional firing mode, the α_3-motoneurons were recruited before the α_2-motoneurons [33], whereas physiologically (including the brain-dead individual) the α_2-motoneurons are recruited before the α_3-motoneurons [32]. Therefore, also the coordination between different kinds of α-motoneuron neuronal network systems has changed following spinal cord injury. The coordination between the different α-motoneuron neuronal networks can be trained by changing the frequency of exercising and activating in this way the slow, the intermediate and the fast neuronal networks differently. It is likely that also the coordination between the α-motoneuron neuronal network systems and the parasympathetic neuronal network system underwent changes. Based on this, detrusor-sphincteric dyssynergia in paraplegics is likely to occur, because the timing between α_2 and α_3-motoneuron systems changed following injury as probably also did the timing between the α-motoneuron systems and the parasympathetic neuronal network system.

3.4.5. Reasons for Pathologic Changes

The question is now, why do the neuronal networks of the sacral micturition center organize themselves in a false way? What changes occur in the isolated spinal cord when it is disconnected from the supraspinal control that is exerted by impulse patterns and trophic effects? Interneuron cell death, sprouting, changes in synapse efficacy (see below), and reorganizations of the automatized spinal cord are possible reasons for change. Obvious interneuron cell death has not been found so far. For cats, it has been reported that there is no sprouting following spinal cord transection [24], even though it is generally believed that there is sprouting [23]. One is therefore mainly left with the changes in the synapses, partial denervation of neurons, lost supraspinal control and the reorganization of the micturition reflex (automatism) for the understanding of dysfunctions in the sacral micturition centre. Since humans with partial cervical spinal cord injury (tetraparetics) can reduce their spasticity and repair bladder functions when undergoing coordination dynamics therapy [55], dysfunctions in the organization of the neuronal network of the sacral micturition centre are at least partly reversible (tetraparetics with some volitional bladder innervation left have bladder synergia). It seems possible to modulate the efficacy of synapses by natural impulse patterns from the periphery and from supraspinal centres to make the organization of the neuronal networks turn more physiologic again (plasticity). Since there is limited regeneration in motoric complete cervical spinal cord injury [56], pathologic changes of the sacral micturition centre are reversible [60]. But still we need to understand how this repair takes place.

Long-term changes in efficacy of synapses, as terminal sprouting, new synapse formation and nerve terminal retraction occur in untreated vertebrate muscle under physiologic conditions [66,67]. The physiological remodelling allows the change of the architecture of synapses in response to different physiological and pathophysiological conditions. Physiological plasticity that operates in the millisecond or second range is facilitation, augmentation, potentiation, depression and desensitization [19]. For neural models of plasticity, see [10]. Efficacy can be functionally defined as synaptic safety margin or the likelihood that postsynaptic potentials will exceed threshold for generation of axon AP. Presynaptic remodelling might be caused by presynaptic axon activity, which means by training. For patients with complete spinal cord injuries, some supraspinal control has to be reconstructed to allow voluntary micturition and voluntary training of the organization of spinal cord networks.

3.4.6. Phase Escape of the Somatic Neuronal Networks as a Possible Cause for Detrusor-Sphincteric Dyssynergia

It was shown that with the activation of the parasympathetic division in paraplegic 9, three phase relations occurred in the functional organization of the units driving the α_2 and α_3-motoneurons (Figures 7B,9). Since phase relations indicate information transfer, it is concluded that the parasympathetic neuronal network interacted with the somatic ones, driving the sphincteric motoneurons. In the brain-dead human HT6 (Figure 5), the phase relations were lost in the functional units consisting of the α and γ-motoneurons and the secondary muscle spindle afferent fibres when the parasympathetic division was activated. In the previous part [58], it was shown that upon the activation of the parasympathetic division, the α_2-motoneuron O2 strongly reduced its activity level by stopping firing oscillatory; in other words, the parasympathetic division inhibited the sphincteric motoneurons. In the paraplegic patients, the phase relations were not lost in the two functional units (Figures 7B,9). Only the number of phase relations changed from two to three, and the phases changed. In the previous part [58], it was shown that the sphincteric motoneuron did not strongly reduce its activity upon the activation of the parasympathetic division. The sphincteric motoneuron in the paraplegic was not inhibited upon the activation of the parasympathetic division. Upon the activation of the parasympathetic division in the paraplegic patient, parasympathetic neuronal networks interacted with the two functional units driving sphincteric motoneurons, since an additional phase relation was built up (Figure

9). However, as phase relations in the somatic units still existed and the activity of the sphincteric motoneurons was not reduced, it is possible that the somatic functional units escaped the inhibitory action of the parasympathetic division by changing their phases (phase escape). The phase escape of the somatic neuronal networks driving the sphincteric motoneurons from the parasympathetic inhibitory action could be a reason for the detrusor-sphincteric dyssynergia at the level of the sacral micturition centre, in terms of human neurophysiology.

3.4.7. Detrusor-Sphincteric Synergia and Dyssynergia in the Framework of System Theory of Pattern Formation – The Solution of the Seemingly Unsolvable Urinary Bladder Coordination Problem

A completely different understanding of detrusor-sphincteric synergia and dyssynergia is obtained in the framework of System Theory of Pattern Formation for Repair [55]. The synergia and the dyssynergia are neuronal network patterns (attractors) with different stabilities. Under physiologic conditions, the stability of the attractor "synergia" is high, and the stability of the attractor "dyssynergia" is low. By intention, a healthy subject or a patient with a repaired bladder can switch between both patterns [60]. Such a schematic continence attractor layout [60] shows some similarity to the attractor layout for the jumping on springboard [55]. The attractor for "jumping in-phase" has a higher stability (deep potential well) than the "jumping in anti-phase" (more shallow potential well). Even more similar is the antagonism property of antagonistic muscles (extensor and flexor), as, for example, the gastrocnemius and the tibialis anterior muscles. During walking, the tibialis anterior is inhibited (not activated) upon contraction of the gastrocnemius muscle and vice versa. The motor programs show antagonistic action. But during standing, both muscles are activated. The motor programs show co-contraction. There are therefore attractors for antagonistic action and co-contraction. A healthy subject can choose by intention between the walking and the standing pattern, which means between the pattern of antagonistic action and co-contraction.

Following spinal cord injury, the patient can have the antagonistic action or the co-contraction pattern preserved. Intermediate, pathologic patterns do occur. When the patient's CNS can partly activate the stepping automatism, often also the pathologic extensor spasticity occurs, which is some kind of co-contraction with emphasis on the extension. During supported treadmill walking, the patient's CNS can improve the walking pattern in the short-term memory. The stability of the antagonistic pattern is increasing, and the stability of the co-contraction pattern is decreasing. The patient can walk better, and the spasticity reduces. The impairment of antagonistic muscle action is therefore a similar problem as the dyssynergia of the urinary bladder but can be measured by electromyography [55]. The important difference between the antagonistic muscle action and the synergia of the bladder is that the antagonistic muscle action can be trained and relearned directly by performing coordinated movements like walking, jumping (Figure 11A) or exercising on special devices for turning (Figure 11B). But for the improvement of the stability of the synergia pattern of the bladder, learning transfer from coordinated movements is needed.

During physiologic micturition, the detrusor contracts and the external sphincter relaxes. But the pattern that the detrusor and the external bladder sphincter contract simultaneously also exists; only its stability is lower. A healthy person can stop (interrupt) the micturition (for continence reason) by activating on volition the external sphincter. In this situation, the detrusor and the external sphincter contract simultaneously. That means by intention, a healthy person can also activate the dyssynergia pattern. The healthy person can choose between the patterns continence and voiding and can even switch between both patterns. But with the loss of supraspinal control, a subject with a complete spinal cord injury cannot choose anymore between both patterns. According to the special injury situation and the stage of disorganization of the neuronal networks of the sacral micturition centre, both patterns (and intermediate patterns) do occur, and the patient cannot choose anymore by intention. The fundamental progress is now that urinary bladder function can be repaired in severe cervical spinal cord injury (motoric complete injury, sensory incomplete) by learning and learning transfer from special movements and some reinnervation of the caudal spinal cord for volitional control [56,60]. Such structural repair can be achieved by a partial regeneration of the spinal cord and/or an establishment of connections via the plexus hypogastricus.

In the reported case, two years of optimal coordination dynamics therapy were needed for bladder repair [56,60]. The bladder repair included the reconstruction of some supraspinal control, which seems to be necessary for physiologic bladder functioning. In the last stage of bladder repair, the female patient felt the synergia and the dyssynergia patterns. For emptying the bladder, she started the voiding with the synergia pattern, which means the detrusor (bladder muscle) contracted and the external sphincter relaxed. Fluid was leaving through the urethra. But when she pressed to reduce residual urine (and generated stress in her CNS, which induces spasticity), she felt that the external sphincter co-contracted with the detrusor. No fluid was leaving the bladder anymore. The organizational state of the CNS had switched from the synergia pattern to the dyssynergia pattern (Figure 5 of Chapter II). But with intention and touch afferent input (behavioural information in the equation of motion of the collective variables, see Ref.55) from the Head's zone of the bladder (the patient touched the skin on top of the bladder), the pattern state switched back to the synergia pattern. This means that the external sphincter relaxed (was inhibited) and the detrusor could push further fluid out of the bladder by continuous contraction to reduce further residual urine. For further discussion and details, see the following part on the cure of urinary bladder [60].

In conclusion, urinary bladder dysfunctions can be cured, including the dyssynergia of the bladder. Destructive operations are therefore not justified anymore, especially in young patients who will have a further life expectation time of 50 years and who want to train hard for a better future. Until there is a cure of the bladder in severe cervical spinal cord injury, catheters, urinals, and diapers have to be used.

By using human electrophysiology and integrative pattern formation strategies, it is possible to understand and cure urinary bladder dysfunction. The understanding of bladder functions under physiologic and pathophysiologic conditions alone is not sufficient for cure; the system theory of pattern formation for repair is needed. On the other hand, with an understanding of the integrative CNS functions alone, one cannot even understand the problem of bladder dyssynergia and cannot develop behavioural information (treatment) to increase for example the stability of the synergia pattern. Human neurophysiology and a clinical understanding of the problem are also necessary. Already Eccles argued: "From mere behavioural descriptions we shall never come to understand how the brain is effective in the learning of motor skills" [70]. With the knowledge of central pattern generators [62], we also cannot understand learning transfer, which is important not only for curing autonomic CNS functions but also for improving higher mental functions in cerebral palsy [48] by learning in the motor system (Chapter II). In incomplete cervical spinal cord injury, a bladder repair is not that difficult, because a partial regeneration of the spinal cord is not needed [55]. Already a functional reorganization can cure bladder (and motor) functions. By using coordination dynamics therapy, the author himself cured urinary bladder functions in four patients with incomplete spinal cord injuries (of different degree) in the cervical, thoracic, and lumbar regions [41,42,55,56,60].

3.4.8. Learning Transfer

The most important step in the development of treatment (on the basis of human repair physiology) to cure human CNS functioning is to measure human CNS self-organization under physiologic (in brain-dead and healthy humans if possible) and pathophysiologic conditions (in patients) to know what became pathologic and what has to be repaired. The simultaneous recording of natural impulse patterns from several single human neurons was the key tool for that diagnostic, which has also been shown in this part. But further knowledge is needed of such natural afferent and efferent impulse traffic of single neurons to improve such learning transfer, achieved by the exercising of certain motor patterns, to further develop this movement-based learning therapy for repair called "coordination dynamics therapy." Before coming to part 5 on the cure of the bladder, therefore, the learning transfer is further analyzed on the single neuron level in human because this learning transfer is not only necessary to cure bladder and other autonomic functions but also to improve higher mental functions, including speech and memory for example in cerebral palsy [48] (Chapter II). Learning transfer is a general property of CNS functioning and is probably also involved in memory, intelligence, and creativity. The property of learning would be limited, if we could not generalize a problem and could not use learned knowledge in other circumstances or disciplines.

3.4.9. Spared Fibre Counts

The relationship between the number of axons spared by an incomplete spinal cord injury and the locomotor efficacy of the animal's hind limb has been studied. It was found that residual fibre counts in excess of 5,000 are associated with useful hind limbs to assist locomotion [12,20]. How many fibres have to spared (or repaired) to cure urinary bladder function is unclear. An early treatment (six to eight hours) with methylprednisolone hemisuccinate may improve the neurological outcome after acute spinal cord injury [2], so that approximately 80% of the new spinal cord injuries are incomplete.

3.4.10. Need for Qualified Human Research

It could be argued that the few measurements of this and the previous part [58] offer not enough safety to draw conclusions concerning the functions of the human CNS; especially since there is only one direct control measurement (HT6). Firstly, one has to remember that it is extremely difficult to get hold of such human measurements (humans cannot be taken from the stock for measurements) and that this kind of research is not organized. Secondly, the functional correlations between neurons, immanent in the sets of natural impulse patterns, may give more insight into the function of the human CNS than hundreds of animal experiments, partly summarized in recent review articles [72-75]. The simultaneous measurement of further parameters, which can clinically be tested, could further improve the safety of the measured parameter dependences. Thirdly, since each spinal cord injury is different from the other one, statistical evaluations are doubtful. The task is not to destroy a developing human neurophysiology (or ignore it) but to complete with it by offering more and better human data. The fact is that these few measurements on the self-organization of parts of the human CNS have direct clinical consequences in the treatment after stroke [43], traumatic brain injury [44,50], hypoxic brain injury [49], cerebellar injury [51,52], spinal cord injury [42,45,55,56,60], in cerebral palsy [48], and in Parkinson's disease [46,47].

References

[1] Abeles, M., *Local cortical circuits. An electrophysiological study*, Springer, Berlin, 1982.

[2] Bracken, M.B., Shepard, M.J., Collins M.D. et al., A randomized controlled trial of methylprednisolone or naloxone in the treatment of acute spinal-cord injury; results of the second national spinal-cord injury study, *N. Engl. J. Med.*, 322, 20 (1990) 1405-1411.

[3] Basany, M., ATPase activity of myosin correlation with speed of muscle shortening, *J. Gen. Physiol.*, 50, Suppl. 6, Part 2 (1967) 197-218.

[4] Barnard, R.J., Edgerton, V.R., Furukawa, T. and Peter, J.B., Histochemical, biochemical and contractile properties of red, white and intermediate fibres, *Am. J. Physiol.*, 220 (1971) 410-414.

[5] Bigland-Ritchie, B., Thomas, C.K., Westling, G. and Johannson, R.S., Classification of human motor units, *Soc. Neurosci*. Abstr., 1988.

[6] Burke, R.E., Levine, D.N., Tsairis, P. and Zajac, F.E.III, Physiological types and histochemical profiles in motor units of the cat gastrocnemius, *J. Physiol.* (London), 234 (1973) 723-748.

[7] Burke, D., Skuse, N.F. and Lethlean, A.K., Isometric contraction of the abductor digiti minimi muscle in man, *J. Neurol. Neurosurg. Psychiatry*, (1974) 825-834.

[8] Burke, R.E., Motor units: Anatomy, physiology, and functional organization. In V.B. Brooks (Ed.), *Handbook of Physiology, Sect. 1, Vol. II, Am. Physiol. Soc. Washington,* D.C., 1981, pp. 351-364.

[9] Burke, D. and Candevia, S.C. (1980), Peripheral motor units. In G. Praxinos (Ed.), *The Human Nervous System*, Academic Press, 1990, pp. 125-145.

[10] Byrne, J.H. and Berry W.O., *Neural Models of Plasticity, Experimental and Theoretical Approaches*, Academic Press, Inc., San Diego, 1989.

[11] Eberstein, A. and Goodgold, J., Slow and fast twitch fibres in human skeletal muscle, *Am. J. Physiol.*, 215 (1968) 535-541.

[12] Eidelberg, E., Straeley, D., Erspamer, R. and Watkins, C.J., Relationship between residual hindlimb-assisted locomotion and surviving axons after incomplete spinal cord injuries, *Exp. Neurol.*, 56 (1977) 312-322.

[13] Flohr, H., *Post-Lesion Neural Plasticity*, Springer-Verlag, Berlin, 1988.

[14] Fowler, C.J. and Kirby, R.S., Abdominal electromyographic activity (decelerating burst and complex repetitive discharges) in the striated muscle of the urethral sphincter in five women with persisting urinary retention, *Br. J. Urol.*, 57 (1985) 67-70.

[15] Fowler, C.J., Kirby, R.S. and Harrison, M.J.G., Decelerating burst and complex repetitive discharges in the striated muscle of the urethral sphincter, associated with urinary retention in women. *J. Neurol. Neurosurg. Psychiatry*, 48 (1985) 1004-1009.

[16] Garnett, R.A.F., O'Donovan, M.J. Stephens, J.A. and Taylor, A., Motor unit organization of human medial gastrocnemius, *J. Physiol.* (London), 287 (1978) 33-43.

[17] Holst, E.v., Die relative Koordination als Phänomen und als Methode zentralnervöser Funktionsanalyse, *Erg. Physiol.*, 42 (1939) 228-306.

[18] Holst, E.v. and Mittelstedt, H., Das Reafferenzprinzip (Wechselwirkungen zwischen Zentralnervensystem und Peripherie), *Die Naturwissenschaften*, 20 (1950) 464-476.

[19] Kandel, E.R., Calcium and the control of synaptic strength by learning, *Nature*, 293 (1981) 697-700.

[20] Kao, C.C., Bunge, R.B. and Reier, P.J., *Spinal Cord Reconstruction*, Raven Press, New York, 1983.

[21] Lambertz, M., Rittweger, J.E., Kluge, W. and Langhorst, P., The common brain stem system (CBS): Dynamic properties of the functional organization and their influence on signal processing, *Wiss. Zeitschrift der Humboldt-Universität zu Berlin, R. Medizin*, 41 (1992) 83-84.

[22] Milner-Brown, H.S., Stein, R.B. and Yemm, R., The orderly recruitment of human motor units during voluntary isometric contractions, *J. Physiol.* (London), 230 (1973) 359-370.

[23] Murray, M. and Goldberger, M.E., Restitution of function and collateral sprouting in the cat spinal cord: the partially hemisected animal, *J. Comp. Neurol.*, 158 (1974) 19-36.

[24] Nacimiento, W., Mautes, A., Töpper, R., Oestreicher, A.B., Gispen, W.H., Nacimiento, A.C., Noth, J. and Kreutzberg, G.W., B-50 (GAP-43) in the spinal cord caudal to hemisection: Indication for lack of intraspinal sprouting in dorsal root axons, *Journal of Neuroscience Research*, 35 (1993) 603-617.

[25] Peter, J.B., Barnard, R.J., Edgerton, V.R., Gillespie, C.A. and Stempel, K.E., Metabolic profiles of three fibre types of skeletal muscles in guinea pig and rabbits, *Biochemistry,* 11 (1972) 2627.

[26] Rittweger, J.E., Kluge, W., Langhorst, P. and Lambertz, M., Rhythmical properties of the common brain stem system (CBS), *Wiss. Zeitschrift der Humboldt-Universität zu Berlin, R. Medizin*, 41 (1992), 91-94.

[27] Schalow, G., Conduction velocities and nerve fibre diameters of touch, pain, urinary bladder and anal canal afferents and α and γ-motoneurons in human dorsal sacral roots, *Electromyogr. Clin. Neurophysiol.*, 31 (1991) 265-296.

[28] Schalow, G., Oscillatory firing of single human sphincteric α_2 and α_3-motoneurons reflexly activated for the continence of urinary bladder and rectum - restoration of urinary bladder function in paraplegia. *Electromyogr. Clin. Neurophysiol.*, 31 (1991) 323-355.

[29] Schalow, G., Recruitment within the groups of γ_1, α_2 and α_3-motoneurons in dogs and humans following bladder and anal catheter pulling, *Gen. Physiol. Biophys.*, 11 (1992) 101-121.

[30] Schalow. G., Aho, A. and Lang, G., Microanatomy and number of nerve fibres of the lower intercostal nerves with respect to a nerve anastomosis. Donor nerve analysis. I (IV). *Electromyogr. Clin. Neurophysiol.*, 32 (1992) 171-185.

[31] Schalow, G., Number of fibres and fibre diameter distributions of nerves innervating the urinary bladder in humans - Acceptor nerve analysis. II (IV), *Electromyogr. Clin. Neurophysiol.,* 32 (1992) 187-196.

[32] Schalow, G. and Wattig, B., Recruitment of α and γ-motoneurons in rats, dogs and humans, *Electromyogr. Clin. Neurophysiol.*, 33 (1993) 387-400.

[33] Schalow, G., Recruitment of motoneurons in the occasional firing mode in paraplegic, *Electromyogr. Clin. Neurophysiol.*, 33 (1993) 401-408.

[34] Schalow, G., Spinal oscillators in man under normal and pathologic conditions, *Electromyogr. Clin. Neurophysiol.*, 33 (1993) 409-426.

[35] Schalow, G., Phase correlated adequate afferent action potentials as a drive of human spinal oscillators, *Electromyogr. Clin. Neurophysiol.*, 33 (1993) 465-476.

[36] Schalow, G., Bersch, U. Göcking, K. and Zäch, G.A., Interspike intervals of secondary muscle spindle and urinary bladder afferents in relation to the oscillation periods of sacral spinal oscillators for continence in man, *Physiol. Res.*, 43 (1994) 7-18.

[37] Schalow, G. Action potential patterns of intrafusal γ and parasympathetic motoneurons, secondary muscle spindle afferents and an oscillatory firing α_2-motoneuron, and the phase relations among them in humans. *Electromyogr. Clin. Neurophysiol.* 33: 477-503, 1993.

[38] Schalow, G., Zäch, G.A. and Warzok, R., Classification of human peripheral nerve fibre groups by conduction velocity and nerve fibre diameter is preserved following spinal cord lesion. *J. Auton. Nerv. Syst.* 52: 125-150, 1995.

[39] Schalow, G., Bersch, U., Göcking, K. and Zäch, G.A., Detrusor-sphincteric dyssynergia in paraplegics compared with the synergia in a brain-dead human by using the single-fibre action potential recording method. *J. Auton. Nerv. Syst.* 52: 151-180, 1995.

[40] Schalow, G. and Zäch, G.A. Mono and polysynaptic drive of oscillatory firing α_1 (FF) and α_2-motoneurons (FR) in a patient with spinal cord lesion. *Gen. Physiol. Biophys.* 15, Suppl.1: 57-74, 1996.

[41] Schalow, G. and Zäch, G.A. Reorganization of the Human CNS, Neurophysiologic measurements on the coordination dynamics of the lesioned human brain and spinal cord. Theory for modern neurorehabilitation (31 case reports). *Gen. Physiol. Biophys.,* 19, Suppl. 1: 1-244, 2000.

[42] Schalow, G. Recovery from spinal cord injury achieved by three months of coordination dynamic therapy. *Electromyogr. Clin. Neurophysiol.* 42: 367-376, 2002.

[43] Schalow, G.: Stroke recovery induced by coordination dynamics therapy and quantified by the coordination dynamics recording method. *Electromyogr. Clin. Neurophysiol.* 42: 85-104, 2002.

[44] Schalow, G.: Improvement after traumatic brain injury achieved by coordination dynamic therapy. *Electromyogr. Clin. Neurophysiol.*, 42: 195-203, 2002.

[45] Schalow, G.: Partial cure of spinal cord injury achieved by 6 to 13 months of coordination dynamic therapy. *Electromyogr. Clin. Neurophysiol.*, 43: 281-292, 2003.

[46] Schalow, G., Pääsuke, M., Ereline, J. and Gapeyeva, H.: Improvement in Parkinson's disease patients achieved by coordination dynamics therapy. *Electromyogr. Clin. Neurophysiol.*, 44: 67-73, 2004.

[47] Schalow, G., Pääsuke, M., Jaigma, P.: Integrative reorganization mechanism for reducing tremor in Parkinson's disease patients. *Electromyogr. Clin. Neurophysiol.*, 45: 407-415, 2005.

[48] Schalow, G., Jaigma, P. Cerebral palsy improvement achieved by coordination dynamics therapy. *Electromyogr. Clin. Neurophysiol.*, 45: 433-445, 2005.

[49] Schalow, G.: Hypoxic brain injury improvement induced by coordination dynamics therapy in comparison to CNS development. *Electromyogr. Clin. Neurophysiol.*, 46: 171-183, 2006.

[50] Schalow, G. and Jaigma, P.: Improvement in severe traumatic brain injury induced by coordination dynamics therapy in comparison to physiologic CNS development. *Electromyogr. Clin. Neurophysiol.*, 46: 195-209, 2006.

[51] Schalow, G.: Symmetry diagnosis and treatment in coordination dynamics therapy. *Electromyogr. Clin. Neurophysiol.*, 46: 421-431, 2006.

[52] Schalow, G.: Cerebellar injury improvement achieved by coordination dynamics therapy. *Electromyogr. Clin. Neurophysiol.*, 46: 433-439, 2006.

[53] Schalow, G., Vaher, I. and Jaigma, P.: Overreaching in coordination dynamics therapy in an athlete with a spinal cord injury. *Electromyogr. Clin. Neurophysiol.* 48: 83-95, 2008.

[54] Schalow, G.: Stem cell therapy and Coordination dynamics therapy improve Spinal cord injury. *Electromyogr. Clin. Neurophysiol.*, 48: 233-253, 2008.

[55] Schalow, G., Jaigma, P., Belle, V.K.: Near-total functional recovery achieved in partial cervical spinal cord injury (50% injury) after three years of coordination dynamics therapy. *Electromyogr. Clin. Neurophysiol.*, 49: 67-91, 2009.

[56] Schalow, G.: Partial cure achieved in a patient with near-complete cervical spinal cord injury (95% injury) after three years of coordination dynamics therapy. *Electromyogr. Clin. Neurophysiol.*, 49: 199-221, 2009.

[57] Schalow, G.: The classification and identification of human somatic and parasympathetic nerve fibres including urinary bladder afferents is preserved following spinal cord injury. *Electromyogr. Clin. Neurophysiol.*, 49: 263-286, 2009.

[58] Schalow, G.: Coordination impairment between the somatic and parasympathetic nervous system divisions in the human sacral micturition centre following spinal cord injury. *Electromyogr. Clin. Neurophysiol.*, 49: 337-367, 2009.

[59] Schalow, G.: Scientific basis for learning transfer from movements to urinary bladder functions in human. *Electromyogr. Clin. Neurophysiol.*, 50: 339-395, 2010.

[60] Schalow, G.: Cure or urinary bladder function in severe cervical spinal cord injury (95% injury, motoric complete). *Electromyogr. Clin. Neurophysiol.*, 50: 155-179, 2010.

[61] Schulz, G., Lambertz, M., Schulz, B., Langhorst, P. and Krienke, B., Reticular formation of the lower brain stem. A common system for cardio-respiratory and somatomotor function: discharge pattern of neighboring neurons influenced by somatosensory afferents, *J. Auton. Nerv.* Syst., 9 (1985) 433-449.

[62] Shimamura, M., Grillner, S., Edgerton, V.G., *Neurobiological Basis of Human Locomotion*, Japan Scientific Societies Press, Tokyo, 1991.

[63] Sica, R.E.P. and McComas, A.J., Fast and slow twitch units in a human muscle, *J. Neurol. Neurosurg. Psychiatry*, 34 (1971) 113-120.
[64] Thomas, C.K., Ross, B.H. and Stein, R.B. (1986). Motor-unit recruitment in human first dorsal interosseous muscle for static contraction in three different directions, *J. Neurophysiol.*, 55 (1986) 1017-1029.
[65] von der Malsburg, C., Nervous structures with dynamic links, *Ber. Bunsen-Ges. Phys.* Chem., 89 (1985) 703-710.
[66] Wernig, A. and Herrera, A.A., Sprouting and remodelling at the nerve-muscle junction, *Progress in Neurobiology*, Vol. 27, 1986, pp. 251-291.
[67] Wernig, A., Anzil, A.P. and Bieser, A., Formation and regression of synaptic contacts in the adult muscle, In: H. Flohr and W. Precht (Eds.), *Lesion-Induced Neuronal Plasticity in Sensorymotor Systems*, Springer Verlag, Berlin, 1981, pp. 38-50.
[68] Windhorst, U., *How Brain-like is the Spinal Cord*, Springer Verlag, Berlin, 1988.
[69] Brindley, G.S., Polkey, C.E., Rushton, D.N. and Cardozzo, L.: Sacral anterior root stimulation for bladder control in paraplegia: the first 50 cases. *J. Neurol. Neurosurg.* Psychiat., 49: 1104-1114, 1986.
[70] Eccles, J.C.: Learning in the motor system. *Progress in Brain Research* 64: 3-18,1986.
[71] Kelso J.A.S.(1995): Dynamic Patterns. *The Self-Organization of Brain and Behavior.* MIT Press, Cambridge.
[72] Martino, G. and Pluchino, S.: The therapeutic potential of neural stem cells. *Nat. Rev. Neurosci.*, 7: 395-406, 2006.
[73] Harel, N.Y. and Strittmatter, S.M.: Can regenerating axons recapitulate developmental guidance during recovery from spinal cord injury. *Nat. Rev. Neurosci.*, 7: 603-616, 2006.
[74] Thuret, S., Moon, L.D.F. and Gage, Gage, F.H.: Therapeutic interventions after spinal cord injury. *Nat. Rev. Neurosci.*, 7: 628-643, 2006.
[75] Bradbury, E.J. and McMahon, S.B.: Spinal cord repair strategies: why do they work? *Nat. Rev. Neurosci.*, 7: 644-653, 2006.

4. Scientific Basis for Learning Transfer from Movements to Urinary Bladder Functions for Bladder Repair in Human Patients with CNS Injury

Coordination Dynamics Therapy (CDT) has been shown to be able to partly repair CNS injury. The repair is based on a movement-based relearning theory, which requires at least three levels of description: the movement or pattern (and anamnesis) level, the collective variable level, and the neuron level. Upon CDT, not only the performed movement pattern itself is repaired, but the entire dynamics of CNS organization is improved, which is the theoretical basis for (re-)learning transfer. The transfer of learning for repair from jumping on springboard and exercising on a special CDT and recording device to urinary bladder functions is investigated at the neuron level. At the movement or pattern level, the improvement of central nervous system (CNS) functioning in human patients can be seen (or partly measured) by the improvement of the performance of the pattern. At the collective variable level, coordination tendencies can be measured by the so-called "coordination dynamics" before, during and after treatment. At the neuron level, relearning can additionally be assessed by surface electromyography (sEMG) as alterations of single motor unit firings and motor programs. But to express the ongoing interaction between the numerous neural, muscular, and metabolic elements involved in perception and action, it is relevant to inquire how the individual afferent and efferent neurons adjust their phase and frequency coordination to other neurons to satisfy learning task requirements. With the single-nerve fibre action potential recording method, it was possible to measure that distributed single neurons communicate by phase and frequency coordination. It is shown that this timed firing of neurons is impaired upon injury and has to be improved by learning. The stability of phase and frequency coordination among afferent and efferent neuron firings can be related to pattern stability. The stability of phase and frequency coordination at the neuron level can therefore be assessed integratively at the (non-invasive) collective variable level by the arrhythmicity of turning (coordination dynamics) when a patient is exercising on a special CDT device. Upon jumping on springboard and exercising on the special CDT device, the intertwined neuronal networks, subserving movements (somatic) and urinary bladder functions (autonomic and somatic) in the sacral spinal cord are synchronously activated and entrained to give rise to learning transfer from movements to bladder functions. Jumping on springboard and other movements primarily repair the pattern dynamics, whereas the exactly coordinated performed movements, performed on the special CDT device for turning, primarily improve the preciseness of the timed firing of neurons. The synchronous learning of perceptuomotor and perceptuo-bladder functioning from a dynamical perspective (giving rise to learning transfer) can be understood at the neuron level. The activated phase and frequency coordination upon natural stimulation under physiologic and pathophysiologic conditions among α and γ-motoneurons, muscle spindle afferents, touch and pain afferents, and urinary bladder stretch and tension receptor afferents in the human sacral spinal cord make it understandable that somatic and parasympathetic functions are integrated in their functioning and give rise to learning transfer from movements to bladder functions. The power of this human treatment research project lies in the unit of theory, diagnostic/measurement, and praxis, namely that CNS injury can partly be repaired, including urinary bladder functions, and the repair can partially be understood even at the neuron level of description in human.

4.1. Data Summary

4.1.1. Neural Network Changes for Understanding Learning Transfer and Pattern Stability Impairment

Single-nerve fibre action potentials (APs) were recorded extracellularly from skin and bladder afferents, α and γ-motoneurons, and secondary muscle spindle afferents from thin sacral nerve roots mainly in a patient with traumatic spinal cord injury sub TH1 and a brain-dead human. Simultaneous single-nerve fibre impulse patterns of skin and urinary bladder afferents and α and γ-motoneurons and secondary muscle spindle afferents were analyzed upon natural stimulation. Phase and frequency coordination for CNS self-organization of neuronal network patterns is analyzed in a brain-dead human (rather physiologic) and patients with spinal cord injury (pathologic case) [1] to explore what has changed in the CNS organization following spinal cord injury [2], to find a scientific basis for learning transfer from movements to urinary bladder functions upon administering a movement and learning therapy called "coordination dynamics therapy," and [3] to understand why integrative patterns do loose stability following injury.

4.1.2. Pattern Change (100ms) Needs More Time Than Pattern Enhancement (6ms)

The time for pattern change was measured in a patient at the neuron level. Identified α and γ-motoneurons fired oscillatory, due to the sustained stretch reflex of the external sphincters (as a part of the continence automatism) induced by an anal catheter (and maybe the bladder catheter). The motoneurons and the secondary muscle spindle afferents transiently synchronized their firing upon repetitive touch and pinprick inside the anal reflex area by reducing the duration of their oscillation period to a resetting of the oscillation cycle. The responses to pinprick stimulation were different from those to touch stimulation in three aspects.

Firstly, the response time to the shortening of the oscillation period was longer than the oscillation period (≈100 ms) for pinprick and shorter for touch. Second, the response to pain application was longer (shortening of several oscillation periods) and stronger than for touch stimulation. Pinprick stimulation reduced the oscillation period to between 5 and 40 ms (mean = 18 ms), and touch stimulation to between 8 and 28 ms (mean = 15 ms). Third, transient synchronization of afferents and efferents was most pronounced for pinprick stimulation.

The shortest latency, however, following touch was approx. 10 ms when measuring from the afferent volley running in the direction of the spinal cord and 110 ms following pinprick. It is discussed that repetitive touch stimulation reinforced the sustained stretch reflex of the anal sphincter, which is possible with no network reorganization (variation of the same network state) and therefore fast, whereas repetitive pinprick stimulation replaced the sustained stretch reflex by the protection reaction of the anal sphincter (switching from one network state to another), which made time-consuming network reorganization necessary. Taking the conduction velocities of skin and pain afferents into consideration, the reinforcement of the continence pattern needed approximately 6ms, and the pattern change from the continence to the protection pattern needed approximately 100ms.

4.1.3. Exact Phase and Frequency Coordination in the Brain-Dead Human

In the brain-dead human, natural simultaneous firing patterns of secondary muscle spindle afferents and α and γ-motoneurons showed relative phase and frequency coordination. The correlation between the firing of a motoneuron and a spindle afferent fibre was in the range of 3 and 5ms. The "Eigenfrequencies" of α_2-oscillators changed only little with time, and the phase relations between the α and γ-motoneurons and the secondary muscle spindle afferents were rather stable. The overlap of the interspike interval distributions of the secondary spindle and bladder stretch receptor afferents with the "Eigenfrequency" ranges (and their halves) of the oscillators showed what afferents were driving what motoneurons with respect to frequency coordination.

Relative phase and frequency coordination was rather stable and specific for the physiologic CNS organization. In a patient with a spinal cord injury, this phase and frequency coordination was analyzed to study CNS self-organization. Simultaneous interspike intervals (IIs) of and phases between the APs of five nerve fibres were recorded, and distributions were constructed. The interspike interval (II) distributions were of a broad peak type. Phase distributions showed one to three peaks interpreted as phase relations between the firings of the nerve fibres.

4.1.4. Loop of Premotor Spinal Oscillators to the Periphery

Following repetitive touch and pinprick stimulation (frequency $\approx$ 1Hz) in- and outside the anal reflex area, the II distributions of α and γ-motoneurons and of secondary spindle afferents assimilated partly or fully, while preserving their phase relations. This coordinated firing is interpreted by the oscillatory firing α neuronal networks building up an external loop to the periphery via the γ-loop.

4.1.5. Change of Phase Relations

Upon touch, pinprick, and anal reflex stimulation, and anal and bladder catheter pulling, the values and the number of the phase relations changed. Two phase relations per oscillation cycle were observed. Two phase relations probably represent the physiologic case for the somatic nervous system. Only one phase relation was found when full synchronization of all units occurred. Three phase relations were found when the parasympathetic nervous system division interacted with the somatic one.

4.1.6. Impairment of Phase and Frequency Coordination Following Spinal Cord Injury

According to the exact phase and frequency coordination among neuron firing obtained from the brain-dead individual in comparison with the impaired coordination in spinal cord injury patients, it is likely that the increased synchronization tendency and instability in the number and the values of phase relations suggest pathologic functioning of the caudal functionally disconnected spinal cord in patients with spinal cord injuries: Oscillatory firing neuronal networks, which lost their specific Eigenfrequencies, interact more easily and randomly with other oscillatory firing networks.

4.1.7. Building of Oscillator Loops to the Periphery for Relearning Patterns and Phase and Frequency Coordination

Since spinal oscillators built up external loops to the periphery, it becomes possible in humans with spinal cord injury to re-preformate oscillatory firing neuronal networks (and the networks they are integrated in) in the sacral micturition centre by coordinated, rhythmic, dynamic, stereotyped movements.

This results in reduced spasticity, repair of movements and repair of urinary bladder functions by learning transfer from movements to bladder functions. The results of this and the following part [43] support my key tenet that learning not only stabilizes the yet to-be-learned patterns (like jumping on springboard or walking on treadmill) but causes also specific modifications of the entire coordination dynamics in the direction of the task requirement and gives rise to transfer of learning.

Further, the task requirements when exercising on the special CDT device is to relearn phase and frequency coordination among neuron firings for that specific movement pattern but causing also long-lasting alterations (improvements) of the phase and frequency coordination of the underlying coordination dynamics of millions of neurons and contributing in this way substantially to a transfer of abstract learning. Learning transfer from movements to urinary bladder functions is also in accordance with functional anatomy, since the somatic and the autonomic nervous systems dovetail with one another in the sacral micturition centre. Even some muscle spindles of the external sphincters are innervated by the somatic and parasympathetic divisions.

4.2. Introduction

4.2.1. General

Coordination Dynamics Therapy has been shown to be able to improve the central nervous system (CNS) functioning after stroke [25], traumatic brain injury [26,33], cerebellar injury [35], spinal cord injury [22,27,37-39,43], cerebral palsy [31], in Parkinson's disease [28,30], and after hypoxic brain injury [32]. This movement-based learning therapy could not only improve the trained movements, but other untrained patterns could also be improved by transfer of learning. In cerebral palsy, the higher mental functions [31] and in spinal cord injury, the autonomic functions [39], including urinary bladder functions [43], could be improved by this movement-based learning therapy. In this article, the mechanisms of the transfer of learning will be tried to clarify to further increase learning transfer and to show on what level of human research urinary bladder functions can be cured in motoric complete cervical spinal cord injury [39,43]. Upon measuring natural communication among single neurons in the human nervous system, insight will be gained into why some trained movements substantially contribute to transfer of (re-)learning (jumping on springboard, exercising on a special device) and others are not or only little (passive movements of arms and legs).

I will concentrate on the learning transfer from movements to urinary bladder functions, because, firstly, this learning transfer is very important, and secondly, high-quality natural impulse patterns have been recorded by me to be involved in single-nerve cell communication related to movement (somatic division) and continence functions (parasympathetic and somatic divisions). A learning transfer from movements to higher mental functions cannot easily be analyzed at the single-neuron level, since it would be difficult to record single-cell communication related to thinking. Phase memory has been measured between single motor units upon repeated muscle activation [19].

But to measure longer acting memory of integrative functions is a different task. It will be tried in this article therefore to understand the "learning transfer" from movements (jumping on springboard (Figure 1C)) to bladder functions (continence and micturition) on the basis of human electrophysiology at the neuron level in the human sacral spinal cord. I will further try to show that the induction of learning transfer in the injured CNS needs more than understanding the relearning of movements. The impaired phase and frequency coordination for CNS self-organization has also to be improved by phase and frequency relearning (when exercising on special devices (Figure 1D)) to have a neuronal network available that has the capability of learning.

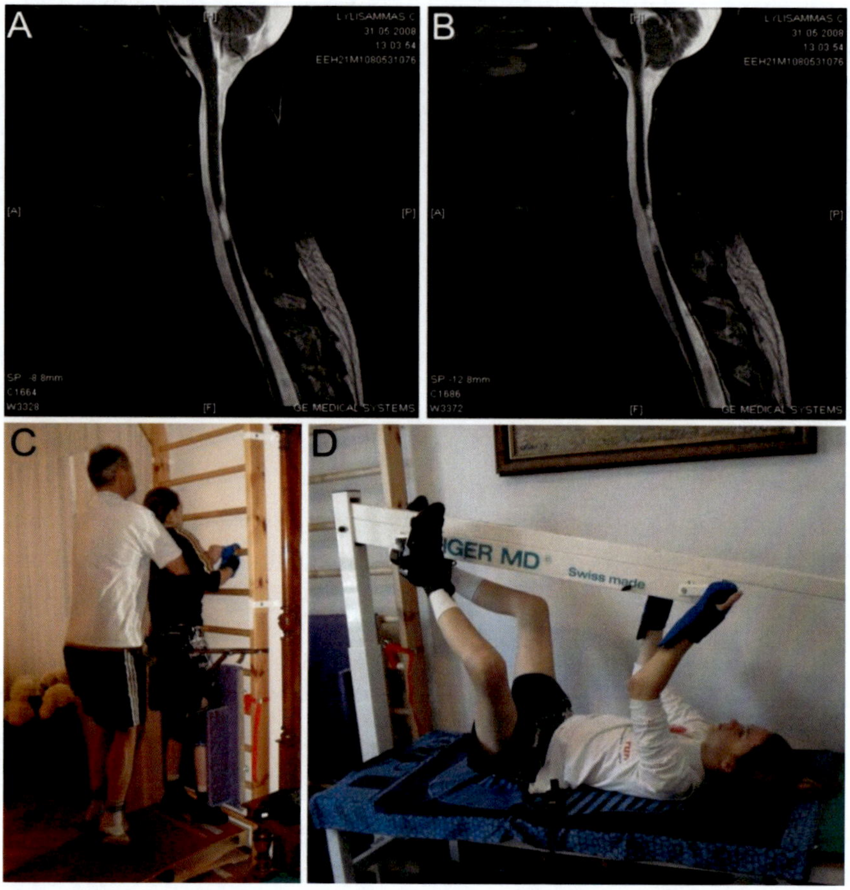

Figure 1. A,B. Magnetic resonance images of a severe spinal cord injury at C5/6 levels of a 20-year-old female patient after three years of optimal coordination dynamics therapy. Both sagittal T2-weighted images show connection between the rostral and caudal spinal cord. C. Jumping on springboard of a 20-year-old patient with a severe spinal cord injury at C5/6 levels (Figure 1A,B) (upon no weight support) after three years of optimal coordination dynamics therapy. The father is jumping behind in interpersonal coordination to increase the jumping amplitude. D. Same patient as in Figure 1C upon exercising on a special coordination dynamics therapy device.

4.2.2. Learning Transfer

The injured, malformed or degenerating human CNS can partly be repaired by learning [21-33,35-39,43]. The training of those movements or functions that are impaired is a too simple approach and only partly successful to repair the CNS integratively and the functions that cannot be trained, like continence or intelligence (cerebral palsy), or cannot be repaired or improved. A scientific approach to CNS repair is to train and relearn those organizational functions of the CNS that are impaired. By comparing the physiologic with the pathologic CNS functions, it was discovered what urinary bladder functions became pathologic and have to be repaired or improved [40-42].

Following severe cervical spinal cord injury (Figure 1A,B), it was shown that motor functions could partially repaired [39] and urinary bladder functions fully repaired [43]. By comparing the natural impulse patterns of single neurons between a brain-dead human and patients with spinal cord injury, it was shown that with respect to bladder dysfunction, mainly the self-organization of neuronal networks of the sacral micturition centre went pathologic ([40-42], Figures 20-22) because of the disconnection from supraspinal centres.

This part concentrates on the "learning transfer" (in severe cervical spinal cord injury) from movements to urinary bladder functions for a repair of bladder functions, consisting of the learning transfer and the improvement of the self-organization of the neuronal networks of the spinal cord and supraspinal centres. In incomplete spinal cord injury, the spared tract fibres can be used for a functional reorganization including a reconnection of the sacral with the pontine micturition centre [38]. In almost complete spinal cord injury [39,43] with nearly no spared tract fibres, remained, however, some connections to the pontine micturition that have to be (and can be) reconstructed by a limited regeneration of the cord [39,43]. This article gives the scientific basis for those movements that are most efficient in inducing learning transfer for bladder repair.

4.2.3. Relearning of Phase and Frequency Coordination and Relearning of Bladder Functions via Learning Transfer from Coordinated Movements

Since following CNS injury the phase and frequency coordination of neuron firing for neuronal network organization is impaired (Figures 20-22), this phase and frequency coordination has efficiently to be improved by learning. This can be achieved when the patient is exercising on the special coordination dynamics therapy device for turning movements (Figure 1D). From all movements that are performed in a coordinated manner, the patient's CNS can relearn the coordinated firing of neurons. However, such relearning of coordinated firing is most efficient if the movement is most integratively exact and utmost coordinatedly performed. A relearned higher precision and specificity of neuronal network organization (a kind of abstract learning) allows higher learning (and information) transfer.

The induction of learning transfer from movements to bladder functions can efficiently be induced when the patient is jumping on springboard, as will be shown in this article, since movement and bladder functions are synchronously activated during the jumping: The networks for generating motor patterns and the networks for generating autonomic and somatic bladder patterns are synchronously activated and entrained in the way that the neurons of the activated networks improve the coordination of their natural impulse traffic. The learning transfer involves two properties to be improved. Firstly, the phase and frequency coordination of neuron firing has to be improvement so that the injured CNS can generate a specific landscape of pattern formation (attractor layout) again with specific stabilities of the attractors and the fluctuation of the pattern states is reduced (Figure 23). Secondly, the geographical landscape of attractors has to be changed specifically by learning and learning transfer so that the repaired CNS can generate the physiologic bladder functions with high pattern stability again [43].

4.2.4. Natural Impulse Patterns of Single Neurons Can Explain Integrative Patterns

The CNS can partially repair itself by the interplay of the body with the external world via the activation of receptors in the different organs and the transmission of the specific information, encoded in natural impulse patterns, in afferents to the neuronal networks and in efferents to the different organs and limbs. Additionally, premotor spinal oscillators (network oscillators) can build up an external loop to the periphery (Figure 15) [18] to include the periphery more directly in the network organization and reorganize the networks more efficiently by entrainment. Recording the natural afferent and efferent impulse traffic in nerve roots, leading to neuronal network organizations (Figure 3) during the organization of urinary bladder and motor patterns upon natural stimulation, and changes in the natural impulse traffic (Figure 8) will reveal changes in the network organizations of continence and motor patterns (Figures 5-9). The organization of bladder function patterns is stimulated (besides intention) by exciting bladder receptors (Figures 4E,G,5G,15h) and the organization of motor patterns is stimulated by exciting somatic receptors (especially pain receptors of the skin (Figures 5,15)).

Since the natural afferent and efferent impulse traffic is part of CNS organization, it should be possible to explain and calculate integrative patterns by using natural impulse patterns of single neurons. It will be shown that the pattern change between continence and protection can be calculated (Figures 7-9) based on human neurophysiologic data at the neuron level. It will therefore be shown that it is possible to measure and understand macroscopic patterns by changes of the natural impulse traffic upon different natural stimulations. Such understanding of integrative functions by impulse pattern at the single-neuron level is a first step in the direction of understanding "I think, therefore I am" (Descartes) by the communication among neurons. The integrative patterns are made visible by an attractor layout, defined in the framework of "System Theory of Pattern Formation for the Injured CNS." Such attractor layouts for continence functions will be used in the following part to understand urinary bladder repair [43].

Knowing that natural impulse patterns can explain behavioural patterns, understanding the learning transfer between movement and continence patterns will be attempted. For a deep understanding of CNS self-organization and reorganization upon natural stimulation during pattern change, we have to go into the complexity of CNS self-organization, which means we have to go into the complexity of phase and frequency coordination among neuron firings [42] (Figures 11,12,14). The exploration of phase and frequency coordination in subnetworks shows some similarity to the analysis of electrical circuits by means of the "rules of Kirchhoff." In human electrophysiology, phase and frequency changes among neurological elements and in electronics currents and voltages among electronic elements are measured. But in both cases a subnetwork is analyzed as a part of a larger network.

4.2.5. Instability of Phase and Frequency Coordination at the Neuron Level Causes Instability and Variability of Movement and Continence Pattern States

A substantial difference between electrical and neuronal networks is the stability of functioning. Instability of macroscopic pattern states is caused by the instability of phase and frequency coordination at the neuron level. For a repair of the injured CNS by learning and learning transfer, the variability of phase and frequency coordination (Figures 20-22) has to be improved to generate more specific and exact coordination changes again (Figures 17-19). In an attractor layout for movements (Figure 9Ab) or bladder functions (Figure 9Bb), the stability of an attractor is given by the depth of the potential well. The basin of attraction for jumping in-phase, for example, is deeper than the one for jumping anti-phase (Figure 9Ab). The jumping in-phase has therefore a higher stability than jumping anti-phase, if the ball (the network state) lies stable in the depth of the basin of attraction, which means if the pattern state shows only little fluctuation (Figure 23A). If the phase and frequency coordination, however, became impaired (high variability of coordination), the attractor layout will be deformed and the ball (the network state) will move about in the attractor basin (or even jumps out of the basin), which means that the network state becomes instable and varies very much (or is even lost) (Figure 23B,C). The performance of jumping on springboard would become poor (changing of stride length, knee flexion, foot positioning, etc.).

There are therefore two kinds of stability of network states. For movement performance with high quality (high quality of CNS organization (healthy CNS)), there are differences in the stability among movement patterns. The jumping in-phase on springboard has a higher stability than jumping anti-phase, probably because the jumping in-phase is easier to generate by the CNS (easier coordinations are only needed). The second kind of stability is related to the precision (fluctuation) of phase and frequency coordination for CNS organization. Following injury, the phase and frequency coordination becomes poor, and more or less all movements can be performed by the patients only with poor performance (if at all). However, some movements become more impaired than others. In stroke patients, for example, the performance of arm movements on the injured side is mostly more impaired than the one of the leg movements, because the networks for generating the arm movements have a higher complexity, and higher precision of phase and frequency coordination is needed for the organization of the arm movement. The networks for organizing the more stereotyped leg movements are not that complex and are also therefore only slightly disturbed by the injury. To repair the injured CNS by learning, therefore, not only the stability of the physiologic attractors has to be increased (deepening of the attractor basin), the stability of pathologic attractors, like spasticity, also has to be decreased (reducing the depths of the potential well). Also the fluctuation of phase and frequency coordination has to be reduced so that the CNS can generate specific geographical landscapes of attractors and movement and continence patterns with little variability again. Without specific phase and frequency changes, there is also no learning transfer possible anymore because of missing specific network organization necessary for learning.

4.2.6. Need for a Regeneration of the Spinal Cord

For the repair of a severe spinal cord injury, a functional repair is not sufficient; the caudal disconnected spinal cord has to be brought under supraspinal control again and at important network sites, new neurons and new connections have to be built. For directing intention downstream to induce pattern change in complete spinal cord injury, some regeneration of tract fibres is needed.

4.2.7. Specific: System Theory of Pattern Formation (Collective Variable Level of Description)

In recent years, several researchers have examined the issue of perceptuomotor learning from a dynamical perspective without, however, providing any operational or formal treatment of individual spontaneous coordination tendencies that exist before, during, and after learning at the collective variable level (the variable is the relative phase between the legs when jumping on springboard, Figure 9Ab) and partly at the neuron level (phase and frequency coordination dynamics, Figures 11,12,14,21,22) to show a self-consistency between the collective and neuron level of description.

By cooperative and competitive interplay, the many billions of neurons of the human CNS generate dynamics in self-organization that can, for special patterns, be described by collective variables or order parameters. Specific equations of motion (the dynamics) of these collective variables generate the time course of organizational states. Coordination patterns are not only determined by the task or biological function, but the patterns adjust continuously to the requirements from the environment, memory, intention, and support given by a therapist. All the specific requirements are captured by the concept of behavioural information and are made part of a vector field that attracts toward required patterns. The coordination pattern dynamics, characterized by equations of motion of the collective variables, takes the following form [44]:

$$dX/dt = F_{intr}(X) + \sum c_{inf} F_{inf}(X,t) \qquad (2)$$

where F_{intr} designates the intrinsic dynamics, and the sum is over different types of so-called behavioural information, F_{inf}, as environmental, memorized, or intended behavioural information. The relative strength of the different influences is parameterized by c_{inf}. For a mathematical solution to equation (2) for the special movement "jumping on the springboard" in the Haken-Kelso-Bunz model [5], see Refs. [33,38]. Since it is extremely difficult or not at all possible to find a mathematical solution to the equations of motion of the collective variables formula (2) (even using many approximations), a substantial progress in evaluating the quality of CNS organization was therefore achieved when it became possible to assess partly the dynamics of CNS organization experimentally (by diagnostic). The dynamics of CNS self-organization at the neuron level by phase and frequency coordination is partly reflected in the temporal stability of coordination patterns, which can be assessed through a process of pattern change. When a subject exercises on the special coordination dynamics therapy and recording device for turning (Figure 1D), the device is imposing all the continuously changing coordinations of arm and leg movements between pace and trot gait. Pattern stability is assessed by measuring the temporal stability of exercising (df/dt; f = frequency) at an own optimal frequency, e.g., 1Hz, upon the pattern change of arm and leg movements. There is only one collective parameter describing the coordination tendencies, which is the temporal stability of exercising. Attractor layouts of pattern stability can be recorded [34].

The equations of motion of the collective variables (formula 2) have important clinical implications for treatment. The behavioural requirements F_{inf} (like intention, support, and instruction) affect the whole coordination dynamics, including stability, rather than certain coordination pattern itself only. If one pattern is trained and improved upon movement-based learning, the not trained patterns also change and improve. There is learning transfer. This learning transfer from movements to urinary bladder functions is understood (and developed further) in this part with data obtained from human electrophysiology.

4.2.8. Magnet Effect

Intrinsic coordination tendencies captured by the intrinsic dynamics influence systematically trained patterns because the degree to which intrinsic tendencies conflict or agree with the required patterns determines the variability of the performed coordination pattern. Such attraction to certain sets of phase relations (of hundreds to millions of phases, necessary to generate a pattern by cooperative and competitive stability) among coordinating neurons was called by von Holst "Magnet effect" [7,8]. The rate of learning transfer is therefore higher if a trained coordination pattern is close to a pattern of the intrinsic coordination tendencies. The trained patterns of nearby phasings are biased or attracted toward the stable intrinsic patterns. In reverse, if the exercised pattern (with the support of therapists and devices) is functionally close to the former physiologic pattern, the "Magnet effect" will "pull" (attract) the deteriorated pattern (due to CNS injury) in the direction of the performed physiologic pattern. With respect to urinary bladder repair, the rate of learning transfer is highest if the activated pattern is close to the coordination tendencies of the coordinating components of the former micturition and continence patterns. It will be shown in this article that during jumping on springboard, the coordinating components for continence and micturition are activated similarly as during physiologic functioning. The "Magnet effect" will attract the neuron firings by entrainment of the neuronal networks in the direction of the coordination pattern "physiologic continence and micturition."

In conclusion, the equations of motion of the order parameters (collective variables) (formula 2) provide us with treatment principles but offer no information about the specific behavioural information (F_{inf}) with the use of which the CNS can efficiently be repaired including learning transfer in the given patient. The human neurophysiology data of this paper, on the other hand, lead to a scientific and practical basis to induce learning transfer from movements to urinary bladder functions for bladder repair. In the previous three parts, it was concentrated on the differences between physiologic and pathologic functioning of the human bladder [40-42], especially on the synergia and the dyssynergia. In this part, the "repair physiology" for bladder repair is presented. Network organizations and changes will be measured in the parasympathetic and somatic nervous system divisions. The stability of network states will be related to the stability of phase and frequency coordination between single neuron firings. It will be shown that the stability of movement and continence patterns is related to the stability of phase and frequency coordinations of neuron firings involved in movement and continence functions. The improvement of the stability of the phase and frequency coordination among millions of neurons and the change of phase and frequency coordination to generate physiologic patterns is the basis for functional repair including learning transfer. The deterioration of phase and frequency coordination upon neuron damage due to injury or degeneration will affect nearly all the coordination relations related to the cooperative and competitive interplay among neurons for generating a pattern. During functional repair, it is tried to establish physiologic stable sets of phase relations again. The jumping on springboard uses mainly the Magnet effect between closely related network states for repair besides the repair of the "Eigenfrequencies" of premotor spinal oscillators. This article is tackling the communication among single neurons and the entrainment of human neuronal networks for repair.

4.2.9. Electrophysiology (Neuron Level of Description)

I have shown in previous parts that the classification scheme for the human peripheral nervous system [12] is preserved following spinal cord injury [40] and can be used as a basis for a comparison of the natural impulse patterns between patients with spinal cord injury and brain-dead humans to clarify bladder functioning. I have analyzed the detrusor-sphincteric dyssynergia (pathologic co-contraction of detrusor and external bladder sphincter) of the urinary bladder [41,42]. Parasympathetically evoked impulse patterns of secondary muscle spindle afferents and sphincteric motoneuron activity were compared between paraplegics and a brain-dead human. It was shown that inflammation, hypertrophy and overactivated bladder stretch (S1) and tension receptors (ST) contributed to bladder dysfunction [40]. The main reason for the detrusor-sphincteric dyssynergia of the bladder, however, seemed to originate from the CNS organization itself. Following spinal cord injury, with the reorganization of the sacral micturition centre, the mutual inhibition of the detrusor and the external bladder sphincter is sometimes lost [41]. The mechanism underlying the loss of the mutual inhibition of detrusor and external sphincter may be due to, besides other reasons, a false sacral reflex reorganization or disconnection from the pontine micturition centre, which is believed to coordinate the activation of detrusor and external bladder sphincter. Intermediate stages of the coordination of detrusor and external bladder sphincter are conceivable. Since in a brain-dead human the parasympathetic division inhibited sphincteric motoneurons with detrusor contraction, pathologic reorganization and adaptation of the CNS, caudal to the level of injury, probably takes place already in the spinal shock phase and thereafter. It was further partly previously shown that the phase and frequency coordination among neurons and neural assemblies in the sacral micturition centre of the human CNS was impaired [42]. The "Eigenfrequencies" of neuronal assemblies (premotor spinal oscillators) became partially lost (Figure 20), and the phase coordination between the firings of motoneurons and muscle spindle afferents became unstable in patients with spinal cord injury (Figure 22) in comparison to the coordination among neurons in the brain-dead human (Figure 21). The instability of the "Eigenfrequencies" of premotor spinal oscillators and the coordination instabilities among the firings of premotor spinal oscillators and spindle afferents lead to the conclusion that the self-organization of the neuronal networks of the sacral micturition centre become unstable following spinal cord injury. The integrative functions of the sacral micturition centre therefore get unstable, and the network patterns probably change their stability values according to the actual spinal cord injury. According to the actual injury, some patients have a synergia of the bladder, others have a dyssynergia and still others have a mixture of both patterns. Now the step from human neurophysiology [40-42] to the System Theory of Pattern Formation [9,33,34,38,44,47] is possible. Since healthy humans have the synergia pattern and the dyssyergia pattern (to stop micturition) and can switch among them by intention, the sacral micturition centre can be repaired by changing the stabilities of these patterns.

The problem of synergia and dyssynergia is not a problem of fixed pathologic circuitries following injury anymore but is more a question of stability of patterns that can be changed by learning. Since patients with severe spinal cord injury cannot train directly urinary bladder functions (or only little), movements have to be trained with which urinary bladder functions can be improved by transfer of learning.

This research reports about the measurements in the human CNS that lead to the movements with which urinary bladder functions can be cured [43] by learning transfer (first fundamental step). To get the urinary bladder under volitional control and to enhance further the stability of micturition and continence functions, some regeneration in patients with spinal cord injury is necessary. The second fundamental step in repair is that this necessary regeneration to cure urinary bladder functions is possible to achieve even in severe cervical spinal cord injury. The following part 5 will report about it [43]. The evolutionary principle of levels of function, which implies that rostral segments of the brain have become dominant over caudal, is restored.

The century step is that the urinary bladder functions can be cured and destructive operations are no longer necessary in patients to whom the movement-based learning therapy "coordination dynamics therapy" can be administered. A drawback is still that the learning effort is enormous in severe cervical spinal cord injury [36], but the effort is much less in incomplete spinal cord injury [38] and in children with spinal cord injury [27]. Nearly 10% of the society suffers from incontinence, including young women after the birth of the first child (stress incontinence). There are no problems for the young mothers to jump on springboard, to walk or run, or to exercise on the special coordination dynamics therapy device for turning.

4.3. Methods

4.3.1. Electrophysiology

Single-nerve fibre action potentials were recorded extracellularly (Figure 3A) from undissected nerve roots (Figure 3C) with two pairs of platinum wire electrode pairs (electrode pair distance 10 mm; electrode distance in each pair 4 mm) at two sites (Figure 3A,B), preamplified (x1000), filtered (RC-filter, passing frequency 100 Hz - 10 kHz), and displayed on a digital storage oscilloscope (Vuko Vks 22-16), and also stored using a PCM-processor (Digital Audio Processor PCM-501ES) and a video recorder. Conduction velocity distributions of afferents and efferents were constructed and calibrated, and group conduction velocities were identified [12].

From multi-unit impulse patterns, simultaneous patterns of several single afferent and efferent nerve fibres were extracted (Figure 3D) by shifting a window through the original trace, taken from the tape, and measuring up the action potential (AP) occurrence times from a counter from the scope.

APs of a single-nerve fibre were identified with the naked eye by picturing first the AP waveforms (templates) with their variations and then identifying the AP occurrence times based on the following criteria: a. AP shape on trace "a," b. AP shape on trace "b," c. reoccurrence of conduction time, d. AP amplitude. Since the adding up of different AP waveforms or the adding up with artefacts can be distinguished with the naked eye, manual identification is superior to computer analyses unless having a computer programmed for all possible adding up of the different potentials and artefacts. Interspike interval (II) and phase distributions were constructed (Figures 11,12). The intraoperative recordings were mainly performed in a brain-dead human and in a patient with a spinal cord lesion sub TH1 (for one year (paraplegic 9)) during the implantation of an electrical anterior root stimulator for urinary bladder control.

4.3.2. Relearning at the Movement, Collective Variable, and Neuron Levels

CNS repair is based on a pattern relearning theory, the theory of coordination dynamics, which requires at least three levels of description: the movement or pattern level (walking, micturition); the coordinative, collective variable level (relative phase between the legs, when jumping on springboard); and the neuron level (phase and frequency coordination among neuron firings).

The nervous system supplies coordination dynamics at these three levels of description rather than particular patterns. At the movement level, existing coordination tendencies are evaluated by the performance of movements and can partly be quantified, for example, by the speed of the performed movement (speed of crawling).

At the collective variable level, existing coordination pattern tendencies are treated in terms of dynamics, i.e., as equations of motion (if possible) of a relevant collective variable that changes during the learning process ($dX/dt = F_{intr}(X) + \sum c_{inf}F_{inf}(X,t)$, see Introduction). By defining a potential function, behavioural changes can be represented by the overdamped movement of a rolling ball (the state of the system) in the potential "landscape." For jumping on springboard and bladder continence, such attractor layouts are shown in Figure 9. For the bladder function, no collective variable could be given.

At the neuron level, existing coordination tendencies are treated in terms of phase and frequency changes among single neuron firings (α and γ-motoneurons and muscle spindle, skin and bladder afferents). Changes of coordination tendencies are measured, invasively, before injury (physiologic, brain-dead human) and after injury (pathologic, patient with spinal cord injury) and their temporal evolution (non-invasively). Non-invasively, single motoneuron firing can be measured by surface electromyography (sEMG) when recording single motor unit firing, for example, in spinal cord injury patients [29].

The repair of the human CNS by learning can be assessed by measuring the existing inner coordination tendencies and their evolution before, during, and after (movement-based) learning. The improvement of CNS self-organization is quantified at the movement level by the performance, at the collective variable level by the arrhythmicity of turning (df/dt, f = frequency) upon exercising on the special coordination dynamics therapy device for turning (Figure 1D), and at the neuron level by improvements of motor programs and single motor unit firings (sEMG).

Information relevant for repair are the existing CNS organizations in patients following CNS injury, here operationalized in terms of behavioural patterns and conceptualized as an attractor layout or coordination dynamics. Cooperative and competitive mechanisms can be evaluated only in relation to existing coordination tendencies, making the individual learner or patient the significant unit, not the group or the species.

Since relearning of patterns involves alterations of the entire attractor layout of the coordination dynamics of movements, autonomic, higher mental, and other functions (even though the dynamics cannot be described until now), learning transfer is also occurring between movement and urinary bladder functions. In the result section, impulse patterns at the neuron level in human are presented, with which insight will be gained into why some excessively trained movements contribute to relearning and learning transfer (synchronous activation of receptors for movement and continence functions upon jumping on springboard (Figure 15), frequency coordination between afferents and motoneurons (Figures 18,19)) and others are not (passive movement with no or only little activation of the neuronal networks). In the Discussion, it is attempted to clarify the learning process in terms of movements, dynamics of collective variables, and dynamics of phase and frequency coordination among single neurons.

4.3.3. Definition of the Term "Coordination Dynamics" to Clarify Unfortunate Connotations

By using the movement, performed on the special CDT and recording device (Figure 1D), as behavioural information in the equation of motion, the collective variable of the corresponding coordination dynamics is the arrhythmicity of turning (temporal stability of this specific coordination pattern range) with pattern change from trot gait (anti-phase) to pace gait (in-phase coordination between arm and legs) and backwards. The arrhythmicity of turning, with ongoing time, was termed "coordination (pattern) dynamics" in former publications.

The mean arrhythmicity of turning per minute was called coordination dynamics value, and the measured change of the arrhythmicity of turning for the pattern range from trot to pace gait and back to trot gait was termed attractor layout of the collective variable "df/dt" or "df/dt/f" (f = frequency). The coordination dynamics values for different loads, ranging between 20 and 200N, are the backbone of my approach to quantify the quality of CNS organization during repair. Therefore, this special coordination dynamics therapy device offers the possibility to measure certain aspects of CNS organization non-invasively to probe CNS organization with ongoing therapy, and it can be used to relearn phase and frequency coordination at the neuron level, at the collective variable level, and at the movement and pattern level. However, the term "coordination dynamics" is also used more generally in this article to describe the self-organization of the neuronal networks of the CNS and its temporal evolution, especially during relearning.

4.4. Results

4.4.1. Clinical Urinary Bladder Function Test (Urodynamics)

To understand learning transfer, defined within the framework of System Theory of Pattern Formation for Repair, from movements to urinary bladder functions with human electrophysiology on the single-neuron level, it is started with the clinical diagnostic of the urinary bladder: urodynamics. The functioning of the urinary bladder can be evaluated by measuring the pressure in the bladder and in the abdomen (colon) and the electromyographic (EMG) activity of the external sphincters and/or functionally associated pelvic floor muscles [45]. Such bladder diagnosis is called urodynamics (Figure 2). Especially the simultaneous activation of the detrusor (line detrusor pressure in Figure 2B) and the sphincteric and/or pelvic floor muscles (line pelvic floor EMG in Figure 2B), the so-called detrusor-sphincter-dyssynergy, can be measured, which may destroy the kidneys in the long term.

Upon retrograde filling of the bladder in a patient with a spinal cord injury, the pressure in the bladder and colon is measured, and the electromyographic activity (EMG) of the pelvic floor is recorded with surface electrodes (Figure 2). The detrusor pressure is obtained by subtracting the abdominal pressure from the bladder pressure. The continence status of the patient is diagnosed by the reports of the first feeling of bladder fullness, the desire to void, and the leaving of fluid out of the bladder. The EMG activity of the pelvic floor informs when the external bladder and anal sphincters are activated. Upon knocking, pressing, coughing, and stimulating bladder reflexes, the bladder status is obtained. This patient of Figure 2 with a spinal cord injury subTh12 had a detrusor-sphincteric dyssynergy of the bladder, because the EMG activity of the external sphincters increased (the sphincters became activated) with the increase of the detrusor pressure (activation of the detrusor). Improvement of urinary bladder function mainly due to therapy could be quantified by repeated urodynamics measurements (Figure 2A,B).

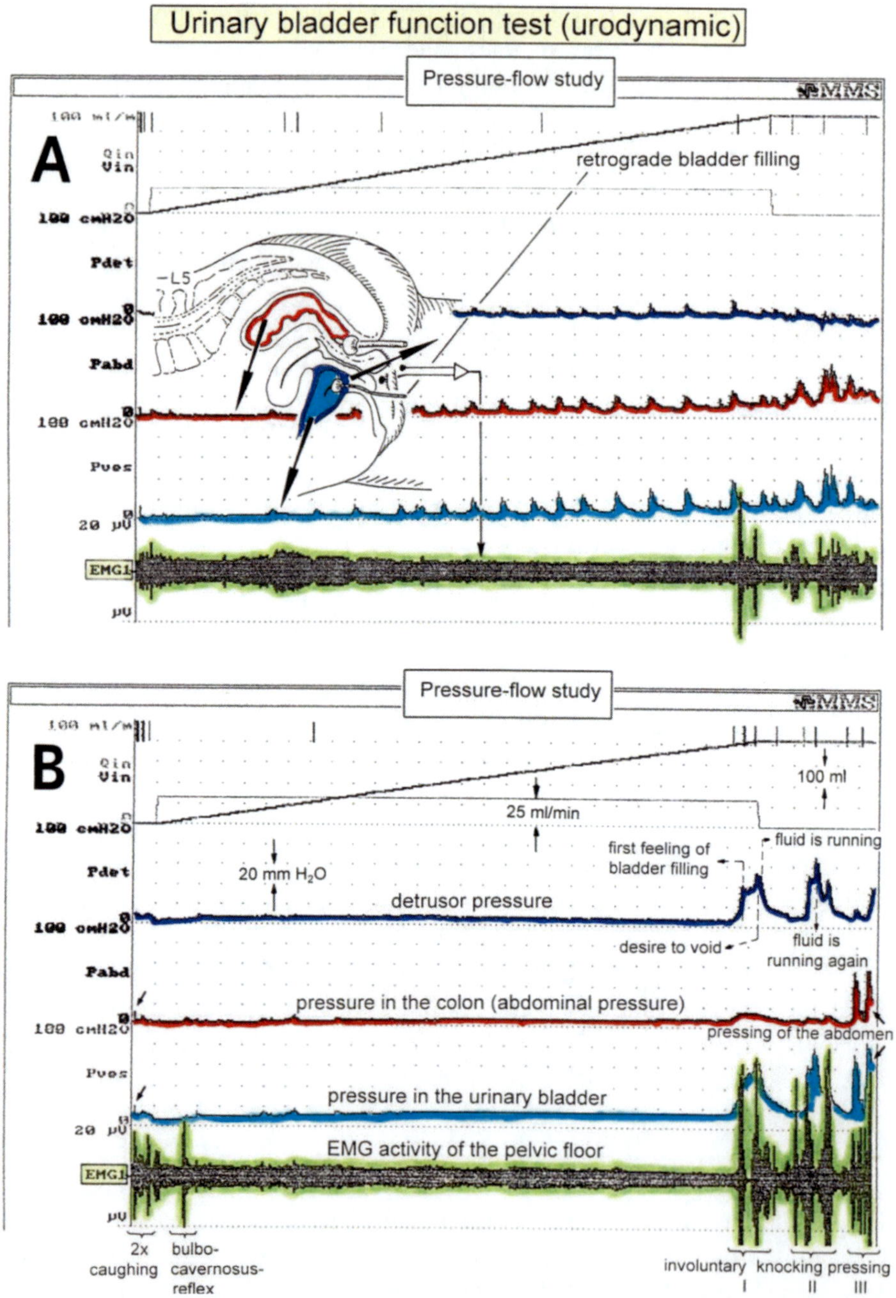

Figure 2. Improvement of the urinary bladder functions mainly due to therapy, quantified by urodynamics in a 30-year-old female patient. A. Three months after the accident in the mountains with resulting paraplegia sub Th12 following spinal cord lesion. B. 12 months after the accident (lesion level lowered to sub L3). In A, the detrusor pressure (P_{det}) is generated by the contracture of the bladder wall, as the pressure difference between abdominal pressure (P_{abd}, measured in the colon) and the bladder pressure (P_{ves}, measured in the bladder). Electromyographic recording obtained with surface electrodes from the sphincters and the pelvic floor (EMG) is shown; the external sphincters and the functionally correlated pelvic floor muscles show similar EMG activity (the rhythmic pressure peaks in A do not originate in the bladder). In A, the detrusor shows nearly no activity with retrograde bladder filling at 25 ml/min; in B, the detrusor shows first activity at 360 ml bladder filling. A detrusor-sphincteric-dyssynergy occurs, because the detrusor pressure peaks occur at the same time as the sphincter EMG activity peaks (B) (bladder and sphincter contract at the same time, so

that fluid can only emerge from the bladder at high bladder pressure; there is a danger of reflux through the ureter into the kidneys). The EMG peaks are a bit irregular, probably because the fluid, leaving the bladder, shunts transiently the EMG electrodes. Exact functional description of B: 2x coughing (B below) increases the EMG activity and passively the pressure in the abdomen and in the bladder (marked by the small arrows, physiologic). The bulbocavernosus reflex (induced by pressure applied to the clitoris) increased the EMG activity of the sphincters (physiologic). Conclusion: The reflex arch is in order; sacral nerve roots and nerves have not been damaged in the accident. I (bottom right): The patient feels an increase of unvolitional detrusor pressure (first feeling of bladder pressure at 360 ml). She tries to contract the sphincters to stop the bladder emptying. Shortly after the desire to empty the bladder, as the detrusor pressure decreases, fluid is leaving the bladder. II: Due to tapping onto the bladder, the bladder reflex is activated (detrusor activated, nearly no abdominal pressure); fluid is leaving the bladder. III: Due to the abdominal muscular pressure, the pressure in the abdomen increases as does passively the pressure in the bladder (the detrusor is not activated); fluid left the bladder. With a delay, the detrusor was activated by the bladder reflex. - The urinary bladder of the patient is partly functioning. It has to be further improved by therapy induced reorganization of the CNS: (1) An earlier feeling of bladder filling, (2) an increase of the time difference between the feeling of the first bladder filling and the unvolitionally emptying of the bladder (for the time being, approx. 10 min, in dependence on whether the patient is physically active (such as walking) or not, (3) further learning how to activate the detrusor on volition, and (4) the physiologic coordination between the bladder and the external sphincter functioning (to stop the detrusor-sphincter-dyssynergy). In cooperation with U. Bersch, Urology, SPZ Nottwil.

4.4.2. Limitation of Urodynamics and Need for Human Electrophysiology for Causal Repair

The evaluation of bladder functioning by means of urodynamics gave the information that the patient had a dyssynergy of the bladder. Repeated testing informed about changes in bladder functioning. But such bladder function tests are giving no information on the pathology of the CNS organization and how to repair urinary bladder functioning causally.

By performing similar bladder tests under operational conditions and recording single-nerve fibre action potentials from sacral nerve roots (Figure 3), the pathology of CNS organization can be explored in patients with spinal cord injury [40-42] and ideas can be found how to repair the injured CNS causally. Since humans have a high capability for learning, it may be possible to repair the injured networks by learning. But we have to "tell" the CNS what it has to relearn for repair. By comparing bladder and motor functions at the neuron level, it may be understandable why learning transfer from movements to bladder functions is possible.

4.4.3. Bladder Functioning At the Neuron level

With the single-nerve fibre action potential recording method [12], it has so far been possible to record single-nerve fibre action potentials from nerve fibres down to a diameter of approximately 3.5 μm in undissected thin long nerve root fascicles. It is therefore possible to record natural impulse patterns from parasympathetic efferents (par), urinary bladder stretch and tension receptor afferents (S1, ST), mucosa afferents from mechanoreceptors of the bladder, the urethra and the anal canal (M), from afferents responding to fluid movement (S2), and from α_2, α_3 and γ-motoneurons and muscle spindle afferents innervating the external striated urinary bladder and anal sphincters (or functionally associated pelvic floor muscles), and to analyze regulatory and organizational mechanisms of parasympathetic neurons and motoneurons in the human CNS.

During a laminectomy, natural impulse traffic to and from the CNS can be recorded from sacral nerve roots (Figure 3A-C). The summed impulse traffic can be split into the natural impulse patterns of single afferent and efferent fibres (Figure 3D).

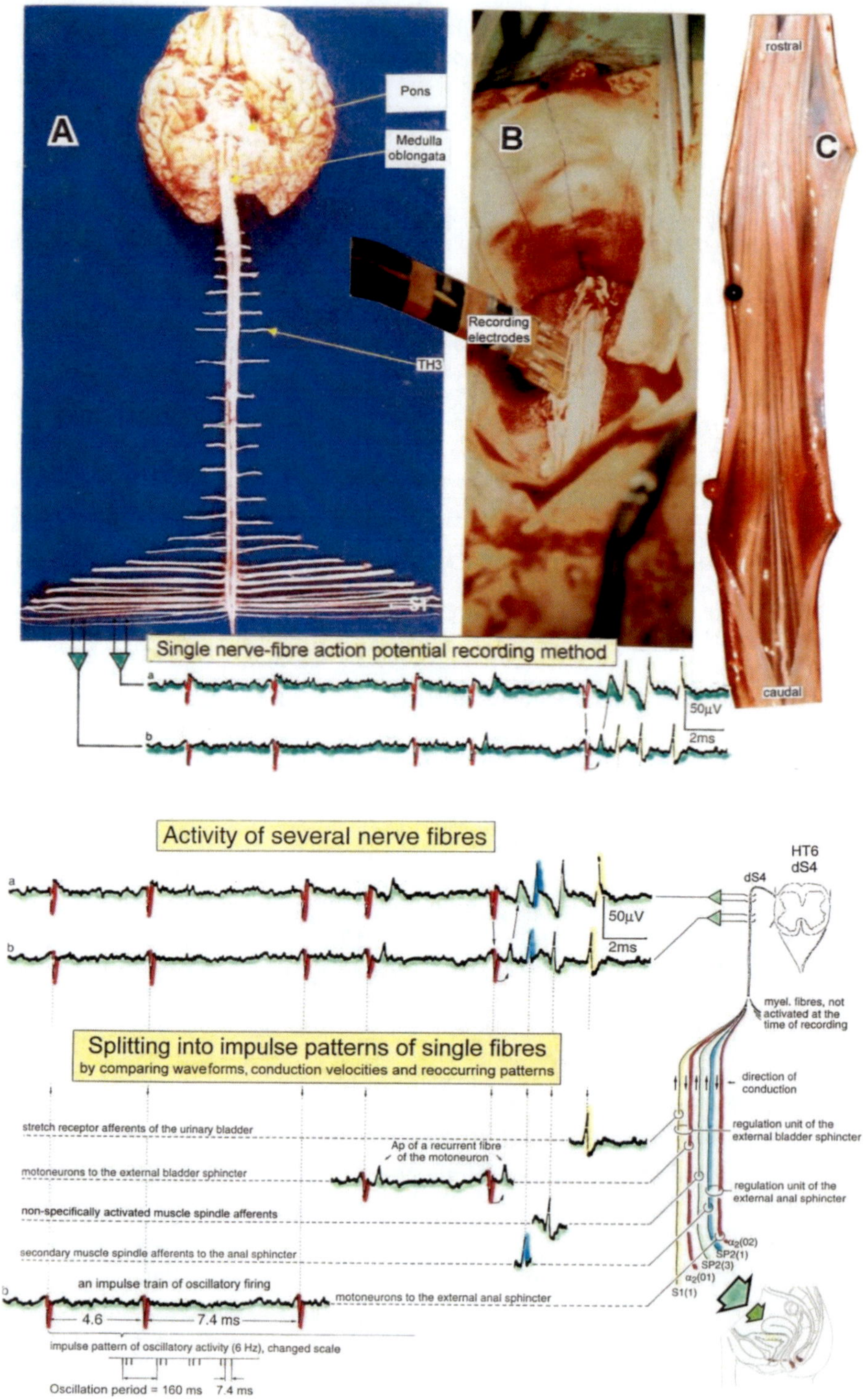

Figure 3. Layout of the recording of single-nerve fibre action potentials to analyze the self-organization of neuronal networks of the human CNS under physiologic and pathophysiologic conditions. A,B,C. By recording with two pairs of platinum wire electrodes (B) from sacral nerve roots (cauda equina, C) containing between 200 and 500 myelinated nerve fibres, records were obtained in which single-nerve fibre action

potentials (APs) were identified from motoneurons (main AP phase downwards) and afferents (main AP phase upwards). By measuring the conduction times and with the known electrode pair distance (10 mm), conduction velocity distribution histograms were constructed in which the myelinated nerve fibre groups larger than 4μm could be characterized by group conduction velocity values. After the recording, morphometry was performed. Distributions of nerve fibre diameters were constructed. and nerve fibre groups were characterized by the peak values of asymmetrical distributions. By correlating the peak values of the velocity distributions with those of the diameter distributions obtained for the same root, a classification scheme was constructed of the human peripheral nervous system. It became thus possible to record natural impulse patterns simultaneously from identified single afferent and efferent nerve fibres and to analyze self-organizing mechanisms of the human CNS under physiologic and pathologic conditions. A. Human CNS with the schematic illustration of the recording layout and original record of single-nerve fibre action potentials. Note the time calibration of 2ms. B. Intraoperative recording layout (when implanting a bladder stimulator) with two pairs of wire electrodes and one temperature sensor. A thin nerve root is positioned over the platinum wire electrodes. C. Dissection of the human cauda equina; at the caudal end of the filum. terminale with thin nerve roots are seen. D. Schematic splitting of the activity of several nerve fibres into simultaneous impulse patterns of single fibres by comparing waveforms, conduction velocities and reoccurring characteristic impulse patterns (rhythmic firing of sphincteric motoneurons). The different conduction times and waveforms were recognized on an expanded time scale. Stretch receptor and secondary muscle spindle afferents contribute to the drive of sphincteric motoneurons and form, together with other afferents, regulation units.

Upon retrograde bladder filling (as in the clinical diagnostic, Figure 2) and the identification of the neuron type, with the use of the classification scheme for human nerve fibres (partly shown in Figure 4G), the natural impulse patterns of identified afferent and efferent fibres can be obtained (Figure 3D) and analyzed. The obtained natural impulse patterns answered an old question: Is it the firing rate of a neuron that codes the information transmitted and processed or does the precise timing of cell discharge code information? Figure 4B,D shows that the information is coded by specific impulse patterns, including the precise timing and the firing rate.

In response to retrograde bladder filling of a brain-dead human, the self-organization of a premotor spinal α_2-oscillator innervating the external striated urinary bladder sphincter is shown in Figure 4D. Because the motoneuron axon of O1 had a recurrent fibre (Figure 4A) at the recording site, each single AP of this motoneuron could be identified safely by the AP of the recurrent fibre. The function of the motoneuron was to secure bladder continence. The activity from urinary bladder receptors, i.e.. the activity of bladder stretch (S1), tension (ST) and flow receptor afferents (S2) (Figure 4E), was an adequate afferent input to the motoneuron. Phase relations between the firing patterns of bladder afferent fibre S1(1) and oscillatory firing urinary bladder sphincteric motoneuron O1 can be seen in the schematized firing patterns in Figure 4B. They were an indicator that the natural impulse pattern of the S1(1) fibre was an adequate drive of the sphincteric motoneuron O1. For retrograde bladder filling up to 550 ml, motoneuron O1 only fired occasionally (Figure 4D,F). This was the storage phase of the bladder, during which the intravesical bladder pressure increased only little. For higher bladder filling volumes, the motoneuron switched via the transient oscillatory firing mode to the continuous oscillatory firing mode (Figure 4D) to generate a higher activity (Figure 4F) for a stronger drive of the urethral sphincter to more strongly secure continence when the storage phase was nearly passed and the bladder pressure increased. For bladder filling volumes higher than 800 ml, the activity of the spinal oscillator decreased again; probably, the oscillator became inhibited (Figure 4D). Pain fibres (not shown in Figure 4G) may have inhibited the oscillatory firing to protect the bladder from mechanical damage. The overflow mechanism was initiated. Probably, fluid entered the trigonum vesicae to activate flow receptors, so that the flow receptor activity (S2) increased strongly (Figure 4E).

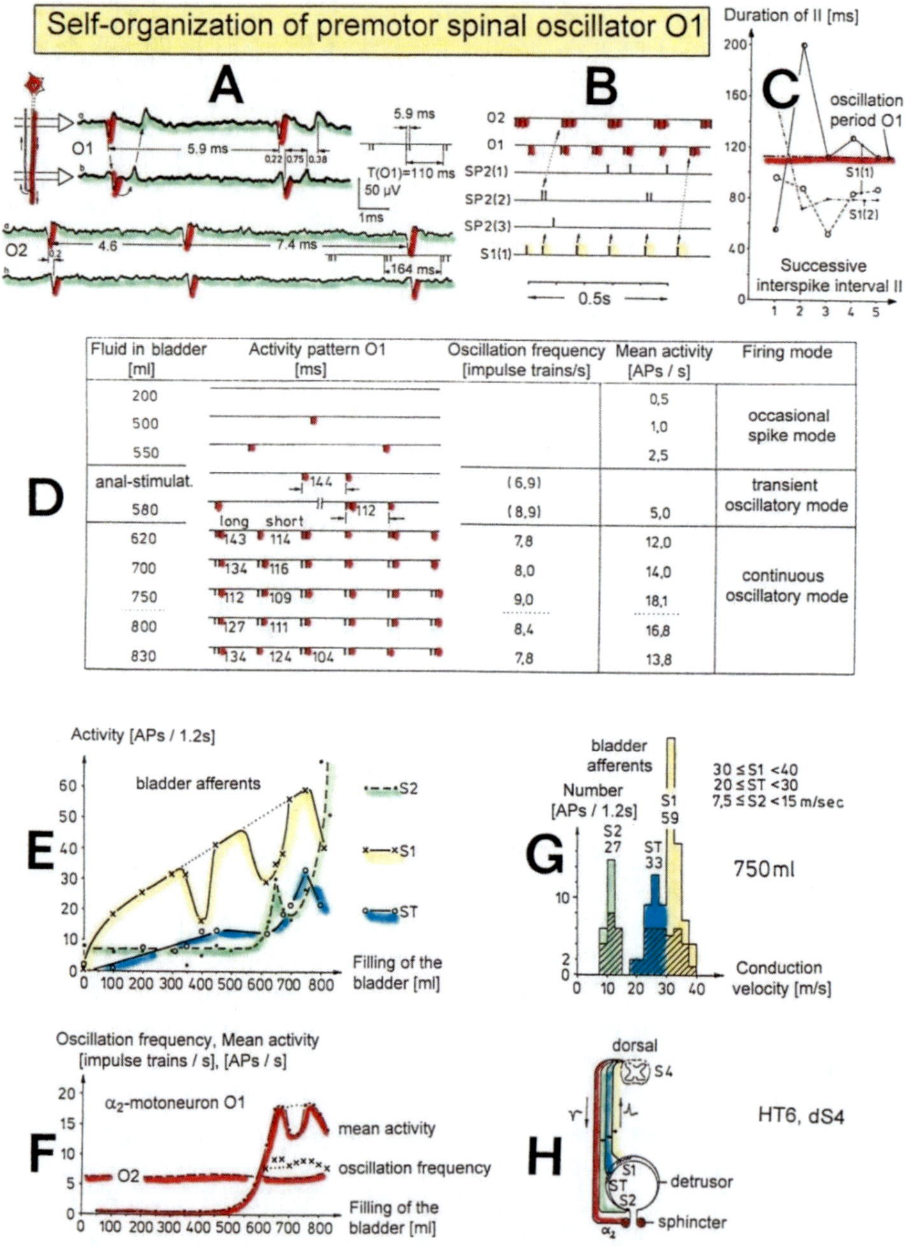

Figure 4. Self-organization of premotor spinal α_2-oscillator O1, which innervates the external urinary bladder sphincter (skeletal muscle). Brain-dead human HT6; recording from a dorsal S4 nerve root. A. Recordings from α_2-motoneurons O_1 and O_2, firing in the oscillatory mode with impulse trains of two (upper recording) and three (lower recording) action potentials (APs). The durations of the oscillation periods were 110 (O_1) and 164 ms (O_2). The interspike intervals of the impulse trains were 5.9 ms (O_1) and 4.6 and 7.4 ms (O_2). Motoneuron O_1 conducted at 36 m/s; its recurrent fibre conducted at 21 m/s. The measurement layout is shown schematically. The inserts show the oscillatory firing modes; they have not been drawn to scale. B. Impulse patterns of oscillatory firing α_2-motoneuron O_2 innervating the external anal sphincter, in relation to

the muscle spindle afferent activity SP2(1 to 3), activated by the stretch of the anal sphincter by the anal catheter, and impulse patterns of oscillatory firing α_2-motoneuron O1 innervating the external urethral sphincter, in relation to the stretch receptor afferent activity (S1(1)) of the urinary bladder, activated by 750 ml bladder filling. Phase relations between APs of SP2(2) and O_2 and between APs of S1(1) and O_1 are indicated by the small arrows. C. Three series of successive interspike intervals of the two stretch receptor afferent fibres S1(1) and S1(2) activated by retrograde bladder filling. The oscillation period of oscillatory firing motoneuron O1, activated only by bladder filling. is shown. D. The firing in the occasional spike mode, the transient and the constant oscillatory firing mode of α_2-motoneuron O_1 in response to filling of the bladder. In the "activity pattern" column. changing durations of oscillation periods are given. The oscillation frequencies in the brackets give the frequencies at the moment of oscillation for the transient oscillatory mode. Downward deflections are schematized APs. Interspike intervals of the close APs ≈ 6.0 ms (A). E. Activity levels of stretch (S1) and tension (ST) and flow receptor afferents (S2) (E) and of sphincteric α_2-motoneuron O_1 (F) in response to retrograde filling of the bladder. The activity values of the S1, ST and S2 afferents are taken from histograms like the one in G. Filling of the bladder was stopped once between 600 and 650 ml. F. The small dotted lines represent mean activity (APs/s) and oscillation frequency (impulse trains/s) of α_2-motoneuron O_1 if bladder filling were not stopped in between. Note that the mean activity increases continuously with the filling of the bladder from 550 to 650 ml, even though motoneuron O_1 started to fire in the oscillatory mode from 620 ml on (D). Note further that the oscillatory firing motoneuron O_2 (frequency of firing with impulse trains is shown) is nearly not affected by the filling of the bladder and by the start of the oscillatory firing of motoneuron O_1. G. Conduction velocity frequency distribution histogram of stretch, tension and flow receptor afferent activity at 750 ml. The activities of afferents S1, ST and S2 are quantified by counting the afferent conduction velocities under the peaks (open plus hatched part), with the conduction velocity limits given in the insert. The counts (27, 33, 59) are given below the peak labelled S1, ST and S2 and plotted into E for the afferent activity at 750 ml. H. Schematic drawing of the anatomical arrangement of the afferents and the motoneuron O_1.

The premotor spinal oscillator O1, of which motoneuron O1 is probably a part [16], was organized by the adequate afferent input (S1, ST, S2, ...) induced by bladder filling. For the oscillator to be formed in the premotor neuronal network in the spinal cord, consisting of motoneurons and interneurons, needs a certain preformation of the networks (connectivity, synapse efficacies, membrane properties of neurites, ...) and adequate afferent input patterns. In modelling such networks, the organization of the premotor spinal oscillators in the spinal cord neuronal networks cannot be separated from its space-time distributed adequate afferent input patterns, giving rise to self-organization. From Figure 4B. it can be seen that the stretch receptor afferent APs (S1(1)) show a relative phase correlation to the impulse trains of the oscillatory firing motoneuron O1. The successive interspike intervals (IIs) of the S1(1) afferents have, on the average, a value similar to that of the oscillation period of oscillator O1 (Figure 4C).

Other premotor spinal oscillators, activating the external striated bladder sphincter not shown here, will also have self-organized themselves by the same or similar afferent input. However, these oscillators subserving the same function will be correlated in their firing in the way that they do not fire in synchrony but distributed in their phases with respect to one another (probably by relative inhibition with respect to each other) to secure that the sphincter muscle does not show rhythmic movements (tremor). Such non-synchrony correlated oscillatory firing of several oscillators has in a small time window been measured [16]. In the case of a non-injured CNS, this premotor spinal oscillator O1 could also be activated volitionally from supraspinal centres. It would be very interesting to see how the afferent input patterns from supraspinal centres to the spinal premotor network would look.

By recording from the dorsal S4 root of the brain-dead human HT6, impulse trains from another oscillatory firing motoneuron and its driving afferents were measured, which served quite a different function (Figure 4F). The α_2-oscillator O2 (Figure 4A,B) innervating the striated external anal sphincter was activated by secondary muscle spindle afferent activity, induced by the anal catheter-stretched muscle spindles, probably located in the anal sphincter or functionally associated pelvic floor muscles. Also, mucosal and skin receptors within the anal reflex area will induce self-organization of premotor oscillators activating the external anal sphincter to secure anal continence. It is evident from the impulse patterns shown schematically in Figure 4B that the impulse trains of this oscillator O2 show a phase relation and an interspike interval relation to its driving spindle afferent APs. But no synchronized firing can be seen between the sphincteric α_2-motoneurons, innervating the external bladder sphincter (O1) and anal sphincter (O2), probably the CNS networks tried to avoid an increased physiologic tremor (macroscopic rhythmic activation of muscles, due to suboptimal regulation). Important for the application of human neurophysiology to neurotherapy is the duality of the functions of the sphincteric motoneurons and secondary muscle spindle afferents, subserving somatosensory and autonomic (parasympathetic) functions. In animals. it was found that also sympathetic fibres innervate muscle spindles [10]. It could therefore be (and seems to be so) that also parasympathetic fibres innervate muscle spindles in the parasympathetic innervation area (S2-S5).

Motoneurons innervating the external sphincters of the bladder and the anal canal subserve somatic functions (contraction of the sphincters on volition or for protection reaction) and parasympathetic functions for the coordination of the detrusor function (parasympathetic) and the external sphincter function. This duality of the sphincteric motoneurons (and spindle afferents in the parasympathetic domain) makes the pattern change from continence to protection reaction understandable and makes learning transfer between somatic and parasympathetic patterns likely. The motoneurons build up two phase relations per oscillation cycle with other motoneurons and secondary spindle afferents for somatic activation (Figures 3,7 of [18]), and build up three phase relations per oscillation cycle when also the parasympathetic division is activated (Figures 4,7 of [18], see also below). The neuronal networks of the somatic and the parasympathetic nervous systems are interlaced and interact with each other. It should therefore be possible to improve parasympathetic functions when improving somatic functions by coordination dynamics therapy, especially as there is indication that also parasympathetic efferents fire rhythmically [41].

The repair of the urinary bladder functioning by reorganization of networks seems to be most difficult because there is false neuronal network organization in the parasympathetic nervous system (overactive (spastic) detrusor) and the somatic division (spastic external (striated) bladder sphincter), and there is false interaction of the interlacing somatic and parasympathetic networks (detrusor-sphincteric dyssynergy: when the detrusor contracts, the external sphincter is also contracting instead of opening (relaxing)) [40-42].

But how can continence pattern organizations be stimulated by movements (jumping on springboard) via bladder afferents?

4.4.4. Movement Stimulation from Natural Afferent Impulse Patterns from the Skin (and Bladder)

When stimulating the skin (Figure 5G,H) under the same recording conditions as for bladder afferents and efferents (Figures 3,4), the afferent input from the skin can nicely be recorded (Figure 5A-E). By waveform comparisons on a stretched time scale (Figure 5B-D), sets of natural impulse patterns of different skin mechanoreceptors can be recorded (Figure 5Eb,F). The activity patterns (Figure 5A) show a touching and a releasing part of activity as can be seen from the action potential appearance of large amplitude units (T1(Parcinian Corpuscle) and T2) and from the schematic activity pattern of the $T1_1$ unit (Figures 5Eb,F). All the 4 touch (T1 until T4) and 1 pain groups, recorded here, seem to have their own touching and releasing part, but according to their latencies and conduction velocities, at different times. It can be seen from Figure 5 that the touching activity part of the slower conducting T3 and T4 fibres with their smaller action potential amplitudes (Figures 5A,C) started about in the middle of the T1 and T2 activity part and lay mainly between the touching and the releasing parts of the T1 and T2 units. The releasing activity of the T3 and T4 fibres appeared after that of the T1 fibres. The activity of the pain fibres (here the slowest conducting fibres) appeared even later at the recording electrodes and stopped appearing after the end of the pinpricking (recorded at the proximal cauda equina, Figure 3).

The very late pain after pinpricking that one normally feels is also probably conducted by very thin myelinated or non-myelinated fibres and cannot be recorded under such conditions.

The different delays and impulse train lengths of the T1 units are interpreted with the spatial distribution of the T1 receptors and the skin indentation upon touch and pinpricking (Figure 5H). The $T1_1$ unit, for example, had its receptor nearest to the needle and responded fastest and longest since the skin indentation first reached the $T1_1$ receptor, gave rise to the largest skin shift and turned away last. The $T1_2$ receptor, being further away from the centre of skin touch, was reached later by the skin indentation, the skin shift was not so large and the skin turned back to the former position more quickly. The action potentials in the $T1_2$ unit therefore appeared later, showed only two (extracellular) action potentials in the impulse train and the release activity part also appeared earlier.

Such natural impulse patterns in many skin afferents inform the CNS about changes in the periphery of the body and give rise to certain network organizations. The neuronal networks of the spinal cord (and supraspinal centres) do self-organize by the natural impulse patterns from the receptors in the periphery (in spinal cord injury) and from supraspinal centres (physiologically both contributions). It is difficult to say whether the networks adapted their self-organization to the impulse patterns of the receptors of the periphery or whether, during phylogeny or ontogeny, the receptors adapted their firing to the needs of the neuronal network. Nevertheless, it is obvious that the natural firing patterns of the afferents and their space-time distributions give rise to the self-organization of the neuronal networks and are therefore of utmost importance for the understanding of the functioning of the CNS. Such parameters are difficult to measure.

To make a conclusion on the organization and reorganization of neuronal networks from the wiring of the neurons (see in [1]) is risky, because one would need to know, for example in humans, the efficacies of the excitatory and inhibitory synapses (LTP, LTD), the membrane properties of local critical areas of neurites, and how membrane properties change with learning and reorganization following CNS injury. Two of the original recordings of simultaneous skin afferent patterns of 14 and 22 receptors are shown in Figure 5. The summed natural firing patterns upon touch (Figure 5Ea) are split and are schematically drawn in Figure 5Eb, and those upon pinprick (Figure 5A-D) can be seen in Figure 5F [12,62]. Electrostimulation of nerves in patients cannot mimic such spatio-temporal impulse codes elicited by natural stimulation and will give rise to non-natural neuronal network organization of the CNS.

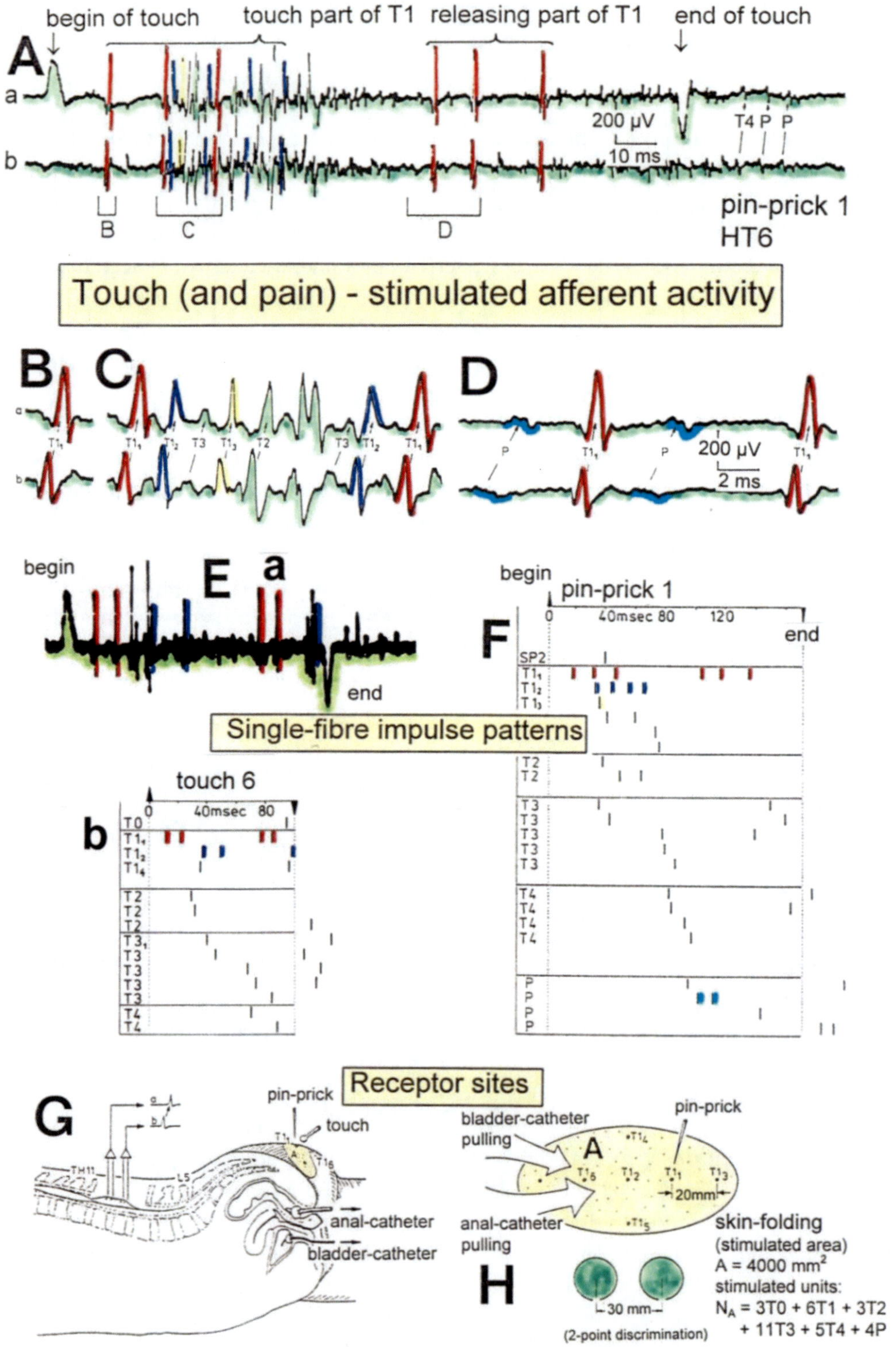

Figure 5. Touch (and pain)-stimulated afferent activity. Touch and pain activity stimulated by pinpricking (A) and touching (Ea) at S5 or Co dermatomes and recorded extracellularly from a dorsal coccygeal root (brain-dead human HT6). T1, T2, T3, T4, P = mark action potentials (APs) from single touch and pain fibres.

Subscripts 1, 2, 3 mark single fibres. A. Whole sweep following pinprick 1 shown at a slow time base. The large upward artifact on trace "a" marks electronically the beginning of the pinprick. The large downward artifact on trace "a" marks the end of the pinprick. Note that two intervals of high activity of large APs occur, one after the beginning of the pinprick with one AP in front, and a second before the end of the pinprick; potentials with small amplitude follow potentials of large amplitude. Time intervals B, C and D are shown in a time-expanded form in Figures B, C and D. B, C, D. Time expanded sweep pieces of A. Identified APs are indicated. Note that the APs from the $T1_1$ touch unit can be safely identified by the waveforms in B, C, D. Eb, F. AP occurrence patterns of single touch and pain fibres following short touch 6 and pinprick 1. No pain afferents are stimulated upon touch 6. Upon pinprick 1, the single-fibre AP activity of the different touch and pain groups is identified by the AP waveforms on traces "a" and "b," and by the conduction times. The single touch afferents of the T1 group are marked with subscripts. One active secondary muscle spindle afferent fibre (SP2) could always be identified in F. Note that for pinprick 1, touch and pain afferents are stimulated whereas for touch 6 only touch afferents. G. Recording and stimulation arrangement for simultaneous recording of several single touch and pain units. A = area stimulated by skin folding, drawn in H in more detail. $T1_1$, $T1_6$ = suggested touch points of the $T1_1$ and $T1_6$-units. H. Drawing of the very approximate skin area stimulated by skin folding. $T1_{1\text{-}6}$ = suggested focal T1 touch points. Two-point discrimination indicated for the sake of comparison. N_A = number of stimulated units in the dorsal coccygeal root. Skin tractions evoked by anal and bladder-catheter pulling are indicated by the large open arrows.

When performing electrostimulation in patients, it should be therefore clear what network states are about to be activated in the CNS and how the artificial patterns differ from the natural pattern states. In basic research, it should also be borne in mind that artificial electrostimulation patterns give rise to artificial neuronal network organization of the CNS. Electrostimulation is therefore not very suitable for analysis of natural integrative functions of the CNS. Below, I shall discuss how the spatio-temporal touch afferent patterns (like touch 6 in Figure 5Eb) give rise to a self-organized network state subserving anal continence, whereas additional activated pain afferents (like pinpricking 1 in Figure 5F) stimulate a network state inducing an escape and protection response.

Due to catheters positioned in the anal canal and the urethra (approximately isometric stretch) (Figure 5G), α and γ-motoneurons (partly) fired oscillatory (Figure 3) as a result of the sustained stretch reflex of the external sphincters (not clear for the urethra) as a part of the continence pattern. These impulse patterns were more distributed and only seldom added up with the afferent input from the mechanoreceptors of the skin (schematically pictured in Figure 6B, SP2(2)-trace), conducted in different fibres but through the same nerve root.

4.4.5. Natural Firing Patterns of Proprioceptive Afferents and α and γ-Motoneurons Measured Simultaneously, and the Phase (and Frequency) Relations between Them

Original recordings of the firing of α-motoneurons (oscillatory), proprioceptive and bladder afferents are shown in Figure 3. Due to catheters positioned in the anal canal and urethra, continence pattern states are activated. These continence states of the CNS are secured by the ongoing communication with the bladder and anal canal via the fibres in the S2-S5 nerve roots. Touch, pinprick, anal and bladder catheter pulling, and filling of the continence organs will change the natural impulse traffic in those fibres.

Figure 6 shows such schematic natural simultaneous impulse patterns of a static and a dynamic γ-motoneuron, two secondary muscle spindle afferents and an oscillatory firing α_2-motoneuron O2 in a dorsal S4 nerve root (there are afferents and efferents in dorsal or ventral lower sacral roots) during continence pattern changes. The small arrows and the dotted lines indicate existing relative phase coordinations between the static and the dynamic γ-motoneurons and motoneuron O2, and between γ-motoneurons and secondary muscle spindle afferents. The dashed-circle line indicates a phase relation between the APs of the static γ-motoneuron (γ_1) and the cross-correlation between SP2(2)-fibre (single ending one of the mother fibre) and SP2(5)-muscle spindle afferent fibre.

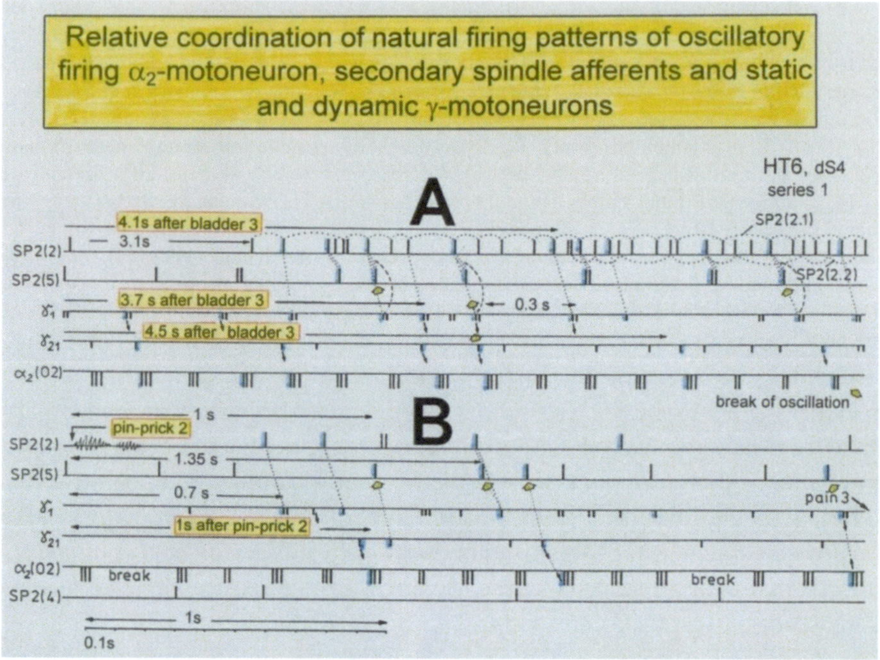

Figure 6. Impulse patterns of simultaneously recorded γ-motoneurons (γ_1 and γ_{21}), secondary spindle afferent fibres (SP2(2), SP2(4), SP2(5)) and oscillatory firing α_2-motoneuron O2 following bladder catheter pulling (bladder 3) (A) and pinprick 2 (B). B was recorded before A. In A, the impulse patterns of the two encoding sites SP2(2.1) and SP2(2.2) of the single parent fibre SP2(2) are indicated by the dotted curves. Times to the activity increases of γ-motoneurons and secondary spindle afferents following stimulation are indicated. Similar time intervals of the occurrence of γ-motoneuron APs and SP2(5) fibre APs (phase coordination) are indicated by the open arrows, and the similar time intervals of γ-motoneuron APs and α-motoneuron APs are indicated by small arrows. Similar time intervals of the APs of fibres SP2(2) and SP2(5) are indicated by the double dotted lines, those of γ_1-APs and the SP2(2) fibre APs by a dotted line, and those of γ_1-APs and the SP2(2)-SP2(5) correlation by a curved dashed line. HT6; dS4.

Including the phase relations between the firings of secondary muscle spindle afferents and the oscillatory firing motoneuron O2, we obtain interlaced loops of coordinations between the firings of γ-motoneurons and secondary muscle spindle afferents, and between secondary spindle afferents and α-motoneurons and between α-motoneurons and γ-motoneurons (co-activity of α and γ-motoneurons) (Figure 13). It becomes obvious from the correlations between the natural impulse patterns (including those of single encoding sites of spindle afferents) that the γ-loop is not a single loop, but a network of loops, because of the divergent projections of γ-motoneurons onto muscle spindles and the probably divergent and convergent projections of secondary muscle spindle afferents into the neuronal network of the spinal cord, consisting of α and γ-motoneurons and interneurons.

More general, phase and frequency coordination can be seen among the natural firing patterns in the afferent and efferent fibres; this means that upstream in the CNS, there should also be phase and frequency coordination among neuron firing. More precise measurements will be presented below. Two phase relations have been observed to occur mostly between the APs of the secondary muscle spindle afferents and the oscillatory firing motoneuron per one oscillation period (Figures 11,12,14) (for somatic activation) in accordance with the “in phase” and “anti-phase” jumping on springboard and crawling. With this coordinated natural impulse traffic to and from the periphery, the change of integrative pattern states can also partly be understood from bladder to movement states, as will be shown.

4.4.6. Response of an Oscillatory Firing α-Motoneuron to Touch and Painful Pinprick Stimulation (Neuronal Network Pattern Change)

I will show now that a small change in the natural impulse patterns travelling into the sacral micturition centre will change there the neuronal network pattern organization. Ten schematically drawn sweep pieces of simultaneous impulse patterns of α_2, α_3 and γ_1-motoneurons and the secondary muscle spindle afferent fibres SP2(1) and SP2(2) are shown in Figure 7. Following a sequence of ten touch stimulations (Figure 7, upper right insert), the rhythmically firing motoneurons responded by a change in the duration of the oscillation period. Of interest is here how fast the α_3-motoneuron reduced its oscillation period following the start of the touch stimulation (measured electrically) or following the volley of skin-afferent APs running through the ventral root skin afferent fibres. The delay of the first reduction of the oscillation period is defined in Figure 7 following touch 7. Beginning and end of the stimulations are marked. There is a possibility that the α_3-motoneuron (S) is actually an α_2-motoneuron (FR), and the α_2 is an α_3-motoneuron. Because of the pathologic oscillatory firing, the type of the motoneuron can no longer be safely identified from the firing pattern, and identification on the basis of conduction velocity is not safe as the distributions of conduction velocities of α_2 and α_3-motoneurons (and of other nerve fibre groups) overlap.

It can be seen in Figure 7 that the α_3-motoneuron mostly reduced its oscillation period after a delay (measured from the beginning of the afferent volley) shorter than the duration of the oscillation period. The first reduction of the oscillation period is marked by a small arrow. The response times are analyzed quantitatively in Figure 8. Following touches 6 and 7, a partial transient synchronization of the firings of afferents and efferents occurred, as can be seen by the similar occurrence times of the impulses. Also, a slight partial synchronization followed touches 2 and 3. Since on the average, the delay in the shortening of the oscillation period occurred after a time interval shorter than the oscillation period, it is of interest to see whether the response times were similar to those for pinpricking. It has been established in humans that the difference between touch and pinprick stimulation is that for pinpricking, additional pain fibres are activated (Figure 5), and pain fibre activity may facilitate different pathways to the motoneural networks.

The touch sites 1, 2, 6 and 7 were inside the anal reflex area and the sites 4,5,9 and 10 were outside of it. The anal reflex area extended approx. 6 cm laterally from the anus.

After touching experiments with the touch stimulation at pen-marked sites 1 to 10 close to the anus (Figure 7), the touching sites were now pinpricked in a similar fashion approx.

every 0.8s to 1s. After a delay, there again was a reduction in the duration of the oscillation periods.

The responses to pinpricking differed from those following touch stimulation in three aspects. Firstly, the response delay times were longer than the oscillation periods.

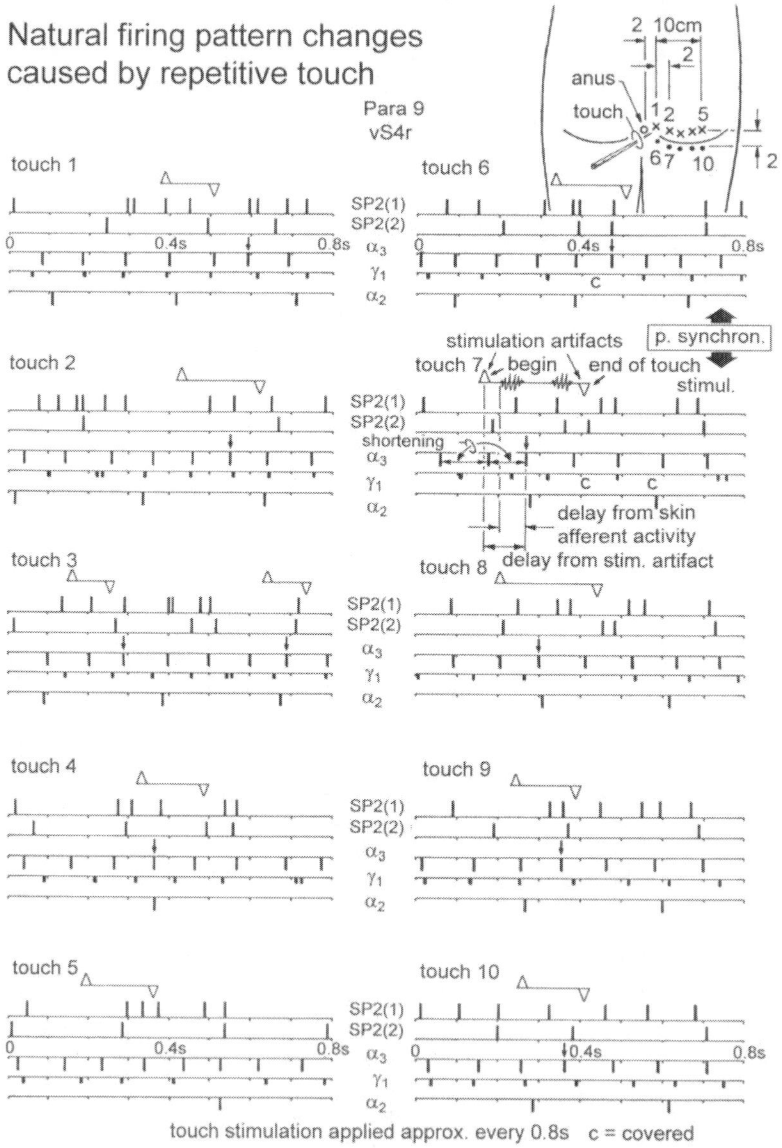

Figure 7. Changes of natural impulse patterns of α_2, α_3 and γ-motoneurons and secondary muscle spindle afferents (SP2(1) and SP2(2)) in paraplegic 9 upon touching sites 1 to 5 (x) and 6 to 10 (•) (insert). The small arrows indicate the first shortening of the oscillation period of the α_3-motoneuron following the beginning of the touch stimulation (marked by an artificially generated artifact). Note that partial synchronization (p. synchron.) of the firing of the α_2 and α_3-motoneurons and secondary muscle spindle afferents occurred upon touching sites 6,7. The α_2 and α_3-motoneurons were only identified by their conduction velocities; it is therefore possible that the α_3-motoneuron was actually an α_2-motoneuron and the α_2-motoneuron an α_3-motoneuron (overlap of velocity distributions); the firing patterns were very pathologic and could not be used for the motoneuron identification.

Secondly, the response to pain application had a longer duration and was stronger; mostly, the duration of two oscillation periods was reduced following pinprick stimulation and the reduction of the duration was larger. Following pinpricks 3, 7 and 8, a very short oscillation period occurred, which resembled a resetting of the oscillation cycle. Thirdly, the transient synchronizations of afferents and efferents following pinpricks 6 were more pronounced.

4.4.7. Latency Difference in the Reduction of the Duration of the Oscillation Period Following Touch and Pinprick Stimulation

It can be seen from Figure 8A that in nine out of ten cases, the latency of the oscillation period shortening following touch stimulation of the perianal skin (see insert in Figure 7) was shorter in duration than the oscillation period (≈100 ms), represented by the dotted line. The shortest latency was approx. 10 ms, when measuring from the afferent volley running in direction of the spinal cord. The latency of the first reduction of the oscillation period, measured from the skin touch stimulation artefact, was longer or shorter than the duration of the oscillation period (Figure 8B). The shortest latency was approx. 30 ms. The difference of the latency, i.e., the time from the stimulation artefact (touching of the skin) to the moment when the skin afferent volley passed the recording electrodes, was approx. 20 ms. According to the group conduction velocities of skin afferents (T1 skin afferents (PC) ≈ 44 m/s at 36 °C), most of the time was lost in the touch receptors and the thin nerve fibres connected to them. With a conduction distance of 0.3 m, a conduction time of t_{cond} = s/v = 300 mm/ 44 mm(ms)$^{-1}$ = 7 ms results. The latency until the reduction of the first oscillation period duration following pinpricking the sites 1 to 10 was mostly longer than the oscillation period itself, independent of whether measured from the skin afferent volley (Figure 8E) or from the stimulation artefact (Figure 8F). Since not every pinprick was strong enough to generate pain (stimulation of pain receptors), it is concluded that the latency from the pinpricking to the first reduction of the oscillation period was longer than the oscillation period. By comparing the latencies to the first reduction of the oscillation period following repetitive touch and pinprick stimulation of sites 1 to 10, it turns out that following touch stimulation the delay is shorter than the oscillation period, whereas it is longer following pinprick stimulation. It seems therefore that touching reinforced the sustained stretch reflex of the anal sphincter while pinprick did not.

The first shortening of the oscillation period in comparison to two preceding periods is plotted in Figure 8D,C (touch) and in Figure 8H,G (pinprick). The reduction of the oscillation period for touch stimulation was between 8 and 28 ms and between 5 and 40 ms for pinpricking. The pinprick stimulation generated a larger transient reduction of the oscillation period. When correlating the latency of shortening to the extent of shortening for pinpricking (Figure 8H), it seems as if the shorter latency correlated with a smaller reduction of the oscillation period (more touch-like), and the long latencies correlated with the greater reduction of the oscillation period (more pain-like). A similar relation was not found following touch stimulation (Figure 8D). It is most likely that the painful pinprick stimulation transiently replaced the sustained stretch reflex of the anal sphincter by a protective reaction of the anal sphincter against pain application; the time needed for the reorganization of the oscillatory firing neuronal network was at least one oscillation period long.

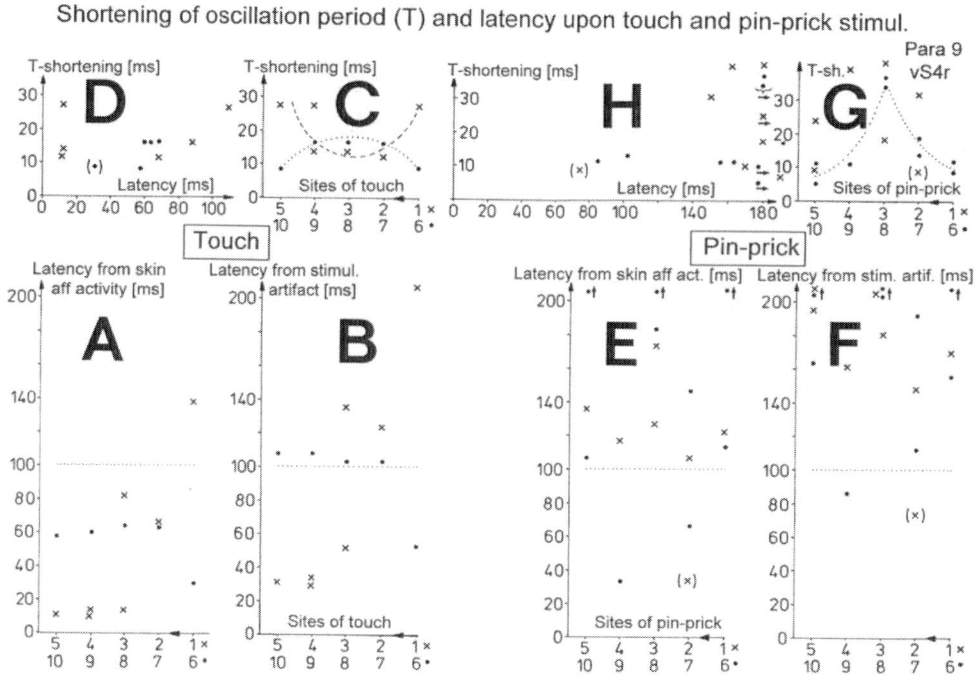

Figure 8. Shortening of the oscillation period T of the α_3-motoneuron (perhaps an α_2-motoneuron upon touch and pinprick stimulation). A, B. Latency of the shortening of the oscillation period (for definition, see Figure 3) measured from the skin afferent activity (A) and from the stimulation artifact (B) with successive touching of sites 1 to 5 (x) and 6 to 10 (•) as indicated in Figure 3. Note that the delay is often shorter than the oscillation period. D,C. Shortening of the oscillation period (T-shortening), with respect to two preceding oscillation periods, in relation to the latency of shortening (D) and in relation to successive touching of sites 1 to 10 (C). E.-H. Same description as for A-D, only with respect to pinpricking. Note that touch stimulation induced a shorter latency for the reduction of the oscillation period than did pinpricking.

The relation between the sites of pinprick (Figure 8G) or touch stimulation (Figure 8C) and the oscillation period shortening shows no clear correlation. The dotted and the dashed lines may show similar latencies. Perhaps the protective reaction against pinpricking of the anal canal and the pelvic floor is highest at sites 3 and 8 at the border of the anal reflex region.

4.4.8. Time Needed for the Change from the Attractor State "Continence" to the Attractor State "Protection"

By comparing the latencies from the moment when the skin afferent volley passed the recording electrodes to the first reduction of the oscillation period following repetitive touch and pinprick stimulation of the sites 1 to 10, it turns out that the delay following touch stimulation (≈ 10ms) is much shorter than for pinprick stimulation (≈ 110ms). It seems therefore that touching reinforced the sustained stretch reflex of the anal sphincter while pinprick stimulated the protection reaction against pinprick (pain). It can be evaluated how much time is needed for the reinforcement of the sustained stretch reflex and for the pattern change from the continence state (sustained stretch reflex) to the protection reaction. With a conduction distance of 0.1 m (distance between the recording electrodes and spinal cord), a conduction time for the fastest skin afferents (group conduction velocity of T1 skin afferents

(Parcinian Corpuscle) ≈ 44 m/s at 36 °C) of t_{cond} = s/v = 100 mm/ 44 mm(ms)$^{-1}$ = 2.3ms is resulting. The conduction time of the action potentials of the motoneurons to cover the distance from the spinal cord to the recording electrodes is resulting also to approximately 2.3ms (the group conduction velocities of the α_2 and α_3-motoneurons are 37 and 50m/s, respectively).

The processing time to reinforce the continence pattern of the anal sphincter was therefore approximately 5.4ms (10ms-4.6ms = 5.4ms). If we assume a synaptic transmission time of 2.5ms, the processing time in the spinal cord is sufficient for the transmission time of two synapses. The reinforcement of the sustained stretch reflex was may be just a di-synaptic pathway.

The conduction times to and from the spinal cord of the protection reaction are approximately 10ms (7.7 + 2.3), because the fastest pain afferent action potentials needed a bit more time than the touch afferent action potentials (group conduction velocity of the fastest pain afferents = 13m/s; t_{cond} = s/v = 100 mm/ 13 mm(ms)$^{-1}$ = 7.7ms). The processing time in the CNS to change from the sustained continence pattern to the protection reaction pattern is therefore approximately 100ms. This time would be enough to involve 40 synapse transmissions. For a reorganization of the neuronal networks for pattern change, conduction times of intraspinal axons and dendrites would also have to be taken into account. But it is obvious that the pain application induced a completely different CNS response.

That pinpricking resulted in a longer response time of motoneurons than touching of the perianal skin is not obvious and cannot be explained based on the reflex theory. Being a stronger stimulus, pinpricking (touch plus pain afferent activity, Figure 5) would be expected to have the same or slightly shorter rather than a much longer response time than touching of the perianal skin (touch afferent activity only, Figure 5).

It is concluded that the change from one network state to another one, restricted to the spinal cord (measurement in a paraplegic), needed approximately 100ms.

With respect to movement learning and learning transfer, strong pain has to be avoided during exercising, because a certain pattern (like jumping on springboard) has to be trained (and neuronal networks entrained) and not a mixture of a movement pattern and a protection reaction.

4.4.9. Attractor Layouts for Jumping on Springboard and Switching between Continence and Protection

For jumping on springboard (Figure 9Aa), the equation of motion of the collective variable φ (angle between the legs) could be solved in the HKB model (5) and an attractor layout constructed for the jumping in in-phase and in anti-phase (Figure 9Ab). In analogy, an attractor layout for bladder continence and protection reaction (Figure 9Bb) can be drawn, even though the anal continence was measured. The enhancement of the continence safety (deepening of the potential well) needed approximately 6ms, and the change from the pattern continence to protection reaction needed approximately 100ms.

It is therefore possible to analyze integrative patterns of the human CNS with natural impulse pattern of single-nerve fibres recorded from sacral nerve roots. Natural impulse pattern can explain integrative functions of the human CNS.

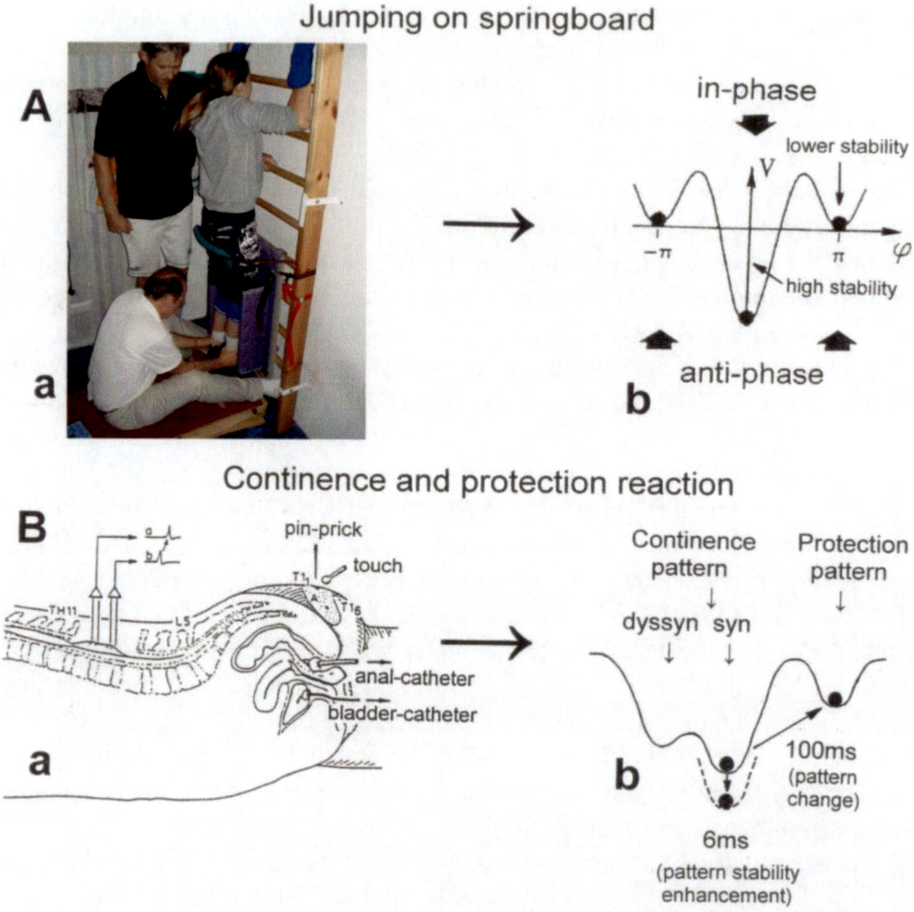

Figure 9. Attractor layout for jumping on springboard in comparison to that of continence and protection.

4.4.10. Relative Phase and Frequency Coordination between the APs of α and γ-Motoneurons and Secondary Muscle Spindle Afferents with No Additional Stimulation and upon Touch, Pinprick, and Bladder Catheter Pulling

Above, it was shown that natural impulse pattern can explain pattern change, which means the generation of integrative patterns can partly be understood at the single neuron level. Natural impulse patterns were related to integrative CNS organizations. How can this aid in understanding why jumping on springboard contributes substantially to the repair of bladder functioning? A better understanding can be achieved to go deeper into the complexity of the cooperative and competitive interplay among neurons that means going deeper into the complexity of phase and frequency coordination of CNS self-organization. Changes of natural firing patterns have roughly been shown in Figure 6. The phase and frequency coordination of coordinated firing will now be analyzed more in detail.

Time correlations of afferent and efferent impulse patterns are easy to detect in the brain-dead individual as the oscillatory firing of α_2-motoneuron (O2) was regularly like an inner clock (Figures 3,4; see below).

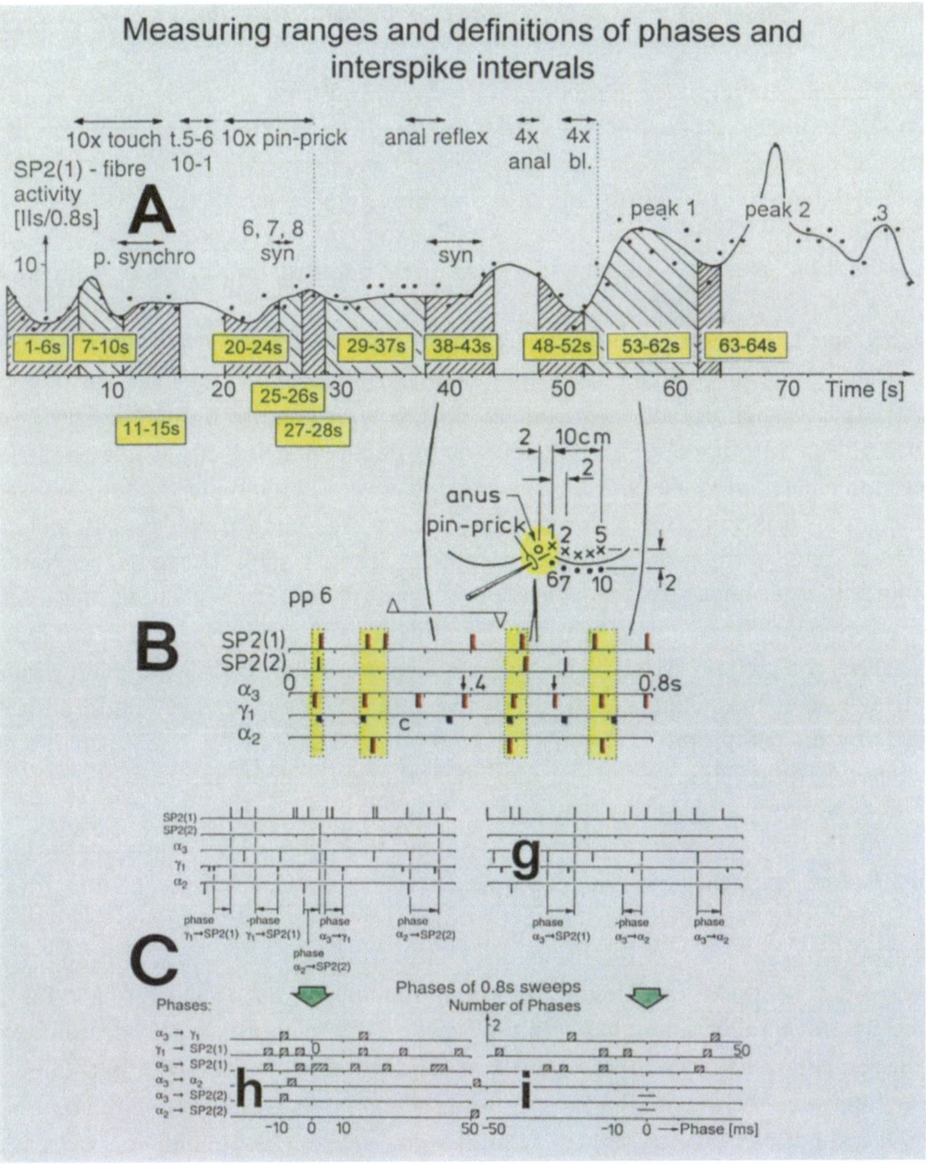

Figure 10. Measurement ranges and definitions of phases for the analysis of phase and frequency coordinations between motoneurons and spindle afferents. Coordination (synchronization) between firing patterns can directly be seen in B. (A) Activity level of secondary muscle spindle afferent fibre SP2(1) in dependence on time. 10x touch = touching sites 1 to 10 shown in B; t. 5-6 = touching alongside the skin from site 5 to site 6; 10x pinprick = pinpricking sites 1 to 10; anal reflex = anal reflex stimulation; 4x anal = fourfold anal catheter pulling; 4x bladder = fourfold bladder catheter pulling; peak 1- peak 2 - 3 = first, second and third peak of spindle afferent activity due to parasympathetic activation; p. synchro = partial synchronization; syn = synchronization of α and γ-motoneurons and secondary muscle spindle afferents. Note the synchronization of the firing patterns following pinprick 6 inside the anal reflex area. (B) A set of single impulse patterns of secondary muscle spindle afferents (SP2(1,2)) and α and γ (intrafusal)-motoneurons and sites of stimulation. The small arrows in the impulse pattern of α_3-motoneuron (S) point to a shortening of the oscillation period following pinprick 6 (pp6). The triangles indicate the beginning and the end of pinpricking. (C)(g,h,i) Definitions of the phases between the different motoneurons and spindle afferents in two sets of impulse patterns (g), and the corresponding sets of phase relation distributions (h,i). Para 9; vS4.

The phases of fusimotors and spindle afferent APs could be defined with respect to the impulses of that inner clock. In the paraplegic, the rhythmic firing was rather irregular. The motoneuron firing therefore cannot be used anymore as a time reference basis. More phases between the extracellular action potentials of the different fibres are necessary to fully describe the correlation between the simultaneous impulse patterns. In Figure 10Cg, the mutual phases between the APs of the different fibres are defined. Figure 10Ch,i shows the corresponding phase distribution histograms. Since too few phases occurred in a sweep piece of 0.8 s duration, phases occurring in certain time intervals were pooled and plotted in Figures. 11 and 12.

In this special pathologic case, the α_2 and α_3-motoneurons fired rhythmically with impulse trains consisting of one action potential (AP), in contrast to the physiologic firing patterns [14], in which α_2 and α_3-motoneurons fire with impulse trains consisting of more than one AP. The identification of motoneurons by conduction velocity is not absolutely safe since group conduction velocity ranges overlap. It is very unlikely, nevertheless, that one of the motoneurons was an α_1-motoneuron (FF) (missing high conduction velocity and high AP amplitude and no correlation to primary spindle afferent fibres), even though they fire physiologically with impulse trains consisting of one AP. It has been shown that oscillatory firing α_1-motoneurons are mainly driven by time locked primary spindle afferent fibres [24,27]. The firing patterns of the α_2 and α_3-motoneurons are strongly pathologic with respect to the length of the oscillation period [14,16] and the impulse train length so that it is impossible in this paraplegic to identify the kind of motoneurons by their discharge patterns of oscillatory firing; this would be possible were the neuronal network driving the motoneurons to fire in a physiologic manner.

In Figure 11, the interspike intervals (IIs) and the phases are shown for similar time intervals. Before stimulation, within the time interval 1-6 s, the α_3-motoneuron fired every 100 ms, the γ_1-motoneuron every 100 to 130 ms, and the SP2(1) fibre every 80 to 150 ms (Figure 11Aa).

The α_2-motoneuron mostly fired every 300 ms and the SP2(2) fibre every 250 ms. At that particular time interval, similar phases (phase relation of broad peak type) occurred twice per α_3-oscillation period between the APs of the α_3 and γ_1 axons, between the γ_1 and the SP2(1) fibres, and between the α_3 and the SP2(1) fibres (Figure 11Ba). One phase relation occurred between the impulses of the α_3 and α_2-motoneurons, and two between the α_3 and the SP2(2) fibres. The broad phase relations between discharge patterns are interpreted as interactions between populations of neurons. Following different stimulations, interspike interval (II) distributions and phase relations changed with time.

Upon touching sites 1 - 5, the IIs of the almost oscillatory firing γ_1-motoneuron reduced in size to be more similar to those of the oscillatory firing α_3-motoneuron (Figure 11Ab). The changing of the different phase relations indicates changes in the interactions between neuronal subnetworks (Figure 11Bb). Upon touching sites 6 to10 (especially sites 6, and 7 (inside anal reflex area)), the IIs of the almost oscillatory firing γ_1-motoneuron increased again (Figure 11Ac). A transient partial synchronization occurred between the different nerve fibres (Figures 11Bc, 14c). Upon pinpricking sites 1 - 5, the IIs of the almost oscillatory firing γ_1-motoneuron reduced again to have a similar II distribution as the α_3-motoneuron.

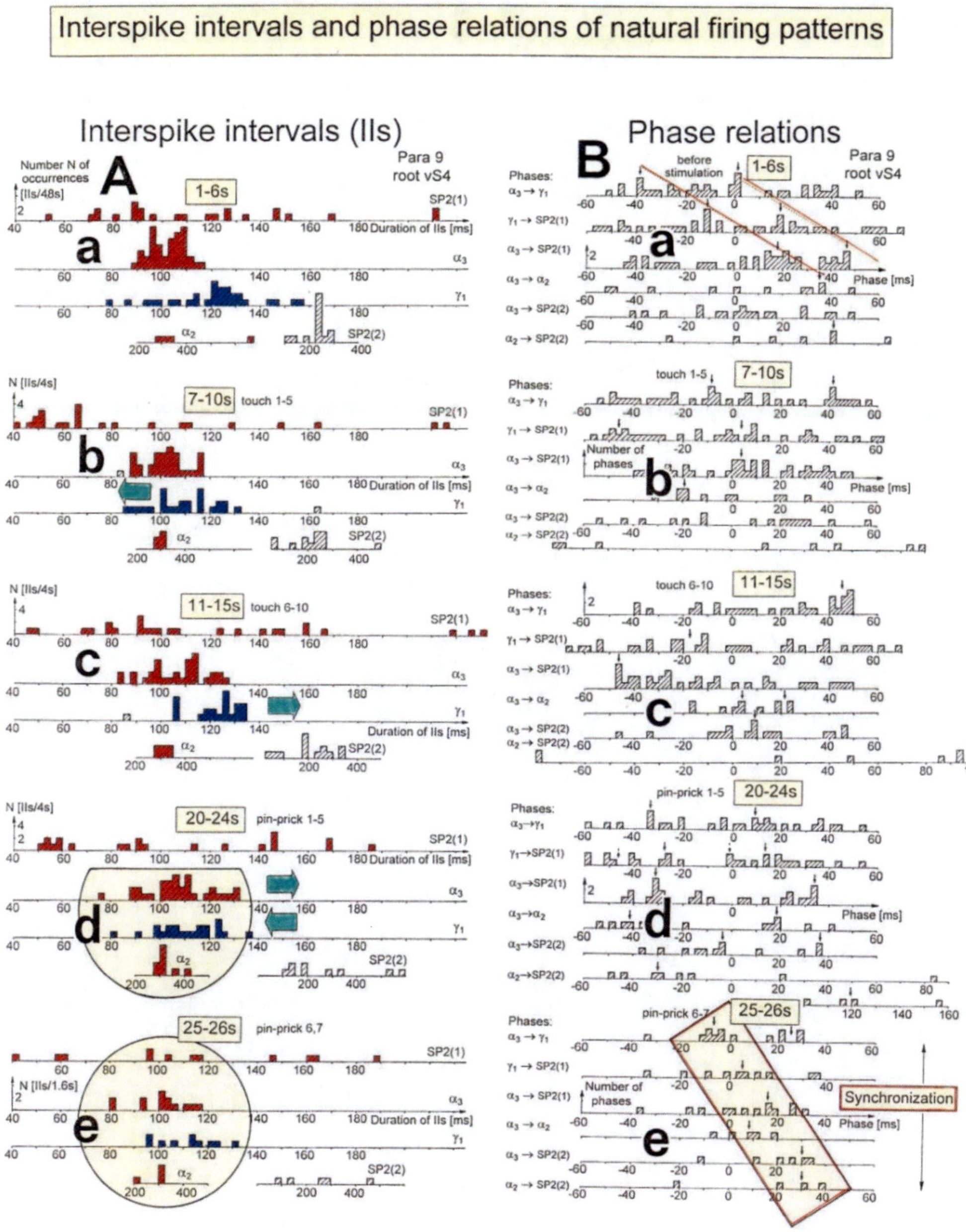

Figure 11. Relative phase and frequency coordination between α and γ-motoneurons and secondary muscle spindle afferents due to touching and pinpricking sacral dermatomes as in Figure 5G. (A) Interspike interval distribution of spindle afferents SP2(1) and SP2(2), α_2 (FR) and α_3-motoneurons (S) and the dynamic fusimotor γ_1 for different time intervals upon touch, pinprick and anal catheter pulling. Interspike intervals (IIs) were collected from several sweeps of 0.8 s duration per second. External loop generation and frequency coordination of α and γ-motoneurons and secondary muscle spindle afferents are marked by the semi-circle and the full circle. The large arrows point to the increase and decrease of the mean II of the distribution. Unsafe identification of α_2 and α_3-motoneurons (or vice versa) because of loss of specific oscillator properties. (B) Histograms of the phases between afferent and efferent fibres for the time intervals indicated, upon different stimulation. Phases were collected from several sweeps of 0.8 s duration per second. The small arrows indicate phase relations. Phase coordination is indicated in a,e. Para 9; vS4.

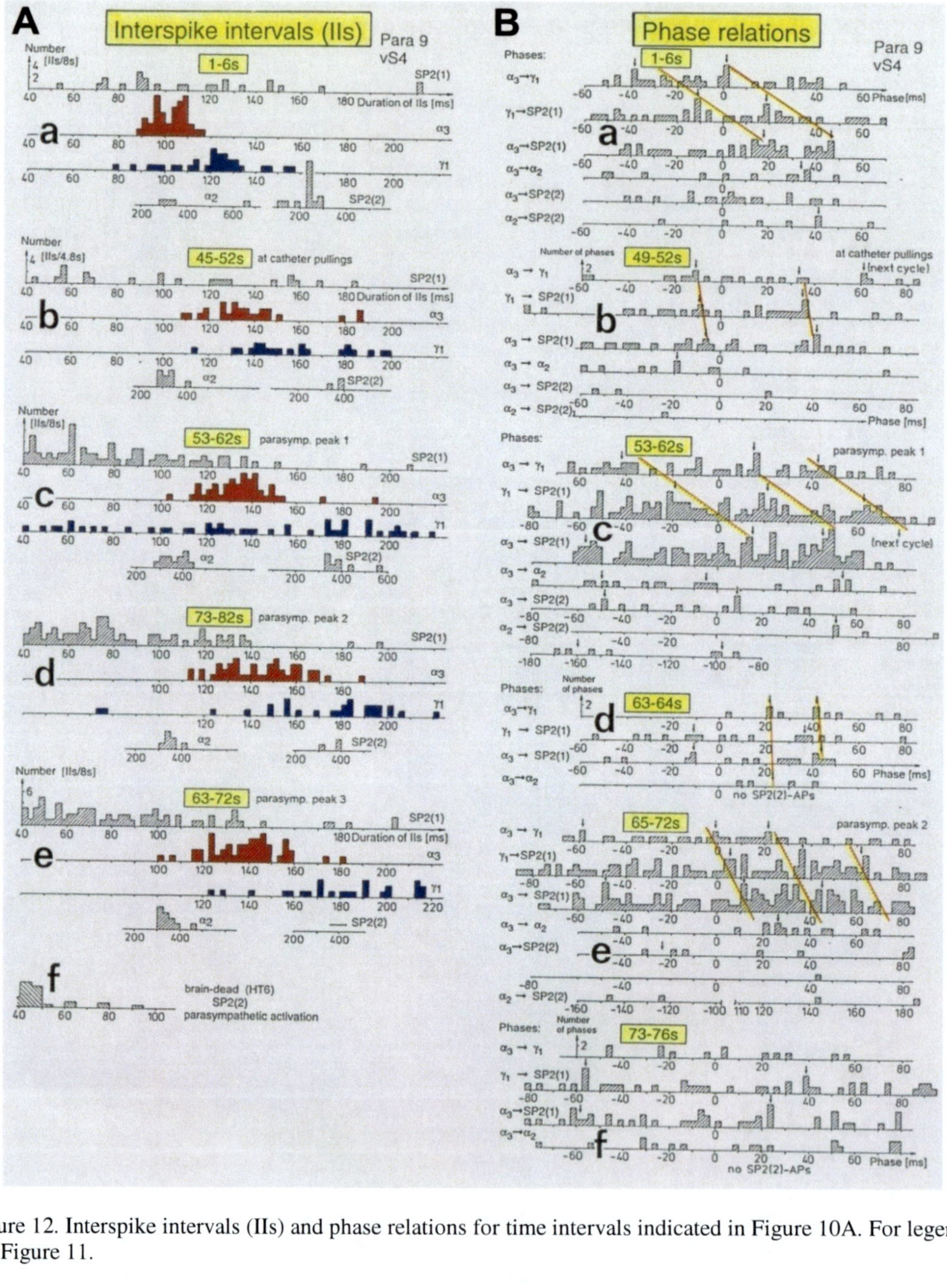

Figure 12. Interspike intervals (IIs) and phase relations for time intervals indicated in Figure 10A. For legend, see Figure 11.

The α_3-motoneuron even slightly increased its IIs (decrease of activity), so that the II distribution of the oscillatory firing α_3-motoneuron and the now oscillatory firing γ_1-motoneuron became very similar (Figure 11Ad). Upon pinpricking sites 6 and 7 (inside the anal reflex area), α and γ-motoneurons and secondary muscle spindle afferents showed similar II distributions (Figure 11Ae). Only one phase relation was organized per oscillation cycle between the different nerve fibres (Figures 11Be, Figure 14e). A synchronization between the APs of the different nerve fibres occurred as can be seen from the direct impulse patterns (Figure 10B). The occurrence of similar II distributions of, and transient constant phases between, the APs of the α_3, γ_1 and SP2(1) fibres are interpreted in the way that, in its oscillatory firing the oscillatory firing, α_3-motoneuron built up an external loop to the muscle spindle innervated by the γ_1 and SP2(1) fibres. The γ-loop became integrated into the oscillatory firing of the α_3-motoneuronal network. Before pinpricking, the γ-loop, consisting partly of the γ_1 and SP2(1) fibres, also contributed to the oscillatory firing, since on the average there existed phase relations. With the pinpricking, however, also the II distributions assimilated, so that this γ-loop was directly included into the oscillatory firing of the α_3-network rather than only contributing to the drive of it. The building up of an external loop to the periphery by spinal oscillators is substantially used when a patient with a spinal cord injury is jumping on springboard (see Figure 15, below). Upon pinpricking sites 8, 9 and 10 (outside of the anal reflex area) and following pinpricking of site 10, the II distribution of the SP2(1) fibre shifted away from those of the α_3 and γ_1 axons. The oscillatory firing α_3-motoneuronal network had abolished its external loop, even though still getting drive from it. Upon anal reflex stimulation and catheter pulling, the external loop was not built up again (Figure 12). Following touch, pinprick and anal reflex stimulation, but not painful catheter pulling, mostly two phase relations existed in paraplegic 9 between the activity of the α_3, γ_1 and SP2(1) fibres per oscillation period (100-140 ms) of the α_3-motoneuron (Figures 11,12), but the phase relations changed with ongoing time. With the activation of the parasympathetic division, upon bladder catheter pulling (Figure 10A), three phase relations occurred per α_3-motoneuron oscillation period (Figure 12Bc,e). At the peaks of parasympathetic activation, three phase relations occurred (Figure 12Bc,e), and only two phase relations were present with little parasympathetic activation (times between the peaks 1 and 2) (Figures 10A,12Bd). Even though the functional units, consisting of fusimotor and α-motoneuron neuronal networks and spindle afferents fibres, were rather unstable in paraplegic 9, in comparison to the brain-dead human (see below), an important difference between skin (somatic) and bladder (parasympathetic) stimulation occurred. Another (third) phase relation per α_3-motoneuron oscillation period occurred with the activation of the parasympathetic division. The activated parasympathetic neuronal network of the sacral micturition and defecation centre seems to have channelled input to the oscillatory firing somatic neuronal network.

4.4.11. Phase Relation Changes between the Action Potentials of the α and γ-Motoneurons and Secondary Muscle Spindle Afferents in Paraplegic 9 upon Somatic and Parasympathetic Activation of the Sacral Micturition Centre

As shown in Figures 11,12, the number (and the values) of phase relations changed between the firings of the different nerve fibres upon different stimulations. In the brain-dead human HT6, two phase relations were found between the α_2-motoneuron and the secondary muscle spindle afferent fibre SP2(2) and the α_2 and the γ_1-motoneuron (Figure 5 of Ref.42 (part 3)). Also in the paraplegic, two phase relations often existed between the firings of the different nerve fibres. It was suggested above that probably a third phase relation occurred when the activated parasympathetic division channelled an additional input to the oscillatory firing somatic neuronal network.

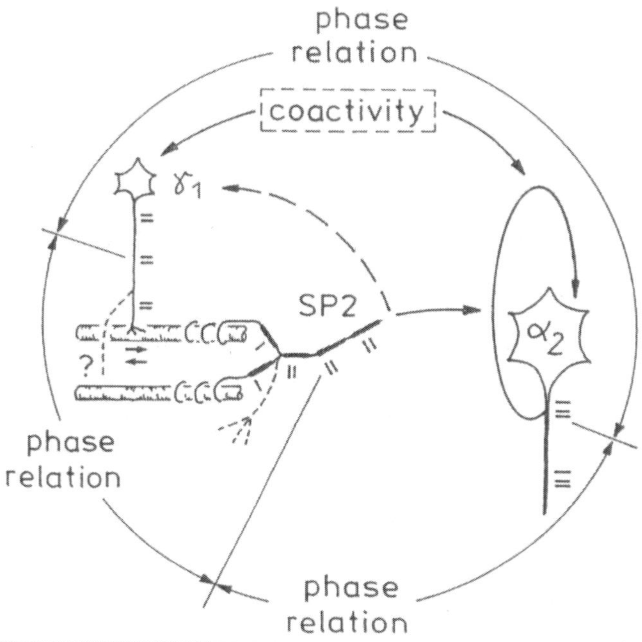

Figure 13. Schematized existing phase relation between α_2 and γ_1-motoneurons and a secondary muscle spindle afferent fibre (SP2). Parallel existing phase relations between other parent afferents and the α_2-motoneuron and between parent secondary spindle afferents are not shown. Phase relation means the increased occurrence of phases in ms in a certain phase range between the action potentials (APs) of the two compared nerve fibres. The complex afferent and efferent muscle spindle innervation was not attempted to be shown. Small arrows at intrafusal muscle fibre indicate local contraction, which is in nuclear chain fibres readily transmitted to the place of afferent innervation. A possible reason of the doublet firing of the SP2 fibre is pictured to occur from single APs (schematized by bars) of two myelinated endings, not necessarily from pacemaker switching. More endings of the parent SP2 fibre and γ_1-motoneurons are indicated by dashed line branches. "Coactivity" indicates a correlation between γ and α-motoneuron spinal cord circuitries for higher activations.

It may therefore be worthwhile to further analyse the number of occurring phase relations per oscillation cycle upon different somatic and parasympathetic stimulations.

Since two phase relation occurred per oscillation cycle between the α_3 and γ_1-motoneurons and the SP2(1) fibre (Figure 11Ba) in paraplegic 9, and also their IIs were rather similar, it is concluded that the neuronal networks of the α_3 and γ_1-motoneurons formed together with the spindle afferent fibre SP2(1) a part of a functional unit. The neural ensemble is built by efficiencies of synapses and projections between the convergence of several fusimotors on one muscle spindle and by the divergence of muscle spindle projections onto several rhythmically firing populations of neurons driving α and γ-motoneurons. Such a functional unit is partly pictured in Figure13 and schematized drawn by three circles in Figure 14. The α_2-motoneuron and the SP2(2) fibre belonged to another functional unit (another ensemble) (longer IIs and the existence of only one phase relation). The two functional units (ensembles) are characterized in Figure 14 by two sets of three circles each. The two functional units interacted with each other, as there existed a phase relation between the α_2 and α_3-motoneurons (Figure 14).

Before stimulation (but with the anal and bladder catheters positioned), there were two phase relations in unit 1 (Figure 14a).

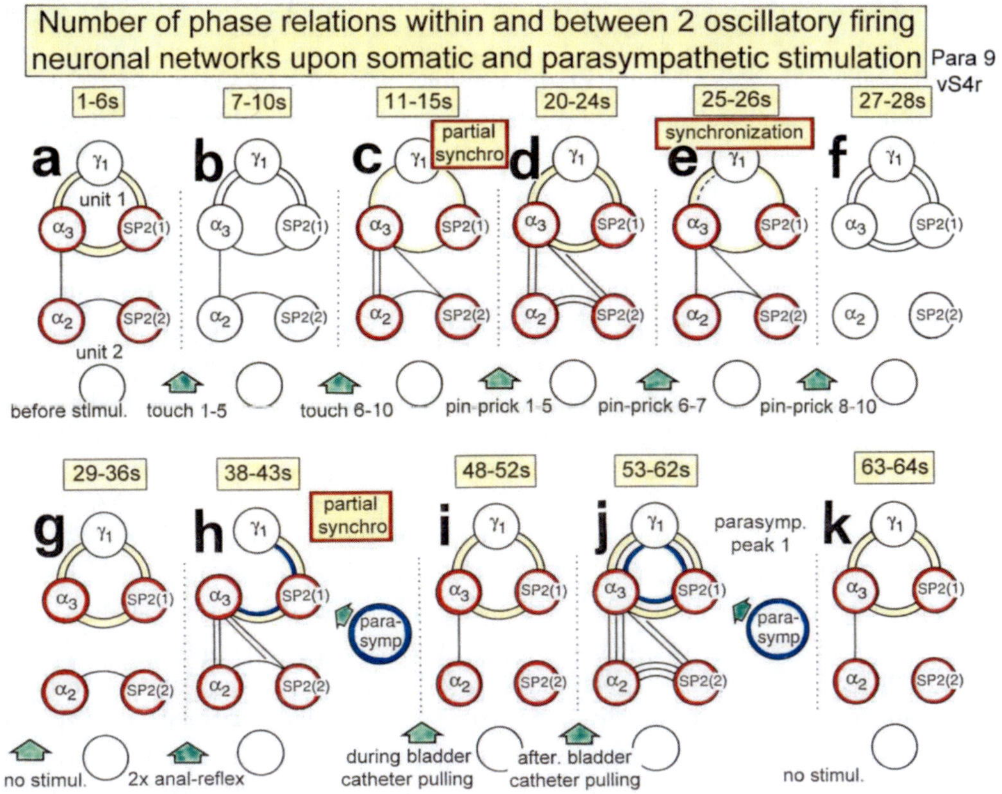

Figure 14. Number of phase relations within and between the two functional units α_3/γ_1/SP2(1) and α_2/-/SP2(2). Time intervals are those of Figure 2A. Note that in "a," the functional unit 1 is with two phase relations per oscillation period in a stage similar to those seen in the brain-dead individual; with synchronization, only one phase relation occurred (e) and the parasympathetic division channeled an extra phase relation to interact with the somatic division (j).

When touching sites 1 to 5 (Figure 7), only slight changes occurred in the two units with respect to the number of phase relations (Figure 14b). But when touching sites 6 to 10, a partial synchronization occurred (Figure 14c), and functional unit 1 reduced the number of phase relations to one. When pinpricking sites 1 to 5, two phase relations occurred again in unit 1 (Figure 14d). Upon pinpricking sites 6 and 7, the number of phase relations between all the components of the two units dropped to one (Figure 14e), and synchronization occurred between the firing patterns (Figure 10B). Since in the brain-dead human two phase relations per oscillation cycle were observed in the functional units, it is possible that synchronization and the existence of only one phase relation for two to three seconds reflected a slight pathologic organization of the networks. Even though upon touching sites 6 to 10 (Figure 14c) or upon pinpricking sites 6 to 7 (Figure 14e) only one phase relation existed in unit 1, and synchronization occurred with both stimulations, it was shown above (Figures 8,9) that the touch afferent input organized a different functional state of unit 1 than pinpricking. The response time until the shortening of the oscillation period was longer than the oscillation period (≈ 100 ms) for pinprick and shorter for touch. It was shown above (Figure 9) that repetitive touch stimulation (most effective inside the anal reflex area) reinforced the sustained stretch reflex of the anal sphincter (continence pattern), and repetitive pinprick stimulation replaced the continence pattern by the protection reaction of the anal sphincter. The number of phase relations alone therefore only provides limited information on the functional state of the organization of the neuronal networks of the human spinal cord. Measurements of a number of parameters are necessary to yield a rather complete description of the functional state of neuronal networks. Following pinprick 8 and 10 and with no stimulation, two phase relations existed again in functional unit 1 (Figure 14f,g), in some similarity to pre-stimulation status (Figure 14a). Following two anal reflex stimulations, partial synchronization occurred in the components of the two units (Figure 5 of [17]) and mainly two phase relations existed (Figure 14h). But the organizational state was still not very similar to the pre- (Figure 14a) or post-stimulation state in unit 1 (Figure 14g), since the parasympathetic division was slightly activated following anal reflex stimulation, as was measured by the impulse pattern (increase of doublet activity) of the secondary muscle spindle afferent fibre SP2(1) (Figures 5,7 of [41] (part 2)). Therefore, probably one phase relation was due to the somatic activation in similarity to Figure 14c,e, and the other phase relation was due to the activation by the parasympathetic division. During bladder catheter pulling (Figure 14i) and with no stimulation (Figure 14k), the number of phase relations and possibly the functional organization was again similar to the pre-stimulation state (Figure 14a). Following strong (painful) bladder catheter pulling with a strong activation of the parasympathetic division (time interval 53-62s (Figure 14j)), measured by the increased doublet firing (see Figure 5 of [41] (part 2)) of the SP2(1) fibre, the functional organization of the sacral micturition centre of the disconnected spinal cord changed completely. Functional unit 1 was now correlated by three phase relations per α_3-oscillation cycle. The functional unit 2 also showed three phase relations per an α_2-oscillation cycle, and interacted with functional unit 1 by three phase relations as well (between the α_3 and α_2-motoneurons; Figure 14j [53-62s]). Between the first and second parasympathetic peak (Figure 10A) at the time interval 63-64s (Figure 14k), the organization form of the two functional units was similar to that before the first parasympathetic activation [49-52s] (Figures 14), only the values of the phase relations changed (Figure 12Bd). With the second strong activation of the parasympathetic division (parasymp. peak 2, time interval 65-72s), the functional unit 1 was bound together again by three phase relations (Figure 10A), in similarity to the first strong activation of the parasympathetic division [42] (part 3), measured by the burst firing of the secondary muscle spindle afferent fibre SP2(1) (Figure 8 of Ref.41 (part 2)) and the increased doublet firing of the SP2(1) fibre (Figure 5 of Ref.41 (part 2)). The functional unit 2 was disorganized, but phase relations still occurred between the α_3 and the α_2-motoneurons and the SP2(2) fibre [42] (Part 3). The α_2-neuronal network and the fusimotor networks, driving the SP2(2) spindle afferent fibre, were integrated differently. After the second strong parasympathetic activation, in the time interval 73-76s (Figure 10A), the functional organization of the two functional units in the spinal cord was similar to that before the activation of the parasympathetic division. Functional unit 2 was slightly disorganized as the SP2(2) fibre strongly reduced its firing [42] (part 3).

4.4.12. Building Up of External Loops to the Periphery by Premotor Spinal Oscillators

It will be shown now that with the building up of simultaneous phase relations between α, γ and SP2 fibres and the assimilation of II distributions (coordinations of rhythms), an external loop of premotor spinal oscillators is built up to the periphery, which makes it possible to directly influence the firing of spinal oscillators by a rhythm training.

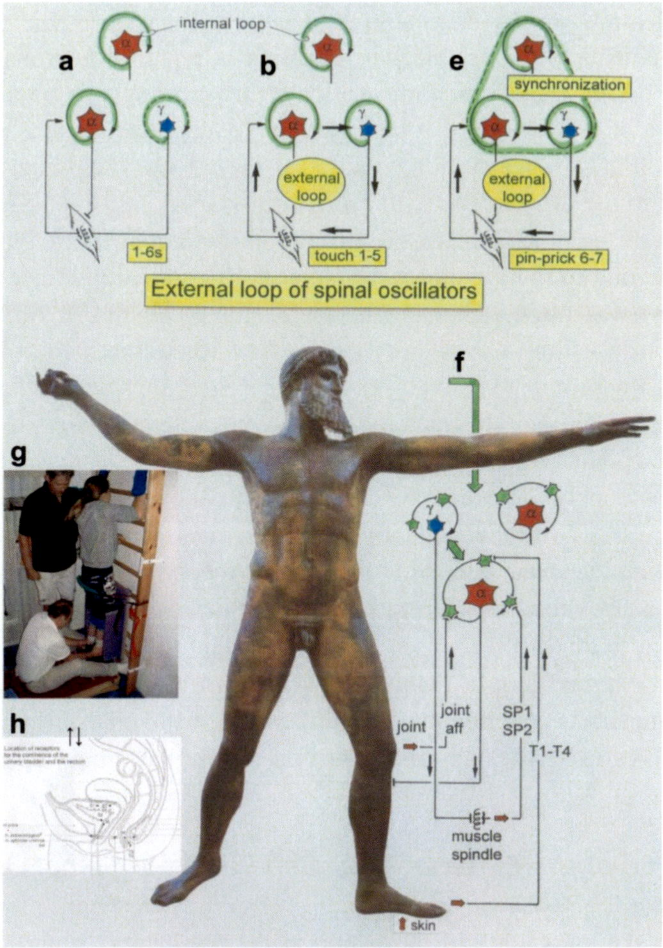

Figure 15. Spreading of oscillatory firing from α-motoneuron neuronal network to include muscle spindles (periphery) and synchronization of different α and γ-motoneuron neuronal networks caused by touch and pinprick stimulation. (a) α-motoneuron neuronal networks fired oscillatory (solid line loop), γ-motoneuron neuronal network did not or did only partly (dashed line loop), upon no additional stimulation; taken from Figure 3Aa,Ba of (8). (b) Oscillatory firing α and γ-motoneuron neuronal networks built up a phase relation with muscle spindle afferents and efferents (external loop to the periphery, indicated by thick arrows) upon touch 1-5, taken from Figure 1Ab, Bb (8). (e) Oscillatory firing α (internal circuitry loop) and γ-motoneuron neuronal networks (external loop) synchronized (broad peak phase relation) upon pinpricks 6-7, taken from Figure 1Ae,Be of (8). The dashed line loop represents synchronization. (f) Oscillatory firing α (internal circuitry loop) and γ-motoneuron neuronal networks (external loop) are extended by analogy from the continence muscles to muscles for locomotion. The open arrows indicate that it may be possible to synchronize spinal oscillators by rhythmic afferent input, generated by rhythmic movements (such as jumping on a springboard or running), and to re-preformate the neuronal circuitry by synapse remodeling to fire more physiologically oscillatory to reduce spasticity and improve locomotion. Extensive pathologic movement like tremor may entrain neuronal circuitry to increase tremor movement. The Greek good is a bronze statue of Zeus found close to the cape of Artemision 460 BC. (g) Supported jumping on the springboard in anti-phase. The patient with the severe cervical spinal cord injury is supported by the author. (h) Location of receptors for the continence of urinary bladder and rectum stimulated upon jumping on springboard. For further details, see Figure 10 of [42].

The somatic and parasympathetic pattern organizations in the sacral micturition centre can simultaneously be entrained by jumping on springboard (Figure 15g) (including the stimulation of movement (Figure 15f) and bladder receptors (Figure 15h)) to allow movement-based learning in the continence and movement patterns and to induce learning transfer from movements to urinary bladder functions. A repair of neuronal network patterns of the functionally disconnected sacral micturition centre in spinal cord injury seems possible.

In Figure 12Ba, it can be seen that there existed two phase relations between the firings of the α_3, γ_1 and SP2(1) fibres, which means that the γ-loop, including the γ_1 and SP2(1) fibres, contributed to the drive of the α_3-oscillator (probably of α_2-type). However, since the II distributions are different (Figure 11Aa), the γ-loop was not a part of this spinal oscillator; it was only contributing to the drive of it, as pictured in Figure 15a. Since the II distribution of the fusimotor γ_1 was often rather broad (for example, Figures 11Aa,12Ac), its driving network was not or only almost oscillatory firing (the dashed line loop in Figure 15a). Upon touching sites 1 to 5 approximately every second, phase relations occurred between the α_3, γ_1 and SP2(1) fibres, even though reaching a different value (Figure 11Bb,c) and the II distributions of the α_3 and γ fibres (Figure 11Ab) and the SP2(1) fibre (Figure 11Ac) became assimilated. The γ-loop became directly connected to the oscillatory firing network: the premotor spinal oscillator (the network driving the α_3-motoneuron; the α_3-motoneuron is a part of the network) built up an external loop to the periphery (Figure 15b). Upon pinpricking sites 6 and 7, one phase relation occurred between the firings of the α_3, γ_1 and SP2(1) fibres (Figure 11Be) and also the II distributions assimilated (Figure 11Ae). The spinal α_3-oscillator had built up a full external loop to include the γ-loop in its oscillatory firing (Figure 15e). Since there was transient synchronous firing of unit 1 with unit 2 (including the α_2 and SP2(2) fibres (Figures 11Be,14e)), probably also the α_2-oscillator built up an external loop to the periphery.

4.4.13. Extension of the External Loop Generation of Spinal Oscillators to Non-Continence Muscles

If one extends the integration of the γ-loop in the oscillatory firing of spinal oscillators innervating striated continence muscles to muscles activated for locomotion (Figure 15f), then, for example, during jumping without weight support (Figure 15g), sky-walking without weight support (Figure 17D of Chapter II), running (Figure 40A of Chapter II) or other coordinated, rhythmic, stereotyped, dynamic movements, at least oscillatory firing α_2 or α_3-oscillators build up external loops to the periphery. It seems partly possible during jumping on springboard to synchronize spinal oscillators with the jumping rhythm, mainly given by the properties (Eigenfrequency) of the springboard. Especially the α_3-motoneuron networks can be entrained efficiently by jumping on springboard, since the "Eigenfrequency" of the springboard and that of the α_3-oscillator network are both in the range of 1Hz (for entrainment, see below). The extension of premotor network organizations serving bladder functions to premotor network organizations serving movement functions is most likely justified, since rhythmic firing of single FF and FR-type motor units, which are innervated by α_1 and α_2-motoneurons, has been recorded by surface electromyography (sEMG) during rhythmic, coordinated movements (Eigenfrequency around 1 Hz) in patients with spinal cord injury (Figures 4,5 of Ref. [29]). Upon exercising on the special coordination dynamics therapy device for turning (Figure 1D), single FF-type motor units in arm and leg muscles coordinated their rhythmic firing with respect to phase and frequency. These single motor unit recordings performed with sEMG support the recordings from single motoneuron axons.

4.4.14. External Loop of Premotor Spinal Oscillators and Rhythmic, Dynamic Stimulation of Motor and Bladder Functions

Upon jumping on springboard (Figures 1C,15g) (and other rhythmic movements like sky-walking or running), premotor spinal oscillators organize themselves to fire transiently oscillatory according to the motor pattern and build up an external loop to the periphery (Figure 15). If the frequency of the rhythmic movement has an integer relationship to the Eigenfrequencies of the premotor networks and more rostral networks, these premotor networks get entrained for more specific self-organization (see below). When jumping on springboard (Figure 15g), not only the motor networks get activated, but also the external sphincteric motoneurons, innervating the external bladder and anal sphincter, as a part of the pelvic floor, get rhythmically activated to counteract the rhythmic weight changes of the intestine. Further, the rhythmic, dynamic, stereotyped up and down movements do stimulate stretch, tension, flow, and mechanical receptors of the bladder (detrusor and proximal urethra) (Figure 15h). This rhythmic movement-related sensory input with ≈1Hz bears similarity to the sensory input stimulated by bladder and anal catheter pulling during the measurements (Figures 10,12). Repetitive phase relation changes in and among neural ensembles will occur in some similarity to the changes following catheter pulling (Figure 12B). Since the neurons involved in the generation of movement and continence (and micturition) patterns (especially if the neurons serve both functions at the same time) are synchronously, rhythmically activated, the pathologic bladder patterns get entrained from the rather physiologic jumping movement, in some similarity to co-movement. The synchronized activation of the somatic and parasympathetic networks allow efficient learning transfer, since the neurons work as coincidence or more general as coordination detector [30]. If there is fluid in the bladder and material in the bowel and rectum, the continence stimulus is stronger. Also walking and running will stimulate and change the intrinsic dynamics of the micturition and defecation centres but not as strongly as the jumping on springboard. The walking and running on a treadmill in severe cervical spinal cord injury is performed under weight support, whereas during jumping on springboard, no weight support is given or needed (Figure 15g). The sympathetic nervous system division, probably innervating the internal urinary bladder sphincter (smooth muscle), will also be activated synchronously. Since the frequency of jumping is with around 1 Hz similar to the "Eigenfrequency" of the α_3-motoneuron oscillators, these oscillator networks should be entrained most efficiently (see below).

4.4.15. Entrainment of Premotor Spinal Oscillator Networks by Rhythmic Movement-Induced Afferent Input and Inputs from Supraspinal Centres

If one approximates for high activation spinal neuronal networks into premotor spinal oscillators (driving the motoneuron) and propriospinal oscillators (generating by coupling with one another movement patterns), then premotor spinal oscillators can be handled in a first approximation as single linear oscillators. The premotor spinal oscillators and the spinal pattern generating networks are self-organized and driven by afferent and supraspinal inputs. When training dynamic, rhythmic, stereotyped movements, the premotor spinal oscillators approximated as linear oscillators are driven by movement-induced afferent input from the periphery (mainly the legs) and surrounding pattern generating networks and possibly supraspinal inputs. These spinal oscillators and most likely their neuronal network can be entrained at least by use of the external loop for a better self-organization. But the functional repair becomes only physiologic if volitional control (by ascending and descending tracts) is intertwined.

If one assumes that loop circuits do not only exist between the premotor spinal oscillators and the periphery, but are a general structure in the CNS, then motor learning involves the formation of loop circuits (or better loop network circuits) between the cortex and the periphery involving the sensory cortex and the thalamus. When a linear oscillatory system is driven by an external periodic input its response contains both frequency components. This is also, in general, true with nonlinear oscillators. However, in this case, if the external frequency is close to the Eigenfrequency of the oscillator itself, then it is possible to have a response at the external frequency only. This phenomenon is known as entrainment or synchronization. It is of paramount importance with respect to biological oscillators because it allows them to "latch on" to the environment. Thus, a rhythm with a free-running period of 24.7 hours may be synchronized to 24 hours when exposed to the natural sequence of day and night.

An oscillator with one degree of freedom can be described by the equation:

$$d^2x/dt^2 + f(x, dx/dt) + \omega_0 x = E \cos\omega_1 t$$

(x = variable, t = time, f = nonlinear term, $\omega_0 = 2\pi f_0$ = frequency of the oscillator, ω_1 = entrainment frequency, E = amplitude)

There exists a set of paired values of amplitude E, and the absolute difference in frequencies $\Delta\omega = |\omega_0 - \omega_1|$ such that the output of the system only contains the frequency ω_1. Figure 16 shows a typical example. Entrainment occurs in the shaded part of the plane. If E and $\Delta\omega$ are below the curve (outside the shaded area), usually the frequencies ω_0 and ω_1 are present. In the case that $\Delta\omega = 0$ but E is very small, the phase of oscillations is not influenced by the input. For further details, including van der Pol oscillator, see [11].

Upon jumping on springboard, the entrainment frequency ω_1 (the jumping frequency) is with 0.9 to 1Hz close to the Eigenfrequency of the premotor spinal α_3-oscillators ($\approx$ 1Hz). The α_3-oscillators and the networks they are integrated in are entrained directly. For the entrainment of the α_2-oscillators (Eigenfrequency $\approx$ 6Hz) and α_1-oscillators (Eigenfrequency $\approx$ 10Hz), subharmonic and super-harmonic entrainment has to be considered. When a nonlinear oscillatory system is driven by an external periodic input z_k, the entrainment can be harmonic (z_k itself has the oscillation period T of the oscillator; case of the α_3-oscillators), subharmonic (z_k has a period, which is an integer multiple of T, mT; case of the α_2 and α_1-oscillators) or super-harmonic (z_k has a period, which is an integer fraction of T, T/m) [11]. With the increasing order of subharmonic entrainment, the entrainment strength reduces for the same coupling strength or entrainment amplitude E. For the α_2 and α_1-oscillators, the entrainment is subharmonic, and the entrainment strength is therefore reduced. In α_2-oscillators, two entrainment phases per oscillation period were mostly observed, which correspond to "in-phase" and "anti-phase" coordination of arms and legs, which enhances entrainment.

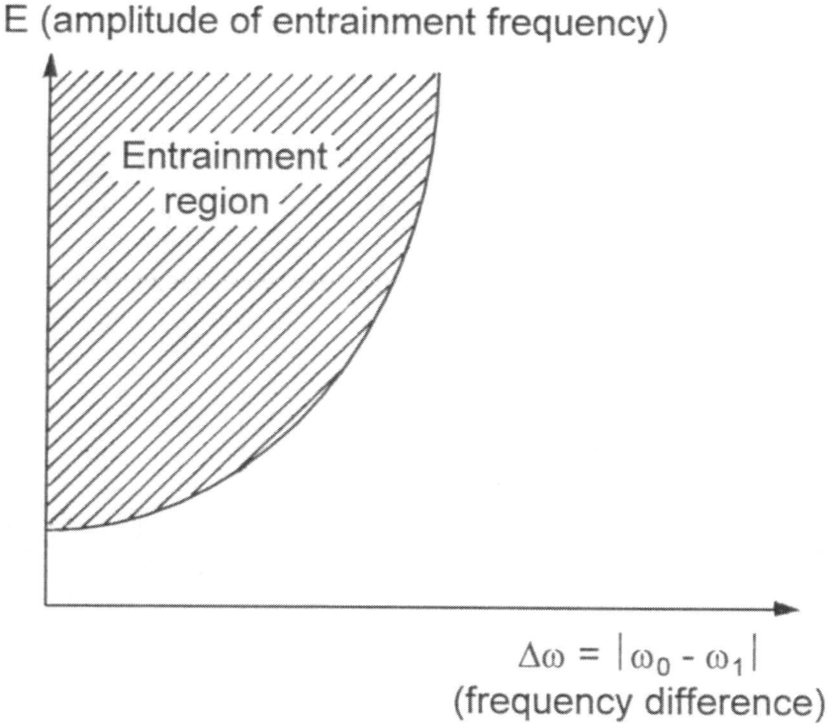

Figure 16. Illustration of the relation between the amplitude E of the external input and the difference in frequencies $\Delta\omega$, which produces entrainment of a nonlinear oscillator.

Also, the change of the number of phase relations between the neural assemblies in the sacral micturition centre may indicate changing entrainment or coupling communication between the somatic and parasympathetic nervous system divisions. However, oscillator models are still far away from human network properties. The jumping on springboard is very rhythmic and stereotyped (frequency of jumping ≈1Hz). But firstly, the movement-induced afferent input enters the network at different levels (premotor neuronal network, propriospinal oscillatory network, brain stem network and higher centres); secondly, oscillators can be driven from different sources; and thirdly, often the rhythmic input patterns consist of impulse trains with increasing interspike intervals and with delays between the responses from different receptors and receptor types. For natural sets of simultaneous impulse patterns of numerous skin receptors induced by touch or pinprick, see Figure 5, and for natural impulse patterns of secondary muscle spindle afferents, see Figure 6. In simulations of networks consisting of populations of interacting oscillators, the known natural afferent input patterns have to be used, and oscillator network structures have to be used, which give the measured output patterns under both physiologic and pathophysiologic conditions. Even though the oscillator models used are far away from human self-organization network forms, their results are still very useful for the interpretation, under certain conditions, of measured data and to understand better the jumping on springboard training (as a part of the coordination dynamics therapy), which contributes substantially to the cure of urinary bladder functions [43]. The consideration about phase and frequency coordination of Figures 11 and 12 when going deeply into the self-organization of human neural networks are quite delicate, but they are necessary if we want to understand the functioning and the repair of the human CNS.

4.4.16. Need for Improving the Stability of Phase and Frequency Coordination to Allow Specific Pattern Formation and Learning Transfer

A young mother, with stress incontinence after giving birth to the first child, could well improve her continence status upon jumping on springboard in addition to other training. But her CNS is not injured; just the periphery has to be repaired by means of changing the CNS.

In severe cervical spinal cord injury (Figure 1A,B), solely jumping on springboard (Figure 1C) is not sufficient for bladder repair. First, of course, the patient has to regain movement functions back (especially the trunk stability) to be able to perform the jumping on springboard. Further, the self-organization of CNS networks by phase and frequency coordination has to be improved to make learning transfer from movements to bladder functions possible, since in every CNS injury, the phase and frequency coordination is impaired. Large instabilities in phase and frequency coordination will not allow specific pattern formation as a basis for learning transfer. However, the stability of phase and frequency coordination can be improved when the patient is exercising on special coordination dynamics therapy devices, especially the one shown in Figure 1D. In the following, measurements are presented to understand the necessity of improving the stability of phase and frequency coordination for functional repair.

The importance of stable phase and frequency coordination to allow specific pattern formation and in consequence learning transfer to other patterns can be understood at the collective variable level and the neuron level. The behavioural information F_{inf} of the coordination pattern dynamics, characterized by equations of motion of collective variables, $dX/dt = F_{intr}(X) + \sum c_{inf} F_{inf}(X,t)$, affect the whole coordination pattern dynamics, including stability, rather than only certain coordination patterns. If the behavioural information includes the exercising of extremely coordinated, integrative movements, like exercising on the special CDT device for turning, then the quality of the CNS self-organization can be enhanced by improving the exactness of self-organization, namely the precision of phase and frequency coordination between neuron and neural assembly firing. By improving the precision of organization of the intrinsic dynamics $F_{intr}(X)$, that is, the specific variability of the injured networks, certain patterns do then already reappear (Figure 23C through Figure 23A).

Neurons often serve more than one network pattern at the same time by time sharing of neuron firing and in this way give rise to learning transfer among the activated patterns. If subnetworks are improved in the organization of one pattern, the organization of the other pattern will also improve.

Neurons involved in the organization of breathing and activating intercostal muscles, for example, are also involved in the organization of trunk stability. By reducing spasticity of the trunk (in patients with Parkinson), the breathing will also improve. Similarly, sphincteric motoneurons are involved in continence and pelvic floor weight bearing. If during pregnancy the (the non-trained) pelvic floor is not trained, sometimes incontinence occurs and after birth stress incontinence.

4.4.17. Phase and Frequency Coordination between Oscillatory Firing α_2-Motoneurons and Their Adequate Afferent Drive in Brain-Dead Human

The relative phase and frequency coordination between the APs of the oscillatory firing α_2-motoneuron O2 and the secondary muscle spindle afferent fibre SP2(1) has partly been shown in Figure 6 and can directly be seen in the original recordings in Figure 17 (brain-dead human).

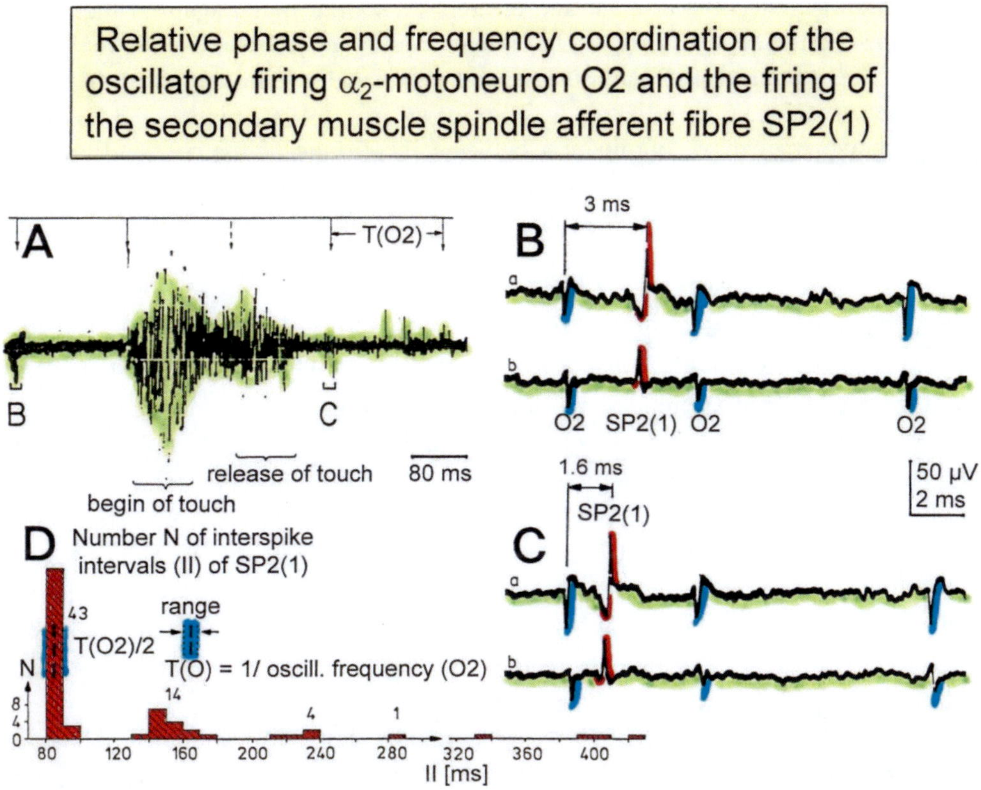

Figure 17. Time relation between the occurrence of the action potentials (APs) of oscillatory firing α_2-motoneuron O2 and the firing of the secondary muscle spindle afferent fibre SP2(1). HT6. S4 dorsal root recording. A. Overall view of the used sweep piece; only trace "a" shown. Four oscillation cycle periods of motoneuron O2 are indicated (T(O2)). The APs of the impulse trains can be recognized only partly, because of the slow time base and poor digitalization. One impulse train (dashed arrow) is lost in the touch stimulated activity, which consists of a touch (large overall activity) and a release part (lower overall amplitude). B,C. Sweep pieces from A, time stretched. In B, motoneuron impulse train APs is marked O2, spindle afferent APs are marked SP2(1). Note that the APs of the spindle afferent fibre are not time-locked to the first AP of the impulse train of the rhythmically firing motoneuron (relative phase coordination). Digitalization four times better than in A, but still rather poor, as can be seen from the low amplitudes of the motoneuron APs on trace "b" in C. D. Occurrence of interspike intervals of the secondary muscle spindle afferent fibre SP2(1). The numbers give the amount of IIs in each distribution peak. The oscillation period of motoneuron O2 (and the range of variation) and the half period are indicated by short dashed lines. Note that the IIs of fibre SP2(1) are very similar to the oscillation period (or the half of it) of α_2-motoneuron O_2 (relative frequency coordination).

The firing of the oscillator and the sweep pieces that are shown time-expanded are indicated at the summary trace (Figure 17A). Figure 17B,C shows the action potential (AP) impulse train of oscillator O2 in connection with one of its driving spindle afferent AP.

Because of the duration of the phase relation of around zero milliseconds between the firing of the driving SP2(1)-fibre and the impulse train of the oscillatory firing motoneuron O2, the SP2(1)-fibre AP (every second AP of the SP2(1)-fibre) appeared at a similar time as the impulse train. Because the AP of the spindle afferent fibre had a characteristic waveform, it was easy to extract its impulse pattern from the summed impulse traffic of this S4 dorsal root. During touch-induced skin afferent activity (Figure 17A), the activities of the motoneuron and the spindle afferent fibre were covered by the skin afferent activity. After the cessation of the skin afferent activity, the afferent and efferent APs were found again at their expected time positions of the regular firings. The phase coordination between the firings of the oscillatory firing motoneuron O2 and the secondary muscle spindle afferent fibre SP2(1) at the time when records B,C were taken, was 1.6 ms (3 ms - 1.4 ms, Figure 17B,C). In Figure 17D, the relative frequency coordination between the firings of the SP2(1)-fibre and the impulse train of the oscillator is indicated. For the time period evaluated, the correlation between the firing of the motoneuron and the spindle afferent fibre was in the range of between 3 and 5ms (Figure 17D).

4.4.18. Relative Frequency Coordination

In Figure 18, considerations concerning the relative frequency coordination are extended to the activity of further afferent fibres and γ-motoneurons of the same root. Figure 18G shows sweep pieces of the original recordings; A through F show the interspike interval distributions of spindle afferents and γ-motoneurons. It can be seen from the overlapping of the oscillator frequency distribution ranges (and the half of it) and from the interspike interval distributions of the afferents that from the viewpoint of frequency coordination, fibre SP2(1) contributed strongly to the drive of oscillator O2, whereas there was a weaker contribution from other afferents (less overlapping between the distributions of the afferents and the range of the basic frequency or the first harmonic of the oscillator). Also, γ-motoneurons showed only little frequency correlation at that time period.

Figure 19 shows the interspike interval distributions of more afferents (including the afferents for bladder filling; stretch receptor afferents S1(1), S1(2)) of another root, together with the oscillation period range (and the half of it) of a second α_2-oscillator (O1). By comparing the oscillation periods (and their halves) and their ranges with the interspike interval distributions of the afferents, it can be suggested which afferents made a (frequency coordination) contribution to the drive of what oscillator at that time interval. For example, the S1(1) urinary bladder stretch afferent fibre activity contributed to the drive of oscillator O1 (activating the external bladder sphincter) because its interspike interval distribution overlaps strongly with the range of the oscillation periods of O1. But the S1(1) distribution does not overlap with the range of the oscillation periods of oscillator O2 or with their halves or quarters. The S1(1) afferent fibre will therefore not have made a substantial contribution to the drive of oscillator O2. On the other hand, the secondary muscle spindle afferent fibre SP2(12) activated oscillator O2 innervating the external anal sphincter, since its interspike interval distribution overlaps with the range of O2 oscillation periods. But the secondary muscle spindle afferent fibre SP2(12) did not activate oscillator O1, as its interspike interval distribution does not overlap with its oscillation period range or the half of it (Figure 19).

By comparing interspike interval distributions of afferent fibres with oscillation period distributions, it can be estimated what afferents made a (frequency coordination) contribution to the drive of the spinal oscillators. These considerations need no knowledge of the connectivity of the neuronal networks. In the frequency coordination between the firings of afferents and oscillators and among oscillators, entrainment or coordination may occur sub- or super-harmonically (see above). The energy transfer, and therefore the coupling strength, will be smaller if the APs coincide in their firing less often.

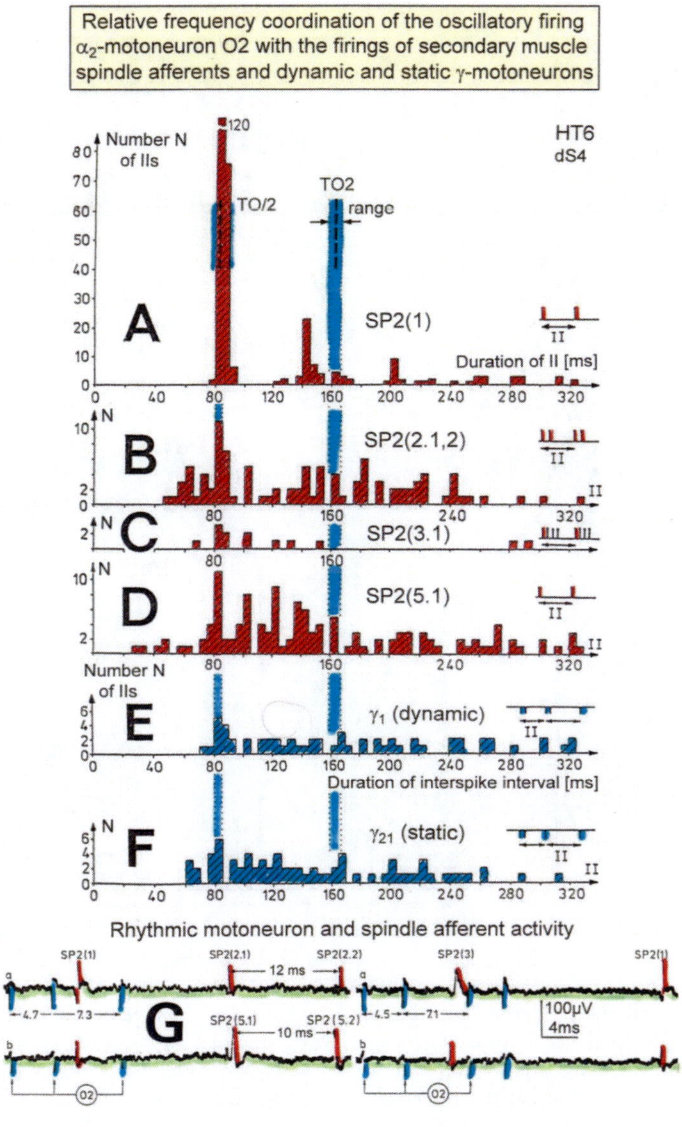

Figure 18. Interspike interval distributions of single endings of four secondary muscle spindle afferents (SP2) and two γ-motoneurons, recorded simultaneously. In A, the oscillation period TO2 (impulse train length = 3 APs) with its range of simultaneously recorded oscillatory firing α_2-motoneuron O2 (see G) is drawn for comparison; also, the halves of the oscillation period TO2/2 are indicated. Note that the interspike interval distributions of spindle afferents and γ-motoneurons have shortest interspike interval, nearly identical to the half of the oscillation period (relative frequency coordination). The schematic impulse pattern in A to F shows the procedure for measuring the interspike intervals. Original records of the firing patterns of α_2-motoneuron O2 and the secondary muscle spindle afferents SP2(1), SP2(2), SP2(3) and SP2(5) are shown in G. Brain-dead human HT6, dS4 root.

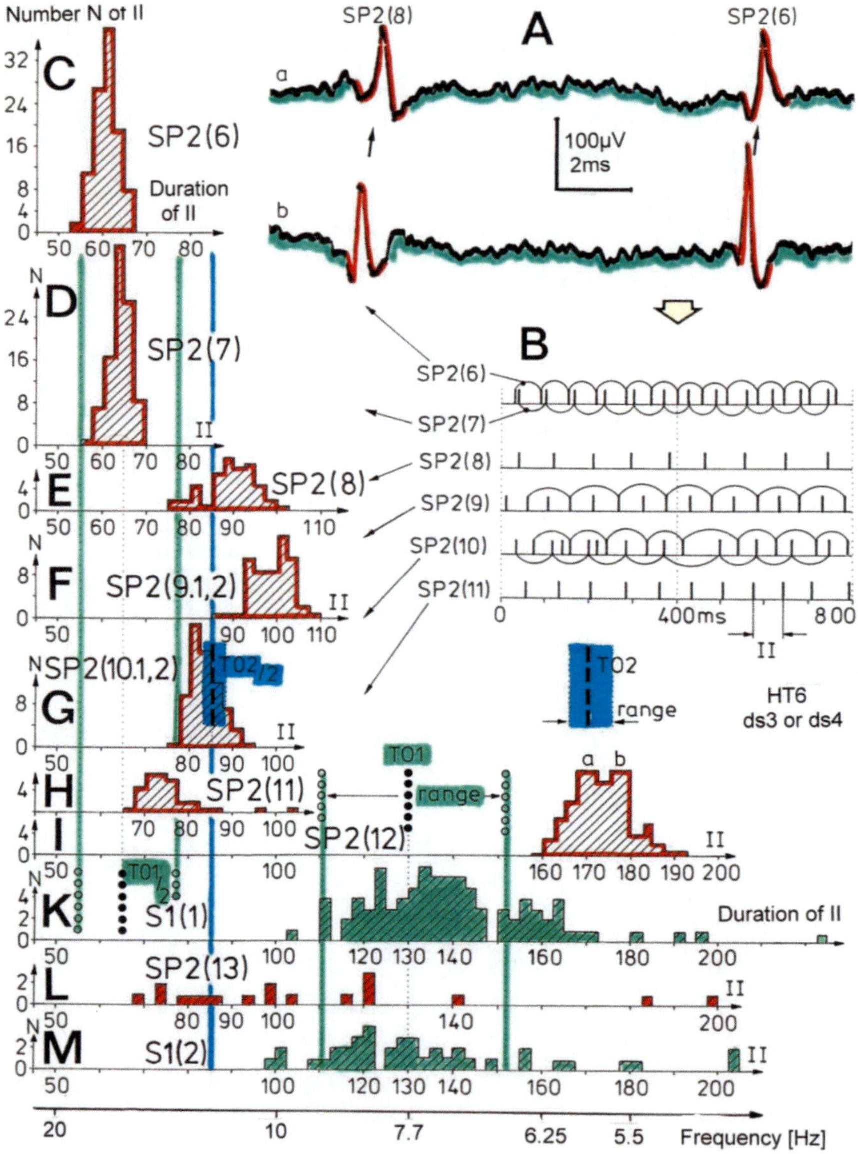

Figure 19. Measurements from brain-dead human HT6 from different spinal cord segments after retrograde bladder filling (700 to 800 ml), with the exception of "I," which was obtained before filling. A. Sweep piece of a recording from a dorsal S3 or S2 root filament. It can be seen that the secondary muscle spindle afferent SP2(6) AP can be distinguished by the waveform on the two traces from the secondary spindle afferent fibre SP2(8) AP (different amplitude of the three phases of the triphasic APs). B. Simultaneously recorded impulse patterns of the six parent secondary spindle afferents SP2(6) through SP2(11) obtained from dS3 or dS2 root recordings. The impulse patterns of SP2(6) and SP2(7) fibres are not separated to show the similarity of the patterns. The impulse patterns of the parent spindle afferents SP2(9) and SP2(10) are split into patterns of the single endings (single ending activity partly connected by circle lines) with the assumption that single endings of parent secondary muscle spindle afferents should have interspike intervals of duration longer than 50 ms. C to H. Interspike interval distributions of six simultaneously recorded single secondary spindle afferent endings. F, G. Interspike interval distributions of parent fibres, which are the sums of the distributions from the two activated endings. I. Interspike interval distributions of a secondary spindle afferent fibre (SP2(12)) of a coccygeal root. K, L, M. Interspike interval distributions of single-fibre afferent activity from a lower sacral dorsal root. In L, most likely the activity from a secondary spindle afferent fibre is shown. In K and M, most likely the interspike intervals from afferents (S1(1) and S1(2)), innervating

stretch receptors of the urinary bladder wall, are shown. In G, H and K, the durations of the oscillation periods (mean and range) of the oscillatory firing α_2-motoneurons are indicated by thick dashed and dotted lines; the motoneurons innervate the external anal sphincter (TO2) and the external bladder sphincter (TO1). The sites of innervation of the oscillatory firing motoneurons are identified (and distinguished from each other) by anal reflex stimulation, bladder filling and catheter pulling. Note that the TO1 and TO2 ranges and their halves overlap with the interspike interval distributions of the secondary spindle and stretch receptor afferents (relative frequency coordination).

As indicated by my measurements, the coupling and the relative coordination during the self-organization of the neuronal networks of the human spinal cord are of an enormous complexity; this self-organization is induced by sets of mutual impulse patterns from stimulated receptors, which are ordered, in time and space (for skin receptors see Figure 5), so as to reflect, in the spinal cord and higher centres, the interplay of the body with the external world.

4.4.19. Impaired Organization of Premotor Spinal Oscillators Following Spinal Cord Injury as an Indicator for Pathologic Network Organization

Following a spinal cord injury, the oscillatory firing networks lose specific properties. The Eigenfrequencies of the premotor spinal oscillators change from a narrow to a broad frequency band (Figure 20).

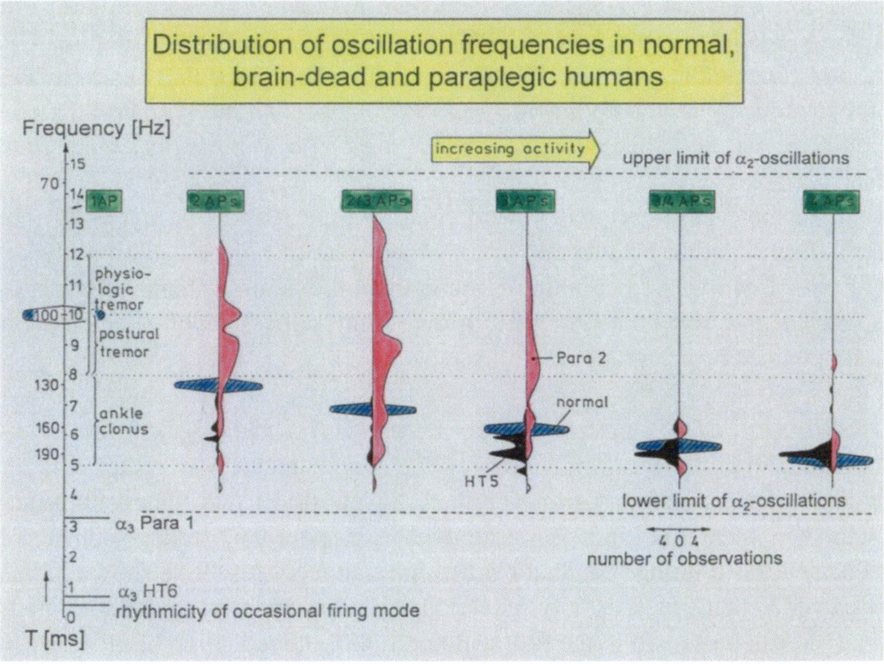

Figure 20. Frequency distributions of oscillation frequencies of continuously oscillatory firing α_2-motoneurons with increasing number of APs per impulse train (increased activity) in paraplegic 2 (open), in brain-dead HT5 (filled), and probably normal human (crosshatched). Frequencies and rhythmic activity changes in the occasional and oscillatory firing mode are indicated. Ranges of physiologic tremor, postural tremor and ankle clonus are also drawn. Note that frequencies for the brain-dead HT5 are too low, and the oscillation frequencies of the spinal cord isolated for a long time (Para 2) are too high and too spread as compared to the theoretically predicted frequency ranges (crosshatched). T = oscillation frequency.

Self-organized α_2-oscillators fire physiologically at an Eigenfrequency (varying within a small frequency band as probably indicated with the hatched distributions in Figure 20) with impulse trains consisting of two to three action potentials. Following brain death, this Eigenfrequency band enlarges (black area in Figure 20). Following a spinal cord injury, the Eigenfrequency band enlarges strongly and includes in this case the frequencies between 4 and 14Hz for firing with two or three action potentials per impulse train (Figure 20). The premotor spinal oscillators have lost their specific properties and could now be excited at frequencies at which they physiologically would not be excited (see above).

4.4.20. Explanation for a Spastic External Bladder Sphincter

In the brain-dead human HT6 of Figure 19, the premotor spinal oscillator TO1 could, according to the overlap of the narrow frequency band harmonically (for simplicity only this case of activation is considered), only be driven by the activity of the bladder stretch receptor afferents S1(1) and S1(2). This oscillatory firing α_2-motoneuron, innervating the FR-type muscle fibres of the external bladder sphincter, is getting activated with the activity of the stretch receptor afferents, when the urinary bladder is filled to secure continence. All the secondary muscle spindle afferents cannot activate the external bladder sphincter harmonically, because their frequency bands did not overlap (frequency band overlap in the frequency range 6.5 until 9Hz). Only the stretch receptor bladder afferents can activate the external bladder sphincter to secure continence. Following a spinal cord injury, the Eigenfrequency band broadens from 6.5-9Hz to 4-14Hz. According to Figure 19, also the secondary muscle spindle afferents SP2(8), SP2(9.1,2), SP2(10.1,2), and SP2(12) could activate the motoneuron innervating the external bladder sphincter. The striated external bladder sphincter can be activated by secondary muscle (and probably other) afferents not involved in continence.

This broadening of the frequency band could be one reason for spasticity (continuous contraction) of the external bladder sphincter. Our patient of Figure 1, with a severe cervical spinal cord injury, had at the beginning of coordination dynamics therapy a spastic external bladder sphincter. She was continent but could not empty the bladder – the emptying of the bladder was achieved by a suprapubic catheter.

4.4.21. Reduction of Spasticity of the External Bladder Sphincter

If one could improve the functioning of the sacral micturition centre and reduce the frequency band of the oscillatory firing premotor spinal oscillators, innervating the external bladder sphincter, then this kind of spasticity of the external bladder sphincter could be reduced. I named this training "oscillator formation training." It will be shown in this and the following part [43] that this spasticity reduction could be achieved by learning and learning transfer. But, as will be shown in the following part [43], the reduction of the spasticity of the external bladder sphincter is only a first step in curing urinary bladder functions. Following CNS injury (including spinal cord injury), not only the frequency coordination, but also the phase coordination among neuron firing, becomes impaired.

4.4.22. Stable Phase Coordination in the Brain-Dead Individual

In a paraplegic, existing relative phase relations among α and γ-motoneurons and secondary muscle spindle afferents are indicated in Figures 11,12 by small arrows. In Figure

14, the natural impulse patterns were used to show the interaction between the somatic and parasympathetic nervous system divisions in the sacral micturition centre. Now I change the representation of the phase relations (shown in Figures 11,12) in order to recognize phase stability, to find reasons for the pathologic organization of neuronal networks in paraplegics. I will start with the physiologic case, namely the brain-dead individual.

To make the phase relation changes better recognizable with time, a representation of phase relations is used, which comes from the measuring of the speed (frequency) of rotation.

The speed of rotation of a turning cylinder with a spot on its surface can be measured with a stroboscope.

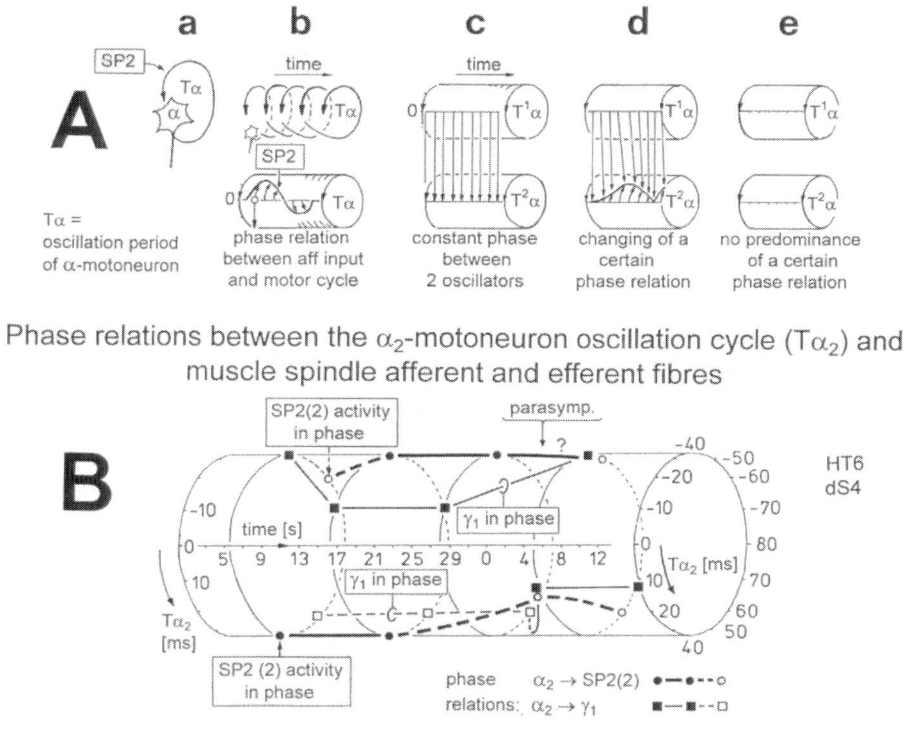

Figure 21. (A) Derivation of the simultaneous description of interspike intervals and phase relations. (a,b) The oscillation period of an oscillatory firing α-motoneuron is schematically characterized by the length of the loop (perimeter). Successive oscillation periods with ongoing time yield a cylinder. Flashing with a stroboscope on such a cylinder with the same frequency as that of the rotation of the cylinder would make a black spot on the turning cylinder not move up or down. If the frequency of the cylinder or the stroboscope changed slowly, the black spot would move up or down. If the black spot moves from left to right with ongoing time, a curve is obtained. By replacing the flashing of the stroboscope by the occurrence of the APs of the spindle afferent fibre (or another oscillatory firing motoneuron) with respect to the APs of the oscillatory firing motoneuron, phase relation changes are made visible in the lower part of "b" for a constant oscillation period (cylinder with no diameter changes). (c) A constant phase between two oscillatory firing motoneurons results to a constant line on the cylinder with ongoing time. (d) A changing phase gives a curve on the cylinder circumference. (e) If there is a loss of predominance of a certain phase between two motoneurons (the black spot gets diffused with ongoing time and is then lost), there is no line or curve. (B) Interspike interval and phase data from the brain-dead human HT6 (root dS4) are plotted in the representation of A. Filled dots and squares represent average phases (phase relations); thick and thin lines connect the dots to show trends. Note that the phase relations change only little; the frequency of the sphincteric α_2-motoneuron ($1/T\alpha_2$) changes only little - the cylinder does not change its diameter.

If the stroboscope flashes light with the same frequency as the cylinder is turning, the spot on the circumference seems to stand still. There is a constant phase between the two frequencies (frequencies are same or multiples of each other). If the phase relation changes, the spot will move.

If no phase relation exists between the turning of the cylinder and the flashing of the light, no spot will be seen. In similarity to stroboscopic measurement of frequencies of turning cylinders, the phase relation between two oscillatory firing spinal oscillators is pictured in Figure 21A. A time axis is introduced on the horizontal line to make phase relation changes visible in dependence on time.

In Figure 21Aa, the loop excitation is pictured for this oscillator model. In Figure 21Ab, the phase relation between the SP2 fibre activity incidence and the oscillatory firing is pictured on the circumference of the oscillation period cylinder of the oscillator. Figure 21Ac,d,e shows different phase relations, namely a constant phase relation (c), a changing phase relation (d), and no phase relation (e). In Figure 21B, phase relation changes are plotted between an α_2-motoneuron and the activity of a secondary muscle spindle afferent fibre and between an α_2 and a γ_1-motoneuron. The data were taken from Figures 4,5 of [42] of a brain-dead individual (probably normal with respect to the number of phases per oscillation cycle and with respect to phase changes). It can be seen that there were two phase relations per α_2-oscillation cycle and that the phase relation changed only little with time. The phase coordination between the firings of the α_2 and a γ_1-motoneuron and the secondary muscle spindle afferent fibre was stable.

4.4.23. Unstable Phase Coordination in the Patient with a Spinal Cord Injury

In Figure 22A,B, different phase relation changes are plotted from Figures 11,12 with respect to the α_3-oscillation cycle (A) and the α_2-oscillation cycle (B). It can be seen that the different phase relations changed strongly in value over time (upon different stimulation) and that also the number of phase relations per oscillation cycle changed. The phase stability of the cooperative and competitive interplay among neurons became impaired. Whether the change of the number of phase relations from two to three following the activation of the parasympathetic nervous system in the sacral micturition centre (Figures 12,14) is physiologic or not is not clear.

4.4.24. Difference of Phase Stability between a Brain-Dead Human and a Paraplegic

The most obvious difference of the phase relation changes between the above-mentioned brain-dead human and the paraplegic was that in the paraplegic, the phase relations varied very much, whereas they changed only little in the brain-dead human. The strong phase relation changes in the paraplegic can be interpreted as instability in the organization of neuronal networks. The correlation of neuronal subnetworks was instable in relation to those of the brain-dead human. Assuming that the neuronal network organization and functioning was rather physiologic in the brain-dead with respect to the firing patterns of the premotor spinal oscillators, the functioning of the networks became instable following spinal cord injury.

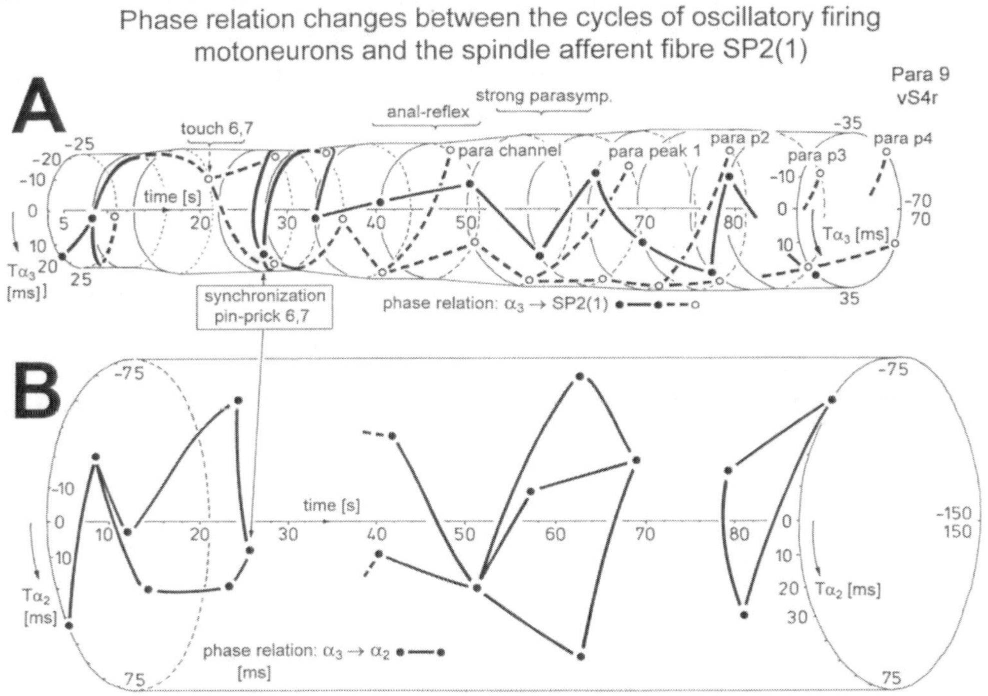

Figure 22. (A) Phase relations between the secondary muscle spindle afferent fibre SP2(1) and the oscillatory firing α_3-motoneuron, taken from Figures 10-12 (and additional data), are plotted on the oscillation period cylinder $T\alpha_3$ (mean oscillation periods are taken from Figures 11A,12A) according to Figure 21A. The cylinder is changing its diameter (perimeter) because the oscillation period changes. Phase changes in ms are scaled on the cylinder circumference. The ongoing time (to the right) is scaled on the axis of the cylinder (time intervals are taken from Figure 10A). Existing phase relations are represented by dots (filled and open (back-side)); lines (filled and dashed (back-side)) only connect the phase relations to show trends. para peak 1, para p2, para p3, para p4 = activity peaks of the SP2(1) fibre due to parasympathetic activation (see Figure 10A right). (B) Phase relations between the α_3 and α_2-motoneurons plotted onto the oscillation period cylinder of the α_2-motoneuron. Dots represent phase relations, taken from Figures 11B,12B. Note that the phase relations of the paraplegic 9 are much more variable than those of the brain-dead human HT6 (Figure 21B); also, the number of phase relations changes.

The more frequent occurrences of changes of phase relations between the different nerve fibres in combination with the changing number of phase relations per oscillation may mean that subnetworks reacted and interacted more quickly and easily with others according to the afferent input. Especially because the oscillatory firing networks lost specific properties, their resonance frequencies changed from a narrow to a broad oscillator frequency band, which means that the oscillators were not excited at a certain frequency anymore but by a broad frequency band. They could now be excited at frequencies at which they physiologically would not be excited. Over-activation and mass effects could be the result. On the other hand, certain networks could escape from driving afferent influence by changing their phase by phase escape to avoid interaction. Functionally far away networks are not reached anymore, which also would result in a loss of specific properties. Therefore, because of the loss of specific properties, some interactions could have occurred more easily and other ones not at all.

W.R. Hess tried, in 1944, to compare biological order and human society [6]. In a society, the upper behaviour of spinal oscillators could be called "putting its flag to the wind." There could be similarities between the organizations of the human nervous system and the organizations between very many individual nervous systems.

4.4.25. Change of the Neuronal Network Organization Following Spinal Cord Injury - Pathologic Network Organization

Following spinal cord injury, the spinal cord neuronal networks have been observed to change their organization, and this can be quantified by six changes of organization.

1. Following spinal cord injury, the spinal oscillators strongly widen their oscillation frequency band [20], which means that their firing becomes unrhythmical (Figure 20). An increase of the more irregular oscillator firing can also be seen in the broadened distributions of the interspike intervals of the impulse trains of oscillatory firing α_2 and α_3-motoneurons [16]. The loss of specificity of oscillatory properties will at least partly be due to the loss of supraspinal inhibition because muscles below the paretic spinal cord injury are over-activated.
2. Following the injury, the phase relations among the oscillators and between oscillators and their driving afferents (for example, secondary muscle spindle afferents) become very instable (Figure 22).
3. Because of the widening of the frequency bands (Figure 20) and the instability in phase coupling (Figure 22), the oscillators lose partly their rhythmicity and their coordination and cooperation properties. The rhythmicity of movements and their coordination and coupling of arms and legs is reduced. The loss of rhythmicity and the loss of coordination can be observed easily when measuring the coordination dynamics.
4. The spinal oscillators are not under full volitional control by the patient anymore (Figure 48 of Ref.21). Sometimes, paraparetic or tetraparetic patients can switch on motoneurons to fire oscillatory, but they cannot switch them off again.
5. Following natural stimulation, the recruitment of motoneurons in the occasional firing mode (low activity mode of motoneuron firing; see Figure 4D for bladder filling up to 550ml) according to the size principle in each nerve fibre group changes (14,15). The level of motoneuron activation increases following spinal cord injury (probably due to loss of inhibition) and the slowly conducting α_3-motoneurons (S) are recruited before the faster conducting α_2-motoneurons (FR), which is pathologic.
6. The coupling of α and γ-motoneurons changes in strength following natural stimulation (15).
7. The loss of rhythmic oscillator properties and also the change of motoneuron firing in the occasional firing mode (when the oscillators have not been self-organized because of low activation) indicate deterioration in the self-organization of the spinal cord neuronal networks below the injury and can at least partly explain the pathologic CNS network organization called spasticity and increased clonus.

The impaired phase and frequency coordinations, measured among α and γ-motoneurons and secondary muscle spindle afferents by an increased variability of coordination, can be introduced in a first approximation in the equation of motion at the collective variable level of description for jumping on springboard ($d\varphi/dt = - dV(\varphi)/d\varphi + (Q\xi_t)^{1/2}$) by an variability increase of strength Q (see Discussion). This increased variability can be pictured by the increased fluctuation of pattern states (pictured by a ball) in an attractor basin (Figure 23B).

4.4.26 .Re-Learning of Phase and Frequency Coordination

This partly impaired phase and frequency coordination between the firings of single neurons and neuron assemblies (Figures 20,22) can be expected to have consequences in the coordination between arm and leg movements because motoneurons innervate muscle fibres, and rhythmic coordinated firing of single motor units has been measured electromyographically with surface electrodes [29] (Figures 79,80 of Chapter III). Indeed, the coordination between arm and leg movements is partly or fully lost following CNS (brain and spinal cord) injury and is often not taking place in the malfunctioning CNS.

This partly impaired phase and frequency coordination at the single neuron level, the assembly level and the macroscopic level can be measured macroscopically when the patient is exercising on a special coordination dynamic therapy device (Figure 1D) on which arms and legs turn with a slightly different frequency (transmission 19 (arms) : 18 (legs)). The phase coordination between arms and legs is imposed by the device. The loss of phase and frequency coordination between arm and leg movements becomes visible and measurable by the arrhythmicity of turning when the patient is exercising on the device. During a turning cycle and during the change of coordination between arms and legs, the turning frequency increases and decreases. This frequency variation (df/dt; f = frequency) can be recorded, quantified and displayed on a computer screen.

During the functional reorganization of the injured CNS of patients, the relative phase and frequency coordination of neuron firing has to be entrained as exactly as possible by the movement induced afferent impulse patterns from the receptors (learning through feedback information) to restore coordination in the range between 3 and 5 milliseconds (Figure 17). The device has therefore to impose the exercising patient a coordination in the millisecond range for the different coordinations of arms and leg movements between pace gait and trot gait. The easy pace and trot gait coordinations, but not the difficult intermediate coordinations, can often be performed by the patient easily. Therefore, the continuous change from the easy to the difficult coordinations and backwards diagnoses the capability of the CNS to organize easy and difficult organizational states. If the movement state can be easily generated by the neuronal networks of the CNS, then the frequency variation of turning is small during the turning cycle, and if the movement state is difficult to organize by the CNS, then the frequency variation is large.

Since in the injured pathologically functioning CNS of patients often coordinations other than pace and trot gait are the easy coordinations, we need a coordination coordinate to judge upon the coordination dynamics. The pace and trot gait coordinations are used for the calibration, since both coordinations between arm and leg movements occur naturally during rhythmic coordinated (automatic) movements like creeping, crawling, walking and running.

The imposed coordination of arm, leg and trunk movements, upon exercising on the special coordination dynamic therapy and measuring device, is in accordance with the coupling possibilities of premotor α_1 (8-12Hz), α_2 (6-9Hz) and α_3-oscillators (0.4 (maybe down to 0.1) -4Hz), even though the frequencies are only a relative coordination parameter, whereas phase is an absolute coordination parameter.

When the hand levers are turned at between ≈ 0.4-1.5 Hz, the resulting frequency difference in turning between arms and legs is approx. 8.5 Hz (low α_1-oscillator frequency or high α_2-frequency) for low hand-turning frequency of 0.5Hz (low α_3-frequency). A slower turning of the hand levers would train directly more the premotor α_2-oscillators ($f < 8.5$Hz). Faster turning of the hand levers (higher α_3-frequency) would train directly the α_1-oscillators in the higher frequency range ($f > 8.5$Hz). Therefore, similar frequencies appear with respect to the frequency of turning on the device for measuring CNS organization and reorganization as have been measured for premotor spinal oscillators.

4.4.27. Improvement of the Organization of Motoneurons in the Occasional and Oscillatory Firing Mode

When turning the levers steadily with medium or high strength, the premotor spinal oscillators in the patient's CNS are entrained because premotor spinal oscillators self-organize themselves for high activation (Figure 4D). The turning of the levers is therefore an oscillator (or assembly) formation therapy. The members of premotor oscillator assemblies (most likely the motoneuron and interneurons) are entrained to improve activation and inhibition by adjusting, for example, the efficacies of the corresponding activated synapses. By turning the levers with little strength at approx. 0.4Hz (releasing the power (load)-setting knob), the motoneurons get only partly organized into premotor spinal oscillators. The motoneurons are firing mainly in the occasional firing mode and are trained for a better recruitment according to the size principle (rhythmicity of repeated recruitment ≈ 0.4Hz (Figure 15 of Ref.21). Between the motoneuron firings in the occasional and oscillatory firing modes, a better coordination of both firing modes is entrained (Figures 15,16 of Ref.21).

4.5. Discussion

4.5.1. Correlation between Natural Impulse Patterns of Single Neurons and Integrative Patterns

Starting from the clinical evaluation of urinary bladder functioning (Figure 2, urodynamic), it was shown that bladder functions can be measured at the single-neuron level by recording single-nerve fibre action potentials from afferent and efferent nerve fibres in sacral nerve roots of the cauda equina during an operation upon stimulating the bladder with bladder filling, anal and bladder catheter pulling, and touching and pinpricking sacral dermatomes (Figures 3-5). Natural impulse pattern changes could simultaneously be recorded from α and γ-motoneurons and secondary muscle spindle afferents (Figures 6-8). Using group conduction velocity values and nerve distances of the human body, conduction times in touch and pain afferents and α-motoneuron axons could be calculated and the times for pattern enhancement and pattern change calculated in the sacral micturition centre. The change from the continence to the protection pattern, measured at the single-neuron level, is pictured in Figure 9, according to the System Theory of Pattern Formation, as a ball (network state) traversing from one basin of attraction into another. The deepness of the potential wells (basins of attraction) indicates the stability of the integrative patterns continence and protection automatism (Figure 9B). It is shown therefore that from evaluations of changes of natural impulse patterns upon natural stimulation integrative CNS functions can be characterized.

The argument of synergetic that the amount of information necessary to describe individual states of neurons is very large and ways must be found to select relevant variables to compress the amount of information (in equations of motion of order parameters (collective variables)), to describe CNS functions, is only partially correct. Sets of natural impulse patterns of relevant neurons also provide deep insight into CNS functioning, because in the process of self-organization of the CNS, many neurons that communicate with each other are involved. The measuring of natural impulse pattern is therefore something like a probing of CNS functioning.

4.5.2. From Natural Impulse Patterns to Coordination Dynamics

To show that learning transfer from movements to bladder functions in the neuronal networks of the sacral micturition centre can be analyzed, phase relations between the firings of different neurons were defined (Figure 10) and their changes plotted (Figures 11-14). It is shown that the cooperative and competitive interplay among neurons serving somatic and continence functions, quantified by phase relation and interspike interval changes, was intertwined. The integrative network organizations for continence (and micturition) and movements were generated by the same networks, including time sharing of neuron firing. Therefore by training movements, generated in the network of the lower sacral spinal cord, also the bladder functions are influenced. This is just what follows out of the equations of motions of the collective variables ($dX/dt = F_{intr}(X) + \sum c_{inf}F_{inf}(X,t)$) of the pattern dynamics: Behavioural information F_{inf} (exercising a movement) affects the whole intrinsic coordination patterns dynamics $F_{intr}(X)$, including stability, rather than certain coordination patterns itself only.

4.5.3. Coherent Activation of Bladder and Motor Functions and Entrainment

Upon jumping rhythmically on springboard (Figure 15g), in addition to the stimulation of mechanoreceptors for movement control, also mechanoreceptors for bladder and rectum control are synchronously activated with the movement (Figure 15h). Continence functions are synchronously activated with the jumping (coherent activation of bladder and movement patterns). Since, additionally, for high activation premotor spinal oscillators build up an external loop to the periphery, neural assemblies are directly entrained to improve their "Eigenfrequencies" and to coordinate their firing with other oscillators (Figure 15). The springboard has an Eigenfrequency ($f \approx 1Hz$; $\omega=2\pi f$), which makes a training in the entrainment region possible (Figure 16). A jumping frequency of 1 Hz is especially efficient for the entrainment of α_3-oscillators because they have an "Eigenfrequency" also around 1 Hz.

4.5.4. Repair of the Phase and Frequency Coordination among Neuron Firings

A completely different contribution to learning and learning transfer is induced if the patient is exercising on the special coordination dynamics therapy device for turning (Figure

1D). This training of very coordinated arm, leg, and trunk movements, imposed by the device, is especially improving the impaired phase and frequency coordination of neuron firing. Following spinal cord injury, the phase and frequency coordination becomes impaired, as turns out when comparing the coordinated firing of neurons in a brain-dead human (physiologic) with that of a patient with a spinal cord injury (Figure 17-22). The network organization becomes deteriorated and escapes entrainment, learning, and learning transfer. The variability of the coordinated firing of neurons becomes too large to enable efficient and specific learning and learning transfer. But when patients exercise very coordinated arm, leg, and trunk movements on the special coordination dynamics therapy device for turning (on which a precise coordination between arm and leg movements is imposed by the device, to which the training individual has to adapt), the CNS of patients can relearn specific self-organization, which means can relearn more exact phase and frequency coordination among neuron firings from the device. Since the CNS network is an open system, many different very coordinated movements should be trained, not to allow the pathologic organization to escape from the repair.

4.5.5. Including of the Measured Increased Variability of Phase and Frequency Coordination among Neuron Firings into the Coordination Dynamics at the Collective Variable Level

Depending on the relationship between the initial coordination pattern dynamics (so-called intrinsic dynamics, $F_{intr}(X)$, depending strongly on the severity of the injury) and the patterns to be relearned (termed behavioural information, $\sum c_{inf}F_{inf}(X,t)$, which act as attractors of the coordination dynamics toward the required patterns), qualitative changes in the attractor layout occur with learning, accompanied by qualitative evidence for loss of stability. The nature of change due to learning (e.g., abrupt versus gradual) arises from the cooperative and competitive interplay between the behavioural information (supported jumping or walking of the patient) and the intrinsic dynamics.

A completely different, additional nature of necessary learning is needed in the repair of CNS injury. The impaired phase and frequency coordination among neuron firing (Figures 20-22) have to be repaired by relearning for proper CNS self-organization. This perturbation of CNS self-organization produces deviations from the attractor states and changes the attractor layout because of altered hard-wiring due to injury. In a first approximation, this tremendously increased variability of phase and frequency coordination can be included into the equations of motion of the collective variables and gives further understanding of pattern change in patients with CNS injury, as, for example, the switch from a movement pattern to a spastic pattern (Figure 23B).

In the Haken-Kelso-Bunz model, the jumping on springboard (Figure 1C) can be described in terms of relative phase between the rhythmically moving legs. Without specific behavioral information, the dynamical description is defined by a vector field (a differential equation) expressing the rate of change in relative phase, $d\varphi/dt$, as a function of the derivative of its potential, $V(\varphi)$:

$$d\varphi/dt = -\,dV(\varphi)/d\varphi + (Q\xi_t)^{1/2} \qquad (1)$$

where $V(\varphi) = -a\cos(\varphi) - b\cos(2\varphi)$ and $(Q\xi_t)^{1/2}$ is the phase and frequency variability of strength Q (where ξ_t is Gaussian white noise of unit variance). Zanone and Kelso (49) introduced noise in Equation 1 (from a logic point of view), because all real systems described by low-dimensional dynamics are coupled to many subsystems at a more microscopic level. One may view noise as a continuously applied perturbation that produces deviations from the attractor state. Such fluctuations are conceptually important in dynamical modelling of phase transition or bifurcation phenomena and are essential in effecting transitions.

I included noise in Equation 1 (from the experimental point of view) because of the measured increased variability of phase and frequency coordination among the coordinated firing of neurons in the human CNS. This, at the neuron level, measured fluctuation of phase and frequency coordination is giving rise to phase transitions or bifurcation phenomena and is essential in causing transitions among attractor states under physiologic (small fluctuation; Figure 20 (HT5 or normal Eigenfrequency distributions); Figure 21B) and pathologic conditions (large variability of phase and frequency coordination; Figure 20 (Para 2 distribution); Figure 22). The relative stability of an attractor states is, therefore, reflected by the depth of each potential well and the strength Q of the variability of phase and frequency coordination, and the attraction of attractor states is reflected by the slope at each point of the potential curve.

The behavioural changes when jumping on springboard (Figure 1C) are represented by the overdamped movement of a rolling ball in the potential landscape for the physiologic (Figure 23A, Q small = little fluctuation of phase and frequency coordination) and the pathologic case (Figure 23B,C, Q large = large variability).

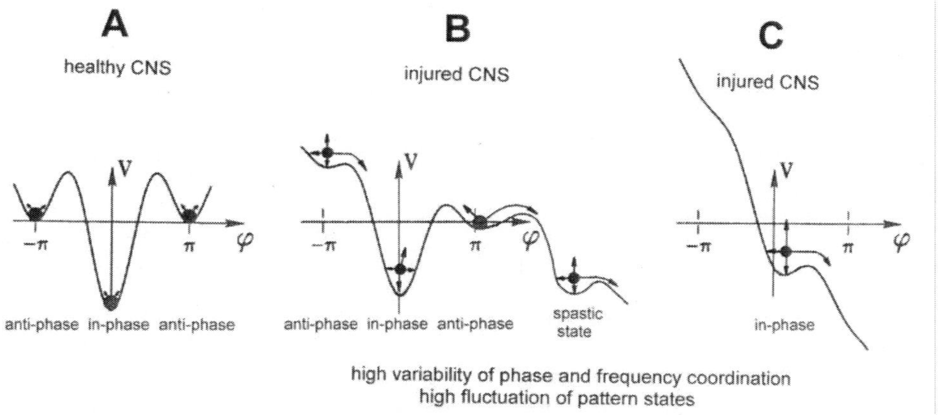

Figure 23. The potential, V(φ), of the coordination dynamics for jumping on springboard of a healthy (A) and injured CNS (B,C). The region around each local minimum acts like a well that weakly traps the system into a coordinated state. Behavioural changes are represented by the over-damped movement of a rolling ball in the potential "landscape." High fluctuations in the stable state, due to the high variability of phase and frequency coordination (in the injured case), will have a greater probability of "kicking" the system out of the basins of attraction (B,C) than for low fluctuations, due to small variability of phase and frequency coordination (in the healthy case (A)). In B, only the in-phase jumping is stable, even though the fluctuation is high. In C, there is only an attractor basin for the in-phase jumping, but the fluctuation is so high that there is high probability that the system is kicked out of the basin of attraction. The patient cannot jump anymore in anti-phase and has problems jumping in-phase. The stability of jumping depends on the motor program (deepness of basin of attraction) and the fluctuation of the pattern state caused by the increased variability of phase and frequency coordination due to the injury.

The increased fluctuation in the rather stable state, due to increased variability of phase and frequency coordination, will have greater probability of "kicking" the system out of the attractor basin (Figure 23B,C), especially in the asymmetric case.

In the healthy CNS, the phase and frequency variability is small and the jumping in-phase and anti-phase are stable (Figure 23A). Following injury, the potential landscape is deformed, and the fluctuation of the jumping network states is high (Figure 23B). The in-phase jumping is still stable in spite of the increased fluctuation, because the basin of attraction is deep. The jumping anti-phase became unstable because the basin of attraction is shallow and the increased fluctuation in the state has a greater probability of "kicking" the system out of the basin. A switch into a spastic state is also possible. In severe CNS injury (Figure 23C), the patient can no longer jump in anti-phase because of the missing of attractors for anti-phase jumping. The jumping in-phase is still possible but unstable.

Upon performing very exact coordinated movements, imposed by devices (as for example shown in Figure 1D), the nervous system of the patient learns to reduce the variability of phase and frequency coordination and achieves in this way a small fluctuation of the network states again as shown in Figure 23A. The treatment (learning) progress is that the in-phase jumping in Figure 23C and the anti-phase jumping in Figure 23B become stable (Figure 23A) again. Also the potential landscape will change due to the reduction of the phase and frequency variability. The important consequence for treatment is that when exercising on special devices and reducing in this way the variability of phase and frequency coordination, the patient can induce the formation of patterns again, without having trained them (learning transfer)! Upon improving the coordinated firing of neurons, a cerebral palsy child may become able to speak or may develop social behaviours (Chapter II).

In conclusion, with respect to bladder repair, the impairment of phase and frequency coordination, measured at the neuron level in human, can be included in the coordination dynamics at the collective variable level. The decrease of the variability of phase and frequency coordination is an essential part of urinary bladder repair.

4.5.6. Improvement of Phase and Frequency Coordination in the Short-Term Memory and the Urge to Void

The effect of improving the phase and frequency coordination in the short-term memory (or short working memory [4]) is significant in the case of the autonomic nervous system. As soon as patients with bladder dysfunction start to exercise on the special device, the urge to void appears, even though often there is only little fluid in the bladder. Following bladder function repair, this exaggerated urge to void, upon exercising on that special device, is strongly reduced. Also in healthy patients, the exercising on the special device improves autonomic functions in the short-terrn memory. In patients with spinal cord injury, also the defecation can be improved upon exercising on that device to stop or reduce medication.

4.5.7. Extended Equations of Motion for Learning

It was shown that in the equations of motion of the collective variables (or order parameters) ($dX/dt = F_{intr}(X) + \sum c_{inf}F_{inf}(X,t)$) that a term can be added to introduce learning in the equation of motion [44,46]. These extended equations were not used by me until now. Emphasis of this research project was to cure CNS injury (including spinal cord injury) and to understand it on the basis of human neurophysiology at the neuron level. Further, the System

Theory of Pattern Formation is only a theory and can explain only a certain aspect of human CNS functioning, especially if the CNS is injured. Already the fluctuation of phase and frequency coordination among neuron firing was only formally included in the equations of motion [49].

4.5.8. Correspondence between Movement Pattern Coordination Dynamics and Phase and Frequency Coordination Dynamics

With the presented measurements of phase and frequency coordination of neuron firings with ongoing time (Figures 8,10-12,14-19,17-22), I introduce a new, lower level of description to the process of learning and pattern change by considering how changes in coordination dynamics at the collective variable and movement levels (Figures 9,23) correspond to phase and frequency coordination dynamics at the neuron level in the human nervous system. Such correspondence between movement coordination pattern dynamics and phase and frequency coordination dynamics constitutes a hallmark of the dynamic pattern approach. By improving the coordination pattern dynamics (by improving the attractor layout (Figure 23) (by improving the specifics of pattern organization)) the accuracy of phase and frequency coordination is increased (the variability of phase and frequency coordination is reduced) and the sets phase and frequency coordination get more characteristic to generate more specific neural network organizations.

4.5.9. Different Routes of Learning in the Healthy CNS: Qualitative and Quantitative Network Changes

In the healthy CNS, there are two different routes of learning depending on the initial attractor landscape: Learning may lead to qualitative changes or to quantitative changes in the initial coordination dynamics. In the former case, the attractor layout evolves (for example) from a bi-stable to a multi-stable structure, whereas in the latter case, the attractor layout is already multi-stable to begin with, but is transformed to a different multi-stable structure. Practice stabilizes the pattern to be learned and causes specific modifications of the entire coordination dynamics in the direction of the task requirement. Such modifications are essentially non-equilibrium phenomena in which previously stable patterns may destabilize temporarily or permanently. How such alterations actually unfold depends on the individual coordination tendencies existing before learning [46].

4.5.10. Different Kinds of Relearning in the Injured CNS, Depending on the Individual Coordination Tendencies Following Injury

Pattern change and pattern relearning involves alterations of the phase and frequency coordination among the neurons involved in the changes of self-organization. Pattern change may thus provide information on the alterations of the phase and frequency coordination taking place during the relearning of patterns. In the injured CNS, there are also the two upper routes of relearning: Pattern relearning may lead to qualitative changes or to quantitative changes in the coordination dynamics. In the former case, the number of phase relations may change between neuron firings (pattern change; Figure 12Bc,e) and the attractor layout evolves from a mono-stable to a bi-stable or from a bi-stable to a multi-stable structure (re-learning; Figure 3 of Ref.43, Figure 3 of part 5), whereas in the latter case, only the values of the phase relations change (Figure 12Ba,b) and the attractor layout is already multi-stable to begin with but is transformed to a different multi-stable structure. The relearning of the multi-stable continence and micturition structure with the subattractors "bladder dyssynergia" and "synergia" needed approximately 18 months upon coordination dynamics therapy, whereas the relearning of the proper relative stabilities of the synergia and dysynergia for physiologic bladder control needed only two months (Figure 3 of Ref.43, Figure 3 of part 5). The difference between qualitative and quantitative network changes cannot only be seen during relearning, but the difference can also be quantified by the times needed for pattern enhancement and pattern change. The enhancement of the pattern "continence" (quantitative change) needed approximately 6ms, whereas the change from the "continence" to the "protection pattern" (qualitative change) needed 100ms (Figure 9Bb). In comparison to the learning in the healthy CNS, there are more learning possibilities needed in the injured CNS, and they do exist. There is not only learning transfer from one movement to another one (CNS symmetries, co-movement) or learning in the vegetative nervous system, but there is also learning transfer between the somatic and vegetative nervous systems, as is shown in the following part [43].

The qualitative change in relearning the proper coordination dynamics includes also the attractor layouts of the somatic and vegetative nervous system divisions and their intermingled structure. Such learning brings about one part of the bladder repair. But the other part of bladder repair is the most qualitative relearning of CNS repair, namely the improvement of the precision of the impaired phase and frequency coordination of neuron firing, which is partly achieved upon exercising very coordinated movements imposed by devices (Figure 1D) and stimulating very coordinated movement-induced afferent input for network entrainment. Since in every injury, malformation, or degeneration of the CNS this phase and frequency coordination for CNS self-organization is impaired, an improvement of the coordinated firing of neurons will improve CNS functioning including those in Alzheimer's and Parkinson's diseases [28]. In limited cell death, the CNS can repair itself by learning. But in substantial neuron cell death, therapy for relearning is needed. Drugs may shift equilibriums in CNS organization for better functioning but cannot repair proper phase and frequency coordination for physiologic functioning.

4.5.11. Learning for Functional and Structural Repair

In psychology, learning has been studied through different windows, such as verbal learning [52], perceptual learning [51], and motor skill learning [50]. Theories of learning have similarly taken many different lines of attack. Although, early on, general laws of learning were sought on the behavioural level [54,55], a shift occurred later to neurophysiological theories of learning [53]. I define relearning in patients as the temporal evolution of the pattern dynamics in the presence of environmental information (therapy) with the intrinsic dynamics mainly depending on the severance of the CNS injury and partly on the repair stage.

Where does this relearning of urinary bladder function stand in the spectrum of psychophysiology when I study learning transfer from coordinated movements to autonomic functions at the pattern level, collective variable level, and neuron level? First of all, such learning is, of course, important in its own right, since 10% of the society suffers from incontinence (as cancer). In aging, coordination dynamics therapy can also be used to improve CNS functioning (Chapter II). I believe that relearning of motor coordination and the transfer of learning to autonomic and higher mental functions provides the possibility to take the study of learning further, particularly in the old quest for laws of learning.

Firstly, motor coordination is an overt behaviour that can be measured as a graded response (rather than categorically). This is important because graded measures of behaviour may provide a natural metric for evaluating the amount of behavioural change during learning. Secondly, coordinated movement can be measured continuously in time, which opens the analysis of behaviour to causality and dynamics. The arrhythmicity of turning, when exercising on a special device (termed coordination dynamics), could be measured (by the coordination dynamics value) during three years of learning, and causality could be drawn to the improvement of motor coordination and to the step-by-step repair of hard-wired neural circuitry (Figure 3 of Ref.39 or Figure 2 of part 5; most likely due to a limited regeneration of the human spinal cord). Thirdly, healthy and pathologic motor patterns can be generated without specific stimulation from the outside and can be recorded by surface EMG. Support by therapists, during treatment, can shift pathologic movements in the direction of physiologic ones and improve the motor patterns by the improved movement-induced afferent input. Therefore, motor patterns are not mere transformation of patterns of stimulation. Together, the three points make it possible to discover intrinsic tendencies of coordination. The interplay of such intrinsic tendencies and learned and relearned coordination patterns may lead to constraints on learning and, hence, lawful aspects of learning. Fourthly, transfer of learning is not only taking place to another untrained movement according to symmetry properties of the CNS [58], but also to autonomic (urinary bladder) and higher mental functions (speech). Fifthly, not only can patterns be learned or relearned, but also an organization principle of the human CNS, namely the phase and frequency coordination among neuron firing, can be improved or relearned. The neuronal networks of the CNS can learn to function more specific in general. Timed firing of neurons and stability of patterns can be altered by learning. With an improved phase and frequency coordination, the patient with CNS injury can reach by intention (supported by movement-related sensory feedback or neural feedback) or spontaneously, patterns organized deeper in the complexity of CNS organization. Sixthly, the hard-wired circuitry can be altered by learning to a limited extent. It was shown that learning is so powerful that the injured [26,27,33], malfunctioning [31], or degenerating [28] CNS can partly be repaired. Learning at a time scale of years (not days as in bimanual coordination tasks [9,46]) may partly repeat development [32,33]. For the repair of urinary bladder functions two to three years of movement-based relearning were needed [43]. The establishment of continence in children during development needs a similar time period.

Seventhly, since movement-based learning and structural repair are intertwined, a mechanism for limited structural repair could well be the stimulation of the endogenous stem cell reserves by movement-based learning. At the physical limit (range of overreaching [36]), performed coordinated movements may stimulate stem cells in the spinal cord to replace lost key neurons for functioning or stimulate the endogenous stem cells in the hippocampal dentate gyrus and olfactory bulb. "Adult born" neurons may migrate to the precise areas vacated by lost neurons (these neuronal networks are most overloaded in activity during the self-organization of physiologic patterns), presumably along processes formed by astrocytes and other structures. They then extend projections toward the denervated targets of original neurons. The migration of the "adult-born" neurons is guided by target-derived growth factors. Their secretion is stimulated by the repeated physiologic activation over months to years. If the pre- and postsynaptic activity of the original neurons and “adult-born” neurons in the injured area have the most important role in guiding axons (and dendrites) that extend from the different neurons (and this is probably the case), then it is likely that the used endogenous stem cell reserves are located in the human spinal cord, since it would be a long distance for migration from the hippocampal dentate gyrus or olfactory bulb to the injury site in the spinal cord. For further considerations, see Chapter VI.

It has been reported that neocortical neurogenesis in humans is mainly restricted to development [60]. The turning over of the neuron population in the adult human neocortex during the last five years was less than 0.07%. We can expect to find even less neurogenesis in the human spinal cord, because of less plasticity. But plasticity (including learning) is an important aspect of the CNS and is necessary for the integration of new memories and newly learned patterns. It is also easy to see the importance of stability for the maintenance of memories (including pattern and phase and frequency coordination stability). There must be, therefore, a delicate balance between plasticity and stability.

These measurements of very little neurogenesis in the adult human brain are in accordance with the very limited structural repair induced by coordination dynamics therapy. The improvements of motor functions at and below the spinal cord injury level in motoric complete cervical spinal cord injury [65,39], however, support the view that new neurons can be built in the adult CNS but only to a very limited extent. After one year of training at physical limits (with more than 30 hours therapy per week), motor functions below the injury level occurred. The consequences for therapy are that the very limited “adult-born” neurons have to be used optimally by therapy, and the self-renewal of the available stem cells has to be stimulated continuously; the patient should not stop the therapy for longer than two days! The optimal use of “adult-born” and functionally integrated neurons means that all newly learned functions should be trained immediately; the therapy has to be updated continuously. The power of this learning therapy is that learning, as a part of the adaptation to the environment, is a very important function for the survival of species including human.

To optimize the structural repair, especially at the injury site, the environment of the repair site has to be conductive for continuous repair. The strategy, to trigger the cells at the injured area to express or re-express endogenous neurotrophins at appropriate levels and locations, is what the therapist has to think of. Automatisms like walking or running are inborn and will get most (epi)genetic support for repair. The relearning of fast walking may, therefore, induce the expression of repair-associated genes most efficiently. Upon walking on volition (including intention), the patient is activating the functioning uninjured network parts. The therapist adds the necessary support to bring the patient into a physiologic performance of walking and inducing thus a rather physiologic movement induced afferent input, propagating to the injury site. The natural afferent and efferent impulse patterns in the original and “adult-born” neurons during walking are the stimulus for generating the geographical landscape of neurotrophins conductive for further structural (and functional) repair, which means self-renewing of multi-potent cells, proliferation and integration. An additional application of stem cells may enhance structural repair upon coordination dynamics therapy. The patients’ CNS has to be taught by learning to repair itself.

An application of stem/progenitor cells close to the injured spinal cord (exogenous stem cell therapy) is unlikely to work, since the movement-based learning is missing to “tell” the stem/progenitor cells what to do. For the homing of these cells, proper information is needed. An additional application of neurotrophins may not be sufficient for the induction of structural repair, if mechanical aspects are involved in migration, proliferation, and integration and neurotrophins are working physiologically only in the μm range [37]. Growth factors would have to be delivered in sufficient quantities to have an effect, but their distribution must be restricted to the site at which they are needed. One of the difficulties in recapitulating the developmental guidance factor milieu will be, therefore, to mimic the floor plate's function in the adult CNS. Further, growth cone attraction and repulsion probably depends on receptor subtypes and intrinsic states of neurons (see also Chapter I). In the case of learning and relearning (upon coordination dynamics therapy), the (expression of) repair-associated genes take care of all these problems at the cellular level.

Eighthly, a repeated expression of the regeneration-associated genes over years at the physical limits of exercising may influence the self-organization of the DNA macromolecule and a window may be opened to view the nature of life from the learning perspective. Movement-based learning may not only influence ontogeny (the development of children with cerebral palsy [31]) but also the phylogeny.

The most important issue in the theory of relearning of motor, vegetative, and higher mental functions following CNS injury is the abstract nature of what has to be relearned: Patterns have to be relearned, and the precise timed firing of neurons has to be relearned. The jumping on springboard and other movements are exercised to relearn movement and continence patterns (and other autonomic functions) including the transfer of learning between them. The turning on the special coordination dynamics therapy and recording device was exercised to improve the impaired timed firing of neurons.

Skill has been defined as a highly organized behaviour in both space and time, and the central problem of skill learning as how such organization of patterning comes about [50]. The understanding that spatiotemporal patterns of behaviour of many interacting neurons needs exact timing of neuron firing and stability of the organizational patterns is foreign to that definition. Also, in the concept that movement is based on a movement plan or score, called motor program, which is established, at least in part, prior to the onset of overt movement [56], the preciseness of timed firing of neurons is missing.

The movements performed during coordination dynamics therapy are designed to repair programs, patterns, and schemata and to improve the timed firing of neurons. But the relearning of patterns and timed firing of neurons cannot be separated. A ten-year-old boy, who suffered a severe brain injury in a car accident, was exercised daily for two hours for a year on the special device for turning, without being able to improve the coordination dynamics values, especially for high loads. But when backward crawling, walking and running were included in the training program, the coordination dynamics values improved strongly [33] (see Chapter II).

4.5.12. What Is Actually Learned upon Therapy

The occurrence of transfer of learning poses the question of what is actually learned upon movement-based learning. At the movement level, patterns and pattern stabilities are relearned. At the collective variable level, phasing patterns and stability are learned. At the neuron level, stability and certain sets (to generate attractors) of phase and frequency coordination are relearned; the variability of coordinated firing of neurons is reduced. Since new neurons and connections are built by neurogenesis, a specific rearrangement of phase and frequency coordination is additionally performed, to integrate the "adult-born" neurons functionally. With respect to stability of patterns and the stability of phase and frequency coordination between neuron firings, giving rise to the patterns, there is self-consistency between the collective and neuron levels of description.

Motor learning and transfer of learning in the injured CNS:

1. The amount of motor learning transfer depends on the similarity between the tasks (old standpoint). But when exercising specific movement patterns, dramatic learning transfer can be achieved from movements to bladder functioning patterns (no similar patterns) over three years, even though the rate of learning is probably small.
2. Learning and learning transfer are alterations in the layout of underlying coordination dynamics. Learning is a fine-tuning, selective process operating on the existing functional organization. Movement-based learning at the limits induces on the long-term structural repair probably by the activation of endogenous stem cell reserves.
3. The learning situation is entered with a certain degree of preorganization of the CNS due to predisposition and injury that clearly constraints the learning process. The rate of learning will decrease with the increase of the injury, which is mainly induced by the increase of the impairment of phase and frequency coordination among neuron firing.
4. Interval training is beneficial, since phase and frequency memory is building up in quality from one series of movements to the next one. This was shown with repeated volitional activation of the same motor units in a patient with spinal cord injury, quantified by sEMG: The phase and frequency coordination did not change from one activation to the next [19]. It seems that the neuronal networks remember the phase and frequency coordinations during interval training. The term short-working memory [4] may characterize more specifically this kind of short-term memory.

4.5.13. Theory of Transfer of Learning

It has long been known that practice of one part of the body in performing a skilled act increases the ability of the bilaterally symmetrical part in the same act [47]. In coordination dynamics theory, it was found that patterns of coordination between symmetrical parts transfer spontaneously. The spontaneous transfer of learning points to an abstract nature of learning [46]. The present knowledge of transfer of learning so far is the following: (a) The amount of transfer seems to be small and positive unless the tasks are practically identical and (b) the amount of transfer depends on the similarity between the tasks [48]. The relearning of urinary bladder functions upon exercising certain movements (jumping on springboard, exercising on a special device) [43] (part 5) indicates that the amount of possible transfer of learning has by far been underestimated. Further, the study of patients with CNS injury reveals more knowledge on CNS functioning, adaptation, and learning than studying only healthy individuals (similar to in animal research).

4.5.14. Influence of Movement-induced Afferent Input on Non-Equilibrium Phase Transitions and Fluctuation Changes of the Collective Variable upon Pattern Shift from Anti-Phase to in-Phase Movement

Learning a bimanual coordination task was studied as a dynamical process [49]. Depending on the relationship between the initial coordination dynamics (so-called intrinsic dynamics) and the pattern to be learned (termed behavioural information, which acts as an attractor of the coordination dynamics toward the required phasing), qualitative changes in the phase diagram occurred with learning, accompanied by qualitative evidence for loss of stability (phase transitions). The nature of change due to learning (e.g., abrupt versus gradual) was shown to arise from the cooperative and competitive interplay between behavioural information and the intrinsic dynamics. Theoretically predicted features of non-equilibrium phase transitions, indicative of loss of stability, accompanied the shift from one pattern (anti-phase) to the other (in-phase), including enhanced fluctuation of relative phase [49].

Upon exercising on the special coordination dynamics therapy and recording device (Figure 1D), the individual is performing (imposed by the device) continuous coordinate changes of arm and leg movements between anti-phase (trot gait) and in-phase (pace gait). Loss of stability is occurring during pattern change from one pattern (anti-phase) to the other (in-phase), quantified by increased arrhythmicity of exercising (df/dt, f = frequency) in accordance with the enhanced fluctuation in the bimanual coordination task (Figure 1 of Ref.61).

But the coordination dynamics (arrhythmicity of exercising) of a top athlete did not show loss of stability of pattern formation when changing from the anti-phase to the in-phase pattern, and healthy individuals can learn to keep the stability during pattern change, if the pattern change is imposed by a device, which means if the movement-related sensory feedback stabilizes the changing self-organization.

A ten-year-old boy suffered a severe brain injury in a car accident [26], which could partially be repaired by coordination dynamics therapy. First, the boy relearned (besides other movements) the in-phase crawling (pace gait) and then the more difficult anti-phase crawling (trot gait). When the patient was able to perform both crawling patterns, he was asked to crawl in anti-phase and should increase the speed (the frequency) of crawling. With increasing crawling frequency, the patient switched spontaneously from the anti-phase pattern to the easier in-phase pattern, as can be observed in patients during jumping on springboard or during the performance of the bimanual coordination task.

There are two conclusions. Firstly, the device imposed extremely coordinated movements and qualitatively increased the learning capacity. Secondly, physiologic movement induced afferent input is very beneficial in learning and relearning and can stabilize unstable patterns. More generally, coordinated bio- or neuro-feedback, especially via the visual system, is increasing the rate of learning qualitatively or even quantitatively.

4.5.15. Frequency Coordination and Motor Learning

Since the motoneurons of the spinal cord networks can build up an external loop to the periphery, it seems likely that also loop circuits exist between different brain parts and the periphery or other brain sites [2] for efficient entrainment and functional reorganization of the CNS. The successful brain repair upon coordination dynamics therapy supports this assumption. But frequency coordination between different brain parts is more than just synchronization in information processing [3]. Synchronization was also observed in my measurements upon rhythmic stimulation (Figures 10B,11,14,15), but this synchronization was only transient. The frequency coordination between the motoneurons and their driving muscle spindle and bladder stretch receptor afferents (Figures 17,18,19) indicates that frequency coordination is also exerted via parts or multiples of a driving frequency (subharmonic or super-harmonic coordination). Coordinated motor unit firing was recorded in patients with spinal cord injury. The precise timing between the spikes was not a synchrony, but a coordination with different phases and different (Eigen) frequencies (Figures 4,5 of Ref.29). The firing of neurons in the occasional firing mode [14,15] has not been touched in this discussion.

4.5.16. Further Research

The substantial progress in understanding learning and transfer of learning with respect to former research (44,46) is that the results of this part are giving functional insights into the process of learning at the neural level. A better understanding of learning and memory at the neuron and functional anatomy levels may allow the application of movement-based learning to repair other CNS injuries, malformations or degenerations. The repair of the urinary bladder in severe cervical spinal cord injury is shown in the following part [43]. Since the spinal cord is part of the CNS, the physiological, pathological, and repair physiological mechanisms should also be valid in other parts of the human CNS. The applied movement-based learning therapy "coordination dynamics therapy" was able to partly repair the injured cerebellum and brain (Chapter II).

But since a limited regeneration of the spinal cord could even be achieved in severe cervical spinal cord injury [39] that means that other tracts in the human CNS may also have the capability for limited regeneration following injury. If in very severe brain injury, the patient cannot recover from the coma anymore, because the ascending reticular activating system (ARAS) is too damaged (possibly following late relief of high brain pressure), a partial regeneration of the ARAS and a functional reorganization may bring some patients out of the permanent coma upon administering efficient treatment for three years. It is mainly from CNS repair in human patients that the discipline of "human repair physiology" gets substantial knowledge.

4.5.17. Ethics of CNS Injury and Clinical Research

In review articles in the journal *Nature* on the repair of spinal cord injury in humans [57,59,63,64], it has been emphasized that there is as yet no cure (57, page 1; 59, page 639), even though there is [22,27,37-39,43] and that the mechanisms that underlie recovery have rarely been definitely established, which is wrong (see this book and Refs.21,34,36-39,43). Human neurophysiology and clinical research in general and learning in specific are foreign to these articles.

Such research politics ruin the progress in human repair physiology and clinical research and steals the future from patients with CNS injury or disease. Clinicians have argued that "withholding a potential therapy from a patient with spinal cord injury (in connection with endogenous stem cell therapy) is in itself unethical" (cited in 59, page 634). This argument is right (especially if we think of rehabilitation centres that do not want to make the step from care to cure), but proper diagnostics are necessary before, during, and after treatment. On the other hand, performing unqualified human research under the name of therapy trials is also unethical.

There must be a balance between progress and safety. Financial interests have to be taken into consideration. Two of the main problems in the cure of CNS injury or disease are that human research is not properly organized worldwide and only animal research is getting financial support (see Epilogue).

References

[1] Arbib M.A.: *The Handbook of Brain Theory and Neural Networks*, MIT Press, Cambridge, 1995.

[2] Asanuma, H. and Pavlidis, C.: Neurobiological basis of motor learning in mammals. *NeuroReport,* 8: I-VI, 1997.

[3] Baker, S.N., Kilner, J.M., Pinches, E.M., Lemon, R.N.: The role of synchrony and oscillations in the motor output. *Exp. Brain Res*, 128: 109-117, 1999.

[4] Basar, E.: Memory as the "whole brain work." A large-scale model based on "oscillations in super-synergy." *International Journal of Psychophysiology*, 58: 199-226, 2005.

[5] Haken, H., Kelso, J.A., and Bunz, H.: A theoretical model of phase transitions in human hand movements. *Biological Cybernetics* 39: 139-156, 1985.

[6] Hess, W.R.: Biological order and human society. In: Akert, K. (Ed.), *Biological Order and Brain Organization - Selected Works of W.R. Hess*, pp. 3-15, Springer-Verlag, Berlin, 1981.

[7] Holst, E.v.: Die relative Koordination als Phänomen und als Methode zentralnervöser Funktionsanalyse. *Erg. Physiol.*, 42: 228-306, 1939.

[8] Holst, E.v.: Relative coordination as a phenomenon and as a method of analysis of central nervous function. In R. Martin (Ed.), *The collected papers of Erich von Holst* (pp. 33-125). Coral Gables, FL: University of Miami , 1973.

[9] Kelso J.A.S.: Dynamic Patterns. *The Self-Organization of Brain and Behavior.* MIT Press, Cambridge, 1995.

[10] Passatore G.M., Fillipi G.M., Grassi G.: Cervical sympathetic nerve stimulation can induce an intrafusal muscle fibre contraction in the rabbit. In: *The Muscle Spindle* (I.A. Boyd and M.H. Gladden, Eds.) pp. 221-226, Stockton Press, New York, 1985.

[11] Pavlidis, T.: *Biological Oscillators: Their Mathematical Analysis.* Academic Press, New York, 1973.

[12] Schalow, G.: Conduction velocities and nerve fibre diameters of touch, pain, urinary bladder and anal canal afferents and □ and □-motoneurons in human dorsal sacral roots, *Electromyogr. Clin. Neurophysiol.*, 31: 265-296, 1991.

[13] Schalow, G.: Oscillatory firing of single human sphincteric α□ and α□-motoneurons reflexly activated for the continence of urinary bladder and rectum - restoration of urinary bladder function in paraplegia. *Electromyogr. Clin. Neurophysiol.*, 31: 323-355, 1991.

[14] Schalow, G.: Recruitment within the groups of γ_1, α_2 and α_3-motoneurons in dogs and humans following bladder and anal catheter pulling. *Gen. Physiol. Biophys.*, 11: 101-121, 1992.

[15] Schalow, G. and Wattig, B., Recruitment of α and γ-motoneurons in rats, dogs and humans, *Electromyogr. Clin. Neurophysiol.*, 33: 387-400, 1993.

[16] Schalow, G.: Spinal oscillators in man under normal and pathologic conditions, *Electromyogr. Clin. Neurophysiol.*, 33: 409-426, 1993.

[17] Schalow, G. and Zäch, G.A.: Reflex stimulation of continuously oscillatory firing α and γ-motoneurons in patients with spinal cord lesion. *Gen. Physiol. Biophys.* 15, Suppl.1: 75-93, 1996.

[18] Schalow, G. and Zäch, G.A.: External loops of human premotor spinal oscillators identified by simultaneous measurements of interspike intervals and phase relations. *Gen. Physiol. Biophys.* 15, Suppl.1: 95-119, 1996.

[19] Schalow, G., Blanc, Y., Jeltsch, W. and Zäch, G.A.: Electromyographic identification of spinal oscillator patterns and recouplings in a patient with incomplete spinal cord lesion: oscillator-formation training as a method to improve motor activities. *Gen. Physiol. Biophys.* 15, Suppl.1: 121-220, 1996.

[20] Schalow, G. and Zäch, G.A.: Neuronal reorganization through oscillator formation training in patients with CNS lesions. *J. Periph. Nerv. Syst.*, 3: 1-24, 1998.

[21] Schalow, G. and Zäch, G.A.: Reorganization of the Human CNS, Neurophysiologic measurements on the coordination dynamics of the lesioned human brain and spinal cord. Theory for modern neurorehabilitation (31 case reports). *Gen. Physiol. Biophys.*, 19, Suppl. 1: 1-244, 2000.

[22] Schalow, G.: Recovery from spinal cord injury achieved by three months of coordination dynamic therapy. *Electromyogr. Clin. Neurophysiol.* 42: 367-376, 2002.

[23] Schalow, G.: Online measurement of human CNS organization. *Electromyogr. Clin. Neurophysiol.* 41: 225-242, 2001.

[24] Schalow, G.: Time axis calibration in human CNS organization for judging dysfunction. *Electromyogr. Clin. Neurophysiol.* 41: 485-505, 2001.

[25] Schalow, G. and Pääsuke, M.: Low-load coordination dynamics in athletes, physiotherapists, gymnasts, musicians and patients with spinal cord injury, after stroke, traumatic brain lesion and with cerebral palsy. *Electromyogr. Clin. Neurophysiol.* 43: 195-201, 2003.

[26] Schalow, G., Pääsuke, M and Kolts, I.: High-load coordination dynamics in athletes, physiotherapists, gymnasts, musicians and patients with CNS injury. *Electromyogr. Clin. Neurophysiol.* 43: 353-365, 2003.

[27] Schalow, G.: Stroke recovery induced by coordination dynamics therapy and quantified by the coordination dynamics recording method. *Electromyogr. Clin. Neurophysiol.* 42: 85-104, 2002.

[28] Schalow, G.: Improvement after traumatic brain injury achieved by coordination dynamic therapy. *Electromyogr. Clin. Neurophysiol.*, 42: 195-203, 2002.

[29] Schalow, G.: Partial cure of spinal cord injury achieved by 6 to 13 months of coordination dynamic therapy. *Electromyogr. Clin. Neurophysiol.*, 43: 281-292, 2003.

[30] Schalow, G., Pääsuke, M., Ereline, J. and Gapeyeva, H.: Improvement in Parkinson's disease patients achieved by coordination dynamics therapy. *Electromyogr. Clin. Neurophysiol.*, 44: 67-73, 2004.

[31] Schalow, G.: Phase and frequency coordination between neuron firing as an integrative mechanism of human CNS self-organization. *Electromyogr. Clin. Neurophysiol.*, 45: 369-383, 2005.

[32] Schalow, G., Pääsuke, M., Jaigma, P.: Integrative reorganization mechanism for reducing tremor in Parkinson's disease patients. *Electromyogr. Clin. Neurophysiol.*, 45: 407-415, 2005.

[33] Schalow, G., Jaigma, P. Cerebral palsy improvement achieved by coordination dynamics therapy. *Electromyogr. Clin. Neurophysiol.*, 45: 433-445, 2005.

[34] Schalow, G.: Hypoxic brain injury improvement induced by coordination dynamics therapy in comparison to CNS development. *Electromyogr. Clin. Neurophysiol.*, 46: 171-183, 2006.

[35] Schalow, G. and Jaigma, P.: Improvement after severe traumatic brain injury induced by coordination dynamics therapy: Comparison with physiologic CNS development. *Electromyogr. Clin. Neurophysiol.*, 46: 195-209, 2006.

[36] Schalow, G.: Symmetry diagnosis and treatment in coordination dynamics therapy. *Electromyogr. Clin. Neurophysiol.*, 46: 421-431, 2006.

[37] Schalow, G.: Cerebellar injury improvement achieved by coordination dynamics therapy. *Electromyogr. Clin. Neurophysiol.*, 46: 433-439, 2006.

[38] Schalow, G., Vaher, I. and Jaigma, P.: Overreaching in coordination dynamics therapy in an athlete with a spinal cord injury. *Electromyogr. Clin. Neurophysiol.* 48: 83-95,2008.

[39] Schalow, G.: Stem cell therapy and Coordination dynamics therapy improve Spinal cord injury. *Electromyogr. Clin. Neurophysiol.*, 48: 233-253, 2008.

[40] Schalow, G., Jaigma, P., Belle, V.K.: Near-total functional recovery achieved in partial cervical spinal cord injury (50% injury) after three years of coordination dynamics therapy. *Electromyogr. Clin. Neurophysiol.*, 49: 67-91, 2009.

[41] Schalow, G.: Partial cure achieved in a patient with near-complete cervical spinal cord injury (95% injury) after three years of coordination dynamics therapy. *Electromyogr. Clin. Neurophysiol.*, in press, 2009.

[42] Schalow, G.: The classification and identification of human somatic and parasympathetic nerve fibres including urinary bladder afferents is preserved following spinal cord injury. *Clin. Neurophysiol.*, in press, 2009.

[43] Schalow, G.: Coordination impairment between the somatic and parasympathetic nervous system divisions in the human sacral micturition centre following spinal cord injury. *Clin. Neurophysiol.*, in press, 2009.

[44] Schalow, G.: Phase relation changes between the firings of α and γ-motoneurons and muscle spindle afferents in the sacral micturition centre during continence functions in brain-dead human and patients with spinal cord injury. *Clin. Neurophysiol.*, in press, 2009.

[45] Schalow, G.: Cure or urinary bladder function in severe cervical spinal cord injury (95% injury, motoric complete). *Clin. Neurophysiol.*, in preparation, 2009.

[46] Schöner, G., Zanone, P.G. and Kelso, J.A.S.: Learning as change of coordination dynamics. *Journal of Motor Behavior*, 24: 29-48, 1992.

[47] Torrens M., Morrison J.F.B.: *The Physiology of the Lower Urinary Tract*, Springer Verlag, 1987.

[48] Zanone, P.G. and Kelso, J.A.S.: Coordination dynamics of Learning and transfer: Collective and component levels. *Journal of Experimental Psychology: Human Perception and Performance*, 23: 1454-1480, 1997.

[49] Bray, C.W.: Transfer of learning. *Journal of Experimental Psychology*, 11: 443-467, 1928.

[50] Schmidt, R.A.: *Motor control and learning*. Champaign, IJ: Human Kinetics, 1988.

[51] Zanone, P.G. and Kelso, J.A.S.: Evolution of behavioural attractors with learning: Nonequilibrium phase transition. *Journal of Experimental Psychology: Human perception and Performance*, 18 (2): 403-421, 1992.

[52] Fitts, P.M.: Perceptual-motor skill learning. In A.W. Melton (Ed.), *Categories of human learning* (pp. 243-285). New York: Academic Press, 1964.
[53] Gibson, J.J.: *Principles of perceptual learning and development*. New York: Appleton, 1969.
[54] Hall, J.F.: *Verbal learning and retention*. Philadelphia: Lippincott, 1971.
[55] Hebb, D.O.: *The organization of behaviour*. New York: Wiley, 1949.
[56] Humphrey, R.: *The nature of learning in its relation to the living system*. London: Kegan,1933.
[57] Koffka, K.: *Principles of Gestalt psychology*. New York: Harcourt, Brace and World,1953.
[58] Schmidt, R.A.: *Motor control and learning*. Champaign, IL: Human Kinetics.
[59] Bradbury, E.J. and McMahon, S.B.: Spinal cord repair strategies: why do they work? *Nat. Rev. Neurosci.*, 7: 644-653, 2006.
[60] Jeka, J.J. and Kelso, J.A.S.: Manipulating symmetry in the coordination dynamics of human movements. *Journal of Experimental Psychology*, 21 (2): 360-374, 1995.
[61] Thuret, S., Moon, L.D.F. and Gage, F.H.: Therapeutic interventions after spinal cord injury. *Nat. Rev. Neurosci.*, 7: 628-643, 2006.
[62] Bhardwj, R.D. et al.: Neocortical neurogenesis in humans is restricted to development. *PNAS,* 103 (33): 12564-12568, 2006.
[63] Schalow, G.: Functional development of the CNS in pupils aged 7 to 19 years. *Electromyogr. Clin. Neurophysiol.*, 46: 159-169, 2006.
[64] Schalow, G. Impulse pattern, innervation densities and two point discrimination of skin and mucosal afferents in humans. Consideration for a sensory reinnervation of urinary bladder and anal canal in spinal cord lesions. IV (IV). *Electromyogr. Clin. Neurophysiol.* 32: 259-285, 1992.
[65] Martino, G. and Pluchino, S.: The therapeutic potential of neural stem cells. *Nat. Rev. Neurosci.,* 7: 395-406, 2006.
[66] Harel, N.Y. and Strittmatter, S.M.: Can regenerating axons recapitulate developmental guidance during recovery from spinal cord injury. *Nat. Rev. Neurosci.*, 7: 603-616, 2006.
[67] Schalow, G.: Building of New Motoneurons in the Human Spinal Cord upon Coordination Dynamics Therapy to Improve Finger Functions in Motoric Complete Cervical Spinal Cord Injury. In: Berkovsky, T.C. (Ed.), *Handbook of Spinal Cord Injuries*, Chapter 4. pp. 231-264, Nova Science Publishers, 2009.

5. Cure of Urinary Bladder Functions in Severe (95%) Motoric Complete Cervical Spinal Cord Injury in Humans

Summary

Severe cervical Spinal Cord Injury (SCI) leads to quadriplegia and autonomic dysfunctions. Bladder/bowel continence, cardiovascular performance, and breathing are impaired besides movements. Even though there are no fully restorative treatments for SCI, I report about a patient who suffered a severe cervical, motoric complete SCI, in whom urinary bladder functions were fully repaired by functional and structural repair (limited regeneration of the cord) upon 2.5 years of Coordination Dynamics Therapy (CDT). On the repair of the blood circulation (no occurrence of pressure ulcers anymore), breathing and motor functions was reported earlier. The mechanism that underlies this important repair of urinary bladder functions is the learning transfer from movements to bladder functions. The human bladder repair is analyzed at the neuron level, the collective variable level (System Theory of Pattern Formation), the movement, and the clinical diagnostic level.

5.1. Introduction

A movement and learning therapy, called coordination dynamics therapy (CDT), has been developed and shown to be able to improve central nervous system (CNS) functioning after stroke [39], traumatic brain injury [40,48], hypoxic brain injury [47], cerebellar injury [49-51], spinal cord injury (SCI) (41,42,52-54,59), in cerebral palsy [45], and in Parkinson's disease [43,44].

The current research in basic clinical neurophysiology and human repair physiology is aimed at developing treatment for CNS injury, especially SCI, to improve movements, autonomic, and higher mental functions.

In Chapter II, it was shown that if 50% (according to MRI) of a cervical spinal cord was spared following injury, the patient could relearn walking, running, and jumping (Figure 40 of Chapter II) and could become continent within two to three years. The administered CDT could partly cure this incomplete SCI at C5/6 levels mainly by a functional reorganization [53]. The question now is can the urinary bladder functions still be repaired in a patient with a ten times more severe SCI, if the approximately spared 5% tract fibres of the spinal cord white matter do not lead to the sacral micturition centre (no spared sacral range)? A functional reorganization is most likely not sufficient for the bladder repair. A regeneration of the spinal cord would be needed.

It will be shown here that in this patient with a severe (motoric complete) cervical SCI, the urinary bladder functions could be cured upon 2.5 years of optimal and intensive CDT. In part 1 through 4, the human neurophysiologic and clinical knowledge was built up to understand the dysfunction of the bladder by comparing the organization of the neuronal networks of the sacral micturition centre between patients with SCI and a brain-dead human [55-58]. The data of the brain-dead human are supposed to be rather physiologic in this respect. It was shown that the natural firing patterns of single somatic and parasympathetic neurons gave rise to the knowledge [55-58] to understand learning transfer from trained movements to urinary bladder functions [58].

It will be shown in this report that the partial solution of the century problem "SCI repair in humans," namely the repair of urinary bladder functions, is not a miracle or genius finding but is the consequence of continuous human neurophysiologic and clinical research. The knowledge to repair urinary bladder functions by CDT cannot only be used in patients with SCI but also in elderly people and young women with stress incontinence.

5.2. Method

5.2.1. Coordination Dynamics Therapy

Schalow Coordination Dynamics Therapy (CDT) is a movement-based learning therapy. Following SCI, the self-organization of the CNS by phase and frequency coordination is impaired, and movements, urinary bladder, bowel, and other functions are impaired, pathologic or lost. For the repair of the impaired phase and frequency coordination among the firings of neurons and neural assemblies, the patient is training on special CDT devices. For the repair of the impaired or lost movements, the patient trains the innate automatisms creeping, crawling, walking and running in the forward and backward direction (for CNS symmetry [50] repair), old-learned movements (because they are deeply built into the CNS and have a high stability) and rhythmic, dynamic, stereotyped movements to repair premotor spinal oscillators. But urinary bladder and bowel functions cannot directly be repaired by training and learning. The repeated retrograde filling of the bladder was not successful in the past (private information). But within the framework of System Theory of Pattern Formation, applied to the injured CNS, urinary bladder functions can be repaired by learning transfer from coordinated movements [58].

5.2.2. Collective Variable Level of Description of CNS Organization

The System Theory of Pattern Formation for Repair gives a theory at hand to partly understand the repair of the injured human CNS by learning and learning transfer. Its application to the somatic and autonomic nervous system divisions extends the theory to understand also the repair of urinary bladder functions. By increasing the stabilities of physiologic urinary bladder patterns (bladder synergia) and decreasing the stabilities of pathophysiologic bladder patterns (bladder dyssynergia, spasticity), continence and physiologic micturition can be achieved. The idea to see deterioration of neuronal network organization from the aspect of pattern stability has its origin in the stability changes of phase and frequency coordination following CNS injury, measured with the single-nerve fibre action potential recording method in human [35-38].

By cooperative and competitive interplay, the many billions of neurons of the human CNS generate dynamics in self-organization in communication with receptors in the periphery that can explain urinary bladder functions. Schematically, collective variables or order parameters capture the collective coordinated activity of the neurons of the CNS. Specific equations of motion (the dynamics) of these collective variables generate the time course of organizational states. If the collective variables are designated by the vector X, the coordination pattern dynamics can abstractly be formulated by the equations of motion (11,61):

$$dX/dt = F_{intr}(X) + \sum c_{inf} F_{inf}(X,t) \qquad (1)$$

where F_{intr} designates the intrinsic dynamics. The sum is over different types of so-called behavioural information, F_{inf}, as environmental, memorized, or intended behavioural information. The relative strength of the different influences is parameterized by c_{inf}. With the behavioural information $\sum c_{inf} F_{inf}(X,t)$ (intention, instruction, treadmill walking, exercising on special devices, crawling, ...), the intrinsic dynamics $F_{intr}(X)$ of neuronal network organization can be changed, including the stability of different patterns, rather than certain coordination patterns itself only [61]. Learning a new pattern involves alterations of the entire layout upon CDT; attractive states are formed anew (Figure 3, continence and micturition states) or previously attractive states vanish (Figure 3, spastic state (spasticity of the external bladder sphincter)). The change of the whole coordination dynamics by the behavioural information is the scientific basis for learning transfer between different coordinated movements and urinary bladder functions for urinary bladder repair.

Since I measured an increased variability of phase and frequency coordination among neuron firing following spinal cord injury in the human spinal cord [58], such fluctuation of coordinated firing has to be included in the equations of motion because it alters the attractor layout and reduces pattern stability. In the Haken-Kelso-Bunz model (and it will be used again for explanation), jumping on springboard can be described in terms of relative phase between the rhythmically moving legs. Without specific behavioral information, the dynamical description is defined by a vector field expressing the rate of change in relative phase, $d\varphi/dt$, as a function of the derivative of its potential, $V(\varphi)$:

$$d\varphi/dt = -\,dV(\varphi)/d\varphi + (Q\xi_t)^{1/2} \qquad (2)$$

where $V(\varphi) = -\,a\cos(\varphi) - b\cos(2\varphi)$ and $(Q\xi_t)^{1/2}$ is the term representing the phase and frequency variability of strength Q. To reduce the fluctuation of phase and frequency among neuron firing to increase pattern stability (reduction of pattern fluctuation (ball moves less about) from Figure 3C to Figure 3F), necessary for urinary bladder repair, the patients have to exercise on special instruments to relearn (improve) the coordinated firing of neurons. The exercising on special devices has to be included as behavioral information in the equations of motion (formula 2).

Since learning and learning transfer during CDT may need a few years, it is of high importance to use such behavioural information, which can really repair the urinary bladder. In previous publications [55-58], the proper behavioural information for learning transfer was developed by using the single-nerve fibre action potential recording method. By analyzing natural impulse patterns in sacral nerve roots upon urinary bladder and bowel stimulations, intravesical stimulation, and movements, it was found that jumping on springboard (Figure 1D) and exercising on the special CDT device (Figure 5C,D) are especially efficient to induce learning transfer from movements to urinary bladder functions [58]. The jumping on springboard was designed to especially repair patterns, including transfer of learning from somatic to autonomic functions (Figure 15 of [58]); and the exercising on the special CDT device was primarily used to reduce the variability of phase and frequency coordination and to increase thus the stability of network states by decreasing pattern fluctuation. Movements are exercised to generate a physiologic attractor layout of patterns again (Figure 3), and the coordinated firing of neurons is trained to reduce fluctuation of pattern states (to increase pattern stability, shown in Figure 3).

Functional and structural repair are intertwined. To induce and enhance the structural repair in the human CNS (necessary in severe SCI), a different strategy is used than in animals. In rats, for example, one is trying to enhance growth of nerve fibres over the injury site. In humans, I induce structural repair by learning in the long-term memory and learning transfer, for which a high complexity and variability of CNS neuronal networks is needed. To induce structural repair (including network rewiring and neurogenesis) in the spinal cord and supraspinal centres by learning in the long-term memory is in accordance with current thinking about long-term memory in the cortex [75] and a relation between neurogenesis and learning [76].

5.2.3. No Indication for Spontaneous Regeneration in the Human Spinal Cord

In lower vertebrates, the CNS fully regenerates following injury [4]. Since ancient times, it has been recognized that injury to the spinal cord is a devastating condition and that the human spinal cord does not regenerate spontaneously even after 20 to 30 years, only the complications increase. An evolutionary explanation for the "loss" of CNS regenerative capacity of human against animals, especially lower vertebrates, is that this is an unselected by-product of gaining the increasingly complex nervous systems that selection pressures have favoured over time [7]. CDT is using this complexity (and variability) of the human nervous system (namely the learning) to partly overcome the "loss" of CNS regenerative capacity and is not going along the line of the nerve fibre growing strategy.

5.2.4. Structural Repair in Animals: The Nerve Fibre Growing Strategy

If the spinal cord of a rat is injured, nerve fibres try to grow over the injury site. Since the distance of growing is too short and the number of growing fibres is too small, it is tried to enhance the power of regeneration to get more fibres growing and over longer distances. Nerve growth factors are used. Applied stem cells may proliferate to neurons and their neurits may cover the injured site. If one assumes that the regeneration over the injury site is inhibited, then one tries to inhibit this inhibition. If scar tissue formation inhibits the regeneration, then the scar tissue is removed. This nerve fibre growing strategy is logical with respect to animals, because in less complex nervous systems, it is often a question of growing, since nerve fibres partly regenerate to specific sites they innervated before the injury. The learning capacity on the other hand is limited. A rat cannot relearn if the nerves to extensors and flexor are transposed [62]. The monkey can relearn [63], and the human can relearn quickly [68]. Further, it took more than two years to cure urinary bladder function by learning in severe cervical SCI (see Results). The rat is not living much longer than the time needed for a repair in humans. A learning strategy would not be very helpful for the rat CNS, unless it is assumed that the rate of learning in rat is higher than in human.

5.2.5. Structural Repair in Human: The Neuronal Network Learning Strategy

The learning capacity of the human CNS is outstanding among the different species, and there is more time for learning and relearning than in the rat. To simplify the explanation, I assume an incomplete thoracic SCI, and for the repair, I concentrate the strategy on the intrinsic spinal cord networks consisting of the intumescentia cervicales, the intumescentia lumbo-sacralis and the connecting spinal cord parts. For repair, the whole subnetwork over the injury site is substantially activated (including the injured networks and the functionally surrounding not injured networks) by exercising rhythmic, dynamic, stereotyped, coordinated arm (mainly located in the networks of the intumescentia cervicalis) and leg movements (mainly located in the intumescentia lumbo-sacralis). To activate many of the spared fibres over the injury site, different coordinated arm and leg movements have to be exercised [54] such as walking (Figure 1C), running, crawling and exercising on the special CDT device.

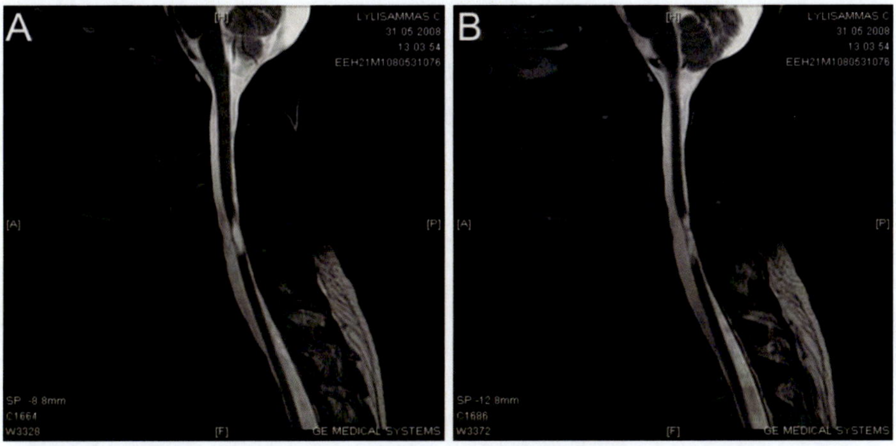

Figure 1A,B. Magnetic resonance images of a severe spinal cord injury at C5/6 levels of a 20-year-old female patient after three years of optimal coordination dynamics therapy. Both sagittal T2-weighted images seem to show connection between the rostral and caudal spinal cord. In comparison to the images made approximately two years earlier, there seem to be more connections between the disconnected spinal cord parts, because at the beginning of therapy connections could be seen only on one sagittal T2-weighted image.

Figure 1C. Fast forward walking on treadmill of a 20-year-old patient with a severe cervical spinal cord injury at C5/6 levels (Figure 1A,B) after three years of optimal coordination dynamics therapy (CDT). Leg support is given by two therapists. The patient is inducing the leg movement with the arms; in this moment only with the left arm. Author is on the right side of the patient.

The crawling, walking, and running are innate automatisms, which may strongly induce genetic support for repair (by changing the expression of regeneration-associated genes), and the exercising on the special device (Figure 5C,D) is primarily improving the stability of the impaired phase and frequency coordination (due to the injury). At the beginning of therapy, the repair is mainly a functional one. But when activating the spared fibres at the limits (range of overreaching [52]), a structural repair seems to be induced.

Figure 1D. Same patient as in Figure 1C upon jumping on springboard. The father is jumping behind in interpersonal coordination to increase the jumping amplitude.

The stimulus for improving the network connectivity over the injury site (not necessarily in the spinal cord) is the specific motor learning process. Those connections should be enhanced or newly built (by sprouting, building of new nerve cells from adult-born stem cells, change of efficiencies of synapses), which are needed during that motor learning process. Ongoing pre- and postsynaptic activity in the injured area during the learning-induced loop circuit formation has probably an important role in guiding the axons that extend from adult-born neurons. Local target-derived growth factors could attract regenerating fibres, or facilitate growth of regenerating fibres along highly active spared fibres, stimulated upon movement-based learning (range of overreaching [52]). Just placing stem or progenitor cells to the injury site seems not to be helpful to enhance repair [59]. The more coordinated the movements are performed, the more the coincidence principle of neurons [44] is used and the more of the relevant fibres and synapses can be activated at their limits at the injury site or those functionally connected to it. To enhance specifically the communication over the injury site, the patient has to concentrate (for example) on the automatic walking (Figure 1C). He/she substitutes partly missing connections in the network activation of the automatic walking by using volitional connections in the learning process (volitional support in the cranial → caudal direction). The therapist, supporting the legs during treadmill walking, brings the patient into the intended automatic walking pattern (Figure 1C) by adding power specifically at moments when additional power is needed during the walking cycle. In this way, massive rather natural afferent impulse patterns (see Figures 5,15 of [58]) enter the spinal cord for processing in the caudal → cranial direction to keep the walking pattern activated and to give information to higher centres. In this way, it is tried to stimulate the functional and structural repair of the injured circuits without disrupting the consolidation of the spared intact neuronal networks. The functionally recruited spared and overloaded intact fibres (and the synapses activated in (network) loop circuits) are supposed to give the stimulus for specific structural repair. The automatic walking pattern the patient intends to perform activates preferentially intrinsic spinal neuronal networks. Patients with incomplete SCI, with some sensory functions left, sometimes feel the activation of the intraspinal circuitry: "When I get into the rhythm during fast walking or running, then the moving gets easier as if I am getting wings."

According to this reasoning, as fewer fibres are spared, according to the injury, fewer fibres can be activated for functional and structural repair. In other words, the more severe the injury is, the more difficult the structural repair will be.

By administering the neuronal network learning strategy, it is an attempt to "teach" the injured nervous system of the patient to repair itself by learning.

In the process of organization and integration of different parts of the spinal cord during the relearning of movement and continence patterns, the incoming natural impulse patterns from the receptors of the periphery (bottom-up activity) and supraspinal centres (top-down activity) are integrated with the endogenous activity of the spinal cord by relative phase and frequency coordination. For the integration of newly built circuits, the coordinated firing of neurons has to be continuously relearned by exercising on special CDT devices (Figures 2;5C,D).

Since neurogenesis in the human CNS seems to be extremely limited [71], the movement-based learning therapy (CDT) has to be optimal and continuous to induce the building of new cells from stem cells and to induce the self-renewal of the "few" available stem cells. A break of CDT for a few months will stop the structural repair and a restart of the neurogenesis may need at least a few months.

5.2.6. Genotype-to-Phenotype Matching during an Individual's Lifetime

The nervous system and other phenotypic traits form during the ontogeny based on the genetic information stored in the cells (genotype). The information specified in the genotype determines aspects of the nervous system that are expressed as innate behavioural tendencies and predispositions to learn. The inherited genotype can completely specify the phenotypic network; i.e., both the network's architecture and connection weights are genetically determined. In this case, the behaviour of the network is entirely innate, and there is no learning [29]. Or the genotype specifies the network's architecture, but the weights are learned [28]. In still other cases, what have been selected during the evolution are good initial weights for learning or good learning rates and momentums [2]. In lampreys and frogs, the behaviour of the network will be mainly innate. In humans, the networks will have good initial weights for learning.

A direct genotype-to-phenotype mapping is implausible. In real life, we cannot predict which phenotype will emerge from a given genotype because of the large nonlinearities present in the mapping process. If the genotype is viewed as a set of instructions, it is not the case that each of these instructions will result in a single network property. Rather, the properties of the network emerge as the result of many interactions among the various instructions and their products [18].

Important processes involved in the genotype-to-phenotype mapping include cell division, cell proliferation from stem cells, and migration of neurons to reach their terminal position, especially because these processes may take place in the adult human CNS. A crucial property of the genotype-to-phenotype mapping in individuals is its temporal character. Development (or maturation) does not simply yield an "individual"; rather, the phenotypic individual is a succession of different phenotypic forms that are sequentially generated by the genotype in interaction with the environment. A model in which the genotype/phenotype mapping (i.e., ontogeny) takes place during the individual's lifetime and is influenced both by the genotype and by the external environment has been described [19].

5.2.7. Activating Genotype-to-Phenotype Mapping for Repair

If there is some similarity between development and repair, the genotype-to-phenotype mapping with its temporal character has to be induced in the patient. The phenotypic repair in the patient is a succession of different phenotypic forms that are sequentially generated by the genotype in interaction with the performed different movements, instructions and concentrations.

To repeat in the patient for repair the succession of different phenotypic forms macroscopically and on the cellular level, the movements attempted to administer to the patient in the sequence creeping, crawling, up-righting, walking and running. Automatisms have to be relearning before volitional movements. In very severe brain injuries or high tetras (C3/5 SCI), the automatisms swallowing and breathing have to be trained from the beginning of therapy. In animal repair research, it is tried to recapitulate the developmental guidance factor milieu [7]; in human repair research (CDT), it is tried to induce the expression of regeneration and learning-associated genes to generate different phenotypes of neurons [70] and to repair the networks.

5.2.8. Judging the Functional Stage of CNS Repair for Therapy Updating

The stage of repair in patients is quantified by different means to understand how the patient's CNS is functioning and how to adapt the therapy for optimal repair. Objectively, the quality of CNS self-organization is measured by recording the coordination dynamics when the patient is exercising on a special device (Figure 5C,D). The stability of coordinated arm and leg movements is measured by the arrhythmicity of exercising on that device upon the continuously imposed changes of the pattern of coordination between arm and leg movements between pace (in-phase) and trot gait (anti-phase).

The performance of other movements, as, for example, walking, is judged visually for optimal performance. Further, the supervising author is also performing supported treadmill walking to feel the volitional and automatic contributions of the patient's CNS to the walking performance and to give instructions to the therapists how to assist the walking with respect to achieve optimal performance (Figure 1C). Detailed anamneses of the patient's feelings with respect to the working of new muscles and further reappearance of sensitivity gives a clinical understanding of what is changing in the patient's CNS during the therapy-induced repair.

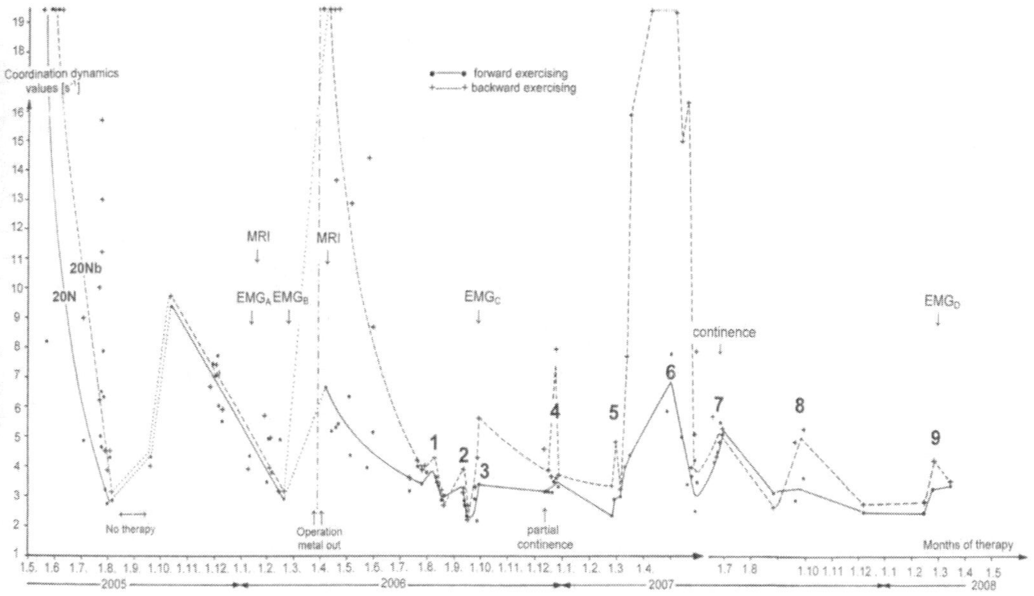

Figure 2. Relation of coordination dynamics values to therapy duration for a load of 20N and for exercising in the forward (lines and dots) and backward direction (20Nb; dashed line and crosses). Note that with no therapy, the coordination dynamics values got worse (increased) and upon therapy, they improved again. Upon metal removal, the coordination dynamics values increased strongly. The transient coordination dynamics value increases (peaks) "1" through "9" fall together with the reappearance of certain muscle functions or specific improvements of motor and autonomic functions and indicate therefore most likely small bits of regeneration. After the large peak "6" of transient coordination dynamics value increase urinary bladder functioning was relearned.

By comparing the actual knowledge obtained from the patient with the continuously measured and graphically pictured parameters (as, for example, shown in Figure 2) and the research knowledge and practical knowledge of similar patients in mind, the movements, performed during therapy, are updated.

5.2.9. Practical Aspects

To get real treatment quickly started in severe cervical spinal cord injury, one has to stop as quickly as possible the urinary bladder infections. They make the patient week and the treatment inefficient. The blood supply of the lower body has to be improved as quickly as possible, to avoid pressure ulcers and make the skin resistant to allow movements, otherwise already strong touching of the patient's skin will lead to bleeding. The special coordination dynamics therapy devices in the sitting and lying position (Figure 5D; help muscles can be used) are especially useful for starting the treatment and to improve the autonomic functions to improve the blood supply of the lower body (especially the skin) and to get sufficient blood supply to the brain for keeping the patient conscious. The start of the treatment is especially difficult in severe cervical spinal cord injury, since nearly no motor functions are spared for training. The earlier the treatment is started, the less the still innervated and not innervated muscles become atrophied. There even may be a critical period for regeneration in the spinal shock phase (see Discussion). For organizational reasons, the treatment is always started too late.

5.3. Results

A 17-year-old female patient suffered a severe cervical spinal cord injury in a car accident. No motor functions remained below the injury level of C5/6, and the patient had impaired feelings. From the MR images, the author estimated that approximately 5% of the spinal cord matter was spared. When the spinal shock faded away, it became obvious that no motor functions remained below the injury level but spotwise sensitivity remained more or less all over the lower body. Two months after the accident, CDT was started. Upon 2.5 years of CDT, the sensitivity improved and some motor functions returned below the injury level, indicating that some regeneration of the spinal cord had occurred. The connectivity over the injury site, according to the magnetic resonance imaging (MRI), may have increased to 8% (Figure 1A,B). In a previous publication, I reported about the improvement of the motor functions of this patient [54]; the improvement was compared to a patient in whom 50% of the spinal cord matter was spared [53]. This part reports on the cure of the urinary bladder functions.

In parts 1 through 3, I reported about the pathologic changes in the sacral micturition centre following a SCI by using the single-nerve fibre action potential recording method [55-57]. In a fourth part, the treatment to cure urinary bladder functions was developed by studying the induction of learning transfer from motor to urinary bladder functions [58].

5.3.1. Transient Increases of the Coordination Dynamics Values Fall Together with the Recovery of Motor and Urinary Bladder Functions

The changes of the coordination dynamics values are shown in Figure 2 for 2.5 years of therapy. Transient worsening of the coordination dynamics (higher values) fall together with the improvement of motor functions and probably indicate successful regeneration of small amounts of nerve fibres. Following the largest peak of transient increase of the coordination dynamics values (indicating the strongest transient regeneration), the urinary bladder started to function rather physiologically again upon 2.5 years of CDT.

5.3.2. Time Course of the Improvement of Urinary Bladder Functions upon CDT

I report about the stages of bladder repair upon two years of therapy. Changes of functions of the detrusor (bladder) and the external and internal bladder sphincters are extracted from a detailed anamnesis and are pictured by an evolution of the attractor layout with the relearning of bladder functions (Figure 3).

1. After the operation (fixation and alignment of the broken cervical spine), a lying catheter was installed in the patient. The patient was suffering continuous infections and fever.

It is understood that the bacteria are “creeping up” the lying catheter into the bladder (especially in female, because of the anatomy of the urethra) to give rise to ongoing infections in spite of antibiotic therapy. Before World War II (time of no antibiotics), patients were dying of such infections.

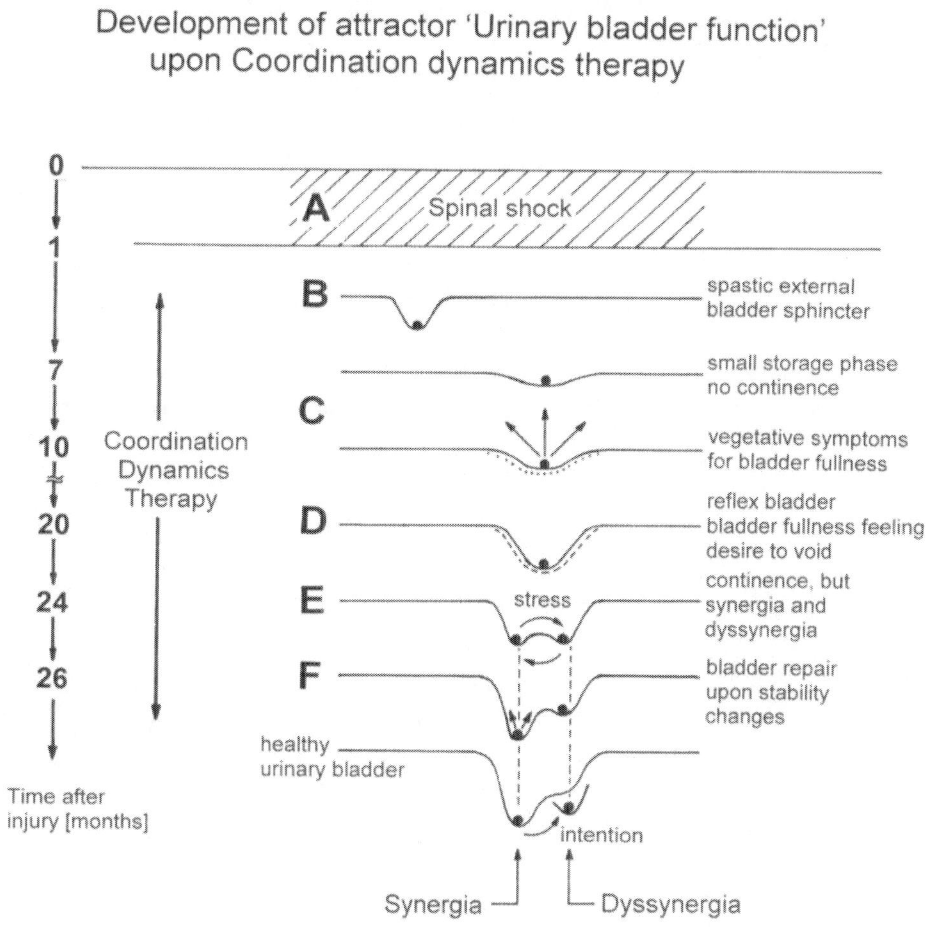

Figure 3. Evolution of the attractor layout of bladder functioning induced by learning transfer from movements to bladder functions upon CDT. The region around each local minimum of the potential landscape acts like a well that weekly traps the system into a coordinated state. Black balls correspond to stable minima of the potential. With learning, the pattern "spasticity of the external bladder sphincter" vanishes and the patterns for bladder functioning ("synergia" and "dyssynergia") appear anew and gain their physiologic stability (physiologic deepness of each basin of attraction). The corresponding attractor layout for physiologic bladder functioning is given. Fluctuation of pattern state (the black ball) (C), and their decrease (F), due to the impairment of phase and frequency coordination of neuron firing, is pictured in "C" and "F" by long and short arrows. Dotted and dashed lines indicate the reoccurrence of bladder sensation. Note that more than two years of optimal CDT were needed for bladder repair.

It is the benefit of Sir Guttmann from Breslau [6] who stopped or reduced these infections by introducing sterile intermittent catheterization. The introduction of antibiotic therapy helped further.

2. One month later (at a time when the spinal shock weaned), a suprapubic catheter was installed, and no more infections occurred. But the bladder did not show any physiologic functions. The patient had no feeling of bladder fullness, no desire to void and did not feel when the fluid was leaving the bladder. The catheter was used for emptying (when opened) and storing (when closed) the urine. Since no fluid was leaving the bladder through the sphincters, the external striated sphincter was spastic

(continuous contraction), and the internal sphincter (smooth muscle), as a part of the detrusor, was probably not working.

Physiologically, the internal sphincter (smooth muscle, slowly acting, probably a part of the detrusor) is keeping the continence for low and medium bladder pressure. For high bladder pressure and sudden bladder pressure increases (as for example during coughing), the fast external sphincter (innervated by α_2-motoneurons and consisting of fast fatigue resistant muscle fibres (FR)) is contracting to secure continence. If the striated external sphincter is not working properly, patients suffer the so-called "stress incontinence."

3. Seven to eight months after the accident (end of 2005), the fluid was leaving the bladder by itself after a small storage phase. This means that therapy had reduced the spasticity of the external sphincter. The patient was now incontinent. So far, the spastic external sphincter had mainly stopped the fluid from leaving the bladder by its spasticity. The internal sphincter started to work a bit to allow a small storage phase. When the fluid was leaving the bladder, there was first no feeling of fluid movement. Later on, the patient had some feeling of fluid movement. Probably flow receptor afferents (S2) started to work. Three months later, the suprapubic catheter was taken out. The patient started to use diapers.
4. Twenty months after the accident (beginning of 2007, upon 18 months of therapy), the patient felt bladder fullness and the desire to void. Probably stretch (S1) and tension receptor connections (ST) started to work again. The vegetative symptoms of bladder fullness information (sweating and sudden heard rate increases, probably transmitted by plexus connections) were replaced by bladder fullness feeling and the desire to void.
5. The patient became able to press the fluid out of the bladder. To get all fluid out, the reflex bladder had to be activated a bit, by tapping, touching or massaging the skin above the urinary bladder, which is the reflex skin area for the bladder reflex. Sometimes body positioning was used to influence the bladder reflex. With these manoeuvres, the desire to void reappeared and the patient could empty the bladder further.

Often patients (to whom no CDT is administered) are training the bladder reflex for emptying. The reflex is stimulated by tapping the skin above the bladder. But if, for example, the external sphincter is spastic (as in our patient), it may not be possible to generate a good functioning reflex bladder.

6. After the appearance of the desire to void, the patient became able to hold the fluid for 30s to 1min. Sometimes, she could keep the continence better and sometimes not so good. This means that the external bladder sphincter (which can be controlled volitionally) started slowly to work, but irregularly. The feeling of bladder emptying became similar to those before the spinal cord injury.
7. Twenty-four months after the accident (spring 2007), the bladder started to function rather physiologically again. After a storage phase, the fluid came out on volition. The detrusor started to work fully. But if the patient was pressing too much at the end of bladder emptying to get all fluid out, the external sphincter contracted. The

external sphincter co-contracted with the detrusor. Detrusor-sphincteric dyssynergia of the urinary bladder occurred. But when she then activated the reflex bladder by tapping or touching the reflex bladder area, the desire to void reappeared, and she could empty the bladder fully. The residual urine was not measured.

At this stage of bladder repair, two patterns existed: the synergia of the bladder, in which the detrusor and external sphincter contracted antagonistically, and the dyssynergia of the bladder, in which the detrusor and the external sphincter co-contracted. The synergia pattern was for emptying the bladder, and the dyssynergia pattern was the pathologic pattern. The pathologic co-contraction of the external sphincter with the detrusor occurred more easily when there was less fluid in the bladder and the patient had to press more (inducing stress to the CNS).

Physiologically, both bladder emptying patterns do exist. The synergia pattern is for emptying the bladder, and the dyssynergia function is for stopping the micturition. But both patterns are under volitional control.

8. Upon two years of coordination dynamics therapy (26 months after the accident), the patient was fully continent again and could empty the bladder on volition. The time interval from the first feeling of the desire to void to the situation that the fluid was coming out by itself (including four times of occurrence of the desire to void) was one hour. The patient did not use diapers anymore. The patient had never used drugs that are supposed to improve urinary bladder functions. The repair of urinary bladder functions was achieved by the relearning of urinary bladder functions including transfer of learning from the movements jumping on springboard, treadmill walking and exercising on the special CDT device for turning (Figure 5D). The strong improvement of urinary bladder functions occurred when the coordination dynamics values strongly increased (got worse) (Figure 2), indicating regeneration.

For patients with incomplete spinal cord injuries, it is very important how long they can hold the urine from the first desire to void until the moment when the fluid comes spontaneously. Can they safely reach the WC or not? The improvement of bladder functioning can also be judged by how long the patient can hold the fluid following the first desire to void. In this case, it was one hour after two years of CDT.

The feeling of the lower abdomen, which was poor after the accident, improved also strongly at the time of nearly full bladder repair. The patient felt again the lower abdomen very good (inside and outside as the patient reported) and felt also again the working of the abdominal muscles. At that time, also the finger functions got a tiny bit better and her supported treadmill walking improved. During walking on treadmill, and during other movements, the patient got goose-pimples all over the body. It seems that an overall improvement of vegetative and somatic functions occurred at the time of full bladder repair.

5.3.3. Attractor Layout Changes during Urinary Bladder Repair

Within the framework of System Theory of Pattern Formation, the repair of the urinary bladder functions can be understood and pictured by the changes in the attractor layout.

One month after the injury, when the spinal shock faded away (Figure 3A), only the pathologic bladder pattern “spasticity of the external sphincter” was present (Figure 3B). Six to ten months later, the spasticity of the external sphincter reduced, and a small storage phase of the bladder reappeared as a first sign of bladder repair (Figure 3C). Twenty months after the operation, the reflex bladder pattern organized itself with bladder fullness feeling and the desire to void (Figure 3D). Twenty-four months after the accident, the attractor layout showed two attractors, the bladder synergia (the detrusor action inhibits the external sphincter) and the dyssynergia (co-contraction of detrusor and external sphincter) (Figure 3E). The state of the system (pictured by the ball) switched easily from the attractor synergia to the attractor dyssynergia. Twenty-six months after the accident, the stability of the attractor synergia had increased and the stability of the attractor dyssynergia had decreased (Figure 3F). On volition (intention), the micturition could be induced and stopped as in healthy individuals.

5.3.4. Conclusion

Upon 2.5 years of coordination dynamics therapy, urinary bladder functions could be cured in severe cervical SCI. Since the injury was motoric complete and the cord was "free" in the spinal canal (Figure 1A,B), some regeneration in the human spinal cord should have occurred.

5.4. Discussion

5.4.1. Integrative Functions and Central Pattern Generators for Urinary Bladder Repair

It was shown in this patient that urinary bladder functions could be cured upon 2.5 years of coordination dynamics therapy (CDT). Three important steps were achieved. First, the patient got the bladder under volitional control again. Second, a physiologic attractor layout for bladder patterns could be generated; and third, the dyssynergia of the urinary bladder could be cured by increasing the stability of the synergia pattern and decreasing the stability of the dyssynergia pattern. The stability changes of the two bladder functioning patterns can be understood within the framework of system theory of pattern formation and human neurophysiology [55-58] but not in the framework of central pattern generators (CPGs).

5.4.2. Similarity of Development between the Ontogenetic Landscape for Locomotion and the Evolution of the Attractor Layout for Urinary Bladder Functions upon Learning

The development of the potential landscape of urinary bladder functions upon repair (Figure 3) shows some similarity to the development of the ontogenetic landscape for locomotion (Figure 4). Also, the attractor layout for locomotor functions develops from a monostable to a bistable or multistable structure. With learning not only attractors appear anew but also attractors vanish. During the relearning of bladder functions, the attractor "spasticity of the external bladder sphincter" vanished. There is some similarity between the repair by relearning and the development of sensory-motor functions during ontogeny. Following spinal cord injury, the stepping automatism and the urinary bladder reflex reappear upon learning. During ontogeny and repair, the automatic stepping is changed by supraspinal control into walking and running patterns, and the bladder reflex (or automatism) is altered into proper micturition and continence functions.

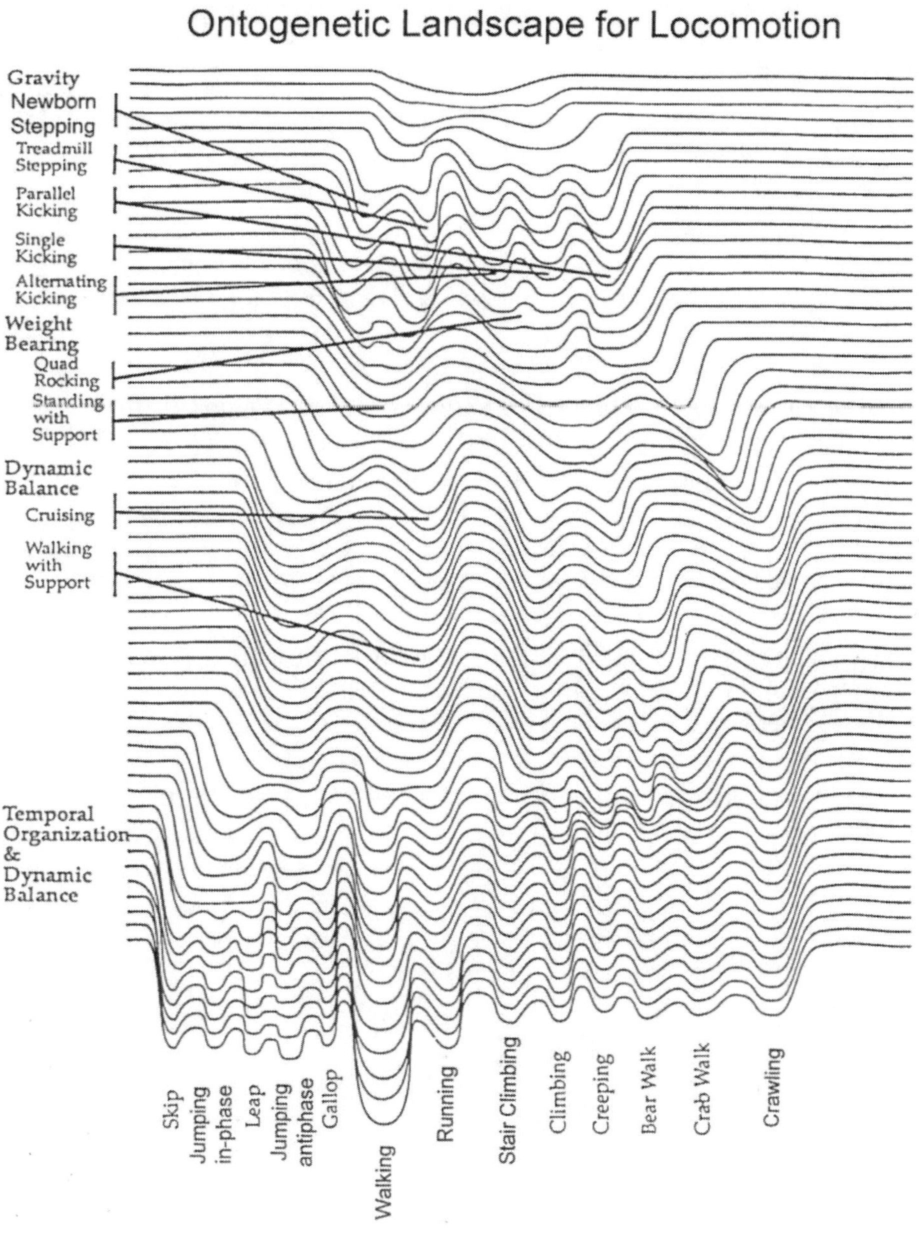

Figure 4. Ontogenetic landscape for locomotion. The evolution of the attractor layout for the different movements develops similarly as the relearning of urinary bladder functioning upon CDT. With permission of Esther Thelen [72].

5.4.3. Vegetative Symptoms Are Not Autonomic Dysreflexia

Normally, exercising does not pose special risks to people with SCI, apart from the starting period. Fractures or muscular injuries occur more often in sports than in patients with SCI. It is argued that people with SCI have atypical physiological responses to exercise (for example, abnormal heart rates and sudden strong sweating), which can limit their ability to sustain intensive exercise [31]. When a patient cannot feel the fullness of the urinary bladder and has no desire to void, then the bladder can inform the mind only by vegetative symptoms, namely sudden heart rate increases and sweating.

Most patients know their vegetative symptoms and empty the bladder before stronger exercises. But if patients do not know their vegetative symptoms for body control or ignore them, then even small blood vessels can tear in the brain during sudden blood pressure increases. Patients with SCI have to relearn their body functions. A big progress for the patient of this report upon 20 months of CDT was when the vegetative symptoms where replaced by the bladder fullness feeling and the desire to void. To treat a patient with severe cervical SCI, theoretical and practical knowledge is needed for the repair of motor and autonomic functions.

5.4.4. Urinary Bladder Repair in Children with Spinal Cord Injury

Nervous system complexity increases not just across phylogeny but also across ontogeny. The loss of regenerative capability with age demonstrates the cost of the complexity [7]. In children with spinal cord injury, it should therefore be easier to repair urinary bladder functions, since the nervous system is less complex and the rate of learning higher.

A four-year-old boy suffered a flaccid paralysis due to a SCI sub Th12/L2 in a car accident. At an age of six years, he underwent CDT for six months. The boy came to the therapy centre in a wheelchair and left it by walking with sticks [42]. The functioning of the urinary bladder improved. Before the therapy, the bladder was emptied by intermittent catheterization. After six months of therapy, the patient was continent for at least three hours and could empty the bladder on volition with little residual urine.

A more likely explanation that children can relearn lost functions faster may not be the not fully achieved nervous system complexity but that the repair during development is partly due to a correction of the development "en route." The increase of the nervous system complexity, including neurogenesis and growing of axons and dendrites, can be used by movement-based learning for functional and structural repair. Young cerebral palsy patients have therefore a good prognosis for repair upon CDT [45].

5.4.5. Continence in Elderly People

Continence problems in elderly people can be due to several reasons. But a generally unknown reason is that the sacral micturition centre is getting an insufficient blood supply through the radicular-medullary arteries. The thickest feeder artery to the sacral micturition centre, the A. spinalis magna (Artery of Adamkiewicz), is supplying 68% of the blood [34] (Figures 12,14,15 of Chapter III) and is often very atherosclerotic. A combination of CDT and a reduction of arteriosclerosis may improve continence in elderly people.

5.4.6. Therapy-induced Regeneration

It has been reported that SCI is a devastating condition for which there is as yet no cure (3,65), even though it has been published that SCI can partly be cured by CDT [41,42,53]. Probably, in most of the patients, the repair was mainly a functional one. The century step forward is that also in a motoric complete cervical SCI, which is really a devastating condition when there are nearly no motor functions remained, the most important urinary bladder function can be cured, and this cure includes some of regeneration of the human spinal cord besides a functional repair.

In experimental (animal) SCI, the spinal cord is only partly injured (spinal hemisection; see discussion of the pictures in the Refs.3,65). But a partial SCI is not the problem to be solved anymore, because it has been shown that a partial SCI can be cured [42,53] (Chapter II). The challenge is the severe and real complete cervical SCI (according to MRI); even though in nearly every injury, a few fibres are spared. But researchers in the field of animal research seem to hesitate to handle urinary bladder problems, which they would get if they would transect the whole spinal cord in rat, cat or dog. Emptying the urinary bladder of animals a few times per day seems not to be an attractive part of research work. Also, regeneration will be much more difficult to achieve in complete SCI.

5.4.7. Extent of Regeneration of the Human Spinal Cord

An important question now is how many regenerating axons do support the functional improvement obtained in this patient with a motoric complete cervical SCI. When in adult rats as little as 1-2% of the corticospinal tracts was spared following electrolytic lesion, performance of forepaw reaching recovered [20,21]. These observations imply that regeneration of a small percentage of axons and functional reorganization of the CNS, including the intrinsic spinal cord networks, could lead to a significant recovery of function. In adult cats as little as 10% of spinal white matter tracts were sufficient to permit spontaneous walking without external support [69]. When in patients large proportions of spinal tracts were transacted owing to intractable pain problems, disturbances in motility were small [32].

Since during CDT those fibres were stimulated to regenerate by coordinated sensory-motor learning in the patient of this report, which were really needed for function, maybe 1 to 3% of the white matter regenerated. Since the trunk muscles and also the intercostal muscles started to function again (no paradox breathing anymore, repair of respiratory function) besides some upper leg muscles, probably less than 1% of the white matter regenerated for the activation of the neuronal networks for driving the pelvic floor (of which the external bladder and anal sphincters are a part), the detrusor and the bowel. It seems that only a very small percentage of fibres need to be successfully reconnected with their targets to mediate urinary bladder recovery. Since the spinal cord was strongly injured only over one vertebra length (Figure 1A,B), axonal grows and elongation will have been shorter than one vertebra length, depending whether atypical synaptic relay circuits were formed at the injury site. This additional small amount of regenerating ascending and descending fibres may not be visible in the MRIs of patients with approximately 5% (sensory) spared white matter.

That these established connections are mainly activated ectopic pathways through root connections, sympathetic chain or plexus connections (misdirected haphazardly during development) is unlikely, since the functions of trunk and leg muscles reappeared from the rostral (below injury level) to caudal direction, indicating a regeneration in general.

5.4.8. Endogenous Adult Neural Stem/Precursor Cells for Spinal Cord Repair

The partial structural repair in this patient was probably achieved by activating intrinsic self-repair processes. The efficient and intensive CDT may have generated a local microenvironment in which endogenous adult stem/precursor cells self-renewed, proliferated, differentiated and migrated to contribute to the CNS repair. But what are the explanations why this intrinsic repair mechanism is so slow (more than two years) and so limited?

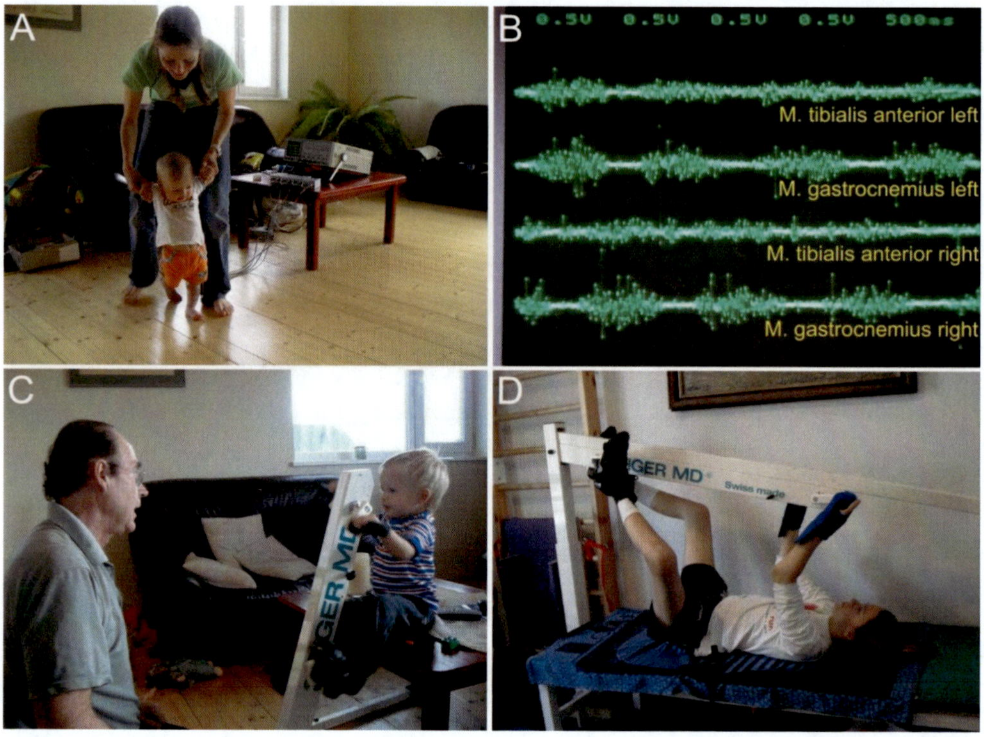

Figure 5. A. Recording of surface EMG during walking of eight-months-old "Jürgen," supported by the mother. B. Surface EMG motor programs of Jürgen show no antagonistic action between the tibialis anterior and gastrocnemius muscles. C. Two-year-old Jürgen becomes able to exercise with difficulties on the special CDT device for babies (the coordination dynamics are given in Figure 6A). D. Patient with severe cervical spinal cord injury during exercising on the special CDT device in the lying position. Note, in this exercise position, patients with spinal cord injury could train already a few days after the spine fixation, since the back would be nicely supported during exercising.

Firstly, if the sciatic nerve in human is injured, the repair needs also more than two years. Secondly, among other reasons (possibly inflammation), the therapy should have started already in the spinal shock phase. If the patient would have started already to exercise one week after the accident (or earlier) on the special CDT device in the lying position (safe position for the reconstructed spine for moving arms and legs (Figure 5D)), the blood supply and the circulation of the cerebrospinal fluid would have reoccurred/improved much faster after the injury. Damage through active secondary processes would have been reduced, and adult neural stem/precursor cells could have reached the injury site earlier. Thirdly, if there is similarity between development and repair (see below), a critical window for more efficient repair may exist. A "plasticity window" could have already been closed before optimal CDT was started. Fourthly, the complexity of the human CNS may not allow a faster regeneration to avoid instability in CNS self-organization (see below). It has been reported that without stimulation, there is only undetectable neurogenesis in the adult neocortex [71].

5.4.9. Combination Therapies to Enhance Regeneration

It is supposed to be widely accepted that no single approach will prove sufficient for successful regeneration [3,7]. A methodology combining the most effective individual therapies is required. But it is shown in this article that CDT alone could already cure urinary bladder functions. This single human therapy was able to induce successful regeneration. I also believe that a combination of therapies may act synergistically to enhance regeneration. But first, it is not enough to ask for combinational treatments, one also has to support them; and second, each single approach must have a solid scientific basis and must be a safe treatment without unacceptable side effects. A combination of stem cell and CDT could enhance regeneration, since the neuronal network repair strategy used in CDT can generate a microenvironment at the injury site for the self-renewal, proliferation, differentiation and migration of neural stem/precursor cells, even though a first therapy trial was not successful [59]. This movement-based learning therapy (CDT) probably stimulates the endogenous stem cell reserves for structural improvement. Additional administration of embryonic or adult neural stem/precursor cells may then enhance the rate of structural repair.

5.4.10. Scientific Basis for Therapies to Repair the Human Spinal Cord

Basic scientific research should provide a rational basis for tailoring specific combinations of therapies [65]. It has been reported that neuro-rehabilitation should have a scientific basis. But CDT has a theoretical basis (this book and Refs.38,50,52-59). Reviewers cite human work without a scientific basis and argue then that neuro-rehabilitation methods should have a scientific basis. It has been measured in human with the single-nerve fibre action potential recording method that the phase and frequency coordination among neuron firings is getting impaired following SCI [58] (part 4). One important consequence of this impairment is that the variability of phase and frequency coordination increases and CNS organization becomes unstable. A macroscopic consequence of this instability of cell communication for self-organization is that movements and other patterns get changed and unstable. This instability of movement patterns can be measured macroscopically (Figure 6C,D) and improved when the patient was exercising on the special CDT and recording device (Figure 5D).

But not only will the injury induce instability of CNS organization. An uncontrolled building of new nerve cells and growing of neurits will also induce instability of phase and frequency coordination of CNS self-organization if the newly formed connections are integrated in the existing injured networks. This instability of CNS functioning will lead to an increase of the instability of physiologic patterns and an increase of the stability of pathologic patterns like the different kinds of spasticity in SCI and epileptic seizures in brain injury. A structural repair of the spinal cord (or CNS) in humans must therefore be a strongly controlled specific mechanism to maintain the delicate balance between plasticity and stability. In Figure 2, it can be seen that with every bit of reoccurrence of muscle function (successful regeneration of a few fibres), the coordination dynamics values got transiently worse (the movement patterns got transiently more instable), until the movement-based learning therapy reestablished stability again. These transient instabilities are marked in Figure 2 with numbers. Following the largest peak 6 of transient network instability, the main urinary bladder repair was achieved.

The use of methods that are sensitive enough to detect potentially small increments in function has been asked for [65]. The coordination dynamics measurements can quantify small increments in function and regeneration (Figure 2). It is not enough to ask for sensitive methods, one also has to use them.

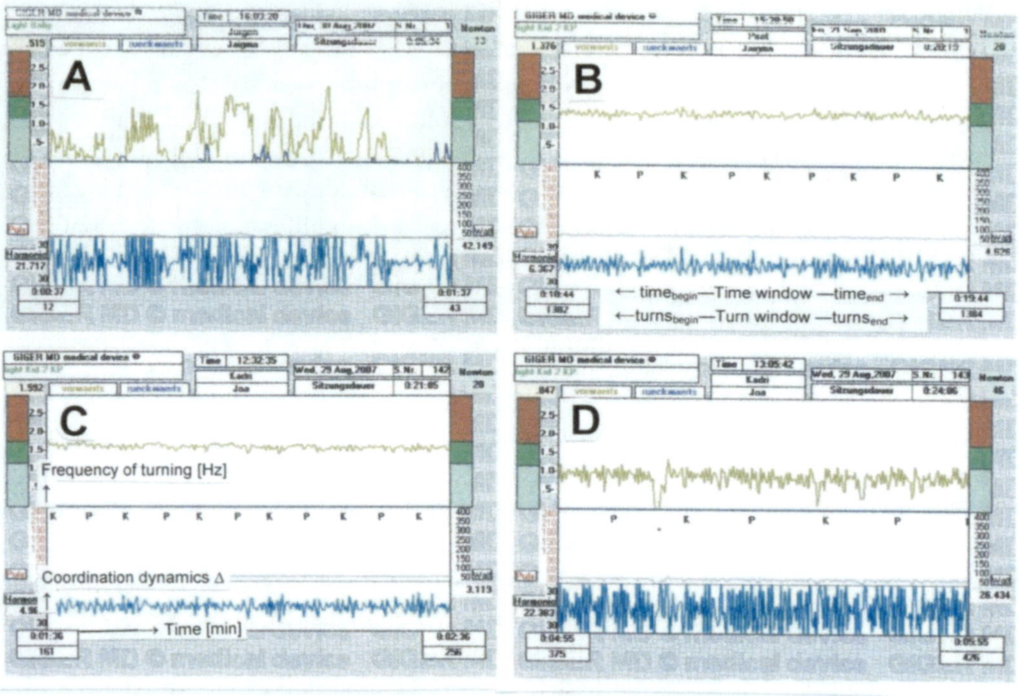

Figure 6. Coordination dynamics recordings from a healthy two-year-old child (Jürgen) (A), his healthy mother (B), and a patient with a formally motoric complete SCI for low (C) and high load (D). Upper trace - frequency; lower trace - arrhythmicity of exercising = coordination dynamics.

The structural repair has therefore to be highly specific and controlled at the injury site to avoid instability, which means the growing of nerve fibres has to be enhanced and inhibited according to the functional needs of the network repair. If neurobiologists really succeed in inducing massive growth of nerve fibre (by enhancing growth or inhibit the inhibition of growth) and if these drugs are applied to humans, then uncontrolled nerve fibre growth may occur and the human CNS probably may get instable in its self-organization; different instabilities may occur as, for example, epileptic seizures and mass contractions of muscles. Upon administering successfully stem cells and nerve growth factors to patients, there is therefore not only the risk of teratocarcinoma and the toxicity of side effects but also the risk of inducing extensive instabilities in the self-organization of the human CNS. To say it with the poet Goethe: “Die Geister die ich rief (neurogenesis, gliogenesis and neurit growth), werd ich nun nicht los” (uncontrolled growing and generation of CNS instabilities) (Zauberlehrling).

The risk of rising instability of CNS organization is much bigger in the human than in the animal CNS due to the much higher complexity and variability of the human CNS organization, necessary for learning.

Brain imaging research focused on detecting the brain centres involved in various sensomotor or cognitive tasks, but CNS organization research attempts to understand the self-organization of distributed local networks and their integration and coordination to “higher” states of processing [66].

Extensive learning potential needs specific variability in the system. For not getting instable in the changing self-organization, there must be stability control. Putting this stability control partly out of order may be dangerous for CNS functioning. In young patients with severe brain injuries, strong instabilities can be observed and measured [47].

If in lower animals the circuitry is rather fixed and the nerve cells are rather specifically connected, then the repair needs only the appropriate cells substituted and the connections to be repaired. But if in the human CNS the specific, fixed wiring is rather limited, then the repair has to be controlled by network functions; an uncontrolled neurogenesis and growing of nerve fibres may be dangerous.

5.4.11. Similarity between Development and CNS Repair

Nerve Fibre Growing Strategy in Animals

It has been shown in frog that there is close similarity between the innervation of muscles during development [33] and reinnervation following denervation [26,27] (Chapter I). Two types of motoneurons with thin and thick axons communicate with slow and twitch muscle fibres to achieve the proper structure, namely that small neurons with thin axons innervate the slow muscle fibres and the large neurons with thick fast-conducting axons innervate the twitch muscle fibres. The repair (the reinnervation) is recapitulating the development. Regenerating axons recapitulate developmental guidance during recovery from the denervation injury.

Two approaches have been emphasized to recapitulating development in the injured CNS: reestablishing crucial developmental cues in the correct pattern to guide regenerating axons and maximizing the sprouting and plasticity of intact fibres through sensory feedback rehabilitation techniques [7]. A faithful recapitulation of developmental path finding and circuit-refining mechanism is likely to be beneficial in the repair of SCI in animals but may only partially hold in humans (see below). By measuring the self-organization of neuronal networks upon pattern change (by using the coordination dynamics assessment) during human development [46] (Chapter II) and human CNS repair [47] (Chapter II), similarities were found between the attractor layouts. More human research has to be performed to learn how during development the human CNS is correcting/optimizing its functioning “en route.”

Here I report that by learning, learning transfer [58], and strong motivation for training, urinary bladder functions could be repaired, including some structural repair. More generally, I report how the mind is taking part in the functional and structural repair of its neuronal networks by which it is generated.

Almost as soon as the first computers were made, the analogy with the brain was accepted. The model of the brain as a computer, with hardwired connections that are adaptive and flexible mostly with respect to the software and the information that is distributed over the hardwired network still dominates. The genius property of the human CNS seems to be that by excessive learning in the long-term memory, the network circuitry can be changed and is not fixed as in computers. The spirit can rise above the physis.

I believe that the property of improving and repairing the CNS structure by the mind is unique to humans. Therefore, not all repair mechanism of the human CNS can be explored by animal research. Even monkey experiments cannot substitute for human CNS research. If nerves, tendons or muscles are transposed, the rat cannot relearn [62], the monkey after several trials [63] and the human immediately [68]. The relearning capacity in the human

CNS with respect to functional reorganization to compensate for changes in the periphery is used in neurosurgery, reconstructive surgery and microsurgery.

Many patients are trying to use their minds to repair their SCI. They concentrate on the movement and performing the movement in the mind again and again, and the natural impulse patterns, generated in the patients' CNS, are running against the injury site in the rostral-caudal direction. But I have not heard of any success with respect to repair in complete SCI by using only the mind. First, at least some fibres probably have to be spared to secrete target-derived growth factors and second, very thoughtful movements have to be performed repeatedly for a long time (in this case two years or more) in cooperation with the concentration on the task. Further, to activate substantially the injured neuronal subnetworks for repair, also coordinated movement induced afferent input is needed (see Method). For network repair, the networks have to be intrinsically activated integratively and at physical limits. One kind of input to the injured network is insufficient for repair.

5.4.12. CNS Development and Repair in Animal and Human: Growth, Guidance and Learning - Why Animal Research Alone Cannot Cure SCI in Human

Two approaches have been proposed to recapitulating development in the injured CNS: Reestablishing crucial developmental cues in the correct pattern to guide regenerating axons and maximizing the sprouting and plasticity of intact fibres through sensory feedback rehabilitation techniques [7]. Reestablishing at least part of the developmental pattern of guidance molecule expression would contribute to any regenerative approach for treating CNS injury in adult animals [7]. Secreted and membrane-bound isoforms of guidance molecules interact with receptor complexes. A recurring theme shared by many of the ligand-receptor combinations directing tract guidance is that interactions can result in growth cone attraction or repulsion, depending on the receptor subtype and intrinsic state of the neuron. This intrinsic state differs among not only neuronal subtypes but also developing neurons, adult neurons and regenerating neurons. Generally, developing neurons possess the intrinsic state most suited for rapid axonal elongation and target finding.

Circuit Connections and Plasticity

In many developing neurons, a surplus of axons initially reaches their targets. Many preliminary synapses form only to retract soon after. Winners of this game are determined by competition for limiting target-derived growth factors. Eventually, excess neuritic branches are pruned, successful neuritic branches stabilize, and many cell bodies that lack victorious nerve terminals undergo apoptosis.

The Flexibility of Immature Circuits

Once axon path-finding and the pruning process have selected for the formation of appropriate synaptic connections, further plasticity occurs in an activity-dependent manner. This results in the potentiation of some connections and inhibition of others. Through learning, neuronal networks become progressively more organized.

Disruption of sensory experiences (behavioural information) within certain age windows results in circuit reorganization, allowing the organism to adapt within the new sensory information including the information generated by sensory-motor coordination. In barn owls, over several weeks they adapt not only regaining auditory-visual coordination but also visual-motor coordination [13]. The ability to adapt correlates negatively with age – owls over 200 days old are unable to regain auditory-visual coordination [14]. Critical periods define the age windows during which various circuits retain the plasticity to sensory deprivation.

Generating Spinal Cord Plasticity

Plasticity applies also to the intrinsic spinal cord circuits that mediate locomotion. Spinal cord neuronal networks generate coordinated activity between groups of agonist and antagonist limb muscles on opposite sides. The networks allow ambulation to become a nearly automatic neural program. Newborn babies (38 (Figure 43)) (Figure 2 of Chapter II) and anencephalic human infants can step automatically. After birth, the boy Jürgen (Figure 5A,C) could step automatically forward but not backwards. This automatic stepping automatism and walking automatism (including arm and leg movements) in the forward and backward direction is used excessively in CDT. The role of axon guidance molecules in coordinating proper neuronal network connections in animals has been emphasized [7]. But the coordination in human infants is very poor [46]. Its improvement depends crucially on the training of coordinated movements. Eight-month-old Jürgen (Figure 5A) has very poor coordination between the motor programs of tibialis anterior and gastrocnemius muscles (Figure 5B, missing antagonicity), even though he is healthy and quite early for his age. At an age of two years, the coordination between arm and leg movements is still poor (similar to in other healthy children), as is measured by the coordination dynamics in Figure 6A. Temporary deprivation of sensory feedback to the rat neuronal networks in the postnatal period results in permanent walking and swimming deficits, defining a critical period for neuronal network plasticity [67]. Humans can relearn walking and other movements throughout life [41,42] following SCI.

Inhibitory neurotransmitters have a crucial role in plasticity. The transition from synchronous (in-phase, pace gait) to alternating bilateral rhythmic limb movement (anti-phase, trot gait) late in gestation depends on glycinergic signalling [16]. Serotonin, acting through 5-hydroxy-tryptamine (5-HT) receptors, delays the maturation of GABA-mediated inputs into lumbar versus brachial spinal circuits [1]. Human babies learn first the pace-gait and then the trot-gait crawling. Human infants who left out the crawling period may have coordination problems throughout life unless CDT is administered. The crawling automatism is used in CDT.

Closing the Plasticity Window and Learning Plasticity

Understanding the mechanism underlying critical period closure will form the basis for approaches to reestablishing plasticity in regenerating adult nervous system. Not only neurotransmitters contribute to the definition of plasticity windows. Extracellular chondroitin sulphate proteglycans (CSPGs), produced mainly by astrocytes, for perineural nets around inhibitory interneurons in the visual cortex, coincident with the closure of the critical period window [17,25,30]. CSPGs inhibit neurite outgrowth, probably blocking the synaptic structural dynamics that partly underlie plasticity [30]. Enzymatic CSPG removal reopens the critical period [30]. Therefore, maturing astrocytes and oligodendrocytes have a role in consolidating neuronal plasticity.

The mechanisms that evolved to consolidate plastic circuitry directly contribute to the inability to regenerate after CNS injury. Hence, plasticity and regeneration are intertwined. The most effective therapeutic approach in SCI repair in animals will untangle these pathways to allow the regeneration of injured circuits without disrupting the consolidation of existing intact circuits. In humans, the mechanisms that consolidate the plasticity in the circuitry may contribute to the control mechanism to avoid instability of physiologic patterns and to avoid the development of pathologic self-organizations as, for example, epileptic seizures. But the learning process offers a well-balanced mechanism for functional and structural repair without inducing large instabilities in the CNS self-organization. Small instabilities do occur during the regeneration of the human spinal cord, as can be seen from Figure 2. These reoccurring small instabilities were resolved upon proper CDT.

Neural Response to Injury and Differences in Mind-Structure Communication for Repair in Animals and Humans

CNS injury, disease, and malformation wreak havoc on the intricately choreographed neural circuitry. Unlike the many similarities across species in patterns of neural development, responses to injury and repair differ greatly. Responses to injury also vary in individual organisms, depending on time, anatomical location and type of injury. The most effective spontaneous injury responses occur in lower vertebrate species, at younger developmental ages, and in the peripheral nervous system (PNS) rather than the CNS. A detailed understanding of the factors responsible for these differences provides insight into the challenge of improving treatments for adult mammalian and human CNS injury. The main difference for repair between animals and humans is that the repair in animals occurs mainly spontaneously and in humans mainly volitionally. The mind-structure communication in animals for inducing non-spontaneous repair is very limited. Also devices for training are missing. Rats, frogs, and lampreys cannot turn on the special CDT device to improve the phase and frequency coordination for the improvement of CNS self-organization (Figure 5CD), cannot jump on springboard for rhythmic natural stimulation of the sacral micturition centre to improve continence [58], cannot perform supported treadmill walking like in Figure 1C, and cannot be instructed and motivated during therapy. It is impossible to explain to a rat that it has to train certain movements for one or two years excessively or should concentrate on certain muscle activations for repairing its injury. If a limb hinders locomotion too much, the rat will eat it.

Translating Animal Data to Human Data

To make use of animal research for the improvement of human diseases, a correlation has to be established between animal and human data. For correlating nerve fibre groups for example between rats and humans, there must be a detailed classification scheme for rat nerve fibre groups and human nerve fibre groups [35,38]. In Chapter IV, single-nerve fibre APs were recorded in rats, and a rat model was introduced with which regeneration can be measured. For a close comparison, a translation from rat data to human data, an up-to-date classification for the rat has to be made. The existing classification schemes [5,8,9,22] are too inaccurate for a comparison to human data. Animal research needs to improve its basic data. To make use of the regeneration data obtained from animals, also regeneration data of humans are needed for a comparison. Such data do not exist, apart from a few exceptions.

To make use of animal data to cure diseases, parallel studies have to be performed in animals and humans, and similar treatments and measurements for quantifying progress have to be used. The switch from animal experiments to human trials is a too simple thinking and too dangerous for patients.

5.4.13. For Repair Functions Also Functions Have to Be Measured (Best with Basic Methods) and They Differ Strongly between Animals and Humans

As emphasized above, in many developing neurons, a surplus of axons initially reach their targets. Winners of this game are determined by competition for limiting target-derived growth factors. Once axon path-finding and the pruning process have selected for the formation of appropriate synaptic connections, further plasticity occurs in a pattern-activity-dependent manner. This results in the potentiation of some connections and inhibition of others. Through learning, neuronal networks become progressively more organized.

These steps also hold during the development of the frog (tadpole stage), when the faster developing large neurons innervate slow and twitch muscle fibres first (Chapter I). The slow muscle fibres also have fast contracting properties, and the membranes are generating action potentials, as the twitch muscle fibres do, even though the morphology is different (no M-line). When the slower developing small motoneurons reach the slow muscle fibres, they "push" the synapses of the large motoneurons away and change membrane and contracture properties (but do not change morphologic properties). The membrane cannot generate an action potential anymore, and the muscle holds a contracture upon depolarization. The formation of appropriate synaptic connections occurs therefore two times (first with the arrival of the fast and then with the arrival of the slowly conducting axons), and the specificity is only relative in that frog model (Chapter I). In human, there are three kinds of muscle fibres, innervated by three kinds of motoneurons [36,55]. The specificity of correlation between muscle fibres and motoneurons is unclear.

The intricate changes during development can only be explored using at least morphology and electrophysiology. The regeneration following a section of the sciatic nerve in the adult frog is following the same scheme, as can be measured with intracellular recording techniques and Ca transient measurements [26,27]. To follow up the morphological changes during reinnervation would be difficult, because the denervated synapses are resolved at different times or speed, which would make it difficult to distinguish between newly built synapses and old denervated synapses. Only single-cell electrophysiology can follow up the reinnervation processes safely.

In humans, the measuring methods are quite different for comparing development and repair of SCI. As non-invasive methods, electromyography and coordination dynamics measurements can be used, beside movement performance measurements. Also, the times for development (Chapter II) and regeneration [57] in humans are quite different from those of frogs, lampreys and rats (Chapter IV).

Figure 5B shows the EMG motor programs of the right and left tibialis anterior and gastrocnemius muscles in the eight-month-old healthy Jürgen (Figure 5A). The single muscles can generate motor programs. But the important antagonistic action between the tibialis anterior and gastrocnemius muscles is still missing. There is still much more co-contraction than antagonistic action between the muscle activations. Also, in a patient with a severe cerebellum injury, the antagonistic action between the tibilais anterior and gastrocnemius muscle was impaired [49] (Figures 67-69 of Chapter II). In the patient of this report (Figures 1C,5D), the generation of motor programs was the biggest problem so far [54].

At an age of two years, Jürgen became able to exercise on the special coordination dynamics therapy device (Figure 5C). Continuously, he could perform approximately only five turnings. The frequency of turning and the arrhythmicity of exercising changed very much (Figure 6A). The mother (Figure 5A) could exercise smoothly (Figure 6B). The patient with the severe cervical spinal cord injury could turn quite smoothly for a low load of 20N (Figure 6C). But she managed to turn for higher load (50N) only with high arrhythmicity (Figure 6D).

The research examples of development and regeneration in animals and humans indicate that they nearly belong to different worlds with respect to CNS organization, measurements of the organization, and the repair following injury. The development of the organization of the human CNS needs different movements to be trained for a proper development and is not so much a “correction en route” but an “improvement of the direction,” because of higher variability of CNS organization. Two-year-old Jürgen can crawl, walk and climb, but his CNS is unable to generate the complicated coordinated movements between arms and legs when exercising on the special device upon changing the patterns continuously between pace and trot gait (Figure 6A). The CNS of the sporty mother can easily generate these intermediate coordinations between pace and trot gait (Figure 6B). Therefore, there is a world of difference between the complexity of CNS organization between Jürgen and his mother with respect to coordinated movements. In the patient with the cervical spinal cord injury (SCI), the supraspinal CNS is working fine; the main problem after some regeneration of the cord is the still missing muscle power.

5.4.14. False Hope

Instead of raising false hope that the animal data can be easily used to cure diseases of the human CNS, the universities and research institutions should first do their duty, namely organizing qualified human neurophysiologic research and support also human research.

5.4.15. Ethics of Spinal Cord Injury (SCI) Research and Clinical Trials

“Transplantation of both embryonic stem cells and embryonic stem cell-derived neural (neuronal or glial) progenitors is able to efficiently promote CNS regeneration in preclinical models of stroke, myelin deficiency, acute spinal cord injury [10,24] and Parkinson’s disease [23].” This sentence of an article makes the reader believe that the research made already the step from the animal research to the human research and treatment is already administered to humans. Looking up the references one finds the title: *Human embryonic stem cell-derived oligodendrocyte progenitor cell transplants remyelinate and restore locomotion after spinal cord injury*. Patients with SCI or journalists think immediately that the SCI problem is already solved! I have learned that it has to be written in the title on what species the research has been performed. In this case, it was the rat. Animal researchers often choose on purpose the words in a way to make the reader believe that this research has already solved the SCI problem in humans. A well-known journal in medicine is even redefining medicine. It makes the reader believe that measurements in mouse are a part of medicine [12]. I have learned in my medical study that medicine has something to do with the cure of diseases in human.

Speciality journals and general audience media need to set reasonable expectations of the safety and efficacy of potential therapies to avoid raising and then dashing the hopes of those living with SCI or those in government, those carrying out research, or the general public [65].

High ethical standards are required by researchers, clinicians and journalists to ensure that results are communicated to the general public in a manner that reflects honestly the safety and efficacy of a potential therapy [65].

Western scientist and clinicians would argue that controls are necessary to identify unambiguously whether a therapy is safe and effective; some clinicians have claimed that withholding a potential therapy from a patient with SCI is in itself unethical [65]. Many SCI units know about CDT. But the patients are not informed that such a therapy exists! A real change from care to cure seems not to be in the interests of the rehabilitation centres, according to my experience.

There is too much false hope generated by animal physiologists, neurobiologists, and researchers working in the field of genetics. Without detailed knowledge of the human physiology and pathophysiology, the animal knowledge is only of limited help to cure diseases in human. False hope also stops the patient from fighting for improvements that are needed for everyday life. Why should a patient fight for two years or more with a movement-based learning therapy to get urinary bladder and some motor functions back, if in a few years time walking can be easily gained with miracle cells or pills? Such false hope is coming from qualified researchers, when they make believe that the animal data can be easily used to repair the human CNS. Qualified human research is needed and has to be organized. Only if there is overlap between animal and human research is there the possibility of rather safe transfer of knowledge from animals to humans. As long as human research is not organized properly, the patients have to suffer or even to die.

Authors of recent review articles presented approximately 1,000 citations on the repair of the human spinal cord following injury [3,7,23,65]. Less than 10% of the citations were from human research. Interesting is that the author was not cited, even though students, young researchers, some physicians and even many journals find the author easily on the internet! I learned a lot from these brilliant research articles with respect to animal research. But with respect to human research and applicability to human patients, they were out of date by 20 years.

"Miracle" operations are not the only danger for patients and the freedom in research. It is the world wide research, treatment, and teaching system that does not allow qualified human research, which is urgently needed to cure diseases and make humans live longer with a higher quality of life. And if there are really operations with unbelievable success, then the treatment before those operations was wrong. Here is an example: A paraplegic patient came by wheelchair from a well-known German rehabilitation centre to a neurosurgery department, was operated on and walked two weeks later out of the neurosurgery department. A second patient was cured in the same way.

The reason for this progress was caused by a mistake of the rehabilitation centre. Bones from the spinal canal caused pressure on the spinal cord. The spinal cord stopped working but was damaged only little. A laminectomy freed the spinal cord, and the spinal cord started to work again. Follow up MRIs are needed.

References

[1] Allein, A.E., Meyrand, P. and Branchereau, P.: Ontogentic changes of the spinal GABAergic cell population are controlled by the serotonin (5-HT) system: implication of 5-HT , receptor family. *J. Neurosci.* 25: 8714-8724, 2005.

[2] Belew, R.K., McInerney, J., Schraudolph, N.N.: Evolving networks: Using the genetic algorithm with connectionist learning. In: *Artificial Life II* (C.G. Langton, Ed.), pp. 511-547 Reading, MA: Addison-Wesley, 1992.

[3] Bradbury, E.J. and McMahon, S.B.: Spinal cord repair strategies: why do they work? *Nat. Rev. Neurosci.*, 7: 644-653, 2006.

[4] Ferretti, P., Zhang, F. and O'Neill, P.: Changes in spinal cord regenerative ability through phylogenesis and development: lessons to be learned. *Dev. Dyn.* 226: 245-256, 2003.

[5] Gasser H.S., Grundfest H.: Axon diameter in relation to the spike dimension and the conduction velocity in mammalian A fibres. *Amer. J. Physiol.* 127: 393-414, 1939.

[6] Guttmann, L.: *Spinal cord injuries. Comprehensive management and research.* Blackwell, Oxford, 1973.

[7] Harel, N.Y. and Strittmatter, S.M.: Can regenerating axons recapitulate developmental guidance during recovery from spinal cord injury. *Nat. Rev. Neurosci.*, 7: 603-616, 2006.

[8] Hunt, C.C.: Relation of function to diameter in afferent fibres of muscle nerve. *J. Gen. Physiol.* 38: 117-131, 1954.

[9] Hursh, J.B.: Conduction velocity and diameter of nerve fibres. *Amer. J. Physiol.* 127: 131-139, 1939.

[10] Keirstead, H.S. et al.: Human embryonic stem cell derived oligodendrocyte progenitor cell transplants remyelinate and restore locomotion after spinal cord injury. *J.Neurosci.* 25: 4694-4705, 2005. (rat).

[11] Kelso, J.A.S.: Dynamic Patterns. *The Self-Organization of Brain and Behavior.* MIT Press, Cambridge, 1995.

[12] Kerschensteiner, M. et al.: In vivo imaging of axonal degeneration and regeneration in the injured spinal cord. *Nature Med.* 11: 572-577, 2005. (rat).

[13] Knudsen, E.I.: Instructed learning in the auditory localization pathway of the barn owl. *Nature* 417: 322-328, 2002.

[14] Knudson. E.I.: Capacity for plasticity in the adult owl auditory system expanded by juvenile experience. *Science* 279: 1531-1533, 1998.

[15] Koepchen, H.P. et al.: "Medullary apneusis" by lesions and cooling in the ventrolateral solitary tract and genesis of respiratory rhythm. *J. Autonom. Nerv. Syst. Suppl.*: 63-69, 1986.

[16] Kudo, N., Nishimaru, H. and Nakayama, K.: Developmental changes in rhythmic spinal neuronal activity in the rat fetus. *Prog. Brain Res.* 143: 49-55 (2004).

[17] Lander, C. et al.: A family of activity-dependent neuronal cell-surface chondroitin sulphate proteoglycans in cat visual cortex. *J. Neurosc.* 17: 1928-1939, 1997.

[18] Nolfi, S., Parisi, D.: "Genotypes" for neural networks. In: *The Handbook of Brain Theory and Neural Networks* (M.A. Arbib, Ed.) pp.431-434, MIT Press, 1995.

[19] Nolfi, S., Parisi, D.: Learning to adapt to changing environments in evolving neuronal networks. *Adaptive Behav.* 1(5): 75-98, 1997.

[20] Li, Y., Field, P.M. and Raisman, G.: Repair of adult rat corticospinal tract by transplants of olfactory ensheathing cells. *Science* 277: 2000-2002, 1997.

[21] Li, Y., Field, P.M. and Raisman, G.: Regeneration of adult rat corticospinal axons induced by transplanted olfactory ensheathing cells. *J. Neurosci.* 18: 10514-10524, 1998.

[22] Lloyd, D.P.C., Chang, H.T.: Afferent fibres in muscle nerves. *J. Neurophysiol.* 11: 199-207, 1948.

[23] Martino, G. and Pluchino, S.: The therapeutic potential of neural stem cells. *Nat. Rev. Neurosci.*, 7: 395-406, 2006.

[24] McDonald, J.W. et al.: Transplanted embryonic stem cells survive, differentiate and promote recovery in injured rat spinal cord. *Nature Med.* 5: 1410-1412, 1999.

[25] McGee, A.W. et al.: Experience-driven plasticity of visual cortex limited by myelin and Nogo receptor. *Science* 309: 2222-2226, 2005.

[26] Miledi, R., Parker, I. and Schalow, G.: Calcium transients in frog slow muscle fibres, in frog skeletal muscle fibres using arsenazo III. *Nature*, 268 No. 5622: 750-752, 1977.

[27] Miledi, R., Parker, I. and Schalow, G.: Calcium transients in normal and denervated slow muscle fibres of the frog. *J. Physiol.*, 318: 191-206, 1981.

[28] Milner P.M.: The cell assembly: Mark II, *Psych. Rev.* 64: 242-252, 1957.

[29] Parisi, D., Cecconi, F., Nolfi, S.: Econets: Neuronal networks that learn in the environment. *Network* 1: 149-168, 1990.

[30] Pizzorusso, T. et al.: Reactivation of ocular dominance plasticity in the adult visual cortex. *Science* 298: 1248-1251, 2002.

[31] Nash, M.S.: Exercise as a health-promoting activity following spinal cord injury. *J.Neurol. Phys. Ther.* 29: 87-103, 2005.

[32] Nathan, P.W.: Effects on movement of surgical incisions into the human spinal cord. *Brain* 117: 337-346, 1994.

[33] Schalow, G. Elektrophysiologische und morphologische Untersuchungen zur Differenzierung der Tonusfasern des Frosches. Inauguraldissertation, Universität des Saarlandes (1983).

[34] Schalow, G. Feeder arteries, longitudinal arterial trunks and arterial anastomoses of the lower human spinal cord. *Zbl. Neurochir.* 51: 181-184, 1990.

[35] Schalow, G.: Conduction velocities and nerve fibre diameters of touch, pain, urinary bladder and anal canal afferents and α and γ-motoneurons in human dorsal sacral nerve roots. *Electromyogr. clin. Neurophysiol.*, 31: 265-296, 1991.

[36] Schalow, G.: Oscillatory firing of single human sphincteric α_2 and α_3-motoneurons reflexly activated for the continence of urinary bladder and rectum. Restoration of bladder function in paraplegia. *Electromyogr. clin. Neurophysiol.*, 31: 323-355, 1991.

[37] Schalow, G. Spinal oscillators in man under normal and pathologic conditions. *Electromyogr. Clin. Neurophysiol.* 33: 409-426, 1993.

[38] Schalow, G. and Zäch, G.A.: Reorganization of the Human CNS, Neurophysiologic measurements on the coordination dynamics of the lesioned human brain and spinal cord. Theory for modern neurorehabilitation (31 case reports). *Gen. Physiol. Biophys.*, 19, Suppl. 1: 1-244, 2000.

[39] Schalow, G.: Stroke recovery induced by coordination dynamic therapy and quantified by the coordination dynamic recording method. *Electromyogr. Clin. Neurophysiol.*, 42: 85-104, 2002.
[40] Schalow, G.: Improvement after traumatic brain injury achieved by coordination dynamic therapy. *Electromyogr. Clin. Neurophysiol.*, 42: 195-203, 2002.
[41] Schalow, G.: Recovery from spinal cord injury achieved by three months of coordination dynamic therapy. *Electromyogr. Clin. Neurophysiol.*, 42: 367-376, 2002.
[42] Schalow, G.: Partial cure of spinal cord injury achieved by 6 to 13 months of coordination dynamic therapy. *Electromyogr. Clin. Neurophysiol.*, 43: 281-292, 2003.
[43] Schalow, G., Pääsuke, M., Ereline, J. and Gapeyeva, H.: Improvement in Parkinson's disease patients achieved by coordination dynamics therapy. *Electromyogr. Clin. Neurophysiol.*, 44: 67-73, 2004.
[44] Schalow, G., Pääsuke, M., Jaigma, P.: Integrative reorganization mechanism for reducing tremor in Parkinson's disease patients. *Electromyogr. Clin. Neurophysiol.*, 45: 407-415, 2005.
[45] Schalow, G., Jaigma, P.: Cerebral palsy improvement achieved by coordination dynamics therapy. *Electromyogr. Clin. Neurophysiol.*, 45: 433-445, 2005.
[46] Schalow, G.: Functional development of the CNS in pupils between 7 and 19 years. *Electromyogr. Clin. Neurophysiol.*, 46: 159-169, 2006.
[47] Schalow, G.: Hypoxic brain injury improvement induced by coordination dynamics therapy in comparison to CNS development. *Electromyogr. Clin. Neurophysiol.*, 46: 171-183, 2006.
[48] Schalow, G. and Jaigma, P.: Improvement in severe traumatic brain injury induced by coordination dynamics therapy in comparison to physiologic CNS development. *Electromyogr. Clin. Neurophysiol.*, 46: 195-209, 2006.
[49] Schalow, G.: Surface EMG- and coordination dynamics measurements-assisted cerebellar diagnosis in a patient with cerebellar injury. *Electromyogr. Clin. Neurophysiol.* 46: 371-384, 2006.
[50] Schalow, G.: Symmetry diagnosis and treatment in coordination dynamics therapy. *Electromyogr. Clin. Neurophysiol.*, 46: 421-431, 2006.
[51] Schalow, G.: Cerebellar injury improvement achieved by coordination dynamics therapy. *Electromyogr. Clin. Neurophysiol.*, 46: 433-439, 2006.
[52] Schalow, G., Vaher, I. and Jaigma, P.: Overreaching in coordination dynamics therapy in an athlete with spinal cord injury. *Electromyogr. Clin. Neurophysiol.*, 48: 83-95, 2008.
[53] Schalow, G.: Near-total functional recovery in partial cervical spinal cord injury (50% injury) after three years of coordination dynamics therapy. *Electromyogr. Clin. Neurophysiol.*, 49: 67-91, 2009.
[54] Schalow, G.: Partial cure achieved in a patient with near-complete cervical spinal cord injury (95% injury) after three years of coordination dynamics therapy. *Electromyogr. Clin. Neurophysiol.*, 49: 199-221, 2009.
[55] Schalow, G.: Unchanged somatic and parasympathetic peripheral nervous system following spinal cord injury. *Electromyogr. Clin. Neurophysiol.*, 49: 263-286, 2009.
[56] Schalow, G.: Coordination between the somatic and parasympathetic nervous systems in the human sacral micturition centre. *Electromyogr. Clin. Neurophysiol.*, 49: 337-367, 2009.

[57] Schalow, G.: Building of New Motoneurons in the Human Spinal Cord upon Coordination Dynamics Therapy to Improve Finger Functions in Motoric Complete Cervical Spinal Cord Injury. In: Berkovsky, T.C. (Ed.), *Handbook of Spinal Cord Injuries*, Chapter 4. pp. 231-264, Nova Science Publishers, 2009.

[58] Schalow, G.: Scientific basis for learning transfer from movements to urinary bladder functions in human. *Electromyogr. Clin. Neurophysiol.*, 50: 339-395, 2010.

[59] Schalow, G.: Stem cell therapy and Coordination dynamics therapy improve Spinal cord injury. *Electromyogr. Clin. Neurophysiol.*, 48: 233-253, 2008.

[60] Schmidt, H., Schalow, G. and Stefani, E.: Trophic influence of small motor axons on the membrane of frog slow muscle fibres. *Exp. Brain Res.,* Suppl. 183, 1975.

[61] Schöner, G., Zanone, P.G. and Kelso, J.A.S.: Learning as a change of coordination dynamics: Theory and Experiment. *J. Motor Behavior* 24: 29-48, 1992.

[62] Sperry, R.W.: The problem of central nervous reorganization after nerve regeneration and muscle transposition. *Quart. Rev. Biol.*, 20: 311-369, 1945.

[63] Sperry, R.W.: Effect of crossing nerves to antagonistic limb muscles in the monkey. *Arch. Neurol. Psychiat.* Chicago, 58: 452-473, 1947.

[64] Thelen, E., Smith, L.B.: *A Dynamic Approach to the Development of Cognition and Action*. MIT Press, Cambridge, 1994.

[65] Thuret, S., Moon, L.D.F. and Gage, F.H.: Therapeutic interventions after spinal cord injury. *Nat. Rev. Neurosci.*, 7: 628-643, 2006.

[66] Varela, F., Lachaux, J.-P., Rodriguez, E. and Martinerie, J.: The brainweb: Phase synchronization and large-scale integration. *Nature Reviews/Neuroscience* 2: 229-239, 2001.

[67] Walton, K.D. et al.: Identification of a critical period for motor development in neonatal rats. *Neuroscience* 51: 763-767, 1992.

[68] Weiss, P. and Brown, P.F.: Electromyographic study on recoordination of leg movements in poliomyelitis patients with transposed tendons. Proceedings of the Society for Exper. *Biology and Medicine*, 48: 384-387, 1941.

[69] Windle, W.F., Smart. J.O. and Beers, J.J.: Residual function after subtotal spinal cord transaction in adult cats. *Neurology* 8: 518-521, 1958.

[70] Kaufmann, S.A.: *The Origin of Order: Self-Organization and Selection in Evolution.* Oxford University Press, New York, 1993.

[71] Bhardwj, R.D. et al.: Neocortical neurogenesis in humans is restricted to development. *PNAS*, 103 (33): 12564-12568, 2006.

[72] Thelen, E. and Smith, L.B.: *A dynamic approach to the development of cognition and action*. MIT Press, Cambridge, 1994.

[73] Chklovskii, D.B., Mel, B.W. and Svoboda, K.: Cortical rewiring and information storage. *Nature* 431: 782-787, 2004.

[74] Nottebohm, F.: Why are some neurons replaced in adult brain? *The Journal of Neuroscience* 22: 624-628, 2002.

Chapter VI

Movement-Based Learning in Humans for CNS Repair Compared with Hippocampus Learning in Animals Quantified at the Cellular, Molecular and Epigenetic Levels

1. An Attempt to Correlate Animal Data on Hippocampus Learning to Treatment Data in Humans

1.1. Adaptation and Repair by Learning

If only genomes were important, humans would not have developed more than mice, and plants would have the same chance as ruling the world as human, mice, and arabidopsis genome would share the same number of genes, approximately 25,000. In fact, it is the epigenome that has evolved tremendously since the appearance of the first multicellular organisms [3]. The term epigenetic, used to describe the interplay between genes and environment resulting in phenotypes, and epigenetic landscape, used to describe the differentiation potential of any cell, are not fully correct.

Simple organisms, such as bacteria, increase their rate of spontaneous mutations to enable the survival of species in a changing environment. Multicellular organisms use complex mechanisms coordinated by the central nervous system (CNS) to behaviourally adapt to changing environments without paying the high price of mutating their genome [4]. Their behavioural adaptation depends on learning, subserved by epigenetic mechanisms. Humans, different from all other species, have the possibility to partially repair their severely injured nervous system by learning. In Chapter II, it was shown that in a patient with cerebral palsy speech could be induced by learning transfer. When the patient learned to exercise by herself on the special coordination dynamics therapy (CDT) device, the patient's CNS partly learned from the device the complex coordinations between pace and trot gait. Her CNS became able to generate the speech pattern, even though it was predicted by different speech therapists that she would never learn to speak. Cognitive function also improved by learning transfer, and continence, which is impaired in very severe brain injury, was fully achieved.

The potential for learning is so high in humans that it can be used to repair basic functions of CNS organization. Animals have a high capacity for spontaneous repair of their nervous system (Chapters I, IV). Humans, on the other hand, have a high capacity for repair by learning, including relearning of principle CNS organization mechanisms as “phase and frequency coordination” among nerve cell firings. A limiting factor for movement-based learning in humans is the structural repair. However, it was measured that there is very limited neurogenesis in the human CNS (spinal cord). It was shown that a few motoneurons could be built anew in a patient with a severe cervical spinal cord injury (SCI) (Chapters II, V). It is also possible that other neurons (interneurons) were built anew upon exercising at the limits. This was indicated by the regeneration of the spinal cord from rostral to caudal. Surface electromyography showed with small motor unit potential recordings that the partial spinal cord repair was probably achieved by neural network repair and not by the regeneration of long tracts like the pyramidal tract.

To enhance the rate of repair of the human CNS by learning, ways must be found to enhance the structural repair, especially the neurogenesis. There seem to be obvious possibilities – the adding of stem/progenitor cells (NPCs) or influencing the epigenetic by learning. In the hippocampus, it was shown that the neuronal network activity itself increased the neurogenesis [6]. If we assume that cells with neurogenic potential exist elsewhere in the brain, most likely to a smaller extent, then the activity itself guides neurogenesis for repair, in accordance with the principle that local cellular environments are important in controlling neurogenesis.

1.2. Excitation-Neurogenesis Coupling Compared with Neural Network-Repair Coupling in Humans

Even though the dentate gyrus of the hippocampus is an exception to the very limited neurogenesis of the adult CNS, it attempts to compare human neural repair data with measurements and considerations in animals on excitation-neurogenesis coupling in adult neural NPCs.

1. Adult neural stem/progenitor cells (NPCs) can intrinsically sense excitatory neural activity and thereby implement a direct coupling between excitation and neurogenesis [6]. – New motoneurons could be built in the human spinal cord, activating finger function. The dorsal flexion of the finger (extension) was generated by dorsal flexion neural networks at or in the SCI (C5/6) spared neural networks for dorsal flexion of the hand. The excitatory neural activity of the remained changed network seems to have sensed the repair of the network including neurogenesis.
2. The activation is sensed via Ca_v 1.2/1.3 (L-type) Ca^{2+} channels and NMDA receptors on the proliferating precursors. The Ca^{2+} channels are opened via excitatory stimuli. Therefore, NPCs themselves can act as the signal detection and processing elements mediating adult excitation-neurogenesis coupling [6]. The impact of excitation could also be indirectly on the surrounding mature hippocampal cells. – There is no

information from treatment (Chapter II) or human neurophysiology (Chapter III) how neuron production is achieved. If there are no NPCs in the human spinal cord, the NPCs would have to be stimulated in other locations as the subventricular zone of the lateral ventricle and transported via the liquor to the spinal cord. To induce migration of NPCs and direct them to the site where they are needed, integrative and specific movements would have to be trained.

3. The excitation through Ca^{2+} channels and NMDA receptors act to inhibit expression of the glial fate genes Hes1 and Id2 and increase expression of NeuroD, a positive regulator of neuronal differentiation. The response in phenotype-determining genes occurs soon after the initiation of the excitatory stimulation [6]. – There is no information from treatment and neurophysiology. The index finger started to work upon three years of treatment. The growing of the motoneuron axon from the spinal cord to the muscle activating the index finger needs at least two years. The building of new motoneurons (and interneurons) may need 0.5 to 1 year in humans.
4. The positioning of the NPCs within the axon-rich environment of the hippocampal subgranular zone seems an ideal setting. Here, activity-dependent depolarizing mechanisms could directly influence proliferating NPCs and trigger activity-dependent gene expression programs [6]. – No information from treatment.
5. Many newborn neurons die within the first two weeks in vivo. – If CDT is stopped for more than three days, the progress is going backwards. Gained functions start to decline. An incomplete SCI patient (C5/6) trained 0.5 years and got a bit better. Then he stopped treatment for 0.5 years and started therapy again. It seemed that he had lost all gained functions in the therapy break. He somehow started CDT from scratch again.
6. The complex question of whether insertion of a new neuron can help store the memory of the same activity that led to creation of the new neuron was not explored [6]. – With respect to movement-based learning, can the activated networks create the structural repair that is needed for the repair of the injured subnetwork? The training of the finger function was included in most of the integrative and coordinated trained movements (training of arms, legs, trunk and other body parts), especially when exercising on the special CDT device. The neuronal subnetworks for activating hand and finger function were activated in coordination with other neural subnetworks. This means the repair (including the creation of new neurons) took place in the networks for finger and hand function and the other activated networks. Repair (change of efficiencies of synapses, synaptogenesis, axon and dendrite growth, neurogenesis) takes place in the activated neural networks, which seems logical. The learning transfer, supported by the "System Theory of Pattern Formation" (Chapter II) and neural network complexity (Chapter V), show that the activated subnetworks are repaired.
7. How neurogenesis allows the network to adapt to different levels of memory storage demands, which in turn are correlated with the activity level in the network. In this view, activity-dependent neurogenesis may affect the future memories or the degradation and clearance of previously stored memories [6]. – If we see memory in the form of patterns of neural activity, activity-dependent neurogenesis supports the relearning of lost pattern functions and supports clearance of post-injury developed pathologic patterns, like spacticity. Therefore, neurogenesis helps the degradation of

pathologic activity patterns. The degradation of pathologic activity patterns includes those of higher mental functions. Following severe brain injury, a 25-year-old male patient became aggressive. Upon three months of CDT, the aggressive behaviour disappeared completely. He even became very happy. Because of the short therapy time, the improvement of the character was probably mainly due to a functional repair by CNS reorganization, excluding substantial neurogenesis. In a 25-year-old patient with cerebral palsy, the aggressiveness declined upon 0.5 years of CDT. Neurogenesis may have contributed to the reduction of the aggressiveness. With CDT, the movement, vegetative and higher mental functions always went into the direction of physiologic functioning or back to the character before the CNS injury. Epigenetic seems to be involved in the repair in the way of control that no pathologic functioning occurs.

8. Neurogenesis elicited a more rapid loss or clearance of previously stored old memory (a) and the newest memories were recalled at a higher fidelity (b) (conclusions from a model [6]). – New neurons with new dendrite and axonal trees can contribute to the reduction of spasticity, which is probably mainly due to the network impairment and the loss of neural network complexity due to the injury. This gives an explanation as to why with CDT spasticity reduces in the long-term memory. But it is possible that physiologic patterns may also loose stability. Physiologic stereotyped patterns like the trot gait coordination and walking also have to be trained, besides complicated coordinations to enhance the complexity and accuracy of neural networks. After repeated performance of the high-load test exercises on the special CDT device, a healthy pupil of the developmental study (Chapter II) felt that his trot gait coordination (for walking) weakened. With new neurons, and in turn with higher network complexity, relearned patterns can be performed more accurately and the variability of the pattern is higher. In poor network complexity, the possibility of movement-based learning is limited. New neurons will improve the accuracy and variability of relearned patterns. But first, as will be analysed below, network instability will occur.
9. That the newest memories were recalled at a higher fidelity, despite the fact that network size was not allowed to increase in the network model, can be intuitively understood from the lack of environment of new neurons in old memories. The synapses of the new neurons can be devoted more fully to the newer memories (patterns), which are then more accurately stored. This advantage that the new neurons with their new synapses enhanced the accuracy of the stored memories (patterns) was dramatically greater in networks that had been more active and had been required to store many memories (patterns) [6]. – This result from a neural network model offers one understanding of why the patient has to train at the limit (more network activity) and has to train different movements (the networks are required to store more memories in the form of patterns). Therefore, training at the limit different movements will increase the quality of the relearned movement patterns and will in turn give rise to learning transfer, so that other functions (patterns) are repaired as well, which cannot be trained, for example, continence and induction of speech. New neurons open more possibilities of learning and relearning.
10. It is of interest to determine the factors that permit excitation-neurogenesis coupling to occur. Cells with neurogenic potential exist elsewhere in the brain, including the

neocortex, but contribute to neurogenesis only in the subventricular zone and the dentate gyrus subgranular zone. However, when cells from non-neurogenic areas are removed and transplanted into neurogenic areas, neuronal progeny results [6]. There are a number of factors that may contribute to the neurogenic potential to induce competence for excitation-neurogenesis coupling in the resident population of NPCs:

a. Type of (natural) local neuronal activity.
b. Access of the NPCs to the local activity.
c. Ability of the local environment to induce activity-sensing competence in the NPCs.
d. Factors intrinsic to the resident NPCs themselves. –

Clinical research gives no direct answer. The naturally activated local neural networks surrounding the injury site and the cells with neurogenic potential will be important for the growing of axonal and dendrite trees and neurogenesis. It is conceivable that progenitor cells are able to identify natural activity related to repair, which would trigger neurogenesis and proliferation. The frog experiments of Chapter I on development and repair show how complicated and differentiated the establishment of a seemingly simple innervation patterns is.

11. Connection of NPCs to the activated networks. Ca_v 1.2/1.3 channels on adult NPCs potently influenced neurogenesis. The channels open primarily in response to the depolarization provided by excitatory synaptic inputs. But proliferating NPCs in vivo are presumable not synaptically connected [6]. – The mechanisms for the activation of the NPCs are not clear. There is no answer from movement-based learning therapy regarding how the NPCs are stimulated. Generally, it is known from the improvement in patients that the repair, including structural repair, is constantly adapting to the needs of the specifically activated circuitry. During treatment, the CNS networks are activated integratively, but specifically those subnetworks are included in the activation that have to be repaired.
12. Critical period plasticity. Adult-born neurons exhibit the same critical classic period plasticity as neurons in the developing nervous system [5]. – When patients have a break in their treatment at the limit, their functions are going backwards, as can be quantified by measuring the coordination dynamics. This experience in patients is in accordance with the transient nature of plasticity. Not to leave out the enhanced synaptic plasticity during the critical period in newly generated neurons of the adult brain, the treatment should not be stopped for longer than three days, especially when we do not know when the possible critical periods appear.
13. Neural network stability. A complete missing aspect in animal research is that neural network patterns have to be stable. Such network stability becomes impaired if the structure of networks is changed. Kelso stated that the stability property of network functions is "foreign" in research [9]. Recently, it was stated that "the transient nature of such enhanced plasticity may provide a fundamental mechanism allowing adult-born neurons within the critical period to serve as major mediators of experience-induced plasticity while maintaining stability of the mature circuitry" [5]. – In humans with every bit of structural change stability of CNS, self-organization is

impaired, which has to then be reestablished by movement-based learning. Chapter III showed that with CNS damage, neuronal network deterioration can be best measured on the premotor spinal oscillators. During repair, instabilities do also occur. With every bit of regained function instabilities occurred, which were quantified by coordination dynamics values and repaired by movement-based learning (Fig.50 of Chapter II). The occurring of instabilities with network changes also makes sense. If one changes an electronic circuitry by adding a new connection, the circuitry function will change and instabilities may also occur. For example, the network starts to oscillate.

Within the framework of System Theory of Pattern Formation movement patterns are actually characterized by their stability, namely how deep potential wells are. The landscapes of potential wells characterize network functions (Chapter II). Upon changing neural networks, instabilities and uncontrolled growing may occur. Two children with cerebral palsy went from Greece to Germany for stem cell therapy. One child got convulsions (network instability) and the other developed cancer (uncontrolled growing). The mother of the child with cerebral palsy, in whom speech was induced by CDT, said: "Our cerebral palsy children are used for research like laboratory animals"! A therapy trial is only justified if the potential treatment is thought completely through the whole treatment (see below).

The brilliant article on excitation-neurogenesis coupling [6] analysed in detail how network excitation can stimulate neurogenesis. The authors stated easily "it is conceivable that pharmacological interventions or manipulations of neural activity could enhance the ability of either native or transplanted NPCs to restore function to damaged areas of the CNS" [6]. – With the knowledge presented in their paper, they should immediately ask the question how the activity in the network can reach the transplanted NPCs: a. what kind of (natural) local neural activity do the transplanted NPCs need? b. how is the access of the NPCs to the local activity? c. how is the ability at the injury site to induce activity-sensing competence in the NPCs? d. what factors intrinsic to the transplanted NPCs influence the success of the transplantation?

The transplanted NPCs are most likely too far away from targeted networks to be activated for proliferation. If some NPCs accidentally become connected to the adult damaged network, instabilities will occur. Generally, when in a highly qualified article the possible clinical applications are suggested, the quality of the article drops dramatically!

14. Proliferating cells and putative progenitors in both subgranular (SGZ) and subventricular (SVG) are closely associated with the vasculature, indicating that factors released from the blood vessels may have a direct impact on adult progenitors [1, 10]. "Infusion of vascular endothelial growth factor (VEGF) promotes cell proliferation in the SVG and SGZ, which can be blocked by a dominant-negative VEGF receptor 2" [2]. In addition, VEGF is required for increased neurogenesis in adult mice exposed to an enriched environment or given the opportunity of voluntary exercise, which are both known to enhance adult neurogenesis. – Following human SCI autonomic innervation, especially of the skin blood vessels, is often impaired. Patients get very vulnerable skin and easily get pressure ulcers, which are slowly healing because of the poor blood supply. Intensive CDT repairs this impaired blood supply within six to nine months so that the skin is no longer vulnerable and pressure

ulcer occurrence reduces. That means movement-based learning is repairing the blood supply and the microcirculation and may have, therefore, an impact on adult progenitors in human.

15. Neural progenitors can be indirectly influenced by neurons outside the microenvironment that are connected to neurons within the neurogenic niche through neural circuits. Both local and distal neurons can exert direct influences on neural progenitors through the ambient levels of neurotransmitters in the neurogenic niche, or even through synaptic contact with neural progenitors. Therefore, adult neurogenesis is subject to complex extrinsic regulation [15]. – CDT tries to activate the different kinds of neural subnetworks in close proximity to the injury site by performing integrative movements and tries thus to influence neurons (networks) outside of the different microenvironment that are connected to neurons within the neurogenic niche through neural circuits. In Chapter II, the movement-based learning of integrative movements with specific concentration on functions surrounding the injury site (where the structural repair is needed) was seen with respect to the System Theory of Pattern Formation and learning transfer.
16. It is learning and not simple training that elicits the survival effects of new neurons in the hippocampus. Learning appears to promote the survival of newborn neurons in cognitively unimpaired aged rats [7]. Learning elicits different influences on neural precursors at different developmental stages. The regulation of SGZ neurogenesis by hippocampus-dependent learning is complicated and can be affected by factors such as the age of the newborn neurons, the stage of learning and specific learning protocols [15]. – When the patient is exercising on the special CDT device, he should not just turn but should try to turn more smoothly. He should try to reduce the arrhythmicity of exercising. The patient should try to improve the performance of the movement by learning. In this learning of human skill, it is essential to have mental concentration with a planned strategy of action and a subsequent evaluation (looking at the coordination dynamics curves on the computer screen) and correction of errors in successive attempts. When well practiced at exercising smoothly, the skill can be accomplished without voluntary attention, as in walking, swimming, cycling or skiing. But simple exercising will also improve CNS functioning; only the rate of learning becomes smaller. This is the learning of automatic movements, in which the process is subconscious. A problem in some patients is (as, for example, in the cerebral palsy patient of Chapter II) that the cognitive functions are so low that they cannot understand that they have to learn to improve their nervous system functioning. The hope in such cases is that the simple training improves their cognitive functions so much that the patients do understand that they have to improve the performance of the trained movements to improve their CNS functioning. The patients should not stop the movement-based learning longer than three days (because relearned functions decline) to influence structural repair at different repair stages. Different movements with different strategies have to be learned or relearned to support different stages of learning.

2. Why an Exogenous Stem Cell Therapy is Unlikely to be Effective

2.1. Excitation-Repair Coupling in the Physiological Context

Above, coupling of excitation to neurogenesis in proliferation adult-derived NPCs was explored, and attempts were made to relate excitation-neurogenesis coupling to treatment of human patients by CDT. Adult neurogenesis is potentially enhanced by excitatory stimuli and involves Ca_v 1.2/1.3 channels and NMDA receptors. But there is indication that this excitation has to be administered in the physiological context. Physiological context means that the neuronal networks functionally and anatomical surrounding the injury site have to be activated in the natural way and probably at the limit. In Chapters III and V, natural impulse patterns have been measured in humans. It was shown into what natural impulse patterns the environmental information is transformed and how CNS subnetworks, like premotor spinal oscillators, self-organize or change the self-organization according to these natural impulse patterns. It has further been shown how the natural impulse patterns do change following CNS injury.

The excitation-repair coupling of neural networks can therefore probably only be achieved by changing the pathologic natural impulse patterns of CNS organization back into the physiologic ones upon movement-based learning. Important neural networks are the premotor spinal oscillators of Chapter III. If we want to repair those neural subnetworks, including structural repair, the excitation-repair coupling has to be physiological. But knowledge of human functional anatomy and neurophysiology is needed for that to stimulate excitation-repair coupling in the physiological context. Then the repair will take place by changing synapse efficiencies, synaptogenesis, axonal and dendrite grows and neurogenesis. Occurring network instabilities are repaired by learning immediately when they occur, as has been shown in a patient with motoric complete SCI (injury level C5/6). In patients with Parkinson's disease, these premotor spinal oscillator networks do coordinate their firing pathologically [14]. The oscillators synchronize their firing and generate motor patterns with tremor. By enhancing the lateral field inhibition upon movement-based learning, the premotor subnetworks would fire more physiologic and at least the tremor can be reduced.

Therefore, to induce excitation-repair coupling, we need to know the natural firing patterns of the different neural subnetworks in the human CNS. The natural firing patterns can be explored by measuring simultaneously natural impulse pattern of several neurons, as has been shown in Chapters III and V.

2.2. Inefficient Excitation-Repair Coupling in Clinical Trials

There are clinical trials performed, in which NPCs are administered to human patients for 6 to 12 months simultaneously with five to six hours supported treadmill walking per day, five times per week. It will be shown now on the basis of the knowledge presented in this book that such therapy will not work, which means that the patients will not substantially benefit from such a therapy.

2.3. Exogenously Applied Stem/Progenitor Cells (NPCs)

The repair at the injury site with respect to NPCs is probably regulated at several steps: (1) proliferation of neural stem/progenitor cells (2) survival of newborn neurons and (3) maturation and integration of newly generated neurons. Firstly, if NPCs are injected into the injury site (the cavity of Fig.36J,K of Chapter II), then some spared fibres will be damaged. Secondly, the excitation-proliferation coupling of above will not take place, because the NPCs are too far away from the spared (hopefully activated) networks and tract fibres. Even if there would be no scar or other tissue, the NPCs will not be stimulated. For an analysis, the in vivo frog data of Chapter I will be used, in which two motoneurons were competing for the innervation of two target cells (two kinds of muscle fibres). To build up the innervation pattern, the two motoneurons competed by membrane contact (distance ≈ 0.1μm). For the communication between the motoneuron and the target (muscle) cells, the distance was a bit longer because there was a basal lamina between the cells. The motoneurons, innervating the slow muscle fibres in the adult frog, exerted their influence onto the target (change of membrane properties and excitation-contraction coupling [16]) by a neurotrophin stored in the axon, since following denervation membrane properties change later when nerve stump is longer [12]. This means the neurotrophin is transported in microtubules in the axon to the synapse. With a longer remaining axon piece, the reservoir of remained neurotrophin is larger and the muscle fibre properties changed later. It is difficult to see how such intricate four-cell communication can be achieved by placing some NPCs to the injury site. Already, the short distances for membrane contact and neurotrophin diffusion will not be achieved. If some cells come close to the activated networks and a few NPCs are activated and proliferated, their number is probably smaller than those that are damaged upon the injection of the cells. Practical knowledge of neurosurgeons is that you should not manipulate in the spinal cord injury site, because large damage will take place. Because of no functional integration due to missing activation, the injected cells will probably die after some time. One can assume that the many cell communications in the human CNS will be as intricate and complex as the four-cell communication for building up the muscle innervation pattern in the frog. When the leg muscles of the tadpole become innervated, the tadpole is using them at a developmental stage when the front limbs have not broken through the skin. When the motoneurons, innervating in adult the frog the slow muscle fibres, selectively innervate their slow muscle fibres and push away the synapses of the other kind of motoneurons (innervating in adult frogs the twitch muscle fibres), they were physiologically activated. Only the motoneurons did not exert the right influence. That means, during building up of the adult innervation pattern, the functioning cells were activated as physiologically as possible according to the stage of development or repair.

Further, with only treadmill walking, the neural subnetworks at the injury site are not sufficiently activated. To achieve a rather optimal activity-neurogenesis coupling, or more generally an activity-repair coupling, those subnetworks have to be activated where at least the repair is needed. Many different science-based movements (Chapter II) have to be performed to reach the complex neural subnetworks for physiologic activation and repair. The exercising on the special CDT device is substantially contributing to the activation of complex subsystems. But no single movement is sufficient to activate all subsystems for repair and improvement. Even in sport, in which the improvement of CNS network functioning is not so far taken into severe consideration, one is also training other movements than those in which the athlete wants to win the competition. Somehow, the functional and anatomical surrounding area to be improved has to be brought into a stage permissive for functional improvement or repair. In the hippocampus, one is speaking of neurogenic areas and non-neurogenic areas. Cells with neurogenic potential do contribute only to neurogenesis if they are stimulated or placed into neurogenic areas. Properly applied treadmill walking, including the activation of the stepping automatism in the forward and backward direction, is not sufficient to activate the networks surrounding the injury site for repair. According to my first experience, the generation of new cells needs 6 to 12 months. The building up of the axonal and dendrite trees may need another two years. A treatment time of 0.5 to 1 year is too short for the repair of the human CNS. The development of the human CNS needs at least 17 years (Chapter II), probably because the network complexity has to become very high. Animals with less network complexity and smaller size may need only one year. For the repair of the human CNS, human data on function and repair are needed. Applying only animal data to human is too risky for the patient.

2.4. Improvement of CNS Functioning in Parkinson's Disease Can Be Objectively Quantified by Measuring the Coordination Dynamics

Above, it was analysed that it is unlikely that a clinical trial in SCI including stem cell therapy and treadmill walking will succeed, because the stem cells will not be integrated in the adult neural networks and the movement-based learning is insufficient to activate the neural networks for excitation-neurogenesis coupling. But the excitation-neurogenesis coupling for the integration of newborn cells into adult damaged networks is also not working in other diseases like Parkinson's. It was reported how messy the treatment situation is in Parkinson's disease [8]. In this disease, the placebo effect is especially strong since the patient's expectations that they will benefit from a treatment induce the release of dopamine, which is also lacking in the disease. The author was emphasizing that there are no objective measures for how well a patient is doing [8]. I have clearly shown in patients with Parkinson's disease that the quality of CNS functioning can be measured by the coordination dynamics when the patient is exercising on the special CDT device. I have further shown that CNS functioning could be improved and quantified by CD values, upon coordination dynamics therapy [13, 14]. The tremor was analysed in detail [14]. The patient with the severe cerebellum injury (Chapter II) also thought that his drug improved his CNS functioning in the short-term memory. The coordination dynamics measurements before and after taking the drug clearly showed that his CNS functioning got transiently worse (Fig.60 of Chapter II).

2.5. Microenvironment (Neurogenic Niche) Permissive for the Differentiation and Integration of New Neurons

In Chapter I, it was shown that the development and the repair were very similar when the two kinds of motoneurons innervated two kinds of target cells (muscle fibres). In adult frogs, a slow muscle fibre can be innervated by two motoneurons. By cutting one axon, the membrane changes its properties in the denervated part. It can now generate action potentials that cannot be generated in the still innervated part of the muscle fibre (Fig. 86 of Chapter III). By partial denervation [11], therefore, we obtained innervation and denervation membrane properties in the same cell but with more than one cell nucleus. In the developing frog, the muscle fibre has approximately a length of 1mm. The distance of action of the neurotrophin is shorter than 0.5mm. Not to influence neighbouring muscle fibres (Ø ≈ 15μm), the distance of action of the neurotrophin, secreted from the motoneuron, must be shorter than 15μm for building up the specific physiologic innervation pattern. Probably the distance of action of the neurotrophin is in the range of 1μm according to the distance between the neuron and the muscle fibre.

One may assume that a similar detailed neurogenic landscape is present in the neural networks of the human CNS. It is difficult to see how such microenvironment can be generated during the administration of neural stem/progenitor cells for treating patients with SCI or Parkinson's disease. Before treating human patients, the microenvironment should be clarified in humans on a scientific basis. To induce competence for excitation-neurogenesis and excitation-repair coupling, the following should be explored in humans:

a. What kind of local neural activity is needed?
b. How can the access of the NPCs to the local activity be achieved?
c. What is the ability of the local environment to induce activity-sensing competence in the NPCs?
d. What is the intrinsic state of the transplanted NPCs after injection?
e. Proliferating cells and putative neural progenitors in both subgranular zone of the dentate gyrus and subventricular zone of the lateral ventricles are closely associated with the vasculature, indicating that factors released from the blood vessels may have a direct impact on adult neural progenitors (1). How can blood vessel supply be achieved for the administrated stem/progenitor cells?
f. Astroglia is supposed to induce neurogenesis from adult neural stem cells (17). During stem cell therapy in spinal cord injury, the astrocytes will not contribute to the stimulation of neurogenesis, because outside the spinal cord matter there is no astroglia. How can the potentially contribution from the astrocytes be simulated in the physiological context?

3. Epigenetic Modification for Repair by Movement-Based Learning

3.1. Overloaded Networks are Required for the Repair of the Human CNS

New neurons can be built in the animal and human CNS (Chapter II). Electrical activity has been shown to regulate development in a variety of species and in various structures (20), including the spinal cord and cortex. Within the mammalian cortex specifically, the development of dendrites and commissural axons in pyramidal cells is activity-dependent (21). Excitatory stimuli act directly on adult hippocampal neural stem/progenitor cells to favour neuron production (6).

In Chapter I it was shown that there was a close similarity between development and repair in the frog peripheral nervous system. The case reports and group studies of Chapter II indicate, on the other hand, that there is only some similarity between development and repair in the human CNS.

Assuming that principles governing the animal CNS development and learning and memory formation in the animal hippocampus partly also hold in human development and repair, it may be understood how, by movement-based learning induced, natural impulse patterns can change the neural networks for repair.

If it is with the activity overloaded networks, which induces CNS repair, then those networks have to be overloaded which have to be repaired and which should take function over. The load in neural networks is increased by going to the limits of exercising and by improving the coordinated firing of neurons via improving the coordination of arm and leg movements, since neurons work as coincidence or better as coordination detectors (Fig.1). Further, by training complicated coordinated movements, neural network patterns can be reached, which lie deeply in the complexity of CNS organisation.

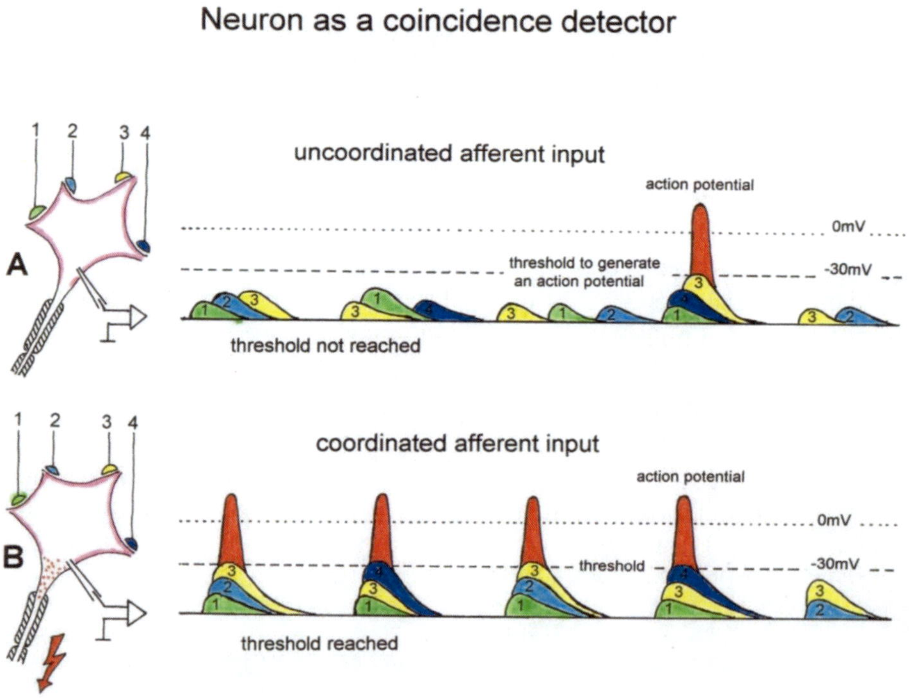

Figure 1. Neuron operating as a coincidence detector. A. Afferent input is reaching rather uncoordinated the cell soma. Only sometimes an action potential is generated, because the threshold of action potential generation is mostly not achieved. B. The action potentials in fibres 1 through 4 are reaching time-coordinated dendrites or the cell soma. The postsynaptic potentials add up and the threshold is achieved at approximately –30mV, and action potentials are generated time-coordinated at the axon hillock. In the real CNS mostly many more smaller postsynaptic potentials will contribute to the generation of an action potential and passive conduction from the dendrites to the cell soma has to be taken into account. Coordinated afferent input may thus induce or enhance (coordinated) communication between neuronal network parts following CNS injury.

For functional repair especially in spinal cord injury the use of tracts has to be changed. By training different movements, different kinds of tract will become overloaded when going to the limits of these coordinated movements. By training impaired functions (with support), the corresponding damaged circuits and the functionally connected networks will become overloaded and get the stimulus for repair. As the building of new motoneurons in a human patient indicates, specifically overloaded networks seem to be also the stimulus for neurogenesis. The building of new neurons is a powerful repair strategy, especially when certain kinds of neurons are missing. But as the human research indicates, the neurogenesis is very limited in the human CNS and the benefit from such repair strategy needs more than a year of intensive training to contribute to functional recovery.

3.2. Selective Requirement for Natural Activity in Specific Neurogenesis and in Shaping the Integration of Specific Neurons into Damaged Adult Neural Networks for Repair

It was discovered that 30% of all cortical interneurons arise from a relatively novel source within the ventral telencephalon, the caudal ganglionic eminence (19). Owing to their late birth, these interneurons populate the cortex only after the majority of other interneurons and pyramidal cells are already in place and have started to functionally integrate. In mice it was shown that activity is essential before postnatal day 3 for correct migration, and that after postnatal day 3, glutamate-mediated activity controls the development of their axons and dendrites (18).

The treatment of patients (Chapter II) and mice data (18) indicate that during development and repair selective activity is necessary for activity-dependent neuron migration and neuron axonal arborisation and neuron dendrite trees besides weight changes of synapses. Further, functional and structural repair can lead to the repair of physiologic CNS functioning, but it can also lead to pathologic functioning like epilepsy and cancer. It seems therefore that for a repair or building up of physiologic CNS patterns, natural physiologic activity patterns are needed, which are generated by movements, vegetative or physiologic cognitive function patterns. Some natural impulse patterns in afferents, efferents and sub-neural networks (network oscillators) in humans were identified in Chapters III and V. To reach and train natural organisational patterns in the deepness of phase and frequency coordination, extremely coordinated and complex arm, leg and trunk movements have to be trained like the movement patterns as performed in Figure 2. For selective activation and repair of the human CNS, it has to be known how the human CNS is functioning and how its patterns are changed by the natural impulse patterns of receptors, informing the CNS about changes in the outside world. With functional anatomy alone it will never be clarified how the human CNS is functioning. Already the functioning of the human muscle spindle with its motor, sympathetic and parasympathetic innervation (Chapter V) is not fully clarified. How can the tremendous functional complexity of the human CNS be explored, if it is not even tried to understand fully the functioning of the human muscle spindle, which is strongly involved in motor, continence functions (Chapter V) and other functions. The present research, putting emphasis on animal research, is bypassing the interests of the human patients, who want to get their CNS repaired or its functioning improved.

The animal research has reached now the point to find out that specific activity is needed for development (and repair) (6,18-21), what was partly published already 20 years ago, based on human neurophysiology (22-24, Chapters III and V).

3.3. A gene Expression Pattern Triggered by Excitation in Proliferating Adult NPCs

For weight changes of synapses, growing of axons and dendrites, neurogenesis and building of glia cells during the process of activity induced structural repair, proteins are needed, which have to be generated by appropriate transcription from genes. It seems that natural patterned activity-dependent depolarization of the plasma membrane triggers activity-dependent gene expression programs (6). Ca^{2+} signalling through Cav1.2/1.3 channels and NMDA receptors can activate a broad array of rapidly responsive transcription factors. Moreover, excitation could in principle induce release of autocrine factors from the NPCs themselves, leading to recruitment of a host of additional signalling pathways to nuclear transcription factors (6).

3.4. Regulation of Epigenetic Modification by Movement-Based Learning

To generate repair in the nervous system it is likely that permanent changes in gene expression patterns are achieved through permanent changes in chromatin remodelling without changes in DNA sequence. The concept of chromatin remodelling addresses a key challenge of how stable changes in gene expression are induced (4) in neural networks to produce long-lasting changes in repair. DNA methylation is one of the many epigenetic modifications that can alter gene expression. Dynamic and reversible DNA methylation may be essential for learning and memory formation and could transmit repair influences onto adult neurogenesis.

Understanding the complex epigenetic regulation of neural activity and adult neurogenesis is integral to designing therapeutic approaches to restore neurogenesis and cognitive functions. It will also give a tremendous insight into understanding how certain environmental or pathological influences, such as stress, physical activity, depression and epilepsy regulate adult neurogenesis (3).

Research in movement-based learning, on the other hand, has to find out how epigenetic mechanisms can efficiently be modified by what movements or learning processes to improve the development (correction en route, especially in cerebral palsy) and the repair of the human CNS and to avoid pathologic CNS changes like epilepsy and cancer.

3.5. Movement-Based Learning in the Postnatal Period

The establishment of adult patterns of DNA methylation during the postnatal period suggests that critical periods exist after which levels of gene expression are relatively stable (4). However, movement-based learning in the adulthood is capable of targeting the epigenome and altering gene expression and hence repair; but the efficiency of repair may be low. CDT should therefore start in CNS injury or malformation as early as possible.

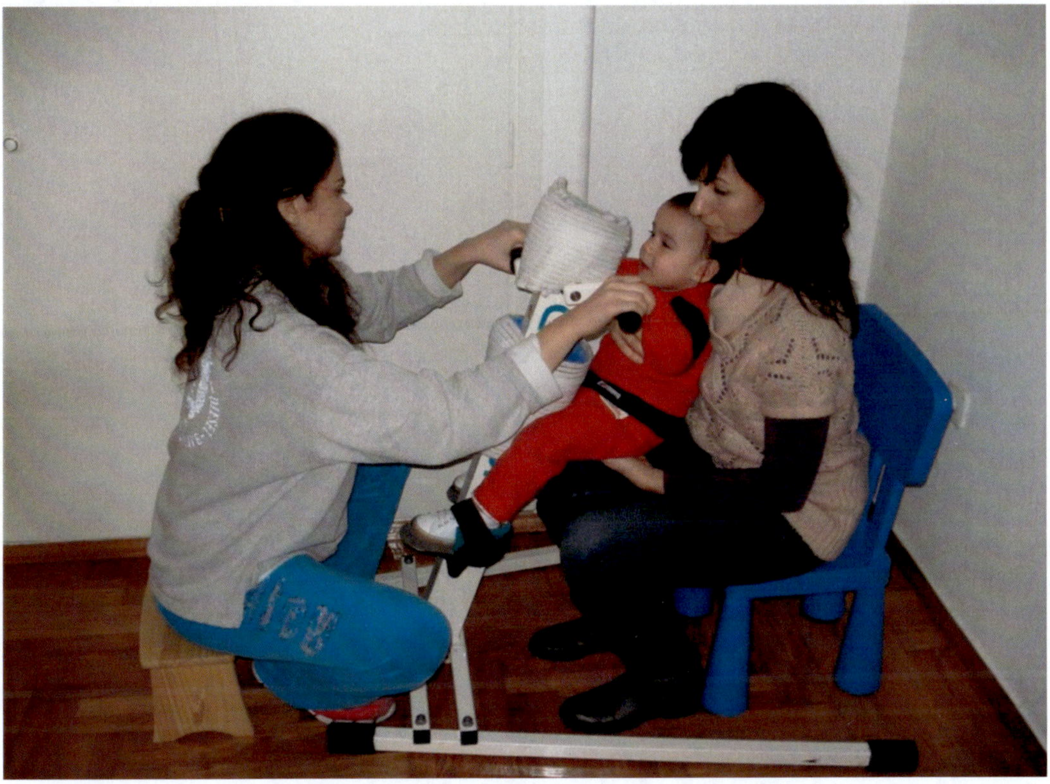

Figure 2. Administration of movement-based learning therapy with coordinated arm and leg movements to a patient with cerebral palsy. The 19-month-old boy is supported by the mother (right side) and the therapist (left side) is turning. The movements are passively administered, because the boy cannot exercise by himself. Also a healthy boy would not be able to exercise by himself, because the movement is too difficult for that stage of development (Chapter II); the CNS has not developed the sufficient neural network complexity to be able to generate the complicated arm and leg movements between pace and trot gait. Still, this movement induced afferent input from the coordinated arm and leg movements helps the injured and developing CNS to repair itself by learning and to improve its development. The main improvement of CNS organisation is caused by the movement in general (often such a movement is the only coordinated one to be able to be administered) and the improvement of phase and frequency coordination of neuron firing for improved neural network organisation.

Figure 2 shows the movement-based learning administered to a 19-months-old child with severe CNS injury. Since the boy can see only extremely little, because of the impairments of the processing of vision, he is also afraid of moving and moves only little while crying. But the exercising on the special CDT device with the mother and physiotherapist is possible. With this administered movement he is crying only sometimes, especially when being moved longer than 10min continuously. As if his CNS is benefiting from this passive exercising and he is feeling good with the improvement of CNS organisation in the short-term memory. Such an infant can laugh and cry at the same time. Laughing because he likes the improvement of CNS organisation and crying because he/she is getting exhausted. Interval training is needed.

3.6. Movement-based learning in the prenatal period

The early environment, consisting of both the prenatal and postnatal periods, has a profound effect on gene expression and adult patterns of behaviour (4). Movement-based learning upon exercising on the special CDT device is normally not possible in the prenatal period. But in premature born babies, the CNS is probably in the stage of prenatal development. A movement-based learning therapy should in principle be therefore beneficial for the repair of the premature CNS, injured for example by hypoxia. Figure 3 shows the special CDT device to administer movement-based learning to a premature born infant.

3.7. Early treatment in CNS injury

If there is some similarity between development and repair and the early environment following injury has a profound effect on gene expression and patterns of behaviour, then early treatment of CNS injury should also be very beneficial for repair.

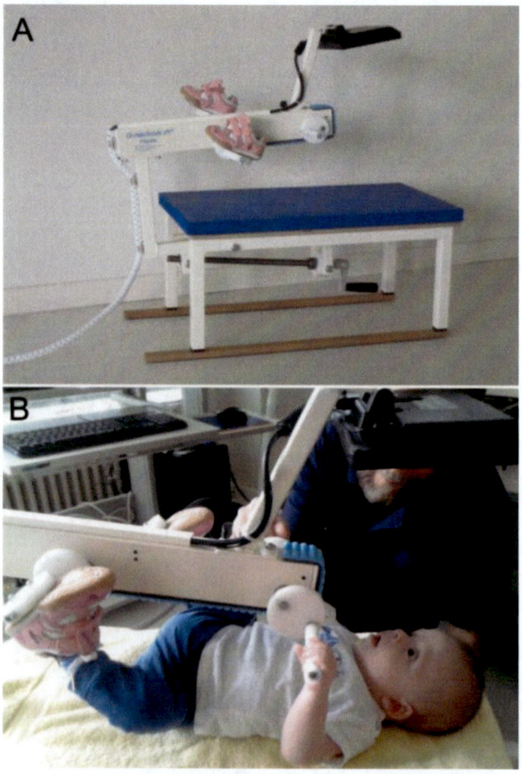

Figure 3. Exercising of a 6 months old baby on the special Coordination Dynamics Therapy device in the lying position. A. Special Coordination Dynamics Therapy (CDT) device for small baby's. B. Exercising of a 6-month-old infant on this special CDT device in the lying position. The infant holds the left handle by itself. The baby is not crying and can turn a few turns by itself. It also watches the computer display of its turning. Note that the instrument can be used also for much smaller baby's including premature born baby's. – This special CDT device for measuring and therapy (int.pat.) is produced by the firm: Giger Engineering, Martin Giger dipl.Ing.ETH/SIA, Herrenweg 1, 4500 Solothurn, Schweiz, www.g-medicals.ch.

In spinal cord injury, the treatment should be started in the spinal shock phase; the special device in the lying position is as safe as lying in bed if the trunk is not made rotating. In stroke patients the therapy should also start as early as possible, especially to improve the blood supply of the brain. To put a patient with severe brain injury into artificial sleep to reduce oxygen consumption is doubtful. The improvement of blood supply by early movement based learning therapy may be more successful then trying to reduce oxygen consumption by putting the patient into artificial sleep. Also, the therapist is loosing partly control over the patient. He cannot speak to the patient any more and cannot become informed about his consciousness and brain functioning. I got a patient with very severe brain injury. After the bicycle accident, the patient could speak and informed the ambulance about his name and address. He was put to artificial sleep. When the artificial sleep was terminated, the patient could not speak any more, indicating that the injury got worse. In Chapter II it was reported that children got earlier out of the coma after severe brain injury when CDT was administered to them in the coma stage.

3.8. Can CDT influence the epigenome to functionally repair genetic defects?

Interesting are two facts concerning movement-based learning. CDT has never caused pathologic CNS changes on the long-term and even in patients with Down's syndrome (trisomy 21) physiologic CNS functioning could be improved. It seems as if CDT is even partly able to repair genetic defects by influencing the epigenome of the patient. A possible mechanism of such a repair could be that the epigenome is modified by movement-based learning in the way that the gene with the defect is not recruited by the epigenome. It seems worthwhile to correlate movement-based learning with epigenome changes. Can the modification of the epigenome by movement-based learning block defect genes and recruit compensatory other genes?

3.9. Dynamic and reversible modification of the epigenetic landscape in comparison to modification of the landscape of pattern formation

The way of dynamic and reversible modification of the epigenetic landscape is probably essential for learning, memory formation and repair of the functional organisation of the injured CNS. Movement-based learning is one of the ways of changing epigenetic landscapes, because during development movement-based learning is used for a "correction en route" of the development and because movements like running, jumping and training balance will have influenced the human genome and epigenome in the last 5 million years, since these movements were essential for getting food and to escape from enemies.

It seems likely to change the epigenetic landscape for repair partly even in defects of the DNA sequence, but not in all cases in some similarity to the repair of the human CNS.

In Chapter II it was shown that speech could be induced in a patient with severe brain injury (cerebral palsy) by improving the phase and frequency coordination of neuron firing, upon exercising on the special CDT device. But the Wernicke's and Broca's areas (speech centres) were not destroyed. These special neural networks for speech were probably only damaged functionally. If these special neural networks for speech were completely destroyed by injury or malformation, the repair of phase and frequency coordination would not have been sufficient for the induction of speech. A terrible lot of new neurons and their functional integration into the existing networks would have been necessary for repair. Such a repair is for the time being out-of-scope for this movement-based learning therapy.

If there is similarity between the modification of the landscape of pattern formation of movements, according to the system theory of pattern formation, and the modification of the epigenetic landscape, then changes in the epigenetic landscape can be repaired, but defects of DNA sequences can only be functionally repaired by epigenetic modification (recruitment of genes) in some cases, especially when the epigenome can substitute other DNA sequences for repair. There seems to be some similarity between the organisation of the genes, accomplished by the epigenome and organisation of the CNS according to the system theory of pattern formation. In genetics exists the problem of how the dynamic organisation is accomplished. Who controls the organisation? In human neurophysiology there is a similar problem. We know a lot about single neurons. But little is known how the tremendously complex CNS organisation of the many billions of neurons is accomplished and how it is controlled to function physiologically.

3.10. Interaction of neural activity and genetic programs during development and repair

The traditional model of brain development includes two phases during which a coarse wiring of the nervous system is laid out, and a later phase during which the coarse connections are refined. In this model, the developmental events that underlie the coarse wiring are the result of predetermined genetic programmes and occur independent of neural activity, whereas the refinement is a result of interactions between the nervous system and the outside world. In this traditional model movement-based learning can be seen as a refinement of impaired development. But the repair of urinary bladder functions and the induction of speech in a 5-year-old girl with a malformation of the cerebrum and thalamus damage upon CDT by learning transfer (Chapter II) cannot just be seen as a refinement of the coarse wiring of the CNS. Therefore, this traditional model has to be modified to accommodate that neural activity and genetic programs interact to specify the composition and organisation of neural networks during all stages of development (20) and repair.

Neural networks are quite resistant to perturbation, suggesting that redundancy is built into neural circuits to ensure that the natural activity in the networks is maintained. Pharmacological or genetic disruptions of crucial network components lead to the expression of alternative circuit mechanisms that generate activity similar to the natural endogenous pattern (20). But in severe brain injury such redundancy is not sufficient anymore and movement-based learning by the training of coordinated movements is necessary to support the building and maintenance of natural activity in the circuits. Spasmolytic drugs alter the natural activity in the developing and repairing networks and may not be helpful for improving development and repair.

3.11. Relative contribution of cell intrinsic versus non-intrinsic fate determinants

Even though many of the components of cell identity, such as general intrinsic physiological properties and cell position, are likely determined at the progenitor level, other components such as choice of synaptic partners are likely influenced by local environmental cues. Therefore, these two mechanisms may contribute to different aspects of cell specification and integration. While earlier decisions, such as the decision to be a certain neuron are likely established at the progenitor level, later restrictions occur over time as they migrate, integrate into the neural networks, and establish the first contacts with other cells. Genetic profiling of cells at critical developmental stages will bring about new insight into interneuron development and subtype specification (22). A complex set of developmental steps are involved in the integration of interneurons into cortical network. These require the interplay of intrinsic genetic programs and their modulation by cell-cell and matrix-cell interactions.

In Chapter I it was shown that the target slow muscle fibre was chosen two times to be a synaptic partner. The target muscle cell was first innervated by fast growing thick axons of large motoneurons and then slowly growing thin axons of small motoneurons (a second group of motoneurons) migrated to the synapses of the fast conducting axons, pushed their synapses away and established its own synapse of different structure and changed membrane properties and Ca^{2+} secretion properties from the sarcoplasmic reticulum of the target slow muscle fibre. There may be, therefore, several critical periods of innervation and cell intrinsic and non-intrinsic determinants are intermingled.

Abnormal development of cortical interneuron subtypes during late embryogenesis due to environmental perturbation coupled with genetic abnormalities might represent a primary cause for many neurodevelopment disorders (22). But the networks can be changed by training certain movements. A so-called autistic child can walk, run and jump, but does not perform a certain movement for a longer period of time than a few seconds. The CNS property to stay in a certain organisational pattern for longer period of time is missing. By forcing the child to stay in the movement pattern "exercising on the special CDT device" upon fixed feet and holding the hands during turning (so that the patient cannot take the hands from the handles) in some similarity to Figure 2 and giving suggestive (with the movement coordinated) instructions, one can often see in eyes and face of the child that one is getting into the deepness of CNS organisation and improves its organisation in the short-term memory. The child is not getting angry of being trapped physically and psychologically, probably because it feels better in his CNS with the improved organisation. To stay longer in a movement or higher mental function pattern can be learned by the child. One of the problems is that it is physically and mentally very hard work for the therapist.

Subtle imbalance in the ratio of excitatory versus inhibitory levels within the cerebral cortex may underlie many neurological and neurodevelopmental disorders such as epilepsy, autism spectrum disorders and schizophrenia (22). CDT can improve CNS functioning including cases of neurological and neurodevelopmental disorders. In Chapter II it was reported about a young woman who improved its epilepsy upon exercising on the special CDT device (in connection with Fig.33 of Chapter II). In a young boy the convulsions reduced (after a transient increase) upon exercising on the special CDT device (23). Before the discovery of drugs against schizophrenia, working therapy was administered to patients, which was helping a bit. However, CDT is much more effective in repair c than the working therapy.

4. Epigenetic Modification for Anticancer Treatment by Movement-Based Learning

4.1. Activation of tumour-suppressor gene by exercise

If movement-based learning (as for example CDT) can suppress genes by modifying the epigenome, it may also be able to activate genes with anticancer effect.

On the long term at the limit performed CDT CNS functioning always changed in the way to become more physiologic and more specific in functioning. Unspecific pathophysiologic CNS functions like spasticity reduced. In the framework of the system theory of pattern formation, the stability of physiologic functions increased (the potential well became deeper) and the stability of pathologic states (like spasticity) decreased (the potential well became less deep) (Chapter II, Fig.4). Since the nervous system is involved in nearly all body functions, it could well be that CDT enhances specific physiologic functions and inhibits non-specific pathologic functions. Cancer is an uncontrolled unspecific growing of cells. It may therefore be that movement-based learning has an anticancer effect.

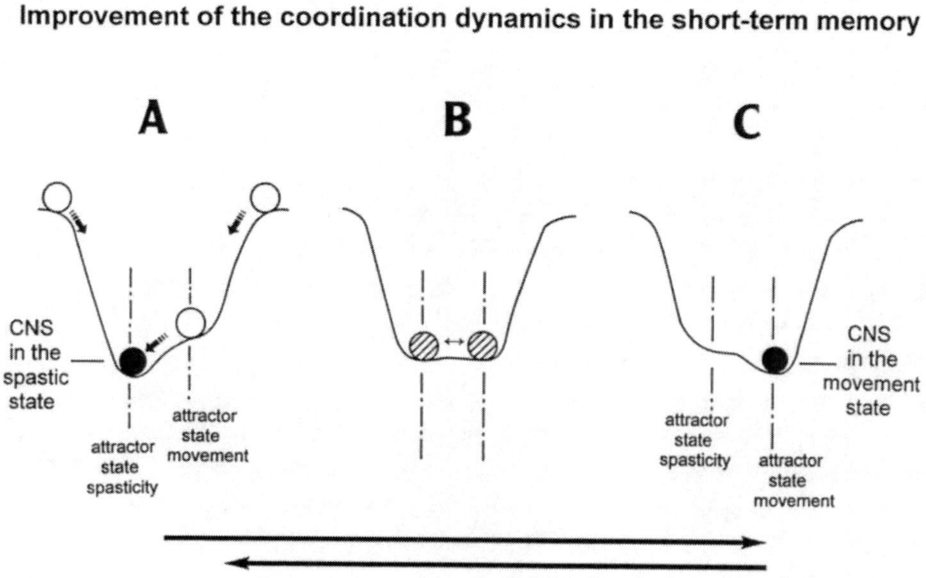

Figure 4. Therapy-induced spasticity (pathologic pattern organisation) reduction in the short-term memory. The position of the ball represents the state of the system and the potential well the attractor. The ball is attracted to the stable position in the deepness of the whole, called attractor state. The attractor layout, consisting of two attractive states of different stability, is changing upon exercising very coordinated rhythmic movements. Black ball = stable state, open ball = very unstable state, hatched ball = spasticity and movement co-exist.

It was reported that exercise might exert an anticancer effect by turning on a tumour-suppressor gene (24). Significant changes in the DNA methylation status of 43 genes occurred in women who engaged, after having completed treatment for breast cancer, in a 6 months moderate-intensity aerobic exercise programme. Association between gene expression level and overall breast cancer survival was found in three. One of these three was L3MBTL1, a putative tumour-suppressor gene. In the exercising patients, L3MBTL1 was demethylated, indicating an increase in gene expression. Elevated expression of this gene may be associated with a lower risk of recurrence and improved survival (24).

4.2. Authors own experience with anticancer effect and body function repair upon Coordination Dynamics Therapy (CDT)

The Author suffered a 'squamous cell carcinoma (epithelioma)' (a malign tumour) in the maxilla (stage between 1 and 2) 4 years ago. The tumour was removed by surgery and a neck dissection was performed: Two lymph nodes with formation of metastases were removed and two stages of further lymph nodes and lymph vessels were taken out. Radiation therapy to the tumour area and chemotherapy was administered to reduce the risk of tumour recurrence from 30% to 15%. To repair the side effects of the anti-cancer treatment, the Author performed CDT at approximately 15 hours per week. Three biopsy's of a strange growing in the following 4 years and other diagnostics showed no sign of malign growing. The exercising on the special CDT device involved coordinated arm, leg and trunk movements and additionally coordinated neck, tong and lip movements.

One year after the tumour extirpation and neck dissection, a reconstruction of the maxilla was performed. The Author went top-fit into the fibula-transplantation operation. In this 9 hour lasting operation pieces of the fibula and the skin with its supplying artery and vein were removed from the right leg. Bone and skin were placed to the former tumour region and artery and vein were connected with the temporalis blood vessels. The transplant had to get blood supply within 7 hours for not being rejected. The Author was informed that during the long-lasting operation he will get oedema and big breathing problems because of the, with the former neck dissection, removed lymph vessels. But only little neck oedema occurred during the operation because the performed CDT had partly repaired the lymph system. One day after the reconstructive operation parts of the transplant was blue, indicating too poor blood supply, even though the transplanted artery was open and supplied blood. The surgeon informed the Author that he will lose some tissue because of poor blood supply. The Author, having brought the special CDT device into the hospital, exercised on the special device, in spite of the many tubes he was connected to, to improve the blood supply of the head including the transplant. As the result of the enhancement of blood circulation, the transplant got enough blood and healed fully in. No transplanted tissue was lost. Later on the surgeon was amazed that his complicated work was fully successful and he said that the transplantation was successful to 100%. Therefore CDT has helped that no big neck oedema occurred in the operation and a tracheotomy (which does not heal very good) could be avoided and has improved the blood supply also in the head so that the transplant was fully accepted. The Author had trained even 1 hour before the operation to be optimally fit for this long-lasting difficult operation. Later on in small operations (in local anaesthesia) part of the fibula bone had to be removed, because it was too big for the implantation of teeth, a vestibule had to be reconstructed and implants for the teeth made. In conclusion, all surgery went fine and all was good because of the continuously performed CDT. CDT helped to repair the body after cancer treatment and may help to exert anticancer effect.

Interesting is that the skin from the leg changed into mucosa in the mouth after transplantation and the hair of the skin did not grow any more in the mouth, even though the hair follicle seem to be still there. The prosthetist argued that nearby mucosa cells migrate into the transplanted skin, because the changes from skin to mucosa started from the sites of healthy mucosa. But it may also be that the change from skin to mucosa cells was induced by gene expression changes induced by function and/or surrounding changes.

Very important is further that with the neck dissection (and tumour resection, radiation and chemotherapy) lost or impaired automatisms like taste, swallowing, coughing, hiccup and others partly reappeared by itself but were enhanced upon CDT. I believe that the exercising of the complicated coordinations between pace and trot gait when exercising on the special device, which stimulate the autonomic nervous system (Chapters III, V), helped to achieve better repair. Even though the eyes are a part of the nervous system and should be reached by CDT, the improvement of the eye functions (wetting of the cornea, opacity of the vitreous body) seems to be most difficult to repair by CDT. The further repair of mouth, throat and neck functions and automatisms improved my quality of life dramatically – many of these important automatisms one only realises when one is losing them.

I saw in TV a lady after several cancer operations in the head. She asked for 'euthanasia' even though her head still did not look too bad. But the lost functions and automatisms, essential (necessary) for quality of life, one cannot see. To report about repair of head functions after cancer treatment is to show that CDT can even help in difficult situations of life.

References

[1] Alvarez-Buylla, A. and Lim, D.A.: For the long run maintaining germinal niches in the adult brain. *Neuron*, 41: 683-686, 2004.

[2] Cao, L., Jiao, X., Zuzga, D.S. et al.: VEGF links hippocampal activity with neurogenesis, learning and memory. *Nat. Genet.*, 36: 827-835, 2004.

[3] Covic, M., Karaca, E. and Lie, D.C.: Epigenetic regulation of neurogenesis in the adult hippocampus. *Heredity*, 105: 122-134, 2010.

[4] Colvis, C.M., Pollok, J.D., Goodman, R.H. et al.: Epigenetic mechanisms and gene networks in the nervous system. *J. Neuroscience*, 25(45): 10379-10379, 2005.

[5] Ge, S., Yang, C., Hsu, K., Ming, G. and Song, H.: A critical period for enhanced synaptic plasticity in newly generated neurons of the adult brain. *Neuron*, 54: 559-566, 2007.

[6] Deisseroth, K., Singla, S., Toda, H. et al.: Excitation-neurogenesis coupling in adult neural stem/progenitor cells. *Neuron*, 42: 535-552, 2004.

[7] Drapeau, E., Montaron, M.F., Aguerre, S. et al.: Learning-induced survival of new neurons depends on the cognitive status of aged rats. *J. Neurophysiol.*, 27: 6037-6044, 2007.

[8] Katnelson, A.: How "sham" brain surgery could be killing off valuable therapies for Parkinson's disease. *Nature*, 476: 143-144, 2011.

[9] Kelso, J.A.S.: *Dynamic Patterns. The Self-Organization of Brain and Behavior*. MIT Press, Cambridge, 1995.

[10] Palmer, T.D., Willhoite, A.R. and Gage, F.H.: Vascular niche for adult hippocampal neurogenesis. *J. Comp. Neurol.,* 425: 479-494, 2000.

[11] Schalow, G. and Schmidt, H.: Local development of action potentials in slow muscle fibres after complete or partial denervation. *Proc. R. Soc*. B 203: 445-457, 1979.

[12] Schalow, G. and Schmidt, H.: Effect of length and temperature on the induction of action potentials in denervated slow muscle fibres of the frog. *Pflügers Arch*. 372: 17-22, 1977.

[13] Schalow, G., Pääsuke, M., Ereline, J. and Gapeyeva, H.: Improvement in Parkinson's disease patients achieved by coordination dynamics therapy. *Electromyogr. Clin. Neurophysiol.* 44: 67-73, 2004.

[14] Schalow, G.: Tremor in Parkinson's disease patients can be induced by uncontrolled activation and uninhibited synchronization of α_2-motoneuron firing to which α_1-motoneuron firing synchronizes. *Electromyogr. Clin. Neurophysiol*., 45: 393-406, 2005.

[15] Zhao, C., Deng, W. and Gage, F.H.: Mechanisms and functional implications of adult neurogenesis. *Cell*, 132: 645-660, 2008.

[16] Miledi, R., Parker, I. and Schalow, G.: Calcium transients in normal and denervated slow muscle fibres of the frog. *J. Physiol*. 318: 191-206, 1981.

[17] Song, H., Stevens, C.F., Gage, F.H.: Astroglia induce neurogenesis from adult neural stem cells. *Nature,* 417: 39-44, 2002.

[18] Marco Garcia, N.V., Karayannis, T. & Fishell, G.: Natural activity is required for the development of specific cortical interneuron subtypes. *Nature*, 472: 351-355, 2011.

[19] Batista-Brito, R. & Fishell, G.: The developmental integration of cortical interneurons into a functional network. *Curr. Top. Dev. Biol.*, 87: 81-118, 2009.

[20] Blankenship, A.G. & Feller, M.B.: Mechanisms underlying spontaneous patterned activity in developing neural circuits. *Nature Rev. Neurosci.*, 11: 18-29, 2010.

[21] Wang, C.L. et al.: Activity-dependent development of callosal projections in the somatosensory cortex. *J. Neuroscience*, 27: 11334-11342, 2007.

[22] Batista-Brito, R. and Fishell, G.: The developmental integration of cortical interneurons into a functional network. *Current. Top. Dev. Biol.,* 87: 87-118, 2009.

[23] Schalow, G. and Nyffeler, T.: Koordinationsdynamik-Therapie: Myelomeningozele (Spina bifida). *Physiotherapie*: 5-19, 2003.

[24] Corless, J.: Powering up. *Nature,* 485: S62-S63, 2012

Epilogue

The Adventure of Performing Medical (Human) Research

1. Funding problems published in 1993

This research has primarily been done on money saved personally, including the equipment (300 000 DM during six years, all the saved money), to bring this clinical research up to the level presented. In Germany, every second day, a paraplegic dies because of bladder infections or by suicide. The author qualified in technical, natural and medical sciences and worked together with R. Miledi at the institute of Biophysics of Sir Bernard Katz. Instead of continuing to work on the trophic effect or on Ca channels, he changed to basic clinical research. He was advised from well-known basic research workers and neurosurgeons that clinical research is worthy of being done if something can be done. At the beginning, reviewers argued it was not possible to record single-nerve fiber action potentials with wire electrodes in humans. Then the argument was that it should only be possible to record from thick nerve fibers, which are of no interest anymore. The author had to use a computer; he has to use average values. A basic animal reviewer could not see the clinical implications and asked then for the ethics of this research, without giving reasons for his ethical considerations. Some American reviewers argued also that if it is possible to record afferents and efferents simultaneously in humans, then this will have consequences. Initially, the Deutsche Forschungsgemeinschaft (DFG) supported the research but refused to accept the application for further money because of organizational reasons. After an intervention by R. v. Weizsäcker, the DFG accepted the application but refused to allocate money because this research project has been judged unqualified and ethically unjustified; they did not want to change the decision even after more than ten papers had appeared in different journals. Now, the DFG raised the organizational argument again and argued that the author cannot apply for money for the same research project again; obviously, the DFG made a faulty decision, and the research worker and the patients have to suffer. The Max Planck Institute for Clinical Research was not interested in the research project, arguing that it exceeded the scope of the field covered by them. Approximately 20 other institutions refused financial or other assistance for various reasons. The "Ministerium für Forschung und Technologie, Forschung im Dienste der Gesundheit" promised financial aid for several years by establishing in the future a Neurotraumatology Research Project, but money is refused for the time being because it is needed for the reunification of Germany. An institution from Switzerland and from the USA refused money basically with the argument that it is better to keep the money in their own countries, even though paraplegics would use the possibility to get treatment in other countries, if available. The trial to allocate funding from private persons by an advertisement in the "Züricher Zeitung" was stopped by an ethical committee. At the rehabilitation conference IRMA VI in Madrid, it was reported that in Amsterdam, it is practically impossible to get funding for research in Rehabilitation. The questions remain: how many paraplegics have still to die until the world's moral responsibility gets sensitized so that the necessary research is organized, which has to cover the range from the basic research up to the patient (question of the Spinal Cord Society, U.S.A.); and, what is the justification for killing animals in basic animal research if no clinical application is intended? This funding problem of human and clinical research was published in 1993 (Schalow G.: Electromyography and Clinical Neurophysiology 33: 477-503, 1993).

2. History of the Human Research Project I Published in 2010

Based on the Author's background of being an engineer in electronics, a theoretical physicist (Dr.rer.nat.), a physician (Dr.med.), and post-doc with Sir Bernard Katz, this research project to cure CNS injuries was started 25 years ago (in 1983) in Finland. It was first supported for 2 years by the 'Deutsche Forschungsgemeinschaft' (DFG). Further funding was refused. Ethical and quality objections were raised. The support of the former Federal President Richard von Weizsäcker could not overthrow the decision of the DFG. The DFG did not want to confirm that the reviewers of the application for funding had a comparable background as the Author. The Author (from West-Berlin) continued the research on personally saved money. He worked together with Departments of Neurosurgery, Neurology, Pathology, and Forensic Medicine of the University of Greifswald in the former German Democratic Republic (GDR), even though this was officially not possible. Since measurements were also performed on brain-dead humans, an ethics committee was established in the former GDR, which decided that the research project was ethically justified. The Author got partially funding from the GDR, even though this was officially not possible for a subject from West-Berlin. Before the re-unification of the GFR and GDR, he could not get funding any more because of shortage of money. The habilitation procedure (Dr.med.habil.) based on the development of the single-nerve fibre action potential recording method in humans was started before the fall of the Berlin Wall and was completed thereafter. When working for 6 years in a rehabilitation centre in Switzerland, the Author applied two times for support by the Swiss National Fund (SNF). Financial support for the research was refused. In recent times, this research project of CNS injury repair has been ignored or there were attempts to destroy it in several ways, including exercising pressure on editors to prevent the publication of papers, exercising pressure on organizers of conferences to prevent presentations, and writing letters to patients and institutions that the method is not working. No real theoretical or clinical scientific discussion could ever be achieved by the Author on the basis of human neurophysiology, even though the results of this research have tremendous clinical consequences. The health care system in middle Europe and North America prefers to let patients with very severe CNS injuries without any qualified treatment instead of inventing and proving new treatments. To find a way to bring qualified movement treatments for CNS repair to patients, the Author has been working for 25 years as a human neurophysiologist and clinical research worker in several countries of Europe, mainly on personally saved money. Animal physiologists and clinicians have not accepted human neurophysiology and the author has been excluded from the research society (congresses, research funding, academic position, publications).

Max Planck argued that it is not possible to convince other people, one has to wait until they die. Einstein was arguing, that intelligence 'is starved out' (New York Times, 12.6.1953). (Schalow G.: *Electromyography and Clinical Neurophysiology* 50, 155-179, 2010).

3. Research across the Berlin Wall

This research project was started in 1983 in Finland. Some pre-experiments were done before to see whether the single-nerve fibre action potential method was working. At the Neuropathology Department of the Turku University the anatomy of the spinal cord was clarified within 2 years. When there was nobody who could be interested in measuring single-nerve fibre action potentials in Finland, the Author (G. Schalow) changed the country for continuing the research project of repairing the human CNS. I thought neurosurgeons of the "Deutsche Demokratische Republik" (GDR) could be interested in doing with me such a research, because researchers and clinicians from countries across Eastern Europe wanted to have contacts to the West. Living in West-Berlin, I had to pass via certain roots (transit ways) through the GDR to reach West-Germany or other European countries. Once I left the transit route to the North (forbidden area for transit) and went to Rostock University and asked for the Chief of the Neurosurgery Department. I was told that there was only a Neurosurgery Department at Greifswald University situated in the North of West-Berlin. I wrote to the Chief of the Neurosurgery Department (Prof. Lang), and he was interested. Subsequently, research cooperation was started to repair human CNS injury between a subject from West-Berlin and a university in the GDR.

On the way from West-Berlin to Greifswald I had to cross the boarder (the Berlin Wall) for leaving West-Berlin. I used the boarder cross at Stolpe (not Checkpoint Charlie). Especially the wall at the 'Brandenburger Tor' (Fig.1) was nasty to see for me because before the building of the wall I used to walk through it to reach the 'Staatsoper unter den Linden' or the 'Komische Oper'. Because of lack of money I changed 1 DM (Westmark) into approximately 4 Mark of the DDR (Ostmark) und could then afford to visit quite often the opera.

On the way to the North I passed two Russian barracks and a Stalin monument. Sometimes I was stopped by Russian military police and I was asked whether I had seen any escaped Russian soldiers. But I was not inspected. There was supposed to be also a concentration camp close to Neubrandenburg where members of the "Hitler Jugend" were put to. Nobody had told me more details because of fear. If denunciated, one was also put to that concentration camp and this was a way of no return. In West-Berlin I lived close to an American army barrack (Finkensteinallee) and sometimes I saw American military police. I felt safe when I saw American soldiers, because when I was in an orphanage, the American soldiers helped the orphanage and especially at Christmas we got chocolate and other nice things to eat. My first ice cream I got was when I became 14 years old and was invited to the American barrack.

Figure 1. Berlin Wall and Brandenburg Gate.

At the beginning of this cooperation between West-Berlin and the GDR, I had to pay 25 "Mark der DDR" per day for using the research facilities. Then later on I had to change 25DM in 25 Mark der DDR (Zwangsumtausch). Even later on I got for one year research money from Ernst-Moritz-Arndt University of Greifswald (900Mark der DDR). The cooperation and the funding was not only helpful in order to progress the research project but was also politically interesting, since the GDR wanted that West-Berlin was seen as an independent country (which the administration of West-Berlin and West-Germany did not want). For sure, the Germans did not want to have their country split into different parts and get their culture partly destroyed. Therefore, it was officially not possible to have research cooperation between a researcher from West-Berlin and a professor from the GDR. Such cooperation was not possible, therefore, and the minister did not officially know anything about it, even though he knew about me. Because of this situation some persons I worked together with in Greifswald thought that I am supposed to be an unofficial member of the secret police (inoffizieller Mitarbeiter der Staatssicherheit (Stasi)), which they told me after the fall of the Berlin Wall. There was not any political pressure put on me, even though I felt some pressure from the surrounding. A very interesting fact was that I had been told that there were reports written about me and for me it was clear that not only my research work was judged.

Since at the beginning I had to pay for every day being in the GDR, I had travelled quite much from West-Berlin to Greifswald and was of course inspected every time, which was unpleasant. With time the border police had got to know me quite well and instead of inspecting me I explained to them what research I had been doing in Greifswald University. Most impressive for them was to look at cross-sections of nerve roots of about $1m^2$ size (for example Fig.22 of Chapter III) to see what a terrible lot of axons of different sizes are contained in nerve roots. But still the crossing of the border was stressful, because sometimes I was examined seriously to see whether I was transporting forbidden things. And of course there were small necessary things I brought to Greifswald. Since certain things had been missing in the former socialistic countries I took small things with me to help my friends who helped me to organize my research project, as for example a pocket calculator, glutaraldehyde and Epon for working with the electron microscope. Because some administration of the GDR thought of this, I sometimes was inspected thoroughly and this was nasty. I had the feeling of loosing all human rights. But to have friends who helped me was important to realise the research project. Already for measuring single-nerve fibre action potentials from brain-dead human cadavers (HTs), I had needed for the start an anaesthetist for sufficient ventilation and medication, then I needed a neurosurgeon to do the laminectomy (not simple in HTs) and I needed a physician or medical student who helped me with the measuring. When working on HTs, the mental load is high and one needs a lot of motivation: the ratio tells you that you are working on a cadaver, but the feeling tells you that you may kill a human being, because the heart is beating. When the artificial ventilation is terminated of an HT, the heart may beat for another one or two hours (the heart is not using only ATP; it can also work with ADP for safety reasons). A pathologist would not start an autopsy on a cadaver in which the heart was still beating. I got a lot of help in Greifswald to realise the research project when no politics were coming in. The measurements in HTs were an essential step to make a repair of the injured human CNS possible. Even though many other scientists and clinicians expected that the success could come only on the long term I got a lot of support. The only country in the world which really supported this research project was the former GDR (Deutsche Demokratische Republik). But all decisions were legally. The DFG from West-Germany decided that this research project had not been ethically justified, most likely without using any ethical committee. In Greifswald an ethical committee was formed to decide whether the measuring in HTs was justified. The committee decided that such measuring was ethically justified. Since there was a lack of dialyse filters (for financial reasons), kidneys had been explanted excessively from HTs in the GDR with permission of the relations. I got those HTs which could not be used for kidney explantation as for example that the former patient had an infection. Further, an explanted kidney can help one patient. Qualified medical research on HTs may help millions of patients. Also, when I tried to further clarify the innervation of the human urinary bladder everything was on legal terms. For further clarifying the anatomy (under the microscope) I needed fresh cadavers from young persons and those ones one can get in the forensic medicine. A forensic autopsy was accompanied by a lawyer to see whether the autopsy was legally performed. After the termination of the autopsy the professor of the forensic medicine explained the lawyer my research project and asked whether it was allowed that I could do a further dissection. The lawyers gave the permission. Also, the urinary bladder was not involved in the death of the persons.

However, when politics were coming in, the situation changed. Being depressed, a friend of mine from the Radiology department showed me a Roentgen picture of a gunshot wound into the pelvis. I had problems to identify the seemingly exploded pelvis. What had happened with that patient? He was an inhabitant from the GDR and wanted to leave the country. He went with the cars from the West and the last bit for border crossing he tried to cover by running. A boarder policeman was running behind him, reached him, and shot him with the machine-gun and destroyed also the pelvis.

4. Back to research

The scientific problem in this research project was now, if it was possible to measure at the neuron level human CNS functioning? Before the Author left the Physiology department of the University of Saarland, he had tried out on different animals that the cauda equina nerve roots might have been suitable for measuring single-nerve fibre action potentials in humans. But to be able to record single-nerve fibre action potentials one needed equipment for that. Since the DFG had not wanted to support the research project any more (see above), equipment was built up on a private basis. Then the success came, namely that it was possible to measure single-nerve fibre action potentials in human during an operation. It had mostly been so in the history of research that if a new qualified measuring method was developed progress in science was achieved. And it was clear to me that a substantial step forward in medicine would be possible. However, the journal 'Nature' was not interested in publishing about this new electrophysiological recording method in human in spite of the fact that there had been not so many methods in human to measure CNS functioning at the neuron level. Journals from the GDR published about the recording technique. With the fall of the Berlin Wall, the journals were taken over by the West. A sad feeling came when an editor said goodbye, because we had been fighting for progress under not very suitable conditions.

The measurements from HTs had brought necessary knowledge for judging the self-organisation of human CNS neural networks: phase and frequency coordination of neuron firing and the stability of networks pattern (Chapters III and V). Probably Desmedt and Eccles liked that research and the Editor of the journal Electromyography and clinical Neurophysiology (Roselle) had given plenty of space for publishing.

The clinical application was not reached in the GDR. When I applied for further funding, I was told that this was not possible, because there was no money any more. The system was going to end because of lack of money; there were no possibilities any more to steer the country. Once I was invited to the Town hall. Probably the administration wanted to see how a private researcher looks like; no political pressure was put on me. But a few times I was denunciated. Because I was living in a student place, the students wanted to discuss with me the political situation. Some of the medical students had been from the army. An army student was really convinced that it was justified to shoot somebody who had been climbing over the Berlin Wall. When I said there was no justification for killing any Germans who wanted to go from one part to the other and what had been the values you defended for which you killed other ones and you risked your own life, I was denunciated for that. I was not afraid to be put into prison, but I was afraid to loose the research condition. I was convinced that such research condition would not come back again and it was so.

The research project ended somehow before the fall of the Berlin Wall because of struggle among professors and not because of political reasons. I habilitated at Greifswald University in medicine (human neurophysiology), which is the highest qualification in medicine in Germany. I started the habilitation with professors from the GDR and ended it with professors from the West, which took nearly all professorships over after the fall of the Berlin Wall. Not all professors coming from the West were good. But the GDR was destroyed (‘wurde abgewickelt’). Some of the medical students got aggressive against some professors from the West as I experienced when I was doing some teaching during my habilitation.

5. Switzerland

The research project was continued in Switzerland in a neuro-rehabilitation centre and the clinical application achieved. The practised physiotherapy methods in the clinic were mainly Bobath and Vojta. These methods were good some 30 years ago, but now out-of-date because of low rate of repair. After 6 years I was thrown out of the rehabilitation centre by mobbing. The reputation and the nice looking clinic seemed to be more important than the progress of the patient, as it is in most rehabilitation centres. When I criticised the low level of neuro-rehabilitation at Zürich University, the professors argued that loud that one could not hear my arguments. I had known that students sometimes had argued that loud for not hearing the argument of the professor (in the 1960ies at the Free University of Berlin). But that professors also were using this strategy when they had no rational arguments was new to me.

6. Back to Finland

I went to a small rehabilitation conference in Helsinki. When I realized that the level was too low, I got angry and went to a neurology conference held simultaneously in Turku (Finland). There, a chief of neurology and neuro-rehabilitation departments got interested in coordination dynamics therapy (CDT). The building up of an intensive care room for treating stroke and other patients with the special CDT device in the lying position was not successful in the Neurology Department, because the staff mainly argued that they were overloaded already with the present work. For additional staff there was no money. But CDT was used in the Neuro-rehabilitation Department. The physiotherapists liked to do CDT. However, the patients could get only 2 hours therapy per day (10 hours per week) for 6 weeks. Such a therapy time is insufficient for the repair of a severely injured CNS. It was not allowed that the patients could pay for additional treatment. Only 10 to 20 years later it became possible to pay privately for additional treatment. I left the university hospital because under these conditions the repair of the CNS did not seem possible to me. Two physicians in Finland became interested in CDT, built up a private therapy place, employed physiotherapists and I supervised the CDT. Patients with CNS injury improved with CDT for 3 months and at least 20 hours per week. TV broadcastings supported this new treatment. But when patients went to the state rehabilitation centre close to Helsinki and argued that they preferred to get therapy in the private place (and that the state is paying for it), because there they would get better and longer treatment for the same amount of money, the administration withdraw the permission for running the place. To put pressure on the physicians, the letters from the administration were brought by the police in a police car. The argument was somehow that CDT is not a physiotherapy method and also not a movement therapy. These arguments were wrong, but the administration was and is always in the power position. They even wanted to cancel the permission to be a physician. The physicians went to court and lost at the Highest Court, even though the reviewer, informing the Court about CDT, was in favour of CDT. The Author was not asked to inform the Court about CDT. On the documentation CD of the Court the private communication between the administration and the Court was deleted, as the physicians told me. It seemed that high positions in the rehabilitation were given to members of political parties (early pension). By trying to compete with the state rehabilitation system, one was attacking members of political parties and they found ways to stop new treatment. If courts are not independent, then the progress in medicine is on the looser side. Before the therapy place in Finland was closed down, the physicians had built up a small therapy place in Spain. Even though there were a lot of small administrative problems, there were no problems with the administration. But Finish patients could not come from Finland, unless they paid privately. According to EU regulation a patient has the right to go to another EU country for a certain treatment if he cannot get it in his own country. Still the Finnish patients could not come to the therapy place in Spain, even though the sun was good for motivation. The Finish administration ignored the EU regulation. Now the physicians are fighting before an EU court. The EU seems to be on their side, because the EU administration wants competition. Even if the physicians would win after 10 to 20 years, the private therapy places had been destroyed and patients lost the possibility of better treatment. One patient, who wanted strongly CDT, and waited for a reopening of the therapy places, died in the meantime.

7. Slovakia

At an age of 9, the son of a friend of mine was overdriven by a car when playing in the street. He also suffered a severe brain injury. Since my friend had been doing the English correction of my papers for several years, he knew about CDT and asked me to come to treat his son. In a rehabilitation clinic CDT was administered to him under my supervision. The boy Benjamin (see Chapter II) got better and the parents of other patients also wanted CDT for their children. The neuro-paediatrician, the physiotherapist and of course the children with severe brain injuries and scoliosis liked CDT. The members of the rehabilitation clinic wanted me there. Suddenly the Chief of the Clinic did not want me there any more and pressure was put on the neuro-paediatrician to stop cooperation with me.

A reason was not given. Then parents and relations of children wanted to build up a small private therapy place for children. My medical certificates were checked thoroughly, but still we did not get the permission to start a small therapy place for children. Even though my friend had several friends in high positions in universities from his medical studies, we could not find out who was blocking us to bring help to the children. Also, a political reason did not exist. My friend was punished during socialistic times because he did not want to become a member of the secret police.

8. USA

There are patients in the USA who would like to come for treatment to Europe. But because of the missing social system, they have no money to come and care and treatment of CNS injuries becomes expensive on the long-term. Christopher Reeve could have afforded to come to Europe. Friends of mine pushed me to apply for money from the Christopher Reeve Foundation and to contact Christopher Reeve. But the application for funding did not reach the 25 reviewers; it had been stopped before. Probably two professors from Switzerland had been also involved to protect him from progress. A patient with a high cervical spinal cord injury (range C3 or C4) has nearly no movement functions left, is dependent on the surrounding and can easily be manipulated.

It seems that Christopher Reeve had been used to get founding for biological research and not for medical research. In an open meeting of Christopher Reeve with politicians of the USA a patient with a thoracic spinal cord injury wanted to push Christopher Reeve to also fight for medical research so that also human patients benefit from the research. In the moment when this person wanted to speak to Christopher Reeve, some persons hold him to put the hand before his mouth so that he could not speak. When Christopher Reeve left the meeting, the patient became free. The patient to whom this happened told me that. As long as the world research society is not informed on what Christopher Reeve really died there are doubts on his death. If a patient gets too many drugs and toxins, the kidneys become overloaded. The death of Michel Jackson was analysed, the one of Christopher Reeve not. With standard care a patient with spinal cord injury cannot die easily.

9. Italy

A rehabilitation physician from North Italy performed something like oscillator formation training and became interested in CDT. He invited me for a 3-days-course including theory and praxis. After the course he wanted to treat his patients with CDT. He applied for instruments.

The hospital administration promised that he would get them, but he did not get them. Then the motivated rehabilitation physician bought a special CDT device by himself and started to treat his patients with it. But then the administration forbid him to administer CDT to protect the patients from becoming irritated by a treatment being different than the normally used one, namely passive movement therapy. Physiotherapists are used to do these movements and many patients like it, because they do not need to work hard. The only disadvantage of this treatment is that the patients practically do not get better.

10. Estonia: the golden years of CDT

A friend of mine from Finland suggested many times to try out Tartu University in Estonia whether progress in neuro-rehabilitation would be possible there. Since there existed a faculty for physiotherapy and sport sciences from socialistic times, there was really interest in movement sciences and research. Further, the university tried to help small firms to be created to build up the country. So it was possible to establish a small therapy place in Tallinn which worked nicely for a few years. Patients and research benefited from it. When the Author left the country, the system switched mainly back to the out-of-date situation as in other countries.

The unit of theory and praxis, which was very beneficial for medicine, lasted only transiently. Especially since neurosurgeons argued that there was nothing on with CDT, the effect on the whole system was minimal. Also, some Master Degree physiotherapists argued, what was the sense of learning CDT if I had a patient for 20min one or two times per week? I believe in the units of theory and praxis and teaching and research. But there must be also the spiritual surrounding allowing for it.

Conclusion

There are more examples to show that it is difficult to bring progress in neuro-rehabilitation. I have been attacked personally many times but I have never been attacked scientifically. It would be very interesting and of great help for medicine if we could get more knowledge concerning the self-organisation of the human CNS. The human genetics (epigenome) was strongly influenced in the last 5 million years and movements were an important part of behaviour. Changes of behaviour in the last 10 thousand years will have not affected the genetics that much. Movements and movement-based learning may therefore be an important part of the health of the human being and to live longer with a higher quality of life. For the time being, progress in medicine is mainly coming from devices and diagnostic and not so much from new methods of repair or healing. It is important, but difficult, to organise qualified medical research. But to replace medical research by biological research is not the solution.

Giselher Schalow
Dr.med.habil.,Dr.rer.nat.,Dipl.Ing.
Untere Kirchmatte 6
CH-6207 Nottwil
Switzerland
www.cdt.host.sk
g_schalow@hotmail.com

Index

A

B

C

D

E

F

G

H

I

J

K

L

M

N

O

P

Q

R

S

T

U

V

W

Y